STP 1405

Effects of Radiation on Materials: 20th International Symposium

Stan T. Rosinski, Martin L. Grossbeck, Todd R. Allen, and Arvind S. Kumar, editors

ASTM Stock Number: STP1405

ASTM
100 Barr Harbor Drive
P.O. Box C700
West Conshohocken, PA 19428–2959

Printed in the U.S.A.

Library of Congress

ISBN: 0-8031-2878-9

Peer Review Policy

Each paper published in this volume was evaluated by two peer reviewers and at least one editor. The authors addressed all of the reviewers' comments to the satisfaction of both the technical editor(s) and the ASTM Committee on Publications.

To make technical information available as quickly as possible, the peer-reviewed papers in this publication were prepared "camera-ready" as submitted by the authors.

The quality of the papers in this publication reflects not only the obvious efforts of the authors and the technical editor(s), but also the work of the peer reviewers. In keeping with long-standing publication practices, ASTM maintains the anonymity of the peer reviewers. The ASTM Committee on Publications acknowledges with appreciation their dedication and contribution of time and effort on behalf of ASTM.

Printed in Bridgeport, NJ
July 2001

Dedication

ARTHUR L. LOWE, JR.
1927-1999

rt Lowe's career spanned 48 years, starting in 1951. This included nearly 37 years with the Bab-ock & Wilcox Company in Lynchburg, VA. He had a leadership role in the application of materials r nuclear fuel cladding and internals, the development of the reactor pressure vessel surveillance rogram, and in the evaluation of radiation effects on materials. Art shared his knowledge and expe-ence for over 30 years as an active member of Subcommittee E10.02 on Behavior and Use of Nu-lear Structural Materials. He has served on numerous other ASTM subcommittees concerned with e testing and evaluation of reactor pressure vessels, fuel cladding, and reactor internals. In 1997 Art as presented with the Peter Hedgecock Award in recognition of his dedication to the activities of ommittee E10. His efforts led to the development and refinement of numerous standards, to the resentation of many technical papers in the Effects of Radiation on Materials symposia series and ther international forums, and to advancement of the materials technology used in nuclear power lants. He has been a mentor to many, and a valued advisor to all. He will be missed greatly.

Foreword

This publication, *Effects of Radiation on Materials: 20th International Symposium,* contains selecte
papers presented at the 20th Symposium on Effects of Radiation on Materials, held 6–8 June, 200
in Williamsburg, Virginia. The symposium was sponsored by ASTM Committee E10 on Nuclea
Technology and Applications. The symposium chairman was Stan T. Rosinski, Electric Power Re
search Institute. Martin L. Grossbeck, Oak Ridge National Laboratory, Todd R. Allen, Argonne Na
tional Laboratory, and Arvind S. Kumar, University of Missouri-Rolla served as co-chairmen.

Contents

Overview

ASTM Committee E10 on Nuclear Technology and Applications sponsors a biennial series of symposia on the effects of radiation on materials. The first symposium was held in 1960 and followed an earlier series begun in 1956 by Committee E10, then called the Committee on Radioisotopes and Radiation Effects. The meetings continue to be a major international forum for the presentation and discussion of research on the influence of radiation on the microstructure and mechanical properties of structural materials. In recent symposia, emphasis has also been placed on the advancement of emerging technologies to improve integrity assessment of major nuclear plant components (e.g., Master Curve fracture toughness methodology) in support of extended plant operation. The proceedings of the Twentieth International Symposium on the Effects of Radiation on Materials are published in this ASTM Special Technical Publication (STP) 1405. The Symposium was held 6–8 June, 2000, in Williamsburg, Virginia.

As demonstrated in previous symposia in this series, the Twentieth Symposium continued the tradition of strong international participation. The 98 registrants for the Twentieth Symposium represented 15 countries and 53% of the participants were from countries other than the U.S. One hundred thirteen presentations were scheduled to be discussed during the Twentieth Symposium and, of the 57 papers in this STP, 67% have corresponding authors from countries other than the U.S. The high level of international collaboration experienced in previous symposia was also apparent in this symposium. Committee E10 considers such international participation important to the success of this series of symposia and to standards development in support of nuclear technology and applications. The organizers extend their appreciation of those international researchers for their participation and the excellent quality of presentations.

The papers presented in this STP are organized in nine sections, four on reactor pressure vessel (RPV) steels, one each on austenitic and ferritic/martensitic alloys, one on proton and spallation neutron sources, one on radiation damage fundamentals, and the final section on other materials. As typical for recent symposia in this series, a large percentage of the papers in this STP (44%) deal with RPV steels and RPV integrity.

The first four sections contain 25 papers that discuss various aspects of radiation effects on RPV plate, forging, and weld materials, including radiation-induced changes on mechanical properties, radiation damage mechanisms and associated microstructural changes, and the modeling of radiation embrittlement mechanisms. A section is also devoted to discussion of the Master Curve fracture toughness methodology for application to RPV integrity assessment.

The section on austenitic alloys contains eight papers on various aspects of radiation-induced segregation, creep, irradiation-creep, swelling, and the associated changes in mechanical properties.

Four papers are presented in the section of ferritic/martensitic alloys. Topics discussed include characterization of fatigue properites and evaluation of the miscrostructural stability in reduced activation ferritic/martensitic steels, the characterization of post-irradiation deformation microstructures in ferritic Fe-Cr alloys, and the evaluation of thermal creep and irradiation creep properties of HT9 materials.

The section on proton and spallation neutron sources contains six papers that examine the effects of high energy proton and spallation neutrons on the mechanical properties and corrosion behavior of various structural alloys.

Four papers are presented in the section on radiation damage fundamentals. Topics discussed include microstructural evolution in martensitic steels and Fe-Cu alloys, the characterization of displacement cascades in austenitic steels, and phase boundary microstructure in nickel and manganese containing stainless steel alloys.

The STP concludes with a section containing ten papers covering a broad range of materials including vanadium alloys, tungsten rods used in the production of tritium, silicon film, silicon carbide and silicon carbide composites, shape-memory alloys, zirconium-niobium pressure tube materials, and other metals and alloys.

The diversity in subject material and the technical quality of the research presented in this STP reflect the continuing importance of the study of radiation effects on materials to the international scientific community and to the nuclear industry. This series of symposia will continue to provide a key international forum for the discussion of radiation effects on materials. On behalf of the editors, I extend my appreciation to all authors for their contributions and to the technical reviewers whose diligence helped maintain the tradition of excellence shown in this series of symposia and contributed significantly to the timely publication of this STP.

Stan T. Rosinski
Symposium Chair and Editor
Electric Power Research Institute
Charlotte, North Carolina

Pressure Vessel Steels—General

L. Myrddin Davies,[1] William L. Server,[2] Vjacheslav Lyssakov,[3] and Stan T. Rosinski[4]

Review of Current Recommendations from the Recent IAEA Specialists Meeting on Irradiation Effects on Pressure Vessel Steels and its Mitigation

---

Reference: Davies, L. M., Server, W. L., Lyssakov, V., and Rosinski, S. T., "**Review of Current Recommendations from the Recent IAEA Specialists Meeting on Irradiation Effects on Reactor Pressure Vessel Steels and its Mitigation,**" *Effects of Radiation on Materials: 20th International Symposium, ASTM STP 1405,* S. T. Rosinski, M. L. Grossbeck, T. R. Allen, and A. S. Kumar, Eds., American Society for Testing and Materials, West Conshohocken, PA, 2001.

Abstract: Under the auspices of the International Atomic Energy Agency (IAEA), the International Working Group on Life Management of Nuclear Power Plants sponsors specialists meetings on radiation embrittlement and mitigation. The purpose of these international meetings is to present state-of-the-art research results in the area of radiation embrittlement, identify critical technical issues that require resolution in the international community, and officially to recommend research activities for IAEA member states in order to resolve these issues. The most recent of these international meetings was held April 1999 in Madrid. This paper discusses the specific conclusions and recommendations drawn at this most recent specialists meeting. Critical technical issues in the area of radiation embrittlement and mitigation, as identified through these specialists meetings, are discussed and summarised.

Keywords: International Atomic Energy Agency, International Working Group on Plant Lifetime Management, specialists meetings, radiation embrittlement and mitigation

Introduction

Under the aegis of the International Atomic Energy Agency (IAEA), the International Working Group on the Life Management of Nuclear Power Plants (IWG-LMNPP) being composed of participants from 27 countries and 2 international organizations, proposes a number of future international specialist meetings at their

---

[1]Chairman of IWG-LMNPP, LMD Consultancy, 176 Cumner Hill, Oxford OX29PJ, UK.
[2]President, ATI Consulting, P.O. Box 5769, Pinehurst, NC 28374, USA.
[3]Scientific Secretary, IWG-LMNPP, Division of Nuclear Energy, IAEA, Wagramerstrasse 5, Vienna A-1400, Austria.
[4]Project Manager, EPRI, 1300 Harris Blvd, Charlotte, NC 28262, USA.

regular meetings as part of its remit. For the past decade specialist meetings on radiation effects and its mitigation on nuclear pressure vessel steels have been held at two-year intervals in a number of locations which have included Balatonfured [*1*], Paris [*2*], Espoo [*3*], Vladimir [*4*], Madrid [*5*]. The next meeting in this series is being organized at the present time for June 2001 in Gloucester. Other specialist meetings on other topics are also held. The program and work of the IWG-LMNPP has been described recently elsewhere [*6*].

The recent process for summarizing conclusions and recommendations from the specialists meetings has utilized a rapporteur selected from the attendees. It is hoped that these summaries can impact future research direction in the IAEA member states and within the international community in general. However, the proceedings from the specialists meetings do not have a wide distribution beyond the government representatives of the member states on the IWG-LMNPP. This paper provides a forum to disseminate the conclusions and recommendations to a wider audience.

Methodology

A list was prepared highlighting technical topics raised in the various meeting sessions at the Madrid meeting. The topics were organized corresponding to 'previous topics' and 'new themes'. This list is shown in Table 1.

Recommendations

Following the summary and review session (which included a discussion of the previous Vladimir [*4*] meeting Recommendations), the following list of key items were identified to be disseminated and carried forward to the next specialist meeting to be held on this topic.

1) Continued support for the IAEA international database(s) on reactor pressure vessel (RPV) toughness was endorsed.
2) Measurement of copper in solid solution should be continued for RPV materials containing significant levels of bulk copper.
3) The role of phosphorus in creating damage during irradiation needs better understanding in terms of both hardening and non-hardening mechanisms.
4) Considering potential inter-granular fracture, the effect of thermal annealing needs further understanding and quantification.
5) The role of high nickel in promoting radiation embrittlement needs further understanding in terms of mechanisms and response at higher fluences; conflicting results have indicated different responses for similar high nickel steels. A new Coordinated Research Programme (CRP) on this topic is being initiated.
6) Further knowledge on the availability and use of reference materials (such as JRQ and HSST plates) for current and future surveillance programmes is still needed.
7) The importance of consistent reporting of fluences was emphasised, and participants were requested to report all fluences in terms of $E > 1MeV$ and $E > 0.5MeV$. To alleviate some concerns involving differences in neutron energy spectrum, the use of dpa is encouraged, although it is recognised that there is still debate on the best measure of dose.

Table 1 – *Topics, sessions and comments from Madrid meeting.*

Item	Previous Topics	Madrid Session No.						Comments
		1	2	3	4	5	6	
1	Databases	X			X			IAEA and others
2	Measure Cu in solution		(X)					(Indirectly determined)
3	Effect of P	X	X	X	X	X		Both for IGF and hardening mechanisms, plus annealing effect
4	Effects of high Ni	X	X	X	X	X		Mixed effects; mechanism not clearly known
5	Reference materials	(X)	X	X	X			Amount available? Enough for the future?
6	Fluence reporting	(X)	X				X	E > 1 MeV; 0.5 MeV; dpa; spectrum
7	Next Specialists Meeting in 2001							UK?
8	Directly measure fracture toughness (FT)	X	X	X	X	X		Master Curve (CRP) emphasis; some J-R curves
9	Dynamic fracture toughness		X	X				Crack arrest data; instrumented Charpy V-notch (CVN)
10	CVN shifts vs. FT shifts	(X)	X	X	X			More data needed
11	Thru-thickness change in toughness				X	X		NVNP-2 sampled at small depth / Magnox microstructure evaluation
12	Irradiation facility catalog							IAEA identified
13	Neutron flux effects	X		X	X	X		Spectrum effects, gradients, thermal vs. irradiation embrittlement

Item	New Themes Presented	Madrid Session No.						Comments
		1	2	3	4	5	6	
14	Specimen reconstitution	X	X					CVN & FT (compact tension)
15	Re-embrittlement after annealing	X	X		X			Model and mechanisms
16	Subsize specimens		X	X	X			Including small punch; based on correlations
17	Material variability and sampling			X				Interpretation of data
18	Effect of prestrain on toughness				X			Sizewell data
19	Relationship to NDT inspection programs				X			Based on integrity assessment
20	Indexing methods		X	X	X			Accuracy needs to be determined

8) Direct measurement of fracture toughness is emphasized along with analysis methods employing the master curve approach; the ongoing IAEA Coordinated Research Programme (CRP) activities are strongly endorsed.
9) More dynamic fracture toughness data are needed.
10) More data should be generated for comparisons between Charpy V-notch and measured fracture toughness shifts; indexing methods in use currently should be re-evaluated. The use of sub-size specimens is required in some instances and appropriate correlations are needed.
11) Through-thickness material properties after irradiation exposure are needed to support structural integrity analyses.
12) The role of neutron flux effects is an area that continues to be debated; resolution of flux effects as a function of fluence, spectrum, temperature and other operating parameters is needed.
13) Efforts to promote better and novel specimen reconstitution techniques are encouraged.
14) Re-embrittlement after thermal annealing is an area of continued interest, especially for WWER steels. Models and mechanistic understanding are necessary.
15) Material variability and sampling issues continue to be areas needing further study. Other effects such as pre-strain can also be important.
16) New advances in computer modelling of radiation damage from first principles are encouraged.

These key items or themes should be used as guidance for RPV embrittlement research topics in the future. Two additional recommendations were considered. Another specialists meeting on irradiation effects should be considered and also a catalogue of world-wide irradiation facilities should be developed. Note that a specialists meeting has been scheduled for June 2001 in Gloucester, UK.

Progress on some of these recommendations can be reported, as indicated earlier. The IAEA has already initiated two major CRPs involving a large number of organisations from member states to investigate: the "Mechanisms of Nickel Content Effect on the Irradiation Embrittlement of RPV Steels" and "Surveillance Programmes Results Application to RPV Integrity Assessment." The latter CRP will further assess the master curve methodology and application of fracture toughness testing. For the irradiations to be conducted during these programmes, the intention is for neutron fluence to be simultaneously reported as E > 0.5 and E > 1.0 MeV and also as dpa.

References

[1] IAEA Specialists' Meeting, held at Balatonfured, Hungary, 1990, 'Radiation embrittlement of nuclear reactor Pressure Vessel Steels', ASTM STP 1170.
[2] IAEA Specialists' Meeting , held at Paris, France, September 1993, IAEA Publication, IWG-LMNPP-93/03.
[3] IAEA Specialists' Meeting, held at Espoo, Finland, October 1995, IAEA Publication, IWG-LMNPP-95/5.
[4] IAEA Specialists' Meeting, held in Vladimir, Russia, September 1997, IAEA Publication, IWG-LMNPP-97/2.

[5] IAEA Specialists' Meeting, held at Madrid, Spain, April 1999. IAEA Publication, IWG-LMNPP-99/2.

[6] Davies, L. M., Gueorguiev, B., Lyssakov, V., 'IAEA Activities on NPPLM programmes', Proceedings of PLIM/PLEX meeting in Madrid, Nuclear Engineering International, Nov. 1999.

Timothy J. Williams[1] and David Ellis[1]

A Mechanistically-Based Model of Irradiation Damage in Low Alloy Steel Submerged Arc Welds

---

Reference: Williams, T. J., and Ellis, D., **"A Mechanistically-Based Model of Irradiation Damage in Low Alloy Steel Submerged Arc Welds,"** *Effects of Radiation on Materials: 20th International Symposium, ASTM STP 1405,* S. T. Rosinski, M. L. Grossbeck, T. R. Allen, and A. S. Kumar, Eds., American Society for Testing and Materials, West Conshohocken, PA, 2001.

Abstract: A model of irradiation damage in low alloy steel submerged arc welds has been developed by fitting test reactor data to physically reasonable mathematical equations. The model is underpinned by a substantial amount of microstructural data obtained using small angle neutron scattering, atom probe microscopy and other techniques. These data have been used qualitatively to show that the basic assumptions of the model are reasonable, and assessed quantitatively using the Russell and Brown modulus hardening model.

The model predicts hardness change in welds irradiated at high neutron dose rates as a function of irradiation dose, irradiation temperature and chemical composition. Correlations have been developed to enable Charpy or fracture toughness shift to be predicted from hardness change. In addition, a simple modification to the basic model allows estimation of damage at power reactor dose rates. This has enabled comparison with US surveillance programme data. Possible reasons for the observed differences in data and predictions are discussed. Plans to validate the model by testing material from a decommissioned RPV are outlined.

Keywords: Irradiation embrittlement, pressure vessel steels, submerged arc weld metal, modelling, copper, nickel, hardness, Charpy shift, fracture toughness shift

Introduction

The mechanisms of irradiation embrittlement of low alloy steels are now fairly well understood. Nevertheless there are a number of significant uncertainties and subtleties of behaviour which remain elusive. These, in conjunction with the common inadequacies of most databases (measurement inaccuracy and lack of values for all the variables that may be important), have been obstacles to the development of a general predictive model, accurately applicable to a wide range of materials and irradiation conditions. Instead

---

[1] Consultant and Senior Engineer, respectively, Rolls-Royce plc, PO Box 2000, Derby, DE21 7XX, United Kingdom.

there exist a number of models describing the effects of irradiation on the mechanical properties of low alloy steels. Each is fitted to a database representing somewhat different material and irradiation conditions from the others, and is typically identified to a specific country or reactor design. This paper presents one such model. It draws heavily on work by recent pioneers in the field, including Odette, Lucas, Fisher, Buswell, Jones and Stoller [*1-7*]. Previous evolutions of the model are described in [*8-11*].

The model has been developed to provide as accurate a description as possible of the particular database it represents, and to try to integrate a range of mechanical and microstructural data. It is hoped that the model will provide a useful reference point for the development of further understanding of the mechanisms of irradiation damage.

Experimental Details

Experimental details have been given previously in [*8-11*]. The following sections summarize and update the data relevant to this paper.

Materials and Specimens

The database used for fitting the model comprised 11 submerged arc welds in 150mm thick ASTM SA533B plate. The welds were austenitized at 920°C for six hours and water quenched, then tempered/stress relieved for 42 hours at 600°C, followed by 6 hours at 650°C. Chemical composition values are shown in Table 1. Several determinations were made for each weld and the maximum value for manganese (1.74%), for example, is the highest mean manganese content of any weld and 1.53% is the mean of the means.

TABLE 1 – *Chemical compositions.*

	Mn	Mo	Ni	Cu	C	Si	S	P
Maximum	1.74	0.54	0.390	0.60	0.062	0.60	0.027	0.035
Mean	1.53	0.50	0.159	0.27	0.053	0.43	0.018	0.023
Minimum	1.42	0.47	0.043	0.07	0.045	0.37	0.012	0.016

The specimens used were standard Charpy V-notched specimens to the ASTM Standard Test Methods for Notched Bar Impact Testing of Metallic Materials (E 23), half Charpy size blocks for hardness testing and "squat" compact tension (CT) specimens. The latter were to the design and dimensions of a standard 25 mm CT specimen (ASTM E399, Standard Test Method for Plane-Strain Fracture Toughness of Metallic Materials), except that the thickness (B) was 40 mm. The Charpy and toughness specimens were in the LT orientation. In the great majority of cases specimens were taken between 3T/4 and T/4, where T is the thickness of the weld. Broken test-pieces were used to provide material for microstructural examination.

Irradiation Conditions

Specimens were irradiated in materials test reactors: the Herald reactor at Aldermaston in the UK, and the Osiris reactor at Saclay in France. Neutron exposures were determined by in-house procedures that generally meet or exceed the requirements of the relevant ASTM Standards (E 693, E 944, E 181, E 261, and E 706). Most irradiations were carried out at a "high" dose rate (about 7 x 10^{-9} dpa/s), but a few were done at a "medium" dose rate (about 7 x 10^{-10} dpa/s) in Herald, and there was one "low" dose rate (about 9 x 10^{-11} dpa/s) irradiation in Osiris. Irradiation temperatures were 225 °C, 255°C, 295 °C and 315 °C. Figure 1 is a plot of dose versus dose-rate for each data set compared with U.S. surveillance data for low nickel welds (discussed below).

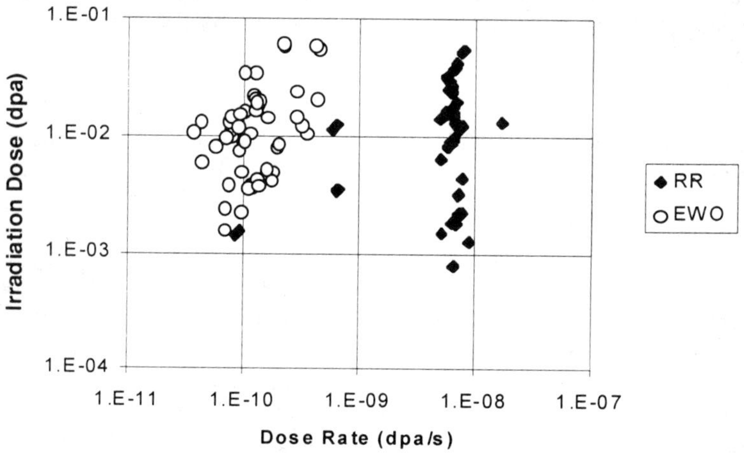

Figure 1 – *Irradiation dose versus dose-rate for each data set. The Rolls-Royce database values are compared with U.S. surveillance conditions for low nickel welds from the Eason Wright and Odette database [14].*

Mechanical Testing

Hardness testing was done using a Vickers Diamond Pyramid machine with a 20kg load. Specimen surfaces were carefully machined to avoid surface hardening and to provide clear edges to the hardness impressions. Most hardness changes were based on six hardness impressions before irradiation, and four after, on each of the Charpy specimens within the data set (normally fifteen but varying in number between one and sixteen). In the case of the half Charpy sized hardness blocks, the number of hardness impressions pre- and post-irradiation was much more limited. Assessment of unirradiated data obtained over several batches spread over a long period of time, suggested that the accuracy of hardness *change* determination (from the Charpy specimen sets) was 2 to 3 VPN. The estimated accuracy of each determination was used to weight the fitting process. Charpy and toughness testing were carried out to the standards cited above (ASTM E 23 and ASTM E 399). The Charpy data were fitted to tanh curves. The toughness data were analysed generally to the ASTM Standard Test Method for Determination of Reference Temperature, T_0, for Ferritic Steels in the Transition Range (E 1921), the Master Curve method, except that test temperature varied.

Microstructural Characterization

A small number of samples were examined using a range of advanced metallographic techniques. Field emission gun scanning transmission electron microscopy (FEGSTEM) was done at AEA Technology Harwell, UK. Small angle neutron scattering (SANS) was done on the D11 spectrometer at the high flux beam reactor of the Institute Laue Langevin, Grenoble, France. Field ion microscopy was carried out in the energy compensated position sensitive atom probe (ECoPoSAP) instrument at Oxford University, UK. Techniques have been documented previously [*11*] and are also described elsewhere in this volume [*12*].

Data Bases and Analysis Techniques

The Rolls-Royce (RR) mechanical property database for this investigation comprised 2 Charpy sets with no hardness measurements, 73 sets with both Charpy and hardness measurements, 32 hardness only specimens and 20 transition regions toughness specimens. The final fitting for the main predictive models was done using in-house software based on a procedure by Britt and Leuke [*13*], which uses error estimates for the dependent and independent variable to avoid bias. Initial assessment of the data, and the fitting of subsidiary models, for example fits to the microstructural data, were done using spreadsheet databases and fitting tools.

The comparisons with published data described below were with the database used by Eason, Wright and Odette to derive improved embrittlement correlations for RPV steels [*14*]. This is based on, but not identical to the U.S. Power Reactor Embrittlement Data Base (PR-EDB), and will be referred to herein as the EWO database.

Hardness Change Model

Base Model – High Dose Rate

In the initial phase of the work the hardness data obtained at the high dose-rate at an irradiation temperature of 255°C (48 data sets) were fitted to provide a base model. Many different model forms were investigated but the following model (Eqs. 1-5) was established as the preferred candidate. This was on the basis of the statistical quality of the fit and conformance to the prevailing concept [e.g. *1-6*] that irradiation damage comprises two components: matrix and precipitation damage. (A third component due to grain boundary embrittlement may be important in some cases [*15*], but there was little evidence for it in these materials.) The detail of the model is also owed in part to the empirical processes of fitting, and in part to the physical insights afforded by the damage characterization work described below and similar work by other investigators.

The parameters underlined were obtained by fitting to the 48 data sets available. The parameters (in Eq 2) defining the temperature dependence of matrix damage were obtained from work by Jones on other materials and reported in [*16*]. The effectiveness of this term is shown in Figure 2, which compares the measured and predicted hardness changes for data at other irradiation temperatures, which were not used in fitting Eqs 1 to 5.

$$\Delta H = \Delta Mtx + \Delta Ppt \tag{1}$$

$$\Delta Mtx = \underline{8.65} \times (2.66 - 0.0065 \times T_{irr}) \times (dose)^{\underline{0.42}} \tag{2}$$

$$\Delta Ppt = (\underline{41} + \underline{1850} \times P) \times (Cu_{mtx} - \underline{0.164})^{\underline{0.398}} \times \tanh\left[\frac{dose}{(\underline{15.7} - \underline{32.2} \times Cu_{mtx})}\right] \tag{3}$$

$$Cu_{mtx} = \begin{cases} Cu, Cu \le 0.35 \\ 0.35, Cu > 0.35 \end{cases} \tag{4}$$

$$\Delta Ppt = 0, \text{ when } Cu_{mtx} \le 0.164 \tag{5}$$

where

ΔH = total hardness change due to irradiation (VPN)

ΔMtx = matrix damage component of hardness change (VPN)

ΔPpt = precipitation component of hardness change (VPN)

T_{irr} = irradiation temperature (°C)

$dose$ = irradiation dose (dpa)

P = phosphorus content (weight %)

Cu_{mtx} = matrix copper content (copper in solid solution at start of irradiation) (weight %)

Cu = total copper content (weight %)

The upper limit to Cu_{mtx} (representing the effect of copper precipitation during heat treatment) was fixed on the basis of FEGSTEM measurements of matrix copper (Figure 3). This value is high relative to other work (for example Eason, Wright and Odette use 0.3% in [*14*]). However this can be explained by the relatively high final heat treatment temperature (650 °C). The fitted value of the copper threshold i.e. the copper content below which hardening is entirely due to matrix hardening, is also high (0.164%) relative to most other work on low alloy steels (estimates range typically from 0.08 to 0.1%). This can not be readily explained. The data underlying this value are shown in Figure 4. Although the effect is not shown clearly (total hardness change is plotted rather than just hardness change due to precipitation), the evidence is reasonably strong. The presumption is that, with a higher copper content, precipitate growth is more rapid.

The dependence of hardness change on phosphorus in Eq 3 was introduced following assessment of residuals. Figure 5 shows the residuals of the model as defined in Eqs 1 to 5, but with the phosphorus value for all data sets set at the mean value for phosphorus for these sets (0.025%). Only those sets for which the model predicts a significant effect of phosphorus (i.e. with $Cu_{mtx} > 0.164\%$ and dose > 0.010 dpa) are included on the plot. Other data would scatter about the zero residual line, obscuring the trend. The overall quality of the fit is illustrated in Figure 6. Figures 7a and 7b show examples of the fits to individual welds. In each case the upper data points are the measured total hardness change value, while the lower data points are the deduced values of precipitation damage, being the total hardness change minus the model prediction for matrix damage.

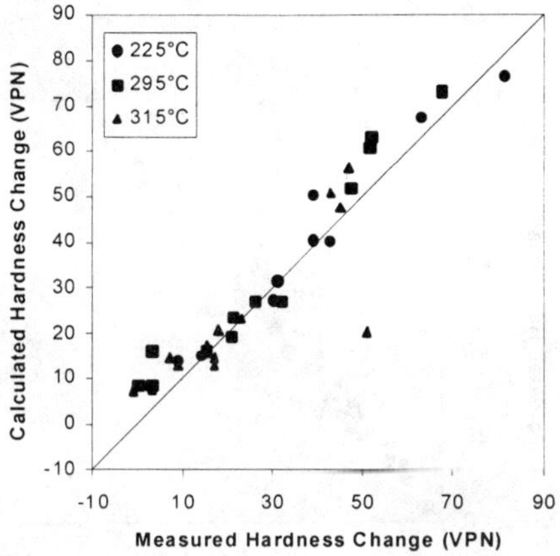

Figure 2 – *Calculated versus measured hardness change for data obtained for irradiation temperatures of 225, 295 and 315 °C. The model was not fitted to these data.*

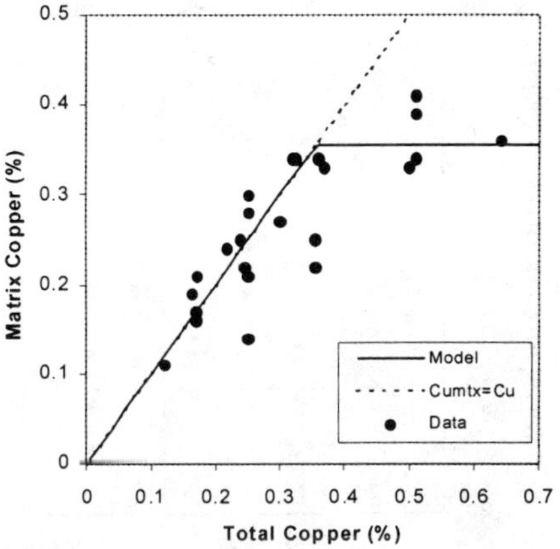

Figure 3 – *Measured matrix copper versus total (bulk) copper. The model assumes a maximum matrix copper content of 0.35%.*

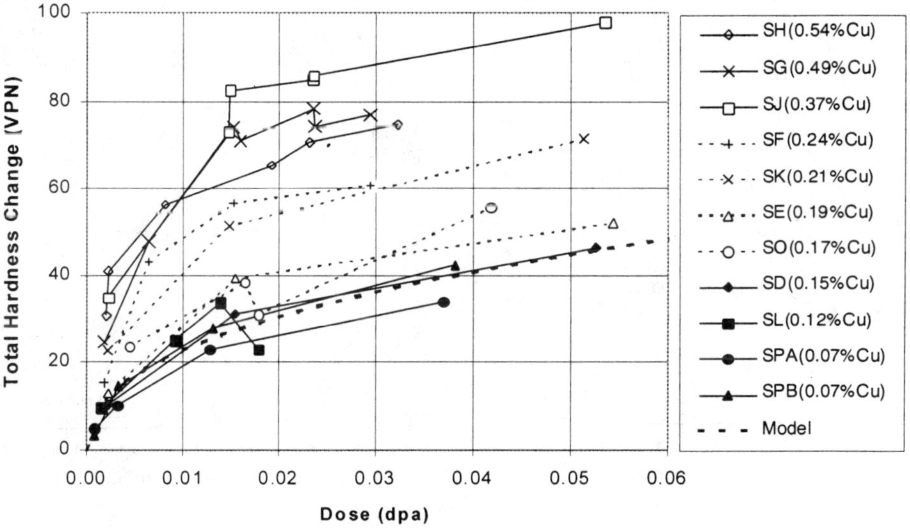

Figure 4 – *Hardness versus dose trend curves for the fitted data; the legend gives weld identities and bulk copper contents. The curves illustrate the apparent insensitivity to copper for copper contents of 0.15% and below. There is a rapid increase in sensitivity between about 0.18% and 0.21% copper. The fitting algorithm produced a threshold of 0.164%. The "model" curve is Eq. 2.*

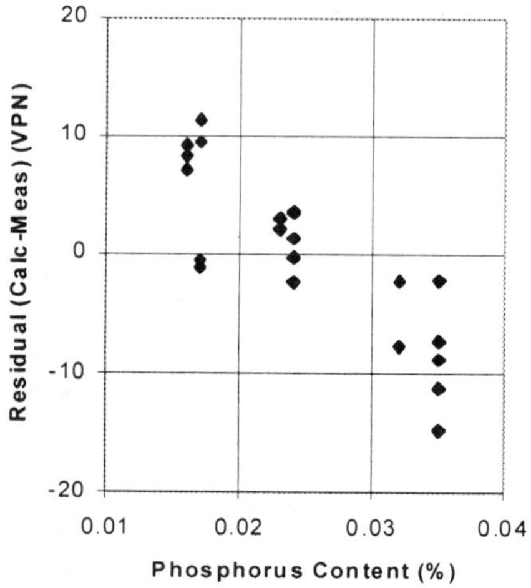

Figure 5 – *Residual vs. phosphorus content for predictions with phosphorus set to 0.025% for all sets, showing the need to include a phosphorus dependency.*

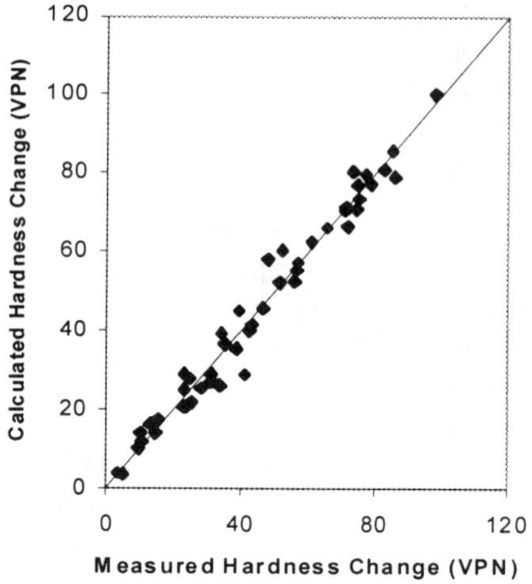

Figure 6 – *Calculated versus measured hardness change for the 48 data sets fitted. The standard deviation for the fit is 4.0 VPN.*

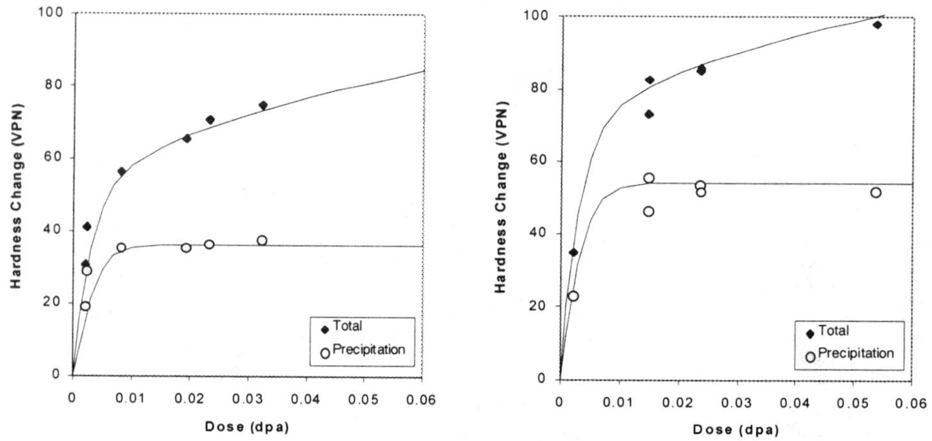

Figures 7a and b – *Examples of model fits: a), left, for weld SH (0.54% Cu); b), right, for weld SJ (0.37% Cu). The upper points are total measured hardness change and the curve is the model fit (Eqs 1 to 5). The lower points are deduced precipitation hardness change (i.e. total measured hardness change minus the model prediction for matrix shift, Eq. 2). The curve through them is the model fit for precipitation hardness change (Eqs 3 to 5).*

Model Extension to Low Dose Rates

The model defined by Eqs 1 to 5 is limited to prediction at the high dose rate (approximately 6.6×10^{-9} dpa/sec). It is now reasonably widely accepted that reducing dose rate may result in an acceleration of precipitation relative to dose. (This is perhaps better seen as an effect of increasing dose rate accelerating precipitation relative to time.) A useful discussion of the phenomenon is given by Odette et al in [*17*]. It is assumed that dose rate does not affect matrix damage or the plateau value of precipitation hardening. The insensitivity of matrix damage to dose rate is reasonably well established by other workers. It is also shown in our data - though not brought out here. The lack of an effect of dose rate on the plateau is an assumption, although limited data do suggest that the assumption is correct. On this basis, Eq. 3 of the above model can be modified to predict damage at low dose rates.

$$\Delta Ppt^{plat} = (\underline{41} + \underline{1850} \times P) \times (Cu_{mtx} - \underline{0.164})^{\underline{0.398}} \tag{6}$$

where

ΔPpt^{plat} = plateau value of precipitation damage (VPN)

Equation 6 is conservative at pre-plateau doses (approximately 0.01 to 0.02 dpa for these welds, depending on copper content) but is accurate once the plateau has been reached. In order to provide accurate estimates of shift for all doses and dose-rates an initial attempt has been made to modify Eq 3 to take account of dose rate by introducing an acceleration factor, *J*. Following Odette [*1*], Fisher et al [*2*], Stoller [*7*] and others, the effect of dose rate is seen as the result of increasing vacancy concentration (hence diffusivity) as dose rate increases. For reasons of pragmatism, *J* is defined as the ratio between the vacancy concentration at the high dose rate, n_v^H, and that at a different dose rate, n_v^i. Thus J is equal to unity for the high dose rate irradiations in the RR database, and increases as dose rate reduces. The modification to Eq. 3 is simply

$$\Delta Ppt = (\underline{41} + \underline{1850} \times P) \times (Cu_{mtx} - \underline{0.164})^{\underline{0.398}} \times \tanh\left[\frac{J \times dose}{(\underline{15.7} - \underline{32.2} \times Cu_{mtx})}\right] \tag{7}$$

$$J = n_v^H / n_v^i \tag{8}$$

where

J = acceleration factor
n_v^H = total vacancy concentration at high dose rate (i.e. 6.6E-9 dpa/s)
n_v^i = total vacancy concentration under irradiation at other dose rates

Indicated acceleration factors were determined for each of the low dose rate irradiations by finding the J values that produced model predictions equal to the observed total hardness change. This was done only for those sets of data with copper contents above the threshold and for which the amount of precipitation predicted from the model with $J = 1$ was less than 95% complete. In order to estimate the uncertainty in the J values, the exercise was repeated for the high dose rate irradiations meeting the same exclusion criteria. The results are shown in Figure 8. There is a considerable amount of scatter, but by comparison with the error bar for the "mean" data point representing the high dose rate data, the evidence for a dose rate effect is clear. One point is off scale. This is a point with a small (2 VPN) predicted hardness change due to precipitation for $J = 1$ and a small (12 VPN) predicted ΔPpt^{plat} and the uncertainty in the observed J is large.

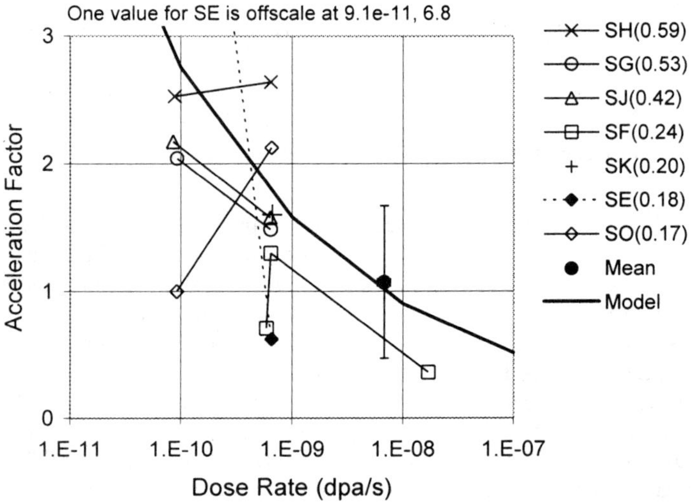

Figure 8 – *Acceleration factor versus dose rate for the medium and low dose rate data. The high dose rate point ("mean") provides an approximate estimate of the uncertainty in the values (error bars are 1 standard deviation). The "model" curve is an approximate fit to the data (Eqs. 8 and 9). Figures given in the key are mean Cu levels for the reduced data set for which low dose rate data exist.*

Conceptually $n_v{}'$ increases from the thermal equilibrium value as dose rate increases from zero. It passes through a point of inflection where the rate of increase of the vacancy concentration is proportional to the rate of increase of dose rate (where the rate of precipitation relative to dose is independent of dose rate), and ultimately saturates. It was beyond the scope of this work to attempt to model this explicitly, but the following equation provides a crude simulation , which has been approximately fitted to the data shown in Figure 8.

$$n_v = \exp\{\ln(\alpha) + \ln(\beta) \times \tanh[(\ln(d') - \ln(d'_0))/\gamma]\} \quad (9)$$

where

d' = dose rate (dpa/s)
α, β, d'_0 and γ are fitted parameters
(α/β corresponds to the thermal equilibrium vacancy density)
($\alpha \times \beta$ corresponds to the saturation vacancy density at high dose rate)
(d'_0 is the dose rate at the point of inflexion)
(γ defines the width of the transition between thermal and saturated conditions)

It was not possible to produce a reliable converged fit (due to the mathematical form of the above equation and narrow range of dose-rates available). However, the following parameter values were selected because they give a reasonable fit to the data and were reasonably consistent with Stoller, Figure 7 of [*18*].

$\alpha = 5.9 \times 10^{-9}$
$\beta = 4.0 \times 10^{4}$
$d'_0 = 4.25 \times 10^{-9}$
$\gamma = 1 \times 10^{19}$

The time to reach the plateau shift at the high dose rate is about 400 hours. This, using the values of α and β above, produces an estimate of the time to reach the plateau for thermal ageing of 2000 years, which is reasonably consistent with other estimates.

Eqs 8 and 9 with the parameters given above provide a reasonable fit to the data (Figure 8) and are not inconsistent with other evidence. However, other possible model forms would give equally good descriptions of the rather limited and scattered data available.

Eason, Wright and Odette [*14*] include time as a variable in their improved embrittlement correlations for the U.S. surveillance database. The dose rate dependency so produced is substantially weaker than Eqs. 7-9 above.

Hardness Change versus Toughness and Charpy Shift Correlations

A correlation was developed between toughness shift and hardness change as shown in Figure 9. The regression fit is given in Eq. 10.

$$\Delta T_0 = 2.68 \times \Delta H \tag{10}$$

where

ΔT_0 = shift in the Master Curve T_0 (reference temperature) ($^\circ$C)

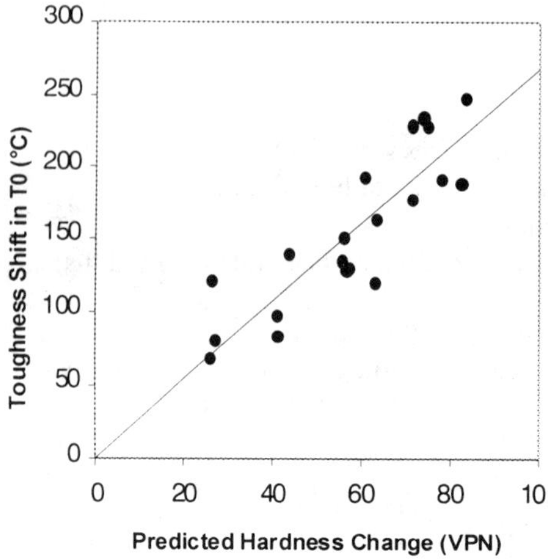

Figure 9 – *Toughness shift in T_0 versus hardness change correlation.*

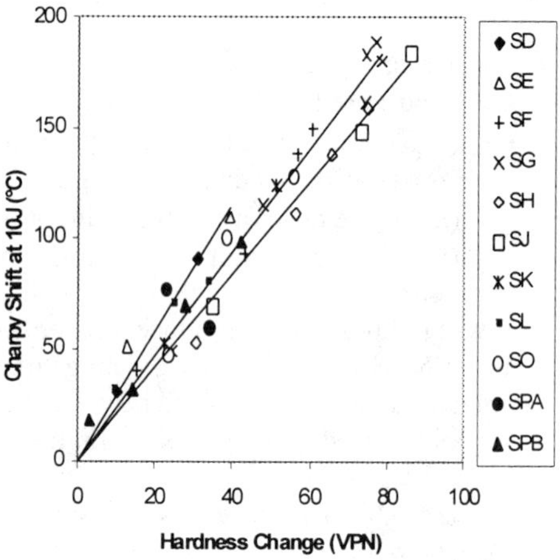

Figure 10 – *Charpy shift at T_{10} versus hardness change correlation. Trend lines shown for welds SE, SG and SJ.*

The Charpy shift versus hardness correlation, shown in Figure 10, was based on work by Odette, Lombrozo and Wullaert [*19*]:

$$\Delta T_{41} = k_{TH} \times \Delta H + \frac{C}{2} \times \ln\left[\frac{(U^I - 10) \times (U^U - 41)}{(U^I - 41) \times (U^U - 10)}\right] \tag{11}$$

where

ΔT_{41} = Charpy shift at the 41J index level (°C)
k_{TH} = correlation coefficient (fitted for each weld individually) (°C.VPN)
C = transition curve width parameter from the Charpy tanh fit (= 60 °C)
U^I = irradiated Charpy upper shelf toughness (J)
U^U = unirradiated Charpy upper shelf toughness

The value of C used was the mean for all Charpy curve fits. To ensure that this could be treated as a constant, the values from the individual tanh fits were examined to determine whether C was correlated to weld heat, copper, nickel, shift, dose or irradiation temperature. No correlation was found. The value of k_{TH} was fitted to each weld individually. The mean k_{TH} over 9 welds was 2.44 (°C/VPN) ± 0.29 (one standard deviation). Taken with Eq 10 this implies about 10% greater toughness shift than Charpy shift (at the 10J index), whereas most published data suggests that this difference exists in plates, but not welds. The welds in this study had a similar heat treatment to plate, and it is possible that this is a significant factor in the correlation.

Any correlation of shift to hardening would break down if significant non-hardening embrittlement (NHE), for example from phosphorus segregation, were to occur. The Charpy shift versus hardness change data were examined for evidence of non-linearity or non-zero intercepts, which might indicate NHE effects. Also selected specimens were fractographically examined. There was no evidence of a significant effect of segregation on mechanical properties within the range of the database, though there was a tendency for the proportion of the boundaries failing intergranularly to increase with irradiation.

Comparison with U.S. Surveillance Data

The model predictions were compared with results from the EWO database described earlier. Data sets were selected which were for weld metal, with nickel contents less than or equal to 0.4% and which had values for all the variables required for prediction and had values of Charpy shift. In total this amounted to 63 sets. The Charpy shift data were converted to hardness change using initially the parameters found in the RR fits above. This however resulted in significant overprediction. The predictions were recalculated using the Odette et al [*19*] estimate of *C* (50°C) and by refitting the value of k_{TH} to the EWO data. The latter gave a value of 1.64 (°C/VPN) (c.f. RR 2.44). In [*20*] Odette and Lucas give $\Delta T/\Delta\sigma = 0.65$ (°C/MPa), where $\Delta\sigma$ is the yield stress change, and $\Delta H/\Delta\sigma =$ 1/3.6 (VPN/MPa); i.e. $\Delta T/\Delta H = 2.34$ (°C/VPN). However in Ref.[*19*] it is noted that $\Delta T/\Delta\sigma$ is non-linear and values of 0.5 (°C/MPa) are evident for lower shifts, giving $\Delta T/\Delta H = 1.8$ (°C/MPa), only 10% greater than the fitted value. Figure 11a shows the basic model (Eqs 1-5) fit to the EWO data using a k_{TH} value of 2.44 (°C/VPN). Figure 11b shows the revised model (incorporating Eqs. 7-9) with the re-fitted k_{TH} of 1.64 (°C/VPN). The open symbols are those for which a dose rate effect would not be expected (i.e. low copper, or full precipitation under high dose rate conditions).

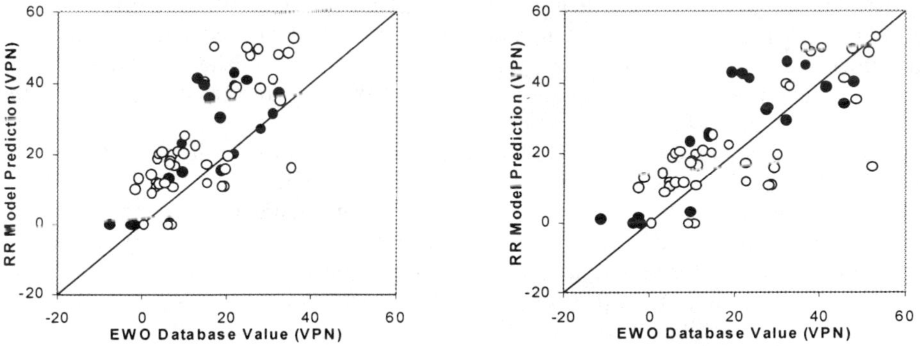

Figure 11a and b – *Calculated versus predicted hardness change for low nickel welds in the EWO database. In a) the calculated values are nominal and assume high dose rate. In b) the acceleration factor is calculated according to dose rate and the Charpy hardness relationship has been refitted in order to scale the results to the 1:1 line. The open circles are sets that are not changed by the acceleration factor.*

The above manipulation shows that there is the potential to reconcile the apparent differences in irradiation sensitivity of the low nickel welds in the two databases. It may be that these stem from differences in the mechanical property correlations rather than from differences in the underlying damage (hardening, yield stress increase) mechanisms. A candidate for the source of differences in the correlations is the heat treatment difference (quenched and tempered versus stress relieved), or, more particularly, the resulting differences in tensile properties. Odette et al [*19*] discuss the effect of unirradiated transition temperature on the yield stress shift correlation and suggest that the $\Delta T/\Delta\sigma$ is greater for higher initial values of Charpy transition temperature. There are similar indications in the RR data base. In the current study the mean start of life 41J temperature for the RR data is –23.5 °C; and that for the EWO data is –37.5 °C.

Microstructural Data

The results of the work to characterize the microstructure of irradiated materials has been used both to aid model development and, at a later stage, to underpin the model. Figure 12 shows an ECoPoSAP map of a copper rich precipitate (CRP) in weld SH. It will be noted that the precipitate contains a small amount of phosphorus which may indicate the need to include of phosphorus in the model. This may relate to work by Odette, Lucas and Klingensmith [*21*] showing that phosphorus significantly increases precipitate number density.

The ECoPoSAP results were used to help estimating the CRP compositions in order to determine volume fraction and precipitate number density from the SANS data. The values used in the current analysis were (in atomic percent): Cu 19; Mn 4.6; Ni 0.7; Si 0.7 and Fe 75. The SANS data are shown in Figures 13 a, b and c. The curves shown are fitted to the data, but in terms of form and fixed parameters are mainly speculative. The number density curve is derived mathematically from the other two.

The data are generally consistent with the model showing clearly the effect of increasing copper in producing precipitation and also the effect of reducing dose rate, although data at low dose rates are only available at lower doses. The increase of precipitate diameter with reducing dose rate was not expected and has implications to the method of accounting for dose rate effects.

The microstructural data were fitted to the Russell and Brown modulus hardening model [*22*]. This followed Odette and Lucas [*20*] and previous work [*11*]. However, in accordance with Stoller [*7*] the Russell and Brown cut off radius was reduced to the mean dislocation spacing. In addition the Russell and Brown inner cut off radius was fitted (this improved the fit for very small particles and was considered valid because particle size has limited meaning) and the Taylor Factor was used rather than the Schmidt Factor. The parameters used/fitted are shown below.

Burger's vector – 0.248 nm
Inner cut off radius (fitted) 0.575 nm
Outer cut off radius – 56 nm
CRP shear modulus (fitted) 44810 MPa
Iron shear modulus - 49000 MPa
Taylor Factor 3.06

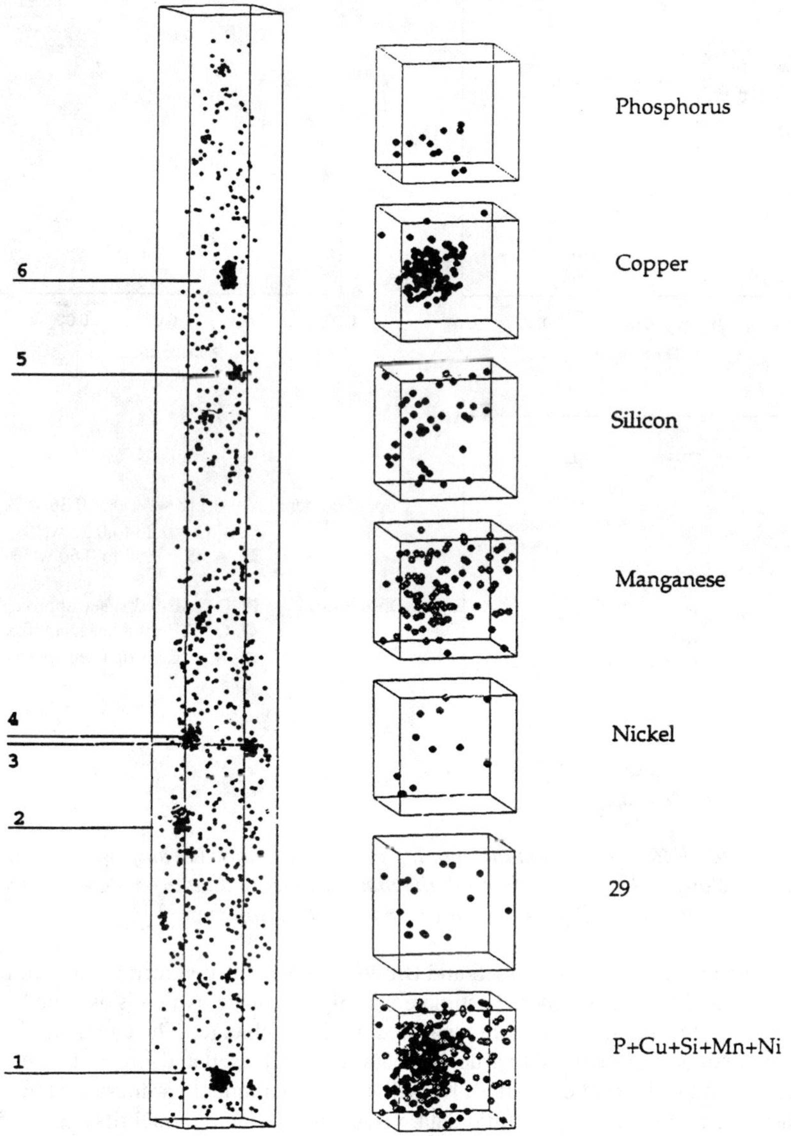

Figure 12 – *(Left) distribution of copper atoms in a 10 x 10 x 110 nm volume of weld SH, and (right) magnified view of the distribution of elements in 4 x 4 x 4 nm volume centered on cluster 4. Element 29 may be iron or (more probably) nickel.*

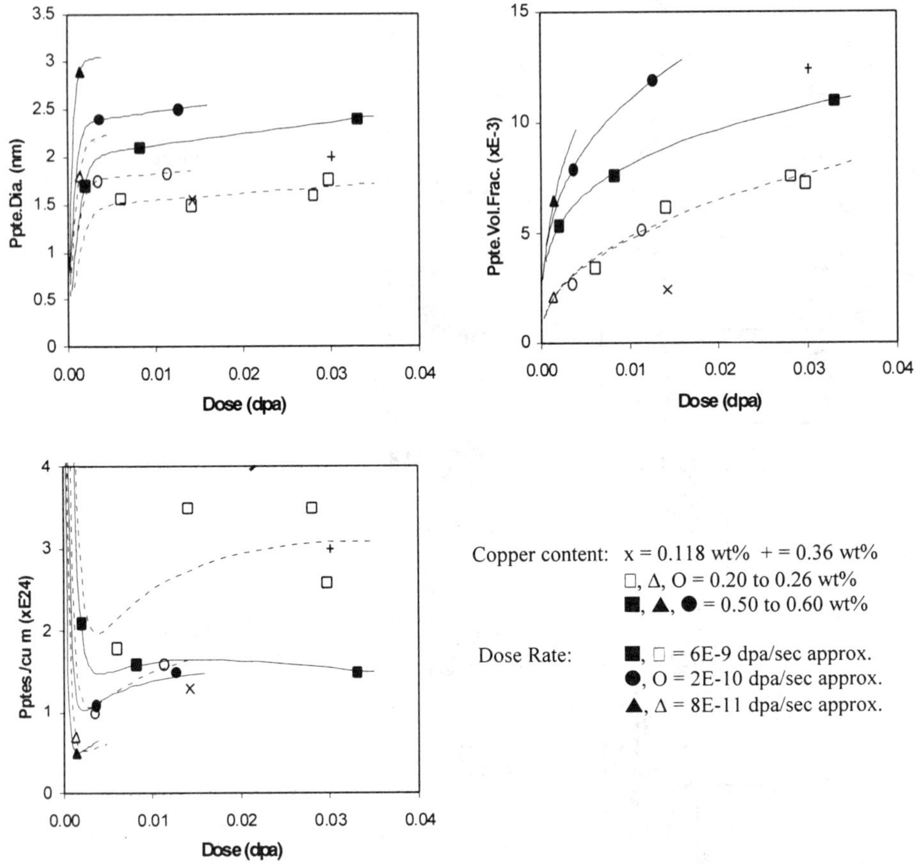

Figure 13 – *SANS data. The curves through the precipitate diameter and volume fraction data are speculative. The curves through the number density data are calculated from the precipitate diameter and volume fraction curves.*

Figure 14 shows the fit to the Russell and Brown model. The deduced values were obtained by subtracting the RR model estimate of matrix damage, which is assumed independent of dose-rate, from the total measured hardness change. The calculated values were obtained from the microstructural data via the Russell and Brown model.

Figures 15a and b show (the re-fitted) Russell and Brown model hardness change predictions, based on the microstructural data, versus the hardness model fits (Eqs. 3-5). The copper remaining in solution is also shown. Although the curves provide quantitatively similar results (which they must, since they were fitted to the same data) the curve shapes are somewhat different. There are a number of possible reasons for this. For example, the Russell and Brown model may not be wholly applicable or accurate in this application; the effective precipitate size may be mis-estimated, or the effective strength of the precipitate may change with size due to composition/modulus changes.

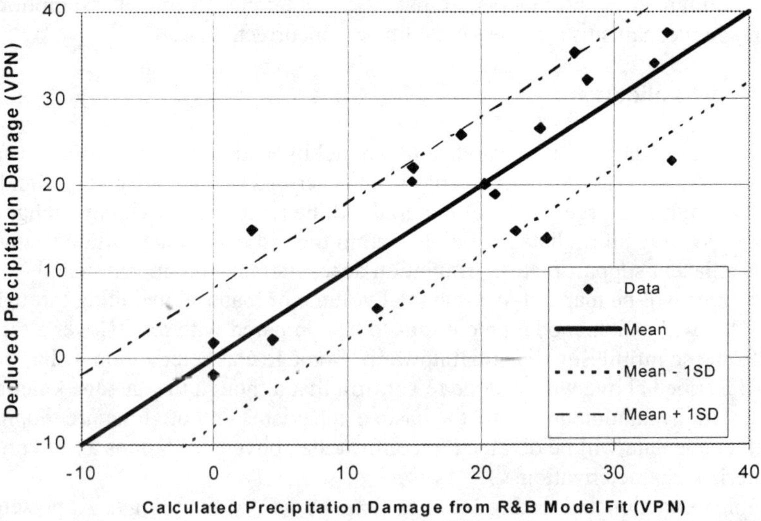

Figure 14 – *Calculated versus measured hardness change from fitting to the Russell and Brown model.*

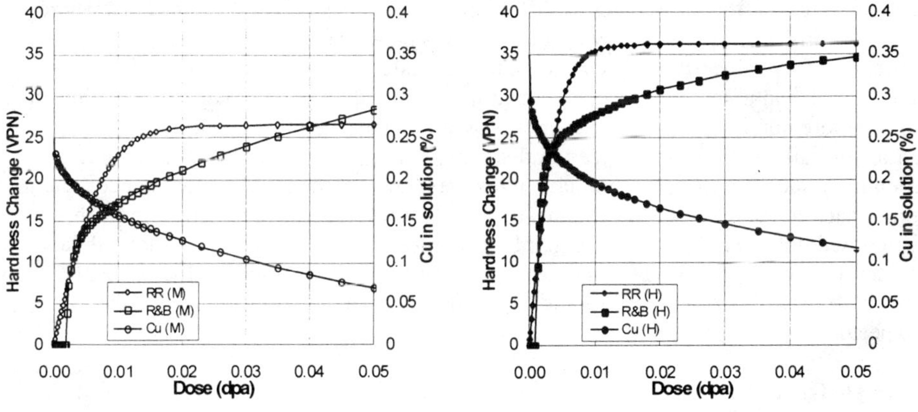

Figure 15a and b – *Comparison between the predictions for precipitation damage based on the RR model (RR) and based on microstructural data via the Russell and Brown model (R&B). The matrix copper is also shown, a) for a medium copper weld and b) for a high copper weld.*

It is evident from Figures 14 and 15 that further work is needed to converge the hardening and microstructural models. It may also be that the RR model assumption that the two damage mechanisms are linearly additive is incorrect.

Plans for Model Validation

Work is currently under way to validate the model by testing material extracted from a decommissioned RPV. The programme has been described in outline by Taylor and Williams [*23*]. In brief, large blocks of material will be removed, providing a range of doses from effectively unirradiated to values within the test data range, within a single weld and its adjacent support plates. Irradiation dose, chemical composition and hardness changes will be mapped over the total volume of material including through thickness. This will enable model predictions to be compared with reality over a range of dose rates from the infinitesimally small upward. Damage characterization, using the techniques described above will be done to confirm that although the damage kinetics may change with irradiation dose rate, the basic mechanisms do not. Fracture toughness, Charpy and tensile data will be obtained to confirm the above correlations and to provide general materials characterization.

The sample extraction and testing are to be done by AEA Technology. At present work is underway to prepare the RPV for sampling and first results are expected towards the end of 2001.

Concluding Comments

The process of developing irradiation damage models on the basis of mechanistic understanding, and the advantages of this approach, have been well described elsewhere [*14, 24*]. Confirmation is scarcely needed, but this paper has provided a further illustration of its value. It has been shown that a model, fitted to a specific limited region of data space, can potentially be used in extrapolation with a greater degree of confidence than if it were purely empirical. It cannot be claimed that the model is yet sufficiently accurate or validated. Several uncertainties remain, in particular the microstructural hardening model, the superposition law, the dependence of the shift/hardening coefficient on material and the detail of the kinetics model. However, the mechanistic approach, coupled with the availability of a range of mechanical property and microstructural data, provides a sound basis for resolving these uncertainties. These things will also be of great value for re-interpretation, should the work to validate the model provide the unexpected.

Acknowledgements

We are greatly indebted to AEA Technology, UK, who carried out or organized all the experimental work, and to their partners and colleagues in Oxford, Grenoble and Saclay. The development of the programme and the assessment of the data have greatly benefited from discussions with many individuals in particular members of the IGRDM (International group for Radiation Damage Mechanisms in RPV Steels). Finally, many thanks to Rosemary Fogg, who is responsible for the databases and fitting processes and who played a vital role in assembling the data and figures for this manuscript.

References

[1] Odette, G. R., "On the Dominant Mechanism of Irradiation Embrittlement of Reactor Pressure Vessel Steels," *Scripta Metallurgica*, Vol. 17, 1983 pp. 1183-1188.

[2] Fisher, S. B., Harbottle, J. E., and Aldridge, N. B., "Radiation Hardening in Magnox Pressure Vessel Steels," *Philosophical Transactions of the Royal Society of London*, Vol. A315, 1985, p.310.

[3] Lucas, G. E., Odette, G. R., Lombrozo, P. M., and Sheckherd, J. W., "Effects of Composition, Microstructure and Temperature on Irradiation Hardening of Pressure Vessel Steels," *Effects of Radiation on Materials: Twelfth International Symposium,ASTM STP 870*, F. A. Garner and J. S. Perrin, Eds., American Society for Testing and Materials, Philadelphia, 1985, pp. 900-930.

[4] Odette, G. R., and Lucas, G. E., "Irradiation Embrittlement of Reactor Pressure Vessel Steels: Mechanisms, Models and Data Correlations," *Proceedings of the Second IAEA* Specialists' *Meeting on Irradiation Embrittlement*, Vienna, Austria, October 1984, ASTM 1985.

[5] Fisher, S. B., and Buswell, J. T., " A Model for PWR Pressure Vessel Embrittlement," *International Journal of Pressure Vessels and Piping*, Vol 27, 1987, p.91.

[6] Buswell, J. T. and Jones, R. B., "The Modelling of Radiation Hardening and Embrittlement in Magnox Mild Steel Submerged-Arc Welds," *Effects of Radiation on Materials: 16th International Symposium, ASTM STP 1175*, A. S. Kumar, D. S. Gelles, R. K. Nanstad, and E. A. Little, Eds., American Society for Testing and Materials, Philadelphia, 1993, pp. 424-443.

[7] Stoller, R. E., "Pressure Vessel Embrittlement Predictions Based on a Composite Model of Copper Precipitation and Point Defect Clustering," *Effects of Radiation on Materials: 17th International Symposium, ASTM STP 1270*, D. S. Gelles, R. K. Nanstad, A. S. Kumar, and E. A. Little, Eds., American Society for Testing and Materials, West Conshohocken, PA, 1996, pp. 25-58.

[8] Williams, T. J., Thomas, A. F., Berrisford, R. A., Austin, M., Squires, R. L., and Venables, J. H., "The Influence of Neutron Exposure, Chemical Composition, and Metallurgical Condition on the Irradiation Shift of Reactor Pressure Vessel Steels," *Effects of Radiation on Materials: Eleventh International Symposium, ASTM STP 782*, H. R. Brager and J. S. Perrin, Eds., American Society for Testing and Materials, Philadelphia, PA, 1982, pp. 343-374.

[9] Williams, T. J., Ellis, D., Swan, D. I., McGuire, J., Whalley, S. P., English, C. A., Venables, J. H., and de la cour Ray, P. H. N., "The Influence of Copper, Nickel and Irradiation Temperature on the Irradiation Shift of Low Alloy Steels," *Proceedings of the Second International Symposium on Environmental Degradation of Materials in Nuclear Power Systems - Water Reactors*, TMS AIME/ANS/NACE, September 1985, ANS Illinois, 1986, pp. 393-399.

[10] Williams, T. J., Burch, P. R., English, C. A., and de la cour Ray, P. H. N., "The Effect of Irradiation Dose Rate and Temperature, and Copper and Nickel Content, on the Irradiation Shift of Low Alloy Submerged Arc Welds," *Proceedings of the Third International Symposium on Environmental Degradation of Materials in Nuclear Power Systems - Water Reactors*, TMS/ASM/ANS/NACE, G.J. Theus, J.R. Weeks Eds., August/September 1987, Michigan, 1998, pp.121-131.

[11] Williams, T. J., and Phythian, W. J., "Electron Microscopy and Small Angle Neutron Scattering Study of Precipitation in Low Alloy Steel Submerged-Arc Welds," *Effects of Radiation on Materials, 17th International Symposium, ASTM STP 1270*, D. S. Gelles, R. K. Nanstad, A. S. Kumar, and E. A. Little, Eds., American Society for Testing of Materials, West Conshohocken, PA, 1996, pp. 191-205.

[12] Williams, T. J., Ellis, D., Hyde, J., and English, C., "Microstructural Evolution in High Nickel Submerged Arc Welds," *Effects of Radiation on Materials, 20th International Symposium, ASTM STP 1405*, S. T. Rosinski, M. L. Grossbeck, T. R. Allen, and A. S. Kumar, Eds., American Society for Testing and Materials, West Conshohocken, PA, 2002 (this volume).

[13] Britt and Leuke, "The Estimation of Parameters in Non-Linear Implicit Models," *Technometrics*, Vol. 15, No. 2, 1973.

[14] Eason, E. D., Wright, J. E., and Odette, G. R., "Improved Embrittlement Correlations for Reactor Pressure Vessel Steels," NUREG/CR-6551, MCS 970501, November 1998.

[15] Bolton, C. J., Buswell, J. T., Jones, R. B., Moskovic, R., and Priest, R. H., "The Modelling of Irradiation Embrittlement in Submerged-Arc Welds," *Effects of Radiation on Materials, 17th International Symposium, ASTM STP 1270*, D. S. Gelles, R. K. Nanstad, A. S. Kumar, and E. A. Little, Eds., American Society for Testing and Materials, West Conshohocken, PA, 1996, pp. 103-118.

[16] Jones, R. B., and Williams, T. J., "The Dependence of Radiation Hardening and Embrittlement on Irradiation Temperature," *Effects of Radiation on Materials, 17th International Symposium, ASTM STP 1270*, D. S. Gelles, R. K. Nanstad, A. S. Kumar, and E. A. Little, Eds., American Society for Testing and Materials, West Conshohocken, PA, 1996, pp. 569-590.

[17] Odette, G. R., Lucas, G. E., and Klingensmith, D., "Anomalous Hardening in Model Alloys and Steels Thermally Aged at 290°C and 350°C: Implications to Low Flux Irradiation Embrittlement," *Effects of Irradiation on Materials:18th International Symposium, ASTM STP 1325*, R. K. Nanstad, M. L. Hamilton, F. A. Garner and A. S. Kumar, Eds., American Society for Testing and Materials, West Conshohocken, PA, 1999, pp. 88-101.

[18] Stoller, R. E., "Modelling the Influence of Irradiation Temperature and Displacement Rate on Hardening Due to Point Defect Clusters in Ferritic Steels," *Effects of Radiation on Materials, 16th International Symposium, ASTM STP 1175*, A. S. Kumar, D. S. Gelles, R. K. Nanstad, and E. A. Little, Eds., American Society for Testing and Materials, Philadelphia, PA, 1994, pp. 394-423.

[19] Odette, G. R., Lombrozo, P. M., Wullaert, R. A., "Relationship Between Irradiation Hardening and Embrittlement of Pressure Vessel Steels," *Effects of Radiation on Materials: Twelfth International Symposium, Twelfth International Symposium ASTM STP 870*, F. A. Garner and J. S. Perrin, Eds., American Society for Testing and Materials, Philadelphia, 1985, pp. 840-860.

[20] Odette, G. R., and Lucas, G. E., "Irradiation Embrittlement of Reactor Pressure Vessel Steels: Mechanisms, Models and Data Correlations," *Radiation Embrittlement of Nuclear Reactor Pressure Vessel Steels: An International Review (Second Volume), ASTM STP 909*, L. E. Steele, Ed., American Society for Testing and Materials, Philadelphia 1986, pp. 206-241.

[21] Odette, G. R., Lucas, G. E., and Klingensmith, R. D., "The Influence of Metallurgical Variables on the Temperature Dependence of Irradiation Hardening in Pressure Vessel Steels," *Effects of Radiation on Materials: 17th International Symposium, ASTM STP 1270*, D. S. Gelles, R. K. Nanstad, A. S. Kumar, and E. A. Little, Eds., American Society for Testing and Materials, West Conshohocken, PA, 1996, pp 606-622.

[22] Russell, K. C., and Brown, L. M., *Acta Metallurgica*, Vol. 20, 1972, pp. 969 -974.

[23] Taylor, L. G., and Williams, T. J., "Validation of Through Life Mechanical Properties of Nuclear Pressure Vessel Steels," *Lifetime Management and Evaluation of Plant, Structures and Components,* Dynamic Testing Agency's Fourth International Conference on "Engineering Structural Integrity Assessment," J. H. Edwards, P. E. J. Flewitt, B. C. Gasper, K. A. McLarty, P. Stanley and B. Tomkins, Eds., Cambridge, UK, September 1998, p 389.

[24] English, C. A., IG-RDM: "History and Purpose and Recent Technical Highlights," *IG-RDM8 Open Workshop, Fracture Toughness Master Curve for Reactor Pressure Vessel Steels*, Nashville, 1999.

Christian Brillaud,[1] Yves Grandjean,[1] and Sébastien Saillet[1]

Vessel Investigation Program of "CHOOZ A" PWR Reactor after Shutdown

---

Reference: Brillaud, C., Grandjean, Y., Saillet, S., **"Vessel Investigation Program of "CHOOZ A" PWR Reactor after Shutdown,"** *Effects of Radiation on Materials: 20th International Symposium, ASTM STP 1405,* S. T. Rosinski, M. L. Grossbeck, T. R. Allen, and A. S. Kumar, Eds., American Society for Testing and Materials, West Conshohocken, PA, 2001.

Abstract: In October 1991 the CHOOZ A PWR plant, operated by the French and Belgian company SENA, was shut down after 24 years. It was the first PWR shutdown in France. According to the EDF life duration project it was decided to perform an examination program on samples taken from the wall of the vessel in order to complete the feedback.

The results obtained show the following:

- The embrittlement measured by the pressure vessel surveillance program is representative of that of the Shell determined at different axial and angular locations for the ¼ of the internal thickness. This observation is consistent with the relative homogeneity of the embrittling elements (Cu, P, Ni) throughout the entire Shell,
- The mechanical properties evolution through the wall thickness of the Shell measured after irradiation is interpreted in relation with fluence variation and unirradiated mechanical properties profile through the thickness,
- The comparison of direct toughness measurement with the French code RCC-M evaluation confirms the safety nature of the conventional approach. For the weld, after an irradiation damage of 2 10^{19} $n.cm^{-2}$ (E>1MeV), a margin of the order of the irradiation effect seems to exist.

Keywords: fracture toughness, Charpy, transition, reactor pressure vessel steel, reference temperature, master curve, irradiation, reactor shutdown

---

[1] Engineers, Electricité de France, Industry Branch, Nuclear Production Division, Corporate Laboratories, GDL Chinon, B.P. 23, Avoine, France.

Introduction

In October 1991 the CHOOZ A PWR plant, operated by the French and Belgian company SENA, was shut down after 24 years. It was the first PWR shutdown in France. According to the EDF life duration project it was decided to perform an examination program on samples taken from the wall of the vessel in order to complete the feedback.

The first objective of the program was to confirm the validity of the irradiation embrittlement monitoring procedure followed in the Pressure Vessel Surveillance Program (PVSP).

The next part of this program was aimed at collecting additional information on material behavior under irradiation.

Two objectives were focused on :

- evaluation of the embrittlement of the material through the thickness and interpretation of the results and,
- toughness tests on the weld metal of joint B/C in order to evaluate the procedure employed in the French RCC-M code. This procedure is based on the use of a reference crack initiation toughness curve indexed on the RT_{NDT} parameter (K_{1C}=f(T-RT_{NDT})).

Chooz A Reactor Vessel and Materials

Chooz A operating conditions

Chooz A was the first 4-loop PWR plant built in France. It was operated from 1967 to 1991 by the Franco-Belgian company SENA. Its net power rating on commissioning was 242 MWe and this was subsequently raised to 305 MWe. From 1967 to 1975, the irradiation temperature was 255 °C (cold leg temperature) and, after the increase of power, was raised to 265 °C until shutdown in 1991. The temperature is about 25°C lower than in the 900 PWR units presently operated by Electricité de France.

Manufacture of the CHOOZ A reactor vessel

The reactor vessel was manufactured by Société des Forges et Ateliers du Creusot (SFAC) from grade A 336 (1.2 MD07) low alloy steel forged Shells. Some characteristics of the Shell manufacture are given in Table 1, where they are compared to the CP1 and CP2 series 900 MWe reactor vessels. In spite of certain differences, these parts are generally quite comparable.

The B/C joint was fabricated by the submerged arc weld process.

Chooz A Materials

Base Metal - Table 2 gives heat chemical analysis of Chooz A Shells C and B, where they are compared to the heat chemical specifications for the first five CP1 900 MWe french units. For this standpoint, and in spite of certain differences (phosphorus of Shell B slightly higher than 900 MWe units), chemical composition of Chooz A Shells B and C are quite comparable to the first five 900 MWe EDF units.

Table 1 - *CHOOZ A manufacture Shells characteristics - Comparison with EDF operating 900 MWe units.*

		Ingot						Shell		
		Weight	Diameter	Crop				Thickness	Diameter	Length
		(t)	mean (mm)	Top (%)	Bottom (%)	Bore (mm)		(mm[1])	inner (mm)	(mm)
CHOOZ A	C	106	2150	20	10	820		205	3206	2245
	B	113.5	2150	20	10	820		230	3206	2430
900 MWe	C	129.8	2400	17	10	820		200	3988	2552

[1] thickness before machining

Table 2 - *Chooz A heat chemical analysis for Shells C and B. Comparison with the specifications for the first five CP1 EDF operating 900 MWe unit (results are in wt %).*

	C	S	P	Si	Cr	Mo	Mn	Ni	Cu
Chooz A Shell C	0.165	0.009	0.010	0.30	0.16	0.38	1.26	0.61	0.10
Chooz A Shell B	0.163	0.012	0.015	0.29	0.14	0.42	1.22	0.59	0.08
900 MWe	≤0.20	≤0.015	≤0.010	0.15-0.35	≤0.30	0.45-0.60	1.2-1.5	0.40-0.80	≤0.010

Material of Chooz A Shells C and B was austenitized 1 hour at 910°C before quenching then was tempered 4 hours at 660°C- 670°C. Total stress relief heat treatment is equivalent to 32 hours at 600°C for Shell B and 26 hours at 600°C for Shell C.

Weld Metal - Available data for weld metal of B/C joint are given in Table 3, where they are compared to the chemical specifications for the EDF 900 MWe units. For available data, chemical composition of Chooz A B/C welded joint is comparable to the 900 MWe units. Because the lack of information about nickel and copper contents, it is not possible to give a trend for the behavior of B/C joint material under irradiation.

Table 3 - *Chooz A chemical analysis for welded joint B/C. Comparison with the specifications for the EDF operating 900 MWe units (results are in wt %).*

	C	S	P	Si	Cr	Mo	Mn	Ni	Cu
Chooz A B/C	0.073	0.012	0.016	0.036	NA	0.44	1.27	NA	NA
900 MWe	≤0.100	≤0.025	≤0.015	0.15-0.60	≤0.30	0.35-0.65	0.80-1.80	≤1.20	≤0.10

The stress relief heat treatment of B/C welded joint is about 7 hours at 600°C.

Chooz A Pressure Vessel Surveillance Program (PVSP) results

Base Metal - For Shell C, specimens used for the PVSP were machined from the acceptance ring, coming from the bottom of the Shell, at the ¼ internal thickness.
For Shell B, specimens were reconstituted with material cut at a tube mouth and irradiated in PVSP capsules. Base metal material was coming from broken half Charpy V HAZ specimens (the portion away from the HAZ was used). Orientation of the specimens is LT. Transition temperature shifts are evaluated for a level of absorbed energy of 56 Joules. The simulated stress relief heat treatment of the PVSP specimens is 24 hours at 600°C. Results for fluence, temperature transition shifts (ΔTT) and upper shelf energy (USE) drop (when available) are given in Table 4.

On the basis of PVSP results, formula giving the temperature transition shift evolution versus fluence was established for Shell C:

$$\Delta TT\ (°C) = 48.39\ (\phi/10^{19})^{0.48} \qquad (1)$$

where ΔTT is the temperature transition shift at 56 J level of energy and ϕ the fluence ($n.cm^{-2}$) of neutrons (E>1MeV). The exponent of this formula (0.48) is greater than the one of the FIS formulae (0.35 [2]) used to evaluate the temperature transition shift in the French RCC-M code. The main reason is the lower irradiation temperature in CHOOZ A compared to the 900 CPY units which probably raised the embrittlement kinetics.

Weld Metal - The wire/flux combination used to fabricate the PVSP welded joint and the B/C joint are not the same. For this reason, results of the PVSP were not directly exploited in this program. Nevertheless, the unirradiated RT_{NDT} of the B/C joint was estimated to be -20 °C.

Table 4 - *Pressure vessel surveillance program results for Chooz A Shells C and B base metal. Due to a reduced number of specimens, it was not always possible to determine the USE drop. These cases are labeled with NA (not available) in the table.*

Material	Fluence (E>1MeV, $n.cm^{-2}$)	ΔTT (°C)	USE drop (J)
Shell C	$0.54\ 10^{19}$	39	2
	$1.21\ 10^{19}$	48	3
	$2.27\ 10^{19}$	66	2
	$3.96\ 10^{19}$	106	7
	$6.13\ 10^{19}$	113	NA
	$6.94\ 10^{19}$	120	6
	$10.84\ 10^{19}$	156	7.6
	$11.61\ 10^{19}$	151	9
Shell B	$2.14\ 10^{19}$	53	NA
	$3.24\ 10^{19}$	58	NA
	$4.94\ 10^{19}$	88	NA
	$6.74\ 10^{19}$	105	NA
	$11.30\ 10^{19}$	148	NA

Chooz A Vessel Sampling and Testing

Ten cylindrical-type trepans of 120 mm diameter were extracted from the full vessel wall thickness (175 mm) in the irradiated core zone (beltline level). The program has been performed on materials from Shells C and B and from the B/C welded joint. Different angular locations were considered. Due to the geometry of the core, the fluence damage for these trepans may be different. Fig. 1 gives the different locations in the vessel where 6 of the ten trepans were extracted. The test program includes chemical analysis, dosimetry, Charpy impact tests, tensile tests and fracture toughness tests. For the latter, the CT12.5 geometry was selected.

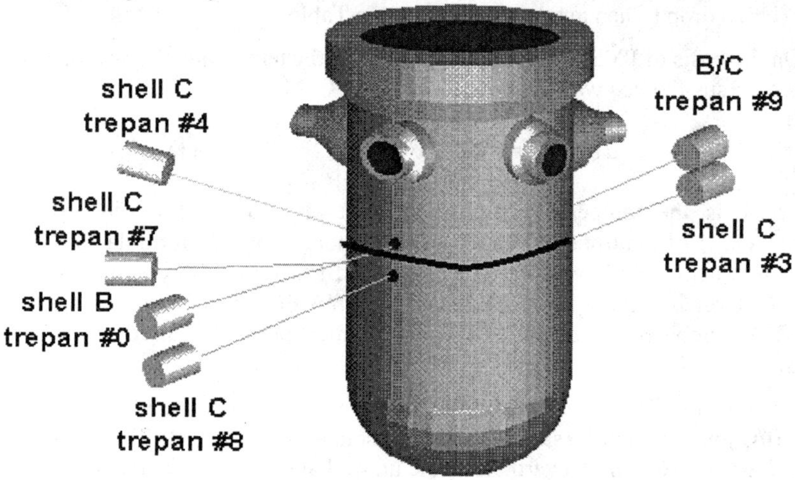

Figure 1 - *Location and identification of the trepans extracted from the Chooz A vessel.*

Table 5 gives the details of the program, including the trepans identification and location (axial and angular) and the type of the specimens sampled.

Table 5 - *Chooz A vessel sampling (KCV is Charpy V notch specimen).*

Trepan Id. #	Material	Location Axial	Location Azimuth	Specimen	Program
0	Shell B	top	135 °	KCV	validation of PVSP
3	Shell C	mid-height	270 °	KCV	validation of PVSP
4	Shell C	mid-height	180 °	KCV	embrittlement through thickness
7	Shell C	mid-height	135°	KCV	validation of PVSP
8	Shell C	top	90°	KCV	validation of PVSP
9	B/C joint	-	270°	CT12.5 tensile	fracture toughness determination

The impact tests were performed on test machines (300 J and 360 J) with an instrumented hammer equipped with an ISO tup. Fracture toughness tests were performed on CT12.5 specimens that were precracked at ambient temperature and 20% side-grooved prior to testing. The final K_{max} during precracking was maintained below 25MPa√m. The tests were conducted under displacement control at a constant crosshead speed of 0.2 mm/min. The specimen notch was machined to allow fixation of the clip gage to measure crack mouth opening in the load line. The applied load and the crack mouth opening displacement were continuously recorded up to fracture.

The fluence integrated by the specimens was determined by the Commissariat à l'Energie Atomique (CEA) based on the activity measured using ^{54}Mn isotope on the specimen themselves (self dosimetry).

Copper, phosphorus and nickel contents were determinate by optical emission spectrometry in an inductively coupled plasma (ICP).

Results

Chemical Analysis

For base metal, Shell C, measurement results of Cu, P, and Ni through the thickness and at different axial locations (top, mid-height and bottom of the Shell) are presented in Figure 2. For top and mid-height locations, samples were machined in the trepans (respectively #8 and #4). For bottom location, the results were obtained from PVSP specimens.

No significant variation of these embrittling elements can be seen. Moreover, we can also notice on Figure 2 that the values measured on PVSP specimen (cross symbol), coming from the bottom of the Shell, are close to those obtained at the top and mid-height location throughout the Shell (diamond and square symbols).

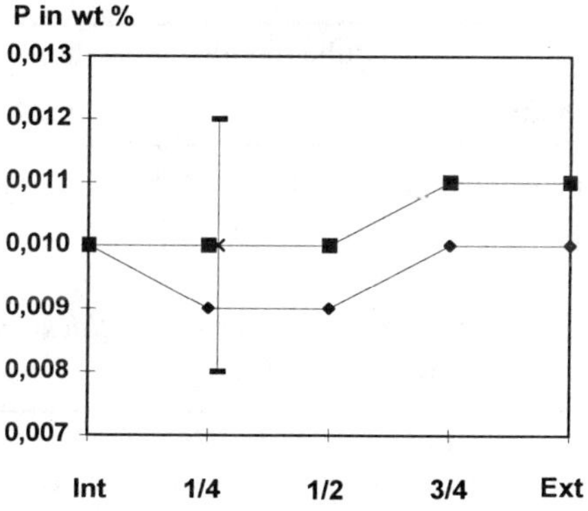

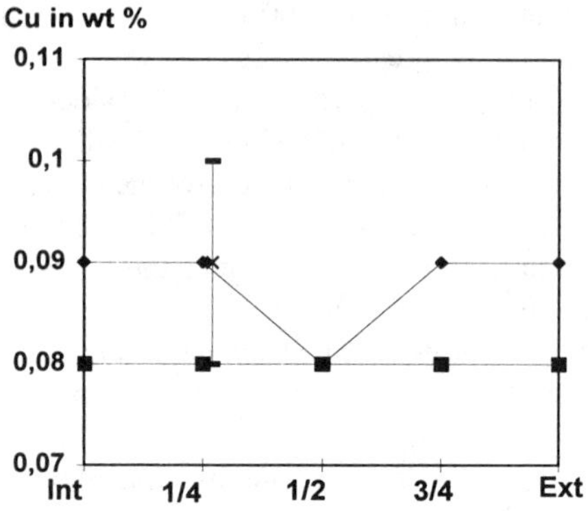

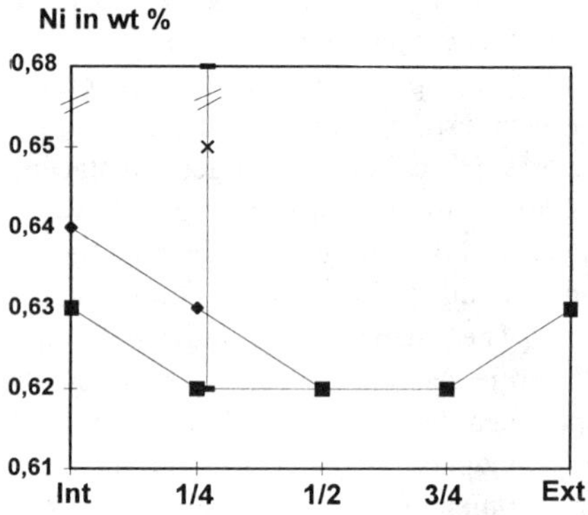

Figure 2 - *Shell C base metal: Phosphorus, Copper and Nickel content profile through the thickness at three axial locations: top (diamond - trepan #8), mid-height (square - trepan #4) and bottom (cross - PVSP specimen). Uncertainty is represented by a bar on the cross symbol (int : internal surface, ext : external surface).*

For Shell B, the same conclusion can be made, when values of Cu, P and Ni measured on PVSP specimen and at the top of the Shell are compared (Table 6).

Table 6 - *Variation of Cu, P and Ni contents over the height of the CHOOZ A reactor B Shell. All measurements were made at the ¼ internal thickness.*

Material	Analysis Origin	Analysis Position (1/4 int. thickness)	Cu wt %	P wt %	Ni wt %
Shell B	PVSP	tube mouth	0.087	0.017	0.64
	trepan #0	top	0.07	0.015	0.61

For welded joint B/C, measurements were made at different thickness on the trepan #9. The values, reported in Table 7, do not show significant variation of Cu, P and Ni elements.

Table 7 - *Variation of Cu, P and Ni contents across the thickness of the CHOOZ A B/C welded joint (trepan #9).*

Analysis position distance from internal surface	Cu wt %	P wt %	Ni wt %
17 mm	0.08	0.014	0.15
33 mm	0.09	0.014	0.21
49 mm	0.09	0.014	0.13

Charpy Impact Data - Part I - Measurement at ¼ Internal Thickness From trepans

The Charpy V specimens from the trepans were sampled in the same orientation as the PVSP (LT orientation). For Shell C, three angular locations (90° #8, 270° #3, 135° #7) and two axial locations (top # 8 and mid-height #3 & #7) are considered.

The transition temperature shifts (ΔTT) measured at 56 J and the drop of the upper shelf energy (USE) level are given in Table 8. These shifts are relative to the results before irradiation in the pressure vessel surveillance program.

Table 8 - *Shells C and B - Transition Temperature Shifts (ΔTT measured at 56 J) and Drop of Upper Shelf Energy (USE) Level Evaluated on Irradiated Material at ¼ Internal Thickness.*

Shell	Azimuth	Axial Position (1/4 int. thick.)	Fluence (E>1MeV, $n.cm^{-2}$)	ΔTT (°C)	USE Drop (J)
B	90 °	top #0	$1.74\ 10^{19}$	43	not evaluated
	90 °	top #8	$2.11\ 10^{19}$	51	24
C	270 °	mid-height #3	$1.94\ 10^{19}$	41	20.8
	135 °	mid-height #7	$2.09\ 10^{19}$	52	3.2

The variation in fluence damage may be due to the different locations (axial and angular) of the sampling in relation with the core geometry. At a quarter thickness from the inside, and for the different locations considered, the transition temperature shifts measured on the base metal of Shell C are very similar. Moreover, the degrees of embrittlement measured on Shells B and C are practically identical.

Charpy Impact Data - Part II - Measurements Through the Thickness

Charpy V transition curves were determined at mid-height location at five positions through the thickness of Shell C (trepan #4). Orientation of the specimens is TS to avoid fluence variation along the notch root. The impact transition temperatures measured at 56 J are given in Table 9.

Table 9 - *Variation of transition temperature (TT at 56 J) through the thickness of Shell C measured on irradiated material (trepan #4, TS orientation).*

Position	Internal	¼ Thick.	½ Thick.	¾ Thick.	External
TT (°C)	36	14	31	6	-9

Dosimetry

The fluence integrated by the specimens through the thickness was evaluated by the Commissariat à l'Energie Atomique (CEA) based on the activity measured using ^{54}Mn isotope on the specimen themselves (self dosimetry). The fitting curve of the individual results is shown on Figure 3.

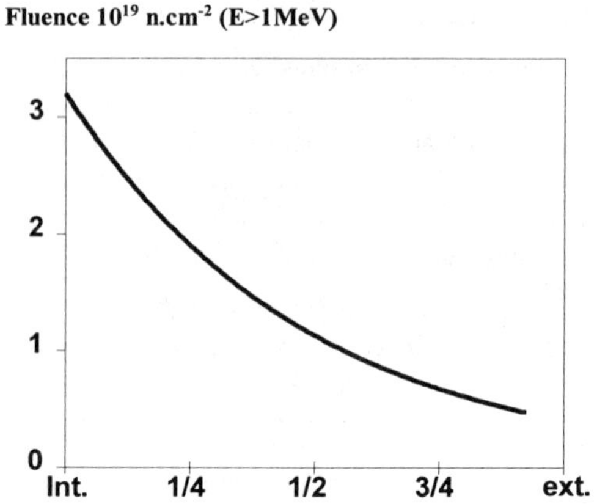

Figure 3 - *Neutron fluence profile through the wall thickness (E>1 MeV, in n.cm^{-2}) - int.: internal surface, ext.: external surface (total thickness = 175 mm).*

The evolution of fluence across the thickness of the Shell C is given by the formula:

$$\Phi_x = \Phi_{surface} \cdot e^{-\alpha x} \qquad (2)$$

where Φ_x is the fluence at x mm from the internal surface, $\Phi_{surface}$ the fluence at the internal surface, x the distance in mm from the internal surface and α a factor given by the fitting curve.

Fracture Toughness Tests on the Weld Metal of Joint B/C (trepan #9)

The toughness tests were carried out on 12.5 mm thick CT specimens (trepan #9). Different testing temperatures were chosen to have a description of the lower part of the transition curve. 12 specimens were tested at 4 temperatures (2 at -100°C, 2 at -50°C, 2 at -25°C and 6 at 0°C). Because the specimens were machined at different thickness, the fluence is ranging from 1.90 10^{19}n.cm^{-2} to 3.00 10^{19}n.cm^{-2}.

The analysis of the fracture toughness data was performed according to E1921 standard and future development (multitemperature determination). The data were normalized to 1T size and the reference temperature, T_0, was determined. Two specimens show a ductile stable growth of the crack before unstable initiation. For one specimen, the Kjc value exceeded the Kjc_{limit} and the censoring procedure of the standard was applied (specimen tested at 0°C). For comparison, T_0 determination was also made by excluding this data.

Different sets of data were used to determine the T_0 parameter :

1) all data (12) and multitemperature determination,
2) all data excepted the specimen with Kjc>Kjc_{limit} (11),
3) six data at 0°C with Kjc values around 100 MPa.√m,
4) six data at 0°C excepted the specimen with Kjc>Kjc_{limit},
5) six data below 0 °C at -25 °C, -50 °C and -100 °C,

6) six data with the higher fluence : 2.45 and 3.00 10^{19} $n.cm^{-2}$ and,
7) six data with the lower fluence : 1.52 and 1.95 10^{19} $n.cm^{-2}$.

The T_0 values are summarized in Table 10. For each set of data, the T_0 values are in very good agreement.

Table 10 - *Determination of T_0 Parameter With the Different Set of Data for Irradiated B/C Welded Joint Material.*

Data Set	Number of Specimen	T_0 (°C)	Remark
all data	12	-19	"reference" T_0
all data excepted UAD98	11	-14	evaluation of "censoring" procedure with 12 specimens
specimens tested at 0°C	6	-21	evaluation of testing temperature and number of specimen on T_0 determination
specimens tested at 0°C excepted UAD98	5	-10	evaluation of "censoring" procedure with 6 specimens
specimens tested below 0°C	6	-17	evaluation of testing temperature and number of specimen on T_0 determination
specimens with the higher fluence	6	-18	evaluation of fluence on T_0 determination
specimens with the lower fluence	6	-20	evaluation of fluence on T_0 determination

The censoring procedure seems to have a low effect on T_0 results depending on the number of specimens. If we consider the uncertainty given in the ASTM standard for 12 specimens for a confidence level of 95%, the scale covered by the "reference" T_0 value is -19 °C ± 14 °C. In this case, the 95 % confidence level seems to cover the effect of the testing temperature, the variation of fluence, the censoring procedure and the number of specimen on T_0 determination

Discussion

The first objective of the program was to confirm the validity of the irradiation embrittlement monitoring procedure followed in the Pressure Vessel Surveillance Program (PVSP). For this, we evaluated the variability in the embrittling elements, copper, phosphorus and nickel throughout the Shells and the welded joint. The results show that, for products with low level of embrittling residual elements, as is the case for the French reactor vessel materials, their concentration does not vary significantly as a function of elevation (for Shell) and through the thickness (for Shell and welded joint). This result confirm that the PVSP specimens are representative with regard to the element content (Cu, P, Ni) involved in irradiation damage [1].

The approach followed to confirm this conclusion was to compare the PVSP transition temperature shift (ΔTT) with the values obtained on specimens machined in the trepans at the ¼ internal thickness. The embrittlement results obtained on samples taken from each Shell (trepan) are compared to those deduced from the pressure vessel surveillance program for the same irradiation fluence in Fig. 4. All specimens were machined at the ¼ internal thickness and have the same LT orientation. For a dose of 1.74 x 10^{19} $n.cm^{-2}$ (E>1MeV), the embrittlement obtained in the monitoring program is the same as that measured on the sample taken from ring B : 43°C. The embrittlement of ring C is found to be about 20°C less than in the monitoring program.

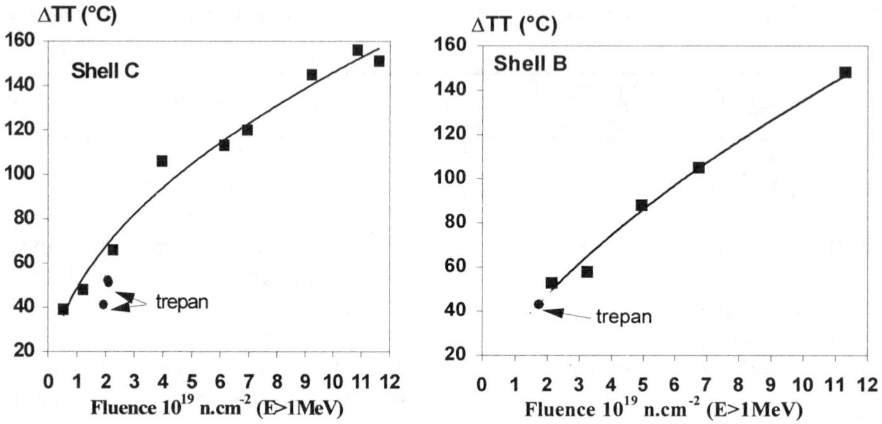

Figure 4 - *Base metal of Shells C and B - Comparison of transition temperature shifts (ΔTT at 56 J) obtained on samples taken from each Shell (trepan, round symbol) with those deduced from the pressure vessel surveillance program (square symbols). All specimens are coming from ¼ internal thickness with the same orientation (LT).*

These results confirm that the embrittlement measured by the PVSP is representative of that of the Shell determined at different axial and angular locations for the ¼ internal thickness. This observation is consistent with the relative homogeneity of the embrittling elements (Cu, P, Ni) throughout the entire Shell.

The next part of this program was aimed at collecting additional informations on material behavior under irradiation. Two objectives were focused on :

- evaluation of the embrittlement of the material through the thickness and interpretation of the results and,
- toughness tests on the weld metal of joint B/C in order to evaluate the procedure employed in the French RCC-M code. This procedure is based on the use of a reference crack initiation toughness curve indexed on the RT_{NDT} parameter (K_{IC}=f(T-RT_{NDT})).

To interpret the mechanical properties variations across the thickness with regard to the fluence values, the knowledge of the unirradiated impact transition temperature profile of Shell C is required. To achieve this task, mechanical test results obtained on a Shell used to qualify the process of elaboration of CHOOZ A Shells (same grade - same manufacture) were used. For these Shell, Charpy V results were available through the thickness and were interpreted in term of transition temperature at 56 J. This profile is then translated to fit the C Shell PVSP unirradiated transition temperature (TT) result measured at ¼ internal thickness to generate the unirradiated transition temperature profile through the C Shell. To calculate the irradiated profile, we added to the as-generated unirradiated profile the transition temperature shift given by the formula (1) elaborated on the basis of C Shell base metal PVSP results. The fluence variation through the thickness, involved in (1), is given by the formula (2) evaluated on the basis of self dosimetry measurement. The Fig. 5 summarizes this work and allows to compare the calculated irradiated profile with the individual transition temperature results obtained with specimens sampled in the trepan (#4).

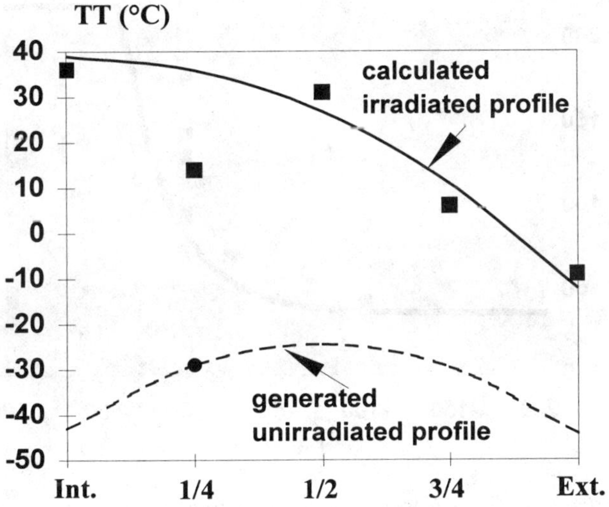

Figure 5 - *Transition temperature (TT °C) evolution through the thickness. Comparison of experimental irradiated results (square symbol trepan#4) with the calculated irradiated profile (continuous line). The experimental unirradiated results at ¼ internal thickness (round symbol) is used to generate unirradiated profile (doted line); the irradiation effect on unirradiated profile is evaluated by (1) and (2).*

The shape of the unirradiated profile shows lower values of unirradiated transition temperature on the surface when compared to the center of the Shell. This evolution of mechanical properties through the thickness is probably associated with temperature gradient during quenching. Excepted for one value (at ¼ internal thickness), the calculated irradiated profile fit well the individual experimental irradiated results obtained with material of the trepan (#4).

Even though this approach remains qualitative to define the unirradiated profile and to calculate the irradiated profile, it allows to interpret the measured mechanical property variations through the thickness with regard to the fluence values.

Fracture toughness tests has been performed on the B/C welded joint material. The size adjusted fracture toughness data of the ½ CT specimens are shown in Fig. 6. In order to accurately evaluate the conservatism of the RCC-M reference fracture toughness curve

(K1c), it is necessary to determine the RT_{NDT} for the irradiated material. Because the material in the PVSP is not the same as the weld, it was decided to compare the toughness results for the irradiated material with the RCC-M curve based on the RT_{NDT} for the unirradiated B/C weld metal (-20 °C). This curve is shown on the figure 6 as a bold line.

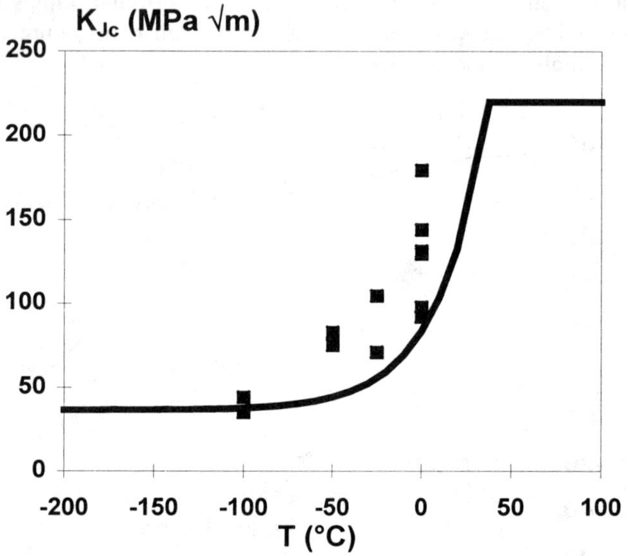

Figure 6 - *Welded joint B/C - Normalized fracture toughness results as a function of test temperature. The individual results are normalized to 1T reference size. The specimens are irradiated at fluence between 1.52 10^{19} n.cm^{-2} et 3.00 10^{19} n.cm^{-2}. The RCC-M French code reference fracture toughness curve K1c evaluated with the unirradiated RT_{NDT} (-20°C) is shown as a bold line.*

All the test results with fluence ranging between 1.52 and 3 10^{19} n.cm^{-2} are above the K_{IC} RCC-M curve evaluated with the unirradiated RT_{NDT}. This result indicates that a margin of the order of the irradiation effect on weld material seems to exist [1].

Conclusion

An extensive investigation program has been carried out on the materials from the CHOOZ A reactor vessel, shutdown after 24 years of operation. The results of the program presented in this paper give the following conclusions:

- There is no major difference in embrittlement elements composition (copper, phosphorus, nickel) between samples from various axial (top, bottom, mid-height), angular and thickness locations.
- For the ¼ internal thickness, the temperature transition shift measured with the PVSP is similar to the temperature transition shifts determined at different axial and angular

locations in the Shell (for an irradiation temperature of 265°C and a fluence of $\approx 2\ 10^{19}$ n.cm^{-2}).

- Mechanical properties evolution through the wall thickness of Shell C measured after irradiation is interpreted in relation with fluence variation and unirradiated mechanical properties profile through the thickness.
- For weld B/C, the reference temperature T_0 determination according to ASTM 1921 standard (with multitemperature determination) for a 95 % confidence level seems to cover the influence of test temperature, fluence variation, number of specimen within the range considered for each parameter.
- The comparison of direct toughness measurement (1/2 CT specimens) with the French code RCC-M evaluation (RT_{NDT}) confirm the safety nature of the conventional approach. For the weld B/C, after an irradiation damage of $\approx 2\ 10^{19}$ n.cm^{-2} (E>1MeV), a margin of the order of the irradiation effect seems to exist.

References

[1] Brillaud, C., Saillet, S., Grandjean, Y. and Houssin, B., "Evaluation of RT_{NDT} Toughness Determination Methodology Using a New Approach," *to be published in Pressure Vessel and Piping, Seatle, 2000.*

[2] Brillaud, C., Hedin, F. and Houssin, B., "A Comparison Between French Surveillance Program Results and Prediction of Irradiation Embrittlement," *Influence of Radiation on Material Properties: 13th International Symposium, ASTM STP 956,* F. A. Garner, C. H. Henager, Jr., and N. Igata, Eds., American Society for Testing and Materials, Philadelphia, 1987, pp. 420-447.

[3] Chaouadi, R., Fabry, A., Van Walle, E. and Van de Velde, J., "Fracture Toughness of the Ni-modified A302-B Plate of the BR3 Reactor Pressure Vessel," *Effect of Radiation on Materials: 19th International Symposium, ASTM STP 1366*, M. L. Hamilton, A.S. Kumar, S. T. Rosinski, and M. L. Grossbeck, Eds., American Society for Testing and Materials, West Conshohocken, PA, 2000, pp. 162-176.

[4] Chaouadi, R., Van Walle, E., Fabry, A., Scibetta, M., and Van de Velde, J., "Fracture Toughness Test Results of Precracked Charpy Specimens of Irradiated 73W Weld Material," *Effect of Radiation on Materials: 19th International Symposium, ASTM STP 1366*, M. L. Hamilton, A.S. Kumar, S. T. Rosinski, and M. L. Grossbeck, Eds., American Society for Testing and Materials, West Conshohocken, PA, 2000, pp. 279-294.

Shigeki Kataoka,[1] Naoto Kato,[1] Kosei Taguchi,[2] Michiyoshi Yamamoto,[3] and Yoshiaki Oka[4]

Development of Reconstitution Technology for Surveillance Specimens in Japan Power Engineering and Inspection Corporation

Reference: Kataoka, S., Kato, N., Taguchi, K., Yamamoto, M., and Oka, Y., **"Development of Reconstitution Technology for Surveillance Specimens in Japan Power Engineering and Inspection Corporation,"** *Effects of Radiation on Materials: 20th International Symposium, ASTM STP 1405,* S. T. Rosinski, M. L. Grossbeck, T. R. Allen, and A. S. Kumar, Eds., American Society for Testing and Materials, West Conshohocken, PA, 2001.

Abstract: In order to establish reconstitution technology for the surveillance specimens of RPV steels, a project was started in 1996FY. We focus on a correlation between the reduction of absorbed energy and the interaction of heat affected zone (HAZ) and plastic zone, to establish applicable reconstitution conditions for Charpy specimens. The plastic zone width was estimated from the hardness distribution of the Charpy specimens, and then the correlation between the plastic zone width and absorbed energy was obtained. Within the limits of 100J, there was no reduction of absorbed energy for the Charpy specimens reconstituted with 10 mm insert length using surface a ctivated joining method (SAJ). By comparing the length of insert material with the sum of HAZ width and plastic zone width, it is clear that the interaction causes the reduction of absorbed energy. The applicable conditions of reconstituted Charpy specimens could be assessed by comparing the insert length with HAZ width + plastic zone width.

Keywords: surveillance specimen, low alloy steels, reconstitution technology

[1]Manager, Tokyo Engineering Center, JAPEIC, Business Court Shin-Urayasu Bldg., 9-2, Mihama 1-Chome, Urayasu-Shi, Chiba 279-0011, Japan

[2]Dr. Eng., Senior Specialist, Mechanical Technology & Design Dept., Isosgo Nuclear Engineering Center, Toshiba Co., 8, Shinsugita-Cho, Isogo-Ku, Yokohama 235-8523, Japan

[3]Senior Engineer, Nuclear Plant Service Engineering Dept., Hitachi Ltd., 3-1-1, Saiwai-Cho, Hitachi-Shi, Ibaragi 317-8511, Japan

[4]Professor, Graduate School of Engineering, University of Tokyo, 7-3-1, Hongo, Bunkyo-Ku, Tokyo 113-8656, Japan

Introduction

Surveillance specimens such as tensile specimens, Charpy specimens and compact tension specimens have been installed in Reactor Pressure Vessels (RPV) to monitor the extent of irradiation embrittlement on RPV steels. The number of the surveillance specimens is insufficient in the case of long-term operation beyond the design life, although the monitoring of the irradiation embrittlement is regulated throughout the plant operational life. In order to maintain the monitoring of the irradiation embrittlement, a project was started in 1996FY entrusted by the Ministry of International Trade and Industry (MITI). The target of the project is to establish the technical standard of reconstitution technology including the reconstitution of the Charpy specimens with 10 mm insert length, which enables the conversion from LT-direction to TL-direction.

Concept of Test Scheme

It is known that the expansion of plastic deformation produced by an impact test is restricted by a hardened zone such as a weld joint, which causes the reduction of absorbed energy [*1,2*]. In case of 100-200 Nm of absorbed energy, plastic deformation would expand by 8-12 mm around the notch tip of a Charpy specimen. In this case, the reduction of absorbed energy would occur for a reconstituted specimen with 10 mm insert length because of the interaction between a heat affected zone (HAZ) at weld joint and plastic zone. Figure 1 shows the schematic image of the correlation between the interaction and reduction of absorbed energy. If the plastic zone width (W_P) + heat affected zone width (W_{HAZ}) was greater than insert length (W_L) represented as the curve (a), the reduction of absorbed energy would occur. The test program for this project was constructed based on this concept.

A HAZ width would be constant if a welding method and welding conditions were fixed. The welding method and conditions, which can minimize the HAZ width, should be selected. On the other hand, a plastic zone width depends on fracture toughness or impact test temperature. The degree of the interaction between the HAZ and plastic zone depends on an insert length, fracture toughness, and test temperature. In order to identify the proper combinations of these parameters to make no interaction, it is important to understand the correlation between the reduction of absorbed energy and the parameters.

Moreover, it is known that the toughness recovery (annealing) of irradiated RPV steels occurs at a relatively higher temperature than the irradiation temperature. However, the annealing does not occur on heat treatment at 450 °C +few minutes or 500 °C +1 minute, according to earlier work on annealing using irradiated RPV steels [*3*]. Therefore, it is possible to avoid annealing by selecting the proper welding method. Nevertheless, the possibility of the effect of annealing was to be confirmed in this project.

Considering the above issues for the reconstitution technology of the Charpy specimens, the following items were studied in this project:

1) Establishment of the correlation between absorbed energy and plastic zone width,
2) Thermal history during reconstitution, and
3) Mechanical properties of reconstituted specimens.

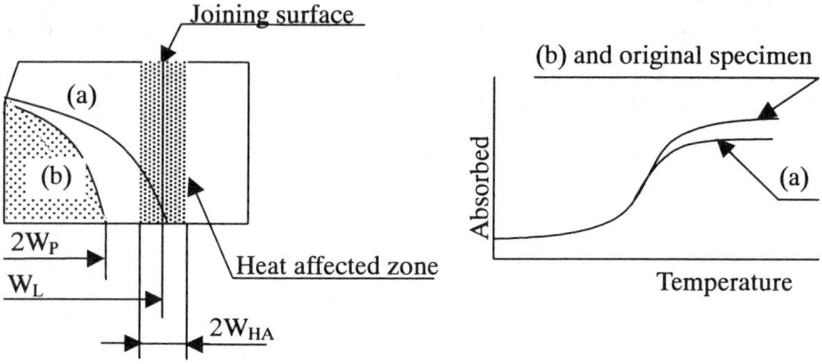

Figure 1-*Schematic description of correlation between interaction and reduction of absorbed energy*

Selection of Joining Method

The major welding methods which have been applied to the reconstitution such as arc-stud welding, friction welding, electron beam welding, laser welding and surface activated joining method (SAJ) [*4*] were studied. Among these welding methods, SAJ has been selected because of its low heat input and narrow HAZ width. SAJ can minimize the HAZ width to approximately 2 mm as shown in Figure 2.

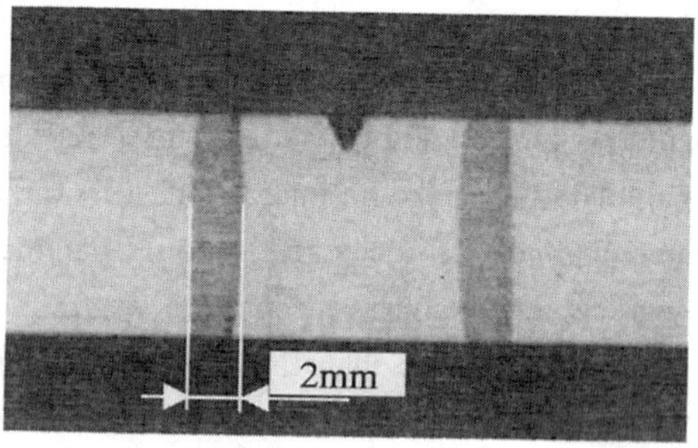

Figure 2-*Reconstituted specimen by SAJ*

Test Procedure

Test Material

Test materials used in this project are shown in Table 1. Low and high toughness RPV steels and weld metals were used which simulated material used for earlier RPV and those for later RPV respectively to cover the whole type of LWR plants in Japan.

Table 1-*Test materials*

Materials	Direction	Sy (MPa)	Su (MPa)	RT_{NDT} (°C)	USE (J)
Low toughness base metal	T	487	637	-8	82
	L	—	—	—	159
High toughness base metal	T	473	618	-40	223
Low toughness weld metal	*1	—	—	-60	195
High toughness weld metal	*1	551	632	-55	222

*1:parallel to welding direction

Measurement of Thermal History during Reconstitution

In order to evaluate the possibility of annealing due to the high temperature greater than irradiation temperature during reconstitution, the temperature profile around joining area was measured using thermo couples. Figure 3 shows the position of thermo couples.

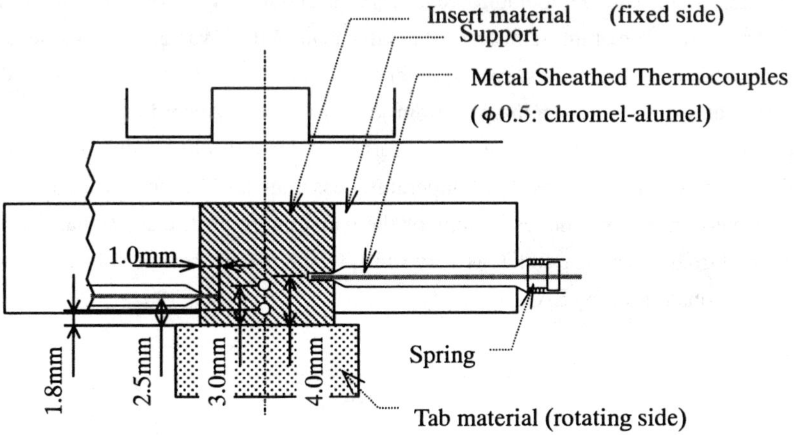

Figure 3-*Measurement of thermal history during reconstitution*

Measurement of Plastic Zone

The profile of a plastic zone produced around a notch tip due to the impact test is identified in this test. The measurement of the plastic zone was performed for the materials listed in Table 1. The profile of the plastic zone on surface along the axial direction of the specimen would be different from that on the inner section because of the different strain field around the notch tip under loading. The plane strain field would be dominant throughout the thickness of the specimen, although the plastic zone size on the surface is greater than that of the inner section. Therefore, measurement was performed on the inner section after impact test. In order to measure the hardened zone, a micro-Vickers test was used.

Impact Test of Reconstituted Specimen

Impact tests of original and reconstituted specimens were performed to determine the mechanical properties of reconstituted specimens. Impact tests were performed for the materials listed in Table 1.

Test Results

Figure 4 shows the temperature profile around the joining area during reconstitution. An insert and tab material was contacted for about 3 seconds with an initial pressing load and rotation speed. Then, the rotation was stopped and pressure increased to join both materials. The maximum temperature reached was about 500°C when the rotation was stopped. The duration of high temperature greater than 400 °C was about 2 seconds and 3 seconds for 300 °C at the nearest measurement point. The maximum temperature did not reach 400 °C at a distance 2.4 mm apart from joining surface. According to other work on the low temperature annealing of low alloy steels [*3,4*], the annealing would not occur until about 500 °C for several minutes. For temperature less than 450 °C, the annealing would not occur for several minutes. The test results of the temperature profile show that the time of high temperature greater than 400 °C is very short, which means that the annealing could be ignored for reconstitution by SAJ.

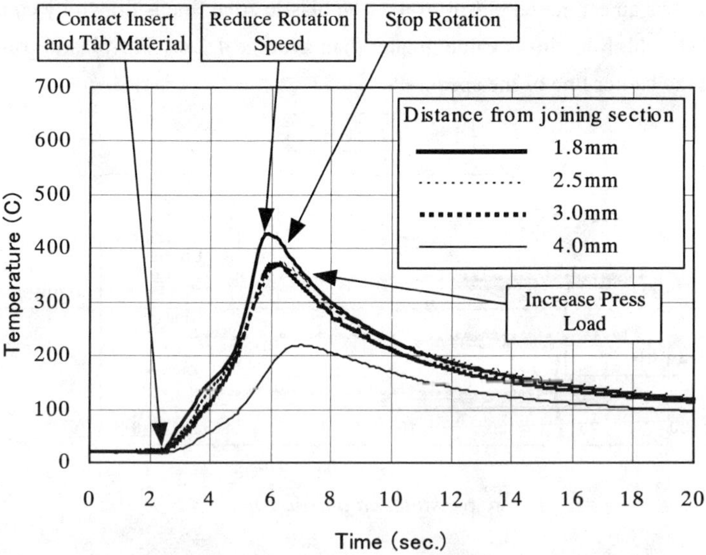

Figure 4-*Heat history around joining section*

Figure 5 shows an example of hardness distribution after impact testing. The average value of hardness at the non-hardened zone of base metal was calculated using the test points within 17.5mm from the edge of the specimens. The hardness in the non-hardened zone was scattered almost within +/- 2σ. In this test, the plastic zone is defined as the high hardness region greater than average + 2σ.

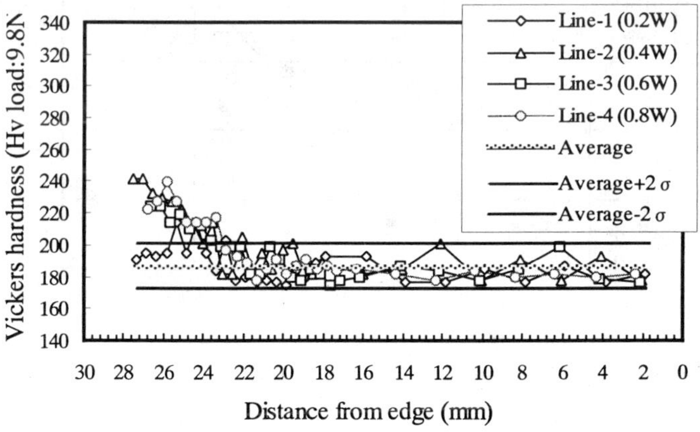

Figure 5-*Distribution of hardness*

Figure 6 shows an estimated plastic zone. The plastic zone line is drawn by connecting the point which has high hardness value greater than average + 2 σ. The plastic zone width is measured on the center line of the specimen.

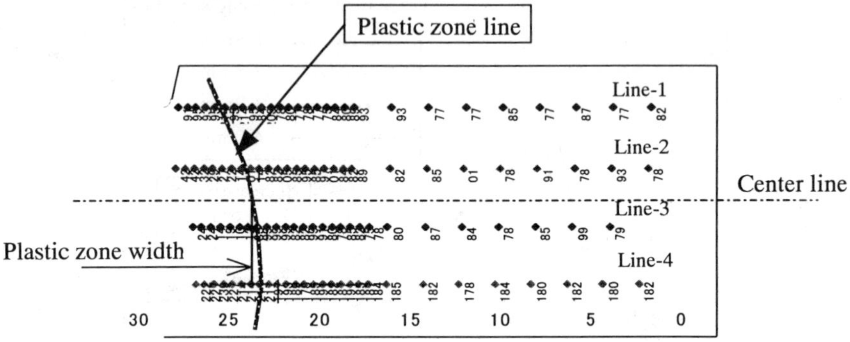

Figure 6-*Estimated plastic zone*

Figure 7 shows the correlation between absorbed energy and plastic zone width obtained by the above method. This figure includes data both from high and low toughness base metal, high and low toughness weld metal. Most data is scattered within +/- σ. This result shows that the correlation between absorbed energy and plastic zone width can be applied to the RPV base metals and weld metals as a common evaluation curve.

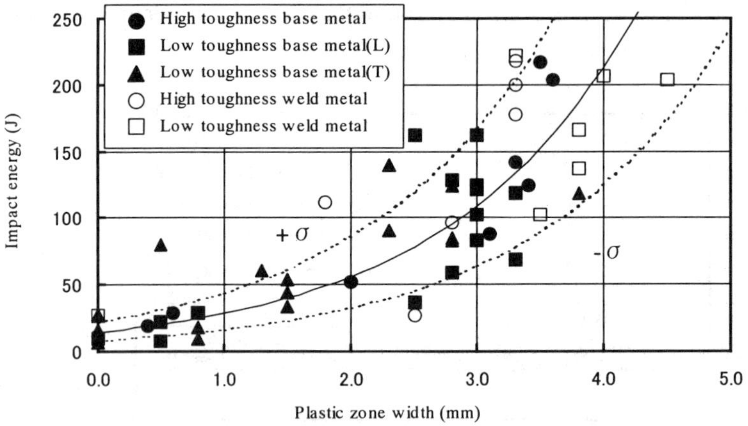

Figure 7-*Correlation between absorbed energy and plastic zone width*

Figure 8 shows the comparison of absorbed energy between original specimens and reconstituted specimens for both high and low toughness base metals. The reduction of absorbed energy occurs especially at high absorbed energy region. On the other hand, for low toughness base metal, there is no reduction of absorbed energy.

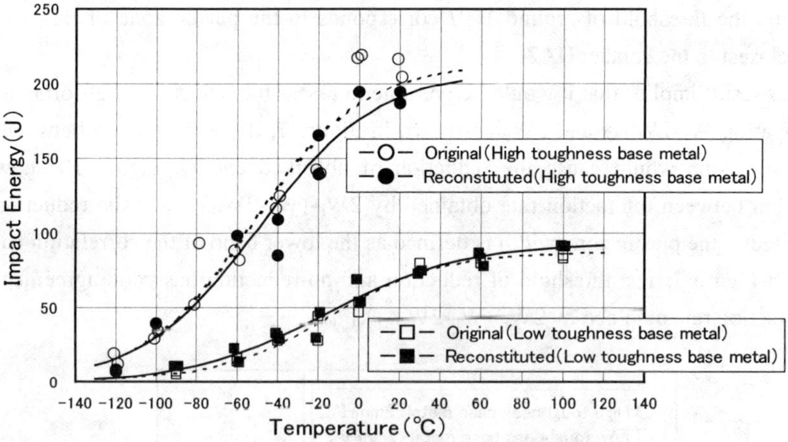

Figure 8-*Comparison of absorbed energy between original and reconstituted specimens*

In order to understand the correlation between the reduction of absorbed energy and original absorbed energy clearly, the reduction rate obtained by subtracting absorbed energy of reconstituted specimens from that of the original specimens at each test temperature which is then plotted against the absorbed energy of the original specimens (Figure 9). The reduction rate is scattered within –10~+10J for the absorbed energy range between 0J to 100J. And for the absorbed energy greater than 100J, the plots are distributed on the plus side, which means the reduction has occurred. This figure represents clearly that a threshold exists at around 100J for 10 mm insert reconstituted Charpy specimens.

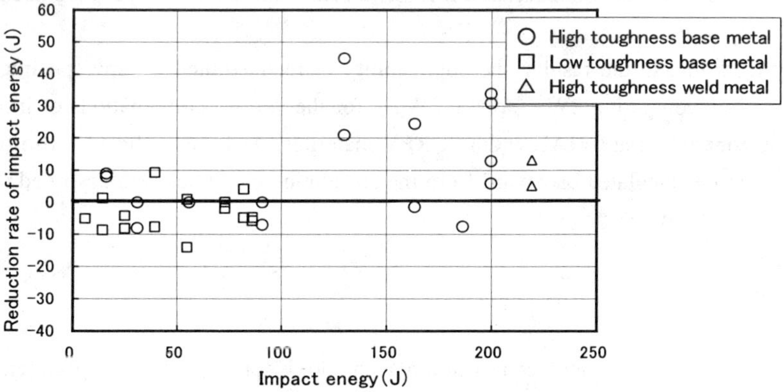

Figure 9-*Reduction rate of absorbed energy for reconstituted specimens*

The plastic zone width at around 100J absorbed energy is obtained as approximately 4 mm on one side using the lower curve of Figure 7. HAZ width is around 2 mm on one side. Therefore, the threshold of around 100J corresponds to the plastic zone of this absorbed energy closest to the contact HAZ.

This result implies that it would be possible to assess the reduction of absorbed energy by comparing W_L-$2W_{HAZ}$ and $2W_P$ as shown in Figure 1, if the interaction between HAZ and plastic zone would cause the reduction of absorbed energy. Figure 10 shows the correlation between interaction rate obtained by $2W_P$-(W_L-$2W_{HAZ}$) and the reduction rate. In this figure, the plastic zone width is defined as the lower band of the correlation curve as shown in Figure 7. The threshold of reduction and no-reduction has good agreement with the interaction rate obtained by $2W_P$-(W_L-$2W_{HAZ}$).

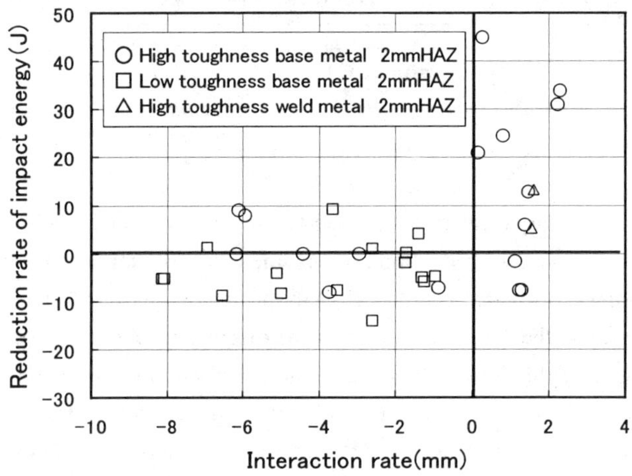

Figure 10-*Correlation between Interaction rate and reduction of absorbed energy*

According to the results, the applicability of reconstituted Charpy specimens can be assessed by comparing $2W_P$ and W_L-$2W_{HAZ}$ for the various combinations of insert length, plastic zone width and HAZ width of RPV materials. And the applicable absorbed energy range can be calculated backward from the correlation curve between absorbed energy and plastic zone width.

Conclusions

In order to establish reconstitution technology for surveillance specimens of RPV steels, basic data was obtained using un-irradiated materials as follows.

(1) The correlation between absorbed energy and plastic zone width was obtained.

(2) The possibility of annealing due to reconstitution could be neglected because of relatively low maximum temperature and short duration time during reconstitution by SAJ method, but this will have to be confirmed.

(3) The applicability of reconstitution could be assessed by comparing an insert length and HAZ width + plastic zone width.

Acknowledgments

The authors are grateful to the Agency of Natural Resources and Energy, the Ministry of International Trade and Industry, and the JAPEIC PLIM committee members for their supports.

References

[1] E. Walle, "Reconstitution : Where do we stand ?," *Effects of Radiation on Materials, 17th Vol. ASTM STP 1270*, American Society for Testing and Materials, West Conshohocken, PA, 1996, pp. 415-447

[2] "Pressure Vessel Steel Research, Belgian activities," SCK/CEN, IAEA Specialist Meeting, 1994

[3] M. Tomimatsu, S. Kawaguchi, M. Iida, "Reconstitution of Fracture Toughness Specimen for Surveillance Test," *3rd Symposium on Small Specimen Test Technique, ASTM STP 1329*, American Society for Testing and Materials, West Conshohocken, PA, 1997

[4] K. Onizawa et al., "Development of a Reconstitution Technique for Charpy Impact Specimens by Surface Activated Joining for Reactor Pressure Vessel Surveillance," *International Journal of Pressure Vessel & Piping*, 70, 1997, pp. 201-207

Pressure Vessel Steels—Master Curve Approach

Bong-Sang Lee,[1] Won-Jon Yang,[2] Moo-Young Huh,[2] Se-Hwan Chi,[1] and Jun-Hwa Hong[1]

Master Curve Characterization of Irradiation Embrittlement Using Standard and 1/3-Sized Precracked Charpy Specimens

---

Reference: Lee, B. S., Yang, W. J., Huh, M. Y., Chi, S. H., and Hong, J. H., **"Master Curve Characterization of Irradiation Embrittlement Using Standard and 1/3-Sized Precracked Charpy Specimens,"** *Effects of Radiation on Materials: 20th International Symposium, ASTM STP 1405,* S. T. Rosinski, M. L. Grossbeck, T. R. Allen, and A. S. Kumar, Eds., American Society for Testing and Materials, West Conshohocken, PA, 2001.

Abstract: Cleavage initiation fracture toughness (K_{JC}) tests have been performed in the transition temperature ranges of five different heats of reactor pressure vessel steels. The main purpose of this study is to investigate the applicability of the ASTM Standard Test Method for Determination of Reference Temperature, T_O, for Ferritic Steels in the Transition Range (E1921-97) to small specimens available in normal surveillance capsules. Precracked Charpy-type specimens with the standard size (10×10×55mm, PCVN) and 1/3-size (3.3×3.3×18.3mm, 1/3-PCVN) are mainly investigated. Results from more than 300 fracture toughness test data are presented including 120 unirradiated PCVN specimens and 100 unirradiated 1/3-PCVN specimens in addition to some CT specimen data and two sets of irradiated PCVN data. The test materials are the IAEA reference steels, so-called JRQ and JFL, and three Korean SA508-3 forging steels that were made by different refining processes. JRQ and a Korean steel, SKA-1, are also evaluated after neutron irradiation by PCVN specimens as well as Charpy impact tests. Fracture toughness data showed that the master curve and its probabilistic bound lines represent the population of the test data measured from different materials and specimens with good accuracy. However, there are still the effects of specimen geometry and size on the measured T_O values. These effects seem to be more significant in microscopically less homogeneous materials. Smaller bend type specimens may give slightly less conservative results than the larger CT type specimens. One of the important findings is that ultra small 1/3-PCVN specimens are also valuable as the standard size PCVN specimens for determining reference temperatures since the differences are rather systematic and predictable. The T_{28J} correlation, based on the PCVN data only, works well for the current data sets including irradiated materials.

---

[1] Principal Researchers and Director, Reactor Materials Dept., Korea Atomic Energy Research Institute, P.O.Box 105, Yusong, Taejon 305-600, Korea.

[2] Graduate Student and Professor, Dept. of Metallurgy Engineering, Korea University, Seoul 136-701, Korea.

Keywords: master curve, fracture toughness, irradiation embrittlement, PCVN, precracked Charpy, reference temperature, T_O

Introduction

Irradiation embrittlement of reactor pressure vessel steels has been extensively studied as an important issue for the structural integrity of nuclear power plants. In spite of the advancement in steel making processes to improve irradiation resistance, many old reactors of radiation-sensitive materials are still being operated. Therefore, from a practical point of view, the most important thing for reactor integrity is to evaluate the degree of irradiation embrittlement, quantitatively as well as accurately.

The current ASME procedure for this purpose is based on qualitative notch-impact testing since quantitative fracture mechanics testing may not be suitable for surveillance programs with a limited volume of specimens. Even though there are experimental relationships between notch toughness and fracture toughness, the fracture initiation on a microscopic scale occurs by different controlling factors in the two different tests [*1*]. Therefore, a more credible procedure for evaluating irradiation embrittlement should be based on actual fracture mechanics testing.

A new standard test method, ASTM E1921-97 for "Determination of Reference Temperature, T_O, for Ferritic Steels in the Transition Range," has been focused as an alternative for indexing irradiation embrittlement [*2*]. Precracked Charpy (PCVN) specimens may be used for this application since surveillance capsules usually contain this type of specimens. Currently, many research groups are investigating the applicability of PCVN specimens to surveillance programs. Studies on developing the code application method are also ongoing projects worldwide. These works are definitely based on a sufficient amount of test data sets.

The purpose of this paper is to present the fracture toughness test results from five different RPV steels, which are representative of old and modern commercial steels. A relatively large number of PCVN specimens have been tested in the transition temperature range following the ASTM E1921-97 procedure. Influences of the experimental factors that may affect the test results are to be discussed. These factors are specimen geometry, test temperature, side-grooving, and so on.

Experimental

Materials

Five different RPV steels are included in this study, which are four SA508-3 steel forgings and an SA533-B1 plate. The SA508-3 forgings consist of three Korean commercial steels and an IAEA reference material (JFL), which have been produced by modern steel making processes. The SA533-B1 plate, which is the IAEA correlation monitor plate (JRQ), may be representative of conventional steels with relatively high residual elements. The Korean SA508-3 steels, VCD-1, VCDA, and SKA-1, are produced by the Korea Heavy Industries and Construction Co. Ltd. (HANJUNG) using different steel making processes, which are vacuum-carbon-deoxidization, vacuum-carbon-deoxidization with aluminum addition, and silicon-killing with aluminum addition processes, respectively. Aluminum was added mainly to reduce the grain size by precipitating aluminum nitrides. The chemical compositions of the test materials are given in Table 1.

The test materials are all quench-and tempered steels and have typical bainitic microstructure. The grain sizes are 20~30μm except for VCD-1, in which the grain size, about 80μm, is much larger than the others. Note that the JRQ steel contains much more residual copper and phosphorous than the others.

Table 1 - *Chemical Composition of Test Materials*

Material	Chemical composition (wt.%)										
	C	Si	Mn	P	S	Ni	Cr	Mo	Al	Cu	V
VCD-1	0.18	0.08	1.43	0.008	0.003	0.78	0.04	0.52	0.005	0.06	0.005
VCDA	0.18	0.10	1.46	0.006	0.003	0.86	0.15	0.51	0.018	0.03	0.004
SKA-1	0.21	0.24	1.36	0.007	0.002	0.92	0.21	0.49	0.022	0.03	0.005
JFL	0.17	0.25	1.44	0.004	0.002	0.75	0.20	0.51	0.016	0.01	0.004
JRQ	0.18	0.24	1.42	0.017	0.004	0.84	0.12	0.51	0.014	0.14	0.002

Specimens

Test material blocks are all actual RPV scale forging or plates thicker than 200 mm. The specimens were sampled only at 1/4T and 3/4T locations of each block. The crack orientation of the specimens is the T-L direction. Two different sizes of bend specimens were mostly investigated; the standard size (10x10x55mm) and 1/3 size (3.3x3.3x18.3 mm) PCVN specimens. Initial notch slots were machined by an electro-discharging wire of ϕ 0.2 mm. Side-grooves were not machined in general. However, some specimens were 20% side-grooved to see a possible constraint effect. In the case of the side-grooved specimens, the notch root radius was 0.25 mm for standard size and 0.1 mm for 1/3-size PCVN specimens. In addition, a number of CT specimens (1/2T-CT, 1T-CT) were tested for comparing the results.

Specimen Irradiation

The standard size PCVN fracture toughness specimens and Charpy impact specimens of two steels (JRQ and SKA-1) were neutron irradiated to a fluence level of about 4.5×10^{19} n/cm^2 (E>1MeV) at 288°C using the LVR-15 research reactor (NRI, Czech Rep.) [*3*]. In order to reduce the irradiation volume, only the notched or cracked parts (10×10×14mm) of the specimens were inserted in the irradiation capsule and the remaining spans of the specimens were electron-beam welded after irradiation.

Fracture toughness test

Fracture toughness testing was conducted following the ASTM E1921-97 standard procedure. Tests were performed by displacement control in an MTS servo-hydraulic test machine equipped with a high-resolution data acquisition system. Fatigue precracking was performed at room temperature. The amount of crack extension from the EDM slot was about 1.5 mm. The crack length was monitored at the surfaces and the applied load was decreased in 3~4 stages with crack extension. The typical stress intensity factor during precracking is estimated at less than 20MPa√m.

The test temperatures were acquired in an isopentane fluid bath or an insulated chamber which are cooled by circulating or injecting liquid nitrogen, for un-irradiated and irradiated specimens, respectively. The temperature was controlled within ± 1°C. After obtaining a sufficient number of valid toughness data at a single temperature, more tests were performed at other temperatures to find the actual temperature dependence of the fracture toughness.

The standard requires the plastic parts of load-line displacement in specimens to calculate the absorbed energy. This can easily be obtained from the total displacement of a machine stroke or a specially designed fixture by assuming elastic machine compliance due to the loading pin indentation and the fixture compliance as well as the elastic displacement of the specimens. The COD gage was not used for the bend tests.

The yield strengths in the test temperature range were measured by 1-inch gage length rod-type specimens and ABI tests [*4*]. Young's modulus was determined by the following experimental equation as a function of temperature:

$$E = 207200 - 57.09 \times T \quad (^\circ\mathrm{C},\ \mathrm{MPa})$$

Load-line displacement rates were 0.5, 0.1, 0.5, 1.0 mm/min for PCVN, 1/3-PCVN, 1/2T-CT, and 1T-CT specimens, respectively.

Test Results

The measured T_O values at various test conditions for the five different steels are summarized in Table 2. Each test condition contains more than six valid data based on the ASTM E1921-97 standard. In some conditions, the number of test specimens is bigger than the one required by the standard. However, in most cases, the determined T_O values did not depend on the number of specimens if the requirement for the minimum number of valid data was met. The reference temperature of each material was determined as the average of two or three T_O values measured at different temperatures using PCVN specimens without side-grooves.

Figure 1 represents the fracture toughness test results from unirradiated PCVN specimens without side-grooves, where the measured values were converted to the equivalent to 1T specimens. Test temperatures were normalized by the reference temperature ($T_{O,\ PCVN}$) of each material. The ASTM validity limit for PCVN specimen size is also indicated in the plot. Figure 1 contains more data points than listed in Table 2, which are spread out over the temperature range from -196°C to -15°C. It shows that the master curve and probabilistic lines predict the actual shape of the curve and the probabilistic occurrence of the measured data reasonably well. It is clear that the shape of the master curve is acceptable in spite that the data points higher than 120 MPa√m are generally not valid by the standard, and some data points above 200 MPa√m may imply small amounts of stable crack growth before cleavage. The scatter of the data points may be well predicted by the theoretical lower- and upper bound lines.

Note that some data are below the 5% lower bound lines. The occurrence of data points below the 5% line may be substantial only in the two steels, VCD-1 and JRQ, which showed relatively poorer fracture properties than the others. This fact might have resulted from the inferiority of the two materials from a microscopic point of view. The microstructural characteristics of the materials are examined in other literatures [*5,6*]. The JRQ steel contains many inclusions as well as phosphorus segregation due to its lack of cleanness. The carbides of VCD-1 are larger than the other steels and are not distributed homogeneously in the matrix. The microstructural inhomogeneity may be responsible for the relatively bigger scatter. However, the 1% probabilistic line bounded most of the test data sets.

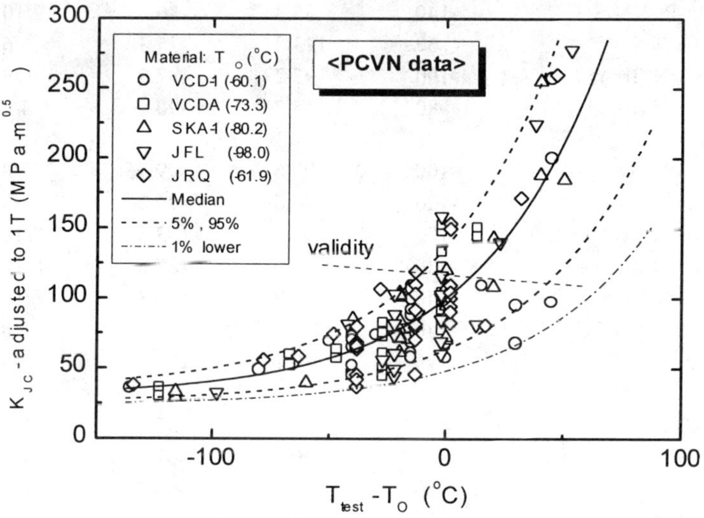

Figure 1 - *Fracture Toughness K_{JC} Test Results Measured by PCVN Specimens*

Table 2 - *Summary of Elastic-Plastic Cleavage Fracture Toughness Test Results*

Material	Specimen	Test Temperature (°C)	$K_{JC(med)}$ for 1T (MPa√m)	T_O (°C)	No. of data (valid/total)
VCD-1	PCVN	-100	60.5	-56.2	8/8
	PCVN	-75	86.8	-64.0	8/8
	1/3-PCVN	-140	49.8	-73.0	7/8
	1/3-PCVN-SG	-140	44.0	-55.2	8/8
VCDA	PCVN	-100	67.9	-67.8	7/7
	PCVN	-75	105.2	-78.8	6/10
	1/3-PCVN	-140	56.4	-88.7	7/9
SKA-1	PCVN	-100	82.6	-85.0	9/9
	PCVN	-80	94.1	-75.4	6/7
	1/3-PCVN	-140	55.1	-86.0	6/8
	1/3-PCVN-SG	-140	55.3	-86.5	7/10
	1T-CT	-60	123.8	-75.3	6/6
	PCVN (Irr.)*	-100	97.2	-97.9	6/9
	PCVN (Irr.)*	-80	115.7	-90.6	4/7
JFL	PCVN	-100	99.6	-99.7	7/8
	PCVN	-120	-74.6	-96.2	8/8
	PCVN-SG	-100	104.9	-103.6	7/10
	PCVN-SG	-120	67.9	-87.8	10/10
	1/3-PCVN	-140	62.9	-100.3	8/15
	1/3-PCVN-SG	-140	-	-	3/8
	1/2T-CT	-90	120.1	-103.3	6/6
JRQ	PCVN	-100	61.2	-57.4	8/8
	PCVN	-75	90.7	-67.5	7/8
	PCVN	-60	101.0	-60.8	6/8
	PCVN-SG	-75	92.1	-68.7	7/8
	PCVN-SG	-60	101.1	-60.8	6/10
	1/3-PCVN	-140	48.0	-68.5	7/8
	1/3-PCVN-SG	-140	55.7	-87.2	6/8
	1/2T-CT	-60	86.9	-49.1	10/10
	PCVN-SG (Irr.)*	60	91.6	66.8	7/8

* : 4.5×10^{19} n/cm^2 (E>1MeV), 288°C

Discussion

Effect of Specimen Size (PCVN, 1/3-PCVN and 1/2T-, 1T-CT specimens)

Figure 2 shows a normalized plot of the 1/3-PCVN data, where the T_O value of each steel is that determined from the PCVN specimens. Actually, the test temperature, -140°C, is the same for all 1/3-PCVN data. Different T_O values of the materials spread out the data sets on the normalized temperature axis. The scatter of 1/3-PCVN data is much larger than that of the PCVN specimens. The main source of the large scatter may be invalid data due to very small specimen size. Another reason is number of test specimens, since more specimens were required to get an accurate T_O value at lower temperatures with smaller specimens. The ASTM valid limit for this specimen is indicated in Figure 2. In fact, it is very difficult to meet all the validity requirements of the ASTM standard using 1/3-sized PCVN specimens. The test temperature is very low in order to obtain a sufficient number of valid data while many of data usually do not meet the size limit. However, even though the scatters in 1/3-PCVN data are much larger than in PCVN data, the median values of the measured toughness data may give reliable information. This will be discussed later.

Figure 3 shows the results from CT specimens. It consists of three sets of complete 1/2T-CT or 1T-CT data and two sets of 1T-CT data spread out on a temperature range. As in Figures 1 and 2, the test temperature was normalized by the T_O values obtained from the PCVN specimens. It shows that the master curve obtained from the PCVN specimens is still valid with the CT specimens. However, two steels, VCD-1 and JRQ, which are less homogeneous microscopically than the others, showed a tendency to produce data near to the lower bound line. This means that the specimen geometry effect on T_O may be significant in less homogeneous materials where larger specimens give more conservative results. This fact is possibly related to the cleanness of the material and correspondingly the inhomogeneous distribution of the cleavage initiation particles [*7*]. Further investigations are needed.

Figure 4 compares the T_O values determined from the PCVN specimens with those from different types of specimens. This clearly shows that the T_O values determined from the 1/3-PCVN specimens are about 10°C lower than those from the PCVN specimens. Note that the T_O values from 1/3-PCVN specimens were not completely validated by the standard requirements. However, the difference was rather systematic and the correlation is also reasonably good. The result from the CT specimens is not clear and the difference may be small except in JRQ steel. Note again that the size effect may be significant in some materials as shown in Figures 1, 3, and 4.

Side-Groove Effect

Up to this point, the results only from the flat type specimens without side-grooves have been presented. One of the important judgements for performing fracture toughness tests is side-grooving. Side-grooves are known to be beneficial to make a crack front straight in elastic-plastic fracture tests [*8*]. In the ASTM E1921-97, side-grooving is only an option. Figure 5 summarizes the test results performed in this study for investigating the side-groove effect. The effect of side-grooves was negligible in the PCVN specimens.

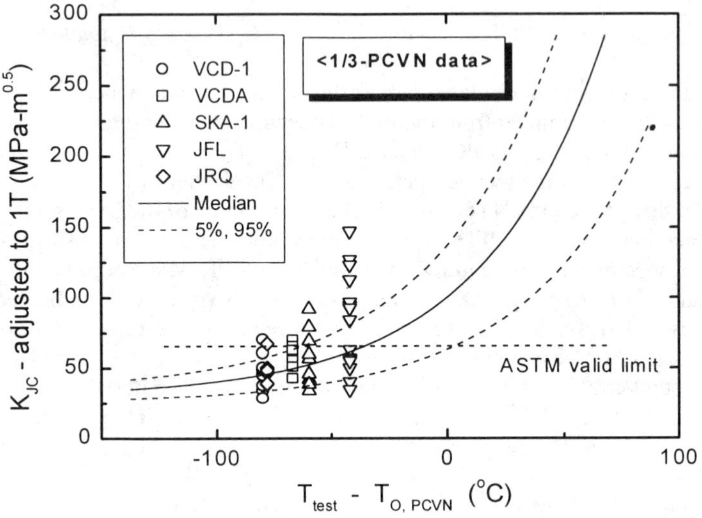

Figure 2 - *Fracture Toughness K_{JC} Results Obtained from 1/3-PCVN Specimens with the Master Curve Determined by PCVN Specimens*

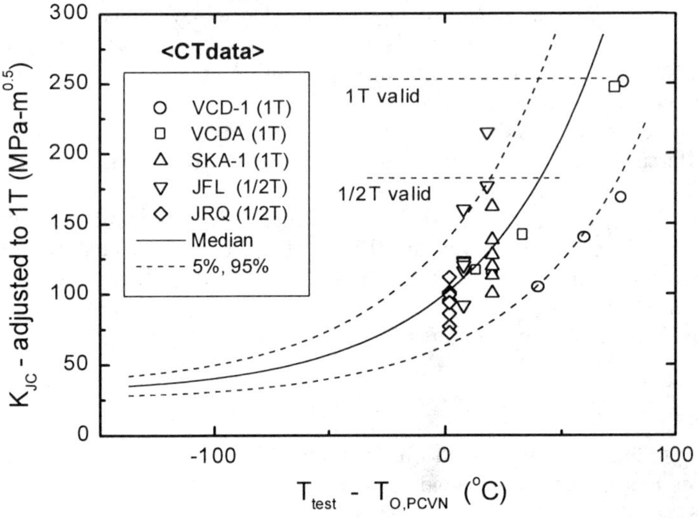

Figure 3 - *Fracture Toughness K_{JC} Results Obtained from CT Specimens with the Master Curve Determined by PCVN Specimens*

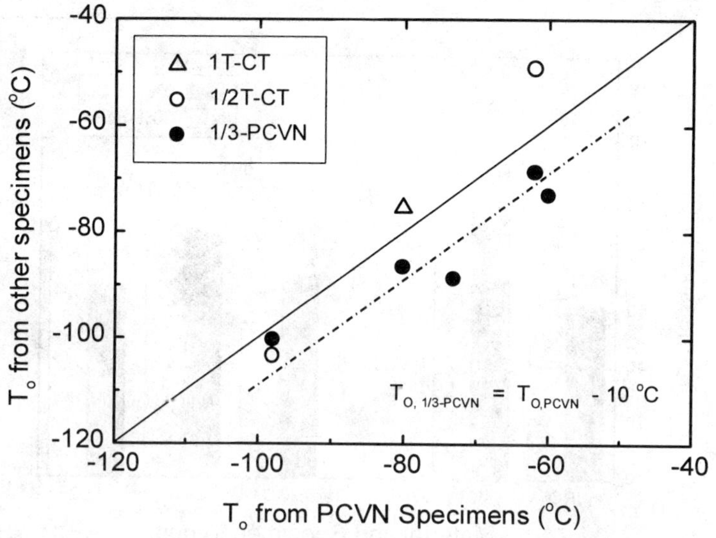

Figure 4 - *Dependence of the Determined T_O Values on Specimen Size*

In smaller 1/3-PCVN specimens, the result is rather confusing as there is no clear tendency. This may come from a practical difficulty to keep the narrow groove tip coinciding with the sharp crack line. Even though the side-groove seems not to affect the mean value of T_O [*8*], it produced a larger scatter in the data sets, as shown in Table 2 and Figure 5. Therefore, side-grooving may be not beneficial in the transition range testing.

Deviation from Different Test Temperatures

Another important point is the selection of the test temperature. Initially, the test temperature is recommended from Charpy impact energy curves, but the temperature is not always revealed as a good selection in practice. In order to see the effect of the test temperature on the determined T_O values, the deviations of the measured T_O from the reference temperature of the material were plotted with respect to the deviations of the test temperatures from the measured reference temperature. Figure 6 contains all the unirradiated data sets obtained from the current study. From the given PCVN data sets, the dependence of the measured value of T_O on the test temperature may not be significant and the standard deviation is estimated at about 5.5°C. The deviations are also similar in 1/3-PCVN and CT specimens except two extreme values from the side-grooved 1/3-PCVN specimens.

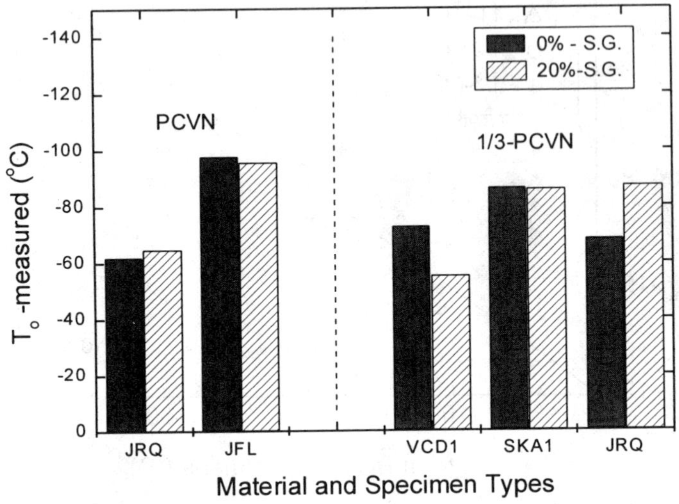

Figure 5 - *Effect of Side-Grooves on the Determined To Values in PCVN and 1/3-PCVN Specimens*

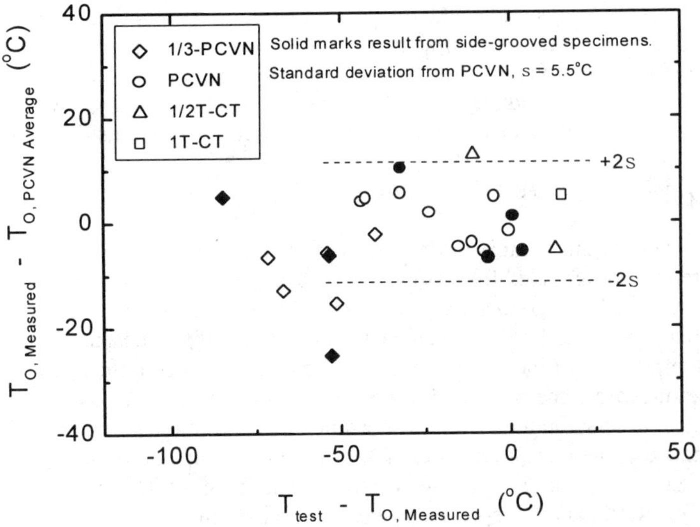

Figure 6 - *Deviation of the Measured To Values from the Reference Temperature of the Corresponding Material at Different Test Conditions*

Charpy Correlation for Evaluating Irradiation Embrittlement

The current practice of indexing radiation induced embrittlement is based on ΔRT_{NDT}, which is determined by either Charpy tests or the chemistry factor of US NRC Regulatory Guide 1.99, Revision 2. Table 3 summarizes the results of two irradiated steels, SKA-1 and JRQ, evaluated by different indexing parameters. Conventional 41J temperature shifts from Charpy curves were determined as 116°C and 6°C for JRQ and SKA-1, respectively. These values are slightly different from the shifts in the reference temperature. ΔT_O was estimated as 134°C for JRQ, which is slightly larger than ΔT_{41J}. Abnormal behavior was observed in the PCVN data of SKA-1, where ΔT_O is a negative number, -18°C. This means that irradiation toughening might have occurred in SKA-1 steel. More experimental data are needed to explain the abnormal behavior in SKA-1 steel.

Wallin [9] proposed a linear relationship between T_O and T_{28J} from analysis with a great number of data sets. The proposed relation was $T_O = T_{28J}$ - 18°C with a standard deviation of 15°C. Figure 7 presents the correlation between the T_{28J} values from Charpy tests and T_O values measured by only Charpy sized specimens in this study. It contains two irradiated data of which one point is plotted at a much higher temperature range. The figure shows that a rather good correlation exists between the two parameters from the same sized specimens. The mean trend line was determined as $T_{O,\,PCVN} = T_{28J}$ - 30°C with a standard deviation of only 5.5°C. In spite of considering a small sized database in this study, selecting the data sets from only the same sized specimens may give more reliable results. The difference of the correlation between Wallin's and in this study, about 12°C, may have come from the different specimen sizes used in the analyses. Wallin's data sets includes a big range of specimen sizes from 10 mm up to 200 mm in thickness, averaged at about 35 mm. This size could give a 12°C more conservative result than the 10 mm thick PCVN specimens as shown in Figure 4.

Table 3 - *Irradiation Shift in Indexing Temperatures of JRQ and SKA-1 (°C)*

Material	Fluence*	Charpy			PCVN
		T_{28J}	T_{41J}	R.G. 1.99 Rev. 2	T_O
JRQ	0	-29.1	-19.6		-62
	4.5	80.5	96.1		67
	Δ	109.6	115.7	80.9	129
SKA-1	0	-59.3	-51.5		-80
	4.5	-55.5	-45.9		-98
	Δ	3.8	5.6	15.3	-18

* : $\times 10^{19}$ n/cm^2 (E>1MeV), 288°C

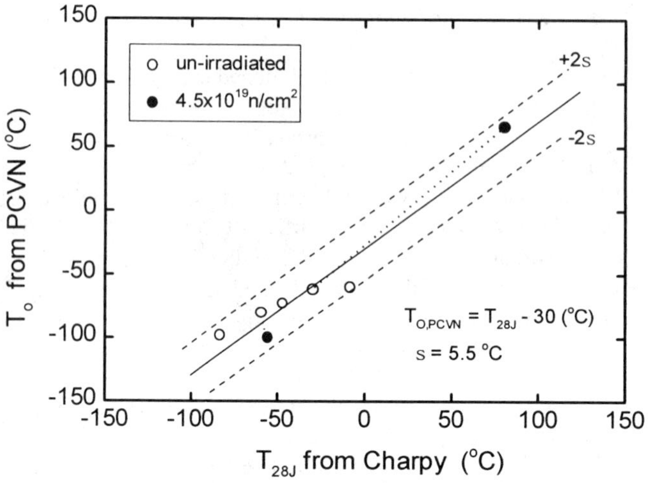

Figure 7 - *Relationship Between Charpy 28J Temperature and the Reference Temperature Determined by PCVN Specimens*

Conclusion

The following conclusions were obtained from this study.

1. Precracked Charpy size (PCVN) specimens could easily be used for evaluating fracture toughness transition characteristics in several unirradiated and irradiated steels and the occurrence populations of the test data were well predicted by probabilistic bound lines.
2. T_O values determined by 1/3-sized PCVN specimens were about 10°C lower than those by PCVN specimens. The scatter was also relatively large in this type of ultra small specimens.
3. The specimen geometry effect may be substantial in lower toughness materials.
4. Side-grooving does not affect the measured T_O values but makes the scatter bigger. Therefore, it is not beneficial for the transition range testing.
5. Test temperature may not affect the measured T_O values significantly with a given specimen size.
6. The correlation between Charpy tests and fracture toughness tests was reasonably good when based on the indexing temperatures rather than the temperature shifts. This implies a possible experimental scatter from both tests. A correlation based on the limited PCVN data from the present study only, $T_{O,\,PCVN} = T_{28J}$ - 30°C, is recommended by considering a possible size effect.

Acknowledgement

This work has been carried out as a part of the Reactor Pressure Boundary Materials Project under the Nuclear R&D Program by MOST in Korea. We also acknowledge Dr. Milan Brumovsky for supporting the irradiation of specimens in NRI, Czech Rep., under a co-operation contract.

References

[1] Chen, J. H., Wang, G. Z., Yan, C., Ma, H., and Zhu, L., "Advances in the Mechanisms of Cleavage Fracture of Low Alloy Steel at Low Temperature: Part I, II, III," *International Journal of Fracture*, Vol. 83, 1997, pp. 105-157.

[2] Yoon, K. K., "Alternative Method of RT_{NDT} Determination for Some Reactor Vessel Weld Metals Validated by Fracture Toughness Data," *Transactions of the ASME: Journal of Pressure Vessel Technology*, Vol. 117, 1995, pp. 378-382.

[3] Brumovsky, M., "*KAERI-NRI Joint Programme on Evaluation of Neutron Irradiation Embrittlement of Reactor Pressure Vessel Steels*," Final Report for NRI Activities, Nuclear Research Institute Rez plc, 1999.

[4] Byun, T. S., Kim, S. H., Lee, B. S., Kim, I. S., and Hong, J. H., "Estimation of Fracture Toughness Transition Curves of RPV Steels from Ball Indentation and Tensile Test Data," *Journal of Nuclear Materials*, Vol. 277, 2000, pp. 263-273.

[5] Kim, J. T., Kwon, H. K., Kim, K. C., Kim, J. M., "Improved Mechanical Properties of the A 508 Class 3 Steel for Nuclear Pressure Vessel Through Steelmaking," *Steel Forgings. Second Volume, ASTM STP 1259*, E. G. Nisbell and A. S. Melilli, Eds., American Society for Testing and Materials, 1997, pp. 18-32.

[6] Viehrig, H., Boehmert, J., "Some Issues by Using the Master Curve Concept," *Transactions of the SMiRT-15*, Seoul, Korea, August 15-20, 1999, pp. V-383-390.

[7] Wallin, K., "The Scatter in K_{IC} - Results," *Engineering Fracture Mechanics*, Vol. 19, No. 6, 1984, pp. 1085-1093.

[8] Wallin, K., "Validity of Small Specimen Fracture Toughness Estimates Neglecting Constraint Corrections," *Constraint Effects in Fracture Theory and Applications: Second Volume, ASTM STP 1244*, Mark Kirk and Ad Bakker, Eds., American Society for Testing and Materials, Philadelphia, 1995, pp. 519-537.

[9] Wallin, K., "A Simple Theoretical Charpy-V - K_{IC} Correlation for Irradiation Embrittlement," *Innovative Approaches to Irradiation Damage and Fracture Analysis, ASME PVP-Vol.170*, 1989, pp. 93-100.

Enrico Lucon[1] and Rachid Chaouadi[1]

Radiation Damage Assessment by the Use of Dynamic Toughness Measurements on Pre-Cracked Charpy-V Specimens

---

Reference: Lucon, E., and Chaouadi, R., "**Radiation Damage Assessment by the Use of Dynamic Toughness Measurements on Pre-Cracked Charpy-V Specimens**," *Effects of Radiation on Materials: 20th International Symposium, ASTM STP 1405,* S. T. Rosinski, M. L. Grossbeck, T. R. Allen, and A. S. Kumar, Eds., American Society for Testing and Materials, West Conshohocken, PA, 2001.

Abstract: The small size of Charpy-type surveillance specimens can make it difficult to obtain fracture toughness measurements which are directly transferable to real-life structures. FEM calculations show that the pre-cracked Charpy specimen (PCCv) tends to lose constraint before the onset of cleavage. However, if the PCCv specimen is loaded under impact conditions, even for moderately strain-rate sensitive materials the deformation level at which loss of constraint occurs is significantly higher, due to the increase in flow properties. This implies that the ductile-to-brittle transition temperature measured under high rate conditions is shifted to higher temperatures with respect to quasi-static toughness tests. This circumstance has been verified through instrumented impact tests on PCCv specimens of two reactor pressure vessel steels, in both baseline and irradiated conditions. The dynamic reference temperature shift due to irradiation has been evaluated and compared with its static counterpart and with index temperature shifts based on conventional impact parameters. Furthermore, the possibility of predicting the dynamic T_0 from the knowledge of the static values, the loading rate and the material's yield strength, has been investigated.

Keywords: dynamic fracture toughness, precracked Charpy-V specimens, impact loading rate, loss of constraint, reference temperature, irradiation-induced embrittlement.

Although nowadays specimen size effects can be effectively modelled by using the widely popular Master Curve approach, incorporated in the ASTM Test Method for Determination of Refernence Temperature, T_0, for Ferritic Steels in the Transition Range (E 1921-97), the small size of Charpy-type surveillance specimens can provide additional uncertainties in the evaluation of the actual toughness properties of a real-life structure or component.

---

[1] Research scientists, SCK•CEN, Reactor Materials Research, Boeretang 200, 2400 Mol, Belgium.

In the nuclear field, for the typical levels of material strength and toughness found in pressure vessel steels, FEM calculations show that the pre-cracked Charpy specimen (PCCv) tends to lose constraint before the onset of cleavage; static test results could therefore provide non-conservative estimates of the conditions of cleavage initiation in the actual pressure vessel. Three-dimensional finite element analyses performed by Nevalainen and Dodds [*1*] have shown that for moderate strain hardening materials, the PCCv specimen might or might not demonstrate adequate constraint, depending on the toughness and strength of the vessel material. Figure 1 shows that, for six different reactor pressure vessel steels, the reference temperature T_o measured from PCCv specimens underestimates in four cases the value calculated with C(T) specimens by approximately 10 °C.

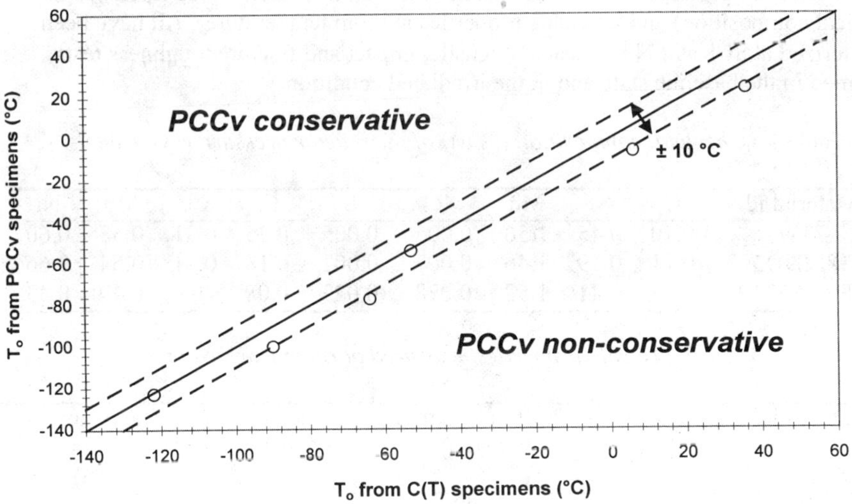

Figure 1 - *Comparison between PCCv and C(T) specimens, in terms of reference temperature, for six pressure vessel steels.*

Previous computational studies by Koppenhoefer and Dodds [*2*], however, have suggested that if the PCCv specimen is loaded under impact conditions (i.e. with a pendulum), even for moderately rate-sensitive materials the deformation level at which loss of constraint occurs is significantly elevated. The reason for this is to be found in the presence of modest viscoplasticity: consequently, impact loading of PCCv samples often enables the determination of fracture toughness under conditions of small-scale yielding for pressure vessel steels, while at the same temperature quasi-statically loaded specimens experience large-scale yielding and loss of constraint [*3*].

In an attempt at experimentally verifying these assumptions, instrumented impact tests have been performed on PCCv specimens in the ductile-to-brittle transition region, in order to characterize the dynamic toughness properties of two reactor pressure vessel steels, in both baseline (unirradiated) and irradiated conditions.

The radiation-induced shift in the dynamic reference temperature $\Delta T_{o,dyn}$ (corresponding to K_{Jd} = 100 MPa√m) has been evaluated and compared with temperature shifts measured both in terms of static fracture toughness and conventional impact parameters.

Finally, the possibility of predicting $T_{o,dyn}$ from the knowledge of $T_{o,st}$, loading rate and yield strength, based on a recently proposed theoretical model [*4*], has been addressed; predictions have been verified against measured values for different pressure vessel steels.

Materials

The reactor pressure vessels considered in this study are described in Tables 1 (chemical composition) and 2 (tensile properties at room temperature). All have been characterized at SCK•CEN by means of tensile, impact and fracture toughness tests performed in the baseline state and in the irradiated condition.

Table 1 - *Chemical composition (% wt) of the reactor pressure vessel steels.*

Material id	C	Si	Mn	P	S	Cr	Cu	Mo	Ni
73W	0.10	0.45	1.56	0.005	0.005	0.25	0.31	0.58	0.60
18MND5	0.21	0.19	1.46	0.007	0.002	0.18	0.11	0.54	0.66
JSPS (A533B)	0.24	0.41	1.52	0.028	0.023	0.08	-	0.49	0.43

Table 2 - *Tensile properties measured at room temperature.*

Material id	Yield stress (MPa)	Ultimate stress (MPa)	Total elongation (%)	Reduction of area (%)
73W	490	600	22	68
18MND5	525	655	24	73
JSPS (A533B)	461	638	18	60

The 73W reference weld material has been extensively characterized at ORNL in the baseline and irradiated conditions [*5*]. Further irradiation campaigns have been carried out in the Belgian BR2 reactor [*6*].

The second steel is a 18MND5 forging of a commercial steam generator, and was provided by SCK•CEN as a contribution to the project "*Relation between different measures of exposure-induced shifts in ductile-brittle transition temperatures*" (REFEREE) for the European Atomic Energy Community [*7*].

The irradiation conditions for these two steels are given in Table 3.

Table 3 - *Irradiation conditions for the two steels investigated.*

Material id	Place of irradiation	Irradiation temperature (°C)	Fast neutron fluence (n/cm²)
73W	USA	288	1.5×10^{19}
	Belgium	290	2.9×10^{19}
18MND5	Holland	260	0.8×10^{19}

Finally, the A533B steel was the object of a round-robin exercise organized by the Japan Society for the Promotion of Science (JSPS), resulting in the obtainment of a large fracture toughness database [*8*].

Experimental Procedure and Fracture Toughness Analyses

Dynamic toughness tests have been carried out using two instrumented 300 J pendula (one for unirradiated and one for irradiated samples), on Charpy-V specimens which had been previously fatigue precracked up to a crack length to width ratio of approximately 0.5 and finally 20% side-grooved. Tests have been performed at different temperatures, following the indications given in the ASTM Test Method for Notched Bar Impact Testing of Metallic Materials (E 23), concerning temperature control and transfer time from the conditioning medium to the impact position.

Force values yielded by the instrumented tup are corrected in order to obtain equivalence between total energy values given by the pendulum encoder and calculated by integrating force and displacement data; this "dynamic calibration" procedure [*9*] has been previously validated by comparison between dynamic and quasi-static maximum force values for a quasi-strain rate insensitive aluminum alloy (6061-T651) [*10*].

As explicitly allowed by the standard Steel – Charpy V-notch Pendulum Impact Test – Instrumented Test Method (ISO 14556:2000), force and time measurements are used to calculate displacements and thereafter evaluate the work spent to fracture the specimen, accounting for the compliance of the impact machine; the critical fracture toughness K_{Jd} to be used in the analyses is evaluated from the value of J-integral corresponding to cleavage instability. The formulas given in the ASTM Test Method for Determination of Reference Temperature, T_o, for Ferritic Steels in the Transition Range (E 1921) for the SE(B) configuration are used.

Replicate tests have been performed at a temperature close to the reference temperature $T_{o,dyn}$, which corresponds to a median dynamic toughness of 100 MPa$\sqrt{m}$ for the reference specimen type (1TCT); the analytical steps prescribed by ASTM E 1921-97 have been followed for the determination of $T_{o,dyn}$ and the dynamic Master Curve with relevant 5% and 95% confidence bounds.

In the analyses, the increased crack-front constraint induced by dynamic loading conditions has been accounted for by using the dynamic yield strength (σ_{yd}) in the establishment of the maximum measuring capacity of the PCCv specimen

$$K_{Jd(\lim)} = \sqrt{\frac{Eb_o\sigma_{yd}}{30}} \tag{1}$$

where

E = Young's modulus at test temperature and
b_o = length of remaining ligament.

The dynamic yield strength is inferred from the values of force at general yield (F_{gy}) measured from instrumented Charpy tests conducted at the same temperature, using [*11*]

$$\sigma_{yd} = 45.8 \cdot F_{gy} \tag{2}$$

Test Results

The results of the dynamic toughness tests performed on precracked Charpy specimens of the pressure vessel steels investigated are given in Table 4, where the following quantities are reported:

- test temperature(s);
- average loading rate calculated at specimen cleavage, in terms of dK/dt;
- total number of impact tests performed (N);
- total number of valid data according to eq.1 (r);
- dynamic reference temperature ($T_{o,dyn}$), corresponding to a median dynamic toughness (normalized to the reference 1TCT size) $K_{med,1TCT}$ = 100 MPa$\sqrt{}$m;
- associated standard deviation, evaluated in accordance with Annex 3 of the ASTM E 1921 standard.

The Master Curve calculations were performed for the different materials and conditions as follows:

- using the single-temperature approach, in accordance with the present version of the ASTM E 1921 standard;
- when tests at different temperatures were available, using the multi-temperature approach [*10*], recently proposed for inclusion in the next revision of the standard.

The values of dynamic reference temperature given in Table 4 have been calculated as the average of all valid single and multi-temperature determinations.

Table 4 - *Dynamic toughness test results.*

Material id	Fast fluence (n/cm²)	T (°C)	dK/dt (MPa√m/s)	N	r	$T_{o,dyn}$ (°C)	σ (°C)	NOTE
73W	-	-50,-40,-30,-20	3.3×10^5	12	8	-29	6.4	(1)
	1.5×10^{19}	0,20,50,80,100	3.6×10^5	14	8	46	6.4	(1)
	2.9×10^{19}	50	5.0×10^5	6	6	78	7.7	(2)
18MND5	-	-90,-70,-50	3.5×10^5	10	6	-71	7.3	(2)
	0.8×10^{19}	-40,-20,25	3.3×10^5	8	5	-26	6.8	(3)
JSPS	-	22	4.8×10^5	9	8	32	6.4	(2)

(1) Only multi-temperature analysis (single-temperature invalid)
(2) Only single-temperature analysis
(3) Average of single and multi-temperature analyses; reference temperature invalid according to E 1921 ($r < 6$)

Comparison between Static and Dynamic Reference Temperatures

For the materials and conditions studied, reference temperatures calculated from static and dynamic toughness tests are shown in Table 5.

Table 5 - *Comparison between static and dynamic reference temperatures.*

Material id	Fast fluence (n/cm²)	dK/dt (MPa√m/s)	$T_{o,st}$ (°C)	$T_{o,dyn}$ (°C)	$\Delta T_{st\rightarrow dyn}$ (°C)
73W	-	3.3×10^5	-77	-29	48
	1.5×10^{19}	3.6×10^5	24	46	22
	2.9×10^{19}	5.0×10^5	39	78	39
18MND5	-	3.5×10^5	-119	-71	48
	0.8×10^{19}	3.3×10^5	-85	-26	59
JSPS	-	4.8×10^5	-6	32	37

The measured increments of T_o due to dynamic loading conditions ($\Delta T_{st\rightarrow dyn}$) are between 22 and 59 °C; no significant influence was found of the value of loading rate in the range investigated (3.3 to 5×10^5 MPa√m/s), corresponding to impact velocities from 1 to 1.5 m/s.

It can be observed that the consistency is particularly remarkable for the steels in unirradiated condition ($\Delta T_{st\rightarrow dyn} = 37 \div 48$ °C).

Assessment of Different Types of Irradiation-Induced Temperature Shifts

The conventional approach to the surveillance of nuclear pressure vessel steels relies on the shift of transition curves obtained from standard Charpy tests, indexed to temperatures associated to the following predefined values:

- 41 J or 68 J absorbed energy (ΔT_{41J} or ΔT_{68J});
- 0.89 mm lateral expansion ($\Delta T_{0.89mm}$);
- 50% shear fracture appearance ($\Delta FATT_{50}$).

More recently, the tendency to quantify the effects of neutron irradiation in terms of actual degradation in the material's fracture toughness has favoured the use of T_o as the temperature to which such effects should be indexed.

Generally, the value calculated from static fracture toughness tests ($T_{o,st}$) is considered; however, the use of $T_{o,dyn}$ could profit from the increased crack-tip constraint induced by dynamic loading conditions.

Irradiation-induced temperature shifts are reported in Table 6 for the materials and conditions investigated in this study.

Table 6 - *Index temperature shifts induced by neutron exposure.*

Material id	Fast fluence (n/cm²)	$\Delta T_{o,st}$ (°C)	$\Delta T_{o,dyn}$ (°C)	ΔT_{41J} (°C)	ΔT_{68J} (°C)	$\Delta T_{0.89mm}$ (°C)	$\Delta FATT_{50}$ (°C)
73W	1.5×10^{19}	101	75	82	104	134	73
	2.9×10^{19}	116	107	105	126	123	115
18MND5	2×10^{19}	34	45	79	78	77	59

Dynamic reference temperature shifts appear in agreement with their static counterpart within the limits inferred from the associated 95% confidence limits, which are approximately equal to ±15 °C.

With respect to the quantities measured from conventional Charpy tests, the shift of fracture toughness (static and dynamic) appears clearly smaller for 18MND5 and of the same order of magnitude for 73W (except with respect to ΔT_{68J}); a reasonable consistency with the shift of the 50% shear fracture appearance ($\Delta FATT_{50}$) can be observed for both steels.

Prediction of Dynamic Reference Temperature

It has been shown [*3,4*] that the Master Curve methodology is fully applicable to dynamic fracture toughness measurements obtained from impact tests on precracked Charpy specimens; more specifically, the shape of the Master Curve is essentially unaffected by the increase in the loading rate. The increase in crack-tip constraint is however translated in a upward shift in the reference temperature T_o.

A rationalization of this loading rate dependence has been recently proposed by Wallin [*4*], based on the application of the Zener-Holloman parameter which relates yield strength (σ_y), temperature (T) and strain rate ($d\varepsilon/dt$) [*13*]

$$\sigma_y = f\left\{T \cdot \ln\left(\frac{A}{d\varepsilon/dt}\right)\right\} \tag{3}$$

where A is the strain rate parameter, which is a function of the activation energy of the yield process.

If the same concept is extended to the case of a fracture toughness test, it is more convenient to express the strain rate in terms of variation of the stress intensity factor with time (dK/dt); therefore, the Zener-Holloman parameter can be applied directly to the reference temperature

$$T_o \cdot \ln\left(\frac{A'}{dK/dt}\right) = const \tag{4}$$

The constant A' in eq.(4) can be defined by associating the quasi-static loading rate (dK/dt = 1 MPa√m/s) to the quasi-static value of the reference temperature ($T_{o,st}$). Renaming log A' as Γ gives the following expression for the reference temperature associated to a generic value of strain rate

$$T_o = \frac{T_{o,st} \cdot \Gamma}{\Gamma - \ln(dK/dt)} \tag{5}$$

By analyzing the loading rate dependence of 59 steels, 15 of which taken from the EPRI pressure vessel material data base [*14*], Wallin [*4*] has proposed the following expression for Γ, as a function of $T_{o,st}$ and $\sigma_{y(To)}$ (yield strength value at T_o)

$$\Gamma = 9.9 \cdot \exp\left\{\left(\frac{T_{o,st}}{190}\right)^{1.66} + \left(\frac{\sigma_{y(To)}}{722}\right)^{1.09}\right\} \tag{6}$$

with a standard deviation of 19.4%. In terms of reference temperature shift, the accuracy is estimated to be of the order of ±20 °C.

If the model previously described is applied to the materials analyzed in this study, the results given in Table 7 and shown in Figure 2 are obtained.

Table 7 - *Comparison between predicted and measured values of $T_{o,dyn}$.*

Material id	Fluence (n/cm²)	dK/dt (MPa√m/s)	$T_{o,st}$ (°C)	$\sigma_{ys(To)}$ (MPa)	$T_{o,dyn}$ (meas,°C)	$T_{o,dyn}$ (pred,°C)	$\Delta T_{o,meas/pred}$ (°C)
73W	-	3.3×10^5	-77	565	-29	-26	-3
	1.5×10^{19}	3.6×10^5	24	708	46	43	3
	2.9×10^{19}	5.0×10^5	39	642	78	58	20
18MND5	-	3.5×10^5	-119	694	-71	-69	2
	0.8×10^{19}	3.3×10^5	-85	744	-26	-46	20
JSPS	-	4.8×10^5	-6	468	32	31	1

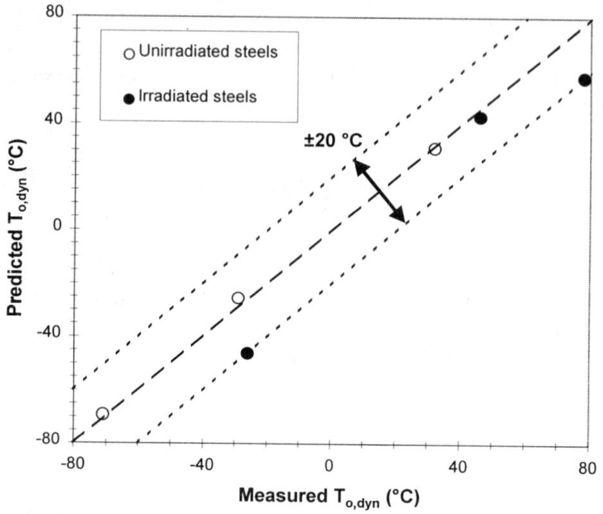

Figure 2 - *Comparison between measured and predicted values of dynamic reference temperature, using Wallin's approach [4].*

The agreement between measured and predicted values is indeed within ±20 °C for all the conditions examined.

The accuracy of the predictions is particularly significant for the steels in unirradiated condition, whereas for the irradiated steels the combined effect of loading rate and neutron irradiation might be somewhat underestimated by the model.

From a practical point of view, it would probably be more useful to obtain a rough estimate the static reference temperature based on dynamic toughness tests; rather than avoiding to carry out static tests (which appears not advisable), this approach could be used as an alternative to estimating the static reference temperature using simple Charpy parameters (such as ΔT_{41J} or ΔT_{28J}), as presently recommended by the ASTM E 1921-97 standard.

This "reversed" approach implies solving eq.5 for $T_{o,st}$

$$T_{o,st} = \frac{T_{o,dyn}\left[\Gamma - \ln(dK/dt)\right]}{\Gamma} \tag{7}$$

and establishing an alternative formulation to eq.6 for calculating Γ, based on $T_{o,dyn}$ and dynamic yield strength (evaluated at $T_{o,dyn}$). This could be accomplished by analyzing the same database as in [*4*].

In any case, Wallin's approach represents in the present formulation an effective tool for estimating $T_{o,dyn}$ and can be consequently used for choosing the test temperature when dynamic fracture toughness tests have to be performed.

Conclusions

The Master Curve analysis has been applied to dynamic toughness measurements performed on precracked Charpy specimens, impact tested on a pendulum; under these conditions, more lenient loss-of-constraint limitations with respect to quasi-static conditions have been shown to apply [*2,3*].

Data obtained on a number of reactor pressure vessel steels, both in the unirradiated and irradiated conditions, show that the shift of the 100 MPa√m reference temperature due to the increase in the loading rate can be somewhat rationalized using a model recently proposed by Wallin [*4*]. This represents also an effective way for choosing the temperature for performing dynamic tests.

Finally, dynamic reference temperature shifts caused by neutron irradiation have been found in general agreement (within ±25 °C) with shifts of both the statically-measured T_o and conventional Charpy index temperatures (T_{41J}, T_{68J}, $T_{0.89mm}$, $FATT_{50}$).

References

[1] Nevalainen, M., and Dodds, R. H., Jr., "Numerical Investigations of 3-D Constraint Effects on Brittle Fracture in SE(B) and C(T) Specimens," UILU-ENG-95-2001, University of Illinois, Department of Civil Engineering, Feb. 1995.

[2] Koppenhoefer, K. C., and Dodds, R. H., Jr., "Constraint Effects on Fracture of Impact Loaded, Pre-cracked Charpy Specimens," UIUL-ENG-95-2007, University of Illinois Department of Civil Engineering, June 1995.

[3] Joyce, J. A., "On the Utilization of High Rate Pre-cracked Charpy Test Results and the Master Curve to Obtain Accurate Lower Bound Toughness Prediction in the Ductile-to-Brittle Transition," *Small Specimen Test Techniques, ASTM STP 1329*, W. R. Corwin, S. T. Rosinski, and E. van Walle, Eds., American Society for Testing and Materials, West Conshohocken, PA, 1998, pp. 253-273.

[4] Wallin, K., "Effect of Strain Rate on the Fracture Toughness Reference Temperature T_o for Ferritic Steels," *Recent Advances in Fracture*, R. K. Mahidhara, A. B. Geltmacher, P. Matic and K. Sadananda, Eds., The Minerals, Metals & Materials Society, 1997.

[5] Nanstad, R. K., Haggag, F.M., McCabe, D.E., Iskander, S.K., Bowman, K.O. and Menke, B.H., "Irradiation effects on Fracture Toughness of Two High-Copper Submerged-Arc Welds, HSSI Series 5," NUREG/CR-5913, ORNL/TM-12156/V1, 1992.

[6] Lucon, E., Chaouadi, R. and Scibetta, M., "Test Results and Analysis of the 73W Weld Material Irradiated in the BR2 Reactor," SCK•CEN Report R-3358, Oct. 1999 (Restricted).

[7] Chaouadi, R., van Walle, E., Fabry A., and Puzzolante, J.-L., "SCK•CEN Contribution to the "Relation between Different Measures of Exposure-Induced Shifts in Ductile-Brittle Transition Temperatures" (REFEREE)," SCK•CEN Report BLG-805, Aug. 1998.

[8] "Standard Test Method for Fracture Toughness within Ductile-Brittle Transition Range," *Standard of the 129th Committee*, Japan Society for the Promotion of Science, 1995 (in Japanese).

[9] Winkler, S., Michael, A. and Lenkey, G.B., "Analysis of the Low-Blow Charpy Test," *Evaluating Material Properties by Dynamic Testing*, ESIS 20, E. van Walle, Ed., MEP London, 1996, pp. 207-233.

[10] Lucon, E., Scibetta, M., Puzzolante, J.-L., Chaouadi, R. and van Walle, E., "Qualification of Force Values Measured with Instrumented Impact Strikers", SCK•CEN Report R-3418, April 2000 (restricted).

[11] Server, W.L., "General Yielding of Charpy V-Notch and Precracked Charpy Specimens," *Journal of Engineering Materials and Technology*, Vol. 100, April 1978, pp. 183-188.

[12] Wallin, K., "Recommendations for the Application of Fracture Toughness Data for Structural Integrity Assessments, " *Proceedings of the IAEA/CSNI Specialists' Meeting on Fracture Mechanics Verification by Large Scale Testing*, Oak Ridge, TN, NUREG/CP-0131 (ORNL/TM-12413), Oct. 26-29, 1992, pp. 465-494.

[13] Priest, A.H., "Influence of Strain Rate and Temperature on the Fracture and Tensile Properties of Several Metallic Materials," *Dynamic Fracture Toughness*, Abington, Cambridge, UK: The Welding Institute, 1977, pp. 95-111.

[14] Server, W.L. and Oldfield, W., "Nuclear Pressure Vessel Steel Data Base," EPRI NP-933, Palo Alto, CA, 1978.

Kunio Onizawa[1] and Masahide Suzuki[1]

Comparison of Transition Temperature Shifts Between Static Fracture Toughness and Charpy-v Impact Properties Due to Irradiation and Post-Irradiation Annealing for Japanese A533B-1 Steels

Reference: Onizawa, K. and Suzuki, M., **"Comparison of Transition Temperature Shifts Between Static Fracture Toughness and Charpy-v Impact Properties Due to Irradiation and Post-Irradiation Annealing for Japanese A533B-1 Steels,"** *Effects of Radiation on Materials: 20th International Symposium, ASTM STP 1405,* S. T. Rosinski, M. L. Grossbeck, T. R. Allen, and A. S. Kumar, Eds., American Society for Testing and Materials, West Conshohocken, PA, 2001.

Abstract: It is assumed in the integrity analysis of reactor pressure vessel (RPV) that the irradiation-induced shift of fracture toughness in the ductile-brittle transition region is the same as the Charpy transition temperature shift. To confirm this assumption, both shifts are compared using irradiated and post-irradiation annealed RPV steels made by Japanese manufacturers. Five ASTM A533B class 1 plates having high- and low-level impurities, which correspond to first generation and modern Japanese reactors are used in this study. Neutron irradiation of precracked Charpy-v (PCCv) specimens as well as standard Charpy-v specimens was carried out at the Japan Materials Testing Reactor (JMTR). The values of fast neutron fluence for this study are 2 to 13×10^{19} (n/cm^2, E>1MeV) considering typical fluence values at the EOL and extended operation of Japanese PWRs. The irradiation temperature was controlled within the range of 290±10°C. Thermal annealing treatments at 350°C and 450°C for 100 hours were performed for irradiated PCCv and Charpy-v specimens. The master curve approach according to ASTM Test Method for Determination of Reference Temperature, T_0, for Ferritic Steels in the Transition Range (E1921) was applied using PCCv specimens. The irradiation-induced shifts of the reference temperature, ΔT_0, obtained in this study were spread in the range of 10 to 200°C. The annealing recoveries of the reference temperature were compared with those of the Charpy transition temperature. Although large scatter was seen in the relation between ΔT_0 values and Charpy 41J shifts, both shifts were almost the same in the average.

Keywords: reactor pressure vessel, surveillance, structural integrity, irradiation embrittlement, fracture toughness, precracked Charpy, master curve, lower bound

[1] Senior Engineer and Head, respectively, Reactor Component Reliability Laboratory, Japan Atomic Energy Research Institute, Tokai-mura, Ibaraki-ken, 319-1195, Japan.

Introduction

To assure RPV structural integrity throughout its operational life, fracture toughness after neutron irradiation must be determined. To do this, the surveillance program for the RPV is performed according to regulations. The results obtained from the Charpy impact tests in the surveillance program are used for the estimation of fracture toughness after irradiation [*1, 2*]. Most codes related to the surveillance program assume that the degree of irradiation embrittlement obtained from Charpy impact tests is equivalent to that of fracture toughness for the material concerned. Recent research, however, has indicated that the correlation is not always conservative in predicting the degree of irradiation embrittlement in terms of fracture toughness [*3, 4*]. Fracture mechanics methodology is, therefore, necessary for the precise evaluation of irradiation embrittlement to assure the RPV structural integrity.

Since fracture toughness values from test specimens of RPV steels show a large scatter in the transition temperature range, fracture toughness must be evaluated statistically. Although intensive research [*e.g. 4*] has been conducted to establish the statistical treatment and the evaluation method of specimen size effect, the previously existing methods need a large number of specimens and/or relatively large size specimens [*5*]. Since only a small number of small size specimens are available from a surveillance capsule, further research is desired to establish an improved test and evaluation method for surveillance program specimens to evaluate the irradiation effect on fracture toughness. A methodology to evaluate fracture toughness of RPV steels has not been established completely, particularly in the transition temperature range. Recently ASTM established a standard test method to determine the reference temperature of fracture toughness in the transition range (E1921-97), commonly called the master curve method. Some efforts on application of the standard to ASME Code Case have also been published [*6-8*]. Further research is still necessary to verify the test method and the applicability to the Japanese RPV steels. The international programs related to this are actually underway in the U.S.A. and the IAEA.

The objectives of this paper are to apply the master curve method to evaluate fracture toughness of Japanese RPV steels using precracked Charpy-V (PCCv) specimens, and to compare the irradiation-induced shift between Charpy impact and static fracture toughness. Based on the results of the fracture toughness tests by PCCv, 1T-CT and 4T-CT specimens, the master curve approach were evaluated on the size effect and the selection of test temperature in the transition temperature range. The shifts of fracture toughness curves due to neutron irradiation and post-irradiation annealing were then compared with Charpy transition temperature shifts. Some of the results were obtained at JAERI in the framework of the IAEA Coordinated Research Project [*9*].

Experiments

Materials

Five kinds of RPV steels of ASTM A533B class 1 made by Japanese fabricators were used in this study. Chemical compositions of the materials are summarized in Table 1. One is designated as JRQ, which was used as a correlation monitor material in the IAEA CRP [*10, 11*]. Steel A and Steel B [*12, 13*] correspond to RPV steels of the 1970's and the late 1980's, respectively. Steel L is a similar A533B-1 steel with a very small amount of impurities which was used in the corrosion fatigue round robin in Japan. The material designated as JSPS A533B-1 was used in the round robin study organized by the Japan Society for Promoting of Science (JSPS) [*14*] and also presented in reference [*15*]. The last material was chosen because of its large database of cleavage fracture toughness for both small and large specimens. However, it has not been used for neutron irradiation study yet. Mechanical properties at room temperature for the materials are listed in Table 2.

Table 1--*Chemical compositions of materials used in this study.*

(Unit in wt.%)

Material	C	Si	Mn	P	S	Ni	Cr	Cu	Mo
JRQ	0.18	0.24	1.42	0.017	0.004	0.84	0.12	0.14	0.51
Steel A	0.19	0.30	1.30	0.015	0.010	0.68	0.17	0.16	0.53
Steel B	0.19	0.19	1.43	0.004	0.001	0.65	0.13	0.04	0.50
Steel L	0.17	0.24	1.36	0.003	0.003	0.61	0.07	0.02	0.47
JSPS A533B-1	0.24	0.41	1.52	0.028	0.023	0.43	0.08	0.19	0.49

Table 2--*Mechanical properties of materials used in this study.*

Material	σ_{ys} (MPa)	σ_{uts} (MPa)	Elong. (%)	T_{41J} (°C)	USE (J)	RT_{NDT} (°C)
JRQ	477	615	22.2	-28	201	-15
Steel A	469	612	26.4	-42	151	-35
Steel B	462	597	24.9	-61	207	-45
Steel L	461	603	24.8	-66	189	-
JSPS A533B-1	461	639	18.5	52	72	-

Fracture Toughness Tests

Four types of specimens were used in this study, namely PCCv, 0.5T-DCT, 1T-CT and 4T-CT [*12*]. In most cases, the specimens were side-grooved by 10% on each side of the specimen after precracking. The notch orientation is T-L direction for all specimens. Each specimen was machined from the approximately quarter thickness position of the material. Fracture toughness tests were performed at mainly lower transition temperatures where a cleavage fracture was observed before the maximum load of the specimen.

Testing and evaluation for fracture toughness were performed according to mainly E1921-97 and in some cases the ASTM Standard Test Method for Plane-Strain Fracture Toughness of Metallic Materials (E399). A valid fracture toughness, K_{IC}, was obtained only from some 4T-CT specimens. The J-integral-based evaluation method following ASTM E1921 was applied to obtain fracture toughness values from the other specimens. Fracture toughness values at cleavage fracture, J_C, were calculated based on the area under the load-displacement curve up to the sudden load drop and the cleavage fracture load (see eq. 1). Fracture toughness K_{JC} values were then converted from J_C by the equation (2). Young's modulus, E was fixed to 206 GPa.

$$J_C = J_{el} + J_{pl} = \frac{K_C^{\ 2}}{E} + \frac{\eta A_{pl}}{B_N b_0} \tag{1}$$

$$K_{JC} = \sqrt{E \cdot J_C} \tag{2}$$

where K_C: Stress intensity factor at the cleavage fracture,
η=2+0.522(a_0/W) for compact tension specimens,
η=1.9 for PCCv specimens,
A_{pl}: Plastic part of the area under load-displacement curve,
B_N: Net thickness, b_0: Initial ligament (=W-a_0),
a_0: Initial crack length, W: Specimen width.

The specimen size effect adjustment was applied to the results from other than 1T-CT specimens. The adjusted fracture toughness K_{JC} values to 1T thickness (25 mm in this study) were expressed as K_{JC_1T} or K_{JC_25mm}.

Irradiation

PCCv specimens of JRQ, Steel A, Steel B and Steel L were irradiated in Japan Materials Testing Reactor (JMTR, thermal output 50MW). The irradiation capsule includes some fluence monitors of pure iron and aluminum-cobalt for fast and thermal neutrons, respectively. Fast neutron fluence values for the specimens were calculated based on the activation measurements of the monitors. The target values for fast neutron fluence were 2 to 13 x 10^{19} n/cm^2 (E>1 MeV) considering typical fluence values at the EOL and extended operation of Japanese PWRs. The irradiation time was varied corresponding to the target fluence, from 590h to 1770h. Consequently fast neutron flux was 1.3 to 2 x 10^{13} n/cm^{2}s (E>1 MeV). The displacements per atom (dpa) for iron by the irradiation to 2 x 10^{19} n/cm^2 (E>1 MeV) was about 3 x 10^{-2}. Temperature for the specimens during irradiation was controlled in the range from 280°C to 300°C and measured at several locations. Charpy-V and tensile specimens were irradiated at the same time and almost the same condition. The shifts of Charpy transition temperatures due to neutron irradiation are summarized in Table 3 together with irradiation conditions. Charpy transition temperature was defined by fitting the data to tanh curve.

Annealing

To evaluate the effectiveness of thermal annealing on fracture toughness, a set of PCCv specimens with some Charpy-v specimens were irradiated and annealed. To determine the annealing response, hardness recovery was measured a series of specimens. The results of the annealing study will be reported elsewhere.

Thermal annealing to RPV is limited to some non-commercial plants and Russian type reactors and has not applied the western LWR type commercial plants. In the annealing feasibility study in USA, two ways of annealing have been suggested; dry (454°C) and wet (344°C) for one week. From the annealing experiments in this study, typical annealing conditions for fracture toughness evaluation were determined. Those are 350°C, 400°C and 450°C for 100 hours. The annealing conditions correspond to dry and wet annealing and an intermediate condition. The measurements of hardness recovery for these conditions to Steel A indicated the recovery rates of 22%, 68% and 98%, respectively. Charpy transition temperature shifts were also evaluated after annealing as listed in Table 3. In this paper, the annealing treatments at 350°C and 450°C for 100 h are presented.

Table 3--*Charpy Transition Temperature Shifts by Neutron Irradiation at JMTR*

Material	Neutron fluence (n/cm^2, E>1MeV)	Irradiation temp. (°C)	ΔT_{41J} (°C)	$\Delta T_{0.9mm}$ (°C)	$\Delta T_{50\%}$ (°C)
JRQ	2.3×10^{19}	290	88	87	81
	2.4×10^{19}	290	80	69	59
	12.1×10^{19}	289	190	176	141
Steel A	7.5×10^{19}	290	138	122	102
	7.7×10^{19}+ A350	289	89*	87*	76*
	7.6×10^{19}+ A450	289	24**	16**	9**
	12.9×10^{19}	287	155	144	123
Steel B	3.4×10^{19}	290	24	17	22
	11.3×10^{19}	288	71	66	65
Steel L	2.9×10^{19}	286	78	69	59

*: Residual shift after post-irradiation annealing at 350°C for 100h

**: Residual shift after post-irradiation annealing at 450°C for 100h

Results and Discussion

Fracture Toughness Test Results

Before neutron irradiation, the reference temperatures, T_0, were evaluated at several temperatures, and by PCCv and 1T-CT. The base test temperature was selected for each material according to E1921 procedure. In addition to the test temperature estimated by E1921, higher temperatures were chosen to check the test temperature dependence on the master curve method. The deviation of each T_0 measurement from the average of T_0 values determined by PCCv specimens is shown in Figure 1. The data points in the lower portion of the figure mean lower T_0 values than the average values, which result in non-conservative evaluations. In the figure, filled symbols indicate T_0 values obtained from data sets including an invalid data point or points. Therefore, those data were determined by the censoring scheme. Testing of PCCv specimens at higher temperature than the average T_0 necessitates a data censoring and leads to a lower, non-conservative T_0 value as illustrated in Figure 1. On the other hand, testing of PCCv specimens below T_0-50°C makes the uncertainty to determine T_0 value larger and requires more specimens. Therefore, it is recommended that testing around (T_0-20°C) can give a comparative result with the testing of 1T-CT. It is noted from this observation that the testing after irradiation should not be performed at too high temperature. To preclude this, before fracture toughness tests, Charpy impact tests were performed to determine the transition temperature at 28J level, T_{28J}. Fracture toughness tests were then performed around (T_{28J}-30~50°C).

The PCCv and 1T-CT data was also analysed using the multi-temperature method [*16*] for T_0 determination. The T_0 values determined using the multi-temperature method were compared to the average of the single temperature T_0 determinations. When the analysis included some invalid data at high temperature, the average T_0 was higher than the T_0 from the multi-temperature analysis. When no invalid data exists (e.g., the 1T-CT data), the T_0 values were approximately the same. This suggests that the data censoring procedure in the multi-temperature analysis should be improved.

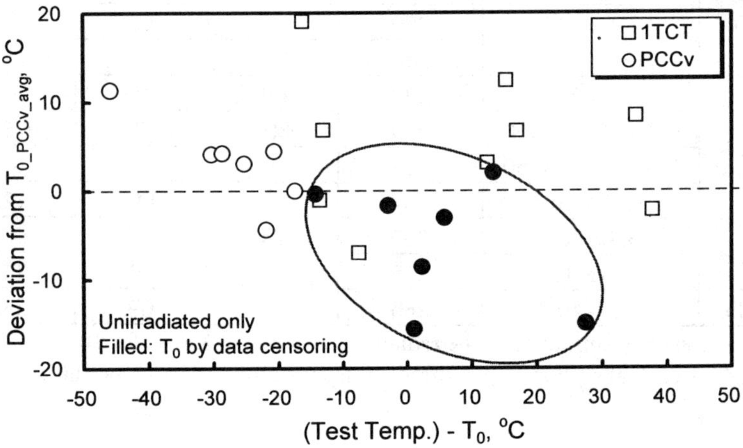

Figure 1—*Relation between test temperature and T_0.*

Using the median value of 1T-equivalent (25 mm was used in this study) adjusted data, the temperature dependence of cleavage fracture toughness within the transition range is expressed by the following equation.

$$K_{JC\_1T(med)} = 30 + 70 \cdot \exp\{0.019 \cdot (T - T_0)\} \qquad (3)$$

This equation has been established by statistical analyses using a huge database including unirradiated and irradiated RPV steels. This equation, which is called a master curve, uniquely defines the cleavage fracture toughness transition curve by a reference temperature, T_0. Around 500 specimens were tested in the transition temperature range as shown in Figure 2. The T_0 value for each material condition was defined by the average of T_0 by PCCv specimens. The master curve and the 5% and 95% tolerance bounds are also illustrated in the figure. Except for the data above T_0+50°C where data are sparse, fracture toughness data are well agreed with the master curve and the tolerance bounds. When the data below T_0+50°C are fitted to the similar type of equation (3) with a fixed constant of 30, almost the same fitting constants were obtained as indicated in Figure 3. Although the fitting includes some invalid data of PCCv specimens, the temperature dependence on fracture toughness in the transition range seems to follow the master curve equation (3).

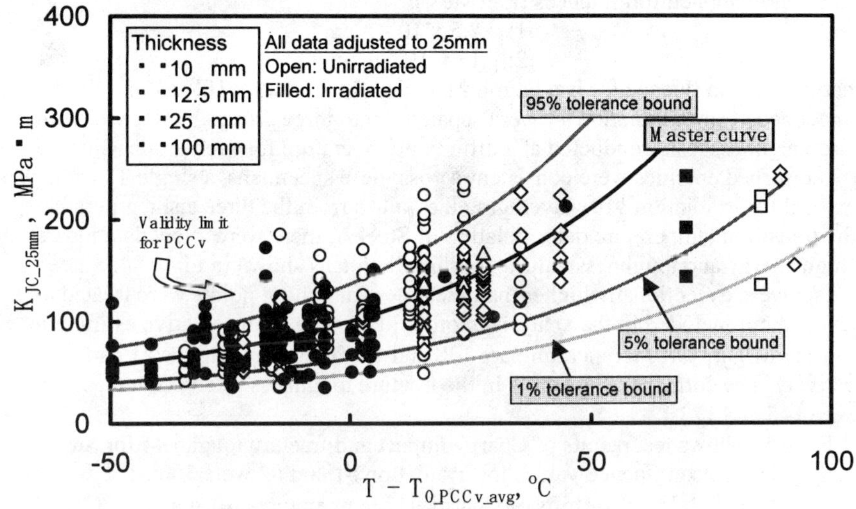

Figure 2—*Adjusted data of fracture toughness to specimen thickness of 25mm and the master curve with tolerance bounds.*

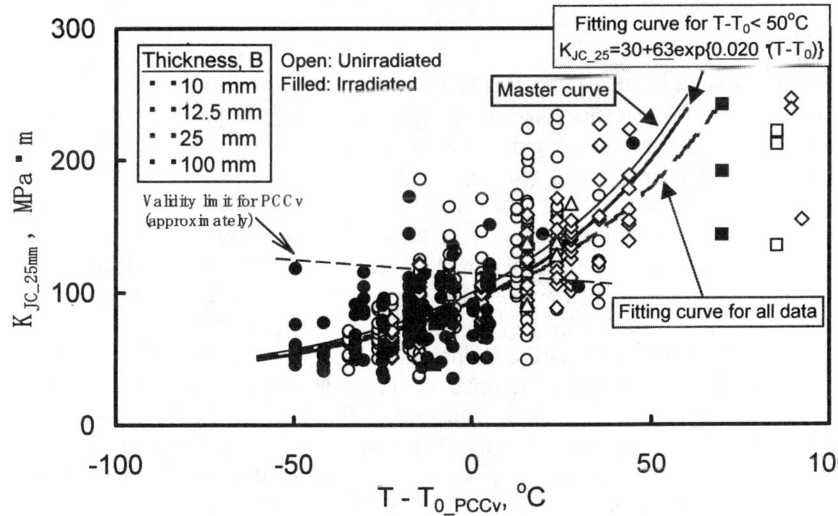

Figure 3—*Comparison of fitting curve for experimental data and the master curve.*

Effect of Irradiation on Charpy Transition Curve and Fracture Toughness

Charpy impact and fracture toughness test results for Steel A before and after neutron irradiation are shown in Figure 4. Two levels of irradiation were completed, with the following neutron fluences (E>1MeV):

#1: 7.5×10^{19} n/cm^2

#2: 13×10^{19} n/cm^2 .

The actual neutron fluence for irradiation #1 varied from 6 to 9×10^{19} n/cm^2 (E>1MeV). The specimens from irradiation #1 were separated into three groups by fluence, with fracture toughness tests conducted at a different temperature for each group. Since Charpy absorbed energies were consistent across these specimens, a single T_0 value was determined for irradiation #1 by averaging the values from the three test temperatures. For the transition temperature determination of Steel A, there were a few data indicating very high energy and toughness values after irradiation as shown in Figure 4. Since those data were exceptionally higher than the other data, those points were treated as outliers and not included in the analysis in order to provide a conservative estimation. The shifts of Charpy 41J temperature are 138 and 155°C for irradiations #1 and #2, respectively. The corresponding shifts in the fracture toughness reference temperature, T_0, were 115 and 173°C.

Figure 5 shows test results of Charpy impact and fracture toughness for Steel B. In this case, total neutron fluence values for irradiation #1 and #2 were 3.4×10^{19} and 11.3 $\times 10^{19}$ n/cm^2 (E>1MeV). The shifts of Charpy 41J temperature are 24 and 71°C, and those of the reference temperature are 14 and 71°C, for irradiation #1 and #2 respectively. Table 4 summarizes the irradiation-induced shifts on the reference temperature T_0 obtained in this study.

Effect of Post-irradiation Annealing on Charpy Transition Curve and Fracture Toughness

The response of Steel A to post-irradiation annealing treatments treatments of 350°C and 450°C for 100 h was determined using both Charpy impact and PCCv specimens The specimens were originally irradiated to a nominal fluence of 7×10^{19} n/cm^2 (E>1MeV). The extents of annealing recovery in the measured properties are illustrated in Figure 6, where the original unirradiated and irradiated results are indicated by dotted lines. After annealing at 350°C, there was only limited recovery in the Charpy 41J transition temperature. On the other hand, annealing at 450°C produced significant recovery in the Charpy transition temperature, with only small residual shift. The measured annealing recoveries for T_{41J} were 35% and 82% for 350°C and 450°C respectively. Test temperatures for the T_0 measurements were based on the Charpy results. The measured T_0 values are summarized in Table 4 and the fracture toughness recovery is illustrated in Figure 6(b). The measured recoveries for T_0 were 32% and 86% for 350°C and 450°C, respectively. For these annealing conditions, The measured recoveries in T_{41J} and T_0 were nearly equivalent. The measured recoveries in hardness after annealing at 350°C and 450°C were 22% and 98%. This hardness recovery exceeded the recoveries in T_{41J} and T_0 at 450°C, but the trend was reversed at 350°C.

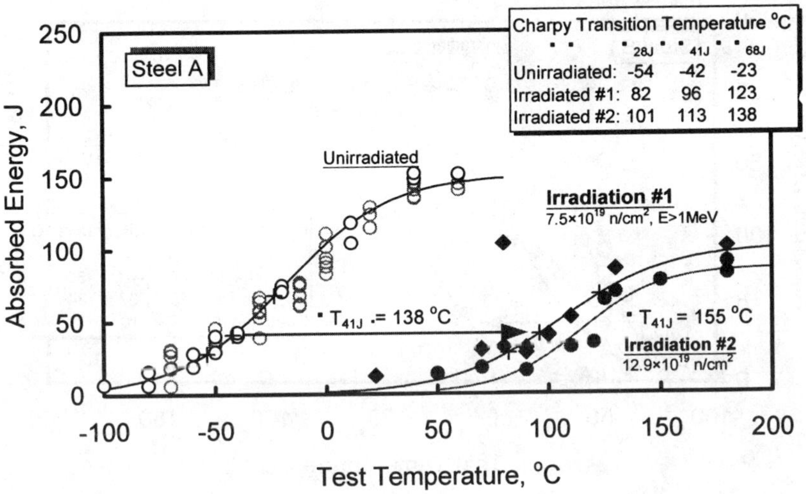

(a) *Charpy impact test results*

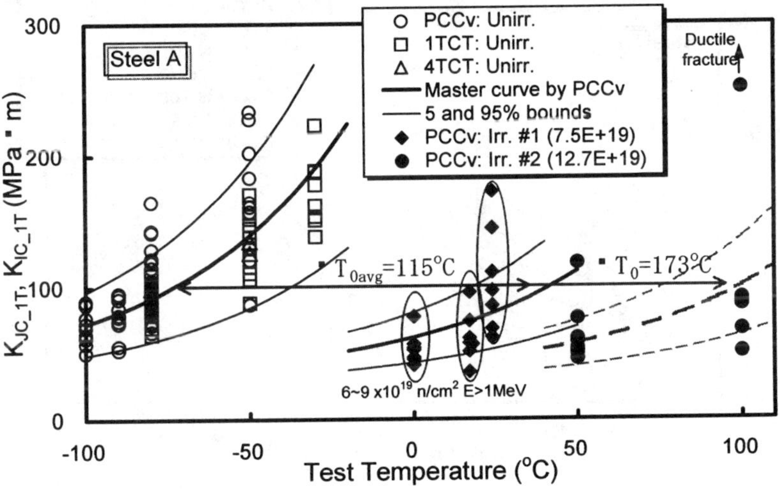

(b) *Fracture toughness data and the master curves*

Figure 4— *Results of Charpy impact test and fracture toughness test for Steel A before and after irradiation.*

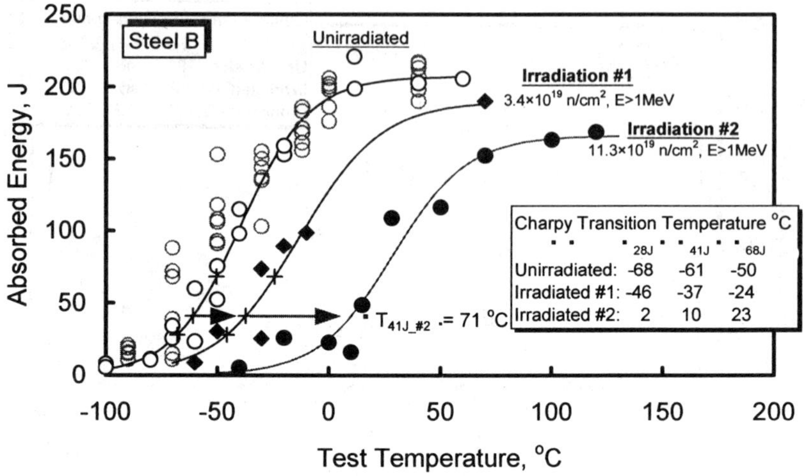

(a) *Charpy impact test results*

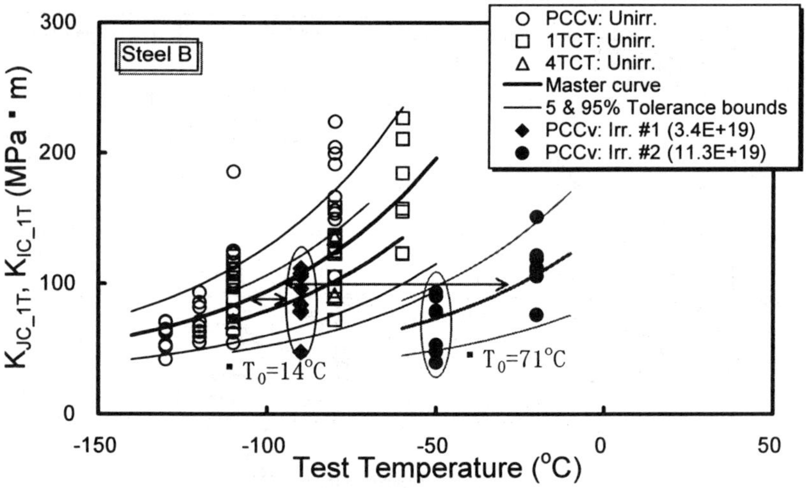

(b) *Fracture toughness data and the master curves*

Figure 5—*Results of Charpy impact test and fracture toughness test for Steel B before and after irradiation.*

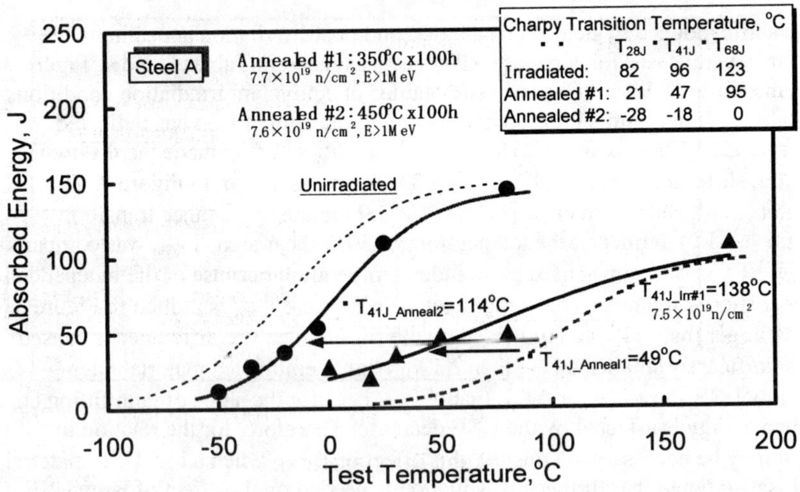

(a) *Annealing recoveries on Charpy transition curves.*

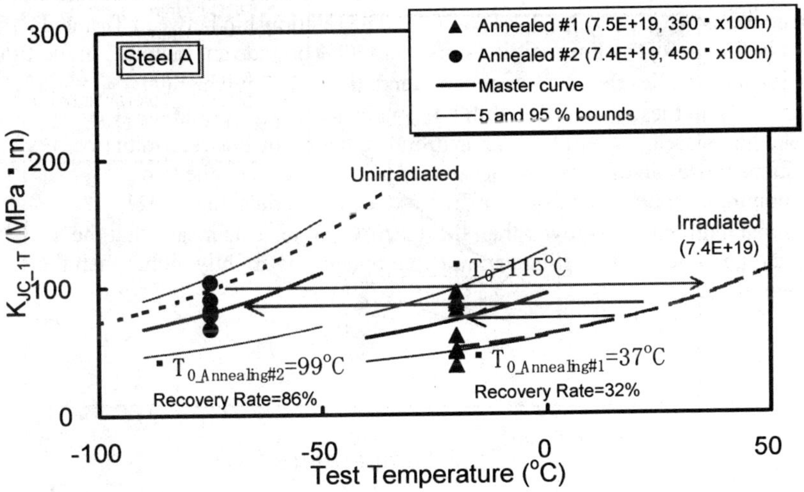

(b) *Annealing recoveries on fracture toughness*

Figure 6—*Test results for Steel A after post-irradiation annealing at 350℃ and 450℃ for 100h.*

Comparison of the Shifts between Charpy Properties and T_0

The shifts induced by neutron irradiation and post-irradiation annealing were measured in several cases for Japanese RPV steels as listed in Tables 3 and 4. Figure 7 compares the shifts of T_0 and Charpy T_{41J} obtained at equivalent irradiation conditions in this study. In previous results in references [*13, 17*], the comparison has indicated slightly larger shifts in T_0 than T_{41J}. However, the additional data made the difference between both shifts smaller; now ΔT_0 is only 3% larger, as shown in Figure 7. The scatter is not small; a standard error is about 15°C. Fracture appearance transition temperature (FATT) defined at the temperature of 50% shear area, $T_{50\%}$, was compared with ΔT_0. FATT is sometimes used as an index temperature because of the independence on the upper shelf toughness. The comparison of ΔT_0 and $\Delta T_{50\%}$ is plotted in Figure 8. ΔT_0 is 30% larger than $\Delta T_{50\%}$ as indicated by the fitting line. The difference is caused from the significantly smaller $\Delta T_{50\%}$ than ΔT_{41J}. Highly embrittled material having relatively low USE shows larger ΔT_{41J} than $\Delta T_{50\%}$ because the slope of tanh fitting curve for absorbed energy is affected by the USE decrease. Therefore, for the relation in Figure 7, it may be necessary to consider this effect on ΔT_{41J} when a low USE material is used. It is also noted that further investigation is needed on the effect of large USE decrease and/or low USE material on the shape of the master curve after irradiation.

ASTM E1921 recommends testing PCCv specimens at T_{28J}-50°C, where T_{28J} is the Charpy 28J transition temperature. Analysis of the data provided in Figure 1 indicates that the optimum test temperature for PCCv specimens is T_0-20°C. These two recommendations are identical if $T_0=T_{28J}$-30°C. The relationship between T_0 and T_{28J} is illustrated in Figure 9. The linear fitting curve and 90% bounds are included in the figure. This relationship implies that the two recommendations are equivalent at T_{28J} = -57°C. Prior to irradiation, these materials exhibit T_{28J} values near -57°C and the two recommendations are in agreement. As neutron irradiation increases, irradiation embrittlement causes an increase in T_{28J} and a deviation between the two recommendations occurs. Therefore for the irradiated materials, the E1921 recommended temperature is lower than the T_0-20°C value. This means that the optimum temperature for T_0 testing of irradiated material is slightly higher than the ASTM recommendation.

Table 4—*Summary of shifts of the reference temperature T_0.*

Material	Neutron fluence (n/cm^2, E>1MeV)	Irradiation temp. (°C)	ΔT_0 (°C)	Note
JRQ	2.1 x 10^{19}	290	105	
	2.4 x 10^{19}	290	76	
	2.9 x 10^{19}	290	105	
	12.2 x 10^{19}	289	198	
Steel A	6.2 x 10^{19}	290	127	Average 115
	7.5 x 10^{19}	288	95	
	9.0 x 10^{19}	291	123	
	7.5 x 10^{19}+ A350	288	78*	
	7.4 x 10^{19}+ A450	288	16**	
	12.7 x 10^{19}	287	173	
Steel B	3.4 x 10^{19}	291	14	
	11.3 x 10^{19}	288	71	
Steel L	2.8 x 10^{19}	287	87	

*: Residual shift after post-irradiation annealing at 350°C for 100h
**: Residual shift after post-irradiation annealing at 450°C for 100h

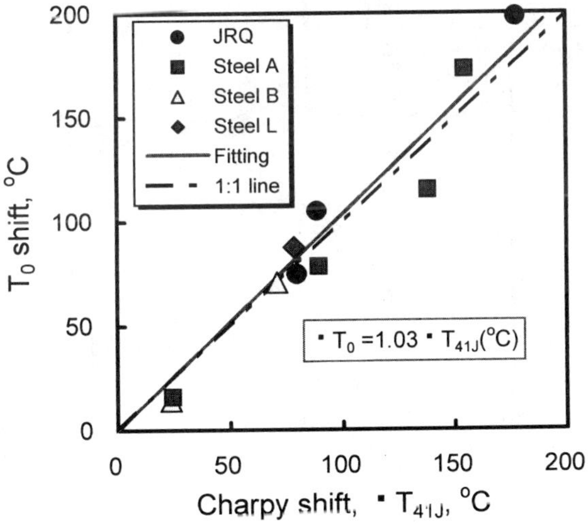

Figure 7—*Comparison of shifts of reference temperature T_0 and Charpy 41J temperature.*

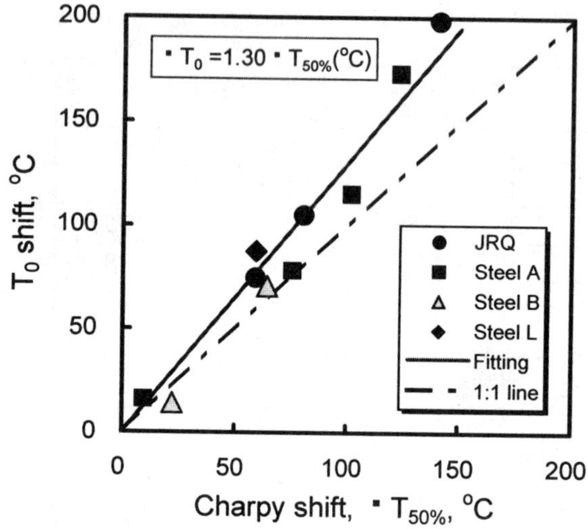

Figure 8—*Comparison of shifts of reference temperature T_0 and Charpy fracture appearance 50% temperature.*

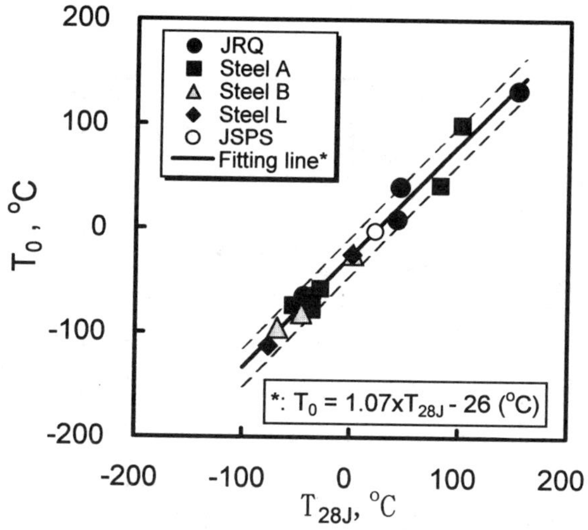

Figure 9—*Comparison of reference temperature T_0 and Charpy 28J temperature.*

Comparison of Lower Bound Fracture Toughness Curve

Recently the following equation (4) has been proposed as the ASME Code Case N-629 [*18*] based on the reference temperature T_0 to alternate the fracture toughness K_{IC} curve of RPV steels.

$$K_{IC} = 36.5 + 22.8 \cdot \exp\{0.036 \cdot (T - RT_{T0})\} \qquad (4)$$

where $RT_{T0} = T_0 + 19.4°C$.

The master curve approach uses 1T-thickness equivalent curve for the median fracture toughness. The current ASME K_{IC} curve corresponds to the lower bound of measured K_{IC} data. For the integrity assessment purpose, the size adjustment may not be applied or should be used for an appropriate defect size and the geometry. The proposed K_{IC}-RT_{T0} curve and as-measured fracture toughness data are compared in Figure 10. Except for the data in higher temperature range ($T-T_0 > 80°C$), the K_{IC}-RT_{T0} curve bounds all experimental data. For comparison, the 1% tolerance bound curve is also shown in Figure 10. The bound curve is found to fit the data points and be more conservative in higher temperature range than the K_{IC}-RT_{T0} curve. The merit of the use of these T_0-based curves is that the fracture toughness can be measured directly by irradiated specimens without any uncertainty of the initial property.

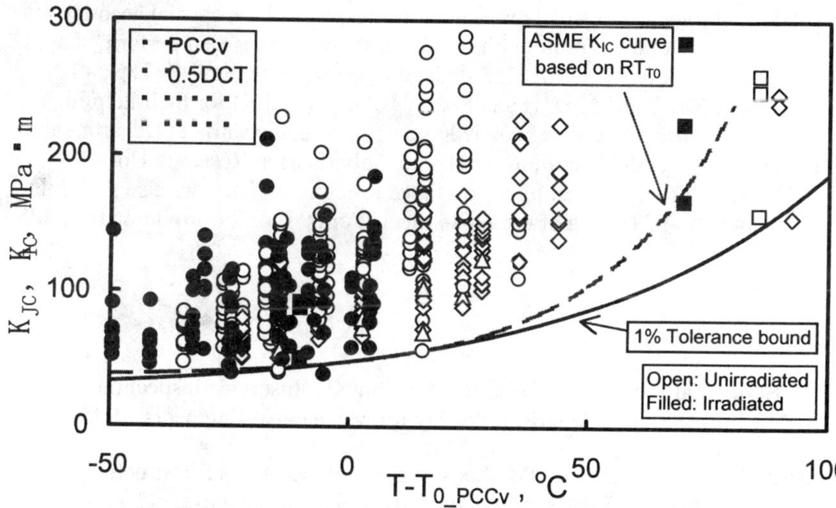

Figure 10—*Comparison of as-measured fracture toughness data with the proposed ASME T_0-base K_{IC} and lower bound curves.*

Summary and Conclusions

Applying the master curve method, fracture toughness tests and the analyses were performed using mainly PCCv specimens in the transition temperature range for Japanese RPV steels. Based on JMTR irradiation and post-irradiation annealing for the specimens of steels, the following conclusions were drawn for the applicability of the master curve method and the irradiation induced shifts between fracture toughness and Charpy properties;

(1) Data sets for PCCv specimens that contain invalid censored data tend to underestimate T_0 as determined by data sets that contain only valid data. Therefore, the test temperature for PCCv specimens should not be higher than T_0 in order to obtain all valid data.
(2) The temperature dependence of experimental fracture toughness data adjusted to 25mm thickness agreed well with the master curve.
(3) The shifts of T_0 by neutron irradiation and post-irradiation annealing were evaluated for Japanese RPV steels and were almost equivalent with Charpy 41J shift.
(4) When the fracture appearance transition temperature (FATT) is used for Charpy shift, it can be 30% smaller than the shift of T_0.

Acknowledgments

A part of this work was carried out under the contract between JAERI and the Science and Technology Agency of Japan under the auspices of the special account law for electric power development promotion. A part of the work was also performed at JAERI in the framework of the IAEA Coordinated Research Program [*9*]. The authors would like to thank Mr. Y. Nishiyama and Mr. T. Tobita of JAERI for their helpful discussion and assistance. Special acknowledgments go to many staffs at Department of JMTR for capsule design and operation, and some individuals at Research Hot Laboratory Division for careful testing of irradiated specimens. They are also grateful to Dr. T. Iwadate and Dr. Y. Tanaka of the Japan Steel Works, Ltd. for providing the JSPS A533B-1 material.

References

[1] ASME Boiler and Pressure Vessel Code, Section XI, "Inservice Inspection of Nuclear Power Plant," American Society for Mechanical Engineers, 1995.

[2] Japanese Electric Association, "Methods of Surveillance Tests for Structural Materials of Nuclear Reactors," JEAC 4201-1991. (in Japanese)

[3] Nanstad, R. K., Haggag, F. M., McCabe, D. E., Iskander, S. K., Bowman, K. O., and Menke, B. H., "Irradiation Effects on Fracture Toughness of Two High-Copper Submerged-arc Welds, HSSI Series 5," NUREG/CR-5913 Vol.1, U. S. NRC, October 1992.

[4] Miglin, M. T., Oberjohn, L. A., and Van Der Sluys, W. A., "Analysis of Results from the MPC/JSPS Round Robin Testing Program in the Ductile-to-Brittle Transition Region," *Fracture Mechanics: Twenty-fourth Volume, ASTM STP 1207*, J.D. Landes, D. E. McCabe and J. A. M. Boulet, Eds., American Society for Testing and Materials, Philadelphia, 1994, pp. 342-354.

[5] Iwadate, T., Tanaka, Y., Ono, S., and Watanabe, J., "An Analysis of Elastic-Plastic Fracture Toughness Behavior for J_{IC} Measurement in the Transition Region," *Elastic-Plastic Fracture: Second Symposium, Volume II-Fracture Resistance Curves and Engineering Applications, ASTM STP 803*, C. F. Shi and J. P. Gudas, Eds., American Society for Testing and Materials, 1983, pp. II-531-II-561.

[6] Server, W. et al., "Application of Master Curve Fracture Toughness Methodology for Ferritic Steels," EPRI TR-108390, Electric Power Research Institute, April 1998.

[7] Server, W. L. and Rosinski, S. T., "Technical Basis for Application of the Master Curve Approach to Reactor Pressure Vessel Integrity Assessment," *Effects of Radiation on Materials: 19th International Symposium, ASTM STP 1366*, M. L. Hamilton, A. S. Kumar, S. T. Rosinski, and M. L. Grossbeck, Eds., American Society for Testing and Materials, West Conshohocken, PA, 2000.

[8] Kirk, M. et al., "Bias and Precision of T_0 Values Determined Using ASTM Standard E1921-97 for Nuclear Reactor Pressure Vessel Steels," *Effects of Radiation on Materials: 19th International Symposium, ASTM STP 1366*, M. L. Hamilton, A. S. Kumar, S. T. Rosinski, and M. L. Grossbeck, Eds., American Society for Testing and Materials, West Conshohocken, PA, 2000.

[9] Onizawa, K. and Suzuki, M., "JAERI's Contribution to the IAEA Coordinated Research Programme on 'Assuring Structural Integrity of Reactor Pressure Vessels' (CRP-IV), - Final Report -," JAERI-Research 2000-022, Japan Atomic Energy Research Institute, May 2000.

[10] CRP Sub-Committee, "Manufacturing History and Mechanical Properties of Japanese Materials Provided for The International Atomic Energy Agency," Japan Welding Engineering Society, Oct. 1986.

[11] Onizawa, K., Suzuki, M., and Ueda, S., "JAERI Contribution for the IAEA Coordinated Research Program Phase III on Optimizing of Reactor Pressure Vessel Surveillance Programmes and Their Analysis," JAERI-M 93-201, Japan Atomic Energy Research Institute, October 1993.

[12] Onizawa, K., Tobita, T., and Suzuki, M., "Investigation on the Evaluation of Cleavage Fracture Toughness Using PCCv Specimens in the Ductile-Brittle Transition Range of Reactor Pressure Vessel Steels," JAERI-Research 97-081, Japan Atomic Energy Research Institute, November 1997.

[13] Onizawa, K. et al., "Effect of Irradiation on Fracture Toughness in the Transition Range of RPV Steels," *Effects of Radiation on Materials: 19th International Symposium, ASTM STP 1366*, M. L. Hamilton, A. S. Kumar, S. T. Rosinski, and M. L. Grossbeck, Eds., American Society for Testing and Materials, West Conshohocken, PA, 2000.

[14] JSPS 129 Committee, "Standard Test Method for Fracture Toughness within Ductile-Brittle Transition Range," Japan Society for Promoting Science, 1995.

[15] Chaouadi, R., "Fracture Toughness Measurements in the Transition Regime Using Small Size Samples," *Small Specimen Test Techniques, ASTM STP 1329*, W. R. Corwin, S. T. Rosinski, and E. van Walle, Eds., American Society for Testing and Materials, 1998, pp. 214-237.

[16] Wallin, K., "Validity of Small Specimen Fracture Toughness Estimates Neglecting Constraint Corrections," *Constraint Effects in Fracture Theory and Applications: Second Volume, ASTM STP 1244*, M. Kirk and Ad Bakker, Eds. American Society for Testing and Materials, 1995.

[17] Onizawa, K., Tobita, T., and Suzuki, M., "Fracture Toughness Evaluation by Precracked Charpy Specimens in the Transition Temperature Range of RPV Steels," Transactions of the 15th International Conference on Structural Mechanics in Reactor Technology, Korea, August 1999, IV-137.

[18] ASME Code Case N-629, "Use of Fracture Toughness Test Data to Establish Reference Temperature for Pressure Retaining Materials for Section XI."

Matthew Wagenhofer[1], Heshan P. Gunawardane[1], and MarjorieAnn E. Natishan[2]

Yield and Toughness Transition Predictions for Irradiated Steels Based on Dislocation Mechanics

Reference: Wagenhofer, M., Gunawardane, H. P., and Natishan, M. E., **"Yield and Toughness Transition Predictions for Irradiated Steels Based on Dislocation Mechanics,"** *Effects of Radiation on Materials: 20th International Symposium, ASTM STP 1405,* S. T. Rosinski, M. L. Grossbeck, T. R. Allen, and A. S. Kumar, Eds., American Society for Testing and Materials, West Conshohocken, PA, 2001.

Abstract: An abundance of empirical data supports the use of the Master Curve, as proposed by Wallin, Saario and Törrönen, to describe the fracture toughness transition behavior of ferritic steels, particularly the notion of a curve shape that is invariant with steel microstructure (other than lattice structure). However, nuclear surveillance programs do not always contain samples of the steel that most limits reactor operations, making direct measurement of fracture toughness impossible. This suggests that a purely empirical argument cannot define the limits of applicability of the Master Curve or validate its use for all conditions of interest. In previous papers a microstructural basis for the existence of a single "Master" fracture toughness transition curve for all ferritic steels was established and limits of applicability have begun to be explored from a theoretical viewpoint. These previous papers established that all steels with the same lattice structure and cleavage fracture mechanism should be expected to adhere to transition behavior that can be defined by a single curve shape with variations in microstructure accounting only for a shift in the transition temperature. In this paper we explore the basis for "Master Curve" validity for irradiated steels by exploring how irradiation affects the microstructure and fracture mode and using the Zerilli-Armstrong constitutive model as the basis for predictions of irradiated steel behavior.

Keywords: Master Curve, fracture toughness transition, ferritic steels, Zerilli-Armstrong constitutive model, irradiation effects

[1]Graduate Students, Department of Mechanical Engineering, University of Maryland, College Park, MD 20742.

[2]Senior Research Scientist, Phoenix Engineering Associates Inc., 3300 Royale Glen Avenue, Davidsonville, MD 21035.

Introduction

There are several methods of determining the ductile-to-brittle transition temperature of a metal. The underlying principle behind all of them is that some measurable quantity, the energy absorbed in fracture (CVN) or fracture toughness (K_{Ic}), is temperature dependent. Two of these methods, the Charpy V-notch test, included in the ASTM Test Methods for Notched Bar Impact Testing of Metallic Materials (E23-92) and the ASTM Test Method for Conducting Drop-Weight Test to Determine Nil-Ductility Transition Temperature of Ferritic Steels (E208-91), yield qualitative results that are not directly related to material properties such as fracture toughness. Thus, any scheme used to correlate these results to fracture toughness contains much uncertainty, necessitating conservative values to be used in any analysis. The third method available, the ASTM Master Curve, detailed in the ASTM Test Method for Determination of Reference Temperature, T_0, for Ferritic Steels in the Transition Range (E1921-97), as proposed by Wallin, Saario and Törrönen [*1*], directly predicts fracture toughness values at any temperature within the transition region based on a minimum of six tests at one temperature. It is derived from the observation that many ferritic steels exhibit a single curve shape for their fracture toughness-temperature behavior. Because this method uses actual fracture toughness tests, the inherent uncertainty is reduced and therefore the method should be more accurate with less conservatism required.

Currently the Master Curve draws upon an empirically derived expression for its temperature dependence. This method works well for steels represented in the data set used to derive this expression, but it cannot predict how the shape of the curve may or may not change for steels not represented. Fracture toughness tests must be performed over a range of temperatures for every significant composition change in order to experimentally validate that the material behavior follows the Master Curve. This can become costly and time consuming, especially in the case of irradiated materials, and it does not address why the curve retains the same shape for so many different steels. As a result, the empirical basis of the current Master Curve lacks the ability to define limits of applicability. By understanding the underlying microstructural material behavior, it should be possible to derive a quantitative, physically-based expression for the temperature dependence of the Master Curve. This would then make it possible to predict how changes in microstructure due to irradiation affect the reference transition temperature without having to perform a multitude of fracture toughness tests.

In previous papers a microstructural basis for the existence of a single "Master" fracture toughness transition curve for all ferritic steels was established [*2,3*] and limits of applicability have begun to be explored from a theoretical viewpoint [*4,5*]. These papers used Zerilli-Armstrong constitutive models of metal flow behavior to establish that all steels with the same lattice structure and cleavage fracture mechanism should be expected to adhere to transition behavior that can be defined by a single curve shape. Variations in microstructure then account only for a shift in the transition temperature. In this paper we explore the basis for "Master Curve validity" for irradiated steels by explaining how irradiation affects the microstructure and fracture mode and using the Zerilli-Armstrong constitutive model as the basis for predictions of irradiated steel behavior.

Zerilli-Armstrong (ZA) Equation and Irradiation Effects, Part I

The ZA constitutive equation [*6*] uses an activation area for dislocation motion as a basis to describe the thermal portion of the flow stress:

$$\sigma_{ZA} = c_0 + K\varepsilon^n + B_0 e^{-\beta T}, \tag{1a}$$

where

$$\begin{aligned} \beta &= \beta_0 - \beta_1 \ln \dot{\varepsilon}, \\ c_0 &= \sigma_G + kd^{-1/2}, \end{aligned} \tag{1b}$$

ε = strain
$\dot{\varepsilon}$ = strain rate
T = absolute temperature
d = average grain diameter
σ_G = contribution to flow stress from solutes and initial dislocation density
k, β_0, β_1, B_0, K and n = constants specific to a material.

In Eqn. (1a), athermal contributions to the flow stress are added linearly with the thermal term, and include, for BCC materials, grain size strengthening, strain hardening, solute strengthening and precipitate strengthening. This linear combination of effects implies independence of thermal, short range, and athermal, long range, barriers to dislocation motion. The athermal, long range barriers have interbarrier spacing orders of magnitude larger than that of the thermal, short range barriers. Thus, an input of thermal energy that allows a dislocation to move past a short range barrier has no effect on a dislocation attempting to move past a long range barrier. In the case of BCC materials, these short range barriers correspond to the atoms of the iron lattice. Consistency with the form of the ZA equation dictates that the contribution of radiation damage to the flow stress be added linearly with the existing terms. This, however, does not determine whether the radiation damage term, hereafter designated σ_{irr}, is temperature dependent or independent.

In a recent publication, Merkle et al. [*7*] demonstrated that neutron irradiation does not affect the fracture toughness temperature dependence. Figure 1 is their log-log plot of fracture toughness versus yield strength for irradiated and unirradiated A533B steel. The slope of the curve for irradiated material is seen to be much steeper than that of the curve for unirradiated material. However, a plot of fracture toughness versus temperature, as shown in Figure 2, results in a similar curve shape for both irradiated and unirradiated materials. Therefore the fracture toughness temperature dependence of the irradiated material should be the same as that of the unirradiated material. A log-log plot of the same fracture toughness values versus an offset yield strength corresponding to the thermal portion, Figure 3, confirms the conclusion of Figure 2 and suggests that the effects of radiation and, correspondingly σ_{irr}, are athermal in nature. In order to quantify this phenomenon for use in the ZA equation, the physical damage mechanisms affecting flow behavior must be reviewed.

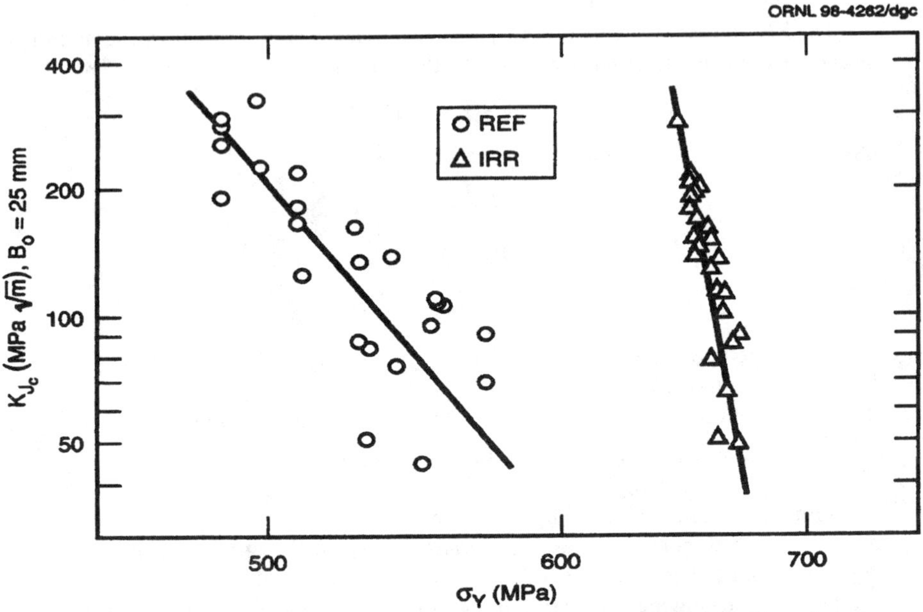

Figure 1 – *Fracture toughness vs. yield strength for unirradiated and irradiated material. A533B C1.1 (JRQ) 12.5mm RCT (after Merkle et al. [7]).*

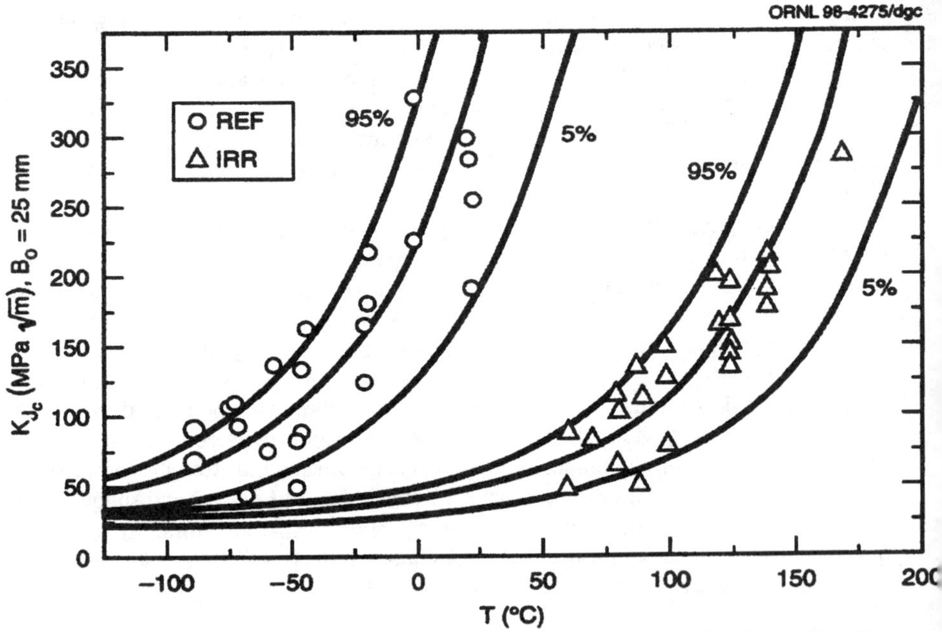

Figure 2 – *Fracture Toughness vs. temperature for unirradiated and irradiated material. A533B C1.1 (JRQ) 12.5mm RCT (after Merkle et al. [7]).*

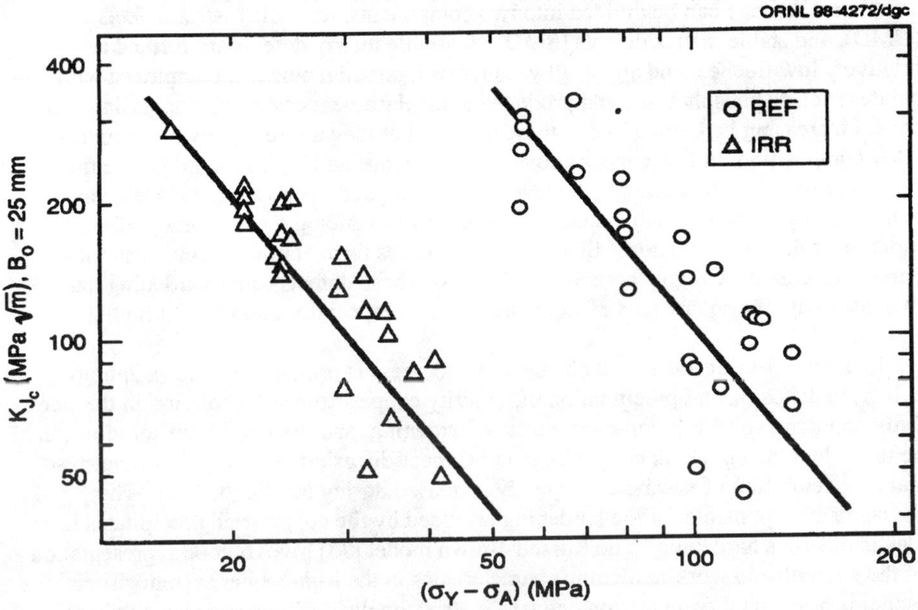

Figure 3 – *Fracture toughness vs. thermal component of yield strength for unirradiated and irradiated material. A533B C1.1 (JRQ) 12.5mm RCT (after Merkle et al. [7]).*

Influence of Irradiated Microstructure on Form of σ_{irr}

Neutron irradiation of RPV steels causes embrittlement effects marked by an increase in yield strength and a decrease in toughness. The increase in yield strength can be related quantitatively to the shift in transition temperature [*8*], and is due to the irradiation-induced fine scale microstructures that obstruct dislocation motion. These microstructures include precipitates, vacancy clusters (microvoids), and interstitial clusters (dislocation loops). Each of these microstructures has been shown to contribute to the three basic mechanisms that control RPV embrittlement [*9*]. They are listed below:

1. Matrix hardening due to radiation-produced point defect clusters and dislocation loops, referred to as the matrix damage contribution;
2. Irradiation-enhanced formation of copper-rich precipitates (age hardening);
3. Irradiation-caused grain boundary segregation of embrittling elements such as phosphorous.

The first two mechanisms cause an increase in yield strength, while the third mechanism causes grain boundary embrittlement without an increase in hardness. Matrix damage has been found to develop continuously during irradiation, exhibiting hardening that has a square root of fluence dependence. Copper-rich precipitates cause hardening which rises to a peak value and is unaffected by subsequent irradiation [*9*]. These two mechanisms are generally considered to be dominant in radiation embrittlement [*10*].

Matrix damage can be divided into two components, unstable matrix defects (UMD), and stable matrix defects (SMD). Unstable matrix defects are formed at relatively low fluence, and are small vacancy or interstitial clusters, complexed with solutes such as phosphorous, and produced in displacement cascades. Increasing flux causes increasing hardening due to these defects, but they are relatively independent of alloy composition. In low copper alloys, at low fluence and high flux, UMD are the dominant source of hardening. However, in high copper steels, these defects delay the copper-rich precipitate contribution to hardening by reducing the efficiency of radiation-enhanced diffusion. Stable matrix features form at high fluence, and include nanovoids and more highly complexed clusters. These defects cause hardening that increases with the square root of exposure which is especially important at high fluences [*11*].

In copper bearing steels, the change in yield stress is primarily due to radiation-enhanced diffusion and precipitation of impurity copper from solid solution in the steel. Only copper in solid solution contributes to hardening, and the fraction in solution can be much less than the bulk copper content. Copper diffusion occurs because the steady state concentration of vacancies is greatly increased during irradiation [*12*]. The mechanism responsible for the hardening provided by the copper-rich precipitates is elastic modulus hardening. The Russell-Brown model [*13*] gives a good representation of the strength and work-hardening characteristics in the iron-copper system. Their model is based on the interaction between matrix slip dislocations and second phase particles, in particular, copper precipitates, which have a shear modulus lower than the shear modulus of the Fe matrix. The strengthening effect of copper precipitates is shown to be the stress required to move dislocations large distances through an array of obstacles [*14*].

$$\Delta\tau = 0.8\frac{\mu b}{L}\left[1-\frac{U_1^2}{U_2^2}\right]^{1/2} \quad ; \quad \sin^{-1}\frac{U_1}{U_2} \le 50^\circ$$
$$\Delta\tau = \frac{\mu b}{L}\left[1-\frac{U_1^2}{U_2^2}\right]^{3/4} \quad ; \quad \sin^{-1}\frac{U_1}{U_2} \ge 50^\circ \tag{2a}$$

$$\frac{U_1}{U_2} = \frac{E_1^\infty \log\frac{r}{r_0}}{E_2^\infty \log\frac{R}{r_0}} + \frac{\log\frac{R}{r}}{\log\frac{R}{r_0}} \tag{2b}$$

where

$\Delta\tau$ = shear stress increase
μ = shear modulus of iron
b = magnitude of the Burger's vector
L = precipitate spacing
U_1 = energy of dislocation in copper
U_2 = energy of dislocation in iron

E_1^∞ / E_2^∞ = ratio of shear modulus of copper precipitate to shear modulus of iron for screw dislocation
R = outer cut-off radius used to calculate the energy of the dislocation ≈ 1000 r_0
r_0 = inner cut-off radius ≈ 2.5b
r = radius of copper precipitate

Models for Radiation Embrittlement

Existing models for irradiation embrittlement combine the two components of hardening. The Stoller model [*13*] computes a change in the shear stress $\Delta\tau$ via a root-square summation of the matrix damage and copper-rich precipitate contributions.

$$\Delta\tau = (\Delta\tau_{Cu}^2 + \Delta\tau_{ICL}^2 + \Delta\tau_{VCL}^2)^{1/2} \tag{3a}$$

where

$\Delta\tau_{Cu}$ = component due to copper precipitation
$\Delta\tau_{ICL}$ = component due to interstitial clusters
$\Delta\tau_{VCL}$ = component due to vacancy clusters

Matrix damage is attributed to small clusters of point defects – vacancies and interstitials, and each term contributes a change in shear stress, $\Delta\tau_{VCL}$ and $\Delta\tau_{ICL}$, respectively. A dispersed barrier model is used to calculate hardening due to these defects.

$$\Delta\tau = \frac{\mu b}{\alpha \ell} \tag{3b}$$

where

μ = shear modulus of iron
b = Burger's vector
$\ell = (Nd)^{-1/2}$ is computed from the number density (N) and diameter (d) of obstacles
α = measure of barrier strength

The contribution to hardening due to copper-rich precipitates is modeled using the Russell-Brown model shown earlier. The incremental increase in shear stress is given by Eqns. (2).

A Taylor factor is used to convert the calculated change in shear stress to a change in the uniaxial yield strength,

$$\Delta\sigma_y \approx 3.1\Delta\tau \tag{3c}$$

McElroy and Lowe [*15*] use the Fisher Model [*16*] of irradiation embrittlement in their study of Linde 80 weld metal. The Fisher Model is empirically derived, and is well established and widely accepted. The matrix damage component of the Fisher Model, in particular, relies heavily on empirical data. The increase in yield strength, $\Delta\sigma_y$, is a linear superposition of the two hardening components after irradiation time, t, and is given by

$$\Delta\sigma_y = \Delta\sigma_{dam} + \Delta\sigma_{Cu} \tag{4a}$$

where

$\Delta\sigma_{dam}$ = component due to matrix damage
$\Delta\sigma_{Cu}$ = component due to copper precipitation.

The matrix damage term $\Delta\sigma_{dam}$ is dependent only on the total accumulated fast neutron fluence, ϕt, and not neutron flux, ϕ.

$$\Delta\sigma_{dam} = A\, F_T(\phi t)^{1/2} \tag{4b}$$

$$F_T = 1.869 - 4.57\times10^{-3}T \tag{4c}$$

where

A = Barton coefficient
F_T = temperature correction factor; 1 at 190°C
T = temperature

The second component, $\Delta\sigma_{Cu}$, is the hardening due to copper-rich precipitates, and can be calculated from the maximum hardening for the given copper content, peak hardening time, and time dependence of hardening derived from empirical fitting to experimental thermal ageing data. Russell and Brown showed that this peak hardening value is proportional to the root of the volume fraction of copper, V_{Cu}, in the alloy [*16*]. The maximum contribution to hardening of the matrix due to copper precipitation is given by

$$\Delta\sigma_{Cu}^{max} = 4.5\times10^{3}V_{Cu}^{1/2} - 90 \text{ (MPa)} \tag{4d}$$

where

V_{Cu} = volume fraction of copper

McElroy and Lowe note that a study of model steels using the Field Emission Gun Scanning Electron Microscope (FEGSTEM) produced an inference based on copper solubility at temperatures from 500°C to 650°C that only 0.23wt% copper was in solid solution at the nominal post-weld heat treatment (PWHT) temperature of 1125°F (607°C) for the weld metals in their database. They established 0.23wt% copper as an upper bound in the model, assigning this value to weld metals having bulk copper contents more than this amount, and assigning the measured bulk values for metals containing less than the upper bound. By using this upper bound of solubility for copper in Linde 80 weld metals, they showed that the predicted and measured yield strength increases agree extremely well [*15*].

Both the Stoller and Fisher models can be used to determine the σ_{irr} term. McElroy and Lowe's success with using the Fisher model, as well as the model's dependence on fluence and the volume fraction of precipitates, makes it the more attractive model. Furthermore, data on these variables can be found in the literature. Work is ongoing to compile this data. The Fisher model is a pseudo-predictive model and the change in yield strength determined from the model can be inserted into the ZA equation. The yield strength of the irradiated material calculated through ZA should be the same as that determined experimentally.

ZA Equation and Irradiation Effects, Part II

The microstructures involved in the hardening mechanisms discussed above, precipitates and vacancy and interstitial clusters, are long-range barriers to dislocation motion in BCC metals [*4*]. As such, these barriers can then be accounted for in the athermal portion of Eqns. (1) by the aforementioned σ_{irr} term:

$$\sigma_{ZA} = d_0 + K\varepsilon^n + B_0 e^{-\beta T}, \tag{5a}$$

where

$$\beta = \beta_0 - \beta_1 \ln \dot{\varepsilon}, \qquad d_0 = \sigma_{irr} + \sigma_G + kd^{-1/2}. \tag{5b}$$

Ultimately this would result in a shift of the fracture toughness versus temperature curve along the temperature axis, but the curve shape would be unaffected, as shown by Merkle et al. [*7*]. Figure 4 is a plot of unirradiated and irradiated yield strength data with ZA predictions superposed over them. In the case of the irradiated data, the σ_{irr} term is approximated with the measured shift in yield strength. The resulting shape of the ZA prediction does not change with this addition, as is expected from the data. The corresponding fracture toughness values shown in Figure 5 demonstrate the consistency of the curve shape from the unirradiated to the irradiated state.

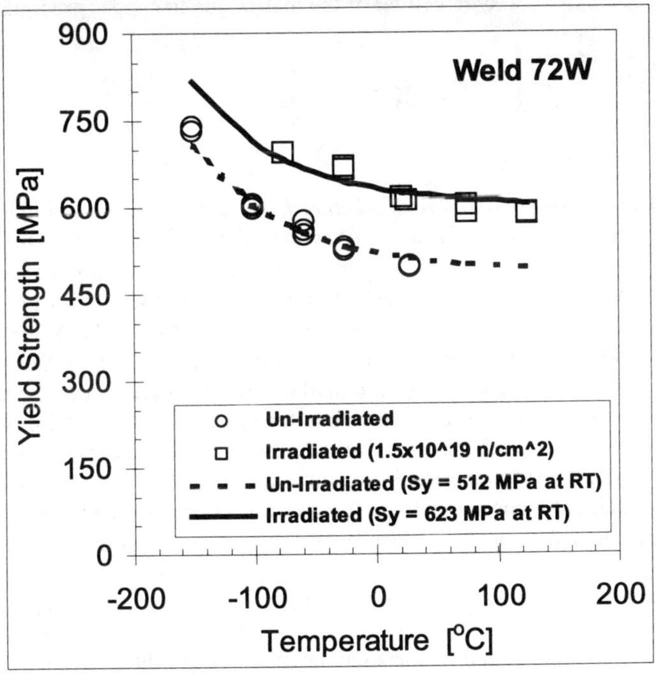

Figure 4 – *Irradiated and unirradiated yield strength data for weld 72W with ZA predictions for both conditions.*

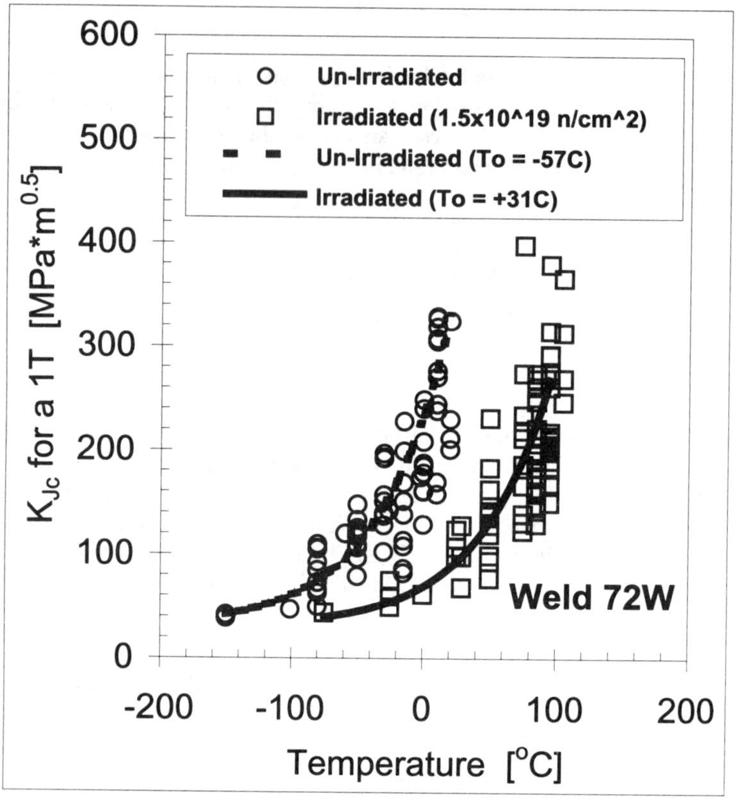

Figure 5 – *Irradiated and unirradiated fracture toughness data for weld 72W showing consistency of curve shape.*

Conclusions

Irradiation damage in RPV steels can be accounted for in the ZA constitutive model by including a new term, σ_{irr}. Only the athermal portion of the ZA flow stress is affected by irradiation. This is because irradiation-induced fine scale microstructures such as vacancy and interstitial clusters, and copper-rich precipitates, are long-range barriers to dislocation motion, and can be accounted for in the athermal portion of the yield stress. Models have been created to describe the matrix hardening due to these microstructures, and can be used to determine σ_{irr}.

Acknowledgements

Support of this project by the Electric Power Research Institute (EPRI) is gratefully acknowledged. This work was funded by EPRI under the Materials Reliability Program (MRP), Contract Number WO6542-05.

References

[1] Wallin, K., Saario, T., and Törrönen K., "Statistical Model for Carbide Induced Brittle Fracture in Steel," *Metal Science*, Vol. 18, January 1984, pp. 13-16.

[2] Natishan, M. E., and Kirk, M. T., "A Physical Basis for the Master Curve," *ASME Pressure Vessel and Piping Symposium*, American Society of Mechanical Engineers, New York, 1999.

[3] Natishan, M. E., and Kirk, M. T., "A Micromechanical Evaluation of the Master Curve," *Fatigue and Fracture Mechanics, Vol. 30, ASTM STP-1360,* K. L. Jerina and P. C. Paris, Eds., American Society for Testing and Materials, Philadelphia, 1998, pp. 51-60.

[4] Natishan, M. E., Wagenhofer, M., and Kirk, M. T., "Dislocation Mechanics Basis and Stress State Dependency of the Master Curve," *Fatigue and Fracture Mechanics, Vol. 31, ASTM STP-1389*, Halford and Gallagher, Eds., American Society for Testing and Materials, Philadelphia, 2000.

[5] Wagenhofer, M., Natishan, M. E., and Gunawardane, H. P., "A Physically Based Model to Predict the Fracture Toughness Transition Behavior of Ferritic Steels," *Engineering Fracture Mechanics*, submitted October 2000.

[6] Zerilli, F. J., and Armstrong R. W., "Dislocation-Mechanics-Based Constitutive Relations for Material Dynamics Calculations," *Journal of Applied Physics*, Vol. 65, No. 5, March 1987, pp. 1816-1825.

[7] Merkle, J. G., Wallin, K., and McCabe, D. E., *Technical Basis for an ASTM Standard on Determining the Reference Temperature, T_0, for Ferritic Steels in the Transition Range*. NUREG/CR-5504, US Nuclear Regulatory Commission, November 1998.

[8] Odette, G. R. and Griesbach, T. J., "Radiation Embrittlement Trend Curves and the Impact on Nuclear Plant Operating Criteria," *Radiation Embrittlement of Nuclear Reactor Pressure Vessels: An International Review (Third Volume), ASTM STP-1101*, L. E. Steele, Ed., American Society for Testing and Materials, Philadelphia, 1989, pp. 174-187.

[9] English, C. A., Phythian, W. J., and McElroy, R. J., "Microstructure and Modeling of RPV Embrittlement," *Microstructure Evolution During Irradiation, Vol. 439,* I. M. Robertson, G. S. Was, L. W. Hobbs, and T. D. de la Rubia, Eds., Materials Research Society, Pittsburgh, 1997, pp. 471-482.

[10] Pavinich, W. A., Griesbach, T. J., and Server, W. L., "An Overview of Radiation Embrittlement Modeling for Reactor Pressure Vessels," *Radiation Embrittlement of Nuclear Reactor Pressure Vessels: An International Review*

(Fourth Volume), ASTM STP-1170, L. E. Steele, Ed., American Society for Testing and Materials, Philadelphia, 1993, pp. 99-117.

[11] Odette, G. R., Mader, E. V., Lucas, G. E., Phythian, W. J., and English, C. A., "The Effect of Flux on the Irradiation Hardening of Pressure Vessel Steels," *Effects of Radiation on Materials: 16th International Symposium, ASTM STP-1175*, A. S. Kumar, D. S. Gelles, R. K. Nanstad, and E. A. Little, Eds., American Society for Testing and Materials, Philadelphia, 1993, pp. 373-393.

[12] Fabry, A., Van de Velde, J., Puzzolante, J. L., Van Ransbeeck, T., Verstrepen, A., Biemiller, E. C., Carter, R. G., and Petrova, T., "Research to Understand the Embrittlement Behavior of Yankee/BR3 Surveillance Plate and Other Outlier RPV Steels," *Effects of Radiation on Materials: 17th International Symposium, ASTM STP-1270*, D. S. Gelles, R. K. Nanstad, A. S. Kumar, and E. A. Little, Eds., American Society for Testing and Materials, Philadelphia, 1996, pp. 138-187.

[13] Stoller, R. E., "Pressure Vessel Embrittlement Predictions Based on a Composite Model of Copper Precipitation and Point Defect Clustering," *Effects of Radiation on Materials: 17th International Symposium, ASTM STP-1270*, D. S. Gelles, R. K. Nanstad, A. S. Kumar, and E. A. Little, Eds., American Society for Testing and Materials, Philadelphia, 1996, pp. 25-58.

[14] Russell, K. C., and Brown, L. M., "A Dispersion Strengthening Model Based on Differing Elastic Moduli Applied to the Iron-Copper System," *Acta Metallurgica*, Vol. 20, July 1972, pp. 969-974.

[15] McElroy, R. J., and Lowe, Jr., A. L., "Irradiation Embrittlement Modeling of Linde 80 Weld Metals," *Effects of Radiation on Materials: 17th International Symposium, ASTM STP-1270*, D. S. Gelles, R. K. Nanstad, A. S. Kumar, and E. A. Little, Eds., American Society for Testing and Materials, Philadelphia, 1996, pp. 68-85.

[16] Fisher, S. B., Harbottle, J. E., and Aldridge, N., "Radiation Hardening in Magnox Pressure-Vessel Steels," *Philosophical Transactions of the Royal Society of London. Series A, Mathematical and Physical Sciences*, Vol. 315, No. 1532, August 1985, pp. 301-332.

Hans-W. Viehrig,[1] Juergen Boehmert,[2] Jan Dzugan[3], and Holger Richter[4]

Master Curve Evaluation of Irradiated Russian VVER Type Reactor Pressure Vessel Steels

---

Reference: Viehrig, H.-W., Boehmert, J., Dzugan, J., and Richter, H., "**Master Curve Evaluation of Irradiated Russian VVER Type Reactor Pressure Vessel Steels,**" *Effects of Radiation on Materials: 20th International Symposium, ASTM STP 1405,* S. T. Rosinski, M. L. Grossbeck, T. R. Allen, and A. S. Kumar, Eds., American Society for Testing and Materials, West Conshohocken, PA, 2001.

Abstract: Results of a joint German/Russian irradiation program performed on the prototype pressurized water reactor VVER-2 of the Rheinsberg nuclear power plant (Germany) are summarized. The experiment comprises Charpy V-notch (CVN), precracked Charpy size (SENB) and compact tension (CT) specimens made of different heats of Russian VVER type reactor pressure vessel (RPV) base and weld metals. Reference temperatures, T_0, were evaluated according to the Master Curve (MC) concept using the multi temperature method. Neutron irradiation induced ductile-to-brittle transition temperature (DBTT) shifts determined on the basis of CVN and SENB tests are compared. On the base of the DBTT the neutron embrittlement sensitivity and the annealing behavior of tested RPV steels are evaluated. Different heats of the same VVER-RPV steel exhibit different neutron induced embrittlement and annealing behavior. The determined CVN transition temperatures correlates to the T_0 temperatures evaluated by the MC concept.

Keywords: reactor pressure vessel steel, integrity assessment, Charpy-V test, transition temperature, fracture toughness, reference temperature, predicting formulas, radiation embrittlement

Introduction

In the fracture mechanics integrity analysis of Russian VVER-type reactor pressure vessels (RPVs) the initiation fracture toughness curve is used, according to the VVER

---

[1] Senior Scientist, [2] Head of Department, [3] Research Scientist, Department of Material Behaviour and Component Safety, Institute for Safety Research, Forschungszentrum Rossendorf e.V., P.O.Box 510119, D-01314 Dresden, Germany.

[4] Research Scientist, now BHG Edelstahl GmbH Freital, P.O.Box 1-64, D-01691 Freital, Germany.

Code for Strength Evaluation of Component and Piping of Nuclear Power Plants (PNAE G-7-002-86). This code is applied for the normal operating and emergency conditions (pressurized thermal shock events) of RPVs. The original evaluation is modified in the IAEA Guideline (IAEA-EBP-WWER-08) on Pressurized Thermal Shock Analysis for VVER Nuclear Power Plants. The generalised lower bound fracture toughness (K_{IC}) curve equation (1) in the IAEA Guideline utilizes material normalizing and indexing parameters.

$$K_{IC} = 26+36\ \exp((0.02\cdot(T-T_K)) \qquad (1)$$

These parameters are the critical temperatures of brittleness, T_{K0} for the initial state and T_K after an operational time. Values of T_{K0} are determined from acceptance tests of vessel materials. The effects of neutron irradiation on toughness are assessed by shifting the curves upward in temperature by an amount which is equal to the ductile-to-brittle transition temperature (DBTT) shift of Charpy-V impact energy curve at 48-J level ($\Delta TT_{48\text{-}J}$) without changing the shape of the curve. This approach is similar to the American Society of Mechanical Engineers (ASME) Boiler and Vessel Code (ASME Code) procedure. The difference is in the determination of the indexing parameter. ASME Code uses a reference temperature, RT_{NDT}, based on a combination of Charpy-V and drop-weight nil-ductility transition (NDT) temperature tests. On the other hand, the T_K defined in the VVER Code PNAE G-7-002-86 is based only on Charpy-V impact tests. Generally, the attention is focused on the use of fracture toughness parameters in the RPV integrity assessment [*1*].

For structural steels, a "master curve" (MC) describes the temperature dependence of the fracture toughness within the brittle-to-ductile transition region. The MC concept proposed by Wallin [*2, 3*] considers size effect and statistical variability in a simple way. This MC concept is adopted in the Standard Test Method for Determination of Reference Temperature, T_0, for Ferritic Steels in the Transition Range (ASTM E 1921-97). Reference temperature T_0 is the temperature at which the value of the median fracture toughness, $K_{JC(med)}$, is equal to 100 MPam$^{1/2}$. Examples of the MC for RPV steels are described elsewhere [*4, 5*]. An actual objective is the development of a method using T_0 as a fracture toughness indexing parameter instead of RT_{NDT} or T_K.

This report presents results of a joint German/Russian research program performed in the VVER-2 reactor of the nuclear power plant Rheinsberg (Germany). The irradiation program was dedicated to investigate the influence of deleterious elements on the embrittlement of VVER type RPV steels due to neutron irradiation. The results are used for validity evaluation of the predictive formulas given in the Russian guide. A further target was the investigation of the annealing behavior of different Russian VVER type RPV steels. The irradiated specimens of this program were tested at Forschungszentrum Rossendorf (FZR) [*6*], MPA Stuttgart (MPA) [*7*] and FZ Jülich (FZJ) [*8*]. The emphasis of this contribution is the application of the master curve concept to the fracture mechanical characterization of different VVER-RPV steels investigated at FZR. A further target is the evaluation of irradiation sensitivity on the basis of Charpy-V and T_0 DBTTs.

Irradiation

About 800 specimens made from 24 different heats of Russian VVER-type RPV were irradiated. Base and weld metals were considered. Different irradiation rigs were loaded in the high flux irradiation positions of the Rheinsberg VVER-2 reactor for one reactor cycle each in the time period from 1984 to 1988. The VVER 2 is a prototype pressurized water reactor with 70 MW electric power and a mean temperature of the cooling water in the irradiation positions of 255°C. The VVER-2 of Rheinsberg NPP was chosen as irradiation facility because the large section of the high flux channel enables irradiation of large and numerous specimens. Additionally, the irradiation conditions are comparable with the VVER-440 conditions regarding neutron flux, neutron spectrum and irradiation temperature. In order to ensure intense cooling the specimens were covered with an electroplated nickel layer and placed in open irradiation rigs and, thus, in direct contact with the coolant. The specimens were irradiated by a medium flux rate of about $1.03 \cdot 10^{12}$ n.cm^{-2} s^{-1} (E>1MeV) .

The irradiation conditions and the detailed results of fast neutron flux and fluence calculations are presented in a research report [*9*]. The neutron fluences of different CVN and SENB specimens of individual test series vary up to the factor of about 2.

Materials and Specimens

A survey of all heats of RPV base (BM) and weld metals (WM) irradiated within the joint irradiation program is given in [*7,12*]. The material investigated by FZR contains heats of VVER-440 BM (15Kh2MFA), VVER-1000 BM (15Kh2NMFAA) and VVER-1000 WM (10KhGMAA). The chemical compositions and microstructures are summarized in Table 1. Table 2 provides the number of tested Charpy V-notched (CVN), precracked Charpy size (SENB) and precracked and side-grooved 1T-CT specimens.

According to the Russian specification the specimens were taken from 1/4 to 3/4 thickness positions and the V-notches are oriented as follows [*10*]:

- specimens of base metal: L-S,
- specimens of weld metal: T-S.

Metallographic investigations were executed to characterize the microstructure of different heats of VVER type RPV steels and to check the orientation of the specimens.
The results of microstructural investigations can be summarized as follows:

- The microstructure of all base metals is mainly bainitic. Heats of 15Kh2MFA base metal apart from heat R3 contain small volume fractions of proeutectoid ferrite (≤3%). Heats of 15Kh2NMFAA base metal do not contain ferrite.
- Prior austenitic grain size of 15Kh2MFA base metal (86 - 111 μm) is clearly larger than the grain size of 15Kh2NMFAA base metal (46 - 82 μm).
- The basic structure of the weld metal 10KhGMAA is bainitic-martensitic with columnar crystals.

Table 1 - *Ni, P and Cu content and microstructure of the investigated VVER-type RPV steels*

Material	Code (no. of heat)	Ni	P	Cu	Microstructure
VVER-440 BM 15Kh2MFA	R 1 (103405)	0.27	0.011	0.10	bainite,< 3% pro-eutectoid ferrite, MnS, MeO.
	R2 (106649)	0.12	0.014	0.12	bainite,< 3% pro-eutectoid ferrite, MnS, MeO; prior austenit grain size: 111 µm
	R3 (109868)	0.12	0.024	0.12	bainite; MnS, MeO; prior austenitgrain size: 86 µm
	D25	0.10	0.017	0.11	bainite,< 3% pro-eutectoid ferrite, MnS, MeO; prior austenit grain size: 93 µm
VVER-1000 BM 15Kh2NMFAA	R16 (KZY 5, 444807)	1.11	0.012	0.07	bainite; prior austenit grain size: 46 µm
	R17 (YU-1, fl, 435593)	1.30	0.013	0.13	bainite; prior austenit grain size: 111 µm
VVER-1000 WM 10KhGNMFAA	R19 (YU-1)	1.71	0.012	0.04	bainite

The volume fraction of inclusions was low for all heats. The shape of the inclusions does not give a hint on the orientation of the specimens in the forging in every case. Table 2 contains the presumed orientation of the specimens. It differs in some cases from the orientation according to the Russian specification.

After the irradiation the specimens were stored several years in a fuel storage pool. During the long storage period the nickel layer was destroyed and the surface of the specimens was corroded. Before testing the surface of each irradiated specimen was cleaned by a rotary iron brush. To ensure reproducible conditions the notch of the CVN specimens was finished by a special grinding dish and the fatigue crack in the SENB specimens was extended to the ratio a/W of about 0.5.

A part of the irradiated CVN specimens (Table 2) of all heats were annealed for 100 hours at 475°C in Argon atmosphere to evaluate the embrittlement mitigation by annealing.

Table 2 - *Charpy-V and SENB specimen characterization*

Material	Code	Number of specimen unirradiated	Number of specimen irradiated	Notch plane orientation [1)]
VVER-440 BM 15Kh2MFA	R1	14 CVN [2)] 10 1T-CT 12 [3)] SENB	30 CVN	unknown
	R2	12 CVN 12+6 [4)] SENB	20 CVN 9 SENB	L-S
	R3	17 CVN 12+6 [4)] SENB	20 CVN 10 SENB	T-S (S-T)
	D25	10 CVN	20 CVN	unknown
VVER-1000 BM 15Kh2NMFAA	R16	12 CVN 10+6 [4)] SENB	30 CVN 10 SENB	L-T
	R17	8 [4)] CVN 11+6 [4)] SENB	30 CVN 10 SENB	L-T
VVER-1000 WM 10KhGMAA	R19	12 CVN 11+6[4)] SENB	30 CVN 10 CVN	T-S

1) according to metallographic investigations
2) tested at the FZJ [*9*]
3) reconstituted specimens: inserts from broken unirradiated 1T-CT specimens
4) reconstituted: inserts from broken unirradiated CVN and SENB specimens

Test Methods

Charpy-V impact tests were carried out on an instrumented 300 J impact-pendulum with a DIN tup. Fracture toughness measurements were performed in accordance with ASTM E 1921-97. SENB and 1T-CT specimens were monotonously loaded by servohydraulic test system MTS 810 - Test Star. The measured J-integral values at brittle failure were transformed into elastic plastic stress intensities factors K_{JC}. The K_{JC} data of each test series were evaluated according to ASTM E1921-97 in the following sequence:

- Check of the measured K_{JC} whether they comply with the validity criteria. K_{JC} values above the validity criterion were censored on the validity limit.
- Adjustment of the measured K_{JC} values on a specimen size of 1T thickness (25.4 mm). K_{JC} values below 50 MPam$^{1/2}$ were not size adjusted.
- Calculation of the reference temperature, T_0 , according to the multiple temperature method [*2, 3*], including all valid and censored values.
- Calculation of MC and tolerance bounds.

Results and Discussion

The transition temperatures (TT) and upper shelf energies (USE) measured by the instrumented Charpy-V impact tests are summarized in Table 3.

Table 3 - *Results of the Charpy-V tests of unirradiated (U), Irradiated (I) and annealed (IA) specimens*

Code	Condition	Φ^{mean}	TT_{48J}	ΔTT_{48J}	R^{TT}	USE	ΔUSE	R^{USE}
		10^{18}	°C	K	%	J	J	%
R 1	U	0	18	-	-	155	-	-
	I	43.6	31	13	-	164	9	-
	IA		11	-7	154	171	16	-
R 2	U	0	-55	-	-	211	-	-
	I	80.7	-23	32	-	178	-33	-
	IA		-57	-2	107	227	16	148
R 3	U	0	7	-	-	171	-	-
	I	45.7	54	47	-	110	-61	-
	IA		14	7	87	138	-33	46
D 25	U	0	-70	-	-	160	-	-
	I	127.6	6	76	-	100	-50	-
	IA		-44	26	66	137	-23	62
R16	U	0	-81	-	-	196	-	-
	I	46.0	-16	65	-	125	-71	-
	IA		-39	42	31	188	-8	89
R17	U	0	-48	-	-	196	-	-
	I	71.8	76	124	-	149	-47	-
	IA		-46	-3	98	202	6	112
R19	U	0	-16	-	-	110	-	-
	I	65.1	175	191	-	60	-50	-
	IA		-1	15	92	120	10	120

Φ^{mean} mean fluence of the test series in $n.cm^{-2}$ (E > 0.5 MeV)

Apart from the heats R1 and R3 of 15Kh2MFA steel the RPV steels exhibit sufficient Charpy-V toughness parameters in the unirradiated condition. The Charpy-V TT of both heats (R1, R3) are above 0°C and so outside the specification for materials in core shells (IAEA-EBP-WWER-08). Chemical composition, metallographic structure and pretreatment do not explain the reason of this behavior. Perhaps, different orientation (probably T) is a possible but not sufficient explanation. The initial toughness properties of the weld metal 10KhGNMFAA are clearly worse than those of the VVER-1000 base metal but they are within the range of the specification.

The results of the reference temperature, T_0, evaluation according to ASTM E1921-97 and the multiple temperature procedure are summarized in Table 4.

Table 4 - *Reference temperatures T_0 determined on unirradiated (U) and irradiated (I) specimens*

Code	cond.	Φ^{mean} / 10^{18}	N	Σn_i	T_0 / °C	ΔT_0 / K	remarks
R1	U		12	1.2	-50		R-specimens
R1	U		10	1.0	-34		1T-CT-specimens
R2	U		8	1.0	-142		C- and R-specimens
R2	I	82.2	9	1.0	-104	38	C-specimens
R3	U		10	1.2	-84		C- and R-specimens
R3	I	51.4	7	1.0	11	95	C-specimens
D25	U		8	1.0	-131		R-specimens
R16	U		10	1.2	-150		R-specimens
R16	I	40.9	9	1.3	-93	57	C-specimens
R17	U		10	1.2	-107		C- and R-specimens
R17	I	73.7	10	1.5	2	109	C-specimens
R19	U		15	1.8	-105		C- and R-specimens
R19	I	61.9	7	0.8	45	150	C-specimens

C-specimen: compact specimen, R-specimen: reconstituted specimen
N: number of evaluated K_{JC}- values

Figure 1 shows all K_{JC} values obtained in this study and the MC together with the 5 and 95% tolerance bounds as a function of the difference between test temperature, T, and reference temperature, T_0. The K_{JC} values were size-corrected to 1T thickness according to ASTM E1921-97. In the range $T_0 \pm 50K$ the Master Curve agrees with the measured K_{JC} values. It is evident that the K_{JC} values of the different heats strongly scatter especially close to the T_0 temperature. Many K_{JC} values are outside the tolerance bounds. The observed scatter indicates inhomogeneities in the structure of the investigated VVER type RPV steels in accordance to metallographic examinations which reveal strong segregation effects.

For the evaluation of the irradiation sensitivity of the tested RPV steels the predicting formula of the Russian code was applied. It predicts the effect of the irradiation on the upper bound of the TT shift (IAEA-EBP-WWER-08) [*11, 12*]:

$$\Delta TT_F = A_F \cdot \sqrt[3]{\Phi} \qquad (2)$$

where

ΔTT_F predicted TT shift,

A_F irradiation embrittlement coefficient and

Φ neutron fluence in 10^{18} n/cm^2 (E > 0.5 MeV).

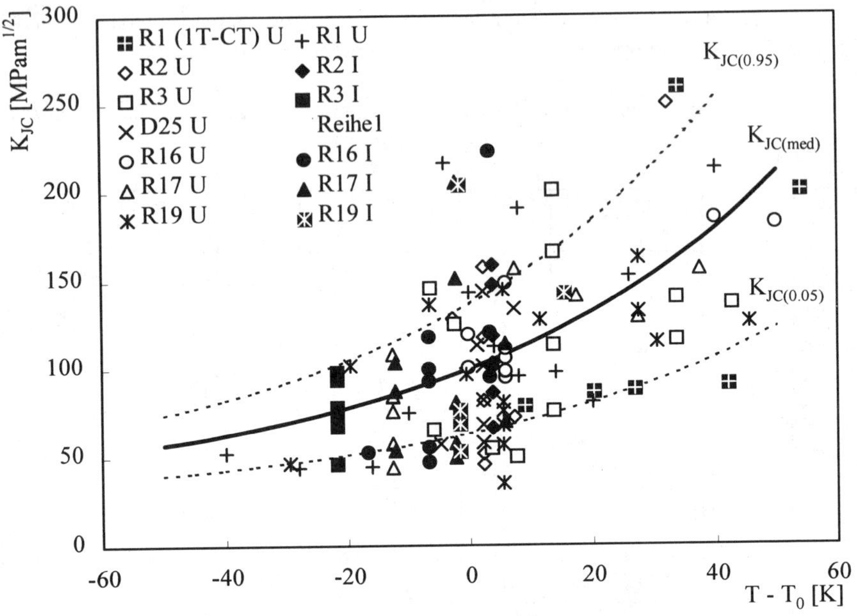

Figure 1 - *Fracture toughness values corrected to 1T and Master Curve*

The irradiation embrittlement coefficient depends on material composition and irradiation temperature. For the VVER-RPV materials following irradiation embrittlement coefficients (A_F) are applied (IAEA-EBP-WWER-08):

- VVER-440 base metal: 18 weld metal: 15 and
- VVER-1000 base metal: 23 weld metal: 20.

These embrittlement coefficients are valid for irradiation temperatures of 270°C (VVER-440) and 290°C (VVER-1000). The TT shift is conservatively estimated with these values for VVER-steels of a chemical composition within the specification. A_F can be adapted to other irradiation temperatures by equation 3 [*11, 13*].

$$A_F^{T_{irr}} = A_F^{T_V} + K \cdot (T_V - T_{irr}) \tag{3}$$

where

T_V temperature for which the factor is valid,

T_{irr} irradiation temperature and

K 0.2 for base and 0.4 for weld metal.

To evaluate the irradiation sensitivity, the irradiation embrittlement coefficient (A_F^m) was calculated from the measured TT and T_0 shift and the mean fluence (Tables 3 and 4) using equation 2. The results are given in Table 5. Furthermore, Table 5 contains the A_F values evaluated according to Eq. 2 and the constant maximum allowed A_F values. These A_F

values are related to the irradiation temperature of 255°C (A_F^{255}) according to equation 3.

Table 5 - *Evaluated coefficients of embrittlement and DBTT*

code	A_F^{255}	Charpy-V Φ^{mean} 10^{18} ncm^{-2}	Charpy-V ΔTT_F^{255} K	Charpy-V ΔTT_{48J} K	Charpy-V A_F^m	Master Curve Concept Φ^{mean} 10^{18} ncm^{-2}	Master Curve Concept ΔTT_F^{255} K	Master Curve Concept ΔT_0 K	Master Curve Concept A_F^m
R1	21	43.6	74	12	3.4				
R2	21	80.7	91	32	7.4	82.2	91	38	8.7
D25	21	127.6	106	76	15.1				
R3	21	45.7	75	47	13.1	51.4	78	95	25.6
R16	30	46	107	65	18.1	40.9	103	57	16.5
R17	30	72.7	125	119	29.7	73.7	126	109	26.2
R19	34	65.1	137	191	47.5	61.9	134	150	37.9
R19 1)	34	35	111	125 2)	38.2				

A_F^{255} constant irradiation embrittlement coefficient for an irradiation temperature of 255°C

ΔTT_F^{255} transition temperature shift calculated with A_F^{255}

1) tested by MPA [*7*] with sub-size Charpy-V specimens machined from 1T-CT specimens

2) ΔTT evaluated with TT_{41J} of the Charpy-V specimens (FZR) and $TT_{1,9J}$ + 65K of the sub-size Charpy-V specimens tested by MPA [*7*]

Figure 2 shows the predicted DBTT shift calculated by use of A_F^{255} values according to equation 2 in comparison with the measured TT shift.

The effect of annealing can be described by a recovery parameter R, defined as change of TT or USE according to equation 4 [*14*]:

$$R^{CVP} = \frac{CVP_I - CVP_{IA}}{CVP_I - CVP_U} \cdot 100 \quad [\%] \tag{4}$$

where

R^{CVP} is the recovery regarding a Charpy-V parameter (TT, USE),

$CVP_{U,I,IA}$ Charpy-V parameter (TT, USE) in the unirradiated (U), irradiated (I) and irradiated and post-irradiation annealed (IA) state.

Table 3 contains the R parameter for the Charpy-V parameter, TT_{48J} and USE, of the investigated RPV steels. In Fig. 3 the R parameters are compared.

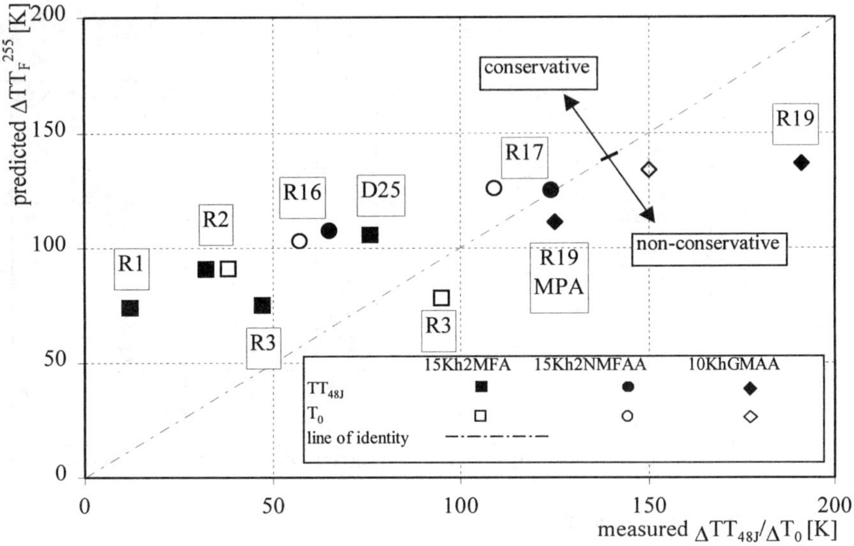

Figure 2 - *Comparison between predicted and measured DBTT shifts*

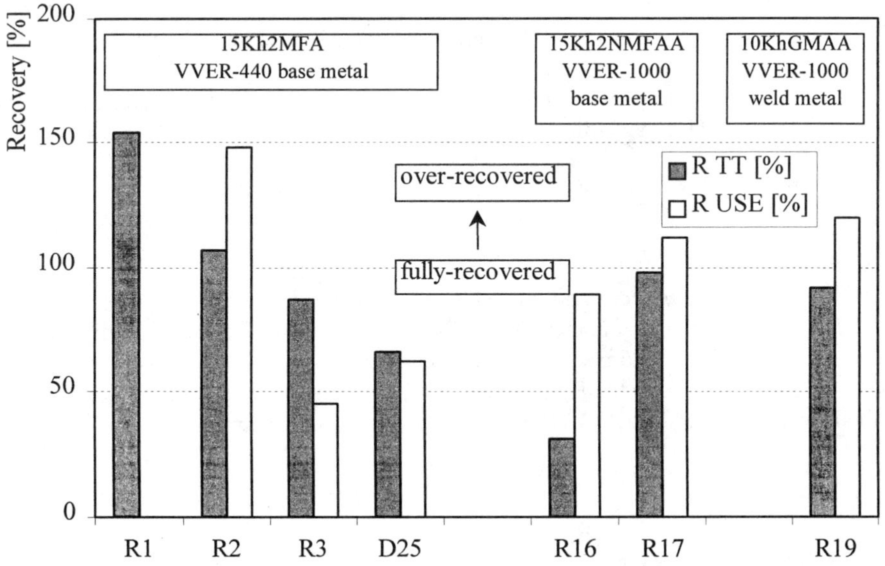

Figure 3 - *Recovery of Charpy-V parameters*

The irradiation and annealing behavior of investigated RPV steels can be assessed on the basis of the available results as follows:

VVER-440 base metal 15Kh2MFA (heats R1, R2, R3 and D25)

- The different heats have nearly the same Cu content but there is difference in the P content (Table 1). A clear dependence of the A_F^m values on the phosphorus content could not be proved by the results. It is shown that heat D25 has the highest irradiation sensitivity A_F^m although the P content is lower than that of heat R3. The mean fluence of heat D25 is about 3 times higher than for heat R3, possibly the fluence dependence does not correspond to the 3rd root rule (Equation 2).
- The TT shift due to neutron irradiation can be conservatively estimated by equation 2 using the maximum limit of A_F^{255}as Figure 2 shows.
- The shift of the T_0 temperature due to neutron irradiation of heat R3 is about twice as high as the Charpy-V transition temperature shift (Table 5).
- In general, the TT and USE are reduced as expected after the irradiation and recovered by subsequent annealing (Table 3). The heats differ in the recovery behavior. Heats R1 and R3 are "over-recovered", i.e. a lower TT and a higher USE are reached than in the unirradiated state (Figure 3). This effect is also known from the literature [*14*]. The USE over-recovers stronger than the TT.

VVER-1000 base metal 15Kh2NMFAA (heats R16 and R17)

- The content of deleterious elements is out of the limits for core shell material [*12*] for both heats in the case of P and for one heat (R17) in the case of Cu. As Figure 2 shows, the TT and T_0 shifts are, nevertheless, conservatively estimated with A_F^{255}.
- Heat R17 with the highest Cu content and neutron fluence provides the largest TT shift (ΔTT_{48J} and ΔT_0) and USE reduction (Tables 3 and 4).
- Both heats exhibit differences in the recovery. While the USE recovers nearly completely for both heats the TT_{48J} recovery is only complete for heat R17 and low for heat R16. The different recovery behaviour is also reflected in Fig. 3.

VVER-1000 weld metal 10KhGNMFAA (heat R19)

- The weld metal has a relatively high Ni content and P content above the limit for core shell material [*12*] and the A_F^m value measured is clearly higher than the A_F^{255}.
- Irradiation induced shift of TT_{48J} of 191 K and T_0 shift of 150 K were measured. Both DBTT shifts are higher than predicted (Table 5).
- Annealing causes almost complete recovery in the TT shift and an over-recovery in USE (Figure 3).

The measured 48-J Charpy-V DBTT (TT_{48J}) of the irradiated VVER base metals are below the critical temperatures of brittleness, T_K wich are determined on the basis of an PTS analysis for VVER-440 and VVER-1000 PWRs [*15*]. Table 3 shows for the VVER-1000 weld metal (code R19) that a neutron irradiation up to about the end of life (EOL) fluence [*15, 16*] caused a TT_{48J} of 175°C. This is above the critical T_K temperature for VVER-1000 weld metal [*15*]. For the assessment of the results it must be noted that the irradiation temperature amounted to 255°C and thus, 15 or 35 K lower than the inlet temperature of cooling water of the VVER-440 or VVER-1000 reactors. In Reference [*16*] a difference in the DBTT value of 85 K is indicated between an irradiation at 250°C and 290°C by a fluence of $6{\cdot}10^{19}$ n/cm^2 (E>0.5 MeV). In the case of a conversion of the measured TT_{48J} shift to an irradiation temperature of 290°C according to equation (3) the value of 191 K is reduced to 135K. The adjustment by 85 K results in a shift of 90 K.

Regarding the influence of nickel on the embrittlement of VVER-1000 RPV steels there are different statements in the literature. According to results of the Kurchatov Institute Moscow, nickel content up to 1.5% does not significantly increase the embrittlement due to neutron irradiation [*16*]. Although the results of the KI indicate, that the fluence dependency on the DBTT shift does not follow the third root law of equation (2), the shift of the DBTT according to this equation and the A_F coefficients can be conservatively predicted up to the EOL fluence. Further investigations, however, revealed that the embrittlement of VVER-1000 weld metals rise more with nickel contents greater 1.2%, than predicted in the Russian Code. Prokovsky [*16*] reports a very strong rise of the DBTT shift for steels with nickel contents greater 1.5% and fluencies above $6{\cdot}10^{19}$ n.cm^{-2} (E>0.5 MeV). The Ni-content (1.7%) and the neutron fluence of the present VVER-1000 weld (code R19) corresponds to these conditions. A comparison of the TT shift determined on sub size Charpy specimens from MPA Stuttgart confirmes the strong DBTT shift of this heat (Figure 3, Table 5) [*7*].

As expected, the fracture toughness based T_0 temperatures of the tested RPV steels are below the Charpy-V transition temperatures TT_{48J}. For the irradiated VVER-1000 weld metal (code R19) a T_0 temperature of 45°C was measured. The ASME Section XI Code Case includes a T_0-based reference temperature according to the following relationship:

$$RT_{T_0} \equiv T_0 + 19K \tag{5}$$

Using this Code Case the RT_{T0} temperature of the base metals are far away from critical temperatures of brittleness, apart from VVER-1000 weld metal heat R19, which is close to the critical temperatures of brittleness. The RT_{T0} temperature of 64°C for the VVER-1000 weld metal (code R19) can be conservatively estimated by the predicting formula of equation (2).

Figure 4 shows the T_0 temperatures of the unirradiated and irradiated specimens in comparison to the Charpy-V transition temperature TT_{48J}. The R19 heat is not included in the correlation because the weight factor according to ASTM E1921 is not fullfiled (Table 4). Both DBTTs indicate a good correlation. The slope well agrees with correlation between Charpy-V derived DBTTs and T_0 presented in References [*18, 19, 20*], but the offset is different. This fact could be caused by scatter of the values and different Charpy-V parameters (TT_{48J}, TT_{41J} and TT_{28J}) in the published correlations. Moreover, the Charpy-V test results could be influenced by different impact tups (ASTM, ISO).

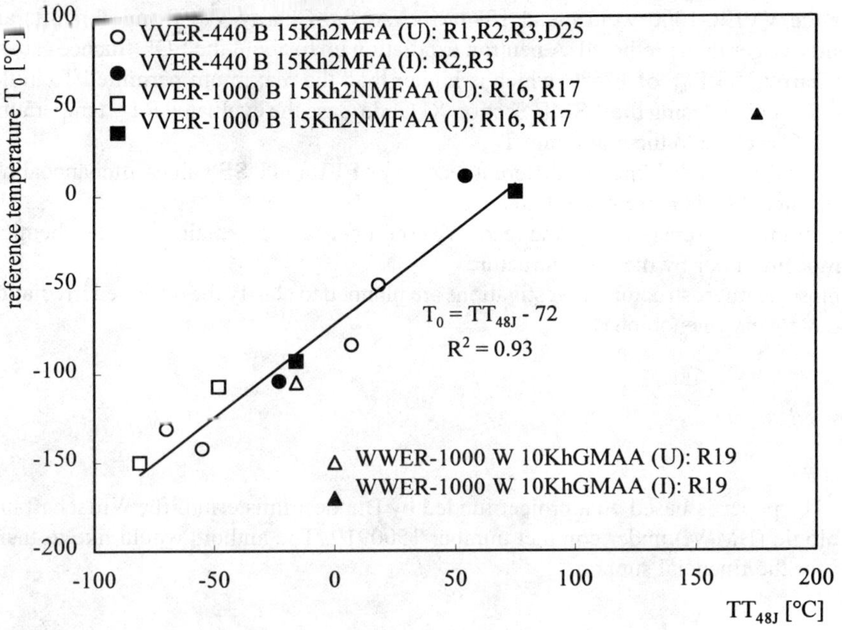

Figure 4 - *Correlation between TT_{48J} and T_0 DBTTs*

Conclusions

CVN and SENB specimens of 7 different VVER-440 and VVER-1000 type RPV steels were tested. Instrumented Charpy V impact tests and static fracture toughness test were performed. On the basis of the test results the shifts of TT, USE and T_0 were determined with respect to the irradiation and annealing treatment. The evaluation of the irradiation behavior regarding individual parameters was very complicated because of the variable chemical composition, microstructure, orientation of the specimens and neutron fluence.

The following conclusions can be stated from the FZR results:

1. The heats of VVER-440 and VVER-1000 base metal exhibit different irradiation sensivity. Generally, the TT shifts due to neutron irradiation could be conservatively estimated by the predicting formula and the maximum irradiation embrittlement coefficients.
2. The Charpy-V transition temperature shifts due to neutron irradiation could be confirmed by the reference temperatures T_0 shifts. For one heat of WWER-440 base metal, the T_0 shift is about two times higher than the Charpy-V transition temperature shift.

3. For the VVER-1000 weld metal 10KhMNAA, the Charpy-V TT and T_0 shifts are clearly larger than predicted. A neutron irradiation up to about the EOL fluence caused a Charpy-V TT_{48J} of 175°C which is higher than the maximum permitted T_K in the Russian guide. Using the ASME Section XI Code Case the evaluated RT_{T0} temperature of 62°C is close to the maximum T_K.
4. The RPV materials indicated different recovery of TT and USE values after annealing. As a rule, USE is more strongly recovered.
5. The irradiation sensivity and the recovery could neither be explained by the chemical composition nor by the microstructure.
6. Nanoscale microstructural investigations are planned to clarify the observed irradiation and recovery phenomena.

Acknowledgments

The paper is based on a project funded by Bundesministerium für Wirtschaft und Technologie (BMWi) under contract number 1500919. The authors would like to thank BMWi for the financial support.

References

[1] Brumovsky, M., Falcnik, M., Kytka, M., Malek, J., and Novosad, P., "Comparison of Static and Dynamic Transition Temperature Shifts in VVER Reactor Pressure Vessel Steels," *Effects of Radiation on Materials: 19th International Symposium, ASTM STP 1366*, M. L. Hamilton, A. S. Kulmar, S. T. Rosinsky, and M. L. Grossbeck, Eds., American Society for Testing and Materials, West Conshohocken, PA, 2000, pp. 56-68.

[2] Wallin, K., "Small Specimen Fracture Toughness Characterisation-State of the Art and Beyond," *Proceedings of the Advances in Fracture Research, ICF 9,* B. L. Karihaloo, Y. W. Mai, M. I. Ripley, R. O. Ritchie, Eds., Amsterdam, 1997, pp. 2333-2344.

[3] Wallin, K., "Master Curve Analysis of Ductile to Brittle Transition Region Fracture Toughness Round Robin Data - the Euro Fracture Toughness Curve,"*VTT Publications 367*, VTT Espoo, Finland.

[4] Server, W. L. and Rosinsky, S. T., "Technical Basis for Application of the Master Curve Approach to Reactor Pressure Vessel Integrity Assessment," *Effects of Radiation on Materials: 19th International Symposium, ASTM STP 1366*, M. L. Hamilton, A. S. Kulmar, S. T. Rosinsky, and M. L. Grossbeck, Eds., American Society for Testing and Materials, West Conshohocken, PA, 2000, pp. 127-142.

[5] Kirk, M., Lott, R., Server, W. L., Hardies, R. and Rosinsky, S., "Bias and Precision of T0 Values determined Using ASTM Standard E1921-97 for Nuclear Pressure Vessel Steels," *Effects of Radiation on Materials: 19th International Symposium, ASTM STP 1366*, M. L. Hamilton, A. S. Kulmar, S. T. Rosinsky, and M. L. Grossbeck, Eds., American Society for Testing and Materials, West Conshohocken, PA, 2000, pp. 142-161.

[6] Viehrig, H.-W., Boehmert, J., Richter, H., "Untersuchung des Zusammenhanges zwischen bruchmechanischen und mechanisch-technologischen Kennwerten bei bestrahlten Reaktordruckbehälterstählen", FZR, *Reaktorsicherheitsforschung Vorhaben 1500919, Abschlussbericht, Berichts-Nr.: FZR-FWSM 6/2000*, Rossendorf, April 2000.

[7] Föhl, J., Weißenberg, T., "Einfluß der Neutronenbestrahlung auf die Werkstoffeigenschaften sowjetischer RDB-Stähle," Staatliche Materialprüfungsanstalt (MPA) Universität Stuttgart, *Reaktorsicherheitsforschung Vorhaben 1500892, Abschlussbericht, Berichts-Nr.: 872701005*, Stuttgart, März 1999.

[8] Gottschalk, H., Derz, H., Pachur, D., "Präzisierung der Aussagen zur Sprödbruchsicherheit von WWER-Reaktordruckbehältern," *Abschlußbericht zum Fördervorhaben 1500869/9 Reaktorsicherheitsforschung*, Forschungszentrum Jülich, Berichts-Nr. KFA-IRW-IB-8/92, Oktober 1992.

[9] Barz, H.-U., Boehmer, B., Konheiser, J. and Stephan, I., "Ermittlung der Neutronendosis von bestrahlten WWER-Reaktormaterialien," *Report Forschungszentrum Rossendorf e.V. FZR-87*, Juni 1995.

[10] Kryukov, A., "Unirradiated Data to the Material of Irradiation Programme Rheinsberg Within the Scientific Technical Co-operation," *Internal Report of the Russian Research Centre Kurchatov Institute*, January 1994.

[11] Petrequin, P., "A Review of Formulas for Predicting Irradiation Embrittlement of Reactor Vessel Materials", *AMES Report No. 6, Catalogue Number: CD-NA-16305 EN-C*, Brussels-Luxembourg, 1996.

[12] Davis, L.M., "A Comparison of Western and Eastern Nuclear Reactor Pressure Vessel Steels," *AMES Report No. 10, Catalogue Number: CD-NA-17327 EN-C*, Brussels-Luxembourg, 1997.

[13] Alekseenko, N. N, Amaev,A. D., Gorynin, I. V., Nikolaev, V. A., "Radiation Damage of Water-Water Reactor Steel," *Energoizdat*, Moscow 1981.

[14] Iskander,S. K., Sokolov,M. A. and Nanstad, R. K., "Comparison of Different Experimental and Analytical Measures of Thermal Annealing Response of Neutron-Irradiated RPV Steels," *Effects of Radiation on Materials: 18th International Symposium, ASTM STP 1325*, R.K. Nanstad, M. L. Hamilton, F. A. Garner, and A. S. Kumar, Eds., American Society for Testing and Materials, West Conshohocken, PA, 1999, pp. 403-420.

[15] Akbashev, I. E., Maximov, Yu. M., Piminov, V. A., "Status of the Problem on the Analysis of WWER Vessel Integrity in Russia," *Workshop on "VVER-RPV Integrity Assessment - Methods and Applications"*, February 23-25 2000 at GRS in Colone, Germany.

[16] Prokovsky, V., "Problems of Assessment of Pressure Vessel VVER-1000 Radiation Lifetime," *Workshop on VVER-RPV Integrity Assessment - Methods and Applications*, February 23-25 2000 at GRS in Colone (Germany)

[17] Amaev, A. D., Erak, D. Yu., Kryukov, A. M., "Radiation Embrittlement of Wwer-1000 Pressure Vessel Materials," *Proceedings of the IAEA Specialists Meeting held in Madrid, Spain 26-29 Aprill 1999, IAEA Working Document IWG-LMNPP-99/2*, Vienna, Austria, pp. 374-385.

[18] Sokolov, M., Nanstad, R.K., "Comparison of Irradiation-Induced Shifts of KJC and Charpy Impact Toughness for Reactor Pressure Vessel Steels," *Effects of Radiation on Materials: 18th International Symposium, ASTM STP 1325*, R. K. Nanstad, M. L. Hamilton, F. A. Garner, and A. S. Kumar, Eds., American Society for Testing and Materials, American Society for Testing and Materials, West Conshohocken, PA, 1999, pp. 167-190

[19] Lee, B. S., Yang, W. J., Huh, M. Y., Chi, S.-H., Hong, J. H., "Master Curve Characterization of Irradiation Embrittlement Using Standard and 1/3-Size precracked Charpy Specimens," *Effects of Radiation on Materials: 20th International Symposium, ASTM STP 1405*, S. T. Rosinski, M. L. Grossbeck, T. R. Allen and A. S. Kulmar, Eds., American Society for Testing and Materials, West Conshohocken, PA, 2002

[20] Onizawa, K., Suzuki, M., "Comparision of Transition Temperature Shifts Between Static Fracture Toughness and Charpy-V Impact Properties due to Irradiation and Post-Irradiation Annealing for Japanese A533B1-1 Steels," *Effects of Radiation on Materials: 20th International Symposium, ASTM STP 1405*, S. T. Rosinski, M. L. Grossbeck, T. R. Allen and A. S. Kulmar, Eds., American Society for Testing and Materials, West Conshohocken, PA, 2002

M. A. Sokolov,[1] J. P. Robertson,[1] L. L. Snead,[1] D. J. Alexander,[2] P. Ferguson,[2] M. R. James,[2] S. A. Maloy,[2] W. Sommer,[2] G. Willcutt,[2] and M. R. Louthan[3]

Fracture Toughness Characterization of 304L and 316L Austenitic Stainless Steels and Alloy 718 After Irradiation in High-Energy, Mixed Proton/Neutron Spectrum

---

Reference: Sokolov, M. A., Robertson, J. P., Snead, L. L., Alexander, D. J., Ferguson, P., James, M. R., Maloy, S. A., Sommer, W. F., Willcutt, G., and Louthan, M. R., **"Fracture Toughness Characterization of 304L and 316L Austenitic Stainless Steels and Alloy 718 After Irradiation in High-Energy, Mixed Proton/Neutron Spectrum,"** *Effects of Radiation on Materials: 20th International Symposium, ASTM STP 1405,* S. T. Rosinski, M. L. Grossbeck, T. R. Allen, and A. S. Kumar, Eds., American Society for Testing and Materials, West Conshohocken, PA, 2001.

Abstract: This paper describes the fracture toughness characterization of annealed 304L and 316L stainless steels and precipitation hardened Alloy 718, performed at the Oak Ridge National Laboratory as a part of the experimental design and development for the Accelerator Production of Tritium (APT) target/blanket system. Materials were irradiated at 25 to 200°C by high-energy protons and neutrons from an 800-MeV, 1-mA proton beam at the Los Alamos Neutron Science Center (LANSCE). The proton flux produced in LANSCE is nearly prototypic of anticipated conditions for significant portions of the APT target/blanket system. The objective of this testing program was to determine the change in crack-extension resistance of candidate APT materials from irradiation at prototypic APT temperatures and proton and neutron fluxes. J-integral-resistance (J-R) curve toughness tests were conducted in general accordance with the American Society for Testing and Materials Standard Test Method for Measurement of Fracture Toughness, E 1820-99, with a computer-controlled test and data acquisition system. J-R curves were obtained from subsize disk-shaped compact tension specimens (12.5 mm in diameter) with thicknesses of 4 mm or 2 mm. Irradiation up to 12 dpa significantly reduced the fracture toughness of these materials. Alloy 718 had the lowest fracture toughness in both the unirradiated and irradiated conditions. All irradiated specimens of Alloy 718 failed by sudden unstable crack extension regardless of dose or test temperature. Type 304L and 316L stainless steels had a high level of fracture toughness in the unirradiated condition and exhibited reduction in fracture toughness to saturation levels of 65 to 100 MPa$\sqrt{m}$. The present reduction in fracture toughness is

---

[1]Oak Ridge National Laboratory, Oak Ridge, TN 37831-6151.
[2]Los Alamos National Laboratory, Los Alamos, NM 87545.
[3]Westinghouse Savannah River Technology Center, Aiken, SC 29808.

similar to changes reported from fission reactor studies. However, the currently observed losses in toughness appear to saturate at doses slightly lower than the dose required for saturation in reactor-irradiated steels. This difference might be attributed to the increased helium and hydrogen production associated with irradiation in the high-energy, mixed proton/neutron spectrum.

Keywords: stainless steel, spallation neutron source, fracture toughness

Introduction

This paper describes the fracture toughness characterization of 304L and 316L stainless steels, and nickel base Alloy 718 performed at the Oak Ridge National Laboratory (ORNL) as a part of the Accelerator Production of Tritium (APT) Project. The APT facility will use an accelerator to provide a spallation neutron source capable of producing tritium (^3H) through the $^3He(n,p)^3H$ reaction. Consequently, the target/blanket (T/B) components of the APT system will be exposed to unique conditions, including a high-energy mixed proton and neutron spectra. Only a limited data base is available for material applications in such an environment. Thus, a comprehensive material testing program was undertaken within the APT Project. This program includes the irradiation of candidate APT T/B system component materials at 25 to 200°C by exposure to high-energy protons and neutrons at the Los Alamos Neutron Science Center (LANSCE) at 800 MeV. The proton flux produced in LANSCE is nearly prototypic of anticipated conditions for significant portions of the APT T/B system.

Background

The APT Project will produce tritium through the $^3He(n,p)^3H$ reaction. The ^3He used for ^3H production will be contained, at approximately 120 psi, in aluminum alloy tubes that will be connected to a $^3H/^3He$ separation facility through a Type 316L stainless steel manifold/piping system. Neutrons for the $^3He(n,p)^3H$ reaction will be produced by proton induced spallation of a tungsten target. Lead blanket assemblies will moderate and multiply the spallation neutrons. The high-energy proton beam that induces spallation in the tungsten target will move from the accelerator portion of the APT system into the T/B system by passing through an Alloy 718 window. Design concepts for the target/blanket system are summarized in Ref. [*1*] and include cladding of the tungsten target elements and containing these elements and the target cooling water with Alloy 718 and providing containment of the lead blanket components with 6061-T6 aluminum extrusions and tubing. The containment vessel for the T/B system will be fabricated from Type 304L stainless steel, and Type 316L stainless steel will be used to fabricate much of the primary cooling water system (internals). The materials irradiation program developed to support the emerging T/B design is outlined in Ref. [*2*] and, for convenience, is summarized in the next few paragraphs.

Models of the tungsten neutron source have been irradiated with 800 MeV protons at power levels prototypic of the power levels anticipated in the APT target. The proton beam had a Gaussian-like intensity profile with a diameter of approximately

$2\sigma = 3.0$ mm, where σ is the standard deviation. The beam profile has two basic effects on the irradiation and specimen design:

1. The specimen had to be small to assure reasonable uniformity in dose within a single sample, and
2. Placement of several specimens within a sample capsule provided the opportunity to determine the effects of dose on a given material.

The specimen also had to be relatively thin, 0.25 to 2.0 mm range, to assure that the energy deposited by the 800-MeV proton beam was properly transferred to the coolant. Details of irradiation procedures and temperature and dose calculations are described in Refs. [*3*,*4*].

Materials Specifications and Heat treatment

Certified mill test reports were obtained for each alloy; the chemical composition, from these test reports, for each material is summarized in Table 1.

Alloy 718 specimens were machined from the as-received material (annealed condition), wrapped in Nb foil and encapsulated in a quartz tube (evacuated and back-filled with Ar). The encapsulated specimens were heat treated through the following steps:

1. Solution anneal at 1065°C for 30 minutes and air cool,
2. Age at 760°C for 10 hours,
3. Furnace cooled from 760 to 650°C and hold for a total aging time of 20 hours,
4. Air cool to room temperature.

The 316L and 304L stainless steel samples were electrodischarge machined (EDM) from as-received (annealed) material and irradiated in the as-machined and cleaned condition and tested after irradiation.

Testing Procedure

The J-integral-resistance (J-R) curve toughness tests were conducted in general accordance with the American Society for Testing and Materials (ASTM) E 1820-99 [*5*] Standard Test Method for Measurement of Fracture Toughness, with a computer-controlled test and data acquisition system [*6*]. In many cases, selection of a specimen design for an irradiation study is a compromise between a desire to test large size specimens to satisfy rigorous validity requirements of the ASTM standard, and necessity to use small specimens due to a variety of limitations caused by imposed irradiation conditions, such as restricted capsule space, difficulty of proper heat removal from large volumes of metal, etc. This APT irradiation project was no exception. As a result of many considerations, disk-shaped compact tension [DC(T)] specimens, 12.5 mm in diameter with thicknesses of 4 mm (0.16T) and 2 mm (0.08T), were selected for this study to develop J-R curve. This small DC(T) specimen has been previously developed at ORNL [*7*] for testing of irradiated materials. Available experience suggested that even such small specimens could provide valuable information on effects of irradiation on a material's ability to resist crack extension, particularly if the specimen thickness matches the component thickness. It was also anticipated that the minimum thickness requirement

Table 1. Chemical composition (wt %) of materials tested

Material	Al	C	Cr	Cu	Fe	Mn	Mo	Ni	P	S	Si	Ti	Others
Alloy 718	0.47	0.04	17.71	0.15	BAL	0.12	3.00	54.79	0.013	0.001	0.13	0.98	Nb and Ta-4.98; Co-0.19; B-0.002
316L	n/a*	0.010	17.33	0.18	BAL	1.61	2.09	10.62	0.024	0.019	0.43	n/a	Co-0.21; N-0.060
304L	n/a	0.013	18.15	0.23	BAL	1.80	0.18	8.35	0.025	0.010	0.43	n/a	Co-0.17, N-0.085

n/a* = Not available.

would decrease for irradiated materials because the yield strength tends to increase and the fracture toughness decreases with irradiation. A thickness of 2 mm was chosen for the compact specimens for the irradiation experiments in the proton beam so that specimens could be cooled to the prototypical irradiation temperatures. Coincidently, the 2 mm thickness was similar to the thickness for some of the materials used in the T/B system. To evaluate a potential dependence of fracture toughness on thickness, J-R curve tests in the unirradiated condition were performed with both 2- and 4-mm-thick specimens. Only a limited number of 4-mm-thick specimens were irradiated.

The specimens were fatigue precracked before irradiation to a ratio of the crack length to specimen width (a/W) of about 0.5, and then side-grooved by 20% (20% SG) of their thickness (10% from each side). The unloading compliance method used for measuring the J-integral using these specimens is outlined in Ref. [7]. Tests were conducted at room temperature, 50, 80, and 164°C. The temperatures were maintained within ±2°C during the tests. Unirradiated specimens were tested in the laboratory on a 98-kN (22-kip) capacity servohydraulic machine, and irradiated specimens were tested in a hot cell with a 490-kN (110-kip) capacity servohydraulic machine with a 22-kN (5-kip) load cell. All tests were conducted in strain control, with an outboard clip gage having a central flexural beam that was instrumented with four strain gages in a full-bridge configuration. After testing, the crack front was marked by cyclically loading the specimen at room temperature. The specimens were then broken open. The unirradiated specimens were examined with a calibrated measuring optical microscope to determine the initial and final crack lengths. The irradiated specimens were photographed, and enlarged prints of the fracture surfaces were fastened to a digitizing tablet to allow the crack length to be measured. Finally, the J-integral was determined and plotted against crack extension using the load/displacement data. From this plot, the provisional critical J-integral, J_Q, was determined as the intersection of the power law regression line with the 0.2-mm crack extension offset line. J_Q values were converted to their equivalent values in terms of stress intensity K_{JQ}:

$$K_{J_Q} = \sqrt{E J_Q} \tag{1}$$

where E is Young's modulus. These K_{JQ} values characterize the toughness of materials near the onset of crack extension. The ASTM standard test method E1820-99 [5] specifies numerous requirements for qualifying the J-R curve and for qualifying J_Q as the critical J-integral at the onset of stable crack extension, J_{Ic}. In general, those requirements can be separated into two categories. The first category contains the requirements that are common for all tests. It includes requirements on the test equipment, machining tolerance, fixture alignment, test rate, temperature stability, rectilinearity of the original and final crack fronts, crack extension prediction, etc. Despite some difficulties with the use of small size specimens (and especially in the irradiated condition), this category of requirements was met in all tests. The second category of requirements is related to the measuring capacity of the specimen. The standard specifies the validity box on the J-R curve, the upper boundary for which is defined by the maximum J-integral capacity for a specimen by the smaller of the following:

$$J_{max} = \frac{b_o \sigma_Y}{20}, \quad or \quad J_{max} = \frac{B \sigma_Y}{20} \tag{2}$$

where b_0 is the original remaining ligament and B is the specimen thickness. The boundary on the right side of the validity box is defined by the maximum crack extension capacity:

$$\Delta a_{max} = 0.25 b_o \tag{3}$$

For the specimens in the present study, the nominal value of b_o is about 4.6 mm. Thus, J_{max} is limited by the specimen thickness, while the nominal value of Δa_{max} is about 1.2 mm. Due to the small size of the specimens, this requirement category was not met in the present study.

Results

Type 304L and 316L Stainless Steels

Both alloys exhibit very high toughness in the unirradiated condition. However, the 4-mm-thick specimens of both alloys exhibit slightly higher (about 10 to15%) toughness than the 2-mm-thick specimens (See Figs. 1 and 2). Figure 3 typifies the positioning of the J-R curve relative to the validity box for a 4-mm-thick specimen in the unirradiated condition for the stainless steels in the present study. The J-R curve for both stainless steels derived from either thickness of specimen cannot be qualified as an E1820 valid curve. Consequently, the J_Q value, denoted by a triangle on Fig. 3, cannot be qualified as J_{Ic}. Such deviation of the J-R curve from the validity box, as shown in Fig. 3 is responsible (at least in part) for the observed size effects on J_Q.

Two 4-mm-thick and two 2-mm-thick specimens of each material were irradiated to a relatively low dose (~0.1 dpa). Although irradiation slightly reduced the fracture toughness of each alloy, the relationship between fracture toughness values derived from 2- and 4-mm-thick specimens remained approximately the same.

It was anticipated that any size dependence would have a propensity to diminish with dose since irradiation reduces the J-integral values of these steels and increases the yield strength resulting in an increase in the J_{max} value, i.e., the difference between the J_Q values measured on 2 and 4 mm thick specimens after irradiation should be less than the difference measured in the unirradiated state. Figure 4 shows the J-R curve derived from testing a 2-mm-thick DC(T) specimen of 304L stainless steel irradiated up to 7.2 dpa and tested at 164°C. The J-R curve for this specimen is still outside the validity box; however, it is much closer to the box than for the unirradiated specimens.

Summarizing the size-related observations in the present study for the stainless steels, the following can be concluded. The K_{JQ} values measured from 2-mm-thick specimens are slightly less (220 to 250 MPa√m) than those obtained from 4-mm-thick specimens (270 to 280 MPa√m) for unirradiated material tested at room temperature. The observed size dependence is consistent with literature results for type 304 and 316 stainless steels in Ref. [*8*]. According to a statistical analysis of those fracture toughness results [*8*], valid K_{JIC} values derived from large specimens should be about 360 MPa√m on average. Thus, the K_{JQ} values obtained from small specimens in this project are conservative relative to the valid, size-independent values. Therefore, the data generated in the present study can be recommended for initial design calculations, especially for the

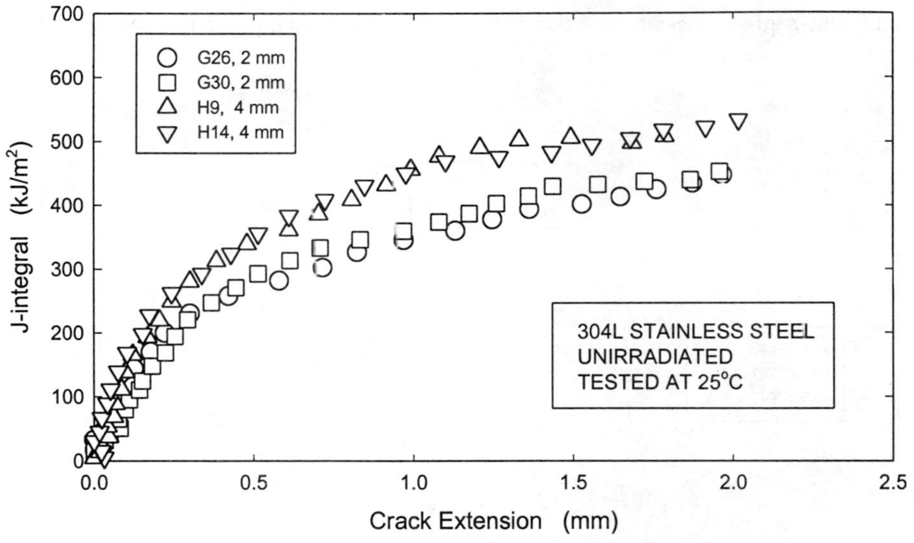

Figure 1. The J-R curves of unirradiated annealed 304L stainless steel measured with 2-mm- and 4-mm-thick specimens at room temperature.

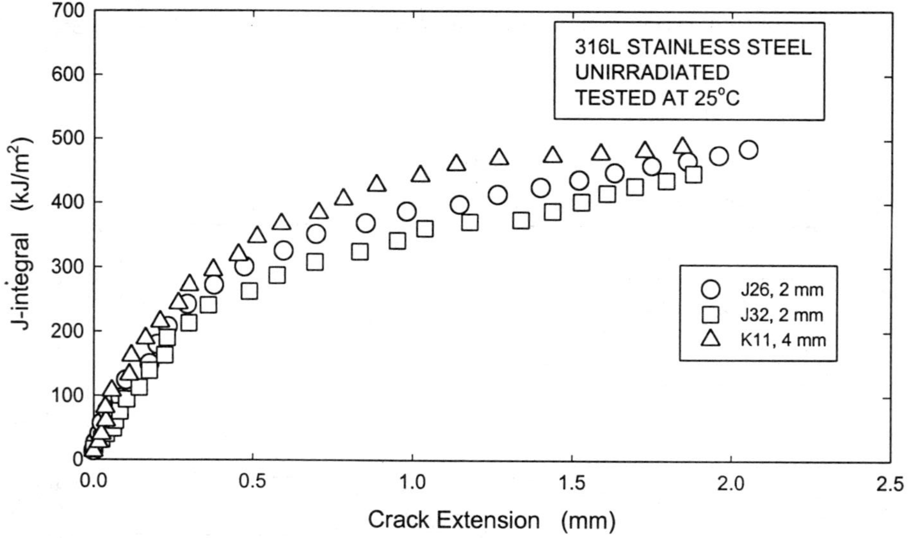

Figure 2. The J-R curves of unirradiated annealed 316L stainless steel measured with 2-mm- and 4-mm-thick specimens at room temperature.

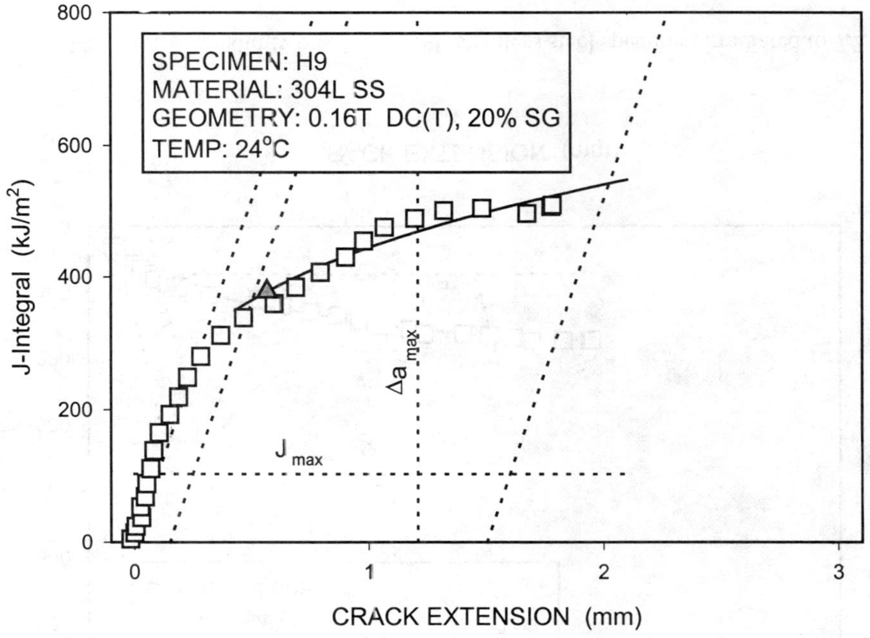

Figure 3. Position of the J-R curve of unirradiated Type 304L stainless steel 4-mm-thick specimen relative to the ASTM E1820 validity box.

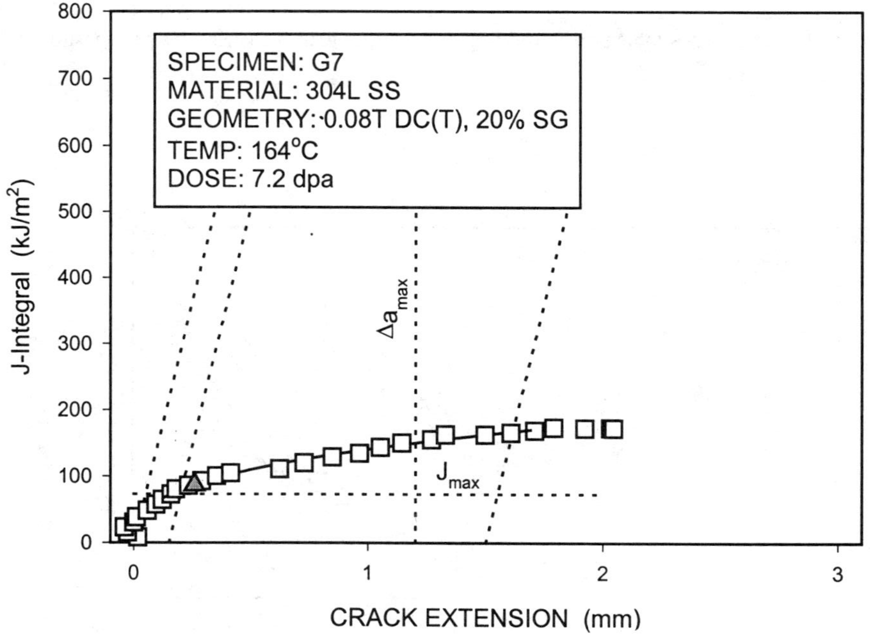

Figure 4. The J-R curve and validity box for a 304L stainless steel specimen irradiated to 7.2 dpa.

end-of-life conditions. This size dependence has a propensity to remain similar in nature for both stainless steels but to diminish in magnitude with irradiation dose because the J-R curves for the irradiated specimens become much closer to the validity box than the J-R curves for the unirradiated specimens.

The 4-mm-thick DC(T) specimens of both stainless steels were tested at room temperature, 50 and 164°C. The toughness decreases slightly from ~275 MPa$\sqrt{m}$ to ~210 MPa$\sqrt{m}$ as the test temperature was increased from 25 to 164°C (Fig. 5), which is a common temperature dependence for the critical J-integral for these type of stainless steels.

Specimens were irradiated to a wide range of doses from 0.1 dpa to approximately 7 to 9 dpa. This irradiation resulted in significant decreases in the toughness of both alloys, although appreciable levels of toughness (~65 to 100 MPa$\sqrt{m}$) were still retained (Figs. 6 and 7). Even after a dose of only 0.1 dpa, fracture toughness decreased more than 25% (see Figs. 6 and 7). The current data indicate that the decrease in the fracture toughness of 304L stainless steel reaches a lower plateau after dose of 3 to 4 dpa at a level of about 100 MPa$\sqrt{m}$. On the other hand, the fracture toughness of 316L stainless steel continues to decrease until it saturates at a level of approximately 60 to 70 MPa$\sqrt{m}$. However, another important observation is that at doses above 4 dpa 316L stainless steel has a tendency to intermittently lose resistance to stable crack growth. Specimens tested after doses of 4.6, 9.3, and 9.4 dpa exhibited local ductile instability (fast crack propagation in a ductile mode for a constant loading rate) after achieving the maximum load. Figure 8 illustrates the load-displacement traces of specimens J4 and J20. Specimen J4 was irradiated up to 9.3 dpa and exhibited ductile instability, while specimen J20 was irradiated up to 3.11dpa and demonstrated a typical load-displacement trace for a specimen exhibiting stable crack growth. Examination of the fracture surfaces of some of the irradiated 316L specimens using a scanning electron microscope (SEM) Philips XL30 did not reveal anything other than ductile tearing modes of crack extension. Figure 9 is a SEM image of the fracture surface of highly irradiated (9.3 dpa) specimen J4 and it shows only ductile dimples. The fact that the crack would propagate for a short distance in an unstable fashion and then arrest, resume propagation in a stable fashion and then repeat the process, may indicate that after irradiation at low temperature in the high energy, mixed proton/neutron spectrum to doses above 4.5 dpa, 316L steel has a propensity to form local zones with very low resistance to stable ductile crack growth.

There is a small amount of fracture toughness data available regarding the behavior of similar austenitic steels after fission neutron irradiation at such low irradiation temperatures. The present data showed that fracture toughness saturates at about the same toughness level as that for material after irradiation in a fission reactor. It appears, however, that the fracture toughness values at low doses for similar, fission neutron-irradiated austenitic stainless steels [*8*] decrease at a slower rate with dose than that observed in the present study. Taking into account that the helium and hydrogen production rates are extremely high under spallation conditions, this difference is consistent with the observation that both hydrogen and hydrogen/helium accumulation lower the room temperature fracture toughness of austenitic stainless steels [*9*]. However due to very limited data available for low temperature fission reactor irradiation of these steels, additional microstructural studies, as well as direct experiments with fission

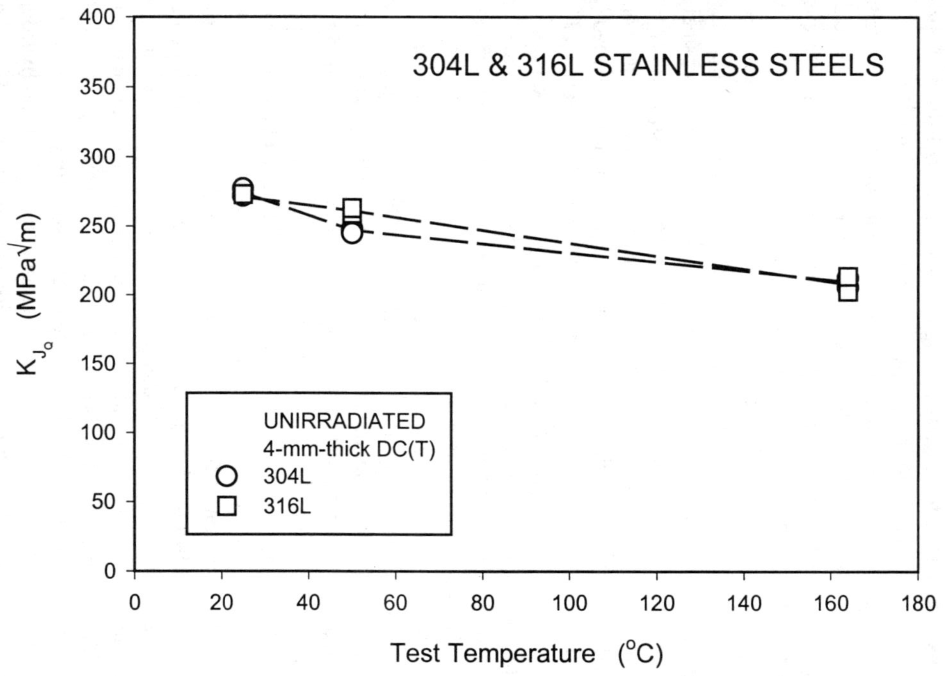

Figure 5. Temperature dependence of fracture toughness for unirradiated 304L and 316L stainless steels.

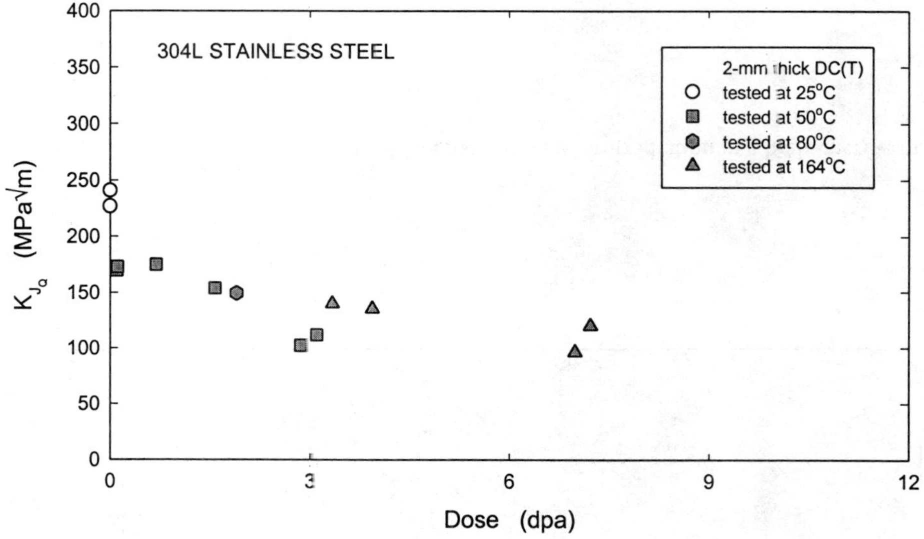

Figure 6. Fracture toughness of annealed 304L stainless steel after irradiation in a high-energy, mixed proton/neutron spectrum.

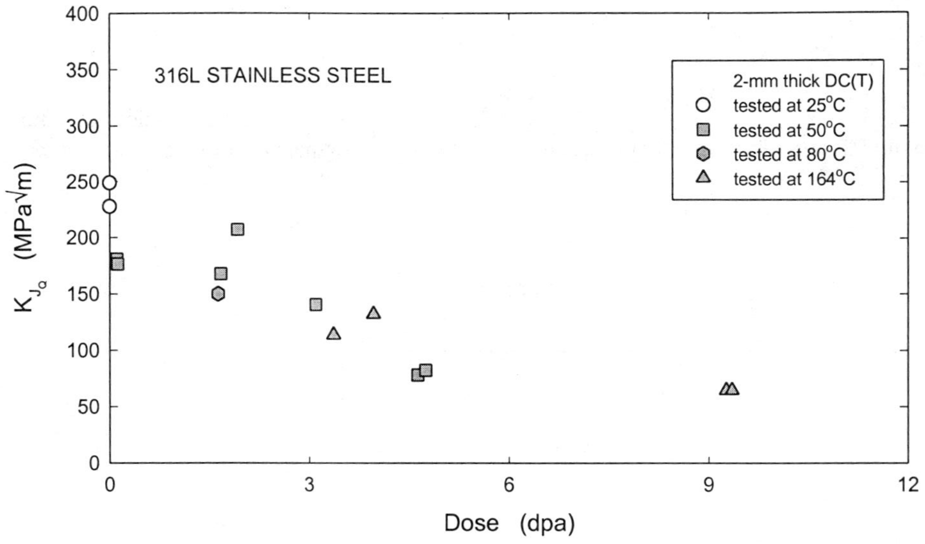

Figure 7. Fracture toughness of annealed 316L stainless steel after irradiation in a high-energy, mixed proton/neutron spectrum.

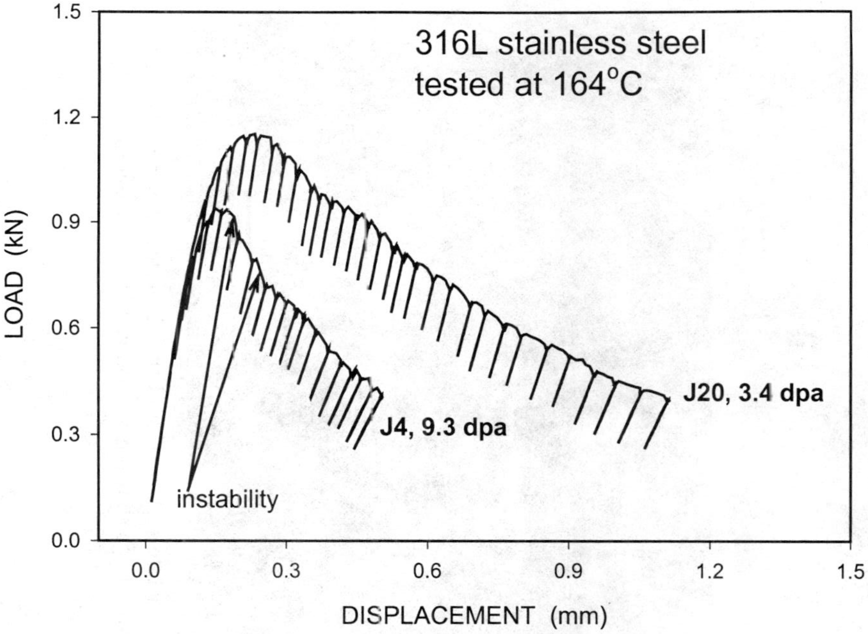

Figure 8. Load-displacement traces of two irradiated 316L stainless steel 2-mm-thick specimens tested at 164°C.

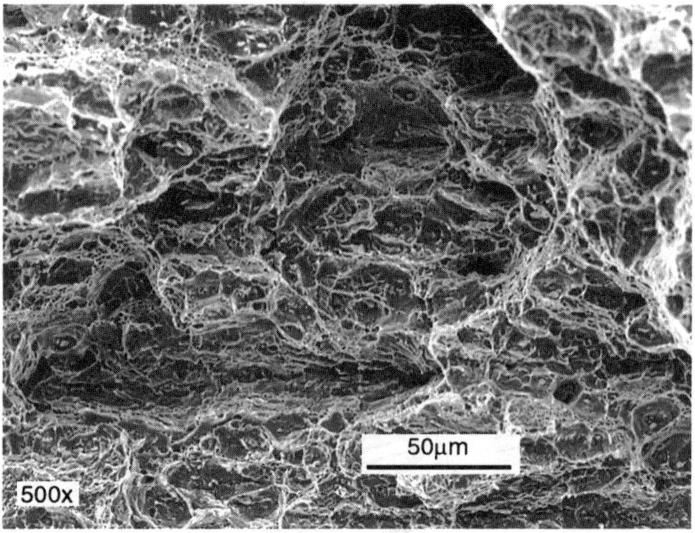

Figure 9. Fracture surface of 304L stainless steel specimen J4 irradiated to 9.3 dpa showing path of ductile crack extension.

reactor irradiation of the same materials at low temperatures are needed to confirm this statement.

Alloy 718

The crack-extension resistance of the precipitation-hardened Alloy 718 is much different from that of the stainless steels. All of the 4-mm-thick specimens failed by sudden, unstable crack extension. The unirradiated 2-mm-thick specimens have an initiation toughness similar to those of the 4-mm-thick specimens, but show more stable crack extension (with some elements of short term instabilities) than the 4-mm-thick specimens (See Fig. 10). After irradiation, only one 2-mm-thick specimen, E14, irradiated to the lowest dose of ~0.1 dpa showed some stable crack extension. All other irradiated specimens failed by sudden unstable crack extension regardless of the dose. For material exhibiting such crack-extension resistance, the J-integral value which corresponds to the point of failure by sudden unstable crack extension, J_c, is reported as the critical J-integral value. For a limited number of 2-mm-thick specimens that demonstrated a mixture of stable and unstable crack propagation, the J-integral at the point of maximum load, J_{Pmax}, is reported as a conservative estimate of the onset of stable crack extension.

In the unirradiated condition, the Alloy 718 exhibited moderate toughness (~120 to 160 MPa$\sqrt{m}$) at test temperatures from 25 to 164°C (See Fig. 11). Despite the scatter in the data at 50°C, there is a slight decrease in toughness as the test temperature increases (See Fig. 11).

The fracture toughness of Alloy 718 decreases steadily after irradiation up to ~ 4 to 5 dpa. After that, the fracture toughness saturates and the dose dependence diminishes (See Fig. 12). Specimens tested after irradiation up to ~ 12 dpa showed fracture toughness values of about 50 MPa$\sqrt{m}$. This reduction in fracture toughness is similar to that reported from fission neutron reactor studies [*10,11*]. However, the neutron irradiation-induced decreases in toughness of heat-treated Alloy 718 after irradiation in fission reactors appeared to saturate after about 8-10 dpa [*10,11*]. The present data reveal that saturation after mixed proton/neutron beam irradiation occurs at smaller doses than that observed after fission neutron-irradiated Alloy 718. This trend is also similar to that observed for stainless steels in the present study.

SEM images (Fig. 13a,b) of specimen E9 did not reveal any stable ductile crack extension prior to brittle instability. This specimen was irradiated to the highest dose, 12.35 dpa. A higher magnification image (Fig. 13b) suggests intergranular fracture as the main mode of crack propagation for this specimen.

Summary

The data presented in this paper summarize the fracture toughness results obtained for type 304L and 316L stainless steels and Alloy 718 after exposure in an environment typical of that for the spallation neutron source proposed for the Accelerator Production of Tritium (APT) Project. These materials were tested in the unirradiated condition and after irradiation at 25 to 200°C by high-energy protons and neutrons at the Los Alamos

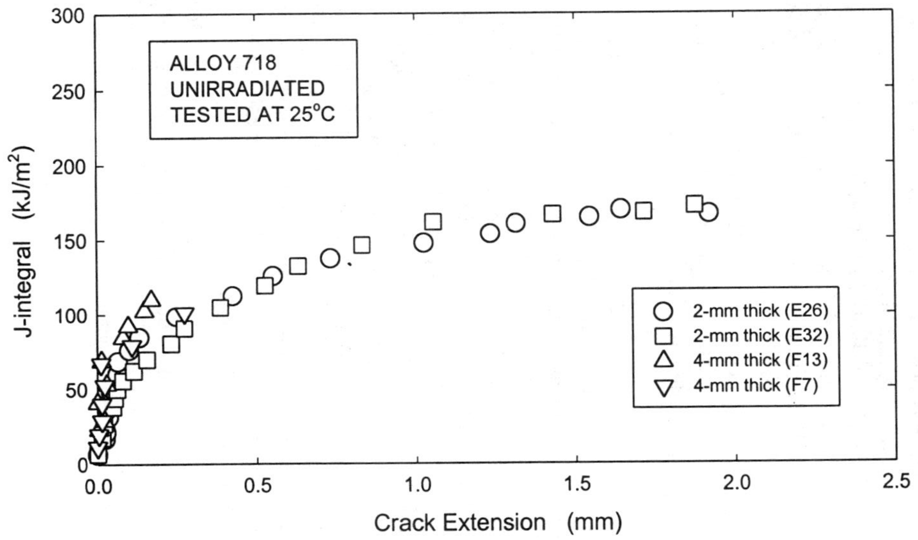

Figure 10. The J-R curves of unirradiated precipitation-hardened Alloy 718 measured with 2-mm- and 4-mm-thick specimens.

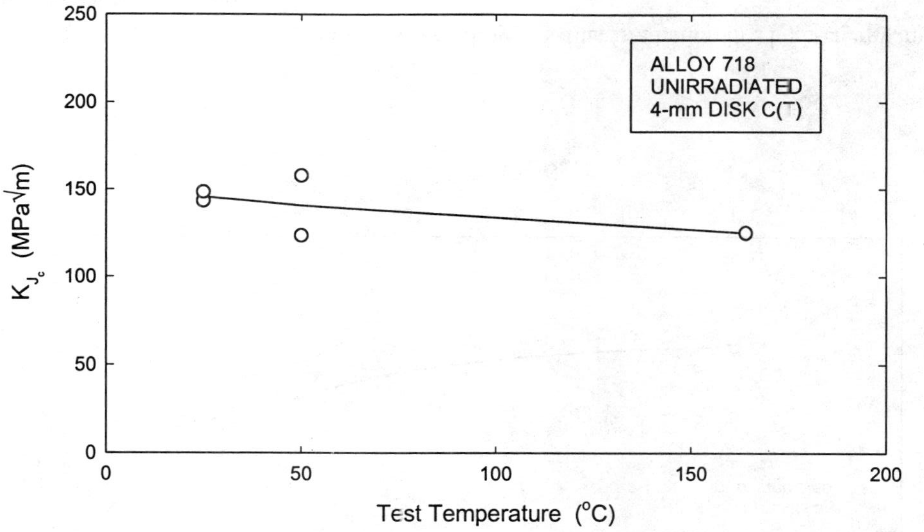

Figure 11. Temperature dependence for fracture toughness for the unirradiated precipitation-hardened Alloy 718.

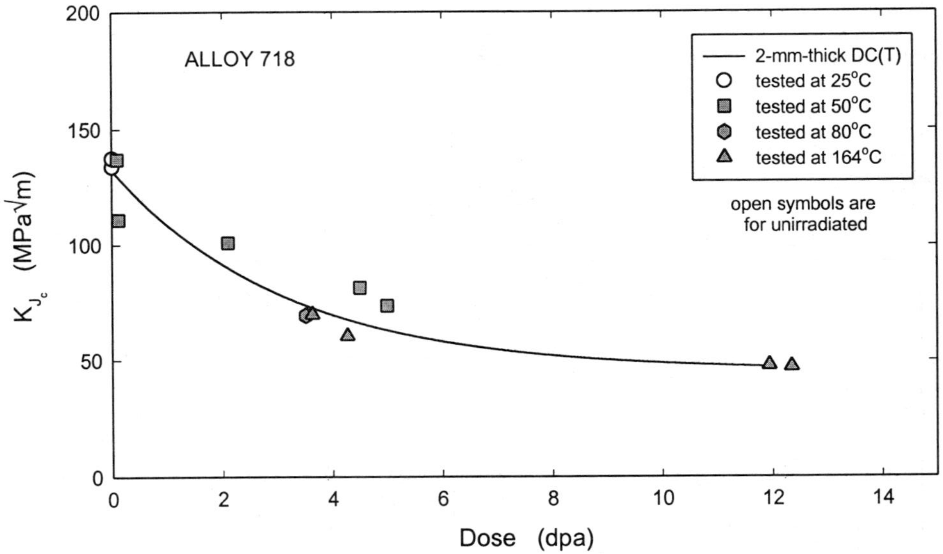

Figure 12. Fracture toughness of precipitation-hardened Alloy 718 after irradiation in a high-energy, mixed proton/neutron spectrum.

(a)

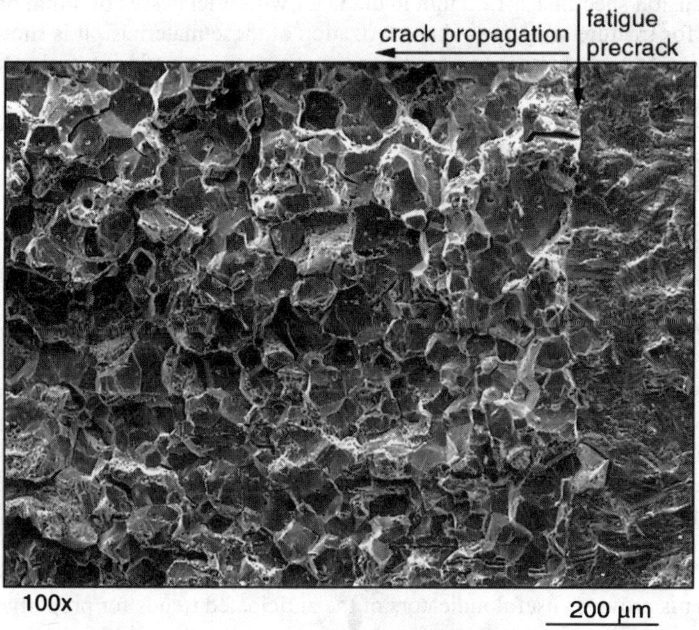

(b)

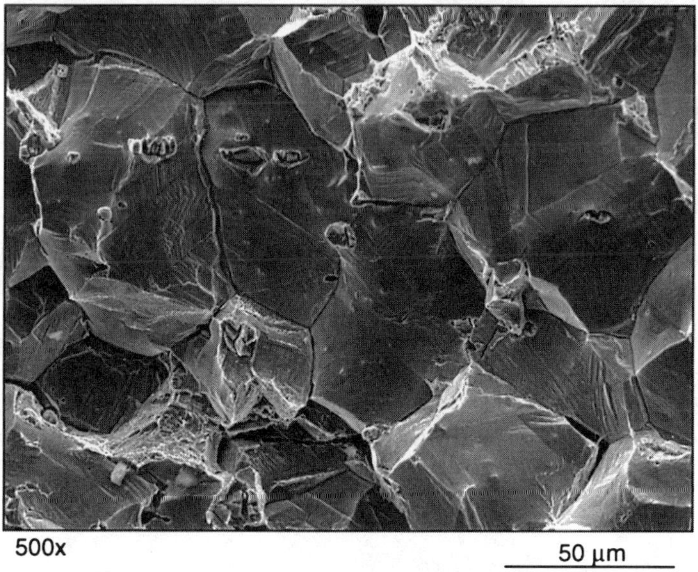

Figure 13. Fragment of the fracture surface of Alloy 718 specimen E9 irradiated to 12.35 dpa at low (a) and high (b) magnification.

Neutron Science Center (LANSCE) to doses up to ~12 dpa. Subsize disk-shaped compact tension specimens, 12.5 mm in diameter with thicknesses of 4 mm or 2 mm, were used for fracture toughness characterization of these materials. It is shown that the smallest specimen geometry provides a conservative estimate of J-integral values at the onset of stable crack extension.

Both 304L and 316L stainless steels exhibited very high toughness in the unirradiated condition. Specimens were irradiated in a wide range of doses from 0.1 up to approximately 7 to 9 dpa. This irradiation resulted in significant decreases in the toughness of both alloys, although appreciable levels of toughness (~65 to100 $MPa\sqrt{m}$) were retained. It was also observed that at doses above 4 dpa 316L stainless steel has a tendency to lose resistance to stable crack growth.

Precipitation-hardened Alloy 718 exhibited intermediate toughness (~120 to 160 $MPa\sqrt{m}$) in the unirradiated condition. This material exhibits a propensity toward unstable crack propagation once tearing is initiated. This phenomenon becomes more pronounced after irradiation. The fracture toughness of Alloy 718 decreases steadily with dose up to ~4 to 5 dpa. After that, the fracture toughness has a tendency to saturate at levels of about 50 $MPa\sqrt{m}$. The SEM examination of a fracture surface of one of the highly irradiated specimens revealed a brittle intergranular mode of crack propagation without any evidence of stable crack growth.

The data presented in this paper demonstrate that the available data base of radiation effects on the fracture toughness of structural metals and alloys exposed to fission reactor environments provides useful indicators of the anticipated trends for property/dose predictions in spallation neutron environments. However, the enhanced gas production associated with the high-energy proton/neutron spectra in spallation neutron sources may cause enhanced degradation of fracture toughness at low doses. Hydrogen production, and a susceptibility to hydrogen embrittlement, may play key roles in the low temperature, low dose degradation. Clearly, some additional microstructural studies and direct experiments with low temperature fission reactor irradiation of these materials are essential to clarify this observation.

Acknowledgments

This research is sponsored by the Accelerator Production of Tritium Program, U.S. Department of Energy, under contract DE-AC05-00OR22725 with UT-Battelle, LLC. The authors also acknowledge and appreciate W. Johnson of General Atomic for a thorough review and J. L. Bishop for preparation of the manuscript.

References

[*1*] M. W. Cappiello and A. M. Baxter, "APT Target/Blanket Design," *AccApp'98*, pp. 289-296, American Nuclear Society, La Grange, IL, 1998.

[2] S. A. Maloy, W. F. Sommer, R. D. Brown, J. E. Roberts, J. Eddleman, E. Zimmermann and G. Willcutt, "Progress Report on the Accelerator Production of Tritium Materials Irradiation Program," *Materials for Spallation Neutron Sources*, edited by M. S. Wechsler, L. K. Mansur, C. L. Snead and W. F. Sommer, p. 131-138, TMS, Warrendale, PA, 1998.

[3] G.J. Willcutt, S.A. Maloy, M.R. James, J. Teague, D.A. Siebe, W.F. Sommer, P.D. Ferguson, "Thermal Analysis of the APT Materials Irradiation Samples," *2nd International Topical Meeting on Nuclear Applications of Accelerator Technology*, Gatlinburg, TN, Sept. 20-23, 1998, pp. 254-259.

[4] M.R. James, S.A. Maloy, W.F. Sommer, P. Ferguson, M.M. Fowler, K. Corzine, "Determination of Mixed Proton/Neutron Fluences in the LANSCE Irradiation Environment," *2nd International Topical Meeting on Nuclear Applications of Accelerator Technology*, Gatlinburg, TN, Sept. 20-23, 1998, pp. 605-608.

[5] Standard Test Method for J-integral Characterization of Fracture Toughness, Designation E1737-96. *Annual Book of ASTM Standards*, Vol. 03.01.

[6] R.K. Nanstad, D.J. Alexander, R.L. Swain, J.T. Hutton, and D.L. Thomas, "A Computer Controlled Automated Test System for Fatigue and Fracture Testing," in *Applications of Automation Technology to Fatigue and Fracture Testing, ASTM STP 1092*, A.A. Braun, N.E. Ashbaugh, and F.M. Smith, Eds., American Society for Testing and Materials, Philadelphia, 1990, pp.7-20.

[7] D.J. Alexander, "Fracture Toughness Measurements with Subsize Disk Compact Specimens," *Small Specimen Test Techniques Applied to Nuclear Reactor Vessel Thermal Annealing and Plant Life Extension, ASTM STP 1204*, W.R.Corwin, F.M.Haggag, and W.L.Server, Eds., American Society for Testing and Materials, Philadelphia, 1993, pp. 130-142.

[8] W.J. Mills, "Fracture Toughness of Types 304 and 316 Stainless Steels and Their Welds," *International Materials Reviews*, 1997, pp. 45-82

[9] M.J.Morgan and M.A.Tosten, "The Effect of Tritium Decay Helium on the Fracture Toughness of Austenitic Stainless Steels," *Hydrogen Effects in Materials*, A.W.Thompson and N.R.Moody, Eds., TMS, Warrendale, PA, 1996.

[10] W.J. Mills, "Effect of Microstructural Variations on the Tensile and Fracture Toughness Properties of Inconel 718 Welds," *Superalloys 718, 625, 706 and Various Derivatives*, E.A.Loria, Ed., TMS, Warrendale, PA, 1994, pp. 845-858.

[11] D.J. Michel and R.A.Gray, "Effects of Irradiation on the Fracture toughness of FBR Structural Materials," *Journal of Nuclear Materials*, Vol. 148, 1987, pp. 194-203.

Pressure Vessel Steels—Microstructure and Modeling

Colin A. English,[1] Susan R. Ortner,[1] Gareth Gage,[1] William L. Server,[2] and Stan T. Rosinski[3]

Review of Phosphorus Segregation and Intergranular Embrittlement in Reactor Pressure Vessel Steels

Reference: English, C. A., Ortner, S. R., Gage, G., Server, W. L., and Rosinski, S. T., **"Review of Phosphorus Segregation and Intergranular Embrittlement in Reactor Pressure Vessel Steels,"** *Effects of Radiation on Materials: 20th International Symposium, ASTM STP 1405,* S. T. Rosinski, M. L. Grossbeck, T. R. Allen and A. S. Kumar, Eds., American Society for Testing and Materials, West Conshohocken, PA, 2001.

Abstract: This paper presents a systematic review of the behavior of phosphorus (P), highlighting the implications of P segregation to grain boundaries under neutron irradiation. The review focuses on Mn-Mo-Ni steels employed in US pressurized water reactors (PWRs), and other PWRs worldwide. Segregation of P to grain boundaries in reactor pressure vessel (RPV) steels can occur during fabrication (especially during the slow cooling stage of a post-weld heat treatment), and as a result of in-service exposure to high operating temperature and irradiation. This segregation of P to grain boundaries can promote a change in the brittle fracture mode from transgranular (TGF) to intergranular (IGF), and a degradation in the mechanical properties. In US RPV steels, most data are on thermal aging of the heat-affected zone (HAZ). Studies in coarse-grained HAZ have shown that the embrittlement arising from segregation of P to grain boundaries is approximately linearly related to the proportion of the brittle fracture that is IGF, and/or the P concentration at the grain boundary. Data are sparse on the effect of irradiation at 288°C on P segregation, and on the contribution of IGF to the total shift in the 41J transition temperature, T_{41J}. In general, the bulk P content appears to be less than about 0.028 wt% P, with base metals having lower levels than weldments. In addition, the consequences of vessel annealing are considered at temperatures around 475°C. It is certain that the annealing treatment will have the consequence of reducing the irradiation hardening, but may significantly increase the grain boundary phosphorus coverage and the likelihood of IGF.

Keywords: phosphorus segregation, intergranular embrittlement, grain boundaries, HAZ, thermal anneal heat treatment

[1] Technical Staff, AEA Technology, Nuclear Science, Harwell, Didcot, Oxfordshire OX11 0RA, United Kingdom.

[2] President, ATI Consulting, 6773-C Sierra Court, Dublin, CA 94568.

[3] Project Manager, EPRI, 1300 Harris Boulevard, Charlotte, NC 28262.

Introduction

This paper reviews the incidence and consequence of phosphorous (P) segregation in steels employed in the pressure vessels of light water reactors (LWRs). It is well known that P can segregate to grain boundaries in steels during service at elevated temperatures [*1*]. The presence of P in the boundary can then lower its cohesive energy, causing the material to fail through intergranular embrittlement. A body of data has been established on the behavior of P in a variety of reactor pressure vessel (RPV) steels used in a number of different reactor designs [*2*]. The purpose of this paper is to present a systematic review of those data for MnMoNi steels employed in US LWRs and, most important, to highlight the implications of the likely P segregation to be found under the service conditions of greatest relevance to US plants.

The RPV steels to be considered are the MnMoNi steels employed in US PWRs (and other LWRs worldwide), namely A302B, A508, A533B. We exclude high-nickel steels. The service conditions relevant to this review are those which involve long-term exposure at temperatures sufficiently high for thermal aging effects to occur, irradiation under conditions in which irradiation-induced segregation occurs, and short-term vessel annealing undertaken to remove the hardening effects of irradiation. P segregation and its embrittlement consequence are discussed separately.

The incidence of P segregation is discussed in terms of the level of P segregation measured at internal grain boundaries. Here, a review of experimental observations is undertaken to highlight the differences between the different classes of steels, emphasizing the importance of microstructure (plate, weld, or heat-affected zone (HAZ)). The levels of P on the grain boundaries at the start-of-life (SOL) will be reviewed, followed by effects of exposure in increasing segregation.

The discussion of embrittlement consequence is concerned with experimental observations of changes in fracture mode, and of changes in impact and fracture toughness properties during service. A key feature of the mechanical property changes induced by P segregation is that they do not include increases in yield stress or hardness, although increases in hardness may affect the embrittlement consequence of a given level of P segregation.

Micromechanisms of P Embrittlement

The presence of P in a predominantly Fe grain boundary is thought to reduce the grain boundary cohesion. Macroscopically, as the grain boundary P level increases, this can appear as an increase in the ductile-to-brittle transition temperature or a decrease in the lower shelf toughness.

Similar or opposite effects are caused by other segregants. In RPV steels the most effective of the other segregants tend to be Mn, B and C. Mn appears to enhance embrittlement [*3*]. Conversely, B and C both enhance the grain boundary cohesion.

Three basic micromechanisms of irradiation embrittlement have been identified and agreed worldwide to control RPV embrittlement:

1. Irradiation enhanced formation of copper-enriched precipitates;

2. Matrix damage due to radiation produced point defect clusters and dislocation loops;
3. Irradiation induced/enhanced grain boundary segregation of embrittling elements such as P.

The first two mechanisms serve to harden the material and increase the yield strength σ_y, whilst the third mechanism causes a drop in the fracture stress, σ_F. The effects of these changes on the fracture behavior may be rationalised within the framework shown in Figure 1, in which the temperature dependences of the yield stress, σ_y, and the fracture stress, σ_F, are plotted. It can be seen that the effect of irradiation in causing hardening or a change in σ_F is to cause a change in the transition temperature, ΔTT. In Figure 1 ΔTT_1 is caused by an increase in σ_y, whilst segregation of P to grain boundaries can lower the fracture stress and results in a shift ΔTT_2. If both mechanisms are operative then a combined shift of ΔTT_3 occurs, whereby the two components are combined by linear addition, i.e.

$$\Delta TT_1 + \Delta TT_2 = \Delta TT_3 \tag{1}$$

This methodology is frequently adopted in the literature for the analysis of irradiated material data in P-containing material [*4*]. The magnitude of the non-hardening term is assumed to be linearly dependent on the increase in grain boundary P during irradiation.

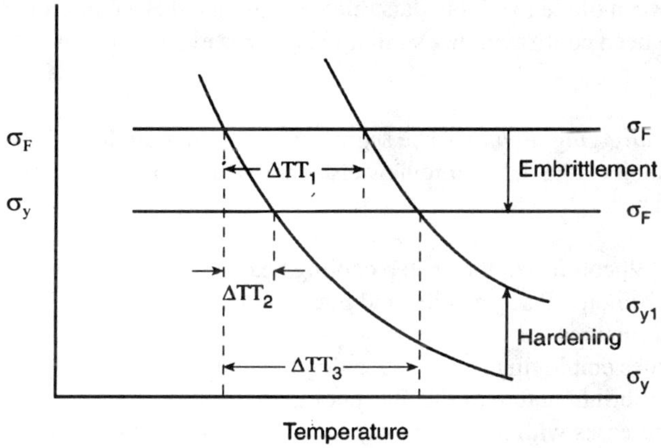

Figure 1 - *Variation of yield stress, σ_y, and fracture stress, σ_F, with temperature.*

This approach to the transition shift describes a shift in the temperature at which a given Charpy impact energy or fracture toughness may be achieved. It contains no explicit reference to the shape of the transition curve. Once sufficient P segregation occurs, however, an additional low-temperature fracture mode is introduced, and the shape of the curve may change. Once intergranular failure is introduced its effects on crack nucleation and growth may not be equivalent, i.e. the relation between fracture toughness tests and Charpy impact tests may also change.

Behavior Of Mn-Mo-Ni Steels

Trend curves describing the irradiation embrittlement of US RPV steels do frequently contain terms in P [*5*, *6*] and an increased sensitivity to irradiation embrittlement with increasing P content has been observed in A533B and A302B [*7*]. Since P can also contribute to hardening, however [*8*], it is not always clear if the effect of P on embrittlement is due to hardening or segregation or both. Where segregation does cause embrittlement, studies have shown that different microstructures exhibit different levels of susceptibility [*1*, *9*]. The review will therefore describe segregation effects in the different microstructures, starting with the most susceptible - the coarse grained heat affected zone (CGHAZ).

Heat Affected Zone Material

In a weldment the weld bead and HAZ structure are very complex, with different regions/zones produced depending on the thermal cycle (columnar, equiaxed and polygonal ferrite in the weld, fine- and coarse-grained regions in the HAZ, grains which have seen full, partial, repeated or no austenitization .....) [*10*]. Such a variety of microstructures is not found within base materials. Increasing the austenitizing temperature promotes grain growth, which is evident in the CGHAZ of weldments. The grain size varies depending on the welding procedure, notably heat input, but can be as large as 100 μm. Much of the available data relate to simulated HAZ microstructures although it has also been confirmed that similar behavior is evident in real HAZ material.

Trends In Thermal Aging Studies - The segregation of P to grain boundaries in CGHAZ has been subject to a detailed review elsewhere [*2*] and the main conclusions will be summarized here.

1. Early studies [*1*, *9*] confirmed that a step cooling heat treatment or an isothermal heat treatment promoted P segregation and intergranular failure (IGF) in simulated CGHAZ microstructures.
2. Subsequent studies employing isothermal aging treatments [*11*, *12*, *13*] were able to confirm the equilibrium nature of the phosphorus segregation; i.e. grain boundary concentration increases with decreasing aging temperature while kinetics of segregation decrease with increasing aging temperature (Figure 2).
3. Several studies [*2*, *12*, *13*] have shown that above a threshold grain boundary P level, the ductile-brittle transition temperature (DBTT) increases linearly with increasing P segregation (Figure 3).
4. CGHAZ from both commercial and model Mn-Mo-Ni steels show the same characteristics [*1*].
5. For a given segregation level in CGHAZ the shift in fracture toughness transition temperature is significantly smaller than that in the Charpy transition temperature [*14*].

Some of these effects are illustrated with reference to the International Atomic Energy Agency Coordinated Research Programme (IAEA CRP), which included a small study examining the effect of thermal aging on simulated CGHAZ material of three model steels based on LWR RPV A533B with varying copper and phosphorus contents [*15*] (see Table 1 for composition). These steels are termed model because they were prepared in relatively small melts to a commercial composition specification. They may not have impurity contents or the microstructural variation typical of actual RPV steels.

Table 1 - *Composition of JPC, JPB, and JPG (in weight percent)*

Alloy	**C**	**Si**	**Mn**	**Ni**	**Mo**	**Cr**	**P**	**Cu**	**S**
JPC	0.18	0.27	1.45	0.81	0.54	0.15	0.007	0.01	0.002
JPB	0.18	0.26	1.42	0.3	0.54	0.15	0.017	0.01	0.001
JPG	0.18	0.27	1.45	0.82	0.55	0.15	0.017	0.16	0.001

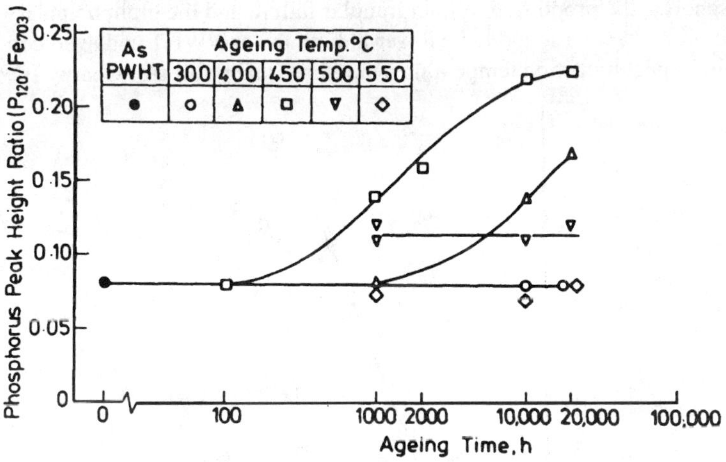

Figure 2 - *Plot of phosphorus peak height ratio versus aging time.*

The CGHAZ microstructure was simulated by a heat treatment at 1200°C for 30 minutes followed by oil quenching and subsequent post weld heat treatment for 25 hours at 615°C. The hardness of the resulting microstructure ranged from 250-268 Vickers Pyramid Number (VPN), and showed no clear effect of copper or phosphorus content. These values are probably marginally above that likely to be present in commercial heat affected zones, which may have an effect on their susceptibility to IGF.

In the as-PWHT (post-weld heat-treated) condition the simulated CGHAZ material of all three alloys was observed to have phosphorus segregation to grain boundaries and an intergranular component in the brittle fracture mode. The magnitudes of both features increased on aging for 2000 hours at 450°C (Table 2).

Table 2 - *Intergranular Content, Grain Boundary Phosphorus Coverage, and Embrittlement of Thermally Aged JPC, JPB, and JPG*

Alloy	Intergranular Content (%)		Grain boundary phosphorus (%)		Embrittlement due to aging (ΔDBTT - °C)
	As-PWHT	Aged	As-PWHT	Aged	
JPC (0.007P)	8	86	<5	15	118
JPB (0.017P)	35	98	10	35	174
JPG (0.017P/0.16Cu)	20	96	8	35	209

The characteristics of the two model steels containing 0.017 wt% phosphorus are shown to be very similar, particularly after aging. In the model steel containing 0.007 wt% phosphorus, the proportion of intergranular failure and the grain boundary phosphorus content were found to be lower in both the as-PWHT and aged conditions. The shift in impact transition temperature (measured using full size Charpy specimens)

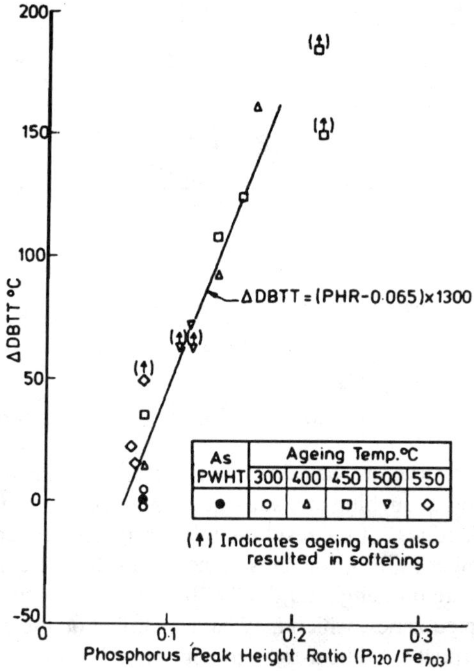

Figure 3 - *Embrittlement as a function of P segregation in A533B simulated CGHAZ.*

due to aging was also much greater in the two higher phosphorus containing alloys, and correlated better with the grain boundary P content than with the overall proportion of intergranular failure observed on the fracture surfaces.

McCabe [*16*] and McCabe et al [*17*] confirmed that increases in P levels occur at the grain boundaries during the aging of coarse-grained regions of five commercially made RPV steels. In these studies a Gleeble simulation of an HAZ weld cycle was used to simulate the metallurgical conditions occurring during welding. After aging at 450°C and 482°C for 168 h, significant transition-temperature shifts were observed (though no post-embrittlement transition temperatures were observed to be above room temperature). Selected steels that had been Gleeble-austenitised to ASTM grain size number 4 and then aged 450°C for 2000 h were examined in the FEGSTEM (field emission gun scanning transmission electron microscope). The McLean equation [*18*] reasonably predicted the observed P segregation of all the modified A302B specimens, but overestimated the segregation in A533B. In addition to the increases in P levels at the grain boundary, increases in Mn, Mo and Ni were also observed, but no increases in Cu Si or Cr were detected.

Effect Of Irradiation - In the IAEA program, the as-PWHT and aged simulated CGHAZ material were also subjected to irradiation at approximately 290°C to average doses of 17.4, 7.3 and 11.5 mdpa for steels JPC, JPB and JPG respectively. This irradiation promoted an increase in both the percentage of intergranular facets (%IGF) on the fracture surface and the grain boundary phosphorus concentration in both the as-PWHT and aged conditions. These data are presented in Table 3. The effect of the irradiation on the intergranular phosphorus concentration was shown to be in reasonable agreement with predictions using the Minor Element Kinetic Segregation Model [*19*] as shown in Figure 4.

Table 3 - *Percentage IGF and Grain Boundary Phosphorus Monolayer Coverage in IAEA Alloys, as a Function of Alloy Composition, Heat Treatment and Irradiation*

Alloy	Condition[1]	Cu, wt%	P, wt%	Dose mdpa	%IGF Unirr.	%IGF Irr.	Monolayer Coverage Unirr. (%)	Monolayer Coverage Irr. (%)
JPC	AR	0.01	0.007	7.69	< 1	< 1	<5	-
	HT			7.26	8	5	<5	6
	HTA			7.52	86	86	15	16
JPB	AR	0.01	0.017	18.22	< 1	< 1	5	-
	HT			18.61	35	83	10	16
	HTA			16.67	98	94	35	44
JPG	AR	0.16	0.017	12.24	1	<1	8	-
	HT			11.69	20	25	8	17
	HTA			11.83	96	90	35	47

[1]AR = as-received plate, HT = heat treated to simulate CGHAZ, and HTA = heat treated and aged under conditions typical of a post-weld stress-relief treatment.

Corresponding increases in the impact transition temperature (Table 4) were shown to be very similar for the as-PWHT and aged conditions of the respective alloys; typically 20°C for the low phosphorus containing steel (JPC) and 50°C for the two high phosphorus containing steels (JPB and JPG). In the high copper steel (JPG) the increase was greater in as-PWHT material than in aged material (69°C compared with 50°C) but this was rationalized in terms of the greater hardening observed in the former; a consequence of the variable irradiation.

Post-Irradiation Annealing Studies - Post-irradiation annealing (PIA) experiments were also conducted on the irradiated, as-PWHT simulated CGHAZ material in model steel JPG. The effect of PIA at 475°C for 168 hours was shown to increase the intergranular content to 95% compared with 25% in the as-irradiated and 20% in the unirradiated as-PWHT condition. The grain boundary phosphorus concentration increased to 29.4% compared with 17% after irradiation and 8% in the unirradiated as-PWHT condition (Table 3). The effect of the irradiation on the intergranular phosphorus concentration was again shown to be in good agreement with predictions using the Minor Element Kinetic Segregation Model [*19*] as shown in Figure 5.

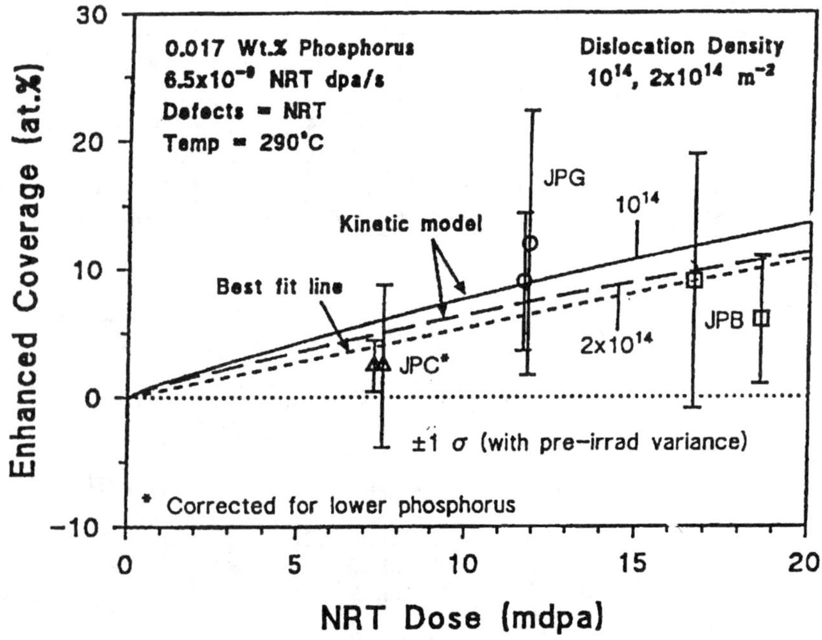

Figure 4 - *Comparison of experimental measurements of phosphorus segregation with predictions of Kinetic Model.*

Table 4 - *Summary of Initial Hardness (H_V) and Transition Temperatures (T_{41J}) in IAEA Alloys, Plus Transition Temperature Shifts (ΔT_{41J}) Due to Thermal Aging or Irradiation, in Different Heat Treatment Conditions*

Alloy	Condition	Cu, wt%	P, wt%	H_V, VPN	T_{41J}, °C	ΔT_{41J}, °C (thermal aging)	ΔT_{41J}, °C (irradiation)
JPC	AR	0.01	0.007	204	-76		15
	HT			259	-125		20
	HTA			268	-7	118	25
JPB	AR	0.01	0.017	206	-82		22
	HT			265	-87		51
	HTA			261	87	174	54
JPG	AR	0.16	0.017	204	-69		33
	HT			250	-120		69
	HTA			260	89	209	50

As a result of this segregation, although the hardness of the material was effectively fully recovered after the anneal, the impact transition temperature actually increased relative to the irradiated condition rather than decreased. Indeed the shift of 155°C

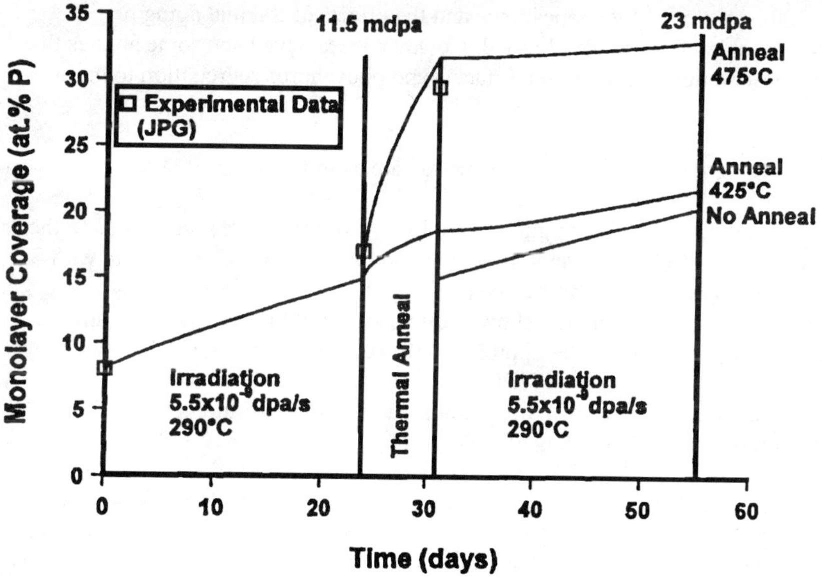

Figure 5 - *Comparison of predictions (solid lines) of the kinetic segregation model for irradiation and post-irradiation annealed JPG with experimental data (open squares). Predicted behavior for continued irradiation with and without annealing at 425°C is also shown.*

relative to the unirradiated as-PWHT condition was significantly greater than that of 69°C produced on irradiation and approaching that induced by thermal aging at 450°C for 2000 hours. This is consistent with the observation that the grain boundary phosphorus concentration was 29.4% after the post irradiation anneal compared with 35% after thermal aging for 2000 hours at 450°C.

The IAEA work described above referred to studies of material with simulated CGHAZ. Gage et al. [*20*] studied whether the brittle fracture mode of HAZ material from the Palisades surveillance specimens included any intergranular component after an anneal for 168h at 454°C. Evidence of IGF was observed, and from the size of the intergranular facets it was estimated that the grain size associated with such facets could be as high as 50-60μm. This indicates that it is probable that IGF is located in material of the CGHAZ. It was observed that the intergranular facets were located in bands, which ran perpendicular to the notch root and parallel to the direction of crack propagation. This is consistent with their being associated with a particular part of the HAZ. Some variability was found between specimens irradiated to different fluences. This was not attributed to the effect of fluence but was more probably associated with the location of the notch in the HAZ.

Weldmetals

Data on the brittle fracture appearance and the effects of thermal aging or irradiation for weldmetals are very limited. However there have been some studies that reported the observation of intergranular facets and phosphorus segregation to grain boundaries.

Thermal Aging - The major points to emerge are as follows:

1. A study of weldmetal (160 wt ppm P) found very little P (<10%) segregated to the grain boundaries after approximately 100000 hours at 285°C. Competition with S for grain boundary sites was suspected [*13*].
2. Fracture surfaces of this material showed only isolated IG facets, though a similar weldmetal (170 wt ppm P) after approximately 100000 hours at 285°C showed significant columnar IG fracture [*13*].
3. A number of studies have shown only small amounts of embrittlement, typically < 20°C, as a result of thermal aging of welds [*21*].

Effect Of Irradiation - Atom probe studies have shown that phosphorus is segregated to grain boundaries in A533B submerged arc weldmetal following neutron irradiation at 288°C for approximately 8750 hours [*22*]. The grain boundary enrichment factor was determined to be approximately 12.5 times that of the bulk value (actual not reported), although this may have arisen as much due to the accumulated thermal history as to the irradiation.

The concentration of phosphorus at grain boundaries in two high copper steel welds characteristic of Linde 80 Flux weld (BW-2) and Linde 1092 flux weld (CE-1) has been measured both before and after irradiation, by FEGSTEM measurements [2]. Measured values of the phosphorus concentration at the grain boundaries are summarized in Table 5. Some increase in monolayer coverage was observed in both irradiated welds although the increase was less than 10% of a monolayer.

Table 5 - *Grain Boundary P Levels in CE-1 and BW-2*

Weld	Start-of-Life (% monolayer)	Irradiated[1] (% monolayer)	Bulk P (wt%)
CE-1	2	5.5	0.014
BW-2	6	14	0.017

[1] 2.74×10^{19} ncm^{-2}; E > 1 MeV at approximately 6×10^{12} n cm^{-2}sec^{-1} and 290°C

The most significant observations of intergranular failure in US RPV weldments were made on irradiated Charpy specimens [*23*] by Gurovich and co-workers. The major results on weldments are:

1. In A533B weldmetal containing 0.009% P, irradiated at ~300°C to 10×10^{19} n cm^{-2} (E > 0.5 MeV), no brittle intergranular fracture was observed on the lower shelf, although 5% ductile intergranular fracture was observed on the upper shelf.
2. In A508 weldmetal containing 0.026% P, irradiated at ~300°C to 10×10^{19} n cm^{-2} (E > 0.5 MeV), 20% brittle intergranular fracture was observed on the lower shelf and in the mid-transition region, and 15% ductile intergranular failure on the upper shelf. In this case no observations were reported on the percentage of IGF in the unirradiated steel.

Base Material

Thermal Aging - The main points to emerge from thermal aging studies of base metal are:

1. P levels at the boundaries were close to the limits of detectability by Auger in a number of A533B and A508 steels after either long-term exposure at typical RPV operating temperatures, or shorter exposures at annealing temperatures. Nonetheless a small number of IG facets were observed [*13*].
2. Significant amounts of IGF have been observed in doped steels [*24*], in forgings subjected to step cooling [*25*], and in forgings exhibiting regions of macro-segregation (ghost lines [*26*]). This has been associated with an increased DBTT [*25, 26*].

Behavior under irradiation - Forging material containing regions of macro-segregation was irradiated to 4.4×10^{23} ncm^{-2}; E > 1 MeV at 285°C [*26*]. No information on the brittle fracture mode or grain boundary chemistry was reported. The irradiation-induced increase in the impact transition temperature for the macrosegregated material was, however, approximately the same if not slightly less than that for the bulk of the material. The actual transition temperature was, of course, still higher due to the higher initial transition temperature prior to irradiation.

A model steel (0.017 wt% P) was irradiated in the normalized and tempered, and step-cooled conditions to 5.35 x 10^{19} n cm^{-2}, E > 1 MeV [*25*]. Step cooling was intended to increase the grain boundary P level, though this was not verified experimentally. It was observed that the irradiation-induced shift in the impact transition temperature was smaller for the step-cooled material than for the normalised and tempered material; 45°C compared with 75°C. However, irradiation-induced increases in yield strength were very similar in both cases; 112 and 118 MPa respectively. A characteristic of the step-cooled material was a lower upper shelf energy compared with the normalized and tempered material in both the unirradiated and irradiated conditions.

The concentration of phosphorus at grain boundaries in two low copper plate (EP2) and forging steels (EF2) has been measured after irradiation by FEGSTEM measurements [2]. The bulk P content was 0.008 wt%. Material was irradiated at a dose rate of approximately 6x10^{12} n$cm^{-2}$$sec^{-1}$; E > 1 MeV at 290°C. Measured values of the monolayer coverage concentration at the grain boundaries are summarized in Table 6.

Table 6 - *Summary of Microstructural Data Obtained from Low Copper Plate and Forging Steels*

Material	**Plate EP2**	**Plate EP2**	**Forging EF1**
Dose (n cm^{-2}; E > 1 MeV)	1.13x10^{18}	2.74x10^{19}	1.13x10^{18}
P Segregation (%monolayer)	7	5	4

Only low levels of P segregation were observed on the grain boundaries. In the plate material no substantial differences were observed between boundaries in the two irradiated materials suggesting that the radiation enhanced segregation is not significant.

In the IAEA CRP3 programme [*15*] the effect of irradiation on the grain boundary P levels was not measured in the irradiated as-received model steels based on A533B with varying copper and phosphorus contents. However, the level of IGF was both very low (<1%) and not affected by irradiation (Table 3) suggesting that irradiation of steels containing up to 0.017 wt% P to doses of < 19 mdpa had not induced damaging levels of grain boundary P.

The most significant observations of intergranular failure in US RPV steels were made on irradiated Charpy specimens [*23*] by Gurovich and co-workers. The major results are:

1. Unirradiated A533B base metal, containing 0.017% P, exhibited neither brittle nor ductile intergranular failure.

2. When material was irradiated to 2.6×10^{19} n cm^{-2} E > 0.5 MeV, and broken in the mid-transition region, 10% of the fracture surface consisted of brittle intergranular facets, and a further 10% of ductile intergranular regions.
3. Neither subsequent annealing nor extension of the irradiation to 10×10^{19} n cm^{-2} E > 0.5 MeV, increased the proportion of intergranular fracture.

Summary of Behavior

Observations of P Segregation and Intergranular Fracture in Different Microstructures

The segregation of phosphorus to grain boundaries has been observed in base material, weldmetal and heat affected zone material of Mn-Ni-Mo steels. Furthermore it has been observed in as-PWHT, thermally aged or irradiated microstructures, although normally significantly enhanced by aging or irradiation. By far the most work has been done on CGHAZ microstructures since these are the most susceptible to intergranular embrittlement, but what limited data there are on the segregation of phosphorus to grain boundaries in base material and weldmetals appears to be in reasonable agreement with that on CGHAZ material.

Segregation can occur during fabrication (especially during the slow cooling stage of a post weld heat treatment), and as a result of in-service exposure to high operating temperature and irradiation. Observations suggest that the typical levels at SOL are 5-8% in base metal, weld and HAZ. In terms of exposure effects, studies suggest the following:

1. P segregation is likely under the conditions considered applicable for vessel annealing, i.e. 1 week at 450-475°C.
2. Long-term aging at temperatures above 300°C may give rise to an increase in P on the grain boundaries.
3. There are limited data at high irradiation doses from accelerated tests in Materials Test Reactors on the P segregation increase after irradiation. The available results are in keeping with predictions of kinetic models.
4. Kinetic models, which predict the results of accelerated irradiations, further suggest that, at low dose rates, reduced point defect recombination will permit the segregation of more P to traps such as grain boundaries. This prediction needs to be validated by low dose rate irradiations.

Observations of Embrittlement Due to P Segregation and Intergranular Fracture in Different Microstructures

The embrittlement arising from segregation of P to grain boundaries can be expressed as an increase in the DBTT, which is approximately linearly related to intergranular content of the brittle fracture and/or the phosphorus concentration at the grain boundary. The correlations obtained are dependent on the product form and heat treatment/microstructure (possibly related to grain size and hardness). In view of the variability it is not feasible to model the changes in mechanical properties to establish a predictive capability, which can be applied generically. However, reasonable trends can be developed for different microstructures. Such trends have been established for CGHAZ in MnMoNi steels, but only indicative data exist for quenched and tempered plate or forging. What limited data there are indicate that the shift in the static fracture toughness transition curve due to segregation is less than that in the impact transition curve. Upper shelf properties can also be affected, although the magnitude of the reduction in impact properties is, by comparison, much less pronounced.

Comparison with Other RPV Steels

The pressure vessels of Magnox reactors, found in the UK, are manufactured using C-Mn steels. The C, Mn, Si and Cu levels in the plate are similar to those found in A533B/A508 steels; the S and P levels are higher; and the Ni and Mo levels lower. A high strength heat resistant Cr/Mo/V steel was developed for VVER-440 RPVs. This ferritic steel containing 0.13-0.18 wt% C and 2.50-3.0 wt% Cr was designated 15Kh2MF. There are additional requirements on the acceptable levels of impurity for materials in core shells subjected to neutron irradiation. The elements so controlled are P, S, Cu, As, Sb, Sn, and Co. In Eastern type steels the relative content of phosphorus is higher, 0.026 – 0.052 wt%.

The observations in the literature have been fully reviewed elsewhere [*2*], and the main features are summarized here:

1. In common with MnMoNi steels, P segregation and IGF are observed in high P steels after either thermal aging or irradiation. The extent of the data is uneven over the different microstructures, e.g. there are few observations of thermal aging in VVER steels.
2. Data on the levels of segregation after irradiation are less complete that in MnMoNi steels but in all cases increases have been reported in one or more microstructures. The data are summarized in Table 7.
3. Ductile IGF has been observed by Gurovich et al in a Russian thermally aged and irradiated pressure vessel steel (25Kh3NM, (1 wt% Ni, 0.024 wt% P). Gurovich et al. [*23*, *27*,] consider that for Eastern steels hardening effects are less relevant to embrittlement than segregation effects, and that the increase in DBTT due to intergranular segregation is less important than that due to segregation to precipitate:matrix interfaces.
4. Nikolaeva et al. [*28*] reported that, in a high-Cr, low alloy VVER RPV steel containing 0.018 wt% P irradiated to high doses at 275°C, the levels of grain boundary P increased with increasing fluence, in agreement with the observations by McElroy et al. [*15*] in irradiated simulated CGHAZ in MnMoNi steels.

Overall the observations in different steels support the observations in MnMoNi steels that an increase in P segregation and/or the level of IGF may occur after thermal aging or irradiation. However, the precise changes are very dependent on the steel, product form and heat treatment/microstructure.

All the data reviewed have concerned the observation of P to grain boundaries. It is important to note a recent study on Fe-P-C alloys, which indicates the potentially important role of carbon [*29*]. A ternary Fe-1200ppmP-30ppmC alloy was examined by Jones and co-workers after aging and irradiation exposure at 240°C. Auger spectroscopy showed that P grain boundary monolayer coverage was virtually unchanged for all SOL, aging and irradiation conditions. In contrast the C coverage was reduced by 50% during the initial aging or irradiation but changed little after longer irradiation. In contrast to studies on MnMoNi steels there was no evidence to support the occurrence of irradiation-induced grain boundary P segregation. The authors therefore considered the irradiation- or thermally-induced embrittlement to be caused by the loss of C from the grain boundary.

Table 7 - *Summary of Data Available on P Segregation in Other RPV Steels*

RPV type	**Start of Life (SOL)**	**Irradiated**	**Thermally Aged**
Mild Steels	Up to 20% in submerged-arc weldmetal, and probably in base metal also	Data reported for observations at a range of temperatures and irradiation doses. Limited data on plate and base metal. Fine grained welds are shown to be more susceptible than coarse grained welds.	Some data on welds, no data on HAZ.
Eastern Steels	Likely to be sensitive to exact heat treatment	Limited Observations Increase of 10% to 20% APFIM study of an irradiated weld irradiated (10^{19} n cm^{-2} (E > 1 MeV) at a flux of 4×10^{11} n cm^{-2} s^{-1} and at 290°C	No Data

Bischler et al. [*30*] examined through-thickness weld samples from C-Mn ferrite steel plates in a decommissioned RPV. They reported that the levels of P and C segregated to the grain boundaries did not vary through the thickness of the vessel. More importantly for this review, the authors considered that the levels of P and C observed were similar to the levels found at the start-of-life in materials of similar thermal history, suggesting that, in these steels at least, low dose rate irradiation had not caused any significant net segregation. This contrasts with the results reported for MnMoNi steels and Eastern steels.

Implications for Vessel Embrittlement

The data described in this review show that P-related embrittlement can be achieved in materials of the kind used in RPVs. The question is, whether it will occur under anticipated operating conditions. This involves two aspects:

1. How much P will be present on the grain boundaries?
2. Will this be sufficient to cause intergranular failure?

Even then this may still not be of any significance with regard to the structural integrity of the vessel. This will depend on (1) the extent of embrittlement, and (2) the consequence of embrittlement for vessel integrity.

The status of each of these is discussed next for MnMoNi steels typical of U.S. production.

How much P will be present on the grain boundaries? - Defining how much phosphorus will be at the grain boundaries appears relatively straightforward. Measurements may be made on archive material, or simulated archive material, to give an indication of SOL levels. The increase in grain boundary P can be calculated, for a range of service conditions, using the appropriate models described. At present, it is considered that the state of knowledge should enable estimates of grain boundary phosphorus levels to be made with reasonable reliability for a range of conditions; e.g. current or end of life (EOL). Further data are necessary, in particular the bench marking of low dose-rate calculations with direct measurements on surveillance samples.

Will this be sufficient to cause intergranular failure? - Whether the phosphorus concentration on the grain boundaries is sufficient to cause IGF is a more difficult question to answer. Clearly this depends on other details of the microstructure (e.g. grain size, other segregants particularly C, hardness) and, until IGF has been observed, that trigger level cannot be measured. Thus, for CGHAZ at SOL, the level can be estimated reasonably well (for a range of hardness values and grain sizes at least). There is insufficient information to show what the trigger levels might be for weldmetal and base metal, though what data exists suggest that the trigger levels may be higher for base metal than for CGHAZ. Measurements of grain boundary P concentrations in these materials when intergranular failure is observed not to occur can, however, be used to define at least a lower bound to the critical levels.

How much is the material embrittled? - With the exception of CGHAZ, the degree of embrittlement resulting from phosphorus segregation and IGF is difficult to predict. Even for CGHAZ the degree of embrittlement measured has been shown to be affected by the choice of test techniques, and to be dependent on a number of parameters related to its original heat treatment/fabrication; e.g. grain size and strength. In real weldments this is compounded by the distribution of coarse-grained zones which will be a function of the weld procedure. Clarification of the microstructure adjacent to the fusion boundary and the likely variations to be expected in reactor pressure vessels would provide some insight into the potential for embrittlement.

What is the consequence of the embrittlement for vessel integrity? - With regard to the structural significance of the embrittled material, changes in parent plate or weldmetal properties clearly have a direct bearing on a structure but these seem to be degraded by relatively small amounts. The major issue that has to be discussed is the consequence of the degradation of narrow zones such as the HAZ in a weldment, particularly when the embrittlement may be even more localised in the CGHAZ.

This issue really concerns the consequence of local brittle zones. There appear to be no directly relevant data on pressure vessel steels in the literature. Thermal aging programmes, designed to encourage P segregation in a range of US RPV steels, followed by mechanical testing will expand the information on typical threshold P levels for embrittlement. The information from such programmes will be useful, but may not be sufficient to determine the structural integrity consequence of embrittled CGHAZ.

However, an interesting parallel is provided in the as-welded steels, used typically in offshore structures. The brittle HAZ in these structures produces poor results in small-scale laboratory tests and large-scale wide test plates and yet in-service experience indicates that it does not cause overall structural integrity problems. If the advantage experienced by the large-scale structure is due to containment of the brittle zone, then P segregation to the CGHAZ in an RPV may be similarly without deleterious consequence. If it is due, to e.g., differences between the fast fracture laboratory tests, and the actual in-service loading conditions for the offshore structures, then the degree of similarity would vary with the anticipated RPV operational or accident scenarios.

The relevance of these observations on local brittle zones is that, although this deleterious effect has been demonstrated in large scale laboratory tests, in-service experience on large scale structures indicates that they do not present an integrity issue. In connection with this last point, it should be noted as well, that no instances of cracking initiating at HAZs have yet been found in the US RPV surveillance programme.

Conclusions on the likelihood of embrittlement in irradiated US RPVs - The P distributions in US plate, forgings, and welds are illustrated in Figure 6*. The bulk P content appears to be less than ~ 0.028 wt% P, with base metal having lower levels than weldments (<0.025 wt%). Data are sparse but in base metal and weld the data strongly suggest that there is only concern for bulk P levels close to the upper limit and irradiation doses that are close to end of life (EOL), as otherwise the observed segregation is low. It is not possible from the data reviewed here to remove the possibility of a contribution from IGF at EOL doses for base metal and welds containing high levels of bulk P. Embrittlement of CGHAZ through P segregation during irradiation requires further work, as appreciable segregation of P to grain boundaries in simulated CGHAZ has been observed under relevant irradiation conditions.

If vessel annealing is considered at temperatures around 475°C, then the consequences of this aging treatment on P segregation in plate, weld and HAZ have to be considered. It is certain that the annealing treatment will have the consequence of reducing the irradiation hardening, but significantly increasing the grain boundary phosphorus coverage and the likelihood of intergranular failure. Comparison between the calculated grain boundary P levels and the estimates of the critical levels can be used

---

* Griesbach, T., private communication.

to identify if embrittlement is likely after annealing or subsequent re-irradiation. Less segregation is expected for an annealing temperature of 454°C for a typical annealing time of 1week, as compared to 475°C. It may be necessary to undertake experiments to determine the trigger level in the steels concerned.

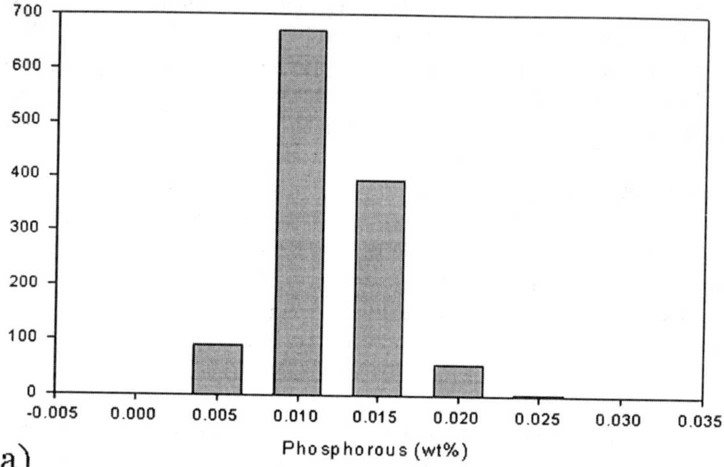

a)

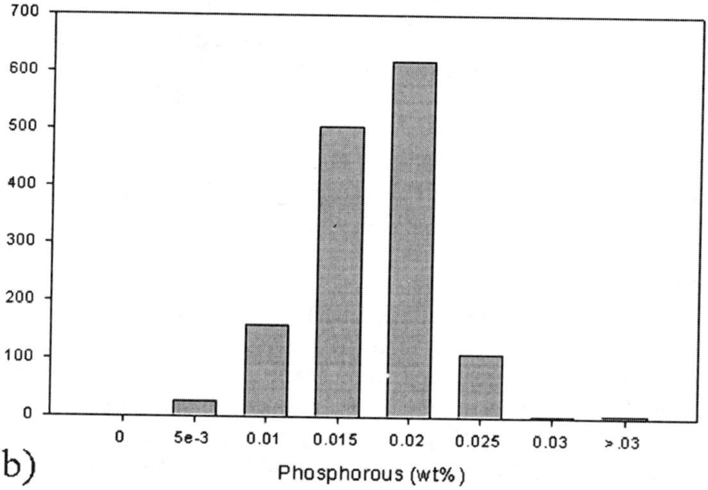

b)

Figure 6 - *Bulk P distributions in US steels a) plates and forgings, and b) welds.*

Conclusions

This review has established the following points:

1. Segregation of phosphorus to grain boundaries in pressure vessel steels is a

relatively common occurrence, and has been observed in base material, weld metal and heat affected zones. Such segregation can occur during fabrication (especially during the slow cooling stage of a post weld heat treatment), and as a result of in-service exposure to high operating temperature and irradiation.

2. The availability of segregation data varies between the different classes of steels. In US RPV steels most information is on thermal aging of HAZ. Extensive observations of phosphorus segregation under thermal aging, together with determination of the relevant activation energies, has enabled segregation at grain boundaries to be well-characterized for CGHAZ material.
3. The segregation of phosphorus to grain boundaries promotes a change in the brittle fracture mode from transgranular to intergranular, and a degradation in the mechanical properties. This is characterized most frequently in terms of the shift in the impact transition curve and an increase in the ductile to brittle transition temperature. However, the measured fracture toughness for the steel can be similarly degraded, although data are limited and, even when available, involve only static rather than dynamic measurements.
4. Studies in CGHAZ have shown that the embrittlement arising due to segregation of P to grain boundaries is approximately linearly related to intergranular component of the brittle fracture and/or the phosphorus concentration at the grain boundary. The correlations obtained are very dependent on the steel, product form and heat treatment/microstructure, and have not been determined for the relevant microstructures in the different classes of steels. In the literature such trends have only been established for CGHAZ in MnMoNi steels. In view of the variability it is not feasible to model the changes in mechanical properties to establish a predictive capability which can be applied generically.
5. The data described in this review show that P-related embrittlement can be achieved in materials of the kind used in RPVs. The question is, whether it will occur under anticipated operating conditions. This involves several aspects:
    - How much P will be present on the grain boundaries?
    - Will this be sufficient to cause intergranular failure?
    - How much is the material embrittled?
    - What is the consequence of the embrittlement for vessel integrity?
6. It is concluded that there is a reasonable capability to estimate the likely levels of grain boundary P coverage which would occur during service. Further data would be welcomed, in particular the benchmarking of calculations with direct measurements on surveillance samples.
7. The trigger level of phosphorus concentration on the grain boundaries that causes IGF can be estimated reasonably well for CGHAZ at SOL (for a range of hardness values and grain sizes at least). There is insufficient information to show what the trigger levels might be for weldmetal and base metal.
8. With the exception of CGHAZ material, the degree of embrittlement resulting from phosphorus segregation and IGF is difficult to predict from current data.
9. With regard to the structural significance of the embrittled material, the major issue

is the consequence of the degradation of narrow zones such as the HAZ in a weldment, particularly when the embrittlement may be even more localized in the CGHAZ. There would appear to be no directly relevant data on pressure vessel steels in the literature on the effect of local brittle zones. The experience in the offshore industry is relevant. Although the deleterious effect of local brittle zones has been demonstrated in large-scale laboratory tests, in-service experience on large-scale structures indicates that they do not necessarily present an integrity issue.

10. The likelihood of embrittlement in irradiated US RPVs was considered. In general, the bulk P content appears to be less than ~ 0.028 wt% P, with base metal having lower levels than weldments. Data are sparse on the effect of irradiation at 288°C on P segregation and the contribution of IGF to the total shift in T_{41J}. In base metal and weld the data strongly suggest that there is only concern for bulk P levels close to the upper limit and irradiation doses that are close to end of life, as otherwise the observed segregation is low. It is not possible from the data reviewed here to remove the possibility of a contribution from IGF at EOL doses for base metal and welds containing high levels of bulk P. Embrittlement of CGHAZ through P segregation during irradiation requires further work, as appreciable segregation of P to grain boundaries in simulated CGHAZ has been observed under relevant irradiation conditions.
11. If vessel annealing is considered at temperatures around 475°C, then the consequences of this aging treatment on P segregation in plate, weld and HAZ have to be considered. It is certain that the annealing treatment will have the consequence of reducing the irradiation hardening, but significantly increasing the grain boundary phosphorus coverage and the likelihood of intergranular failure. It may be necessary to undertake specific validation experiments to determine the trigger level in the steels being considered.

References

[1] Druce, S. G., Gage, G., and Jordan, G., *Acta Metallugica*, Vol 34, 1986, pp. 641-652.

[2] Review of Phosphorus Segregation and Intergranular Embrittlement in Reactor Pressure Vessel Steels (PWRMRP-19): PWR Materials Reliability Project (PWRMRP), EPRI, Palo Alto, CA: 2000. TR-114783.

[3] Grabke, H. J., Hennesen, K., Moller R., and Wei, W., "Effects Of Manganese On The Grain Boundary Segregation, Bulk and Grain Boundary Diffusivity Of Phosphorus In Ferrite", *Scripta Metallurgica*, Vol. 21, 1987, pp. 1329-1334.

[4] Jones R. B., and Bolton C. J., "Neutron Radiation Embrittlement Studies In Support Of Continued Operation And Validation By Sampling Of Magnox Reactor Steel Pressure Vessels And Components", in Proc. 24th Water Reactor Safety Meeting organised USNRC October 1996.

[5] Eason, E. D., Wright J. E., and Odette, G. R., "Improved Embrittlement Correlations for Reactor Pressure Vessel Steels," NUREG/CR-6551 MCS/070501 1998.

[6] Wang, J.A., "Development of Embrittlement Prediction Models for U.S. Power Reactors and the Impact of Heat-Affected Zone to Thermal Annealing", *Effects Of Radiation On Materials: 18th International Symposium, ASTM STP 1325*, R.K.Nanstad, M.L.Hamilton, F.A.Garner and A.S.Kumar, Eds., American Society for Testing and Materials, West Conshohocken, PA, 1999, pp 525-540.

[7] Rosinski, S. T., Hawthorne, J. R., and Rocheau, G. E., "Experimental Tests of Irradiation-Anneal-Reirradiated Effects on Mechanical Properties of RPV Plate and Weld Materials." *Effects Of Radiation On Materials: 18th International Symposium, ASTM STP 1325*, R.K.Nanstad, M.L.Hamilton, F.A.Garner and A.S.Kumar, Eds., American Society for Testing and Materials, West Conshohocken, PA, 1999, pp 388-402.

[8] Odette, G. R., Lucas, G., and Klingensmith, D., "On the Effect of flux and Composition on Irradiation Hardening at 60°C" *Effects Of Radiation On Materials: 17th International Symposium, ASTM STP 1270*, D. S. Gelles, et al., Eds., American Society for Testing and Materials, West Conshohocken, PA, 1996 pp 547-568.

[9] Tavassoli, A., Bougault, A., and Bisson, A., Proc Int Conf on the effects of Residual impurity on Weldability and Weld Properties. London, 15-17 Nov. 1983, pp43-68.

[10] Lancaster, J. F., "Metallurgy of Welding", pub. George Allen & Unwin, London, 1980.

[11] Gage, G., Druce, S. G., and Popkiss, E.W., Proceedings of Intl Conference on Nuclear Power Plant Life Extension, July 31-August 2, 1988, Vol. 1, p 86.

[12] Vatter, I. A., Hippsley, C. A., Druce, S. G., *International Journal of Pressure Vessel and Piping*, Vol 54, 1993, pp31-48.

[13] Vatter, I. A., Hippsley, C. A., "Quantification of Microstructural Effects on Thermal Aging of Induced Embrittlement in Pressure Vessel Steels." AEA Technology Report AEA-RS-4464, 1993.

[14] Jordan, G., Wigham, D. R., and Hippsley, C. A., AEA TSD-2206.

[15] McElroy, R. J., English, C. A., Foreman, A. J. E., Gage, G., Hyde, J.M., Ray, P.H.N., and Vatter, I.A., "Temper Embrittlement, Irradiation Induced Phosphorus Segregation and Implications for Post-Irradiation Annealing of Reactor Pressure Vessels," *Effects Of Radiation On Materials: 18th International Symposium, ASTM STP 1325*, R.K.Nanstad, M.L.Hamilton, F.A.Garner and A.S.Kumar, Eds., American Society for Testing and Materials, West Conshohocken, PA, 1999, pp296-316.

[16] McCabe, D. E., "Initial Evaluation of the Heat-Affected-Zone, Local Embrittlement Phenomenon As It Applies To Nuclear Reactor Vessels." Letter report, September 1999,ORNL/NRC/LTR-99/10.

[17] McCabe D. E., Nanstad R K., English C A., Ortner S R., "Investigation of Temper Embrittlement in Reactor Pressure Vessel Steels Following Thermal Aging, Irradiation, and Thermal Annealing," *Effects of radiation on Materials: 20th International Symposium, ASTM STP 1405*, S. T. Rosinski, M.L. Grossbeck, T. R. Allen, and A.S. Kumar, Eds, American Society for Testing and Materials, West Conshohocken, PA, 2002. This volume.

[18] McLean, D., in "Grain Boundaries in Metals", Clarendon Press, Oxford (1957).

[19] Druce, S. G., English, C. A., Foreman, A. J. E., McElroy, R. J., Vatter, I. A., Bolton, C. J., Buswell, J. T., and Jones, R. B., "The Modelling Of Irradiation-Enhanced Phosphorus Segregation In Neutron-Irradiated Reactor Pressure Vessel Submerged Arc Welds", *Effects Of Radiation On Materials:17th International Symposium, ASTM STP 1270*, D. S. Gelles, et al., Eds., American Society for Testing and Materials, West Conshohocken, PA, 1996.pp. 119-137.

[20] Gage, G., Atkins, T., Lane, C., Ray, P. H. N., and Wigham, D., "Palisades Annealing Recovery and Irradiation Program Evaluation of the Recovery of Hardness and the incidence of Intergranular Fracture after Post-Irradiation Annealing". AEA Technology Repórt AEA-T 1696, July 1997.

[21] Swift, R. A., and Gulya, J. A., "Temper Embrittlement of Pressure Vessel Steels" *Welding Research Supplement 57-s*, February 1973.

[22] Miller, M. K., and Burke, G., "An APFIM Survey of Grain Boundary Segregation and Precipitation in Irradiated Pressure Vessel Steels." *Effects Of Radiation On Materials:16th International Symposium, ASTM STP 1175*, A. S. Kumar, et al., Eds., American Society for Testing and Materials, West Conshohocken, PA, 1993, pp. 492-502.

[23] Gurovich, B. A., Korolev, Y. N., Kuleshova, E. A., Nikolaev, Y. A., and Shtrombakh, Y. I., "Irradiation Embrittlement of Reactor Pressure Vessel Steels due to Mechanisms other than Radiation Hardening" *Effects Of Radiation On Materials:18th International Symposium, ASTM STP 1325*, R.K.Nanstad, M.L.Hamilton, F.A.Garner and A.S.Kumar, Eds., American Society for Testing and Materials, West Conshohocken, PA, 1999, pp 271-295.

[24] Hasegawa, M., Nakajima, N., Kusunoki, N., and Suzuki., K., *Trans Japan Institute of Metals*, Vol 6, 1975, pp. 641.

[25] Soulat, P., Miannay, D., and Horowitz, H., "Temper Embrittlement of Phosphorus on thermal and Neutron embrittlement of Pressure vessel steels." *Radiation Embrittlement of Nuclear Pressure Vessel Steels: An International Review (Fourth Volume), ASTM 1170*, Lendell E., Steele, Ed., American Society for Testing and Materials, Philadelphia, 1993, pp. 283-299.

[26] Soulat, P., Houssin, B., Bocquet, P., and Bethmont, M., "Analysis of radiation Embrittlement Results from a French Forging Examined in the Second Phase of an IAEA-Co-ordinated Research Program". *Radiation embrittlement of Nuclear Pressure Vessel Steels: An International Review (Fourth Volume), ASTM 1170*, Lendell E., Steele, Ed., American Society for Testing and Materials, Philadelphia, 1993, pp. 249-265.

[27] Gurovich, B. A., Kuleshova, E. A., Nikolaev, Yu. A., and Shtrombakh, Ya. I., *Journal of Nuclear Materials*, Vol. 246, 1997, pp. 91.

[28] Nikolaeva, A. V., Kevorkyan, Y. R., and Nikolaev, Y. A., "Comparison Of Observed and Predicted Data On Radiation Induced Grain Boundary Phosphorus Segregation In VVER Steels," *Effects of Radiation on Materials, 19th International Symposium, ASTM STP 1366*, M.L.Hamilton, A.S. Kumar, S.T.Rosinski and M.L.Grossbeck, Éds, American Society for Testing and Materials, West Conshohocken, PA, 1999, pp. 399-414.

[29] Jones, R. B., Cowan, J. R.,Corcoran, R. C., and Walmsley, J. C., "Embrittlement, Hardening, and Grain Boundary Composition in an Irradiated and Thermally Aged Fe-P-C Alloy, *Effects of Radiation on Materials, 19th International Symposium, ASTM STP 1366*, M.L.Hamilton, A.S. Kumar, S.T.Rosinski and M.L.Grossbeck, Eds, American Society for Testing and Materials, West Conshohocken, PA, 1999. pp. 473-491.

[30] Bischler, P. J. E., Callen, V.M., Corcoran R.C., and Spellward, P., "A Microstructural Examination Of Irradiated Reactor Pressure Vessel Weld Samples" in IAEA Specialist Meeting held Madrid 1999, proceedings to be published in *International Journal of Pressure Vessel and Piping*.

Valery A. Pechenkin,[1] Igor A. Stepanov,[1] and Yury V. Konobeev[1]

Modeling of Phosphorus Accumulation on Grain Boundaries in Iron Alloys under Irradiation

Reference: Pechenkin, V. A., Stepanov, I. A., and Konobeev, Yu. V., "**Modeling of Phosphorus Accumulation on Grain Boundaries in Iron Alloys under Irradiation,**" *Effects of Radiation on Materials: 20th International Symposium, ASTM STP 1405,* S. T. Rosinski, M. L. Grossbeck, T. R. Allen, and A. S. Kumar, Eds., American Society for Testing and Materials, West Conshohocken, PA, 2001.

Abstract: Kinetics of phosphorus accumulation on grain boundaries (GB) in iron-based alloys is treated theoretically, taking into account both the radiation-induced segregation (RIS) in the matrix and the Gibbsian adsorption (GA) at GB. For steady-state conditions analytical expressions are derived for component profiles near GB and component concentrations on GB. Modeling of phosphorus accumulation at GB in iron alloys is carried out using two different models: 1) the McLean's model generalized to take into account the radiation-enhanced phosphorus diffusion via vacancy and interstitial mechanisms as well as RIS near GB, 2) a model of RIS in a ternary Fe-P-Ni alloy accounting for the binding energy of phosphorus atoms with interstitials as well as a possibility of high phosphorus content near GB at high irradiation doses. Predictions of modeling are analyzed regarding the dependence on temperature, dose and Fe and P diffusion parameters available for iron alloys. It is shown that the GB phosphorus concentration calculated as a function of temperature reveals one or two maxima depending on dose and the choice of material parameters.

Keywords: iron alloys, irradiation, embrittlement, grain boundaries, radiation-induced segregation, Gibbsian adsorption, phosphorus accumulation, modeling

The phosphorus accumulation on grain boundaries (GB) in irradiated iron alloys is being extensively studied relative to the radiation embrittlement (RE) of pressure vessel steels. As has been pointed out in Ref[*1*], RE is caused by a cooperative action of three mechanisms: radiation hardening of the matrix, segregation of impurities (mainly of

[1] Head of Laboratory, Research Scientist and Head of Division, respectively, State Scientific Center of Russian Federation, Institute of Physics and Power Engineering, Bondarenko sq.1, Obninsk, Kaluga region, Russia 249033.

phosphorus) at GB and at precipitate/matrix interfaces. Particularly, for Russian steels the main contribution to RE is related to the phosphorus segregation at intergranular and interphase boundaries [*1,2*]. The phosphorus segregation at point defect sinks can be important for the radiation-induced phosphides formation too [*2,3*].

Two alternative approaches are usually used for modeling impurity atoms and alloying element accumulation on GB in alloys under irradiation [*1,4-6*]. One of them is the McLean's kinetic model for a component accumulation on GB due to Gibbsian adsorption (GA) modified to account for the radiation-enhanced diffusion via the vacancy mechanism. Such an approach has several limitations: 1) radiation enhancement of diffusion via interstitial mechanism is not considered, 2) radiation-induced segregation (RIS) of alloy components near GB is ignored, 3) Guttmann's equilibrium GB component concentrations in thermal annealing conditions [*7*] are used as an upper limit of ones in irradiation conditions. In an alternative approach, RIS near GB in dilute binary alloys is considered to neglect the GA of components on GB. Phosphorus concentration on GB is calculated simply by doubling the concentration at the nearest to GB lattice plane.

The aims of the present paper are as follows: 1) to study relative contributions of RIS and GA to the component segregation at GB in multicomponent alloys under irradiation with application to phosphorus and nickel accumulation on GB in iron alloys, 2) to generalize McLean's model for irradiation conditions and to model the phosphorus accumulation on GB in a wide range of temperatures and doses, 3) to calculate the phosphorus segregation near GB using the model of RIS in ternary Fe-P-Ni alloys, 4) to evaluate a sensitivity of model predictions to a variation of diffusion parameters of Fe and P in iron alloys.

For this purpose, the systems of equations for component and point defect fluxes from the matrix to GB and in the matrix of multicomponent alloys under irradiation is formulated on the base of linear non-equilibrium thermodynamics. Differences in component diffusivitics both via vacancy and interstitial mechanisms in the matrix as well as GA of components on GB are considered. A physical criterion permitting to divide the temperature scale into two regions with different mechanisms of GB component accumulation is formulated. In low temperature region a "kinetic" equilibrium is established between matrix and GB component concentrations due to point defect fluxes. In high temperature region both RIS and GA are responsible for a "thermodynamic" equilibrium between these concentrations. Analytical expressions for steady state component profiles near GB and for component concentrations on GB are derived for a wide range of temperatures. These expressions allow one to treat the competition of components (in particular, phosphorus and nickel in pressure vessel steels) both in equilibrium segregation on GB and in RIS near GB. In the absence of irradiation the expressions are reduced to the well-known Guttmann's expressions for equilibrium segregation on GB.

In the present paper modeling of phosphorus accumulation at GB in iron alloys is carried out using two different kinetic models: 1) the McLean's model, generalized to take into account the radiation-enhanced phosphorus diffusion via vacancy and interstitial mechanisms as well as RIS near GB, and 2) a model of RIS in a ternary Fe-P-Ni alloy accounting for a binding energy of undersized phosphorus atoms with interstitials as well as a possibility of high phosphorus content near GB at large irradiation doses.

Predictions of these models regarding the dependence on temperature, dose and material diffusion parameters available are analyzed. It is shown that the GB phosphorus concentration calculated as a function of temperature reveals one or two maxima depending on dose and on the choice of material parameters. The results point to a high sensitivity of model predictions to the scatter in experimental data available on diffusion parameters of Fe and P in iron alloys.

Steady-State Segregation

In order to apply the McLean's approach to the component accumulation on GB one should know limiting concentrations on GB at a large time. In thermal annealing conditions those are given by the Guttmann's expressions [*8*]. Under irradiation these concentrations correspond to the steady-state segregation, when component fluxes in the matrix J_k and ones from the matrix to GB j_k are equal to zero while point defect (PD) fluxes to GB j_α ($\alpha = v$ for vacancies and $\alpha = i$ for interstitials) are constant. The steady-state component profiles $C_k(r)$, where r is the distance from GB, in the matrix as well as the steady-state component concentrations C_k^g on GB are established at a sufficiently large irradiation time (see below). Under the steady-state

$$J_k = j_k = 0, \quad j_v = j_{v0} = j_i = j_{i0} = \varepsilon K R_g, \tag{1}$$

where ε is the cascade efficiency, K is the PD generation rate, index "0" denotes an alloy with no segregation, R_g is the size of a cell, from which PD are absorbed on GB (see [*9*]), which is of the order of PD free path length. In Eq (1) biasing of PD sinks caused by their stress fields as well as by RIS [*10*] were neglected.

Steady-State Component Profiles near Grain Boundary

By estimates, under irradiation the steady-state PD concentrations are established during the time of the order of $1/(D_v \rho_s)$, where ρ_s is the PD sink strength. The characteristic time needed for establishing the steady-state component profiles near PD sinks is larger by several orders of magnitude. This is due to a large difference in diffusion coefficients of PD, $D_\alpha = \sum_j d_{j\alpha} C_j$, and components $D_k = d_{ki} C_i + d_{kv} C_v$. PD profiles near sinks change during establishing the component profiles. However, such a change is insignificant. Therefore, the steady-state vacancy concentration profiles $C_{v0}(r)$ in an alloy with no segregation [*9*] as well as the condition $J_\alpha(r) = J_{\alpha 0}(r)$ $(\alpha = i, v)$ will be used below to obtain the steady-state component profiles near GB.

To derive analytical expressions for steady-state component profiles the following fluxes of components and PD in a concentrated substitutional alloy will be used

$$J_k = -D_k \alpha_k \nabla C_k + C_k (d_{kv} \nabla C_v - d_{ki} \nabla C_i) = 0, \tag{2}$$

$$J_{v0} = J_v = -D_v \nabla C_v + C_v \sum_k d_{kv} \alpha_k \nabla C_k ,$$

$$J_{i0} = J_i = -D_i \nabla C_i - C_i \sum_k d_{ki} \alpha_k \nabla C_k , \tag{3}$$

where α_k is the thermodynamic factor.

Any correlation effects, external forces and the dependence of PD activation energies on alloy composition [*1,11*] are ignored.

Eq (2) can be written as follows

$$\nabla C_k = \frac{C_k d_{kv}}{\alpha_k D_k} \nabla C_v - \frac{d_{ki} C_k}{\alpha_k D_k} \nabla C_i . \tag{4}$$

Eq (4) can be solved analytically taking $\sum_k C_k = 1$ (in this case $\sum_k \nabla C_k = 0$) and using the additional simplifications: 1) $\alpha_j = 1$ and 2) PD profiles C_v and C_i near GB can be replaced by C_{v0} and C_{i0} (then $D_k \approx D_{v0} C_{v0} (d_{kv} / D_{v0} + d_{ki} / D_{i0})$). In this case the solution of Eq (4) has the following form [*12*]

$$C_k = \frac{C_{k0} (C_{v0} / < C_{v0} >)^{g^0_{kN}}}{\sum_j C_{j0} (C_{v0} / < C_{v0} >)^{g^0_{jN}}}, \tag{5}$$

where $g^0_{kN} = g_{kN} / (D_{v0} D_{i0} f_0)$, $g_{kN} = d_{kv} d_{Ni} - d_{ki} d_{Nv}$,

$$f_0 = \left(\frac{d_{kv}}{D_{v0}} + \frac{d_{ki}}{D_{i0}} \right) \left(\frac{d_{Nv}}{D_{v0}} + \frac{d_{Ni}}{D_{i0}} \right) \cdot \sum_j \frac{C_{j0}}{(d_{jn} / D_{v0} + d_{ji} / D_{i0})} \cdot \frac{d_{ji}}{D_{i0}},$$

$< C_{v0} >$ is the average vacancy concentration in the matrix [*10*], D_{v0} and D_{i0} are the vacancy and interstitial diffusion coefficients in an alloy without segregation. The index "*N*" stands for one alloy component picked out from major alloy components (for iron in a Fe-P-Ni alloy below).

Steady-State Component Concentrations on GB

Since an atomistic model of GB in alloys is absent at present, for a crude estimation we will use below a model developed in Refs[*8,13*] together with linear non-equilibrium thermodynamics approach for point defect and component fluxes in the matrix and from the matrix to GB at times when the system is approaching to the steady-state. Following Guttmann [*8*] and Defai and Prigogine [*13*], GB will be considered as a flat layer of the width d in which chemical potentials of components μ^g_k and thermal equilibrium point defect concentrations $C^g_{\alpha e}$ (α=*v* and *i* for vacancies and interstitials, respectively) differ from those in the matrix, μ^b_k and $C^b_{\alpha e}$. GB is considered as a perfect sink or source of PD ($\mu^g_v = \mu^g_i = 0$) separated from the matrix by a distance *b* of the order of interatomic spacing.

PD and component fluxes *J* in the matrix are proportional to gradients of chemical potentials if correlation factors and external forces are ignored (see a comprehensive review by Allnatt and Lidiard [*14*]). PD and component fluxes *j* from matrix to GB will be constructed as a discrete analog of *J* by replacing the gradients with differences of chemical potentials divided by *b*. The flux j_k of a component "*k*" between GB and the nearest crystallographic plane in the matrix can be written as

$$j_k = j_k^k + j_k^i - j_k^v, \tag{6}$$

where j_k^k is the term determined by the difference in chemical potentials of "k" atoms on GB and in the matrix, j_k^i and j_k^v are the terms determined by the fluxes of interstitials and vacancies to GB, respectively.

Accounting for these terms one can write the fluxes j_k and PD fluxes j_α to GB as follows

$$j_k = \frac{D_k^b C_k^b}{kT}\frac{(\mu_k^b - \mu_k^g)}{b} + \frac{d_{ki}^b C_i^b C_k^b}{kT}\frac{(\mu_i^b - \mu_i^g)}{b} - \frac{d_{kv}^g C_v^b C_k^g}{kT}\frac{(\mu_v^b - \mu_v^g)}{b}, \tag{7}$$

$$j_v = -\frac{C_{ve}^g}{kT}\sum_k d_{kv}^b C_k^b \frac{(\mu_k^b - \mu_k^g)}{b} - \frac{C_v^b}{kT}\sum_k d_{kv}^g C_k^g \frac{(\mu_v^g - \mu_v^b)}{b}, \tag{8}$$

$$j_i = \frac{C_i^b}{kT}\sum_k d_{ki}^b C_k^b \frac{(\mu_k^b - \mu_k^g)}{b} + \frac{C_i^b}{kT}\sum_k d_{ki}^b C_k^b \frac{(\mu_i^b - \mu_i^g)}{b}, \tag{9}$$

where "g" and "b" indexes denote GB and its nearest plane in the matrix, C_k and C_α are the component and PD concentrations, respectively, d_{ka}^b and d_{ka}^g are the component "k" diffusivities corresponding to atomic jumps from the matrix to GB and back, respectively, $D_k^b = d_{ki}^b C_i^b + d_{kv}^b C_{ve}^g$. In Eqs (7) to (9) it was taken into account that PD migration energies E_α^{mb} between GB and the nearest plane can differ from those E_α^m at large distances from GB.

In steady-state conditions one can obtain from Eqs (7) and (1)

$$\mu_k^g = \mu_k^b + \left(\frac{d_{ki}^b C_i^b}{D_k^b}\mu_i^b - \frac{d_{kv}^g C_v^b C_k^g}{D_k^b C_k^b}\mu_v^b\right) = \mu_k^b + (\varepsilon_{ki} - \varepsilon_{kv}) = \mu_k^b + \varepsilon_k \tag{10}$$

The "kinetic" term ε_k in Eq (10) disturbs a "thermodynamic" equilibrium between component concentrations on GB and in the matrix. The relation between these concentrations can be obtained by inserting the Guttmann's expression [8] for μ_k^g in Eq (10)

$$\frac{C_k^g}{C_k^b}\exp\left(\beta_{kv}\frac{C_k^g}{C_k^b}\right) = \frac{C_N^g}{C_N^b}\exp\left(\beta_{Nv}\frac{C_N^g}{C_N^b}\right)\exp\left(\frac{\Delta G_{kN}}{kT} + \delta_{kN}\right), \tag{11}$$

where $\beta_{kv} = \varepsilon_{kv}/kT$, $\delta_{kN} = (\varepsilon_{ki} - \varepsilon_{Ni})/kT$, the difference in free enthalpies of "k" and "N" components on GB, ΔG_{kN} was defined elsewhere [7].

Using Eq (1) and the approximation $C_{a0}^b = C_{a0}(b) \approx C_{ae}^b + \varepsilon K R_g b / D_{a0}$ [9], one can obtain the following estimates of β_{kv} and δ_{kN}

$$\beta_{kv} \approx \beta \cdot \frac{D_{Ns}^b}{D_{ks}^b}\cdot\frac{d_{kv}^g}{\langle d_v^g\rangle}, \quad \delta_{kN} \approx \beta \cdot \frac{\left(d_{ki}^b d_{Nv}^b - d_{Ni}^b d_{kv}^b\right)\cdot C_{ve}^g}{\langle d_i^b\rangle D_k^b}, \tag{12}$$

$$\beta \approx \frac{\beta_0}{1+\beta_0}, \quad \beta_0 = \frac{\varepsilon K b R_g}{D_{Ns}^b \cdot C_{ve}^g / C_{ve}^b} \tag{13}$$

In Eq (13) the thermal equilibrium vacancy concentration on GB C_{ve}^{g} can significantly exceed C_{ve}^{b} in the matrix. D_{ks}^{b} is the self-diffusion coefficient of component "k" near GB, $< d_{\alpha}^{b} >$ is the component diffusivity, averaged over the component concentrations on GB ($\alpha = v$) and in the matrix ($\alpha = i$).

The temperature dependence of β_{kv} and δ_{kN} keeps mainly in the strongly temperature dependent parameter β_0 (see Fig. 1 below). Physically, β_0 equals the ratio of the PD flux from the region R_g to GB due to irradiation $\varepsilon K R_g$ (this flux is responsible for RIS) and the flux of thermal vacancies between GB and the matrix $D_{Ns}^{b} C_{ve}^{g} / C_{ve}^{b} b$ (this flux is responsible for thermodynamic equilibrium between GB and the matrix due to thermal vacancies). So, the parameter β_0 divides the temperature scale into two regions of different mechanisms of the GB component accumulation. At low temperatures ($\beta_0 >> 1$) a "kinetic" equilibrium governed by PD fluxes is established between matrix and GB component concentrations. At high temperatures ($\beta_0 << 1$) both RIS in the matrix and GA are responsible for a "thermodynamic" equilibrium between these concentrations.

"Thermodynamic" and "Kinetic" Equilibrium between the Matrix and GB

Since the denominator in Eq (13) increases exponentially with increasing the temperature, $\beta_0 << 1$ and $\beta_{kv} C_k^g / C_k^b << 1$, $\delta_{kN} << 1$ at sufficiently high temperatures. The same inequalities are valid at low PD generation rates. In these cases Eq (10) is reduced to $\mu_k^g = \mu_k^b$ and the solution of Eq (11) with account of Eq (5) can be solved using Guttmann's approach [8]. The solution has the following form

$$C_k^g = \frac{C_{k0} (C_{v0}^b / < C_{v0} >)^{g_{kN}^0} \cdot \exp(\Delta G_{kN} / kT)}{\sum_j^n C_{j0} (C_{v0}^b / < C_{v0} >)^{g_{jN}^0} \cdot \exp(\Delta G_{jN} / kT)} \tag{14}$$

where $C_{v0}^b = C_{v0}(b)$.

The analytical expression (14) allows one to treat the competition and cooperation of components both in the equilibrium adsorption on GB and in RIS near GB. In the absence of RIS (K=0 or $g_{kN} = 0$) these expressions are reduced to the well-known Guttmann's expressions [8] for multicomponent alloys.

At low temperatures (β_O>>1) PD fluxes from the matrix to GB significantly exceed the flux of thermal vacancies between the matrix and GB. This leads to a "kinetic" equilibrium between them. Then Eq (10) is reduced to $\varepsilon_{ki} \approx \varepsilon_{kv}$ and the relation between steady state component concentrations on GB C_k^g and in the nearest to GB crystallographic plane in the matrix C_k^b has the following form

$$C_k^g = \frac{C_k^b d_{ki}^b / d_{kv}^g}{\sum_j C_j^b d_{ji}^b / d_{jv}^g}, \tag{15}$$

where C_k^b should be taken from Eq (5) after replacing C_{v0} with C_{v0}^b.

Component diffusivities d_{kv}^g corresponding to their jumps from GB to the matrix can be expressed in terms of jumps in the opposite direction d_{kv}^b using detailed balance in thermal equilibrium conditions when Guttmann's relations between component concentrations in the matrix C_{k0} and on GB C_{ke}^b are valid. Then Eq (15) takes the form

$$C_k^g = \frac{C_k^b \exp(\Delta G_{kN} / kT) d_{ki}^b / d_{kv}^b}{\sum_j C_j^b \exp(\Delta G_{jN} / kT) d_{ji}^b / d_{jv}^b}. \tag{16}$$

Applying Eqs (14) and (16) to phosphorus in iron-based alloys one can see, that at low temperatures, when phosphorus concentration on GB approaches unity according to Eq (14), the same holds according to Eq (16), i.e. both equations lead to the same result in low temperature region.

Modeling of Phosphorus Segregation on Grain Boundaries in Iron-Based Alloys

Although the equations for PD and component fluxes in the matrix were formulated above for concentrated alloys, they can be used also at low contents of an undersized alloying component "*m*" which forms bounded mixed dumbbells, such as P, Si, S etc. in iron-based alloys. In this case the analysis given above remains valid after replacing of d_{mi} with $\xi \cdot d_{mi}$, where $\xi = \exp(E_m^b / kT)$, E_m^b is the mixed dumbbell binding energy [*15*]. But, if due to RIS the concentration of *m*-component becomes high near GB, more complicated equations should be used [*16*].

For the present the accumulation of phosphorus on GB in ferritic steels at low irradiation temperatures is usually calculated taking into account only RIS near GB [*4,5*]. In Refs[*4,5*] an atomistic model for the migration of substitutional solute atoms in dilute binary f.c.c. alloys under irradiation developed earlier in [*17,18*] was used to calculate the phosphorus accumulation on GB in ferritic steels. Later a similar model for b.c.c. dilute binary alloys was developed in Refs[*19,20*]. The merit of these models is a correct account of all various jumps of solute atoms via vacancy and interstitial mechanisms. The limitations are as follows: these models are applicable only at low solute concentration near GB and do not account for component competition in RIS. In addition, the four-frequency model was choosen in Refs[*19,20*] for the vacancy mechanism of phosphorus migration in α-Fe, assuming that there may be a significant interaction between a solute atom and a vacancy only at the nearest-neighbor separation. This model is a particular case of more general five-frequency model developed in Ref[*21*], where the solute atom-vacancy interaction at larger separations was considered. Molecular dynamics calculations of phosphorus atom-vacancy interactions [*22,23*] and vacancy migration energies [*23*] in the neighborhood of phosphorus atom in α-Fe using interatomic potentials from Ref[*24*] pointed out to a significant binding energy (0.39 eV) between P atom and a vacancy at the second-neighbor separation. So the five-frequency model seems to be more applicable for vacancy mechanism of P migration in α-Fe. Since at low temperatures phosphorus concentrations near GB can be high due to RIS (see below) and a competition of phosphorus with other alloying elements (S, Si, Ni etc.) can influence its segregation near GB, we will use below a model of RIS in ternary alloys developed in Refs[*15,16*] in order to model RIS near GB in a Fe-P-Ni alloy. This model accounts for a binding energy of phosphorus atoms with interstitials as well as a competition of P and Ni in RIS near GB. It has covered RIS arising both due to inverse Kirkendall effect and due to migration of interstitial-solute atom complexes. In the particular case of binary alloys this model is reduced to one developed in Ref[*25*]. The approach in Refs[*15,16,25*] differs from one developed in Refs[*26,27*], where a nonrandom occupation of interstitials by atoms of different components is simply accounted for by modifying the diffusion coefficients (see discussion in Refs[*25*] and [*15*]). The model predictions are discussed below together with ones based on the modified McLean's model.

Generalization of the McLean's Model to Irradiation Conditions

At present for modeling of GB phosphorus accumulation under irradiation at high temperatures the McLean's equilibrium model [*27*] is modified[*1,4,6*]. In this approach only vacancy mechanism of the radiation-enhanced phosphorus diffusion is considered, and the equilibrium GB phosphorus concentration is determined by Guttmann's expressions [*8*]. As shown above, at relatively high temperatures both GA and RIS are of importance and steady state GB component concentrations under irradiation should be determined from Eq (11) instead of Guttmann's expressions. In addition, the radiation-enhanced phosphorus diffusion via interstitial mechanism should be accounted for because this mechanism is thought to be major one under irradiation [*4*]. Thus, the McLean's model modification suggested is as follows:

1) Radiation-enhanced diffusion coefficient of phosphorus D_p is used instead of self-diffusion coefficient

$$D_p = d_{pv}\langle C_{v0}\rangle + \xi_p d_{pi}\langle C_{i0}\rangle \quad (17)$$

where $\langle C_{v0}\rangle$ and $\langle C_{i0}\rangle$ are the average vacancy and interstitial concentrations in the bulk depending on the PD generation rate and sink density.

2) Steady-state GB component concentrations from Eq (14), where phosphorus diffusivity via interstitials d_{pi} should be replaced with $\xi \cdot d_{pi}$ are used instead of Guttmann's equilibrium concentrations in McLean's approach. As was discussed above, these concentrations represent an approximation of correct ones and are valid in the limits of sufficiently low and high temperatures. Probably, these concentrations underestimate real ones at intermediate temperatures where $\beta_0(T) \sim 1$ because of the binding of phosphorus atoms with interstitials (see Eqs (16) and (14)).

3) Radiation-enhanced diffusion of phosphorus from the bulk to GB should provide increased phosphorus concentration not only on GB but due to RIS also in the matrix near GB. Therefore, an effective GB width d_{ef} is used in calculations instead of d

$$d_{ef} = d + \frac{2h}{C_p^{gs}} \cdot \sum_{j=1}^{n} C_p(j \cdot b), \quad (18)$$

where $C_p(r)$ – related to RIS phosphorus concentration profile near GB, n is deduced from the following equation $nb \cong R_g$.

Input Parameters

Modeling of phosphorus accumulation on GB and radiation-induced segregation of phosphorus near GB was carried out using models discussed for the ferritic ternary Fe-1.07Ni-0.018P (wt.%) alloy with the same Ni and P contents as in the pressure vessel steel investigated in Refs[*3,4*]. Experimental data available and estimates of diffusion parameters of Fe and P in iron alloys offer a wide scatter. Therefore two different sets of the parameters were used in calculations. In the Set 1 most of the input parameters were taken from [*3,4*]: phosphorus and iron self-diffusion coefficients are $D_P = 7.12 \cdot 10^{-3} \exp(-2.67eV/kT)\, m^2/s$ and $D_{Fe} = 5 \cdot 10^{-5} \exp(-2.6eV/kT)$ m^2/s, respectively, GB width d=0.33 nm, $b=d$, ΔG_P =0.43 eV, ΔG_{Ni} =0.12 eV. In the calculations the interaction of P and Ni on GB [*3*] as well as a possible alteration of ρ_s during irradiation were ignored. Other parameters were taken as follows: ε=0.1, the recombination coefficient $\mu_R = 8 \cdot 10^{20}\, m^{-2}$, Fe, Ni, and P migration energies $E^m_{Fe,v} = E^m_{Ni,v} = 0.6\, eV$, $E^m_{P,v} = 0.67\, eV$, $E^m_{Fe,i} = E^m_{P,i} = E^m_{Ni,i} = 0.3\, eV$ (see [*28*] for Fe), P-interstitial binding energy $E^b_P = 0.3\, eV$. In order to account for Ni enrichment observed at PD sinks in f.c.c. and b.c.c. iron-based alloys it is suggested that the ratio of Fe and Ni diffusivities via vacancy mechanism is equal to 2.5. The Set 2 of parameters is based on ones used in [*4*] and adopted in [*20*]: $E^m_{Fe,v} = E^m_{Ni,v} = E^m_{P,v} = 1.2\, eV$, $E^m_{Fe,i} = E^m_{Ni,i} = E^m_{P,i} = E^b_P = 0.15\, eV$, $\varepsilon = 1$, $d^v_{Fe,0} = d^v_{P,0} =$ $= 5 \cdot 10^{-6}\, m/s^2$, $d^v_{Ni,0} = 2.5 \cdot 10^{-6}\, m/s^2$. Other parameters are the same as in the Set 1.

Results and Discussion

Calculations were performed for the dose rate K=10^{-9} dpa/s and for two constant values of PD sink strength ρ_s = 10^{14} m^{-2}, 10^{13} m^{-2}. In Fig. 1 the temperature dependence of the parameter β_0 is shown for the Set 1 of input parameters at three values of the effective activation energy $E_s = E_{sd} - (E_v^{f,b} - E_v^{f,g})$, where E_{sd} is the self-diffusion activation energy in α-Fe. Since the value of β_0 is sensitive to the difference in vacancy formation energies in the matrix, $E_v^{f,b}$, and on GB, $E_v^{f,g}$, the calculations were performed taking $E_v^{f,b} - E_v^{f,g} =$ 0, 0.2 and 0.4 eV. It is seen that the temperature dependence is very sharp and the temperature T_0 determined from the equation $\beta_0 = 1$ decreases significantly with decreasing E_s. Since the main temperature dependence of β_0 is related to E_s, a change of the parameter set and ρ_s only slightly influences T_0 due to R_g changing. In all cases considered the conditions $\beta_0 \ll 1$ and $\beta_0 \gg 1$ are valid in wide ranges of temperatures.

Steady-state GB component concentrations in Fe-1.07Ni-0.018P (wt.%) calculated using Eq (14) for Set 1 of parameters, K=10^{-9} dpa/s and ρ_s=10^{14} m^{-2} are shown in Fig. 2(a). The calculation demonstrate that steady-state GB phosphorus concentration

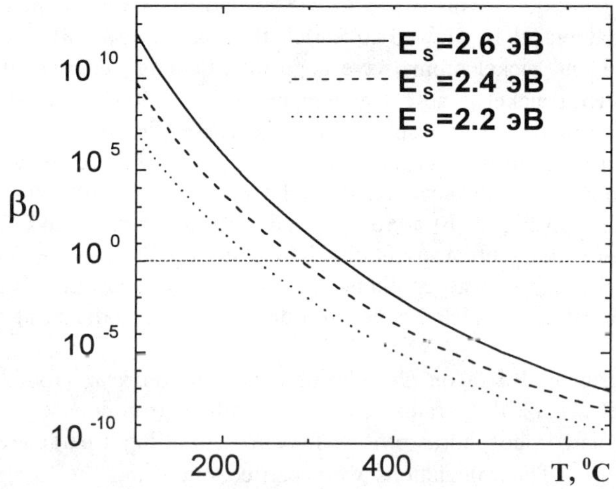

Figure 1 - The *dependence of* β_0 *parameter on the temperature and the effective self diffusion energy* E_s *in the Fe-1.07Ni-0.018P alloy at* $K = 10^{-9}$ *dpa/s.*

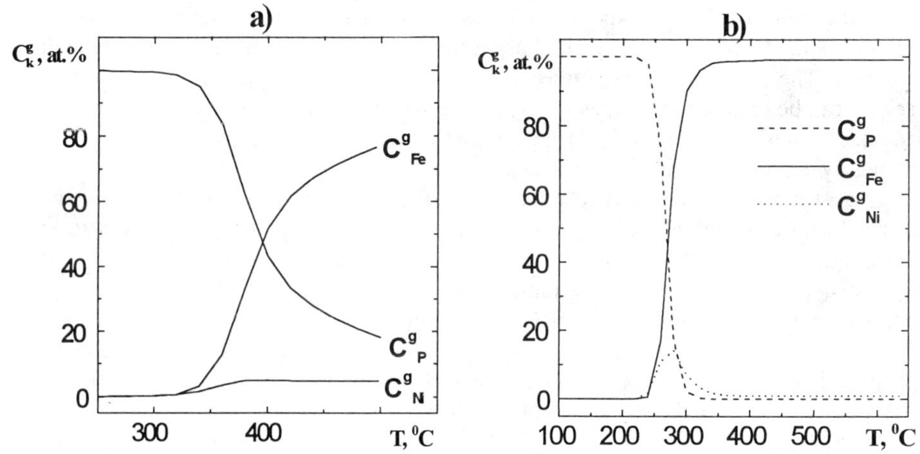

Figure 2 - *Steady-state grain boundary component concentrations in the Fe-1.07Ni-0.018P alloy calculated taking into account both RIS and GA (a) and RIS only (b) for Set 1 of parameters at* $\rho_s = 10^{14} m^{-2}$, $K = 10^{-9}$ *dpa/s.*

is higher than thermal equilibrium one defined by Guttmann's expressions [*8*] in the temperature range 100-500°C. This is due to RIS of phosphorus in the matrix near GB. Despite the fact that nickel is allowed to enrich the matrix near GB due to RIS via vacancy mechanism and nickel atoms have a positive "binding energy" with GB, the enrichment of GB with nickel is absent at temperatures ≤ 350°C. This is related to a competition between nickel and phosphorus both in RIS near GB and GA on GB. It should be noted that in the present calculations a cooperation of these components in GA on GB [*1*,*6*] was not taken into account. Possibly, such an account could increase the nickel content on GB. In Fig. 2 (b) a crude estimate of the steady-state GB component concentration is made taking into account only RIS near GB. Eq (5) at $r = 0$ was used for the estimate, i.e., the matrix was continued up to the GB layer. In this case (in the absence of GA) the calculated nickel concentration on GA is significant at temperatures 250-300°C.

In Fig. 3 experimental data on phosphorus concentration near GB in the Russian pressure vessel steel 15Kh2MFA irradiated at 270°C with neutron fluence of 10^{24} n/m^2 [*1*] along with the calculated phosphorus profiles for two sets of input parameters at $K=10^{-9}$ dpa/s and $\rho_s=10^{14}$ m^{-2}. The calculations were carried out using the model of RIS in ternary alloys developed in Refs[*15*,*16*] and discussed above. Earlier this model was applied for modeling Si segregation near GB in irradiated Fe-Cr-Ni alloys [*16*]. The phosphorus concentration at the distance of 30 nm from GB for the Set 1 of parameters is much lower than one for the Set 2. This is mainly due to a relatively small size of 180 nm of the computational sell in the former case. This case should be recalculated for larger sizes of the cell. Both the experimental data and calculations reveal a significant enrichment of the matrix near GB with phosphorus due to RIS even at this relatively high temperature. The difference in experimental and calculated phosphorus profiles in a near GB region can be related to two reasons: 1) a pure knowledge of the ratios of component diffusivities which determine RIS, 2) an ignored additional RIS of components during sputtering, which was used in the experiment of Ref[*1*] (see Ref[*29*]).

Figure 4 shows the results of calculation of GB phosphorus accumulation using the generalized McLean's model described above. The calculation was made for two sets of parameters, $\rho_s=10^{14}$ m^{-2}, $K=10^{-9}$ dpa/s and four exposures: $t_1=1$ month, $t_2=1$ year, $t_3=10$ years, $t_4=30$years. Additional calculations with the McLean's model in isothermal conditions without irradiation demonstrate an absence of detectable GB phosphorus accumulation for these times at low temperatures up to 200°C due to low vacancy mobility. In irradiation conditions a significant GB phosphorus accumulation reveals at low temperatures due to the radiation enhancement of phosphorus mobility via vacancy and interstitial mechanisms as well as RIS. A comparison of the results for two sets of parameters in Fig. 4 shows that the model predictions are similar at temperatures above 300°C and are qualitatively different at lower temperatures. The maximum at 450-500°C is caused by an increased phosphorus diffusivity and GA. In the low temperature region considered the GB phosphorus concentration decreases with increasing temperature for the Set 1 of parameters and increases for the Set 2 revealing an additional maximum at about 300°C. This difference in phosphorus accumulation is related to the difference in mechanisms governing average PD concentrations in the bulk. These concentrations are determined by the PD sink strength for the Set 1 of parameters and by the PD

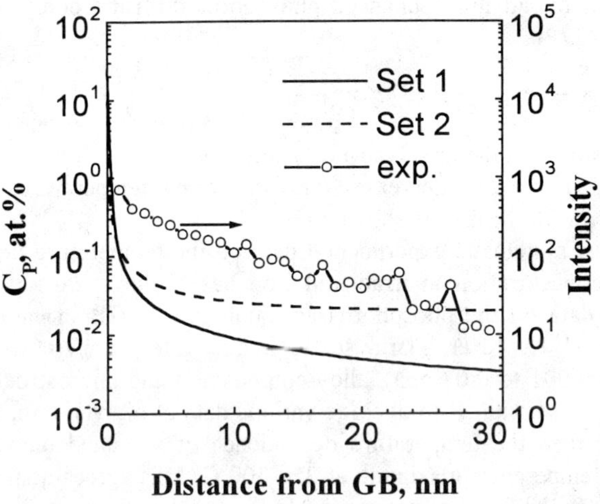

Figure 3 - *Experimental data on the phosphorus concentration near GB in WWER-400 base alloy (SIMS results) irradiated at 270°C [1] and calculated P profiles for two sets of input parameters.*

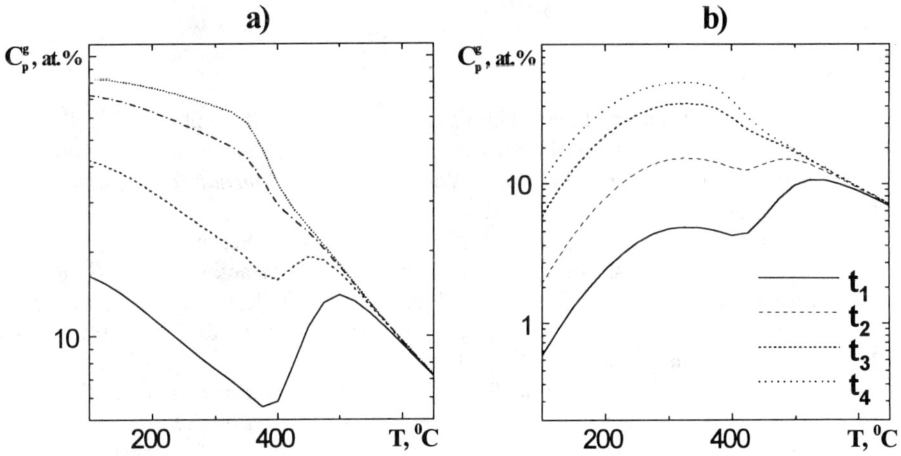

Figure 4 - *GB phosphorus concentration in the Fe-1.07Ni-0.018P alloy calculated using the modified McLean's model for Set 1 (a) and Set 2 (b) of parameters at* $\rho_s = 10^{14}\ m^{-2}$, $K = 10^{-9}$ *dpa/s and four irradiation times:* $t_1 = 1$ *month,* $t_2 = 1$ *year,* $t_3 = 10$ *years,* $t_4 = 30$ *years.*

recombination for the Set 2 [*12*]. Inserting these concentrations in Eq (17) one can obtain the following estimates of radiation enhanced phosphorus diffusion coefficients D_{P1} for Set 1 and D_{P2} for Set 2 [*12*]

$$D_{P1} = \frac{\varepsilon K}{\rho_s} \exp\left(\frac{E_b}{kT}\right), \quad D_{P2} = \sqrt{\frac{\varepsilon K D_0}{\mu_R}} \exp\left(\frac{2E_b - E_v^m}{2kT}\right), \tag{19}$$

where D_0 is the preexponential factor of vacancy diffusivity.
It is seen from Eq (19) that D_{P1} decreases with increasing temperature whereas D_{P2} increases in accordance with the results shown in Fig. 4.

For the present any systematic experimental data on the temperature dependence of the GB phosphorus concentration in irradiated iron based alloys are absent. Only in Ref[*30*] experimental data on the phosphorus accumulation on GB in various pressure vessel steels are collected. However, a large scatter in dose rate (from 10^{-9} to 10^{-12} dpa/s), irradiation dose (from 0.001 to 0.03 dpa), alloy composition and microstructure leads to a significant scatter in these data. If to average various data at a given temperature, as it was done in Ref[30], then the temperature dependence of averaged data reveals two maxima with the low temperature maximum at 250-300°C. This agrees qualitatively with the model predictions for the Set 2 of parameters in Fig. 4. However, more accurate data are needed for a definite conclusion.

Acknowledgment

This work was supported by the Russian Foundation for Basic Research under Project # 98-02-17400.

References

[1] Gurovich, B. A., Kuleshova, E. A., Nikolaev, Yu. A., and Shtrombakh, Ya. I., "Assessment of Relative Contributions from Different Mechanisms to Radiation Embrittlement of Reactor Pressure Vessel Steels," *Journal of Nuclear Materials*, Vol. 246, 1997, pp. 91-120.

[2] Davies, M., Kryukov, A., English, C., Nikolaev, Y., and Server, W. L., "A Comparison of East/West Steels for Pressurized Water Reactors," *Effects of Radiation on Materials, ASTM STP 1366*, M. L. Hamilton, A. S. Kumar, S. T. Rosinski, and M. L. Grossbeck, Eds., American Society for Testing and Materials, West Conshohocken, PA, 2000, pp. 3-15.

[3] Pavinich, W. A., Griesbach, T. J., and Server, W. L., "An Overview of Radiation Embrittlement Modelling for Reactor Vessel Steels," *Radiation Embrittlement of Nuclear Reactor Pressure Vessel Steels, ASTM STP 1170*, L. E. Steele, Ed., American Society for Testing and Materials, Philadelphia, PA, 1993, pp. 99-118.

[4] Druce, S. G., English, C. A., Foreman, A. J. E., McElroy, R. J., Vallet, I. A., Bolton, C. J., Buswell, J. T., and Jones, R. J., "The Modelling of Irradiation-Enhanced Phosphorus Segregation in Neutron Irradiated Reactor Pressure Vessel Submerged-Arc Welds," *Effects of Radiation on Materials, ASTM STP 1270*, D. S. Gelles, R. K. Nanstad, A. R. Kumar, and E. A. Little, Eds.,

American Society for Testing and Materials, West Conshohocken, PA, 1996, pp. 119-137.

[5] McElroy, R. J., English, C. A., Foreman, A. J. E., Gage, G., Hyde, J. M., Ray, P. H. N., and Vatter, I. A., "Temper Embrittlement, Irradiation Induced Phosphorus Segregation and Implications for Post-Irradiation Annealing of Reactor Pressure Vessels," *Effects of Radiation on Materials, ASTM STP 1325*, R. K. Nanstad, L. M. Hamilton, F. A. Garner, and A. S. Kumar, Eds., American Society for Testing and Materials, West Conshohocken, PA, 1999, pp. 296-316.

[6] Nikolaev, Yu. A., Nikolaeva, A. V., Zabusov, O. O., Gurovich, B. A., Kuleshova, E. A., and Chernobaeva, A. A., "Radiation- and Thermal-Induced Adsorption of Phosphorus at Grain Boundaries in Low Alloyed Steel," *Fizica Metallov i Metallovedenie*, Vol. 81, No. 1, 1996, pp. 120-128.

[7] McLean, D., "Grain Boundaries in Metals," Clarendon, Oxford, 1957.

[8] Guttmann, M., "Equilibrium Segregation in a Ternary Solution: a Model for Temper Embrittlement," *Surface Science*, Vol. 53, 1975, pp. 213-227.

[9] Pechenkin, V. A., and Epov, G. A., "Analytical Expressions for Steady-State Component Profiles of Irradiated Substitutional Alloys Near Point Defect Sinks," *Journal of Nuclear Materials*, Vol. 186, 1992, pp. 269-276.

[10] Pechenkin, V. A., Epov, G. A., Stepanov, I. A., and Konobeev, Yu. V., "Effect of Radiation-Induced Segregation on Precipitate Stability and Swelling in Irradiated Alloys," *Effects of Radiation on Materials, ASTM STP 1325*, R. K. Nanstad, L. M. Hamilton, F. A. Garner, and A. S. Kumar, Eds., American Society for Testing and Materials, West Conshohocken, PA, 1999, pp. 850-865.

[11] Perks, J. M., and Murphy, S. M., " Modelling the Major Element Radiation-Induced Segregation in Concentrated Fe-Cr-Ni Alloys," *Materials for Nuclear Reactor Core Applications, British Nuclear Energy Society*, London, 1987, pp. 165-169.

[12] Pechenkin, V. A., "On Segregation at Grain Boundaries in Multicomponent Alloys under Irradiation," Preprint IPPE-2788, 1999.

[13] Defay, R., and Prigogine, I., "Tension Superficielle et Adsorption," Paris, Dunod, 1951.

[14] Allnatt, A. R., and Lidiard, A. B., "Statistical Theories of Atomic Transport in Crystalline Solids," *Rep. Prog. Phys.*, Vol. 50, 1987, p. 373.

[15] Pechenkin, V. A., and Epov, G. A., "The Influence of Radiation-Induced Segregation on Precipitate Stability in Austenitic Steels," *Journal of Nuclear Materials*, Vol. 207, 1993, pp. 303-312.

[16] Pechenkin, V. A., and Stepanov, I. A., "Modeling the Radiation-Induced Segregation of Undersized Solutes Near Grain Boundaries," *Materials Scince Forum*, 1999, Vol. 294-296, pp. 771-774.

[17] Murphy, S. M., "A Model for Segregation in Dilute Alloys During Irradiation," *Journal of Nuclear Materials*, Vol. 168, 1989, pp. 31-42.

[18] Murphy, S. M., and Perks, J. M., "Analysis of Phosphorus Segregation in Ion-Irradiated Nickel," *Journal of Nuclear Materials*, Vol. 171, 1990, pp. 360-372.

[19] Barbu, A., and Lidiard, A. B., "Solute Segregation in Dilute bcc Alloys Under Irradiation," *Phil. Mag. A*, Vol. 74, 1996, pp. 709-722.

[20] Lidiard, A. B., "The Migration of Phosphorus in Ferritic Iron Alloys Under Irradiation," *Phil. Mag. A*, Vol. 79, 1999, pp. 1493-1506.

[21] Le Claire, A. D., "Solute Diffusion in Dilute Alloys," *Journal of Nuclear Materials*, Vol. 69-70, 1978, pp. 79-96.

[22] Vasiliev, A. A., Rybin, V. V., and Zisman, A. A., "The Nature of the Phosphorus Mobility in bcc Iron Irradiated at Low Temperatures," *Journal of Nuclear Materials*, Vol. 231, 1996, pp. 249-253.

[23] Minashin, A. M., and Pechenkin, V. A., "Molecular Dynamics Calculation of Phosphorus Diffusivity via Vacancy Mechanism in α-Fe," Proceedings of the 10th Int. Seminar on "Radiation Physics of Solids, 3-8 July, 2000, Sevastopol, Ukraine, pp 333-336.

[24] Fujiwara, T., and Ishii, Y., "Structural Analysis of Models for the Amorphous Metallic Alloy $Fe_{100-x}P_x$," *J. Phys. F: Metal Phys.*, Vol. 10, 1980, pp. 1901-1911.

[25] Bakai, A. S., and Turkin, A. A., "Radiation-Modified Phase Diagrams of Binary Alloys," *Effects of Radiation on Materials, ASTM STP 1125*, R. E. Stoller, A. S Kumar, and D. S. Gelles, Eds., American Society for Testing and Materials, Philadelphia, PA, 1992, pp. 709-730.

[26] Lam, N. Q., Kumar, A., and Wiedersich, H., *Effects of Radiation on Materials, ASTM STP 782*, H. R. Brager, and J. S. Perrin, Eds., American Society for Testing and Materials, Philadelphia, PA, 1982, p. 985.

[27] Wiedersich, H., Okamoto, P. R., and Lam, N. Q., "A Theory of Radiation-Induced Segregation in Concentrated Alloys," *Journal of Nuclear Materials*, Vol. 83, 1979, pp. 98-108.

[28] Takaki, S., Fuss, J., Kugler, H., Dedek, U., and Schulz, H., "The Resistivity Recovery of High Purity and Carbon Doped Iron Following Low Temperature Electron Irradiation," *Radiation Effects*, Vol. 79, 1983, pp. 87-122.

[29] Yacout, A. M., Lam, N. Q., and Stubbins, J. F., "Near-Surface Compositional Modifications of Ternary Alloys During Sputtering at Elevated Temperatures," *Nucl. Instr. Meth. B*, Vol. 42, 1989, pp. 49-60.

[30] Faulkner, R. G., Song, S., Meade, D., and Goodwin, C. C., "Radiation-Induced Grain Boundary Segregation," *Materials Science Forum*, Vol. 294-296, 1999, pp. 67-74.

Shenhua Song,[1] Roy G. Faulkner[1], and Peter E. J. Flewitt[2]

Grain Boundary Phosphorus Segregation Under Irradiation and Thermal Aging and Its Effect on the Ductile-to-Brittle Transition

Reference: Song, S., Faulkner, R. G., and Flewitt P. E. J., "**Grain Boundary Phosphorus Segregation Under Irradiation and Thermal Aging and Its Effect on the Ductile-to-Brittle Transition,**" *Effects of Irradiation on Materials: 20th International Symposium, ASTM STP 1405,* S. T. Rosinski, M. L. Grossbeck, T. R. Allen, and A. S. Kumar, Eds., American Society for Testing and Materials, West Conshohocken, PA, 2001.

Abstract: Embrittlement of low-alloy steel components used in the nuclear power industry is classified into hardening embrittlement and non-hardening embrittlement. Hardening embrittlement stems from precipitation of carbides or copper-rich compounds. Non-hardening embrittlement results mainly from grain boundary segregation of phosphorus. To examine the non-hardening embrittlement in 2.25Cr 1Mo steel, phosphorus intergranular segregation in samples subjected to neutron irradiation and thermal aging at 270°C and 400°C is determined by modelling and transmission electron microscopy. The ductile-to-brittle transition temperature for the steel is determined by means of small punch testing with 3 mm diameter disc specimens. It is indicated that there is a reasonable agreement between the behavior of intergranular phosphorus segregation and shifts in the ductile-to-brittle transition temperature. Measured and predicted phosphorus intergranular segregation in a range of pressure vessel steels is also discussed.

Keywords: grain boundaries, segregation, neutron irradiation, steels, ductile-to-brittle transition, and analytical electron microscopy

Introduction

The exposure of low alloy steel to higher temperature heat treatment or neutron irradiation is known to cause intergranular segregation of elements such as P, S, and Si [*1*]. This, in addition to yield point increases caused by neutron damage inducing copper-rich precipitates to form, is responsible for toughness reduction in neutron irradiated material. Conventional aging treatments, without irradiation, can also cause

[1] Research Associate and Professor, respectively, IPTME, Loughborough University, Loughborough, Leics, LE11 3TU, UK.
[2] Manager of Structural Integrity, BNFL Magnox Generation, Berkeley Center, Berkeley, Glos, GL13 9PB, UK

P inter-granular segregation in certain critical temperature regimes[*2*]. The decrease in toughness experienced is usually demonstrated by an increase in the ductile to brittle transition temperature in these steels. A series of impact energy tests at different temperature will thus confirm the effect. However, in the neutron irradiated situation it is costly to expose large quantities of material to the neutron flux and to measure their properties because of the need to work with radiological protection. Therefore it is desirable to use small specimens to detect the effect. This paper discusses the use of these small Charpy specimens and the utilization of a ball punching method for small disc-type specimens to study the DBTT shifts as a function of neutron irradiation and temperature. The results are discussed in terms of the observed and modeled inter-granular P segregation behavior.

Materials

Low alloy steel is studied. The chemical composition is given in Table 1.

Table 1 - Steel Alloy Compositions, wt.%

Material	C	Mn	S	P	Si	Cr	Ni	Mo	Sn	As	Sb
Low Alloy	.09	.46	.01	.04	.27	2.3	.15	1.0	.01	.01	.01
Pressure Vessel	.18	1.3	.02	.02	.36	Cu-	0.1	-	-	-	-

The phosphorus content is 0.039 wt.%, equivalent to 0.072 at.%. The material was induction melted in Ar. The resulting ingots were hot-rolled to 10 mm diameter rods and austenitised at 1150°C for 2 hours, furnace cooled to 950°C, held for 1 hour and oil quenched. They were toughened at 650°C for 2 hours, followed by water quenching. The rod samples were cut into quasi-prepared discs 3 mm in diameter and 0.5 mm thickness and also mini-Charpy V-notch specimens 3mm x 4mm x 27mm in size (Fig. 1a and b respectively). After neutron irradiation or thermal treatment at 270° C for 46 days or 400° C for 86 days, the discs were ground to a 0.25 mm nominal thickness for the disc ball punch testing [*3*].

Observed and predicted intergranualr segregation behavior is also discussed for a range of pressure vessel steels. The general composition of such steels is given in Table 1.

Neutron Irradiation Program

Material irradiation was performed in a thermal light water research reactor, named SAPHIR, operated by the Swiss Federal Research Institute for Reactor Research at Villigen, Switzerland. The reactor operates with a thermal power output of 10 MW. The thermal neutron flux is 8×10^{13} n cm^{-2} s^{-1} at the core edge and up to 1.2×10^{14} n cm^{-2} s^{-1} in the central irradiation positions. In this work specimens were irradiated for 46 days with a neutron dose rate of 1.05×10^{-8} dpa s^{-1} at 270°C (42 mdpa) and for 86 days at a neutron dose rate of 1.75×10^{-8} dpa s^{-1} at 400°C (130 mdpa). Thermal control specimens were aged at similar times and temperatures to simulate the heating effects of the irradiation.

Mechanical Testing

The small ball punch test technique has been described in great detail in references 3 and 4. In brief the test jig consists of a disc specimen and a specimen holder. The holder is comprised of upper and lower dies, and four clamping screws. This facilitates prevention of cupping upward during punching and plastic deformation is concentrated in the region immediately below the 1mm diameter ball punch. Fig. 1a gives a schematic picture of the test jig.

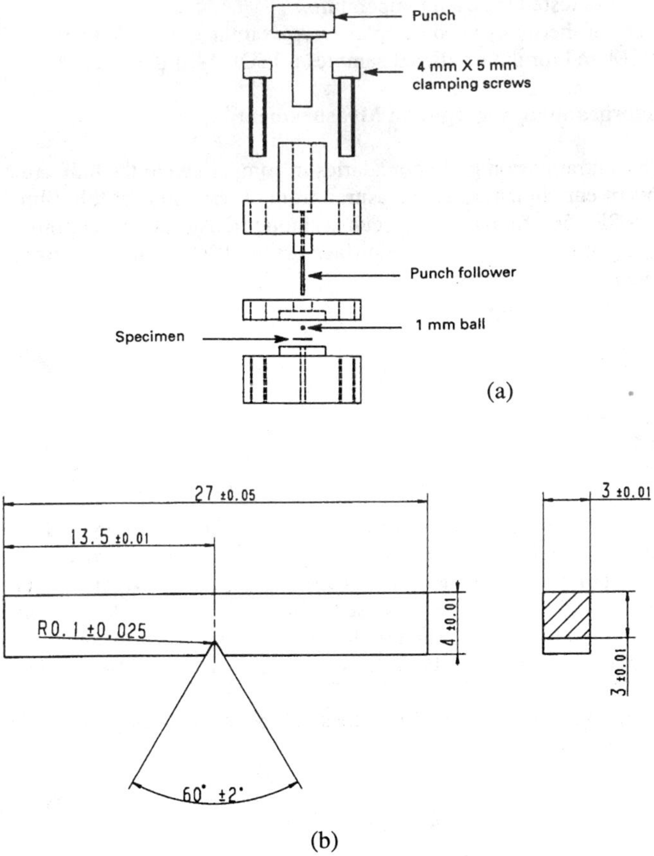

Fig.1 Schematic diagram of: (a) the test jig; (b) mini-Charpy specimen.

The small punch test is performed on a Universal test machine, a Hounsfield H10KM, equipped with an environmental chamber cooled with liquid nitrogen. Each test is carried out with a constant crosshead speed of 0.1 mm min^{-1}. Two identical miniaturized jigs are used for testing. Three to six specimens are used for each test result. The temperature is controlled by a chromel-alumel thermocouple place in the environmental chamber and a second thermocouple is used to measure the actual jig temperature. During testing, the load displacement data are recorded at every 4 μm displacement and stored in a microcomputer that allows the data to be subsequently processed.

Mini-Charpy testing (specimen details are shown in Fig. 1(b)) is accomplished on a machine developed at EPFL-CRP Fusion Technology, Villigen-PSI, Switzerland [*5*]. It has been shown from testing of other ferritic materials that the DBTTs determined from the mini-Charpy test are approximately 50°C lower than those determined by the standard Charpy test. Since only limited material was available, the mini-Charpy test was only conducted on 270°C aged and irradiated samples and only one specimen was tested for each temperature.

Fracture surfaces of thermally aged samples were examined on a Cambridge Stereoscan 360 SEM and for the irradiated samples a JEOL JSM 840A SEM.

Experimental Intergranular Segregation Measurement

Phosphorus concentrations on grain boundaries in comparison to the bulk are expressed in terms of enrichment ratios measured from observations of thin film specimens in the VGHB501 field emission gun scanning transmission electron microscope and on grain boundary fracture surfaces in the JEOL scanning Auger electron spectrometer.

Results

Ball Punch Tests

Typical load displacement curves are shown in Fig. 2. The maximum load increases with decreasing temperature and the elongation to failure decreases. The energy of fracture is normalized to allow for specimen-to-specimen variation. The concept of specific fracture energy (SFE) is used. The SFE is defined as the fracture energy per unit specimen thickness. The fracture energy is measured from the area under the load displacement curves. The SFE as a function of temperature is shown for all samples in Fig. 3.

The yield strength may be evaluated from the small punch test using a formula due to Mao and Takahashi [*6*].

$$\sigma_y = \frac{0.36 P_y}{t^2}$$

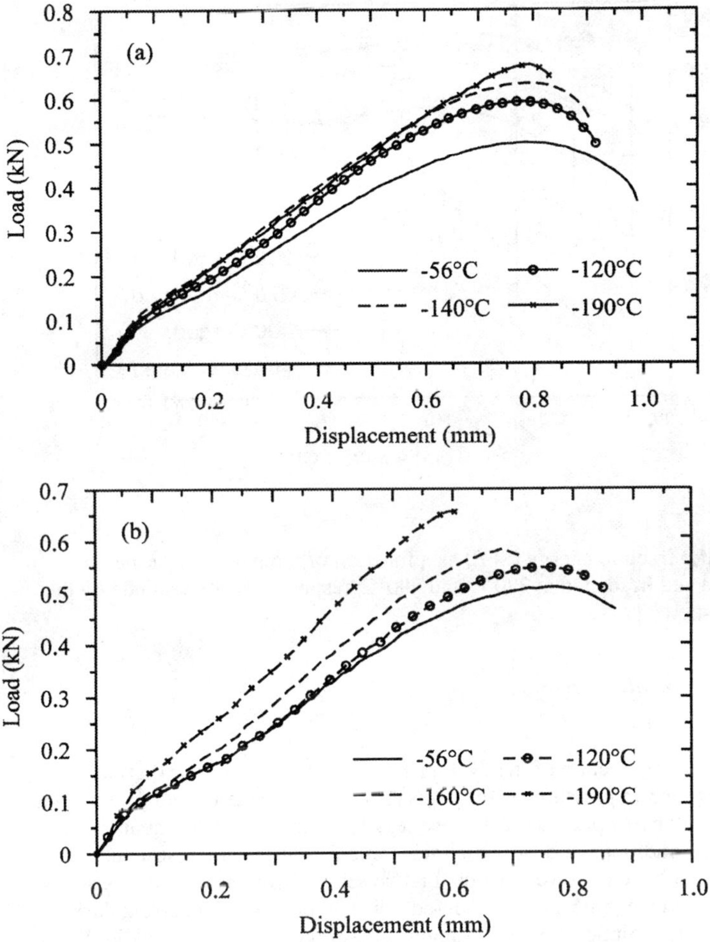

Fig.2 Typical load extension curves determined by the small ball punch test for: (a) irradiated; (b) non-irradiated materials.

where P_y is the yield load at the point where the load-extension curve deviates from linearity in newtons and t is the initial specimen thickness in mm. The yield strength as a function of test temperature is shown in Fig. 4.

Mini-Charpy Tests

The limited mini-Charpy test data from the 270°C aged and irradiated material are shown in Fig. 5.

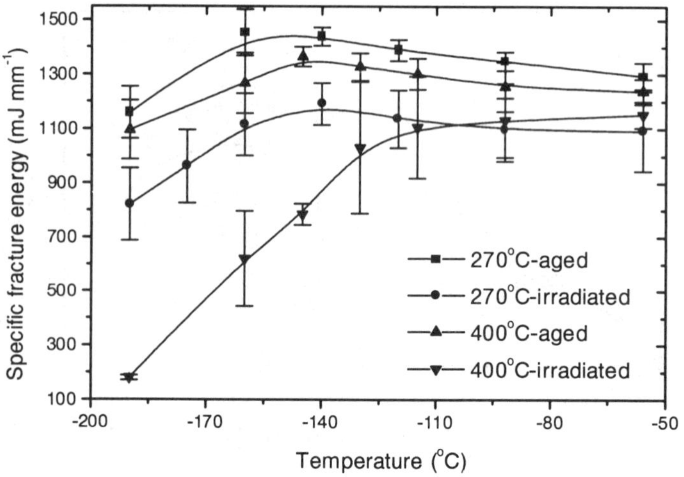

Figure 3. Specific fracture energy (SFE) as a function of temperature for the specimens aged and irradiated at 270°C and 400°C, respectively, determined by the small punch test.

Intergranular Segregation Measurements

A combined survey picture of the experimental results obtained for P grain boundary segregation in pressure vessel steels as a function of temperature is shown in Fig. 6. It should be emphasized that these results come from experiments with widely-differing conditions of dose, dose rate, composition and microstructure and should therefore only be reviewed in a qualitative sense. There is always a base level enrichment caused by equilibrium and non-equilibrium processes occurring during prior heat treatment (solution treatment, aging, stress relief annealing, welding). This contribution is subtracted out from the results quoted in the figure.

Discussion

The majority of atoms apart from carbon and nitrogen are immobile at 270°C. This means that it is unlikely that 270°C ageing will cause any embrittlement. There is no confirmation of this because no archive specimen data are available, but it seems a reasonable assumption. Fig. 3 confirms this, suggesting that the DBTT shifts

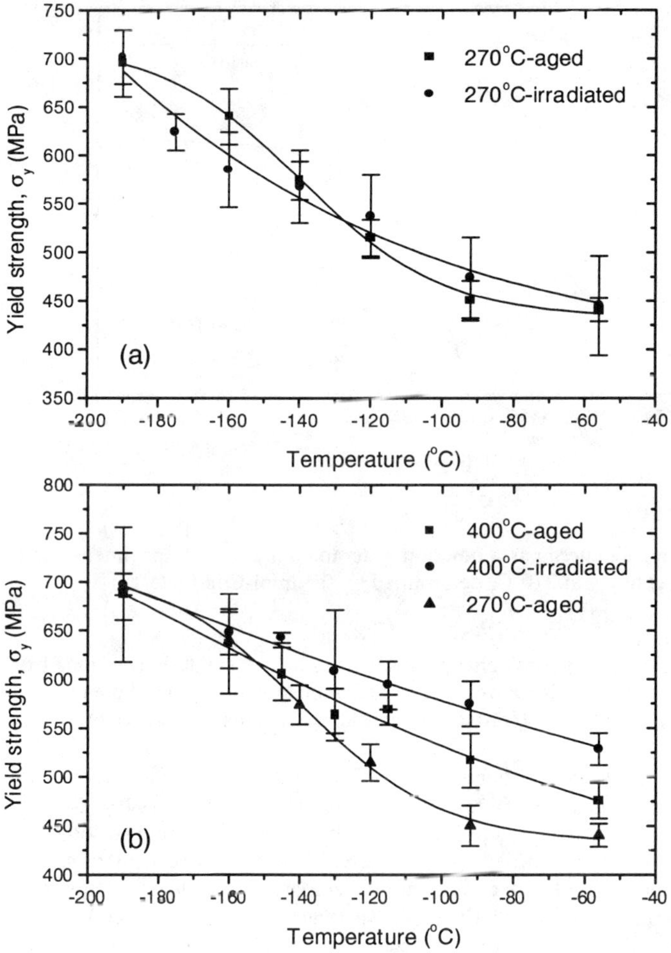

Figure 4. Yield strength as a function of temperature (a) for the specimens aged and irradiated respectively at 270°C, and (b) for the specimens aged and irradiated respectively at 400°C with that for the 270°C-aged specimens plotted for comparison, determined by the small punch test.

Upwards after ageing at 400°C and after irradiation at 270°C and 400°C in increasing amounts. The only curve that can be analyzed to define a transition temperature is that of the 400°C irradiated material. If the temperature at the mean of the maximum and minimum SFE values in horizontal portions of the curve, the DBTT for this material is –155°C. The other treatments do not extend to low enough temperatures to get a sufficiently horizontal portion to the curve to make this determination. The

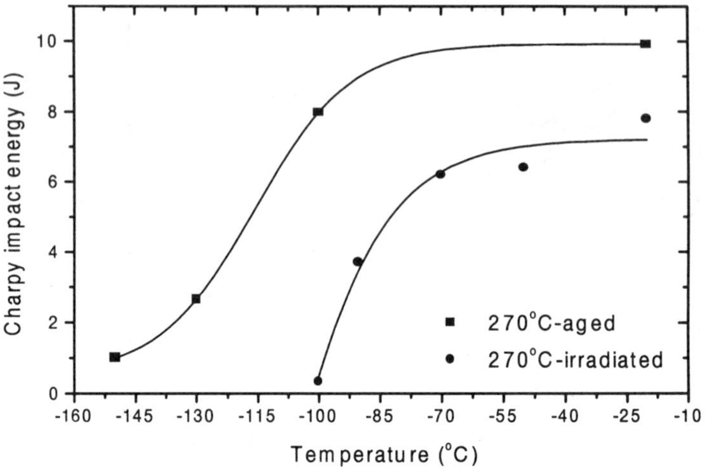

Figure 5. Charpy impact energy as a function of temperature for the specimens aged and irradiated respectively at 270°C, determined by the mini-Charpy test.

fact that all curves in Fig. 3 go though a peak seems to be a characteristic of data from the small punch test [*4,7,8*]. Other workers have demonstrated that small punch testing procedures can be used effectively to predict mechanical properties of austenitic steels [*9*].

The non-hardening effect of irradiation at 270°C is confirmed by Fig. 4a, which shows the change in σ_y measured from the ball punch test. Nevertheless Fig. 5 indicates that there is a drop in the upper shelf energy on irradiation and so some irradiation hardening cannot be ruled out. Certainly, RPV experience suggests that there should be some irradiation hardening at this temperature. A clearer picture is confirmed by Fig. 4b, where some hardening is seen during irradiation at 400°C. This is presumably due to some irradiation-induced carbide precipitation or defect cluster dislocation loop effects. It is also significant that σ_y increases on moving from 270°C aging to 400°C ageing. This means that there is a thermal hardening effect that must be caused by additional carbide precipitation.

The ductile to brittle transition temperature (DBTT) effects seen in Fig. 3 indicate shifts according to the following order: 270°C age - 400°C age - 270°C irradiation - 400°C irradiation. The results indicate that irradiation shifts the transition to higher temperatures more at 400°C than at 270° C, although it should be emphasized that the 400°C irradiations were for three times the dose of the 270°C irradiation. This larger shift at 400°C is supported by irradiation-induced segregation (RIS) solute drag model calculations for P GB segregation in these steels (Fig. 7) showing that at 270°C, P enrichment at GBs caused by irradiation is only likely to be by factors of 1 or 2 (Fig. 7a). At 400°C the enrichment ratio predicted is >10 (Fig.7b). The predictions in Fig.

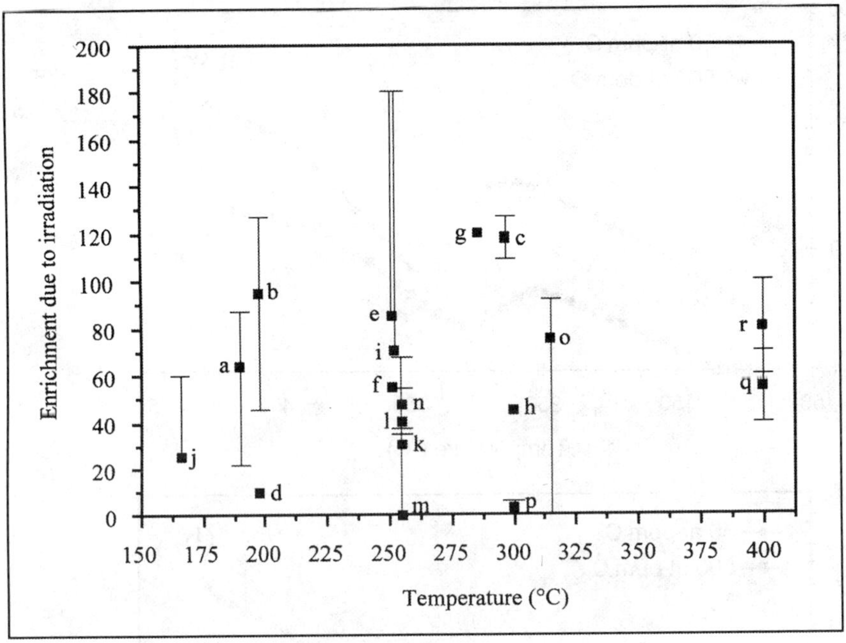

Fig. 6 Experimental source data for Irradiation Induced Intergranular Phosphorus Segregation in Ferritic Steels. Thermal effects are subtracted out. Enrichment is the ratio of the segregated concentration to the bulk concentration. Reference key: a are AES data from Jones[16], dose = 2.287-2.656mdpa, dose rate = 1.7-3.3 $x10^{-12}$ dpa/s; b are AES data from Jones[16], dose = 1.055-1.862mdpa, dose rate = 3.4-$4x10^{-12}$ dpa/s; c are AES data from Jones[16], dose = 1.929mdpa, dose rate = $3x10^{-12}$dpa/s; d are FEGSTEM data from Druce[15], dose = 0.94mdpa, dose rate = 10^{-12}dpa/s; e are FEGSTEM data from Druce[15], dose = 9.2mdpa, dose rate = 10^{-9}dpa/s; f are AES data from Druce[15], dose = 9.2mdpa, dose rate = 10^{-9}dpa/s; g are AES data from Druce[15], dose = 9.5mdpa, dose rate = 10^{-9}dpa/s; h are FEGSTEM data from Druce[15], dose = 1.52mdpa, dose rate = 10^{-12}dpa/s; i are FEGSTEM data from Druce[15], dose = 8.61mdpa, dose rate = 10^{-9}dpa/s; j are FEGSTEM data from Druce[15], dose = 9.11mdpa, dose rate = 10^{-9}dpa/s; k are FEGSTEM data from Meade[17], dose = 1.5mdpa, dose rate = $8.9x10^{-11}$dpa/s; l are FEGSTEM data from Meade[17], dose = 12.34mdpa, dose rate = $6.4x10^{-10}$dpa/s; m are FEGSTEM data from Meade[17], dose = 29mdpa, dose rate = $6x10^{-9}$dpa/s; n are FEGSTEM data from Meade[17], dose = 38mdpa, dose rate = $6x10^{-9}$dpa/s; o are FEGSTEM data from Meade[17], dose = 26.62mdpa, dose rate = $6.3x10^{-9}$dpa/s; p are FEGSTEM data from Faulkner[10] for 2¼Cr-1Mo steel, dose = 0.042dpa, dose rate = $1.05x10^{-8}$ dpa/s; q are FEGSTEM data from Faulkner[10] for 2¼Cr-1Mo steel, dose = 0.13dpa, dose rate = $1.75x10^{-8}$ dpa/s; r are AES data from Kameda and Bevolo [14], dose = 6.6 mdpa, 1.5 x 10^{-8} dpa s^{-1}.

7 are supported by field emission gun transmission electron microscope observations [*10*]. Enrichment ratios of 2-3 for the 270°C exposure (appropriate to Fig. 7b) are observed. For the 400°C exposure enrichment ratios of around 30 (appropriate to Fig

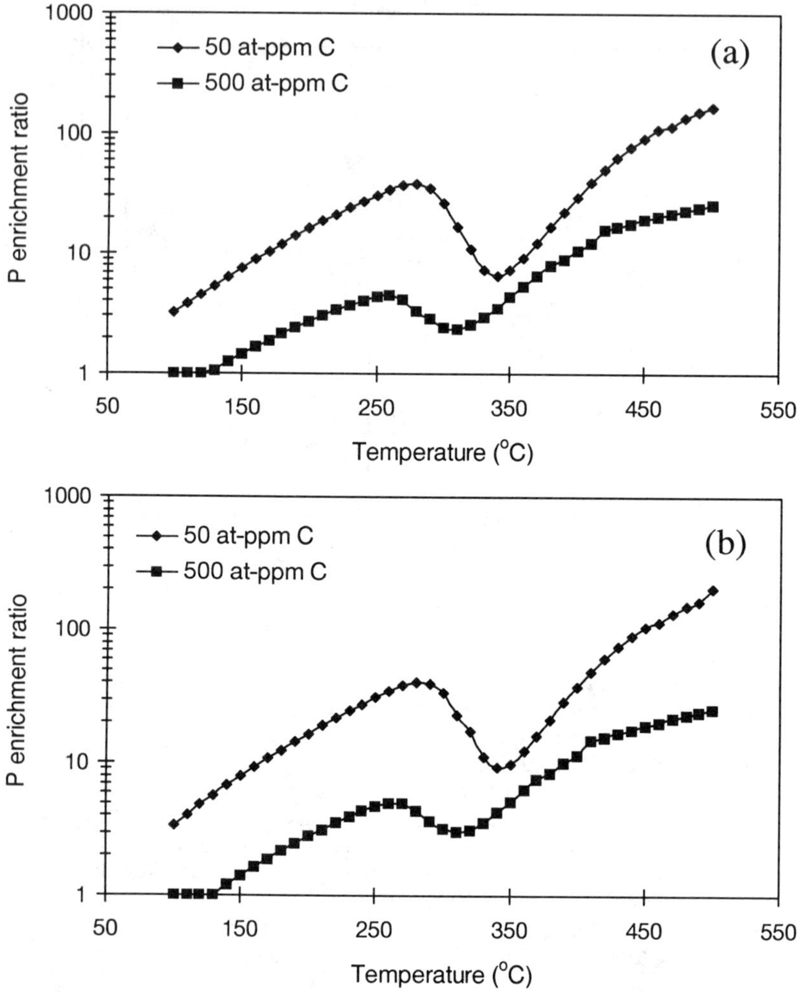

Fig.7 Phosphorus inter-granular segregation as a function of temperature for irradiation conditions appropriate to a) 270°C irradiation and b) 400°C irradiation.

7a) are seen. These observed ratios are corrected by subtracting out the ratio caused by the thermal ageing component measured from the thermal control samples.

The net conclusion is that while there are some small irradiation-induced hardening effects, the majority of the DBTT shifts can be explained by non-hardening behavior, i.e., grain boundary segregation of phosphorus. This is further supported

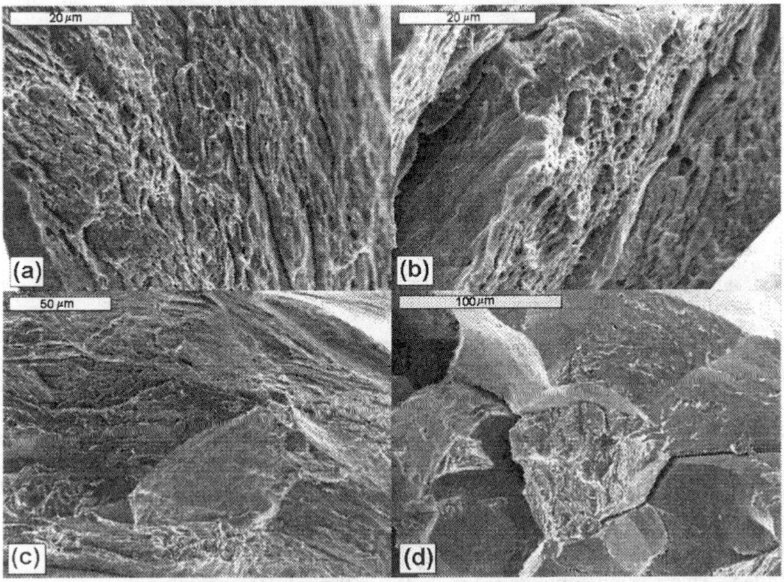

Fig.8 SEM micrographs for the 400°C irradiated samples tested at a) -56°C, b) -92°C, c) -130°C, and d) -145°C. The fractures become progressively more inter-granular as the temperature decreases.

by the SEM pictures shown in Fig. 8. The majority of the mini-Charpy fracture surface in the 400°C irradiated material, the most severely embrittled material, is seen to be inter-granular (Fig.8d).

The mini-Charpy data at 270°C (Fig. 5) are in reasonable agreement with the small specimen test results seen in Fig. 3. If the temperature corresponding to the mean value of the highest and lowest Charpy values is taken as the DBTT, then the DBTT will be about –115°C for the 270°C aged material and about –88°C for the 270°C aged and irradiated material. The ΔDBTT is thus 27°C.

Assuming that the DBTT for mini-Charpy tests is about 50°C lower than for

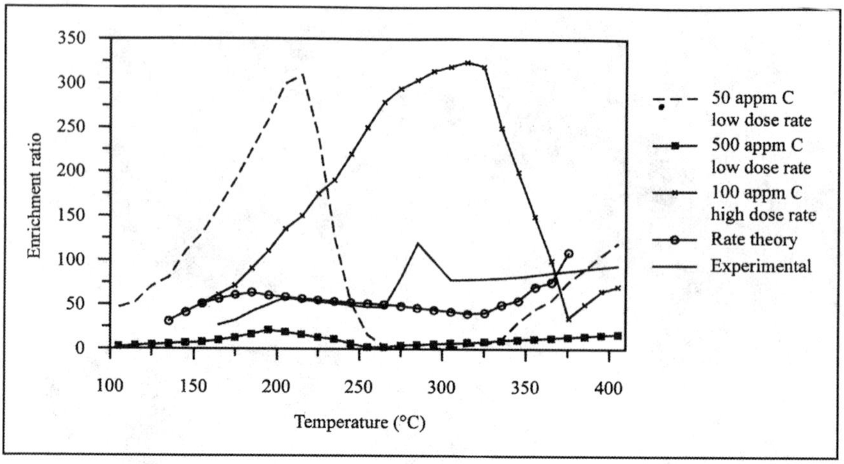

Fig. 9 Theoretical predictions of P Intergranular Segregation in Ferritic Steels as a function of Temperature (dislocation density, 1 x 10^{-15} m^{-2}, grain size 10 µm, dose rate 5 x 10^{-12} dpa s^{-1}, dose 5 mdpa). A summary of the experimental data from Fig. 6 is also included. Two carbon contents are included in the solute drag theory predictions. Another solute drag prediction for a higher dose rate and dose (1 x 10^{-8} dpa s^{-1} and130 mdpa) is included. The rate theory predictions are taken from Druce[15].

conventional tests in high alloy Fe-12 Cr steels [*5*], we obtain DBTTs of –65° C for the 270°C aged material and –38°C for the 270°C aged and irradiated material. The aged material result is consistent with values due to Wada and Hagel [*11*], who obtained similar values with the standard test on similarly heat-treated material.

Kameda and Mao [*7*], working with ferritic steels indicate the following relationship for values from small specimen and standard Charpy tests.

$$DBTT_{spt} = 0.4\ DBTT_{sct}$$

where the temperature is in K. Using the above figures for the 270°C aged and irradiated material (– 65°C and –38°C), the DBTT for the aged material should be around –190°C and for the irradiated material it should be –179°C in the small specimen test. A rough assessment of our small specimen DBTTs from Fig. 3 suggests that the aged material might have a DBTT of around –230°C and the irradiated material might have a value of –220°C. Kameda's results were taken from tests that were operated at much higher strain rates (1.2 – 2.0 mm min^{-1}) than ours (0.1 mm min^{-1}). A theoretical analysis of this [*12*] suggests that the DBTT reduces with decreasing strain rate. As a consequence, the factor of 0.4 in the above equation might be smaller in our work. This, together with the fact that the arbitrary 50°C shift between mini-Charpy and small specimen tests from Marmy [*5*] was estimated for much higher alloy ferritic steels, suggests that our DBTT results from the small specimens treated at 270°C are reasonable.

Figure 9 provides the appropriate modeling predictions of the temperature dependence of phosphorus grain boundary segregation in pressure vessel ferritic steels, and the experimental data taken from Fig. 6 are given for comparison/validation purposes. These experimental data are averaged where data points are clustered at a single temperature, i.e., 200°C, 260°C, 280°C, and 300°C, and points p and q from Fig. 6 are omitted because they are from a 2.25% Cr 1% Mo steel, not a pressure vessel steel.

The important modeling issue is that the solute drag-NES model [2] is very sensitive to the site competition effect when carbon is considered (Fig. 9). Carbon competes for sites on the grain boundary and within the complexes that are transporting the phosphorus to the grain boundary. Beere [*13*] and Kameda and Bevolo [*14*] have also shown that there is a relationship between carbon and phosphorus intergranular segregation in steels. The rate theory predictions of Druce [*15*] give quite good fits with observed data, because they have been curve-fitted. This automatically accommodates for any site competiton within the rate theory model.

Two carbon concentrations are shown for the solute drag model at the lower dose rate (5×10^{-12} dpa s^{-1}) for reasons described earlier. It is seen that the experimental curve fits reasonably well with the solute drag model predictions for free carbon contents within this range. It is possible that the variable carbon solubility in ferritic steel could change the situation. It is estimated from thermodynamic calculations that the solubility of carbon can increase by an order of magnitude over a 100°C range at the temperatures of interest here. This would have the effect of reducing the enrichment of phosphorus from the 50 appm to the 500 appm curve over the range of temperature occupied by the low temperature peaks seen in Fig. 9.

The peak seen in the experimental curve at about 290°C is explained by the solute drag model predictions because higher dose rates (1×10^{-8} dpa s^{-1}) shift the non-equilibrium segregation peak from 210°C to temperatures in excess of 300°C. Since some of the experimental datá were from irradiations at higher dose rates, this peak would seem to position itself somewhere in the 250°C to 300°C regime.

The rate theory predictions give good first with the experimental data but it must be remembered that they were curve-fitted to the available experimental data in the first place.

It is worth commenting further that the experimental data were take from a wide range of dose and dose rates, and that the experimental analysis techniques differed in several cases. Where Auger electron spectroscopy (AES) is used the enrichment ratios are artificially high because the intergranular fracture necessary for these measurements always occurs along the most heavily segregated boundaries. It therefore seems inadvisable to attempt to 'normalise' these data to, say, a fixed fluence, any further.

Conclusions

Thermal and neutron irradiation-induced embrittlement of a 2.25% Cr-1Mo- low alloy ferritic steel studied by small ball specimen punch testing and mini-Charpy testing shows that the effects of 42 mdpa irradiation at 270°C are to increase the DBTT but to maintain, or slightly increase, the yield strength. At 400°C, with 130 mdpa, both the yield strength and the DBTT are increased. The increased hardening is supposed to originate from increased carbide precipitation and cluster damage (which will be variable at each of the two irradiation temperatures because of the different doses used) whilst the DBTT shifts are caused by irradiation-induced phosphorus inter-granular segregation (RIS). This is supported by SEM fractography showing substantial evidence for intergranular failure in irradiated material. The relationships between DBTT results from small specimen ball punch tests, mini-Charpy tests and conventional Charpy tests are discussed and the small specimen and mini-Charpy test results given in this work are shown to comply well with measurements made on similar steels by the conventional method.

Experimental and theoretical predictions of intergranualr phosphorus segregation in pressure vessel steels are summarized and discussed. Both solute drage and rate theory models are shown to adequately predict segregation as a function of the prevbailing experimental parameters. Site competition between carbon and phosphorus is seen as an important issue

Acknowledgements

The work was supported by BNFL Magnox Generation and is published with permission of the Director of Technology and Central Engineering Division of BNFL Magnox Generation. Also we are indebted to Mr. Frank Page of Loughborough University and Mr. Nick Mayley-Jones of BNFL Magnox Generation for help with the SEM.

References

[1] Faulkner, R.G., "Segregation to Boundaries and Interfaces in Solids," *Int Met Rev,* Vol.41, 1996, pp198-208.

[2] Song, S-H., Faulkner, R.G., and Flewitt, P.E.J., "Effect of Boron on Phosphorus-Induced Temper Embrittlement," *J. Mater. Sci.,* Vol.34, 1999, pp 5549-5556.

[3] Vorlicek, V., Exworthy, L.F., and Flewitt, P.E.J., "Evaluation of a Miniaturised Disc Test for Establishing the Mechanical Properties of Low Alloy Ferritic Steels," *J. Mater. Sci.*, Vol 30, 1995, 2936.

[4] Ullmaier, H., Hishinuma, A., Lucas, G.E., and Jung, P., "Recommendation of Miniaturised Techniques for Mechanical Testing of Fusion Materials in an Intense Neutron Source," *J. Nucl. Mat.*, Vol. 232, 1996, pp186.

[5] Marmy, P. EPFL Laussane, Switzerland, Private Communication, 1999.

[6] Mao, X, and Takahashi, H., "Development of a Further-Miniaturized Specimens of 3mm Diameter for the TEM Disk Small Punch Tests," *J. Nucl. Mat*, Vol. 150, 1987, pp 42.

[7] Kameda, J, and Mao, X., "Small Punch and TEM-Disc Testing Techniques and their Application to Characterisation of Radiation Damage,"*J. Mater. Sci.*, Vol. 27, 1992, pp983.

[8] Lucas, G. E., Odette, G. R., and Sheckherd, J. W., "The Use of Small-Scale Specimens for Testing Irradiated Material," *ASTM STP 888,* W. R. Corwin

and G. E. Lucas, Eds., American Society for Testing and Materials, 1986, pp.112-140.

[9] Hankin, G.L., Toloczko, M.B., Hamilton, M.L., and Faulkner, R.G., "Validation of Shear Punch-Tensile Correlation Technique using Irradiated Materials," *J. Nucl. Mater.*, Vol. 258-263, 1998, pp1651-1656

[10] Faulkner, R.G., Song, S-H., Flewitt, P.E.J., Victoria, M., and Marmy, P., "Grain Boundary Segregation under Neutron Irradiation in Dilute Alloys," *J.Nucl. Mater.*,Vol. 255, 1998, pp 189-209

[11] Wada, T., and Hagel, W.C., "Effect of Trace Elements, Molybdenum, and Intercritical Heat Treatment on Temper Embittlement of 2.25Cr1Mo Steel," *Metall Trans,* Vol 7A, 1976, pp 1419.

[12] Kameda, J., "A Kinetic Model for Ductile-Brittle Fracture Mode Transition Behavior," *Acta. Metall,* Vol. 34, 1986, pp 2391.

[13] Beere, W. B., Private Communication, BNFL Magnox Generation, 1988.

[14] Kameda, J. and Bevolo, A. J., "Neutron-Irradiation-Induced Intergranular Solute Segregation in Iron-Based Alloys," *Acta. Metall.,* Vol. 37, 1989, pp 3283.

[15] Druce, S. G., English, C. A., Foreman, A. J. E., McElroy, R. J., Vatter, I. A., Bolton, C. J., Buswell, J. T., and Jones, R. B., "The Modelling of Irradiation-Enhanced Phosphorus Segregation in Neutron Irradiated RPV Submerged Arc Welds," *ASTM STP1270,* Gelles, D. S., Nanstad, R. K., Kumar, A. S., and Little, E. A., Eds., American Society for Testing and Materials, 1996, pp119.

[16] Jones, R. B., Cowan, J. R., Corcoran, R. C., and Walmsley, J. C., BNFL Magnox Generation, Private Communication, 1998.

[17] Meade, D., PhD Thesis, Loughborough University, pp. 238, 1998.

Roger E. Stoller[1] and Lawrence R. Greenwood[2]

An Evaluation of Through-Thickness Changes in Primary Damage Production in Commercial Reactor Pressure Vessels

Reference: Stoller, R. E. and Greenwood, L. R., "**An Evaluation of Through-Thickness Changes in Primary Damage Production in Commercial Reactor Pressure Vessels,**" *Effects of Radiation on Materials: 20th International Symposium, ASTM STP 1405,* S. T. Rosinski, M. L. Grossbeck, T. R. Allen, and A. S. Kumar, Eds., American Society for Testing and Materials, West Conshohocken, PA, 2001.

Abstract: An extensive database of atomic displacement cascades in iron has been developed using the method of molecular dynamics (MD). More than 300 simulations have been completed at 100K with energies between 0.1 and 100 keV. This encompasses nearly all energies relevant to fission reactor irradiation environments since a 100 keV MD cascade corresponds to the average iron cascade following a collision with a 5.1 MeV neutron. Extensive statistical analysis of the database has determined representative average values for several primary damage parameters: the total number of surviving point defects, the fraction of the surviving point defects contained in clusters formed during cascade cooling, and a measure of the size distribution of the in-cascade point defect clusters. The cascade energy dependence of the MD-based primary damage parameters has been used to obtain spectrum-averaged defect production cross sections for typical fission reactor neutron energy spectra as a function of depth through the reactor pressure vessel. The attenuation of the spectrum-averaged cross sections for total point defect survival and the fraction of either interstitials or vacancies in clusters are quite similar to that for the NRT dpa. However, the cross sections derived to account for the energy dependence of the point defect cluster size distributions exhibit a potentially significant variation through the vessel. The production rate of large interstitial clusters decreases more rapidly than dpa whereas the production of large vacancy clusters is slower than dpa.

Keywords: damage attenuation, displacement cascades, ferritic steels, modeling, molecular dynamics, point defects, pressure vessels, radiation damage

Introduction

Current regulatory practice in the United States calls for radiation-induced reactor pressure vessel (RPV) embrittlement to be predicted on the basis of the neutron fluence above 1 MeV [*1,2*]. The correlation of embrittlement with the so-called fast fluence (i.e. neutron fluence above 1.0 MeV) is rationalized by the assumption that most atomic displacements are generated by the high energy portion of the spectrum, and is well

[1] Senior Research Staff, Oak Ridge National Laboratory, P. O. Box 2008, Oak Ridge, TN.
[2] Lead Scientist, Pacific Northwest National Laboratory, P. O. Box 999, Richland, WA.

supported by ongoing data analysis [*3*]. However, it is well known that neutrons of lower energy also contribute to embrittlement and the use of atomic displacements per atom (dpa) has been recommended [*4,5*] as an improved correlation parameter, particularly when differences in neutron energy spectrum must be accounted for. The change in neutron energy spectrum as a function of depth in the RPV is an important example of this situation. Because of this through-thickness variation in the neutron energy spectrum, Revision 2 of Regulatory Guide 1.99 (RG-1.99/2) specifies that either dpa or an empirical, dpa-based exponential be used to obtain an equivalent fast fluence for calculating Charpy shifts at locations within the vessel [*1,2*].

Differences in neutron energy spectra are manifested as differences in the energy spectra of the primary knockon atoms (PKA) produced in elastic collisions with these neutrons. Since low and high energy PKA can produce damage structures that are both qualitatively and quantitatively different from each other, a systematic investigation of the energy dependence of primary damage formation should provide some insight into how this damage may be attenuated through an RPV. The potential impact of differences in PKA energy spectra are amenable to investigation by displacement cascade simulations using the method of molecular dynamics (MD). MD simulations provide a detailed picture of the formation and evolution of displacement cascades, and recent advances in computing equipment permit the simulation of high energy displacement events involving several million atoms [*6-10*]. The results presented below encompass MD cascade simulation energies from near the displacement threshold to as high as 100 keV.

RPV Irradiation Environment

Representative neutron energy spectra were obtained from Ref. [*11*] for typical pressurized water [PWR] and boiling water reactors [BWR]. The analysis described below was carried out for both reactor types, but the results presented here will focus on the PWR. The conclusions reached for the BWR are essentially the same as for the PWR, with the primary difference being that the relative change between the inner and outer diameters of the BWR are reduced because the RPV thickness is ~156 mm whereas the PWR RPV thickness is ~218 mm. Because of other design differences, the fast flux at the inner diameter of the RPV is lower for the BWR, $\sim 1.1 \times 10^{14}$ $n/m^2/s$, than for the PWR, $\sim 7.5 \times 10^{14}$ $n/m^2/s$. A previous analysis demonstrated that differences between the BWR and PWR neutron energy spectra were modest [*12*].

The neutron and PKA energy spectra for the PWR case are shown in Figs. 1 (a) and (b), respectively. To illustrate how the spectrum changes through the RPV, spectra are shown for three locations: 8 mm (the first RPV node from the transport calculations), and 1/4 and 3/4 of the way through the RPV. The range of the MD cascade energies used in this study are included in Fig. 1(a), where the four cascade energies shown represent the average PKAs from the neutron energies indicated by the arrows. An MD energy of 0.1 keV is near the minimum required to yield stable displacements. Neutrons with energies below about 1 keV produce atomic displacements primarily from the recoils generated by neutron capture (n,γ) reactions. Such recoils have energies of a few hundred eV. PKA spectra calculated with the SPECTER code [*13*] for each of the neutron spectra in Fig.

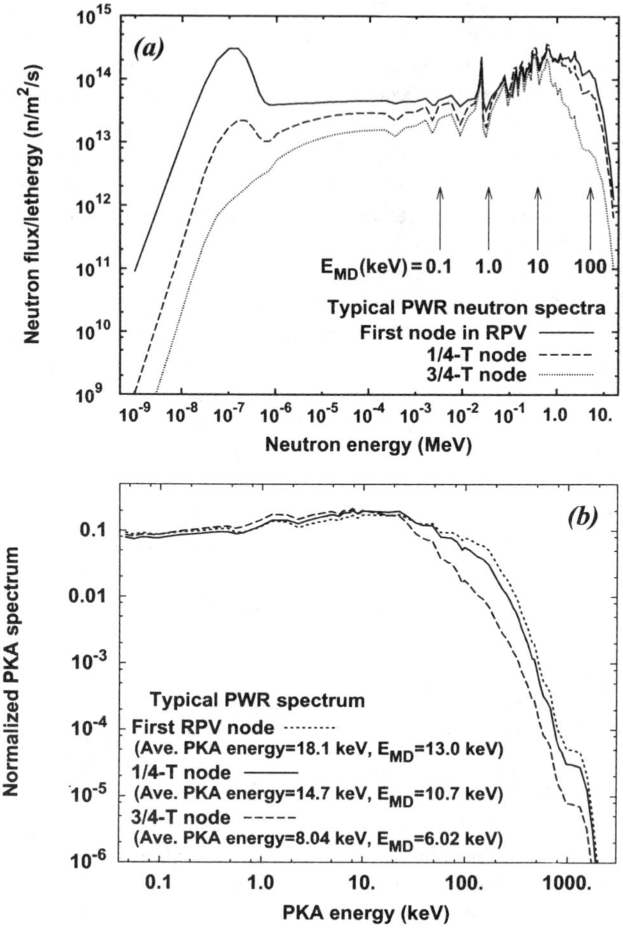

Figure 1 - *Neutron energy spectra (a) and PKA spectra (b) for three locations in a typical pressurized water reactor pressure vessel.*

1(a) are shown in Fig. 1(b). The average PKA energy and its corresponding damage energy are indicated for each PKA spectrum. The difference between PKA and damage energy is discussed in the next section.

MD Cascade Simulations

The MD displacement cascade simulations were carried out using the MOLDY code [*14*], and the interatomic potential for iron that is described in Refs. [*15 and 16*]. The MOLDY code simply describes elastic collisions between atoms and does not account for energy loss mechanisms such as electronic excitation and ionization. Therefore, the cascade energy E_{MD} is analogous to the damage energy in the secondary displacement model by Norgett, Robinson, and Torrens (NRT) [*17*]. PKA energies that correspond to a given E_{MD} can be calculated using the procedure described in Ref. [*17*], and the neutron energy required to generate PKA with a given average energy can be calculated for an elastic collision. For example, the highest MD cascade energy used in this work is 100 keV. A 100 keV damage energy corresponds to an iron PKA energy of ~176 keV, which is the average recoil energy from a collision with a 5.1 MeV neutron. Thus, the simulations encompass most of the neutron energy range experienced by fission reactor components.

The energy dependence of several primary damage parameters have been obtained from the MD simulations. First is the total number of stable displacements created, with vacancies and interstitials being formed in equal numbers. These include all surviving defects present after in-cascade recombination is complete and the simulation cell has returned to thermal equilibrium. A simulation time of about 15 to 20 ps is required to reach this condition at the highest energies. Second, because many of the surviving point defects are contained in small clusters, rather than as isolated defects, the fraction of surviving vacancies and interstitials contained in clusters is determined. It is convenient to express these parameters as a fraction of the displacements predicted by the NRT model [*17*].

The energy dependence observed in the vacancy and interstitial cluster size distributions suggested that some method of accounting for their spectrum dependence should also be included in this analysis. A single parameter was chosen for both types of defect size distribution for this initial investigation, the number of defects in clusters above a specified size. Clusters containing 5 or more vacancies were observed only at cascade energies of 2 keV and higher, and interstitial clusters of 10 or more were observed only at 20 keV and above. These apparent thresholds made the values of 5 and 10 convenient choices to evaluate the effect of energy on the vacancy and interstitial defect cluster size distributions, respectively. In-cascade clustering is significant because such clusters provide nuclei for the growth of larger defects and their formation directly within the cascade means that extended defects can evolve more quickly than if the clusters were formed only by the much slower process of classical nucleation. The presence of relatively large in-cascade clusters could disproportionately contribute to the nucleation of defects such as interstitial loops and microvoids. The number of vacancies and interstitials in "large" clusters is also expressed as a fraction of the NRT displacements at each energy.

An initial statistical analysis of the cascade database up to 50 keV has been published previously [*6*], and the energy dependence of total defect survival and interstitial clustering values are discussed in Ref. [*18*]. Since that time, a series of eight 100 keV simulations have been completed and the vacancy cluster analysis has been carried out on the complete database. The 100 keV results fall on a smooth extrapolation of the lower energy data, as shown in Fig. 2 for the total defect survival fraction. At each energy, the data point is an average of between 8 and 128 cascades, with error bars shown indicating the standard error on the mean. These error bars are almost too small to be observed at the higher energies in Fig. 2, indicating that the minimum observed in the curve near 20 keV is statistically significant. The minimum arises from the extensive subcascade formation

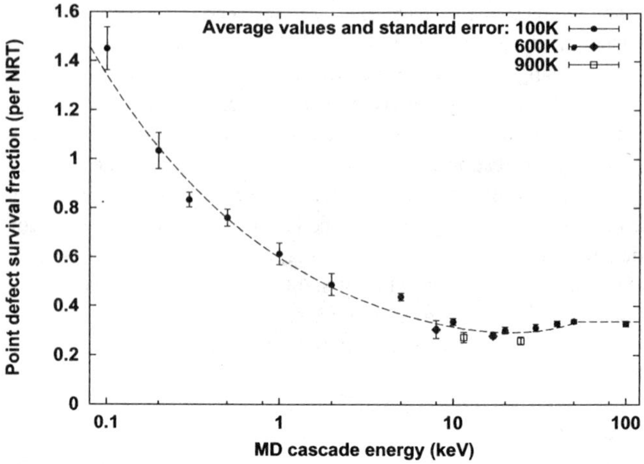

Figure 2 - *Energy dependence of total point defect survival obtained from MD cascade simulations.*

that occurs above 10 keV [*6,8-10*]. Although a detailed comparison of the results of similar work is beyond the scope of this paper, it is worth pointing out that similar observations on the energy dependence of defect formation have been made by others who used a different interatomic potential for iron [*19*].

The interstitial clusters formed in these iron cascade simulations tend to be well defined, with most clusters consisting of nearest neighbor arrangements of <111> or <110> dumbbells, or <111> crowdions. Occasionally, interstitials may be added to a cluster in the second nearest neighbor location or the atomic arrangements can be more complex [*20*]. The use of computer visualization and animation software enables such interstitials to be clearly identified as part of the cluster. However, the vacancy clusters produced in these iron cascades tend to be diffuse, with clearly correlated arrangements out to the fourth nearest neighbor distance [*9,10*]. As a result, two different criteria were employed in the vacancy cluster analysis to define what constitutes a vacancy cluster. The criterion were that all vacancies in the cluster be within: (1) the second (d_{2nn}=0.2867 nm) or, (2) the fourth (d_{4nn}=0.4754 nm) nearest neighbor distance of another vacancy in the cluster. Application of the fourth nearest neighbor criterion nearly doubles the fraction of vacancies in clusters, but is supported by Monte Carlo aging studies of cascade debris [*10, 21*].

The total fraction of vacancies in clusters for both the 2nn and 4nn clustering criteria is shown in Fig. 3(a) and the fraction of vacancies in clusters of 5 or more using the same criteria is shown in Fig. 3(b). For purposes of comparison, the previously derived trend lines for the interstitial clustering parameters from Ref. [*10*] are also shown in Figs. 3(a) and (b). At low energies, the dependence of vacancy clustering on cascade energy is noticeably different from that of the interstitials. The fraction of vacancies (per NRT displacement) continues to increase down to the lowest energies, whereas the interstitial

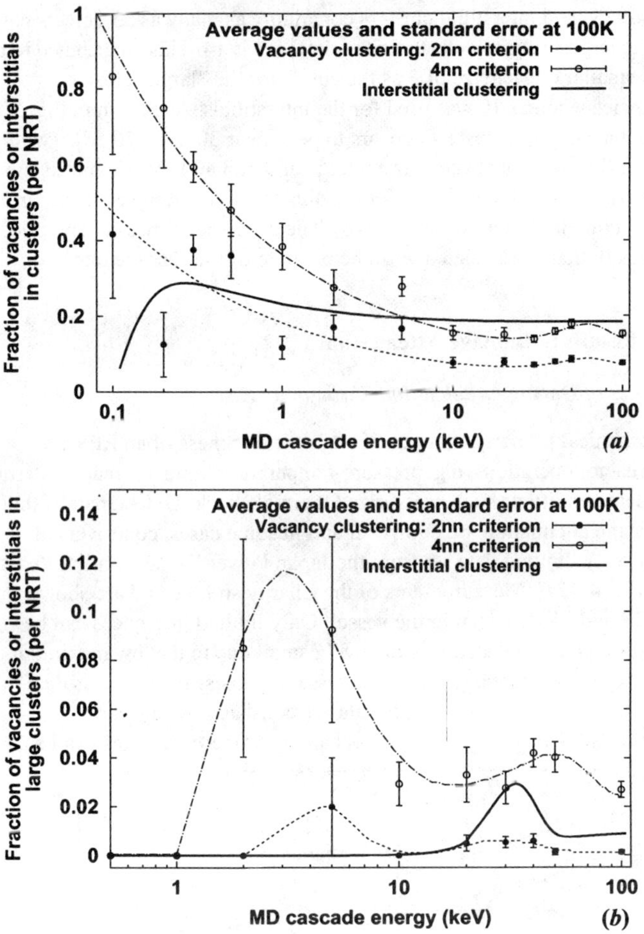

Figure 3 - *Energy dependence of normalized primary damage parameters from MD cascade simulations: (a) total vacancies and interstitial in clusters, and (b) vacancies in clusters of 5 or more and interstitials in clusters of 10 or more.*

clustering fraction decreases below about 300 eV. A much larger fraction of the vacancies than interstitials are in clusters at low energies, and the decrease with energy is much more rapid for the vacancies. At high energies, the clustering fraction for both defect types becomes nearly asymptotic, with some indication of a peak in the vacancy clustering curve near 50 keV. A somewhat larger fraction of the interstitials are in clusters at high energies, even for the less restrictive 4nn vacancy clustering criterion.

The analysis of the cluster size distributions revealed further differences between the clustering behavior of vacancies and interstitials. Fewer large vacancy clusters are observed than interstitials, and interstitial clusters containing as many as 33 defects have been observed, but the largest vacancy cluster in the database is 14. This contributed to the decision to choose a smaller cluster size of 5 as the cutoff for the "large" cluster designation for vacancies whereas 10 was used for the interstitials. As shown in Fig. 3(b), the fraction of interstitials in large clusters appears to peak near 30 keV [*10,18*]. Two peaks are observed for the vacancy cluster parameter near 3 and 40 keV. The peaks are small for the 2nn criterion but well developed for the 4nn case. No simple explanation has been found for the structure in the curves in Fig. 3(b). The high-energy peaks may reflect a maximum energy density that is associated with the breakup of the cascade into subcascades.

Application of MD Results to Damage Attenuation

Conventional Estimates of Damage Attenuation Through an RPV

Prediction of mechanical property changes through the thickness of an RPV are required for routine reactor operations, e.g. pressure-temperature limits for reactor startup, and analysis of accident conditions such pressurized thermal shock. Unfortunately, the available data on damage attenuation are equivocal and in some cases, controversial [*22,23*]. A recent, but very limited data set from the Japan Power Demonstration Reactor (JPDR) is shown in Fig. 4 [*24*]. Measurements of the Charpy shift at 41 J are shown for two locations about 18 and 71 mm deep in the vessel. Only limited inferences can be drawn from this data because: (1) the shift is relatively small due to the low maximum JPDR dose of about $2x10^{22}$ n/m^2, and (2) the total vessel thickness is much less than that of most commercial RPVs. However, the JPDR data are consistent with some interpretations of other data which suggest that mechanical property changes can be

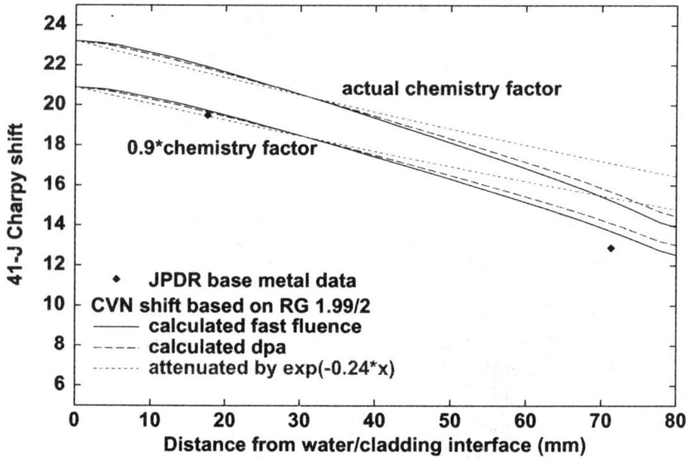

Figure 4 - *Comparison of JPDR base metal data and predictions of RG-1.99/2 with various attenuation models.*

attenuated faster than would be predicted based on either dpa or the exponential attenuation formula employed in RG-1.99/2 [22].

This point is illustrated by the two sets of curves in Fig. 4 which compare calculations of the expected attenuation of the Charpy shift through the vessel using the correlation from RG-1.99/2. The neutron fluence needed to compute the shift at each depth was determined in three ways : (a) the actual calculated fast fluence, (2) proportional to the calculated dpa, and (3) proportional to RG-1.99/2's exponential formula. The upper set of curves was obtained using the chemistry factor calculated from the material's measured composition, and the lower curves were obtained by a simple ratio procedure to obtain agreement with the data at 18 mm. In spite of the limited range of the data, the exponential attenuation is clearly very conservative with respect to the data. Of course, the small shift and known scatter in Charpy data [*25*] mean that this data does not provide a definitive demonstration of faster than dpa attenuation.

A further comparison of different attenuation assumptions is shown in Fig. 5 for the same PWR neutron spectrum discussed above [*11*]. Figure 5(a) compares the RG-1.99/2 attenuation formula with the actual dpa, neutron fluence with E>0.1 MeV, and neutron fluence with E>1.0 MeV. A value for the exposure parameters at the water/RPV interface was obtained by linear interpolation between the values at the last node in the water and the first node in the RPV. The exposure parameters in Fig. 5(a) have been normalized using these interpolated values. For this thicker vessel geometry, deviations between the exponential attenuation formula and dpa are also seen, and the exponential is not conservative at every point. The ratio based on the exponential is nearly the same as fast fluence (E>1.0 MeV) at the 1/4-T position and closer to dpa beyond the 3/4-T position.

The Charpy shifts predicted with this PWR spectrum for a weld with 0.25 wt% Cu and 0.75 wt%Ni at a fast fluence of $2x10^{23}$ n/m^2 are shown in Fig. 5(b). The procedure of RG-1.99/2 was used for each exposure parameter in Fig. 5(a), i.e. the shift was calculated by attenuating the surface fluence with E>1.0 MeV according to the attenuation ratios obtained for dpa, fluence with E>0.1 MeV, or the exponential function. Because of the fluence dependence in the embrittlement correlation, the Charpy shift reduces more slowly through the vessel than does any of the exposure parameters. The relative attenuation in Charpy shift is also a function of the neutron fluence as shown in Table 1, where ratios of the exposure parameters and predicted Charpy shifts are compared at two different fluences. Although the exposure ratios are independent of fluence, the predicted Charpy shift profile becomes flatter as the fluence increases. For this full-thickness PWR RPV, the overall attenuation ratios based on the exponential formula are similar to those based on dpa. The agreement is not as good for a thinner BWR vessel (see for example Figure 4).

Through-Thickness Variation in MD-based Parameters

Energy dependent functions were obtained by fitting the data in Figs. 2 and 3, and these were used to evaluate the effect of the variation in the neutron energy spectrum through an RPV. These energy-dependent functions have a form similar to those derived previously [*26*], and they were employed in the SPECOMP code [*27*] to compute effective production cross sections for point defect survival, total vacancy and interstitial clustering,

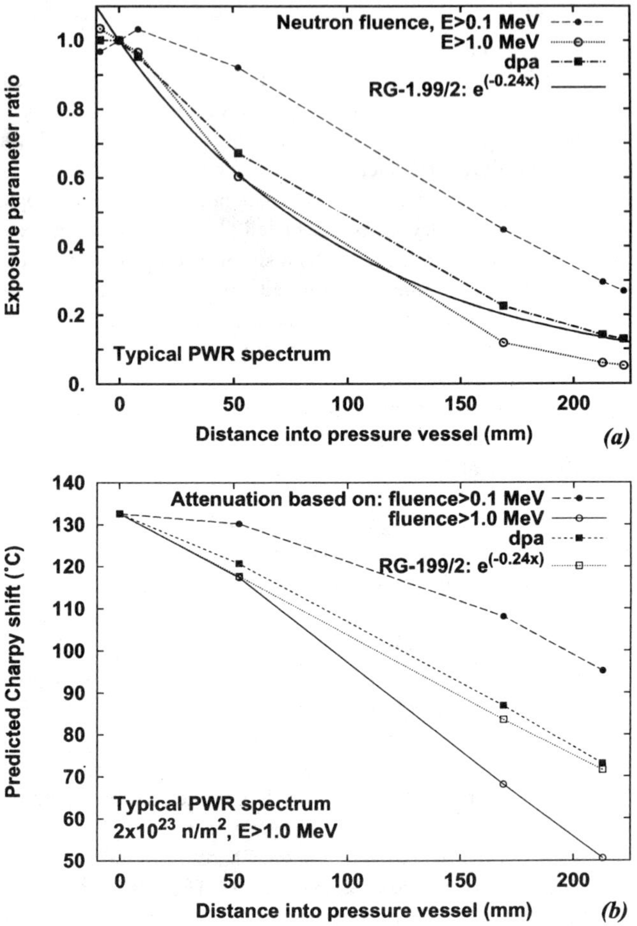

Figure 5 - *Comparison of exposure parameter and 41 J Charpy shift through a typical PWR RPV.*

and vacancies and interstitials in large clusters. The values at 50 keV were used for all higher energy cascades. PKA spectra for iron such as those shown in Fig. 1(b) were obtained from SPECTER [*13*] and used to weight these MD-based cross sections in order to calculate spectrum-averaged values for various locations through the RPV. More details on the SPECOMP/SPECTER calculations can be found in Refs. [*12 and 18*].

A comparison of the effects of neutron energy spectrum changes through a PWR RPV on the MD-based exposure parameters, fast fluence (E>1.0 MeV), and dpa is shown in Fig. 6. Total MD defect survival, total interstitials in clusters, and interstitials in large clusters (≥10) are shown in Fig. 6(a), and total vacancy clustering and vacancies in large

Table 1- *Attenuation of exposure and damage parameters for typical PWR spectrum*

	Ratio: $\frac{\text{Value at outer diameter}}{\text{Value at inner diameter}}$			Ratio: $\frac{\text{Value at 3/4-T}}{\text{Value at 1/4-T}}$		
	Exposure parameter	41-J Charpy shift		Exposure parameter	41-J Charpy shift	
		$5x10^{22}$ n/m^2	$2x10^{23}$ n/m^2		$5x10^{22}$ n/m^2	$2x10^{23}$ n/m^2
Neutron fluence, E>1.0 MeV	0.0599	0.272	0.382	0.196	0.477	0.581
Neutron fluence, E>0.1 MeV	0.296	0.619	0.718	0.487	0.761	0.830
dpa	0.144	0.434	0.551	0.337	0.631	0.720
$e^{-0.24x\ [in]}$	0.134	0.423	0.539	0.331	0.622	0.710

(≥5) clusters are shown in Fig. 6(b). There is essentially no difference between dpa and either total MD defect survival or total interstitials in clusters in Fig. 6(a). However, a potentially significant difference is seen between dpa and the production of large interstitial clusters. Since these clusters are only produced in relatively high-energy cascades, their production is attenuated more rapidly than dpa. This is consistent with the change in the PKA spectrum through the vessel shown in Fig. 1(b).

Attenuation of the total vacancy clustering fraction in Fig. 6(b) also follows dpa quite closely, whereas the fraction of vacancies in large clusters is attenuated somewhat more slowly than dpa. The vacancy cluster parameters in Fig. 6(b) were derived for the 4nn clustering criterion. Results for the 2nn criterion are qualitatively similar, with a larger difference between dpa and vacancies in large clusters. The slower-than-dpa attenuation of large vacancy clusters is a result of the lower energy peak shown in Fig. 3(b). The relative importance of vacancy clusters produced by 1 to 10 keV PKA increases as the higher energy PKA are removed from the spectrum.

Summary

It is now possible to simulate the atomic displacement cascades generated by nearly the complete fission neutron spectrum. The results presented above include cascade energies as high as 100 keV, corresponding to a 176 keV PKA, which is the average iron recoil from a neutron with an energy of 5.1 MeV. Thus, these results are fully relevant to materials irradiated in any fission reactor environment. Note that the highest *average* PKA energy obtained from SPECTER for any of the PWR RPV neutron spectra was only 18 keV. This extensive database of cascade simulations can now be used to derive statistically-relevant primary damage parameters and provide a basis for analyzing

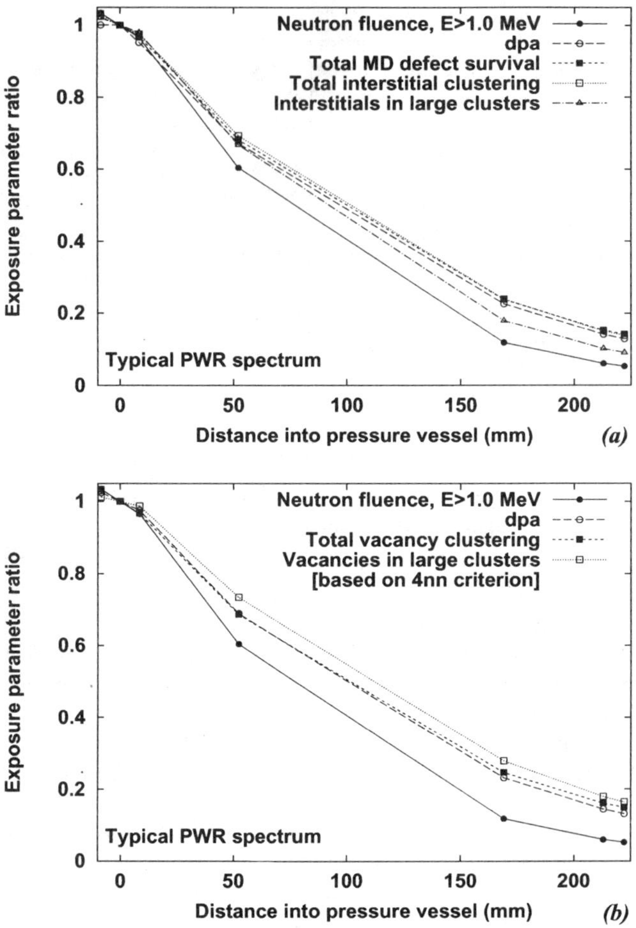

Figure 6 - *Attenuation of MD-based exposure parameters compared with fast fluence and dpa.*

differences in radiation environments.

Primary damage parameters derived from the MD cascade simulations have been used to obtain effective defect production cross sections for several locations through typical PWR and BWR reactor pressure vessels for comparison with conventional exposure parameters such as dpa and fast fluence. The analysis indicates that the deviations from dpa-like attenuation are rather modest. This implies that neutron energy spectrum changes due to attenuation should be well accounted for through the use of dpa as long as total defect production is the parameter controlling mechanical property changes. Although the results were not presented here, the spectrum-averaged defect production cross section derived for PWR and BWR vessels were nearly identical, implying that spectrum differences should not contribute to differences in RPV embrittlement between the two reactor types.

The most significant difference between any of the MD-based parameters and dpa was for the production of large interstitial clusters. If these clusters contribute significantly to embrittlement, attenuation of mechanical property changes could be faster than that predicted by dpa attenuation. This could be offset by the impact of large vacancy clusters since they were attenuated slightly more slowly than dpa. However, it is not possible to predict mechanical property changes simply on the basis of primary damage formation. The spectrum-averaged defect production cross sections will be used to develop an improved description of the radiation damage source terms in a kinetic embrittlement model that has been used to predict radiation-induced mechanical property changes based on microstructural evolution.

Acknowledgments

Research sponsored by the Office of Nuclear Regulatory Research, U.S. Nuclear Regulatory Commission under inter-agency agreement DOE 1886-N695-3W with the U.S. Department of Energy and by the Division of Materials Sciences and Engineering, U.S. Department of Energy under contract DE-AC05-00OR22725 with UT-Battelle, LLC, and the Office of Fusion Energy Sciences, U.S. Department of Energy at the Pacific Northwest National Laboratory.

References

[1] "Radiation Embrittlement of Reactor Vessel Materials," Regulatory Guide 1.99, Revision 2, U.S. Nuclear Regulatory Commission, May 1988, available from the U.S. National Technical Information Service, Springfield, VA.

[2] Randall, P. N., "Basis for Revision 2 of the U.S. Nuclear Regulatory Commission's Regulatory Guide 1.99," *Radiation Embrittlement of Nuclear Reactor Pressure Vessel Steels: An International Review, STP 909*, L.E. Steele, Ed., American Society for Testing and Materials, West Conshohocken, PA, 1986, pp. 149-162.

[3] Eason, E. D., Wright, J. E., and Odette, G. R., "Improved Embrittlement Correlations for Reactor Pressure Vessel Steels," NUREG/CR-6551, MCS 970501, U.S. Nuclear Regulatory Commission, November 1998, available from the U.S. National Technical Information Service, Springfield, VA.

[4] Recommendations of IAEA Specialists Meeting on Radiation Damage Units in Graphite and Ferritic and Austenitic Steels, *Nuclear Engineering and Design*, 1975, 33, p. 48.

[5] Stoller, R. E. and Odette, G. R., "Recommendations on Damage Exposure Units for Ferritic Steel Embrittlement," *Journal of Nuclear Materials* 1992, 186, pp. 203-205.

[6] Stoller, R. E. and Calder, A. F., "Statistical Analysis of a Library of Molecular Dynamics Cascade Simulations in Iron at 100K,"*Journal of Nuclear Materials* 2000, 283-287, 186, pp. 203-205.

[7] Phythian, W. J., Stoller, R.E., Foreman, A. J. E., Calder, A. F., and Bacon, D. J., "A Comparison of Displacement Cascades in Copper and Iron by Molecular Dynamics and Its Application to Microstructural Evolution," *Journal of Nuclear Materials* 1995, 223, pp. 245-261.

[8] Stoller, R.E., "Point Defect Survival and Clustering Fractions Obtained From Molecular Dynamics Simulations of High Energy Cascades," *Journal of Nuclear Materials* 1996, 233-237, pp. 999-1003.

[9] Stoller, R.E., Odette G.R., and Wirth, B. D., "Primary Damage Formation in bcc Iron," *Journal of Nuclear Materials* 1997, 251, pp. 49-69.

[10] Stoller, R.E., "The Role of Cascade Energy and Temperature in Primary Defect Formation in Iron," *Journal of Nuclear Materials* 2000, 276, pp. 22-32.

[11] Albert,T.E., Gritzner,M.L., Simmons, G.L., and Straker, E.A., "PWR and BWR Radiation Environments for Radiation Damage Studies, "EPRI NP-152, Electric Power Research Institute, Palo Alto, CA, September 1977.

[12] Stoller, R.E. and Greenwood, L.R., "An Evaluation of Neutron Energy Spectrum Effects Based on Molecular Dynamics Displacement Cascade Simulations," *Effects of Radiation on Materials, STP 1366*, M. L. Hamilton, A. S. Kumar, S. T. Rosinski, and M. L. Grossbeck, Eds., American Society for Testing and Materials, West Conshohocken, PA, 2000, pp. 548-559.

[13] Greenwood, L.R. and Smither, R.K., "SPECTER: Neutron Damage Calculations for Materials Irradiations," ANL/FPP/TM-197, Argonne National Laboratory, Argonne, IL, January 1985.

[14] Finnis, M.W., "MOLDY6-A Molecular Dynamics Program for Simulation of Pure Metals," AERE R-13182, UKAEA Harwell Laboratory, Harwell, UK, 1988.

[15] Finnis, M.W. and Sinclair, J.E., "A Simple Empirical N-body Potential for Transition Metals," *Philosophical Magazine-A* 1984, 50, pp. 45-55 and Erratum, *Philosophical Magazine-A*, 1986, p. 161.

[16] Calder, A.F. and Bacon, D.J., "A Molecular Dynamics Study of Displacement Cascades in α-iron," *Journal of Nuclear Materials* 1993, 207, pp. 25-45.

[17] Norgett, M.J., Robinson, M.T., and Torrens, I.M., "A Proposed Method of Calculating Displacement Dose Rates," *Nuclear Engineering and Design* 1975, 33, pp. 50-54.

[18] Stoller, R.E. and Greenwood, L.R., "A Comparison of the NRT Displacement Model and Primary Damage Formation Observed in Molecular Dynamics Cascade Simulations," presented at the 10th International Symposium on Reactor Dosimetry, 12-17 September 1999, Osaka, Japan, to be published by the American Society of Testing and Materials in ASTM STP 1398

[19] N. Soneda and T. Diaz de la Rubia, "Defect Production, Annealing Kinetics, and Damage Evolution in α-Fe: An Atomic-scale Computer Simulation," *Philosophical Magazine A* 1998, 78, pp. 995-1019.

[20] Stoller, R.E., "Molecular Dynamics Simulations of High Energy Cascades in Iron," *Microstructure of Irradiated Materials*, I. M. Robertson, L. E. Rehn, S. J. Zinkle, and W. J. Phythian, Eds., Materials Research Society, Pittsburgh, PA, 1995, pp. 21-26.

[21] B. D. Wirth and G. R. Odette, "Kinetic Lattice Monte Carlo Simulations of Cascade Aging in Iron and Dilute Iron-Copper Alloys," *Multiscale Modeling of Materials*, V. V. Butalov, T. Diaz de la Rubia, P. Phillips, E. Kaxiras, and N. Ghoniem, Eds., Materials Research Society, Pittsburgh, PA, 1999, pp. 211-216.

[22] Gold, R. and McElroy, W. N., "Radiation-Induced Embrittlement in Light Water Reactor Pressure Vessels," *Nuclear Engineering and Design* 1987, 104, pp. 155-174.

[23] Kussmaul, K., Föhl, J., and Weissenberg, T., "Assurance of the Pressure Vessel Integrity with Respect to Irradiation Embrittlement: Activities in the Federal Republic of Germany," *Radiation Embrittlement of Nuclear Reactor Pressure Vessel Steels, ASTM STP 1011*, L. E. Steele, Ed., American Society of Testing and Materials, West Conshohocken, PA, 1989, pp. 3-26.

[24] Suzuki, M., "Investigation of In-service Irradiated pressure Vessel Steel from the Decommissioned Reactor, JPDR," ICONE-7411, 7th International Conference on Nuclear Engineering, Tokyo, Japan, April, 1999, pp. 1-8.

[25] Nanstad, R. K., McCabe, D. E., and Swain, R. L., "Evaluation of Variability in Material Properties and Chemical Composition for Midland Reactor Weld WF-70," *Effects of Radiation on Materials, STP 1325*, R. K. Nanstad, M. L. Hamilton, F. A. Garner, and A. S. Kumar, Eds., American Society for Testing and Materials, West Conshohocken, PA, 1999, pp. 125-156.

[26] Stoller, R. E. and Greenwood, L. R., "Subcascade Formation in Displacement Cascade Simulations; Implications for Fusion Reactor Materials," *Journal of Nuclear Materials* 1999, 271&272, pp. 57-62.

[27] Greenwood, L. R., "SPECOMP Calculations of Radiation Damage in Compounds," *Reactor Dosimetry: Methods, Applications, and Standardization, ASTM STP 1001*, H. Farrar IV and E. P. Lippincott, Eds., American Society of Testing and Materials, West Conshohocken, PA, 1989, pp. 598-602.

Hiroshi Kawanishi[1] and Masahide Suzuki[2]

Hardness and Microstructure Changes with Thermal Annealing of Neutron-Irradiated Fe-Cu Alloys

---

Reference: Kawanishi, H. and Suzuki, M., "**Hardness and Microstructure Changes with Thermal Annealing of Neutron-Irradiated Fe-Cu Alloys,**" *Effects of Radiation on Materials: 20th International Symposium, ASTM STP 1405,* S. T. Rosinski, M. L. Grossbeck, T. R. Allen, and A. S. Kumar, Eds., American Society for Testing and Materials, West Conshohocken, PA, 2001.

Abstract: In order to investigate mechanisms in recovery of irradiation-induced hardness by thermal annealing for pressure vessel steels containing copper, Fe-0.6 wt.% Cu model alloys irradiated to a dose of 0.0055 dpa at 290°C in JMTR were isochronally annealed at temperatures ranging from 250 to 600°C ($\Delta T = 50$°C, $\Delta t = 15$ min). Vickers hardness measurements showed that 70% of the irradiation-induced hardness was eliminated by the annealing. TEM examinations revealed that three kinds of precipitates were formed in the specimens; those are small copper precipitates (disk-shaped, 3-8 nm in diameter), ordered structures of possibly Cu_3Fe ($L1_2$-type, $a_o = \sim 0.343$ nm) and Fe_3O_4 particles (structure of spinel-type). Copper precipitates slightly increased the number density and size with the post-irradiation annealing below 400 °C, indicating clearly that copper precipitates contributed to hardening, though the degree may be small. The ordered structure was proven by the occurrence of superlattice reflections in selected area diffraction patterns. The post-irradiation annealing reduced the intensities of the superlattice reflections and the zone of Cu_3Fe-structure. The ordered structure of Cu_3Fe was observed to be destroyed by thermal annealing, resulting in a remarkable decrease in hardness. The hardness during thermal annealing was concluded to be governed by two competitive processes of hardening by the growth of copper precipitates and softening by the order-disorder transformation of Cu_3Fe.

Keywords: ordered structure, superlattice, Cu_3Fe, $L1_2$-type, thermal annealing, hardness, RPV steel, TEM, copper precipitate, Fe-Cu alloy

---

[1]Research Associate, Department of Quantum Engineering and Systems Science, University of Tokyo, 7-3-1, Hongo, Bunkyo-ku, Tokyo 113-8656, Japan.
[2]Senior Engineer, Japan Atomic Energy Research Institute, Tokai-mura, Naka-gun, Ibaraki-ken 319-1195, Japan.

Introduction

Neutron irradiation of reactor pressure vessel (RPV) steels containing a small amount of copper induces an increase in ductile-to-brittle transition temperature (DBTT) and a decrease in upper shelf energy (USE). Heat treatment after neutron irradiation can bring about a significant recovery of the degradation for DBTT and USE. Thermal annealing of the RPV steels is one option to restore an already embrittled RPV for extending the life of a RPV. Two components, which are responsible for irradiation hardening/ embrittlement of RPV steels, have been reported using various up-to date techniques by many investigators [*1-8*]. The first is the formation of three-dimensional or spherical copper rich-clusters having diameters of a few nm where the clusters are suggested to be alloyed with Mn and Ni. The second is microvoid or vacancy cluster-solute atom complex resulting from the clustering or collapse of point defects produced during irradiations. Regarding the response of neutron-irradiated RPV steels with thermal annealing, atom probe field microscope (APFIM) studies have demonstrated that the concentration of copper in solution in the matrix decreases with neutron-irradiations and thermal annealing [*1-3*]. Pareige et al. have reported via small angle neutron scattering (SANS) and tomographic APFIM experiments of neutron-irradiated French Chooz A pressure vessel steels that neutron-induced clusters (Cu, Si, Ni, Mn) are dissolved during thermal annealing, while copper precipitates are formed [*9*].

We have reported through TEM examinations of neutron-irradiated Fe-0.6 wt.% Cu model alloys that hardening by neutron irradiation is considered to be due to the formation of ordered structures, which have been postulated to be of Cu_3Fe-type [*10*]. In addition, small copper precipitates less than several nm in diameter have been found to exist in the irradiated specimens. Since the number density and size of the copper precipitates are roughly similar to those in the unirradiated specimens, the contribution of small copper precipitates to the irradiation hardening has been thought to be small. If the formation of ordered structures is primarily the cause of irradiation hardening, the mechanisms of recovery of hardening may also be different from those reported already. Then, TEM examinations of neutron-irradiated Fe-0.6 wt.% model alloys were performed to investigate the evolution of ordered structures and small copper precipitates with thermal annealing.

The paper will present the results of TEM examinations that thermal annealing induces a slight increase in the size and number density of small copper precipitates, and the dissolution of Cu_3Fe-structure, leading to a conclusion that the recovery of hardness for neutron-irradiated Fe-Cu alloys is controlled by two competitive processes of the growth of copper precipitates and the order-disorder transformation of Cu_3Fe-structure.

Experimental

The work is a continuation of previous investigations; the procedure of specimen preparation, the compositions of the specimens and the details in neutron irradiation were already described [*10, 11*]. 3-mm disks of Fe-0.6 wt.% Cu alloys irradiated in JMTR at 290°C to a dose of 0.0055 dpa were isochronally annealed at temperatures ranging from 250 to 600°C where the temperature was raised at 50°C-interval ($\Delta T = 50°C$) and the holding time (Δt) was 15 min. The annealing was carried out in a vacuum of ≤ 0.5 torr, and the disks were chemically polished with a solution of 2 %-HF + 1 %-H_2O + 97 %-H_2O_2 before hardness measurements. Vickers hardness measurements were done with a load of 1 kg. Specimens thermally annealed at 300, 400, and 600°C were electropolished at -50°C in a mixture of 5 %-$HClO_4$ and 95 %-CH_3OH where the applied voltage was 30 V, followed by TEM examinations using an electron microscope JEM-200CX.

Results

Hardness Changes

Figure 1 shows the recovery of the irradiation-induced hardness for Fe-0.6 wt.% Cu with thermal annealing, indicating that ~70 % of the irradiation-induced hardness is recovered until the 600°C-annealing. Two points are noted in Figure 1. The first is that the recovery takes place at a low temperature (250°C) lower than the irradiation temperature. The second is that the curve characteristically shows a plateau or step around 300-400°C. These are interpreted in Discussion.

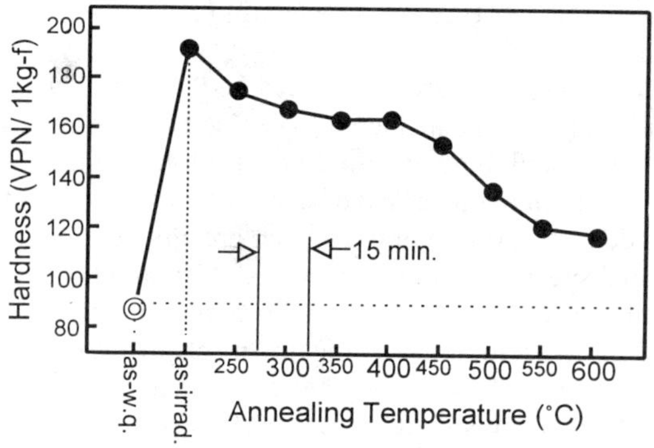

Figure 1— *Hardness changes of Fe-0.6 Cu irradiated in JMTR at 290°C to a dose of 0.0055 dpa with thermal annealing.*

Evolution of Small Copper Precipitates with Thermal Annealing

TEM examinations has shown the presence of small copper precipitate, ordered structure of possibly Cu_3Fe-type and magnetite (Fe_3O_4) in the specimens which are thermally annealed after neutron irradiation. However, descriptions are made on copper precipitate and Cu_3Fe-structure in the paper since magnetites have been observed in unalloyed iron, thermally aged and neutron-irradiated Fe-0.6Cu alloys, and thought to affect hardening little [*10, 11*].

TEM examinations were preferentially done using a dark field imaging technique, which was generally useful in observing small or thin precipitates. To analyze the selected area diffraction pattern (SADP) corresponding to their image or contrast is indispensable to identify precipitates or secondary phases different from the matrix. To find and index the reflections from a precipitate on a SADP is the most generic way to identify precipitates. However, such reflections often become invisible on SADPs due to their low intensities when precipitates are small or thin. It has been empirically known that the reflections of copper precipitates in neutron-irradiated Fe-0.6Cu alloys are too weak to identify [*10*]. However, there is a Kurdjumov-Sachs (K-S) orientation relationship between the matrix and copper precipitates in Fe-Cu alloys, namely,

$$(110)_{Fe} \,//\, (111)_{Cu} \text{ and } [1\bar{1}1]_{Fe} \,//\, [1\bar{1}0]_{Cu}.$$

This means that the vectors of $[110]_{Fe}$ and $[111]_{Cu}$ are parallel to each other in a diffraction plane. Furthermore, the spacings of $(110)_{Fe}$ and $(111)_{Cu}$ planes are 0.2027 and 0.2087 nm,

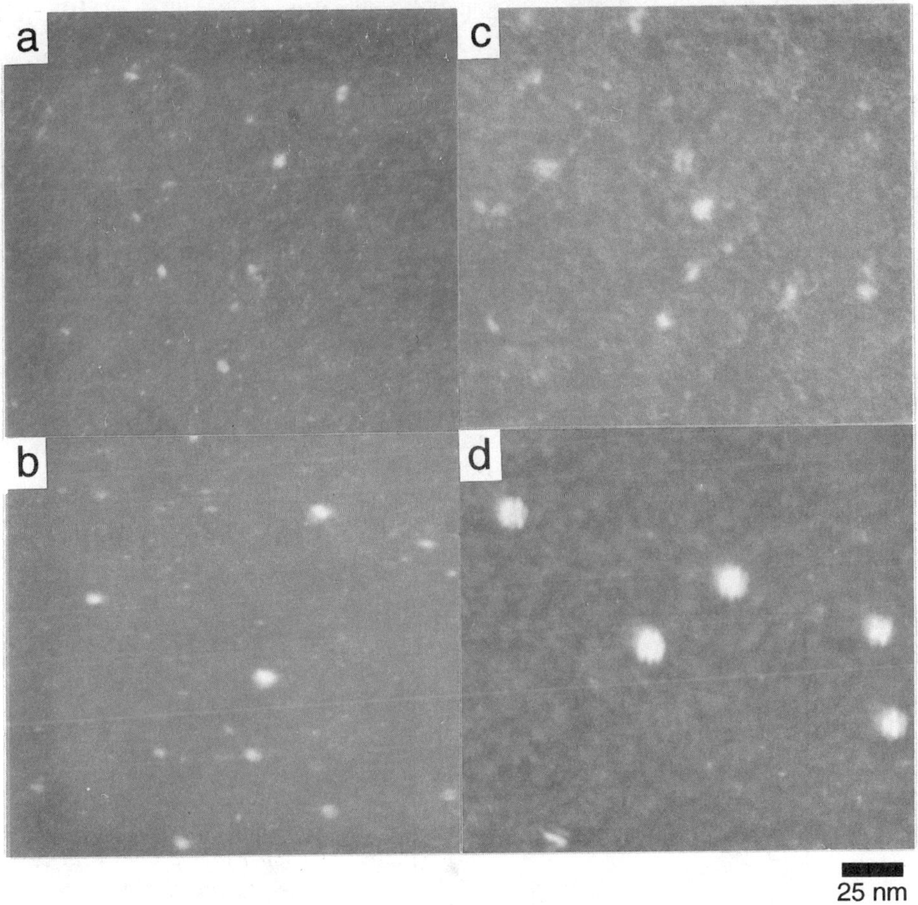

Figure 2 — *Dark field micrographs of copper precipitates taken by setting an objective aperture to a* $(110)_{Fe}$ *reflection for the specimens (a) as-irradiated, (b) 300°C-annealed, (c) 400°C-annealed and (d) 600°C-annealed. The image of copper precipitates is thought to be originated from two* $(111)_{Cu}$ *-reflections overlapped to the* $(110)_{Fe}$ *reflection.*

respectively. Thus, the reciprocal lattice vectors (**g**) of $[110]_{Fe}$ and $[111]_{Cu}$ are understood almost completely to overlap each other in a reciprocal lattice space. In other words, the reflections of $(111)_{Cu}$ always exist at the position of $(110)_{Fe}$ reflections on a SADP as far as copper is precipitated in the matrix, even if the reflections of copper are invisible. In practice, dark field images of copper precipitates can be observed by setting an objective aperture on a given $(110)_{Fe}$-reflection. The validity of this technique has been verified in the previous study [*10*]. As described previously [*11*], there are twelve habit planes (namely, $(112)_{Fe}$) in the matrix for the precipitation of copper. This means that two $(111)_{Cu}$ reflections are always overlapped to a $(110)_{Fe}$ reflection. The total number density of copper precipitates can be calculated by multiplying 6 by that obtained using a $(110)_{Fe}$ reflection.

A sequence of copper precipitates with thermal annealing are shown in Figure 2, which are taken by the technique described above. The SADPs corresponding to each images are different among the four micrographs. It must be noted that the contrast observed in Figure 2 consists of two components originating from copper precipitates and strain field produced by the strain of $(110)_{Fe}$-plane in the matrix around precipitates. The minimum size to be detected was approximately 2 nm in the present experiments. Caution must be paid in determining the sizes of copper precipitates since the strain contrast around them may apparently increase their sizes.

A banded structure is observed in the 600°C-annealed specimen. Though works using high-resolution electron microscope (HREM) have postulated that the banded structure arises from the formation of twinned 9R structure [*12-14*], we consider that the banded structure results from the growth twin of copper precipitates with annealing. The reason is briefly described in Discussion.

Figure 3 shows a sequence of the average size and number density of copper precipitates with thermal annealing. It is clearly indicated in Figure 3 that the sizes increase with the annealing while the number densities also increase up to 400°C, but decrease at 600°C-annealing.

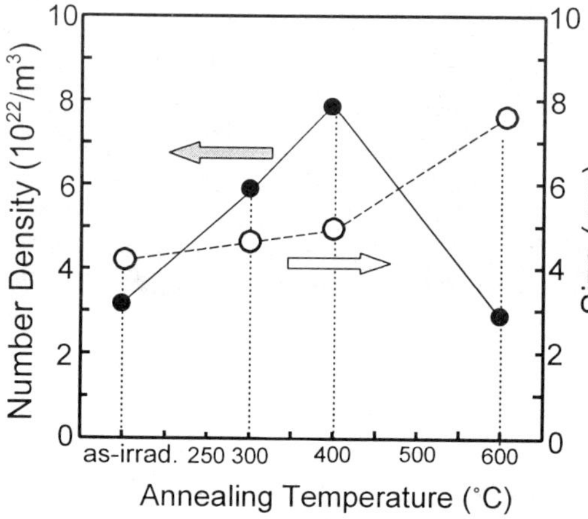

Figure 3— *A sequence of sizes and number densities for copper precipitates with thermal annealing.*

Ordered Structure of Cu_3Fe-Type

In addition to small copper precipitates, the previous paper [*10*] has reported that the ordered structures, which possibly have the composition of Cu_3Fe ($L1_2$-type), are formed in neutron-irradiated Fe-0.6wt.% Cu, indicating that neutron irradiation induces the disorder-order transformation in the matrix. In general, it was rather difficult to identify so-called superlattice reflections from such zone, partially because the SADP analysis was disturbed by lots of reflections of magnetite (Fe_3O_4) and partially because the occurrence of superlattice reflections was more sensitive to the electron beam direction or the specimen orientation during observations. In addition, it is emphasized that the images of the ordered structures may be often superimposed to those of Fe_3O_4 and copper precipitates since those reflections are often laid near or to overlap each other.

It has been continued in the present study to obtain further evidence of the ordered structures in the specimens irradiated. Figure 4 is an instructive example in showing how the diffraction conditions during observations are drastically influenced by the direction of the incident electron beam. Figure 4 (a) is a $(311)_{Fe}$ diffraction pattern for the specimen annealed at 400°C after neutron irradiation, where no reflections showing ordered structures take place though a few reflections of Fe_3O_4 appear. Figure 4 (b) is the SADP obtained when the specimen is tilted a little without varying the position in order to observe the microstructure by a $(\bar{1}21)_{Fe}$ reflection; the tilting angle (θ) is calculated to be about 1.2° using the equation of $2sin\theta/2 = \lambda\, \boldsymbol{g}_{121}$ where λ is the wavelength of electrons at 200 kV and $\boldsymbol{g}_{121}$ the reciprocal lattice vector of $(121)_{Fe}$-plane. A regular array of extra reflections can be seen in Figure 4 (b). These extra reflections are analyzed to form a (310) pattern corresponding to a cubic crystal. The reflections lying on a line of $(10\bar{3})_{Fe}$ and $(\bar{1}21)_{Fe}$ spots in Figure 4 (b) are thought to be due to double diffraction. The lattice parameter of the cubic crystal is determined to be approximately 0.340-0.345 nm from Figure 4 (b), and close to that of Cu_3Fe-structure; the lattice parameter for Cu_3Fe is estimated to be 0.343 nm from Vegard's law regarding the lattice parameter in a binary alloy system of iron and copper. We must pay attention to the occurrence of (001) reflections in Figure 4 (c), indicating that the crystal is neither a fcc nor a bcc structure. This is the ground for postulating the formation of the ordered structure of $L1_2$-type. It might be worthwhile noting that the intensities of extra reflections greatly varied by slightly moving the specimen, for instance, for their focusing, suggesting that the visibility of the ordered structures is also governed by the buckling of the specimen. The image by a $(002)_{Cu3Fe}$ superlattice reflection in Figure 4 (d) is shown in Figure 5 (b).

Figures 5 represent a sequence of the ordered structures of Cu_3Fe-type with thermal annealing where Figures 5 (a), (b) and (c) correspond to the specimens as-irradiated, annealed at 400°C and at 600°C, respectively. Micrographs in Figures 5 are taken by (a) a $(1\bar{1}0)_{Cu3Fe}$ reflection on a $\{(110)_{Fe}+(111)_{Cu3Fe}\}$ pattern, (b) a $(002)_{Cu3Fe}$ reflection on a $\{(311)_{Fe}+(310)_{Cu3Fe}\}$ pattern and (c) a $(1\bar{1}0)_{Cu3Fe}$ reflection on a $\{(135)_{Fe}+(111)_{Cu3Fe}\}$ pattern, where $\{(110)_{Fe}+(111)_{Cu3Fe}\}$, for example, represents the appearance of two patterns of $(110)_{Fe}$ and $(111)_{Cu3Fe}$ on their SADP. Though it was always desirable to use an identical reflection, e.g., $(1\bar{1}0)_{Cu3Fe}$ reflection on $(110)_{Fe}$-pattern, for their imaging among the specimens, it was commonly most difficult to realize it. It was almost impossible to estimate quantitatively the volume of the Cu_3Fe-zone in the specimens

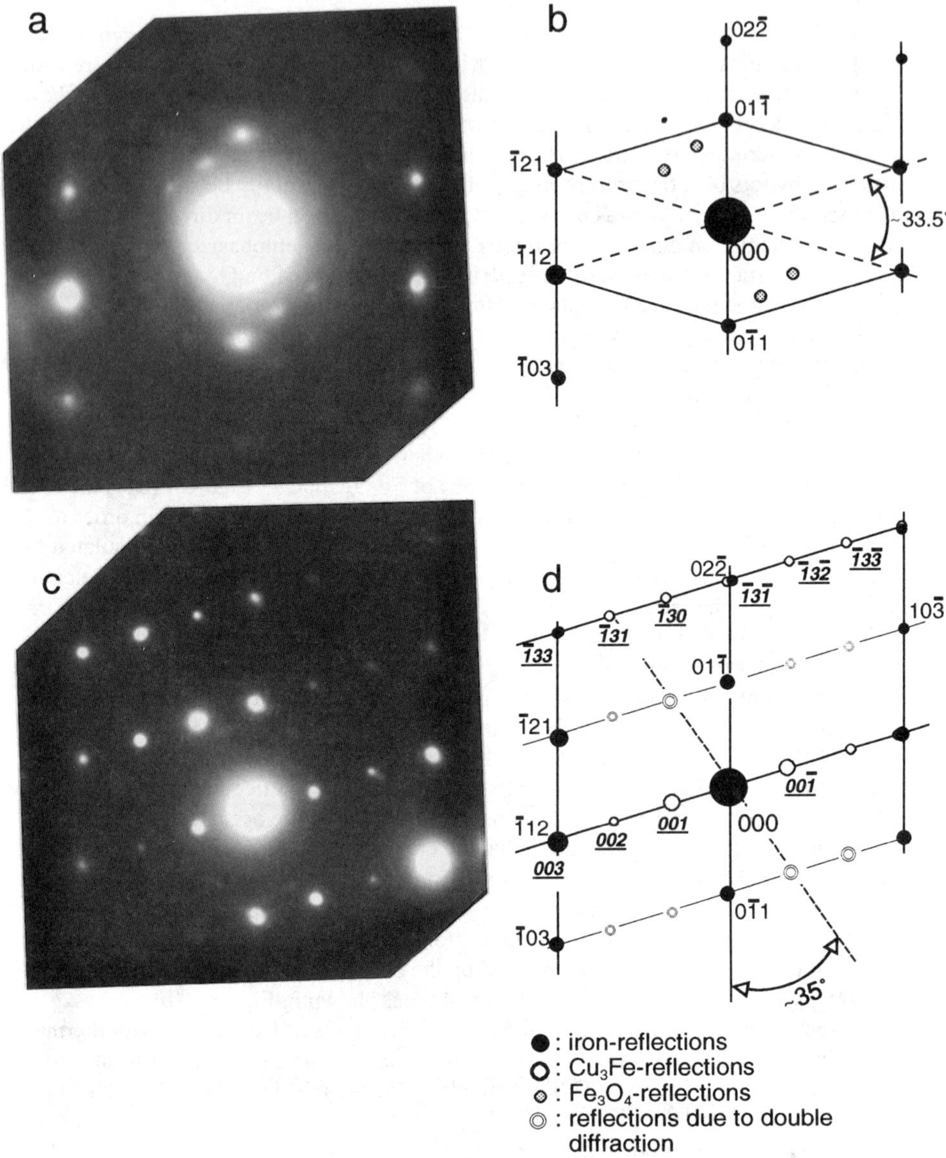

Figure 4 --- *An example to show how the occurrence of superlattice reflections is sensitive to the direction of the incident electron beam. (a)* $(311)_{Fe}$*-SADP for the specimen annealed at 400°C, and (b) the SADP obtained by tilting the specimen by about 1.2° from the situation of Figure 4 (a) without varying their position, resulting in the occurrence of a* $(310)_{Cu3Fe}$*-pattern. Their indices are shown in (b) and (d) corresponding to (a) and (c), respectively.*

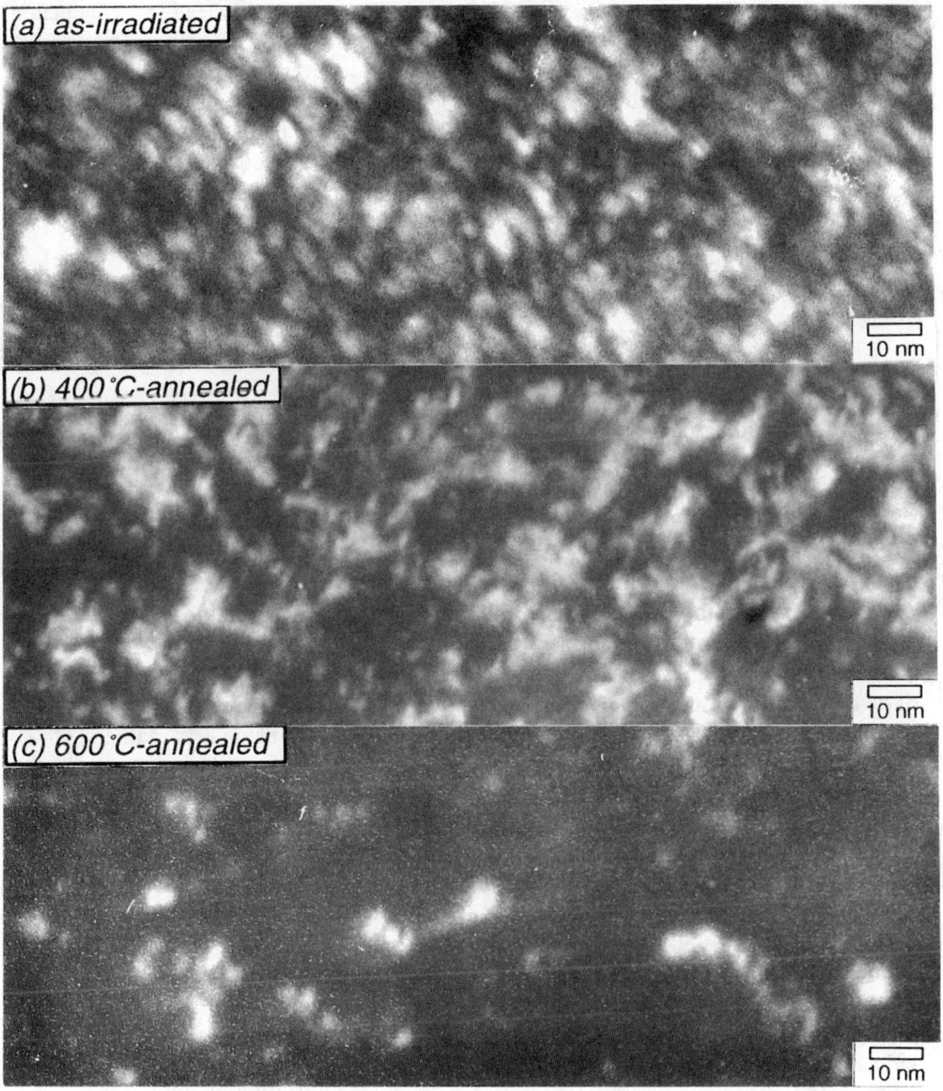

Figure 5 — *A sequence of ordered-structures speculated to be of Cu_3Fe-type with thermal annealing. Images are taken by (a) $(110)_{Cu3Fe}$, (b) $(002)_{Cu3Fe}$ and (c) $(110)_{Cu3Fe}$ reflection in the specimens as-irradiated, annealed at 400°C and annealed at 600°C, respectively.*

because the contrast remarkably varied by a slight change in the electron beam direction. However, it is proven from a series of micrographs in Figure 5 that the Cu_3Fe-zone decreases in size with thermal annealing, and that the intensities of superlattice reflections become lower with temperatures. Thus, the observations reasonably lead to a conclusion that thermal annealing destroyed the ordered-zone formed during neutron irradiations; the order-disorder transformation takes place, resulting in recovery of neutron-induced hardness. On the other hand, information on small copper precipitates, which grow with the annealing, indicates that copper precipitates contribute to hardening. Thus, hardness during thermal annealing is concluded to be determined by two competitive processes of hardening by the growth of small copper precipitates and softening by the dissolution of ordered structures.

Discussion

Contrast and Shape of Copper Precipitates

The shape of precipitates is a key issue to discuss strengthening mechanisms associated with precipitation; the treatments for spherical and plate-shaped precipitates are entirely different. Copper precipitates have been reported to be spherical in previous works [*1,3,5, 15-17*]. In fact, copper precipitates usually show a round shaped contrast in conventional TEM examinations. This does not indicate directly that the precipitate is spherical because a disk-shaped contrast may arise from a thin precipitate lying almost parallel to the specimen surface if the operating reflection (**g**) for imaging is normal to a plane within the precipitate. If we carefully examine the K-S orientation relationship between the matrix and copper precipitate, it is easily understood that there are various copper reflections on or near low order reflections of the matrix like (110) or (002) in a reciprocal lattice space since there are the twelve habit planes of copper precipitate in the matrix, namely, $(112)_{Fe}$ plane. This means that the image of the precipitate readily appears when such a reflection of the matrix is operated, and often makes the interpretation of the contrast complicated. If a thin precipitate exists inclined at some angle to the specimen surface, the possibility of Bragg diffraction by the planes of the precipitate reduces because the respective diffraction planes almost parallel to the direction of the electron beam vary with the depth. Thus, the contrast becomes very faint, and often invisible. This is probably the reason why we rarely identify the elliptical contrast for the precipitate inclined to the specimen surface. Figure 6 is the unusual example of such precipitates by a $(220)_{Fe}$ reflection in Fe-0.6Cu aged at 500°C for 1000 h where the intensity of the reflection from the precipitates is very low and the contrast is mainly produced by strain field around the precipitates, indicating clearly that copper precipitates are of plate-type.

Generally, it is very difficult to characterize the shape of small precipitates having the sizes less than a few nm in diameter. However, the determination of their shape becomes possible occasionally from contrast theory of electron microscopy. Figure 7 is the dark field micrograph taken by a $(1\bar{2}1)_{Fe}$ reflection on the $(135)_{Fe}$-SADP for the specimen thermally annealed at 600°C after the irradiation. Line-of-no-contrast (LNC) is observed in relatively large precipitates denoted A, B and C, but the LNCs as to A and B are not parallel to each other. It has been well known from diffraction theory of electron microscopy that LNC appears when the displacements of crystal lattice due to coherent precipitates are parallel to the reflecting plane of the electron beam. Ashby and Brown

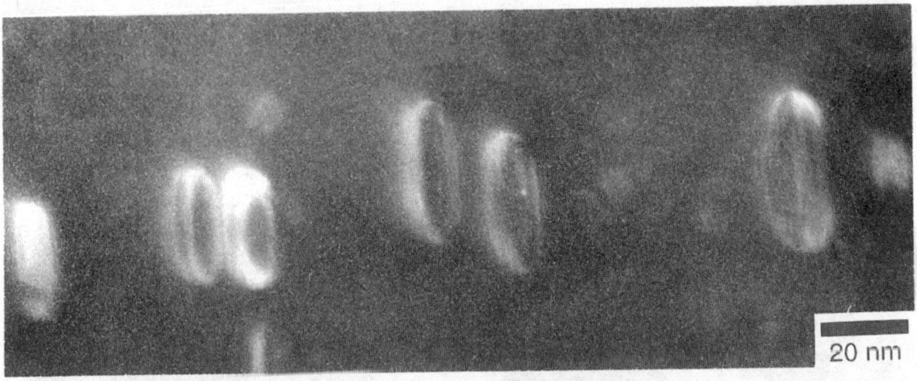

Figure 6 — *Copper precipitates inclined to the specimen surface in Fe-0.6Cu aged at 500°C for 1000 h. The contrast is produced by strain field around the precipitates, indicating clearly that copper precipitate is of plate-type.*

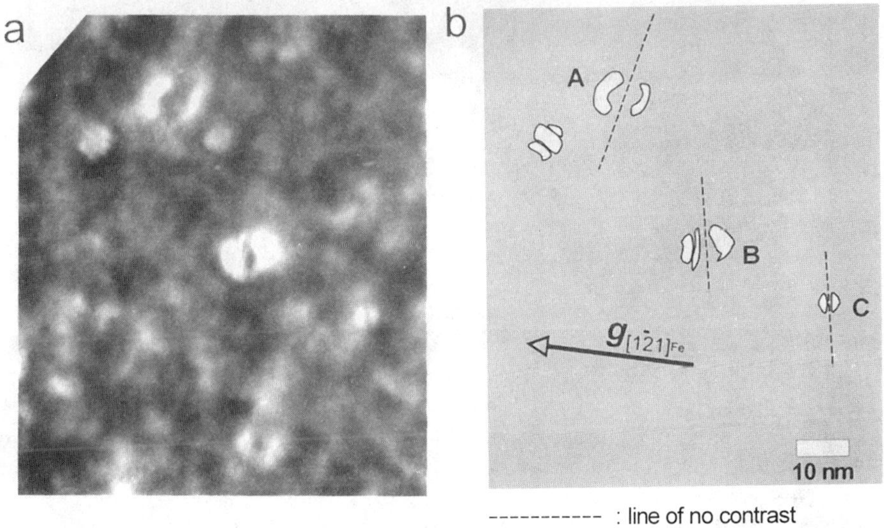

Figure 7 — *Image of copper precipitates with line-of-no-contrast observed in the specimen thermally annealed at 600°C after irradiation. Precipitates of plate-type are evidenced by the fact that two line-no-contrasts (A and B) are neither parallel to each other nor normal to* **g***-vector as sketched in Figure 7 (b).*

have proved that the direction of LNC becomes normal to the **g**-vector operated for their imaging due to symmetrical strain around precipitates/inclusions in the matrix when they are spherical [*18-19*]. When a precipitate is of plate-type, it has been known that LNCs are not always parallel to **g**-vector, and that the direction of LNC varies depending on both ***g***-vector and displacement vector relating to the strain field around precipitates [*20*]. It appears that mathematical analysis have not yet made on the precipitates of plate-type. However, we can paradoxically determine the shape of small precipitates by inquiring the parallelism of two LNCs or the directionality of LNC against ***g***-vector operative in their imaging. If two LNCs are not parallel to each other or there is the LNC being not normal to ***g***-vector, the corresponding precipitate is immediately judged to be of plate-type. Using these criteria, copper precipitates in Figure 7 are easily determined to be of plate-type. It is true that the criteria described above can be applied to the identification of the shape of large sized precipitates, but it is rather difficult practically to select the reflection and the direction of the electron beam for this purpose.

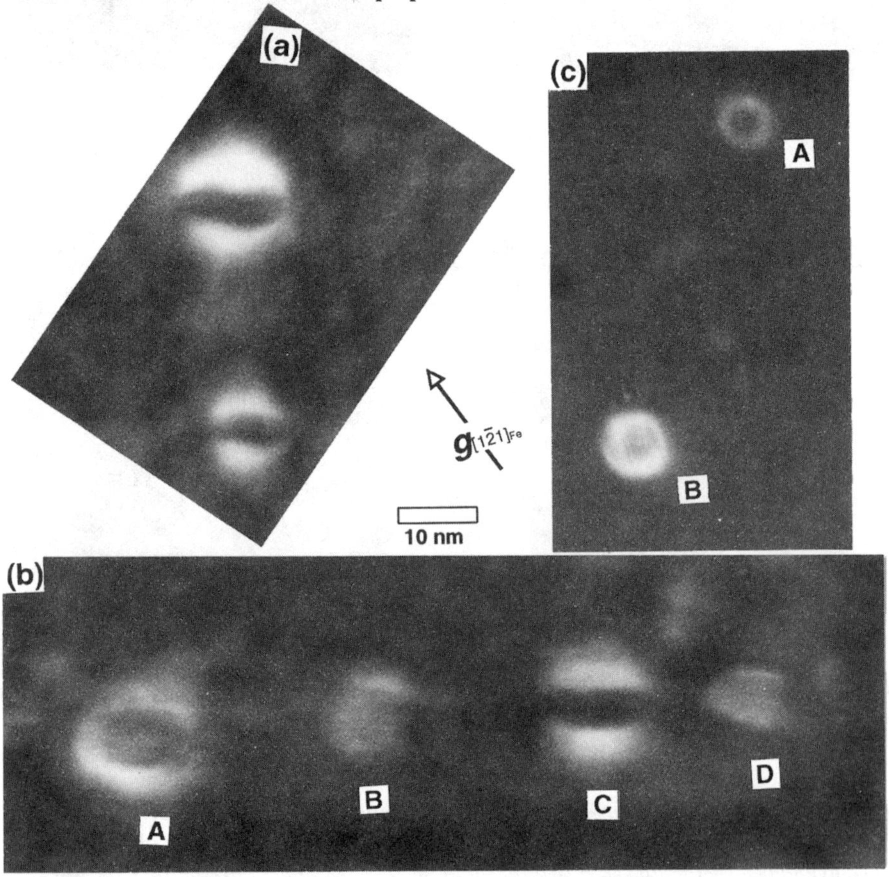

Figure 8 — *Some diffraction contrasts for copper precipitates of plate-type in the specimen thermally annealed at 600°C: (a) Structure factor contrast due to a difference in extinction distance between copper and the matrix, (b) structure factor contrast for the precipitate A and C and precipitate-reflection contrast for those B and D, and (c) ring contrast for A due to the displacement component parallel to the plane of precipitate, similar to prismatic dislocation loop.*

Small copper precipitates can be recognized as white dots (Figure 2) in dark field images using the reflections from copper precipitates. A question might come out; if precipitate is plate-like as described above, a line or string contrast corresponding to the thickness may be observed during TEM observations when the electron beam is parallel to the surface plane of the precipitate. It is the case, but it was unsuccessful in the present study to directly identify such contrasts. The reason for this may be partially due to difficulty to find low intensities of the reflections for small copper precipitates, and partially due to magnetic effects of the specimen. In general, it is rare to regard the images of small copper precipitates using a reflection of the matrix. However, they may appear originating from a relative difference in the extinction distance between the precipitate and the matrix (structure factor contrast) [*21*]. Figures 8 (a) and (b) are the examples of such contrasts taken by a $(1\bar{2}1)_{Fe}$ reflection on the SADP of $(135)_{Fe}$ in the specimen thermally annealed at 600°C. The contrast for the precipitates becomes more visible by the strain field produced around the precipitates and the shapes of two precipitates in Figure 8 (a) are close to rod, and the LNCs are not normal to $\boldsymbol{g}_{[\bar{1}2]Fe}$. Then, these precipitates are understood to be of plate-type and inclined at some angle to the specimen surface by applying the criteria described above. In Figure 8 (b), the contrast for the precipitates denoted A and C is similar to those in Figure 8 (a) while the precipitate A having a wide width is thought to lie on a habit plane differing from the precipitate C, the habit plane of which is steeply inclined to the specimen surface; the contrast for the precipitate A is close to ring contrast described just below. The contrast for the precipitates B and D are thought to occur by the precipitate reflection which exits near or overlaps the $(1\bar{2}1)_{Fe}$ reflection used for their image since they show no contrast of their strain field. Ring contrast similar to a prismatic dislocation loop can be seen in Figure 8 (c), which often takes place by a matrix reflection. Ring contrast is the characteristic of the plate-shaped precipitate since the contrast is caused by the secondary displacements parallel to the precipitate plane; plate-shaped precipitates can be regarded roughly as a dislocation loop. In contrast, no ring contrast generally appears in spherical precipitates since they have the radial and isotropic displacements around the precipitates/inclusions [*18-19,21*].

Figure 9 is the example showing banded structure for the copper precipitate in the unirradiated specimen (aged at 500°C for 1000 h). The image is taken by a $(1\bar{1}1)_{Cu}$ reflection on the SADP of the superimposed $(111)_{Fe}$ and a couple of $(110)_{Cu}$ patterns. The SADP corresponding to the copper precipitate in Figure 9 is analyzed to a (110) pattern of a fcc crystal structure. The interface of the precipitate is parallel to the $[1\bar{1}2]_{Cu}$ direction, which is in agreement with the axis of twin in a fcc crystal. The banded structure of this type can be seen in the copper precipitates as small as ≥2 nm in size in both the aged and irradiated specimens. Copper precipitates are usually observed as a disk without bands. This is due to the fact that the image is produced by a couple of copper reflections diffracted at the adjacent twin bands within a precipitate. In Figure 9, only one side of the adjacent twin bands appears by operating a single reflection of the precipitate. It has been postulated through the studies using a HREM technique [*12-14*] that the banded structure is due to a twinned 9R structure. The question is to identify correctly whether the diffraction pattern corresponding to copper precipitate is due to a fcc Cu or a twinned 9R structure of a bcc. We like to suggest that the banded structure is due to growth twin since the diffraction pattern is analyzed to be due to a fcc structure. The details regarding the analysis of the SADP are described elsewhere.

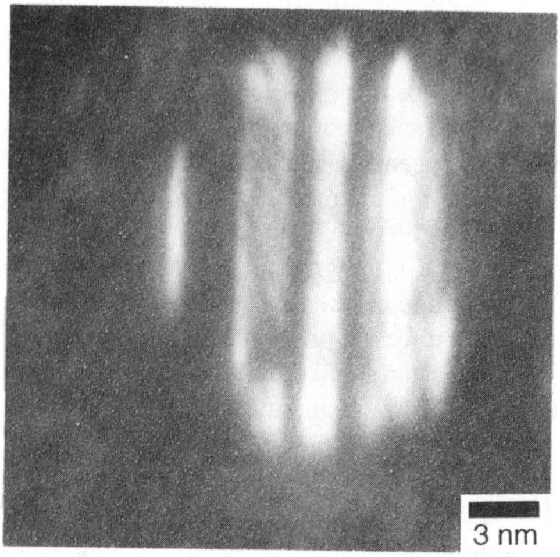

Figure 9 -- *Dark field image of the copper precipitate by a* $(1\bar{1}1)_{Cu}$ *reflection in the unirradiated specimen. Banded structure is suggested to arise from development of growth twin with (111) twin plane.*

The Orientation Relationship between Ordered Structure and the Matrix

We have proposed in our previous paper [10] that there is a orientation relationship (OR) between the ordered structure and the matrix, namely,

$$(011)_{Fe} // (111)_{Cu3Fe} \text{ and } [100]_{Fe} // [\bar{1}\bar{1}2]_{Cu3Fe} \text{ - - - - - - - - - - - - - OR1.}$$

On the other hand, the OR deduced from Figures 4 (c) and (d) is expressed as

$$(311)_{Fe} // (310)_{Cu3Fe}, \text{ and } [01\bar{1}]_{Fe} // [\bar{1}3\bar{1}]_{Cu3Fe} \text{ (or } [\bar{1}12]_{Fe} // [100]_{Cu3Fe} \text{ - - OR2.}$$

At first sight, two ORs seem to be disparate. However, it is revealed from considerations of crystal geometry that OR1 is equivalent to OR2. Figure 10 is the stereographic projection for two cubic crystals on (011) and ($\underline{\mathit{111}}$), where (hkl) and ($\underline{\mathit{hkl}}$) signify the poles of the matrix and the ordered structure, respectively. The [$\bar{1}$00] pole is placed so as to coincide with the [$\underline{\bar{1}\bar{1}2}$] pole in Figure 10; this procedure realizes the OR1. In Figure 10, the ($\bar{2}$11) pole completely coincides with the ($\underline{\mathit{001}}$) pole, while the poles of (101) and (13$\bar{1}$) are situated near to those of ($\underline{31\bar{1}}$) and ($\underline{\bar{1}30}$), respectively (refer to three circles). Calculations shows that the (101) pole makes an angle of ~0.8° with the ($\underline{31\bar{1}}$) pole, and that the angle between the poles of (13$\bar{1}$) and ($\underline{\bar{1}30}$) is ~4.9°. This means that the $(131)_{Fe}$-reciprocal lattice plane is situated so as to make the angle of ~4.9° with that of $(130)_{Cu3Fe}$. This explains that the occurrence of $(310)_{Cu3Fe}$-diffraction pattern on a $(311)_{Fe}$-SADP is very sensitive to the direction of the electron beam as described above. Then, the OR2 should be modified as follows;

$(311)_{Fe}$ is nearly parallel to $(310)_{Cu3Fe}$, and $[01\bar{1}]_{Fe}$ is nearly parallel to $[\bar{1}3\bar{1}]_{Cu3Fe}$

Two points can be reduced from the present considerations. One is that the SADPs are properly analyzed, and the other that the ordered structure is of a cubic structure.

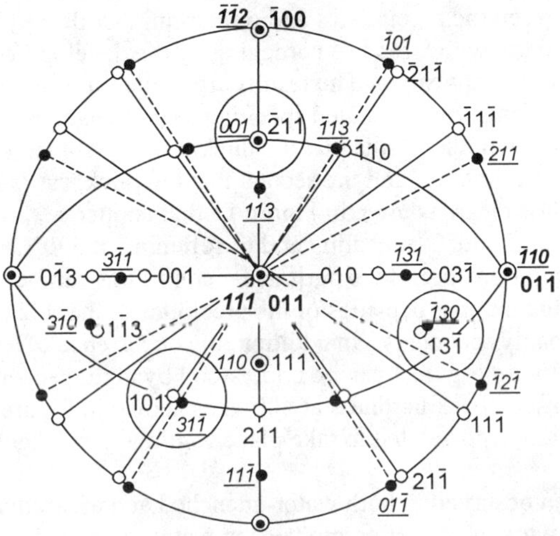

Figure 10 --- *Stereographic projection to show the orientation relationship between the matrix (hkl) and ordered structure ($\underline{hkl}$). The diagram consists of {(011) + ($\underline{111}$)} projection for cubic crystals where $[\bar{1}00]$ // $[\underline{\bar{1}\bar{1}2}]$. The drawing indicates clearly that the OR2 (marked by circles) deduced from Figs. 4 (c) and (d) is equivalent to the OR1 proposed previously [10].*

● ($\underline{hkl}$) : pole for ordered structure
○ (hkl) : pole for the matrix

Hardening due to Copper Precipitates During Thermal Annealing

It is surprising that softening takes place as soon as thermal annealing starts as shown in Figure 1, and that both size and number density increase at a low temperature less than the irradiation temperature in Figure 3. One possible explanation for the results is as follows: In case of thermal aging before irradiation, copper atoms appear to flow into G.P.-zone (probably ~100 % Cu) at the initial stage of the aging, resulting in hardening without the notable growth of copper precipitates until overaging takes place [*10-11*]. When the specimens are neutron-irradiated, it is speculated that a high concentration of vacancies produced in the matrix are thought to enhance diffusion of copper and iron atoms, resulting in the formation of a stoichiometric compound (Cu_3Fe) possibly inside G.P.-zone. On the other hand, it is assumed that the decomposition of Cu_3Fe releases free copper atoms in the matrix, resulting in their growth of copper precipitates. Though the question is to discuss the stability of stoichiometric compound of Cu_3Fe in relation to the vacancy concentration, this remains to be investigated further.

Chemical hardening mechanisms due to plate-shaped copper precipitates, the habit plane of which is $(112)_{Fe}$, has been considered in Ref [11]. Unfortunately, it was not possible to evaluate precisely the contribution of copper precipitates to hardening during thermal annealing since determining their thickness in the present work was unsuccessful. The hardening ($\Delta\sigma_y$) by plate-shaped precipitates is expressed as the sum of three works; those are relating to the energy increase for disordering the precipitate in the slip plane (*Ep*), the interface increase at the interface (*Es*) and the work by the formation of misfit dislocation (*Ei*). *Ep* is the most dominant component to rule the hardening, i.e., $\Delta\sigma_y \propto Ep$ roughly, (refer to Equation 2 and Table 2 in Ref [11]). Therefore, $\Delta\sigma_y$ is considered to be roughly proportional to t/L ($\approx t\cdot\rho^{1/3}$) since $\rho L^3 \approx 1$, where t is the thickness of the precipitates, L the mean distance between the precipitates and ρ their number density. A rough estimation was done in order to know the trend of hardening by the development of small copper precipitates with the thermal annealing. The results are plotted in Figure 11 where data are normalized as $t\cdot\rho^{1/3} = 1$ for the as-irradiated specimens. Two cases are examined since their thickness is unknown; the first is that the thickness is assumed to be 1/10 of the averaged diameter of the precipitates, and the second that the thickness is kept constant during thermal annealing. It is clearly shown in Figure 11 that copper precipitates contribute to hardening until 400°C-annealing and to softening at 600°C-annealing. Furthermore, the hardening in the present experimental conditions can be understood to primarily be ruled by the number densities of the precipitates; the role of the increase in their thickness is probably secondary. In addition, the occurrence of a plateau of hardness curve at 300-400°C in Figure 1 can be understood by taking account of the fact that those precipitates exhibit a peak hardness at 400°C as shown in Figure 11 while the order-disorder transformation is postulated to take place continuously (Figure 12).

Small copper precipitates have been observed in both water-quenched and as-irradiated Fe-Cu model alloys, suggesting that parts of copper precipitates in water-quenched specimens remain during neutron irradiation. In order to examine the influence of the cooling rate during quenching, thin plate (0.15 mm in thickness, the same alloys as those used in the previous studies [*10-11*]) was annealed at 850°C for 2 h and quenched into

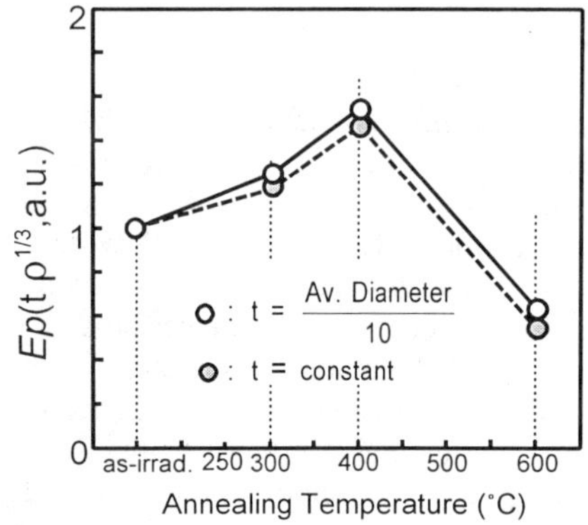

Figure 11 -- *A rough estimation of hardening by small copper precipitates with thermal annealing where their thickness is assumed twofold, namely, 1/10 of the diameter and constant. Data are normalized as* $t\cdot\rho^{1/3} = 1$ *for as-irradiated specimens.*

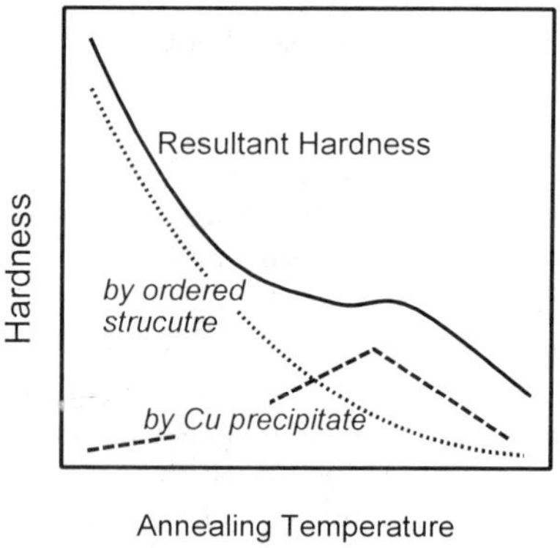

Figure 12 -- *A schematic illustration to interpret the occurrence of a plateau as to hardness curve around 300-400°C in Figure 1 with thermal annealing.*

ice- brine. TEM observations of these specimens showed no evidence of copper precipitates, supporting a hypothesis that copper precipitates observed in the previous studies are formed probably due to their slow cooling rate during water-quenching for bulk specimens (e.g., 3 mm$^{\phi}$-rod).

Present and Previous Data

In the present experiment, the occurrence of (100) reflections has been confirmed (Figure 4). This strongly manifests the formation of ordered structures by neutron irradiation since the reflections of (100)-type do not occur in a simple bcc or fcc crystal. Numerous works [*1-3,5,9*] have reported that spherical or three-dimensional copper-rich precipitates induce irradiation hardening. Though it was impossible to determine the compositions of small precipitates observed in the present study, it appears reasonable that copper precipitates are fcc copper termed ϵ phase (100 %-copper) in binary phase diagrams of Fe-Cu system. Judging from their sizes (< a few nm) reported, their copper-rich precipitates are thought to be the same as copper precipitates observed in our study (Figure 2). If so, the mechanisms proposed regarding irradiation hardening are not in agreement with ours; our data show that the formation of such precipitates is not the dominant mechanism for irradiation hardening because the recovery of hardness takes place with an increase in sizes and number densities of copper precipitates as shown in Figures 2 and 3.

If it is allowed to assume that copper-rich precipitates are identical with the copper precipitates in the present study, it might be beneficial to consider a simple question as to why both results do not coincide. Copper-rich precipitates have been reported using many up-to-date techniques such as SANS and APFIM, besides TEM. Their works have been mainly done on commercial RPV steels while our data are obtained from a binary Fe-Cu system. This may result in different results, but this possibility appears less since their works also addressed Fe-Cu binary alloys in addition to RPV steels, and there are no discrepancy regarding the results between the binary alloys and RPV steels [e.g., *22*]. We have described above that three kinds of precipitations take place in neutron-irradiated Fe-Cu alloys; those are copper precipitates, ordered structures of Cu_3Fe and magnetites (Fe_3O_4). In SANS studies, experimental data have been analyzed by assuming the presence of copper precipitates alone; the analysis, if possible, is desirable, in which the precipitation of Cu_3Fe and Fe_3O_4 are taken into account.

Tomographic APFIM study is a powerful approach since the positions of each atom can be exhibited three-dimensionally. APFIM has essentially the spatial resolution, e.g., $\leq$0.5 nm to x- and y-directions, and $\leq$0.2 nm to z-direction, depending on the respective machines. This means in a three-dimensional representation that each atom may be displayed at the position away within the resolution from their original position in a crystal; precipitates might be seen as spread or swelled, and recognized as a sphere or an ellipsoid even if they are of plate-type. The spatial resolution is atomic especially in the z-direction on many machines. However, if the precipitate is inclined at some angle to the specimen surface and we consider the shift of the atoms within the x-y plane, the precipitate looks swelled in some degree. This effect might strongly influence the results in case of fine precipitates less than 1-2 nm in size.

In addition, it may be reasonable to speculate that atoms (e.g., Fe in Fe-Cu alloys or Fe, Mn, and Ni in RPV steels) around copper precipitate may simultaneously enter the inside of the precipitate in a three-dimensional representation. This suggests that copper precipitates may have the compositions like Cu_XFe_Y in Fe-Cu alloys or $Cu_XMn_YNi_Z$ in RPV steels. If the thickness of the precipitate becomes much larger than the spatial resolution, the concentration of copper in copper precipitates must increase; data shown in Table 5 in Ref [10] can be understood by considering the effects described above.

Another issue is to know how information obtained from Fe-Cu model alloys can be applied to the cases of neutron-irradiated commercial RPV steels like A 533B steel. The studies on microstructural evolutions of these steels during thermal annealing are still limited. Pareige et al. [3] have reported on French Chooz A steels that neutron-induced Cu-Ni-Mn-Si clusters are "dissolved" during annealing while copper precipitates form and develop. Their description as to copper precipitates is in agreement with the present TEM observations. It is of interest to examine whether the clusters have ordered structure or not; further examinations are required for this purpose.

The hardening by neutron irradiations of overaged Fe-Cu alloys has been reported to be low, compared to those as-quenched in the previous work (Figure 1 in Ref [9]), indicating that pre-existing copper precipitates prevent the formation of ordered structures in the matrix. Present data indicate that the precipitation of copper takes place and develops during thermal annealing. This suggests the possibility that the degree of hardening may become lower by re-irradiation of Fe-Cu alloys as well as steels bearing large copper precipitates since copper precipitates, which develop with thermal annealing, are considered to delay the disorder-order transformation in the matrix.

Conclusions

The following conclusions are introduced by Vickers hardness measurements and TEM examinations of Fe-0.6 wt.% Cu alloys irradiated at 290°C to 0.0055 dpa with thermal annealing.

1. Small copper precipitates grow with thermal annealing.
2. Copper precipitates are plate-shaped.
3. Irradiation-induced ordered structures are dissolved with thermal annealing.
4. Ordered structures are postulated to be Cu_3Fe ($L1_2$-type).
5. Recovery of hardness during thermal annealing is determined by two competitive processes of hardening by development of copper precipitates and softening by order-disorder transformation of ordered structures.

References

[1] Buswell, J. T., Phythian, W. J., McElroy, R. J., Dumbill, S., Ray., P. H. N., Mace, J., and Sinclair, R. N., "Irradiation-Induced Microstructural Changes, and Hardening Mechanisms, in Model PWR Reactor Pressure Vessel Steels," *Journal of Nuclear Materials,* vol. 225, 1995, pp. 196-214.

[2] Miller, M. K., and Russel, K. F., "Atom Probe Characterization of Copper Solubility in the Midland Weld after Neutron Irradiation and Thermal Annealing," *Journal of Nuclear Materials,* vol. 250, 1997, pp. 223-228.

[3] Pareige, P., Stoller, R. E., Russel, K. F., and Miller, M. K., "Atom Probe

Characterization of the Microstructure of Nuclear Pressure Vessel Surveillance Materials after Neutron Irradiation and after Annealing Treatments," *Journal of Nuclear Materials*, vol. 249, 1997, pp. 165-174.

[4] Wirth, B. D., Odette, G. R., Pavinich, W. A., Luca, G. E., and Spooner, S. E., "Small Angle Neutron Scattering Study of Linde 80 RPV Welds," *Effects of Radiation on Material: 18th International Symposium, ASTM STP 1325*, R. K. Nanstad, M.. L. Hamilton, F. A. Garner, and A.S. Kumar, Eds., American Society for Testing and Materials, West Conshohocken, PA, 1999, pp. 30-51.

[5] Buswell, J. T., Little, E. A., and Sinclair, R. N., "An Analysis of Radiation Effects in Model A 533B Pressure Vessel Steels Containing Copper, Phosphorus and Nickel Additions," *Effects of Radiation on Material: 18th International Symposium, ASTM STP 1325*, R. K. Nanstad, M. L. Hamilton, F. A. Garner, and A. S. Kumar, Eds., American Society for Testing and Materials, West Conshohocken, PA,1999, pp. 102-121.

[6] Stoller, R. E., "The Impact of Mobile Point Defect Clusters in a Kinetic Model of Pressure Vessel Embrittlement," *Effects of Radiation on Material: 18th International Symposium, ASTM STP 1325*, R. K. Nanstad, M. L. Hamilton, F. A. Garner, and A. S. Kumar, Eds., American Society for Testing and Materials, West Conshohocken, PA, 1999, pp.14-29.

[7] Nanstad, R. K., McCabe, D. E., and Swain, R. L., "Evaluation of Variability in Material Properties and Chemical Composition for Midland Reactor Weld WF-70," *Effects of Radiation on Material: 18th International Symposium, ASTM STP 1325*, R. K. Nanstad, M. L. Hamilton, F. A. Garner, and A. S. Kumar, Eds., American Society for Testing and Materials, West Conshohocken, PA, 1999, pp. 125-156.

[8] Rosinski, S. T., Hawthorne, J. R., and Rochau, G. E., "Experimental Tests of Irradiation-Anneal-Reirradiation Effects on Mechanical Properties of RPV Plate and Welded Materials," *Effects of Radiation on Material: 18th International Symposium, ASTM STP 1325*, R. K. Nanstad, M. L. Hamilton, F. A. Garner, and A. S. Kumar, Eds., American Society for Testing and Materials, West Conshohocken, PA, 1999, pp. 388-402.

[9] Pareige, P., Auger, P., Welzel, S., Van Duysen, J. -C., and Miloudi, S., "Annealing of a Low Copper Steel: Hardness, SANS, Atom Probe, and Thermoelectric Power Investigations," *Effects of Radiation on Materials: 19th Internation Symposium, ASTM STP 1366*, M. L. Hamilton, A. S. Kumar, S. T. Rosinski and M. L. Grossbeck, Eds., American Society for Testing and Materials, West Conshohocken, PA, 2000, pp. 435-447.

[10] Kawanishi, H., and Suzuki, M, "Microstructure and Hardening in Thermally Aged and Neutron-Irradiated Fe-Cu Model Alloy", *Effects of Radiation on Materials, 18th International Symposium, ASTM STP 1366*, M. L. Hamilton, A. S. Kumar, S. T. Rosinski, and M. L. Grossbeck, Eds., American Society for Testing and Materials, West Conshohocken, West Conshohocken, PA, 2000, pp. 492-515.

[11] Kawanishi, H., and Suzuki, M, "Electron Microscopy of Precipitation and Age Hardening in Fe-0.6Cu Model Alloy," *Effects of Radiation on Material: 18th International Symposium, ASTM STP 1325*, R. K. Nanstad, M. L. Hamilton, F. A. Garner, and A. S. Kumar, Eds., American Society for Testing and Materials, West Conshohocken, PA, 1999, pp. 945-969.

[12] Othen, P. J., Jenkins, M. L., Smith, G. D. W., and Phythian, W. J., " Transmission

Electron Microscope Investigations of the Structure of Copper Precipitates in Thermally-Aged Fe-Cu and Fe-Cu-Ni," Phi. Mag. Lett., vol. 64, 1991, pp.383-391.
[13] Hardouin Duparc, H. A., Doole, R. C., Jenkins, M. L., and Barbu, A., "A High-Resolution Electron Microscopy Study of Copper Precipitation in Fe-1.5 wt.% Cu under Electron Irradiation," *Phil. Mag. Lett.*, vol. 71, 1995, pp. 325-333.
[14] Othen, P. J., Jenkins, M. L., and Smith, G. D. W., "High-Resolution Electron Microscopy Studies of the Structure of Cu Precipitate in α-Fe," *Phil. Mag. A*, vol. 70, 1994, pp. 1-24.
[15] Hornbogen, H., and Glenn, R.C., "A Metallographic Study of Precipitation of Copper from Alpha Iron," *Trans. Met. Soc. AIME*, vol. 218, 1960, pp.1064-1070.
[16] Osamura, K., Okuda, H., Takashima, M., Asano, K., and Furusaka, M., "Small-Angle Neutron Scattering of Phase Decomposition in Fe-Cu Binary Alloy," *Met. Trans, JIM*, vol. 34, 1993, pp. 305-311.
[17] Goodman, S. R., Brenner, S. S., and Low, Jr., J. R., " An FIM-Atom Probe Study of the Precipitation of Copper from Iron-1.4 At. Pct Copper," *Met. Trans*. vol. 4, 1973, pp.2363-2369.
[18] Ashby, M. F. and Brown, L. M.,, "Diffraction Contrast from Spherical Symmetrical Coherency Strains," *Philosophical Magazine*, vol. 8, 1963, pp.1083-1103.
[19] Ashby, M. F. and Brown, L. M., "On Diffraction Contrast from Inclusions," *Phil. Mag.*, vol. 8, 1963, pp.1649-1676.
[20] Ruhle, M., "Study of Small Defect Clusters in Irradiated Materials by Means of Transmission Electron Microscopy," Proc. of the Symposium on Radiation Damage in Reactor Materials, vol. 1, Vienna, 1969, pp. 113-157.
[21] Hirsch, P. B., Howie, A., Nicholson, R. B., and Pashley, D. W., "Electron Microscopy of Thin Crystals," Butterworth, London 1965, pp. 336-348.
[22] Auger, P., Pareige, P., Akamatsu, M. and Blavette, D., "APFIM Investigation of Clustering in Neutron-Irradiated Fe-Cu Alloys and Pressure Vessel Steels," *Journal of Nuclear Materials*, vol. 225, 1995, pp. 225-230.

Ryuta Kasada[1], Takeshi Kitao[1], Kazunori Morishita[1], Hideki Matsui[2], and Akihiko Kimura[1]

Effects of Copper Concentration and Neutron Flux on Irradiation Hardening and Microstructure Evolution in Fe-Cu Model Alloys

---

Reference: Kasada, R., Kitao, T., Morishita, K., Matsui, H., and Kimura, A., **"Effects of Copper Concentration and Neutron Flux on Irradiation Hardening and Microstructure Evolution in Fe-Cu Model Alloys,"** *Effects of Radiation on Materials: 20th International Symposium, ASTM STP 1405,* S. T. Rosinski, M. L. Grossbeck, T. R. Allen, and A. S. Kumar, Eds., American Society for Testing and Materials, West Conshohocken, PA, 2001.

Abstract: Irradiation hardening and microstructure evolution under neutron irradiation have been investigated for pure Fe, Fe-0.15Cu, Fe-0.28Cu and Fe-0.46Cu alloys. All the alloys were annealed at 780°C for 20 min and quenched in iced water. Neutron irradiations were performed in the Japan Material Test Reactor (JMTR) up to a fluence of about 1×10^{22}n/m^2 at different fluxes (3.0×10^{16} and 1.5×10^{15} n/m^{2}s) at 290°C utilizing a so-called "multi-division temperature control irradiation rig" to investigate flux effects under controlled irradiation temperature and flux. The irradiation hardening increased with increasing copper concentration. Positron annihilation lifetime spectrometry (PAS) revealed that the second lifetime component (τ_2) was observed only in pure iron and Fe-0.15Cu alloy irradiated at the low flux condition. As for the flux effect, the irradiation hardening was larger at the lower flux condition in all the model alloys and A533B steel. Post-irradiation annealing experiments indicated that there were two recovery stages: the first was above 350°C and the second was around 550°C. The amount of hardening-recovery in the first stage decreased with increasing copper concentration and depended on the flux; a larger recovery was observed at the lower flux condition. In contrast, the amount of recovery in the second stage increased with increasing copper concentration and was independent of the flux. It is considered that the first recovery is related to the annealing out-of-matrix defects and the second one is due to dissolution of copper precipitates. The τ_2 of PAS disappeared after the annealing at 350°C, indicating that microvoids decomposed during the annealing.

Keywords: irradiation hardening, flux effects, hardening-recovery, Fe-Cu alloys, A533B

---

[1]Graduate students, associate professor and professor, respectively, Institute of Advanced Energy, Kyoto University, Gokasho, Uji, Kyoto 611-0011, Japan.

[2]Professor, Institute for Materials Research, Tohoku University, Katahira, Sendai 980-0812, Japan

Introduction

The lifetime extension of nuclear plants has been dictated so as to supply steady electric power in the future. Soundness of reactor pressure vessel steel (RPVS) under neutron irradiation for the long term has been considered to be one of the critical issues for the extended operation of the plant, since the RPVS has been known to suffer irradiation embrittlement. The RPVS of early plants often contains amounts of copper as an impurity that enhances the irradiation embrittlement by the formation of copper-rich precipitates (CRP) [*1-4*]. Precise evaluation of irradiation embrittlement in such a copper-containing RPVS is essential to guarantee the safe operation of the early plant.

Many experimental and theoretical studies have been done so far to establish the evaluation and simulation method of neutron irradiation embrittlement of RPVS, based on the results obtained with experimental reactors, accelerators and high voltage electron microscopes [*5-10*]. Since the irradiation-induced copper-precipitation behavior can be influenced by damage rate, it has been strongly recommended to investigate the damage rate effects on the irradiation embrittlement of copper-containing RPVS.

Recently, a reactor irradiation technique has been improved in irradiation of specimens with controlled irradiation temperature and period. In the Japan Materials Test Reactor (JMTR), a so-called multi-section and multi-division controlled irradiation rig was developed where the specimen temperature was controlled by the heater and a part of specimens was removed from the reactor core position during the reactor operation [*12*].

In this study, the effects of neutron flux on the irradiation hardening of Fe-Cu model alloys as well as A533B RPVS have been investigated in order to understand the mechanism of the effects of damage rate on the irradiation hardening of Fe-Cu alloys utilizing the controlled irradiation rig of JMTR.

Table 1- *Chemical compositions of pure iron and Fe-Cu alloys.*

Chemical Compositions (wt.ppm) *wt.%

Material	Cu*	C	O	N
Pure Fe	0.002	1	26	2
Fe-0.15Cu	0.154	2	27	2
Fe-0.28Cu	0.280	2	15	2
Fe-0.46Cu	0.464	4	23	2

Experimental Procedure

The materials used in this study were pure Fe, Fe-0.15Cu, Fe-0.28Cu and Fe-0.46Cu model alloys for RPVS. The chemical compositions of the alloys are shown in Table 1. These alloys were produced from high purity materials by arc-melting method in high purity Ar gas, resulting the concentration of interstitial impurity that was less than a few wt. ppm for carbon and nitrogen and less than 27 wt. ppm for oxygen. The alloy ingots were cold-rolled to 0.25 mm thickness of sheets, and then punched out into 3 mm diameter disk specimens and 4 × 16 mm sub-sized tensile specimens. The specimens were finally annealed at 780°C for 20 min, followed by quenching in iced water. A533B steel, which contained 0.03 wt. % of Cu, was also used in this study. The details of chemical compositions and heat treatment for the steel were described in this proceeding.

Neutron irradiations were performed in the Japan Material Test Reactor (JMTR) at 290°C up to a dose of 1.05×10^{22} n/m^2 (>1 MeV) with neutron flux of 3.0×10^{16} n/m^{2}s (HDR; higher dose rate), and to a dose of 9.3×10^{21} n/m^2 (>1 MeV) with neutron flux of 1.5×10^{15} n/m^{2}s (LDR; lower dose rate), using a so-called multi-division temperature control irradiation rig that enabled part of the specimens removed during reactor operation so as to investigate flux effects under the controlled irradiation temperature. These fluxes and the irradiation period correspond to displacement dose rates of 4.5×10^{-9} dpa/s (97.4hr) and 2.2×10^{-10} dpa/s (1728hr), respectively. The irradiation conditions are summarized in Table 2.

Tensile tests were carried out at a cross-head speed of 0.2 mm/min at room temperature. Micro-Vickers hardness tests with a load of 0.2 kg and positron annihilation lifetime spectrometry (PAS) were carried out at room temperature, following isochronal (30min) annealing from 350°C to 650°C in 50°C steps in a vacuum of 1.0×10^{-5} Pa.

Table 2- *Irradiation conditions of higher dose rate (HDR) and lower dose rate (LDR) irradiations.*

	Higher Dose Rate (HDR)	Lower Dose Rate (LDR)	(units)
Temperature	290	290	°C
Dose (E>1 MeV)	1.05×10^{22}	9.3×10^{21}	n/m^2
Dose (dpa)	1.58×10^{-3}	1.40×10^{-3}	dpa
Irradiation period	97.4	1728	hour
Dose rate (E>1 MeV)	3.0×10^{16}	1.5×10^{15}	n/m^{2}s
Dose rate (dpa)	4.5×10^{-9}	2.2×10^{-10}	dpa/s

Results

Irradiation Effects

The effects of neutron irradiation at different neutron fluxes• on the irradiation hardening of Fe-Cu model alloys are shown in Fig.1. The yield stress of Fe-Cu alloys increased with copper concentration before and after the irradiation (Fig.1(a)) and the irradiation hardening significantly increased with increasing copper concentration. The effect of neutron flux on irradiation hardening was observed in all the alloys, that is, the LDR irradiation induced a larger hardening than the HDR one. The same tendency was also recognized for A533B, though the amount of hardening was much smaller than in the model alloys. It appears that the effect of neutron flux is independent of copper concentration, which will be discussed later.

PAS results of two-component analysis are shown in Fig. 2. A longer lifetime component that indicates formation of microvoids was only observed in pure Fe and Fe-0.15Cu alloy irradiated at LDR condition. Since the PA spectra of Fe-0.28 and 0.46Cu alloys could not be separated into two components, the average lifetime, τ_m, was shown in the figure. As shown in the figure, the lifetime of longer component decreased with increasing copper concentration, suggesting that copper atoms suppressed growth of microvoids. At HDR condition, no long lifetime component or microvoid was observed in any of the alloys. It is expected that the over-sized copper atoms in the iron lattice trap vacancies and prevent them from aggregating into microvoids. As for the A533B steel, it appears that no microvoid is formed in either of the irradiation conditions.

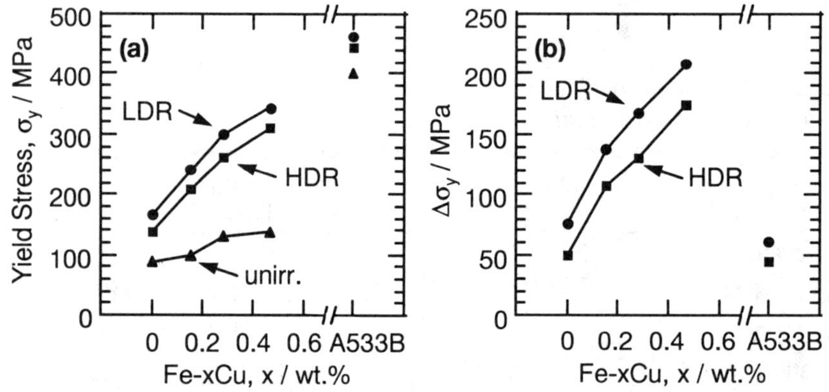

Fig. 1- *Tensile test results for unirradiated, HDR and LDR irradiated Fe-Cu alloys and A533B; (a) yield stress and (b) increase in yield stress after irradiation.*

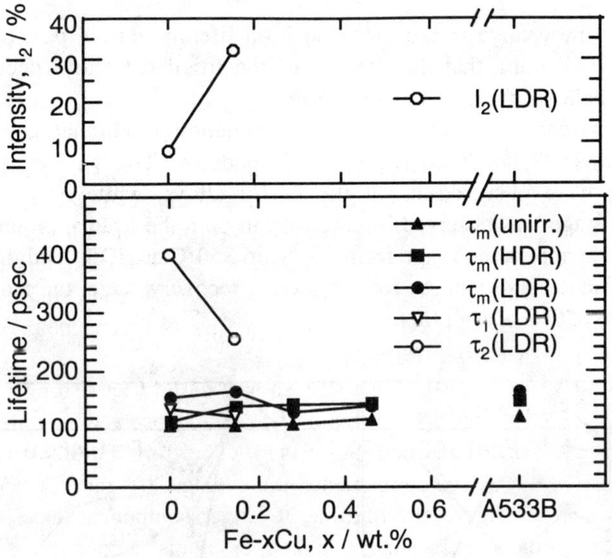

Fig. 2- *PAS results for unirradiated, HDR and LDR irradiated Fe-Cu alloys and A533B.*

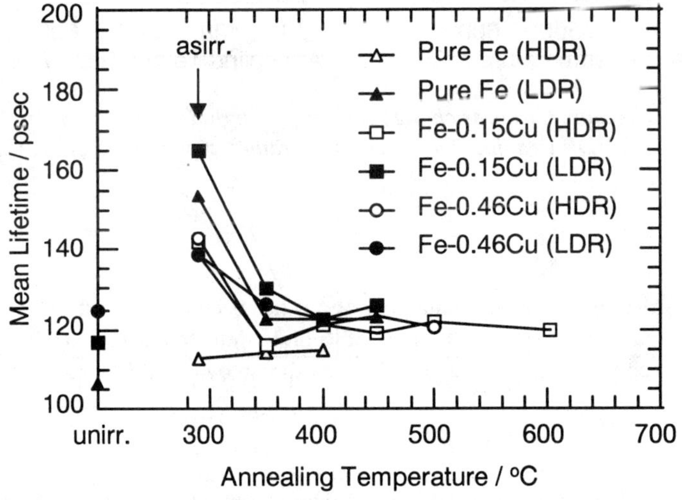

Fig. 3- *Positron lifetime recovery behavior during post-irradiation isochronal annealing, in the HDR and LDR irradiated pure Fe, Fe-0.15Cu and Fe-0.46Cu.*

Post-irradiation isochronal annealing behavior

Figure 3 shows the change in the mean positron lifetime during post-irradiation isochronal annealing, indicating that the lifetime of the irradiated alloys decreased to values for unirradiated alloys after annealing to 350°C.

Micro-Vickers hardness change during post-irradiation isochronal annealing is shown in Fig. 4 for a) HDR condition and b) LDR condition. Two recovery stages of irradiation hardening were observed in all the Fe-Cu alloys, while only the lower temperature recovery stage was observed in pure iron in both the irradiation conditions. The first stage is in the temperature range from 300°C to 350°C at HDR condition, while it is above 350°C at LDR condition. As for the second recovery stage, the temperature range is from 550°C to 650°C of both fluxes.

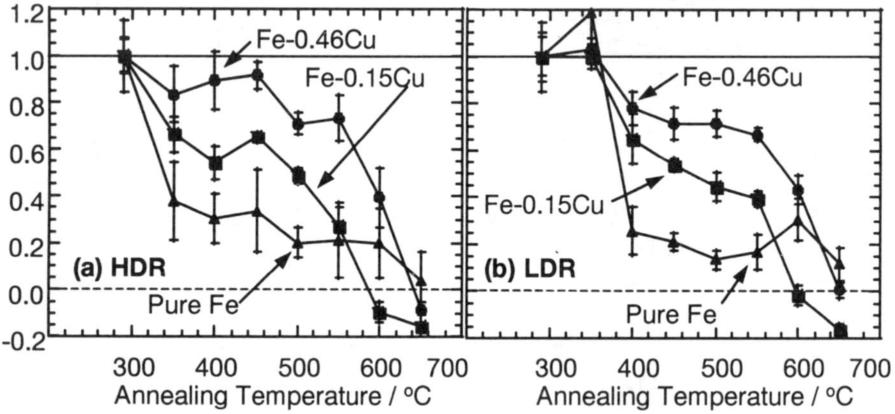

Fig. 4- *Normalized Vickers hardness change during post-irradiation isochronal annealing: (a) HDR irradiated and (b) LDR irradiated pure Fe, Fe-0.15Cu and Fe-0.46Cu.*

Discussion

Irradiation hardening of copper-containing alloys has been considered to consist of two components; namely, matrix damage of point defect clusters (PDC) and irradiation-enhanced copper-rich precipitate (CRP), as shown by the following:

$$\Delta\sigma_{tot(t)} = \Delta\sigma_{PDC(t)} + \Delta\sigma_{CRP(t)}, \tag{1}$$

where $\Delta\sigma_{tot(t)}$, $\Delta\sigma_{PDC(t)}$ and $\Delta\sigma_{CRP(t)}$ are the total hardening, hardening by PDC, hardening by CRP, at an irradiation time of t, respectively[*1*]. Furthermore, $\Delta\sigma_{PDC(t)}$ can be separated into two components; small interstitial loops and microvoids. According to our previous study on recovery behavior of proton irradiation-induced hardening in Fe-Cu model alloys, three recovery stages were observed during isochronal annealing from 80 to 650°C. The second and third stage was observed at around 350 and 600°C, respectively, which were attributed to annealing out of matrix damage and resolution of copper precipitates, respectively [*5*]. Figure 4 indicates that almost the same behavior was observed for neutron irradiated Fe-Cu alloys, and the first and second stage in Figure 4 corresponds to the second and third stage in proton-irradiated Fe-Cu alloys. In proton-irradiated alloys, the first recovery stage appeared at around 200°C, since the irradiation temperature was 60°C [*10*]. In order to estimate the contribution of each component to irradiation hardening in neutron irradiated Fe-Cu alloys, the amount of the hardening recovered at each stage was evaluated from Fig. 4 and shown in Fig. 5, with the assumption that the lower and higher temperature recovery stages of irradiation hardening are due to annealing out of PDC and resolution of CRP, respectively [*8*].

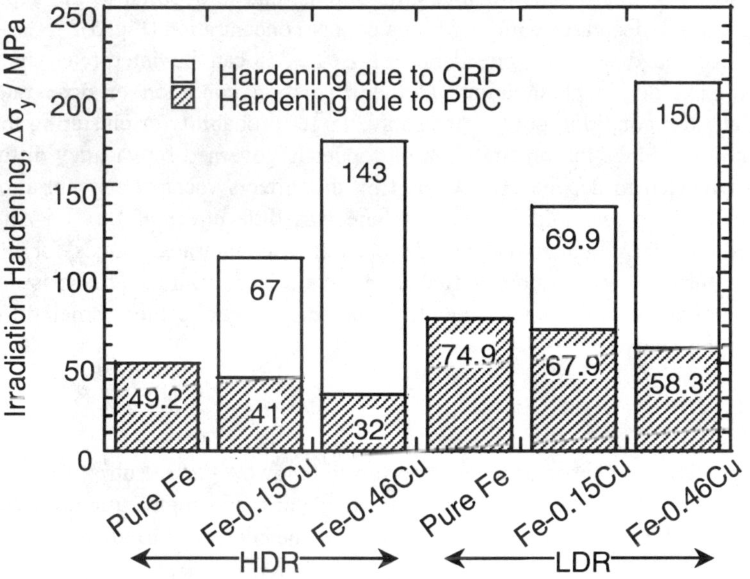

Fig. 5- *Contribution to irradiation hardening of each hardening, Δf_{PDC} and Δf_{CRP} evaluated from Fig. 4.*

From Fig. 5, it is suggested that:

1) Irradiation hardening by PDC significantly depends on dose rate and copper concentration. The hardening is larger at LDR condition and decrease with increasing copper concentration.
2) Irradiation hardening by CRP appears to be independent of dose rate, and increases with increasing copper concentration.

In addition, the contribution of microvoids to irradiation is considered to be very small, because the density and size of microvoids calculated from PAS results is not enough to explain the hardening and there is no recovery of irradiation hardening at 350°C for LDR condition in spite of occurrence of recovery of microvoids as shown in Fig. 3. Therefore, irradiation hardening by PDC is mainly attributed to small interstitial clusters. Although the contribution of microvoids to irradiation hardening is rather small, formation of interstitial clusters is affected by microvoids as well as vacancies. In both the irradiation conditions, $\Delta\sigma_{PDC}$ decreases with increasing copper concentration, which is interpreted as follows. Since copper atoms trap vacancies, as suggested by Fig. 2, and cause recombination with interstitial atoms, the increase in copper concentration increases the density of recombination sites that reduces the number of interstitial atoms to form the dislocation loops. Although it is possible that the aggregation of copper around microvoids increases the density of interstitial atoms because of decreasing in sink efficiency of microvoids for interstitial atoms, that is opposed to the experimental results that $\Delta\sigma_{PDC}$ decreases with increasing copper concentration (Fig. 5).

As for the dose rate dependence of $\Delta\sigma_{PDC}$, it can be interpreted in terms of "recombination dominant situation" at the irradiation condition or dose rate region. Namely, at LDR condition point defects have larger probability of clustering than in the HDR condition. Since the precipitation of copper is governed by vacancy diffusion [*9*], $\Delta\sigma_{CRP}$ is expected to depend on neutron flux that affects vacancy concentration during irradiation [*10*]. In this study, however, there was little effect of flux on $\Delta\sigma_{CRP}$. It is considered that the obtained values of $\Delta\sigma_{CRP}$ are the saturated values for the alloys. Therefore, further investigation of fluence effects as well as flux effects is demanded. Also, it is possible that the weak dependence of $\Delta\sigma_{CRP}$ might be due to small difference in the flux in this study.

Conclusions

The effects of copper concentration (up to 0.46wt.%) and neutron flux (3.0×10^{16} n/m^{2}s and 1.5×10^{15} n/m^{2}s) on irradiation hardening and microstructure evolution in Fe-Cu model alloys were investigated by utilizing the controlled irradiation rig of JMTR. Obtained results are:

1) Irradiation hardening in Fe-Cu model alloys significantly increased with copper concentration, though the yield stress of unirradiated alloys depended weakly on the copper concentration.
2) Irradiation hardening in LDR condition was larger than that in HDR condition. The effect of neutron flux on irradiation hardening was due mainly to the difference in the contribution of matrix defects to the hardening. No effect of neutron flux on precipitation behavior of copper was observed.
3) Dose rate dependence of irradiation hardening was also observed in the A533B steel, although the difference is smaller than for the model alloys.

Acknowledgement

The authors would like to thank Inoue Foundation for Science for the financial support to participate this conference. They are much obliged to the staff of RI facility in I. M. R. of Tohoku University for their help in dealing with the activated specimens.

References

[1] Fisher, S. B., Harbottle, J. E., and Aldridge, N., "Radiation Hardening in Magnox Pressure-Vessel Steels," *Philosophical Transactions of the Royal Society of London*, Vol. A315, 1985, pp. 301-332.

[2] Buswell, J. T., Little, E. A., and Sinclair, R. N., "An Analysis of Radiation Effects in Model A533B Pressure Vessel Steels Containing Copper, Phosphorus and Nickel Additions," *Effects of Radiation on Materials: 18th International Symposium, ASTM STP 1325*, R. K. Nanstad, M. L. Hamilton, F. A. Garner, and A. S. Kumar, Eds., American Society for Testing and Materials, West Conshohocken, PA, 1999, pp. 30-51.

[3] Odette, G. R., Lucas, G. E., and Klingensmith, D., "Anomalous Hardening in Model Alloys and Steels Thermally Aged at 290°C and 350°C: Implications to Low Flux Irradiation Embrittlement," *Effects of Radiation on Materials: 18th International Symposium, ASTM STP 1325*, R. K. Nanstad, M. L. Hamilton, F. A. Garner, and A. S. Kumar, Eds., American Society for Testing and Materials, West Conshohocken, PA, 1999, pp. 88-101.

[4] Odette, G. R., Mader, E. V., Lucas, G. E., Phythian, W. J., and English, C. A., "The Effect of Flux on the Irradiation Hardening of Pressure Vessel Steels," *Effects of Radiation on Materials: 16th International Symposium, ASTM STP 1175*, A. S. Kumar, D. S. Gelles, R. K. Nanstad, and E. A. Little, Eds., American Society for Testing and Materials, Philadelphia, PA, 1993, pp. 373-393.

[5] Suzuki, M., Idei, Y., Kizaki, M., and Onizawa, K, "Effects of Damage Rate on Irradiation Hardening and Post-Irradiaition Annealing Characteristics of Carbon Steel," *Effects of Radiation on Materials: 18th International Symposium, ASTM STP 1325*, R. K. Nanstad, M. L. Hamilton, F. A. Garner, and A. S. Kumar, Eds., American Society for Testing and Materials, West Conshohocken, PA, 1999, pp. 379-387.

[6] Dohi, K., Onchi, T., Kano, F., Fukuya, K., Narui, M., and Kayano, H., "Effect of Neutron Flux on Low Temperature Irradiation Embrittlement of Reactor Pressure Vessel Steel," *Journal of Nuclear Materials*, 1999, Vol. 265, pp. 78-90.

[7] Yanagita, S., Xu, Q, Yoshiie, T., and Ino, H., "Neutron Irradiation Rate Dependence of Damage Structure in Fe-Cu Model Alloys," *Effects of Radiation on Materials: 19th International Symposium, ASTM STP 1366*, M. L. Hamilton, A. S. Kumar, S. T. Rosinski, and M. L. Grossbeck, Eds., American Society for Testing and Materials, West Conshohocken, PA, 2000, pp. 516-524.

[8] Mathon, M. H., Barbu, A., Dunstetter, F., Maury, F., Lorenzelli, N., and De Novian, C. H., "Experimental Study and Modeling of Copper Precipitation under Electron Irradiation in Dilute FeCu Binary Alloys," *Journal of Nuclear Materials*, 1997, Vol. 245, pp. 224-237.

[9] Kimura, A., Shibamoto, H., Yuya, H., Hasegawa, M, Yamaguchi, S., and Matsui, H., "Damage Structure of Proton Irradiated Fe-0.3wt.%Cu Alloy," *Effects of Radiation on Materials: 17th International Symposium, ASTM STP 1270*, D. S. Gelles, R. K. Nanstad, A. S. Kumar, and E. A. Little, Eds., American Society for Testing and Materials, West Conshohocken, PA, 1996, pp. 220-231.

[10] Shibamoto, H., Koyama, K., Yuya, H., Hasegawa, M, Kimura, A., Matsui, H., and Yamaguchi, S., "Effects of Proton Irradiation on Positron Annihilation and Micro-Vickers Hardness of Fe-C-Cu Model Alloys," *Effects of Radiation on Materials: 17th International Symposium, ASTM STP 1270*, D. S. Gelles, R. K. Nanstad, A. S. Kumar, and E. A. Little, Eds., American Society for Testing and Materials, West Conshohocken, PA, 1996, pp. 623-636.

[11] Akamatsu, M., Van Duysen, J. C., Pareige, P., and Auger, P., "Experimental Evidence of Several Contribution to the Radiation Damage in Ferritic Alloys," *Journal of Nuclear Materials*, 1995, Vol. 225, pp. 192-195.

[12] Narui, M., Sagawa, T., Endo, Y., Uramoto, T., Shikama, T., Kayano, H., and Kiritani, M., "Development of Experimental Rigs for Temperature-Controlled Reactor Irradiations for Fundamental Study of Radiation Effects on Fusion Materials," *Journal of Nuclear Materials*, 1994, Vol. 212, Issue Part B, pp. 1645-1648.

Stephen E. Cumblidge,[1] Gary L. Catchen,[1] Arthur T. Motta,[1] Gerhard Brauer,[2] and Jürgen Böhmert[3]

Effects of Neutron Irradiation and Thermal Annealing on Model Alloys Using Positron Annihilation Techniques

Reference: Cumblidge, S. E., Catchen, G. L., Motta, A. T., Brauer, G., and Böhmert, J., **"Effects of Neutron Irradiation and Thermal Annealing on Model Alloys using Positron Annihilation Techniques,"** *Effects of Radiation on Materials: 20th International Symposium, ASTM STP 1405,* S. T. Rosinski, M. L. Grossbeck, T. R. Allen, and A. S. Kumar, Eds., American Society for Testing and Materials, West Conshohocken, PA, 2001.

Abstract: We present the results of a systematic investigation of neutron-irradiated and thermally annealed Fe-Cu-Ni-P model alloys using positron annihilation spectroscopy (PAS), including lifetime and Doppler broadening techniques, and Rockwell hardness. These alloys were examined in the as-fabricated state, after irradiation at 270° C to 1 x 10^{19} n.cm^{-2}, and to 8 x 10^{19} n.cm^{-2}, and after successive post-irradiation isochronal anneals at temperatures from 200 to 600° C. The results can be qualitatively explained by invoking an irradiation-induced microstructure consisting of a combination of small dislocation-type defects or defect clusters (matrix damage) and dense precipitation of fine scale irradiation-induced precipitates. The matrix damage anneals between 350° C and 450° C. The irradiation-induced precipitates also evolve with annealing, but at higher temperatures. The combined effect of high Cu and high Ni concentrations leads to more extensive irradiation-induced precipitation than in cases where either element is missing, whereas the effect of P is less pronounced. We analyze and compare the results with similar measurements performed on irradiated pressure-vessel steels and with other positron measurements on model alloys, reported in the literature.

Keywords: positron annihilation, positron lifetime, Doppler broadening, irradiation hardening, model alloys, neutron irradiation, irradiation induced precipitates, matrix damage.

---

[1] Graduate Student, Professor and Associate Professor, respectively, Department of Mechanical and Nuclear Engineering, The Pennsylvania State University, University Park, PA 16802, USA.

[2] Group Leader, Institut für Ionenstrahlphysik und Materialforschung, Forschungzentrum Rossendorf, PF 510119, D-01314 Dresden, Germany.

[3] Group Leader, Institut für Sicherheitsforschung, Forschungzentrum Rossendorf, PF 510119, D-01314 Dresden, Germany.

Introduction

Positron annihilation spectroscopies have been used with some success to help characterize the damage structure that develops under fast-neutron irradiation and that is responsible for pressure-vessel embrittlement [*1-6*]. At issue is the nature of the fine-scale damage (< 10 nm) that forms under irradiation and that causes the increase in the nil-ductility temperature (NDT) and the decrease in the upper-shelf energy (USE) in the Charpy Test [*7*]. A substantial amount of work has been put into this characterization, using techniques such as small-angle neutron scattering (SANS) [*8, 9*], atom probe field ion microscopy (APFIM) [*10, 11*], transmission electron microscopy (TEM) and a variety of others [*12*], in addition to positron annihilation spectroscopies. In such studies, investigators have identified irradiation-induced precipitates (Cu-rich precipitates) and defect clusters (matrix damage) as the likely candidates for causing irradiation embrittlement [*13*]. However, the complex nature of the irradiation-induced precipitation and defect clustering processes in industrial alloys such as A533B steel, in combination with the fine scale of the damage, make it difficult to determine the exact nature of the mechanisms. Positron annihilation spectroscopies, both in lifetime or Doppler broadening experiments, can provide unique information about the nature of the damage. In combination with other techniques, they can be used to help determine the embrittlement mechanism operative in pressure-vessel steels under irradiation.

The use of model alloys in irradiation testing programs can help eliminate some of the uncertainties related to alloy composition [*4, 14*]. By systematically varying the alloying contents, we can determine the separate effects of specific alloying additions, and combinations of additions on irradiation embrittlement. We present the results of an investigation conducted on a series of model alloys, prepared with systematic variations of Cu, P and Ni. We conducted a series of positron-annihilation-lifetime and Doppler-broadening measurements, complemented by Rockwell hardness measurements on several model alloys that were not irradiated and that were irradiated to 1×10^{19} n.cm^{-2} and 8×10^{19} n.cm^{-2} (E > 0.5 MeV) at 270° C. We performed measurements on both the as-irradiated samples and on the same samples after performing successive isochronal anneals, conducted at temperatures ranging from 200 to 600° C. We analyze the results and compare them with similar measurements on irradiated pressure-vessel steels and with other positron measurements on model alloys reported in the literature.

Experimental Methods

Samples Examined

Table 1 gives the compositions of the eight model alloys used in this study [*15*], which represent the averages of three independent spectroscopy and standard chemical analysis measurements. The alloys were austenitized at 980-1000° C for two hours and oil quenched. They were then tempered at 670° C for ten hours, and cooled in air. The samples were originally in the form of Charpy V-notch test specimens, and after testing were cut into 1 cm^2 x 1 mm slices with a wafering diamond blade, parallel to the fracture surface. The samples used here were taken from specimens tested in the temperature region corresponding to the lower shelf energy. The surface layer was removed by using nitric acid to remove the cold-worked region, created by the cutting of the samples. The amounts of Cu, P and Ni are given in the last column as high, medium or low.

Table 1. *Model alloy chemical composition (wt. ppm)*

Material	C	S	Cu	P	Ni	Si	Mn	Cu/P/Ni
A	0.01	0.004	0.02	0.002	0.01	0.15	0.39	LLL
B	0.01	0.005	0.42	0.012	0.01	0.24	0.49	HML
C	0.01	0.004	0.12	0.010	1.98	0.09	0.35	MMH
D	0.01	0.004	0.12	0.012	1.10	0.12	0.41	MMM
E	0.01	0.004	0.12	0.039	1.13	0.2	0.46	MHM
F	0.01	0.004	0.42	0.012	1.19	0.21	0.47	HMM
G	0.01	0.004	0.11	0.013	0.01	0.37	0.48	MML
H	0.01	0.004	0.11	0.039	0.01	0.24	0.49	MHL

Irradiation and Annealing

Materials A-F were irradiated at surveillance positions to a fluence of 1 x 10^{19} $n.cm^{-2}$ (E > 0.5 MeV) at the ROVNO-1 commercial reactor at a flux of 4 x 10^{11} $n.cm^{-2}s^{-1}$. Materials A, B, G, and H were irradiated at the KOLA-3 commercial reactor to 8 x 10^{19} $n.cm^{-2}$ (E > 0.5 MeV) at a flux of 3 x 10^{12} $n.cm^{-2}s^{-1}$. All alloys were irradiated at 270° C. For the isochronal annealing studies, the samples were annealed at successively increasing temperatures for 30 minutes in sealed quartz tubes filled with ultra-high purity argon. The samples were then etched in hydrochloric acid after each annealing to remove any surface oxidation and to ensure a clean, uniform surface.

Positron Annihilation Spectroscopy

To measure positron lifetime distributions on non-irradiated samples and samples with low ^{60}Co activity, we used a standard BaF_2 two-detector, fast-slow-coincidence positron-lifetime spectrometer. To examine samples with significant ^{60}Co activity, we used a modified three-detector system [*16*]. In the three-detector arrangement, we use a "start" detector to detect the 1.28 MeV γ-ray emitted following the β^+-decay of ^{22}Na, one "stop" detector to detect one of the two 511-keV annihilation quanta, and a third NaI "check" detector, which must detect the second 511-keV annihilation quantum, coincident with the response of the stop detector, for the lifetime to be recorded. This arrangement reduces the interference caused by the two γ-rays emitted in the decays of ^{60}Co by roughly two orders of magnitude, although it reduces the counting rate. Both positron systems have a time resolution of 270 ps. The positron source consisted of 75 μCi of carrier-free ^{22}Na, surrounded by two sheets of Kapton film. The equipment was kept in a room where the temperature was well controlled to avoid electronic drift. The positron lifetime system was frequently calibrated using well-annealed Fe foils to determine the time resolution and the fraction of positrons annihilating in the source.

We used the PATFIT package to perform the deconvolution of the positron lifetime distributions [*17*]. We used the program RESOLUTION to analyze the lifetime distributions corresponding to a Fe foil to determine the resolution function of the system and to determine the fraction of positrons annihilating in the positron source. Using the resolution function generated for a known sample by RESOLUTION, we then used the program POSITRONFIT to perform the deconvolution of the positron lifetime distributions measured on the experimental samples.

For the Doppler-broadening (DB) measurements, we used a Ge x-ray detector with an energy resolution of 1.07 ±0.05 keV at 511 keV. The DB spectrometer was frequently calibrated using a ^{181}Hf 482 keV γ-ray source and a ^{85}Sr 514 keV γ-ray source. We analyzed the energy distribution of 511 kev annihilation quanta using the Doppler lineshape S-parameter, which gives a measure of the "sharpness" of the distribution of 511 keV annihilation quanta. In a metal the S parameter is smaller when the positrons annihilate with higher-momentum electrons, such as those bound to an atom. This will happen more frequently in defect-free crystals. When open-volume defects are present, the S parameter is larger (less variability in the energy of the annihilation quanta), as positrons will tend to annihilate with low-momentum electrons, such as those in the electron gas normally found in higher proportions in open volume defects. As positrons thermalize before annihilation, the positron's momentum is not a significant factor in measuring the S-parameter.

Hardness Testing

For the 15N Rockwell superficial Rockwell hardness tests, we used an American Chain and Cable Company model 6TT Rockwell hardness tester. Before each measurement, the tester was checked for accuracy using a standard test block of known hardness. The calibration anvil was used for the tests, as the samples were very small. In testing each sample, we performed five hardness measurements and used the mean value.

Results

Examination of Samples in the Non-Irradiated and As-Irradiated States

Table 2 summarizes the results obtained for measurements of the average positron lifetime $\bar{\tau}$ and the S-parameter, including both the values obtained before irradiation and those obtained after irradiation but before annealing. Asterisks indicate samples that have not yet been examined because of their high activities, and the numbers in parentheses indicate the experimental error. The average lifetime for the non-irradiated samples varies widely, between 114 and 145 ps, likely indicating differences in the microstructure before irradiation (e.g. higher average lifetimes could indicate a higher dislocation density). The higher S-parameters measured in the samples with high $\bar{\tau}$ (indicating greater percentage of annihilations occurring in well-defined defects such as dislocations) support this interpretation. Some of this sample-to-sample variation could also have been caused by the different compositions of the alloys, but no systematic trend was observed between alloy composition and positron lifetimes. On the other hand, the hardness measurements show that alloys with higher Cu and Ni content tend to exhibit higher hardness values, likely as a result of solid-solution strengthening. Table 3 summarizes the results obtained for the 15N superficial Rockwell hardness tests performed on the irradiated and non-irradiated model alloys. The Rockwell Hardness varies between 32 and 53 for the different alloys.

After neutron irradiation, the positron lifetimes generally increase, with the exception of those for samples C, F and G. These changes are likely caused by a combination of irradiation-induced precipitates and defect clusters formed under irradiation. Longer positron lifetimes (165 ps for dislocations and 175 for vacancies [*18*]) and larger Doppler broadening S-parameters characterize open-volume defects, whereas trapping in irradiation-induced precipitates is associated with shorter lifetimes. Further irradiation of sample B to 8 x 10^{19} n.cm^{-2} causes an additional increase in the average positron lifetime. Finally the average positron lifetime in sample H increases significantly after irradiation to 8 x 10^{19} n.cm^{-2}.

The deconvolution of these positron lifetime distributions using a one or two-lifetime unconstrained fit shows no evidence for the longer lifetimes (>200 ps) usually associated with microvoids, depleted zones, or higher-order vacancy clusters. The distribution could be well fit using two lifetimes: one at about 165 ps and one at about 110 ps or shorter, as predicted by trapping theory [*19*] for a combination of defect trapping and bulk annihilations. This is in agreement with the results of Lopes-Gil [*3*] and Brauer [*2*], who observed no irradiation-induced voids after irradiations conducted at 290° C and 270° C respectively.

Table 2: *Average positron lifetimes $\bar{\tau}$ and S-parameters for the model alloys, before and after irradiation*

Material	$\bar{\tau}$ (ps)			S-Parameter		
	Non-irradiated	1x10^{19} n.cm^{-2}	8x10^{19} n.cm^{-2}	Non-irradiated	1x10^{19} n.cm^{-2}	8x10^{19} n.cm^{-2}
A	116(1)	162(2)	*	0.5126(4)	0.5353(6)	*
B	125(2)	134(2)	145(1)	0.5151(7)	0.5203(5)	0.5258(6)
C	137(2)	120(2)		0.5254(5)	0.5140(5)	
D	145(2)	173(1)		0.5266(3)	0.5452(5)	
E	114(1)	180(1)		0.5105(5)	0.5486(5)	
F	117(2)	116(1)		0.5120(7)	0.5168(7)	
G	143(2)		136(1)	0.5284(4)		*
H	134(2)		165(1)	0.5123(5)		*

Table 3: *Rockwell Hardness, before and after irradiation*

Material	15N Rockwell Hardness		
	Non-irradiated	$1x10^{19}$ n.cm^{-2}	$8x10^{19}$ n.cm^{-2}
A	32 (1)	56 (3)	66 (2)
B	44 (5)	53 (4)	62 (2)
C	50 (4)	54 (1)	
D	53 (2)	64 (1)	
E	51 (1)	65 (1)	
F	49 (1)	62 (1)	
G	44 (3)		59 (1)
H	46 (1)		64 (2)

Successive Isochronal Anneals after Irradiation to $1x10^{19}$ n.cm^{-2}

We present the results for samples A-F, measured after irradiation to 1 x 10^{19} n.cm^{-2} at 270° C, and after successive anneals for 30 minutes at the temperatures indicated. The high activity of samples G and H precluded us from performing the post-annealing positron measurements in this first stage of the study.

Figure 1 shows the average positron lifetime and the relative change in S-parameter and in Rockwell Hardness for Material A (low levels of Cu, P and Ni) and B (high levels of Cu, medium levels of P, and low levels of Ni) after irradiation to $1x10^{19}$ n.cm^{-2} at 270° C and after successive anneals for 30 minutes at the temperatures indicated. The values of these parameters measured after irradiation but before annealing correspond to the values shown at an annealing temperature of 0° C. The non-irradiated values are also indicated. The error bars reported for average positron lifetime and S-parameter originate from the corresponding fitting process, and those associated with the Rockwell Hardness indicate the variability within the measurements conducted. As the annealing temperature increases from 200 to 350° C, little change is observed in $\bar{\tau}$. Between 350 and 450° C, $\bar{\tau}$ decreases abruptly to values close to those for of the non-irradiated material, and remains there up to 600° C. Corresponding decreases in the S-parameter and in the Rockwell hardness values are observed in the temperature range of 350-450° C. This result suggests that the higher positron lifetime observed after irradiation (relative to the non-irradiated value) below 350° C is caused by irradiation-induced open-volume defects; the annealing of these defects between 350 and 450° C causes the concurrent decrease in hardness observed in that temperature range.

For sample B (low Ni content but higher contents of Cu and P), the average positron lifetime starts to decrease at somewhat lower annealing temperatures; and, at 450° C the lifetime values are close to those observed at the same annealing temperature in sample A. The S-parameter also decreases, indicating the annealing of defects in that temperature range. Interestingly, the hardness shows a marked increase, starting at 350°

C and peaking at 500° C. This peak is paralleled by corresponding peaks in $\bar{\tau}$ and S, starting at 450° C.

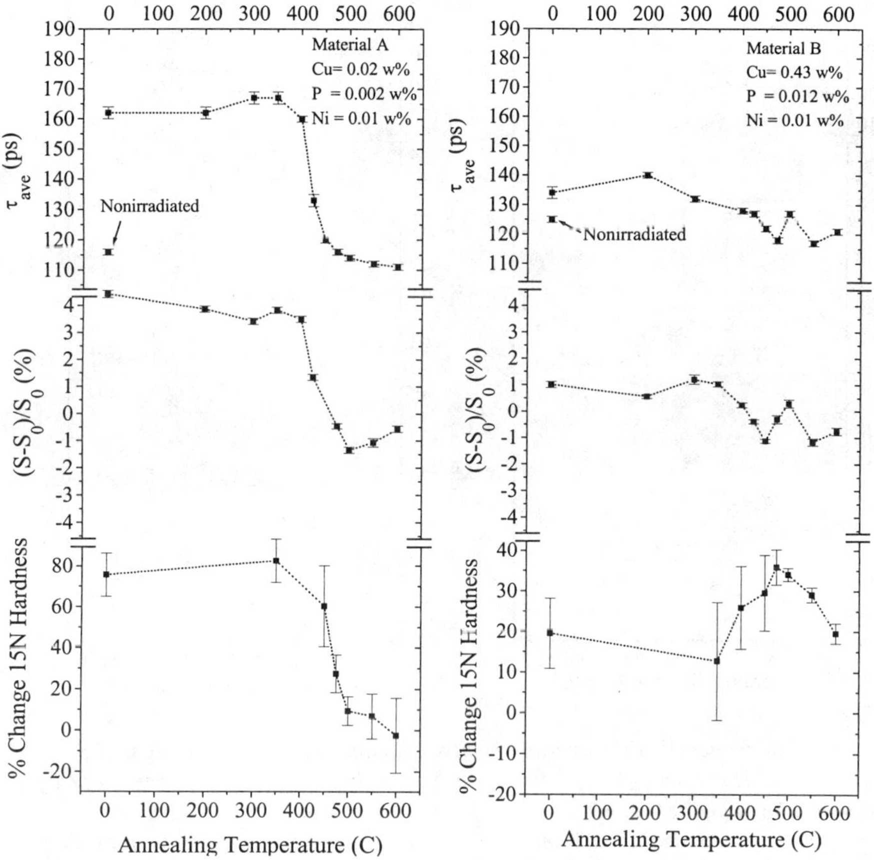

Figure 1: *The dependence of the average positron lifetimes, Doppler-broadening S-parameters relative, and 15N Rockwell superficial hardness values on annealing temperature for neutron-irradiated ($1x10^{19}$ n/cm^2 at 270°C) materials A and B. The S-parameter and 15N Rockwell hardness results are shown as the percent change from the as-fabricated (non-irradiated and non-annealed) state. The samples were successively annealed for 30-minutes at each temperature.*

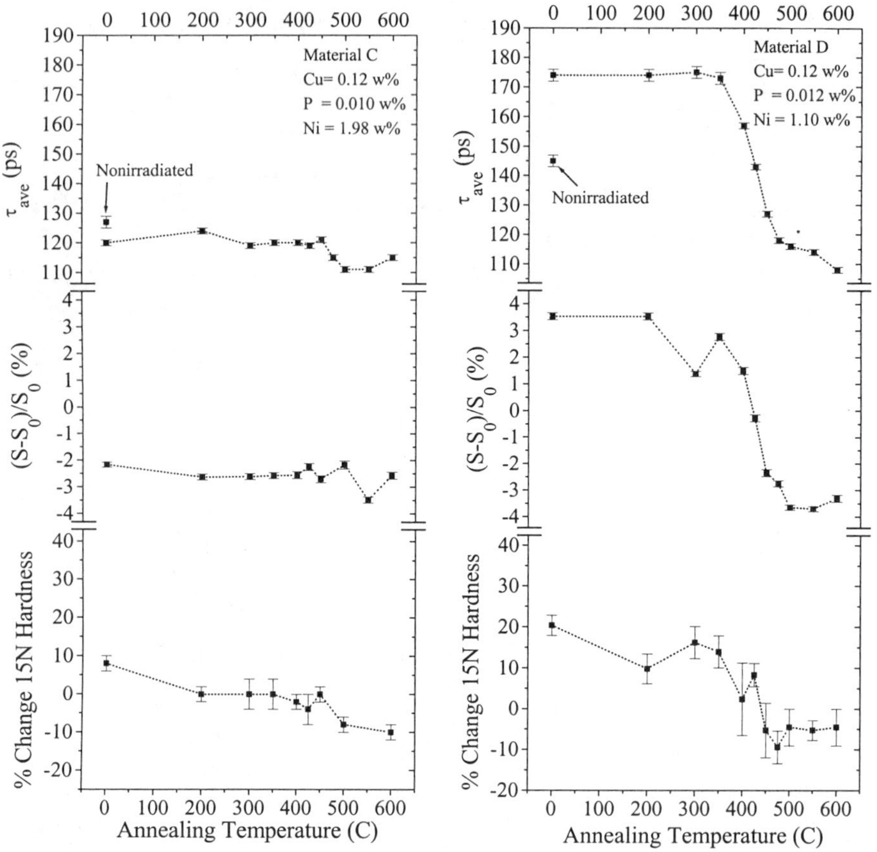

Figure 2: *The dependence of the average positron lifetimes, Doppler-broadening S-parameters, and 15N Rockwell superficial hardness values on annealing temperature for neutron-irradiated ($1x10^{19}$ $n.cm^{-2}$ at 270 °C) materials C and D. The S-parameter and 15N Rockwell hardness results are shown as the percent change from the as-fabricated (non-irradiated and non-annealed) state. The samples were successively annealed for 30-minutes at each temperature.*

Figure 2 shows the average positron lifetime, S-parameter, and Rockwell hardness for sample C, which has the highest Ni content of the annealed samples and medium values of both Cu and P and for sample D, which contains medium levels of Ni, Cu and P. For sample C, the average positron lifetime in the as-fabricated material *decreases* significantly after irradiation, and does not change appreciably with annealing. This behavior is mirrored by the S-parameter, which also decreases with irradiation and remains constant after annealing. The hardness shows a modest increase after irradiation, and exhibits a gradual decrease with increasing annealing temperature.

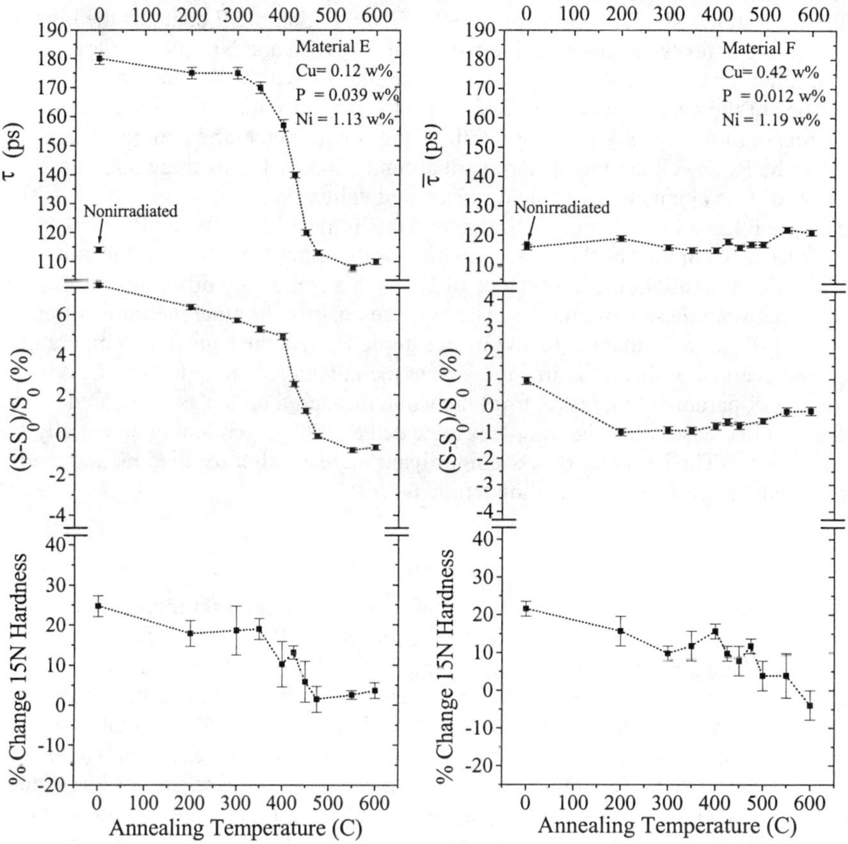

Figure 3: *The dependence of the average positron lifetimes, Doppler-broadening S-parameters, and 15N Rockwell superficial hardness values on annealing temperature for neutron-irradiated ($1x10^{19}$ n/cm^2 at 270° C) materials E and F. The S-parameter and 15N Rockwell hardness results are shown as the percent change from the as-fabricated (non-irradiated and non-annealed) state. The samples were successively annealed for 30-minutes at each temperature.*

For sample D, the average positron lifetime for the as-irradiated sample is the highest among the samples that we examined, and it remains relatively unchanged after annealing up to 350° C. In the same way as in the baseline sample A, the average positron lifetime decreases abruptly between 350 and 450° C, reaching values close to the non-irradiated values at 450° C, with no further evolution up to 600° C. This behavior is also reflected in the S-parameter, which decreases in the same temperature range, and in the hardness, although the scatter in this case is much larger.

Figure 3 shows similar results for samples E and F. Sample E contains the highest value of P for the annealed samples and medium values of Cu and Ni. The average positron lifetime for the as-irradiated sample E is similar to that for D. The same broad characteristics of the average positron lifetime as observed in sample D are seen, i.e., abrupt decreases of the average positron lifetime, the S-parameter, and a more gradual decrease for the Rockwell hardness, starting at around 350° C. The average positron lifetime at 450° C is again close to the non-irradiated value.

Sample F has a high Cu concentration and medium (0.12%) Ni and P concentrations. The contrast of the measurements for this sample with those for sample B (0.01% Ni) should highlight the importance of Ni, as this is the only difference in composition between these two samples. The average positron lifetime measured after irradiation is 116 ps, (as compared to 134 ps in sample B). As the annealing temperature increases, the average positron lifetime remains essentially constant up to 600° C. After irradiation, the S-parameter increases from the value measured on the as-fabricated material, and it decreases after the sample is annealed at 200° C, remaining essentially constant thereafter. The hardness shows a significant increase after irradiation, and then exhibits a steady decrease with annealing temperature.

Discussion

In this work we have relied mostly on the average positron lifetime, as this parameter is not strongly influenced by the fitting procedure. It is possible to perform the deconvolution of the positron spectra to obtain individual lifetimes, and to attempt to compare those with the lifetimes of well-known defects. For the examinations mentioned above, the longest lifetime we have found after deconvolution was 200 ps. If taken at face value, a 200 ps lifetime would correspond to that of a di-vacancy in Fe, which would mean that in addition to dislocation loops, the matrix damage would consist of mono and di-vacancy defects. However, the actual clusters formed may interact with solute atoms such that their compositions could change, thereby affecting their positron lifetimes.

We can propose an interpretation of the results that fits the observations. We emphasize that, because positrons do not provide a direct measure of some of these microstructural features, we need to combine and confirm these measurements by using other experimental techniques. In the baseline material A, no irradiation-induced precipitates are formed because of the low levels of alloying elements. We can attribute the increase in average positron lifetime in sample A, measured after irradiation, to the formation of matrix damage, such as vacancies, di-vacancies, or dislocation loops, which anneal between 350 and 450° C. As the defects anneal, hardness decreases, and the average positron lifetime also decreases, because the fraction of annihilations in defects decreases.

In contrast with sample A, sample B has high Cu and P content. The smaller increase in average positron lifetime measured after irradiation can be attributed to the existence of some irradiation-induced precipitation, in addition to the matrix damage also present in sample A. Because of the greater positron affinity for Cu than for Fe [*20*], these precipitates tend to trap the positrons, increasing the number of annihilations having the lower characteristic lifetime. Because both of these irradiation features trap positrons, the measured value lies in between the characteristic lifetimes for these two traps. However, in the absence of Ni, it is likely that less irradiation-induced precipitation occurs than when Ni is present in significant quantities; thus some of the Cu remains in solution. As the annealing takes place, the matrix damage is annealed, and thermal aging causes more of the Cu to precipitate. This precipitation causes the average positron lifetime to decrease and the hardness to increase. Above 450° C, the precipitation is complete and a ripening process starts to occur, which decreases the precipitate number density. This "Ostwald ripening" process causes the decrease in hardness, observed between 500 and 600° C. This hardening under thermal aging was observed by others: a change in Hv_{20} hardness of 40 was observed in Fe-0.9 % Cu model alloy after 1 hour thermal aging at 550° C [*21*].

By comparing the results for sample B (0.01% Ni) with those for sample F (1.19% Ni), we can assess the effect of Ni. The average lifetime in sample F measured after irradiation is about 120 ps (essentially the same as before irradiation), but we observe a significant increase in hardness, indicating that the microstructure evolution is taking place, although not affecting the average positron lifetime.

For sample F, we can interpret this average positron lifetime as resulting from saturation trapping of positrons in irradiation-induced precipitates. If a combination of high amount of either Cu and a medium amount of Ni, or a high amount of Ni with a medium amount of Cu, would lead to a high number density of irradiation-induced precipitates, then samples C and F would show more extensive irradiation-induced precipitation than samples B, D and E. We would expect an elementally-pure and coherent metal precipitate to show the annihilation lifetime and Doppler-broadening characteristics corresponding to those for the precipitated metal. For pure Cu, we measured a positron annihilation lifetime of 119±1 ps, which is close to the lifetime measured on Fe, 110 ±1 ps.

In this scenario, the matrix defects in sample F are completely obscured by very extensive precipitation that traps all the positrons, so that they annihilate with the lifetimes associated with the precipitates. It has been proposed that Ni acts as a Cu-dispersant, enhancing Cu precipitation and making it finer [*8*]. A fine distribution of irradiation-induced precipitates is more efficient at trapping positrons than is a coarse distribution. Because these precipitates are stable to higher temperatures, the annealing of matrix damage is not observed, and we have no way of assessing its extent in these samples.

A similar scenario is operative for sample C. In that case, the as-fabricated microstructure contains a greater amount of matrix dislocation-type defects, so the as-fabricated average positron lifetime is 137 ps, which is reduced by irradiation to 120 ps. The aforementioned hypothesis would also explain this result. Extensive trapping of positrons at irradiation-induced precipitates would obscure the matrix damage. By contrast, in samples E and D, and to a lesser extent in B, the matrix damage represents a greater percentage of the annihilations; and, thus, the average positron lifetime increases. Also, because a majority of the positrons annihilates in well-defined defects, the S-parameter is higher for samples B, D, E than for C and F.

We note that radiation-induced copper-rich precipitates are not necessarily either pure or coherent. Using atom probe field ion microscopy, Miller et al. report that these precipitates consist a mixture of elements, consisting of Fe, Cu, Mn, and Si, with P segregating to the precipitate boundary [*11*]. Investigations of precipitates in irradiated reactor pressure-vessel steels using SANS show that these precipitates have a magnetic-to-nuclear scattering ratio that is not consistent with that for pure Cu [*22*] but instead is consistent with that for a mixture of Cu, Mn, and Ni [*8*]. Using high-resolution electron microscopy, Othen et al. show that pure Cu precipitates in ferritic steel maintain a b.c.c. structure until they reach a diameter of 5 nm, at which point they transform into a 9R structure [*23*]. This result suggests that, as long as the precipitate size remains around the typical diameter of 1-2 nm, the precipitates could remain coherent, but with some stresses that could cause misfit defects at the matrix-precipitate interface. Brauer et al. have estimated a positron lifetime in irradiation-induced precipitation of 124-145 ps at such a misfit defect located at the Cu-Fe interface [*6*]; the same value was estimated by Phythian et al. to be ~ 130 ps [*21*]. Such lifetimes in irradiation-induced precipitates depend on a variety of characteristics of these precipitates, including their degree of coherency, crystal structure, composition, and element distribution. Given the lifetimes observed of around 120 ps, it is likely that these irradiation-induced precipitates are coherent with the matrix.

We examine the effect of P by comparing specimens D (0.012% P) and E (0.39% P). The results for the average positron lifetime, the S-parameter and the hardness are very similar, indicating that P does not play a significant role in the irradiation-induced microstructural evolution and in the post-irradiation annealing of these model alloys. It is interesting to compare the present results obtained in model alloys with previous results obtained in the examination of neutron-irradiated pressure-vessel steels. The same general irradiation and annealing behavior is observed in the pressure-vessel steels as is observed in the model alloys. After irradiation to 1.5×10^{19} $n.cm^{-2}$, the average positron lifetime in A533B steel (incorrectly reported as A508 in [*24*]) increases to ~ 150 ps. This average positron lifetime decreases between 300 and 450° C, and increases again between 450 and 600° C [*24*]. This behavior was also observed in VVER steels [*2*]. This suggests that a similar interpretation of the results is valid, i.e., that matrix damage (with a characteristic lifetime of about 165-175 ps) develops in parallel with irradiation-induced precipitates, which should be rich in Cu and Ni.

Conclusions

We have used positron annihilation spectroscopies to examine a series of model iron alloys in which we systematically varied the alloying concentrations of Cu, P and Ni. We measured the average positron lifetime, the S-parameter, and the Rockwell hardness, on model alloy samples irradiated to 1×10^{19} $n.cm^{-2}$ and 8×10^{19} $n.cm^{-2}$, and we measured the same quantities after performing successive isochronal anneals of the samples that had been irradiated to 1 x 10^{19} n.cm^{-2}. The main conclusions of this study are as follows:

1. No evidence was observed for the long positron lifetimes (> 250 ps) normally associated with higher-order defect clusters or microvoids. This result is in agreement with the results obtained by several other researchers who did not observe microvoids in complex alloys after high temperature neutron irradiation.

2. The effects of Cu and Ni, especially in combination, on microstructural evolution and on the inferred irradiation-induced precipitation, as measured by the average positron lifetime, S-parameter and hardness, were much more pronounced than the effects of P.

3. In samples without a high concentration of either Cu or Ni, combined with a medium concentration of the other element, the average positron lifetime increases after irradiation to 1 x 10^{19} n.cm^{-2}. The microstructural feature associated with this lifetime anneals between 350° C and 450° C. The S-parameter and the hardness in these samples exhibit similar annealing behavior. This behavior could be caused by the formation of defect clusters produced under irradiation.

4. In samples having either high Cu and medium Ni or high Ni and medium Cu, we observe a low average positron lifetime after irradiation to 1 x 10^{19} n.cm^{-2}, which occurs at the same time that the hardness increases. This behavior changes little with annealing temperature. This low average positron lifetime could result from saturation positron trapping in irradiation-induced precipitates.

5. These results and interpretation are in general agreement with the current view that irradiation embrittlement of ferritic steels at high temperature is caused by fine-scale damage, which consists of a combination of irradiation-induced precipitates and defect clusters, but not microvoids.

We emphasize that these observations need to be combined the observations performed with other complementary experimental techniques to obtain a more complete view of microstructural evolution in pressure vessel steels under irradiation.

Acknowledgments

We thank the FERMI group (research consortium of electrical utilities and fuel vendors, including Pennsylvania Power and Light, Philadelphia Electric Co., General Public Utilities, Commonwealth Edison, Public Services Electric and Gas and Westinghouse Electric Corp.) for sponsoring this research. We thank the Radiation Science and Engineering Center, Breazeale Nuclear Reactor at Penn State for providing use of the facilities and for the use of a high resolution gamma detector in the Doppler broadening experiments.

References

[1] Ghazi-Wakili, K., Zimmerman, U., Brunner, J., Tipping, P., Waeber, W. B., and Heinrich, F., "Positron Annihilation Studies on Neutron Irradiated Pressure Vessel Steels," *Physica Status Solidii (a)*, 1987, 102, pp. 153.

[2] Brauer, G., Liszkay, L., Molnar, B., and Krause, R., "Microstructural Aspects of Neutron Embrittlement of Reactor Pressure Vessel Steels - A View from Positron Annihilation Spectroscopy," *Nuclear Engineering and Design*, 1991, 127, pp. 47-68.

[3] Lopes-Gil, C., De Lima, A. P., Ayres de Campos, N., Fernandes, J. V., Koegel, G., Sperr, P., Trifshaeuser, W., and Pachur, D., "Neutron Irradiated Reactor Pressure Vessel Steels Investigated by Positron Annihilation," *Journal of Nuclear Materials*, 1989, 161, pp. 1-12.

[4] Hempel, A., Saneyasu, M., Tang, Z., Hasegawa, M., Brauer, G., Plazaola, F., and Yamaguchi, S., "Effects of Neutron Irradiation on Fe-Cu model Alloys and RPV Steels Probed by Positron Annihilation and Hardness Measurements," Effects of Radiation on Materials: 19th International Symposium, ASTM STP 1366, 1999, Seattle, pp. 560-578.

[5] Pareja, R., De Diego, N., De La Cruz, R. M., and Del Rio, J., "Postirradiation Recovery of a Reactor Pressure Vessel Steel Investigated by Positron Annihilation and Microhardness measurements," *Nuclear Technology*, 1993, 104, pp. 52-63.

[6] Brauer, G., Puska, M. J., Sob, M., and Korhonen, T., "Positron Affinity for Precipitates in Reactor Pressure Vessel Steels," *Nuclear Engineering and Design*, 1995, 158, pp. 149-156.

[7] Odette, G. R., "On the Dominant Mechanism of Irradiation Embrittlement of Reactor Pressure Vessel Steels," *Scripta Metallurgica*, 1983, 17, pp. 1183-1188.

[8] Wirth, B., "On the Character of Nano-Scale Features in Reactor Pressure Vessel Steels Under Neutron Irradiation," Ph.D. thesis in Mechanical and Environmental Engineering. University of California at Santa Barbara: 1998.

[9] Brauer, G., Eichhorn, F., Frisius, F., and Kampmann, R., "Investigation of Neutron Irradiated Soviet Type Reactor Pressure Vessel Steels by Small Angle Neutron Scattering," Effects of Radiation on Materials: 16th International Symposium, ASTM STP 1175, 1993, pp. 503-515.

[10] Miller, M. K. and Burke, M. G., "An Atom Probe Field Microscopy Study of Neutron Irradiated Pressure Vessel Steels," *Journal of Nuclear Materials*, 1992, 195, pp. 68-82.

[11] Pareige, P., Stoller, R. E., Russel, K. F., and Miller, M. K., "Atom Probe Characterization of the Microstructure of Nuclear Pressure Vessel Surveillance Materials after Neutron Irradiation and after Annealing Treatments," *Journal of Nuclear Materials*, 1997, 249, pp. 165-174.

[12] Williams, T. J. and Phythian, W., "Electron Microscopy and Small Angle Neutron Scattering Study in Low Alloy Steel Submerged-Arc Welds," Effects of Radiation

on Materials: 17th International Symposium, ASTM STP 1270, 1996, pp. 191-205.

[13] Stoller, R. E., "Pressure Vessel Embrittlement Predictions Based on a Composite Model of Copper Precipitation and Point Defect Clustering.," Effects of Radiation on Materials: 17th International Symposium, ASTM STP 1270, 1996, pp. 25-58.

[14] Valo, M., Krause, R., Saarinen, K., Hautojärvi, P., and Hawthorne, J. R., "Irradiation Response and Annealing Behavior of Pressure Vessel Model Steels and Iron Ternary Alloys Measured with Positron Techniques," 15th ASTM International Symposium on Radiation Effects on Materials, ASTM STP 1125, 1992, Nashville, TN, pp. 172-185.

[15] Böhmert, J., Ulbricht, A., Kryukov, A., Nikolaev, Y., and Erak, D., "Composition Effects on the Radiation Embrittlement of Iron Alloys," Effects of Radiation on Materials: 20th International Symposium, ASTM STP 1405, 2000, Williamsburg, VA, pp. submiited.

[16] Van Hoorebeke, L., Fabry, A., Walle, E. v., Velde, J. V. d., Segers, D., and Dorikens-Vanpraet, L., "A Three-detector Positron Lifetime Setup Suited for Measurements on Irradiated Steels," *Nuclear Instruments and Methods in Physics Research A*, 1996, 371, pp. 566-571.

[17] Kirkegaard, P., Eldrup, M., Mogensen, O. E., and Pedersen, N. J., *Computer Physics Communications*, 1981, 23, pp. 307.

[18] Vehanen, A., Hautojärvi, P., Johansson, J., Yli-Kaupila, J., and Moser, P., "Vacancies and Carbon Impurities in Alpha-iron: Electron Irradiation," *Physical Review B*, 1982, 25, pp. 762-780.

[19] Frank, W. and Seeger, A., "Theoretical Foundation and Extension of the Trapping Model," *Applied Physics*, 1974, 3, pp. 61-66.

[20] Puska, M. J., P. Lanki, and Nieminen, R. M., "Positron Affinities for Elemental Metals," *Journal of Physics F: Condensed Matter*, 1989, 1, pp. 6081.

[21] Phythian, W., de Diego, N., Mace, J., and McElroy, R., "Characterization of Model Fe-Cu-Mn-Ti-N Alloys by Electron Microscopy and Positron Annihilation," Effects of Radiation on Materials: 16th International Symposium, ASTM STP 1175, 1993, Denver, pp. 462-491.

[22] Grosse, M., Eichhorn, F., Böhmert, J., and Brauer, G., "Characterization of Irradiation Induced Precipitates by Small Angle X-Ray and Neutron Experiments," Effects of Radiation on Materials: 17th International Symposium, ASTM STP 1270, 1996, Sun Valley, pp. 1123-1133.

[23] Othen, P., Jenkins, M. L., Smith, G., and Phythian, W. J., "Transmission Electron Microscope Investigations of the Structure of Copper Precipitates in Thermally Aged Fe-Cu and Fe-Cu--Ni," *Philosophical Magazine Letters*, 1991, 64, pp. 383.

[24] Cumblidge, S. E., Motta, A. T., and Catchen, G. L., "Examination of Neutron Irradiated Pressure Vessel Steel Using Positron Annihilation Lifetime Spectroscopy," Materials Research Society Fall Meeting 540, 1998, pp. 471-476.

Jonathan M. Hyde,[1] David Ellis,[2] Colin A. English,[1] and Timothy J. Williams[2]

Microstructural Evolution in High Nickel Submerged Arc Welds

Reference: Hyde, J. M., Ellis, D., English, C. A., and Williams, T. J., "**Microstructural Evolution in High Nickel Submerged Arc Welds,**" *Effects of Radiation on Materials: 20th International Symposium, ASTM STP 1405,* S. T. Rosinski, M. L. Grossbeck, T. R. Allen, and A. S. Kumar, Eds., American Society for Testing and Materials, West Conshohocken, PA, 2001.

Abstract: The effect of nickel on irradiation sensitivity in ferritic steels is substantial, complex, and not yet fully understood. In this study, the evolution of irradiation induced microstructures in high nickel submerged arc welds has been examined in a series of small angle neutron scattering (SANS), field emission gun scanning transmission electron microscopy (FEGSTEM) and optical position sensitive atom probe (OPoSAP) experiments. For a low (~0.05%) copper content weld, SANS measurements showed that a high density of small features had been formed during irradiation. It was expected that clustering would be evident from visual examination of the OPoSAP elemental maps, but this was not the case. However, a statistical assessment of the data revealed that the manganese atoms were non-randomly distributed. For a medium copper (0.15%) weld, clusters were again not visible in OPoSAP, but non-randomness of both manganese and nickel could be detected at a low dose; at higher doses copper was non-random also. At high levels of copper (0.24% and 0.56%), clusters microalloyed with nickel, manganese, copper, silicon and iron were visible. The SANS data showed that the feature size tended to increase with copper and dose, but the number density remained fairly constant. The paper discusses how the differences between the SANS and OPoSAP data might be reconciled, and the implications in relation to the mechanisms of irradiation damage.

Keywords: Irradiation embrittlement, pressure vessel steels, submerged arc weld metal, nickel, microstructure, small angle neutron scattering, atom probe field ion microscopy

[1] Senior Research Scientist and Principal Technical Consultant respectively, AEA Technology plc, B220 Harwell, Didcot, Oxon. OX11 0RA, UK.

[2] Senior Engineer and Consultant, respectively, Rolls-Royce plc, PO Box 2000, Derby, DE21 7XX, UK.

Introduction

The characterization of the microstructural changes caused by irradiation damage is a key activity in the development of an understanding of the phenomenon sufficient to enable construction of accurate predictive models. Such models are essential if predictions are to be made for combinations of variables not adequately represented by empirical data. It is widely held that there are three main components of irradiation damage in reactor pressure vessel steels: matrix damage, precipitation damage, and non-hardening embrittlement. This paper focuses primarily on the second of these, the damage caused by clustering, or precipitation, of solute atoms. The third component is not significant in the steels discussed, and imaging techniques for the matrix damage have not yet been developed.

It is unfortunate that the precise nature of matrix damage is not yet established and, until this is achieved, embrittlement modelling will remain incomplete. However, precipitation damage is the most significant contributor to irradiation embrittlement in steels with more than about 0.15% copper. Understanding it is a key to understanding irradiation dose rate (flux) effects, and to understanding the differences in behavior between high nickel (>1.5%) materials (discussed in this paper) and those with more typical (~ 0.7%) nickel contents.

There are a number of techniques available to characterise the very fine scale damage caused by neutron irradiation. None, however, produce complete and unequivocal data. All require considerable care in application and interpretation, and all are costly relative to mechanical testing. The approach taken has been to use selected techniques in combination for a limited range of materials and irradiation conditions.

Overview of Microstructural Techniques Used

Table 1 provides a summary of the techniques used. The FEGSTEM (field emission gun scanning transmission electron microscopy) was performed at AEA Technology Harwell using a Vacuum Generators HB501 instrument, fitted with an Oxford Instruments beryllium-windowed X-ray detector system for EDX analysis. The SANS (small angle neutron scattering) data were obtained using D11 at the Institute Laue Langevin at Grenoble France and using PAXY at Saclay France. The OPoSAP (Optical Position Sensitive Atom Probe) work was performed at the University of Oxford. Details of the techniques are given in [*1, 2, 3*]. Details of the data interpretation are given later.

Materials

The materials used have been previously described [*4, 5, 6*]; a summary is given here for convenience. Those used in this investigation are shown in Table 2. All of the materials are two metre long submerged-arc welds in 150mm thick ASTM A533B plate and were made specifically for irradiation test programmes. They were austenitized at 920°C followed by water quenching, then stress relieved/tempered for 40 hours at 600°C, then 6 hours at 650°C. Chemical analysis was done using wet chemistry methods. The compositions given in Table 2 are averages of the approximately 10 to 20 measurements made at various locations within each weld. In subsequent tables best estimate values are given depending on the location of the individual specimen within the weld. The high copper level in weld WV was achieved by doping the welding flux with copper carbonate.

Table 1 - *Summary of microstructural techniques employed.*

Technique	Information on matrix damage?	Information on irradiation induced clusters?	Information on non-hardening embrittlement?
FEGSTEM	No (although dislocation loops may provide contrast)	Provides approximate concentrations of alloying elements (excluding Fe)	Yes from analysis of grain boundaries. Profile across boundary provides estimate of coverage.
OPoSAP	No unless associated with compositional fluctuations	Direct analysis of clusters - size and composition	Limited since the probability of analysing a grain boundary is low
SANS	Limited - interpretation difficult	Cluster size. A ratio is linked to composition.	No

Table 2 - *Nominal material compositions (wt%).*

Material	C	Mn	Si	S	P	Cr	Ni	Mo	Cu
WF	0.038	1.41	0.53	0.007	0.009	0.095	1.72	0.35	0.023
WG	0.037	1.21	0.60	0.007	0.008	0.094	1.72	0.35	0.240
WK	0.051	1.45	0.48	0.005	0.011	0.043	1.64	0.38	0.147
WL	0.046	1.34	0.40	0.008	0.013	0.055	1.64	0.41	0.034
WV	0.046	1.34	0.40	0.010	0.010	0.055	1.66	0.41	0.560

Test Matrix

Although a large number of weld and plate materials have been used over a period of years for irradiation experiments, only a relatively small number of these have been utilized for damage characterization studies and, of these, only high nickel weld metals are included in this paper.

Specimens have been selected to enable studies to be carried out on the effect of dose on low copper material and the effects of dose and dose rate on medium and high copper materials. The samples examined were cut from broken (at low energy) Charpy or hardness specimens that had been irradiated in the Herald reactor at Aldermaston in the United Kingdom or the Osiris reactor at Saclay in France over a range of irradiation temperatures and doses. Details have been given elsewhere [*4-6*]. Table 3 summarizes the matrix of specimens examined and the techniques that have been applied to them.

Table 3 - *Specimen Exposure and Usage.*

Specimen	Copper[1] (wt%)	Dose (mdpa)	Dose Rate (dpa/sec)	SANS	FEGSTEM	OPoSAP
WF073	0.023	0	0	✓	-	-
WF011	0.022	14.42	5.96E-9	✓	-	✓
4WL743	0.046	0	0	✓	✓	-
WLD330	0.034	0	0	✓	-	-
4WL769	0.038	1.56	9.34E-11	✓	-	-
4WL772	0.038	32.12	5.80E-9	✓	-	-
4WL773	0.052	32.18	5.81E-9	✓	✓	✓
WK086	0.184	0	0	✓	-	-
WK039	0.136	13.25	6.97E-9	✓✓[2]	✓	✓
WK087	0.184	37.85	6.74E-9	✓	✓	✓
WG117	0.240	0	0	✓	-	-
WG114	0.240	1.45	8.68E-11	✓	✓	✓
WG010	0.240	14.05	6.52E-9	✓	✓	✓
WG102	0.240	55.19	8.19E-9	✓	✓	✓
WV227	0.500	0	0	✓	✓	-
WV216	0.500	1.46	5.07E-9	✓	-	-
WV126	0.550	1.48	8.86E-11	✓	-	-
WV127	0.550	3.52	6.59E-10	✓	-	-
WV396	0.540	6.38	5.68E-9	-	-	✓
WV214	0.500	9.29	6.56E-9	✓	-	-
WV400	0.540	12.11	6.30E-10	✓	-	-
WV466	0.530	13.07	4.73E-9	-	✓	✓
WV014	0.620	17.90	6.48E-9	✓	-	-
WV012	0.620	30.10	6.23E-9	✓	✓	✓

1 Best estimate copper content for the individual specimen.
2 Duplicate specimens examined from the same block.

Interpretation of Microstructural Data

In high dose, high Cu steels Cu-enriched clusters can be imaged using both FEGSTEM and OPoSAP techniques. At low doses and in low Cu steels, irradiation-induced clusters are much less well defined. Consequently, different approaches to characterise these features are required.

Analysis of Steels in Which Clusters are Not Well Defined

In the low Cu steels the formation of matrix damage is believed to dominate the irradiation response. In this case techniques such as FEGSTEM are not appropriate since they do not have the resolution required to image these defects. Similarly, the OPoSAP cannot detect the presence of vacancy or interstitial clusters in an irradiated steel. However, irradiation can also lead to the development of composition fluctuations, which may be associated with matrix damage. OPoSAP data can be analyzed using composition frequency distributions [*7*] and contingency table [*8, 9, 10*] analyses to determine if fluctuations exist and are statistically significant. These techniques involve dividing the three-dimensional (3D) reconstructed image of the atomic scale structure of the material into approximately cubic blocks such that each block contains exactly the same number of atoms.

A composition frequency distribution analysis for a particular element A is performed counting the number of blocks with no A atoms, with 1 A atoms, with 2 A atoms etc. and comparing with the number expected from a random distribution. As usual, with the χ^2 statistic, it is necessary to combine classes with small probabilities so that the expected number of observations is greater than 5. The X^2 test statistic can then be evaluated

$$X^2 = \sum_{i=1}^{N} \frac{(O_i - E_i)^2}{E_i}$$

where O_i are the experimental observations, E_i the values expected from a binomial distribution and N is the number of categories (N-1 degrees of freedom).

An example is shown below in Figure 1 for Mn in an irradiated low Cu high Ni weld (4WL773). In this case, a X^2 value of 162 with 3 degrees of freedom is significant at the 1% level, which implies that there is less than a 1% chance that the observed distribution of Mn is random.

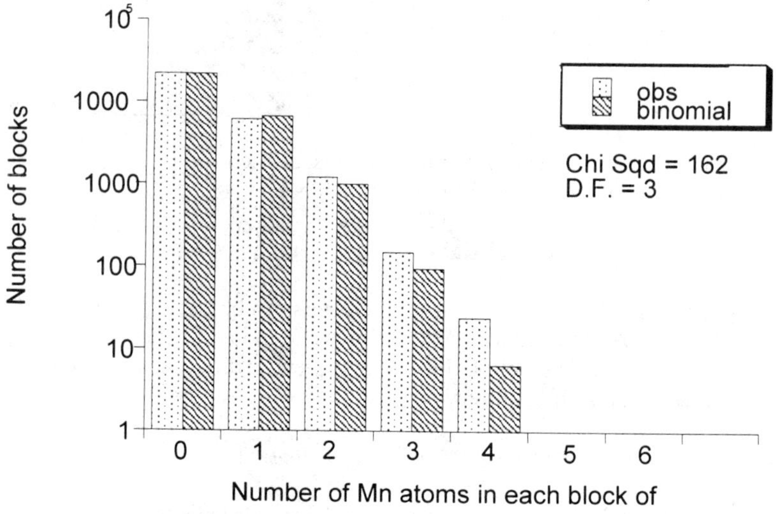

Figure 1 - *Composition frequency distribution for Mn atoms in 4WL773.*

A contingency table analysis is designed to analyze the extent of co-segregation of various elements, for instance, to determine whether the non-random fluctuations in Mn are spatially associated with non-random fluctuations in, say, Cu or Ni. A two-dimensional (2D) table is then drawn up showing the number of experimental observations within each category (e.g. number of Mn atoms per block and number of Ni atoms per block). The null hypothesis is that the two categories are independent. To test the null hypothesis it is necessary to build a second table detailing the 'Estimated Values'. These are simply calculated from the row and column sums of the experimental values. For instance the estimated number of blocks containing 2Mn and 1Ni atom is

$$E_{2Mn+1Ni} = \frac{\textit{Observed blocks with } 2Mn \textit{ atoms} * \textit{Observed blocks with } 1Ni \textit{ atom}}{\textit{Total number of blocks}}$$

The X-squared test statistic can then be evaluated

$$X^2 = \sum_{i=1}^{r}\sum_{j=1}^{c}\frac{\left(N_{ij} - E_{ij}\right)^2}{E_{ij}}$$

where r is the number of rows, c is the number of columns, N_{ij} are the experimental observations and E_{ij} the estimated values. If the two variables (categories) are independent, then X^2 will have a χ^2 distribution with (r-1)(c-1) degrees of freedom. The probability that the calculated X^2 value is consistent with the null hypothesis (i.e. that the categories are independent) can be determined by comparison with tabulated values of the χ^2 distribution for the appropriate number of degrees of freedom.

SANS techniques now have the capability of determining the presence of features that exist in the material on the order of ~1nm in diameter. Furthermore, SANS provides a good average since the volume analyzed is large in comparison to the feature size. However, SANS only unambiguously provides information on feature size and size distribution. To determine absolute irradiation induced volume fraction and hence feature number density it is necessary to make assumptions concerning the composition and magnetic properties of the irradiation induced features. The simplest solution is to choose a representative mean feature composition. This can be based on the observed features in high Cu RPV steels or simply a prediction on the type of features expected in these steels. Provided that all of the SANS data is interpreted in a self consistent manner, then a comparative analysis of the relative irradiation induced volume fraction as a function of dose and dose rate in low Cu steels can be used to show trends explicitly.

SANS data were obtained from both irradiated and equivalent unirradiated samples. For each unirradiated-irradiated pair of materials the differential cross sectional variation with Q obtained from the unirradiated sample was subtracted from the irradiated differential cross sectional variation with Q to determine the irradiation-induced response. A maximum entropy algorithm was used to fit the experimental irradiation-induced differential cross sectional variation with Q to obtain a volume weighted particle size distribution and hence the mean irradiation-induced feature diameter. To obtain absolute volume fractions and hence feature number densities the feature contrast factor was assumed to be the same as that observed in the high Cu welds (see Table 4 below). Plots have been generated to show the trends between the irradiation-induced volume fraction with both dose and copper content.

The FEGSTEM technique can also be used to determine if irradiation induced clusters exist that are too small to generate the black-white strain contrast [*1*] normally associated with irradiation-induced clusters. This is achieved by taking a series of spot measurements where the electron beam is focused on very small regions (~2nm in diameter) of the foil that do not exhibit contrast. The copper associated with each spot measurement is then determined and presented in the form of a bar chart showing the number of spot measurements versus their copper content. A narrow distribution indicates that the copper is randomly distributed whereas a wide distribution indicates that some spot measurements sampled sub-visible irradiation-induced clusters.

Analysis of Steels in Which Clusters are Well Defined

In principle, interpretation of microstructural data is somewhat simpler since both the OPoSAP and FEGSTEM techniques can resolve the Cu-enriched irradiation induced clusters. The methodology is shown schematically in Figure 2. The composition of irradiation-induced clusters can be estimated directly from the OPoSAP data by sampling the atoms within each cluster and calculating their mean composition. The FEGSTEM can also provide information on the cluster composition provided that the features are sufficiently large to show strain contrast. The difference in EDX spectra obtained with an electron beam focused on a cluster and the neighbouring matrix can be used to confirm the relative concentrations of elements such as Cu, Mn, Ni and Si. However, the FEGSTEM has insufficient resolution to determine their Fe content. Relatively few clusters can be examined using these techniques, which limits the accuracy of the estimated mean composition. Furthermore, determining the precise extents of irradiation-induced clusters is difficult since they do not have well defined interfaces with the matrix. Consequently it is possible that the Fe content is slightly overestimated if the sampling volume sampled both cluster and matrix.

As detailed for the low Cu steels, the irradiation-induced size distribution was calculated from the experimental irradiation-induced differential cross sectional variation with Q. The mean feature size can then be compared with the size of the clusters observed by atom probe microanalysis. The feature contrast depends on the cluster composition, the cluster vacancy content and also their magnetic properties. The atom probe and FEGSTEM data directly provide an estimate of the mean cluster composition. However, not all of the specimens examined by SANS have also been examined using OPoSAP and FEGSTEM techniques and so it was necessary to make use of the available data. The mean cluster composition from all of the OPoSAP analyses of high Cu high, high Cu welds was chosen for the analysis of all of the high Cu, high Ni SANS data. The vacancy content and magnetic properties are not well known although studies have shown that the magnetic properties of binary Fe Cu alloys are less than predictions based simply on the Fe content [*11*]. Given these uncertainties and also the uncertainty associated with the precise Fe content of the clusters it was decided to analyze the SANS data using three different scenarios as detailed in Table 4. The feature density was used as a fitting parameter to ensure that the A-ratio of the assumed feature was consistent with the experimentally determined A-ratio (typically 2.4 or 2.5 for these steels). The mean density of bcc Fe is ~7.9g/cm^3. A lower density implies the presence of vacancies associated with the features.

Table 4 - *Assumptions of cluster composition used for the analysis of SANS scattering data.*

Scenario	Cu (at.%)	Mn (at.%)	Ni (at.%)	Si (at.%)	Fe (at.%)	Magnetic moment of Fe in feature	Feature Density (g/cm^3)	Feature A-ratio
1	9.0	11.0	10.0	3.0	67	0.7	6.8	2.5
2	18.0	22.0	21.0	6.0	33	0.7	7.6	2.5
3	27.0	31.0	33.0	9.0	0	0	7.9	2.4

For each of these scenarios, the irradiation-induced volume fraction was calculated. This, together with the mean feature size, was used to determine the feature number density. Self-consistency was determined in two ways: First, by comparing the calculated number density with the estimated number density from the OPoSAP data (where possible) and second, by calculating the reduction in matrix levels of Cu, Mn, Ni and Si and ensuring that they were consistent with the weld composition and matrix levels of these elements measured by both OPoSAP and FEGSTEM techniques.

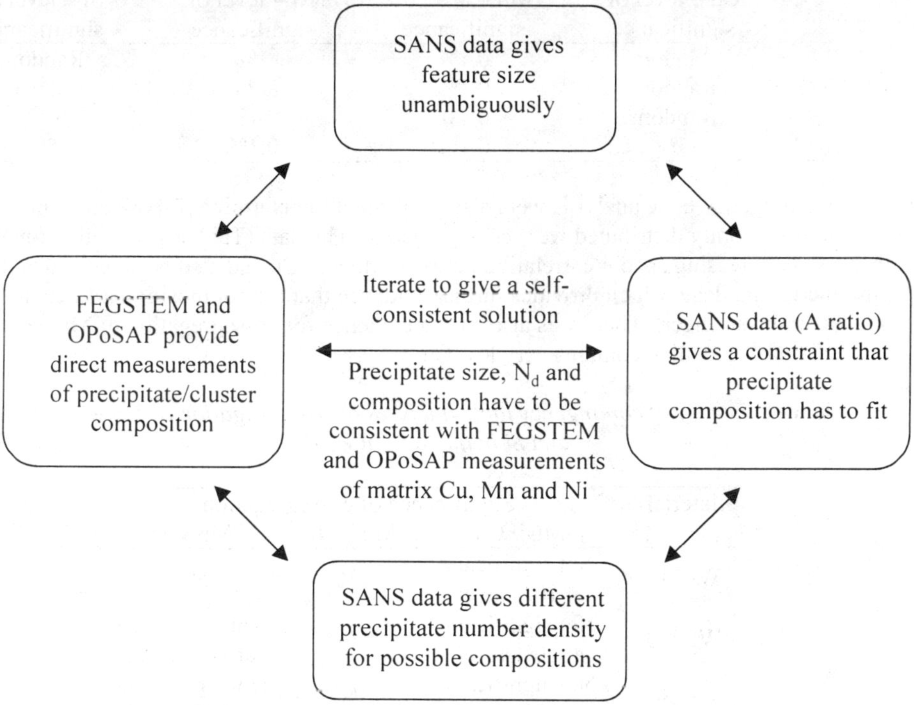

Figure 2 - *Methodology for interpretation of microstructural data.*

Results

Results Obtained from Low/Intermediate Copper Welds (WF, 4WL, WK)

Atom maps showing the distribution of Cu, Mn and Ni atoms in WF11, 4WL773, WK39 and WK87 are shown in Figure 3. Each dot represents an individual atom and so any regions that exhibit a higher density of dots indicate the presence of clustering. In fact, it is quite difficult to judge from a two-dimensional image whether the distributions are random or not. However, by rotating the volume in 3D, it is clear that Cu, Mn and Ni distributions in 4WL and both WK samples are not random. To investigate this further, composition frequency distribution analyses were performed on each data set using a block size of 27 atoms. The results are summarized in Table 5. The analysis showed that the Mn distributions in each material are not random. In the lowest Cu welds, there was less evidence for non-random Ni distributions than in the medium Cu weld (WK) and in the medium Cu weld, the Cu was found to become increasingly non-random with dose.

Table 5 - *Summary of composition frequency distribution analyses of low Cu welds.*

Material	Cu – level of significance	Mn – level of significance	Ni – level of significance	Si – level of significance
WF11	Random	0.1%	1%	Random
4WL773	Random	0.1%	1%	0.1%
WK39	Random	0.1%	0.1%	0.1%
WK87	5%	0.1%	0.1%	5%

Contingency table analyses were also performed to determine if those elements that were non-randomly distributed were co-segregated. The data (Table 6) show that for WK there is an increasing spatial correlation between Mn and Cu and also between Mn and Ni with increasing dose, which provides further evidence that the compositional fluctuations are becoming stronger. There was also strong evidence for co-segregation of Mn and Ni in irradiated WF, which contains very low Cu.

Table 6 - *Summary of contingency table analyses of co-segregation of Mn with Cu, Ni and Si in the low Cu welds.*

Material	Significance of co-segregation		
	Mn v Cu	Mn v Ni	Mn v Si
WF11	Not significant at 10% level	1%	5%
4WL773	Not significant at 10% level	5%	Not significant at 10% level
WK39	Not significant at 10% level	5%	Not significant at 10% level
WK87	10%	1%	1%

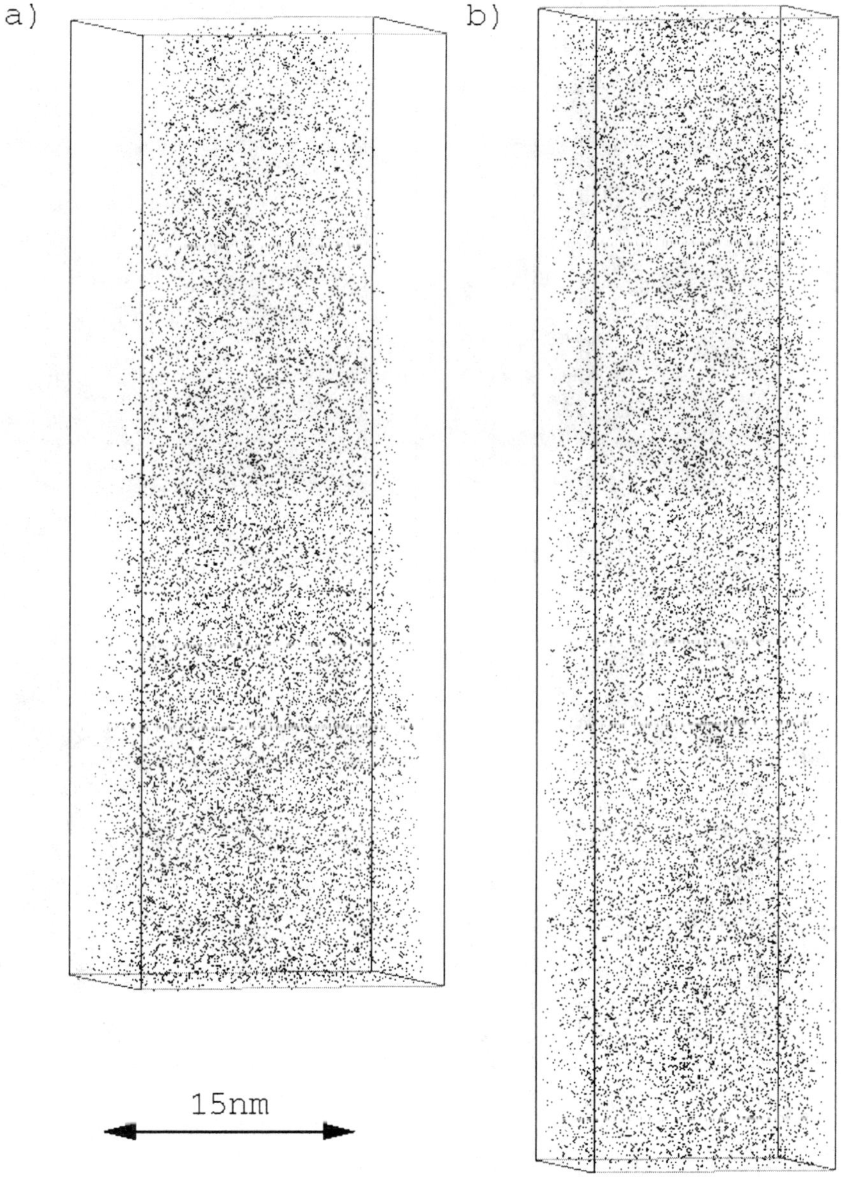

Figure 3 - *Atom map of (A) WF11, (B) 4WL773, (C) WK39, (D) WK87 and (E) WK87 (repeat examination) showing distributions of Cu, Mn, and Ni atoms.*

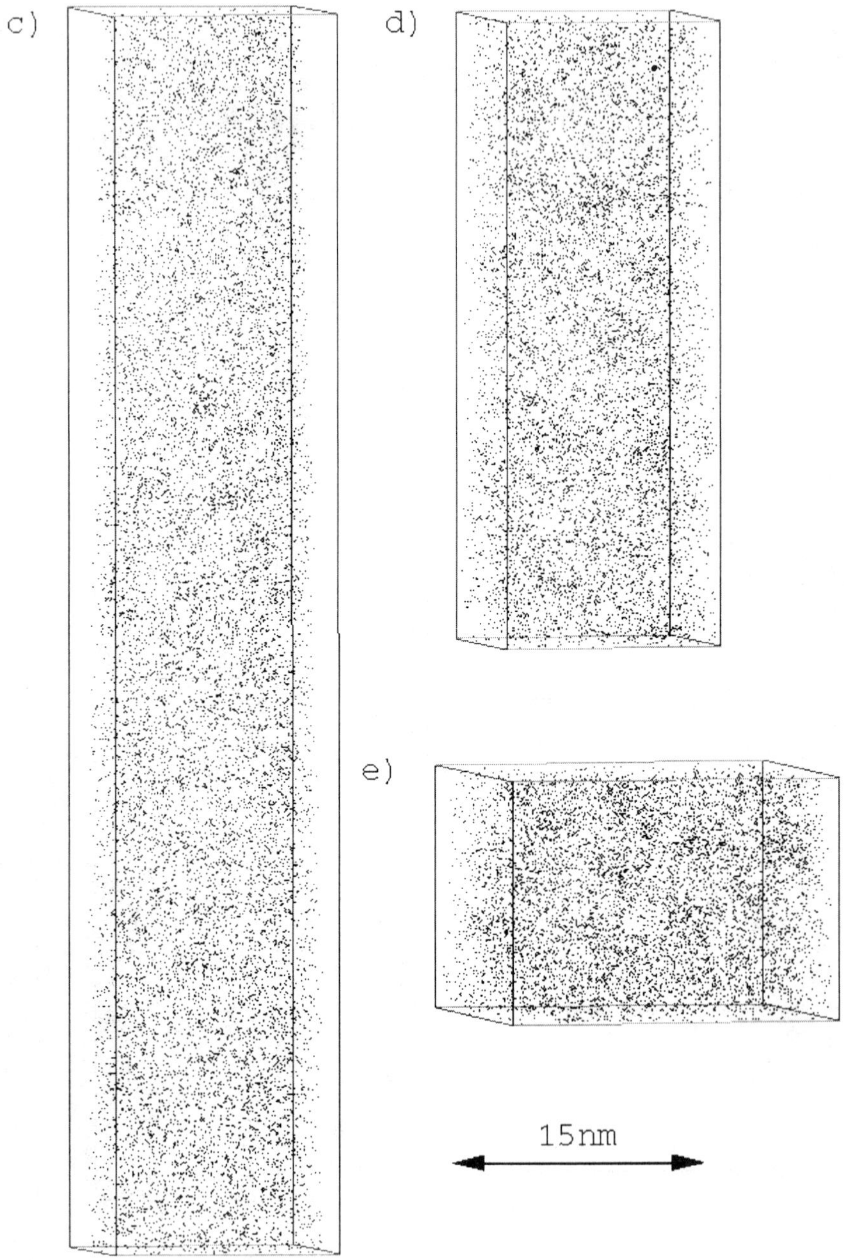

Figure 3 - *Atom map of (A) WF11, (B) 4WL773, (C) WK39, (D) WK87 and (E) WK87 (repeat examination) showing distributions of Cu, Mn, and Ni atoms.*

FEGSTEM analyzes were only performed on 4WL773. No evidence for irradiation-induced clusters was found. This is consistent with the atom probe data since the FEGSTEM does not have sufficient resolution to detect very fine scale composition fluctuations.

Several SANS experiments were performed on the low and intermediate Cu welds. They confirmed that the irradiation induced features are all extremely small, with a mean diameter of ~1.2nm (Table 7). Furthermore, the irradiation-induced volume fraction was found to increase with both the Cu content of the weld and dose. This is consistent with the atom probe observations which showed increasingly non-random distributions of Ni and Cu with dose and an increasing spatial association of both Ni and Cu with Mn with increasing dose and Cu content.

Further interpretation of the SANS data to estimate the number density of features requires knowledge of the irradiation induced feature compositions. This is discussed in the discussion section of this paper.

Table 7 - *Summary of SANS data obtained from low and intermediate Cu welds.*

Material	Dose (mdpa)	Dose rate (dpa/s)	Mean Feature Diameter (nm)	Volume fraction (nominal contrast factor)[1]
WK39	13.32	6.97E-9	1.2	0.082
WF11	14.42	5.96E-9	1.1	0.026
4WL772	32.12	5.80E-9	1.3	0.097
4WL773	32.18	5.81E-9	1.2	0.087
WK87	37.85	6.74E-9	1.3	0.166
4WL769	1.56	9.34E-11	Too small to quantify	

[1] Volume fractions calculations are based on the assumed nominal contrast for feature. The nominal contrast chosen was identical for all materials and conditions.

Results Obtained from Medium Copper Weld WG

OPoSAP analyses were performed on three irradiated samples of WG. A relatively small volume of material (~1400 nm^3) was obtained from the lowest dose/dose rate condition (WG114), which prevented a detailed statistical analysis. However, within the analyzed volume the Cu, Mn and Ni atoms appeared to be randomly distributed. Significantly greater volumes of both WG010 (~3000 nm^3) and WG102 (~5500 nm^3) were analyzed before the specimens failed. The atom maps in Figure 4 show not only that the distributions of Cu, Mn and Ni in both welds are not random, but also that the features are much better defined in the higher dose condition. Qualitatively, the enrichments in Cu, Mn and Ni in WG010 look like fluctuations whereas the equivalent enrichments in WG102 look like clusters.

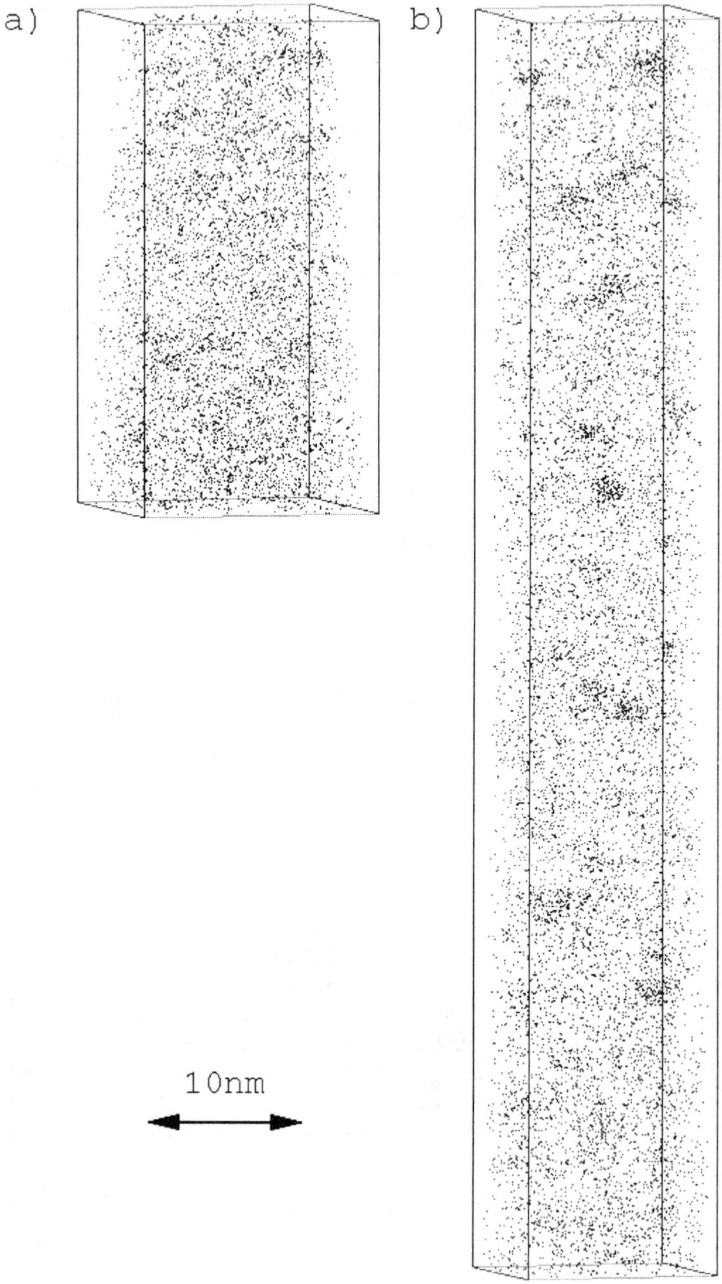

Figure 4 - *Atom map showing distribution of Cu, Mn and Ni in (a) WG010 and (b) WG102. Atom maps drawn to same scale as indicated by the scale marker.*

The compositions of some of these features were examined by locating the regions most highly enriched in Cu+Mn+Ni within a sphere of diameter 1nm (Table 8). The analysis confirms that as the dose increases the copper content of the clusters increases.

Table 8 - *Comparison of Cu, Mn, and Ni enriched regions in WG010 with those observed in WG102. (Concentration, wt%, within sphere of diameter 1nm).*

Sample	Dose (mdpa)	Si	P	Cr	Mn	Fe	Ni	Cu	Mo	Co	O	C
WG010	14	1	0	0	11	67	17	3.6	0.0	0	0	0
WG102	54	4.1	0	0	11	64	13	8.2	0	0	0	0

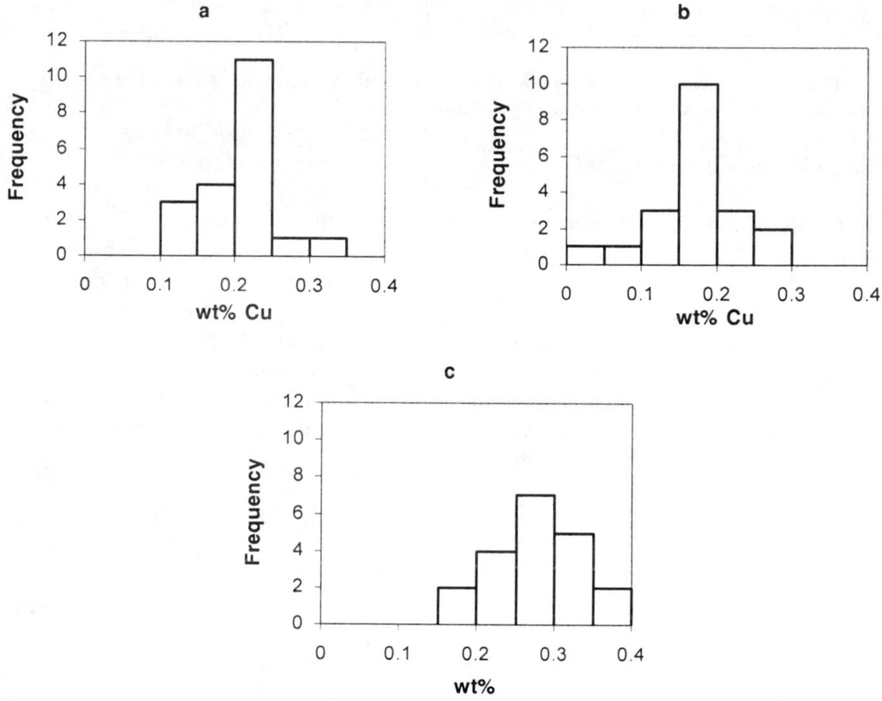

Figure 5 - *Comparison of copper spot distributions in (a) WG114, (b) WG010 and (c) WG102.*

FEGSTEM analyses were also performed on all three irradiated WG samples. The distribution of spot measurements, taken to avoid contrast features, demonstrated that the Cu distribution was becoming increasing non-random with increasing dose (Figure 5). However, individual contrast features were not sufficiently well defined to enable cluster composition estimates to be made in WG010. In WG102, it was possible to show that the clusters were enriched in Cu, Mn and Ni and probably Si confirming the observations made using the OPoSAP.

The SANS data confirmed that the features are extremely small (1.2-1.3nm) in the two lower dose conditions and slightly larger in the higher dose condition (1.5nm) and that the irradiation-induced volume fraction increases with increasing dose (Table 9).

Table 9 - *Summary of SANS data from analysis of WG.*

Material	Dose (mdpa)	Dose rate (dpa/s)	Mean Feature Diameter (nm)	Volume fraction (nominal contrast factor)[1]
WG010	14.05	6.52E-9	1.2	0.110
WG102	55.19	8.19E-9	1.5	0.206
WG114	1.45	8.68E-11	1.3	0.047

[1] Volume fractions calculations are based on the assumed nominal contrast for feature. The nominal contrast chosen was identical for all materials and conditions.

Results Obtained from Weld WV

OPoSAP experiments were performed on WV in three irradiated conditions (6.3, 13 and 30mdpa). Atom maps showing the distributions of Cu, Mn, and Ni are shown in Figure 6. Even in the lowest dose condition, the atom maps clearly show the presence of irradiation-induced clusters. With increasing dose, the clusters become better defined. Figure 7 shows a magnified view of one of the clusters observed in the highest dose condition (WV012). The cluster is clearly ramified and contains a significant Fe content.

The mean cluster composition in each material condition was estimated by locating the regions most highly enriched in Cu, Mn and Ni and calculating the composition within a sphere of diameter 1nm (Table 10).

Table 10 - *Summary of mean composition (wt%) of irradiation-induced features observed in WV.*

Material	Si	P	Mn	Fe	Ni	Cu	Mo
WV396	2.5	0.1	5.9	73	5.3	12.5	0.1
WV466	4.7	0.0	10.9	68	7.8	8.6	0.0
WV012	5.4	0.3	11.2	66	9.7	7.1	0.2

FEGSTEM analyses were also performed on WV466 and WV012. An analysis of contrast features and neighboring matrix confirmed that the irradiation-induced clusters contain Cu, Mn and Ni in approximately equal quantities and also a somewhat lower Si content.

SANS experiments have been performed on samples of WV irradiated over a range of dose rates as well as different doses. The results are summarized in Table 11.

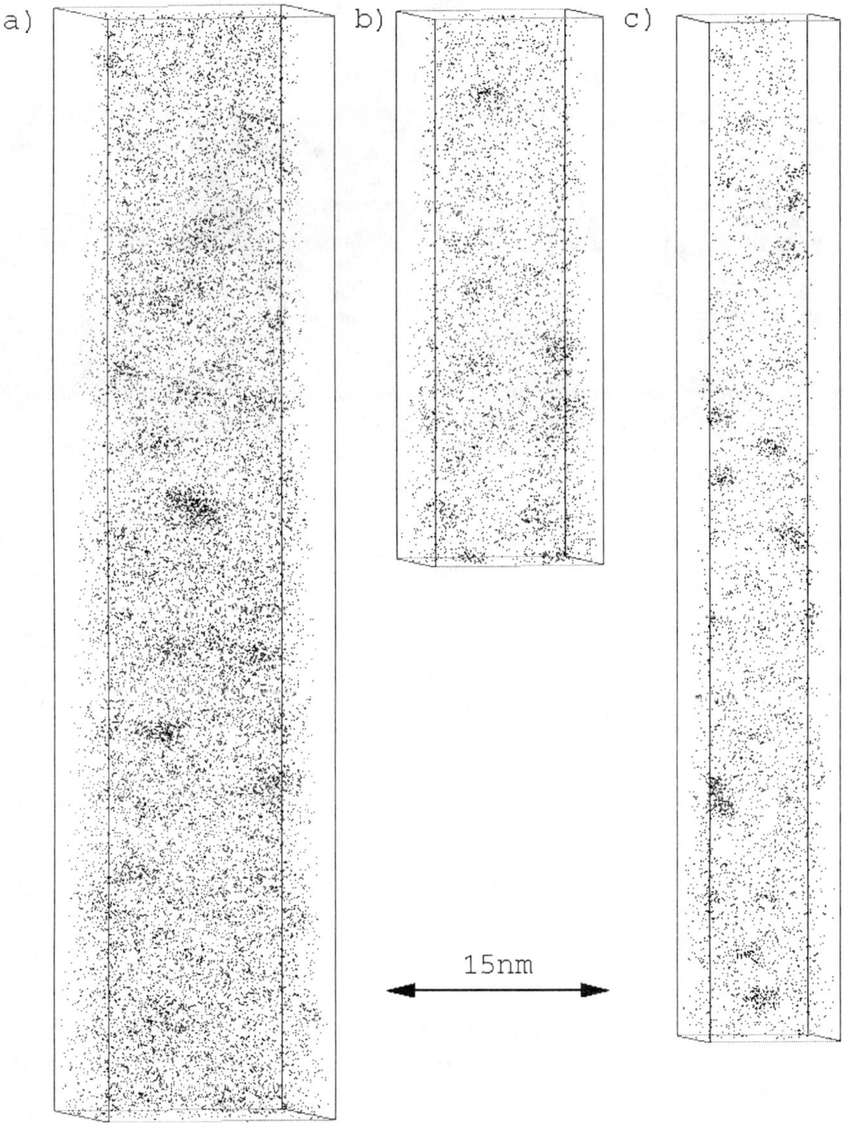

Figure 6 - *Atom maps showing distribution of Cu, Mn and Ni in (a) WV396, (b) WV466 and (c) WV012.*

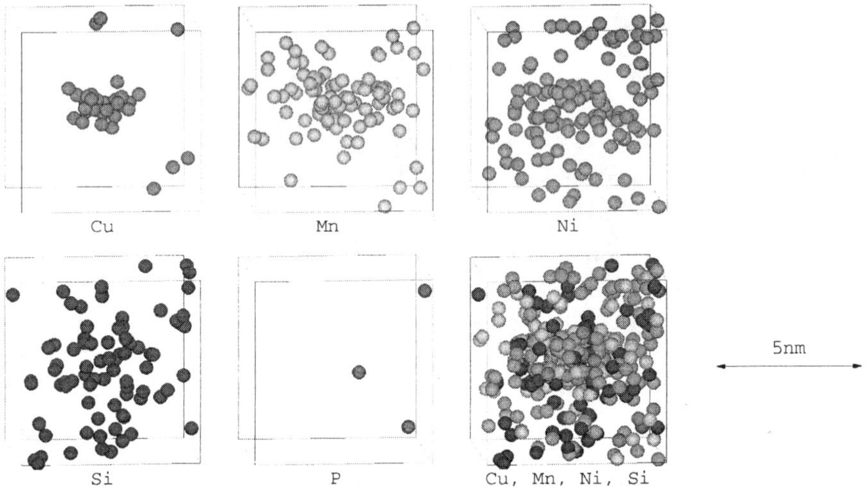

Figure 7 - *Magnified view showing one cluster observed in WV012 (extents of outline box 6x6x6nm³) showing distribution of Cu, Mn, Ni, Si, and P.*

Table 11 - *Summary of SANS data obtained from irradiated WV samples.*

Material	Dose (mdpa)	Dose rate (dpa/s)	Mean Feature Diameter (nm)	Volume fraction (nominal contrast factor)[1]
WV216	1.46	5.07E-9	1.1	0.110
WV214	9.29	6.56E-9	1.6	0.219
WV014	17.90	6.48E-9	1.7	0.261
WV012	30.10	6.23E-9	1.9	0.359
WV127	3.52	6.59E-10	1.7	0.208
WV400	12.11	6.30E-10	2.0	0.341
WV126	1.48	8.86E-11	1.9	0.198

[1] The volume fraction calculations are based on an assumed nominal contrast for feature. The nominal contrast chosen was identical for all materials and conditions.

Discussion

In this discussion we focus on two aspects, firstly, the self-consistency of the data from the different techniques, and secondly, the trends in the data as a function of Cu content, dose, and dose rate.

Self-Consistency of Results

OPoSAP vs. FEGSTEM - Both FEGSTEM and OPoSAP techniques are capable of detecting irradiation induced clusters. If the clusters are sufficiently well defined the OPoSAP can provide good estimates of their compositions. Similarly, the FEGSTEM can provide good estimates of the ratios of all elements except Fe in the clusters. Where both the FEGSTEM and OPoSAP gave evidence for well-defined clusters the estimated compositions from the two techniques were reasonably consistent with each other. In addition, visible clustering was observed in WG010 (14 mdpa) in the OPoSAP, but the clusters were not well defined. These is in agreement with the FEGSTEM observations of this material where no contrast features were observed, but the breadth of the Cu matrix spot measurements provided evidence of a non-random distribution of Cu.

A comparison of cluster compositional data from OPoSAP and FEGSTEM analyses is given in Table 12 below. The data from the OPoSAP have been converted to a ratio of elements ignoring iron in order to compare with FEGSTEM data. Both are presented in at. %. Both techniques showed that the clusters were found to be Mn, Ni and Cu enriched, with the Mn and Ni contents exceeding the Cu content. It is possible that the higher estimated P and Mo contents observed using the FEGSTEM technique may have resulted from coherent Bremstrauhlung radiation [12]. The mean composition of the visible clusters observed in the OPoSAP was ~67%Fe, 11%Mn, 10%Ni, 9%Cu and 3%Si.

Table 12 - *Comparison of cluster compositions determined by FEGSTEM (at.%) and by atom probe (at.%)*

Specimen	Technique	Mn	Ni	Mo	Cu	Si	P
WV466	FEGSTEM	14	36	4	26	7	14
	Atom Probe	34	25	0	27	15	0
WV012	FEGSTEM	20	27	4	20	15	14
	Atom Probe	33	29	1	21	16	1
WG102	FEGSTEM	34	17	0	14	32	2
	Atom Probe	35	25	0	26	13	0

OPoSAP vs. SANS - In the medium and high Cu welds WG and WV, clusters were detected by both techniques in all samples except the low dose sample of WG (1.5mdpa). Here, no visible evidence of Cu clusters was detected in the OPoSAP. In the other cases, the diameters measured by the OPoSAP were broadly consistent with the SANS data.

The contrast factor associated with the mean cluster composition observed from the OPoSAP analyses of the high Cu welds (WV) and high dose intermediate Cu weld (WG102) was used to determine the absolute irradiation-induced volume fraction and hence approximate number density of irradiation-induced features. In addition, two further scenarios were also evaluated, as detailed in Table 4. The results for the three scenarios for the high dose WV and WG samples are summarized below in Table 13. For comparison, the number density of features observed in the OPoSAP data is also presented. As previously stated, it is difficult to identify clusters from a 2D image.

However the clusters, particularly if they exist at a very high number density, can be more easily identified when the data are rotated in 3D on a workstation.

The analyses show that the estimated cluster number density is strongly dependent on the assumed feature composition. If it is assumed that the clusters do not contain Fe, then the estimated number density is lower than that observed by OPoSAP microanalysis. If it is assumed that the clusters contain ~33at.% Fe in which the effective mean magnetisation of each Fe atom is 0.7, then the SANS and OPoSAP number density data correlate reasonably well. However, the atom probe indicated that the clusters contain ~67%Fe. There are several possible explanations, which include the fact that analysis of OPoSAP data may lead to an overestimation of the Fe content of the clusters since the method used may, inadvertently, sample both cluster and a small amount of matrix. Furthermore, the effective magnetisation of Fe atoms within the clusters and the vacancy content of the clusters both strongly affect the scattering response, and neither parameter is known. Significantly, for the three scenarios examined, the resultant reduction in matrix levels is not strongly dependent on the precise details of the clusters.

Table 13 - *Comparison of estimated cluster number densities using the three scenarios detailed in* Table 4 *for high dose WV and WG welds and comparison with OPoSAP estimates.*

Steel	Scenario (see Table 4)	Assumed Fe content of clusters	Cluster V_f	Cluster N_d ($x10^{23}$ m^{-3})	Reduction in matrix levels (at.%) Cu	Mn	Ni	Si	N_d from OPoSAP Data ($x10^{23}$ m^{-3})
WV012	1	67	0.03026	84	0.24	0.29	0.26	0.08	~50
	2	33	0.01703	47	0.30	0.36	0.35	0.10	~50
	3	0	0.00971	27	0.26	0.30	0.32	0.09	~50
WG102	1	67	0.01713	97	0.13	0.16	0.15	0.04	~60
	2	33	0.00967	55	0.17	0.21	0.20	0.06	~60
	3	0	0.00553	31	0.15	0.17	0.18	0.05	~60

All of the SANS data, interpreted using scenario one, are detailed in Table 14. The most important area of disagreement with the OPoSAP data is that the SANS data provided strong evidence for strong scattering centers in cases where FEGSTEM and OPoSAP provided no evidence for visible clusters, but where the OPoSAP provided strong evidence for non-random distributions of solute elements. The result was particularly surprising since the OPoSAP has sufficient resolution to detect clusters that are ~1.2 nm in diameter as proven by the examination of irradiation induced clusters in low Ni welds [*13*].

This is clearly an area requiring further research. In this paper we tentatively suggest that SANS is sensitive to the fluctuations observed in the OPoSAP. Although, speculation, this may be because point defects are associated with the fluctuations. In the discussion below we interpret these features in the low Cu high nickel welds as solute fluctuations rather than clusters.

Table 14 - *Estimated cluster number density (N_d) from SANS analysis of high Ni welds and comparison with OPoSAP estimates.*

Assumed cluster composition – 67Fe-11Mn-10Ni-9Cu-3Si (at.%) (Scenario 1)							
			Reduction in matrix levels (at.%)				
Steel	Cluster V_f	Cluster N_d ($x10^{23}$ m^{-3})	Cu	Mn	Ni	Si	Estimated Cluster N_d from OPoSAP Data ($x10^{23}$ m^{-3})
WF11	0.00217	31.2	0.017	0.021	0.019	0.006	Fluctuations
4WL772	0.00828	81.0	0.064	0.079	0.072	0.021	No Data
4WL773	0.00743	72.6	0.058	0.071	0.064	0.019	Fluctuations
WK39	0.00686	75.9	0.053	0.065	0.059	0.018	Fluctuations
WK87	0.01389	96.7	0.108	0.132	0.120	0.036	Fluctuations
WG114	0.00391	34.0	0.030	0.037	0.034	0.010	Fluctuations
WG010	0.00915	100	0.071	0.087	0.079	0.024	Not determined
WG102	0.01713	96.9	0.133	0.163	0.148	0.044	60
WV012	0.03026	84.3	0.236	0.288	0.262	0.079	50
WV014	0.02200	82.6	0.171	0.209	0.190	0.057	No Data
WV126	0.01669	45.8	0.130	0.159	0.144	0.043	No Data
WV127	0.01754	71.9	0.137	0.167	0.152	0.046	No Data
WV214	0.01846	91.1	0.144	0.176	0.160	0.048	No Data
WV216	0.00927	149	0.072	0.088	0.080	0.024	No Data
WV400	0.02875	68.6	0.224	0.274	0.249	0.075	No Data

Data Trends

The data trends are plotted in Figures 8, 9 and 10 and are discussed for low and medium/high copper welds in the next two sections. The curves drawn in the figures are intended to indicate the possible data trends; clearly they are entirely speculative in the case of the medium and low dose rate data. The number density curves are derived mathematically from the other two. As described earlier, the volume fractions and number densities depend on the assumptions made.

Low Cu welds - A summary is given of the main trends in Table 15. The main feature is that solute fluctuations are observed in all these welds and there is no evidence from the OPoSAP or FEGSTEM of visible clusters. In the very low Cu welds WF and WL (<0.04 Cu wt.%) there is strong evidence for Mn and Ni being both non-random and spatially correlated. In both alloys there was no evidence of a non-random distribution of Cu, this may either reflect that Mn and Ni associate before Cu or alternatively, the low level of Cu makes a statistical detection of a non-random distribution difficult. In WF, Si was found to be non-random, in contrast to the random distribution observed in WL. The

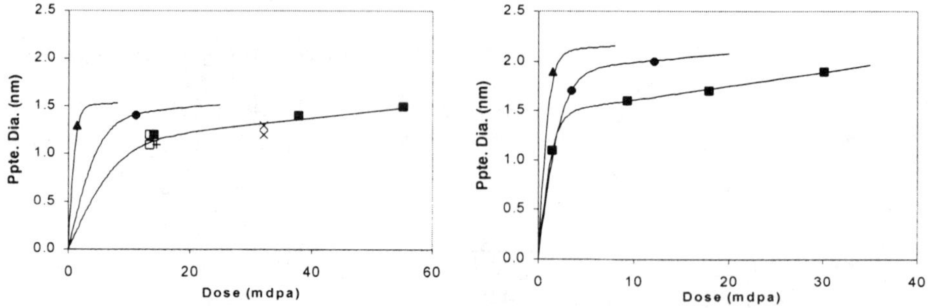

Figure 8 - *Effect of irradiation dose on mean diameter of SANS scattering centers for high nickel welds with low and medium copper (left) and high copper (right)*

Key to Figures 8, 9 and 10:

Approximate dose rates in dpa/sec are: 7E-9 (■), 7E-10 (●), 8E-11 (▲).
Copper levels for right figure are : 0.5 to 0.62 wt% (all data points)
Copper levels for left figure are: 0.240 wt% (filled symbols)
0.136 wt% (open squares)
0.038 to 0.052 wt% (x)
0.022 wt% (+)

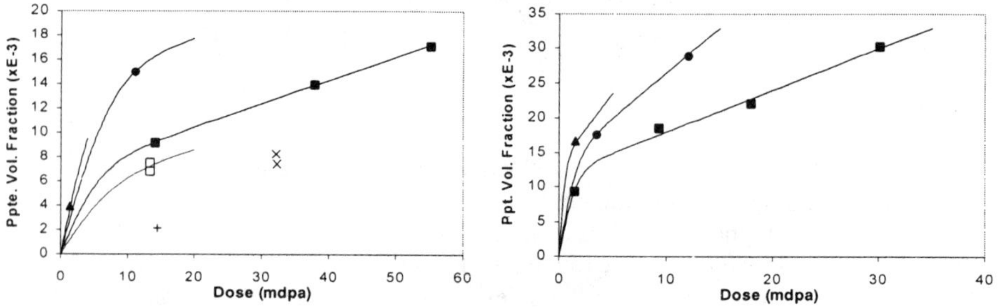

Figure 9- *Effect of irradiation dose on volume fraction of SANS scattering centers for (a) High Nickel Welds with Low and Medium Copper and (b) High Copper.*

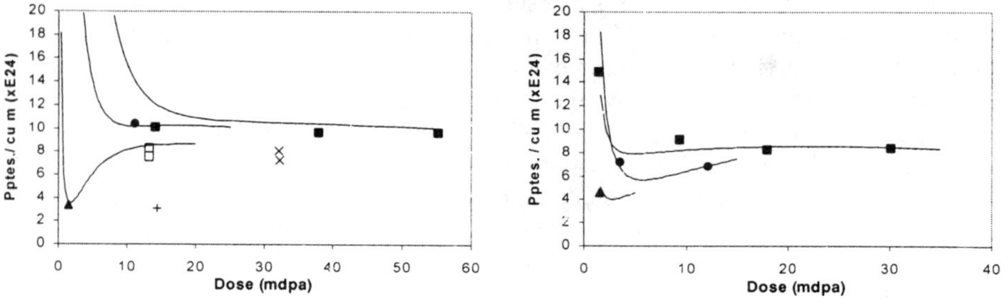

Figure 10 - Effect of irradiation dose on number density of SANS scattering centers for high nickel welds with (a) low and medium copper and (b) high copper.

data on WK demonstrates that with increasing fluence the 'degree' of non-randomness increases. At both doses the Mn, Ni and Si are non-random. At the higher dose Cu is observed to be non- random, and the spatial correlations of the non-random elements become more pronounced. This suggests that the degree of non-randomness increases as the dose increases.

Table 15 - *Summary of low Cu data from the FEGSTEM, OPoSAP and SANS.*

Technique	FEGSTEM	OPoSAP	SANS (Volume fraction, V_f, using scenario 1, D is feature diameter)
WF low Cu	N/A	Mn, Ni non random, Cu Si random Mn and Ni spatially correlated	Scattering centers detected $D = 1.1$ nm $V_f = 0.0022$
4WL	No evidence of clusters	Mn, Ni and Si non-random Cu random Mn and Ni correlated	Scattering centers detected $D = 1.2\text{-}1.3$ nm $V_f = 0.007\text{-}0.008$
WK low dose	No evidence of clusters	Mn, Ni and Si non random No significant co-segregation	Scattering centers detected $D = 1.2$ nm $V_f = 0.007$
WK high dose	No evidence of clusters	Mn, Ni, Si and Cu non random Mn Cu, Mn Ni, and Mn Si correlated	Scattering centers detected $D = 1.3$ nm $V_f = 0.014$

As discussed above it is not possible to unambiguously identify the features giving rise to the SANS response in the low Cu welds. The assumption is that the fluctuations are responsible as discussed above. It is interesting to note that the diameter only increases slowly with dose (Table 15), whilst the volume fraction increases rapidly. Note however, that this analysis is based on the assumption that the features are the same in all material conditions. Since they develop with dose, their scattering contrast will also increase and it is therefore likely that the analysis overestimates the rate of increase in volume fraction.

Table 16 - Summary of medium and high Cu data from the FEGSTEM, OPoSAP, and SANS.

Technique	FEGSTEM	OPoSAP	SANS (Volume fraction, V_f, using scenario 1, D is feature diameter)
WG low dose 1.5 mdpa	No evidence for clusters.	Small volume analyzed Mn, Ni, and Cu all random	Scattering centers detected D = 1.3 nm V_f = 0.0039
WG medium dose 14 mdpa	Matrix Cu distributions broader - consistent with increased clustering of Cu. No visible contrast features detected	Cu enriched regions detected. Irradiation reduced matrix Cu by <0.08wt%	Scattering centers detected D = 1.2 nm V_f = 0.009
WG high dose 54 mdpa	Cu enriched regions detected by direct observation. Matrix Cu distributions in broad agreement with this. Irradiation reduced matrix Cu by 0.07wt%	Cu, Mn, Ni enriched regions detected D ~1.5nm. Irradiation reduced matrix Cu by 0.1wt%	Scattering centers detected D = 1.5 nm V_f = 0.017
WV 6.3 and 9.3 mdpa	N/A	In 6.3 mdpa sample, Cu enriched clusters detected. D ~ 1-2nm	In 9.3 mdpa sample, scattering centers detected D = 1.6 nm V_f = 0.018
WV 13 and 17.9 mdpa	13 mpda sample - Cu enriched regions detected by direct observation. Matrix Cu distributions in agreement with this. Irradiation reduced matrix Cu by 0.13wt%	13 mpda sample - Cu enriched clusters detected. Irradiation reduced matrix Cu by 0.14wt% D ~ 1.5-2nm	17.9 mdpa sample - Scattering centers detected D = 1.7 nm V_f = 0.022
WV 30 mdpa	Cu enriched regions detected by direct observation Matrix Cu distributions broad in agreement with this. Irradiation reduced matrix Cu by 0.1wt%	Cu enriched clusters detected. D ~ 2nm.	Scattering centers detected. D = 1.9 nm V_f = 0.030

Medium to High Cu Welds - The observations are summarized in Table 16. Well-defined clusters form once Cu has increased to 0.25wt%. These clusters evolve as the dose increases. On increasing the Cu concentration there is enhanced precipitation and loss of matrix Cu. In this composition range (intermediate to high Cu), at a sufficiently high dose the Mn/Ni/Cu/Si clusters have a composition that is relatively insensitive to bulk Cu. Most importantly, both the analysis of the SANS data and the FEGSTEM and OPoSAP data indicate that there is still significant Cu to precipitate out. For instance in WG, the matrix copper prior to irradiation is ~0.25wt.%. Following irradiation to 30mdpa only ~0.1% Cu had precipitated out.

The effect of irradiation dose on the scattering center diameter (directly determined from the SANS analysis) is shown in Figure 8. Three dose rates are represented as indicated in the key to the figures. As noted earlier, the lines through the points are speculative. The mean diameter of scattering centers changes only slowly with irradiation dose beyond a few mdpa. The data indicate that the diameters in the high Cu weld are higher than in the medium Cu weld. In addition, both plots show an effect of dose rate, i.e., the scattering center diameter increases more rapidly (in terms of dose) with decreasing dose rate, but may ultimately reach the same maximum value.

The volume fraction of scattering centers is shown in Figure 9. Within the dose range plotted, the volume fractions show no evidence of reaching a plateau. Unlike the plot of diameter versus dose for the medium and low copper materials, the volume fraction is sensitive to the copper levels within this range. For high copper material and for material with 0.24 wt% copper, a dose rate effect is indicated as for diameter. The number of scattering centers remains fairly constant beyond a relatively low dose (around 5 to 10 mdpa).

Overall Data Trends

There appears to be strong dependence on the Cu level in the observed behavior, which is summarized in Table 17 below and shown diagrammatically in Figure 11. With increasing Cu content and dose, the irradiation-induced clustering becomes better defined. At low doses and low copper contents only compositional fluctuations were observed. At high doses and high copper contents well-defined clusters were observed and could be analyzed to determine their size, number density and composition. Between these two categories, a third category, 'cloud' has been identified. Here, solute clustering was clearly visible by eye, but not sufficiently well-defined to be able to assign them a diameter or describe them as real clusters. All data points represent specimens irradiated at high dose rate (approx. 7E-9 dpa/sec) except for the point at 1.45 mdpa for which the dose rate was approximately 8E-11 dpa/sec.

Mechanistic Interpretation

Odette has pointed out the importance of Mn and Ni rich clusters in high Ni alloys [*14*]. He argued that the existence of Mn Ni rich clusters was consistent with the results of thermodynamic calculations, which demonstrated that at relatively low temperatures a new phase field opens in the Mn Ni rich corner of the Fe-Cu-Mn-Ni rich quaternary. Interestingly, the thermodynamic models predict that Mn Ni rich precipitates containing little Cu can contain appreciable levels of Fe, in agreement with the OPoSAP data. Odette also argued that Cu was essential for the formation of Mn Ni rich clusters, and

Table 17. *Summary of microstructural results on high Ni welds.*

Very Low Cu (WF and WL)	Mn and Ni are the first elements to become non-random. In the WL specimens examined the dose is too low for Mn/Ni rich clusters/precipitates to develop.
Low Cu (WK)	As the Cu increases both the Mn and Ni become non-random and, as the dose increases, the Cu becomes non-random as well i.e. the data on WL and WK are consistent with Mn, Ni and Cu fluctuations developing.
Intermediate and high Cu (WG and WV)	Well-defined clusters form once Cu has increased to 0.25wt% . These clusters evolve as the dose increases. On increasing the Cu concentration there is enhanced precipitation and loss of matrix Cu. In this composition range (intermediate to high Cu), at a sufficiently high dose the Mn/Ni/Cu/Si precipitates have a composition that is relatively insensitive to bulk Cu.

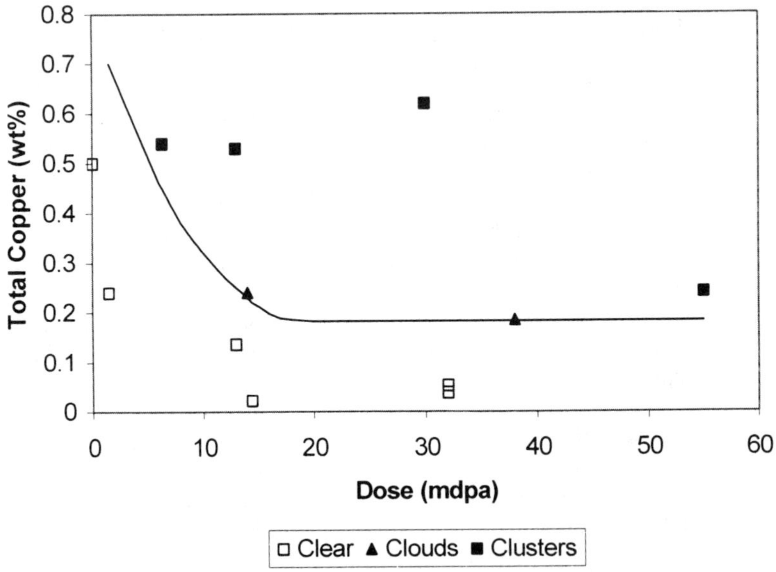

Figure 11 - *Summary of specimens examined by atom probe showing conditions under which visible clusters form*

that cluster formation would occur more readily in alloys containing Cu. This is in agreement with the present observations.

In cases where clustering is not apparent, there is nevertheless significant hardening. In Figure 11 the four specimens with doses between 10 and 20 mdpa are, in ascending order of Cu, WF11, WK39, WG10 and WV466. The irradiation-induced hardness changes for these were 32, 32, 45, and 93 VPN respectively. Clearly, "matrix damage" is substantial. However, it remains to be seen whether matrix damage is related to the fluctuations.

Conclusions

A key conclusion from the studies of the medium to high Cu welds is that there is no evidence for saturation in the precipitation behavior in the dose range examined. In particular in the medium to high Cu steels relatively small amounts of Cu precipitated out by 50-60mdpa. In low Cu alloys, clusters were not observed in the OPoSAP or FEGSTEM, but there was strong evidence of composition fluctuations in Mn and Ni, and also Cu at high fluences. It is most likely that these are a precursor to the clustering observed in the medium to high Cu materials. Overall Cu still appears to have a controlling effect, and it is interesting to note that the clusters observed at high doses in the medium and high Cu welds had similar compositions. Furthermore, there is evidence of flux effects with an increased volume fraction of clusters at the lower dose rates in the high Cu welds.

A number of important technique issues emerged. Firstly, statistical analysis of OPoSAP elemental distributions is potentially very important in giving information in the early stages of cluster formation. Secondly, the analysis of SANS data needs greater refinement, when solute fluctuations are observed in the OPoSAP. It is particularly important to establish a model that is self-consistent with both OPoSAP and SANS observations.

References

[1] Buswell, J.T., English, C.A., Hetherington, M.G., Phythian, W.J., Smith, G.D.W., Worrall G.M., "An Analysis of Small Clusters Formed in Thermally Aged and Irradiated FeCu and FeCuNi Model *Alloys," Effects of Radiation on Materials: 14th International Symposium (Volume II), ASTM STP 1046*, American Society for Testing and Materials, Philadelphia, 1990, pp. 127-153.

[2] Kostorz, G., "Small Angle Scattering and its Applications to Material Science," *Treatise on Materials Science and Technology', Vol. **15** 'Neutron Scattering*, G. Kostorz, Ed, Academic Press, New York, 1979, pp. 227-289.

[3] Cerezo, A., Godfrey, T.J., Sijbrandij, S.J., Smith, G.D.W., Warren, P.J., *Review of Scientific Instruments*, Vol. **59,** 1998, p. 862.

[4] Williams, T. J., Ellis, D., Swan, D. I., McGuire, J., Whalley, S. P., English, C. A., Venables, J. H., and de la cour Ray, P. H. N., "The Influence of Copper, Nickel and Irradiation Temperature on the Irradiation Shift of Low Alloy Steels," *Proceedings of the Second International Symposium on Environmental Degradation of Materials in*

Nuclear Power Systems - Water Reactors, TMS AIME/ANS/NACE, September 1985, ANS Illinois, 1986, pp. 393-399.

[5] Williams, T. J., Burch, P. R., English, C. A., and de la cour Ray, P. H. N., "The Effect of Irradiation Dose Rate and Temperature, and Copper and Nickel Content, on the Irradiation Shift of Low Alloy Submerged Arc Welds," *Proceedings of the Third International Symposium on Environmental Degradation of Materials in Nuclear Power Systems - Water Reactors*, TMS/ASM/ANS/NACE, G. J. Theus and J. R. Weeks, Eds., August/September 1987, Michigan, 1998, pp. 121-131.

[6] Williams, T. J., and Phythian, W. J., "Electron Microscopy and Small Angle Neutron Scattering Study of Precipitation in Low Alloy Steel Submerged-Arc Welds," *Effects of Radiation on Materials, ASTM STP 1270*, D. S. Gelles, R. K. Nanstad, A. S. Kumar, and E. A. Little, Eds., American Society for Testing and Materials, West Conshohocken, PA, 1996, pp. 191-205.

[7] Hyde, J.M., Miller, M.K., Hetherington, M.G., Cerezo, A., Smith, G.D.W. and Elliott, C.M., "Spinodal Decomposition in Fe-Cr alloys : Experimental Study at the Atomic Level and Comparison with Computer Models. II. Development of Scale and Composition Amplitude," *Acta Metallurgica,* Vol. **43**, 1995, p. 3403.

[8] Everitt, B.S., "The Analysis of Contingency Tables," Chapman and Hall, London, 1977.

[9] Hetherington, M.G., Cerezo, A., Hyde, J.M., Smith, G.D.W. and Worrall, G.M., "Statistical Analysis of Atom Probe Data, " J. de *Physique*, Vol. 47-C7, 1986, p. 495.

[10] Miller, M.K. and Smith, G.D.W., "Atom Probe Microanalysis: Principles and Applications," Materials Research Society, 1989.

[11] Sumiyama, K., Yoshitake, T., Nakamura, Y., "Magnetic Properties of Metastable bcc and Fcc Fe-Cu Alloys Produced by Vapor Quenching" *Journal of the Physical Society of Japan*, Vol. **53** 1984 p. 3160.

[12] Spence, J.H.C, Titchmarsh, J.M. and Long, N., *Microbeam Analysis*, J. T. Armstrong, Ed., San Francisco Press, 1985, p. 349.

[13] Unpublished AEA Technology research.

[14] Odette, G. R., Materials Research Symposium Proceedings Vol 373, 1995, p. 137.

Pressure Vessel Steels—Mechanical Properties

Matthew J. DeVan[1] and Wayne A. Pavinich[2]

An Evaluation of the Effect of Radiation Environment on Linde 80 Reactor Vessel Welds

Reference: DeVan, M. J. and Pavinich, W. A., **"An Evaluation of the Effect of Radiation Environment on Linde 80 Reactor Vessel Welds,"** *Effects of Radiation on Materials: 20th International Symposium, ASTM STP 1405,* S. T. Rosinski, M. L. Grossbeck, T. R. Allen, and A. S. Kumar, Eds., American Society for Testing and Materials, West Conshohocken, PA, 2001.

Abstract: The B&W Owners Group (B&WOG) developed the Master Integrated Reactor Vessel Surveillance Program (MIRVP) to encompass all domestic operating pressurized water reactors (PWRs) with reactor vessels that contain Linde 80 submerged-arc welds. These PWRs include the reactor vessels fabricated by Babcock & Wilcox (B&W) that include B&W-designed 177-Fuel Assembly (FA) plants and Westinghouse-designed plants. The B&WOG Reactor Vessel Working Group has sponsored studies over the past several years to develop an understanding of the mechanisms and trends of radiation embrittlement for reactor vessel submerged-arc welds fabricated using Linde 80 weld flux. As part of the MIRVP, a Westinghouse-designed capsule was included to provide a comparison of irradiation data in the Westinghouse neutronic environment with the B&W 177-FA environment. This paper presents the effect of irradiation environment on the embrittlement trends of Linde 80 submerged-arc welds based on mechanical test data available through the B&WOG and the literature, and existing trend curves.

Keywords: submerged-arc welds, reactor vessel, radiation embrittlement, irradiation temperature

Introduction

The Master Integrated Reactor Vessel Surveillance Program (MIRVP) was developed to encompass all domestic operating pressurized water reactors (PWRs) with reactor vessels that contain Linde 80 submerged-arc welds [*1,2,3*]. These PWRs are the reactor vessels fabricated by Babcock & Wilcox (B&W) that include B&W-designed 177-Fuel Assembly (FA) plants and several designs of Westinghouse plants. The operating environments of the Westinghouse-designed and B&W-designed reactors are similar, but not exactly the same. It was not clear if the small environmental differences would cause measurable differences in embrittlement response of the reactor vessels. An evaluation

[1]Supervisory Engineer, Engineering and Licensing, Framatome Technologies, Inc., P. O. Box 10935, Lynchburg, VA 24506.
[2]President, Invex Engineering, 1113 Laurel Hill Road, Knoxville, TN 37923.

was presented in BAW-1543, Revision 3 indicating that these small differences would not significantly affect embrittlement response [*1*]. The implication is that data derived from materials irradiated in B&W-designed and Westinghouse-designed reactors could be used interchangeably in reactor vessel integrity analyses. The U.S. Nuclear Regulatory Commission (NRC) Staff approved the MIRVP concluding: *"The surveillance programs documented in report BAW-1543, Revision 3, should be capable of monitoring the effect of neutron irradiation and thermal environment on the fracture toughness of the ferritic reactor vessel beltline materials in the plants that are participating in the material surveillance program"* [*4*]. As part of the MIRVP, a capsule (designated as "W1") was specifically fabricated, irradiated, and tested to benchmark potential differences in environmental effects between B&W-designed and Westinghouse-designed reactors.

The W1 Capsule was inserted and irradiated in the Surry Unit No. 2 reactor vessel to provide a comparison between the effects of the irradiation environments on mechanical test data between Westinghouse-designed reactor vessels and B&W-designed 177-FA reactor vessels. The W1 Capsule contained compact fracture and Charpy V-notch specimens that were fabricated from three Linde 80 submerged-arc weld metals: SA-1526 (weld wire heat 299L44/flux lot 8596), SA-1585 (weld wire heat 72445/flux lot 8597), and WF-70 (weld wire heat 72105/flux lot 8669). The materials in the W1 Capsule are from the same archive sources as the weld metals included in the MIRVP capsules TMI2-LG1, CR3-LG1, CR3-LG2, and DB1-LG1 that were previously irradiated in B&W 177-FA host reactors (i.e., Three Mile Island Unit 2, Crystal River Unit 3, and Davis-Besse). The chemical compositions of these weld metals are given in Table 1. This test matrix was designed to isolate the radiation environmental effects and to allow direct evaluation of these effects.

Table 1 – *Chemical Composition of Linde 80 Weld Metals Irradiated in MIRVP Capsules W1, TMI2-LG1, CR3-LG1, CR3-LG2, and DB1-LG1*

	Chemical Composition, wt%								
Weld Metal	C	Mn	P	S	Si	Cr	Ni	Mo	Cu
SA-1526	0.09	1.53	0.013	0.017	0.53	0.08	0.70	0.42	0.37
SA-1585	0.08	1.45	0.016	0.016	0.51	0.09	0.59	0.38	0.22
WF-70	0.09	1.63	0.018	0.009	0.54	0.11	0.58	0.40	0.39

Evaluation of the W1 Capsule Results

The parameters that are known to affect radiation embrittlement are neutron flux, neutron energy spectrum, gamma heating, and irradiation temperature. The neutron flux values for the capsules in this evaluation are within one order of magnitude [*1*], and the flux effects are expected to be minimal within this range. It has been demonstrated that radiation temperature has a slight effect on embrittlement and gamma heating may affect the embrittlement of capsule specimens [*5,6*]. The potential effect of gamma heating has been previously evaluated [*7*]; and it was demonstrated that the maximum increase in irradiation temperature caused by gamma heating is approximately 4°C (6°F). Therefore, the W1 Capsule results are evaluated considering the effect of irradiation temperature and the expected data scatter that is inherent in ferritic steels. Irradiation temperatures are defined as the reactor vessel inlet temperature of the respective reactor vessel where the reactor vessel surveillance capsule was irradiated. This paper is based on the results of the W1 Capsule mechanical test data and the data from weld metals fabricated with the same weld wire heats.

Review of W1 Capsule Results

Direct comparison of mechanical test data of the three Linde 80 welds (SA-1526, SA-1585, WF-70) indicate that the upper-shelf fracture toughness and the decrease in upper-shelf Charpy energy are similar for each weld regardless of the irradiation location. However, the 41 J (30 ft-lb) shifts for welds SA-1526 and WF-70 that were irradiated in Surry Unit No. 2 appear to be much larger than those from identical welds irradiated in the B&W-designed reactors. Table 2 shows the comparison of the Charpy 41 J (30 ft-lb) shifts and upper-shelf energy (USE) decreases from the W1 Capsule compared with other MIRVP capsule data. Comparison of the USE data for all three Linde 80 submerged-arc weld metals indicate the decreases are similar regardless of where the material was irradiated. The 41 J (30 ft-lb) shifts are equivalent for weld metal SA-1585 regardless of the irradiation location. However, the disparity in the 41 J (30 ft-lb) shift data for weld metals SA-1526 and WF-70 must be evaluated to determine the factors that affect the disparity.

Table 2 - *Summary of 41 J (30 ft-lb) Shift and Upper-Shelf Energy Decrease Data for the Linde 80 Weld Metals SA-1526, SA-1585, and WF-70*

Material	Capsule	Fluence, n/cm^2 x 10^{19}	Irradiation Temperature, °C (°F)	41 J (30 ft-lb) Shift, °C (°F)	Upper-Shelf Energy Decrease, J (ft-lb)
SA-1526	TMI2-LG1	0.830	291 (556)	120.0 (216)	42.7 (31.5)
SA-1526	W1	0.669	284 (543)	145.6 (262)	49.5 (36.5)
SA-1585	CR3-LG1	0.510	291 (556)	77.2 (139)	31.9 (23.5)
SA-1585	W1	0.660	284 (543)	76.7 (138)	38.0 (28.0)
SA-1585	CR3-LG2	1.670	291 (556)	91.1 (164)	35.3 (26.0)
WF-70	TMI2-LG1	0.585	291 (556)	67.8 (122)	14.9 (11.0)
WF-70	W1	0.644	284 (543)	103.3 (186)	19.7 (14.5)
WF-70	DB1-LG1	0.663	291 (556)	76.1 (137)	20.3 (15.0)
WF-70	CR3-LG2	1.190	291 (556)	70.6 (127)	21.7 (16.0)

Evaluation of 41 J (30 ft-lb) Shift Data

There are two possible explanations for the disparities in the 41 J (30 ft-lb) shift data for the SA-1526 and WF-70 weld metals:

1. There is a significant difference in the irradiation environments between B&W-designed and Westinghouse-designed reactors causing the observed behavior. The only known parameter that could cause the difference in behavior is the different irradiation temperatures of Surry Unit No. 2 reactor and B&W-designed reactors. However, if irradiation temperature is causing the apparent anomalous shift behavior, it is inconsistent that the 41 J (30 ft-lb) shift behavior of weld metal SA-1585 is uniform regardless of irradiation environment? Neutron spectral and flux differences can be ruled out as a source of the difference because a) neutron spectra are similar between B&W-designed and Westinghouse-designed reactors and b) neutron flux values are within one order of magnitude; therefore no significant flux effect will be seen within this range of neutron flux.

2. The data variation is within the expected scatter of the transition region. Previous work sponsored by the B&W Owners Group (B&WOG) and work performed at the Oak Ridge National Laboratory (ORNL) has shown large variations in 41 J (30 ft-lb) shifts, as high as ±18°C (±32°F) [*8*] (Figure 1). The anomalous shift values of weld metals SA-1526 and WF-70 are apparently within this scatter band. However, further analysis is necessary to verify the variations are within the expected scatter.

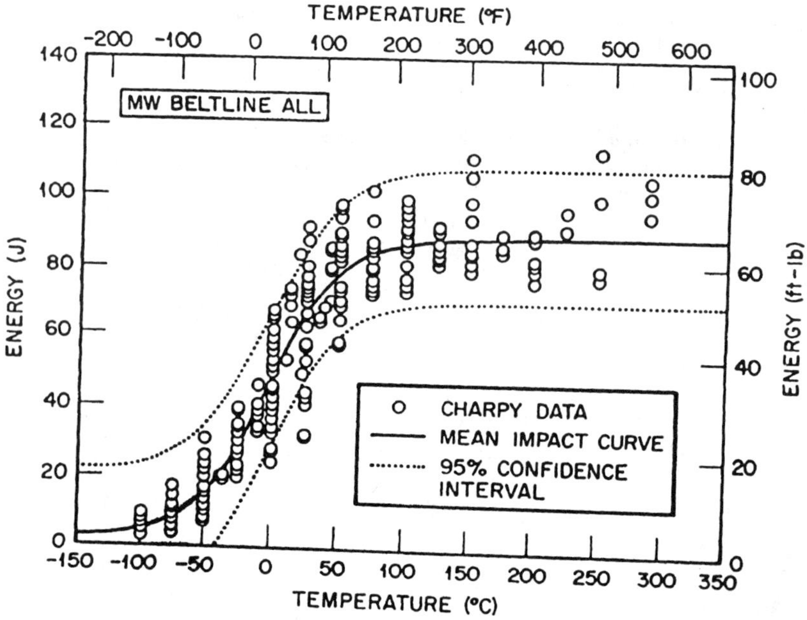

Figure 1 - *Large Scatter of Unirradiated Charpy Data for Linde 80 Weld Metal WF-70*

The Effect of Irradiation Temperature on Embrittlement

Previous work sponsored by the B&WOG indicated that there is a small irradiation temperature effect on embrittlement. This conclusion is based on a theoretical model of radiation embrittlement and a slight improvement in statistical parameters that results when 41 J (30 ft-lb) shift data are adjusted 1-degree for each 1-degree of irradiation temperature difference. These findings are similar to other evaluations by Odette, et al and Stoller [*5,6*]. Eason, et al [*9*] found a small improvement in standard error when an irradiation temperature parameter was added to the correlation of all complete 41 J (30 ft-lb) shift data sets. This parameter effectively adjusts the shift prediction approximately 1-degree for each 1-degree decrease in irradiation temperature. This correlation is based on a large number of 41 J (30 ft-lb) shift data sets (609 total) from all reactor vessel surveillance programs of domestic nuclear power plants. Although it is recognized that there is a small irradiation temperature effect, it is not large enough to account for the apparent disparities observed for the W1 Capsule 41 J (30 ft-lb) shift values when compared with those data from B&WOG supplemental capsules. There is only a 7°C (13°F) difference in irradiation temperature between Surry Unit No. 2 and the B&W-designed 177-FA host reactor vessels, whereas the disparities for weld metals SA-1526 and WF-70 are approximately 26°C (50°F). Therefore, the source of the disparity can not be fully attributed to irradiation temperature difference.

Evaluation of the W1 Capsule 41 J (30 ft-lb) Shift Results

The 41 J (30 ft-lb) shift data are highly scattered which are evident from the W1 Capsule results. Therefore, it must be determined if the observed scatter is expected or is caused by environmental differences between Surry Unit No. 2 and the B&W 177-FA host plants (Three Mile Island Unit 2, Crystal River Unit 3, and Davis-Besse). This is accomplished by comparing the surveillance data with the predictions of trend curves such as that of Regulatory Guide 1.99, Revision 2 [*10*] or those of NUREG/CR-6551 [*9*]. This comparison will assure the scatter is within an expected embrittlement response if the trend curve adequately tracks the data. For a single measurement, there is a 95% probability that the error in prediction will be less than two standard deviations (2σ) of the applicable trend curve. For multiple measurements the data scatter around the mean curve and the average error should decrease as the number of measurements increases. For "n" individual measurements of 41 J (30 ft-lb) shift there is a 95% probability that the average prediction error within $2\sigma/\sqrt{n}$ [*11,12,13*].

$$\frac{-2\sigma}{\sqrt{n}} < \sum \frac{(\Delta T_{41m} - \Delta T_{41p})}{n} < \frac{2\sigma}{\sqrt{n}} \qquad (1)$$

where

ΔT_{41m} = the measured 41 J (30 ft-lb) shift
ΔT_{41p} = the predicted 41 J (30 ft-lb) shift
n = the number of ΔT_{41m} data
σ = the standard deviation of the applicable trend curve

If the previous inequality is satisfied for the SA-1526, SA-1585, and WF-70 weld metal 41 J (30 ft-lb) shift data with the same archive sources, then there is reasonable assurance that the observed scatter in these data is expected and not caused by the difference in irradiation temperature provided that the trend curve tracks the data. The inequality was applied using the Regulatory Guide 1.99, Revision 2, Position 1.1 shift prediction and the transition temperature shift model presented in NUREG/CR-6551. The NUREG/CR-6551 equation inputs, in addition to fluence and copper and nickel contents, are phosphorous content, irradiation temperature and irradiation time. The results are presented in Table 3.

Table 3 - *Evaluation of Residuals for Weld Metals WF-70, SA-1585, and SA-1526*

Material	Capsule	Fluence, n/cm^2 $x10^{19}$	Predicted Shift, °C (°F)	Measured Shift, °C (°F)	Residual	Residual (Irradiation Temperature corrected)
Regulatory Guide 1.99, Revision 2, Position 1.1						
WF-70	TMI2-LG1	0.585	88.3 (190.9)	67.8 (122)	-68.89	-68.89
WF-70	W1	0.644	91.6 (196.9)	103.3 (186)	-10.90	-23.90
WF-70	DB1-LG1	0.663	92.6 (198.7)	76.1 (137)	-61.72	-61.72
WF-70	CR3-LG2	1.190	113.1 (235.5)	70.6 (127)	-108.50	-108.50
SA-1585	CR3-LG1	0.510	56.9 (134.4)	77.2 (139)	4.61	4.61
SA-1585	W1	0.660	63.4 (146.2)	76.7 (138)	-8.22	-21.22
SA-1585	CR3-LG2	1.670	87.2 (188.9)	91.1 (164)	-24.89	-24.89
SA-1526	TMI2-LG1	0.830	105.4 (221.8)	120.0 (216)	-5.77	-5.77
SA-1526	W1	0.669	97.6 (207.6)	145.6 (262)	54.37	41.37
				Average Residual	-25.55	-29.88
				$2\sigma/\sqrt{n}$	18.67	18.67
NUREG/CR-6551						
WF-70	TMI2-LG1	0.585	75.4 (167.8)	67.8 (122)	-45.82	---
WF-70	W1	0.644	81.7 (179.1)	103.3 (186)	6.88	---
WF-70	DB1-LG1	0.663	78.1 (172.5)	76.1 (137)	-35.49	---
WF-70	CR3-LG2	1.190	89.7 (193.4)	70.6 (127)	-66.38	---
SA-1585	CR3-LG1	0.510	52.3 (126.1)	77.2 (139)	12.86	---
SA-1585	W1	0.660	61.1 (142.5)	76.7 (138)	-4.45	---
SA-1585	CR3-LG2	1.670	71.7 (161.0)	91.1 (164)	2.99	---
SA-1526	TMI2-LG1	0.830	94.3 (201.7)	120.0 (216)	14.32	---
SA-1526	W1	0.669	93.1 (199.5)	145.6 (262)	62.49	---
				Average Residual	-5.84	---
				$2\sigma/\sqrt{n}$	15.67	---

It is observed that the average residual is well within the bounds of the inequality when comparison of the data with the NUREG/CR-6551 trend curve is made. However, the inequality is not satisfied when comparison of the data is made with the prediction of Regulatory Guide 1.99, Revision 2, Position 1.1 with and without a correction for the different irradiation temperatures. The reason for the lack of satisfaction of the inequality is that the Regulatory Guide 1.99, Revision 2, Position 1.1 correlation consistently over predicts 41 J (30 ft-lb) shifts for Linde 80 weld metals. The resultant residuals in Table 3 for the Regulatory Guide 1.99, Revision 2, Position 1.1 predictions are large and negative as expected. The results using the NUREG/CR-6551 trend curves are more valid because it was developed using a) a data set of 609 measurements instead of 202 (151 base metal and 51 weld metal) for Regulatory Guide 1.99, Revision 2, b) an improved understanding of radiation embrittlement and c) physically-based functional forms. This evaluation indicates that the scatter of the data is expected and that the difference in the two environments is not causing unexpected results.

Evaluation of W1 Capsule Results Based on Wire Heat Data

The W1 Capsule results were compared with all other reactor vessel surveillance program 41 J (30 ft-lb) shift data with the same weld wire heats (i.e., wire heats 299L44, 72445, and 72105) to determine any environmental effects. The available data for the three different weld wire heats are shown in Table 4.

All 41 J (30 ft-lb) shift data were adjusted by 1-degree for each 1-degree difference in irradiation temperature from 291°C (556°F), so that the scatter in the data could be isolated from irradiation temperature effects. Figures 2 through 4 present the surveillance data for the weld wire heats 299L44, 72445, and 72105 respectively with respect to the transition temperature shift trends calculated in accordance with Regulatory Guide 1.99, Revision 2, Position 2.1. Observation of these plots indicate that the 41 J (30 ft-lb) shift data for the three weld wire heats are within the 2σ bounds for the individual heats including the 41 J (30 ft-lb) shift data from the W1 Capsule. The 41 J (30 ft-lb) shift data from the W1 Capsule are therefore within the scatter expected for the individual heats of material and environmental effects are not responsible for the apparent discrepancies between data from specimens irradiated in B&W-designed and Westinghouse-designed reactors.

Conclusions

Based on the analysis in this paper, it is concluded that the environmental differences in B&W-designed reactors and Westinghouse-designed reactors did not significantly affect the mechanical test data results of the W1 Capsule as compared to other mechanical test data of the MIRVP. The upper-shelf energy data all indicate no environmental effect. There are apparent differences in 41 J (30 ft-lb) transition temperature shift data arising from irradiation in different environments for the weld metals WF-70 and SA-1526. However, these differences were determined to be within the scatter expected for Charpy 41 J (30 ft-lb) transition temperature shift data.

Table 4 - *41 J (30 ft-lb) Shift Data (ΔT_{41}) for Weld Wire Heats 72105, 299L44, and 72445 Fabricated with Linde 80 Flux*

Weld Wire Heat	Capsule	Fluence x 10^{19}, n/cm^2	Irradiation Temperature, °C (°F)	ΔT_{41}, °C (°F) (Determined by TANH Fits)	Irradiation Temperature Adjusted ΔT_{41}, °C (°F)
72105	OC2-C	0.102	291 (556)	25.6 (46)	25.6 (46)
72105	OC2-A	0.337	291 (556)	59.4 (107)	59.4 (107)
72105	OC2-E	1.210	291 (556)	96.7 (174)	96.7 (174)
72105	OC3-A	0.081	291 (556)	8.3 (15)	8.3 (15)
72105	OC3-B	0.312	291 (556)	38.9 (70)	38.9 (70)
72105	OC3-D	1.450	291 (556)	78.9 (142)	78.9 (142)
72105	TMI1-LG1	0.585	291 (556)	67.8 (122)	67.8 (122)
72105	DB1-LG1	0.663	291 (556)	76.1 (137)	76.1 (137)
72105	CR3-LG2	1.190	291 (556)	70.6 (127)	70.6 (127)
72105	Z1-T	0.310	276 (529)	60.0 (108)	45.0 (81)
72105	Z1-U	1.020	276 (529)	105.6 (190)	90.6 (163)
72105	Z1-X	1.260	276 (529)	106.7 (192)	91.7 (165)
72105	Z1-Y	1.560	276 (529)	112.2 (202)	97.2 (175)
72105	Z2-U	0.270	276 (529)	76.7 (138)	61.7 (111)
72105	Z2-T	0.779	276 (529)	99.4 (179)	84.4 (152)
72105	Z2-Y	1.460	276 (529)	123.9 (223)	108.9 (196)
72105	W1	0.644	284 (543)	103.3 (186)	96.3 (173)
299L44	TMI1-E	0.107	291 (556)	41.1 (74)	41.1 (74)
299L44	TMI1-C	0.866	291 (556)	92.2 (166)	92.2 (166)
299L44	TMI1-LG1	0.968	291 (556)	125.6 (226)	125.6 (226)
299L44	TMI1-LG1	0.830	291 (556)	120.0 (216)	120.0 (216)
299L44	CR3-LG1	0.779	291 (556)	112.2 (202)	112.2 (202)
299L44	SU1-T	0.281	281 (538)	95.0 (171)	85.0 (153)
299L44	SU1-V	1.940	281 (538)	138.9 (250)	128.9 (232)
299L44	SU1-X	1.599	281 (538)	130.0 (234)	120.0 (216)
299L44	W1	0.669	284 (543)	145.6 (262)	138.6 (249)
72445	CR3-LG1	0.510	291 (556)	77.2 (139)	77.2 (139)
72445	CR3-LG2	1.670	291 (556)	91.1 (164)	91.1 (164)
72445	PB1-V	0.634	283 (542)	59.4 (107)	51.4 (93)
72445	PB1-S	0.829	283 (542)	91.7 (165)	83.7 (151)
72445	PB1-R	2.190	283 (542)	76.1 (155)	78.1 (141)
72445	PB1-T	2.230	278 (533)	100.6 (181)	87.6 (158)
72445	W1	0.660	284 (543)	76.7 (138)	69.7 (125)

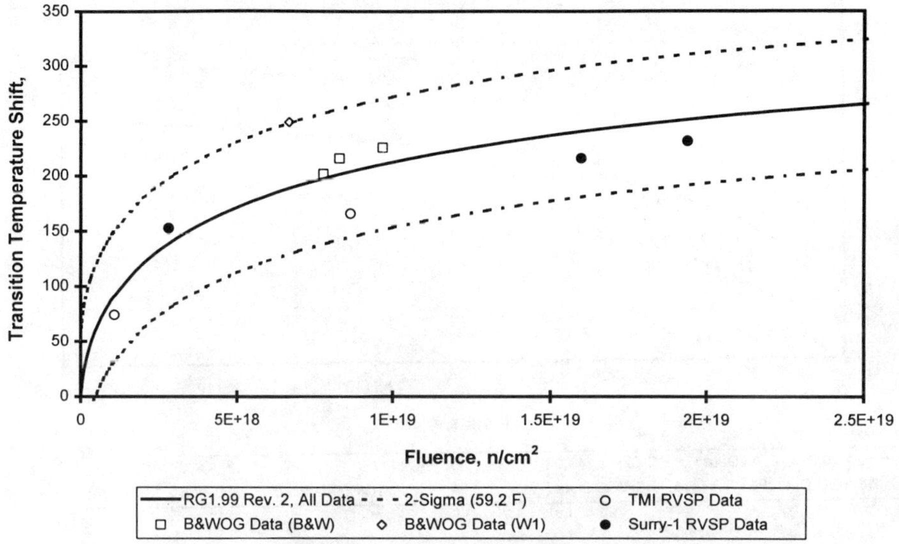

Figure 2 - *Weld Wire Heat 299L44*

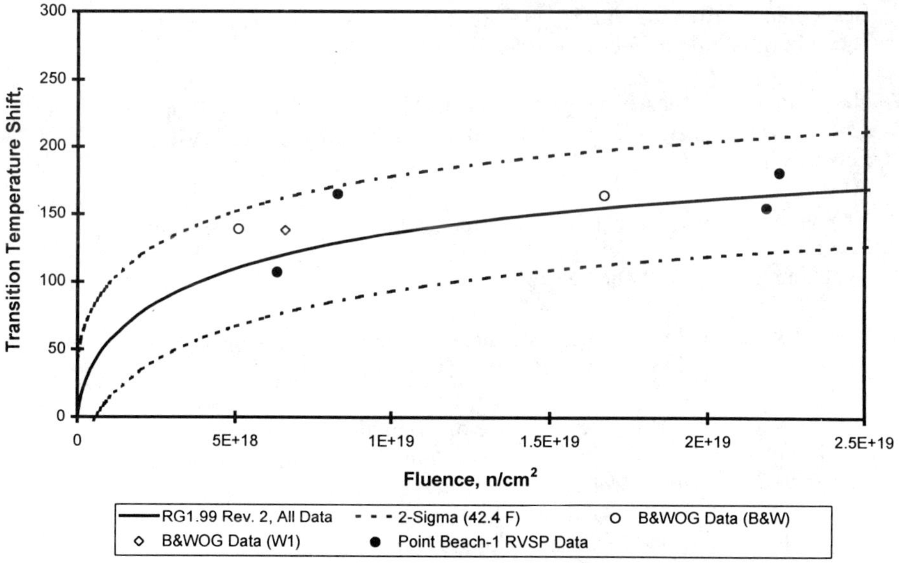

Figure 3 - *Weld Wire Heat 72445*

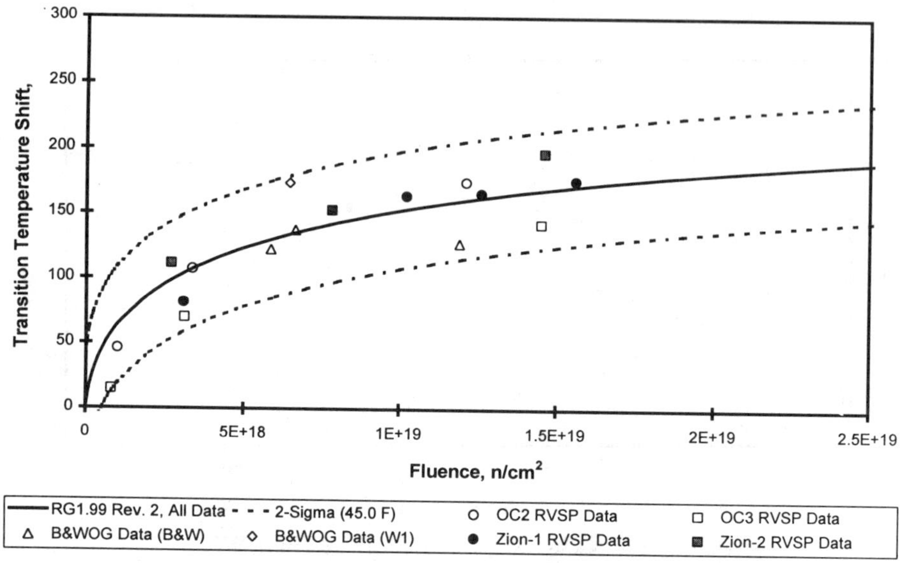

Figure 4 - *Weld Wire Heat 72105*

References

[1] Fyfitch, S., Gross, L. B., and Lowe, A. L. Jr., "Master Integrated Reactor Vessel Surveillance Program," *BAW-1543, Revision 3*, B&W Nuclear Technologies, Inc., Lynchburg, Virginia, February 1989.

[2] Harbison, L. S., "Master Integrated Reactor Vessel Surveillance Program," *BAW-1543, Revision 4*, B&W Nuclear Technologies, Inc., Lynchburg, Virginia, February 1993.

[3] DeVan, M. J. "Supplement to the Master Integrated Reactor Vessel Surveillance Program," *BAW-1543, Revision 4, Supplement 3*, Framatome Technologies, Inc., Lynchburg, Virginia, February 1991.

[4] Safety Evaluation by the Office of Nuclear Reactor Regulation B&W Report BAW 1543, Revision 3 (TAC NO. 75131), June 11, 1991.

[5] Stoller, R. E., "Modeling the Influence of Irradiation Temperature and Displacement Rate on Hardening Due to Point Defect Clusters in Ferritic Steels," *Effects of Radiation on Materials: 16th International Symposium, ASTM STP 1175*, A. S. Kumar, D. S. Gelles, R. K. Nanstad, E. A. Little, Eds., American Society for Testing and Materials, Philadelphia, 1993, pp. 394-423.

[6] Odette, G. R., "Recent Advances in Predicting Embrittlement of Reactor Pressure Vessel Steels," *Proceedings of the Second International Symposium on Environmental Degradation of Materials in Nuclear Power Systems – Water Reactors*, American Nuclear Society, La Grange Park, Illinois, 1986, pp. 121-130.

[7] Lowe, A. L. Jr. et al, "Evaluation of Surveillance Capsule Temperatures," *BAW-2040*, Babcock &Wilcox, Lynchburg, Virginia, March 1989.

[8] McCabe, D. E., Nanstad, R. K., Iskander, S. K., and Swain, R. L., "Unirradiated Materials Properties of Midland Weld WF-70," *NUREG/CR-6249*, United States Nuclear Regulatory Commission, October 1994.

[9] Eason, E. D., Wright, J. E., and Odette, G. R., "Improved Embrittlement Correlations for Reactor Pressure Vessels Steels," *NUREG/CR-6551*, United States Nuclear Regulatory Commission, November 1998.

[10] "Radiation Embrittlement of Reactor Vessel Materials," *Regulatory Guide 1.99, Revision 2*, U. S. Nuclear Regulatory Commission, May 1988.

[11] Draper, N. R. and Smith, H., "Applied Regression Analysis (Second Edition)," John Wiley & Sons, New York, 1981.

[12] Johnston, J., "Econometric Methods (Second Edition)," McGraw-Hill, New York, 1972.

[13] Vinod, H. D., and Ullah, A., "Recent Advances in Regression Methods," Marcel Dekker Inc., New York, 1981.

S. K. Iskander,[1] R. K. Nanstad,[1] C. A. Baldwin,[1] D. W. Heatherly,[2] M. K. Miller,[1] and I. Remec[3]

Reirradiation Response Rate of a High-Copper Reactor Pressure Vessel Weld

---

Reference: Iskander, S. K., Nanstad, R. K., Baldwin, C. A., Heatherly, D. W., Miller, M. K., and Remec, K., **"Reirradiation Response Rate of a High-Copper Reactor Pressure Vessel Weld,"** *Effects of Radiation on Materials: 20th International Symposium, ASTM STP 1405,* S. T. Rosinski, M. L. Grossbeck, T. R. Allen, and A. S. Kumar, Eds., American Society for Testing and Materials, West Conshohocken, PA, 2001.

Abstract: The Charpy impact response of reirradiated Heavy-Section Steel Irradiation (HSSI) Program Weld 73W has been determined at three fluence levels. The Charpy specimens had previously been irradiated at 288°C to 1.8×10^{19} cm^{-2} (E > 1 MeV) and annealed at 454°C for 168 h. The results show that the change in the 41-J Charpy energy level transition temperature ($\Delta TT_{41\text{-}J}$) of the reirradiated specimens is slightly higher than predicted by the vertical shift method, but significantly less than predicted by the lateral shift method. Previous results have also shown that the upper-shelf energy (USE) over-recovers as a consequence of annealing, which may explain why the USE value after a significant amount of reirradiation is approximately equal to the USE value in the unirradiated condition.

Keywords: annealing, reactor pressure vessel, Charpy V-notch, irradiation, neutron, weld metal, copper

Introduction

A consideration in the decision to anneal a reactor pressure vessel (RPV) is the reembrittlement rate after annealing. This response is important in order to assess the cost-to-benefits ratio of restoring fracture toughness properties of RPVs that have been degraded by neutron irradiation. There is a dearth of fracture toughness data on reembrittlement rates, and the use of Charpy data may be a useful indicator of such rates.

---

[1]Metals and Ceramics Division, Oak Ridge National Laboratory, P.O. Box 2008, Oak Ridge, TN 37831

[2]Engineering Technology Division, Oak Ridge National Laboratory, P.O. Box 2009, Oak Ridge, TN 37831

[3]Computational Physics and Engineering Division, Oak Ridge National Laboratory, P.O. Box 2008, Oak Ridge, TN 37831

It is assumed that the rate of toughness degradation upon reirradiation may be assessed by means of the 41-J temperature shift ($\Delta TT_{41\text{-}J}$) and the USE drop (ΔUSE) of the Charpy impact energy curve.

A few years ago, there was much research on the topic of thermally annealing RPVs [*1*],[2],[*3*],[*4*],[*5*]. About a dozen RPVs were annealed in Eastern Europe [*6*], a process which was followed with much interest in the United States, and the principal author of this paper participated in one of the delegations composed of U.S. Nuclear Regulatory Commission (NRC) staff, industry, and research groups that visited the site of the Novovorenzh Nuclear Power Plant, Unit 3, and its annealing [*7*]. During this period an amendment was made to 10CFR50.66, "Thermal Annealing Rule," NRC *Regulatory Guide 1.162* on Annealing [8] was issued, and American Society of Mechanical Engineers (ASME) Code Case N-557 on dry annealing of a nuclear RPV was published [*9*]. Models for the recovery of the transition temperature and USE were published [*10*] with the help of a database assembled at Oak Ridge National Laboratory (ORNL) [*11*]. The American Society for Testing and Materials (ASTM) Standard Guide for In-Service Annealing of Light-Water Cooled Nuclear Reactor Vessels, ASTM 509-97 was revised to address annealing recovery and reembrittlement. Recently this activity has subsided, in part because of economic issues. However, since research with irradiated material takes many years, it would be prudent to continue research on developing an understanding of reembrittlement rates. A major revision of NRC *Regulatory Guide 1.99*, Rev. 2 [*12*] is being considered that would effect the pressurized thermal shock (PTS) screening criteria (RT_{PTS}). Moreover, the methodology for determining RT_{PTS} is being evaluated, and these events could have an impact on whether any plants would ever be candidates for annealing.

Material Used in this Study

High-copper HSSI Weld 73W has been very extensively characterized in the unirradiated and irradiated conditions [*13*]. A relatively large number of specimens were tested in the unirradiated and irradiated conditions such that statistical data on the effects of neutron irradiation on Charpy, tensile, initiation and crack-arrest fracture toughness have been published. These results have established a correlation between the irradiation-induced shifts in fracture toughness and Charpy $\Delta TT_{41\text{-}J}$, and it is assumed that such a correlation continues in the reirradiated case.

For reference, the chemical composition and mechanical properties of Weld 73W are shown in Tables 1 and 2, respectively.

Table 1 - *Chemical composition and the standard deviation of the various elements of HSSI Weld 73W*

	Composition, wt %, and standard deviation									
	C	Mn	P	S	Si	Cr	Ni	Mo	Cu	V
Mean	0.098	1.56	0.005	0.005	0.45	0.25	0.60	0.58	0.31	0.003
Standard Deviation	0.007	0.026	0.0004	0.0006	0.028	0.006	0.006	0.009	0.010	0.0001

Table 2 - *Mechanical properties of HSSI Weld 73W*

Material	CVN impact USE (J)	CVN 41-J transition temperature (°C)	Room temperature tensile strength (MPa)	
			Yield	Ultimate
HSSI weld 73W (undersize specimens)	118	–38	495	603
HSSI weld 73W (full size specimens)	135	–40	495	603

There were a number of slightly undersize Charpy specimens used in the irradiation, annealing, and reirradiation studies. As may be seen from Table 2, there is very little difference between the 41-J transition temperatures $TT_{41\text{-}J}$ of the full-size and undersize specimens and thus in reference to the $TT_{41\text{-}J}$ no distinction will be made regarding specimen size. The results of testing unirradiated, irradiated and annealed Charpy specimens are shown in Fig. 1. An interesting observation is that the USE has over-recovered to levels higher than the original unirradiated value. Unirradiated specimens were aged at 454°C for 168 h and increases in the USE were also observed for these specimens. This has been frequently reported in the literature, but is nonetheless unexpected since HSSI Weld 73W had been stress relieved at 607°C (1125°F) for 40 h, which is typical for welds in commercial RPVs. This has also been confirmed by plotting the percent recovery of the USE versus percent recovery of $TT_{41\text{-}J}$ as shown in Fig. 2, which was prepared from data in Ref. [*11*]. The reason for this over-recovery is still unknown. The relationship shown in Fig. 2 between recovery of the $TT_{41\text{-}J}$ and over-recovery of the USE implies that if the $TT_{41\text{-}J}$ recovers, then the USE will also recover percentage-wise at least as much.

Reirradiation, Dosimetry and Testing of the Charpy Specimens

Three containers, each with ten Charpy specimens, and 0.5-mm-diam (0.020-in.) Fe and 0.1%Co/Al wire dosimeters in the V-notch grooves of the specimens were encapsulated in three containers, as shown in Fig. 3. The three containers were electrodischarge machined from solid aluminum stock. The tolerances on the inside and outside dimensions of the containers were required to be very small. The specimens had previously been irradiated to an average fluence of 1.8×10^{19} cm^{-2} (E > 1 MeV) and then annealed at 454°C for 168 h. The packaging of the irradiated specimens and the dosimeters, using manipulators in the hot cell, required high dexterity of the ORNL operators. The three containers were then loaded with three fission radiometric dosimetric sets (FRDS), in the HSSI Program/University of California, Santa Barbara (UCSB) capsule located at the University of Michigan Ford Nuclear Reactor (FNR) and reirradiated to 0.4, 0.8, and 2.95×10^{19} cm^{-2} (E > 1 MeV) at 288°C. The specimens irradiated to these fluences will be referred to as the low-, medium-, and high-flux Charpy specimens (see Fig. 4).

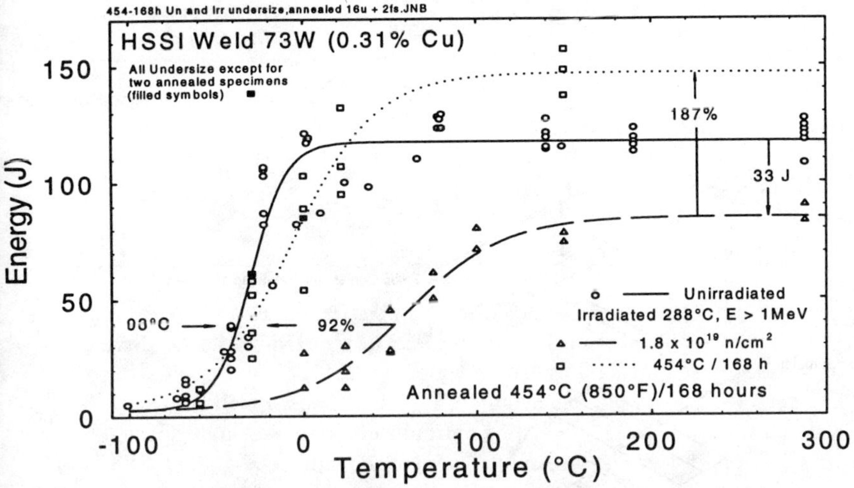

Figure 1 - *The results of testing unirradiated, irradiated, and annealed Charpy specimens.*

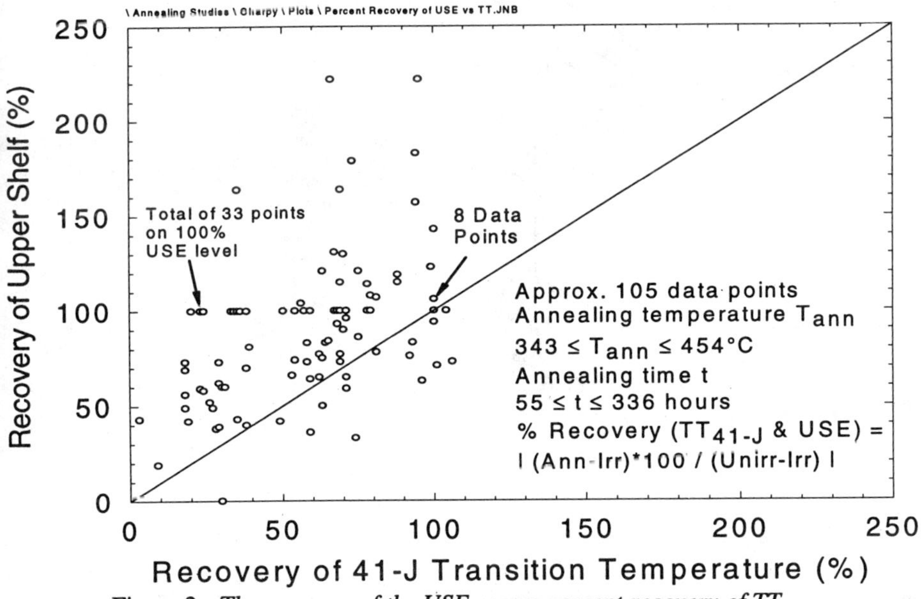

Figure 2 - *The recovery of the USE versus percent recovery of TT_{41-J}.*

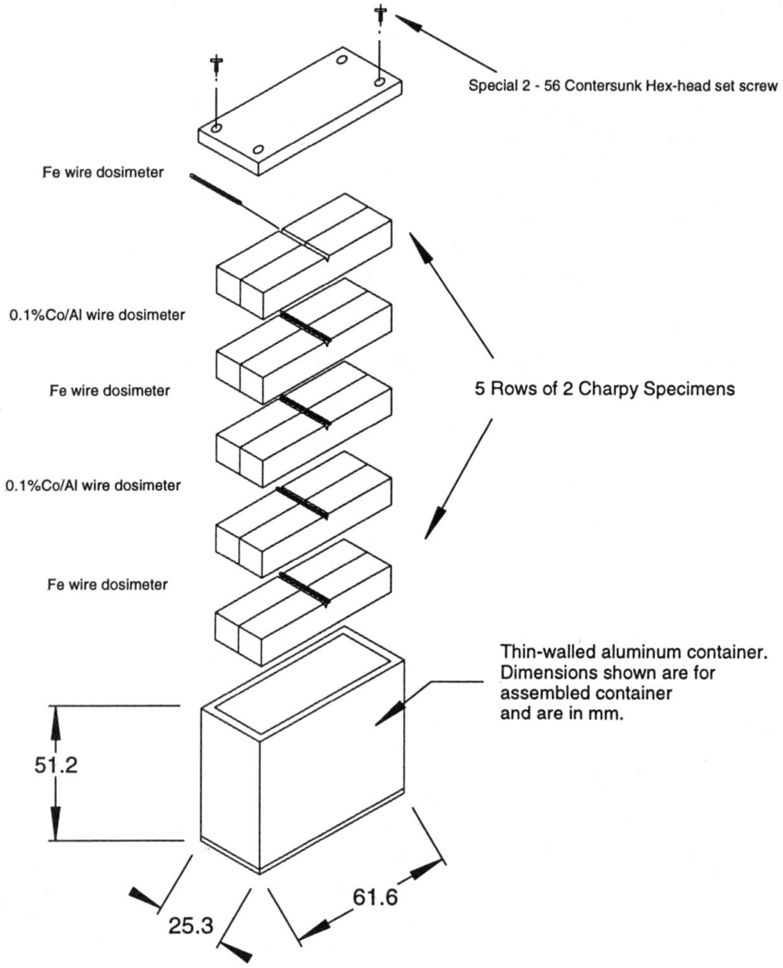

Figure 3 - *Charpy specimens were encapsulated in three containers with 0.5-mm-diam (0.020-in.) Fe 0.1%Co/Al wire dosimeters in the V-notch grooves of the specimens.*

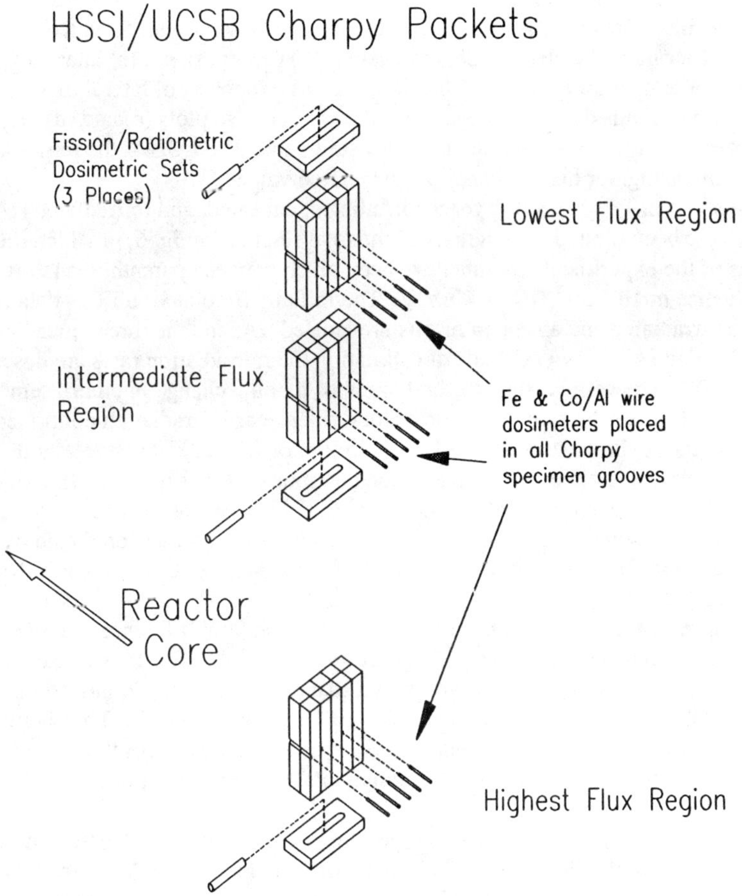

Figure 4 - *Arrangement of the three Charpy specimen containers in the University of Michigan Ford Nuclear Reactor.*

The FRDS were retrieved after about 20 days of irradiation and used to estimate the flux. The duration of irradiation exposure required was then estimated, the irradiation completed and the containers retrieved, and the activity of the wire dosimeters counted. The low-flux containers were located at a higher elevation than the reactor core, and it was necessary to perform three-dimensional dosimetric calculations because preliminary calculations of flux made using an approximate three-dimensional analysis (with an estimated error of less than 10%) showed that the calculated flux was 20% lower than the initial three-dimensional estimates.

The results of testing the Charpy specimens are shown in Fig. 5, which is composed of four plots, labeled (a) Charpy energy, (b) percent shear, (c) lateral expansion as a function of temperature, and (d) Charpy energy as a function of lateral expansion. The lateral expansion is difficult to measure in the hot cell, so plots (c) and (d) helped identify potential error to be corrected by remeasurement. The USE of the reirradiated material is still as high or higher than the unirradiated value.

The shifts in $TT_{41\text{-}J}$, $\Delta TT_{41\text{-}J}$, for the irradiated, annealed, and reirradiated (IAR) specimens have been plotted as functions of the total fluence in Fig. 6, in which the coordinates of the experimental points have been given between parentheses [shifts in K and total fluence in 10^{19} cm^{-2} (E > 1 MeV)]. The initially irradiated $\Delta TT_{41\text{-}J}$ data are labeled I, the irradiated and annealed results are labeled IA, and the three reirradiated values are labeled IAR. Two methods of estimating the reirradiation rates are described in ASTM E 509. The vertical shift method assumes that the change in embrittlement due to reirradiation follows the same function of total fluence as if irradiation continued with no annealing, and is shown as the dotted curve in Fig. 6. The lateral shift method assumes that embrittlement due to reirradiation continues follows the same function of increase in fluence as the unirradiated material and is shown as the dashed curve. It may be seen that the measured values of $\Delta TT_{41\text{-}J}$ are well below the values predicted by the lateral shift method, but are somewhat higher than the values predicted by the vertical shift procedure.

To estimate the rate of reembrittlement beyond the largest fluence, the rate predicted by the fluence factor given in *Regulatory Guide 1.99*, Rev. 2, was used with the chemistry factor adjusted until the curve, shown as a dash-dot in Fig. 7, passed through the $\Delta TT_{41\text{-}J}$ at the accumulated fluence of 4.8×10^{19} cm^{-2} (E > 1 MeV). The chemistry factor obtained implies an effective copper content of about 0.02%, for the 0.6% Ni of HSSI Weld 73. It may be recalled that the actual total copper content of this weld is 0.31%.

A curve of the $\Delta TT_{41\text{-}J}$ vs $\Delta\Phi$ was regression fit through the four data values associated with IA, and IAR; this has also been plotted in Fig. 7 as a dash-dot-dot curve. This curve can be used to predict the $\Delta TT_{41\text{-}J}$ for a set of ten specimens now being reirradiated at the FNR to an accumulated total fluence of about 6×10^{19} cm^{-2} (E > 1 MeV).

The Kurchatov Institute in Moscow, Russia, has performed similar experiments, and their results have also been plotted in Fig. 6 [*14*]. The lack of agreement between the Kurchatov and ORNL results at approximately the same fluence is obvious, but the reasons for this difference are not known.

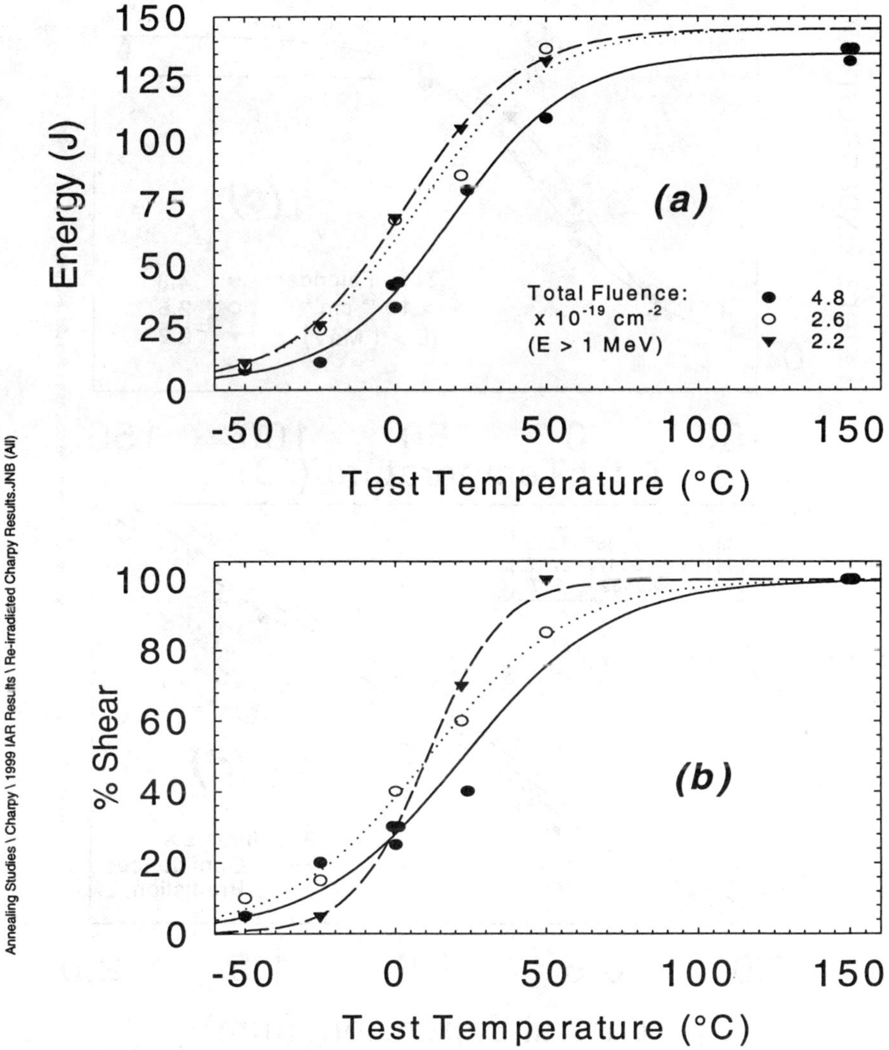

Figure 5 - *Results of Charpy specimen testing: (a) Charpy energy, (b) percent shear, (c) lateral expansion as a function of temperature, and (d) Charpy energy as a function of lateral expansion.*

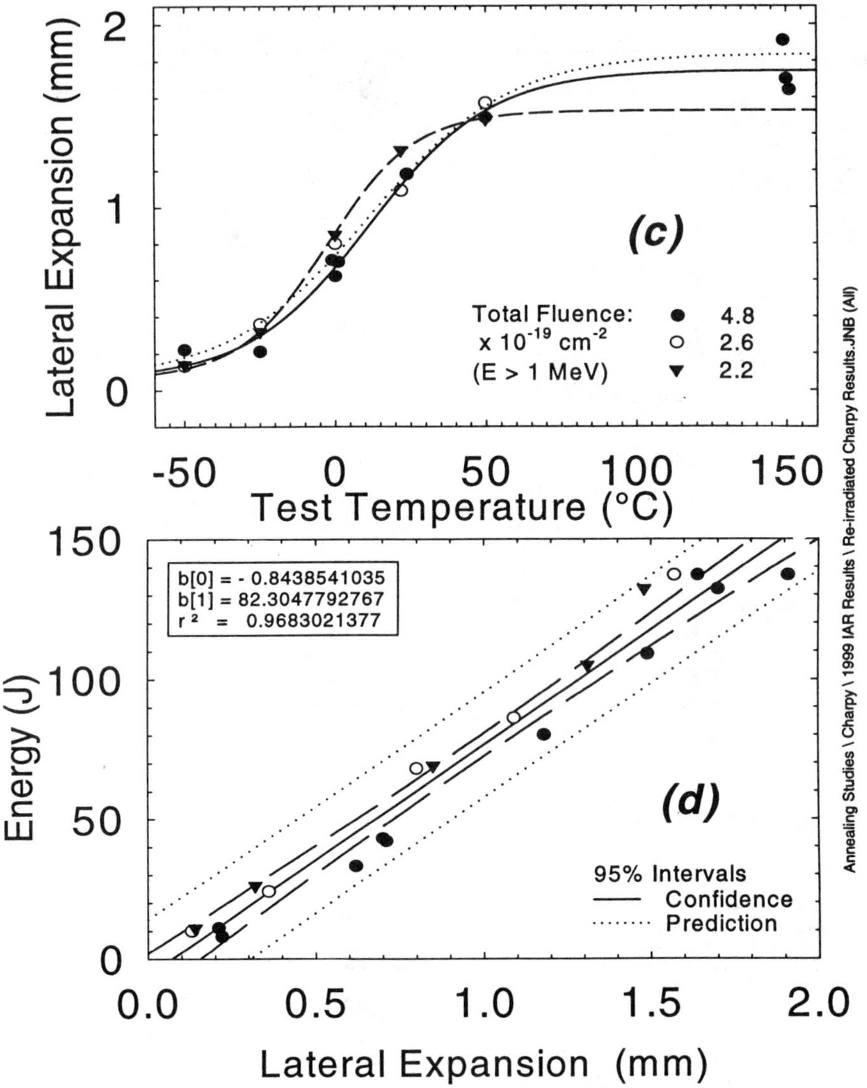

Figure 5 - *(Continued) - Results of testing Charpy specimens: (a) Charpy energy, (b) percent shear, (c) lateral expansion as a function of temperature, and (d) Charpy energy as a function of lateral expansion.*

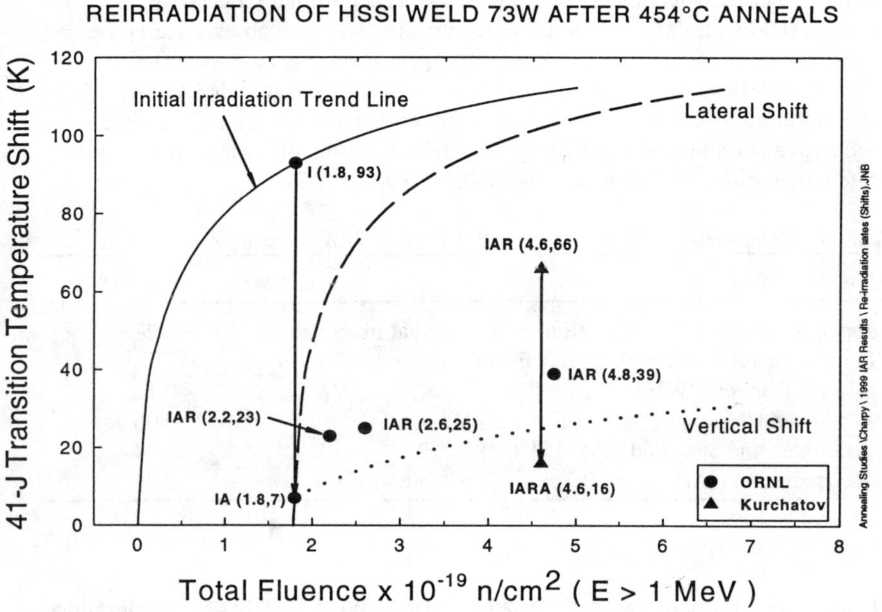

Figure 6 - *Shifts in 41-J energy level transition temperature of irradiated, annealed, and reirradiated Charpy specimens versus total fluence.*

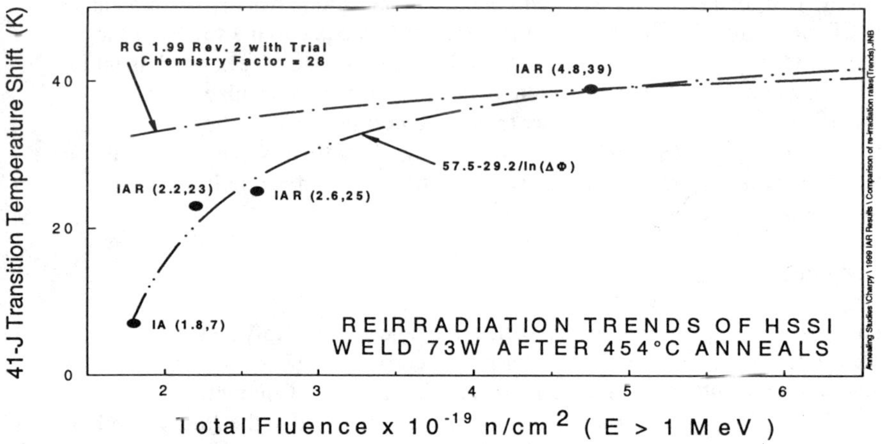

Figure 7 - *Reirradiation trends obtained using expression given in RG 1.99, Rev 2, and a regression fit through the data.*

Research in progress at ORNL has given indications copper precipitates out of the matrix and is unavailable to cause further radiation embrittlement [*15*],[*16*]. Table 3 compares the amounts of copper under various conditions. The 0.12% at. wt Cu in the ferrite matrix that contributed to the embrittlement has been further reduced by precipitation to 0.06% at. wt. A small decrease in copper was also observed after the 168-h anneal at 454°C. No change in the matrix copper level was observed after reirradiation. The 0.04 % at. wt. agrees reasonably in order of magnitude with the amount 0.02 inferred by curve fitting discussed above. The matrix composition was determined with a combination of an energy-compensated atom probe and an energy-compensated three-dimensional atom probe [*17*].

Table 3 - *Comparison of bulk copper in HSSI Weld 73W in various conditions*

Condition	Atomic %
Bulk copper, at end of weld fabrication, postweld heat treatment (PWHT) or stress relief at 607°C (1125°F) for 40 h	0.27
Matrix Copper, PWHT	0.12
Irradiated (I), 2×10^{19} cm^{-2} (E > 1 MeV)	0.06
Irradiated and annealed (IA), 454°C (850°F)	0.04
Reirradiation (IAR), 1×10^{19} cm^{-2} (E > 1 MeV)	0.04

Summary and Conclusions

The Charpy impact response of reirradiated HSSI Weld 73W has been determined at three fluence levels. The Charpy specimens had previously been irradiated and annealed at 454 °C for 168 h, and the results have shown that the $\Delta TT_{41\text{-}J}$ of the reirradiated specimens is slightly higher than predicted by the vertical shift method, but significantly less than predicted by the lateral shift method. Previous results have also shown that the USE over-recovers as a consequence of annealing, which may explain why the USE value after reirradiation is still approximately equal to the values of the unirradiated condition. One explanation of the relatively low rate of reembrittlement of HSSI Weld 73W compared to the initial rate may be the decreased availability of radiation sensitive elements such as copper to cause embrittlement. If the reembrittlement rate of an RPV can be shown to be significantly less than the initial one, then annealing may be a technically attractive consideration even before it becomes absolutely necessary.

Acknowledgments

This research is sponsored by the Office of Nuclear Regulatory Research, U.S. Nuclear Regulatory Commission, under Interagency Agreement DOE 1886-N695-3W with the U.S. Department of Energy under Contract DE-AC05-00OR22725 with UT-Battelle, LLC. It was performed in the Heavy-Section Steel Irradiation Program, managed by T. M. Rosseel. The authors appreciate the financial support from the U.S. Nuclear Regulatory Commission. The authors also appreciate the technical review by J. G. Merkle, manuscript preparation by J. L. Bishop, and many helpful suggestions from M. A. Sokolov and the efforts of the hot cell staff: L. J. Turner, P. S. Bishop, G. W. Parks, and the Charpy testing by R. L. Swain. The authors also acknowledge P. Pareige for measurements of the matrix copper compositions in the PWHT, I, and IA conditions shown in Table 3.

References

[*1*] Oland, C. B., Bass, B. R., Bryson, J. W., Ott, L. J., and Crabtree, J. A., *Marble Hill Annealing Demonstration Evaluation*, NUREG/CR-6552 (ORNL/TM-13446), Oak Ridge National Laboratory, Oak Ridge, Tenn., February 1998.

[*2*] Iskander, S. K., Sokolov, M. A., and Nanstad R. K., "Comparison of Different Experimental and Analytical Measures of the Thermal Annealing Response of Neutron-Irradiated RPV Steels," *Effects of Radiation on Materials: 18th International Symposium, ASTM STP 1325,* R. K. Nanstad, M. L. Hamilton, F. A. Garner, and A. S. Kumar., American Society for Testing and Materials, West Conshohocken, PA, 1999.

[*3*] Iskander, S. K., Sokolov, M. A., and Nanstad, R. K., "A Perspective on Thermal Annealing of Reactor Pressure Vessel Materials from the Viewpoint of Experimental Results," *ICONE-4, Proceedings of the ASME-JSME 4th International Conference on Nuclear Engineering, New Orleans, Louisiana, March 10-14, 1996,* Vol. 1, Part A, ASME International, New York, 1996.

[*4*] Iskander, S. K., Sokolov, M. A., and Nanstad, R. K., "Effects of Annealing Time on the Recovery of Charpy V-Notch Properties of Irradiated High-Copper Weld Metal," *Effects of Radiation on Materials: 17th Volume, STP 1270,* D. S. Gelles, R. K. Nanstad, A. S. Kumar, and E. A. Little, Eds., American Society for Testing and Materials, Philadelphia, 1996, pp.277-293.

[*5*] Iskander, S. K., Sokolov, M. A., and Nanstad, R. K., "Some Aspects of the Role of Annealing in Plant Life Extension," *Transactions of the American Nuclear Society,* Vol. 71, La Grange Park, Illinois, 1994, pp. 191-192

[*6*] Kryukov, A. M., and Sokolov, M. A., "Investigation of Material Behavior Under Reirradiation after Annealing Using Subsize Specimens," *Small Specimen Test Techniques Applied to Nuclear Reactor Vessel Thermal Annealing and Plant Life Extension, ASTM STP 1204*, W. R. Corwin, F. M. Haggag, and W. L. Server, Eds., American Society for Testing and Materials, Philadelphia, 1993, pp. 417-423.

[*7*] *Report on Annealing of the Novoronezh Unit 3 Reactor Vessel in the USSR*, MPR-1230, MPR Associates, March 1991.

[*8*] "Format and Content of Report for Thermal Annealing of Reactor Pressure Vessels," *Regulatory Guide 1.162*, U.S. Nuclear Regulatory Commission, February 1996.

[*9*] "In-Place Dry Annealing of a PWR Nuclear Reactor Vessel," Code Case N-557, ASME Boiler and Pressure Vessel Code, Sect. XI, Div. 1, Approval Date March 19, 1996.

[*10*] Eason, E. D., Wright, J. E., Nelson, E. E., Odette, G. R., and Mader, E. V., *Models for Embrittlement Recovery Due to Annealing of Reactor Pressure Vessel Steels,* NUREG/CR-6327 (MCS 950302), Modeling and Computing Services/University of California, Santa Barbara, May 1995.

[*11*] Stallman, F. W., Wang, J. A., and Kam, F. B. K., *TR-EDB: Test Reactor Embrittlement Data Base, Version 1*, NUREG/CR-6076 (ORNL/TM-12415), Martin Marietta Energy Systems, Inc., Oak Ridge National Laboratory, Oak Ridge, Tenn., January 1994.

[*12*] "Radiation Embrittlement of Reactor Pressure Vessel Materials," *Regulatory Guide 1.99, Revision 2*, U.S. Nuclear Regulatory Commission, Washington, May 1988.

[*13*] Nanstad, R. K., McCabe, D. E., Menke, B. H., Iskander, S. K., and Haggag, F. M., "Effects of Irradiation on K_{Ic} Curves for High-Copper Welds," *Effects of Radiation on Materials, 14th International Symposium, (Volume II), ASTM STP 1046,* N. H. Packan, R. E. Stoller, and A. S. Kumar, Eds., American Society for Testing and Materials, Philadelphia, 1990, pp. 214-233.

[*14*] Sokolov, M. A., Chernobaeva, A. A., Nanstad, R, K. Nikolaev, Y. A., and Korolev, Y. N., "Irradiation, Annealing, and Reirradiation Effects on American and Russian Reactor Pressure Vessel Steels,"*Effects of Radiation on Materials: 19th International Symposium, ASTM STP 1366*, M. L. Hamilton, A. S. Kumar, S. T. Rosinski, and M. L. Grossbeck, Eds., American Society for Testing and Materials, West Conshohocken, PA, 2000, pp. 415-434.

[*15*] Pareige, P., Stoller, R. E., Russell, K. F., and Miller, M. K., "Atom Probe Characterization of the Microstructure of Nuclear Pressure Surveillance Materials after Neutron Irradiation and after Annealing Treatments," *Journal of Nuclear Materials*, 249, 165-174 (1997).

[*16*] Miller, M. K., and Russell, K. F., "Atom Probe Characterization of Copper Solubility in the Midland Weld after Neutron Irradiation and Thermal Annealing," *Journal of Nuclear Materials,* 250, 223-228 (1997).

[*17*] Miller, M. K., *Atom Probe Tomography*, Kluwer Academic/Plenum Publisher, 2000.

Ryuta Kasada,[1] Tetsuya Suzuki,[2] Keisuke Itoh,[2] Yoshinobu Naruse,[2] and Akihiko Kimura[1]

Relation Between Resistivity and Mechanical Properties in Heat Affected Zone of Welded Pressure Vessel Steel

Reference: Kasada, R., Suzuki, T., Itoh, K., Naruse, Y., and Kimura, A. **"Relation Between Resistivity and Mechanical Properties in HAZ of Welded Pressure Vessel Steel,"** *Effects of Radiation on Materials: 20th International Symposium, ASTM STP 1405,* S. T. Rosinski, M. L. Grossbeck, T. R. Allen, and A. S. Kumar, Eds., American Society for Testing and Materials, West Conshohocken, PA, 2001.

Abstract: The relation between resistivity and mechanical properties, such as hardness and the ductile-brittle transition temperature (DBTT), in heat-affected zone (HAZ) of welded A533B pressure vessel steel (PVS) has been investigated by means of a small specimen test technique before and after neutron irradiation. A submerged arc welded thick plate of A533B for plant use was supplied by Japan Steel Works, Ltd. Neutron irradiation was performed in the Japanese Materials Test Reactor (JMTR) at 290°C up to 1.3×10^{24} n/m^2. Resistivity measurements with a pair of potential probes, 1 mm distance apart, were performed on the surface of specimens at different locations across the fusion boundary. Before the irradiation, the resistivity of the weld was higher than that of the base metal and the resistivity in the HAZ gradually reduced with increasing distance from the fusion boundary. After the neutron irradiation, the resistivities at all the locations were reduced. Post-irradiation annealing experiments revealed that resistivities of base metal and HAZ almost recovered after annealing to 500°C, while only a part of the recovery was observed for the resistivity of the weld. On the contrary, the recovery of irradiation hardening was faster in the weld than in the base metal. It is considered that the reduction of resistivity is due to precipitation of solute elements, such as manganese, nickel and copper. The correlation between resistivity change and irradiation hardening is also discussed.

Keywords: nondestructive evaluation, resistivity, hardness, ductile-brittle transition temperature, heat-affected zone, pressure vessel steel

[1]Graduate student and professor, respectively, Institute of Advanced Energy, Kyoto University, Gokasho, Uji, Kyoto 611-0011, Japan.
[2]Research scientists, Chubu Electric Power Company Inc., Nagoya 459, Japan.

Introduction

For nuclear plant lifetime extension, the evaluation and prediction of the neutron irradiation-induced deterioration of the reactor pressure vessel (RPV) are critical issues. It has been considered that the evaluation of mechanical properties of welded joints of pressure vessel steel (PVS) is rather difficult because of the complex metallurgical structure depending on the location in the heat-affected zone (HAZ). Recently, the Charpy V-notch (CVN) impact test utilizing sub-sized specimens has been developed as one of the options for resolving the shortage of surveillance test specimens for plant lifetime extension of light water reactors [*1-5*]. In previous studies [*4, 5*], we investigated the dependence of ductile-brittle transition (DBT) behavior on the location of V-notch in HAZ of the welded A533B PVS, using the small specimen test technique (SSTT).

In addition to the SSTT, nondestructive evaluation (NDE) techniques have been developed to survey irradiation embrittlement of structural materials using electric and magnetic property changes such as, electric resistance, Barkhausen noise (BHN), superconducting quantum interference device (SQUID), and positron annihilation spectrometry (PAS)[*6-8*]. It is suggested that such NDE techniques might be used to evaluate the lifetime of PVS, as well as the current destructive evaluation methods. Among these, the electric resistivity measurement was reported to be effective in detecting the irradiation embrittlement of PVS [*6*], where a good correlation between ΔDBTT and neutron irradiation-induced resistivity change has been shown.

In the present study, the electrical resistivity method was applied for the HAZ specimens of welded A533B PVS in order to develop an NDE method to evaluate the irradiation embrittlement of welded steel.

Experimental Procedure

The material used in this study was A533B PVS, for nuclear reactor pressure vessel, which was produced by Japan Steel Works, Ltd. The chemical compositions of base metal and deposit are shown in Table 1 with the heat treatment conditions. The welding conditions for this steel were as follows: 1) 32V of arc voltage, 2) 630A of beam current, and 3) 650mm/min of traveling speed, with pre-heat. Post-welding heat treatment was done at 625°C for 15 hr. All the procedures followed the actual pressure vessel fabrication procedures. Charpy V-notch (CVN) specimens were cut out of the welded joint into 3.3 mm and 1.5 mm size of miniaturized specimens. The detail of this procedure was shown in refs. *4, 5*. The specimen location is identified by the symbols Mx for base metal, Dx for deposit and B for fusion boundary, where x denotes the distance from the fusion boundary in mm. For the resistivity measurement, the 1.5 mm size of CVN specimen that had a V-notch at the M1 position was used. Neutron irradiation was carried out in the Japan Materials Test Reactor (JMTR) at 290°C up to 1.3 × 10^{24} n/m^2 (E>1 MeV) or 0.20 dpa. The irradiation period was 540 hr. Post-irradiation experiments were performed for the 1.5 mm size of CVN specimens, compared with as-received and thermal controlled 3.3 mm size specimens. Micro-Vickers hardness test was carried out with a load of 0.2 kg at room temperature. Resistivity measurement was conducted at room temperature, using conventional four probe with a direct current of 623.8 mA. The resistivity was measured for both directions of the current ten times and the average values were obtained. The measurements were also carried out following isochronal annealing from 325°C to 525°C for 0.5 hr in a vacuum of 1.0×10^{-5} Pa.

Table 1- *Chemical compositions and heat treatments of A533B cl.1 steel and deposit.*

Chemical composition / wt. %

	C	Si	Mn	P	S	Ni	Cr	Cu	Mo	V	Al	N
Base	0.17	0.16	1.42	<0.003	<0.003	0.63	0.13	0.03	0.51	<0.01	0.026	0.0093
Weld	0.10	0.07	1.27	0.008	<0.003	0.82	0.03	0.08	0.44	<0.01	0.006	0.0047

Heat treatments
Quenching : 890°C, 4 hr, water quench. • Tempering: 660°C, 4hr, air cooled.

Results

Irradiation Effects

After the neutron irradiation, Vickers hardness of the HAZ specimen increased at all positions of the welded joints of the steel, as shown in Fig. 1 where the hardness was plotted against the position across the fusion boundary. In order to clarify the thermal effect under the irradiation at 290°C up to 540 hr on the irradiation hardening, hardness test was also carried out for the thermal controlled specimen. There was no significant effect of the aging on the hardness of welded joint.

Resistivity measurements were carried out for as-received, thermal controlled and as-irradiated welded specimens. Fig. 2 shows the dependence of the resistivity on the location in the HAZ of the welded steel. A small increase in the resistivity was observed after the thermal aging at 290°C up to 540 hr, while a significant reduction of the resistivity was induced by the irradiation at 290°C at all the locations from M9 to D7.

The irradiation-induced change in the hardness and resistivity is shown in Fig. 3. There is no apparent correlation between them.

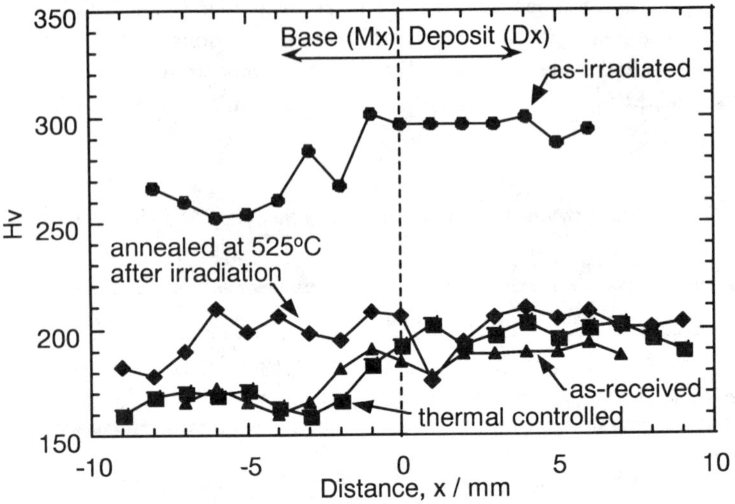

Fig. 1- *Vickers hardness of as-received, as-irradiated, thermal controlled and annealed at 525°C after irradiation.*

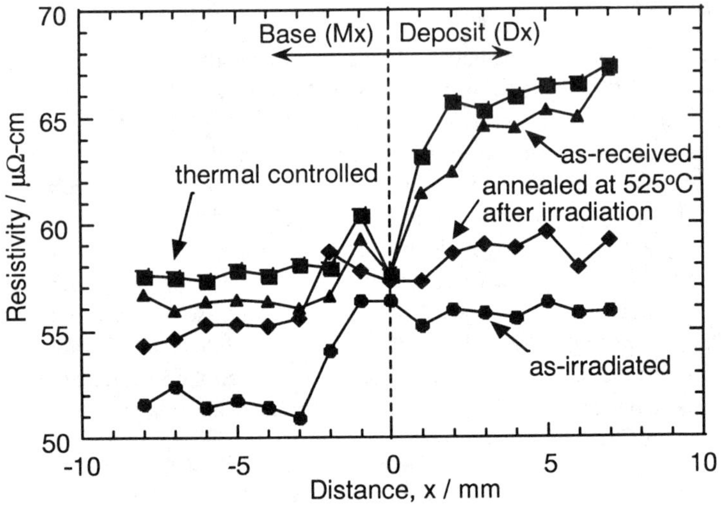

Fig. 2- *Resistivity of as-received, as-irradiated, thermal controlled and annealed at 525°C after irradiation.*

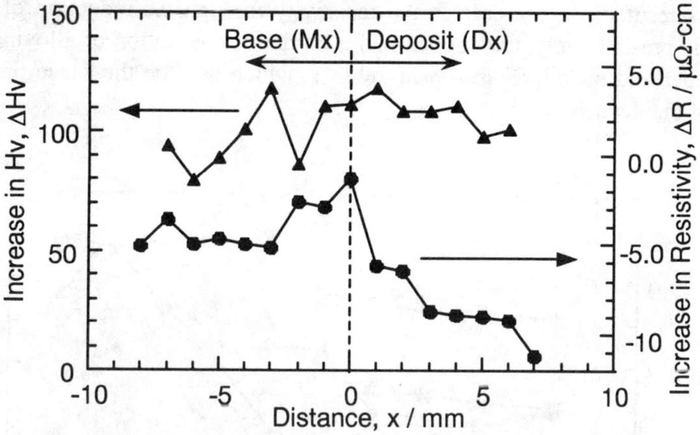

Fig. 3- *Changes in Vickers hardness and resistivity of the as-irradiated specimen.*

Isochronal Annealing Experiment

Post-irradiation isochronal annealing behavior of the resistivity is shown in Fig. 4 for several locations in the HAZ of the steel. The irradiation-induced reduction of the resistivity was recovered by the annealing. Among the HAZ locations, the M1 location showed the smallest reduction of the resistivity and the D4 location showed the largest one. At the M1 location, annealing at 400°C resulted in almost all the recovery of the resistivity change, and the other locations, except for the D4 location, recovered after annealing to 500°C, while the D4 recovered only a half of the irradiation-induced resistivity change. The resistivities of thermal-controlled specimens show no significant change during annealing, as shown in Fig. 5.

Discussion

Difference in Resistivity between Base Metal and Deposit

As shown in Fig. 2, the resistivity of deposit was larger than that of the base metal in the as-received condition. This may be explained by the difference in the microstructure between them, which depends on the chemical compositions and heat treatment conditions. The A533B steel consists of bainitic structure that contains high density of dislocations and precipitates. As shown in the ref. *9*, there was no significant difference in the resistivity between the quenched Fe-Cu alloy and the cold-worked one. Therefore, the difference in the resistivity between the base metal and deposit is considered to be due to a change in the distribution of solute atoms and precipitates. In general, solute atoms increase the resistivity of the lattice, while precipitation of the solute atoms results in a decrease in the resistivity through the reduction of the solute atoms. In the present study, the deposit had a higher concentration of alloying elements such as copper and nickel, as shown in Table 1, which may be the cause of the higher resistivity of the deposit.

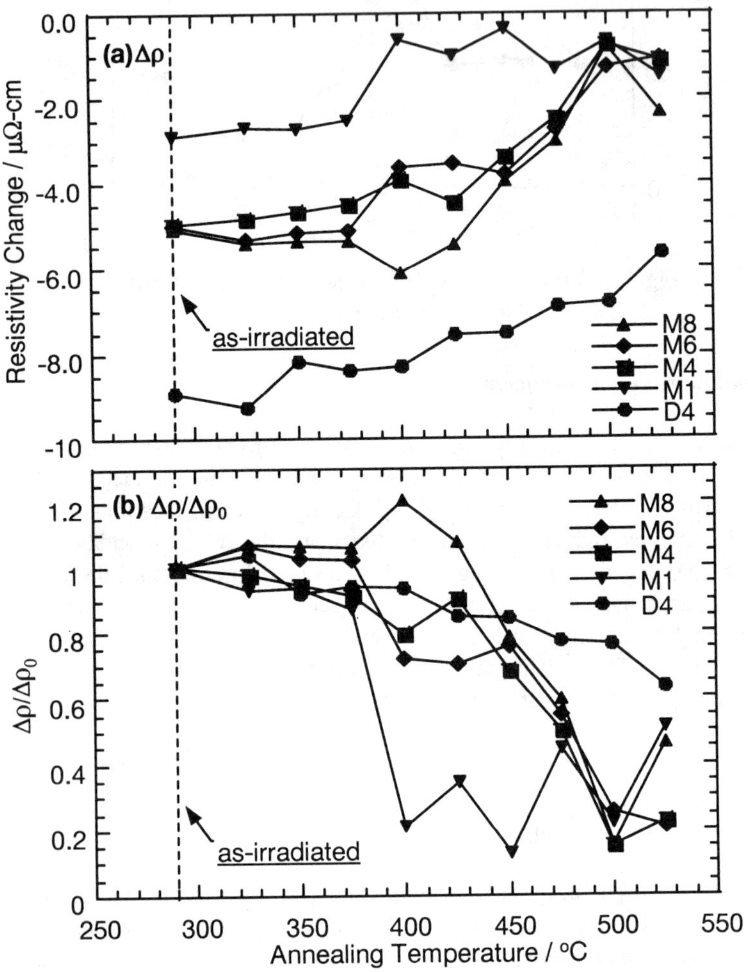

Fig. 4- *(a) Resistivity change and (b) ratio of the resistivity change between the isochronally annealed and as-irradiated specimen.*

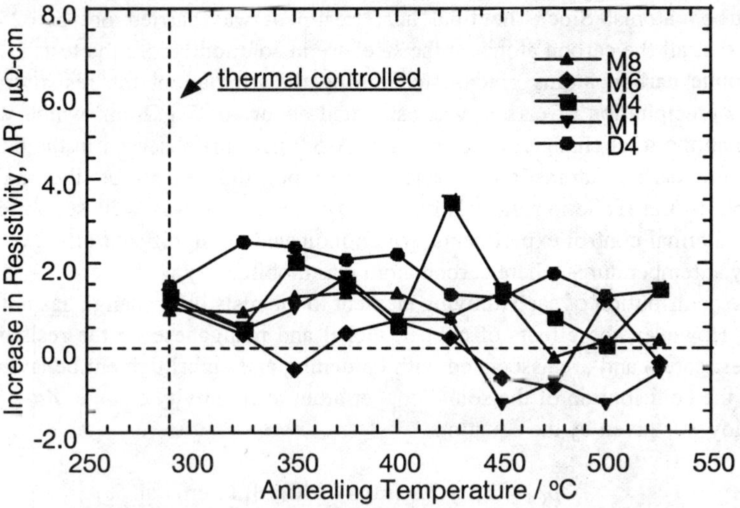

Fig. 5- *Resistivity change of the thermal controlled specimen with isochronal annealing.*

Factor Controlling Irradiation-Induced Resistivity Change

In pure iron, irradiation generally results in an increase in the resistivity, due to irradiation-induced point defects (vacancies and interstitial atoms) and their clusters [*10, 11*]. However, the present study showed an irradiation-induced decrease in the resistivity. Such a case was also observed in Fe-Cu alloys irradiated at 140°C, and Fe-Ni alloys irradiated at 50°C followed by annealing up to 275°C [*11*]. The reduction of resistivity has been explained by the redistribution of alloying elements during irradiation, that is, the decrease in the concentration of solid-solution atoms through their precipitation. However, it is difficult to identify the contribution of each element to the resistivity change, because the A533B commercial steel contains many alloying elements. In the following discussion, it was assumed that each alloying element contributed independently to the resistivity of the steel.

The steel used in this study contains 0.17 wt. % (0.79 at. %) of carbon atoms in the base metal and 0.10 wt. % (0.46 at. %) in the deposit, respectively. According to the work by Takaki et al.[*10*], the contribution of solute carbon to the resistivity change in iron, $\Delta\rho_c$, was given by the following:

$$\Delta\rho_c = 1.5 \times 4.9 \times \Delta C_c \ [\mu\Omega\text{-cm / at. \%}] \quad (1)$$

where the first coefficient, 1.5, is a form factor caused by the difference in specimen between this experiment and the reference, and ΔC_c is a change in the concentration of solute carbon atoms. Since the final heat treatment was carried out at 625°C, it is expected that all the carbon atoms in the steel are in solution before the tests [*12*]. When all the solute carbon atoms precipitated during the irradiation, the resistivity change caused by precipitation of carbon was estimated about –0.37 μΩ-cm, which was much smaller than the total change in the resistivity (-5.0 μΩ-cm) observed in the base metal. Thus, solute carbon atoms and carbides were not thought to be the main factor responsible to the reduction in resistivity of the steel. This was well supported by the results of thermal control experiments, which indicated no significant effect of aging on resistivity at temperatures where carbon atom was mobile.

The contribution of each alloying element to the resistivity change has not yet been revealed. However, the effects of copper, nickel and manganese on the resistivity have been investigated, and are associated with hardening and embrittlement behavior [*9, 11*]. In ref. *9*, the contribution of the solute copper atom to resistivity change, $\Delta\rho_{Cu}$, in Fe-Cu binary alloy was given by the equation,

$$\Delta\rho_{Cu} = 1.5 \times 3.9\ (\pm 0.4) \times \Delta C_{Cu}\ [\mu\Omega\text{-cm / at. \%}] \quad (2)$$

where ΔC_{Cu} was a change in the concentration of solute copper atoms. When all of the 0.08 wt. % of copper atoms in the deposit were in solid solution before irradiation and precipitated during the irradiation, $\Delta\rho_{Cu}$ was estimated to be –0.41 μΩ-cm. This value was also small compared with the total change. After post-irradiation annealing up to 400°C, the resistivity was not recovered but further decreased. It indicated that copper precipitates were thermally stable up to 450°C[9]. This is in agreement with the results that the deposit in this study was not recovered by post-irradiation annealing up to 525°C, as shown in Fig. 4.

A similar effect can be observed in Fe-Ni and Fe-Mn alloy [11] and can be explained as in the case of copper, while the resistivity change was larger than that observed in Fe-Cu alloy because of the larger content than copper in the used steel. Assuming that the contribution of nickel atom to resistivity change in this steel is similar to that of copper, the change in resistivity caused by decreasing in solid solution nickel for the deposit is estimated to be –4.2μΩ-cm, which is a reasonable value to explain the irradiation-induced reduction in the resistivity of deposit. In ref. *11*, the resistivity in Fe-Ni and Fe-Mn was completely recovered by annealing up to 450°C. It was suggested that the resistivity of base metal recovered up to 525°C was corresponding to the resolution of those alloying elements, since both the solubility limit of nickel and manganese are several wt.% at 600°C. Although the precipitation particle is known to contribute to the increase in resistivity, the contribution is thought to be small, because the reduction increases with increasing copper concentration [*11*].

Irradiation-induced point defects and clusters increase the resistivity. Frenkel defects induce the resistivity increase, $\Delta\rho_{FD}$, which was given by the equation,

$$\Delta\rho_{FD} = 1.5 \times 0.3 \times \Delta C_{FD}\ [\mu\Omega\text{-cm / at. \%}] \quad (3)$$

where ΔC_{FD} was change in the density of Frenkel defect [10]. Again, the contribution of irradiation induced-defects to the resistivity is considered to be smaller than that of carbon atoms.

Irradiation-induced resistivity change is given by the equation

$$\Delta\rho_{\text{total}} = \Sigma\Delta\rho_s + \Sigma\Delta\rho_p + \Delta\rho_{FD} \tag{4}$$

where $\Delta\rho_s$ and $\Delta\rho_p$ are the resistivity changes caused by each solute atom and precipitate, respectively. For the irradiation condition in this study, precipitation was thought to be under the early stage of growth and $\Sigma\Delta\rho_p$ and $\Delta\rho_{FD}$ were thought to be negligible.

Correlation between Resistivity and Mechanical Properties

In ref. 12, the increase in hardness, ΔHv, in the as-irradiated specimens was assumed to consist of two components:

$$\Delta \text{Hv} = \Delta \text{Hv}_{\text{PDC}} + \Delta \text{Hv}_{\text{PPT}} \tag{5}$$

where $\Delta \text{Hv}_{\text{PDC}}$ and $\Delta \text{Hv}_{\text{PPT}}$, was the hardening induced by the accumulation of irradiation-induced point defect clusters (PDC) and radiation-enhanced precipitation, respectively. The latter component of Eq. (5), which is proportional to square root of the number density of precipitates, is described by the following equation:

$$\Delta \text{Hv}_{\text{PPT}} = k\,(\text{N}_{\text{unirr}} - \text{N}_{\text{asirr}})^{1/2} \tag{6}$$

where N_{unirr} and N_{asirr} was the number density of alloying atoms in solid solution in unirradiated and as-irradiated specimen, respectively, and k is a constant. On the other hand, the resistivity change of ΔR is written by the following equation as discussed in the previous section:

$$\Delta \text{R} = m\,(\text{N}_{\text{asirr}} - \text{N}_{\text{unirr}}) \tag{7}$$

where m is a constant. From Eqs. (5), (6) and (7), the increase in hardness of as-irradiated specimen is described by the following equation:

$$\Delta \text{Hv} = \Delta \text{Hv}_{\text{PDC}} + \alpha\,(-\Delta \text{R})^{1/2} \tag{8}$$

where α is a constant.

The relation between resistivity change and hardness change is shown in Fig. 6, for each location of the as-irradiated specimens. The fitting curves calculated by applying the Eq. (8) to the results are also shown, where $\Delta \text{Hv}_{\text{PDC}}$ is assumed to be 0, 40, 80 and 100. The irradiation hardening observed in this study was thought to be mainly due to PDC, because the present steel contained less copper and other elements, such as nickel and manganese, are considered to be less hardener at present irradiation conditions. Thus, the assumption that ΔHv_{PDC} is from 80 to 100 was thought to be corresponding to the steel. This also indicates that there is a difficulty to monitor the total hardness change by resistivity measurement in HAZ of welded steel. However, this method is thought to be useful for the estimation of hardness in PVS containing higher concentration of copper where copper precipitates mainly govern the hardening.

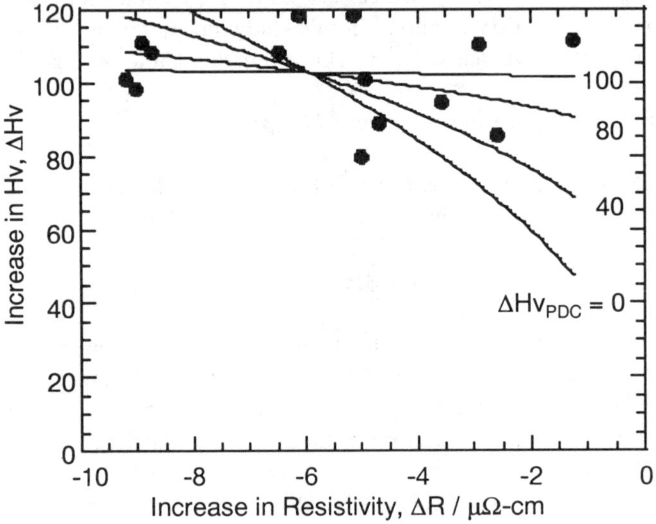

Fig. 6- *Relation between hardness change and resistivity change after neutron irradiation.*

The relation between resistivity change and ΔDBTT that was reported in the previous study [*5*], at each location of weld joint, is shown in Fig. 7. Because the ΔDBTT in the A533B steels can be assumed to be proportional to the irradiation hardening, the relation between resistivity change and ΔDBTT can be fitted in the same way as irradiation hardening. The result also indicates that ΔHv_{PDC} corresponds to a large portion of the total hardening.

Conclusions

The electrical resistivity change induced by neutron irradiation was investigated for the HAZ of welded A533B. The irradiation at 290°C decreased the resistivity at all locations around the fusion boundary. It was considered that the decrease in the resistivity was attributed to the reduction of solid solution atoms, such as copper, nickel and manganese. It appears that the resistivity method is not adequate to evaluate the irradiation embrittlement in HAZ of welded A533B steel having a low copper content.

However, the method will be useful for the estimation of hardness in PVS having higher copper contents where mainly copper precipitates govern the hardening.

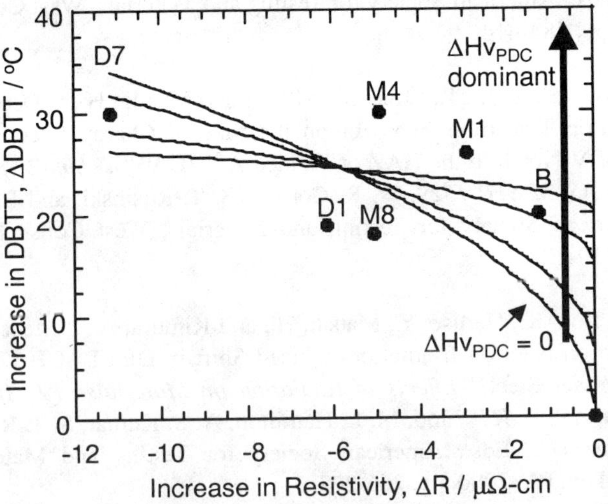

Fig. 7- *Relation between resistivity change and ΔDBTT in as-irradiated specimen.*

Acknowledgment

We would like to thank the Inoue Foundation for Science for the financial support to participate in this conference.

References

[1] Rosinski, S. T. and Corwin, W. R., "ASTM Cross-Comparison Exercise on Determination of Material Properties Through Miniature Sample Testing," *Small Specimen Test Technique, ASTM STP 1329*, W. R. Corwin, S. T. Rosinski, and E. van Walle, Eds., American Society for Testing and Materials, 1998, pp. 3-14.

[2] Kumar, A. S., Louden, B. S., Garner, F. A., and Hamilton, M. L., "Recent Improvements in Size Effects Correlations for DBTT and Upper Shelf Energy of Ferritic Steels," *Small Specimen Test Techniques Applied to Nuclear Reactor Vessel Thermal Annealing and Plant Life Extension, ASTM STP 1204*, W. R. Corwin, F. M. Haggag, and W. L. Server, Eds., American Society for Testing and Materials, Philadelphia, 1993, pp. 47-61.

[3] Sokolov, M. A. and Nanstad, R. K., "On Impact Testing of Subsize Charpy V-Notch Type Specimens," *Effects of Radiation on Materials: 17th International Symposium, ASTM STP 1270*, D. S. Gelles, R. K. Nanstad, A. S. Kumar, and E. A. Little, Eds., American Society for Testing and Materials, West Conshohocken, PA, 1996, pp. 384-414.

[4] Kimura, A., Suzuki, T., Jincho, M., and Matsui, H., "Dependence of Ductile-Brittle Transition Behavior on the Size of Charpy Specimen and the Location of V-Notch in the HAZ of Welded A533B PVS," *Small Specimen Test Technique, ASTM STP 1329*, W. R. Corwin, S. T. Rosinski, and E. van Walle, Eds., American Society for Testing and Materials, West Conshohocken, PA, 1998, pp. 110-122.

[5] Suzuki, T., Itoh, K., Naruse, Y., Matsui, H., and Kimura, A., "Effect of Location of V-Notch on Neutron Irradiation-Induced Shift in DBTT of HAZ of Welded Pressure Vessel Steel," *Effects of Radiation on Materials: 19th International Symposium, ASTM STP 1366*, M. L. Hamilton, A. S. Kumar, S. T. Rosinski, and M. L. Grossbeck, Eds., American Society for Testing and Materials, West Conshohocken, PA, 2000, pp. 266-278.

[6] Goto, T., Kamimura, T., Kumano, S., Takeuchi, I., Maeda, N., and Yamaguchi, A., "Changes in Electromagnetic Properties of a Low-Alloy Steel Caused by Neutron Radiation," *Effects of Radiation on Materials: 18th International Symposium, ASTM STP 1325*, R. K. Nanstad, M. L. Hamilton, F. A. Garner, and A. S. Kumar, Eds., American Society for Testing and Materials, West Conshohocken, PA, 1999, pp. 589-600.

[7] Otaka, M., Shimizu, S., Morinaka, R., Maeda, N., and Yamaguchi, A., "Detection of Radiation Damage in Low Alloy Steel Using a SQUID Sensor," *Effects of Radiation on Materials: 18th International Symposium, ASTM STP 1325*, R. K. Nanstad, M. L. Hamilton, F. A. Garner, and A. S. Kumar, Eds., American Society for Testing and Materials, West Conshohocken, PA, 1999, pp. 568-575.

[8] Hempel, A., Saneyasu, M., Tang, Z., Hasegawa, M., Brauer, G., Plazaola, F., Yamaguchi, S., Kano, F., and Kawai, A., "Effects of Neutron Irradiation on Fe-Cu Model Alloys and RPV Steels Probed by Positron Annihilation and Hardness Measurements," *Effects of Radiation on Materials: 19th International Symposium, ASTM STP 1366*, M. L. Hamilton, A. S. Kumar, S. T. Rosinski, and M. L. Grossbeck, Eds., American Society for Testing and Materials, West Conshohocken, PA, 2000, pp. 560-578.

[9] Mathon, M. H., Barbu, A., Dunstetter, F., Maury, F., Lorenzelli, N., and De Novian, C. H., "Experimental Study and Modeling of Copper Precipitation under Electron Irradiation in Dilute FeCu Binary Alloys," *Journal of Nuclear Materials*, 1997, Vol. 245, pp. 224-237.

[10] Takaki, S., Fuss, J., Kugler, H., Dedek, U., and Schultz, H., "The Resistivity Recovery of High Purity and Carbon Doped Iron Following Low Temperature Electron Irradiation," *Radiation Effects*, 1983, Vol. 79, pp. 87-122.

[11] Nikolaev, V. A., and Rybin, V. V., "Mechanisms Controlling the Composition Influence on Radiation Hardening and Embrittlement of Iron-Base Alloys," *Effects of Radiation on Materials: 17th International Symposium, ASTM STP 1270*, D. S. Gelles, R. K. Nanstad, A. S. Kumar, and E. A. Little, Eds., American Society for Testing and Materials, West Conshohocken, PA, 1996, pp. 3-24.

[12] Fisher, S. B., Harbottle, J. E., and Aldridge, N., "Radiation Hardening in Magnox Pressure-Vessel Steels," *Philosophical Transactions of the Royal Society of London*, 1985, Vol. A315, pp. 301-332.

Marc G. Horsten[1] and William P. A. Belcher[2]

Fracture Toughness and Tensile Properties of Irradiated Reactor Pressure Vessel Cladding Material

---

Reference: Horsten, M. G. and Belcher, W. P. A., "**Fracture Toughness and Tensile Properties of Irradiated Reactor Pressure Vessel Cladding Material,**" *Effects of Radiation on Materials: 20th International Symposium, ASTM STP 1405,* S. T. Rosinski, M. L. Grossbeck, T. R. Allen, and A. S. Kumar, Eds., American Society for Testing and Materials, West Conshohocken, PA, 2001.

Abstract: A comprehensive testing programme was undertaken to evaluate the effects of irradiation, thermal ageing, specimen orientation, and test temperature on the fracture toughness and tensile behaviour of a Type 309L/308L stainless steel strip clad deposit. Specimens were irradiated in the high flux reactor (HFR) Petten to nominal neutron doses of 0.05 dpa and 0.1 dpa at 295°C. Testing was performed at room temperature, 100, 200 and 295°C. The fracture resistance properties of the clad material were unaffected by irradiation to 0.1 dpa, which is approximately twice the predicted end-of-life dose. The same neutron dose resulted in an increase in 0.2% yield stress (15-20 MPa) and a small loss of ductility (3%). Thermal ageing to the simulated pressurised water reactor (PWR) end-of-life thermal condition (1000 hrs at 400°C) had no significant effect on the fracture resistance behaviour or tensile properties of clad material in either irradiated or unirradiated condition. Clad fracture resistance properties were unaffected by orientation in the plane of the clad layer and were only dependent on test temperature.

Keywords: pressure vessel cladding, fracture resistance, neutron irradiation, tensile, thermal ageing, orientation, temperature

Introduction

The inside surface of PWR reactor pressure vessels (RPVs) is covered with a corrosion resistant austenitic stainless steel cladding layer to prevent general corrosion of the low alloy RPV steel and minimise contamination of the coolant with activated corrosion products. In modern RPVs, such as British Energy's Sizewell B PWR, the cladding is applied to the ferritic base material by a submerged-arc, strip clad process, which commonly utilises Type 309L stainless steel for the first layer and Type 308L for subsequent layers.

[1]Materials scientist, NRG, PO Box 25, 1755 ZG Petten, Netherlands.
[2]Metallurgist, British Energy Generation Ltd., Barnwood, Gloucester, UK.

The welding process and associated heat treatment parameters are carefully controlled to prevent underclad cracking. The structural integrity safety case for the Sizewell B PWR RPV includes fracture analysis to assess defect tolerance during both normal service and accident conditions. The growth of defects, postulated to be present at start-of-life, and limiting defect depths are evaluated for surface breaking base metal defects in an unclad vessel. The assumption of surface breaking defects gives a more pessimistic assessment than assuming that the defects are covered by the cladding, provided that the clad layer remains intact. Additional analysis was therefore performed to confirm that, for postulated underclad (buried) defects, the cladding ligament at end-of-life remained intact under the most severe accident conditions. This additional analysis uses the analytical defect model shown in Figure 1.

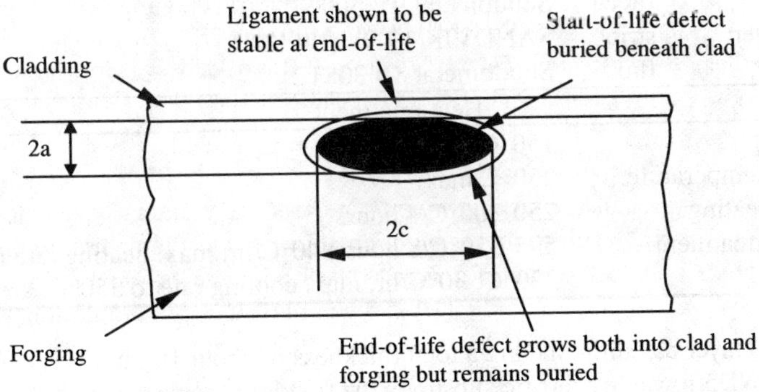

Figure 1 – *Defect assumptions for cladding ligament analysis.*

Fracture toughness properties of irradiated austenitic stainless steel cladding, which were utilised in this analysis for Sizewell B, were a conservative lower bound to the limited data available at the time [1-7]. These data suggested little deleterious effect from thermal ageing at PWR operating temperatures, but a marked degradation in start-of-life fracture toughness was expected due to irradiation. The scarcity of the data, particularly from modern strip clad material, however, made definition of realistic fracture toughness properties uncertain. In order to confirm the fracture toughness data assumed in the fracture analysis, a comprehensive testing programme was undertaken to evaluate the effects of irradiation, thermal ageing, and test temperature on the fracture toughness and tensile behaviour of austenitic stainless steel cladding. This paper presents the results of tensile tests and fracture mechanics experiments with small compact (C(T)) specimens on Type 309L/308L strip clad material, representative of the Sizewell B RPV cladding, after irradiation to target dose levels of 0.05 dpa and 0.1 dpa at 295°C.

Materials and Specimens

Materials

The programme cladding material was produced using consumables, weld procedures, and heat treatments representative of the Sizewell B RPV core shell manufacture, as detailed in Table 1.

Table 1 – *Details of cladding consumables and heat treatment.*

		Strip clad welding consumables
First layer:	strip	SANDVIK 24-13L (ER 309L)
	flux	Soudometal 9V 309T1
Second layer:	strip	SANDVIK 19-9L (ER 308L)
	flux	Soudometal 8B 308T2
		Heat treatment
Preheating:		150°C min.
Inter-pass temperature:		250°C max.
Post clad heating:		250-300°C/4 hours
Final heat treatment:		595-620°C/8 hours; 40°C/hr. max. heating rate from 350°C; 30°C/hr. max. cooling rate to 350°C; air cool.

The two layer cladding, having a total thickness of about 10 mm, was deposited on to an 18MND5 low alloy ferritic substrate, the first layer comprising Type 309L and the second Type 308L weld metal. Microstructural, microhardness and compositional variations across the austenitic/ferritic interface were examined and found to be similar to those for the Sizewell B cladding [8]. Only the 5 mm thick top layer, consisting mainly of Type 308L material, was utilised in the programme. The chemical composition of the top layer of cladding is given in Table 2.

Table 2 - *Chemical composition of the top layer of cladding Type 308L material.*

Element	C	Si	Mn	Cr	Ni	Mo	N_2	Fe
Contents (wt-%)	0.028	0.85	1.10	19.33	10.51	0.06	418 ppm	balance

Specimens

Reduced thickness C(T) specimens with a thickness of 5 mm and tensile test specimens with a 20 mm gauge length and of 4 mm diameter were machined in the plane of the cladding, see Figure 2.

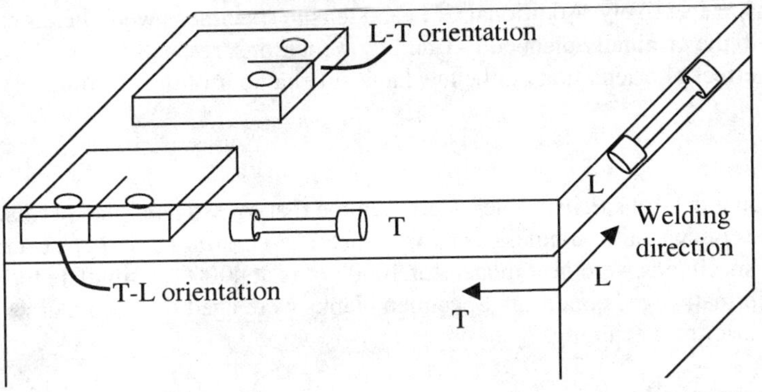

Figure 2 – *C(T) and tensile specimens machined in the plane of the cladding.*

The dimensions of the C(T) and tensile specimens are given in Figure 3. All specimens were taken from the top layer of the cladding to minimise material variability effects.

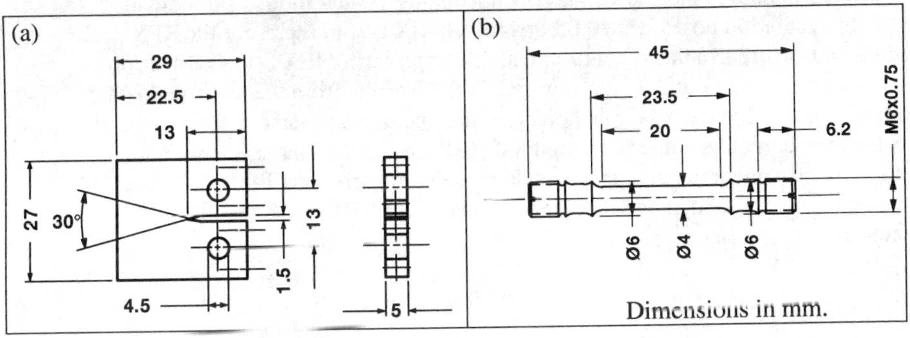

Figure 3 –*C(T) (a) and tensile (b) specimen used in the experimental study.*

Further details of the specimen machining plan and their relative position within the clad block are given in [9]. A total of 75 C(T) and 45 tensile specimens were machined from the cladding.

Orientation

The planned orientation of test specimens was L-T for C(T) specimens and L for tensile specimens to represent the orientation of postulated underclad defects, namely perpendicular to the direction of welding (L). Following manufacture, however, microstructural examination of the heat-affected zone in the base metal and the weld bead geometry in the cladding layer [10] showed that the actual direction of welding was perpendicular to that marked on the clad block by the manufacturer. Due to this reference marking error, the majority of the C(T) and tensile specimens had the T-L and

T orientations, respectively. Additional C(T) and tensile specimens were therefore machined with the originally planned L-T and L orientations, respectively, to determine any possible effect of orientation on the mechanical behaviour of the cladding.

Thermal Ageing

Based on a previous British Energy programme that investigated the thermal ageing kinetics of austenitic stainless steel weld metal and castings [11, 12], several cladding test specimens were heat treated for 1000 hours at 400°C to simulate the RPV end-of-life thermally aged condition. Specimen blanks were aged in an inert gas environment prior to machining.

Specimen Irradiations

Specimens were irradiated in the High Flux Reactor at Petten as part of the SINEXT (R312) experiment. Sample holders, containing specimens stacked on top of each other, were positioned within a threefold thimble device. Irradiation temperature was controlled to within ±15°C of the 295°C target by an inert gas mixture, with C(T) specimen temperatures monitored via thermocouples located in load-pin hole inserts. The target irradiation doses were 0.05 dpa for SINEXT-2, to represent the RPV cladding end-of-life irradiated condition, and 0.1 dpa for SINEXT-1. The neutron exposure times were 17 and 24 days, respectively. Results from the neutron dosimetry monitors contained in each sample holder, which include assessment of the neutron metrology uncertainties, have been reported [13]. The accumulated fluence and resulting displacement damage results, which are summarised in Table 3, indicate average doses of 0.049 dpa and 0.080 dpa achieved for SINEXT-2 and SINEXT-1, respectively.

Table 3 – *Neutron dosimetry results.*

Irradiation experiment		Displacement damage (dpa)	Thermal fluence Φ_0 (10^{23} m^{-2})	Fluence E >1 MeV (10^{23} m^{-2})
SINEXT-2	Min.	0.037	2.88	2.42
target:	Max.	0.061	4.73	4.06
0.05 dpa	Average	0.049	3.78	3.25
SINEXT-1	Min.	0.067	7.64	4.31
target:	Max.	0.093	11.0	6.02
<0.1 dpa	Average	0.080	9.78	5.20

Fluence monitor data have been used to estimate the individual damage dose for each specimen. These values have been utilised in assessing the data.

Testing

All testing was performed in the hot cell laboratory at Petten. The test matrix for C(T) and tensile specimens in the unirradiated, irradiated, and thermally aged and irradiated conditions is given in Table 4. Testing was performed at room temperature (RT), 100°C, 200°C and 295°C. Irradiated tensile specimens were tested only at 100°C and 295°C.

Table 4 – *Test matrix for cladding experiments.*

Specimen condition	Test temperature, °C							
	RT		100		200		295	
	C(T)[1]	T[2]	C(T)[1]	T[2]	C(T)[1]	T[2]	C(T)[1]	T[2]
Unirr. control								
As-received	3+[2]	2	3+[2]	3	6+[2]	3	5+[5]	4+[3]
Thermally aged	2	1	3	2	-	-	2	2
Irradiated								
SINEXT-2	3	-	3	2	3	-	3+[1]	3+[1]
SINEXT-1	3	-	3	2	3	-	3+[1]	3+[1]
Thermally aged[3] and irradiated								
SINEXT-2	-	-	2	2	-	-	2	2
SINEXT-1	-	-	2	2	-	-	2	2

[1] C(T) specimen orientation is T-L, except [ ], which have L-T orientation.
[2] Tensile specimen orientation is T, except [ ] , which have L orientation.
[3] 1000 hours at 400°C to simulate the end-of-life thermally aged condition.

Tensile

Both unirradiated and irradiated tensile specimens were tested at a strain rate of $5x10^{-4}s^{-1}$ in the same electro-mechanical test machine. The test procedure was consistent with ASTM Standard for Tension Testing of Metallic Materials (E8) and ASTM Standard for Elevated Temperature Tension Testing Tests of Metallic Materials (E21). Test temperatures were controlled to an accuracy of ±2°C. Values of 0.2% yield stress (0.2%YS), ultimate tensile strength (UTS) and uniform elongation (UE) were calculated from the engineering stress-strain curve, and total elongation (TE) from the broken specimen halves.

Fracture Resistance

Fatigue precracking of the C(T) specimens was performed with an R-ratio of 0.1 at 10 Hz frequency. The ΔK range was decreased with increasing crack extension. Single specimen fracture resistance testing used the direct current potential drop (DCPD) method for measuring crack extension. The rate of increase in stress intensity factor was in the range of 0.8-1.0 $MPa\sqrt{m}\cdot s^{-1}$ in the elastic regime. Load and DCPD were recorded as a function of load-line displacement. Post-test measurements of fatigue crack length and crack extension (Δa) were used to correlate the measured DCPD signal with actual crack growth, from which Δa values were obtained and plotted against corresponding J data. Fracture parameters $J_{0.2}$, $J_{0.2bl}$, $J_{0.5}$ and $dJ/da_{0.5}$ were determined and initiation fracture toughness, $K_{0.2bl}$, was calculated. Testing and analysis procedures were based on ESIS P2-92 [14].

Tensile Test Results

Tensile properties of clad material in the unirradiated condition are shown in Figures 4(a) and 4(b) as a function of test temperature for the two specimen orientations L and T and both thermally aged and unaged conditions.

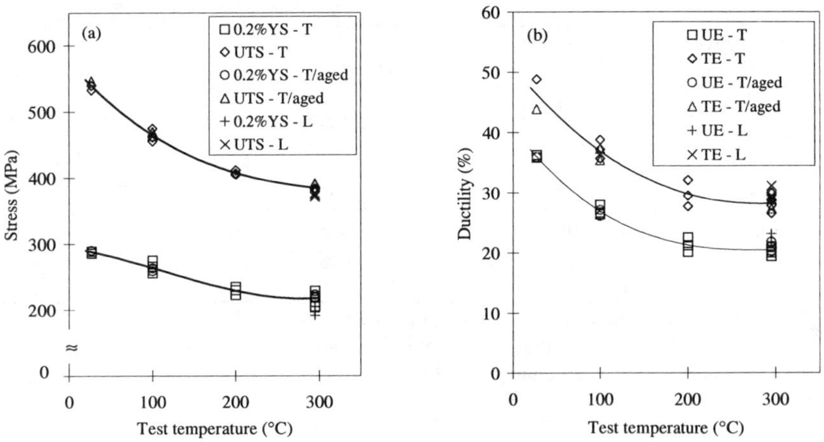

Figure 4 – *Temperature trend curve of (a) 0.2%YS and UTS and (b) tensile ductility of unirradiated material.*

There is no marked influence of ageing for 1000 hours at 400°C on either 0.2%YS, UTS or tensile ductility (TE and UE). Furthermore, although the L and T oriented specimens, tested at 295°C, suggest marginally lower strength values in the L orientation, these differences are not considered to be experimentally significant. No clear evidence of orientation on ductility is evident.

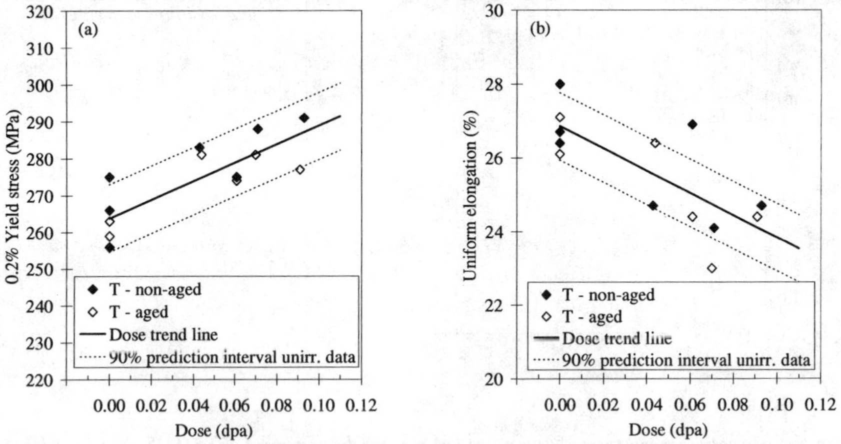

Figure 5 – *Irradiation response of cladding material at 100°C test temperature: (a) hardening and (b) loss of ductility.*

Tensile properties following irradiation suggested a trend of increasing 0.2%YS and decreasing ductility with neutron dose, but in each case, the effect was small. Figures 5(a) and 5(b) indicate these trends at a test temperature of 100°C. At the maximum neutron dose (~0.1 dpa), the irradiation hardening effect was of the order 15-20 MPa at both 100°C and 295°C, while UE values were approximately 3% lower.

Fracture Resistance Test Results

In order to assess the influence of thermal ageing, orientation, and irradiation on cladding fracture resistance, individual J-Δa curves at a particular test temperature were combined.

Influence of Thermal Ageing

J-Δa curves at 100°C from thermally aged specimens are compared with non-aged specimens (all T-L orientation) in Figures 6(a) and 6(b), for the unirradiated and ~0.05 dpa irradiated conditions, respectively. The fracture resistance behaviour is seen to be independent of thermal ageing.

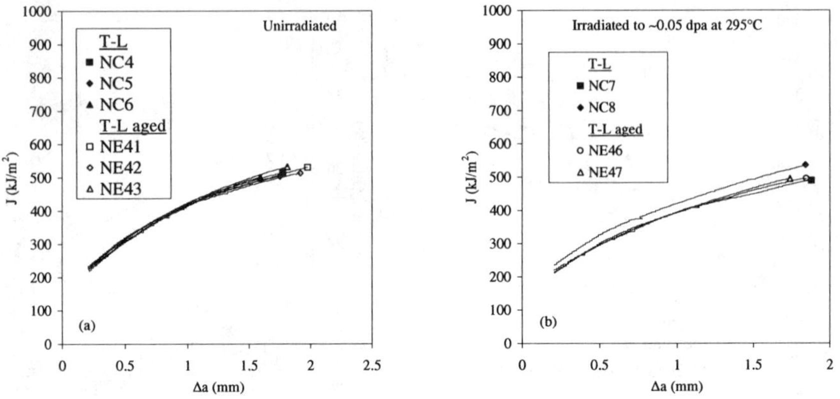

Figure 6 – *Fracture resistance curves of (a) unirradiated, and (b) ~0.05 dpa irradiated material tested at 100°C.*

Within the bounds of normally expected scatter, this pattern was repeated for all test temperatures in each condition and leads to the conclusion that cladding fracture resistance is not significantly affected by thermal ageing to the end-of-life condition.

Influence of Specimen Orientation

The fracture resistance of T-L and L-T orientated specimens at 295°C is shown in Figures 7(a) and 7(b) for the unirradiated and ~0.05 dpa irradiated conditions, respectively. Within typical experimental scatter, the J-Δa curves are essentially unaffected by orientation differences.

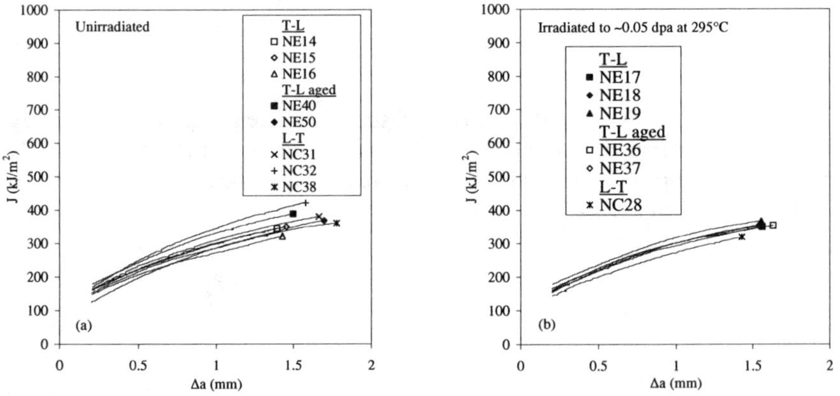

Figure 7 - *Fracture resistance curves of (a) unirradiated and (b) ~0.05 dpa irradiated material tested at 295°C.*

Again, this behaviour was observed at all other test temperatures in the unirradiated and ~0.1 dpa irradiated materials, resulting in the conclusion that the fracture resistance of the cladding can be considered to be orientation independent.

Influence of Neutron Irradiation

On the basis that no clear effect of thermal ageing or specimen orientation on fracture resistance had been observed in any of the specimens, irradiated or not, the J-Δa data from T-L, T-L aged, and L-T specimens were combined for each combination of irradiation condition and test temperature.

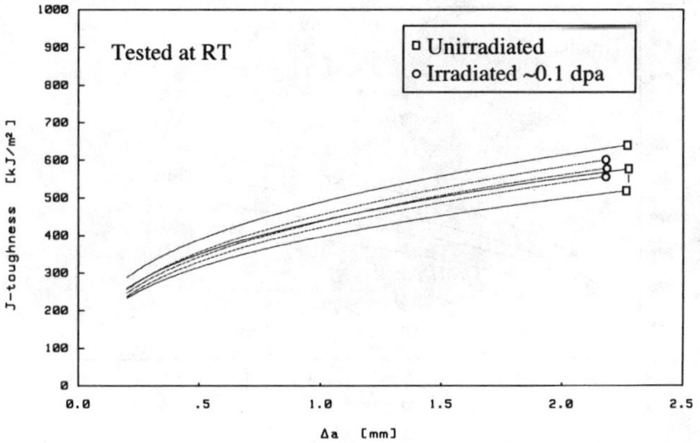

Figure 8 – *Fracture resistance of unirradiated and ~0.1 dpa irradiated material at RT.*

Results indicated that the curves for the irradiated material were within the 90% prediction intervals for the unirradiated material. In other words, there was no evidence of any irradiation effect up to the maximum neutron dose, Figures 8 and 9. Similar observations have been reported previously on modern strip cladding material [15].

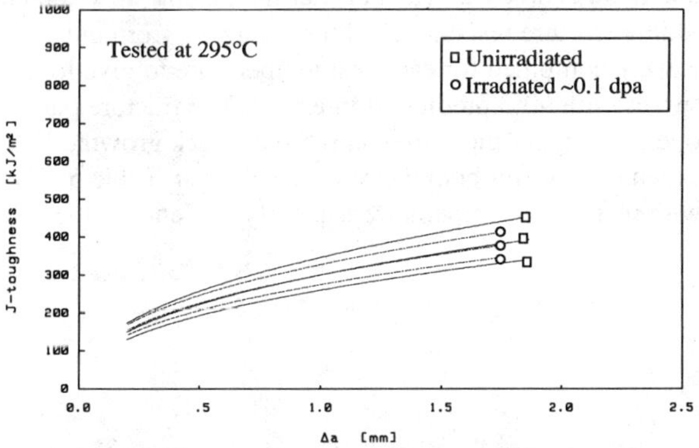

Figure 9 - *Fracture resistance of unirradiated and irradiated material at 295°C.*

The absence of any effect from either thermal ageing, noted previously, or neutron irradiation on fracture resistance behaviour makes it unlikely that any synergistic effect will occur under the combined influence of irradiation and thermal ageing in the RPV cladding, during service. It is however not possible to fully exclude a synergistic effect based on this evidence alone.

Influence of Test Temperature

The mean J-Δa curves for the ~0.05 dpa irradiated material at the four distinct test temperatures RT, 100, 200 and 295°C are shown in Figure 10.

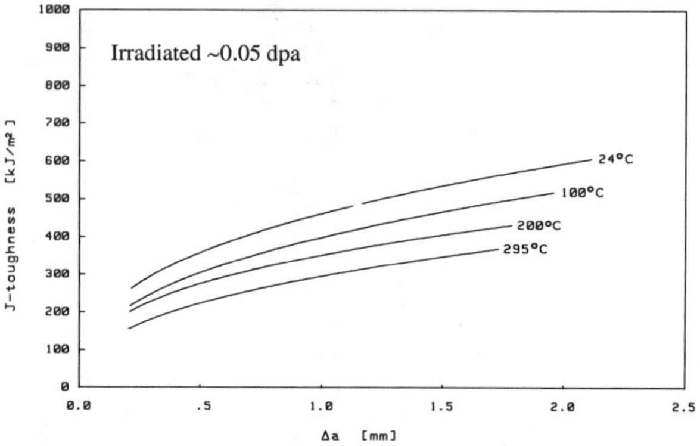

Figure 10 – *Fracture resistance temperature dependence of irradiated material.*

Increasing temperature can be seen to reduce the fracture resistance of the irradiated clad material. A similar fall off in fracture resistance with increasing temperature was observed in unirradiated and ~0.1 dpa irradiated materials. Minor differences are expected to result from material variability.

On the basis that there is no clear effect of thermal ageing, specimen orientation or irradiation on cladding fracture resistance, all the J-Δa data from unirradiated and irradiated specimens were combined for each test temperature to give four mean fracture resistance curves with 90% prediction intervals. The fracture parameters $J_{0.2}$, $J_{0.2bl}$, $J_{0.5}$ and the slope of the resistance curve at 0.5 mm crack growth, $dJ/da_{0.5}$, were obtained from these mean and lower bound curves, as given in Table 5. The decrease in fracture resistance with increasing temperature is clearly evident.

Table 5 – *Fracture parameters from mean and lower bound fitted curves.*

T_{test} (°C)	$J_{0.2}$ (kJ/m^2) mean	$J_{0.2}$ (kJ/m^2) lower	$J_{0.2bl}$ (kJ/m^2) mean	$J_{0.2bl}$ (kJ/m^2) lower	$J_{0.5}$ (kJ/m^2) mean	$J_{0.5}$ (kJ/m^2) lower	$dJ/da_{0.5}$ (kJ/m^3) mean	$dJ/da_{0.5}$ (kJ/m^3) lower
RT	256	232	309	276	350	318	241	218
100	219	202	270	246	312	288	241	223
200	185	163	229	196	266	234	209	184
295	152	135	187	162	224	198	188	166

Engineering Fracture Initiation

Since the fracture resistance data obtained from the 5mm thick C(T) specimens did not meet current J-validity requirements for size-independent fracture behaviour [14], the fracture parameters in Table 5 cannot be considered strictly applicable to larger cladding thicknesses. Nevertheless, the results do characterise the fracture resistance behaviour of the 5 mm top layer of cladding and are relevant to analyses which consider failure in this specific region. In view of the above conclusions on the absence of irradiation, thermal ageing and specimen orientation effects, individual $J_{0.2bl}$ data from all tests have been combined in Figure 11.

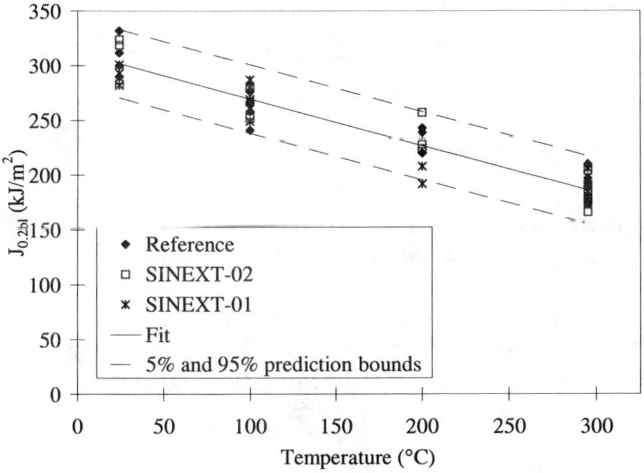

Figure 11 – *The temperature dependence of $J_{0.2bl}$ data.*

This figure can be used to describe the trend with temperature of engineering fracture toughness. The fitted correlation is a simple linear temperature relationship, with the inherent variability, characteristic of weld material inhomogeneity, defined by the 90% prediction interval.

Fractography

Fracture surfaces of C(T) test specimens were examined systematically by scanning electron microscopy to investigate any differences in fracture mechanism. In all cases, fracture was by ductile microvoid coalescence, with no evidence for any change in fracture mechanism due to differences in ageing condition, specimen orientation, irradiation condition or test temperature [10].

Comparison with Fracture Analysis Assumptions

The initiation fracture resistance values, J_{Ic}, for RPV cladding, which were utilised in the Sizewell B RPV fracture analysis are compared in Figure 12 with the lower bound $J_{0.2bl}$ values from the present programme.

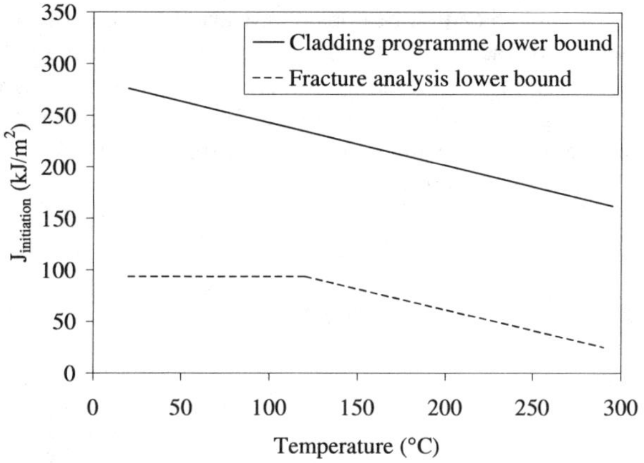

Figure 12 - *Lower bound initiation fracture resistance of cladding used in fracture analysis compared with current programme data.*

Over neutron irradiations of up to twice the predicted end-of-life dose for the Sizewell B RPV, the data assumed in fracture analysis of the cladding ligament (see Figure 1) are seen to be conservative. Results from this programme therefore provide strong support for the Sizewell B RPV safety case.

Conclusions

The following conclusions can be drawn from the experiments on the simulated Sizewell B RPV cladding material.

- The fracture resistance properties of the clad material are unaffected by irradiation to approximately 0.1 dpa at 295°C, which is twice the predicted end-of-life dose.
- Irradiation to approximately 0.1 dpa at 295°C results in an increase in 0.2% yield stress (15-20 MPa) and a small loss of ductility (3%).

- Thermal ageing to the simulated end-of-life thermal condition (1000 hours at 400°C) has no significant effect on the fracture resistance behaviour or tensile properties of the cladding material in either the irradiated (thermal ageing prior to irradiation) or unirradiated condition.
- The cladding fracture resistance properties are unaffected by orientation in the plane of the clad layer.
- The cladding material fracture resistance properties are dependent only upon test temperature.
- Fracture resistance properties used in fracture analysis of the Sizewell B PWR RPV cladding are conservative.

Acknowledgements

The authors acknowledge useful discussions with respective NRG and BEGL colleagues. The work was funded under the HSE Industry Management Committee programme. This paper is published with the permission of British Energy Generation Ltd. and NRG Petten.

References

[1] Ahlstrand, R. and Rajamaki, P., "Toughness and Fatigue Properties of Stainless Steel Submerged Arc Weld Cladding Overlay and Significance of Postulated Flaws in the Cladding Overlay," *Int. J. Pres. Ves. & Piping Vol. 33*, 1988, pp 129-142.

[2] Haggag, F. M., Corwin, W. R. and Nanstad, R. K., "Irradiation Effects on Strength and Toughness of Three-Wire Series-Arc Stainless Steel Weld Overlay Cladding," Oak Ridge National Laboratory Report NUREG/CR-5511, Oak Ridge, Tennessee, February 1990.

[3] Corwin, W. R., "Stainless Steel Cladding Investigations, Heavy-Section Steel Technology Programme Semi-annual Progress Report for October 1985-March 1986," Oak Ridge National Laboratory Report NUREG/CR-4219-Vol.3-No.1, Oak Ridge, Tennessee, June 1986.

[4] Haggag, F. M. and Iskander, S. K., "Results of Irradiated Cladding Tests and Clad Plate Experiments", *Nuclear Engineering and Design Vol. 118*, 1990, pp 297-304.

[5] Haggag, F. M. and Nanstad, R. K., "Effects of Thermal Ageing and Neutron Irradiation on the Mechanical Properties of Stainless Steel Weld Overlay Cladding," *Fifth Int. Symposium on Environmental Degradation of Materials in Nuclear Power Systems – Water Reactors*, 25-29 August 1991, Monterey, California, pp 327-332.

[6] McCabe, D. E., "Fracture Evaluation of Surface Cracks Embedded in Reactor Vessel Cladding: Unirradiated Bend Specimen Results," NUREG/CR-4841 (MEA-2200), Oak Ridge, Tennessee, March 1987.

[7] McCabe, D. E., "Fracture Evaluation of Surface Cracks Embedded in Reactor Vessel Cladding," NUREG/CR-5207, Oak Ridge, Tennessee, September 1988.

[8] Abbott, K. R., Mason, J. E. and Belcher, W. P. A., "Examination of the Clad Ferritic Interface of Sizewell B Representative RPV Material," Nuclear Electric Report EPD/SXB/REP/0020/96, June 1996.

[9] Horsten, M. G., Van den Broek, F. P., Jong, M. and de Vries, M. I., "Fracture Toughness and Tensile Properties of Irradiated Vessel Cladding Material," ECN Report ECN-CX-97-106, December 1997.

[10] Belvroy, R. L. and Horsten, M. G., "Microstructural Investigation of Irradiated Vessel Cladding Material," ECN report ECN-CX-97-131, February 1998.

[11] Pumphrey, P. H., "Ageing Kinetics of CF3 Cast Stainless Steel in the Temperature Range 300°C to 400°C," *Materials Science and Technology Vol. 6(3)*, March 1990, pp 211-219.

[12] Gage G. and Belcher W. P. A., "The Influence of Ageing on the Impact Properties and Fracture Toughness of a Simulated PWR Pump Bowl Repair Weldment," *IAEA Specialists Meeting on Thermal and Mechanical Degradation in Reactor Materials*, Abingdon, England, November 19-21, 1991.

[13] Ketema, D. J., "Neutron Dosimetry Results of SINEXT-01 and 02," Internal ECN note 71070/NUO/DJK/mh/10887, 17th November 1997.

[14] European Structural Integrity Society, "Procedure for Determining the Fracture Behaviour of Materials", ESIS P2-92, January 1992.

[15] J.K. Sharples et al, "Study to generate Information on RPV Cladding Materials Data Including Manufacturing Aspects and Provide State-of-the-Art Report on the Role of Cladding in RPV Safety Analysis – Part 1 (Summary Report – Task 1)," AEAT/NJCB/01339001/R01 Issue 1 (Draft), July 2000.

O. P. Maksimkin[1] and O. V. Tivanova[1]

True Characteristics of Strength and Ductility for Neutron-Irradiated Metals and Alloys

Reference: Maksimkin, O. P. and Tivanova, O. V., **"True Characteristics of Strength and Ductility for Neutron-Irradiated Metals and Alloys,"** *Effects of Radiation on Materials: 20th International Symposium, ASTM STP 1405,* S. T. Rosinski, M. L. Grossbeck, T. R. Allen, and A. S. Kumar, Eds., American Society for Testing and Materials, West Conshohocken, PA, 2001.

Abstract: The results of comparative investigations of variations in engineering and true characteristics of strength and ductility for austenitic stainless steel 12Crl8Nil0Ti, high-nickel alloy 03Cr20Ni45Mo4, and commercially pure molybdenum irradiated by neutrons in the WWR-K reactor core at a temperature lower than 370 K and the fluences ranging from $5 \cdot 10^{22}$ to $2 \cdot 10^{24}$ n/m^2 (E>0.1MeV) are presented.

In tensile experiments at 293 K at the constant strain rate $\sim 10^{-4}$ c^{-1} the curve showing variations in geometrical dimensions of specimens was recorded together with the strengthening diagram. This revealed regularities in deformation-induced change of specimen diameters measured at various cross-sections over the specimen operating length, and allowed us to determine true values of the yield strength and the failure stress, and to find the value of narrowing related to the start of the localized plastic deformation.

For steel specimens, true values of strains and stresses (with the specimen diameter and length variations taken in account), corresponding to onset of recording the martensite ferromagnetic phase, are determined. It has been shown that, under the present test conditions, true stresses of α'–phase formation in deformed steel are found to decrease as the neutron fluence increases.

Keywords: true characteristics of strength, ductility, irradiation, necking, martensite transformation, austenitic stainless alloys, molybdenum

---

[1]Head of division, Dr. Sc. (Solid State Physics), professor, and postgraduate student, respectively, Institute of Atomic Energy, National Nuclear Center, Almaty, 480082, Republic of Kazakhstan.

Introduction

Investigation of regularities in radiation strengthening and embrittlement of reactor structural materials will be more correct if it is possible, in the course of mechanical tests, to provide reliable and accurate determination of "true" values of stresses and strains. In order to find these parameters, one must know current values of cross-section areas in every moment of test (that means the measured values of diameters, thickness and width) and calculated length of a specimen. However, in practice it is extremely difficult to determine continuously varying geometrical sizes of a radioactive specimen directly in a tensile test because the safety-related time of contact with the specimen is very limited. Due to this situation, theoretical calculations persist as a predominant way to define section areas needed for determining the "true" yield stresses in radiation mechanical tests [1]. Numerical inaccuracies seem to be inevitable in such an approach. Besides, considerable troubles occur when determining regularities and character in formation of the neck shape under deformation. In view of solution of the problems like this, we have developed contact-free the remote-controlled optical-electron extensometer (OEE). It is useful in works with thin, fragile and irradiated materials, the contact with those is unwanted in tests.

In this work the results of experiments on determination of true mechanical characteristics by means of the specially developed OEE, the effect of localization of deformation in various materials before and after irradiation are analyzed.

Experimental Technique

As investigation objects, metal materials, applied as the reactor structural materials, have been chosen, such as austenitic stainless steel 12Cr18Ni10Ti, high-nickel alloy 03Cr20Ni45Mo4 and technically pure molybdenum. From these materials the cylindrical specimens have been prepared; their sizes are given in Figure 1a. A part of specimens prior subjected to thermal treatment have been irradiated in the WWR-K reactor core at the temperature not higher than 370 K to the following neutron fluences, as $5 \cdot 10^{22}$, $2 \cdot 10^{23}$, $7 \cdot 10^{23}$ and $2 \cdot 10^{24}$ n/m^2 (E>0.1MeV). Mechanical tests on unirradiated and irradiated materials have been performed at room temperature using the universal testing machine "Instron 1195", provided by the optical electron extensometer and the scanning ferroprobe. Strength and ductility characteristics of investigated materials and fluences of neutrons are given in Table 1.

As special experiments show, the data on variation in geometrical sizes, obtained by means of OEE, coincide, with an accuracy to 1%, with the results of control measurements of the diameter and the length of a specimen, carried out at the measuring microscope.

In view of investigation of the deformation-induced martensitic $\gamma \rightarrow \alpha$ ($\gamma \rightarrow \varepsilon \rightarrow \alpha$) transition in steel 12Cr18Ni10Ti, we have developed a device that allows to judge on formation and distribution of the ferromagnetic phase basing on the length of a deformed irradiated specimen. Its measuring element is the ferroprobe (Ferster 1.053, Germany), and one can determine with it the presence of the magnetic phase of the content from 0.05 to 50% (relatively to a standard specimen) within a local region (1 mm^3) where a test probe contacts with a specimen surface. The probe is mounted in a cramp, normally with respect to a specimen, pressed to it in the course of tensile strain and moves strictly along its axis. The probe motion is carried out by means of a reverse engine providing the velocity within the range from 10^{-3} to 10 mm/s.

A signal generated by the ferroprobe is registered in the recorder; also opportunity for automate collection and processing of the data is provided by means of computer at the on-line/off-line regimes. Capabilities of the developed measuring system are illustrated by typical extensograms (Figure 1c), recorded at various moments of time t_1, t_2,... t_i, corresponding to the points 1, 2,..., i on the tensile diagram, and by typical magnetograms for steel 12Cr18Ni10Ti (Figure 1b).

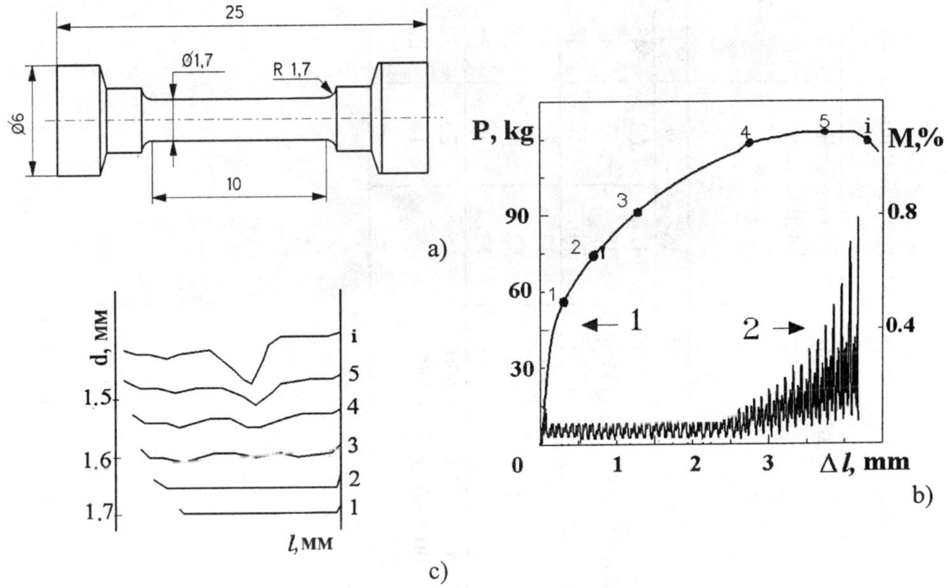

Figure 1–*(a) Configuration and sizes (in mm) of the strained specimen; (b) Tensile diagram (1) and the curve of variation in the amount of the ferromagnetic phase (2); (c) extensograms in the coordinates "the specimen diameter – the length", describing the specimen profile in the moment of deformation 1, 2, 3, 4, 5, i.*

The obtained primary diagrams have been used for determination of characteristics of the strength (yield stress -$\sigma_{0,2}$, ultimate stress- σ_U) and the plasticity (δ), whereas the extensograms have served as a base for determination of the diameter d in any section of the specimen operational length, allowing to determine "true" values of the yield stress (σ_{true}), and such types of deformation as the longitudinal (ε) and the transverse (ψ) ones.

Table1– *Strength and ductility characteristics of investigated materials*

material	Fluence n/m^2 (E>0,1MeV)	Yield stress, MPa	Tensile stress, MPa		Uniform strain, related to tensile stress %		Ultimate strain, %		Strain related to localization recording on the extensogram %	
			eng.	true	ε	ψ	ε	ψ	ε	ψ
12Cr18Ni10Ti	0	205	660	1230	42	63	45	90	37	47
	$5 \cdot 10^{22}$	285	690	1190	38	53	40	70	35	44
	$2 \cdot 10^{24}$	500	760	1250	34	50	38	79	30	43
03Cr20Ni45Mo4	0	260	650	990	33	44	35	61	32	44
	$5 \cdot 10^{22}$	390	670	1000	32	40	38	100	32	40
	$2 \cdot 10^{24}$	515	710	1018	27	36	35	110	27	36
Molybdenum	0	[1]480/440	550	690	15	20	32	119	15	20
	$2 \cdot 10^{23}$	[1]680/640	610	740	8	16	20	93	3	3
	$7 \cdot 10^{23}$	[1]760/680	–	–	–	–	15	141	1	1

[1]–upper/lower yield stress

In view of determining peculiarities in a real pattern of strain localization, the tensile strain is being stopped periodically, a specimen is turned about its axis, and appropriate tensograms, corresponding to various images of the specimen profile, are recorded. Analysis of tensograms has shown that the transverse section of a deformed specimen represents the geometrical figure differing from a circle, obtained as a result of intersection between a cylinder (operational segment of the specimen) and a cone (the yield band). Nevertheless, in subsequent calculations it has been assumed that the specimen's section is a circle of a diameter calculated from the extensogram. The value of the transverse deformation is calculated by the formula $\psi=2 \ln d_0/d_i$.

It has been assumed that the process of localisation is of step-like nature, and as the criterion for step identification , the following condition has been adopted: $d \ln(d_0/d_i)$: $d \ln(l_i/l_0)=const$, where indices 0 and i refer, respectively, to the initial values and to the current ones. In this case the relation ψ–$2 \ln d_0/d_i=f(\varepsilon)$, in log coordinates, is represented by a set of rectilinear sections having various slope angles α_i. with respect to the deformation axis. The combination $tg\alpha_i=n_i$ is used as a parameter for identification of a given stage of localization of deformation.

Experimental Results and Discussion

Steel 12Cr18Ni10Ti – The specimens, subject to thermal treatment, austenization (1423K, 30 min.) and irradiation by neutrons (to maximum fluence 10^{24} n/m^2, E>0.1MeV), have been deformed at room temperature at the tensile rate $0.16 \cdot 10^{-4}$ s^{-1} with on-line recording of the geometrical dimensions of specimen and the distribution of the α'–phase over its length. Analysis of obtained extensograms has shown that already at the early steps of tension (far before achieving the tensile strength) deformation is heterogeneous, and a microneck occurs in a specimen. In some cases, on achieving maximum load value, several micronecks, developing at the same time and competing, are observed. At further strain one of the necks occupies a dominating position, it develops extensively, and destruction occurs in it.

By outcomes of the measurements performed with the extensometer, graphs have been constructed depicting variations in the specimen diameter, marked in several sections (including, in a place of future breakdown) versus the extent of the longitudinal deformation. Figure 2 demonstrates that at early stages of tensile strain values of diameters for every specimen section vary following a certain falling step-like curve. It should be noted that already at early stages of the deformation process variation in specimen transverse size is observed not in all points of the specimen simultaneously, and its rate isn't constant. Under further tensile strain, the section is getting prominent where the rate of variation in the specimen diameter increases gradually, and, on achieving the maximum load, variation in transverse size of the specimen comes about but in this localized region, where the rate of diameter reduction continuously increases.

In the curve of the deformation dependence of necking $\psi=f(\varepsilon)$, corresponding to a place of future breakdown in the unirradiated steel specimen, one can separate three segments of different angles of slope (see Figure 3). The first segment is characterized by the fact that its experimental points, within measurement inaccuracy corridor, at lesser extents of deformation, lie strictly on a straight line, and at larger amounts of deformation, the straight line transforms into a curvilinear step-like curve. The slope angle of the averaged curve for this area, practically, corresponds to the value of the slope angle of the graph $\psi = f(\varepsilon)$ at the rectilinear segment. On extensograms corresponding to the step-like curve, local narrowing (micronecks) are observed.

The second segment of the curve $\psi = f(\varepsilon)$ corresponds to development of the dominating neck and gradual reduction of deformation beyond it. The third segment of evenly marked localization of deformation is characterized by catastrophically fast growth of the neck. Here it should be noted that the coordinates of the point of inflection between the second and third segments corresponds to stop of uniform elongation and maximum load in the engineering tensile diagram.

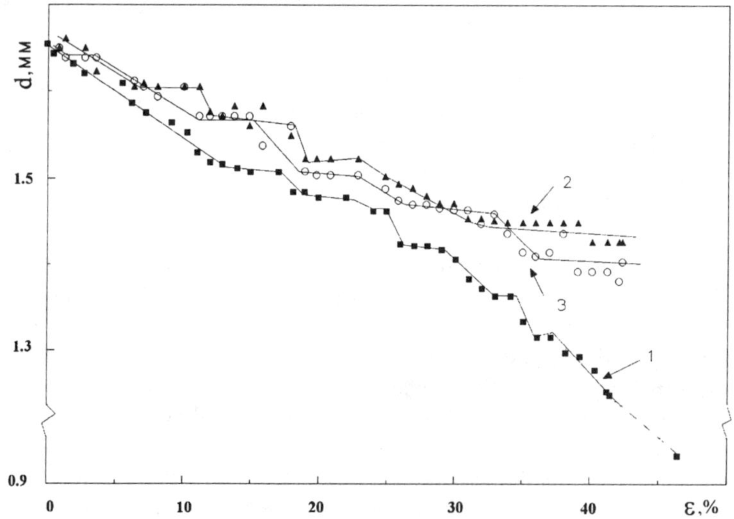

Figure 2– *Variation in a diameter(d) of the specimen made of steel 12Cr18Ni10Ti, measured 1 – in place of future breakdown; 2 – near the localization band; 3–outside the localization band versus the extent of strain.*

As it was mentioned earlier, analysis of extensograms shows that presence of heterogeneous deformation over the entire length of the specimen from steel 12Cr18Ni10Ti is revealed already in the first segment of the curve $\psi = f(\varepsilon)$. It has been established that of the so-called martensite of deformation of the α'–phase is revealed in the region of transition between the rectilinear section and the step-like one in this curve. Combined analysis of extensograms and the appropriate data concerning distribution of the α'–phase over the specimen length has shown that maximum contents of martensite are observed in micronecks [2-4]. As it is known that martensite has higher strength than austenite, then, probably, in the course of deformation local narrowing in a specimen are strengthened by the α'–phase exactly to an extent sufficient for slow evolution of the neck, i.e., strengthening at the expense of martensite compensates geometrical loss of strength. As deforming continues, development of several necks is observed, correspondent to the second segment in the curve $\psi = f(\varepsilon)$. Combined analysis of extensograms for the curves of the distribution of the α′–phase over the specimen length has shown that in competing necks the rates of martensite formation are, practically, the same, whereas in other specimen segments accumulation of martensite gradually disappears. In the neck, where geometrical loss of strength will exceed martensitic strengthening, deformation will continue, and specimen destruction will come.

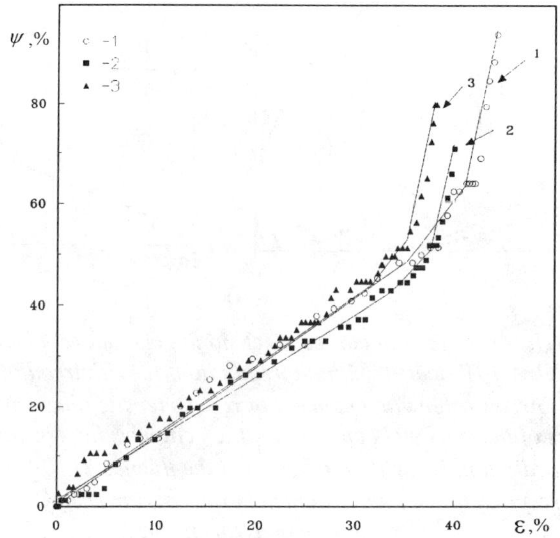

Figure 3– *Variation in the neck of the specimen from the unirradiated steel 12Cr18Ni10Ti (1) and from the same steel irradiated by neutrons (2 – $5 \cdot 10^{22}$, 3 – $2 \cdot 10^{24}$ n/m^2) versus the extent of strain.*

For illustration, in Figure 4(a) two kinetic curves (2 and 3) are given: one of them reflects the martensite phase occurrence and its accumulation in the neck (2), and another is related to the place remote significantly from the neck (3). One can see that kinetics of martensite accumulation is different in two cases: in the developing neck the α′–phase content continuously (exponentially) increases, whereas far from the neck the process of martensite formation comes about with saturation at the final stages of the deforming process.

With the minimal current values of the specimen diameter determined from the extensograms, the machine tensile diagrams were transformed into the yield curves: "true stress (σ) – logarithmic strain (ε)", approximated by the Bolfinger–Hollomon equation $\sigma = C \cdot \varepsilon^n$ where n is the hardening work exponent of the deformation strengthening, and C is the constant. Then the linear anamorphoses of this function was constructed, and in the log σ–log ε graphs the salient points and its coordinates (Figure 4(b)), as well as the angles of the slopes of the appropriate rectilinear sections were found. It has been established that the strain ε_{cr} corresponding to the onset of the α′–phase production coincides with the coordinate of one of the salient points in the $\ln \sigma_{true} - \ln \varepsilon$ graph, where no peculiarities have been observed, similar to the $\ln\sigma_{eng.}$–lnε graph.

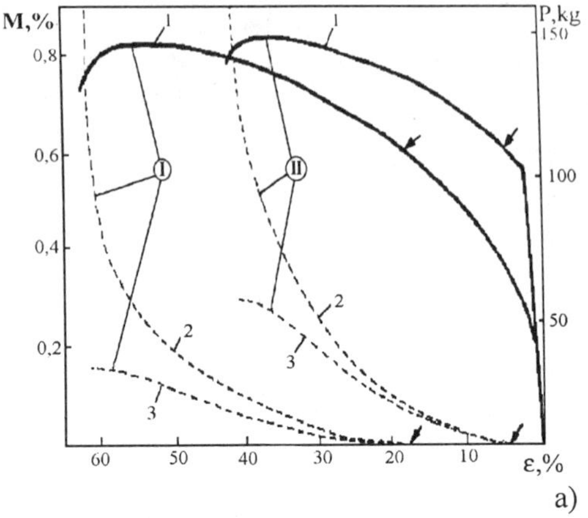

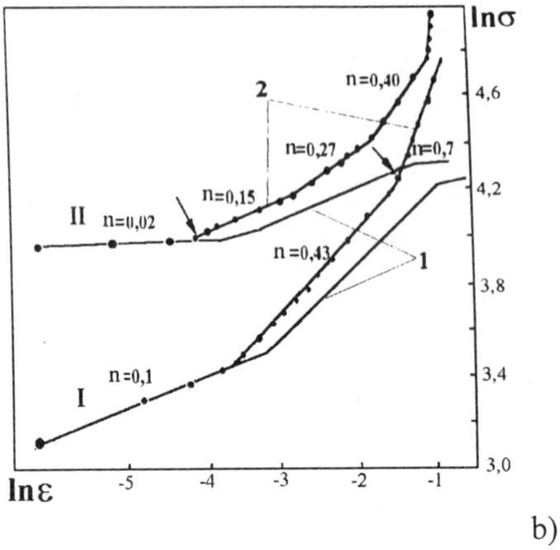

Figure 4– *(a)The experimental curves of tensioning(1) and α′–phase accumulation (2,3) in the 12X18H10T steel: I–without irradiation, II–with irradiation at the fluence* $2{\cdot}10^{24}$ *n/m²*. *Arrows show to a moment of α′–martensite phase arising. (b)The linear anamorphoses of the steel yield curves: 1 – the engineering tensions, 2 - the true ones. I - without irradiation, II - with irradiation at the fluence* $2 \cdot 10^{24}$ *n/m²*. *The dots on the curve correspond to the time instants of extensogram record. Arrows show to α′–martensite phase transformation.*

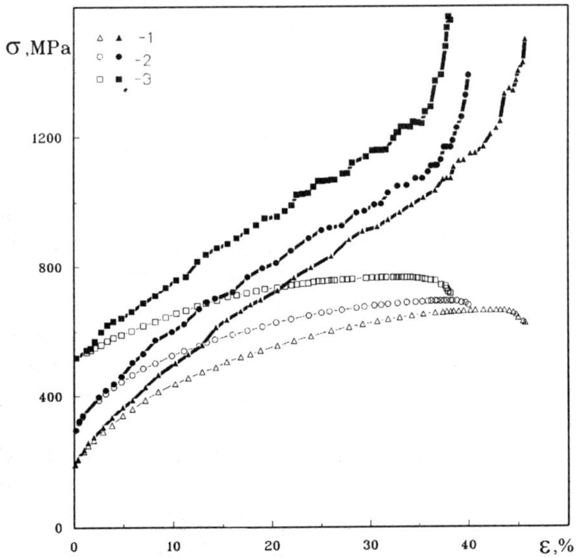

Figure 5 – *Stress–strain curves for specimens of steel 12Cr18Ni10Ti: true curves are depicted by filled symbols, engineering curves– by hollow symbols (1- without irradiation, 2,3- after irradiation to the fluences* $5{\cdot}10^{22}$ *(2),* $2{\cdot}10^{24}$ *(3) n/m²*).

It is of special interest that the experimentally determined values of σ_{eng}^{cr} are close for unirradiated (545MPa) and irradiated (540MPa) steels, whereas the values of σ_{true}^{cr}, calculated by means of the extensograms, are substantially different (700 and 560MPa respectively), the difference increasing with the irradiation dose.

The graphs of the true stresses, calculated for the section of future breakdown, versus deformation in Figure 5. are presented; for comparison, the graphs for conventional stresses are presented here too. One can observe that before destruction the true stresses exceed substantially (~twice) the conventional ones.

High-Nickel Alloy 03Cr20Ni45Mo4 – The specimens were subject to thermal treatment at 1423 K for 30 minutes and cooling in water. The rate of test comprises $0.16{\cdot}10^{-3}\ s^{-1}$. Analysis of extensograms for the deformed specimens of the alloy 03Cr20Ni45Mo4 in the unirradiated/irradiated state, it follows that before the moment of identification of the localized region, variation in a shape of the entire specimen comes about comparatively uniformly. The curves of variation in the diameter over the length of the deformed specimen coincide up to the point correspondent to observation of a neck on the extensograms. After that, variation in the specimen diameter is marked only within the localised region.

In the curve $\psi = f(\varepsilon)$ two clear rectilinear segments are identified, observed both in the irradiated state and in the unirradiated one (Figure 6). The first segment, judging by analysis of the extensogram, corresponds to uniform deformation, the second one is associated with localisation of deformation of necking, and the slope angle for the second segment is larger than for the first one. The curves of necking versus deformation for the primary specimen of alloy and a similar specimen irradiated by fluence $5{\cdot}10^{22}\ n/m^2$, practically coincide. As a result of irradiation by fluence $2{\cdot}10^{24}\ n/m^2$, the second segment of the curve has been shifted toward lesser deformation values.

Compared to steel 12Cr18Ni10Ti, structure of high-nickel alloy 03Cr20Ni45Mo4 is more stable; and, seemingly, owing to this, in the course of deformation any noticeable effects aren't observed. A neck in a specimen occurs in the region corresponding to a maximum load and has evenly observed borders. In-reactor irradiation (fluence $2{\cdot}10^{24}\ n/m^2$) has weakly influenced on the necks in specimens of alloy 03Cr20Ni45Mo4 whereas localizing starts earlier, and the extent of transverse deformation slightly increases.

The values of the stresses, calculated for a section in a place of future breakdown, are presented in Figure 7, demonstrating that at initial stages of strain (up to ε=20%) the stresses in the section of the irradiated ($5{\cdot}10^{22}\ n/m^2$) specimen are somewhat higher than for the primary one. Under further deforming the values of stress, practically, coincide, whereas irradiation with the fluence $2{\cdot}10^{24}\ n/m^2$ has led to a more considerable increase in the stresses at the same extents of deformation. It is of interest that in irradiated specimens of alloy 03Cr20Ni45Mo4 before destruction sharp decrease of values of the true stresses is observed, associated, rather, with inaccurate determination of a section area, related to neglect of cavities in the band of deformation localization.

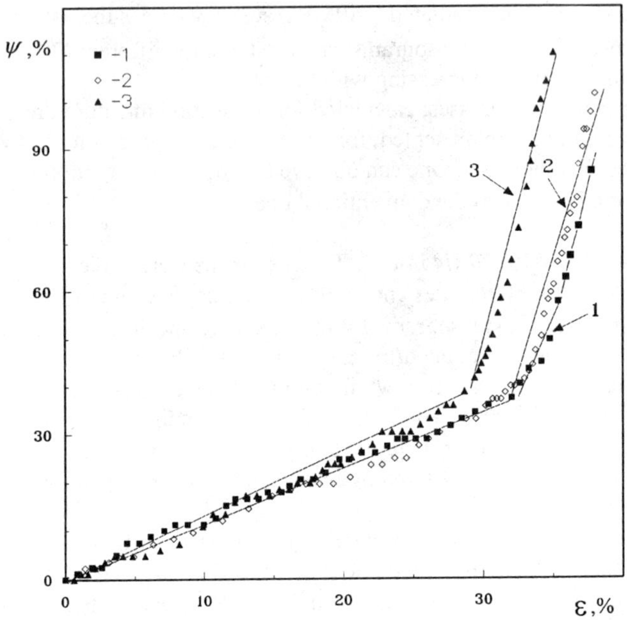

Figure 6 –*Necking in the specimen form the alloy 03Cr20Ni45Mo4, unirradiated (1) and irradiated by neutrons (2– $5{\cdot}10^{22}$, 3 –$2{\cdot}10^{24}$ n/m^2) versus the extent of strain.*

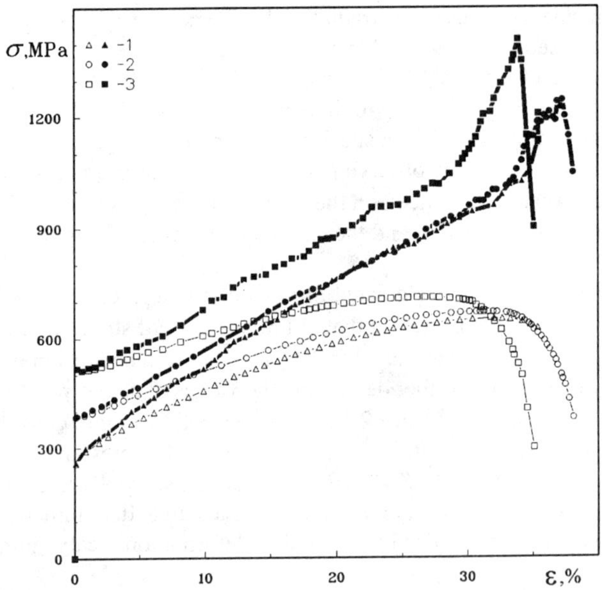

Figure 7 – *Stress–strain curves for specimens of the alloy 03Cr20Ni45Mo4 true curves are depicted by filled symbols, engineering curves– by hollow symbols (1- without irradiation, 2,3- after irradiation to the fluences $5{\cdot}10^{22}$ (2), $2{\cdot}10^{24}$ (3) n/m^2).*

Molybdenum – Thermal treatment at 1423 K for two hours, cooling with a furnace, the strain rate is $0{,}45 \cdot 10^{-3}$ c^{-1}. Extensograms obtained at tests of the molybdenum specimens in unirradiated state witness that strain is almost uniform over the entire length of a specimen to a certain strain level (ε=15 %), when local necking is revealed in a specimen, resulting in its destruction. In irradiated molybdenum a visible neck is recorded much earlier than in unirradiated. The curves of variation in a diameter determined for various places in a deformed specimen, calculated from extensograms, have shown to rather uniform strain up to the moment of localization, corresponding to maximum load. Further variation in a diameter is observed only in a single localised region of a specimen: in its neck.

On the curve ψ=*f*(ε) built for unirradiated molybdenum four rectilinear segments (stages) are observed (see Figure 8). Every subsequent stage differs from a previous one by a larger value of the tangent of slope angle. The first segment is characterized by uniform strain over the entire specimen. Coordinates of the inflection point between the first and second rectilinear segments correspond to the point on the tensile diagram, characterised by the maximum load. In the second segment necking comes about. In the extensograms related to this segment, local narrowing is observed. Tangent of the slope angle for the graph ψ=*f*(ε) in the second segment is higher by a factor of ~1.5 than in the first one. The third segment is characterized by considerable development and deepening the neck and, respectively, by increase in the tangent of the slope angle of the straight line ψ=*f*(ε) by a factor of ~5.5. In the fourth segment (pre-destruction) catastrophically fast development of the neck to the specimen mass is observed. The slope angle of the segment increases, compared to the previous one, by a factor of three. It should be noted that the last three stages correspond to deformation occurring only in a localized volume.

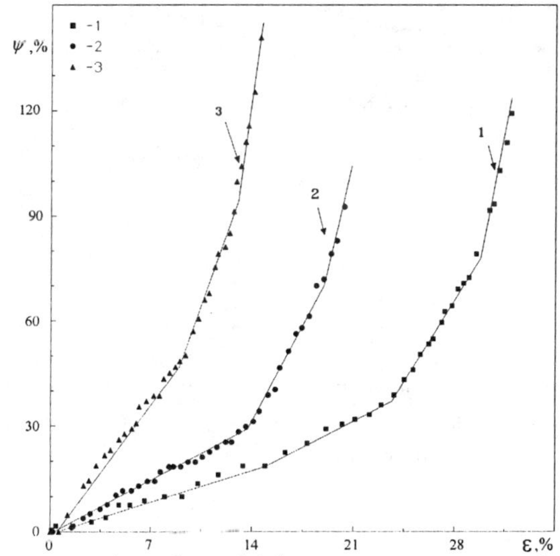

Figure 8 – *Variation in necking the molybdenum specimen, unirradiated (1) and subject to neutron irradiation (2– $2 \cdot 10^{23}$, 3– $7 \cdot 10^{23}$ n/m^2) as the extent of strain increases.*

In the curve $\psi = f(\varepsilon)$, built for the specimen irradiated at the fluence $2 \cdot 10^{23}$ n/m^2, one can separate four segments. The first segment corresponds to deformation over the entire mass of the specimen and synchronous formation of a neck seen in the appropriate extensograms. Due to this, the slope angle for this segment of the curve is greater than for the unirradiated specimen. The next three segments correspond to deformation in a localised volume. And the slope angles for segments are nearly equal to those for the appropriate segments of the curve $\psi = f(\varepsilon)$ for unirradiated specimen.

On the curve $\psi = f(\varepsilon)$, built for the specimen irradiated at the fluence $7 \cdot 10^{23}$ n/m^2, the first (uniform) step is absent: extensograms record local narrowing right after elastic strain.

In Figure 9 the graphs of the conventional stresses, calculated over the initial area of the section, and the true stresses, calculated over varied areas of the section in the place of future breakdown, versus the longitudinal strain are presented. The values of true stresses differ from the conventional ones, and the greater the extent of irradiation, more substantial is the difference.

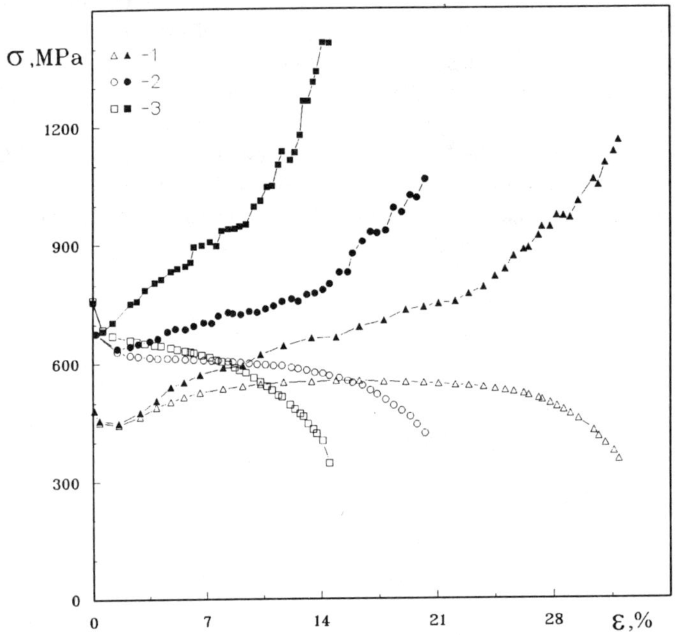

Figure 9 – *Stress–strain curves for the molybdenum specimen: : true curves are depicted by filled symbols, engineering curves– by hollow symbols (1- without irradiation, 2,3- after irradiation to the fluences $2 \cdot 10^{23}$ (2), $7 \cdot 10^{23}$ (3) n/m^2).*

Conclusion

The new experimental complex for mechanical tests including the "Instron" testing machine has been developed. It enables to obtain a simultaneous remote-controlled measuring of variation in the specimen geometrical sizes, magnetization value, and its gradients over the specimen working length.

The true values for both the yield stress and the tensile strain have been determined for unirradiated and neutron-irradiated austenitic stainless steel 12Crl8Nil0Ti, high-nickel alloy 03Cr20Ni45Mo4 and commercially pure molybdenum.

For austenitic steel 12Crl8Nil0Ti the critical values of true stresses have been obtained corresponding to the formation of the deformation-induced α′-martensite.

References

[1] Jitsukawa, S., Grossbeck, M. L., and Hishinuma, A., "Stress–Strain Relations of HFIR–Irradiated Austenitic Stainless Steels," *Journal of Nuclear Materials*, 1991, pp. 563-567.

[2] Ibragımov, Sh., Maksimkin, O.P., and Sadvakasov, D., " Martensitic γ→α Transformation and Mechanical Properties of Steel 12Cr18Ni10Ti Irradiated by Neutrons," *Physica of Metals and Metallography*, Vol. 7, 1990, p 3.

[3] Maksimkin, O.P., "The Deformation Induced Martensitic Transformation in Cr–Ni Stainless Steel Irradiated by Neutrons, " *Physica Status Solidi*, Vol.163, 1997, R7–R8.

[4] Potehin, B.A., " Tensile Peculiarities on Cylindrical Specimen of Metastable Austenitic Stainless Steels," *Physica of Metals and Metallography*, Vol. 7, 1979, pp. 1058-1064.

Randy K. Nanstad,[1] Donald E. McCabe,[2] Mikhail A. Sokolov,[3] Colin A. English,[4] and Susan R. Ortner[4]

Investigation of Temper Embrittlement in Reactor Pressure Vessel Steels Following Thermal Aging, Irradiation, and Thermal Annealing

Reference: Nanstad, R. K., McCabe, D. E., Sokolov, M. A., English, C. A., and Ortner, S. R., **"Investigation of Temper Embrittlement in Reactor Pressure Vessel Steels Following Thermal Aging, Irradiation, and Thermal Annealing,"** *Effects of Radiation on Materials: 20th International Symposium, ASTM STP 1405,* S. T. Rosinski, M. L. Grossbeck, T. R. Allen, and A. S. Kumar, Eds., American Society for Testing and Materials, West Conshohocken, PA, 2001.

Abstract: The Heavy-Section Steel Irradiation Program at Oak Ridge National Laboratory includes a task to investigate the propensity for temper embrittlement in coarse grain regions of heat-affected zones in prototypic reactor pressure vessel (RPV) steel weldments as a consequence of irradiation and thermal annealing. For the present studies, five prototypic RPV steels with specifications of A302 grade B, A302 grade B (modified), A533 grade B class 1, and A508 class 2 were given two different austenitization treatments and various thermal aging treatments. Thermal aging treatments were conducted at 399, 425, 454 and 490°C for times of 168 and 2000 h. Charpy V-notch impact toughness vs temperature curves were developed for each condition with ductile-brittle transition temperatures used as the basis for comparing the effects of the various heat treatments. Very high austenitization heat treatment produced extremely large grains which exhibited a very high propensity for temper embrittlement following thermal aging. Intergranular fracture was the predominant mode of failure in many of the materials and Auger analysis confirmed significant segregation of phosphorus at the grain boundaries. Lower temperature austenitization treatment performed in a super Gleeble to simulate prototypic coarse grain microstructures in submerged-arc weldments produced the expected grain size with varying propensity for temper embrittlement dependent on the material as well as on the thermal aging temperature and time. Although the lower temperature treatment resulted in decreased

[1]Leader, Fracture Mechanics Group, Metals and Ceramics Division, Oak Ridge National Laboratory, Oak Ridge, TN 37831-6151.

[2]Engineer, Fracture Mechanics Group, Metals and Ceramics Division, retired from Oak Ridge National Laboratory, Oak Ridge, TN 37831-6151.

[3]Metallurgist, Fracture Mechanics Group, Metals and Ceramics Division, Oak Ridge National Laboratory, Oak Ridge, TN 37831-6151.

[4]Principal Technical Consultant and Technical Consultant, respectively, AEA-Technology, Harwell, Didcot, Oxfordshire, OX11 ORA, United Kingdom.

propensity for temper embrittlement, the results did provide motivation for the investigation of the potential for phosphorus segregation as a consequence of neutron irradiation and post-irradiation thermal annealing at 454°C. One of the A 302 grade B (modified) steels was given the Gleeble treatment, irradiated at 288°C to about 0.8×10^{19}n/cm (>1 MeV) and given a thermal annealing treatment at 454°C for 168 h. Charpy impact testing was conducted on the material in both the irradiated and irradiated/annealed conditions, as well as in the as-received condition. The results show that, although the material exhibited a relatively small Charpy impact 41-J temperature shift, the heat-affected zone-simulated material did exhibit significant intergranular fracture in the post-irradiation annealed condition.

Keywords: reactor pressure vessel, thermal aging, intergranular fracture base metal, temper embrittlement, thermal annealing

Introduction

In 1989, Combustion Engineering had proposed to the U.S. Nuclear Regulatory Commission (NRC) that consideration be given to replacing the fracture toughness based dynamic K_{IR} curve with the quasi-static K_{Ic} curve for setting allowable temperature/pressure limits during normal upset conditions.[5] The idea was accepted for consideration pending the development of clear evidence that there can be no unanticipated trigger sources that would introduce running cleavage cracks. Such sources could be either crack pop-ins or possibly local brittle zones (LBZs) that reside along the fusion line of multipass weldments. The existence of LBZs was suggested as one of the possibilities for reactor pressure vessel (RPV) weldments. The U.S. Nuclear Regulatory Commission asked the Heavy-Section Steel Irradiation (HSSI) Program at Oak Ridge National Laboratory (ORNL) to review the available LBZ information and determine its applicability to nuclear vessel fabrication and service conditions and conduct experiments in support of the evaluations. A detailed discussion of the LBZ evaluation is presented by McCabe [*1*], from which the unirradiated temper embrittlement results reported here were taken. Part of the overall study was a consideration of the effects of radiation and thermal annealing on the coarse grain region of the heat-affected-zone (HAZ) in a submerged-arc weldment. Research projects designed to evaluate LBZ damage mechanisms generally use electric resistance heating of specimen blanks to simulate material that has been through a weld pass cycle [*2,3,4*]. Brittle zone microstructures are created in sufficient bulk to develop reliable Charpy transition curves that represent the various microstructures identified within the LBZ region.

The thermal history and consequent morphology of HAZ welds are extremely complex. Each individual weld pass produces a four-zone microstructural gradient depicted in Fig. 1. The coarse grain HAZ (CGHAZ), schematic (D), immediately

---

[5]R. K. Nanstad, notes from meeting with the U.S. Nuclear Regulatory Commission, January 9, 1990.

adjacent to the fusion line consists of enlarged prior austenite grains from the subsequent weld passes that grow rapidly due to being at a temperature close to the melting point. Upon cooling, the prior austenite can transform to martensite or a lower bainite/martensite mix, depending on the carbon equivalent (CE) of the base metal. At slow cooling rates, low CE steels will transform to a lower toughness microstructure, pearlite and/or upper bainite. With low CE, the cooling rate must be quite rapid to form martensite and lower bainite. Lower bainite and/or martensitic steel is more brittle as cooled, but these microstructures are converted to high toughness steel after PWHT. The second zone, fine grain HAZ (FGHAZ), schematic (C), from the fusion line in Fig. 1 is fine-grain prior austenite resulting from slower austenitic grain growth at lower temperatures. The third zone, intercritical coarse grain HAZ (ICCGHAZ), schematic (B), is material heated into the A_{c3} and A_{c1} (intercritical) transformation zone. A mixture of ferrite and austenite has precipitated along prior austenite grain boundaries which embrittles the material in this local region upon cooling to room temperature [*2,3*]. These small volumes of material, ICCG HAZ, have been identified as the LBZ trigger point locations. It is understood that LBZ material is created by the reheating of coarse grain prior austenite from the subsequent weld passes that are adjacent to the fusion line. It therefore follows that there can be an appreciable amount of coarse grain prior austenite along a fusion line that is not brittle crack initiation material. The fourth zone is the subcritical reheated coarse grain zone of the (CGHAZ), which undergoes very little changed property. An interesting outcome of the fundamental studies on microalloyed steels has been a demonstration that ICCG HAZ microstructures can cause about 15 to 20°C transition temperature increase over unsensitized material [*3*]. When these same materials are postweld heat treated at 450°C (842°F), the fracture toughness of ICCG HAZ material is restored and, in some cases, slightly improved over that of the unsensitized condition.

RPV weldments are principally multi-pass submerged-arc welds (SAWs) with high heat input that can promote the development of coarse prior austenite grains. Carbon equivalent is important from the standpoint of the transformation products formed on cooldown. Several important alloying elements contribute to CE, as indicated from the following equation:

$$CE = \frac{Mn}{6} + \frac{Cu + Ni}{15} + \frac{Mo + V + Cr}{5}. \tag{1}$$

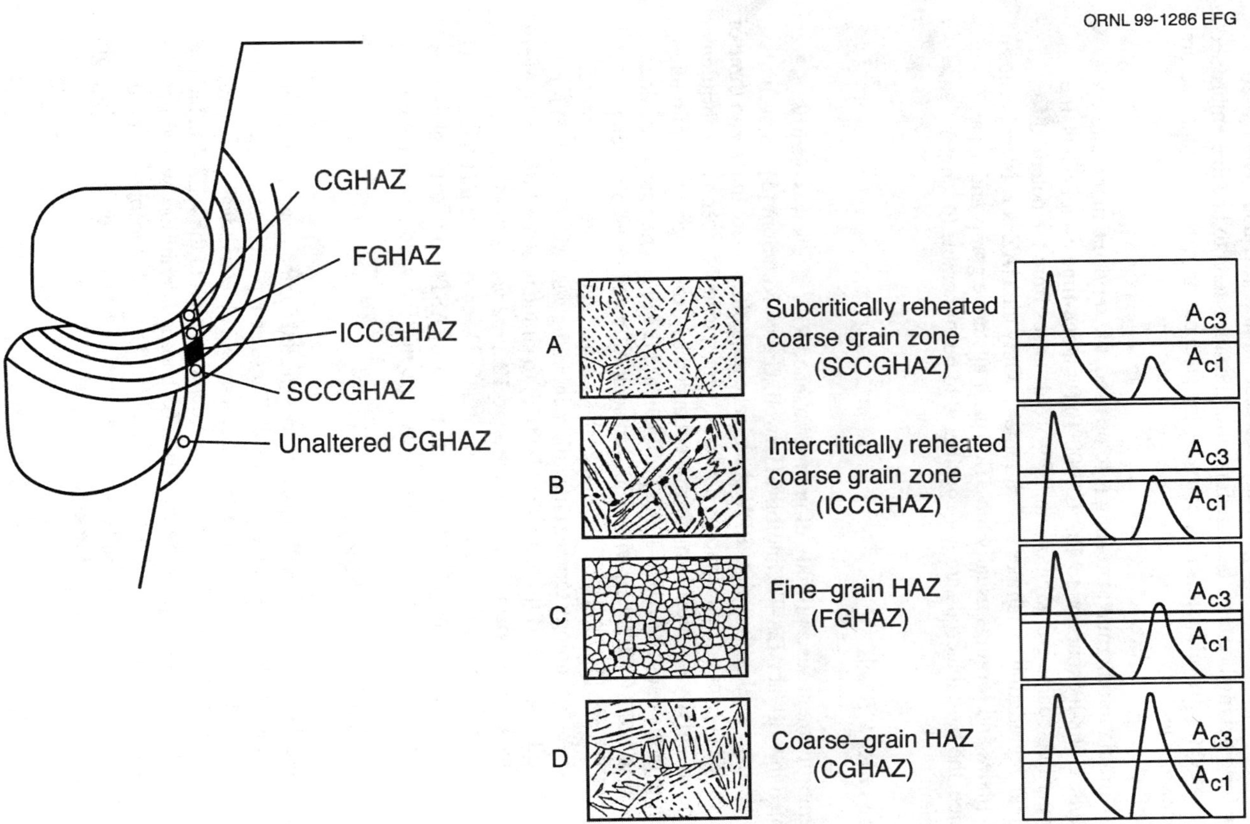

Figure 1. *Schematic representation of microstructures local to the HAZ of multipass welds.*

The CE controls austenite stability and hence promotes transformation to low-temperature transformation constituents, such as lower bainite and martensite. The typical CE in low-carbon microalloyed steel is about 0.37, while the typical RPV steel has a CE of about 0.62. Increased CE generally decreases weldability due to increased propensity for cracking because the transformation products tend to become embrittled as cooled. However, postweld tempering converts the steel to one with excellent fracture toughness properties.

If post-irradiation thermal annealing is to be used, the service temperature cycle could include a 168-h excursion to 482°C (900°F) for a one-time cycle. Thus, the concern regarding coarse grain regions in RPV welds arises from the potential for metallurgical degradation by temper embrittlement. Only the HAZ is at issue because RPV steel plates and forgings usually have fine prior austenite grains and, as a consequence, the base materials are not considered to be vulnerable to temper embrittlement.

Evaluation of RPV Steels

The embrittlement susceptibility of grain-coarsened RPV steels was clearly demonstrated in work at AEA-Technology, Harwell, United Kingdom [*5*]. Eleven laboratory heats of steel with typical RPV chemical composition were made and three of these with varied copper and phosphorus contents were selected for temper embrittlement studies. The sensitivity to embrittlement was enhanced by heat treating to obtain huge prior austenite grain size (about 0 to 1, ASTM Standard E 112). The austenitization temperature was 1200°C (2200°F) with a 30-min soak followed by an oil quench. All were given PWHT at 615°C (1140°F) with a 24-h soak and again followed by an oil quench. The AEA-Technology temper embrittlement aging was 450°C (840°F) for 2000 h for the fully aged condition and 475° (890°F) to simulate an end-of-life recovery anneal cycle. One other exposure was irradiation to 12.2 mdpa at 288°C (550°F). Table 1 is a summary of their findings. It is important to keep in mind that the ΔT_0 values are extremely grain size dependent and with grain sizes of ASTM 0 to 1, the above results may not truly represent the coarse grain microstructure in RPV weld heat-affected zones.

ORNL Temper Embrittlement Studies

Five commercially produced RPV steels were selected for the present program; their reported chemical elements are shown in Table 2. These as-reported analyses served as the basis of selection for the materials to be evaluated. An objective was to include at least one material with the phosphorus level considered as high in the AEA-Technology experiment (Table 1). High copper, by the AEA-Technology definition, was almost automatic in the commercial steels that were available. The high phosphorus, on the other hand, was not as easily obtained because that level of phosphorus is quite high for U.S. RPV base metals, as shown in Fig. 2 [*6*].

Table 1. *Summary of AEA-Technology Temper Embrittlement Results*

Chemistry variable (wt %)		Aging (h/°C)	Charpy impact T_o (°C)		Charpy impact ΔT_o (°C)
Cu	P		Initial	Aged	
0.01	0.007	2000/450	-86	32	118
0.01	0.017	2000/450	-55	119	174
0.16	0.017	2000/450	-95	114	209
0.16	0.017	*a*	-95	-26	69

[a]Irradiated 12.5 m dpa.

Table 2. *Commercial Materials Used for ORNL Temper Embrittlement Studies*

Material	Code	Content (wt %)			
		P	Ni	S	Cu
A 302 grade B	Maine Yankee	0.015	0.2	0.017	0.14
A 508 class 2	Midland	0.010	0.76	0.015	0.02
Modified A 302 grade B	GE (Z5)	0.016	0.62	0.015	0.17
Modified A 302 grade B	GE (Z7)	0.007	0.53	0.014	0.16
A 533 grade B	HSST Plate 01	0.018	0.75	0.013	0.18

Confirmatory chemical analyses and tensile properties were obtained during the course of the present investigation, and these properties are reported along with the as-received values in Table 3. The initially reported chemical compositions were not entirely reliable, as can be seen. One material, A 302 grade B, received in the as-normalized condition, was not available in the heat-treated condition desired. Consequently, the normalized microstructure was converted into a quenched-and-tempered product using a double austenitize, quench, and temper treatment [*1*]. This simulated the heat treatment reported on the low phosphorus modified A 302 grade B steel listed in Table 2. This particular modified A 302 grade B (Z7) came from a commercial reactor nozzle cutout. The end result was suitable from a mechanical properties standpoint but it was later determined that the quenched-and-tempered microstructure had prior austenite grain size of ASTM 5, which was large in comparison with the other commercially made steels (Table 4).

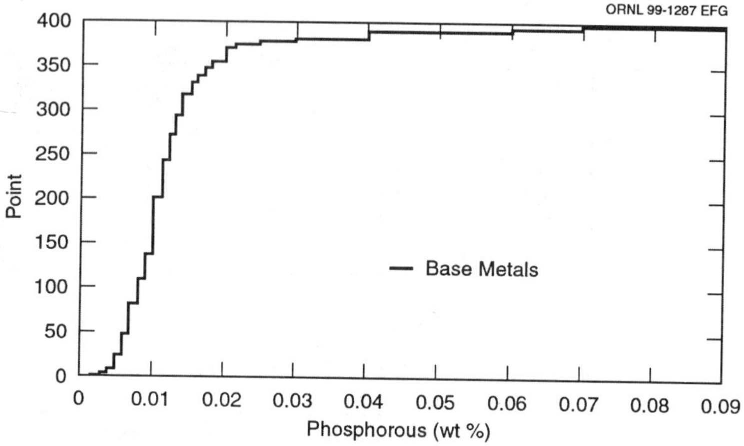

Figure 2. *Cumulative number of chemical samplings made on RPV base metals versus the phosphorus content.*

Figure 3. *Example of the typical microstructure after the AEA-Technology austenitize, quench, and postweld heat treatment, A533 grade B steel, nital etch, 100×.*

Table 3. *Chemical Composition and Tensile Properties of Materials for ORNL Temper Embrittlement Studies*

Material	Condition[a]	Composition (wt %)												
		C	Mn	P	S	Si	Ni	Cr	Mo	V	Nb	Cu	As	Sn
A302B	AR	0.21	1.20	0.015	0.015	0.28	0.19	0.24	0.60	0.004	0.006	0.13	0.015	0.012
	AEA	0.21	1.15	0.014	0.016	0.31	0.21	0.22	0.57	0.005	0.005	0.14	0.018	0.011
	AS	0.22	1.12	0.014	0.015	0.31	0.18	0.21	0.56	0.005	0.003	0.15	0.17	0.010
A508 class 2	AR	0.20	0.63	0.009	0.013	0.19	0.75	0.37	0.60	<0.01	—	0.02	<0.01	<0.01
	AEA	0.22	1.04	0.013	0.010	0.28	0.73	0.21	0.57	0.002	0.004	0.05	0.015	0.006
	AS	0.22	0.66	0.013	0.009	0.27	0.71	0.34	0.58	0.002	0.003	0.05	0.015	0.006
Mod A302B	AR	0.17	1.27	0.014	0.016	0.17	0.61	0.10	0.50	0.002	0.004	0.16	0.013	0.017
	AEA	0.20	1.12	0.015	0.018	0.19	0.61	0.09	0.47	0.003	0.005	0.18	0.018	0.016
	AS	0.19	1.10	0.016	0.017	0.19	0.54	0.09	0.46	0.003	0.003	0.19	0.017	0.015
Mod A302B	AR	0.26	1.47	0.08	0.014	0.15	0.53	0.09	0.52	0.002	0.003	0.16	0.014	0.018
	AEA	0.25	1.42	0.010	0.013	0.18	0.55	0.09	0.48	0.002	0.004	0.17	0.021	0.015
	AS	0.27	1.42	0.011	0.016	0.18	0.49	0.09	0.50	0.002	0.004	0.19	0.021	0.015
A533B	AR	0.24	1.34	0.016	0.013	0.23	0.80	—	0.53	—	—	—	—	—
	AEA	0.23	1.15	0.010	0.012	0.28	0.74	0.10	0.55	0.003	0.004	0.12	0.013	0.010
	AS	0.22	1.49	0.010	0.012	0.27	0.65	0.10	0.55	0.003	0.004	0.20	0.020	0.015

[a]AR = As reported; AEA = AEA + 2000 h at 450°C; AS = as-receivec and aged.

As-received tensile properties

Material	Yield strength		Tenside strength	
	(MPa)	(ksi)	(MPa)	(ksi)
A302B	538	78.1	689	100.1
A508 class 2	394	57.1	569	82.6
A533 grade B	470	68.2	620	89.9
Mod A302B Z5	400	58.1	538	78.1
Mod A302B Z7	472	68.5	632	91.7

Table 4. *Grain Sizes of the Commercially Made Materials in the As-Received Condition*

Material	Size by E 112	Average grain diameter (μm)
A302 grade B[a]	5	65
A508 class 2	8	22
A533 grade B	8.5	19
Mod A302 grade B Z5	10	11
Mod A302 grade B Z7	9	16

[a]Reprocessed from normalized to quenched and tempered.

The general plan to evaluate temper embrittlement in the five RPV steels was structured to be in three phases, starting with the AEA-Technology heat treatment schedule discussed earlier, except that in this case commercially produced RPV steels are used. The three phases are defined as follows:

Phase I - Simulate the AEA-Technology heat treatment. Austenitize at 1200°C (2200°F) for 30 min., oil quench, postweld heat treat at 615°C (1140°F) for 24 h, oil quench, followed by two different aging treatments:

1. Age 2000 h at 450°C (840°F).
2. Age 168 h at 490°C (914°F).

Phase II - Use electric resistance heating (Gleeble) to simulate a typical submerged-arc weld cycle to develop the coarse grain microstructure. Austenitize at 1260°C (2300°F) for 10 s, postweld heat treat at 615°C (1140°F) for 24 h, oil quench, followed by two different aging treatments:

1. Age 2000 h at 450°C (840°F).
2. Age 168 h at 450°C (840°F).
   Note: The rapid cool following PWHT should maximize the temper embrittlement susceptibility.

Phase III - Perform the same heat treatment schedule as in Phase II followed by two different aging cycles, an irradiation cycle, and an irradiation/thermal annealing cycle:

1. 168 h at 482°C (900°F).
2. 168 h at 399°C (750°F).
3. Irradiate to 1×10^{19} n/cm^2 (>1 MeV at 288°C).
4. Irradiate to 1×10^{19} n/cm^2 (1 MeV at 288°C), then 168 h at 450°C.

Additionally, baseline CVN data were generated to provide the unaged condition in each case to quantify transition temperature shifts and to supplement the metallurgical evaluations. In all cases, temper embrittlement is evaluated by CVN transition curve shifts.

Results of Temper Embrittlement Studies

Phase I - All of the commercially produced steels that were given the AEA-Technology heat treatment cycle had martensite to lower bainite microstructures. Grain size by ASTM Standard Test Methods for Determining Average Grain Size (E 112) was generally between 0 and 00; meaning that the typical austenite grain diameter was on the order of 360 to 500 μm. The microstructure after PWHT is shown in Fig. 3.

The CVN energy, E_J, data were least-squares fitted to the following hyperbolic tangent equation:

$$E_J = A + B \tanh[(T - T_o)/C] \,, \qquad (2)$$

where

A = (USE + LSE)/2,
B = (USE - LSE)/2,
C = width of the transition, and
T_o = mid-transition temperature.
USE = upper-shelf energy
LSE = lower-shelf energy

Parameter, T_o, was selected to be the transition curve temperature level of reference. T_o is a point halfway between the lower shelf and upper shelf Charpy energy. This point was selected because it tends to be a position that is least affected by changes in curve shape. The LSE used in fitting data to Eq. (2) was held constant at 2.7 J [*7,8*].

Table 5 lists the CVN T_o temperatures of the five commercial RPV steels in the as-received condition both before and after embrittlement aging. As-received RPV steels with small grains on the order of ASTM 8 to 11 showed insignificant embrittlement. The A302 grade B exception was caused by larger grains on the order of 5. Aging of the materials following austenitization and PWHT to simulate the AEA-Technology heat treatment schedule showed aging-induced transition temperature shifts from 12 to 145°C, with the aging at 450°C/2000 h causing greater embrittlement than those at 490°C/168 h.

In summary, the results from the Phase I study (to simulate the AEA-Technology experiments using commercially made steels) showed that only two of the five steels exhibited temper embrittlement similar to the level of the AEA-Technology experiments. The conversion to large prior austenite grains by the AEA-Technology austenitizing treatment by itself did not always cause significant change in toughness from the as-received toughness. For example, as a result of the AEA-Technology heat treatment, comparing rows 1 and 3 in Table 5, there were small increases in T_o for two materials, while the other three changes in T_o were negative, with two of these being substantial. The dependence of ΔT_o on prior-austenite grain size was clearly demonstrated in four of the five materials tested and weakly demonstrated in the fifth. However, the grain sizes

Table 5. *Effects of Thermal Aging on Five RPV Steels Given the AEA Heat Treatment*

Condition	Aging time (h)	Aging temperature (°C)	Charpy impact transition temperature, T_o (°C)				
			A302B	A533B	A508 class 2	Modified A302B	
						High P	Low P
As received	Initial	—	0	20	0	-9	-9
As received	2000	450	64	38	10	-4	10
AEA austenitize[a] and PWHT[b]	Initial	—	10	-67	4	-23	-73
AEA austenitize[a] and PWHT[b]	2000	450	45	78	41	34	42
AEA austenitize[a] and PWHT[b]	168	490	44	65	16	16	-3

[a]30 min at 1200°C; oil quench.
[b]24 h at 615°; oil quench.

that resulted from the AEA-Technology austenitizing treatment are considerably larger than those expected in actual ICCGHAZ regions, suggesting that these Phase I results do not truly represent the coarse-grain situation of interest in commercial RPV welds. They do, however, demonstrate the susceptibility of these commercial steels to temper embrittlement, providing the motivation for further study with more prototypical microstructures.

Phases II and III - The electrical resistance method of heating and cooling with equipment (Gleeble, Inc.) that is designed specifically for simulation of weld cycles was applied to austenitization of the same five materials tested in Phase I. Because the intercritical reheating cycle, identified as the source of LBZ behavior in low-carbon microalloyed steels, was determined to be of no relevance to RPV steels, only a single simulation cycle was needed in the present experiment to simply create the simulated coarse prior austenite grains in RPV HAZs. A report by Lundin and Mohammed [9] was used to decide upon the most appropriate thermal cycle. Peak temperature and time were selected to develop ASTM grain size in the range of 4 to 5. The peak temperature was limited to 1260°C (2300°F), specifically to be below the burning temperature for steels of typical RPV chemistry [1]. The thermal cycle used is shown in Fig. 4. Specimen blanks prepared for the Gleeble were cylindrical rods about 76 mm long and 14.35 mm in diameter (3 by 0.564 in.). About 5 mm (0.2 in.) of material at the center of each rod was determined to be representative of the weld cycle material. An example photomicrograph of the resulting microstructure appears in Fig. 5, showing a mixture of martensite and lower bainite.

Table 6 lists the Gleeble weld simulation and aging data developed to date. The PWHT cycle used was identical to that used in the AEA-Technology simulation, namely, a 24-h soak time at 615°C (1140°F) followed by an oil quench. Note that the as-tempered toughness reflects a favorable microstructural condition that in all probability resulted from the fast cooling of the weld cycle simulation. Aging time at 168 h represents the expected hold time that would be used in an RPV annealing procedure. There are three aging temperatures, 399, 450, and 482°C, that represent the lowest to highest practical annealing temperatures that have been under consideration. At 399°C (750°F), there appears to be no evidence of temper embrittlement in four materials. This arguably could also be true for the high phosphorus modified A 302 grade B steel (Z5) that for a lack of sufficient material was not included in this aging exposure. At 450 and 482°C (842 and 900°F), however, embrittlement develops and increases with increasing temperature. Hence, these two annealing conditions produce temper embrittlement damage, provided there is coarse grain HAZ material. However, Fig. 6 shows that none of the transition temperatures, after annealing for 168 h, exceeded room temperature. After 2000 h at 450°C, only two of the five materials had T_0 temperatures that exceeded room temperature. Scanning electron fractography was performed on the modified low-phosphorous A302B steel in the Gleeble/PWHT condition after aging at 450°C for 168 and 2000 h, and for the high-phosphorous steel after 2000-h aging only. After 168 h of aging, the Charpy impact fracture surface of the low-phosphorous steel exhibited mixed

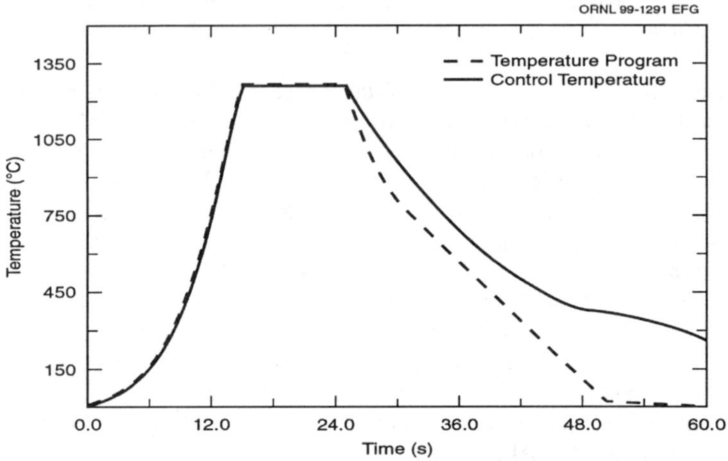

Figure 4. *Gleeble-generated weld thermal cycle to develop ASTM 4.5 grain size.*

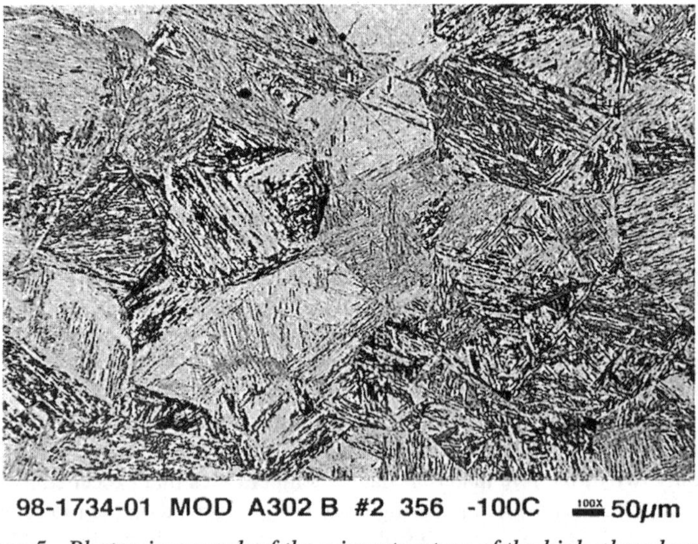

Figure 5. *Photomicrograph of the microstructure of the high-phosphorus modified A302 grade B steel after Gleeble austenitize and postweld heat treatment, nital etch, 100×.*

Table 6. *Charpy Impact Transition Temperatures for ORNL Materials Austenitized by Gleeble, PWHT, and Thermal Aging*

Austenitize method	Aging temperature (°C)	Aging time (h)	Transition temperature, T_o (°C)				
			A302B	A533B	A508 class 2	Modified A302B	
						High P	Low P
Gleeble	Initial	0	-87	-67.0	-102	—	-67
Gleeble	399	168	—	-67.1	-98.6	-89.4	-71.1
Gleeble	450	168	-44	-28	-80	—	-47
Gleeble	482	168	—	7.4	-72.6	-47.4	-33.6
Gleeble	450	2000	43	66	-46	-14	-4

cleavage and intergranular fracture. In this case, intergranular fracture was the predominant mode in the region near the notch tip, but cleavage fracture was the predominant mode for the bulk of the fracture surface away from the near-notch region. However, in the case of the material aged for 2000 h, the fracture surface exhibited predominantly intergranular fracture throughout, and this observation applied to both the low- and high-phosphorous steels. At higher test temperatures, where the Charpy energy is on the upper shelf, fractography showed only ductile void coalescence for both the low- and high-phosphorous steels after 2000 h.

Six commercial welds were examined for maximum grain size along the fusion line (Table 7). The results of this small sampling of actual RPV welds suggested that the presumed coarse-grain size and uniform distribution characteristics do not always exist. The coarse grains that could be found were much smaller than ASTM 4.5, and the percentage of coarse grains along all fusion lines was found to be very low. Example photomicrographs taken at selected large grain locations are provided in Ref. 1.

Table 7. *Largest Coarse-Grain Sizes Found in HAZs of Six Commercially Fabricated RPV Welds*

Weld source	ASTM E 112 grain size
HSSI Fifth Irradiation Series, 72W	6.5
HSSI Fifth Irradiation Series, 73W	9.0
SNUPPS beltline	7.5
SNUPPS longitudinal	7.5
Midland beltline	10
Midland nozzle course	6.5

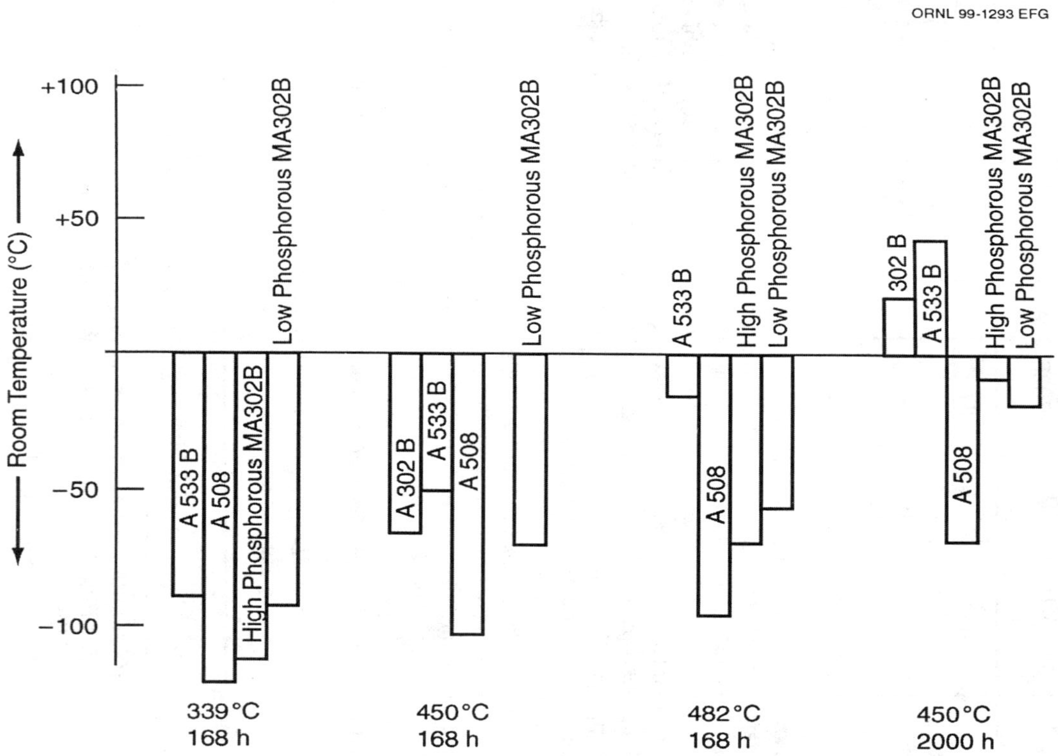

Figure 6. *Charpy impact mid-transition temperatures after aging for 168 h at 399, 450, and 482°C and 2000 h at 450°C (MA302B).*

Measurement of Element Diffusion to Grain Boundaries

Measurement of element diffusion to grain boundaries was conducted as part of a collaboration with AEA-Technology, Harwell, United Kingdom. The segregation of critical elements to grain boundaries on four selected specimens was measured by scanning transmission electron microscopy/energy dispersive X-ray analysis. These results were then compared with theoretical calculations using the McClean [*10*] model as well as the Murphy and Perks [*11*] model. Experimental results are shown in Table 8. The grain-boundary analyses also included Cu, Si, and Cr, with low levels of segregation being observed. Measurement replication was four times, and in some cases the variability was very large. Phosphorous segregation to grain boundaries was also calculated. The McClean equation reasonably predicted segregated concentrations of all the modified A302B specimens but overestimated the A533B segregation. The problem could be that the bulk phosphorous given in Table 2 was used in the calculation, instead of the more recent phosphorous determinations that appear in Table 3.

Table 8. *AEA-Technology Measurements of Grain Boundary Phosphorus Coverage*

Material	Treatments	Equivalent monolayer coverage (%)			
		Mn	Mo	P	Ni
Modified A 302 B (low phosphorous)	Gleeble austenitized and PWHT (unaged)	9	12	1	1
Modified A 302 B (low phosphorous)	Gleeble austenitized and PWHT (aged at 450°C, 2000 h)	19	16	10	1.5
Modified A 302 B (high phosphorous)	Gleeble austenitized and PWHT (aged at 450°C, 2000 h)	14	32	17	8
A 533 grade B (high phosphorous)	Gleeble austenitized and PWHT (aged at 450°C, 2000 h)	19	15	5	10

The same Gleeble-austenitized and 450°C aged A 533 grade B steel analyzed by AEA was also submitted to Auger analysis at ORNL. In this method, a fresh fracture surface is created under vacuum and analyzed. Part of the surface is intergranular, exposing the grain boundaries (GBs), as shown in Fig. 7, and the other part of the surface is transgranular (TG) cleavage. Bulk phosphorous was determined on cleavage surfaces and on grain boundaries after 10 Å were removed by sputtering (SP) (see Fig. 7). On average, the difference between bulk and GB phosphorous is a factor of about 10. A number of other elements were also measured and Table 9 summarizes the observations in terms of the change from the levels measured within the grains. As observed in the AEA-Technology STEM/EDX measurements, molybdenum is a significant segregant primarily because it is strongly attracted to phosphorus. As further shown in Ref. 1, intergranular fracture vanishes at Charpy upper-shelf test temperatures.

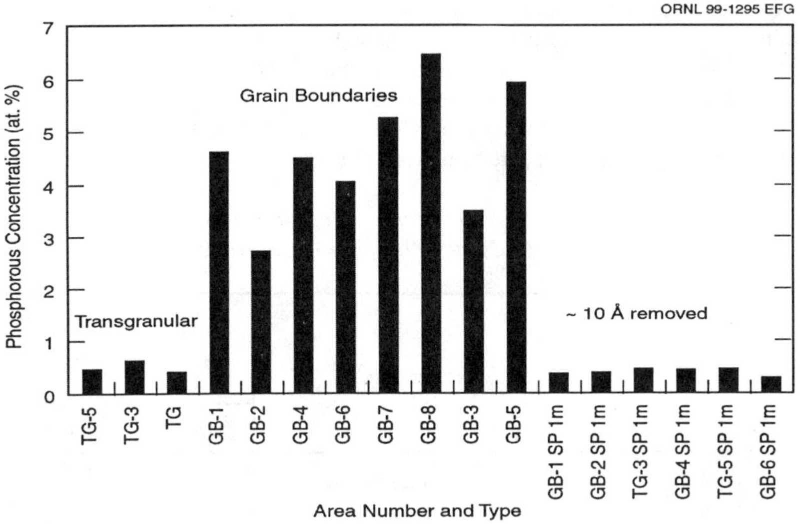

Figure 7. *Auger analysis of phosphorus concentration in A533 grade B class 1 following Gleeble austenitization and thermal aging at 450° for 2000 h.*

Table 9. *Results of Auger Analysis on A533 Steel in the Gleeble/PWHT Condition Aged at 450°C at 2000 h*

	Composition (at. %)						
	P	Mo	C	O	Fe	Ni	Cu
Average intergranular	0.50	x	x	x	x	x	x
Average grain boundaries	4.62	6.3x	2.0x	0.9x	0.8x	1.8x	2.6x

Effects of Irradiation and Thermal Annealing

The low phosphorus modifed A302B steel was selected for the first irradiation and annealing experiment. A total of 30 Charpy impact specimens (3 groups of 10 specimens each) were machined, one group from the as-received plate and two groups from the Gleeble austenitized/PWHT material. The specimens were irradiated in the HSSI Irradiation, Annealing, and Reirradiation (IAR) facility at the University of Michigan Ford Reactor at an irradiation temperature of 288°C ± 4°C. The specimens were exposed to a fluence of 1.0×10^{19} n/cm^2 (1 MeV) at a flux of about 4×10^{11} n/cm^2/s. One group of the Gleeble treated specimens was subsequently annealed at 460°C ± 2°C for 168 h in air in a hot cell at ORNL. The specimens were then tested at various temperatures and scanning electron fractography was performed on selected specimens from each group. Table 10 provides a summary of the CVN results for the three groups and, for comparison, for the Gleeble/PWHT material after thermal aging only. The material is not very radiation sensitive as shown by the relatively low transition temperature shift of 22°C. The unirradiated material was also aged at the irradiation temperature of 288°C for approximately the same time (119 days) as the irradiation exposure; as the results show, there was no effect of such low-temperature aging. The expected effect of post-irradiation thermal annealing at 460°C for 168 h would be nearly full recovery of the transition temperature. In this case, however, the transition temperature did not recover, perhaps indicative of another operative metallurgical mechanism in the material. Although hardness or tensile tests have not yet been performed, the upper-shelf energy increase after annealing to a level greater than that of the unirradiated condition is similar to other observations, at least implying that radiation hardening may have decreased as a consequence of the annealing treatment. Similarly, the material aged at 450°C for 168 h shows the same effect. This further supports the potential for another mechanism of embrittlement.

Table 10. *Summary of Irradiation, Annealing, and Thermal Aging Effects on Simulated HAZ in A302B (Modified) Steel*

Condition	Charpy impact mid-transition (°C)		Charpy upper-shelf energy (J)	
	T_o	ΔT_o	USE	ΔUSE
Gleeble initial	-67	–	129	–
Irradiated	-45	22	121	-8
Irradiated/Annealed	-38	29	140	11
Aged 288°C/119 days	-70	-3	117	-12
Aged 450°C/168 hours	-47	20	141	12
Aged 450°C/2000 hours	-4	63	129	0

Scanning electron fractography was performed on an XL-30 Philips scanning electron microscope in an ORNL hot cell. Figure 8 shows the as-received plate after irradiation and that cleavage was the dominant fracture mechanism; no intergranular fracture was observed. Figure 9 shows the Gleeble/PWHT material after irradiation with cleavage fracture as the dominant mechanism; in this case, however, about 15% intergranular fracture was observed. Figures 10 and 11 show fractographs for two different specimens of the Gleeble/PWHT in the irradiated/annealed condition. Both specimens exhibit significant intergranular fracture, with the fraction of intergranular fracture appearing somewhat greater in the specimen which failed at higher energy. Figures 12(a) and (b) show higher magnification fractographs of the specimen in Fig. 11. It is apparent that intergranular fracture is the predominant mechanism of failure of this material, with the fraction of intergranular fracture being greater than 75%.

Discussion

Experiments at AEA-Technology have clearly demonstrated that steels with the typical RPV chemistry can be made to be highly sensitive to temper embrittlement provided the prior-austenite grain size is enlarged to more than 360 μm in diameter. The SAW process used in RPV fabrication is a high-heat-input welding procedure, and grains of substantial enlargement would seem quite likely; more so than in offshore SMAW weldments. Hence, there was a need to pursue more information on temper embrittlement. The scope of interest was the temper-embrittlement properties of commercially made RPV steels, the influence of grain-size and thermal-cycle effects, and the effects of neutron irradiation and thermal annealing.

The objective of Phase I was to confirm the AEA-Technology findings, except in this case commercially made steels were used. This work clearly showed that the temper embrittlement seen with the AEA-Technology laboratory-made steels did in fact apply to the five selected commercially made steels. However, the CVN ΔT_o magnitude was somewhat less.

Phase II introduced an anticipated more-representative weld HAZ coarse grain size, as well as the heating and cooling rates of a typical weld pass. Most RPV steels have CE of about twice that of microalloyed steels. Consequently, the transformation products formed with the high cooling rates of a weld pass are favorable for the development of high fracture toughness, provided postweld stress relief tempering is used. All of the commercial steels had exceptionally good (low-temperature) preaged transition temperatures after simulated weld cycles. Given a 168-h anneal at 450°C, the fracture toughness was degraded, but not as much as the fracture toughness improved from the austenitization and PWHT cycle. The rapid cool following PWHT should maximize the temper embrittlement sensitivity.

Phase III expanded the study to look at other aging-condition variables. The austenitization, PWHT, and grain size were the same as in Phase II. Aging time was fixed

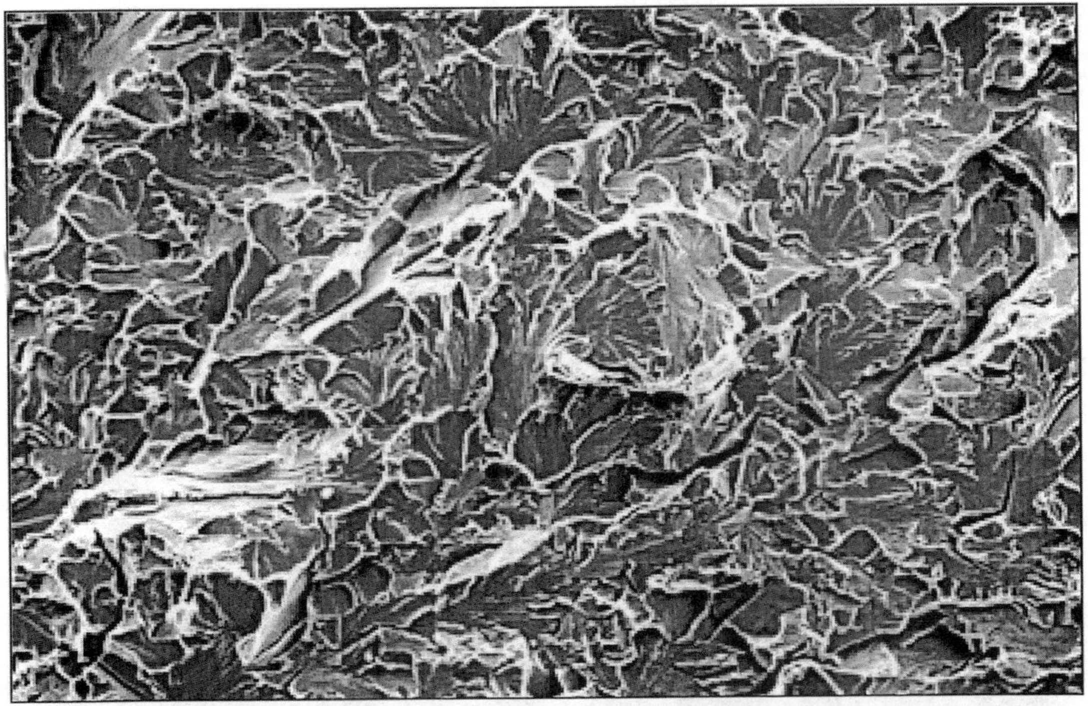

Figure 8. *Scanning electron fractograph of as-received modified A302B steel Charpy impact specimen following irradiation at 288°C to 1 × 10^{19} n/cm^2.*

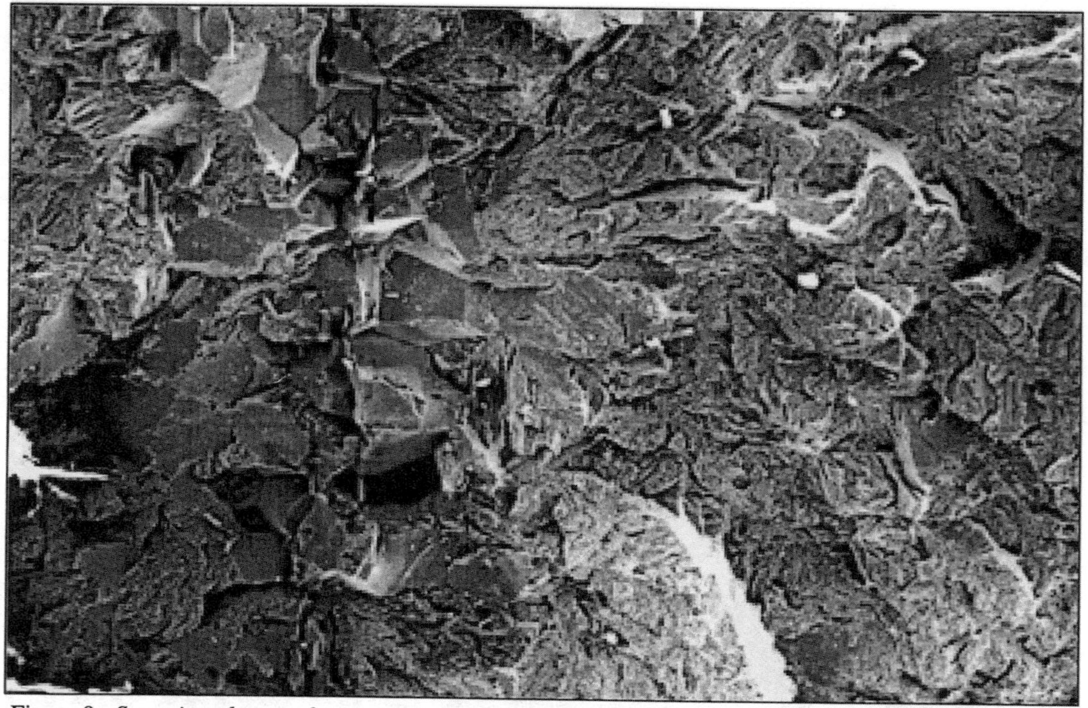

Figure 9. *Scanning electron fractograph of Gleeble austenitized/PWHT modified A302B steel Charpy impact specimen following irradiation at 288°C to 1 × 10^{19} n/cm^2.*

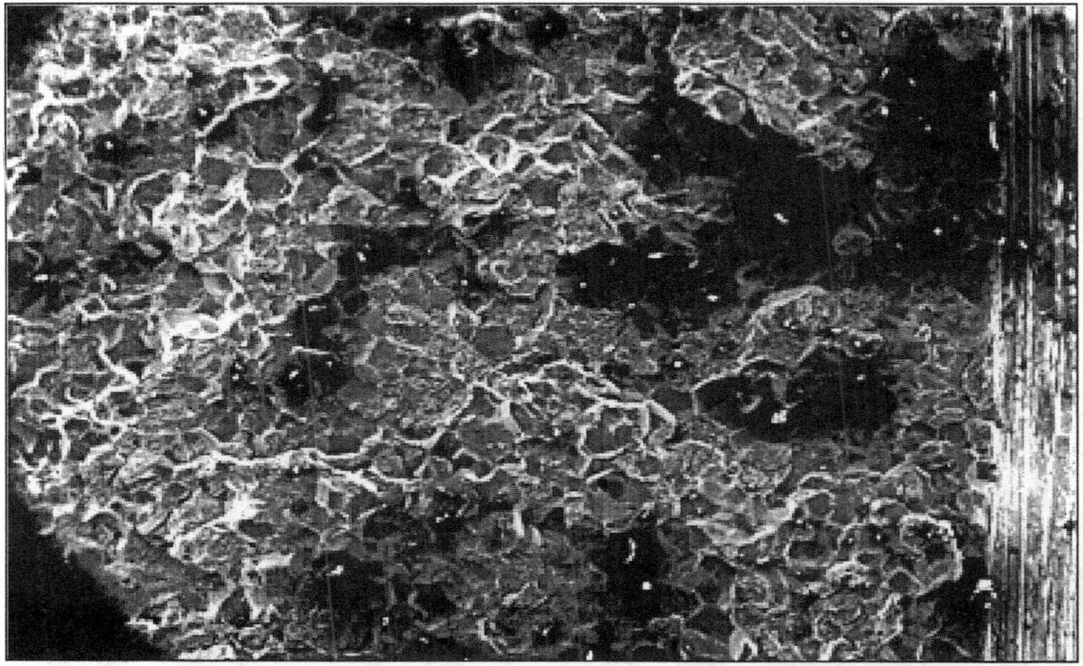

Figure 10. *Scanning electron fractograph of Gleeble austenitized/PWHT modified A302B steel Charpy impact specimen following irradiation at 288°C to 1 × 10^{19} n/cm^2 and thermal annealing at 460°C for 168 h. Specimen failed with 4 J. The notch of the specimen appears at the right side of the fractograph.*

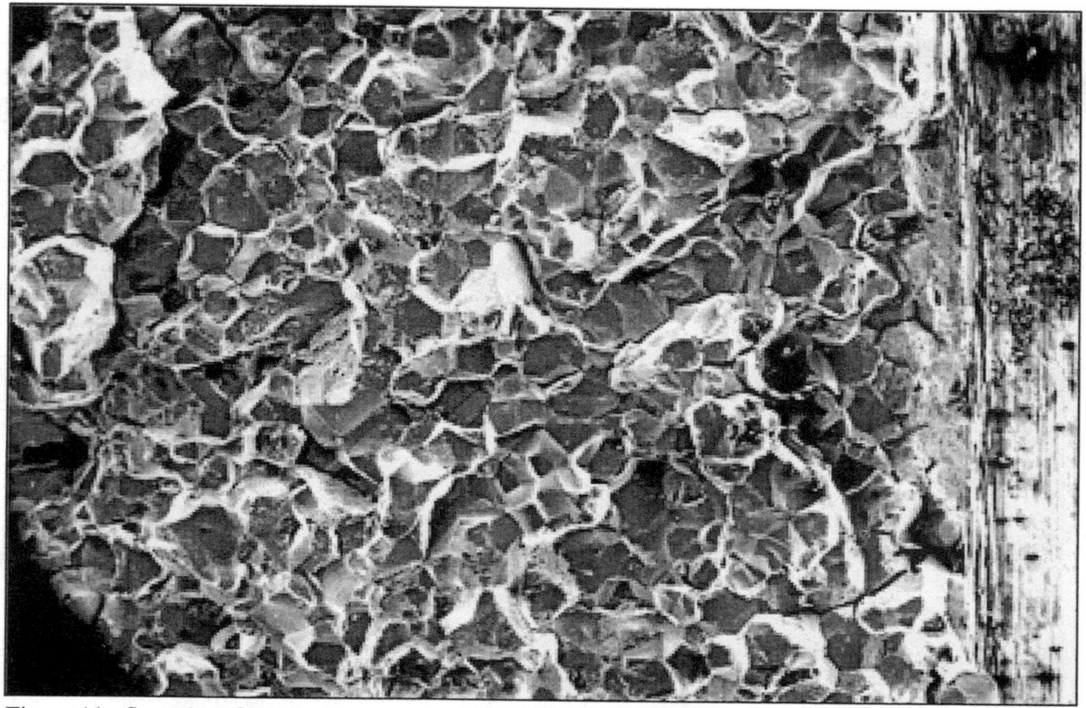

Figure 11. *Scanning electron fractograph of Gleeble austenitized/PWHT modified A302B steel Charpy impact specimen following irradiation at 288°C to 1 × 10^{19} n/cm^2 and thermal annealing at 460°C for 168 h. Specimen failed with about 37 J. The notch of the specimen appears at the right side of the fractograph.*

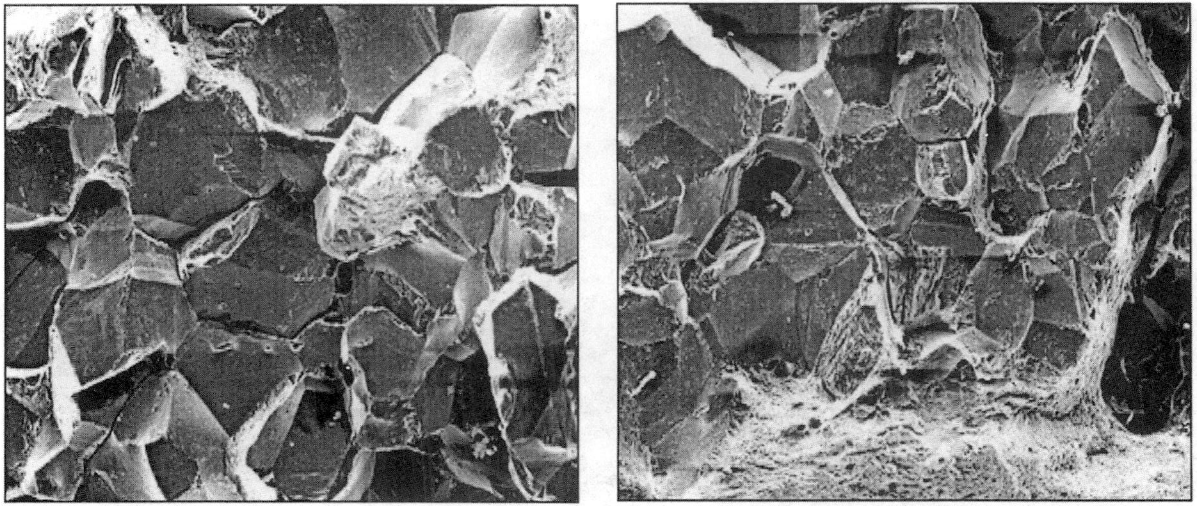

Figure 12. *Scanning electron fractographs of Gleeble austenitized/PWHT modified A302B steel Charpy impact specimen following irradiation at 288°C to 1 × 10^{19} n/cm^2 and thermal annealing at 460°C for 168 h (higher magnification of same specimen as shown in Fig. 11) showing (a) region near crack front and (b) region near one side of the specimen showing shear lip area.*

at 168 h, but aging temperature was varied. Given the aging times used, a lower-bound aging temperature for temper embrittlement was observed. At 399°C (750°F) there was no detectable temper embrittlement in any of the steels. At 450 and 482°C (842 and 900°F), significant transition-temperature shifts could be seen. However, there were no post-embrittlement (168 h) T_0 temperatures observed above room temperature.

Irradiation experiments were performed to determine whether an RPV steel with demonstrated temper-embrittlement insensitivity at temperatures below 400°C can suffer embrittlement at 288°C when the thermal exposure is combined with neutron irradiation or when irradiation is followed by post-irradiation annealing at about 454°C for 168 h , a process intended to recover fracture toughness. The results for one relatively low-phosphorus RPV steel, initially heat-treated to simulate the coarse grain microstructure in a RPV submerged-arc weldment, have indicated a surprising propensity for temper embrittlement following irradiation and thermal annealing. Calculations of irradiation-induced segregation of phosphorus using the Murphy-Perks model were made by AEA-Technology. It was concluded that irradiation appears to have a similar effect to a 482°C/168-h thermal aging treatment, provided the initial segregation level is allowed to be representative of the PWHT condition. Although the results indicate that the monolayer coverage could be on the order of 10%, the post-irradiation annealing treatment would not be expected to deposit sufficient additional phosphorus to result in such a high fraction of intergranular fracture. Additional such calculations will be performed, but measurements of elemental segregation to the grain boundaries in that steel are highly desirable to confirm the nature of the segregation. The role of molybdenum in this case is implicated, as is the potential role of free carbon.

Summary of Observations and Conclusions

Local grain-coarsened regions in simulated RPV weld HAZs of five representative RPV materials have been studied. Two grain-coarsening schedules have been used: (1) the AEA-Technology cycle for prior austenite grain growth creates massive grains to determine if the extremely large ΔTT values observed in the AEA-Technology model alloys are representative of commercial RPV alloys; and (2) the Gleeble cycle for prior austenite grain growth creates coarse grain sizes representative of multipass weld LBZs. The thermal history is prototypic of submerged-arc weld cycles, except that a rapid cool following the PWHT was imposed.

1. No examples of structural failures traced to LBZ in RPV welds were found in the literature.
2. Thermal aging at 450°C for 2000 h of Gleeble simulated HAZs resulted in almost complete intergranular fracture of A533B (P = 0.18%) and modified A302B (P = 0.007%) steels and a factor of 10 higher grain boundary phosphorous content in A533B.
3. The McClean model of phosphorus segregation reasonably predicted the measured results for 4 of the 5 steels.

4. Irradiation of a Gleeble simulated HAZ in modified A302B steel (P = 0.007%) to 1.0×10^{19} n/cm^2 (>1 MeV) at 288°C showed:
   a. The irradiation-induced Charpy shift was 22°C.
   b. Thermal annealing at 450°C for 168 h produced no recovery of transition toughness.
   c. Irradiated specimens exhibited about 10 to 20% intergranular fracture.
   d. After thermal annealing at 460° for 168 h, specimens exhibited more than 75% intergranular fracture.
5. Microscopical examination is required to explain the causes of such extensive intergranular fracture in this particular steel. Further experiments with the same steel given a prototypic cooling rate following the PWHT are recommended.

Acknowledgments

This research sponsored by the Office of Nuclear Regulatory Research, U.S. Nuclear Regulatory Commission, under Interagency Agreement DOE 1886-N695-3W with the U.S. Department of Energy under Contract No. DE-AC05-00OR22725 with UT-Battelle, LLC. The authors are indebted to the late Michael Vassilaros of the NRC who assisted with the evaluations of this work during the aging studies. Assistance at ORNL of the following individuals is also greatly appreciated: Alan Frederick for Gleeble austenitization, Edward Hatfield for heat treatment, Janie Gardner and Jackie Mayotte for metallography, Ronald Swain for mechanical tests, James King for welding technology guidance, and Julia Bishop for preparation of the manuscript.

References

[1] McCabe, D. E., *Initial Evaluation of the Heat-Affected Zone, Local Embrittlement Phenomenon as it Applies to Nuclear Reactor Vessels*, ORNL/NRC/LTR-99/10, September 1999.

[2] Barnes, A. M., *The Effect of Intercritical Thermal Cycles on HAZ Microstructure and Toughness in C-Mn-Al-Nb Steel*, Report 402/1989, The Welding Institute, August 1989.

[3] Arimochi, K., "CTOD Behavior of Steel Weld with 500 MPa Class Strength and HAZ-CTOD Characteristics of Newly Developed Low Al-B-N Steel," *Transactions of the Japan Welding Society*, 20(2), 1989.

[4] Thaulow, C., and Paaw, A. J., "Materials Characterization with Respect to HAZ Local Brittle Zones,"*WELDTECH 88*, London, November 1988, pp. 24-1 to 24-14.

[5] McElroy, R. J., Foreman, A. J. E., Gage, G., Phythian, W. J., Ray, P. H. N., and Vatter, I. A., "Optimization of Reactor Pressure Vessel Surveillance Programmes and Their Analysis," contribution to IAEA CRP 3 Research Program, AEA-RS-2426, December 1993.

[6] J. A. Wang, *Embrittlement Data Base, Version 1*, USNRC Report NUREG/CR-5506 (ORNL/TM-13327), August 1999.

[7] Nanstad, R. K., McCabe, D. E., Swain, R. L., and Miller, M. K., *Chemical Composition and RT_{NDT} Determination for Midland Weld WF-70*, NUREG/CR-5914 (ORNL-6740), 1992.

[8] Nanstad, R. K., McCabe, D. E., and Swain, R. L., "Evaluation of Variability in Material Properties and Chemical Composition for Midland Reactor Weld WF-70," *Effects of Radiation on Materials: 18th International Symposium, ASTM STP 1325*, R. K. Nanstad, M. L. Hamilton, F. A. Garner, and A. S. Kumar, Eds., American Society for Testing and Materials, West Conshohocken, PA, 1999, pp. 125-156.

[9] C. D. Lundin and S. Mohammed, *Effect of Welding Conditions on Transformation and Properties of Heat-Affected Zones in LWR Vessel Steels*, USNRC Report NUREG/CR-3873 (ORNL/SUB/78-7637/1), November 1990.

[10] McClean, D., *Grain Boundaries in Metals*, Clarenden Press, Oxford (1957).

[11] Druce, S. G., Gage, G., and Jordan, G. R., *Acta Metallurgica*, 34, 1986, pp. 641-652.

Jürgen Böhmert,[1] Andreas Ulbricht,[1] Alexander Kryukov,[2] Yuri Nikolaev,[2] and Dmitry Erak[2]

Composition Effects on the Radiation Embrittlement of Iron Alloys

---

Reference: Böhmert, J., Ulbricht, A., Kryukov, A., Nikolaev, Y. and Erak, D., **"Composition Effects on the Radiation Embrittlement of Iron Alloys,"** *Effects of Radiation on Materials: 20th International Symposium, ASTM STP 1405,* S. T. Rosinski, M. L. Grossbeck, T. R. Allen, and A. S. Kumar, Eds., American Society for Testing and Materials, West Conshohocken, PA, 2001.

Abstract: The effects of copper, phosphorus and nickel on radiation embrittlement were investigated using an irradiation experiment at surveillance positions in two Russian VVER-type reactors with 8 iron-base testing alloys. Chemical composition varied between 0.015 to 0.42 % Cu, 0.002 to 0.039 % P, 0.01 to 1.98 % Ni, 0.09 to 0.37 % Si, and 0.35 to 0.49 % Mn. Charpy-V impact tests and tensile tests were performed with specimens machined from these alloys. The specimen were tested in the as-received state, in the irradiated state (fluence: 1×10^{19} and 8×10^{19} cm^{-2} [$E_n > 0.5$ MeV]) and in the post-irradiation annealed state.

Irradiation produces significant hardening and embrittlement. The effect increases with the Cu and P content. Ni causes an additional embrittlement. It does not depend on the Ni content within the range of 1.1 to 2 % Ni. The irradiation effect can be eliminated by annealing at 475 °C / 100 h but not absolutely in the case of high Cu or P content.

Small angle neutron scattering experiments of these alloys show that not only the volume fraction but also the type of the radiation defects vary in dependence on the composition of the alloys.

Keywords: radiation embrittlement, iron alloy, neutron scattering, Charpy impact testing, phosphorus, nickel, copper

[1]Leader and research scientist of the Material and Components Safety Division of the Institute for Safety Research at the Research Center Rossendorf Inc., P.O.Box 510119, 01314 Dresden, Germany.

[2]Research scientists at the Russian Research Center, Kurchatov Institute, Kurchatov Sq. 1, 123182 Moscow, Russia.

Introduction

The sensitivity of a reactor pressure vessel (RPV) steel against irradiation embrittlement is, in large part, controlled by the material composition. One of the most important parameters which influence the level of embrittlement is the copper content as has been recognized for many years [*1,2*]. Thus, the copper effect is already part of the legal guides for predicting the irradiation embrittlement [*3*]. Russian results gained on Cr-Mo-V alloyed RPV steels show a strong harmful effect of phosphorus [*4-6*]. A weak effect of phosphorus was established by Odette and Lucas [*7*] with a comprehensive statistical analysis on ASTM-related RPV steels irradiated at surveillance position in US-American NPPs and in material test reactors. For these types of RPV steels, it has been proven a clear influence of nickel [*7-11*]. The effect depends on the irradiation conditions. It is partly independent on the copper content but also partly enhancing the harmful role of copper in a synergistic manner.

Since the VVER 1000 was set up, for the Russian RPV design, steels of higher nickel content have been used. Unfortunately, mistakes of the surveillance program have prevented up to now that the nickel effect can reliably be estimated from the results of the surveillance programs [*12*]. Nevertheless, there are serious indications that high nickel content of 1.5 to 2 % which is contained in welds of the first VVER 1000/320-type reactors is very detrimental [*13*]. The nickel content is not, however, considered in the Russian guide for predicting the irradiation embrittlement.

Few data have been published about the composition effects on the annealing response. A study of Hawthorne (quoted in [*14*]) on weld and SA-533-B-type RPV steel showed mainly the influence of copper. Systematic experiments were performed on Russian VVER 440-type RPV steels. Above all the phosphorus content defines the level of recovery [*15*]. However, the thermal stability of the irradiation damage seems to increase due to nickel [*12*]. Thus, nickel-containing steels require higher annealing temperatures.

Results so far have mainly been determined by statistical correlation analyses of irradiation experiments and surveillance programs from nuclear power reactors. The large number of parameters of influence reveals the element-specific effect only uncertainly. Therefore, the Russian Research Center Kurchatov Institute Moscow designed a special irradiation program to investigate the effect of copper, phosphorus and nickel on the irradiation embrittlement under constant, VVER-typical irradiation conditions. Specimens made from 19 iron alloys with systematically varied content of copper, phosphorus and nickel were irradiated at surveillance position of VVER 440-type reactors and the mechanical properties were tested after irradiation and post-irradiation annealing. Additionally, microstructural investigations by small angle neutron scattering were planned. The following paper reports on the results of the first step of the irradiation experiment on 8 selected alloys. The selection of the alloys aimed at maximum variation of the three mentioned elements. This work was carried out together with the Forschungszentrum Rossendorf e.V. (Germany). The alloys are carbon poor and, thus, do not correspond to RPV steels. They exhibit a ferritic microstructure that is not so complex and complicated as a real microstructure of a RPV steel. Consequently, the experiments provide a good basis for a physical understanding of composition effects and add to the manifold international efforts on this field [*7,11,16,17*] which partly use simple model alloys as well [*18-20*]. Of course, special care has to be taken to transform the results gained on test alloys to the irradiation behavior of RPV steels. This will be subject of further investigations.

Experimental

Materials

The composition of the 8 investigated iron-base test alloys is given in Table 1. The alloys are carbon poor (C < 0.01 %), free of vanadium and molybdenum and contain manganese in the range of 0.36 to 0.49 % and silicon in the range of 0.11 to 0.37 % without systematic variation. 4 alloys (designated A,B,G,H) are nickel poor (Ni ≤ 0.12 %), 4 other alloys (designated C-F) have a nickel content of approx. 1.2 % or 2 % respectively. The content of copper varies between 0.015 % for the very pure alloy A to 0.42 % (alloy B, F) whereas the content of phosphorus ranges from 0.002 % for alloy A to maximally 0.039 % (alloy E, H).

The alloys were fabricated under the same condition in an industrial small-scale production. The final heat treatment complies with the heat treatment of the VVER 1000-type RPV steel. The alloys exhibit a ferritic microstructure. Their grain size is different. Particularly the nickel-rich alloys have a fine homogeneous grain. Fig. 1 shows the microstructure of alloy H as an example. From these alloys Charpy V-notch standard specimens and small tensile specimens (diameter: 3 mm, measuring length: 16 mm) were machined.

Table 1 - Chemical composition of the test alloys (in wt-%, Fe balance).

Code	C	S	Cu	P	Ni	Si	Mn
A	0.01	0.004	0.015	0.002	0.01	0.15	0.39
B	0.01	0.005	0.42	0.012	0.01	0.24	0.49
C	0.01	0.004	0.12	0.010	1.98	0.09	0.35
D	0.01	0.004	0.12	0.012	1.10	0.12	0.41
E	0.01	0.004	0.12	0.039	1.13	0.20	0.46
F	0.01	0.004	0.42	0.012	1.19	0.21	0.47
G	0.01	0.004	0.11	0.013	0.012	0.37	0.48
H	0.01	0.004	0.11	0.039	0.010	0.24	0.49

Irradiation

Twelve or fifteen Charpy specimens respectively and six or four tensile specimens were positioned at surveillance positions of the VVER 440 reactors ROVNO 1 and KOLA 3 over one reactor cycle. Both reactors differ in the core design and, thus, in the neutron flux (ROVNO 1: 0.4×10^{12} $cm^{-2}\cdot s^{-1}$, KOLA 3: 3.0×10^{12} $cm^{-2}\cdot s^{-1}$ [E_n > 0.5 MeV]). Fluences of 10×10^{18} cm^{-2} (= F_1) or 80×10^{18} cm^{-2} (= F_2) [E_n > 0.5 MeV] respectively were reached. Measurements of the ^{54}Mn-activity of every specimen after irradiation coincide within ±15 % for one specimen. The values between the different alloys show the same scatter.

The irradiation temperature was about 270 °C. After irradiation a part of the specimens was annealed at 475 °C/100 h.

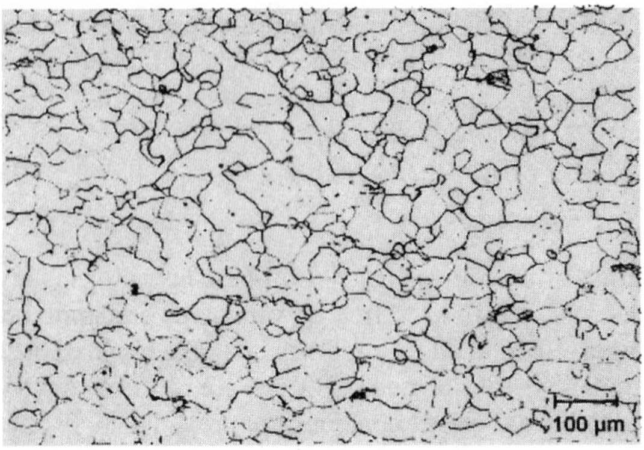

Figure 1 - Microstructure of alloy H.

Testing

The specimens were tested by Charpy-V impact tests and tensile tests in the unirradiated, irradiated and, partly, post-irradiation annealed state. For the Charpy impact tests a instrumented pendulum impact testing machine KMD-30 of a nominal maximum impact energy of 300 J (highest absorbed energy=305.3 J) and an impact velocity of 5.64 m/s and with an automatic temperature condition and specimen manipulation system was available. In some cases the impact energy was not enough to completely break the specimen in the upper shelf energy range. These specimens simplistically obtained an upper shelf energy of 305 J. The ductile-brittle transition temperature TT was determined from the impact energy-temperature curves. The values of TT were calculated from a nonlinear regression fit of a hyperbolic tangent equation to the Charpy. impact energy results. According to the Russian guide the transition temperature is related to consumed impact energy of 47 J. Furthermore, in the following other relations of the Russian guides are used:

$$\Delta TT_{irr} = TT_{irr} - TT_0 = A_{irr} \cdot \sqrt[3]{F} \quad . \tag{1}$$

TT_o and TT_{irr} are the transition temperatures in the initial state and after irradiation to the fluence F in 10^{18} cm^{-2} ($E_n > 0.5$ MeV). A_{irr} characterizes the radiation sensitivity and depends on the content c (in wt %) of copper and phosphorous

$$A_{irr} = 800 \cdot \left(c_P + 0.07 \cdot c_{Cu}\right) \quad . \tag{2}$$

Eq. (2) is valid for weld materials of VVER 440-type reactors and an irradiation temperature of 270 °C.

As the results show, the number of the available specimens for one impact energy-temperature curves does not meet the requirements according to the German rule KTA 3203. However, as the transition temperature range was very small, the determination of TT could be estimated reliably.

Tensile tests were performed at a constant crosshead speed of 1 mm/min and at room temperature and the load versus displacement was recorded. For every state and every alloy two or three specimens were available.

Small angle neutron scattering (SANS) experiments

The SANS experiments were carried out at the BENSC V4 spectrometer of HMI Berlin (Germany) [*21*] and the PAXE spectrometer of LLB CEA Saclay (France) [*22*] under saturating magnetic field H = 1.4 T perpendicular to the neutron beam direction. The performances of both facilities are similar. A 2-dimensional position-sensitive detector consisting of 64×64 cells with a size of 1×1 cm^2 was used. The total width at half maximum of the wavelength distribution is 10% of the nominal 0.6 nm wavelength. Two configurations were used for sample-detector distance (1 and 4 m) and collimation lengths, respectively covering a scattering vector Q range of approx. 0.1 to 3 nm^{-1}. The data analyses were processed by software routine of LLB and BENSC, respectively. A comprehensive description of the raw-data treatment formula, including transmission measurement, background and sensitivity corrections is given in [*23*]. Several methods were applied for the absolute calibration at the facilities. One was performed by measurement of a water standard [*23*] and the other one by direct beam method [*24*]. A sample was measured at both facilities for comparison and a significant difference was not obtained.

Results

Mechanical Testing

In every case, irradiation clearly affects the mechanical properties. The strength parameters (ultimate tensile stress, yield stress) increase, the ductility (uniform elongation, total elongation, upper shelf energy) is reduced and the transition temperature is shifted to higher values. Generally, the effect on the transition temperature is the bigger the higher the content of copper (alloy B, F), phosphorous (alloy E, H) and nickel (alloy C-F) and the higher the fluence.

Table 2 - Results of the Charpy impact tests.

alloy code	unirradiated			irradiated ROVNO-1 F_1= 10×10^{18} cm^{-2}			annealed ROVNO-1			irradiated KOLA-3 F_2= 80×10^{18} cm^{-2}		
	n	TT_0 [°C]	USE [J]	n	TT_{irr} [°C]	USE [J]	n	TT_A [°C]	USE [J]	n	TT_{irr} [°C]	USE [J]
A	7	48	268	8	149	198	7	26	291	12	174	141
B	7	-72	261	8	49	238	7	-42	305	12	49	225
C	6	-66	282	6	96	159	9	-68		-		
D	6	-47	304	8	117	190	7	-52	305	-		
E	7	-51	300	8	166	175	7	-29	305	-		
F	5	-58	245	6	176	160	7	-14	285	-		
G	6	-60	300	-			-			12	-23	300
H	7	-22	292	-			-			12	137	156

n - number of the tested specimen; USE - upper shelf energy; TT_0 - transition temperature, unirradiated; TT_{irr} - transition temperature, irradiated; TT_A - transition temperature, post-irradiated annealed.

As expected, the alloys show already in the unirradiated initial state clear differences of the mechanical properties. Surprisingly, the pure alloy A has a considerably higher transition temperature than the other alloys in the initial state.

Tables 2 and 3 summarize the results of Charpy impact tests and the tensile tests. Fig. 2 shows typical sets of impact energy-temperature curves for the four states tested of alloy A. The dependence of the transition temperature shift ΔTT on the fluence is depicted in Fig. 3. Comparing the shift ΔTT_{irr} of the Ni-poor alloys (A, B, G, H) for the fluence F_1 and F_2, it does not seem that ΔTT_{irr} follows the relation as expected according to (1). This could result from a flux effect. A such effect is described for VVER 440-type weld metal on the base of surveillance results [*25*]. For comparable differences in the flux and similar level of fluence a constant shift of 40 °C to higher temperature is observed if the flux is lower. To evaluate the composition effect the correction was used and all results are related to the fluence level F_1 by means of Eq. (1).

Table 3 - Results of the tensile tests.

alloy code	unirradiated				irradiated ROVNO-1 $F_1 = 10\times10^{18}\ cm^{-2}$				annealed ROVNO-1				irradiated KOLA-3 $F_2 = 80\times10^{18}\ cm^{-2}$			
	$R_{p0.2}$	R_m	A_m	A_5	$R_{p0.2}$	R_m	A_m	A_5	$R_{p0.2}$	R_m	A_m	A_5	$R_{p0.2}$	R_m	A_m	A_5
A	257	361	224	386	453	533	111	245	222	346	252	417	513	538	67	186
B	293	384	180	351	505	584	122	252	348	436	129	306	482	535	118	239
C	338	428	193	359	514	579	121	246	318	421	209	352				
D	317	423	211	376	519	589	119	247	306	421	198	345				
E	351	447	200	342	546	638	124	246	327	440	226	384				
F	388	475	161	296	489	581	124	250	431	526	154	306				
G	234	352	192	399									300	396	154	306
H	260	363	216	404									455	546	127	258

$R_{p0.2}$ - yield stress [MPa], R_m - ultimate tensile stress [MPa], A_m - uniform elongation [%], A_5 - total elongation [%].

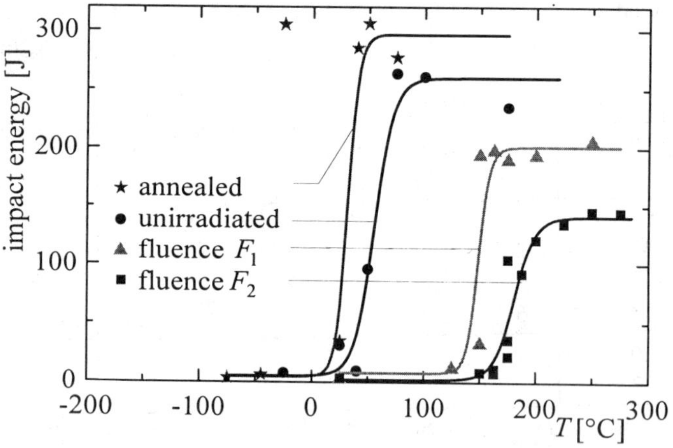

Figure 2 - Charpy impact energy versus temperature for alloy A in different states.

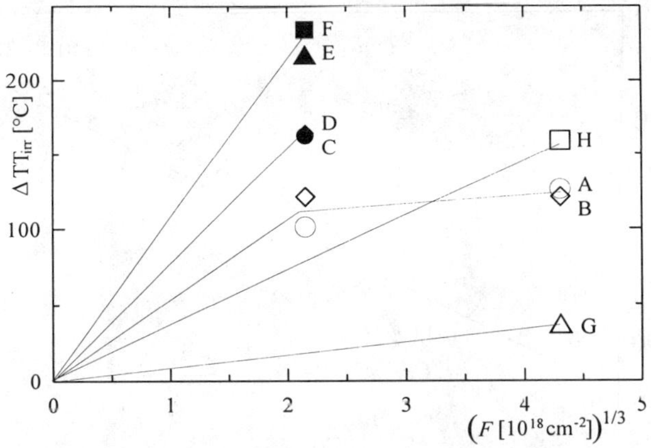

Figure 3 - Irradiation-induced shift ΔTT_{irr} of the transition temperature.

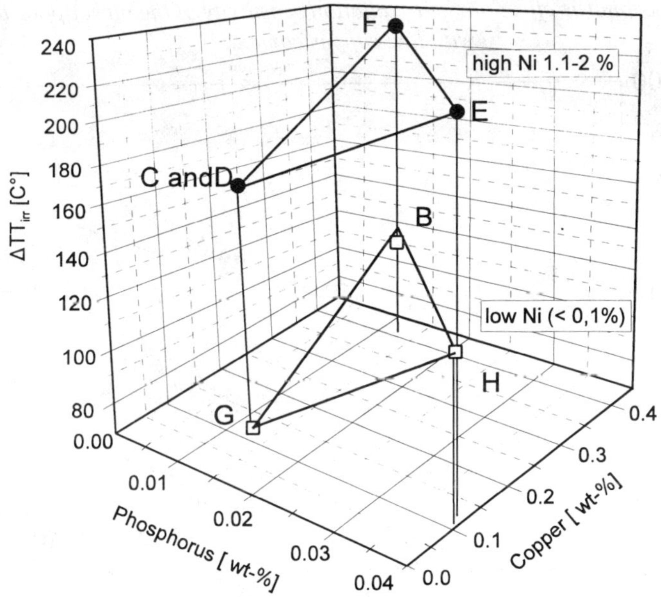

Figure 4 - Chemical composition effect on the irradiation-induced shift ΔTT_{irr} of the transition temperature.

Fig. 4 shows the influence of the composition on the transition temperature shift more clearly. Apart from the expected copper effect, phosphorus and nickel affect the transition temperature in the same way. For Ni-poor alloys the shift seems to follow the prediction of the Russian guide for weld metals as Figs. 5 and 6 illustrate.

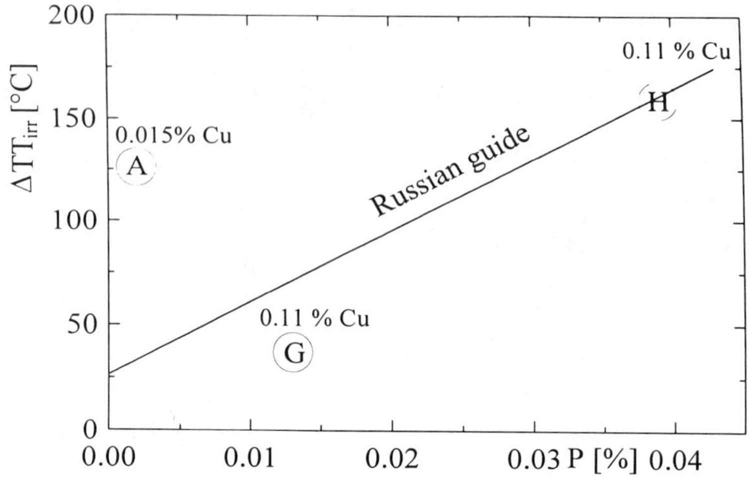

Figure 5 - Phosphorus effect on the irradiation sensitivity of the nickel-poor test alloys with low copper content.

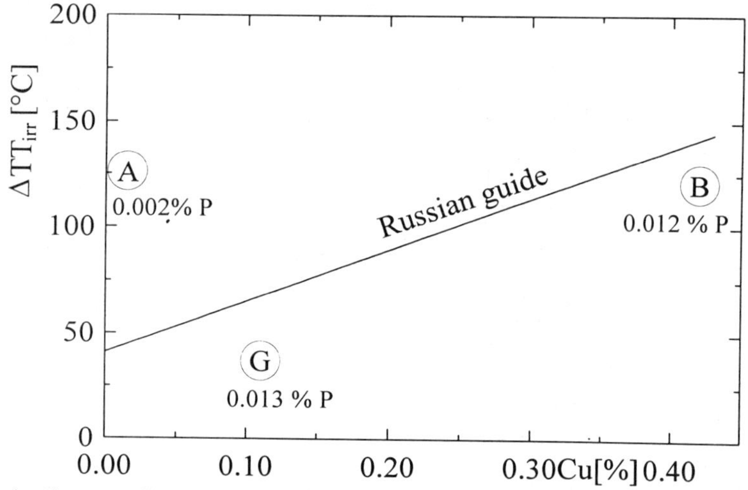

Figure 6 - Copper effect on the irradiation sensitivity of the nickel-poor test alloys with low phosphorus content.

The nickel-rich alloys show an additional shift of the transition temperature that, however, hardly depends on the nickel content in the investigated range between 1.1 and 2 %. It amounts to about 120°C at a fluence of 10×10^{18} cm^{-2} and at a flux of 0.4×10^{12} cm$^{-2} \cdot$s^{-1} [$E_n > 0.5$ MeV].

A surprisingly strong effect occurs at the very pure alloy A containing only manganese and silicon apart from marginal residues of other elements. This is not understandable from the compositional point of view and gives a hint that other irradiation barriers are formed in a pure α-iron matrix.

An influence of the composition on the annealing behavior can also be proven (Fig.7). For low copper or phosphorus content, annealing at 475 °C/100 h removes the irradiation-induced changes of the mechanical properties completely. If the content of these elements, however, is high, then a residue of 20 to 25 % of the changes survives. Nickel has no influence on that. The pure alloy A again drops out of the trend and exhibits an "overrecovery".

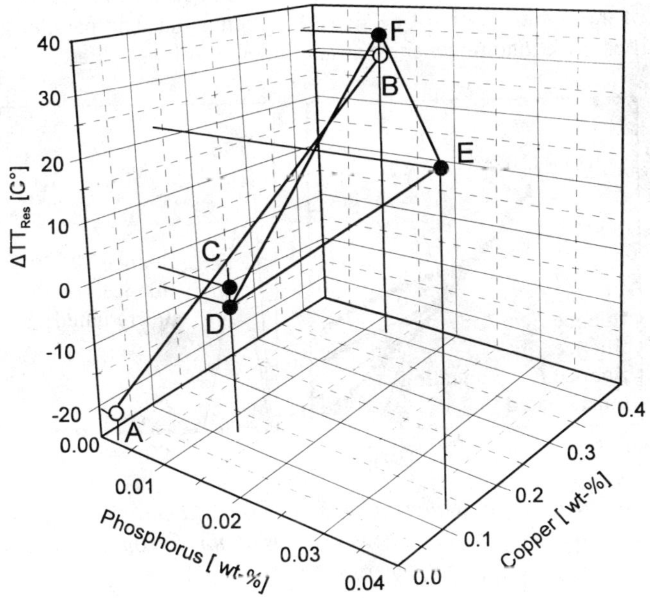

Figure 7 - Chemical composition effects on the residual shift ΔTT_{Res} of the transition temperature (• high Ni, ∘ low Ni).

SANS Experiments

Inhomogeneities of size R (radius) and a volume fraction c_V in the sample result in a course of the macroscopic scattering cross section $d\Sigma/d\Omega(Q)$ as follows

$$\frac{d\Sigma}{d\Omega}(Q,\alpha) = k\cdot\int_0^\infty c_V(R)\cdot R^3\cdot\left(\Delta\eta_{nuc}^2 + \Delta\eta_{mag}^2\cdot\sin^2(\alpha)\right)\cdot S(Q,R)\ dR \quad . \tag{3}$$

In Eq. (3) $Q=|\mathbf{Q}|=4\pi\cdot\sin\theta/\lambda$ is the magnitude of the scattering vector, 2θ is the full scattering angle, λ is the wavelength, k is a constant (in case of spherical particles = $4\pi/3$), $S(Q,R)$ is the so-called shape factor as the Fourier transform of the shape of the inhomogeneities, $\Delta\eta$ is the difference of scattering length densities between the particles and matrix (scattering contrast $\Delta\eta^2$), and α is the angle between the scattering vector and the magnetization of the sample. The product $c_V(R)\cdot\Delta\eta^2$ is established by a Fourier transformation method according to Glatter[*26*]. Under the assumption that the scattering particles are non-ferromagnetic inhomogeneities, the magnitude $\Delta\eta_{mag}$ equals the magnetic scattering length of iron. Thus, the volume fraction of inhomogeneities and the size distribution, here defined as volume distribution function $c_V(R)$, can be estimated from the magnetic scattering cross section.

Fig. 8 shows the measured $d\Sigma/d\Omega(Q)$ and the calculated size distribution function $c_V(R)$ of different states of alloy A. The evidently high volume fraction of irradiation-induced defects correlates with the high susceptibility of this alloy tó neutron embrittlement. An analysis of the A-ratio, ratio of $d\Sigma/d\Omega$(Q) perpendicular and parallel to the magnetization, identifies these inhomogeneities as agglomerates of vacancies.

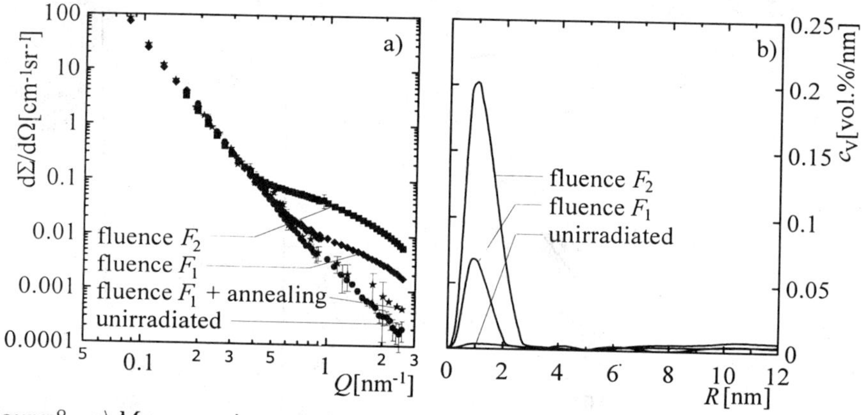

Figure 8 - a) Macroscopic scattering cross sections via magnitude of the scattering vector, b) volume distribution functions via radius of non-ferromagnetic inhomogeneities in different states of alloy A.

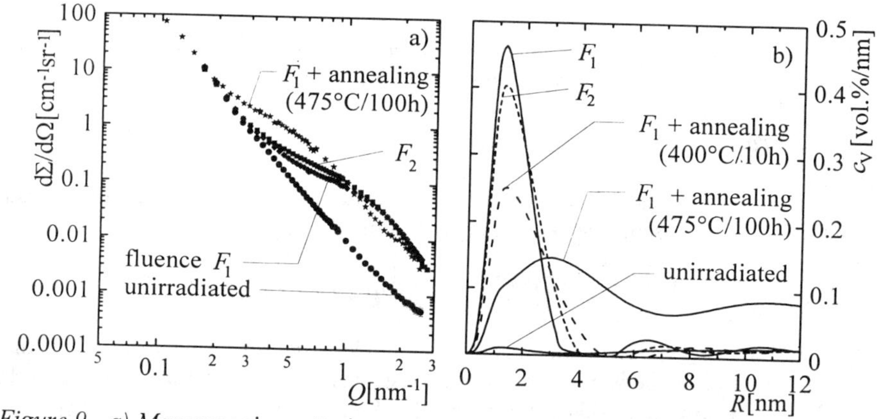

Figure 9 - a) Macroscopic scattering cross sections via magnitude of the scattering vector b) volume distribution function via radius of non-ferromagnetic inhomogeneities in different states of alloy B.

Fig. 9 shows $d\Sigma/d\Omega(Q)$ and $c_V(R)$ of alloy B. Irradiation produces a high fraction of nano-scale defects. However, the changes of the ultra-fine defects due to different fluence levels F_1 and F_2 are irrelevant.

Thermal annealing reduces irradiation defects. Particle growth was observed in the case of alloys with the highest copper-content (alloy B, F), especially shown in Fig. 9.

A comparison of the volume distribution functions of inhomogeneities in nickel-poor and nickel-rich alloys after irradiation is illustrated in Figure 10. The volume fraction of irradiation-induced inhomogeneities strongly depends on the chemical composition. Copper-rich alloys provide the highest contribution, nickel-rich alloys show lower values than the nickel-poor alloys.

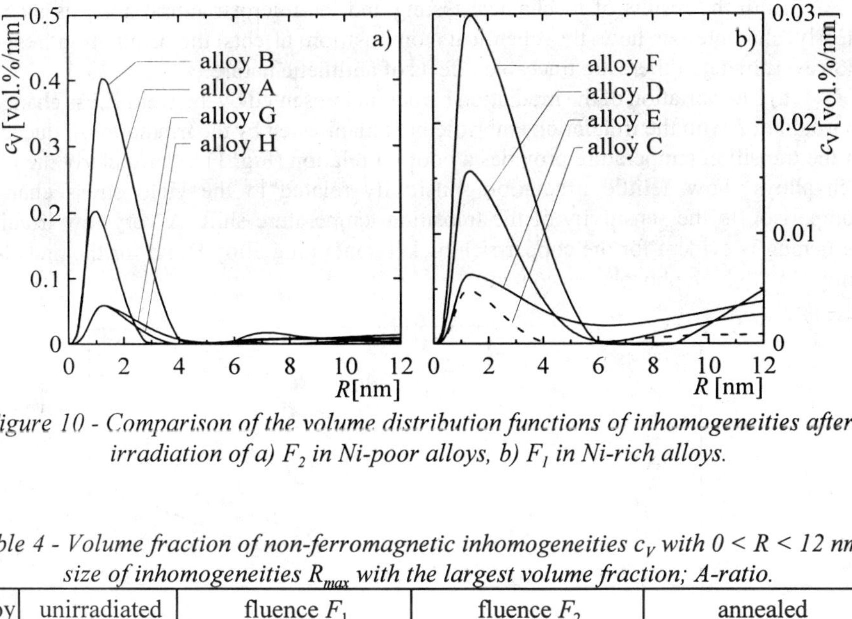

Figure 10 - Comparison of the volume distribution functions of inhomogeneities after irradiation of a) F_2 in Ni-poor alloys, b) F_1 in Ni-rich alloys.

Table 4 - Volume fraction of non-ferromagnetic inhomogeneities c_V with $0 < R < 12$ nm; size of inhomogeneities R_{max} with the largest volume fraction; A-ratio.

alloy	unirradiated		fluence F_1			fluence F_2			annealed		
	c_V [vol.%]	A-ratio	c_V [vol.%]	R_{max} [nm]	A-ratio	c_V [vol.%]	R_{max} [nm]	A-ratio	c_V [vol.%]	R_{max} [nm]	A-ratio
A	0.028	1.6-1.8	0.1	1.0	1.6-1.8	0.26	1.0	1.6-1.8	0.027		1.6-1.8
B	0.10	1.7	0.8	1.3	5 - 6	0.81	1.3	4.5	1.0	3.0	5.5-6.5
C	0.026	2.2	0.013	1.3	1.3-1.4						
D	0.029	2.0	0.048	1.3	2.0						
E	0.031	1.8	0.036	1.4	2.6						
F	0.031	3.2	0.089	1.4	5				0.69	> 4	5 - 6
G	0.080	1.7				0.16	1.3	2.6	0.068		
H	0.093	1.6				0.14	1.3	1.7			1.7

Generally, neutron irradiation produces inhomogeneities in the size range between 0.5 and 6 nm. The size distribution function has a maximum at about 1.2±0.2 nm. The shape of the peak and the position of the maximum do not depend on the fluence. Analyses of the A-ratio show that the type of the irradiation-induced inhomogeneities is not identical for the different alloys. An A-ratio smaller 2 is typical for pure matrix defects whereas an A-ratio higher 3 is characteristic for Cu-containing irradiation defects.

An overview of the estimated microstructure parameters by SANS is given in table 4.

Discussion

Both the results of mechanical testing and the microstructural analysis by SANS clearly characterize how the chemical composition affects the irradiation response. However, in detail there are traceable effects of antithetic nature:

a) The variation of the irradiation hardening presented by the yield stress change due to irradiation with the irradiation embrittlement manifested by the irradiation-induced shift of the transition temperature provides a poor correlation (Fig. 11). Particularly the nickel-rich alloys show a little irradiation sensitivity related to the yield stress change in comparison to the sensitivity of the transition temperature shift. A very low irradiation hardening is evident for the copper-rich nickel-containing alloy F and for the nickel-poor alloy H with high phosphorus content.

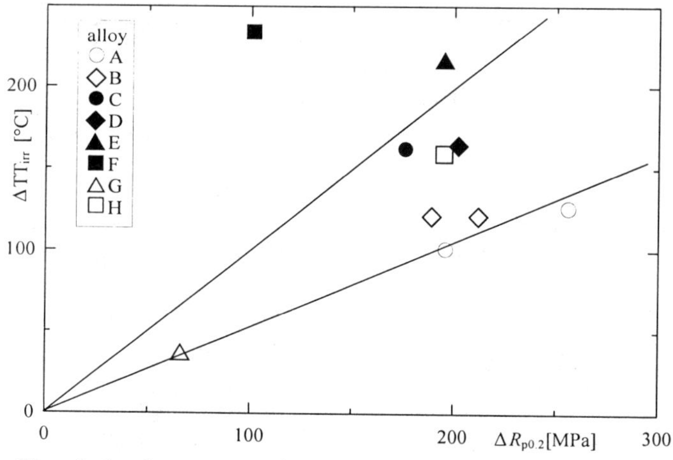

Figure 11 - Correlation between irradiation-induced shift of the transition temperature ΔTT_{irr} and yield stress change $\Delta R_{p0.2}$.

b) The volume fraction of irradiation-induced microstructural defects is considerably lower for the nickel-rich alloys C-F than for the nickel-poor alloys A,B,G, and H (Fig.10) although the irradiation effect on the transition temperature is higher (Fig. 4).

c) Phosphorus leads to an irradiation-induced shift of transition temperature but does not have an increased SANS effect. As the alloys with high phosphorus content (E,H) are accompanied with relatively low irradiation hardening, a special non-hardening irradiation embrittlement mechanism cannot be excluded. A such mechanism caused by grain boundary segregation has repeatedly been suggested for highly phosphorus-containing steels (e.g. [*27,28*] but was not confirmed in every case [*29*].

d) A surprisingly strong effect occurs at the very pure alloy A. Its dependence of the fluence and its A-ratio (but also the annealing behaviour) significantly differ from alloy B. A sophisticated analysis of SANS results [*30*] gives evidence that the microstructural irradiation defects are rather planar, contrary to the other alloys where the shape of defects can be described as spherical.

e) The results of mechanical tests and SANS experiments are quite consistent. High content of copper does not only raise the irradiation embrittlement but also increases the volume fraction of microstructural irradiation defects. The A-ratio is approximately 5 and evidently higher than for the other alloys. Values of about 5 are typical for copper-rich precipitates. For alloy B, the rise of fluence from F_1 to F_2 (factor of 8) does not change both the transition temperature and the yield stress. This corresponds to the SANS curves which also does not show a clear fluence effect for these two fluence levels.

The findings reveal the complicated and complex processes due to irradiation even for the simply structurized iron-base alloys investigated. In the unirradiated initial state, strength and toughness are mainly controlled by the type and amount of the elements in solid solution. Moreover the different grain size between the alloys also influences the mechanical behaviour and for the copper- and nickel-rich alloy F a pre-precipitation of copper cannot be excluded.

In the consequence of primary defect production as a result of neutron-lattice atom interaction, the steady state vacancy concentration is increased and a large population of small point defect clusters is produced. Non-equilibrium point defect concentration, radiation-enhanced diffusion and defect cluster-solute atom interactions lead to complicated microstructural evolution. Obviously, ultra-fine scale irradiation defects are formed in every case. However, their type, their effectivity to impede the dislocation movement, and, perhaps, their distribution can be very different for the different alloys. Apart from that the decrease of residual solute atoms reduces the original solid solution hardening. Thus, further work is required to understand the trends in more detail.

Conclusion

The study presents the first results of an extensive irradiation program to characterize how the chemical composition affects the irradiation embrittlement. The harmful effect could be proven not only for copper but also for phosphorus and nickel. As the alloys exhibit a ferritic microstructure, the effect can mainly result from the interaction between the solute atoms and the primary irradiation defects in the iron matrix. The agreement between the results obtained with the nickel-poor alloys of this study and those achieved by the surveillance programs with weld metal indicates that the irradiation behavior of VVER 440-type weld metals is not essentially influenced by the metallurgical state but by the chemical composition. The recently valid guide seems to provide a useful assessment of the embrittlement for high flux levels. Nickel has a clearly negative effect and must be considered for VVER 1000-type material. However, the results do not suggest an increase of irradiation sensitivity with the nickel content within the range of 1.1. to 2.0 % as often assumed. In this stage of investigation synergistic effects have not yet been proven. The program is to be continued and completed. This is in preparation.

References

[1] Steele, L, E., "Critical Aspects of Neutron Irradiation Embrittlement of Pressure Vessel Steel and Welds," Symposium on the Technology of Pressure Retaining Steel Components, Vail Village, 1970, M.S. Weckler, Ed., *Nuclear Metallurgy,* New York, Vol. 16, 1970, pp. 270.

[2] Hawthorne, J. R., Fortner, E., Grant, S. P., "Radiation Resistant Experimental Weld Metals for Advanced Reactor Vessel Steels," *Welding Journal Research Supplement,* Vol. 35, 1970, pp. 453.

[3] Petrequin, P., "A Review of Formulas for Predicting Irradiation Embrittlement of Reactor Vessel Materials," *AMES Report No. 6*, Paris 1996, EUR 16455 EN.

[4] Amayev, A. D., Kryukov, A. M., Levit, V. I., Sokolov, M. A., "Radiation Stability of WWER-440 Vessel Materials", *Radiation Embrittlement of Nuclear Reactor Pressure Vessel Steels: An International Review, ASTM STP 1170*, L.E. Steele, Ed., American Society of Testing and Materials, Philadelphia, 1993, pp. 9.

[5] Amaev, A. D., Kryukov, A. M., Levit, V. I., Platonow, P. A., Rogov, M. F., Sokolov, M. A., "Radiation Damage and Recovery in WWER-440 Vessel Materials," *PLEX 93*, Zürich, 1993.

[6] Platonov, P. A., Krasikov, E. A., Strombakh, Ya. I., "About Radiation Embrittlement kinetics of Steels for Reactor Vessels," *Irradiation Embrittlement and Mitigation, Proc. of a Specialists Meeting on Irradiation Embrittlement and Mitigation,* Espoo (Finland) IWG-LMNPP-95 Vol. I, 1995.

[7] Odette, G. R., Lucas, G. E., "Irradiation Embrittlement of RPV Steels: Mechanisms, Models and Data Correlations," *Radiation Embrittlement of Nuclear Reactor Pressure Vessel Steels: An International Review, ASTM STP 909*, L.E. Steele, Ed., American Society for Testing and Materials, Philadelphia, 1986, pp. 206.

[8] Guionnet, G., Houssin, B., Brasseur, D., Lefort, A., Gros, D., Pedreau, R., "Radiation Embrittlement of PWR Reactor Vessel Weld Metals: Nickel and Copper Synergism Effects," *Effects of Radiation on Materials: Eleventh Conference, ASTM STP 782*, H.R. Brager, J.B. Perrin, Eds., American Society for Testing and Materials, 1982, pp. 392-411.

[9] Hawthorne, J. R., "Significance of Nickel and Copper Content to Radiation Sensitivity and Postirradiation Heat Treatment Recovery of Reactor Vessel Steels", *Effects of Radiation on Materials: Eleventh Conference, ASTM-STP 782*, H.R. Brager, J.S. Perrin, Eds., American Society for Testing and Materials, 1982, pp. 375.

[10] Davies, L. M., Venables, J. H., Williams, T. J., "Irradiation Embrittlement of a Variety of RPV Steel Plates and Weldments," *IAEA Meeting*, Vienna, 1979.

[11] Odette, G. R., Lucas, G. E., Klingensmith, R. D., "The Influence of Metallurgical Variables on the Temperature Dependence of Irradiation Hardening in Pressure Vessel Steels," *Effects of Radiation on Materials: 17th International Symposium ASTM STP 1270,* D.S. Gelles, R.K. Nanstad, A.S. Kumar, E.A. Löffle, Eds., American Society for Testing and Materials, 1996, pp. 606-627.

[12] Davies, M., Kryukov, A., English, C., Nikolaev, Y., "East/West Steels for RPV," *Irradiation Effects and Mitigation, Proc. of IAEA Specialists Meeting,* Vladimir (Russia), Sept. 1997, IWG-LMNPP-97/2, Vienna, 1997, pp. 42.

[13] Vishkarev, O. M., Zwezdin, Yu. I., Shamardu, V. K., Tulyakov, G. A.,"Radiation Embrittlement of Soviet 1000-MW Pressurized Water Reactor Vessel Steel 15Kh2NMFAA," *Radiation Embrittlement of Nuclear Reactor Pressure Vessel Steels: An International Review, ASTM STP 1170,* L.E. Steele, Ed., American Society for Testing and Materials, Philadelphia, 1993, pp. 218.

[14] Planman, T., Pelli, R., Törrönen, K., "Irradiation Embrittlement Mitigation," *AMER-Report No. 1,* Espoo, EUR 16072 EN, Sept. 1994.

[15] Amaev, A., Kryukov, A., Levit, V., Platonov, P., Sokolov, M., "Mitigation of Radiation Embrittlement of Pressure Vessel Steel by Annealing and Reembrittlement," *Proc. of the 5th German Russian Seminar on Components Safety and Quality Assessment in Reactor Technology*, Stuttgart, 1994.

[16] Taboada, A., Mayfield, M., "USNRC Research on Reactor Pressure Vessel Steel Embrittlement," *Effects of Radiation on Reactor Pressure Vessel Steel Embrittlement: 16th International Symposium, ASTM STP 1175*, A.S. Kumar, D.S. Gelles, R.K. Nanstadt, E.A. Little, Eds., American Society for Testing and Materials, Philadelphia 1993, pp. 14-24.

[17] Williams, T.J., Phythian, W.J., "Electron Microscopy and Small Angle Neutron Scattering Study of Precipitation in Low Alloy Steel Submerged-ARC Welds," *Effects of Radiation on Materials: 17th International Symposium, ASTM STP 1270*, D.S. Gelles, R.K. Nanstadt, A.S. Kumar, E.A. Little, Eds., American Society for Testing and Materials, 1996, pp. 191-205.

[18] Nikolaev, V. A., Rybin, V. V., "Mechanism Controlling the Composition Influence on Radiation Hardening and Embrittlement of Iron-base Alloys," *Effects on Radiation on Materials: 17th International Symposium, ASTM-STP 1270*, D.S. Gelles, R.K. Nanstadt, A.S. Kumar, E.A. Little, Eds., American Society for Testing and Materials, 1996, pp. 3-4.

[19] Odette, G. R., Lucas, G. E., Klingensmith, R. D., Stoller, R. E., "On the Effect of Flux and Composition on Irradiation Hardening at 60 °C," *Effects of Radiation on Materials: 17th International Symposium, ASTM STP 1175*, A.S. Kumar, D.S. Gelles, R.K. Nanstadt, E.A. Little, Eds., American Society for Testing and Materials, Philadelphia, 1996, pp. 547-567.

[20] Phythian, W.J., Diego, N.de, Mace, J., McElroy, R.J., "Characterization of Model Fe-Cu-Mn-Ti-N Alloys by Electron Microscopy and Position Annihilation," *Effects of Radiation on Materials: 16th International Symposium, ASTM STP 1175*, A.S. Kumar, D.S. Gelles, R.K. Nanstadt, E.A. Little, Eds., American Society for Testing and Materials, Philadelphia, 1993, pp. 462-491.

[21] HMI Berlin: "Neutron-Scattering Instrumentation at the Research Reactor BER II," Berlin Neutron Scattering Center - BENSC, May 1996.

[22] Laboratoire Leon Brillouin: "Equipments Experimentaux," Edition LLB, 1995, 69.

[23] Strunz, P., Saroun, J., Keiderling, U., Wiedenmann, A., Przenioslo, R., "General formula for determination of cross-section from measured SANS intensities," *Journal of Applied Crystallography.,* Vol. 33, April 2000.

[24] Cotton, J. P., "Initial Data Treatment," *Neutron, X-Ray and Light Scattering,* P. Lindner, Th. Zemb, Eds., North Holland, 1991, pp. 19-31.

[25] Kryukov, A. M., "The State of the Art of VVER-type RPV: Radiation Embrittlement and Mitigation," *Proc. of the Specialists Meeting on Irradiation Effects and Mitigation,* Vladimir (Russia), 1998.

[26] Glatter, O., "Determination of Particle-Size Distribution Functions from Small-Angle Scattering Data by Means of Indirect Transformation Method," *Journal. of Applied Crystallography,* 13, 1980, 7-11.

[27] Druce, S. G., English, C. A., Foreman, A. J. E., McElroy, R. J., Vatter, I. A., Bolton, C. J., Busewell, J. T. Jones, R. B., "The Modelling of Irradiation-Enhanced Phosphorus Segregation in Neutron Irradiated Reactor Pressure Vessel Submerged-Arc Welds," *Effects of Radiation on Materials: 17th International Symposium, ASTM STP 1270,* D.S.Gelles, R.K.Nanstad, A.S.Kumar, E.A.Little, Eds., American Society for Testing and Materials, 1996, pp 119-137.

[28] Miller, M. K., Burke, M. G., "An APFIM Survey of Grain Boundary Segregation and Precipitation in Irradiated Pressure Vessel Steels," *Effects of Radiation on Materials: 16th International Symposium, ASTM STP 1175,* A.S.Kumar, D.S.Gelles, R.K.Nanstadt, E.A.LIttle, Eds., American Society for Testing and Materials, Philadelphia, 1993, pp. 492-502.

[29] Buswell, J. T., Jones, R. B., "The Modelling of Radiation Hardening and Embrittlement in Magnox Mild Steel Submerged-Arc Welds," *Effects of Radiation on Materials: 16th International Symposium, ASTM STP 1175,* A.S.Kumar, D.S.Gelles, R.K.Nanstadt, E.A.Little, Eds., American Society for Testing and Materials, Philadelphia, 1993, pp. 424-443.

[30] Gokhman, A., Boehmert, J., Ulbricht, A., "Contribution to the Determination of Microstructural Parameters from Small Angle Scattering Experiments at Reactor Pressure Vessel Steels," *Forschungszentrum Rossendorf, Wissenschaftlich-technische Berichte FZR-288,* Febr. 2000.

Austenitic Alloys

Yury V. Konobeev,[1] Valery A. Pechenkin,[1] and Frank A. Garner[2]

The Determination of Bias Factor Stress Dependence from Experimental Data on Irradiation Creep and Stress-Affected Swelling in Austenitic Stainless Steels

---

Reference: Konobeev, Yu. V., Pechenkin, V. A., and Garner, F. A., "**The Determination of Bias Factor Stress Dependence from Experimental Data on Irradiation Creep and Stress-Affected Swelling in Austenitic Stainless Steels,**" *Effects of Radiation on Materials: 20th International Symposium, ASTM STP 1405,* S. T. Rosinski, M. L. Grossbeck, T. R. Allen, and A. S. Kumar, Eds., American Society for Testing and Materials, West Conshohocken, PA, 2001.

Abstract: The measured irradiation creep strain ε in structural materials is usually fitted with the equation: $\varepsilon/\sigma = B_o dpa + DS$, where σ is the applied stress, S the void swelling, and B_o and D are the material parameters. The stress dependence of swelling is sometimes described using the equation $S = S_o(1+M\sigma_{hyd})$, where S_o is the stress-free swelling and σ_{hyd} is the hydrostatic stress component. On the basis of the SIPA mechanism of irradiation creep it is shown in this paper that the parameter D depends on stress as follows: $D = D_o/(1+M\sigma_{hyd})$, where D_o is a constant and the parameter B_o depends on dose. Knowing the constant M, one can determine the constant D_o from experimental data on dose dependence of creep tube diametral strain at several hoop stress levels. This permits the determination of the stress dependence of the bias factor. Such a determination was made in this paper for the austenitic stainless steel 316, for which a complete set of data became available recently. One of the important results of the present paper is the conclusion that to a reasonable approximation the irradiation creep rate $\dot{\varepsilon}^{ic}$ can be determined from data on the stress-free swelling rate and stress-affected swelling.

Keywords: swelling, irradiation creep, neutron irradiation, stress, stainless steel

Both void swelling and irradiation creep phenomena seriously affect the performance of structural materials that will see irradiation service in thermal, fast, and fusion reactor environments as well as in spallation neutron facilities. In a series of papers by Garner et al., it has been well established that these irradiation phenomena are mutually interrelated in both austenitic and ferritic/martensitic steels. A summary of the major insights on the two related

---

[1] Head of Division and Head of Laboratory, respectively, State Scientific Center of Russian Federation, Institute of Physics and Power Engineering, Obninsk, Russia.
[2] Senior Staff Scientist VI, Pacific Northwest National Laboratory, Richland, WA, 91352.

phenomena is presented in Ref. [*1*]. In particular, for structural materials the measured irradiation creep strain ε can be often fitted with the well-known equation: $\varepsilon/\sigma = B_o dpa + DS$, where σ is the applied stress, S is the void swelling, and B_o and D are the material parameters. The dependence of swelling on stress has been often described with the linear equation $S = S_o(1+M\sigma_{hyd})$, where S_o is the stress-free swelling and σ_{hyd} is the hydrostatic component of the applied stress. Several findings were made recently in investigations of the interrelationship between irradiation creep and swelling rates in structural materials. Of particular interest is the role of irradiation-induced swelling to first accelerate irradiation creep at the onset of void growth, but then to throttle back the creep rate as the swelling rate grows larger. This late-term decrease in creep rate has been designated "creep cessation" or "creep disappearance" by Garner and coworkers. In addition, a stress-enhancement of void swelling is somehow involved in the interaction between swelling and creep, such that its operation tends to compensate for the decline in the creep-swelling coupling coefficient D.

The objective of the present paper is first to consider the linear relationship between irradiation creep and swelling rates and to undertake an attempt to determine the stress dependence of the bias factor. This objective will be realized using a complete set of experimental data which recently became available from experiments with pressurized creep tubes made from cold-worked AISI 316 stainless steel irradiated in the Materials Open Test Assembly (MOTA) [*2*].

Irradiation Creep and Swelling Rates

Let's formally divide the strain tensor ε_{ik} in two parts as follows:

$$\varepsilon_{ik} = \frac{1}{3}\delta_{ik}\varepsilon_{ll} + \varepsilon_{ik}^{ic}, \tag{1}$$

where δ_{ik} is the Kronecker's symbol, and

$$\varepsilon_{ik}^{ic} = \varepsilon_{ik} - \frac{1}{3}\delta_{ik}\varepsilon_{ll} \tag{2}$$

is the deviatoric strain tensor. Under irradiation, the first term on the right side of eq. (1) describes strains caused by the voidage V (void volume fraction) in a material, and the second term describes the irradiation creep. The voidage $V=S/(1+S)$, where S is the void swelling, is equal to the sum of diagonal elements of the strain tensor. Thus, the voidage and irradiation creep rates can be found as follows:

$$\frac{dV}{dt} = \frac{d}{dt}\varepsilon_{ll}, \tag{3}$$

$$\frac{d}{dt}\varepsilon_{ik}^{ic} = \frac{d}{dt}\varepsilon_{ik} - \frac{1}{3}\delta_{ik}\frac{d}{dt}V. \tag{4}$$

Following Ref. [*3*], the dislocation structure of a material will be considered as a superposition of dislocations of three orthogonal Burgers vectors directed along coordinate axes 1, 2 and 3. Let us denote through L_i the total length per unit volume of both line dislocations and dislocation loops, the Burgers vector of which is directed along axis i. Further, it will be assumed below that a uniaxial stress σ is applied along axis 1. In this case, obviously, one has L_2=L_3 and $\varepsilon_{11} \neq \varepsilon_{22} = \varepsilon_{33}$. Then $d\varepsilon_i / dt = d\varepsilon_{ii} / dt$ $(i = 1,2,3)$ can be expressed in terms of the climb velocity of dislocations u_i (equal to loop growth rate) as follows:

$$d\varepsilon_i / dt = bL_i u_i \,. \tag{5}$$

where b is the Burgers vector modulus. The climb velocity of line dislocations as well as the interstitial loop growth rate can be expressed through the values $D_v C_v$ and $D_i C_i$, where D_v, C_v and D_i, C_i are the diffusion coefficients and mean concentrations of vacancies and interstitials, respectively, under irradiation, as follows:

$$u_i = \frac{1}{b}(Z_i^{(i)} D_i C_i - Z_v^{(i)} D_v C_v) \,. \tag{6}$$

where $Z_\alpha^{(i)}$ (α=i, v) are the sink efficiencies for point defects of dislocations and dislocation loops having Burgers vector directed along the i-axis. Assuming both the climb velocity of line dislocations and the loop growth rate are identical and combining the eqs. (5)-(6), one obtains

$$d\varepsilon_i / dt = L_i (Z_i^{(i)} D_i C_i - Z_v^{(i)} D_v C_V) \,. \tag{7}$$

Substituting eq. (7) into eqs. (3)-(4) one gets the following expression for the voidage rate

$$dV / dt = (Z_i^{(1)} L_1 + 2Z_i^{(2)} L_2) D_i C_i - (Z_v^{(1)} L_1 + 2Z_v^{(2)} L_2) D_v C_v \,, \tag{8}$$

and for irradiation creep strain rate along axis 1:

$$d\varepsilon_1^{ic} / dt = \frac{2}{3}[(Z_i^{(1)} L_1 - Z_i^{(2)} L_2) D_i C_i - (Z_v^{(1)} L_1 - Z_v^{(2)} L_2) D_v C_v] \,. \tag{9}$$

The values $D_\alpha C_\alpha$ can be obtained from the steady state rate equations. Ignoring all point defect sinks except for dislocations, loops and voids and assuming that the voids are not biased sinks for point defects, one obtains

$$D_\alpha C_\alpha = G\Psi / S_\alpha \,, \tag{10}$$

where G is the dose rate,

$$\Psi = 2/(1 + \sqrt{1 + 4G\mu_r / D_v S_i S_v}) \,, \tag{11}$$

$$S_\alpha = D_{void} + Z_\alpha^{(1)} L_1 + 2Z_\alpha^{(2)} L_2 \,, \tag{12}$$

μ_r is the temperature independent coefficient of point defect recombination, $D_{void} = 4\pi \langle r_v \rangle N_v$ is the sink strength of voids, $\langle r_v \rangle$ the mean void radius, N_v the void number density. In cases where point defect recombination can be ignored, $\Psi = 1$.

Substituting eq. (10) in eqs. (8)-(9) and omitting the subscript "1" for ε^{ic} one can write the expressions for voidage and irradiation creep rates in the following form:

$$\dot{V} = \frac{\Psi D_{void}}{S_v S_i}\left[(Z_i^{(1)} - Z_v^{(1)})L_1 + 2(Z_i^{(2)} - Z_v^{(2)})L_2\right], \tag{13}$$

and

$$\dot{\varepsilon}^{ic} = \frac{2\Psi}{3S_v S_i}\{3L_1 L_2 (Z_i^{(1)} Z_v^{(2)} - Z_i^{(2)} Z_v^{(1)}) + D_{void}\left[(Z_i^{(1)} - Z_v^{(1)})L_1 - (Z_i^{(2)} - Z_v^{(2)})L_2\right]\}. \tag{14}$$

For convenience, in eqs. (13)-(14) and everywhere below, the dose derivative of any function f will be denoted as $\dot{f}$.

Obviously, from eqs. (13)-(14) it is possible to write the following general equation which interrelates the irradiation creep and voidage rates:

$$\dot{\varepsilon}^{ic} = B_o + (2/3)\frac{(Z_i^{(1)} - Z_v^{(1)})L_1 - (Z_i^{(2)} - Z_v^{(2)})L_2}{(Z_i^{(1)} - Z_v^{(1)})L_1 + 2(Z_i^{(2)} - Z_v^{(2)})L_2}\dot{V}, \tag{15}$$

where

$$B_o = \frac{2\Psi}{S_v S_i} Z_v^{(1)} L_1 Z_v^{(2)} L_2 \left(\frac{Z_i^{(1)}}{Z_v^{(1)}} - \frac{Z_i^{(2)}}{Z_v^{(2)}}\right). \tag{16}$$

As seen from eq.(16), the value of B_o is not a constant, but depends on sink strengths of voids and dislocations, which, in turn, are dependent on dose and temperature.

Let's consider separately two well-known proposed mechanisms of irradiation creep: 1) the mechanism of stress-induced preferential nucleation (SIPN), 2) the mechanism of stress - induced absorption (SIPA).

SIPN Mechanism

The mechanism of stress-induced preferential nucleation of loops (below, SIPN mechanism) was initially proposed in Ref. [*4*] as an important mechanism of irradiation creep. The corresponding model was formulated in an attempt to explain the observed anisotropy of Burger's vector orientations in an ensemble of dislocation interstitial loops in irradiated stainless steels. Various aspects of the model have been reviewed in Ref. [*5*]. One of the difficulties of this model is an obvious absence of a nucleation barrier for interstitial clusters larger than di-interstitials. Therefore, one would expect no effect of stress on interstitial loop nucleation.

In the framework of the SIPN mechanism concept, $Z_\alpha^{(1)} = Z_\alpha^{(2)}$ so that $B_o = 0$, and eq. (15) is reduced to the following:

$$\dot{\varepsilon}^{ic} = \frac{2}{3}\frac{L_1 - L_2}{L_1 + 2L_2}\dot{V}. \tag{17}$$

As is seen from eq. (17), the irradiation creep rate is directly proportional to the voidage rate. Therefore, at low doses the creep rate is required to be small since swelling is usually absent for a substantial dose. Such a prediction on the basis of the SIPN mechanism is in serious contradiction with observations of the irradiation creep in cold-worked austenitic stainless steels during the transient regime of void swelling.

The stress dependence of both L_1 and L_2 is unknown for the present. One can expect that the dislocation structure anisotropy should increase with increasing the applied stress. So, in the case of extreme anisotropy, when $L_2=0$, the explicit dependence of irradiation creep rate on stress disappears and creep strain becomes equal to $(2/3)V$. In this case from eq. (1) an obvious result follows: the strain in the direction of the applied stress must be equal to the voidage, $\varepsilon_1 = V$. Unfortunately, available experimental data on irradiation creep and swelling at high neutron doses and high stresses do not allow us to make any definite conclusions regarding the applicability of the SIPN-mechanism. For example, in Fig. 1 the data on the irradiation-induced creep and swelling of annealed AISI 304L irradiated in EBR-II at ~390°C to exposures of ~80 dpa are shown [*6*]. As it follows from these data, at 80 dpa the linear dependence of diametral strain on hoop stress in the creep tubes is maintained up to ~190 MPa, despite the magnitudes of the strains becoming very close to the void swelling level at high hoop stresses. Due to scatter of data on void swelling at high hoop stresses one can not judge conclusively that the diametral strain and swelling exactly coincide, but they do closely approach each other. Another set of experimental data on the stress dependence of diametral strain and swelling in 20% cold-worked 316 stainless steels Core 1 tubes irradiated in FFTF-MOTA at 550°C to 80 dpa [*2*] shown in Fig. 2 demonstrates that such a coincidence, if it exists, will take place only at hoop stresses higher than 250 MPa. Thus, the SIPN mechanism of irradiation creep does not appear to operate in cases where the swelling rate is negligible and its potential contribution is rather questionable at high doses and high applied stresses in structural alloys. However, one can not completely exclude on the basis of only these data the possibility that the SIPN mechanism might be operative in pure metals or dilute alloys.

SIPA Mechanism

The mechanism of stress-induced preferential absorption of interstitials by dislocations was originally proposed in Refs. [*3,7*] and then developed in Refs. [*8,9*]. An original treatment of this mechanism was made in Ref. [*10*]. The basic idea of the SIPA mechanism is that it is dependent on the applied stress inducing a small change of Z_α that is linear with stress. This model takes into account the inhomogeneity interaction between point defects and dislocations. The dislocation structure is considered to be isotropic in this model. The stress induced change is different for dislocations with different orientations of Burgers vector with respect to the external stress direction, so that

$$Z_\alpha^{(i)} = Z_\alpha (1 + \xi_\alpha^{(i)} \sigma). \qquad (18)$$

Considering the point defects as inclusions with elastic moduli different from those of the matrix, in Ref. [*3*] it was estimated that the values of $\xi_\alpha^{(i)}$ as being of order $1/E$, where E is the Young's modulus of a material, and, consequently, are very small. However, the treatment of point defects as inclusions is somewhat questionable. It is quite probable that the

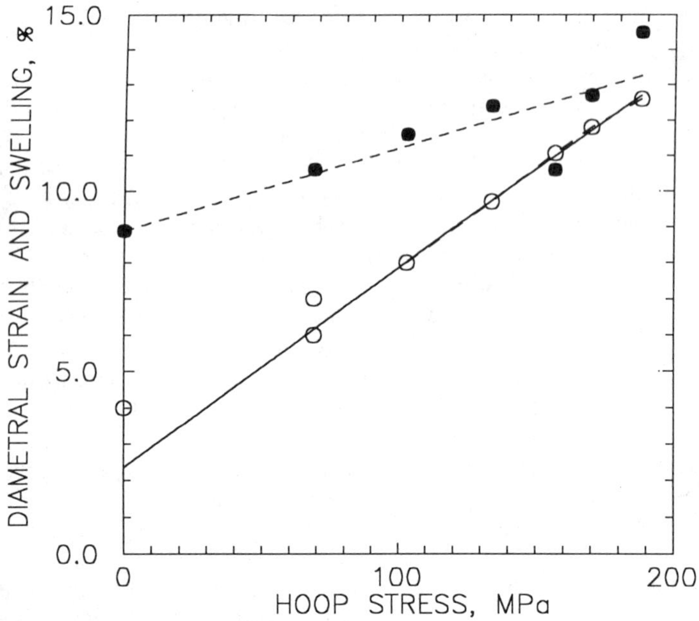

Figure 1 - *Experimental data on stress dependence of diametral strain (open circles) and void swelling (filled circles) in creep tubes made from annealed AISI 304L and irradiated in EBR-II at ~390°C to dose of 80 dpa [6]. The dashed curve is the linear best fit of the swelling data at hoop stresses ranging from 0 to 188 MPa. The solid line is the best fit of diametral strains at stresses ranging from 103 to 188 MPa.*

inhomogeneity interaction inadequately describes the real interaction between point defects and dislocations. From this point forward, a linear dependence of Z_α on stress will be assumed to be valid, but the ξ-values will be treated as adjustable parameters. Just as values of Z_i and Z_v cannot be simultaneously determined in the stress-free case from experimental data on the swelling rate, but only the bias factor

$$B = (Z_i - Z_v) / Z_v, \tag{19}$$

it is also impossible to determine simultaneously four values $\xi_\alpha^{(i)}$ (i=1,2) in the model of dislocation structure under consideration. Therefore, it will be assumed that the bias factors for dislocations of the "1" and "2" orientations depends linearly on stress and for the determination of stress dependence of the bias factors only two values, $\xi^{(1)}$ and $\xi^{(2)}$ are needed:

$$B^{(i)} \equiv (Z_i^{(i)} - Z_v^{(i)}) / Z_v^{(i)} = B + \xi^{(i)}\sigma . \tag{20}$$

Let's take $Z_v^{(1)}L_1 = Z_v^{(2)}L_2$ in eqs. (13) and (15). Due to the weak dependence of $Z_v^{(i)}$ on stress that means, in fact, that $L_1 \approx L_2 = L/3$, where L is the total length of

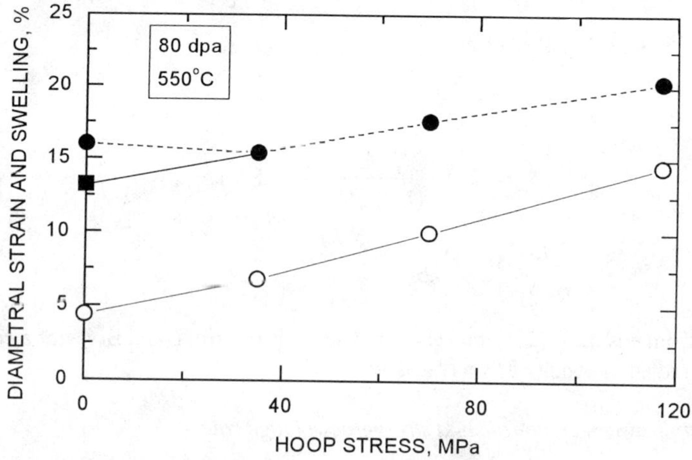

Figure 2 – *Experimental data on stress dependence of diametral strain (open circles) and swelling from density measurements (filled squares) in 20% cold-worked 316 stainless steels (Core 1) tubes irradiated in FFTF-MOTA at 550°C to 80 dpa. The filled squares correspond to swelling obtained from diameter measurement of the unstressed tube [2].*

straight dislocations and loops per unit volume. Then, because the stress dependence of the voidage rate is presumably determined by the stress dependence of the bias factor, from eqs (13) and (20) one obtains that

$$\dot{V} = \dot{V}_o(1 + M\sigma), \tag{21}$$

where $\dot{V}_o$ is the voidage rate in the stress-free case:

$$\dot{V}_o = B\frac{Q}{(1+Q)^2}\Psi, \tag{22}$$

$$Q = Z_v L / D_{void}, \tag{23}$$

and

$$M = (\xi^{(1)} + 2\xi^{(2)})/3B. \tag{24}$$

As seen from eq. (21), the voidage rate increases with increasing applied tension stress and, on the contrary, decreases with increasing compression stress. Such a behavior is predicted in all theoretical approaches [*3,7,8-10*] based on the SIPA mechanism of irradiation creep.

The irradiation creep rate can now be written in several forms.

1. Relationship between creep rate and stress-affected voidage rate

First, from eqs. (15)-(16) one has

$$\dot{\varepsilon}^{ic}/\sigma = B_o + N\dot{V}\,, \tag{25}$$

where

$$B_o = (2/9)\frac{Q^2\Psi}{(1+Q)^2}(\xi^{(1)} - \xi^{(2)})\,. \tag{26}$$

$$N = (2/3)\frac{(\xi^{(1)} - \xi^{(2)})}{B(1+M\sigma)}\,, \tag{27}$$

As seen from eqs. (26)-(27), the value of N depends on stress, and B_o is not a constant but depends on irradiation conditions via Q and Ψ.

2. *Relationship between creep rate and stress-free voidage rate*

When comparing eqs. (25) and (21) one notes that in the approximations made above the irradiation creep rate given by eq. (24) can be expressed in terms of the stress-free voidage rate as follows

$$\dot{\varepsilon}^{ic}/\sigma = B_o + N_o\dot{V}_o\,, \tag{28}$$

where B_o and N_o are :

$$B_o = (1/3)N_oB\frac{Q^2\Psi}{(1+Q)^2} = (1/3)N_oQ\dot{V}_o\,, \tag{29}$$

$$N_o = (2/3)(\xi^{(1)} - \xi^{(2)})/B\,. \tag{30}$$

With the irradiation creep strain rate in the form of eq. (28) and with B_o and N_o values from eqs. (29)-(30), the total strain rate, $\dot{\varepsilon}_1$, in the direction of the applied stress can be expressed in terms of the stress-free voidage as follows:

$$\dot{\varepsilon}_1 = \frac{\dot{V}_o}{3}[1 + \sigma(M + 3N_o + N_oQ)]\,. \tag{31}$$

From eq. (31) it follows that to determine the value of N_o from experimental data on creep strain rate, one needs to know the parameter M and the sink strength ratio Q. With magnitudes of N_o and M known from experimental data the parameters $\xi^{(1)}$ and $\xi^{(2)}$ can be calculated as follows:

$$\begin{aligned} \xi^{(1)} &= B(M + N_o), \\ \xi^{(2)} &= B[M - (N_o/2)]. \end{aligned} \tag{32}$$

3. *The irradiation creep rate in terms of the sink strength ratio Q*

As seen from eq. (24), the ratio $\dot{\varepsilon}^{ic}/\sigma$ can be also expressed in terms of the sink strength ratio Q as follows:

$$\dot{\varepsilon}^{ic}/\sigma = (2/3)(\xi^{(1)} - \xi^{(2)})\Psi\frac{Q(1+Q/3)}{(1+Q)^2}. \qquad (33)$$

Interestingly, in the case where the point defect recombination is an insignificant process, i.e. at $\Psi=1$, the ratio $\dot{\varepsilon}^{ic}/\sigma$ is of maximum value of $(3/8)BN_o$ at $Q=3$. At $Q>3$ the ratio decreases slightly with increasing Q and approaches the level of $(1/3)BN_o$ at $Q>>1$. In the same sink-dominant case the stress-free voidage rate $\dot{V}_o$ as a function of Q is small at $Q<<1$ and $Q>>1$ and reaches its maximum value of $(1/4)\,B$ at $Q=1$. In addition, the values of $\dot{V}_o$ and Q are not independent of each other in this case because $\dot{V}_o/B = Q/(1+Q)^2$ and, consequently, the ratio Q can be expressed in terms of $\dot{V}_o$ as follows:

$$Q = \frac{B}{2\dot{V}_o} - 1 \pm \sqrt{(\frac{B}{2\dot{V}_o} - 1)^2 - 1}. \qquad (34)$$

After substituting the expression (34) in the right side of eq. (31) one finds that the total strain rate is the following complicated function of the stress-free voidage rate:

$$\dot{\varepsilon}_1 = \frac{\dot{V}_o}{3}[1+\sigma(M+3N_o)] + \frac{\sigma}{3}N_o[\frac{B}{2} - \dot{V}_o \pm \sqrt{(\frac{B}{2} - \dot{V}_o)^2 - \dot{V}_o^{\,2}}]. \qquad (35)$$

As seen from eq. (35), $\dot{\varepsilon}_1/\sigma = BN_o/3$ at $\dot{V}_0 = 0$.

Discussion

According to experimental data shown in Fig. 1, both the diametral strain and void swelling in creep tubes made from annealed AISI 304L and irradiated in EBR-II at ~390°C up to doses of 80 dpa [*6*] appear to be linear in stress. For hoop stresses ranging from 0 to 188 MPa. the best fit of swelling data results in $S,\% = 8.90(1 + M_h\sigma_{hoop})$, where $M_h = 2.6\times10^{-3}\ MPa^{-1}$. In this range of hoop stresses, the data on diametral strain can be best fit using the following form: $\Delta D/D,\% = 3.50(1 + 13.5\times10^{-3}\sigma_{hoop})$. Thus, from eq. (31) one obtains $N_{oh}(3+Q) = 10.9\times10^{-3}\ MPa^{-1}$ The dose dependence of the diametral strain and swelling is linear in the range from 10 to 80 dpa [*6*], the stress-free swelling rate is 0.13 %/dpa. Using eq. (34) and taking the value of the bias factor B equal to 2.6% one obtains the ratio Q being equal to 18 (if Q>1) or 1/18 (if Q<1) Microstructural data of Ref. [*11*] give evidence that Q<1 in neutron irradiated annealed AISI 304. This results in $N_{oh}=3.6\times10^{-3}\ MPa^{-1}$. Using this magnitude of N_{oh} and taking $\sigma_{hoop}=(2/\surd3)\sigma$, $B=2.6\%$, $M=(2/\surd3)\ M_h=3.00\times10^{-3}\ MPa^{-1}$, $N_o=(2/\surd3)N_{oh}=4.16\times10^{-3}\ MPa^{-1}$, from eq. (32) one obtains that for the annealed AISI 304L $\xi^{(1)}=1.86\times10^{-4}\ MPa^{-1}$ and $\xi^{(2)}=0.24\times10^{-4}\ MPa^{-}$

^{1} at ~390°C.

Strictly speaking, it is necessary to distinguish the total creep strain ε_1 in eq. (31) from the diametral strain ε_d in creep tubes, which is determined by the initial and final creep tube diameters, d_o and d_f as follows: $\varepsilon_d = (d_f - d_o)/d_o$. Obviously, the creep rate $\dot{\varepsilon}_1$ and diametral strain rate $\dot{\varepsilon}_d$ are simply related to each other: $\dot{\varepsilon}_1 = \dot{\varepsilon}_d/(\varepsilon_d + 1)$. Also, the stress-free voidage and swelling rates, $\dot{V}_o$ and $\dot{S}_o$ are mutually related as follows $\dot{V}_o = \dot{S}_o/(1+S_o)^2$.

Recently, the authors of Ref. [2] investigated the relationship between swelling and irradiation creep using eight sets of pressurized tubes, representing two heats of 20% cold-worked 316 stainless steel irradiated at either 550°C, 575°C, 600°C and 670°C in the FFTF-MOTA to doses of 79-84 dpa. Experimental data on the stress dependence of diametral strain and void swelling in Core 1 tubes irradiated at 550°C to 80 dpa are shown in Fig. 2 [2]. If we take only the stress-free swelling value from diameter measurements, one finds that at 550°C and 80 dpa $S,\% = 13.3\,(1 + M_h \sigma_{hoop})$ with $M_h = 4.5 \times 10^{-3}$ MPa^{-1}. To calculate the dose rates of diametral strain at stress levels of 0, 35, 70 and 117 MPa the experimental data on diametral strain at 550°C were best fitted using the function $A(dpa)^n$.

Experimental data on the diametral strain and results of fitting are shown in Table 1. In an attempt to determine the value of N_o, it was supposed that $Q>1$ up to 80 dpa in 20% cold-worked 316 stainless steel at 550°C, that the function Ψ equals unity, i.e. the point defect recombination is negligible and eq. (35) was used. The stress σ in eq. (35) was taken as the hoop stress.

The value of the bias factor B was taken to be equal to 2.6%. The calculated values of the sink ratios Q are equal to 47.0, 20.8 and 4.46 at 24, 37 and 80 dpa respectively. The calculated values of D_o are shown in Table 2. The average value of N_o is equal to $(1.07 \pm 0.25) \times 10^{-3}$ MPa^{-1}. Using the average value of $N_o = 1.07 \times 10^{-3}\,MPa^{-1}$, from eq. (32) one can calculate the parameters $\xi^{(1)}$ and $\xi^{(2)}$: $\xi^{(1)} = 1.45 \times 10^{-4}\,MPa^{-1}$ and $\xi^{(2)} = 1.03 \times 10^{-4}\,MPa^{-1}$.

Table 1 - *Experimental data on diametral strain in 20% cold-worked 316 stainless steel at 550°C [2] and results of their best fitting with A(dpa)^n*

Parameter	Diametral strain ε_d, %			
	Hoop stress 0 MPa	Hoop stress 35 MPa	Hoop stress 70 MPa	Hoop stress 117 MPa
24 dpa	0.13	0.41	0.68	1.40
37 dpa	0.47	1.22	1.88	3.60
80 dpa	4.42	6.77	10.0	14.5
A, 10^{-5} %	2.136	35.54	69.55	430.6
n	2.793	2.290	2.185	1.855

These magnitudes of $\xi^{(1)}$ and $\xi^{(2)}$ are higher at least by two orders of magnitude as compared with the ones predicted in Refs. [3,10] on the basis of SIPA mechanism of irradiation creep arising from of the inhomogeneity interaction between point defects and dislocations. Such a result is in agreement with the conclusion made in Ref. [12] where irradiation creep due to elastodiffusion has been examined, i.e. anisotropic diffusion induced by applied external stress. Using point-defect data generated by computer simulation, the author of Ref. [*12*] found that creep rates due to elastodiffusion in iron and copper are to be up to thirty times larger than those due to SIPA mechanism treated in Refs. [*3,10*].

It is of interest to point out that eqs. (26) and (29) predict the maximum value of the B_o parameter (in the case when Q>>1 and Ψ=*1*) of 9×10^{-6} (MPa×dpa) $^{-1}$.

Table 2 - *Calculated values of N_o in 20% cold-worked 316 stainless steel that was irradiated in FFTF-MOTA at 550°C* [2]

Dose, dpa	N_o, 10^{-4} MPa^{-1}		
	Hoop stress 35 MPa	Hoop stress 70 MPa	Hoop stress 117 MPa
24	9.17	6.77	8.90
37	12.9	9.15	10.5
80	19.1	11.5	8.61

According to Ref.[*13*] in creep tubes of the Nb-stabilized 1.4981 stainless steel irradiated the fast reactor Rapsodie at hoop stresses from 0 to 200 MPa, the magnitude of M_h in the equation $S=S_o(1+M_h\sigma_{hoop})$ depends on irradiation temperature. At temperatures of 405, 440 and 480°C M_h is equal to 0, 1.8×10^{-3} and 5×10^{-3} MPa^{-1}, respectively. Converting the hoop stresses to the effective ones, one obtains that M ranges from 0 to 5.8 x 10^{-3} MPa^{-1} in the temperature interval of 405 - 480°C. On the basis of eq. (24) one can expect that either the values of ξ increase, and/or the bias factor decreases with increasing irradiation temperature For the present, the origin of such a temperature behavior of these parameters is not clear, however

Conclusions

From the analysis made above one can conclude that both the irradiation creep in austenitic stainlees steels and its interrelationship with void swelling can be appropriately described on the basis of the SIPA mechanism. The predicted linear dependence of swelling on hoop stress is in agreement with available experimental data. The consideration of the widely used relationship $\dot{\varepsilon}^{ic}/\sigma = B_o + D\dot{S}$ describing the relative rates of irradiation creep and swelling leads to a conclusion that the parameter B_o depends on dose. The parameter D depends on applied stress as follows: $D=D_o/(1+M\sigma)$, where M is the parameter which determines the linear dependence of swelling on stress and D_o is the material parameter. This allows us to relate the irradiation creep dose rate to the stress-free swelling dose rate $\dot{S}_o$ as follows: $\dot{\varepsilon}^{ic}/\sigma = B_o + N_o\dot{S}_o/(1+S_o)^2$. A decrease of the parameter D with increasing hoop stress was observed recently by Garner and coworkers in creep tubes made from the 20% cold-worked 316 stainless steel neutron irradiated at 550°C up to 80 dpa [2]. In the case when the point defect recombination is negligible and sink strength of precipitates can be neglected, the dose dependence of B_o as well as of the stress-free swelling is determined only by the ratio of sink strengths of dislocations and voids. In such a case the ratio of the sink strengths can be calculated from experimental data on the dose dependence of stress-free swelling that allows the constant N_o to be obtained if the parameter M is known. With the parameters N_o and M it is possible to determine the stress dependence of the bias factors for dislocations oriented differently in respect to the direction of applied load.

References

[1] Garner, F.A., "Chapter 6, Irradiation Performance of Cladding and Structural Steels in Liquid Metal Reactors", in Vol. 10A, Part 1, *Nuclear Materials, Materials Science and Technology: A Comprehensive Treatment*, R.W. Cahn, P. Haasen, and E.J. Kramer, Eds., VCH Publishers, 1994, pp. 419-534.

[2] Garner, F.A., Toloczko, M.B., and Puigh, R.J., "The Relationship Between Swelling and Irradiation Creep in 20% Cold-Worked 316 Stainless Steel", *Effects of Radiation on Materials, 19th International Symposium, ASTM STP 1366*, M.L. Hamilton, A.S. Kumar, S.T. Rosinski, and M.L. Grossbeck, Eds., American Society for Testing and Materials, West Conshohocken, PA, 1999, pp.667-678.

[3] Heald, P.T. and Speight, M.V., *Philosophical Magazine,* 1974, Vol. 29, p. 1075.

[4] Hesketh, R.V., *Philosophical Magazine*, 1962, Vol. 7, p. 1417.

[5] Ehrlich, K., "Irradiation Creep and Interrelation with Swelling in Austenitic Stainless Steels", *Journal of Nuclear Materials,* 1981, Vol. 100, pp. 149-166.

[6] Garner, F.A., Porter, D.L., and Hudman, G.D., "Irradiation Creep and Swelling of Annealed Type 304L Stainless Steel at ~390°C and High Neutron Fluence", *Radiation Materials Science, Proceedings of the International Conference on Radiation Materials Science*, Alushta, May 22-25, 1990, Vol. 1, Kharkov, 1990, pp. 68-74.

[7] Heald, P.T. and Speight, M.V. "Point Defect Behavior in Irradiated Materials", *Acta Metallurgica*, 1975, Vol. 23, pp. 1389-1399.

[8] Bullough, R. and Willis, J.R. *Philosophical Magazine*, 1975, Vol. 31, p. 855.

[9] Bullough, R. and Hayns, M.R., *Journal of Nuclear Materials,* 1975, Vol. 57, p. 348.

[10] Wolfer, W.G. and Ashkin, M., "Diffusion of Vacancies and Interstitials to Edge Dislocations", *Journal of Applied Physics*, 1976, Vol. 47, No. 3, pp. 791-800.

[11] Michel, D.J., and Smith, H. H., "Fatigue Behavior and Microstructure of Neutron Irradiated Thin Section Type 304 Stainless Steel at Elevated Temperatures", *Properties of Reactor Structural Alloys After Neutron or Particle Irradiation, ASTM STP 570*, American Society for Testing and Materials, Philadelphia, PA, 1975, pp. 156-170.

[12] Woo, C.H., "Irradiation Creep due to Elastodiffusion", *Journal of Nuclear Materials,* 1984, Vol. 120, pp. 55-64.

[13] Schneider, W., Herschbach, K., and Ehrlich, K., "Interdependence of In-Pile Creep and Void Swelling in Ti and Nb-Stabilized Stainless Steels", *Effects of Radiation on Materials, 11th Conference, ASTM STP 782*, H. R. Brager and J. S. Perrin, Eds., American Society for Testing and Materials, Philadelphia, PA, 1982, pp. 30-43.

James I. Cole,[1] Todd R. Allen,[1] Hanchung Tsai,[2] Shigeharu Ukai,[3] Shunji Mizuta,[3] Naoaki Akasaka,[3] Takako Donomae,[3] and Tsunemitsu Yoshitake[3]

Swelling and Microstructural Evolution in 316 Stainless Steel Hexagonal Ducts Following Long-Term Irradiation in EBR-II

Reference: Cole, J. I., Allen, T. R., Tsai, H., Ukai, S., Mizuta, S., Akasaka, N., Donomae, T., and Yoshitake, T., **"Swelling and Microstructural Evolution in 316 Stainless Steel Hexagonal Ducts Following Long-Term Irradiation in EBR-II,"** *Effects of Radiation on Materials: 20th International Symposium, ASTM STP 1405,* S. T. Rosinski, M. L. Grossbeck, T. R. Allen, and A. S. Kumar, Eds., American Society for Testing and Materials, West Conshohocken, PA, 2001.

Abstract: Swelling behavior and microstructural evolution of 12% cold-worked 316 SS hexagonal ducts following irradiation in the outer rows of EBR-II is described. Immersion density measurements and transmission electron microscopy (TEM) examination were performed on a total of seven irradiation conditions. The samples were irradiated to temperatures between 375 and 430°C to doses between 23 and 51 dpa and at dose-rates ranging from 1.3×10^{-7} to 5.8×10^{-7} dpa/s. Dose-rates and temperatures approach conditions experienced by a variety of components in pressurized water reactors (PWR's) and those which may be present in future advanced reactors designs. TEM analysis was employed to elucidate the effect of radiation on the dislocation, void and precipitate structures as a function of irradiation conditions. A moderate dose-rate effect was observed for samples which were irradiated at dose-rates differing by a factor of two. Lower dose-rate samples contained voids of larger diameter and typically swelled more in the bulk. The dislocation and precipitate structure was not visibly influenced by a dose-rate decrease.

Keywords: radiation effects, austenitic stainless steels, swelling, voids, transmission electron microscopy

Introduction

Austenitic stainless steels continue to play a prominent role both as a major structural component in existing light water reactors and in future advanced reactor concepts. Swelling and radiation-induced microstructural changes in austenitic stainless

[1] Argonne National Laboratory-West, Box 2528, Idaho Falls, ID 83403-2528.
[2] Argonne National Laboratory, 9700 South Cass Ave., Argonne, IL 60439-4803.
[3] Japan Nuclear Cycle Development Institute, 4002, Narita-Cho, O-Arai-Machi, Ibaraki-Ken, 311-1393 Japan.

Table 1 - *Independent chemical analysis of base alloy composition.*

C	Mn	P	S	Si	Cr	Ni	Mo	N	Fe
0.03	1.58	0.021	0.018	0.54	17.36	10.98	2.38	0.017	Bal.

steels have been documented extensively over a wide range of dose and temperatures mainly associated with irradiation environments found in fast reactor cores[*1-3*]. Because of the substantial amount of evidence that indicates swelling and microstructural evolution will be dependent on dose-rate, it is essential that the impact of dose-rate be more fully investigated.

With the shutdown of the Experimental Breeder Reactor-II (EBR-II), the opportunity has presented itself to examine the effects of low dose-rate irradiation on a variety of components retrieved from outer rows of the reactor and specifically address swelling and microstructural changes in materials irradiated further from the core center where nearly all previous materials test were performed.

For this study, two hexagonal ducts used to house the reflector subassemblies were removed and sectioned for analysis. Because of gradients in dose, dose-rate, and temperature, a variety of samples experiencing different irradiation conditions can be obtained from a single hex duct. By comparing ducts placed at various distances from the reactor core and samples taken from various positions along the length of an individual duct, the effect of temperature, dose and dose-rate can be explored.

Experimental

TEM microstructural characterization of irradiated 12% cold-worked 316 stainless steel (SS) has been performed on 7 of 40 samples for which density measurements have been reported previously [*4*]. The 40 samples were sectioned from one of two hex ducts (U9861 and U9922) retrieved from EBR-II following long-term irradiation. Samples taken from U9861 (row 8 of EBR-II) experienced a dose-rate of approximately twice that of U9922 (row 9 of EBR-II) for the same axial position.

The base alloy composition as provided by the vendor is shown in Table 1, while a list of the 7 samples examined from the two different hex ducts along with the irradiation parameters is provided in Table 2. Four 3 mm TEM discs were punched from each of seven thinned 3/4 inch density discs sectioned from hexagonal duct flats. The TEM discs were then prepared both as thin foils and extraction replicas and observed in a 200 kV JEOL 2010 TEM equipped with a double tilting specimen stage and energy dispersive x-ray (EDX) detector.

Void diameters were measured for approximately 250 voids from several different regions in each sample in order to obtain overall size distribution profiles. In taking images of the voids, a large deviation from the Bragg condition was used in order to minimize interfering contrast from the dislocation structure. Precipitate identification was performed using extraction replicas. Many of the grain boundary precipitates could not be extracted, so they were identified in the bulk by electron diffraction. It was not possible to obtain compositional information in the bulk because the high radiation fields

Table 2 - *Irradiation parameters, void diameters, densities and swelling data.*

Sample ID	**Temp. (°C)**	**Dose (dpa)**	**Dose Rate (dpa/s)**	**Void Diam., Avg. (nm)**	**Void Density (voids/m^3)**	**Void Swelling $\Delta V/V$**	**Swelling: Immersion Density**
U9861 E2	376	23.2	2.70×10^{-7}	6.1	3.2×10^{21}	0.04%	0.03%
U9922 N3	377	23.7	1.31×10^{-7}	8.8	3.9×10^{21}	0.14%	0.21%
U9861 E4	394	49.5	5.77×10^{-7}	12.9	4.6×10^{21}	0.52%	0.83%
U9922 N8	407	51.1	2.82×10^{-7}	14.1	7.9×10^{21}	1.2%	1.6%
U9861 E14	408	35.7	4.16×10^{-7}	6.1	4.8×10^{21}	0.06%	0.12%
U9922 N15	414	35.7	1.97×10^{-7}	8.8	2.6×10^{21}	0.09%	0.3%
U9922 N9	429	24.7	1.36×10^{-7}	16	7.3×10^{20}	0.19%	0.01%

of the samples would overload the EDX detector. Precipitate identification was performed using techniques of selected area diffraction, microdiffraction and EDX.

Precipitate identification was performed using a commercial software application to match scanned diffraction patterns to computer generated patterns produced from crystallographic data of known phases in both unirradiated and irradiated stainless steels[*5,6*]. Combining crystallographic and composition information allowed a majority of the phases present in the samples to be identified.

Results

Because there were no archival hex can samples available from the same lot of 12% 316 SS as that which had been irradiated, samples from a different lot, but which had undergone similar thermomechanical treatment, were prepared as thin foils to evaluate the pre-irradiated microstructure. Figure 1 provides a TEM image of the unirradiated 316 hex can. The grain size is fairly large, and there is a substantial density of dislocations due to the initial 12% cold working. Additionally, there is a high density of extended defects which were identified as twins using electron diffraction. Twinning at higher levels of strain is a common mode of deformation in low stacking fault materials such as austenitic stainless steel. The grain boundaries are free of second phase precipitation with a few large inclusions scattered randomly throughout the single phase austenite matrix.

A representative lower magnification image of hex duct U9861 E15 following irradiation is shown in Figure 2a. The image reveals an extremely dense population of radiation-produced features. The dense dislocation line networks seen in the unirradiated

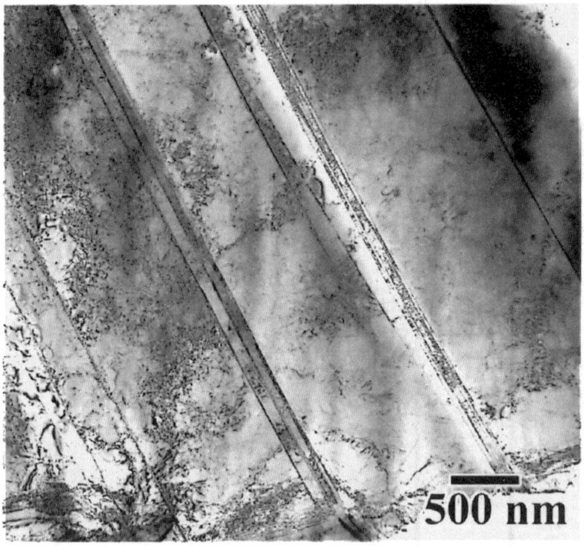

Figure 1 - *Unirradiated 12% CW 316 SS exhibiting high dislocation density and deformation twins.*

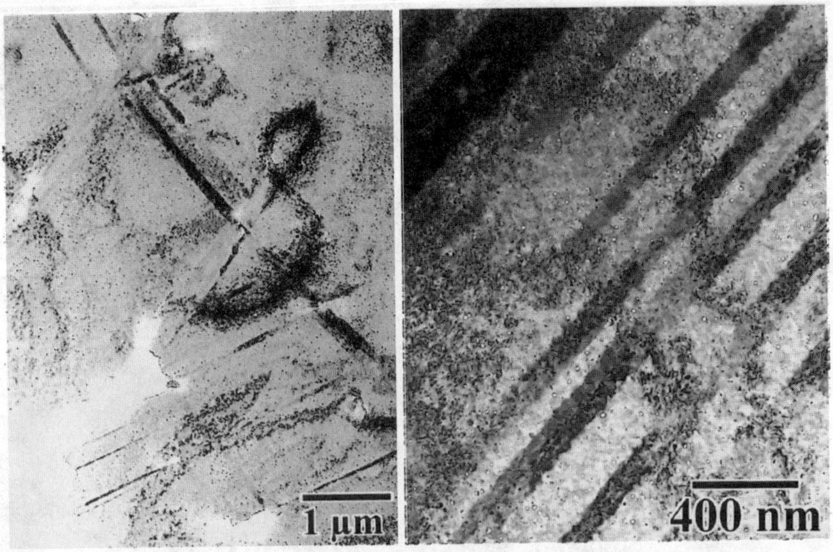

Figure 2 - *Irradiated 12% CW 316 SS, a) Sample U9922 N15 exhibiting high irradiation produced defect densities and the presence of deformation twins, b) higher magnification image of deformation twins in hex duct U9922 N15.*

cold worked sample can no longer be observed. Because of the high image contrast and small size of the voids, they can not be seen at this magnification. The radiation damage appears to be quite homogeneous at this scale. As in the unirradiated case, there is a high density of deformation twins resulting from the initial cold work state (Figure 2b provides a higher magnification image of the twins). There was no clear evidence of regions denuded of voids adjacent to the grain boundaries in any of the samples examined.

Void Structure

Table 2 shows the measured average void diameters and void density for each sample. TEM images of 6 pairs of hex duct samples are provided in Figure 3 going from lower to higher dose (left to right), and higher to lower dose-rate (top to bottom). For all but the highest dose, there appears to be a slight increase in void size with decreasing dose-rate. Void density changes with dose rate are inconsistent. The two samples irradiated at ~376°C to 23 dpa (E2 and N3) show essentially no change in void density with decreasing dose-rate. The void density decreases with decreasing dose-rate at 35 dpa (E14 and N15) and increases with increasing dose-rate at 50 dpa (E4 and N8). Differences in irradiation temperature (although small) among these samples may play a role in these inconsistencies.

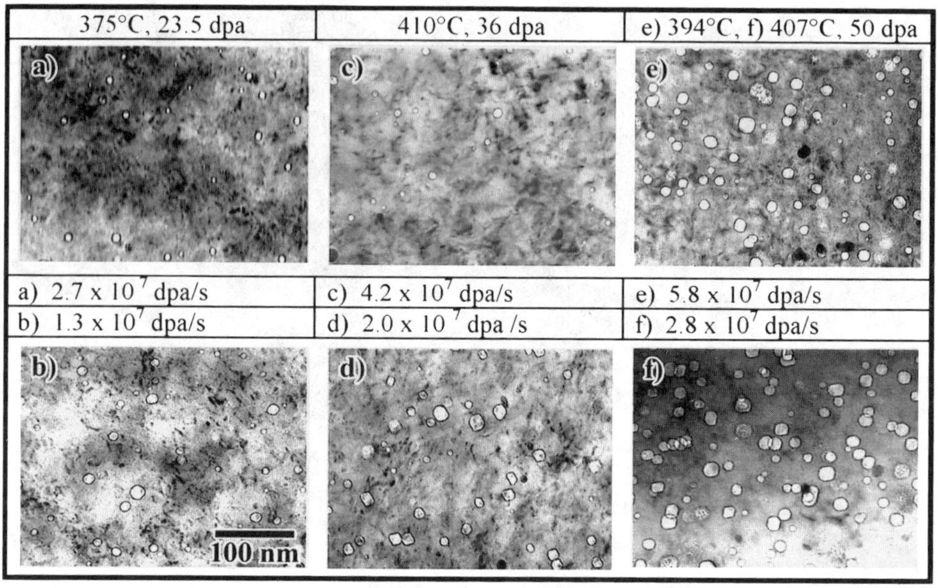

Figure 3 - *Void images in hex ducts a) U9861 E2, b) U9922 N3, c) U9861 E14, d) U9861 N15, e) U9861 E4, and f) U9922 N8.*

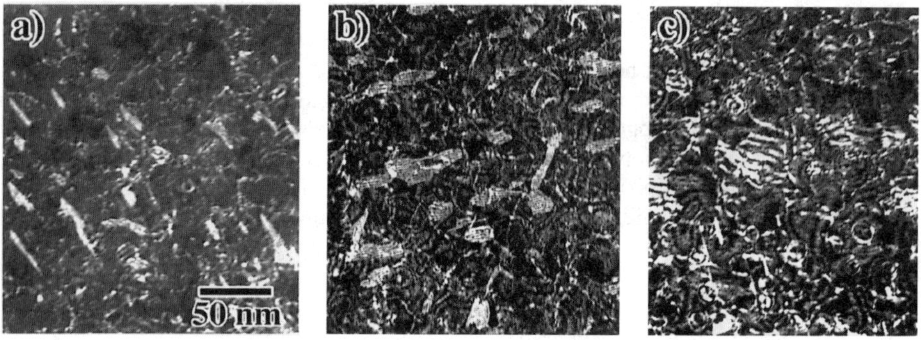

Figure 5 - *WBDF images of the dislocation structures in the 316 Stainless steel hex ducts: a) U9861 E2, 376°C, 23.2 dpa, 2.7 x 10^{-7} dpa/s, b) U9922 N15, 414°C, 35.7 dpa, 1.97 x 10^{-7} dpa/s, c) U9922 N8, 407°C, 51.1 dpa, 2.82 x 10^{-7}dpa/s.*

The data in Table 2 illustrate that the size distributions and densities of voids roughly correspond with swelling calculated from bulk density measurements (immersion density) as described in an earlier study on the same materials[*4*]. The swelling due to voids alone can be calculated using:

$$\frac{\Delta V}{V} = \frac{(4/3)\pi R^3 N}{1-(4/3)\pi R^3 N} \tag{1}$$

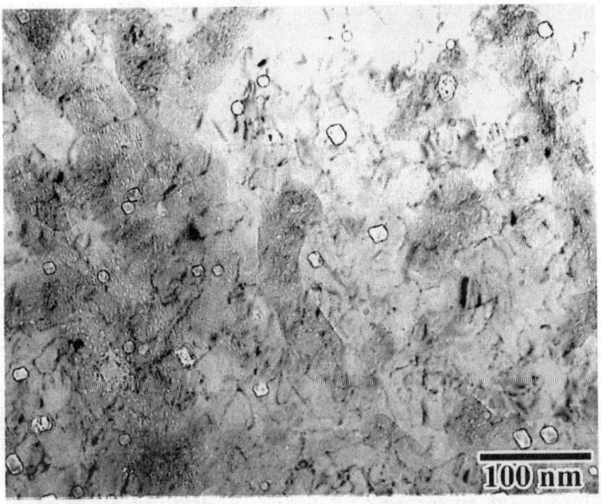

Figure 4 - *Samples irradiated to approximately 25 dpa at 430°C and a dose-rate of 1.4 x 10-7 dpa/sec.*

where R is the average void radius and N is the density of voids. These calculated values are also shown in Table 2 for all the samples examined. For specimen N9, the calculated swelling from the voids is larger than the bulk swelling calculated from immersion density measurements. This is consistent with the bulk densification observed in several of the samples. Densification typically occurred in samples which had been irradiated at temperatures above ~386°C to a dose below 30 dpa.

Figure 4 shows the void microstructure of sample U9922 N9 which was irradiated at 429°C to a dose of 24.7 at a dose-rate of 1.4 x 10^{-7} dpa/sec. Sample U9922 N3 (Figure 3b) was irradiated to the same dose at a comparable dose-rate but at a temperature of ~376°C. Comparing the images and the data in Table 2 reveals that with an increase in temperature of ~55°C, void size nearly doubles and the void density is nearly halved.

Dislocation Structure

WBDF images of the dislocation structures reveal a high density of loops and network dislocations. Representative images of the dislocation structures in samples irradiated from lowest to highest dose (left to right) are provided in Figure 5. The images presented in the figure are taken at an operating g-vector of <111>. The images reveal a high density of faulted Frank dislocation loops which have a Burgers vector of 1/3<111> and are sessile on the slip plane. There is a complete absence of unfaulted perfect dislocation loops (Burgers vector of 1/2<110>). The absence of loops which are unfaulted indicates the high stability of faulted loops in the 316 SS alloy. Detailed quantitative analysis of the loop structure will not be reported at this time but will be reported in a future paper. Significant differences in dislocation structure were not observed as a function of temperature and dose-rate in the ranges investigated. However, with increasing dose, there appeared to be a substantial decrease in the density of faulted loops with a concurrent increase in the dislocation network density. This observation is consistent with the concept of loop growth and unfaulting to form the network with increasing dose[*1-3*].

Precipitate Structure

A substantial amount of small irradiation-induced/enhanced precipitation occurred in the hex ducts both on the grain boundaries and within the grain interiors, while the bulk matrix maintained its austenite crystal structure. Precipitates were present in all of the samples examined and ranged in size from approximately five nanometers to several hundred nanometers.

Many of the larger precipitates (>100 nm), as presented in Figure 6a, were rich in Ti. The vendor provided chemical analysis did not include Ti, and it is present as an impurity. The presence of the such Ti-rich precipitates was confirmed on a bulk TEM disc using wavelength dispersive spectroscopy (WDS) in an SEM. The large Ti-rich precipitates are most likely TiO inclusions often found in commercial purity stainless steels, originating from the steel making process. Another Ti-rich precipitate was also identified which had the crystal structure of TiC, but contained other elements (Cr, Fe, Ni, Mo). Figure 6b provides an image of such a TiC precipitate extracted from the matrix. These precipitates were much smaller (<100 nm) than the former Ti-rich precipitate and also more abundantly distributed throughout the grain interiors.

Additional phases identified in the hex ducts by electron diffraction include M_6C, $M_{23}C_6$ and M_7C_3 (Figures 7 provides an images of an M_6C precipitate on a grain boundary). M_6C precipitates were much more common than $M_{23}C_6$ or M_7C_3 and were frequently found to nearly continuously decorate grain boundaries in the higher dose samples. Other phases infrequently observed were a sulfur-rich precipitate (possibly MnS) and an extremely small (<10 nm) Fe-rich intermetallic phase that was likely chi phase but could not be positively identified because of its small size and low volume fraction. Precipitation in the hex ducts was fairly inhomogeneous, as there were regions that showed a prevalence of one type of precipitate and a complete absence of other types. Intermetallic phases laves and sigma were not observed in the replicas or in the bulk, but have been observed in several studies at higher temperatures[*6-8*].

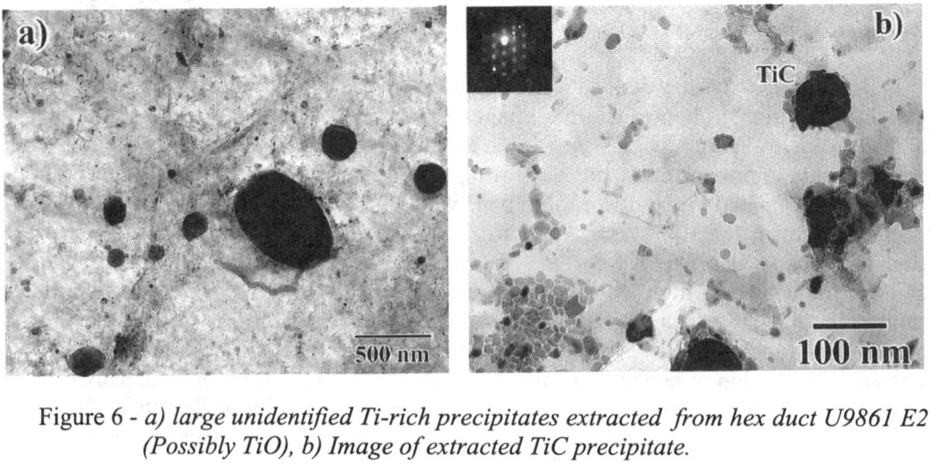

Figure 6 - *a) large unidentified Ti-rich precipitates extracted from hex duct U9861 E2 (Possibly TiO), b) Image of extracted TiC precipitate.*

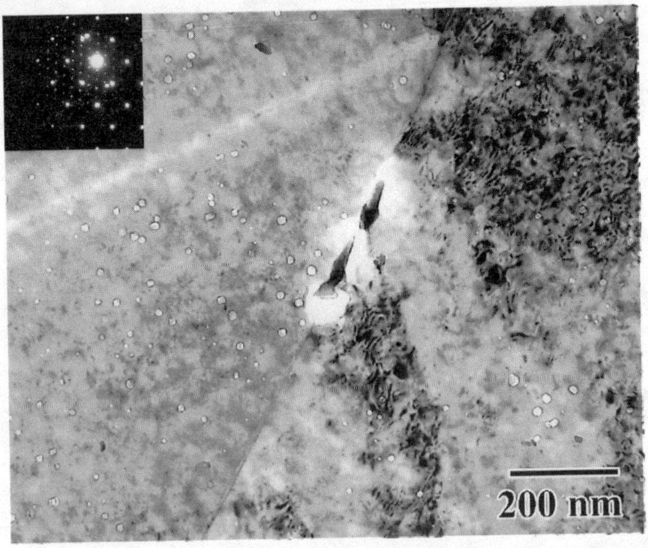

Figure 7 - *Grain boundary M_6C precipitate found in hex duct U9922 N15 (TEM thinfoil image).*

Discussion

Sample pairs were chosen to compare the effect of temperature, dose, and dose rate on microstructural development.

Effect of Temperature

Samples N3 and N9 were irradiated at a 55°C difference in temperature, but with similar dose-rate and dose (24 dpa, 1.3 x 10^{-7} dpa/sec). The swelling for these samples has been calculated from immersion density measurements (see Table 2). Although the bulk swelling of sample N3 is substantially greater than N9 (0.21% and 0.01% respectively), calculated swelling from voids ($\Delta V/V$) reveal that N9's swelling should be slightly greater. As mentioned earlier, analysis of the density measurements of 12% cold-worked 316 stainless steel[4] indicated that at higher temperatures (>386°C) and low dose (<30 dpa), sample density increases (densification). Significantly, densification did not occur at lower temperatures (<386°C) for any of the samples. Densification has been attributed to a lesser extent to the rearrangement of the dislocation structure, and to a greater extent formation of a large volume fraction of precipitates with a consequent decrease in volume[*7,9*]. The irradiation temperature also affected the swelling rate at higher dose. The swelling rate at doses greater than 30 dpa is greater at higher temperature (>386°C) than at lower temperature. The differences in low dose densification and high dose swelling rate are illustrated in the graph shown in Figure 8.

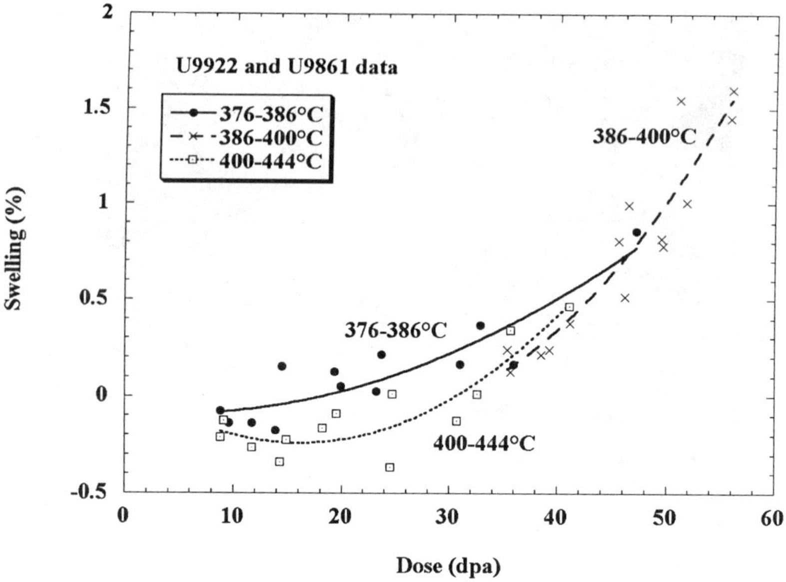

Figure 8 - *Swelling as a function of dose for three temperature regimes. For higher temperatures, the low dose densification and the high dose swelling rate are greater.*

The graph illustrates that at lower dose the swelling rate is diminished in the temperature range between 400 and 444°C, while with increasing dose the swelling rate approaches that of the lower temperature samples (376-386°C) and eventually surpasses it. The void sizes and densities of samples N3 and N9 indicate a possible explanation for why the higher temperature samples have a greater swelling rate at higher dose. Even though the swelling of sample N9 is less than N3 at ~25 dpa, sample N9 has an average void size twice as large as sample N3. This larger void size will promote proportionately greater swelling as the density of the voids increases with increasing dose.

Differences in precipitate distribution and morphology as a function of temperature over the range investigated were not apparent in either the TEM thin foils or the extraction replicas. Compositions for a particular precipitate type varied substantially among and within the replicas and elements detected in an individual precipitate type were not detected in all of the precipitates of that type. For example some of the TiC precipitate contained Mo while others did not. Because of the variability in precipitate composition, a temperature effect on composition could not be discerned.

Effect of Dose

Samples N8 and N15 were irradiated to different doses (50 and 35 respectively), but at similar temperature and only slightly different dose rate (~410°C at dose-rates of ~2.8 x 10^{-7} dpa/sec and 2.0 x 10^{-7} dpa/sec respectively). The measured swelling of sample N8 is 5 times larger than that of N15. At higher dose (sample N8), the void density and size

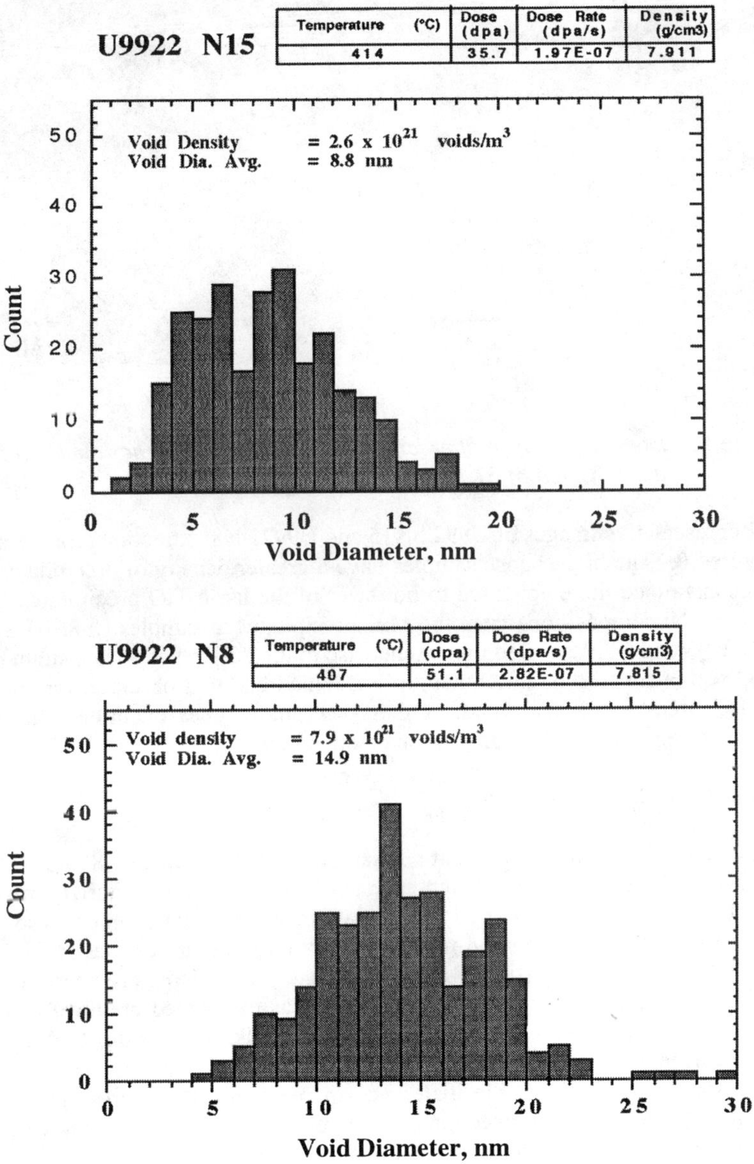

Figure 9 - *Void diameter distributions for samples U9922 N15 and N8.*

are greater, causing the increased swelling. At increasing dose, not only do the voids become larger and more prevalent, but the distribution becomes more sharply peaked. Figure 9 shows the void distributions for samples N15 and N8. The void distribution becomes localized in a narrower range of diameter as the voids grow.

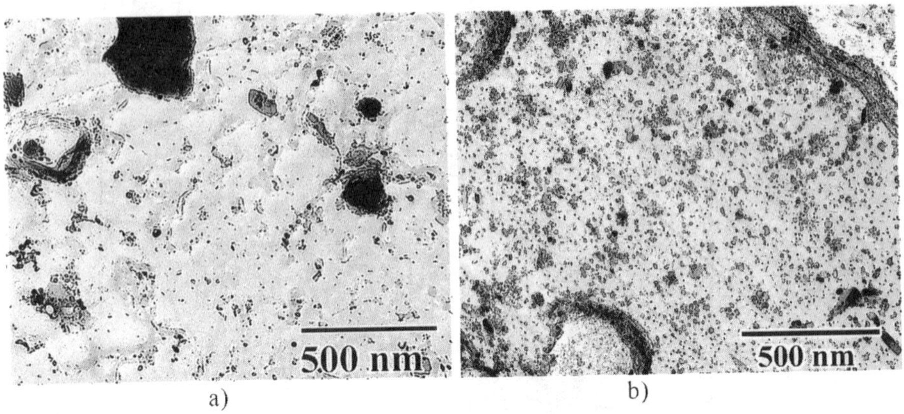

Figure 10 - *Dose comparison of precipitation behavior in 316 hex ducts. a) 35.7 dpa at 414°C and b) 51.1 dpa at 407°C.*

Representative images of U9922 N15 and U9922 N8 extraction replicas are provided in Figure 10. The higher dose samples have a greater density of precipitates, and in the regions examined there appeared to be fewer of the large TiO precipitates. Comparing the Cr and Ni compositions from the M_6C precipitates in samples U9861 E2 and E7 and U9922 N15 and N8 suggested that an increase in the dose resulted in a slight depletion of Cr and with a concurrent enrichment of Ni. This trend was observed for the majority of the M_6C precipitates examined. Other precipitate types examined did not show a statistically significant effect of dose on composition.

Effect of Dose-Rate

Referring to Table 2 and Figure 3, sample pairs E2/N3, E4/N8 and E14/N15 were irradiated at different dose-rates, but each pair was irradiated at similar temperature and dose. The difference in dose-rate is about a factor of two for each of these pairs. Both the calculated void swelling and measured bulk swelling are greater at the lower dose-rate for all three pairs of samples. Void diameters was also consistently greater at the lower dose-rate, while the overall void density, as mentioned previously, varied. The data show that the increases in void diameters and bulk swelling due to decreasing dose-rate were seen at all temperatures and doses examined. However, these differences did not translate into a observable difference in mechanical properties as reported in a companion paper in these proceeding[*10*]. Similar increases in swelling with decreasing dose-rate have been observed in 304 stainless steel[*11,12*].

Conclusions

EBR-II 12% CW 316 stainless steel hexagonal ducts were examined in the transmission electron microscope. Observed swelling reached a maximum of 1.6% at a dose of approximately 50 dpa. Sample densification occurred at lower doses (<30 dpa) and temperatures above approximately 400°C. Void diameters increased with both temperature and dose. Over the dose-rate ranges examined there appeared to be a moderate increase in swelling as dose-rate decreased, corresponding to an increase in void size. The dislocation structure was composed of faulted dislocation loops and dense networks of line dislocations. Loop density decreased and network density increased with increasing dose.

Substantial precipitation was observed in all of the hex ducts examined. Trends that could be discerned as a function of irradiation parameters include an increase in small-precipitate density as a function of dose as well as a moderate depletion of Cr and enrichment of Ni with dose. The majority of the identified precipitates were chromium-rich carbides.

Acknowledgements

Work supported by U.S. Department of Energy, under contract No. W-31-109-ENG-38 with the Japan Nuclear Cycle Development Institute.

References

[1] Maziasz, P. J., "Overview of Microstructural Evolution in Neutron-Irradiated Austenitic Stainless Steels," *Journal of Nuclear Materials*, Vol. 205, 1993, pp. 118-145.

[2] Zinkle, S. J., Maziasz, P. J. and Stoller, R. E., "Dose Dependence of the Microstructural Evolution in Neutron-Irradiated Austenitic Stainless Steel," *Journal of Nuclear Materials* Vol. 206, 1993, pp. 266-286.

[3] Garner, F. A., "Irradiation Performance of Cladding and Structural Steels in Liquid Metal Reactors," *Materials Science and Technology: Nuclear Materials,* ed. B. R. T. Frost, Vol. 10, 1993, New York: VCH. pp. 422-543.

[4] Allen, T. R., Cole, J. I., Tsai, H., Ukai, S., Mizuta, S. and Yoshitake, T., "The Effect of Low Dose Rate Irradiation on the Swelling of 12% Cold-Worked 316 Stainless Steel," *Proceedings of the Ninth International Symposium on Environmental Degradation of Materials in Nuclear Power Systems-Water Reactors*, August 1999, Newport Beach, CA 1999 in press.

[5] Mansfield, J., *Convergent Beam Electron Diffraction of Alloy Phases,* Adam Hilger Ltd., Bristol, England 1984.

[6] Yang, W. J. S., Brager, H. R. and Garner, F. A. "Radiation-induced Phase Development in AISI 316," *Conference Proceedings of Phase Stability During Irradiation,* J. R. Holland, L. K. Mansur, and D. I. Potter, Eds., The Metallurgical Society of AIME, Warrendale, PA, 1981, pp. 257-270.

[7] Brager, H.R. and Straalsund, J.L., "Defect Development in Neutron Irradiated Type 316 Stainless steel," *Journal of Nuclear Materials,* Vol. 46, 1973 pp. 124-158.

[8] Brager, H. R., and Garner, F. A., "Dependence of Void Formation on Phase Stability in Neutron-irradiated Type 316 Stainless Steel," *Effects of Radiation on Structural Materials, ASTM STP 683,* J. A. Sprague and D. Kramer, Eds., American Society for Testing and Materials, 1979, pp. 207-232.

[9] Straalsund, J. L. and Paxton, M. M., "Thermal Densification of Austenitic Stainless Steel," *Nuclear Technology*, Vol. 13, 1972, pp. 99-102.

[10] Cole, J. I. and Allen, T. R., "The Effect of Low Dose-Rate Irradiation on the Microstructure of 304 Stainless Steel," *Materials Research Society Symposium,* Vol. 540, Materials Research Society, Boston, 1999, pp. 433-438.

[11] Yoshitake, T., Donomae, T., Mizuta, S., Tsai, H., Strain, R. V., Allen, T. R. and Cole, J. I., "Tensile Properties of 12% Cold-Worked Type 316 Stainless Steel Irradiated in EBR-II under Lower-Dose-Rate Conditions to High Fluence," in these proceedings.

[12] Porter, D. L. and Garner, F. A., "Swelling of AISI Type 304L Stainless Steel in Response to Simultaneous Variations in Stress and Displacement Rate," *Effects of Radiation on Materials: Twelfth International Symposium, ASTM STP 870,* Garner, F. A. and Perrin, J. S. Eds., American Society for Testing and Materials, Philadelphia, 1985, pp. 212-220.

Todd. R. Allen,[1] James. I. Cole,[1] and Ed. A. Kenik[2]

Radiation-Induced Segregation and Void Swelling in 304 Stainless Steel

Reference: Allen, T. R., Cole, J. I., and Kenik, E. A., **"Radiation-Induced Segregation and Void Swelling in 304 Stainless Steel,"** *Effects of Radiation on Materials: 20th International Symposium, ASTM STP 1405,* S. T. Rosinski, M. L. Grossbeck, T. R. Allen, and A. S. Kumar, Eds., American Society for Testing and Materials, West Conshohocken, PA, 2001.

Abstract: Void swelling and radiation-induced segregation have been measured in 304 stainless steel. Samples were irradiated in the outer regions of the EBR-II reactor where displacement rates of 2.0×10^{-8} and 6.6×10^{-8} dpa/s are comparable to those in pressurized water reactor components. Samples were irradiated at temperatures from 371-390°C to total doses of up to 20 dpa. Void swelling reached a maximum of 2 % at 20 dpa. Nickel enrichment and chromium depletion to of 20 at% and 12 at% respectively were measured. Both void swelling and radiation-induced segregation were dependent on dose rate, increasing as the dose rate decreased. Grain boundary compositions were measured both near and in areas free of precipitates. The presence of a precipitate significantly changes the grain boundary compositions near the precipitate.

Keywords: void swelling, radiation-induced segregation, austenitic steels

Introduction

For operating nuclear reactors, many core components are constructed from 304 stainless steel [*1*]. Performance of 304 stainless steel under long-time, low dose rate irradiation is important to extended life operation. The dose rates for typical pressurized water reactor (PWR) components range from about 1.4×10^{-7} dpa/s for components in the inner core region to 1.4×10^{-10} dpa/s for the core barrel (these displacement rates are

[1]Argonne National Laboratory-West, PO Box 2528, Idaho Falls, ID 83403-2528.
[2]Metals and Ceramics Division, Oak Ridge National Laboratory, Oak Ridge, TN 37831.

calculated using fast neutron fluences from [*2*] converted to dpa using $0.7x10^{21}$ n/cm^2=1 dpa [*3*]), with thicker components reaching temperatures up to 400°C [*4*].

As part of the Experimental Breeder Reactor (EBR-II) reactor materials surveillance (SURV) program [*5-9*], test samples of 304 stainless steel were placed into EBR-II in 1965, with the intention of determining microstructural, corrosion, and mechanical property changes due to irradiation and thermal aging. The peak displacement rate for the materials in the SURV subassemblies was approximately $6.5x10^{-8}$ dpa/s. This displacement rate is about two orders of magnitude lower than used in a typical accelerated reactor materials test but within the range of displacement rates experienced by commercial light-water reactor (LWR) core components [*4*].

Following shutdown of the EBR-II reactor, surveillance test samples constructed of 304 stainless steel were retrieved to determine the effect of low dose rate irradiation on mechanical properties and microstructure. In addition to the SURV specimens, a large quantity of hexagonal duct material, made of 304 stainless steel with a thickness of approximately 1mm, was retrieved from the reflector region of the EBR-II. These EBR-II components were irradiated at temperatures greater than most PWR components. Even so, the material is valuable for understanding life extension problems in PWRs. Test samples can be irradiated in PWRs at operating temperatures, but to obtain the high doses expected at end of life, the samples are irradiated at a dose rate higher than experienced by actual core components. The EBR-II material is irradiated at a representative dose rate, but at a slightly elevated temperature. The scenario is illustrated in Figure 1. As the development of irradiated microstructure is both temperature and dose rate dependent, a high temperature or high dose rate test by itself is not entirely representative. An analysis that includes both high dose rate (operating PWR) and high temperature (EBR-II) data provides a greater opportunity to identify and understand mechanisms of radiation damage.

This work reports on the swelling, microchemical, and microstructural changes that occur in 304 stainless steel irradiated at temperatures from 371-390°C. Samples were irradiated at dose rates ranging from $2.0x10^{-8}$ to $6.6\ x10^{-8}$ dpa/s to doses of 20 dpa.

Experiment

Two analyses were performed on the 304 stainless steel samples: bulk swelling was measured using immersion density and microstructural and microchemical characterization was performed using transmission electron microscopy. Density samples were examined from six different surveillance (SURV) subassemblies irradiated in row 12 of EBR-II and from reflector hex cans irradiated in row 10 and 14 of EBR-II. SURV density specimens were prepared by sectioning an approximately 1 mm thick disk from the 1.11cm SURV hardness/corrosion specimens (see Figure 2). For the reflector subassemblies, three-quarter inch (1.9 cm.) diameter samples were punched at selected locations of the 1mm thick hex cans using a remotely controlled hydraulic punch.

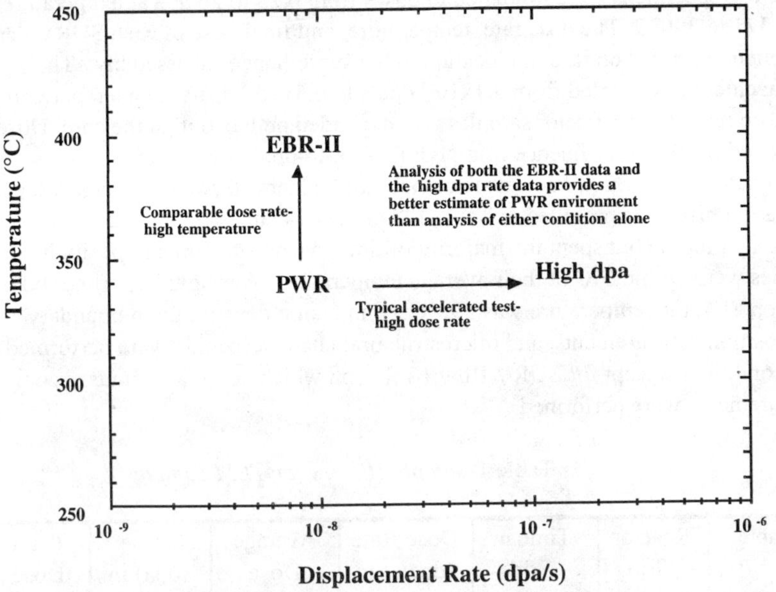

Figure 1-*Analyzing both higher temperature and higher dose rate samples provides a better understanding of actual PWR materials effects.*

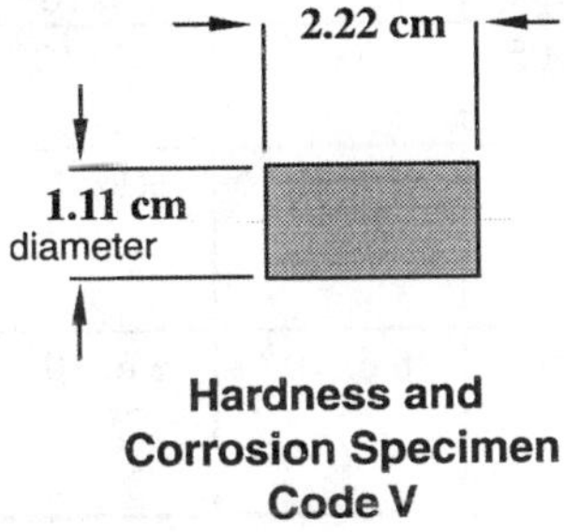

Figure 2-*Dimensions of Surveillance (SURV) hardness specimen*

Samples for microstructural and microchemical analysis were prepared from the density samples. The density disks were thinned in a hot cell to about 250 μm by mounting and grinding using standard metallographic sample preparation techniques. Three millimeter TEM sample blanks were then punched from these thinned disks using a special mechanical punch developed for hot cell use. TEM sample blanks were electropolished at -30°C using a 5% perchloric acid/95% methanol solution until electron transparent. Grain boundary compositions were measured using a Phillips CM200 FEG-STEM with an EMiSPEC X-ray analysis system. Other microstructural characterization was performed using a JEOL 2010 transmission electron microscope.

SURV specimens were irradiated to doses from 0.25 to 20 dpa at temperatures ranging from 371 to 390°C. The dose rate, temperature, and final dose of each SURV density specimen depended on its axial location in the surveillance subassembly. The displacement rates varied from 3.1×10^{-8} dpa/s to 6.5×10^{-8} dpa/s. The temperature, dose, and dose rate of the reflector samples also depended on location in the core. Doses were calculated from reactor fluence using NJOY flux-to-dpa cross sections.

For samples in which microstructure and microchemistry were analyzed, the irradiation history is provided in Table 1. The row 14 subassembly was moved once during its lifetime but spent the majority of time in row 14. While in position 14E10, samples were within 5°C of their average temperature. During the short period of time in position 8F4, the temperature was 390°C. Immersion density, grain boundary composition measurements, and microstructural characterization were performed on all four conditions except for SURV Row 12 K-5 on which no grain boundary composition measurements were performed.

Table 1-*Sample History for TEM Samples*

Sample	Reactor Grid Position	Time in Grid Position (MWD*)	Dose Rate (dpa/s) in Grid Position	Average Dose Rate (dpa/s)	Dose (dpa) in Grid Position	Total Dose (dpa)	Avg Temp (°C)
Refl Row 14	8F4 14E10	5951 348584	2.9×10^{-7} 1.5×10^{-8}	2.0×10^{-8}	2.4 7.6	10	379
Refl Row 10	10C2	187505	4.7×10^{-8}	4.7×10^{-8}	12.2	12.2	378
SURV Row 12 K-5	12E8	215110	3.1×10^{-8}	3.1×10^{-8}	8.9	8.9	375
SURV Row 12 K-6	12E8	215110	6.6×10^{-8}	6.6×10^{-8}	19.6	19.6	389

*MWD=Megawatt-Days

Density and microstructure samples were taken from three different lots of 304 stainless steel. Because the subassemblies were fabricated from different lots of steel, differences in composition may be critical. Compositions from the SURV samples were measured prior to insertion in the reactor. Unirradiated archive material exists for the row 10 subassembly, but not for the row 14 subassembly. Therefore, composition was determined from samples taken from both irradiated hex ducts using Inductively Coupled Plasma-Atomic Emission Spectroscopy (ICP-AES) for major elements and a LECO IR-412 Carbon Determinator for carbon. Table 2 provides the ICP-AES composition measurements for each hex duct and the vendor specification chemistry for the SURV samples. The row 10 subassembly has a significantly smaller concentration of Mo.

Table 2-*Typical Compositions for Microchemistry Samples*

Element	Row 10 Subassembly (at. %)	Row 12 SURV (at. %)	Row 14 Subassembly (at. %)
Cr	19.6	19.4	19.6
Ni	8.5	9.4	9.1
Fe	69.8	68.4	69.0
Mo	<0.02	0.12	0.07
Mn	0.82	0.90	1.01
C	0.4	0.4	0.5
Si	0.92	1.3	0.76

Results

In the results section, the density and void distribution measurements are presented first followed by the grain boundary segregation measurements.

Density and Void Distributions

In Figure 3, swelling is plotted as a function of dose for both the row 12 surveillance samples and the row 10 and 14 subassembly samples. The swelling of the surveillance samples increases continually with increasing dose, reaching 2% by 19.6 dpa. For samples irradiated to doses near 10 dpa, the swelling increases for samples taken farther from the core center (irradiated at lower displacement rate).

Figures 4 and 5 show micrographs and void size distributions for the row 10, 12, and 14 samples irradiated to a dose near 10 dpa. The swelling increases with decreasing dose rate (larger row numbers). The row 14 sample (lowest dose rate) has a higher density of large voids and a symmetrical distribution about the average size. The row 12 and 10 samples have a similar average void size, but the row 12 sample has a larger density of voids. The void size distribution in the row 10 and 12 samples have a positive skew with a tail of large void sizes up to about 30 nm.

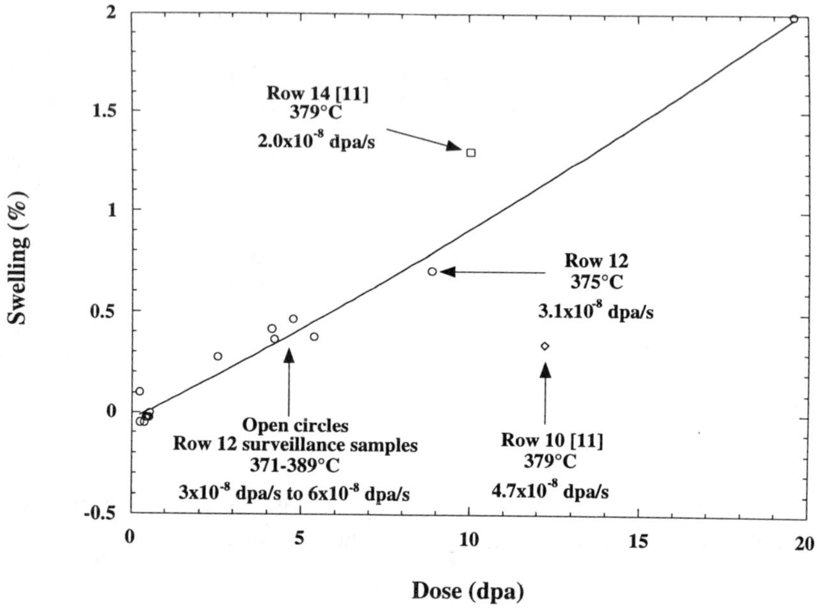

Figure 3-*Swelling as a function of dose for samples taken from Rows 10, 12, and 14.*

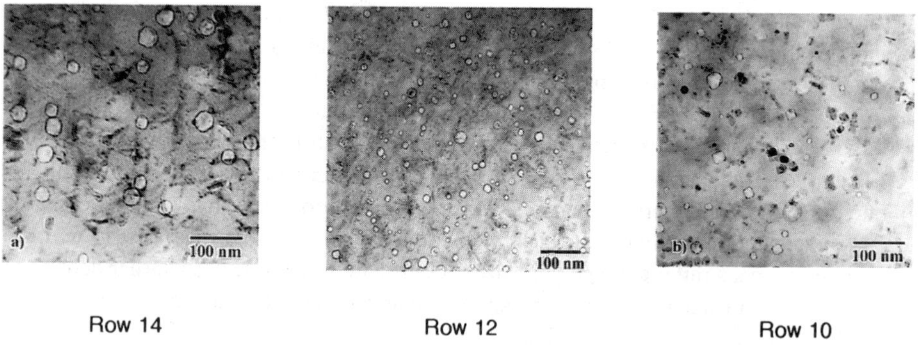

Figure 4-*Micrographs of voids in samples from rows 14, 12, and 10.*

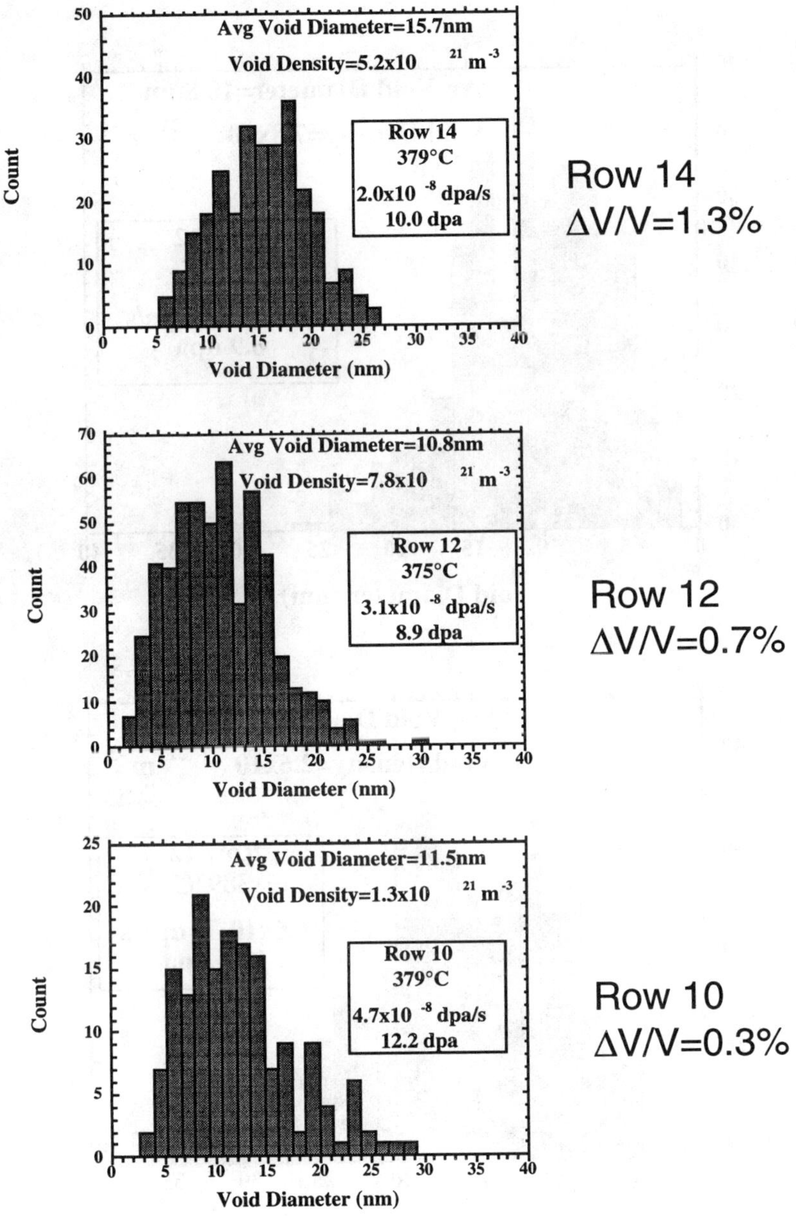

Figure 5-*Void size distributions for samples from row 10, 12, and 14. ΔV/V is the swelling measured using immersion density*

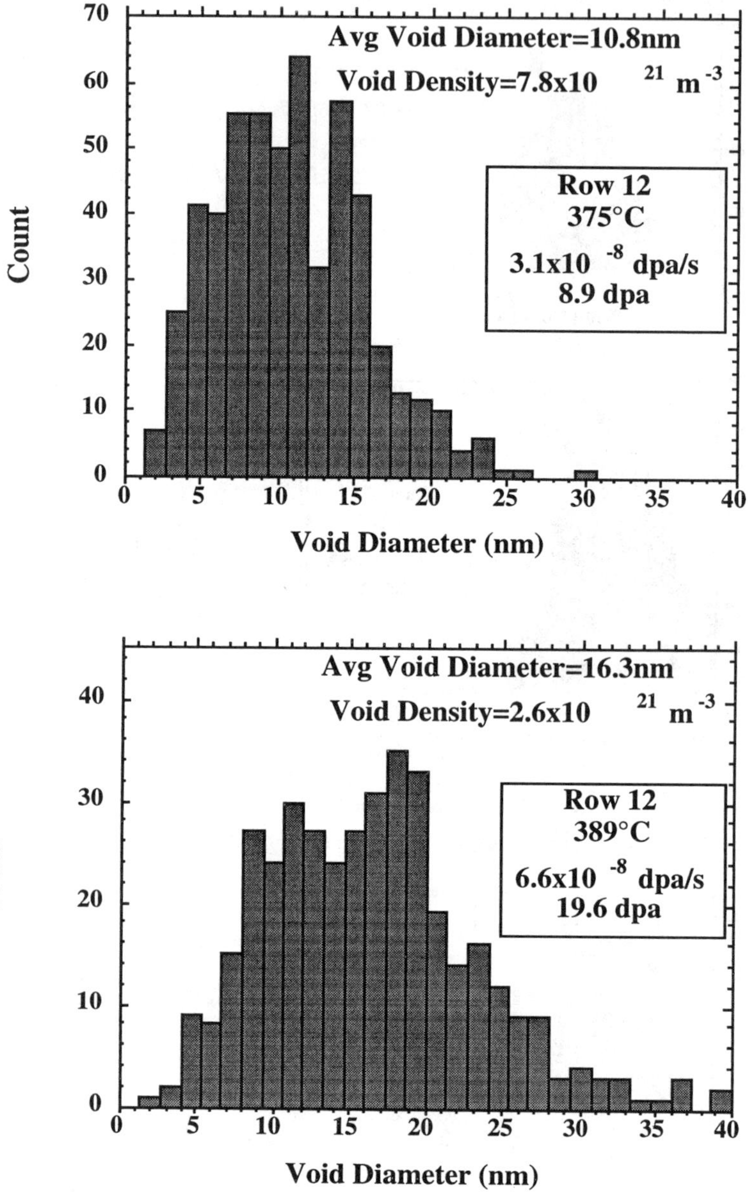

Figure 6-*Void size distributions for SURV samples in row 12.*

Void size distributions for the two SURV samples irradiated in row 12 are shown in Figure 6. These samples are from the same lot of steel and were irradiated in the same experimental subassembly, but at different axial heights. The higher dose sample was irradiated closer to the core centerline and experienced about twice the dose. The higher dose sample was also irradiated at a slightly higher temperature. For the sample irradiated to a larger dose and higher temperature, the average void size is larger and the density of voids is smaller. The void size distribution in the higher dose sample has a significant tail extending to larger sizes.

Radiation-induced Grain Boundary Segregation

Figure 7 indicates the change in grain boundary concentration for row 10 (12.2 dpa), row 12 (19.6 dpa), and row 14 (10.0 dpa) samples. As the displacement rate decreases, the amount of nickel enrichment and chromium depletion increases. Even at half the dose, the row 14 sample has greater radiation-induced segregation (RIS) of Ni and Cr than the row 12 sample.

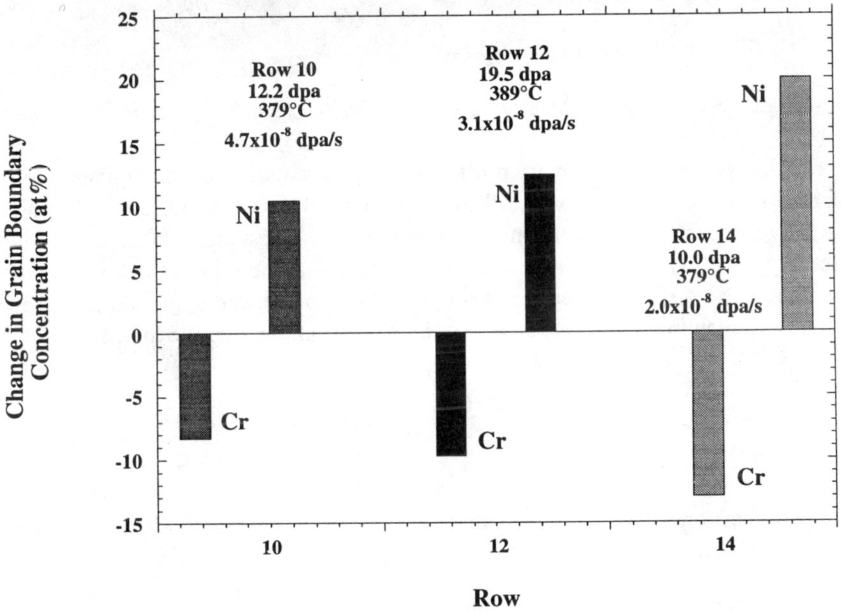

Figure 7-*Radiation-induced grain boundary segregation in samples from rows 10, 12, and 14.*

Figures 8 and 9 show the grain boundary segregation for two different boundaries in the row 12 high dose (19.6 dpa) sample. In Figure 8, the segregation was measured near a precipitate. The precipitate was Cr-enriched with a Cr concentration of 30 at%. Near the precipitate, the Cr concentration is lower than at grain boundary locations farther from the precipitate. Away from the boundary, the Cr concentration is 13-14 at%. The Ni concentration is very high nearest the precipitate and lower away from the precipitate.

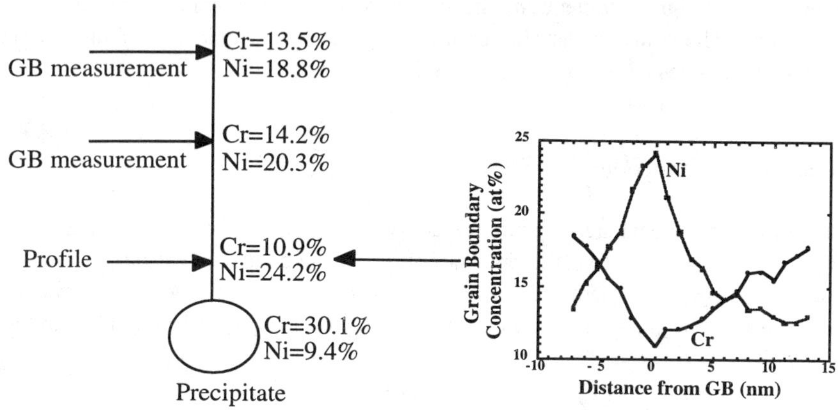

Grain Boundary near Precipitate

Figure 8-*Grain boundary segregation near a precipitate for a row 12 sample*

Figure 9 shows the grain boundary profiles and measurements on a precipitate free boundary in the row 12 high-dose (19.6 dpa) sample. The concentrations of Ni and Cr are relatively constant. Only the measurement in the thickest area of the sample shows noticeably lower Ni and higher Cr concentrations. Using the sum of the x-ray counts under the Fe, Ni, Cr, and Si peaks as a rough measure of thickness, the sample is roughly 1.5 times as thick in the thickest area (Ni-18.1%) as compared to the thinnest area (Ni=23.8%).

Discussion

The implications of irradiation conditions on the swelling and RIS in 304 stainless steel are discussed below.

Swelling

Figure 3 showed the swelling of the EBR-II surveillance material as a function of dose. The swelling reached a maximum of 2% at 19.6 dpa. Austenitic stainless steel alloys eventually reach a terminal swelling rate of about 1%/dpa [*10*]. At the maximum dose analyzed (19.6 dpa), the 304 SURV material is still early in the transient portion of the swelling versus dose with a swelling rate near 0.1%/dpa.

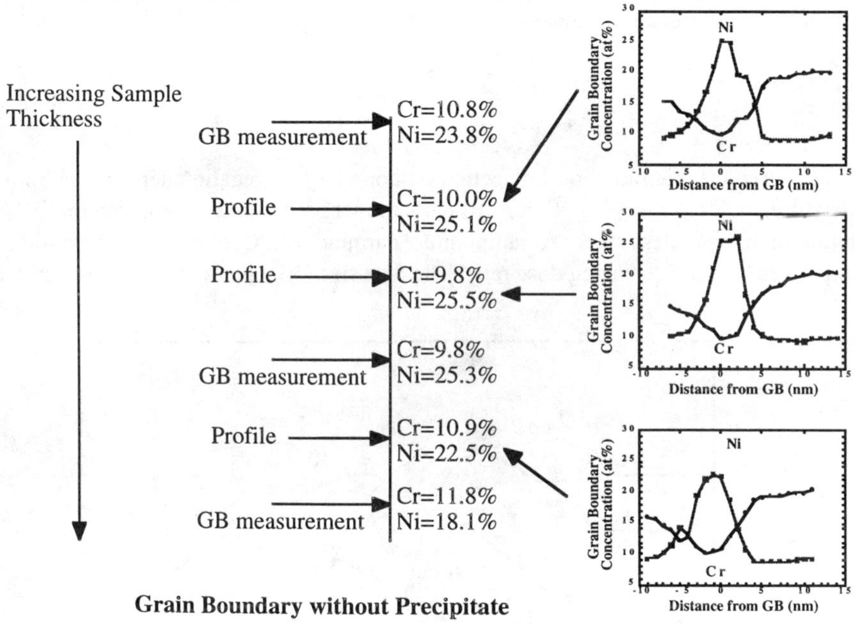

Figure 9-*Grain boundary segregation in a precipitate-free area for a row 12 sample*

The effect of dose rate on swelling can be seen by noting the relative swelling of the row 10, 12, and 14 samples. For samples irradiated near 375°C and 10 dpa, swelling increases with decreasing dose rate. The row 14 and row 10 density measurements have been previously reported by Garner et al. [*11*], who, in a much larger study, showed a clear dependence of swelling on dose rate in 304 stainless steel.

The differences in void size distribution between the two row 12 samples (Figure 6) are likely to be due to differences in both temperature and dose. Void size is known to increase with increasing dose and the larger dose sample has greater swelling. Increasing the temperature is known to increase the void size and decrease the density. The higher temperature sample has larger voids and a smaller density.

Radiation-induced Grain Boundary Segregation

The discussion of radiation-induced segregation is divided into two sections. First, the effect of dose rate on segregation is explained. Next, the effect of grain boundary precipitation on segregation is discussed.

Effect of Dose Rate

Theory predicts that radiation-induced grain boundary segregation depends on both temperature and dose rate [*12*]. Figure 10 plots model predictions for chromium depletion in 304 stainless steel. At temperatures around 375°C, chromium depletion should increase with decreasing dose rate. The data from Figure 7 confirms this trend.

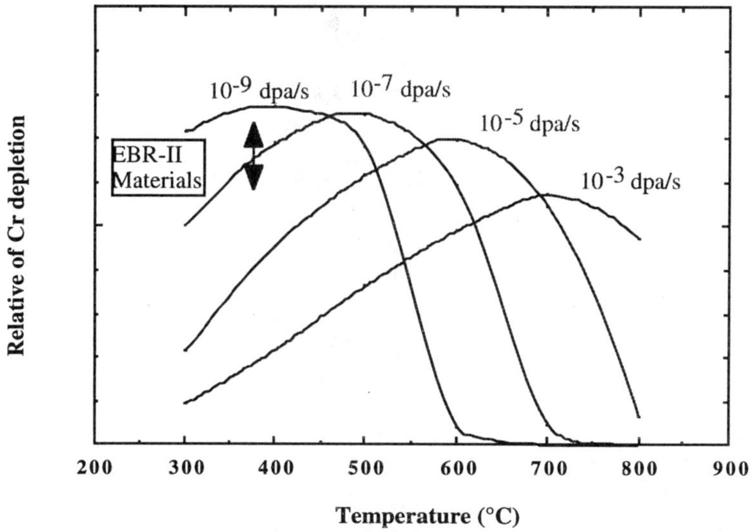

Figure 10-*Relative chromium depletion as a function of temperature and dose rate*

The row 12 and 14 samples have significantly greater molybdenum concentration than the row 10 sample. Segregation measurements in samples from Magnox reactor control rods (a 4 wt. % boron steel) irradiated at temperatures from 290 to 330°C and to doses from 0.04 to 0.35 dpa, indicated that increasing Mo content reduced the Cr depletion [*13*]. In the EBR-II samples, greater grain boundary chromium depletion occurs in the sample with greater bulk molybdenum concentration, contrary to the Magnox measurements. Examining Table 2, the differences in segregation do not correlate with any changes in bulk composition (there is no consistent change in bulk composition moving from row 10 out to row 14 as there is with segregation). Since the differences in chromium depletion in the EBR-II materials do not correlate with differences in bulk composition, they are likely to be caused by dose rate differences.

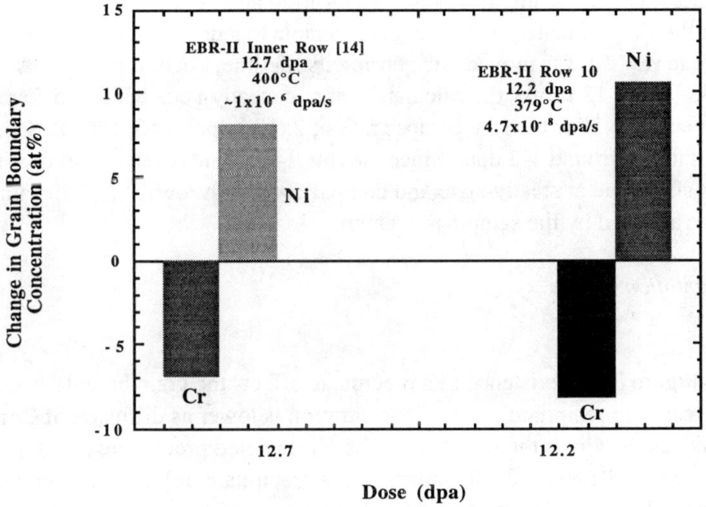

Figure 11-*Comparison of Grain Boundary Segregation between inner row and outer row samples irradiated in EBR-II.*

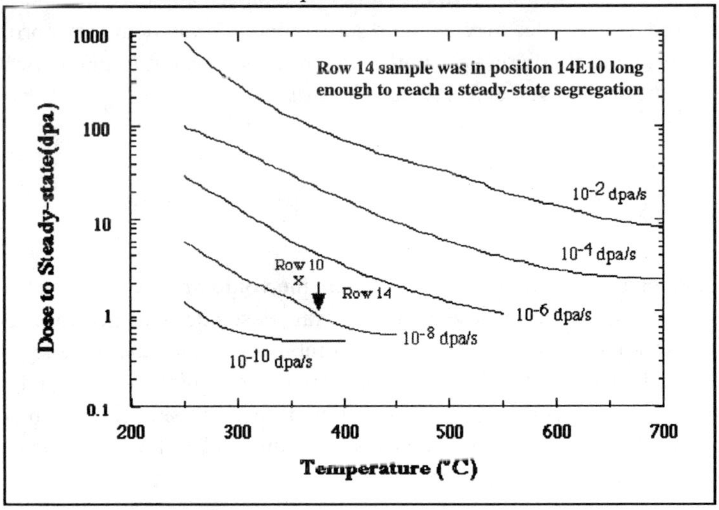

Figure 12-*Calculations indicate that even though the row 14 sample was moved once during its exposure, the RIS should have reached steady-state in the final position*

Dumbill measured segregation in an Fe-18Cr-15Ni alloy irradiated with neutrons at 400°C to 12.7 dpa in EBR-II [*14*]. A dose rate was not reported, but a typical experimental location in EBR-II had dose rates on the order of 10^{-6} dpa/s. The segregation measured by Dumbill is compared to the segregation measured in this study in Figure 11. The segregation is larger in the lower dose rate samples. In the work by Dumbill, the Fe-18Cr-15Ni alloy has a greater bulk nickel concentration and therefore would be expected to exhibit greater nickel enrichment [*15*] for similar irradiation conditions. The greater segregation in the EBR-II materials is likely attributable to a dose rate difference.

The sample in row 14 was moved once during its lifetime, obtaining 7.6 of its total 10 dpa in row 14. Figure 12 shows the calculated time to steady-state RIS for different dose rates and temperatures. For the row 14 dose rate of 2×10^{-8} dpa/s at 375°C, RIS should reach steady-state by around 1-2 dpa. Since the row 14 sample received 7.6 dpa in position 14, it should be at steady-state and comparisons with row 12 and 10 samples are not likely to be affected by the sample movement.

Effect of Precipitation

As seen in Figure 8, the existence of a precipitate affects the grain boundary segregation. Near the precipitate, the Cr concentration is lower as diffusion of Cr to the precipitate reduces the Cr on the boundary. The Ni-depleted precipitate also rejects Ni and locally increases the Ni on the boundary. This precipitate did not show an increased Si concentration. The composition near grain boundary precipitates has a noticeable effect on the localized attack of electropolished samples. The 5% perchloric acid/95% methanol polishing solution tends to etch away precipitates and the surrounding regions. Heavily precipitated samples may be more susceptible to grain boundary corrosion. The precipitation is due to irradiation as similar precipitation was not found in the same 304 stainless steel thermally aged at 371°C for 18 years.

RIS and Void Swelling

For irradiation temperatures near 375°C, both the void swelling and radiation-induced grain boundary segregation increase with decreasing dose rate. The decreased dose rate reduces the fraction of point defects lost to recombination, allowing more to participate in diffusional events that lead to RIS or void formation and growth. Although the enrichment of slow diffusing nickel and depletion of fast diffusing chromium at the void surface reduces the subsequent flux of vacancies to the void surface [*16*], the major factor affecting void growth is the displacement rate.

Conclusion

Both void swelling and radiation-induced segregation have been measured in 304 stainless steel irradiated in the EBR-II reactor. For temperatures near 375°C, a dose of 20 dpa causes 2% swelling. At 20 dpa, the swelling is still in the transient phase with a rate of about 0.1%/dpa. At temperatures near 375°C, both void swelling and RIS increase as the dose rate decreases. Precipitation on the grain boundary can significantly affect the radiation-induced grain boundary segregation profiles in the vicinity of the precipitate.

Acknowledgments

The authors gratefully acknowledge the efforts of M. E. Vaughn, J. P. Webb, E. L. Wood, and the staffs at the Hot Fuels Examination Facility and Analytical Laboratory at ANL-West. Thanks to K. N. Grimm and K. A. Bunde for performing dose calculations and to R. T. Jensen for performing the temperature calculations. Work supported under contract W-31-109-Eng-38 with the Department of Energy. Research at the Oak Ridge National Laboratory ShaRE User Facility was sponsored by the Division of Materials Sciences and Engineering, U. S. Department of Energy under contract DE-AC05-00OR22725 with UT-Batelle, LLC., and through the SHaRE Program under contract DE-AC05-76OR00033 with Oak Ridge Associated Universities. Work supported under contract W-31-109-Eng-38 with the Department of Energy.

References

[1] IAEA-TECDOC-1119

[2] Shah, V. N. and MacDonald, P. E., *Aging and Life Extension of Major Light Water Reactor Components*, Elsevier, Amsterdam, 1993.

[3] Andresen, P.L., Ford, F.P., Murphy, S.M., and Perks, J.M., *Proc. 4th Int. Symp. on Environmental Degradation of Materials in Nuclear Power Systems - Water Reactors,* Jekyll Island, GA, August 1989 NACE, Houston, 1990, p. 1.

[4] Van Duysen, J. C., Todeschini, P., and Zacharie, G., *Effects of Radiation on Materials: 16th International Symposium, ASTM STP 1175*, A. S. Kumar, D. S. Gelles, R. Nanstad, and E. A. Little, Eds., American Society for Testing and Materials, Philadelphia, 1993, p. 747.

[5] Greenberg, S., ANL-7624, Sept 1969.

[6] Greenberg, S., Strain, R. V., and Ebersole, E. R., ANL-7682, June 1970.

[7] Greenberg, S., Strain, R. V., and Ebersole, E. R., ANL-7937, Sept 1972.

[8] Ruther, W. E., Hayner, G. O., Carlson, B. G., Ebersole, E. R., and Allen T. R., "The EBR-II Materials-Surveillance Program: Part IV. Results of SURV-4 and SURV-6," ANL-98/3, January, 1998.

[9] Ruther, W. E., Staffon, J. D., Carlson, B. G., and Allen, T. R., "The EBR-II Materials-Surveillance Program: Part IV. Results of SURV-5," ANL-98/4, March, 1998.

[10] Garner, F. A., Materials Science and Technology: A Comprehensive Treatment, Volume 10a, Nuclear Materials, Part I, pp. 419-543, VCH Publishers 1994.

[11] G.M. Bond, B. H. Sencer, F.A. Garner, M.L. Hamilton, T.R. Allen and D.L. Porter., "Void Swelling of Annealed 304 Stainless Steel at ~370-385°C and PWR-Relevant Displacement Rates", *Proceedings of the Ninth International Symposium on Environmental Degradation of Materials in Nuclear Power Systems-Water Reactors*, S. Bruemmer, P. Ford, G,. Was Eds., Newport Beach, CA, TMS, Warrendale, PA, August 1999, p.1045.

[12] Was, G. S. and T. R. Allen, J. Nucl. Mater, 205, 332 (1993).

[13] Spellward, P., Walmsley, J., Scowen, R., Partridge, N., Stump, J., Corcoran, R., Gilmour, T., and Callen, V., *Proc. 8th Int. Symp. on Environmental Degradation of Materials in Nuclear Power Systems - Water Reactors,* Amelia Island, Fl, August 1997 (American Nuclear Society, LaGrange Park, IL), p. 734.

[14] Dumbill, S. and Williams, T. M., in*: Proceedings of the Conference on Materials for Nuclear Reactor Core Applications,* (BNES London 1987 Vol. 1) p. 119.

[15] Allen, T. R., Busby, J. T., Was, G. S., and Kenik, E. A., " On the Mechanism of Radiation-Induced Segregation in Austenitic Fe-Cr-Ni Alloys," *Journal of Nuclear Material* 255,44 (1998).

[16] Allen, T. R., Busby, J. T., Gan, J., Kenik, E. A., and Was, G.S., "The Correlation Between Swelling and Radiation-Induced Segregation in Fe-Cr-Ni Alloys," *Effects of Radiation on Materials: 19th International Symposium, ASTM STP 1366*, M. L. Hamilton, A. S. Kumar, S. T. Rosinski, and M. L. Grossbeck, Eds., American Society for Testing and Materials, West Conshohocken, PA, 2000, pp. 739-755.

Naoaki Akasaka[1], Ichiro Yamagata[1], and Shigeharu Ukai[1]

Effect of Irradiation Environment of Fast Reactor's Fuel Elements on Void Swelling in P, Ti-Modified 316 Stainless Steel

Reference: Akasaka, N., Yamagata, I., and Ukai, S., **"Effect of Irradiation Environment of Fast Reactor's Fuel Elements on Void Swelling in P, Ti-Modified 316 Stainless Steel,,"** *Effects of Radiation on Materials: 20th International Symposium, ASTM STP 1405,* S. T. Rosinski, M. L. Grossbeck, T. R. Allen, and A. S. Kumar, Eds., American Society for Testing and Materials, West Conshohocken, PA, 2001.

Abstract: The difference in swelling behavior of P, Ti-modified 316 stainless steel (PNC316) between sample irradiation using material irradiation vehicle (e.g. FFTF/MOTA) and fuel assembly irradiation in FBR was investigated. From the evaluation of the dose dependence in the swelling of PNC316, it is clearly shown that swelling of claddings irradiated by fuel assemblies occurred earlier than the swelling of samples irradiated by material irradiation vehicle in the range of irradiation temperatures from 723K to 823K. The microstructural evolution of the claddings irradiated by fuel assemblies, which is represented by void formation, should be mainly affected by the irradiation environments: irradiation temperature history, temperature gradient within the thickness of the claddings, and by the primary and secondary stresses within the cladding during irradiation.

Keywords: 316 stainless steel, neutron irradiation, fast reactor, irradiation history, swelling

Introduction

The core materials for fast breeder reactors require having good mechanical properties at high temperature, high dimensional stability against fast neutron exposure and resistances for other deterioration in the qualities of materials. The most important problem to be solved is the suppression of swelling under the neutron irradiation when the austenitic stainless steels are applied to the core material of fast breeder reactors. It is not too much to say that the key to success for long life fuel assembly of fast breeder reactors made of austenitic stainless steel is the improvement of the swelling resistance of the material.

The first resulting material in our material development program is a 20% cold-

[1] Assistant senior engineer, engineer, and senior researcher, respectively, O-arai Engineering Center, Japan Nuclear Cycle Development Institute, 4002 Narita, O-arai, Ibaraki 311-1393, Japan.

worked P,Ti-modified 316 stainless steel which is optimized within the chemical composition range of the Japanese Industrial Standard (JIS) SUS316 and has a typical composition of Fe-16Cr-14Ni-0.05C-2.5Mo-0.8Si-0.025P-0.004B-0.1Ti-0.1Nb [*1*]. This material has been designated PNC316. The swelling data of this material have been taken until now using sample irradiation vehicle (e.g. FFTF/MOTA) [*2*] and fuel assemblies [*3*]. In recent years the swelling data were taken from high dose irradiated fuel cladding. These data show that the swelling behavior of claddings irradiated by fuel assembly is not like the swelling behavior of samples of cladding irradiated by sample irradiation vehicles.

Recently, many investigations of the effect of irradiation temperature history on microstructure change have been energetically carried out [*e.g. 4-6*] since this effect was pointed out [*e.g. 7,8*]. The irradiation conditions of fuel assembly of fast breeder reactor, such as irradiation temperature, stress state of material and so on, change continuously with burn up of MOX fuel. But the influences of changing these irradiation conditions on swelling behavior are not sufficiently clear. Seran et al. reported that the temperature and swelling gradient in cladding wall led to enhanced swelling [*9*]. Therefore, the assessment of the influence of peculiar irradiation conditions of fuel assembly on irradiation behavior is very important for designing and life estimation of fuel assembly.

In this study, the reason for enhanced swelling on fuel cladding of fast breeder reactor is discussed based on the differences of irradiation environment between the sample irradiation vehicle and fuel assembly.

Experimental Procedure

Cladding samples and fuel claddings used in this experiment are made of PNC316. The chemical composition, solution treatment condition, and cold-working level of claddings are listed in Table 1. Dimension of the claddings is the MONJU type of 6.5 mm in diameter, 0.47 mm in thickness. The samples were irradiated in FFTF/MOTA [*10*] and Phenix P-4 which were sample irradiation vehicles. The irradiation temperature ranged from 405 to 670°C, and the maximum neutron dose was as high as 205 dpa at 405°C. The fuel pins were assembled in two sets of JOYO C-type fuel irradiation vehicles [*11*] and a fuel irradiation assembly in FFTF. The maximum neutron dose was as high as about 105 dpa at core midplane.

Table 2 summarizes the irradiation vehicles and swelling evaluation methods. The samples that were irradiated in FFTF/MOTA were periodically discharged and diameter measurements were made prior to reinsertion into the reactor. The swelling was obtained from the diameter change assuming isotropic deformation. The samples, which completed irradiation in both of Phenix P-4 and FFTF/MOTA up to target neutron dose, were also measured for density. The fuel pins in the fuel assembly, which completed irradiation in Joyo and FFTF up to target burnup in MOX fuel, were cut into about 20 mm lengths, and these specimens were also measured for density after defueling. Some specimens that were irradiated with both sample irradiation vehicles and fuel irradiation assemblies were analyzed for their microstructures using transmission electron microscopy (TEM).

Table 1 - *Chemical composition, solution treatment condition, and cold-working level of claddings.*

Lot.	Chemical Composition (wt%)							
	C	P	Si	Mn	Mo	B	Ti	Nb+Ta
K554	0.052	0.028	0.82	1.83	2.49	0.0031	0.080	0.079
S554	0.047	0.030	0.92	1.74	2.50	0.0039	0.097	0.070
S553	0.048	0.031	0.93	1.78	2.51	0.0044	0.098	0.073
60MK	0.054	0.028	0.78	1.72	2.35	0.0039	0.078	0.080
60MS1	0.057	0.026	0.79	1.85	2.54	0.0036	0.075	.0.095

Lot.	Chemical Composition (wt%)			Solution Treatment Condition	Cold-Working Level (%)
	Ni	Cr	Fe		
K554	13.84	16.52	bal.	1080°C x 2min.	19
S554	13.74	16.50	bal.	1085°C x 1min.	19
S553	13.81	16.60	bal.	1085°C x 1min.	19
60MK	13.45	16.22	bal.	1080°C x 2min.	18
60MS1	13.72	16.33	bal.	1095°C x 1min.	20

Table 2 - *The summary of irradiation test vehicles including irradiation condition and swelling evaluation Methods.*

Irradiation Vehicle	Lot. of Irradiated Cladding	Irradiation Condition		Swelling Evaluation Methods	Remark
		Temprature Range (°C)	Maximum Neutron Dose (dpa)		
Phenix/P-4	K554 S554	420-650	95	Diameter Density	Sample
FFTF/MOTA	K554 S554	405-670	205	Diameter Density	Sample
Joyo/C2M	S553	380-620*	44	Density	Fuel
Joyo/C3M	S553	380-590*	89	Density	Fuel
FFTF/MFA-1	60MK 60MS1	390-620*	107	Density	Fuel

*) Average temperature range of fuel pin during irradiation period

Result

Swelling Behavior

Figure 1(a) shows dose dependence on swelling of samples irradiated by sample irradiation vehicles. Swelling of these samples occurred slightly up to 100 dpa at the nominal temperature range of cladding of fast breeder reactor. This result shows that the cladding tube made of PNC316 has a high swelling resistance in sample irradiation. Over 100 dpa of neutron dose, swelling occurred distinctly at lower irradiation temperature range around 400°C, but very low swelling occurred continuously at higher irradiation temperature ranges around 550°C.

Figure 1(b) shows dose dependence on swelling of cladding specimens irradiated by fuel irradiation assembly. It is a striking difference, compared with the results of swelling behavior of samples irradiated by sample irradiation vehicles, that swelling of fuel claddings at the irradiation temperature ranged from 450°C to 550°C occurred at lower neutron dose and that the swelling rate seems higher. At the irradiation temperature except from 450°C to 550°C, it is not clear that there is a difference of swelling behavior of claddings compared with samples, because neutron dose of claddings irradiated by fuel assembly at this temperature ranges were relatively low at this temperature.

Microstructural Observation

For the purpose of investigating the difference of swelling behavior between sample irradiation and fuel assembly irradiation, microstructural observation of cladding irradiated on both conditions was carried out. The irradiation temperature of observed specimens was selected around 500°C, because there is distinct difference in swelling behavior between samples and fuel claddings. Figures 2 and 3 show microstructures of samples irradiated by sample irradiation vehicles and fuel claddings irradiated by fuel assembly, respectively. These microstructures were observed on thin foil specimens that were sectioned from the midwall region in the direction of

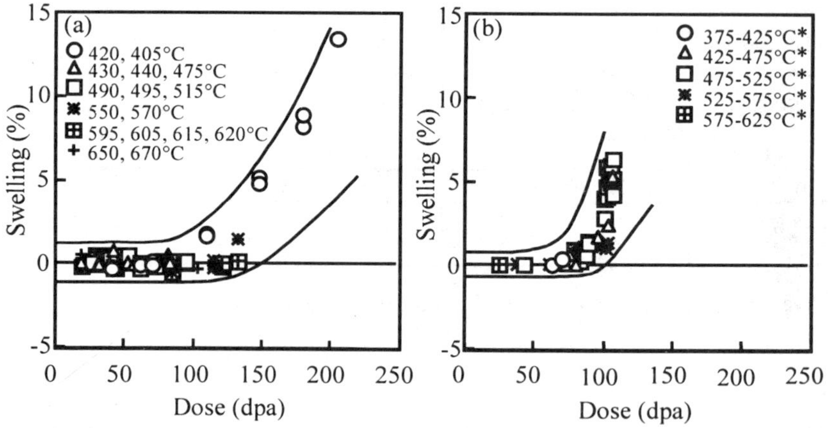

Figure 1 - *Dose dependence of swelling of irradiated claddings at different temperature ranges; (a) Results on sample irradiation test, and (b) Results on fuel assembly irradiation (* Based on average temperature during irradiation period at midwall of claddings).*

thickness. And the irradiation temperature of fuel claddings corresponds to that of the cladding midwall averaged over the entire irradiation period, which was predicted by means of a code calculation based on the operation histories of reactors.

As shown in Figure 2, there are only a few voids, and there are some precipitates in the matrix of the samples. These precipitates were classified into two types of shapes. The first type of precipitate shape is a blocky type. The size and density of this precipitate were relatively large and low, respectively. These precipitates were identified as M_6C by image of electron diffraction pattern and result of composition analysis using EDX. The other type of precipitate shape is a needle

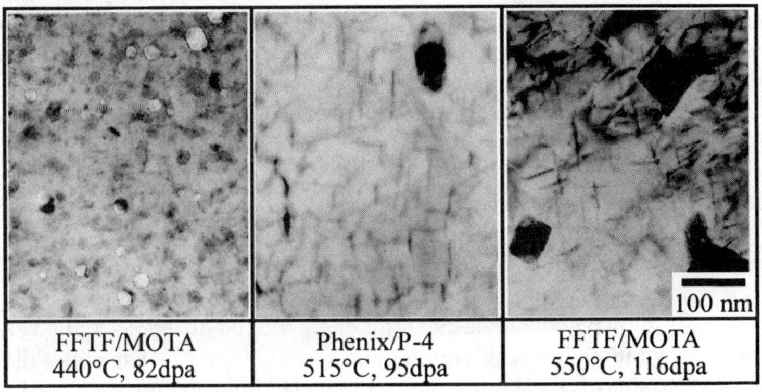

Figure 2 - *Typical microstructures of PNC316 cladding irradiated with sample irradiation vehicles.*

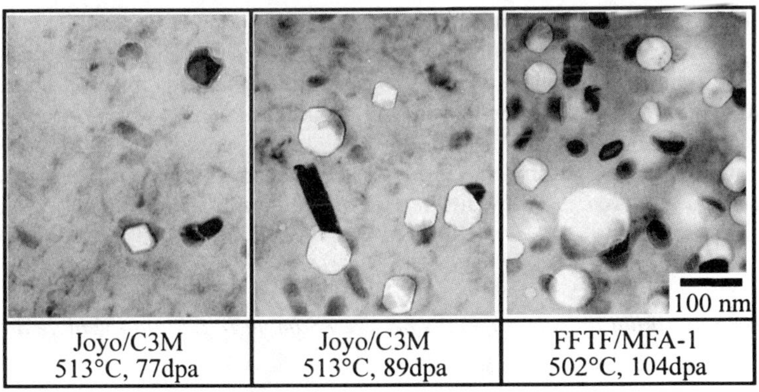

Figure 3 - *Typical microstructures of PNC316 cladding irradiated with fuel assembly (irradiation temperatures are average values during irradiation period).*

type. The density of this precipitate was relatively high, and the formation of this precipitate was remarkable in the matrix of the samples that were irradiated at around 550°C. This needle-like precipitate seems to be phosphide of the type (Fe,Ti)P [*12,13*].

On the other hand, void formation of the fuel claddings occurred at lower neutron dose comparing with samples, as shown in Figure 3. It is clear from the evolution of microstructure that the swelling increases with increasing neutron dose as shown in Figure 1(b) because of increasing void density. And formation of blocky precipitates was recognized in the matrix. This precipitate was identified as M_6C and/or G-phase from electron diffraction pattern and EDX experiment as the same method as identification of precipitate in samples. The needle-like phosphides were also formed in matrix. But it is striking difference comparing with samples that the density and size of blocky precipitate are higher and smaller, respectively. The phosphides, which were formed in matrix at lower neutron dose, were dissolved at higher neutron dose indicating that the phosphides can become unstable under certain conditions, consistent with other such observations [*12*].

Discussion

Microstructural Evolution

As already shown above, it is clear that there are the differences of swelling and microstructural evolution between samples and fuel claddings. The major different points of microstructure between sample and fuel claddings are not only void structure but also precipitate formation and stability. In particular, formation of G-phase and Laves-phase occurred in fuel claddings at lower neutron dose comparing with samples, as shown in Figure 4. The size and density of G-phase and/or M_6C in fuel cladding are finer and higher than samples, respectively. Dissolution of needle-like phosphide also occurred in fuel.

G-phase and needle-like phosphide are radiation-induced precipitates [*14*]. The changing of formation behaviors of radiation-induced precipitates indicates the difference of point defect behavior between fuel claddings and samples. These precipitates are strongly related with swelling behavior [*15-18*]. Therefore, not only point defect behavior but also radiation-induced solute segregation and precipitate behavior affect void formation behavior. It is considered that these differences of irradiation behavior are caused by differences of irradiation environment between fuel claddings and samples.

Differences of Irradiation Environments between Sample Irradiation Vehicle and Fuel Assembly

Usually, samples such as the open tube of cladding has been enclosed in a capsule made of stainless steel, and the capsule has been set in the sample irradiation vehicle like FFTF/MOTA. The irradiation temperature of the sample is determined by heat transfer to coolant from sample and capsule that are uniformly generated heat

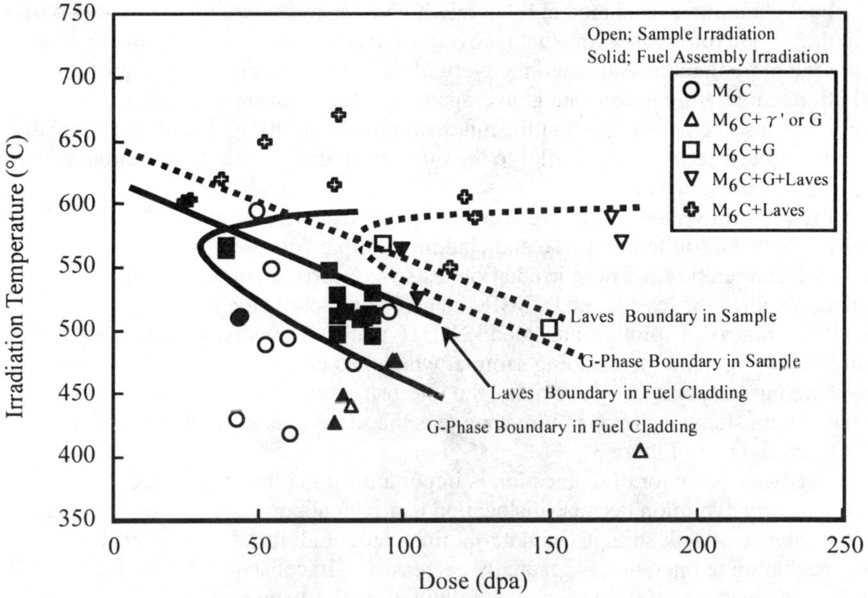

Figure 4 - *Difference of precipitate boundary of G-phase and Laves between sample irradiation condition and fuel assembly irradiation condition*

by gamma-ray. Therefore, it seems that the temperature of samples during irradiation has been kept relatively constant if gamma-ray flux has not remarkably changed during reactor operation. Specially, FFTF/MOTA had control machinery of irradiation temperature with thermocouples and insulation gas controller. The inside of the capsule has been filled with static sodium.

On the other hand, the irradiation environment of the fuel assembly has been changed dynamically compared with sample irradiation vehicle. With burning MOX fuel, the irradiation temperature of cladding at one axial position has been decreasing little by little, and there is some thermal gradient to coolant from MOX. The stress of the cladding has also changed every hour because irradiation creep relaxes the stress, which is caused by thermal condition, release of FP gas from MOX, and so on. There is also flowing of sodium outside the cladding.

The Influence of History of Irradiation Temperature in Fuel Cladding

Figure 5 shows the history of irradiation temperature of midwall of fuel claddings irradiated in Joyo and FFTF at the center plane of reactor core. The difference of temperature between maximum and minimum temperature was about 30°C during irradiation period because the temperature of cladding was gradually decreased with burning of MOX fuel.

Fuel cladding is irradiated at relatively high temperature in the initial period of irradiation. Therefore, it seems that recovery of dislocation of cladding in this case progressed more than the estimated recovery of dislocation of cladding where the irradiation temperature is constant at average temperature during irradiation period. Then, there is some possibility that the microstructural evolution of cladding irradiated with temperature decreases is similar to the microstructural evolution of sample which has low dislocation density.

Figure 6 shows that the microstructures of neutron irradiated solution annealed PNC316, 20% cold-rolled PNC316 and cladding sample with step-wise change of irradiation temperature. These irradiation tests were carried out with sample irradiation vehicle of Joyo named CMIR. There are a lot of fine precipitates and voids in the matrix of solution-annealed PNC316 that had low dislocation density at starting irradiation and the cladding sample, which had also low dislocation density after changing temperature. Specially, the microstructure of cladding samples irradiated with step-wise temperature change is the same as microstructures of fuel cladding as shown in Figure 3.

Recovery behavior of dislocation is important for the investigation of microstructural evolution because dislocation is principal point defect sink. It seems that the balance of sink strength in material fluctuates at all times during irradiation when irradiation temperature is gradually decreased. In consequence, the behavior of radiation-induced segregation fluctuates following with fluctuation of supersaturation of point defect. It seems that the reason of formation of fine G-phase and M_6C in irradiated cladding with irradiation temperature change is the changing of behavior of radiation-induced segregation. It is thought that the formation of void on PNC316 is

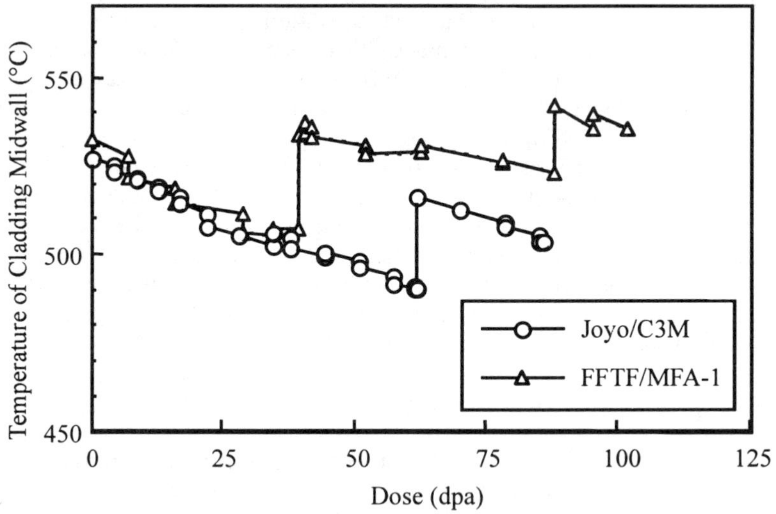

Figure 5 - *Temperature history in midwall of cladding at axial position of core center plane during irradiation in fuel assembly.*

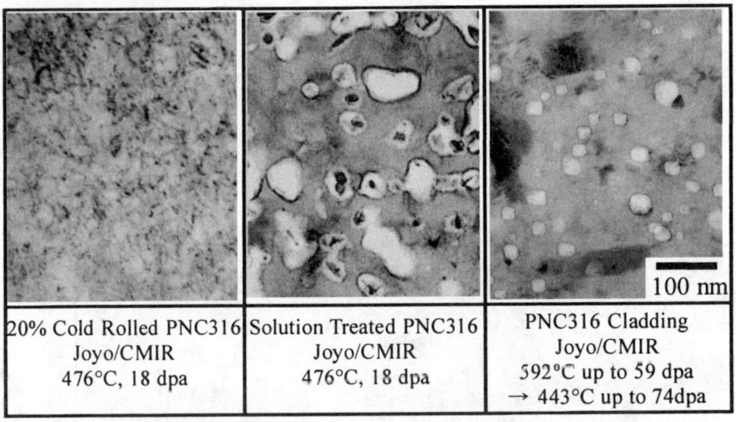

Figure 6 - *Typical microstructures of 20% cold rolled PNC316, solution treated PNC316 and temperature changed PNC316 cladding irradiated with sample irradiation vehicle in Joyo*

relevant to changing of behavior of radiation-induced segregation. Because controlled adding of minor elements cause suppression of swelling of this material [*1*]. Therefore the decrease of irradiation temperature as same as temperature history of fuel assembly seems to affect the swelling behavior in PNC316.

Influences of Other Irradiation Environments

Temperature Gradient in Cladding Wall – The temperature gradient is generated in cladding wall during reactor operation because there is heat flux to sodium coolant from MOX fuel. The temperature gradient was estimated with heat transfer calculation code that the temperature of inner surface of cladding was higher about 40°C than outer surface. Because there is a temperature dependence on void formation, the swelling gradient occurs in the cladding wall depends on the temperature gradient. This swelling gradient leads to secondary stress in the cladding wall, and this stress is possible to enhance swelling. Figure 7 shows microstructure of fuel cladding, which sectioned three part of the wall in direction of thickness. The void formation was not uniform and occurred at side of outer surface of cladding at first. The next region that formed void was side of inner surface. It is reported that secondary stress, which was occurred by nonuniform structure of void in cladding, enhances swelling [*9,19*]. But this mechanism for swelling enhancing needs nonuniform swelling at first. Therefore, the secondary stress seems not to be the origin for swelling occurring in the fuel claddings, because the swelling of the sample, which was irradiated with the sample irradiation vehicle, dose not occur at lower than 100 dpa.

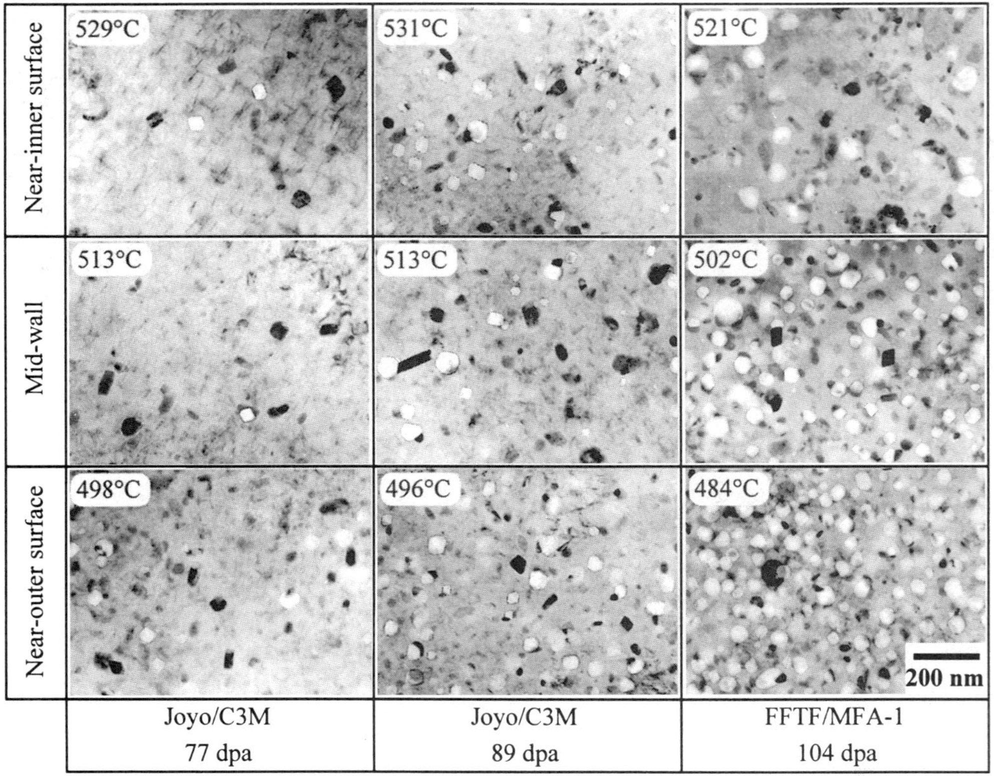

Figure 7 - *Typical microstructure of fuel cladding which sectioned three part of wall in direction of thickness (each temperature indicated in photographes are average temperature at each regions during irradiation pe*

Influence of Other Stress – It seems that other stress that are expected to cause enhanced swelling on the fuel cladding are two stresses, which are residual stress caused by final cold-working of cladding and increasing gas pressure in fuel pin caused by releasing FP gas from MOX fuel. The value of hoop stress on the side of outer surface that was caused by residual stress of cladding was estimated at about 200 MPa. Generally, swelling is enhanced under applying stress compared with swelling under free stress [*20,21*]. But it seems that the residual stress does not mainly lead to enhance swelling in fuel cladding. If the residual stress leads to enhanced swelling, there is no difference in swelling behavior between samples and fuel claddings. The residual stress that is caused by cold-working is generated in all cladding, which is not only fuel claddings but also samples.

Another primary hoop stress that is caused by increasing gas pressure in fuel pin was estimated about 100 MPa at the end of life of fuel assembly by gas analysis. On the other hand, swelling of the pressurized tube, which simulated the influence of primary stress on swelling, were only slightly enhanced compared with the swelling of open tube, as shown in Figure 8. Therefore, it is considered that the primary stresses caused in fuel cladding are not also mainly factor for enhancing swelling.

Influence of Sodium – The outer surface of fuel cladding is exposed to sodium coolant that has a relatively large flow rate. Under this condition, solute elements in

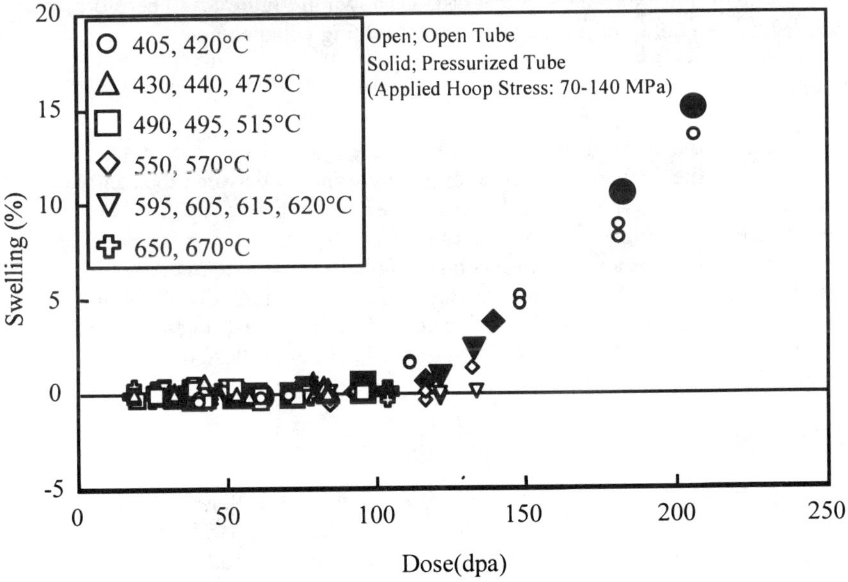

Figure 8 - *Swelling behavior of presurrized tube comparing with swelling of open tube*

the cladding material have flowed out to coolant from outer surface of cladding. Then, it is possible that decreasing solute element concentration in PNC316 affects swelling behavior, because the swelling of PNC316 is suppressed by addition of minor element. On the other hand, it is reported that thickness of lacked zone of solute elements in fuel cladding made of Type 316 stainless steel which irradiated in fuel assembly are about 5μm, and that the thickness of lacked zone tend to saturate depending on irradiation time [*22*]. Therefore, this influence seems to be not major factor for swelling enhancement.

Influence of Helium – Generally, helium produced in stainless steel by transmutation of element has influence on swelling behavior [*14*]. It is also reported that helium concentration inside the inner surface of the cladding is two times higher than inside the outer surface of the cladding because of injection occurring of recoil helium from MOX fuel [*23*]. But the result of helium analysis in cladding irradiated in Joyo/C3M, which were sectioned three region in direction of wall thickness, is that helium concentrations were similar between inside the inner surface and inside the outer surface, namely, about 60 appm inside the inner surface and midwall, and about 55 appm inside the outer surface. Moreover, if the void formation behavior is affected by helium injected to inner surface, it seems that void formation begins at inner surface of cladding. But the void formation of fuel claddings began at outer surface, and formation of void at midwall of cladding which has helium content as higher as in side of inner surface began at last, as shown in Figure 8. Therefore, this influence seems also not to be a major factor on swelling enhancement.

Conclusion

The reason for enhanced swelling of fuel cladding for fast breeder reactor is discussed based on the differences of irradiation environment between the sample irradiation vehicle and the fuel assembly. There are some different points between the sample irradiation vehicle and the fuel assembly. It seems that the most important influence on swelling enhancement in fuel cladding is irradiation temperature history during lifetime, which the irradiation temperature decreased gradually attending with burning of MOX fuel. The individual influences of other irradiation environments are not clear, but there may be some of the synergistic effect of these influences for swelling enhancement.

Acknowledgment

The authors would like to thank Ms. T. Donomae and Mr. S. Onose for useful discussions concerning this study. We would also like to thank Mr. M. Sekine for his assistance in conducting this experiment and discussions of this study.

References

[1] Itaki, T., Yuhara, S., Shibahara, I., Kubota, H., Itoh, M., and Nomura, S., "Effects of Alloying Elements on Mechanical and Swelling Properties of Type 316

Stainless Steel and Advanced Materials for Fast Reactor Application," *Materials for Nuclear Reactor Core Application*, Vol. 1, BNES, London, 1987, pp. 203-210.

[2] Shibahara, I., Akasaka, N., Onose, S., Okada, H., and Ukai, S., "Swelling of Advanced Austenitic Stainless Steels Developed for the Environment of Heavy Neutron Exposure," *Journal of Nuclear Materials*, Vols. 212-215, 1994, pp. 487-491.

[3] Ukai, S., Yoshitake, T., Akasaka, N., Donomae, T., Katsuyama, K., Mitsugi, T., and Asaga, T., "Irradiation Performance of FBR Monju-Type Fuel with Modified Type 316 Stainless Steel at High Burnup," *Transactions of the American Nuclear Society*, Vol. 79, 1998, pp. 115.

[4] Muroga, T., Ohnuki, S., Garner, F. A., and Zinkle, S. J., "The Influence of Neutron Spectrum and Irradiation History on Microstructural Evolution in Fusion Structural Materials," *Journal of Nuclear Materials*, Vols. 258-263, 1998, pp. 130-139.

[5] Akasaka, N., Hattori, K., Onose, S., and Ukai, S., "Effect of Temperature Change on Void Swelling in P, Ti-modified 316 Stainless Steel," *Journal of Nuclear Materials*, Vols. 271&272, 1999, pp. 370-375.

[6] Nita, N., Fukumoto, K., Kimura, A., and Matsui, H., "Effect of Temperature Change on Microstructural Evolution of Vanadium Alloys under Neutron Irradiation in JMTR," *Journal of Nuclear Materials*, Vols. 271&272, 1999, pp. 365-369.

[7] Bates, J. F., and Gelles, D. S., "Increased Swelling in 20% Cold Worked 316 Stainless Steel due to Inreactor Temperature Reductions" *Journal of Nuclear Materials*, vol. 71, 1978, pp. 365-367.

[8] Foster, J. P., and Boltax, A., "Effects of Temperature Changes on the Swelling Behavior of 20% Cold-worked AISI Type 316 Stainless Steel" *Nuclear Technology*, Vol. 47, 1980, pp. 181-188.

[9] Seran, J. –L., Touron, H., Maillard, A., Dubuisson, P., Hugot, J. P., Le Boulbin, E., Blanchard, P., and Pelletier, M., "The Swelling Behavior of Titanium-Stabilized Austenitic Steels Used as Structural Materials of Fissile Subassemblies in Phénix," *Effects of Radiation on Materials: 14th International Symposium (Volume II), ASTM STP 1046*, N. H. Packan, R. E. Stoller, and A. S. Kumar, Eds., American Society for Testing and Materials, Philadelphia, 1990, pp. 736-752.

[10] Puigh, R. J. and Schenter, R. E., "In-Reactor Creep Rupture Experiment in the Materials Open Test Assembly (MOTA)," *Effects of Radiation on Materials: 12th International Symposium, ASTM STP 870*, F. A. Garner and J. S. Perrin, Eds., American Society for Testing and Materials, Philadelphia, 1985, pp. 795-802.

[11] Asakura, F., Mizoo, M., Kinjo, K., Teranuma, S., Nakamura, T., Enokido, Y., Nakashima, Y., and Aoyama, T., "The Experience of Experimental Fast Reactor JOYO Operation and Maintenance," *International Conference on Fast Reactor and Related Fuel Cycles*, Vol. 1, Kyoto, Japan, October, 1991, pp. 2.7.

[12] Maziasz, P.J., "Formation and Stability of Radiation-Induced Phase in Neutron-

Irradiated Austenitic and Ferritic Steels," *Journal of Nuclear Materials*, Vol. 169, 1989, pp. 95-115.

[13] Watanabe, H., Muroga, T., and Yoshida, N., "The Effect of Phosphorus on the Microstructure of an Fe-16Cr-17Ni Austenitic Ternary Induced by Fast Neutron Irradiations," *Effects of Radiation on Materials: 16th International Symposium, ASTM STP 1175*, Arvind S. Kumar, David S. Gelles, Randy K. Nanstad, and Edward A. Little, Eds., American Society for Testing and Materials, Philadelphia, 1993, pp. 951-963.

[14] Maziasz, P. J., "Overview of Microstructural Evolution in Neutron-Irradiated Austenitic Stainless Steels," *Journal of Nuclear Materials*, Vol. 205, 1993, pp.118-145.

[15] Shibahara, I., Ukai, S., Onose, S., and Shikakura, S., "Irradiation Performance of Modified 316 stainless steel for Monju Fuel," *Journal of Nuclear Materials*, Vol. 204, 1993, pp. 131-140.

[16] Mansur, L. K., "Theoretical Evaluation of a Mechanism of Precipitate-Enhanced Cavity Swelling during Irradiation," *Philosophical Magazine A*, Vol. 44, 1981, pp. 867-877.

[17] Brager, H. R., and Garner, F. A., "Microchemical Evolution of Neutron-Irradiated Stainless Steel," *Effects of Radiation on Materials: 10th International Symposium, ASTM STP 725*, D. Kramer, H. R. Brager, and J. S. Perrin, Eds., American Society for Testing and Materials, Philadelphia, 1981, pp. 470-483.

[18] Coghlan, W. A., and Garner, F. A., "Effect of Nickel Content on the Minimum Critical Void Radius in Ternary Austenitic Alloys," *Radiation-Induced Changes in Microstructure: 13th International Symposium, Vol. I, ASTM STP 955*, F. A. Garner, N. H. Packan, and A. S. Kumar, Eds., American Society for Testing and Materials, Philadelphia, 1987, pp.315-329.

[19] Akasaka, N., Yamagata, I., and Ukai, S., "Effect of Temperature Gradients on Void Formation in Modified 316 Stainless Steel Cladding," *Journal of Nuclear Materials*, Vols. 283-287, 2000, pp. 169-173.

[20] Garner, F. A., Gilbert, E. R., and Porter, D. L., "Stress-Enhanced Swelling of Materials during Irradiation," *Effects of Radiation on Materials: 10th Conference, ASTM STP 725*, David Kramer, H. R. Brager, and J. S. Perrin, Eds., American Society for Testing and Materials, 1981, pp. 680-697.

[21] Foster, J. P., and Flinn, J. E., "Residual Stress Behavior in Fast Neutron Irradiated SA AISI 304L Stainless Steel Cylindrical Tubing," *Journal of Nuclear Materials*, vol. 89, 1980, pp. 99-112.

[22] Ukai, S., Yoshida, E., Enokido, Y., and Nihei, I., "Quantitative Characterization on Dissolution and Deposition Behavior of SUS 316 Stainless Steel Cladding Constituents under Flowing Sodium in Fast Reactors," *Materials for Nuclear Reactor Core Application*, Vol. 1, BNES, London, 1987, pp. 341-348.

[23] Garner, F. A., Hunter, C. W., Johnson, G. D., Lippincott, E. P., and Schiffgens, J. O., "The Origin and Consequences of Radial Helium Profiles in Fast Reactor Cladding," *Nuclear Technology*, vol. 58, 1982, pp. 203-217.

Alexandre V. Kozlov[1], Evgeniy A. Kinev, S. V. Bryushkova and Irina A. Portnykh.

The Swelling Dependence of Cold Worked 16Cr-15Ni-2Mo-1Mn Steel on Neutron Irradiation Temperature, Fluence and Damage Rate During Its Use as a Cladding Material in the BN-600 Reactor

Reference: Kozlov, A. V., Kinev, E. A., Bryushkova, S. V., and Portnykh, I. A., **"The Swelling Dependence of Cold Worked 16Cr-15Ni-2Mo-1Mn Steel on Neutron Irradiation Temperature, Fluence and Damage Rate During Its Use as a Cladding Material in the BN-600 Reactor,"** *Effects of Radiation on Materials: 20th International Symposium, ASTM STP 1405,* S. T. Rosinski, M. L. Grossbeck, T. R. Allen, and A. S. Kumar, Eds., American Society for Testing and Materials, West Conshohocken, PA, 2001.

Abstract: The radiation swelling of samples of fuel element cladding made of steel 16Cr-15Ni-2Mo-1Mn after its use up to a maximum damage dose 73 - 93 dpa in the reactor BN -600 was studied. The samples were cut out from areas of different location in the core, so the irradiation temperature of various samples varied from 390 to 610 °C. The damage dose varied from 40 -92 dpa and the radiation damage rate varied from $0.9 \cdot 10^{-6}$ to $2 \cdot 10^{-6}$ dpa/s.

The conditions necessary to have the radiation swelling occurred were considered theoretically within the statistical thermodynamics of solids. An analytical expression connecting the incubation period of swelling with a temperature, fluence, radiation damage rate, and characteristics of material microstructure was obtained. The swelling rate at the post-incubation stage at various irradiation temperatures was estimated. The character of swelling dependencies on temperature, dose, and radiation damage rate obtained from the experimental data agreed well with the results of the theoretical analysis. In the range of 390 to 600 °C the nucleation damage dose initiating swelling increased with irradiation temperature. The swelling rate in the post-incubation phase also increased. Decreasing radiation damage rate shifted the fluence dependence of swelling into a region of lower fluences and temperatures.

Keywords: F02, N01, P04, S05, S10

[1]Doctor, Head of Post-Irradiation Examination Department, SUSE "SB RDIPE", P.O.Box 29, 21ª Klara-Tsetkin St.-60, Zarechny, 624051, Sverdlovsk region, Russia.

Introduction

Radiation swelling occurring under high dose irradiation is one of factors that limits operation duration of reactor structural materials. Swelling essentially influences on fuel lifetime of a reactor. It has been demonstrated most clearly in fast reactors where damage dose accumulation rate is high [*1*]. A great amount of test results has been obtained in the reactors. Tentative representations based on these data have shown that swelling of austenitic steels occurs in a limited temperature range from 420 to 550 °C. At the same time there are test data showing that swelling occurs at lower temperatures [*2*]. Thus it widens the amount of power reactors in which it is necessary to take into account for this phenomenon. There are some difficulties in obtaining such data at low irradiation temperature (< 400 °C), which can be realized only in thermal neutron reactors in which dose accumulation rate is relatively slow. These needs to be a rather long time to accumulate a larger damage dose.

There are many studies involving attempts of theoretical description of swelling and looking for empirical regularity. However, presently there is no theoretical model permitting a reliable prediction of swelling behaviour in a wide radiation damage rate and temperature and dose ranges.

In this paper we tried to find out a connection of swelling value with neutron irradiation temperature and radiation damage rate using a theory of statistical thermodynamics of solid body based on test data of influence of high dose irradiation on a cladding material in the BN-600 reactor.

Experimental Data

Swelling examination of samples from 17 fuel elements selected from 4th fuel assemblies after their operation in the BN-600 reactor up to different burnups was performed. Maximum damage dose of a cladding of cold work 20 % 16Cr-15Ni-2Mo-1Mn steel was 93 dpa.

A diameter *d* was measured every 15 ° in azimuth bearing on 540 samples with an accuracy of 0.01 mm with the following averaging in a test section. As an initial diameter we used the diameter of a cladding in the bottom of the gas cavity for each fuel element. Swelling on 23 samples was determined by a hydrostatic method as well as by geometrical measuring. It has appeared that plastic deformation - ε is proportional to swelling (with correlation coefficient ~ 0.64). As a value of swelling, ξ was submitted as a value based on a diameter changing with taking into account a correction on cladding plastic deformation ($\xi=3(\Delta d/d)-3\varepsilon$). The error of swelling determination was about 1 %.

Further, averaging of swelling value was performed on three to five samples cut out from one and the same cladding area having length 10-15 mm. Figure 1 exhibits these results. One can see that the points are in the area limited by a parabola curve, with the top corresponding to ~ 480 °C. Using such data configuration it seems that 480 °C is the temperature of swelling maximum. However, the points presented in Figure 1 belong to different damage doses and various radiation damage rates. In order to obtain a dependence of swelling from temperature, dose, and radiation damage rate a more careful analysis is needed.

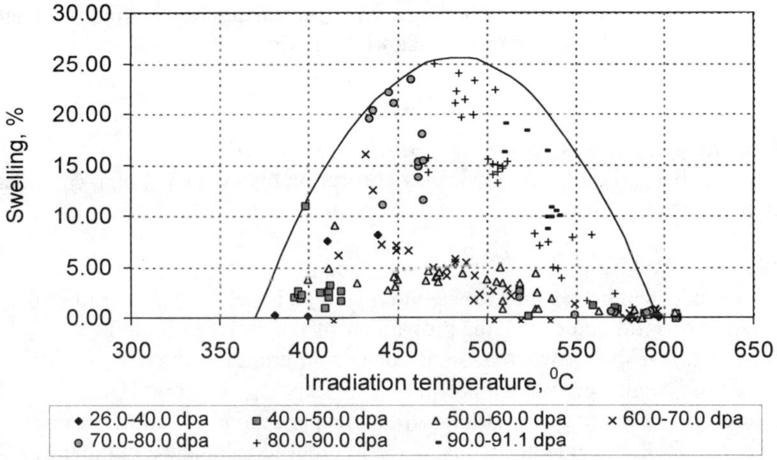

Figure 1 - *Dependence of Swelling from Irradiation Temperature of Samples of Fuel Element Cladding after their Operation up to Different Damage Doses in the BN-600 Reactor*

Theoretical Viewing of Swelling Regularity

Vacancies and interstitials are generated under neutron irradiation so their concentration increase. At the same time there are diffusion processes such as both point defects and impurity atoms interact with them. Absorption of vacancies and interstitials by sinks (grain boundaries, dislocations, twinning dislocation, the second phase precipitates) and also recombination among them cause a reduction of their concentration. The result of such competition processes (defect generation and leaving) is increase in concentration of both type of defects. Concentration changing dependence on time is defined by both point defect mobility (migration energy and temperature) and "power" of sink (sink extend, elastic interaction of the given type defect with sinks).

Interstitials have mobility at operation temperature of BN reactors and reach quasi-equilibrium concentration for a short period of time. Vacancy mobility is lower and changes significantly in a range (370 – 650) °C. In particular, with temperature increasing within this range both dissolution processes of vacancy clusters as non-optimum energy defects are intensified and more energy surviving defects as dislocation loops and vacancy voids are formed and grow.

Let's consider on what a growth intensity of void formation (or its nucleus) depend. Designate r – void radius; E_v – vacancy formation energy. During elemental action of vacancy absorption by void a void surface and its energy increase and matrix energy decrease (on E_v). Let's calculate increasing of surface energy during elemental action. We consider that the void volume is $n\upsilon$ (where n – a number of vacancies in a void, υ– a volume per atom equal to vacancy volume in accepted approximation), neglecting relaxation of an area occupied by a void. Increasing the void radius under vacancy absorption is defined from the equation in the first approximation.

$$\Delta r = \upsilon / (4 \cdot \pi \cdot r^2) \qquad (1)$$

Surface energy increases on $2\upsilon\tau/r$ *(τ - surface tension coefficient).* Thus, energy is changed on U under vacancy absorption by a void and on $(-U)$ under evaporation of vacancy from a void.

$$U = 2 \cdot \upsilon \cdot \tau / r - E_v \upsilon \qquad (2)$$

Now we can calculate vacancy flows into a void and back. According to statistical thermodynamic of solid bodies [*3*], the probability of vacancy overjumping is proportional to $exp(-E_{mv}/kT)$, where E_{mv} - is vacancy migration energy. A number of overjumpings of vacancy per second is equal to $\nu_0 * exp(-E_{mv}/kT)$, (ν_0 – Debye frequency, describing a number of atom oscillations per second). If all of six directions are equivalent the factor 1/6 is added in order to calculate a number of overjumps in a given direction. If energy is changed during vacancy jumping the probability of such overjump is proportional to $exp(-U/kT)$, under case of equivalent jumps in the five remaining directions (without changing of energy). Therefore the probability of vacancy jumping into a void using requirements of normalization is calculated as follows,

$$\varpi = \frac{\exp(-U / kT)}{5 + \exp(-U / kT)} \qquad (3)$$

Then vacancies flow into a void can be expressed as

$$N_+ = \frac{\exp(-U / kT)}{5 + \exp(-U / kT)} \cdot 4 \cdot \pi \cdot r^2 \cdot c_v / b^2 \cdot \nu_0 \cdot \exp(-E_{mv} / kT) \qquad (4)$$

where

T = irradiation temperature
b = lattice parameter, and
c_v = vacancy concentration in matrix.

Similarly a flow of vacancies evaporated from a void can be expressed by Eq. (4) if their concentration inside voids equal to 1 ($c_v = 1$).

$$N_- = \frac{\exp(U / kT)}{5 + \exp(U / kT)} \cdot 4 \cdot \pi \cdot r^2 / b^2 \cdot \nu_0 \cdot \exp(-E_{mv} / kT) \qquad (5)$$

Interstitials also come into a void with the reduction of its size (that is, to operate the same as vacancy evaporation). We can neglect interstitial emission from a void because this process requires a major amount of energy and its probability is practically equal to zero under operation temperatures. Interstitial flow into a void - N_i is expressed by an equation similar to Eq. (4), only instead of vacancy characteristics there are interstitial characteristics.

$$N_i = \frac{\exp(-U_i / kT)}{5 + \exp(-U_i / kT)} \cdot 4 \cdot \pi \cdot r^2 \cdot c_i / b^2 \cdot \nu_0 \cdot \exp(-E_{mi} / kT) \qquad (6)$$

where

E_{mi} = interstitial migration activation energy
E_i = interstitial formation energy, and
U_i = energy of interstitial absorption by a void, by another words, a difference between system energy before and after absorption.

$$U_i = -(2 \cdot \upsilon \cdot \tau / r + E_i) \qquad (7)$$

Note that Eq. (4) and Eq. (5) were obtained in approach when point defect concentration near void surface was constant, i.e. both absorption and evaporation rate of point defects is lower than rate of delivering of the defects to void surface. Condition of void growth is a performance of the relation $N_+/(N_- + N_i) > 1$. It is given in the equation.

$$\frac{c_v \cdot \exp(-E_{mv} / kT) / (5\exp(U / kT) + 1)}{\exp(-E_{mv} / kT) / (5\exp(-U / kT) + 1) + c_i \cdot \exp(-E_{mi} / kT) / (5\exp(U_i / kT) + 1)} > 1 \qquad (8)$$

If we substitute the coefficient of surface tension for γ-Fe, equal $\tau = 2.2$ J/m^2 [*4*], and $Å_v = 1.5$ eV into Eq. (2) so we obtain that $U < 0$ at $r > 2.5*10^{10}$m. When r is equal to lattice period ($3.65*10^{-10}$m) the value (U/kT) is egual (-5) and grows on the module with void radius increasing at 900 K (it is higher than the given range). Using energy characteristics of diffusion in γ-Fe it is possible to estimate the values included in Eq. (8). If addends of the second order smallness are neglected in the sums we receive void growth condition as follows.

$$\frac{c_v}{0.2 \cdot \exp(U / kT)) + c_i \cdot \exp((E_{mv} - E_{mi}) / kT)} > 1 \qquad (9)$$

It is possible to present this equation as expression for threshold vacancy concentration at which swelling begins.

$$c_{vt} = 0.2 \cdot \exp(U / kT) + c_i \cdot \exp((E_{mv} - E_{mi}) / kT) \qquad (10)$$

There are two addends in the Eq. (10). The first decreases vacancy concentration that is necessary for void growth that begins under nuclei size increase (dependence U(r) is shown in Eq. (2)). In particular if there are, for example, gas bubbles, the void growth begins much earlier. If already there are voids but threshold vacancy is not reached they will be dissolved. Temperature reduction causes a decrease of threshold vacancy concentration (because of U has negative value).

When the second addend is compared with the first one the interstitial concentration influences essentially on threshold vacancy concentration; it is higher the more the value of threshold vacancy concentration. In the first approximation it is possible to consider interstitial concentration (quasi-equilibrium under irradiation) proportional to neutron flux density. Increasing temperature decreases this effect because ($Emv - Emi$) has positive value.

Therefore we can imagine the following model of swelling process under irradiation. Vacancy concentration grows slowly under neutron irradiation. Its rate depends on vacancy generation intensity (neutron flax density), mobility (temperature), and an amount of sinks. Sinks are exhausted with fluence accumulation in examination steel it causes in acceleration of vacancy accumulation in a solution. However before reaching threshold vacancy concentration determined by Eq. (10) there is no swelling, it explains a presence of swelling incubation period. After that swelling intensity depends on vacancy concentration, temperature and void size.

Connection of the swelling incubation dose with irradiation parameters can be traced using Eq. (10). As the interstitial concentration is proportional to neutron flux density - φ ($c_i = B(S)\cdot\varphi$, where $B(S)$ is a function from structure characteristics of a material S), and vacancy concentration is proportional to damage dose D ($c_{vi} = A(\varphi, S)\cdot D$, where A(S) also a function from structure characteristics of a material) in order to estimate incubation dose D_0 the following equation can be used

$$D_0 = (0.2\cdot\exp(U/kT) + B(S)\cdot\varphi\cdot\exp((E_{mv} - E_{mi})/kT))/A(\varphi,S) \quad (11)$$

As shown in Eq. (11) incubation dose should be lower at lower both temperature and neutron flux density, it corresponds to [5].

It is difficult to explain quantitatively the swelling kinetics after incubation period due to major amount of unknown parameters included in the equation of balance. Moreover during swelling these parameters change with structural evolution (dislocation annealing and, dissolution and formation of the second phase precipitates) and growth of intensity of vacancy absorption by pores due to increasing of their surface area under growth. The numerous attempts, described in the literature, of approaches and assumptions allow understanding some characteristics of swelling process but not to describe the effect in whole.

Void formation and their growth begins after achieving the threshold vacancy concentration. Vacancy sinks into voids depend on a relation between power of sinks - voids and neutral sinks - interstitials, dislocations, grain boundaries etc. which do not lead to swelling One can see that swelling rate must increase with void growth and also with exhaustion of neutron sink power. However the process becomes quasi-equilibrium in time, as voids size is great enough, absorption of the same specific amount of vacancies resulted in inappreciable change of voids surface, and after annealing of "labile" neutral sinks only stable ones are remained. So flows relationship becomes stationary, and portion of vacancies leaving into voids has not time-depended. Dose swelling rate ($v = d\xi/dD$) becomes practically constant. Quasi-equilibrium processes of such type have the equation as follows,

$$\xi = v\cdot(D - D_0)\cdot(1 - \exp(-p\cdot(D - D_0))) \quad (12)$$

where

v = swelling rate at steady-state (stationary) stage depending both from an amount of stationary sinks and characteristics of point defects diffusion mobility. It is possible to consider that v is the constant for given material characterizing a limit to which a swelling rate approaches asymptotically

D_0 = incubation dose of swelling depends on irradiation temperature (in our case it is the dose corresponding 1 % swelling) radiation damage rate and structural characteristics of a material, and

p = coefficient characterizes an arrival rate into stationary state of swelling. It depends on the same structural and irradiation characteristics as D_0 is.

The equations are used to analyze test data presented in the previous issue.

Results

The complexity of analysis of test data file is connected with differences in irradiation characteristics (T, D, dD/dt) of test samples. Therefore in order to analyze a dependence of swelling from neutron irradiation dose the whole file is divided into groups of 20 degree temperature ranges (and on 10 degree in temperature area of high swelling). Data belong to different radiation damage rates inside group therefore we drew curves average on radiation damage rate. Figure 2 shows the swelling – damage dose dependencies for different irradiation temperature ranges. One can see that groups of points with higher irradiation temperature ranges are shifted to a higher dose area. Figure 2 (a,b) presents dependencies for the highest and the lowest temperature ranges (the middles of the ranges correspond to 390, 450, 470 and 510 °C). The dependencies are described by the Eq. (12) and obtained by selection of optimal values of the coefficients v, D_0 and p. Such curves correspond to 530 and 550 °C in Figure 2c, as at higher temperatures even at 70 – 80 dpa swelling is not observed practically. Values of swelling parameters from Eq. (12) for various irradiation temperatures are exhibited in Table 1.

Table 1 - *Equation Parameters of Dependence of Swelling ξ from Irradiation Characteristics* $\xi = v \cdot (D - D_0) \cdot (1 - \exp(-p \cdot (D - D_0)))$

Temperature of middle interval, °Ñ.	390	450	470	505	530	550
v, %/dpa	3	3	3	3	3	3
D_0, dpa	26	39	42	54	58	68
ð, dpa^{-1}	0.0030	0.0048	0.0051	0.0055	0.0058	0.0075

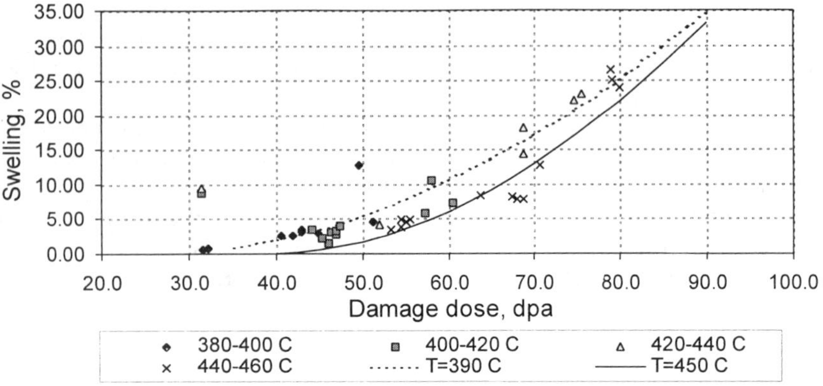

a

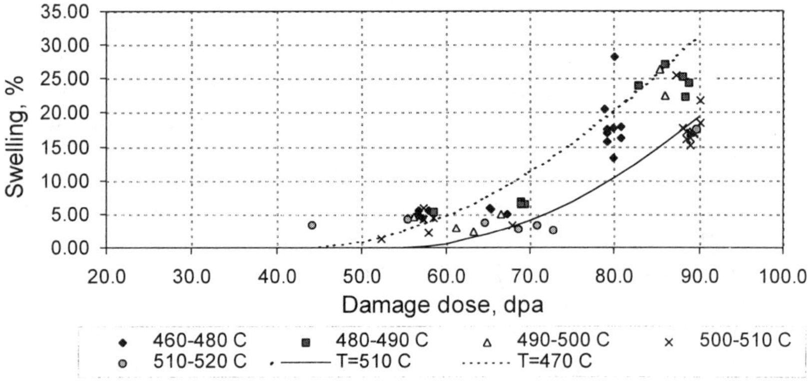

b

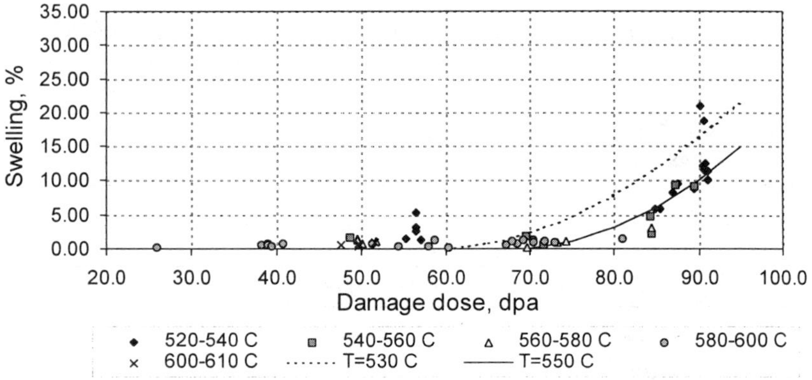

c

Figure 2 - *Dependence of Swelling from Damage Dose for Samples Irradiated within Different Temperature Ranges (a,b,c).*

Note that v obtains the value ~3 %/dpa, D_0 and p decrease under temperature decreasing in a temperature range 390 - 550 °C, Figure 3. If we draw down a linear extrapolation of dependencies $D_0(T)$ and $p(T)$ up to cross with an axis of temperatures we will obtain the values 295 and 261 °C. These values are close not only between themselves but also to temperature of the beginning of vacancy diffusion [*6*]. On physical sense these values should be equal between themselves and coincide with the inferior temperature boundary of void formation (T_H). Vacancies have no sufficient diffusion mobility in order to move at considerable distances at lower temperatures.

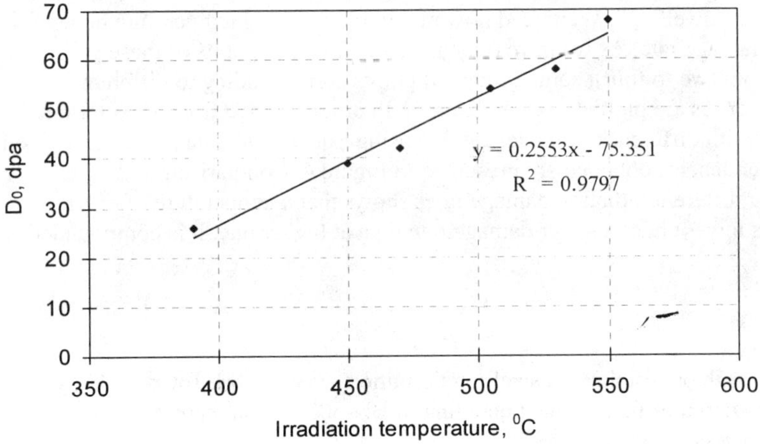

a

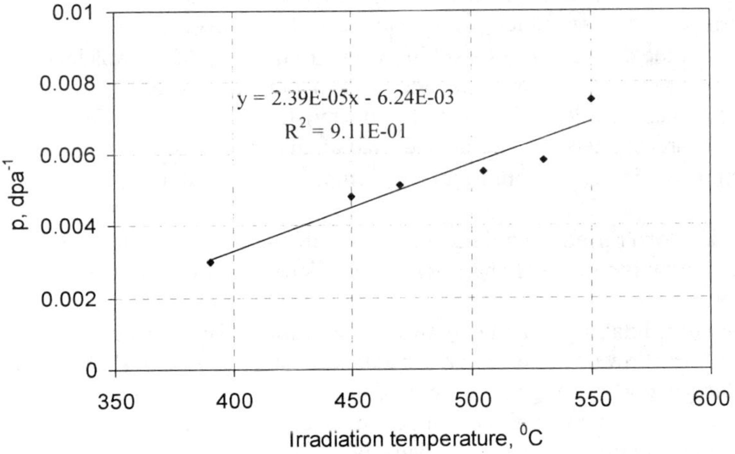

b

Figure 3 - *Dependence of D_0 (a) and ð (b) Parameters from Irradiation Temperature.*

v, $D_0(T)$, and $p(T)$ values, as well as a lower temperature at which swelling does not happen are presented in Table 1. Note that they are estimation values and do not take into account dependencies from radiation damage rate. Obtaining more accurate values requires to measure swelling of a set of samples with given radiation damage rates and irradiation temperatures. Using the described above treatment procedure of results and magnification of statistics it is possible to improve an accuracy.

In order to analyze test dependencies of swelling from radiation damage rate (dD/dt) data groups were selected from the whole file. In the each group the value (dD/dt) does not differ more than 0.5×10^{-8} dpa/c, and damage doses are identical with accuracy 0.5 dpa. These data groups plotted on a diagram in coordinates "irradiation temperature – swelling". We traced how maximum swelling temperature depends on radiation damage rate. As we have only a small amount of points of these groups (usually from two to four), some groups of points corresponding to different irradiation doses are plotted in each diagram. In order to leave from a scattering connected with difference of damage doses using Eq. (12) all data are reduced to one dose. Dependencies obtained are presented in Figure 4. Comparison of diagrams concerning different radiation damage rates shows that a temperature of maximum swelling is lower under a lower damage rate than at higher one. It is compounded with the Eq. (11).

Conclusions

Based both on test data on swelling definition of cold work 16Cr-15Ni-2Mo-1Mn steel as a material of fuel element cladding of BN-600 and on theoretical analysis the following was concluded.

- Incubation swelling dose depends on neutron irradiation temperature and radiation damage rate. The incubation dose decreases with reduction of these characteristics within tested temperature and dose rate interval.
- A view of the dependence of swelling value from damage dose was installed. The equation of this dependence includes three parameters one of them (rate at a stage of stationary swelling) is material characteristics, others depend on both structural characteristics of material and irradiation parameters.
- Swelling rate increases with irradiation temperature growth in post-incubation period.
- There is a lower irradiation temperature at which there is no void formation. The temperature is a material characteristic and depends on vacancy diffusion parameters.
- Quantitative relations obtained between a swelling value and irradiation characteristics allow one to describe test data satisfactorily. Use of the equations obtained to predict swelling in a wide irradiation condition spectrum needs more accurate definition of parameters included in the equations and depended on irradiation temperature and radiation damage rate.

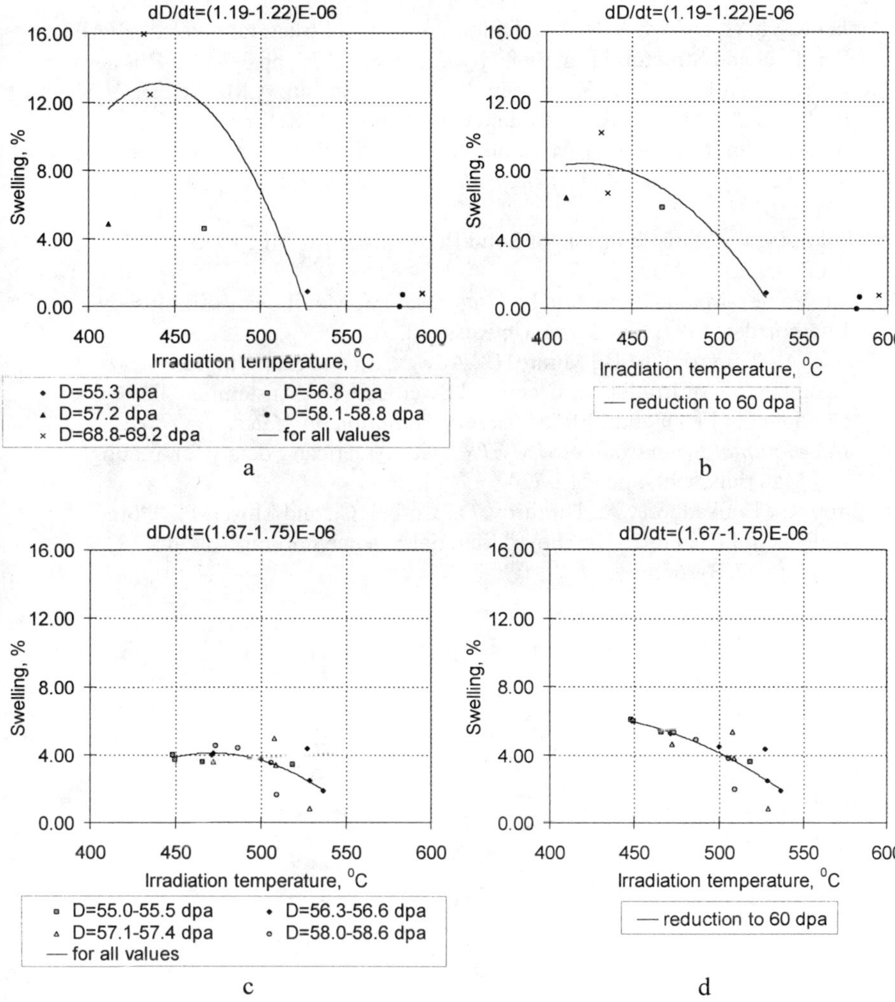

Figure 4 – *Temperature Dependence of Swelling for Samples, Irradiated with Different Radiation Damage Rates, a) Damage Rate dD/dt = (1.19-1.22) E-06, Test Data; b) Damage Rate dD/dt = (1.19-1.22) E-06, Reduction to 60 dpa; c) Damage Rate dD/dt = (1.67-1.75) E-06, Test Data; d) Damage Rate dD/dt = (1.67-1.75) E-06 Reduction to 60 dpa.*

References

[1] Konobeev, Yu.V., Bykov, V.N., "Radiation Porosity in Structural Steels," / VAET, Ser: Fuel and Structural Material, ¹1(b), Ìoscow, 1977, pp.3-13 (in Russian).

[2] Porollo, S.I., Vorobjev, A.N., Konobeev, Yu.V., Dvoriashin, A.M., Krigan, V.M., Budylkin, N.I., Mironova, E.G., and Garner, F.A., "Swelling and Void-Induced Embrittlement of Austenitic Stainless Steel Irradiated to 73-82 dpa at 335-365°C", *Journal of Nuclear Materials*, Vols. 258-263, 1998, pp. 1613-1617.

[3] Zhirifalko, L., Jr., Statistic Physic of Solid Body, Moscow, Mir, 1975, p 382 (in Russian).

[4] Physical Values, Manual reduction by Grigoreva, I.S, Meilihova E.Z., Moscow, Energoizdat, 1991, pp.53 - 54. (in Russian).

[5] Garner, F.A., Toloczko, M.B., Munro, B., Adway, S., and Standring, J., "Comparison of Irradiation Creep and Swelling of an Austenitic Alloy Irradiated in FFTF and PER", *Effects of Radiation on Materials: 18th International Symposium, ASTM STR 1325*, American Society for Testing and Materials, 1999, pp. 713-724.

[6] Dimitrov, C., Benkaddour, A., Dimitrov, O., Corbel, C., and Moser, P., "Point Defects in FCC Fe-Cr-Ni Alloys" Material Science Forum, Vols. 15-18, 1987, pp. 753-765.

Tsunemitsu Yoshitake,[1] Takako Donomae,[1] Shunji Mizuta,[1] Hanchung Tsai,[2] Robert V. Strain,[2] Todd R. Allen,[3] and James I. Cole[3]

Tensile Properties of 12% Cold-Worked Type 316 Stainless Steel Irradiated in EBR-II Under Lower-Dose-Rate Conditions to High Fluence

Reference: Yoshitake, T., Donomae, T., Mizuta, S., Tsai, H., Strain, R. V., Allen, T. R., and Cole, J. I., **"Tensile Properties of 12% Cold-Worked Type 316 Stainless Steel Irradiated in EBR-II Under Lower-Dose-Rate Conditions to High Fluence,"** *Effects of Radiation on Materials: 20th International Symposium, ASTM STP 1405,* S. T. Rosinski, M. L. Grossbeck, T. R. Allen, and A. S. Kumar, Eds., American Society for Testing and Materials, West Conshohocken, PA, 2001.

Abstract: The effects of lower-dose-rate irradiation on the tensile properties of 12% cold-worked type 316 stainless steel hexagonal ducts which were irradiated in the outer rows of EBR-II were investigated. Tension tests were conducted on a total of seventeen irradiation conditions. Fractographic examinations were performed on four irradiation conditions which were selected based on the results of tension tests. The results of the tension tests and fractographic examinations showed that the 316 stainless steel hexagonal ducts exhibited irradiation hardening typical to 316 stainless steels irradiated at similar temperatures and retained respectable ductility after the irradiation. There was no significant effect of dose rate on tensile properties for the range of irradiation conditions of this study.

Keywords: 12% cold-worked type 316 stainless steel, tensile properties, strength, ductility, failure mode, EBR-II reflector duct, lower-dose-rate irradiation, fast reactor core material

Introduction

Studying the effects of long-term, low-dose-rate neutron exposure on core structural materials is important because these conditions exist for many of the fixed or hard-to-replace components in the outer regions of fast reactor and light-water reactor cores. However, because of the difficulty of obtaining suitably irradiated samples for testing and characterization, these effects have not been extensively investigated. The Experimental Breeder Reactor-II (EBR-II) operated for approximately 30 years and has many components that experienced long-term, lower-dose-rate irradiation. During the

[1] Japan Nuclear Cycle Development Institute, 4002 Narita-cho, Oarai-machi, Ibaraki 311-1393, Japan.

[2] Argonne National Laboratory, 9700 South Cass Ave., Argonne, IL 60439.

[3] Argonne National Laboratory-West, P.O.Box 2528, Idaho Falls, ID 83403-2528.

shutdown of EBR-II, some of its peripheral core components were retrieved to evaluate the effects of dose rate on properties especially at low dose rates. The focus of this study is to evaluate the effects of low-dose rate irradiation on mechanical properties, swelling and microstructural evolution in 12% cold-worked type 316 stainless steel irradiated as the hexagonal ducts of reflector subassemblies in EBR-II. In this paper, the tensile properties of such hexagonal ducts are reported.

Experimental Procedure

Material

Materials used in this work are 12% cold-worked type 316 stainless steel, which were irradiated in the EBR-II as the hexagonal ducts of reflector subassemblies. Table 1 shows the chemical composition. These ducts were finally annealed at 1121---1149°C and water-quenched to room temperature. Final cold-working level was 11.7%.

To obtain samples irradiated to high dose at low dose rate, samples were prepared from two reflector subassemblies; U9861 and U9922 which were irradiated in row eight and row nine of the EBR-II core. The dose rate of U9861 is almost twice that of U9922 at same damage level. These two ducts had been irradiated at temperatures ranging from 371 to 444°C depending on axial location. The fast neutron-damage dose was between 0.9 and 40.5 dpa and the dose rate range was from 1.0×10^{-8} to 3.52×10^{-7} dpa/s.

Specimen Preparation

Tensile specimens were cut using a Charmille Andrew Electric Discharge Wire-Cutting Machine after machining rectangular strips with two holes from two hexagonal ducts in parallel with longitudinal axis of them.

The tensile specimen was designed to meet both the American Society of Mechanical Engineers (ASME) and the Japanese Industrial Standards (JIS) specifications, and is shown in Figure 1.

Tension Test

Tension tests were conducted in air using an Instron Model TT screw-driven tension testing machine which was installed in the Clean Transfer Area of the Alpha-

Table 1---*Chemical composition of 12% cold-worked type 316 stainless steel*

Element	% by weight	Element	% by weight
C	0.041	Cr	17.48
Mn	1.19	Ni	11.34
P	0.033	Cu	0.35
S	0.023	Mo	2.61
Si	0.64	Fe	bal.

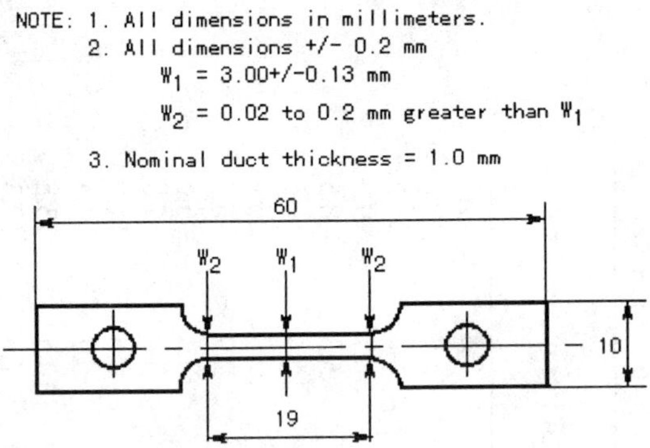

Figure 1---*Dimensions of tension specimen*

Gamma Hot Cell Facility (AGHCF) at Argonne National Laboratory. Test temperature was nominally 380°C or 430°C and selected to correspond to the irradiation temperature. Additionally two specimens were tested at room temperature and 430°C regardless of irradiation temperatures. All tests were performed at a crosshead speed of 1.9×10^{-2} mm/s, which for the gauge length of 19 mm, resulted in a strain rate of $1.0 \times 10^{-3}\ s^{-1}$.

Tensile specimens were carefully mounted onto the testing machine using an unique pair of electromagnetic specimen grips to prevent specimens from being dropped or bent. A thermocouple in contact with the lower specimen grip was used to monitor specimen temperature based on the relation between gauge temperature and the lower grip temperature which was established in separate furnace heating tests. To prevent undesirable specimen straining during heat up to the test temperature, the load was continuously monitored and the position of the actuator was adjusted to keep loads well below the load required for yielding.

Yield strength was determined by the 0.2% offset method. Uniform elongation and total elongation were obtained from an engineering stress-strain curve.

No archive material remains for these samples, but room temperature tensile properties are available from material certifications.

Fractographic Examination

In order to characterize the failure mode, scanning electron microscope (SEM) fractography of the fracture surfaces was conducted on four specimens which were selected after tension tests using an ETEC Model U1 Scanning Electron Microscope in the Electron Beam Laboratory of AGHCF. In addition, optical metallography was performed on the samples.

Table 2---*Results of tension tests*

Specimen number	Irradiation condition			Test temperature, °C	Strength, MPa		Elongation, %	
	Temperature, °C	Dose, dpa	Dose rate, $x10^{-7}$ dpa/s		Yield	Ultimate tensile	Uniform	Total
ET11	438	14.0	1.63	432	583	689	8.8	11.3
NT12	432	19.2	1.06	429	643	750	8.0	10.3
ET10	432	19.5	2.28	430	588	701	8.8	11.5
NT11	422	30.1	1.66	432	683	770	6.7	9.3
ET08	417	30.2	3.52	430	707	770	5.3	8.5
ET15	371	0.9	0.10	383	568	687	11.4	14.5
NT15	371	2.2	0.12	23	792	916	17.8	26.4
ET06	374	9.1	1.06	377	786	856	2.9	6.2
NT01	375	13.6	0.75	380	791	861	3.3	6.6
NT08	409	40.5	2.23	428	709	776	5.4	7.8
ET03	425	30.2	3.52	429	613	719	8.6	11.9
NT04	417	40.5	2.24	428	670	744	6.7	9.4
ET04	435	19.2	2.24	428	568	701	10.5	13.8
NT05	425	30.1	1.66	428	649	739	7.6	10.3
ET05	444	9.1	1.06	429	545	667	10.4	13.9
NT09	436	14.2	0.78	429	610	752	8.4	11.8
ET14	371	0.9	0.10	427	569	689	10.9	14.2

Note1) Crosshead speed : 1.14 mm/min
Note2) ET ; U9861, NT ; U9922 (reflector subassembly number)

Experimental Results

Tensile Properties

The results of the post-irradiation tension tests and the irradiation condition of each specimen are listed in Table 2. Since there were small gradients in dose and irradiation temperature along the specimen axis, the irradiation conditions listed were determined for the specimen midpoint.

Figure 2 is a graph of yield strength as a function of fluence (dose). Yield strength increased with increasing dose and irradiation hardening was more obvious at 380°C than at 430°C. Yield strength at 430°C is represented in two groups in which dose rate is above or below 2.0×10^{-7} dpa/s. No noticeable dose rate effect can be observed in this graph. Yield strength of ET14, which was irradiated at 371°C and tested at 427°C, was comparable with that of ET15 which was irradiated and tested at about 380°C. It seems that strength is more strongly influenced by irradiation temperature than by test temperature in this temperature range. Data from the room temperature, low-dose test (NT15) revealed that significant hardening occurred very early in irradiation when compared with unirradiated tensile data from the inspection certificate of this duct. The trends in ultimate tensile strength appeared to be similar to the yield strength as a function of dose, with UTS having a higher stress level (Fig. 3).

Uniform elongation and total elongation are shown as a function of dose in Figures 4 and 5, respectively. Uniform elongation decreased modestly with dose in relation to irradiation hardening, and a decrease in elongation was more evident at 380°C than 430°C. As with the yield strength, Figure 4 shows that there was no obvious dose rate effect on uniform elongation at 430°C. Total elongation showed comparable trends with both dose and dose rate to uniform elongation. Total elongations ranged from 6.2% to 26.4% in this study, so it seems that all specimens retained respectable ductility and that no brittle failure was indicated under these irradiation conditions.

Figures 6 and 7 show yield strength and uniform elongation as a function of dose rate respectively. In these figures, data at 430°C are represented in two groups, dose levels greater or less than 30 dpa. There is no significant dose rate effect on yield strength and uniform elongation under these conditions. Ultimate tensile strength and total elongation showed similar behavior to yield strength and uniform elongation respectively.

Fractography

SEM examinations were performed on four specimens; ET03, NT04, ET04 and NT05 which were selected as functions of dose and dose rate according to the results of tension tests. The results of the SEM examinations are given in Table 3, and fractographs of fracture surfaces are shown in Figure 8. All specimens showed significant area reduction ranging from 27% to 36% with no noticeable dose effect. Fracture surfaces for all four samples exhibited typical ductile fracture and indicated a ductile mode of failure.

Figure 9 provides optical micrographs of the longitudinal surfaces of the gauge sections near fracture. Samples were electro-etched using 10% oxalic acid after polishing. The micrographs in Figure 9 indicate that twinning is more prominent in the lower dose specimen, ET04.

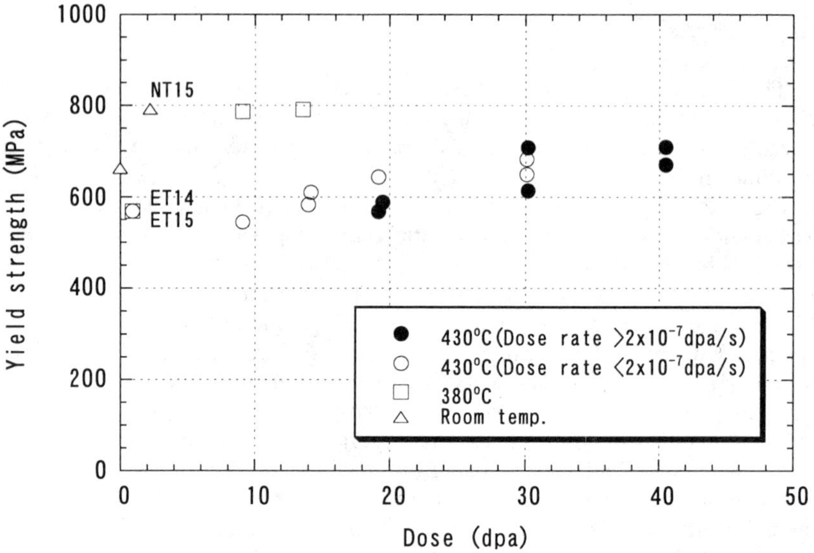

Figure 2---*Yield strength of 12% cold-worked type 316 SS after irradiation in EBR-II (Legend indicates test temperature)*

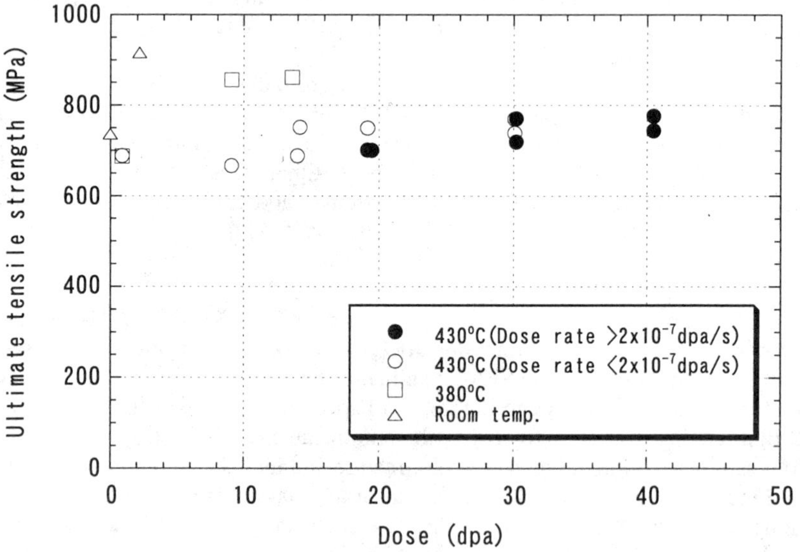

Figure 3---*Ultimate tensile strength of 12% cold-worked type 316 SS after irradiation in EBR-II*

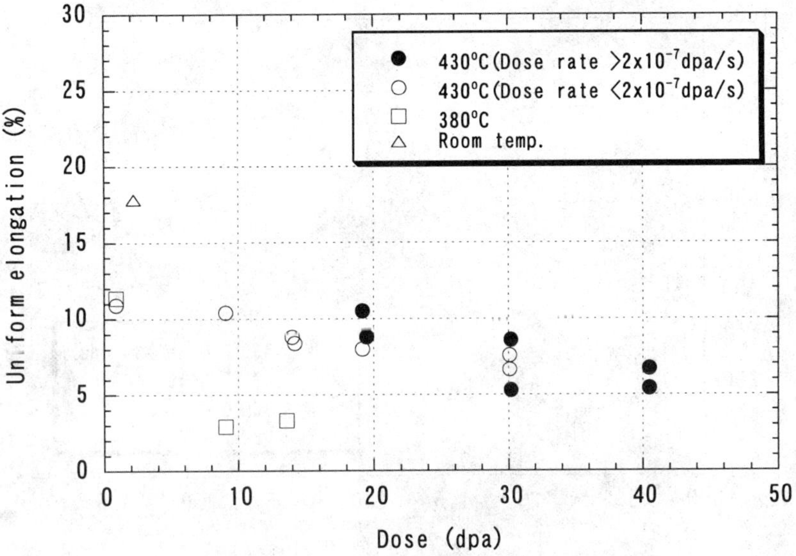

Figure 4---*Uniform elongation of 12% cold-worked type 316 SS after irradiation in EBR-II*

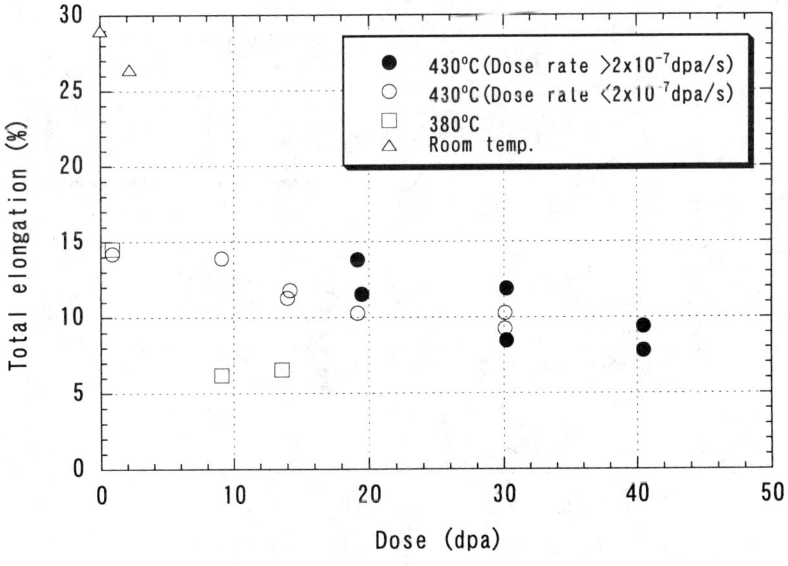

Figure 5---*Total elongation of 12% cold-worked type 316 SS after irradiation in EBR-II*

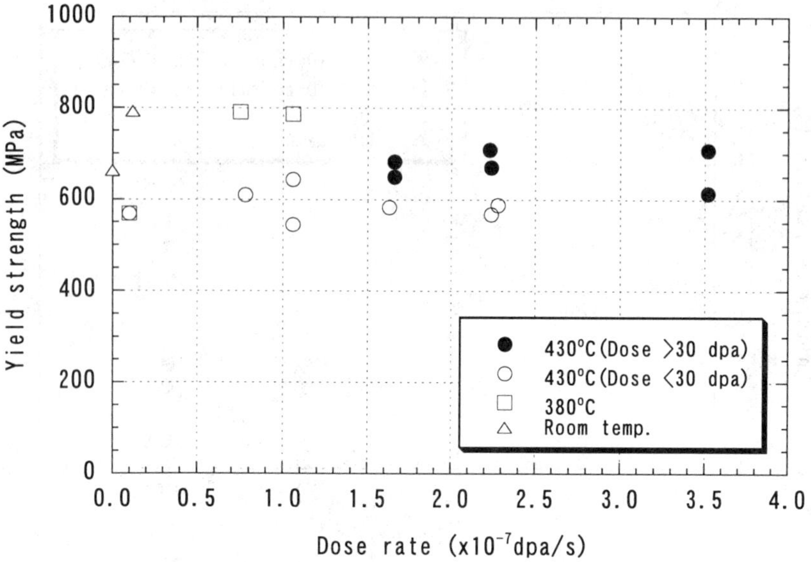

Figure 6---*Dose rate dependence of yield strength of 12% cold-worked type 316 SS after irradiation in EBR-II*

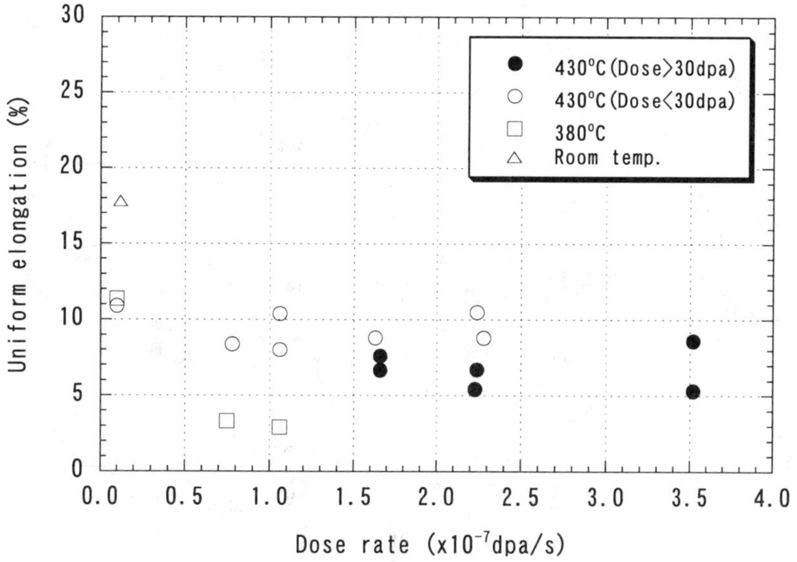

Figure 7---*Dose rate dependence of uniform elongation of 12% cold-worked type 316 SS after irradiation in EBR-II*

Table 3---*SEM results*

Specimen number	Irradiation condition			Test temperature, °C	Fracture mode	Gauge section area, mm^2		Reduction in area, %
	Temperature, °C	Dose, dpa	Dose rate, $x10^{-7}$ dpa/s			Initial	After testing by SEM	
ET03	425	30.2	3.52	429	ductile	3.12	2.29	27
NT04	417	40.5	2.24	428	ductile	3.03	1.95	36
ET04	435	19.2	2.24	428	ductile	3.07	1.96	36
NT05	425	30.1	1.66	428	ductile	3.02	1.95	36

Note 1) ET ; U9861, NT ; U9922 (reflector subassembly number)

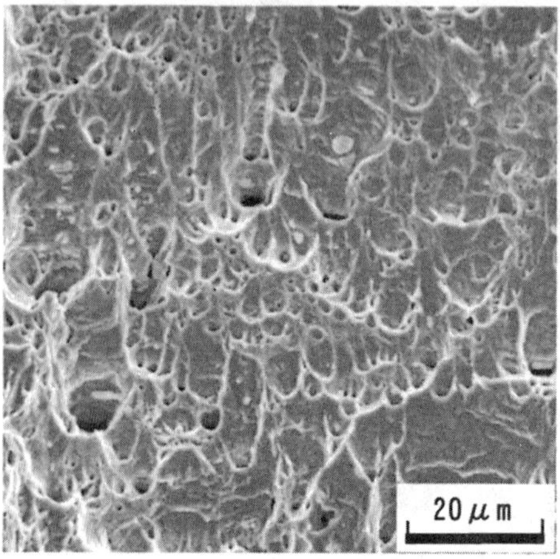

(a) ET04 (19.2dpa, 2.24×10^{-7}dpa/s, 435°C)

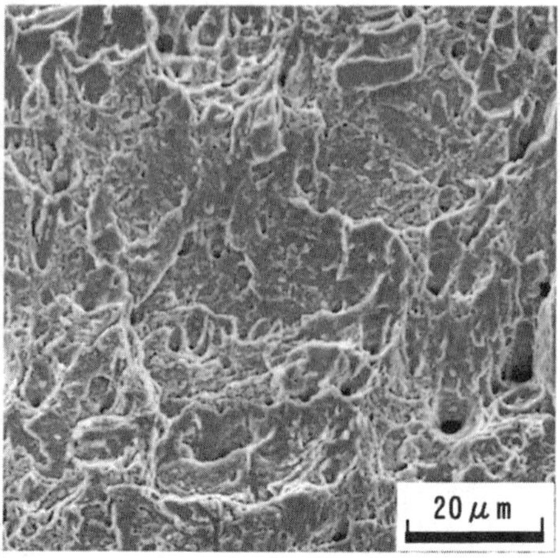

(b) NT04 (40.5dpa, 2.24×10^{-7}dpa/s, 417°C)

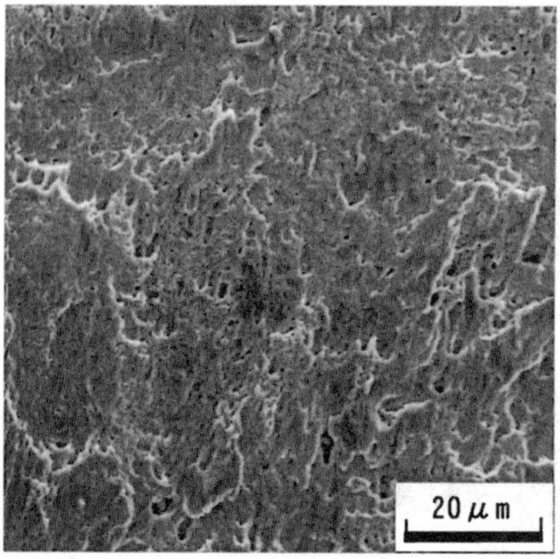

(c) NT05(30.1dpa, 1.66×10^{-7}dpa/s, 425°C)

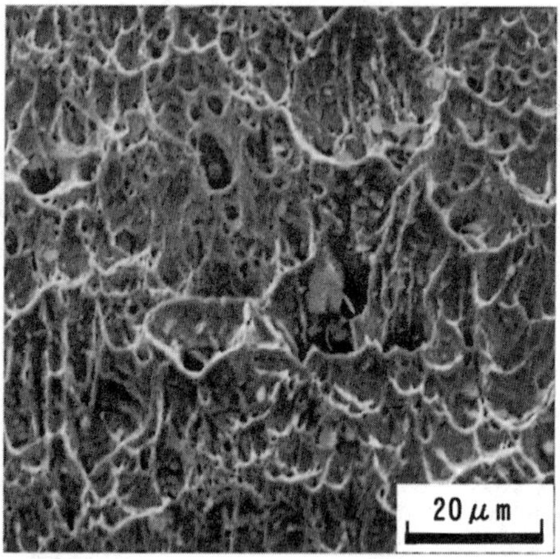

(d) ET03(30.2dpa, 3.52×10^{-7}dpa/s, 425°C)

Figure 8---*Scanning electron micrographs of fracture surfaces in tension specimens of 12% cold-worked type 316 SS irradiated in EBR-II*

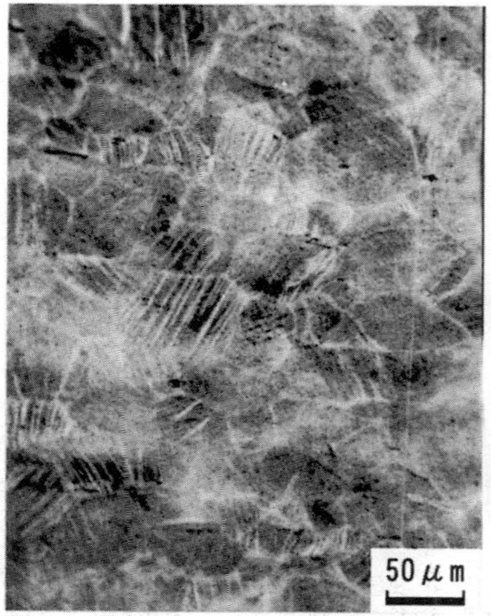

(a) ET04 (19.2dpa, 2.24×10^{-7}dpa/s, 435°C)

(b) NT05 (30.1dpa, 1.66×10^{-7}dpa/s, 425°C)

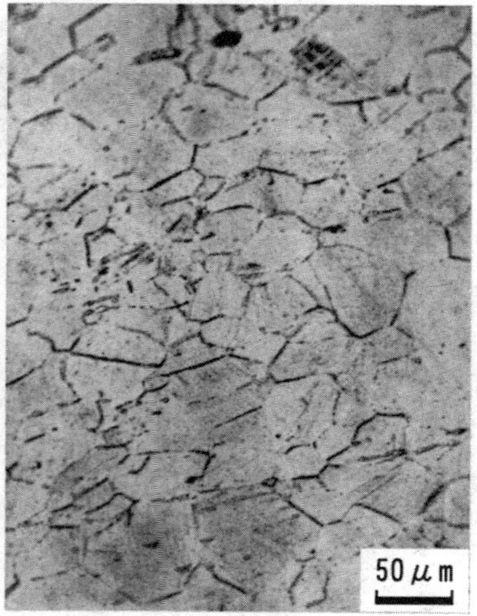

(c) NT04 (40.5dpa, 2.24×10^{-7}dpa/s, 417°C)

Figure 9---*Optical micrographs of longitudinal surface of gauge section near fracture for 12% cold-worked type 316 SS irradiated in EBR-II*

Discussion

Dose Effect on Tensile Properties

Tensile properties of irradiated austenitic stainless steels have been investigated extensively [*e.g. 1*]. In this study, tensile properties of 12% cold-worked type 316 SS showed typical temperature and dose dependence for irradiated austenitic stainless steels and good agreement with that of 20% cold worked AISI 316 SS irradiated in EBR-II as reported by Garner et al. [*2*]. In previous work, yield strength saturated at fast neutron fluences ranging from 3 to 5 x 10^{26} n/m^2 (E>0.1MeV). In this study, yield strength and ultimate tensile strength irradiated and tested at 430°C slightly increased up to 40 dpa, although the strength at 380°C was saturated near 10 dpa. Doses of 10 and 40 dpa correspond to 2 and 8 x 10^{26} n/m^2 (E>0.1MeV) respectively in EBR-II, so it is indicated that saturation of strength at 430°C may be delayed in this case. Dupouy et al. conducted tensile tests on annealed 316 stainless steels after irradiation at 390°C in PHENIX and RAPSODIE. PHENIX operated at a dose rate that was about three times higher than that of RAPSODIE. The test result indicated that the 0.2% proof stress was higher and saturated at a lower damage level under higher dose rate conditions [*3*]. In lower dose rate conditions, irradiation hardening might be delayed because recovery is more effective compared to hardening from irradiation defects in these temperature ranges. These results can be interpreted in an alternative way. The irradiation temperature of tension specimens tested at 430°C ranged from 409 to 444°C. Thus the moderate increase in yield strength with dose is suggested to be strongly reflected by the effect of irradiation temperature. As seen in Table 2, the irradiation temperatures were relatively low (409 ~ 425°C) in the higher dose region, while they were relatively high (432 ~ 444°C) in the lower dose region.

For studying the soundness and behavior of core and structural materials, the irradiation hardening and loss of ductility which results in embrittlement in such materials must be understood. Type 316 stainless steel still retained respectable ductility after the irradiation in this study as shown in Figures 4 and 5. For the irradiated austenitic steels, it is well-known that the yield strength approaches ultimate tensile strength with increasing dose [*4*] and that the total elongation approaches the uniform elongation at high dose condition where there is more than 6% swelling [*5*]. These results indicate concretely that work hardening is small after yielding and that reduction in area hardly occurs beyond the maximum load. The ratio of yield and ultimate tensile strength, YS/UTS, and the ratio of uniform and total elongation, UE/TE, are shown as a function of dose in Figures 10 and 11, respectively. YS/UTS reached about 0.92, although YS/UTS at 430°C increased moderately with dose up to 40 dpa in Figure 10. YS/UTS for unirradiated material tested at room temperature is 0.9 as provided in the vendor inspection certificate. This indicates that there was no significant decrease in work hardening under this irradiation condition and that the material still retained work hardenability after the irradiation. On the other hand, UE/TE was less than 0.8, and it suggested that specimens exhibited a certain reduction in area. But the amount of total elongation at 380°C was decreased to about 6%, while UE/TE was almost 0.5 at around 10 dpa. These results indicate that it is necessary to take into account this phenomenon in order to evaluate the ductility of irradiated material. The results of SEM examination and

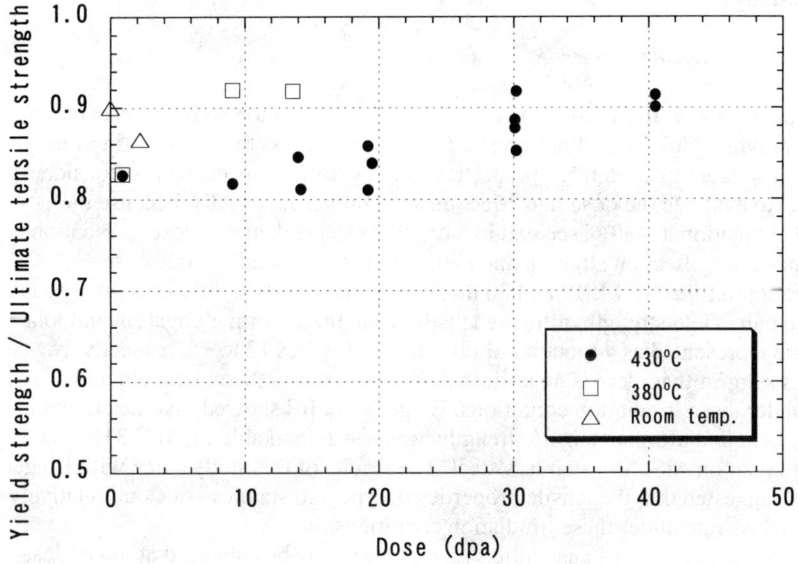

Figure 10---*The ratio of yield and ultimate tensile strength of 12% cold-worked type 316 SS after irradiation in EBR-II*

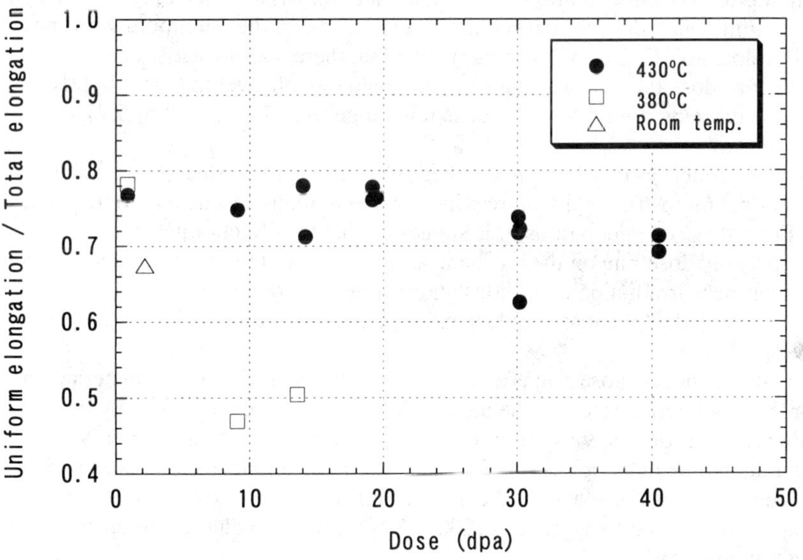

Figure 11---*The ratio of uniform and total elongation of 12% cold-worked type 316 SS after irradiation in EBR-II*

reduction in area also suggest that the failure mode would be nearly ductile failure in this irradiation condition and that there is no significant effect of neutron damage on failure mode in this study.

Dose Rate Effect on Tensile Properties

In Figures 6 and 7, the yield strength and uniform elongation were represented in two groups for which dose was either greater than 30 dpa or less than 30 dpa, in order to evaluate the dose rate effect on these properties roughly. But it is necessary to consider the dose level to evaluate the dose rate effect on such properties strictly, because the strength and elongation at 430°C seemed to have slight dependence of dose as mentioned above. The dose rate effect on strength and elongation is evaluated at three dose conditions of approximately 14, 19 and 30 dpa with a factor of two difference in dose rate for each dose pair. Yield strength, ultimate tensile strength, uniform elongation and total elongation are represented as a function of dose rate in Figures 12 to 15, respectively. These figures suggest that there is no noticeable effect of dose rate on strength and elongation under these irradiation conditions. Brager et al. [*6*] showed that the effect of dose rate on yield and ultimate tensile strengths was less remarkable on AISI 316 stainless steel than on AISI 304 stainless steel. The results of this study agree with Brager et al., so it is suggested that the tensile properties of type 316 stainless steel are relatively insensitive to dose rate under these irradiation conditions.

The swelling behavior of austenitic stainless steel may be enhanced at lower dose rates [*7*]. Thus it is important for evaluation of the soundness and irradiation behavior of core components which are fixed in the outer regions of the core for long periods during their service lives to consider the swelling and the effect of dose rate on material in such a low dose rate conditions. As mentioned in the introduction, the evaluation of swelling behavior and microstructural evolution of this duct material has also been investigated. Analysis of the immersion density measurements indicated that the maximum swelling was 1.6% (51.1 dpa, 407°C, 2.82×10^{-7} dpa/s) , and that there was no statistically significant effect of dose rate on swelling behavior even though swelling was slightly increased with decreasing dose rate in the dose rate range from 1.31×10^{-7} to 5.77×10^{-7} dpa/s [*8*].

The tensile properties of 12% cold-worked type 316 stainless steel in this study seem to reflect this microstructural feature which was essentially insensitive to dose rate in this case. From these considerations, it is suggested that the effect of neutron irradiation, especially dose rate on the mechanical properties of this material seems to be very small under these irradiation conditions and that the soundness of FBR core components such as reflector subassemblies and core vessels will be maintained in their projected lives.

On the other hand, no dose rate effect on the tensile properties of this material was observed for the low temperature and the narrow dose rate range experienced by the samples in this study. From the view point of the evaluation of the irradiation effects on the core material and the development of the future high-performance core material, it is necessary to compare these results with the data at higher dose rates (on the order of 10^{-6} dpa/s) and higher irradiation temperatures (500 ~ 600°C) which would be obtained in the fuel subassembly region.

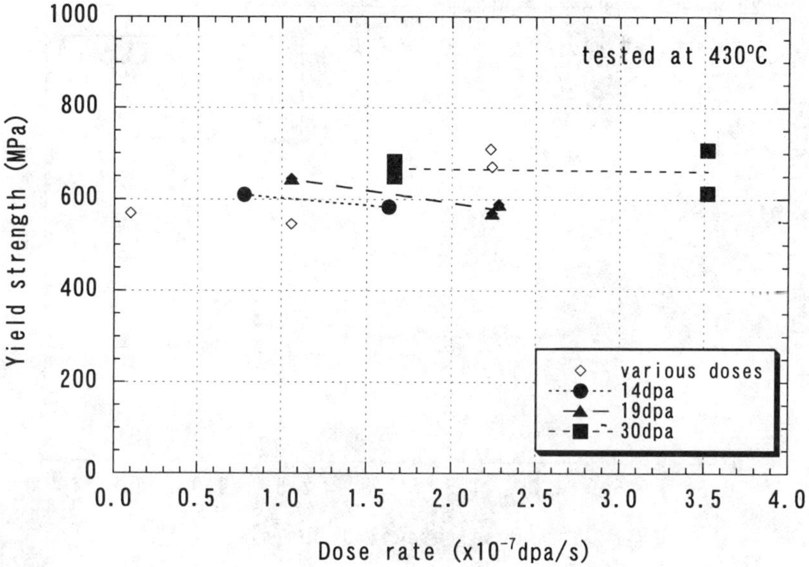

Figure 12---*Dose rate dependence of yield strength of irradiated 12% cold-worked type 316 SS tested at 430°C*

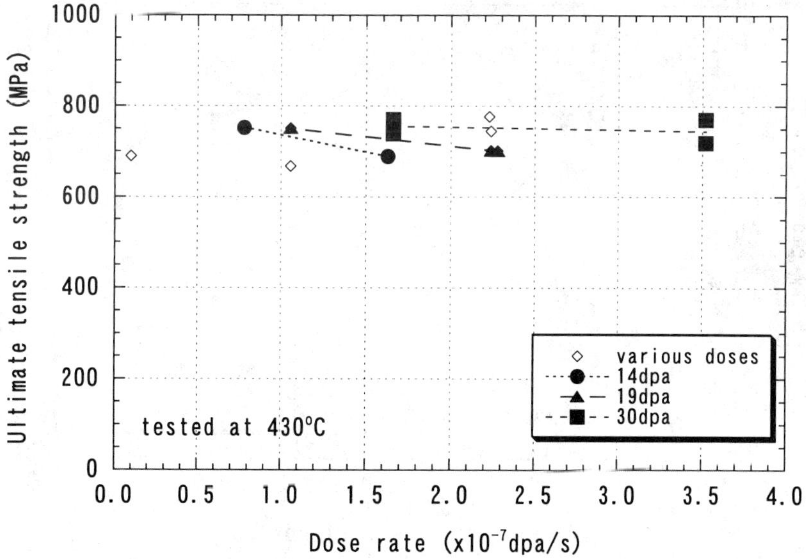

Figure 13---*Dose rate dependence of ultimate tensile strength of irradiated 12% cold-worked type 316 SS tested at 430°C*

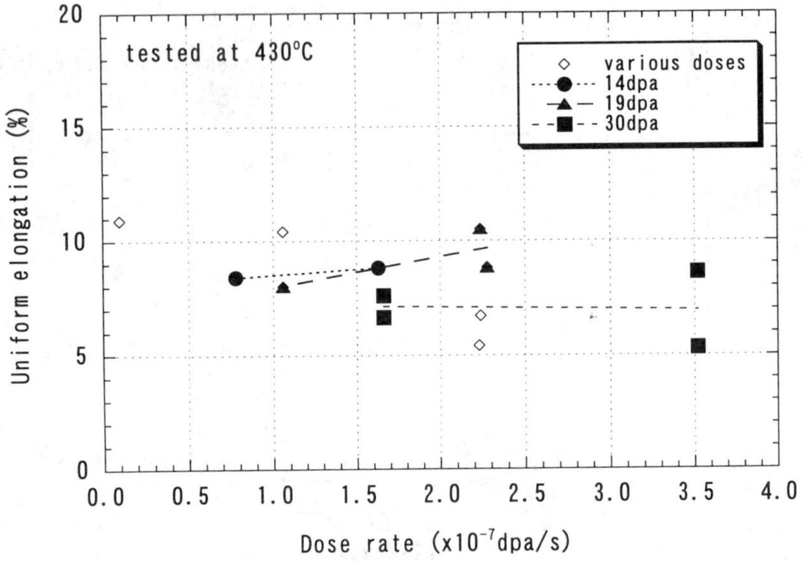

Figure 14---*Dose rate dependence of uniform elongation of irradiated 12% cold-worked type 316 SS tested at 430°C*

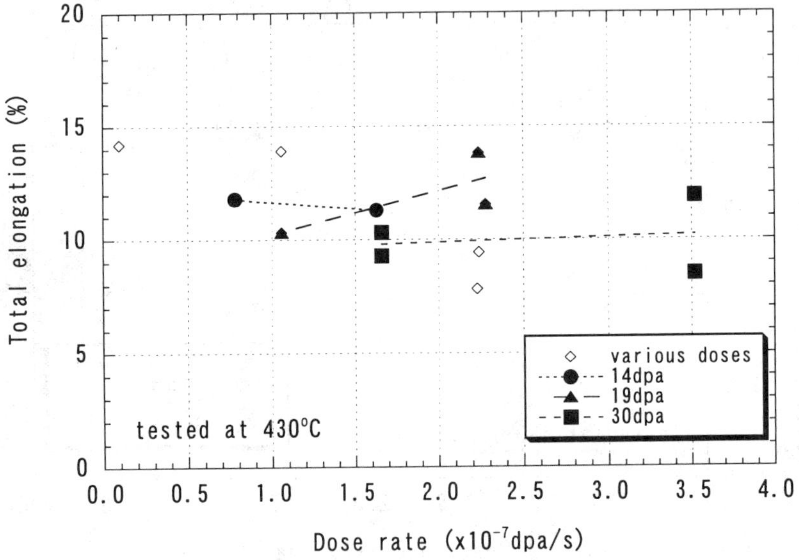

Figure 15---*Dose rate dependence of total elongation of irradiated 12% cold-worked type 316 SS tested at 430°C*

Conclusions

The following conclusions were drawn from this study:

1. The 12% cold-worked type 316 stainless steel showed typical temperature and dose dependence on irradiation hardening, and retained respectable ductility after the irradiation.
2. Fractographic examination indicated a ductile mode of fracture.
3. There was no significant effect of dose rate on tensile properties under the irradiation conditions examined, and this suggested that the soundness of FBR core components such as reflector subassemblies and core vessels would be maintained for their projected lives.

Acknowledgments

The authors wish to thank S. Ukai, Japan Nuclear Cycle Development Institute, for initial planning of the experiment. They also wish to thank K. N. Grimm, Argonne National Laboratory, for the EBR-II radial reflector DPA calculations, and D. O. Pushis, W. C. Kettman and T. S. Bray, Argonne National Laboratory, for specimen preparation and testing.

Work supported by U. S. Department of Energy, under contract No. W-31-109-ENG-38 with the Japan Nuclear Cycle Development Institute.

References

[1] Fish, R. L., Straalsund, J. L., Hunter, C. W., and Holmes, J. J., "Swelling and Tensile Property Evaluations of High-Fluence EBR-II Thimbles," *Proceedings of the Symposium on Effects of Radiation on Substructure and Mechanical Properties of Metals and Alloys, ASTM STP 529,* American Society for Testing and Materials, Philadelphia, 1973, pp. 149-164.

[2] Garner, F. A., Hamilton, M. L., Panayotou, N. F., and Johnson, G. D., "The Microstructural Origins of Yield Strength Changes in AISI 316 During Fission or Fusion Irradiation," *Journal of Nuclear Materials,* 1981, 103-104, pp.803-808.

[3] Dupouy, J. M., Erler, J., and Huillery, R., "Post Irradiation Mechanical Properties of Annealed and Cold Worked 316 Stainless Steel after Irradiation to High Fast Neutron Fluences," *Proceedings of International Conference: Radiation Effects in Breeder Reactor Structural Materials, Scottsdale,* The Metallurgical Society of AIME, New York, 1977, pp. 83-93.

[4] Holmes, J. J., and Straalsund, J. L., "Effects of Fast Reactor Exposure on The Mechanical Properties of Stainless Steels," *Proceedings of International Conference: Radiation Effects in Breeder Reactor Structural Materials, Scottsdale,*

The Metallurgical Society of AIME, New York, 1977, pp. 53-63.

[5] Fissolo, A., Cauvin, R., Hugot, J.-P., and Levy, V., "Influence of Swelling on Irradiated CW Titanium Modified 316 Embrittlement," *Effects of Radiation on Materials: 14th International Symposium, Vol. II, ASTM STP 1046,* American Society for Testing and Materials, Philadelphia, 1990, pp. 700-713.

[6] Brager, H. R., Blackburn, L. D., and Greenslade, D. L., "The dependence on Displacement Rate of Radiation-Induced Changes in Microstructure and Tensile Properties of AISI 304 and 316," *Journal of Nuclear Materials,* 1984, 122-123, pp.332-337.

[7] Seran, J. L., and Dupouy, J. M., "The Swelling of Solution Annealed 316 Cladding in RAPSODIE and PHENIX," *Effects of Radiation on Materials: 11th International Conference, ASTM STP 782,* American Society for Testing and Materials, Philadelphia, 1982, pp. 5-16.

[8] Cole, J. I., Allen, T. R., Tsai, H., Ukai, S., Mizuta, S., Akasaka, N., Donomae, T., and Yoshitake, T., "Swelling and Microstructural Evolution in 316 Stainless Steel Hexagonal Ducts Following Long-Term Irradiation in EBR-II," *Effects of Radiation on Materials: 20th International Symposium, ASTM STP 1405,* American Society for Testing and Materials, West Conshohocken, PA, 2002.

Akihiro Uehira[1], Shigeharu Ukai[2], Syunji Mizuta[2], and Raymond J. Puigh[3]

Irradiation Creep Deformation of Modified 316 and 15Cr-20Ni Base Austenitic Fuel Elements Irradiated in FFTF

Reference: Uehira, A., Ukai, S., Mizuta, S., and Puigh, R. J., **"Irradiation Creep Deformation of Modified 316 and 15Cr-20Ni Base Austenitic Fuel Elements Irradiated in FFTF,"** *Effects of Radiation on Materials: 20th International Symposium, ASTM STP 1405,* S. T. Rosinski, M. L. Grossbeck, T. R. Allen, and A. S. Kumar, Eds., American Society for Testing and Materials, West Conshohocken, PA, 2001.

Abstract: Irradiation creep of modified 316 (PNC316) and 15Cr-20Ni base austenitic stainless steels was determined using the diametral change of the fuel pins, which were irradiated under the fission gas pressurized condition in FFTF to a peak displacement damage of 107 dpa in PNC316 and 120 dpa in 15Cr-20Ni base. The creep compliance B_0 and creep-swelling coupling coefficient D were found to be 1.1 to 3.0 x 10^{-6} $MPa^{-1}dpa^{-1}$ and up to1.3 x 10^{-2} MPa^{-1} for instantaneous volumetric swelling rate, respectively. Those are almost the same level as those derived using the pressurized tubes irradiated in FFTF-MOTA. D values tend to decrease and disappear with developing void swelling.

Keywords : fast breeder reactor, core material, austenitic stainless steel, irradiation creep, fuel pin, FFTF

Core component materials for fast reactors must be highly resistant to radiation-induced swelling and creep and also have adequate high temperature creep strength. Austenitic stainless steels have superior high temperature strength, but less swelling resistance compared with ferritic or martensitic stainless steels. The modified 316 (PNC316) and 15Cr-20Ni base austenitic stainless steels were developed by Japan Nuclear Cycle Development Institute (JNC) as Monju

[1] Researcher, O-arai Engineering Center, Japan Nuclear Cycle Development Institute, 4002 Narita, O-arai, Ibaraki 311-1393, Japan.
Present address: Steel tube works, Sumitomo Metal Industries, 1 Nishino-cho, Higashi-mukojima, Amagasaki 660-8660, Japan.

[2] Senior researcher and Assistant senior researcher, respectively, O-arai Engineering Center, Japan Nuclear Cycle Development Institute, 4002 Narita, O-arai, Ibaraki 311-1393, Japan.

[3] Manager, Pacific Northwest National Laboratory, Richland, WA 99352.

core materials [*1-3*]. In PNC316 and 15Cr-20Ni, void swelling resistance and high temperature creep properties were improved by adjusting the minor elements such as titanium, niobium,phosphorous, boron and the amount of cold working [*4*].

In studies at irradiation creep behavior, it is reported [*5,6*] that the behavior of austenitic claddings irradiated as fuel pins is not the same as for irradiated in pressurized tubes. The measured irradiation creep strains of a cold worked 316 Ti fuel pin irradiated in Phenix were three times higher than those predicted by using pressurized tubes irradiation test results. On the other hand, in the case of 20% cold worked 316 fuel pin irradiated in EBR-II, the observed irradiation creep strains were smaller than expected. Additionally, it was reported [*6-9*] that swelling coupling irradiation creep tended to decline when void swelling rate became higher. There is the possibility that the creep-swelling coupling coefficient D diminishes with increased swelling rate.

In this paper, the irradiation creep of the PNC316 and 15Cr-20Ni base stainless steels were studied using the fuel pins irradiated in FFTF and compared with results of pressurized tubes irradiated in FFTF-MOTA [*10*].

Experimental Procedure

The chemical compositions, solution treatment conditions and cold-worked levels of PNC316 and 15Cr-20Ni base fuel pin claddings irradiated in FFTF are shown in Table 1. In PNC316 (lots: 60MK and 60MS), the minor elements such as titanium, niobium, phosphorous, and boron were adjusted in the range of JIS (Japanese Industrial Standard) 316 and the amount of cold working was optimized in order to improve the void swelling resistance and high temperature creep strength. For further improvement in the swelling resistance, the ratio of nickel to chromium and the amount of titanium were increased beyond JIS 316 specification in 15Cr-20Ni base stainless steels (lots: 60AK and 60AS).

PNC316 and 15Cr-20Ni base fuel pins were irradiated in FFTF as MFA-1 (Monju Fuel Assembly-1) [*11*] and MFA-2 (Monju Fuel Assembly-2), respectively. The difference of MFA-1 and MFA-2 assemblies is only the materials for the components, which are made of PNC316 for cladding, wrapper and wire in MFA-1 and 15Cr-20Ni in MFA-2. The as-fabricated specifications of MFA-1 and MFA-2 are listed in Table 2. The number of pins in the fuel assembly and dimensions of each component are exactly the same as those in the Monju fuel assembly. The irradiation conditions of MFA-1 and MFA-2 in FFTF are shown in Table 3. The peak displacement damage reached as high as 107 and 120 dpa, respectively.

The outer diameters of fuel pins were measured in a post irradiation examination (PIE) with a laser profilometer in order to determine the diametral changes of claddings due to radiation-induced creep and swelling. Density measurements of claddings were also conducted to directly determine the swelling, which was used to subtract from the total diametral change to give the amount of creep strain itself. Both diameter and density measurements were conducted for 6 pins in MFA-1 and 5 pins in MFA-2. Based on these PIE data, the creep compliance B_0 and creep-swelling coupling coefficient D of PNC316 and 15Cr-20Ni were estimated.

Table 1— *The chemical compositions (mass %), solution treatment conditions and cold working levels of PNC316 and 15Cr-20Ni base fuel pin claddings.*

Lots	60MK (PNC316)	60MS (PNC316)	60AK (15Cr-20Ni base)	60AS (15Cr-20Ni base)
C	0.054	0.057	0.057	0.064
Si	0.78	0.79	0.41	0.75
Mn	1.72	1.85	1.72	1.91
P	0.028	0.026	0.026	0.028
S	0.003	0.002	0.005	0.002
Ni	13.45	13.72	19.66	19.73
Cr	16.22	16.33	14.89	15.05
Mo	2.35	2.54	2.55	2.56
B	0.0039	0.0036	0.0044	0.0031
N	0.0090	0.0030	0.002	0.002
Ti	0.078	0.075	0.28	0.25
Nb	0.080	0.095	0.09	0.11
Solution treatment	1353 K x 2 min	1368 K x 1 min	1398 K x 2 min	1353 K x 1 min
Cold work	18.0 %	20.6 %	18.0 %	15.6 %

Table 2— *As-fabricated specifications of MFA-1 and MFA-2 fuel assemblies.*

Number of fuel pins		169
Cladding tube		
Outer diameter	(mm)	6.5
Inner diameter	(mm)	5.56
Wall thickness	(mm)	0.47
Wrapper tube		
Distance from face to face	(mm)	104.6
Distance from corner to corner	(mm)	110.6
Wall thickness	(mm)	3.0
Fuel column length	(mm)	914.4

Table 3 — *Irradiation conditions of MFA-1 and MFA-2.*

		MFA-1	MFA-2
Peak displacement damage	(dpa)	107	120
Peak burn up	(GWd/t)	147	162
Peak linear power	(W/cm)	349	352
Cladding temperature	(K)	633~923	633~913
Effective full power days		903	1004

Results and Discussion

Calculation of Irradiation Creep Strains

The typical examples for measured diametral changes and swelling strains in the representative PNC316 and 15Cr-20Ni base fuel pins are shown in Figs. 1 and 2. At the distance from core bottom (DFCB) of about 300 to 500 mm, the swelling strains contributed mainly to the total diametral changes. The relationship in total diametral change and swelling strain for the PNC316 and 15Cr-20Ni base fuel pins is shown in Figs. 3 and 4. The difference in the diametral change and swelling strain corresponds to the creep strain, and it seems to be almost constant even with increasing swelling strain.

The hoop irradiation creep strain, ε_H, is calculated by using the following correlation:

$$\varepsilon_H = (\Delta D/D_0)_{total} - (\Delta D/D_0)_{swelling} - (\Delta D/D_0)_{thermal} \quad (1)$$

where

$(\Delta D/D_0)_{total}$ = total diametral change
$(\Delta D/D_0)_{swelling}$ = swelling strain determined by density measurement
$(\Delta D/D_0)_{thermal}$ = thermal creep strain

The amount of thermal creep strain was predicted by using the correlation based on PNC316 and 15Cr-20Ni base pressurized tubes irradiated in FFTF-MOTA. The contribution of thermal creep to the total diametral change could not be negligible at the temperature above about 773 K. The hoop creep strain, ε_H, was converted to the equivalent creep strain, by using the correlation, $\varepsilon = 2/\sqrt{3}\ \varepsilon_H$, based on Henckey's flow rule. The following derivation is conducted using this equivalent creep strain and equivalent stress, $\sigma = \sqrt{3}/2\ \sigma_H$.

Determination of Irradiation Creep Compliance: B_0

The irradiation creep is able to be expressed in the general form [6]:

$$\dot{\varepsilon} / \sigma = B_0 + D\dot{S} \quad (2)$$

where

$\dot{\varepsilon}$ = instantaneous irradiation creep strain per displacement damage
σ = stress
B_0 = irradiation creep compliance
D = creep-swelling coupling coefficient
$\dot{S}$ = instantaneous volumetric swelling

The average creep modulus, Bave, is defined as using displacement damage, ϕt,

$$\begin{aligned} \text{Bave} &= B_0 + D\dot{S} \\ &= (d\varepsilon / d\phi t) / \sigma \end{aligned} \quad (3)$$

Stress in a cladding tube of a fuel pin increases almost linearly due to release of generated fission gases from the fuel during irradiation, and the following correlation was determined at each

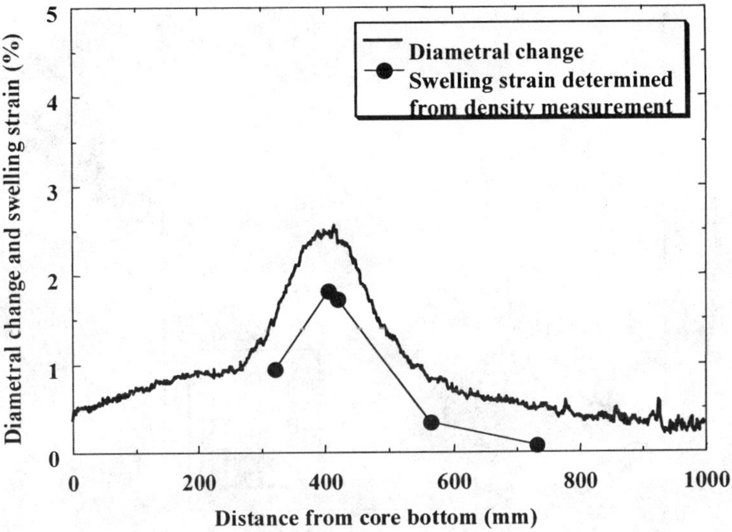

Figure 1 — *Measured diametral change and swelling profiles of PNC316 fuel pin (Pin No.186019)*

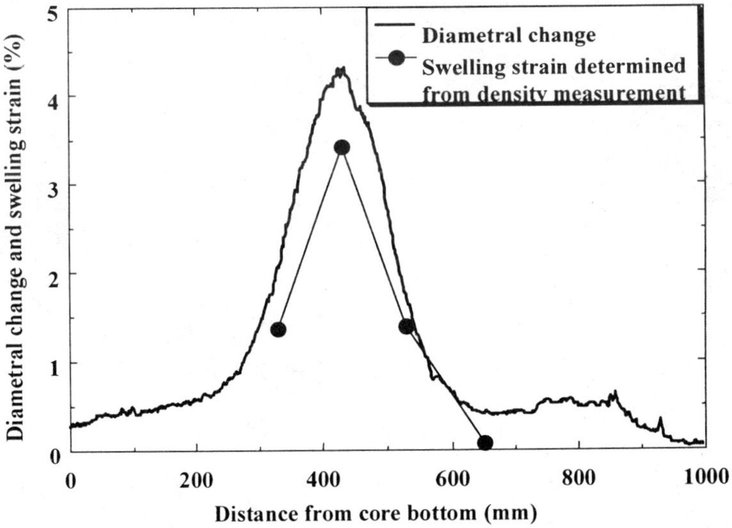

Figure 2 — *Measured diametral change and swelling profiles of 15Cr-20Ni base fuel pin (Pin No.186019)*

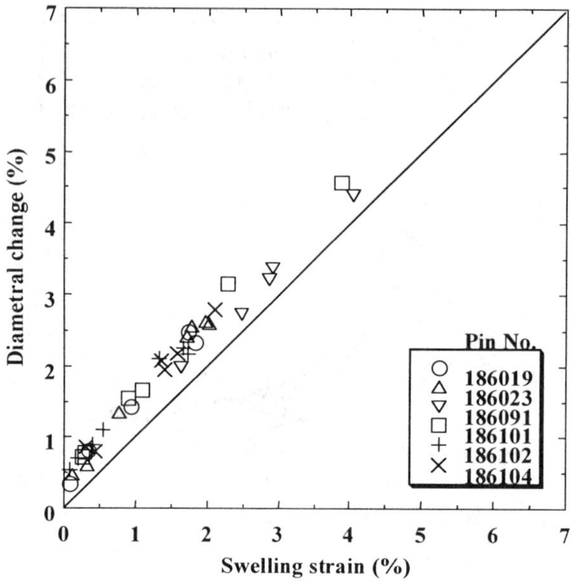

Figure 3 — *Relation between diametral change and swelling strain in PNC316 fuel pins*

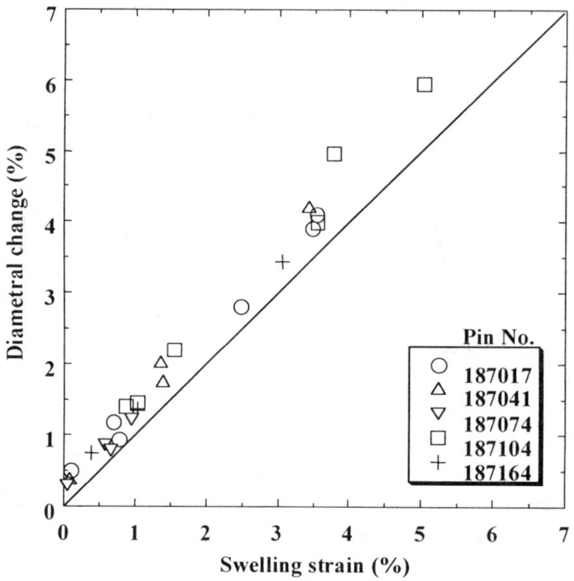

Figure 4 — *Relation between diametral change and swelling strain in 15Cr-20Ni base fuel pins*

DFCB location of every MFA-1 and MFA-2 fuel pins,

$$\sigma = a\ \phi t + b \quad (4)$$

where a and b are constant. Equation (3) is integrated using equation (4), and average creep modulus, Bave, can be expressed as:

$$Bave = \varepsilon' / (1/2\, a\, \phi t'^{2} + b\, \phi t') \quad (5)$$

ε' = irradiation creep strain at the end of irradiation

$\phi t'$ = displacement damage at the end of irradiation

Values of Bave at each DFCB location in all pins were calculated using equation (5). The typical results from fuel pin No. 186019 are shown in Fig. 5. Bave values tend to increase with increasing instantaneous volumetric swelling rate at the end of irradiation, since Bave values include creep-swelling coupling component. From equation (2) the irradiation creep compliance, B_0, can be determined when swelling rate becomes zero at the data nominal line as shown in Fig. 5.

The data ranges of B_0 derived for PNC316 fuel pins in MFA-1 and for 15Cr-20Ni base fuel pins in MFA-2 are shown in Fig. 6, which compares them with PNC316 and 15Cr-20Ni base pressurized tubes irradiated in MOTA [*10*] and other austenitic stainless steels, such as 20% cold-worked 316 and cold-worked 316Ti from the literature [*5,6,12*]. The creep compliances, Bo, of PNC316 and 15Cr-20Ni base fuel pins show similar behavior. The derived range of B_0 for PNC316 and 15Cr-20Ni base fuel pins is 1.1 to 3.0 x 10^{-6} $MPa^{-1}dpa^{-1}$, which is located in the range of other austenitic stainless steels, although it tends to be slightly larger than PNC316 and 15Cr-20Ni irradiated in MOTA.

Determination of Irradiation Creep-Swelling Coupling Coefficient: D

In this study, irradiation creep-swelling coupling coefficient D was estimated, based on equation (2) with the instantaneous swelling rate. The equation (6) can be derived using equation (3) and (4).

$$d\varepsilon = (B_0 + D\dot{S})(a\phi t + b)\, d\phi t \quad (6)$$

The following equations are obtained by integrating Equation (6).

$$\varepsilon' = B_0 (1/2\, a\, \phi t'^2 + b\, \phi t') + aD \int \dot{S}\, \phi t\, d\phi t + bD \int \dot{S}\, d\phi t$$
$$= B_0 (1/2\, a\, \phi t'^2 + b\, \phi t') + aD \int \phi t\, dS + bDS' \quad (7)$$

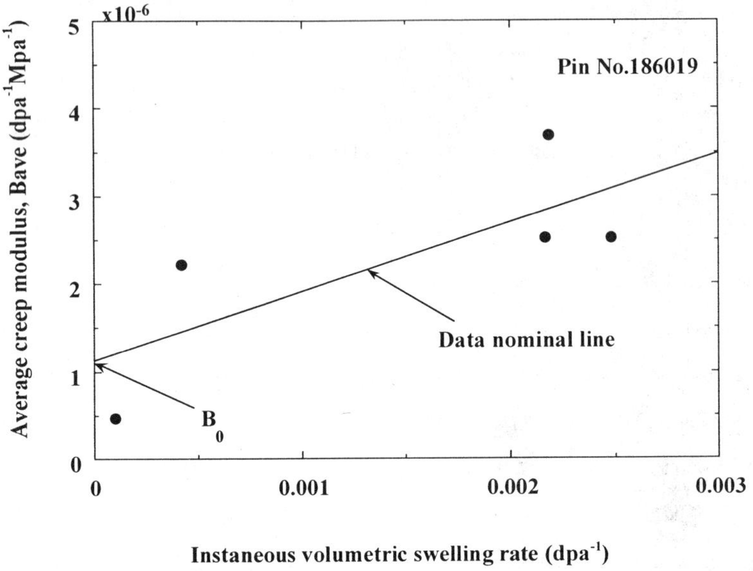

Figure 5 — *Determination of creep compliance Bo*

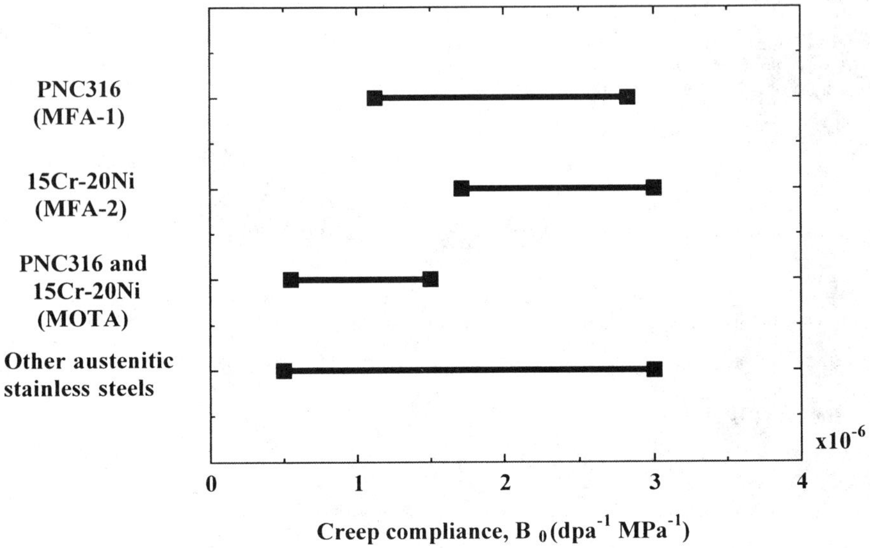

Figure 6 — Comparison of creep compliance of Bo with other data in literatures

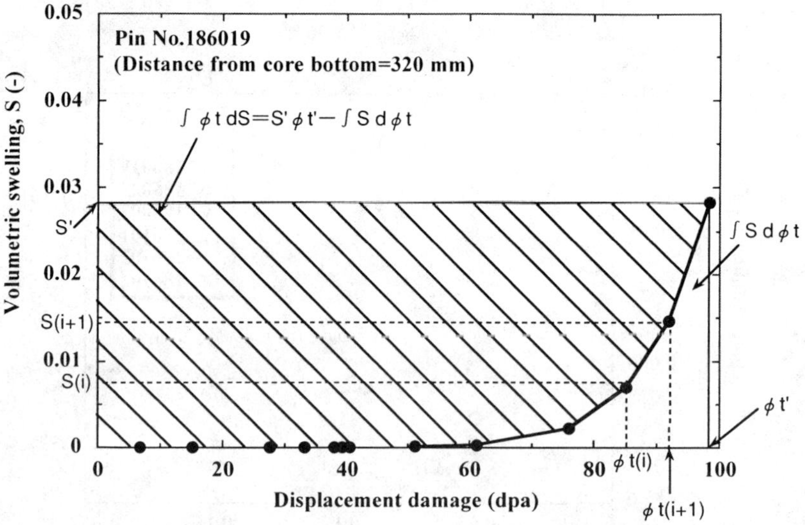

Figure 7 — *Calculation of integral part in equation (8) to determine D from relationship between displacement damage and swelling*

where

S' = volumetric swelling at end of irradiation.

The second integral part of equation (7) can be expressed by the following equation, as shown by the shaded area in Fig.7.

$$\int \phi t\, dS = S' \phi t' - \int S\, d\phi t \tag{8}$$

Equation (8) can be approximately calculated by

$$\int \phi t\, dS = S' \phi t' - \Sigma \{1/2\, [S(i) + S(i+1)]\, [\phi t(i+1) - \phi t(i)]\} \tag{9}$$

where

S (i) = volumetric swelling at (i) irradiation cycle

S (i+1) = volumetric swelling at (i+1) irradiation cycle

ϕt (i) = displacement damage at (i) irradiation cycle

ϕt (i+1) = displacement damage at (i+1) irradiation cycle

The creep-swelling coupling coefficient, D, can be finally determined using equation (10), which was obtained by substituting equation (9) to equation (7):

$$D = \{\varepsilon' - B_0 (1/2\, a\, \phi t'^2 + b\, \phi t')\} / [a\, S' \phi t' - a \Sigma \{1/2\, [S(i) + S(i+1)]\, [\phi t(i+1) - \phi t(i)]\} + b\, S'] \tag{10}$$

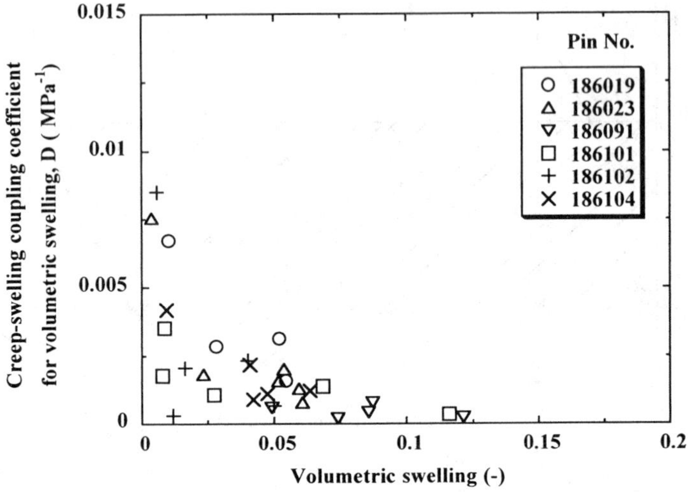

Figure— 8 *Dependence of creep-swelling coupling coefficient on volumetric swelling in MFA-1 PNC316 fuel pins*

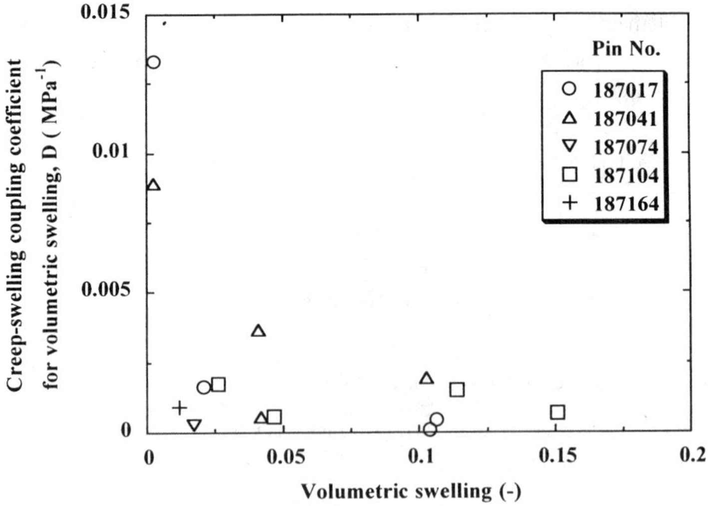

Figure— 9 *Dependence of creep-swelling coupling coefficient on volumetric swelling in MFA-2 15Cr-20Ni fuel pins*

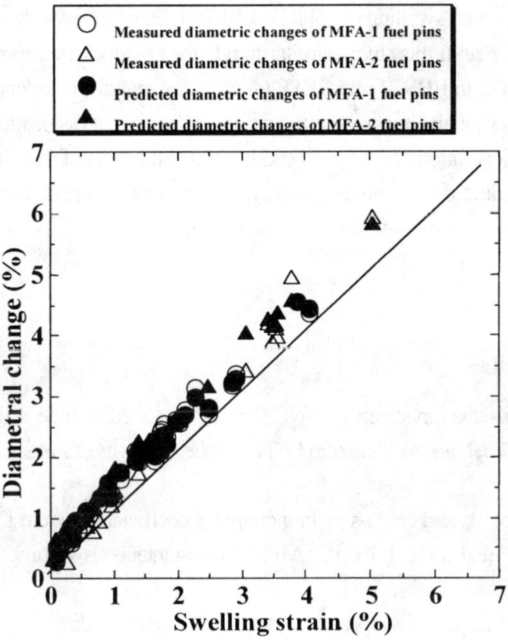

Figure 10— *Comparison of predicted diametral changes based on MFA-1 and MFA-2 fuel pins with measured changes*

The values of D derived for PNC316 and 15Cr-20Ni base fuel pins are shown in Figs.8 and 9, respectively. It appears that the D values tend to diminish with increasing swelling. The swelling-creep coupling coefficient was demonstrated to decline and almost disappear with developing swelling. This disappearance of D values as swelling increases accounts for the measured constant creep strain even in increasing swelling as shown in Figs.3 and 4. Fig. 10 compares the measured diametric changes and predicted ones using the derived Bo and D values. Both agree reasonably well. From these analyses it is suggested that the diametric increase of MFA-1 and MFA-2 fuel pins due to irradiation creep is due mainly to the creep compliance Bo term, and the contribution of the creep-swelling coupling D term is minor at the swelling dominant regime.

In the MOTA irradiation tests with the pressurized tubes of PNC316 and 15Cr-20Ni, D values were also derived using both instantaneous and average swelling rates. It was shown that the D values for instantaneous swelling rates are declined from 9 x 10^{-3} MPa^{-1} to 2 x 10^{-3} MPa^{-1} with increasing swelling [*10*]. Therefore, it is concluded that irradiation creep behavior of PNC316 and 15Cr-20Ni in fuel pin cladding is almost same as that of pressurized tubes irradiation tests.

With regard to the creep-swelling coupling coefficient, D, which was shown to be declined with increasing swelling rate in this study, similar trends for D values are reported in AISI 316 pressurized tubes irradiated in EBR-II and FFTF [*6-9*]. An analytical explanation is provided by Woo and Garner, based on the anisotropic to isotropic transition of dislocation microstructure associated with the void evolution [*13,14*]. It can be said that results of this study substantially support earlier work conducted by Woo and Garner. More exact correlation will be necessary in terms of theoretical-basis interpretation.

Conclusions

In this study, the irradiation creep of PNC316 and 15Cr-20Ni base stainless steels was determined by means of fuel pins irradiated in FFTF. The results are summarized as follows:

(1) The creep compliance B_0 and creep-swelling coupling coefficient D were found to be 1.1 to 3.0 x 10^{-6} $MPa^{-1}dpa^{-1}$ and up to 1.3 x 10^{-2} MPa^{-1} for instantaneous volumetric swelling rate, respectively.
(2) D values tend to decrease and disappear with developing void swelling.
(3) The irradiation creep behavior of Bo and D for fuel pin irradiation tests is the same level derived from the pressurized tubes irradiated in FFTF-MOTA.

Acknowledgement

We wish to gratefully acknowledge Mr.MPaxton of PNNL for Executing cooperation program, and Mr.T.Asaga, Mr.T.Mitsugi, Mr.K.Katsuyama, Mr.T.Omori and other members of Fuels and Materials Division in O-arai Engineering Center for carrying out this experiment.

References

[1]Nomura, S., Shikakura, S., Ukai, S., Seshimo, I., Harada, M., Shibahara, I., and Katsuragawa, M., "Development of Long Life FBR Core Materials," *Proceedings of International Conference on Fast Reactors and Related Fuel Cycles,* 1991, Vol. I, pp. 7.4-1.

[2]Tateishi, Y., "Development of Long Life FBR Fuels with Particular Emphasis on Cladding Material Improvement and Fuel Fabrication," *Journal of Nuclear Science and Technology*, 1989, Vol. 26, No. 1, pp. 132-136.

[3]Ukai, S., Harada, M., Nomura, S., Shikakura, S., and Shibahara, I., "Development of Long Life FBR Core Materials", *Proceedings of the International Symposium on Material Chemistry in Nuclear Environment*, 1992, pp. 347-357.

[4]Ukai, S., Akasaka, N., Hattori, K.and Onose, S., "The Effects of Phase Stability on Void Swelling in P, Ti-Modified 316 Stainless Steels during Neutron Irradiation", *Effects of*

Radiation on Materials: 18 th International Symposium, ASTM STP 1325, R.K.Nanstad, M.L.Hamilton, F.A.Garner and A.S.Kumar, Eds., American Society for Testing and Materials, West Conshohocken, 1999, pp.808-820.

[5]Selan, J., L., Touron, H., Maillard, A., Dubuisson, P., Hugot, J., P., Boulbin, E., Blanchard, P., and Pelletier, M., "The Swelling Behavior of Titanium-Stabilized Austenitic Steels Used as Structural Materials of Fissile Subassemblies in Phenix ," *Effects of Radiation on, Materials: 14 th International Symposium (Volume II), ASTM STP 1046*, N. N. Packan, R. E. Stoller, and A. S. Kumar, Eds., American Society for Testing and Matcrials, West Conshohocken, 1990, pp.739-752.

[6]Garner, F., A., " Irradiation Performance of Cladding and Structural Steels in Liquid Metal Reactors" , *Materials Science and Technology, vol. 10 A*, R. W. Cahn, P. Haasen and E. J. Kramer, 1994, pp.420-543.

[7]Porter, D. L., and Garner, F. A., " Cessation of Irradiation Creep in AISI 316 Concurrent with High Level of Swelling" , *Influence of Radiation on Material Propwrties:13 th International Symposium, Part II, ASTM STP 956*, 1987, pp.11-21.

[8]Garner, F. A , Porter, D.L and Makenas, B. J, "A third Stage of Irradiation Creep Involving Its Cessation at High Neutron Exposures", J. Nuclear Materials 1987, 148, 279-287.

[9]Garner, F.A, Toloczko, M.B and Puigh, R.J, "The Relationship between Swelling and Irradiation Creep in 20% Cold-Worked 316 Stainless Steel", to be published in *Effects of Radiation on, Materials: 19 th International Symposium, ASTM STP 1366*, M.L.Hamilton, A.S.Kumar, S.T.Rosinski and M.L.Grossbeck, Eds., American Society for Testing and Materials, West Conshohocken, PA, 1999.

[10]Uehira, A., Mizuta, S., and Ukai, S., "Irradiation Creep of 11Cr-0.5Mo-2W,V,NbFerritic-Martensitic, Modified 316, and 15Cr-20Ni Austenitic S. S. Irradiated in FFTF to103 - 206 dpa, ", *to be published in Ninth International Conference on Fusion Reactor Materials*, 1999.

[11]Ukai, S., Yoshitake, T., Akasaka, N., Donomae, T., Katsuyama, K., Mitsugi, T., and Asaga, T., "Irradiation Performance of FBR Monju-Type Fuel with Modified Type 316 Stainless Steel at High Burnup", *American Nuclear Society Transactions*, 1998, pp.115.

[12]Maillard, A., Touron, H., Scran, J., L., and Chalony, A., "Swelling and Irradiation Creep of Neutron-Irradiated 316Ti and 15-15Ti Steels," *Effects of Radiation on Materials: 16 th International Symposium, ASTM STP 1175*, A. S. Kumar, D. S. Gelles, R. K. Nanstad and E. A. Little, Eds., American Society for Testing and Materials, West Conshohocken , 1993, pp.824-837.

[13]Woo,C.H., Garner,F.A. and Holt,R.A., "Irradiation Creep due to SIPA under Cascade Damage Conditions", ibid [12], pp.27-37.

[14]Woo,C.H, J.Nuclear Materials, 1995, 225, 8-14.

Jean-Christophe Brachet,[1] Xavier Averty,[2] Pierre Lamagnère,[1] Ana Alamo,[1] Frank Rozenblum,[3] Olivier Raquet,[4] and Jean-Luc Bertin[1]

Behavior of Different Austenitic Stainless Steels, Conventional, Reduced Activation (RA) and ODS Chromium-Rich Ferritic-Martensitic Steels Under Neutron Irradiation at 325°C in PWR Environment

Reference: Brachet, J.-C., Averty, X., Lamagnère, P., Alamo, A., Rozenblum, F., Raquet, O., and Bertin, J.-L., **"Behavior of Different Austenitic Stainless Steels, Conventional, Reduced Activation (RA) and ODS Chromium-Rich Ferritic-Martensitic Steels Under Neutron Irradiation at 325°C in PWR Environment,"** *Effects of Radiation on Materials: 20th International Symposium, ASTM STP 1405,* S. T. Rosinski, M. L. Grossbeck, T. R. Allen, and A. S. Kumar, Eds., American Society for Testing and Materials, West Conshohocken, PA, 2001.

Abstract: The main objective of this paper is to summarize CEA data recently obtained on different kinds of steels irradiated at 325°C in the Osiris experimental reactor in a typical PWR environment, that is, pressurized water – P=155 bars – with controlled PWR type chemistry. The different steels studied can be classified in four groups: 304/316 type austenitic stainless steels, conventional 9-12%Cr(Mo,V,Nb) and reduced activation 7.5-11%Cr(W,V,Ta) martensitic steels, and two ferritic-martensitic alloys strengthened by oxide dispersion (ODS). Some of those steels have been included with different initial metallurgical conditions, that is, (1) for austenitic steels : solution annealed and cold-worked structures; (2) for martensitic steels : tempered, cold-worked and as-quenched martensitic structures. This experimental irradiation, named "Alexandre", has been carried out in the Osiris experimental reactor (under a mixed fast/thermal neutron flux) for different neutron fluence levels with a maximum irradiation damage of ~9dpa. The main results obtained are discussed, with a special emphasis on the chemical composition and initial metallurgical condition effects on the tensile and uniform corrosion properties of both conventional and reduced activation chromium-rich martensitic steels.

Keywords: 304/316, 8-12%Cr martensitic steels, reduced activation, ODS, neutron irradiation at 325°C, tensile properties, uniform corrosion, PWR environment

Commissariat à l'Energie Atomique, CEA-Saclay, 91191 Gif-Sur-Yvette, France:
[1] DTA/CEREM/SRMA
[2] DRN/SEMI
[3] DRN/SIREN
[4] DTA/CEREM/SCECF

Introduction

Type 304/316 austenitic steels are the actual reference materials for different nuclear applications (PWR in-core components, ITER Fusion reactor project, …). Nevertheless, these materials present limitations for in-service conditions at low irradiation temperatures (below 400°C), such as hardening/embrittlement, irradiation assisted stress corrosion cracking (IASCC) and high residual activation due to their Ni (+ Co) content.

As an alternative to conventional austenitic stainlees steels, 8-12% Cr martensitic steels are recognised to have good potential for both fission and fusion reactors structural component applications. These steels have an intrinsic high resistance to swelling under heavy neutron irradiation conditions and they display attractive mechanical properties (strength, toughness, ...) and interesting physical properties (low thermal expansion coefficient, adequate thermal conductivity, ...). However, they are prone to hardening/embrittlement for low in-service temperatures (less than ~ 350-400°C) and, consequently, an important decrease of their toughness is generally observed with the increasing neutron fluence in this temperature range. Furthermore, if they have to be used in an aqueous environment, their corrosion resistance can be suspect when compared to that of austenitic stainless steels.

The aim of this paper is to summarize some recent data obtained at CEA on the behavior of a great variety of steels including 304/316 type stainless steels (which are reference materials), conventional and reduced activation martensitic steels and two Oxide Dispersion Strengthened (ODS) ferritic-martensitic steels, irradiated at 325 +5/-10°C in the experimental Osiris reactor under mixed neutron spectrum and for a typical PWR environment, that is, pressurized water with the relevant controlled chemistry. One of the main objectives of this irradiation called "Alexandre" is to provide a direct comparison of the properties of all these types of steels, some of them under different metallurgical conditions, as a function of the fluence for neutron doses ranging from ~0.9 up to 9 dpa. This last dose is assumed to represent a "saturation" dose for the hardening/embrittlement of both austenitic [1] [2] and martensitic [3] steels. The properties studied are : tensile properties, including Reduction of Area (RA) measurements, in-pile and out-of-pile uniform corrosion and residual activation (dose rate measurements and γ spectrometry). Moreover, Scanning Electron Microscopy (SEM) fractograph analysis, Transmission Electron Microscopy (TEM) and Small Angle Neutron Scattering [4] studies are in progress to understand the irradiation-induced evolution of properties.

Experimental

Materials

Four groups of steels have been studied. Denominations, suppliers, nominal chemical compositions and metallurgical conditions of the alloys are summarized in Table 1.

Table 1 - *Nominal chemical compositions and metallurgical conditions of the different steels*

Alloys	Suppliers	Nominal chemical composition (*)	Initial metallurgical conditions (**)
		Austenitic stainless steels :	
304-SA	Creusot-Loire	18.7Cr-9.8Ni	Solutioned Annealed
316L(N)-SA	"SuperPhenix" grade	17Cr-12.4Ni-2.5Mo	Solutioned Annealed
316-CW	UGINE	17Cr-10.7Ni-2.3Mo	Cold-Worked
316Ti-CW	Creusot-Loire	17.4Cr-12.8Ni-2.4Mo-0.3Ti	Cold-Worked
		Conventional martensitic (+ one bainitic) steels :	
EM10(1)	Aubert & Duval	9Cr1Mo(0.2Ni)	N+T+CW+T, AQ
EM10(2)	Imphy	9Cr1Mo(0.5Ni)	N+T+CW
T91	Marrel-Industrie	9Cr1Mo(Nb,V)	N+T+CW
MANET	Saarstahl	10.5Cr0.6Mo(Nb,V)	N+T
HT9	Sandvik	12Cr1Mo0.5W(V)	N+T, AQ
"M3"-Experimental	Imphy	9Cr0.5Mo	N+T+CW+T
"T5" - Exp.	Imphy	9Cr(Nb,V)	N+T+CW
2 1/4Cr	/	2.4Cr1Mo	Tempered bainite
		RA martensitic steels :	
F82H	JAERI - NKK	7.5Cr2W(V,Ta)	N+T, AQ
LA12LC - Exp.	UKAEA – B.S.	9Cr0.7W(V)	N+T+CW
LA13Ta - Exp.	UKAEA – B.S.	9Cr3W(V,Ta)	N+T+CW
LA4Ta - Exp.	UKAEA – B.S.	11Cr0.7W(V,Ta)	N+T+CW
		ODS ferritic-martensitic steels :	
EM10+Y_2O_3 - Exp.	CEA	9Cr1Mo+Y_2O_3	Extruded+N+R
MA957	Inco Alloys	13Cr0.3Mo0.9Ti +Y_2O_3	As-extruded

For martensitic steels :
(*) Carbon concentration ranges from 0.1 to 0.2 wt%,
(**) N: normalization at 980-1075°C, T : tempering at 740-800°C, CW: cold-working 10-20% and AQ: as-quenched martensite (this last metastable metallurgical condition could simulate welds without post-weld heat treatments, for in-service structures repairing and/or replacement as examples).

Irradiation conditions

The "Alexandre" irradiation started in January 1996 and has been carried out in the "Coralline" high pressure water installation of the OSIRIS reactor at Saclay. Different partial unloadings of samples resulted in different neutron damage levels. The main characteristics of the irradiation conditions are summarized in Table 2.

Post- Irradiation Examinations (PIE)

The doses of ~5.2 and ~9.3 dpa were achieved at the end of 1999. As such, weighing of these samples has already been done but the tensile results are not yet available.

Table 2 – *"Alexandre" irradiation conditions*

Average reactor power	65 MW
Temperature	325 °C +5/-10°C
Fast/thermal neutron flux ratio	1.1
Maximum neutron flux	2 10^{14} n/cm².s - E>1MeV
Achieved doses (estimation)	~ 0.9, 2, 3.5, 5.2 and 9.3 dpa corresponding respectively to exposure times ~ 43, 97, 140, 252 and 361 days
Environment	- Pressurized water (155 bars) - controlled PWR chemistry (O_2<10ppb) - water coolant flow ~ 2-4 l/h

- *Tensile testing* - Tensile tests have been done on plate samples 2mm wide, 1mm thick and 8 mm in gauge length at a conventionnal strain rate of 3 10^{-4} s^{-1}. Accurate and systematic quantifications of the Reduction of Area (RA) to failure have been performed using an image analysis procedure (Figure 1). It is worth pointing out that RA can be correlated to the impact behavior (USE) as shown later, and so, is a ductility parameter of great importance to fully characterize the brittleness (or not) of the samples after thermal ageing [5] [6] and/or after irradiation [7] [8]. This is illustrated in Figure 2 which compares two irradiated martensitic alloys showing nearly the same total elongation (TE) values (obtained from the engineering tensile curves) but very different RA (and UE) values.

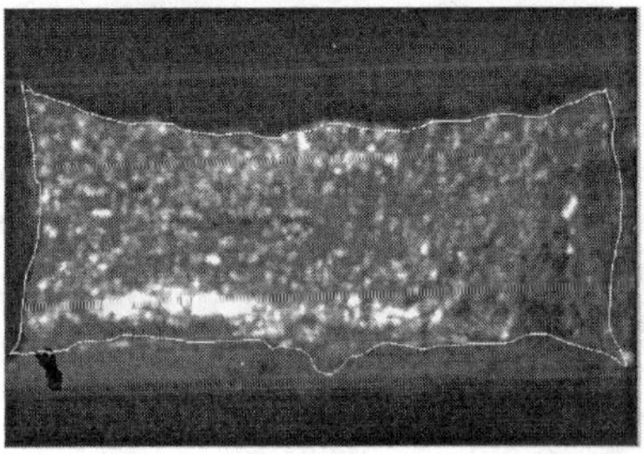

Figure 1 - *304-SA irradiated at 3.5 dpa => Determination of the Reduction of Area by image analysis procedure :*
** RA (325°C - 0 dpa.) ~ 80%*
** RA (325°C – 3.5 dpa) = 52%*

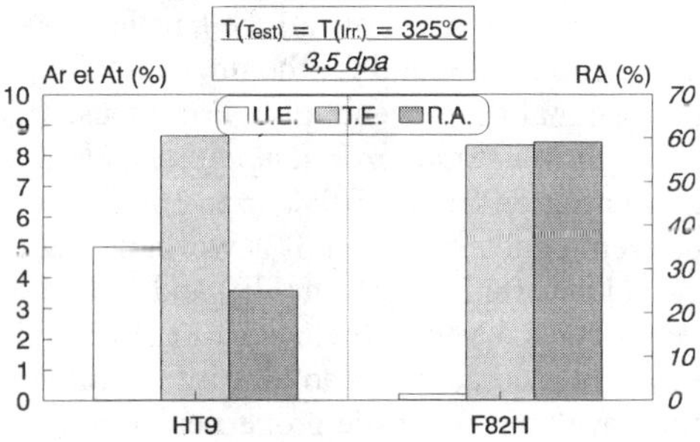

Figure 2 - *Post-irradiation tensile ductility at 325°C for tempered HT9 and F82H steels, showing the same Total Elongation values but very different Reduction of Area values.*

- *Uniform corrosion measurements* - Systematic weighing of the irradiated samples after Ultrasonic (US) cleaning has been performed without descaling. For comparison purposes and to get some insight of a possible irradiation effect on the uniform corrosion kinetics, out-of-pile static autoclave tests have been carried out on the same materials and with the same environmental conditions (same temperature and water chemistry). For these last tests, two types of experiments were used :

(1) 10 000 hours duration autoclave test without partial unloadings;

(2) different autoclave opening times ranging from 500 up to 10 000 hours to provide the corrosion kinetics.

For both experiments, weighing has been done before and after descaling of the oxidized layer. Visual examinations of the corroded samples were done to verify that no significant localized corrosion had occurred (which could make difficult the interpretation of the weight gain results).

- *Residual neutron activity* - Systematic dose-rate measurements have been performed on the first unloaded samples - that is for a dose of ~0.9 dpa - and for different times of cooling ranging from 9 months up to 39 months.

Results and discussion

Post-Irradiation (PI) tensile properties at $T_{Test}=T_{Irr.}=325°C$

Post-Irradiation (PI) tensile properties of some of the alloys studied have already been presented in [9]-[11]. Here, the aim is to synthesize the tensile properties at 3.5 dpa (which is the maximum dose available for the tensile properties up to now) and to focus on the influence of chemical composition and metallurgical conditions for all the ferritic-martensitic steels studied. It must be pointed out that, for all the alloys studied, an important increase of strength and associated ductility loss is observed for doses as low as 0.9 dpa. Figure 3 shows the engineering tensile curves obtained on the EM10 (9Cr1Mo) steel at 325°C. It is obvious from this figure that the initial metallurgical conditions have a major influence on the PI tensile behavior.

Figures 4 to 6 show all the tensile data obtained at 325°C - 3.5 dpa on the martensitic steels and for the three metallurgical conditions considered, as a function of the nominal Cr concentration. From these figures it should be noticed that :

- *for the "standard" tempered martensitic structures (Figure 4)*, increasing Cr concentration from 7.5-9% up to 10-12% is detrimental to the PI ductility parameters, as conventionally observed. It is worth noting that the HT9 - 12%Cr - steel retains quite good Uniform Elongation (UE) and Total Elongation (TE) values at 3.5 dpa but shows very poor RA values (below 30%). As mentioned earlier, this reflects the necessity to measure accurately and to take into account the RA mechanical parameter to be able to fully analyze the tensile properties of these alloys after irradiation.

- *for cold-worked martensitic structures (Figure 5)*, the Cr influence is not obvious and these metallurgical conditions are characterised by a lower strength and a lower PI ductility decrease (especially for LA12LC and "T5" alloys). As already proposed in [11], it could be argued that the initial high number of dislocations of CW structures should acts as effective sinks for defect annihilation, producing a partial recovery of the irradiation damage at lower irradiation temperatures than for conventional tempered martensitic structures. However, further work will be necessary to explain the apparent relative insensivity to the Cr concentration (up to 11% - see LA4Ta) of the CW martensitic structures.

- *for as-quenched martensitic structures (Figure 6)*, very high strength values are achieved at 3.5 dpa, that is, up to 1800 MPa for the HT9 steel. It is quite surprising that, for such hardened structures, UE and TE values remain high. Also, as shown on Figure 6-c, the RA evolution is quite complex as a function of the neutron dose when compared to more conventional tempered structures: a RA decrease is observed for HT9 (showing an evolution very close to the tempered martensitic structure one), nearly constant RA values are observed for EM10 and a slight increase of this parameter is observed for F82H with increasing irradiation damage. One can assume that all these contradictory RA evolutions are the mark of different competitive microstructural evolutions within the metastable as-quenched martensitic structure, that is : inclination to form Cr-C clusters (pre-precipitation of carbides phenomena, eventually assisted by the neutron flux), anhilation and/or interaction of radiation damage (point defects, dislocation loops, cascades, ...) with the initial dislocation network due to the displacive martensitic phase transformation upon cooling, etc... It shows clearly that further fundamental studies must be carried out to get a better insight of these complex as-quenched martensite microstructural evolutions under neutron irradiation at low temperatures. Moreover, one must take into account that RA measurements from rectangular cross-sections are more complex than for cylindrical tensile samples. For this last point, some further finite clement calculations are planned to determine the triaxiality ratio created in the specimens (with their particular geometry) after strain localization.

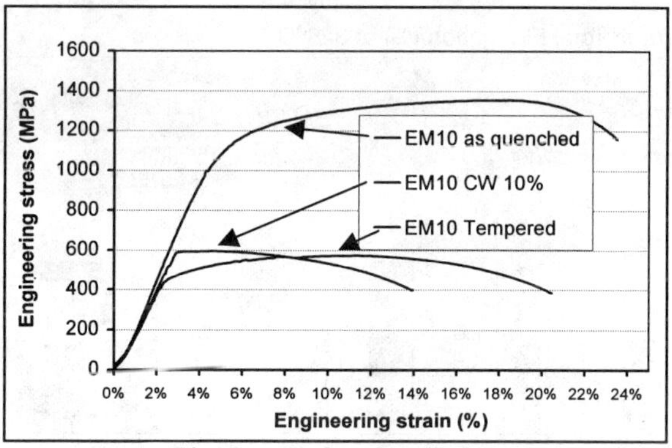

Fig. 3-a: Before irradiation

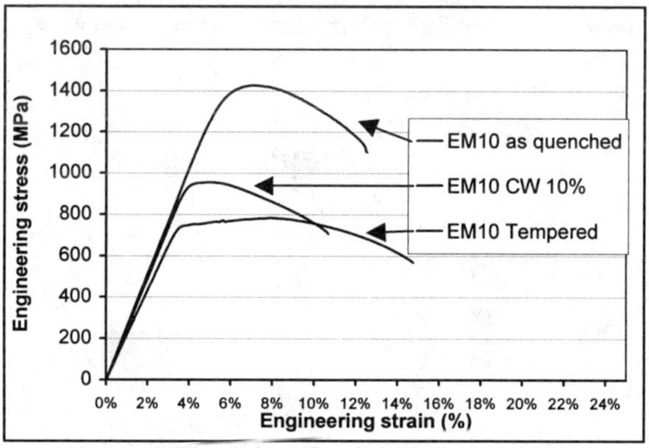

Fig. 3-b: After irradiation

Figure 3 - *Engineering tensile curves of EM10 (9Cr1Mo) for three initial metallurgical conditions, before and after irradiation up to 3,5 dpa* ($T_{test}=T_{irrad.}=325°C$).

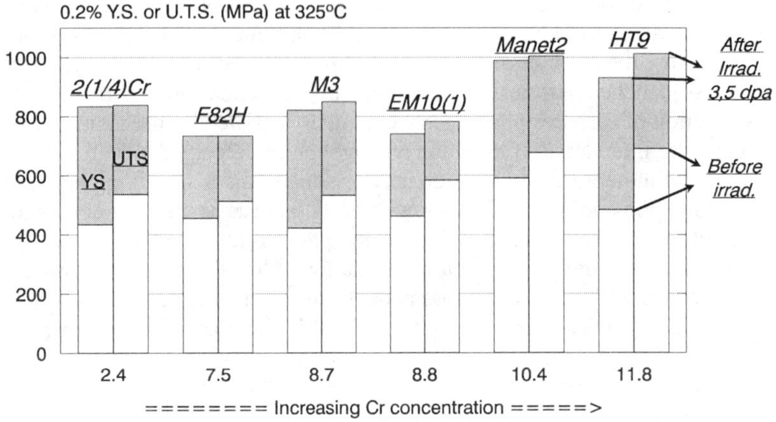

Fig. 4-a

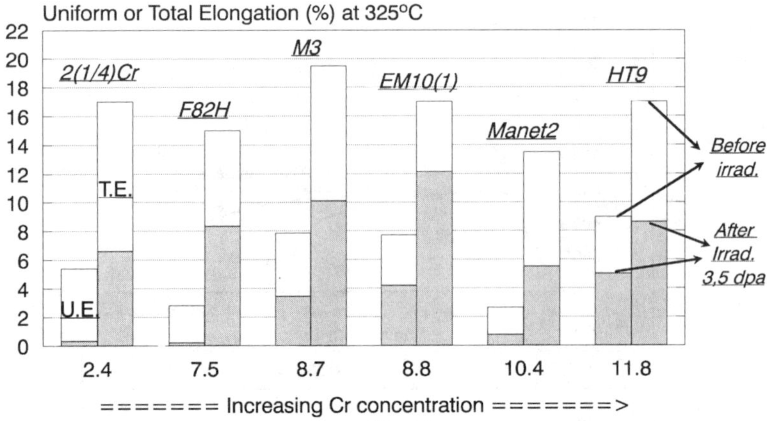

Fig. 4-b

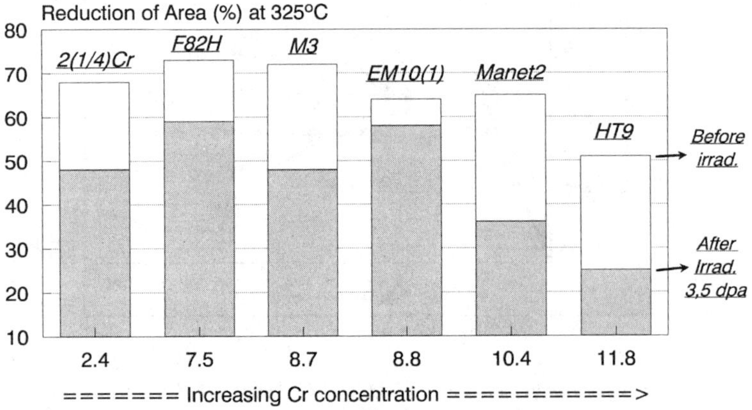

Fig. 4-c

Figure 4 : *Tensile properties at $T_{test}=T_{irrad.}=325°C$, for tempered martensitic structures before and after irradiation up to 3.5 dpa.*

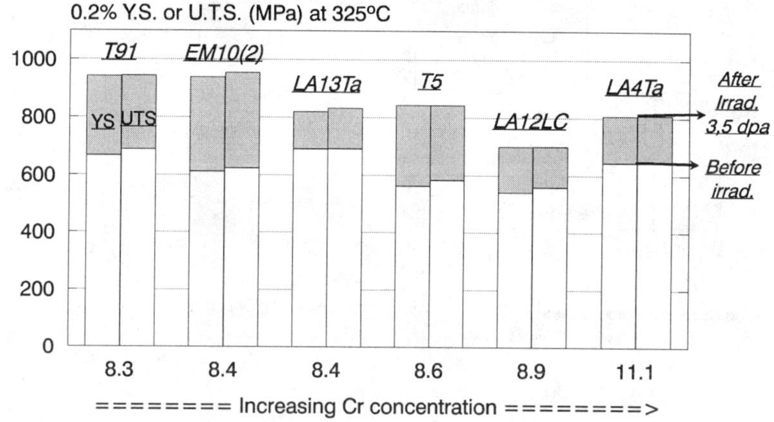

Fig. 5-a

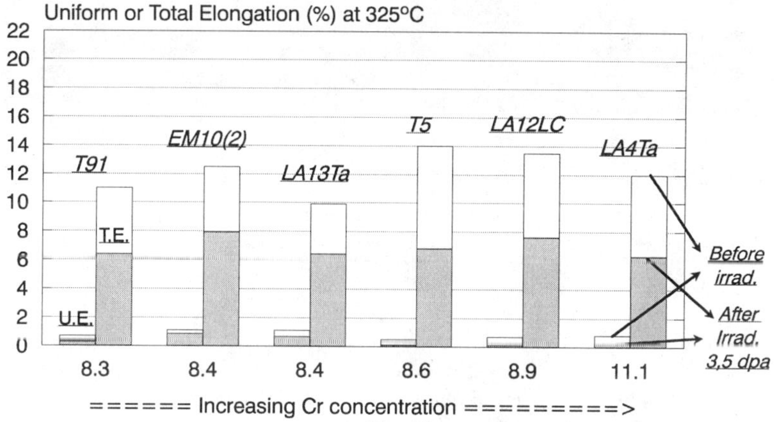

Fig. 5-b

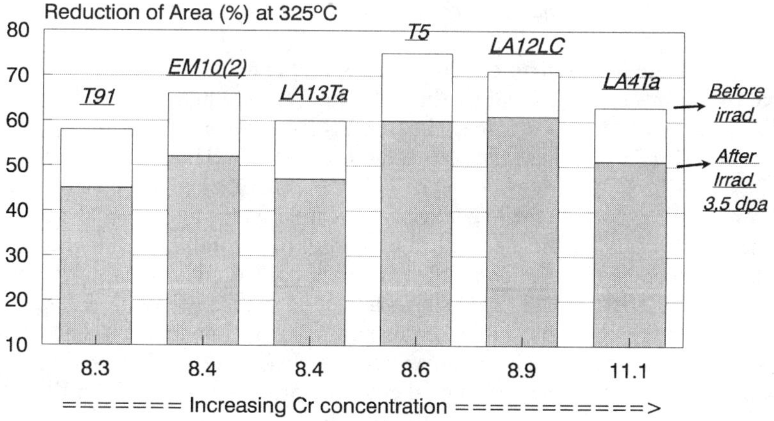

Fig. 5-c

Figure 5 : *Tensile properties at $T_{test}=T_{irrad.}=325°C$ for Cold-Worked (10-20%) martensitic structures before and after irradiation up to 3.5 dpa.*

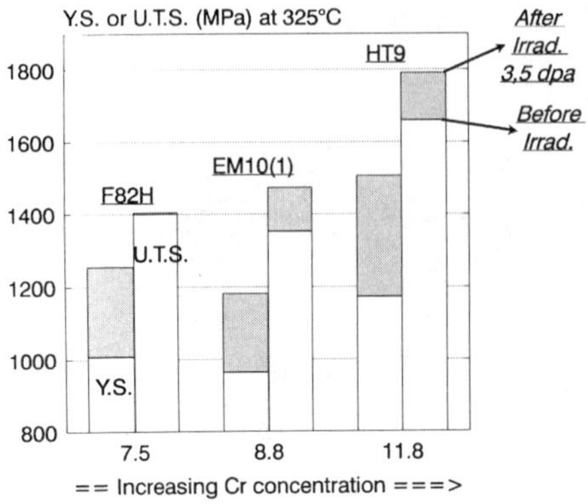

Figure 6-a

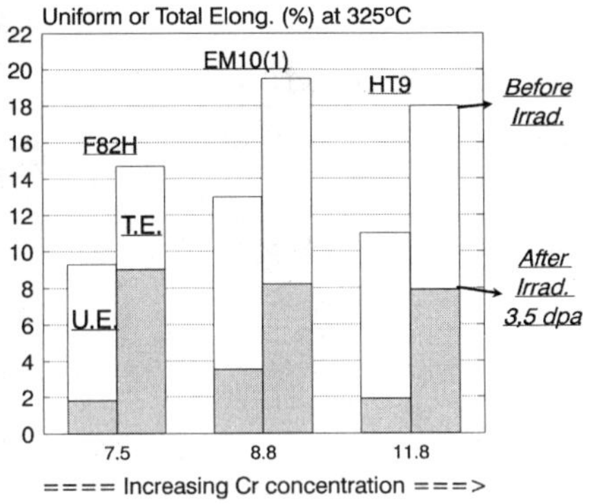

Figure 6-b

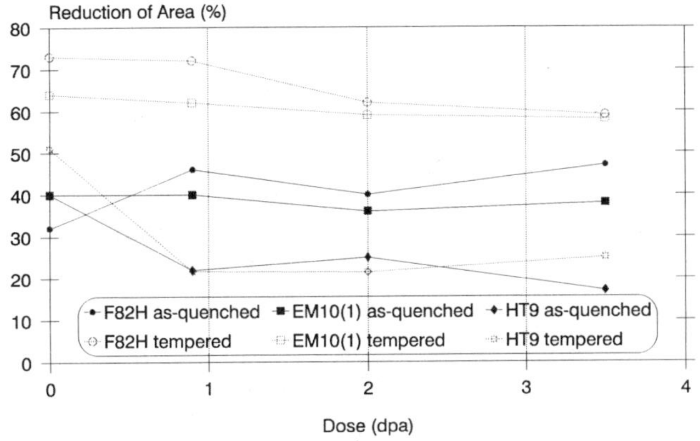

Figure 6-c

Figure 6 : *Tensile properties at T_{test}=$T_{irrad.}$=325°C, for* as-quenched *martensitic structures (6-a, 6-b) and compared to tempered martensite (6-c), before and after irradiation up to 3.5 dpa.*

Finally, to compare the strength/ductility ratio of all the materials and for all the metallurgical conditions studied, Figure 7 shows 0.2% Yield Strength (YS) values as a function of RA values at 325°C - 3.5 dpa. The best YS/RA ratio are observed, respectively, for : 316L(N)-SA austenitic steel, EM10(1) and F82H tempered martensitic steels, LA12LC and T5 CW martensitic steels. This last experimental alloy ,without Mo addition, shows a quite good PI tensile behavior, indicating that, Mo and/or W are not necessarily needed to optimize the mechanical properties of chromium-rich martensitic steels for use under neutron irradiation at low temperatures. Moreover, as already pointed out in [11], the ODS ferritic MA957 steel shows a good PI tensile behavior at 2 dpa (which is the only neutron dose available for this particular steel up to now). This is quite surprising when considering its high Cr concentration - close to 13%. However, this trend has to be confirmed at higher radiation damage before drawing any definitive conclusion.

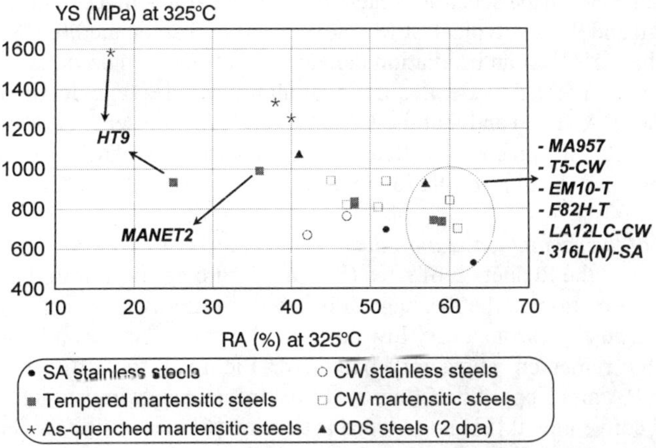

Figure 7 : *Representation of 0.2% Yield Strength (YS) as a function of Reduction of Area (RA) at 325°C for all the steels irradiated at 3.5dpa.*

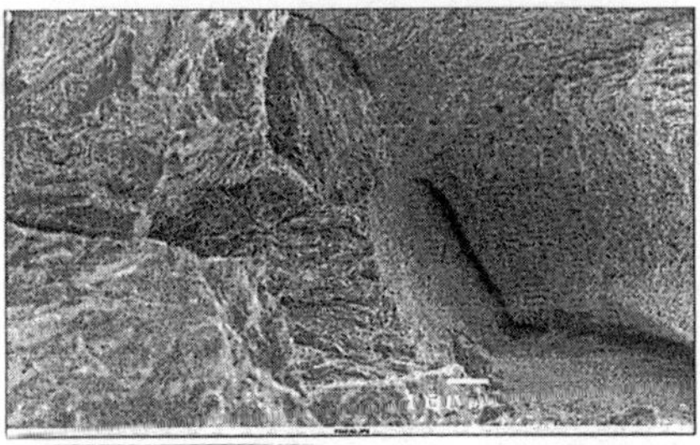

Figure 8-a : *SEM fractograph showing a mixed intergranular / quasi-cleavage / ductile failure mode for HT9-T steel irradiated at 3.5dpa and tested at RT*

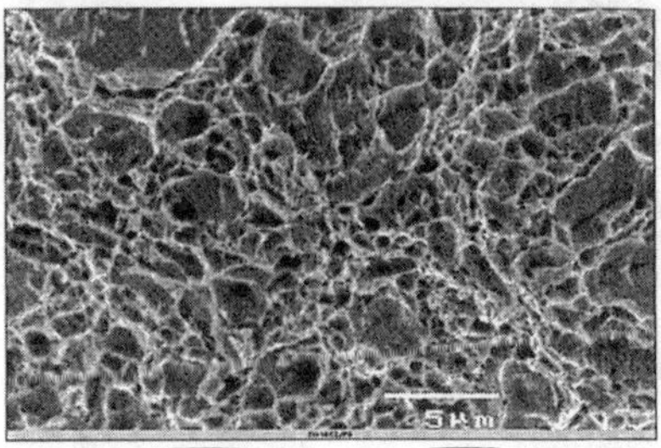

Figure 8-b : *SEM fractograph showing a pure ductile failure mode for LA12LC-CW steel irradiated at 3.5dpa and tested at RT*

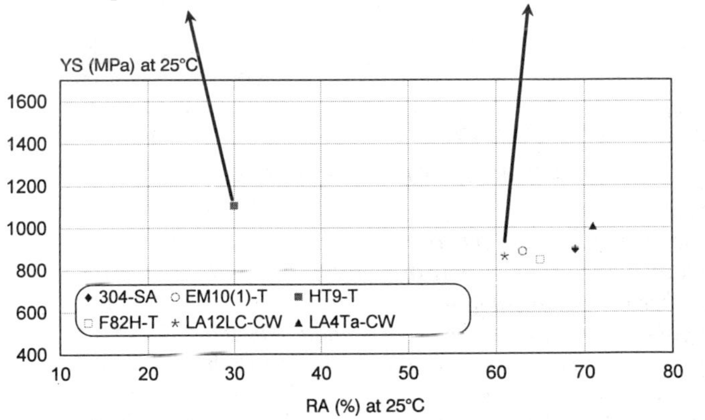

Figure 9 : *0.2% Yield Strength (YS) as a function of Reduction of Area (RA) at Room Temperature for some of the steels irradiated at 3.5 dpa.*

Post-Irradiation (PI) tensile properties at Room Temperature

Some of the irradiated tensile specimens have been tested at Room Temperature (RT). Figures 8-a, 8-b and 9 show typical SEM fractographs and associated 0.2%YS v.s. RA values measured at 25°C, for an irradiation damage of 3.5 dpa. It is obvious that the HT9 steel has failed with a mixed mode, i.e., intergranular/quasi-cleavage/ductile mode (consistent with a poor RA value) and that the LA12LC-CW steel presented a 100% ductile failure mode at RT, in agreement with a limited increase of the DBTT for this steel, as already shown in [11] after irradiation up to 2.5 dpa in HFR at 300°C.

In-situ observations of the strain localization during tensile tests at RT before irradiation

The measurement of the Reduction of Area (RA) at rupture has been found to provide interesting information on the mechanical behavior of irradiated steels. For example, the LA12LC steel presents a very low uniform elongation before and after irradiation while its RA remained satisfactory. Moreover, Figure 10 shows the empirical relationship between RA and Upper Shelf Energy (USE, from Charpy impact test) values for a great variety of conventional [5] and reduced activation [6] martensitic steels before and after thermal ageing. Recently, such a correlation has been observed on irradiated martensitic steels [8].

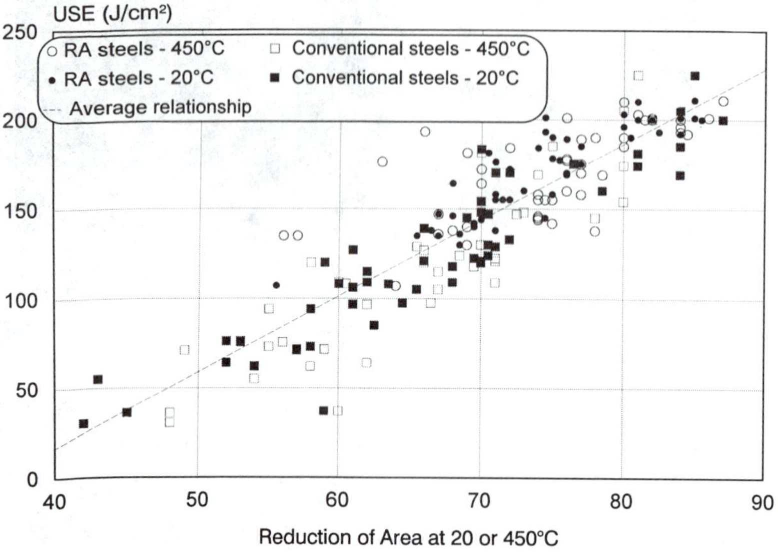

Figure 10 - *Empirical relationship between RA (from tensile tests(*)) and USE(**) parameters obtained on a great variety of conventional and RA martensitic steels before and after thermal ageing at 400-600°C up to 10 000-20 000 hours. Note that each couple of points has required the experimental determination of a full Impact transition curve, that is, for test temperatures ranging from –200 up to +400°C.*

(*) Tensile tests have been performed on cylindrical (diameter = 2mm) samples

*(**) Impact tests have been carried out on non-standard (3.5*10*55mm) V-notched samples*

Therefore, it was decided to complete the study of the mechanical behavior by in-situ observations during tensile tests at room temperature. The first objective was to characterise the strain localization and the ductility. Moreover, image analysis has been performed in order to evaluate the elasto-plastic behavior after strain localization.

As a first step, in-situ observations have been performed before irradiation. In figures 11 and 12 the results obtained on 304-SA austenitic stainless steel and tempered EM10 steel are presented. In each case, image 1 refers to the specimen before tensile test and image 2 represents the maximum load situation. From image 3, we may observe that strain localization appears at elongations higher than the (usually called) uniform elongation (UE). This phenomenon is related to " diffuse necking " [12] leading to a slight decrease in load without strain localization. As a consequence, strain to necking (STN) is significantly higher than UE (for 304-SA steel STN = 85.1 % and UE = 71.6 % and for EM10 steel STN = 12.8 % and UE = 9.0 %). In image 4 (last image before rupture), necking is observed at the centre of the specimen. We may observe that, after "diffuse necking" situation, the localization of the strain is faster for 304-SA than for EM10. Also, when compared to 304-SA tensile behavior, an important RA is reached for EM10 steel before rupture despite a reduced uniform elongation.

Image analysis has been used to evaluate the plastic behavior of the specimen beyond necking. The results are plotted in Figure 13 for tempered EM10 steel. The stress-strain curve deduced from image analysis up to 50% strain differs from the behavior extrapolated from uniform elongation stage using a stress-strain law of the form :

$$\sigma = \sigma_0 \varepsilon^n \text{ (with n= 0.146 and } \sigma_0 = 1091 \text{ MPa for tempered EM10)} \quad (1)$$

This phenomenon can be attributed to the increase of damage after strain localization. A Bridgman correction [13] has been applied in order to estimate the effect of the neck curvature of the specimen on the equivalent stress. This calculation is only valid for a cylindrical specimen on the hypothesis of isotropic necking. Thus, as mentioned above, we intend to perform finite element calculations to determine the triaxiality ratio created in the specimens after strain localization for the specimen geometry used presently (rectangular cross-section).

The analysis presented here will be performed shortly on specimens irradiated at 5.2 and 9.3 dpa. This will enable us the determination of the onset of necking more precisely by taking into account the diffuse necking. Moreover, the stress-strain behavior on irradiated specimens beyond necking will be studied, in particular, when strain localization appears just after the yield stress.

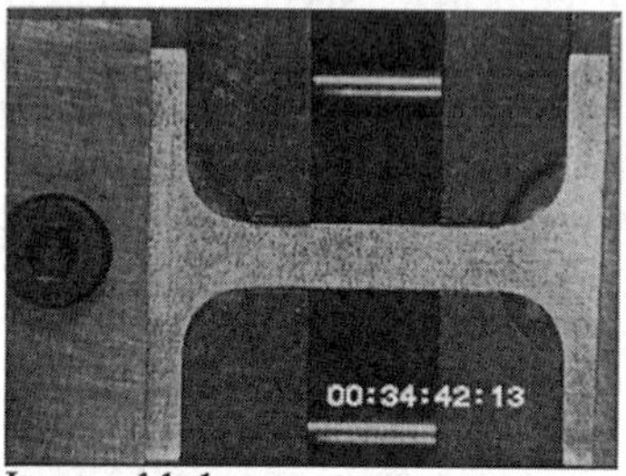

Image 11-1

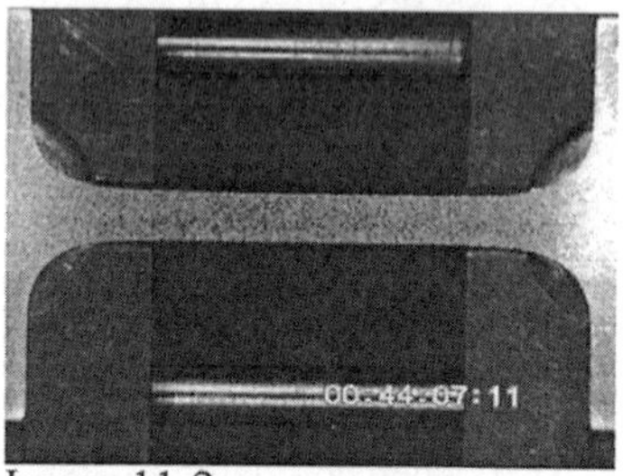

Image 11-2

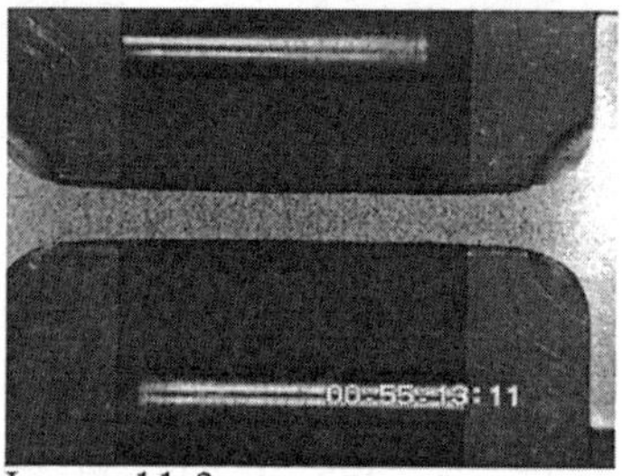

Image 11-3

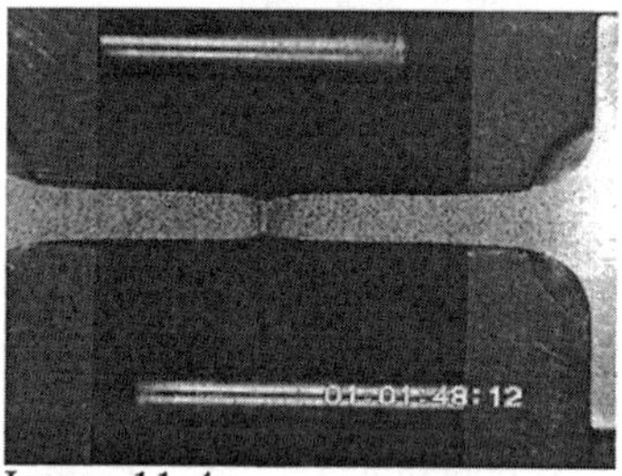

Image 11-4

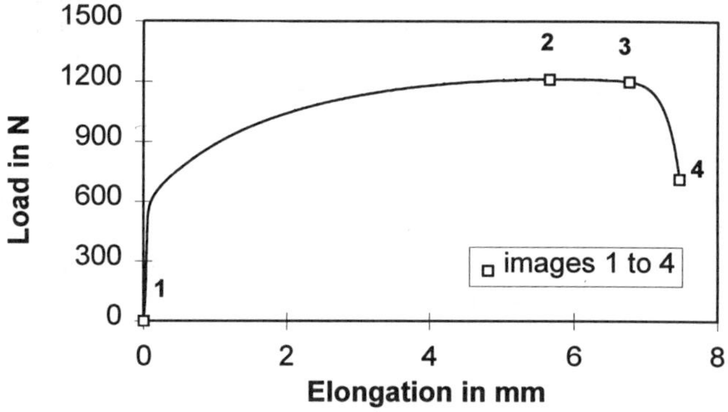

Figure 11 : *In-situ observations and corresponding engineering tensile curve obtained on austenitic 304-SA stainless steel before irradiation.*

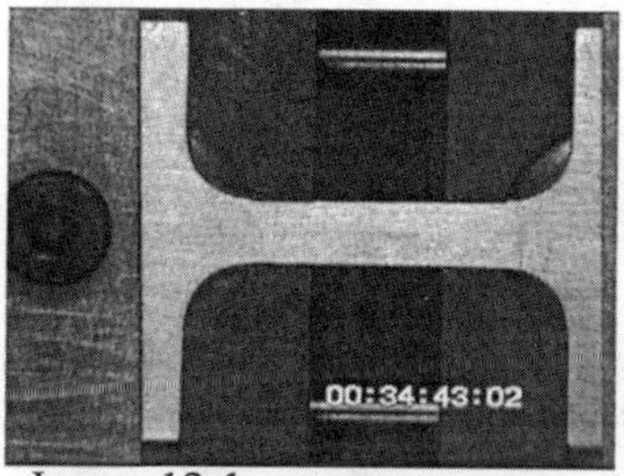

Image 12-1

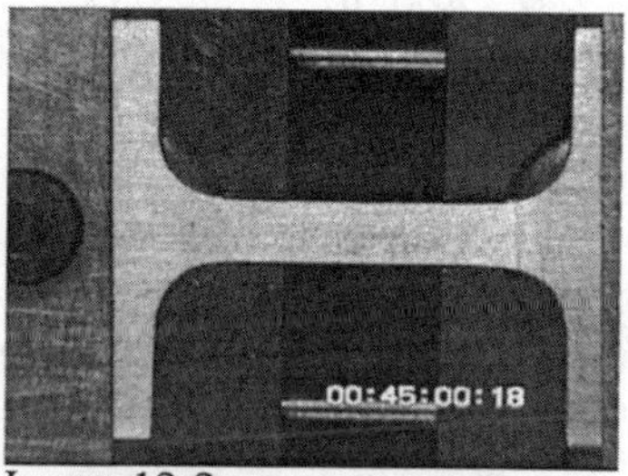

Image 12-2

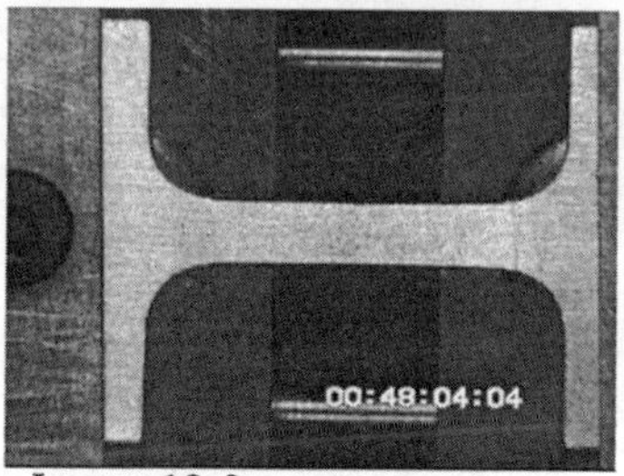

Image 12-3

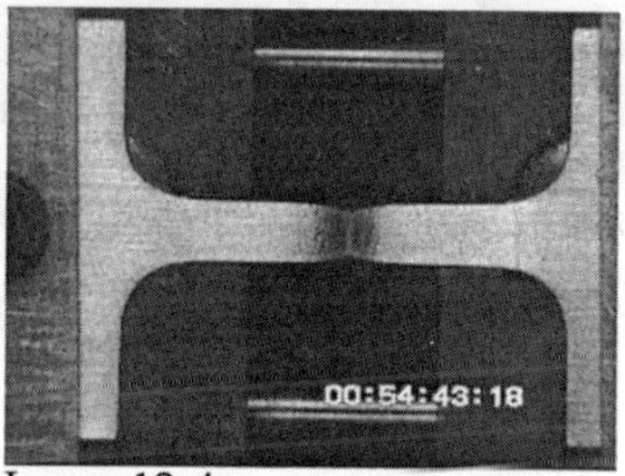

Image 12-4

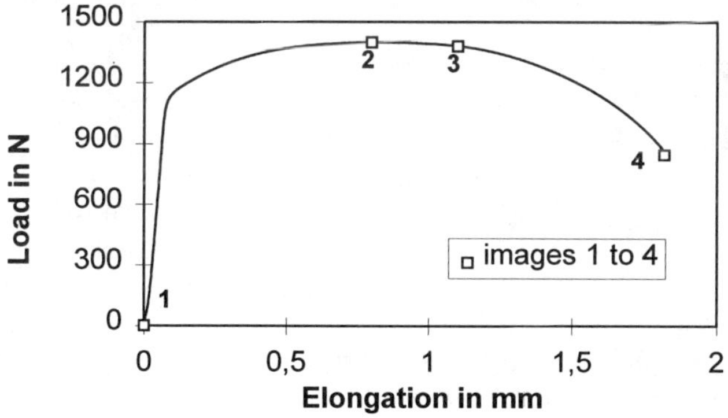

Figure 12 : *In-situ observations and corresponding engineering tensile curve obtained on tempered EM10 martensitic steel before irradiation.*

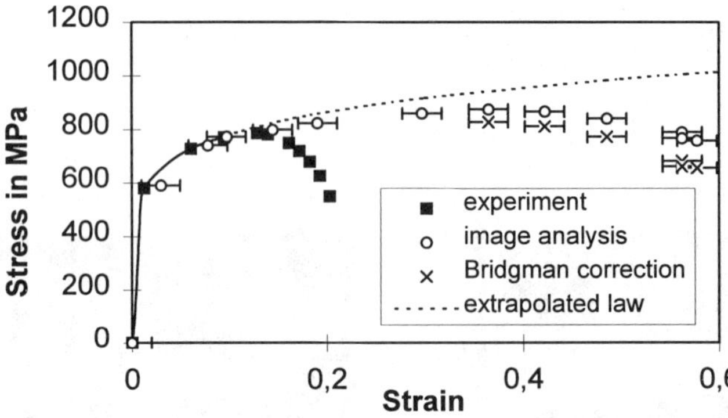

Figure 13 : *Stress-Strain behavior of tempered EM10 steel and calculations from in-situ image analysis results and using a "Brigman" [13] corrected law.*

In-pile and out-of-pile uniform corrosion

- *Uniform corrosion kinetics :* Figure 14 shows the typical weight evolution of F82H steel as a function of the exposure time for both in-pile and out-of-pile conditions.

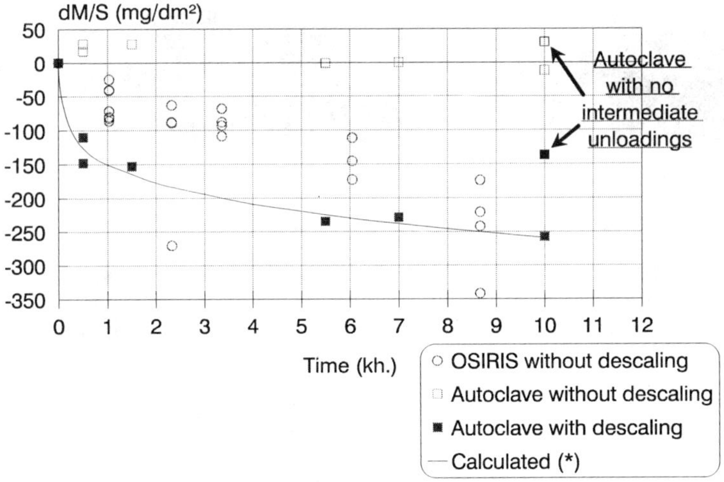

Figure 14 : *Weight gain evolution of F82H as a function of the time at 325°C.*
(*) $dM/S = K.t^n$ with $K=-28,4 mg/dm^2$ and $n=0,24$

All the martensitic, bainitic and ferritic steels studied here behave qualitatively as shown in this figure, that is:

=> a (quasi-linear) progressive and slight weight decrease is observed as a function of the exposure time on samples irradiated in Osiris, without descaling;

=> no significant weight change is observed after out-of-pile autoclave tests without descaling ;

=> after descaling and for a total exposure time of 10000 hours, the measured out-of-pile autoclave weight decrease ranges from ~ - 150 to ~ - 80 mg/dm² for the autoclave experiment (1) without intermediate unloadings, and from ~ - 300 to ~ - 170 mg/dm² for autoclave conditions (2) including intermediate unloadings.

One can make the assumption that the different corrosion behaviors observed between autoclave tests (1) and (2) should be related to the influence of the autoclave opening. This opening could promote some structural modifications of the outer surface of the oxidized layer, with a possible further detrimental effect on the corrosion barrier efficiency of the layer. Also, it is generally assumed that uniform corrosion follows a parabolic kinetics ($dM/S = K t^n$) where the n value is close to 0.4-0.5 [14] [15]. For the present out-of-pile autoclave (2) tests and after descaling, the fitted values of time exponent seem to be systematically much lower (n~0.15-0.3), and their meaning is not yet understood. However, one can refer to the three basic corrosion processes as already proposed in [14] and [15]:

(A) - loss of iron (and other alloying elements ?) into the water (weight reduction) and formation of an inner layer of Fe-Cr spinel oxide ;

(B) - deposition (weight gains) of crystals of magnetite ;

(C) - deposition (weight gains) of "loose crud" from the water.

Considering the antagonistic effects of these different processes on the positive or negative weight gain evolutions as a function of the exposure time, and taking into account that processes (B) and (C) could be partially erratic in nature, then the typical weight gain evolutions shown in Figure 11 can be qualitatively understood. Moreover, referring to [15], it is expected that the observed differences between in-pile Osiris corrosion results and out-of-pile autoclave results (without descaling) arise as a consequence of two possible phenomena, that is, water radiolysis effect under neutron irradiation on one hand, and the influence of the flowing water in the case the of Osiris loop experiment vs. the "static" confined environment for the out-of-pile autoclave testing on the other hand.

- *Alloys chemical composition influence on the uniform corrosion* - Figure 15 shows the weight gain measurements from 10 000 hours out-of-pile autoclave tests and from in-pile Osiris tests for exposure times ranging from ~3355 h. (~3.5 dpa) up to ~8670 h. (~9.3 dpa). This figure shows the influence of the nominal chromium concentration on the corrosion behavior of all the martensitic, bainitic and ferritic steels studied. It can be observed that, despite a significant scatter of the measurements, the Cr influence is slight for Cr concentration ranging from ~7.5 % (F82H) up to ~10.5% (Manet). Only the HT9 (Cr~12%) and the ODS MA957 (Cr~13%) steels display a significant gain on the uniform corrosion resistance. The autoclave results obtained on the 2 1/4Cr bainitic steel do not seem to be obvious when compared to alloys with higher Cr contents. However, preliminary SIMS analysis have shown a quite different outer oxidised layer with a lower adherence for this low Cr steel.

For the other main alloying elements, Mo and/or W additions seem to have only a slight influence on the corrosion resistance of chromium martensitic steels on the one hand. On the other hand, some chemical measurements of the environmental water after corrosion tests gave some indications of preferential release of W from the corroded samples (and, to a less extent, of Mo) for both in-pile and out-of-pile testing conditions.

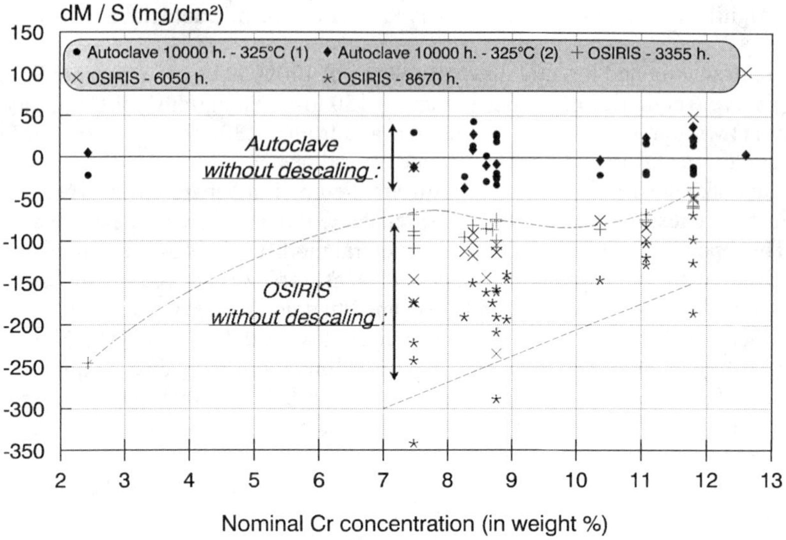

Figure 15-a

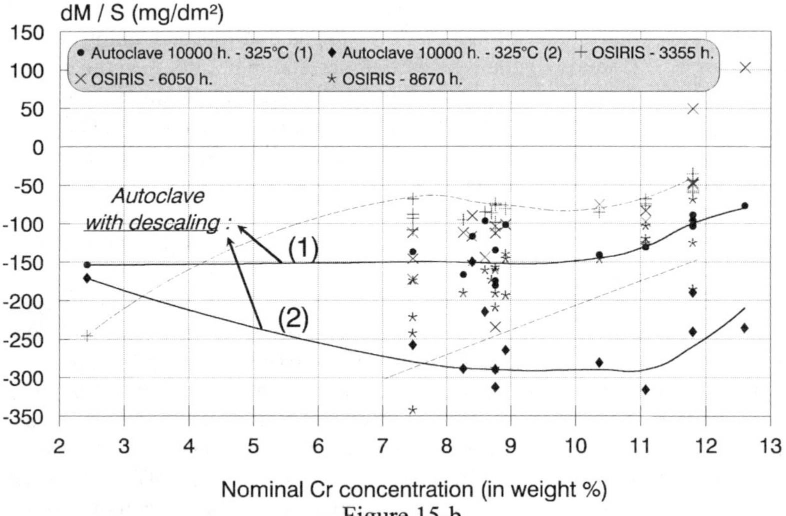

Figure 15-b

(Autoclave (1) and (2) refer respectively to tests without and with unloadings)

Figure 15 : *Evolution of the measured weight gain as a function of the nominal Cr concentration of the alloys:*
(1) after out-of-pile autoclave tests before (15-a) and after (15-b) descaling;
(2) after in-pile (OSIRIS) exposure (without descaling) for different times corresponding respectively to irradiation doses of:
~ 3.5 dpa (3355 h.), ~ 5.2dpa (6050 h.) and ~9.3 dpa (8670 h.)

Such a selective release of some of the alloying elements in the water seems to be of technological importance since activation of these elements could contribute to the "ambiant" dose rate level, in the case of applications of these steels to structural components in an aqueous environment. Consequently, the full analysis of the in-service activation of such components needs to carefully take into account the concentration and the chemical nature of the released elements and not only the nominal alloying elements concentration in the parent material.

Finally, it must be noted that, the initial metallurgical conditions have no significant influence on the corrosion resistance, for the experimental conditions tested here.

Post-Irradiation (PI) dose rate measurements

PI dose rate measurements have been carried out on the first unloaded samples from Osiris for a corresponding dose of 0.9 dpa. These measurements were made for different cooling times. Figure 16 summarizes the dose rate measurements obtained for 39 months after irradiation of all the studied alloys (each bar of the graph corresponds to one sample). It is obvious from this figure that reduced activation martensitic steels display an attractive low level of the residual activity, especially when they are compared to austenitic stainless steels of the 304 and 316 types.

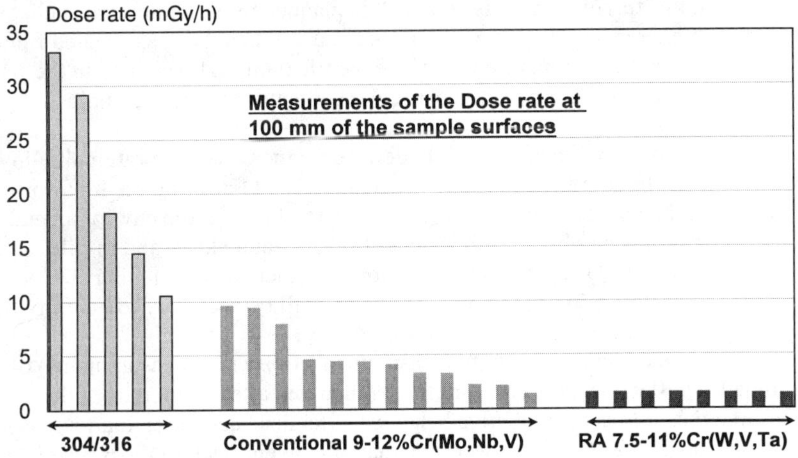

Figure 16 : *Dose rate values for all the steels irradiated at 0.9 dpa measured at 39 months after irradiation (each bar corresponds to one sample).*

Conclusion and further work

The present irradiation, named "Alexandre," has been carried out in the Osiris experimental reactor at 325°C for different neutron damage levels ranging from ~0.9 dpa up to ~9.3 dpa. Post-irradiation (PI) tensile and corrosion properties of 304/316 type and chromium rich ferritic-martensitic steels (including RA and ODS variants) have been established in PWR environmental conditions.

Type 7.5-12%Cr martensitic steels have been studied for different initial metallurgical conditions, that is, tempered, cold-worked and as-quenched martensite conditions. It is then observed that increasing the nominal chromium concentration of the alloys from the 7.5-9% range to 10-12% induces an increasing hardening and, consequently, a more detrimental loss of ductility for steels with an initial tempered condition. For cold-worked initial condition, such behavior is not observed, indicating a strong influence of the initial dislocations network introduced by the cold-working. Moreover, as-quenched martensitic structures show a complex behavior. Competitive evolutions between hardening and recovery processes seem to occur within these metastable structures. All these data highlight the need for further microstructural studies to achieve a better insight of all the related physical phenomena.

Moreover, careful measurements of the reduction of area show that this ductility parameter is of great importance to analyze the PI tensile behavior of the different kinds of steels studied. Then, preliminary in-situ image analysis observations of the strain localization during tensile testing at room temperature have been performed on samples before irradiation. For future work, this approach is planned to be applied on irradiated materials for the higher doses available, that is, 5.2 and 9.3 dpa. Also, associated finite element calculations are in progress to determine the triaxiality ratio created in the specimens after strain localization for the particular geometry of tensile samples presently used.

Uniform corrosion properties are mainly defined by the Cr concentration of alloys. Nevertheless, only a slight effect is observed within the 7.5-11% Cr range. It is worth noting that some indications of selective release of W (and Mo) in the environmental pressurized water have been observed for both out-of-pile (autoclave) and in-pile (pressurized water loop of Osiris reactor) conditions. Further work including microstructural characterization of the oxidized layers will be necessary to quantify these selective alloying element release phenomena.

Finally, the residual neutron activity measurements (dose rate measurements) of the samples unloaded at 0.9 dpa have been briefly summarized after 39 months of decaying. These measurements assess the attractive behavior of reduced activation martensitic steels showing a dose rate 10 times lower than that of conventional 304/316 type stainless steels. Complementary γ-spectrometry measurements and activity calculations are in progress to identify the radionucleides responsible for the observed dose rate evolutions.

Aknowledgements:

The authors would like to acknowledge J.-P. Pizzanelli, A. Castaing, L. Séjourné, P. Coffre, G. Pinte, O. Rabouille, J.-J. Espinasse, for their technical contributions, A. Richard, D. Poquillon, M.-T. Cabrillat for financial support and Dr. P. Yvon for discussion of the paper.

References

[1] Robertson, J.-P., Ioka, I., Rowcliffe, A.F., Grossbeck, M.L., and Jitsukawa, S., "Temperature Dependance of the Deformation Behavior of Type 316 Stainless Steels After Low Temperature Neutron Irradiation," *Effects of Radiation on Materials : 18th Int. Symposium, ASTM STP 1325*, 1999, pp. 671-688.

[2] Horsten, M.G., and De Vries, M.I., "Tensile properties of type 316L(N) stainless steel irradiated to 10 displacements per atom," *Journal of Nuclear Materials, 212-215*, 1994, pp. 514-518.

[3] Klueh, R.L., and Alexander, D. J., "Tensile and Charpy Impact Properties of Irradiated Reduced-Activation Ferritic Steels," *Effects of Radiation on Materials : 18th Int. Symposium, ASTM STP 1325*, 1999, pp. 911-930.

[4] Mathon, M.-H., De Carlan, Y., Geoffroy, G., Averty, X., De Novion, C.H., and Alamo, A., "Study by SANS and TEM of the Microstructural Evolution of Reduced Activation and Conventional Martensitic Steels After Thermal Ageing and Neutron Irradiation," *Effects of Radiation on Materials: 20th Int. Symposium, ASTM STP 1405*, 2002.

[5] Brachet, J.-C., Castaing, A., and Foucher, C., "Effects of the chemical composition on the behavior of 9-12Cr/0.5-1Mo/(W,V,Nb) thermal aged steels," *Proc. of Int. Symposium on "Materials ageing and component life extension," Milan, Italy, Oct. 10-13 1995, Engineering Materials Advisory Services Ltd., UK*, 1995, pp. 75-87.

[6] De Carlan, Y., Alamo, A., Mathon, M.-H., Geoffroy, G., Castaing, A., "Effect of thermal ageing on the microstructure and mechanical properties of 7-11CrW RAF/M steels," *Proc. of ICFRM 9, Colorado Springs, Colorado, USA, Oct. 11-15, 1999, to be published in Journal of Nuclear Materials.*

[7] Lucas, G.E., et al., "Implications of radiation-induced reductions in ductility to the design of austenitic stainless steel structures," *Journal of Nuclear Materials 233-237*, 1996, pp. 207-212.

[8] Horsten, M. G., Van Osch, E. V., Gelles, D. S., and Hamilton, M. L., "Irradiation Behavior of Ferritic-Martensitic 9-12%Cr Steels," *Effects of Radiation on Materials : 19th Int. Symposium, ASTM STP 1366*, 2000, pp. 579-593.

[9] Alamo A., Brachet J.-C., Averty X., "Preliminary results on irradiation experiment performed in Osiris-Reactor at 325°C including RAF/M steels," *Proc. of the IEA Working Group Meeting on Reduced Activation Ferritic/Martensitic Steels, JAERI, Tokyo, Japan*, Oct. 26-31 1997, *Ed. K. Shiba*, p. 256.

[10] Averty, X., Brachet, J.-C., Pizzanelli, J.-P., and Castaing, A., "Tensile properties of austenitic, martensitic and ferritic steels irradiated at 325°C up to 3.4 dpa in the Osiris experimental reactor," *Proc. of Int. Symposium Fontevraud IV : "Contribution of materials investigation to the resolution of problems encountered in PWR,"* Sept. 1998, France, Vol. 1, A74.

[11] Alamo, A., Horsten, M., Averty, X., Materna-Morris, E. I., Rieth, M., and Brachet, J.-C., "Mechanical behavior of reduced activation and conventional martensitic steels after neutron irradiation in the range 250-450°C," *Proc. of 9th Int. Conf. on Fusion*

Reactor Materials - ICFRM 9, Oct. 11-15 1999, *Colorado Spring, USA*, to be published in *Journal of Nuclear Materials*.

[12] Brunet, M., Arrieux, R., Boivin, M., "Détermination par éléments finis en grandes déformations, des courbes limites de formage en contraintes, " *Proc. of Int. Symp. on Plastic Instability, Considère Mémorial, Presses de l'école nationale des Ponts et Chaussées*, 1985, pp.227-233.

[13] Bridgman, P.W., "Studies in large plastic flow and fracture", *Harvard Univ. Press, Cambridge*, 1946.

[14] Gott, K., Lind, A., and Low, J., "The effects of irradiation on corrosion of the 10-12% Cr-MoV-Nb martensitic steel Manet," *Proc. of SOFT Conference, Rome*, Sept. 1992.

[15] Källström, R., "The effects of irradiation and radiolysis on the aqueous corrosion of high chromium martensitic steels," *Fusion final report MANET 4, Studsvik/M-93/106*, March 18th 1994.

Ferritic Martensitic Alloys

David S. Gelles,[1] Margaret L. Hamilton, [1] and Robin Schäublin[2]

Post-Irradiation Deformation Microstructures in Fe-9Cr

Reference: Gelles D. S., Hamilton, M. L., and Schäublin, R., "**Post-Irradiation Deformation Microstuctures in Fe-9Cr,**" *Effects of Radiation on Materials: 20th International Symposium, ASTM STP 1405,* S. T. Rosinski, M. L. Grossbeck, T. R. Allen, and A. S. Kumar, Eds., American Society for Testing and Materials, West Conshohocken, PA, 2001.

Abstract: The deformed microstructures of both irradiated and unirradiated Fe-9Cr uniaxial tensile specimens have been examined to identify controlling mechanisms. Deformation following irradiation is found to occur in poorly defined channels, causing formation of discrete steps at surfaces and delineated by nonuniformly distributed highly elongated voids. Deformation is by motion of $\frac{a}{2}$<111> dislocations, which interact with and decompose irradiation-induced a<100> loops. The structure formed after extensive deformation consists of highly complex cell walls and moderate densities of individual slip dislocations.

Keywords: post-irradiation deformation, ferritic steel, fractography, dislocations, voids

Introduction

Martensitic steels provide excellent radiation damage resistance for fission and fusion reactor applications. Significant hardening results from irradiation at temperatures of 400°C and below, but total elongation remains in the 7% range even in simple Fe-Cr binary alloys[*1*]. Hardening is understood to arise due to the formation of voids, precipitates, and dislocation loops with both a<100> and $\frac{a}{2}$<111> Burgers vectors. However, microstructural understanding of post-irradiation deformation behavior in ferritic/martensitic steels represents an important field for research that has been largely unexplored. This is in part due to limitations imposed by specimen availability, imaging challenges, and as will be shown, localized flow limitations. We have previously attempted to examine post-irradiation deformation in Fe-6Cr and –12Cr specimens,

---

[1] Staff Engineer and Senior Scientist, PNNL Materials Research Department, Richland, WA 99352.
[2] Research Associate, EPFL-CRPP Fusion Technology, Villigen PSI, Switzerland.

without success (as will be shown, because of the limited extent of the behavior)[*1*], but are unaware of any other similar efforts. In the previous attempt, specimens of irradiated and deformed miniature sheet tensile specimens were prepared about 1.5 mm from the fracture surface, but the microstructures found were characteristic of undeformed material. This topic is of importance because post-irradiation deformation in general, and embrittlement in particular, are expected to define the low temperature limits for fusion structural applications. The topic is of particular interest because ferritic steels provide an unusual example where irradiation produces a Burger's vector (a<100>) that is not normally encountered in unirradiated steels, nor is it found in most other body centered cubic metals. Furthermore, it is not clear how slip dislocations of type $\frac{a}{2}$<111> can interact with a<100> Burger's vectors so as to annihilate all a<100> segments.

Experimental Procedures

Miniature sheet tensile specimens of Fe-Cr alloys with nominal gauge dimensions of 5.1 mm x 0.25 mm x 1.0 mm were irradiated in fusion irradiations in the Fast Flux Test Facility (FFTF), Richland, WA, and were tested to provide understanding of post-irradiation deformation response in simple alloys[*1,2*]. Later irradiation experiments duplicated these conditions with the intention of going to higher dose levels, but did not exceed the dose previously attained because FFTF was shut down prematurely. As the conditions being irradiated were duplicates, they were ignored and further deformation testing of these specimens was terminated. From those available specimen conditions, the two listed in Table 1 were easily accessible and were chosen for this study. It was decided to limit the present study to Fe-9Cr in order to best match with the Fusion low activation candidate alloy compositions. This meant that two specimens were available from capsule ZP-1 and four were available from ZR1. Unirradiated control duplicates could not be found, so SS-3 control samples that are somewhat larger with nominal gauge dimensions of 7.62 mm x 0.76 mm x 1.5 mm but with identical heat treatment were substituted in order to provide control condition microstructures. The heat treatment given was intended to duplicate standard normalization and tempering: 1040°C/1h/AC +760°C/2h/AC.

Table 1 - *Irradiation Conditions for Miniature Sheet Tensile Specimens*

Capsule ID	MOTA/Basket	Irr. Temp (°C)	Fluence (n/cm^2)	Dose (dpa)
ZP1	1E/BCB-1	370	2.79×10^{22}	10.4
ZR1	1E/2C-3	403	9.46×10^{22}	40.5

Specimen flats were mechanically polished to Linde 600 grit and deformed at room temperature either to failure on the first test (UE and/or TE given) or to about 2% for the irradiated specimens and 8% for the unirradiated specimens without failure. Testing details are provided in Table 2 with elongation estimated from test data.

Following deformation, specimens were examined by scanning electron microscopy (SEM) to identify regions that were deformed. Unirradiated SS-3 specimens were then ground from 0.030" (0.75 mm) to about the thickness of the miniature specimens: 0.010" (0.25 mm). Disks 1 mm in diameter were then punched from promising areas so that the edge of the specimen was retained on one side to allow determination of the stress axis. Each disk was then mounted in a 3 mm stainless steel disk using recently developed procedures [*3*] and prepared using normal polishing procedures. Transmission electron microscopy (TEM) was performed on a JEOL 1200EX TEM operating at 120 KeV and using a double tilting ±45° goniometer stage. Imaging included procedures for identifying each of the a<100> and $\frac{a}{2}$<111> Burgers vectors in a field of view[*4*]. All micrographs were digitized and stereo images were prepared as anaglyphs.

Table 2 - *Test Details for Deformed Tensile Specimens*

Specimen ID	Condition	Yield (MPa)	Max. load (MPa)	Elongation (%)
6721	unirradiated	194	326	16.4 UE, 27.3 TE
6720	“	139	250	8.5
6724	“	142	239	7.2
ZR01	40.5 dpa at 403	534	534*	4.1 TE
ZR02	“	540	554*	2.3
ZP01	10.4 dpa at 370	619	619*	2.6

*Maximum load response as per reference 2, page 1240.

Results

Surface Features

SEM revealed that surfaces of both irradiated and unirradiated specimens had developed clearly defined steps. The vertical surfaces of these steps were not flat, but instead showed fine structure. However, the irradiated specimens showed no other features whereas the unirradiated specimens developed more rounded step edges and more structure on the original surfaces. Examples are provided in Figures 1 and 2. Figure 1 compares the specimens at low magnification and Figure 2 shows examples of specimen surfaces at higher magnifications.

Figure 1 shows the unirradiated specimens on the left and the irradiated samples on the right at decreasing strain from left to right. From Figure 1 it can be shown that strain is much more localized following irradiation. In fact specimen ZP02, second from right is about to fail in the necked region with little indication of deformation elsewhere. Slip steps can be identified only at the ends of in the gage sections, even in the failed specimen. Horizontal arrows define neck regions and diagonal arrows denote individual channel traces. From this figure, the reason for the previous inability to find deformation in reference 1 studies is apparent; deformation is so localized that material 1.5 mm from the fracture surface shows few slip steps. Deformation in the unirradiated specimens is much more uniform.

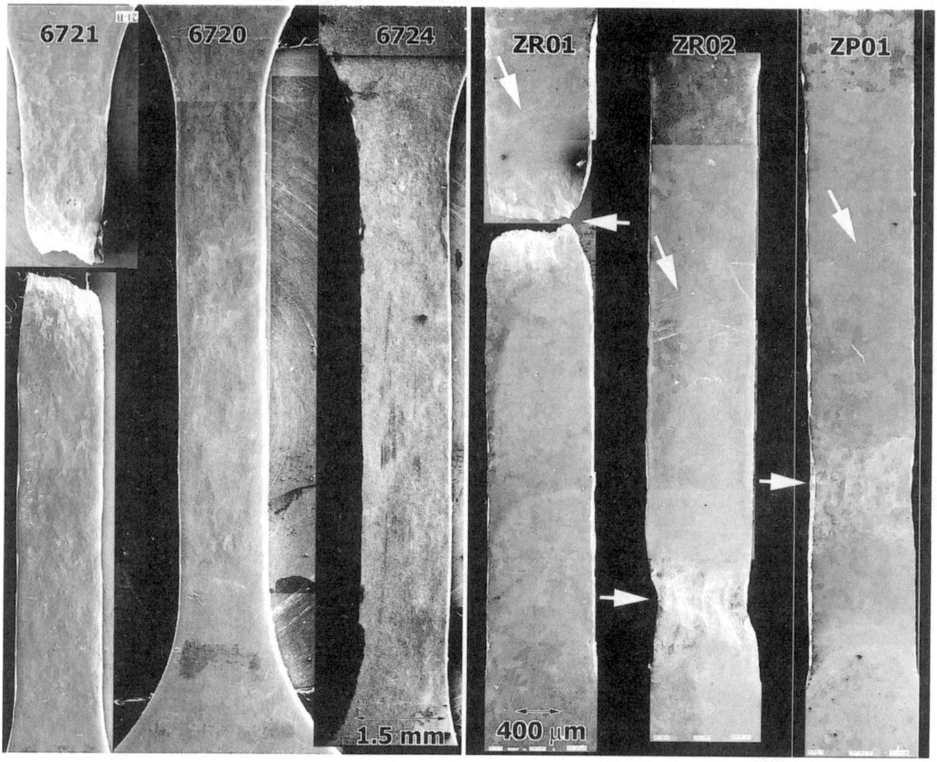

Figure 1 - *Low magnification examples of deformed tensile specimens with unirradiated specimens on the left. Note differences in magnification*

Figure 2 shows the fracture surfaces at higher magnification with unirradiated specimen 6721 on the left and irradiated specimen ZR01 on the right. Dimple rupture is apparent on the fracture surface following irradiation. The grain size is on the order of 200 μm and each grain contains slip steps appearing as wavy lines, but these lines are difficult to resolve in the unirradiated condition at lower magnification. Slip step spacing increases with distance from the fracture surface. At higher magnification in the lower images, it is apparent that in the irradiated case, the surface is stepped with no features on the original surface and fine structure on the step surfaces. In general, the fine structure is ridged, but several examples can be found where steps appear on this surface. Therefore, once deformation initiates on a plane, it tends to continue without change on that plane. In comparison, the unirradiated sample shows undulations on the original surface, step edges that are rounded and step surfaces that are more uneven. (Scratches are on the lower left and should be ignored). Therefore, in the unirradiated case, deformation is less restricted to deformation channels.

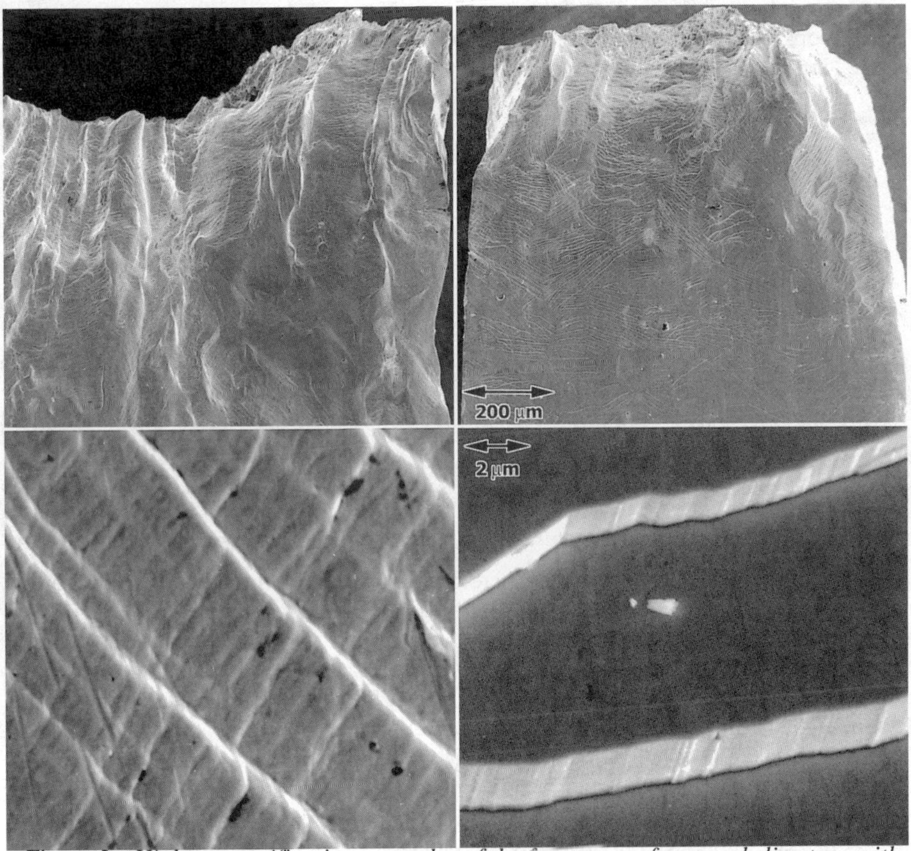

Figure 2 - *Higher magnification examples of the fracture surfaces and slip steps with unirradiated condition 6721 on the left and irradiated condition ZR01 on the right.*

Microstructural Examination

The unirradiated deformed condition was found to contain a moderate density of straight screw $\frac{a}{2}$<111> dislocation line segments, but $\frac{a}{2}$<111> loops produced by deformation were also found. An example demonstrating procedures that allow identification of all $\frac{a}{2}$<111> Burgers vectors, is given in Figure 3 for an area in specimen 6724 deformed approximately 7%. Figure 3 provides comparison of the same area imaged using $\bar{g}=01\bar{1}$ (vertical) in Figure 3a and 200 (horizontal) in Figure 3b taken near (011) orientations and $\bar{g}=10\bar{1}$ (with $\bar{g}$ as shown) in Figure 3c taken near (131)[*4*]. In Figure 3a only $\frac{a}{2}[1\bar{1}1]$ and $\frac{a}{2}[11\bar{1}]$ are in contrast, in Figure 3b, all $\frac{a}{2}$<111> dislocations are in contrast and in Figure 3c, only $\frac{a}{2}[\bar{1}11]$ and $\frac{a}{2}[11\bar{1}]$ are in contrast. Therefore, those dislocations that appear in Figures 3a and 3b, but not in 3c, that tend to lie on the diagonal from lower left to upper right, are of type $\frac{a}{2}[1\bar{1}1]$ and are of screw configuration. A small loop, most visible in Figure 3a just above and to the right of center, appears to be an edge loop with $\frac{a}{2}[1\bar{1}1]$ Burgers vector.

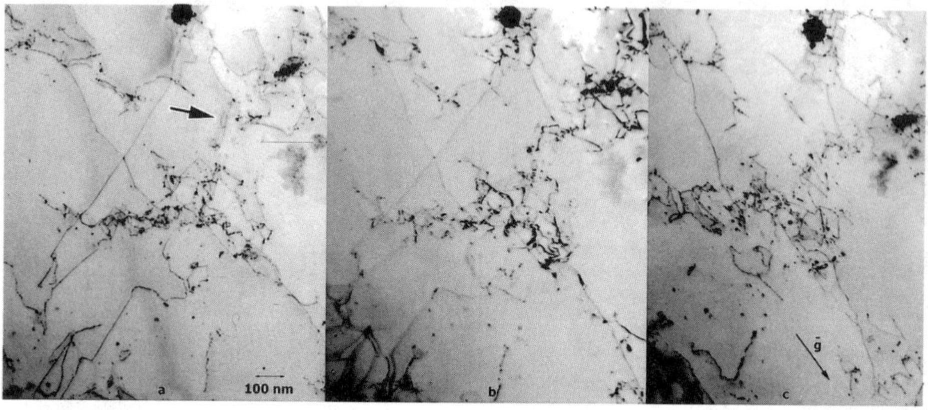

Figure 3 - *Dislocation structure in unirradiated deformed specimen 6724C*

Except in samples that were taken where necking was occurring, the microstructures observed in irradiated samples were typical of the as-irradiated structure, as found in earlier work[*1*]. That structure consisted of equiaxed voids, a<100> loops, $\frac{a}{2}$<111> line segments, and possibly large $\frac{a}{2}$<111> loops that are as large as foil dimensions. Examples are provided in Figure 4 from specimens ZP01 away from the necked area and ZR01 near the fracture surface, showing bright field images with $\bar{g}=01\bar{1}$ (horizontal) and 200 (vertical) taken near (011) orientations, the lower dose and temperature on the left. For Figure 4, it can be noted that in $\bar{g}=200$ contrast, all $\frac{a}{2}$<111> dislocations are weakly visible and one set of a<100> are visible in strong contrast, those on the (200) plane, whereas in $\bar{g}=01\bar{1}$ contrast, only two sets of $\frac{a}{2}$<111> dislocations and the remaining a<100> dislocations are visible (with a[100] invisible)[*1,4*]. Therefore, Figure 4 shows loops on the order of 50 nm predominantly of a<100> type with a few $\frac{a}{2}$<111> dislocation segments. Similar microstructures were reported previously in undeformed specimens[*1,5-8*]. These observations therefore demonstrate that procedures used to prepare areas of interest did not introduce extraneous deformation.

Figure 5 provides an example of the structure found in a deformed region of specimen ZR02, under various imaging conditions: $\bar{g}=01\bar{1}$ (horizontal), 200 (vertical) and in void contrast taken near a (011) orientation. In Figure 5, it can be noted that the contrast from a<100> dislocations is very different from that in Figure 4, in that voids are often linked to dislocations, and many examples of highly elongated voids can be identified. Therefore, post-irradiation deformation effectively removes the a<100> loops created by irradiation but strain is very nonuniform, resulting in severe distortion of many voids but negligible distortion of neighboring voids.

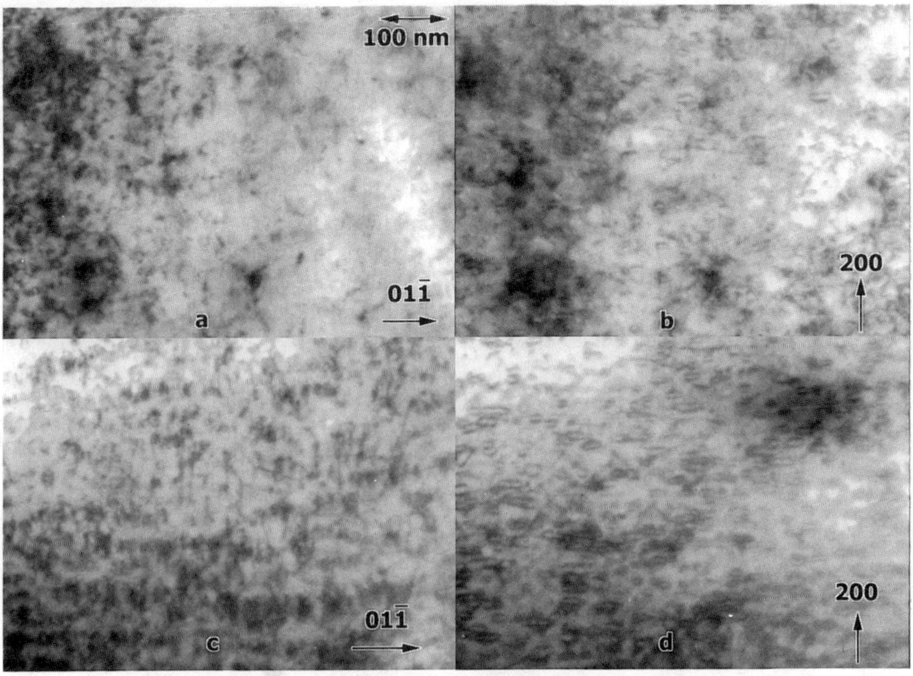

Figure 4 - *Microstructures in apparently undeformed regions for ZP01 in a) and b) and ZR01 adjacent to the fracture surface in c) and d)*

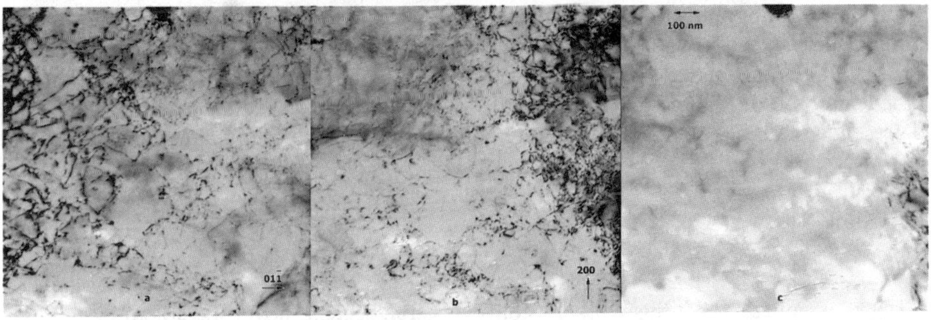

Figure 5 - *Microstructures in a deformed region of ZR02 in $01\bar{1}$ (horizontal), 200 (vertical) and void contrast*

The imaging procedures that allow identification of all $\frac{a}{2}\langle 111\rangle$ Burgers vectors were applied to post-irradiation deformation structures with less success because lattice rotation due to localized deformation was more severe and because a[100] Burgers vectors are present. Examples are given in Figures 6 and 7. Both use the same imaging conditions as those used in Figure 3 with $\bar{g}=01\bar{1}$ horizontal in a and 200 vertical in b taken near (011) orientations but with either $\bar{g}=1\bar{1}0$ or $10\bar{1}$ (with $\bar{g}$ as shown) in c taken near (113) or (131), respectively[*4*]. In Figures 6a and 7a, only $\frac{a}{2}[1\bar{1}1]$, $\frac{a}{2}[11\bar{1}]$, a[010] and a[001] dislocations are in contrast, in Figure 6b, all $\frac{a}{2}$<111> dislocations and a[100] dislocations are in contrast, in Figure 6c, only $\frac{a}{2}[\bar{1}11]$, $\frac{a}{2}[1\bar{1}1]$, a[100] and a[010] dislocations are in contrast and in Figure 7c, only $\frac{a}{2}[\bar{1}11]$, $\frac{a}{2}[11\bar{1}]$, a[100] and a[001] dislocations are in contrast. Figure 6 was selected to show behavior in an early stage of deformation and Figure 7 to show more advanced stages of development. [Note that this is possible because local deformation was much greater in the necked region of specimen ZR02 than for that in specimen ZP01 (see Figure 1).] Examination of Figure 6 demonstrates structure at the lower left with several features similar to that shown in Figure 3, consisting of a<100> loops. However, much of this structure appears different, indicating that many of the a<100> loops have been altered. Other regions of Figure 6 are quite different in that no a<100> loops remain. Therefore, early in the deformation process, it is possible to remove the a{100} loop structure.

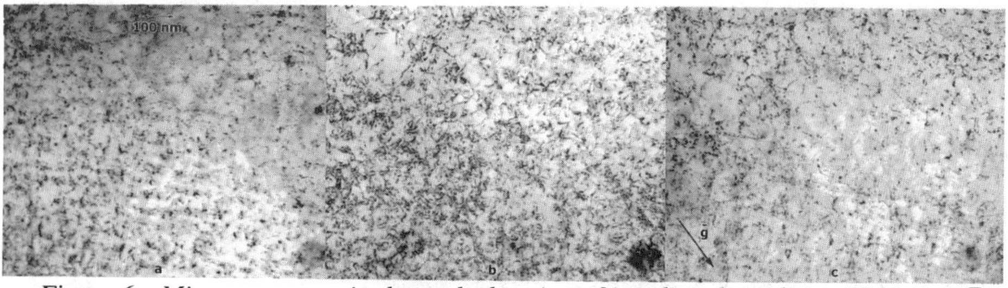

Figure 6 - *Microstructures in the necked region of irradiated condition ZP01 in* $01\bar{1}$ *(horizontal), 200 (vertical) and* $1\bar{1}0$ *(as shown)*

Figure 7 indicates that later in the deformation process the dislocation structure coarsens and develops a subgrain structure. Dislocations are found to be predominantly of $\frac{a}{2}$<111> Burgers vector type and of screw configuration. The subgrain boundaries appear to be linked to voids and are non-planar as indicated by the subgrain structure appearing on the center left of Figure 7a. However, the major deformation channels, as defined by the presence of highly elongated voids, are not parallel to nearby subgrain boundaries.

Discussion

It is now apparent that this work was only possible because novel specimen miniaturization procedures have been developed. Deformation in irradiated specimens is extremely localized, at least for the miniature specimens used in this study. Furthermore, it is straightforwardly possible to relate the dislocation structure available for analysis with reference to the tensile direction. However, no attempt has yet been made to relate Burgers vector anisotropy to the stress state. Instead, the emphasis was placed on understanding the dislocation reactions that lead to alteration of the irradiation-induced dislocation structure and channel deformation.

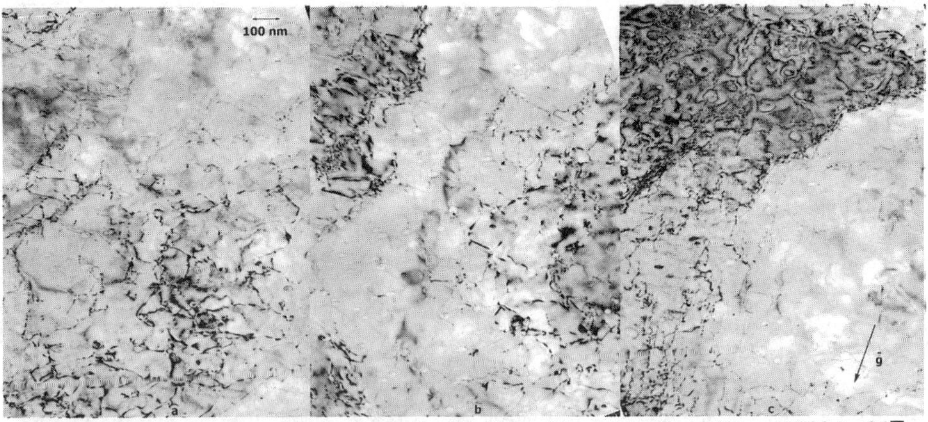

Figure 7 - Microstructures in the necked region of irradiated condition ZR02 in 01$\bar{1}$ (horizontal), 200 (vertical) and 10$\bar{1}$ (as shown)

Deformation is found to remove irradiation-induced a<100> edge dislocation loops. That process can be envisioned as follows. A moving $\frac{a}{2}$<111> dislocation with the correct sense will interact with an a<100> loop according to the following reaction:

$$\tfrac{a}{2}[\bar{1}11] + a[100] = \tfrac{a}{2}[111] \qquad (1)$$

Replacement of a section of the a[100] loop with an $\frac{a}{2}$[111] line segment by this process can take place with significant reduction in energy, and it can be expected to continue rapidly, replacing the entire a[100] loop with an $\frac{a}{2}$[111] loop lying on the (100) plane. This replacement reaction can occur with any moving $\frac{a}{2}$<111> dislocation but its sense will determine which side of the a<100> loop it will be attracted to. Figure 6b provides excellent demonstration that this reaction occurs easily upon post-irradiation deformation. However, stereoscopic examination of features in Figure 6b reveals that most of the $\frac{a}{2}$<111> loops present are not on {100} planes, and therefore, once formed, such loops are able to glide to more favorable configurations.

Reduction in dislocation density then occurs with further interaction between these loops and moving dislocations of similar Burgers vector. However, interactions between $\frac{a}{2}$<111> loops and dislocations of a different Burgers vector will not be energetically favorable, representing the reverse of the above interaction, and therefore, such loops will represent barriers to dislocation slip.

It is apparent that with sufficient deformation, all irradiation-induced loops can be destroyed, leaving a dislocation structure with screw component characteristics similar to those created in deformed unirradiated conditions. Even after the irradiation-induced dislocation structure is removed, localized deformation continues, resulting in a higher density of surface steps and a transition into necking response. This may be a consequence of favorable dislocation nucleation or enhanced dislocation glide because the remaining microstructural obstacles to dislocation glide, the voids, are distorted.

It has been shown that some voids are elongated or, at least, deformed from the original equiaxed and faceted shape. The deformed elongated cavities are located in regions that form narrow bands that are about 100 nm wide as shown in Figure 8. The analysis of the shape of the elongated cavities allowed measurement of the local strain applied. The local strain is given by the ratio of length of the cavity, measured from top to bottom in the direction of the strain, and the original size, as measured by the size of the cavity at mid-height perpendicular to the strain axis. The average strain for highly elongated voids is about 300 %, which represents the passage across a 25 nm cavity of about 100 dislocations.

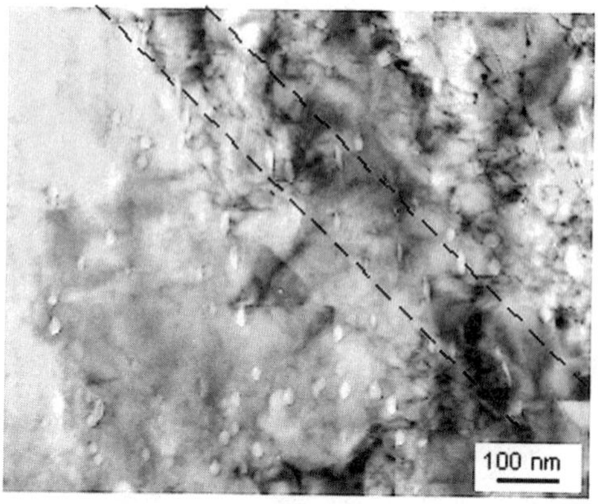

Figure 8 - Microstructure in the necked region of irradiated condition ZR02 showing typical elongated cavities in poorly defined narrow bands. A band is delineated by the dashed lines.

Even though these bands are poorly defined, they clearly indicate deformation of the material occurring by localized shear. When these bands are correlated with the stepped traces revealed by the SEM analysis, it is clear that the irradiated material deforms via a channeling mechanism. Although channeling has been demonstrated in a number of irradiated fcc and ferritic materials[*9-10*], this is to our knowledge the first evidence for this mechanism in a Fe-9Cr type ferritic alloy.

Conclusions

A novel specimen preparation procedure has made it possible to examine post-irradiation deformation in Fe-9Cr miniature sheet tensile specimens. It is found that

1) Deformation in irradiated specimens is highly localized so that specimens prepared more than 1 mm from the fracture surface usually showed no evidence of deformation.
2) Surface slip steps created during post-irradiation deformation are clearly defined with no other apparent surface damage whereas in unirradiated specimens the steps are more poorly defined and other surfaces show evidence of local deformation. The surface steps in irradiated specimens are nonplanar but features created early in the deformation are reproduced during further deformation creating ridges on the steps that are rarely altered with further deformation. Therefore, once localized slip is established in irradiated specimens, it generally continues without change. Localized necking occurs in irradiated specimens as slip steps get larger and slip planes increase in density.
3) Transformation of a<100> loops created during irradiation of ferritic alloys appears to proceed easily based on observed rapid disappearance of such features. It is proposed that this transformation is based on the reaction

$$\frac{a}{2}[\bar{1}11] + a[100] = \frac{a}{2}[111] \quad (1)$$

4) Continued post-irradiation deformation removes all irradiation-induced dislocation structure, replacing it eventually with $\frac{a}{2}[111]$ screw dislocations typical of deformation in unirradiated specimens. However, evidence for highly localized plasticity is found in the void structure so that some voids become highly elongated but nearby voids remain undistorted.
5) As local strains increase during post-irradiation deformation, a highly defined subgrain structure develops, whereas such structure is not found in unirradiated specimens deformed to about 8% uniform elongation.

References

[1] Hamilton M. L., Gelles D. S., and Gardner P. L., "Post-irradiation Deformation Behavior in Ferritic Fe-Cr Alloys," *Effects of Radiation on Materials: 16th International Symposium, ASTM STP 1175*, A. S. Kumar, D. S. Gelles, R. Nanstad, and E. A. Little, Eds., American Society for Testing and Materials, Philadelphia, 1993, pp. 545-557.

[2] Hamilton M. L. and Gelles D. S., "Post-irradiation Strength and Deformation of Ferritic Fe-Cr Binary Alloys," *Effects of Radiation on Materials: 15th International Symposium, ASTM STP* 1125, R. E. Stoller, A. S. Kumar, and D. S. Gelles, Eds., American Society for Testing and Materials, Philadelphia, 1992, pp. 1234-1242.

[3] Gelles D. S., Ohnuki S., Shiba K., Kohno Y., Kohyama A. , Robertson J. P., and Hamilton M. L., "Microstructural Examination of Irradiated Isotopically Tailored Ferritic/Martensitic Steels Following Shear Punch Testing," *Fusion Materials* Semiannual Progress Report for Period Ending December 31, 1998, *DOE/ER-0313/25* Oak Ridge National Laboratory, Oak Ridge, TN, 1999, pp. 143-150.

[4] Gelles D. S., Kimura A. and Shibayama T., "Analysis of Stress-Induced Burgers Vector Anisotropy in Pressurized Tube Specimens of Irradiated Ferritic-Martensitic Steels: JFMS and JLF-1," *Effects of Radiation on Materials: 19th International Symposium, ASTM STP 1366*, M. L. Hamilton, A. S. Kumar, S. T. Rosinski and M. L. Grossbeck, Eds., American Society for Testing and Materials, W. Conshohocken, PA, 2000, pp. 535-547.

[5] Gelles D. S., "Microstructural Examination of Neutron-Irradiated Simple Ferritic Alloys," *Journal of Nuclear Materials*, 1982, 108 - 109, pp. 515-526.

[6] Gelles D. S., "Neutron Irradiation Damage in Ferritic Fe-Cr Alloys," *Effects of Radiation on Materials: 14th International Symposium*, Vol. 1, *ASTM STP 1046*, N. H. Packan, R. E. Stoller and A. S. Kumar, Eds., American Society for Testing and Materials, Philadelphia, 1990, pp. 73-97.

[7] Gelles D. S., "Segregation Effects in Ferritic Fe-Cr Alloys," *Effects of Radiation on Materials: 16th International Symposium, ASTM STP 1175*, A. S. Kumar, D. S Gelles, R. Nanstad, and E. A. Little, Eds., American Society for Testing and Materials, Philadelphia, 1993, pp. 66-93.

[8] Gelles D. S., "Void Swelling in Binary Fe-Cr Alloys at 200 Dpa," *Journal of Nuclear Materials*, 1995, 225, pp. 163-174.

[9] Fish R. L., Straalsund J. L., Hunter C. W., and Holmes J. J., "Swelling and Tensile Property Evaluation of High-Fluence EBR-II Thimbles," *Effects of Radiation on Substructure and Mechanical Properties of Metals and Alloys, ASTM STP 529*, J. Moteff, Ed., American Society for Testing and Materials, 1973, pp. 149-164.

[10] Victoria M., Baluc N., Bailat C., Dai Y., Luppo M. I., Schäublin R., and Singh B.N., "The Microstructure and Associated Tensile Properties of Irradiated FCC and BCC Metals," *Journal of Nuclear Materials*, 2000, 276, pp. 114-122.

Takanori Hirose,[1] Hideo Sakasegawa,[1] Akira Kohyama,[2] Yutai Katoh,[2] and Hiroyasu Tanigawa[3]

Effect of Specimen Size on Fatigue Properties of Reduced Activation Ferritic/Martensitic Steel

Reference: Hirose, T., Sakasegawa, H., Kohyama, A., Katoh, Y. and Tanigawa, H., **"Effect of Specimen Size on Fatigue Properties of Reduced Activation Ferritic/Martensitic Steel,"** *Effects of Radiation on Materials: 20th International Symposium, ASTM STP 1405,* S. T. Rosinski, M. L. Grossbeck, T. R. Allen, and A. S. Kumar, Eds., American Society for Testing and Materials, West Conshohocken, PA, 2001.

Abstract: Small Specimen Testing Technology (SSTT) and related remote-control testing techniques are indispensable for the effective use of the limited volumes of material testing reactors and proposed intense neutron sources. As a part of this work, a new fatigue test machine with a laser extensometer for hot-cell usage has been developed. Materials used in this work were Japanese reduced activation ferritic/martensitic steel, JLF-1 (Fe-9Cr-2W-V-Ta) and its weldment. Correlations between fatigue life characteristics and fracture mechanisms were investigated for full- and mini-sized hourglass type specimens to clarify the effect of specimen size on fatigue properties These tests revealed that there was not significant difference in the number of cycles to failure in both specimens.

Keywords: reduced activation ferritic/martensitic steel, JLF-1, fatigue, TIG welding, small specimen testing technology

Introduction

Many outstanding accomplishments in the areas of fusion science and plasma physics over the past several decades emphasize the urgent needs of R & D activities in the area of fusion engineering/technology. The consensus that materials is the "Queen Technology to Fusion" or that materials are the key to fusion engineering/technology has been widely recognized and many IEA activities on fusion materials have been initiated

[1] Graduate student, Graduate School of Energy Science, Kyoto University, Yoshida Honmachi, Sakyo, Kyoto 606-8501, Japan
[2] Professor and associate professor, respectively, Institute of Advanced Energy, Kyoto University, Gokasho, Uji, Kyoto 611-0011, Japan
[3] Post doctoral researcher, Japan Atomic Energy Research Institute, 2-4 Shirakata, Tokai, Naka, Ibaraki 319-1195, Japan

in the recent years. Presently, reduced activation ferritic/martensitic steels (RAFs), vanadium alloys and SiC/SiC composite materials are considered to be the most promising candidates [*1,2*]. Among them, the RAFs R & D is placed at the highest priority in the Japanese and the European programs [*3,4*]. To accelerate the development of RAFs, a working group under Annex-II of the International Energy Agency (IEA) implementing agreement on a program of research and development on fusion materials has been formed to coordinate a collaborative program between the European Union (EU), the USA and Japan. As one of the steels for the IEA test program, JLF-1 (9Cr-2W-V-Ta) steel has been provided from the Japanese university activity JUPITER program. The second large heat of JLF-1 was produced in 1996. The 1.5 metric ton heat of JLF-1 steel was produced into plates with thickness of 15 and 25 mm. The first large heat was mainly used to obtain basic mechanical properties and neutron and charged particle irradiation experiments including HFIR irradiation [*5,6*]. The emphasis on the second large heat was to investigate joining technology and performance of the welded joints. Fracture toughness tests using 25-mm-thick plate are another important issue for the second heat [*7*]. In fusion applications, structural materials will suffer from cyclic stress caused by thermal cycling during reactor operation [*8*]. Therefore investigation of fatigue properties is very important, although only limited data on fatigue properties of RAFs had been obtained [*9*].

The objectives of this work are to investigate the fatigue properties of the second heat of JLF-1 and its welded joints, and to investigate the effect of specimen size on fatigue properties, in order to establish postirradiation and in-situ fatigue test methods, as an important part of small specimen test technology (SSTT) R & D. For a clear understanding of fatigue properties under the anticipated fusion environment, the effect of specimen size on the fatigue properties was studied utilizing the reduced activation ferritic steel JLF-1 and its weldment. The correlation between fatigue life characteristics and fracture mechanisms was investigated for full- and miniature-sized specimens. Based on the comparison of fatigue characteristics between the two sizes, the effects of specimen size on fracture mechanisms and their consequences on fatigue life times are discussed.

Experimental Procedure

Materials

The materials used were a reduced activation steel (JLF-1) and its weldment fabricated by tungsten inert gas (TIG) arc welding. Table 1 shows the chemical composition and heat treatment conditions.

Table 1 *Chemical composition (mass%) and heat treatment condition of JLF-1*

	C	Si	Mn	P	S	Al	Cr	W	V	Ta	N	Ti	B
BM[1]	0.10	0.05	0.45	0.003	0.002	0.003	8.85	1.99	0.20	0.080	0.0231	—	0.0002
WM[2]	0.061	0.13	0.43	0.005	0.003	0.003	9.16	1.91	0.25	0.081	0.0259	0.019	0.0001
WW[3]	0.061	0.10	0.45	0.003	0.003	0.002	8.96	1.82	0.25	0.084	0.0332	0.028	0.0001

[1] BM: Base metal, [2] WM: Weld metal, [3] WW: Welding wire

Normalizing: 1323K x 1h/A.C., Tempering: 1053K x 1h/A.C.

The TIG welding was performed using wire having a diameter of 1.2mm. The heat input was 14.5~15.8kJ/cm. Post welding heat treatment condition was 1013K x 3h/F.C. More detail on the JLF-1 second large heat are presented elsewhere [*10*].

Specimen

The shape and dimension of the specimens are presented in Figure 1. It is well known that the hourglass type specimen has good resistance to buckling, which is a very important issue in to miniaturized specimens for push-pull tests. Moreover it is desirable for ion irradiation experiments, because fracture initiation site was limited to around the specimen's waist portion. Full-sized and mini-sized hourglass type fatigue specimens were machined from the $25mm^t$ x 320^w x 500^lmm TIG welded plate. The width of the weldment was about 20mm. Both BM and WM specimen were oriented in the transverse direction. As mentioned above, the width of the welding line was narrow. In the case of full-sized WM specimen, the hourglass specimen was fabricated across the weldment. All relevant gauge volume consisted of the weld metal, with the heat affected zone (HAZ) being located outside of the test section. A mini-sized WM specimen made entirely of the weld metal was also prepared. In order to ease stress concentration, the root radius to minimum diameter (R / d) ratio of both specimens was selected to be 8, which is the upper limit of the ASTM E-606 recommendation of 4 to 8. The minimum diameter of full- and mini-sized specimen was 6mm and 1.25mm, respectively. The mini-sized specimens, called SF-1 (Small Fatigue) are proposed to use in accelerator driven D-Li stripping reaction neutron sources, such as IFMIF [*11*]. But also other proposed neutron source like the Gas Dynamic Trap (GDT-ns) requires the application of SSTT [*12*].

Development of Fatigue Test Machine for Hot-cell Usage

To develop mechanical testing machines for irradiated specimens, easy and remote handling process is indispensable. Especially for fatigue tests, gripping of specimens and attaching of extensometers by remote handling is considered to be difficult. In this work, remote handling issues were regarded as the most important. For measuring strain, a laser

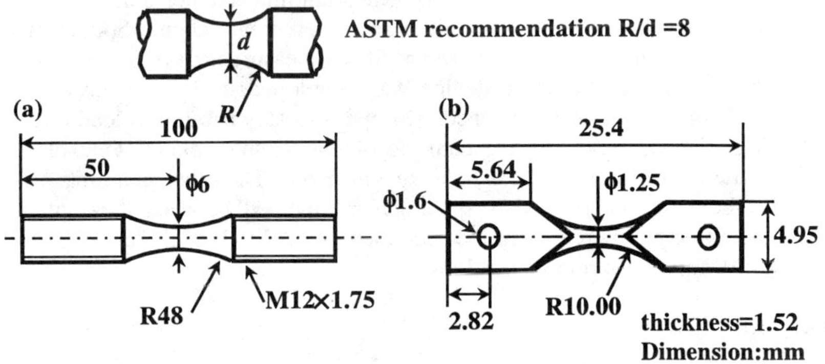

Figure 1 -*Dimensions of fatigue specimens (a) Full-sized specimen, (b) Mini-sized specimen.*

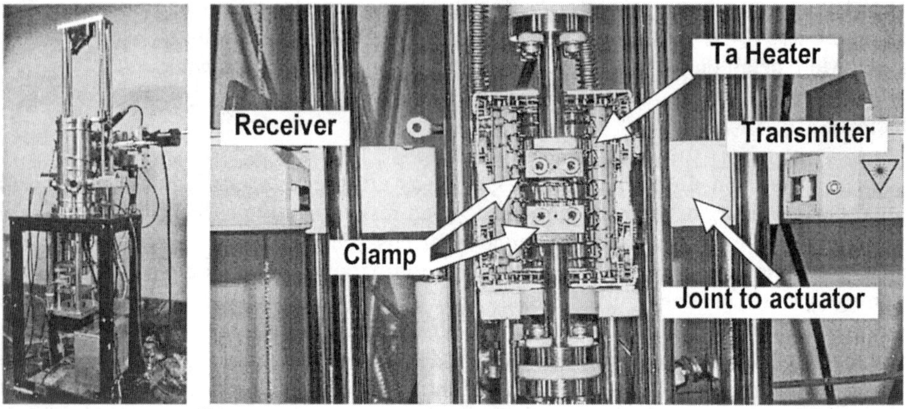

(a) (b)

Figure 2 -*Fatigue test machine for hot-cell usage (a) Fatigue machine, (b) Inside of vacuum chamber*

extensometer was applied, which enables monitoring deformation during the test without any specimen contact [*13*]. Figure 2 shows newly developed fatigue test machine for hot-cell usage. As shown in this figure, the laser extensometer consists of a transmitter and a receiver joined to an actuator. Laser measures change of minimum diameter during tests. Strain measurement was carried out by measuring the deviation from the reference value at the waist portion. Diametral strain, ε_d, was defined as follows

$$\varepsilon_d = (D - D_0) / D_0 \quad (1)$$

where
D = specimen's minimum diameter during test
D_0 = reference diameter, specimen's minimum diameter with unloaded condition

A measurement accuracy of 0.1μm was obtained by averaging the measured values. The measured strain value was also used for controlling the actuator during the test. The upper limit of distance between the transmitter and receiver was about ~500mm, which is enough to install vacuum chamber in between. Strain measurement is carried out outside of the chamber. A unique clamping device was developed in order to grip SF-1 type specimen strictly during the test. The specimen needs to stay parallel to load direction to avoid buckling. The clamping device consists of two counter plates. One plate has a groove along the load direction and a hole for guide pin. The other plate has a serrated surface and a hole for the pin. SF-1 specimen is clamped between these plates. The groove and the guide pin help to keep the specimen parallel to load direction. This device enables easy operation with remote handling.

Test Condition

For full-sized specimens, an electrohydraulic servo-controlled testing machine with a 10kN load-cell was used. For mini-sized specimen, an electromotive testing machine with a 2kN load-cell was used. Load controlled fatigue tests were carried out at 0.1~10Hz, with sinusoidal stress waveform at push-pull condition. Diametral strain controlled fatigue tests of full- and mini-sized specimen were carried out with a triangular stress waveform and a total diametral strain range, $\Delta\varepsilon_d$ of 0.2% ~ 3.0%. The diametral strain rate was 0.01%/s, the stress condition was push-pull. $\Delta\varepsilon_d$ was converted to total axial strain range, $\Delta\varepsilon_a$, by the following formula described in ASTM E-606

$$\Delta\varepsilon_a = (\sigma / E)(1 - \nu_e) - 2\Delta\varepsilon_d \qquad (2)$$

where:
σ = applied stress
E = Elastic modulus
ν_e = Elastic Poisson's ratio

The number of cycles to failure was defined as a point where a tensile peak stress decreased by 25% from the tensile peak stress at the first cycle and as a fracture cycle in the case that the specimen was broken. All tests were performed at an ambient temperature in air.

Results and Discussions

Effect of Heat Treatment

It is well known that fatigue tests are quite sensitive to specimen geometry and surface conditions including residual stress and strain introduced by machining. Although extreme care was taken during specimen preparation, there still remained a concern about

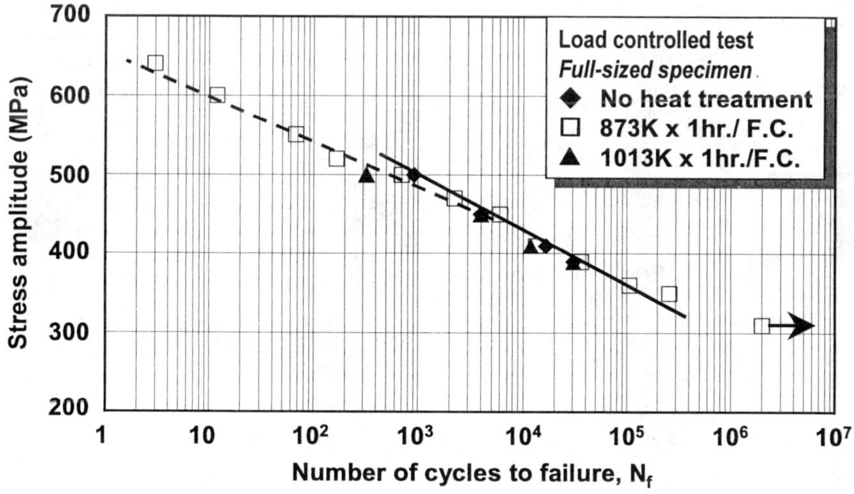

Figure 3 -*Effect of post machining heat treatment (PMHT) on fatigue life of JLF-1.*

this influence on the results. To see the influence, three types of full sized fatigue test specimens; (1) as fabricated, (2) 873K x 1 hr Post Machining Heat Treatment (PMHT) as standard stress relief condition and (3) 1013K x 3 hr PMHT as full stress relief condition, were tested. Figure 3 shows the results of the fatigue test on full sized specimens. The results of the specimens with three types of PMHT were almost identical. This looks sufficient to consider that residual stress and strain induced by machining has no effect on fatigue properties. Still the rest of the fatigue tests in this work have been done with 873K x 1 hr PMHT for the full-sized specimens and without PMHT for the mini-sized specimens.

Prediction of Fatigue Limit by Vickers Hardness Test and Tensile Test

As shown in Figure 4, it was found that fatigue life curve (S-N curve) couldn't be drawn by a single linear line with a slight decline toward a shorter lifetime at larger stress amplitude. The origin of this slight declination was interpreted to be the influence of cyclic creep. Reduction of area was significant in higher stress conditions. Therefore, under the true strain controlled condition, there will be no concern about this out of linearity. Another important information shown in this figure is the effect of welding on fatigue properties. Comparing with fatigue life of base metal (BM) and weld metal (WM), the value for WM is considerably longer than that of BM. It is commonly well known that there are good correlations of fatigue limit with Vickers hardness and tensile strength [14].

$$\sigma_{wo} \cong 1.6\ \mathrm{Hv} \pm 0.1\ \mathrm{Hv} \quad (\mathrm{Hv} \leq 400) \tag{3}$$

$$\sigma_{wo} \cong 0.5\ \sigma_B \tag{4}$$

where

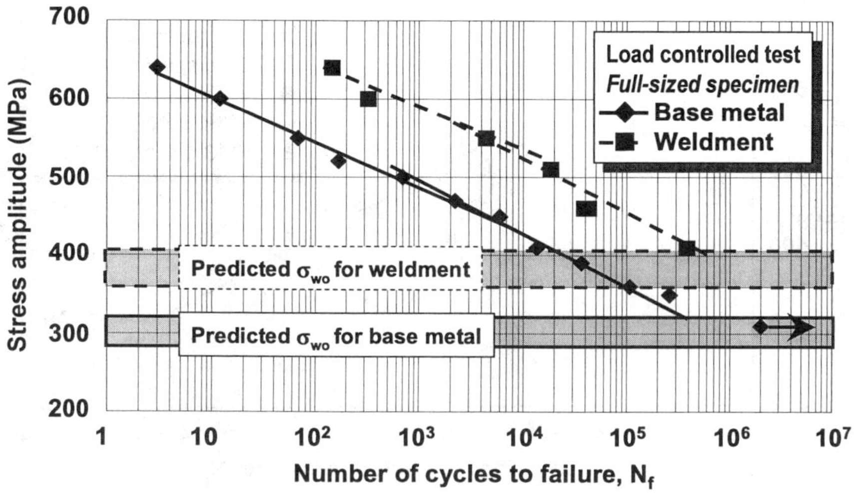

Figure 4 -*Results of load controlled fatigue tests, Estimation of fatigue limit.*

σ_{wo} = fatigue limit (in MPa)
Hv = Vickers hardness
σ_B = tensile strength (in MPa)

According to the results of Vickers hardness test and tensile test [*7,9*], σ_{wo} was predicted to be 304±19 MPa and 384±24 MPa, for BM and WM, respectively. These values appear to successfully predict the result of fatigue test with enough accuracy. This method could be applied to irradiated materials.

Diametral strain controlled tests

All specimens broke in the vicinity of the specimen's waist. Full-sized WM specimens broke in the weld metal. Typical hysteresis loops obtained by laser extensometry are presented in Figure 5. As shown in these figures, the laser extensometer was successfully applied to diametral strain controlled fatigue tests. Furthermore, emphasis has been put on a strictly parallel alignment of the specimen axis to the load direction. The clamping device was successfully applied, because buckling did not occur even at large strain range of $\Delta\varepsilon_a$=6%. The shape of hysteresis loop revealed that stress range and plastic strain range in each cycles did not depend on specimen size. Plastic strain range of BM and WM did not change during tests except for the end of tests. This phenomenon was observed in both full- and mini-sized specimen [*15*]. Significant difference between BM and WM in fracture apparatus was not observed. Defects introduced by welding, such as blowhole, did not affect to fracture apparatus.

The N_f values with the mini-sized specimens were compared with those of full-sized specimens. The relationship between N_f and the converted total axial strain range, $\Delta\varepsilon_a$, is presented in Figure 6. As shown in this figure, fatigue lifetime of WM was slightly shorter than BM. However, the difference between the lifetime of BM and that of

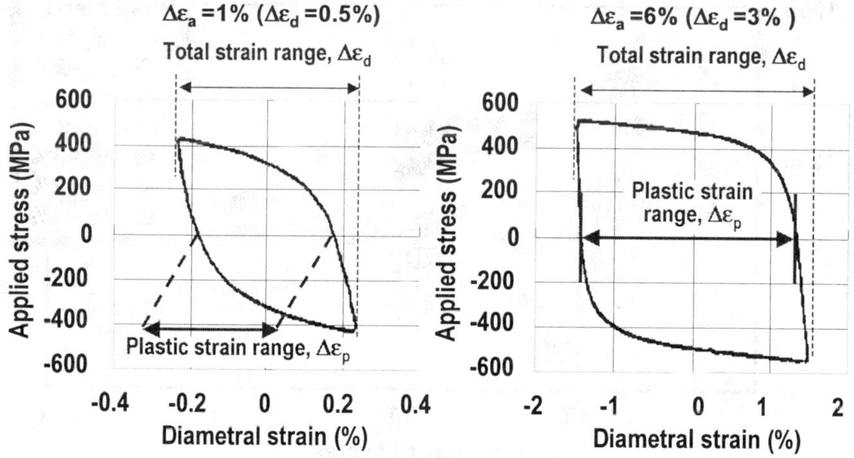

Figure 5 -*Typical hysteresis loops of JLF-1 base metal mini-sized specimen, obtained by laser extensometer. Around* $N_f/2$ *cycle.*

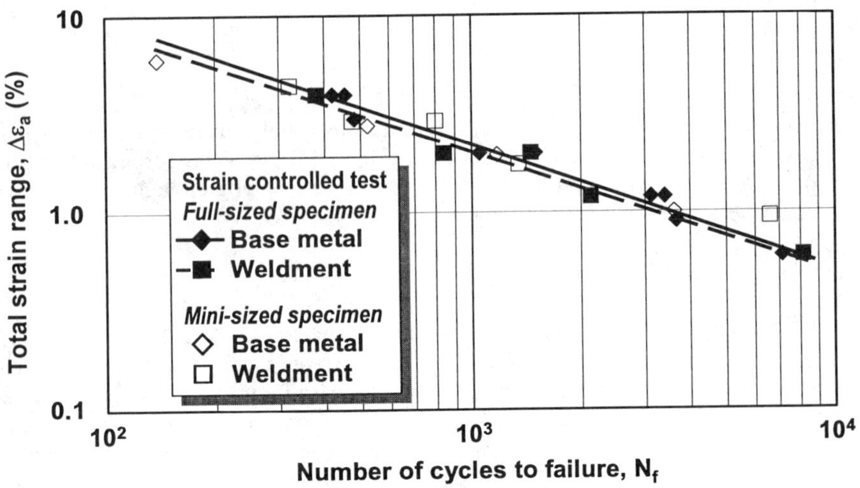

Figure 6 -*Fatigue life curves for JLF-1 and its weldment.*

WM was not significant. Significant size effect on fatigue lifetime was not observed in this work [*16, 17*].

Cyclic stress response curves are presented in Figure 7. As shown in this figure, cyclic softening was observed in all tests. The tensile stress increased, however, at initial stage in several tests performed at higher strain range. It implies that the difference in both stress range and inclination of stress range did not depend on specimen size. The difference in stress range and inclination between specimen sizes was less than 5%.

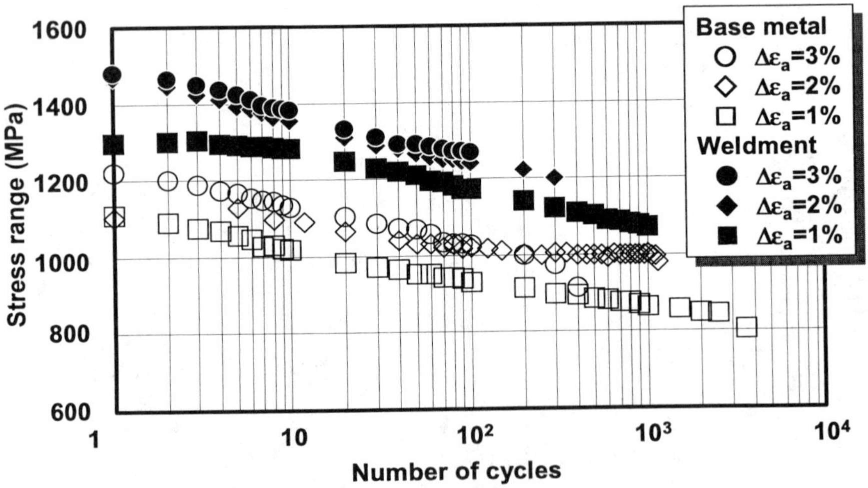

Figure 7 -*Cyclic stress response curves for JLF-1 and its weldment obtained from mini-sized specimen.*

According to these results, fatigue properties are not sensitive to the specimen sizes studied in this work.

Summary

A miniature fatigue specimen test machine for a hot laboratory has been successfully developed and the specimen size effect was studied. According to load controlled fatigue tests, the fatigue limit could be predicted by the well-known correlation between fatigue limit and Vickers hardness, and/or tensile strength. Fatigue life property was not obtained by a single linear line, but was slightly deviated downward with increasing applied stress amplitude. This can be attributed to the influence of cyclic creep.

So far, there is no significant specimen size effect on fatigue properties. Strain controlled fatigue tests on JLF-1 and its weldment were carried out using full-sized and mini-sized specimens. Fatigue life of JLF-1, TIG weldment was longer and the fatigue strength was larger than those of JLF-1 base metal. Number of cycles to failure, plastic strain range and applied stress are independent of specimen size. Cyclic softening was observed in both BM and WM. Fatigue lifetime of WM was shorter than that of BM. According to these results, fatigue properties were not sensitive to specimen size studied in this work. Neutron irradiation of JLF-1steel and its weldment is ongoing in Japanese fission reactors, such as the JMTR and the JOYO. Post-irradiation fatigue tests will be carried out in the near future.

Acknowledge

The authors would like to express their sincere appreciation to Prof. Namba, Prof. Nishimura, Prof. Muroga and Dr. Nagasaka of National Institute of Fusion Science, Dr. Shiba, Dr. Miwa and Dr. Jitsukawa of Japan Atomic Energy Research Institute for their useful help in many aspects to this work.

References

[1] Proceedings of international conference on fusion reactor materials, Journal of Nuclear Materials, Tokyo (1984), Chicago (1985), Kurlsrhue (1987), Kyoto (1989), Clearwater (1991), Stresa (1993), Obninsk (1995).

[2] Kohyama, A., Matsui, H. and Hishinuma, A., Proceedings of 10th Pacific Basin Nuclear Conference, 1996, p. 883.

[3] Proceedings of the IEA Working Group Meeting on Ferritic/Martensitic Steels, Tokyo (1992), Oak Ridge (1993), Sun Valley (1994), Oak Ridge (1995), Baden (1995), Culhum (1996), Tokyo (1997).

[4] Kohyama, A., Hishinuma, A., Gelles, D. S., Klueh, R. L., Dietz, W. and Ehrlich, K., "Low-Activation Ferritic and Martensitic Steels for Fusion Application," *Journal*

of Nuclear Materials, 233-237, 1996, pp. 138-147.

[5] Shiba, K. and Kohyama, A. "The progress of round-robin test in Japan and Japanese program," Proceedings of the IEA Working Group Meeting on Ferritic/Martensitic Steels, JAERI, Tokyo, 1997.

[6] Kohyama, A., Kohno, Y., Kuroda, M., Kimura, A. and Wan, F., "Production of Low Activation Steel; JLF-1, Large Heats –Current Status and Future Plan-," *Journal of Nuclear Materials*, 258-263, 1998, pp. 1319-1323.

[7] Nishimura, A., Inoue, N. and Muroga, T., "Fracture Toughness of Low Activation Ferritic Steel (JLF-1) Weld Joint at Room Temperature," *Journal of Nuclear Materials*, 258-263, 1998, pp. 1242-1247.

[8] Ishii, T., Fukaya, K., Nishiyama, Y., Suzuki, M. and Eto, M., "Low Cycle Fatigue Properties of 8Cr-2WVTa Ferritic Steel at Elevated Temperatures," *Journal of Nuclear Materials*, 258-263, 1998, pp. 1183-1186.

[9] Nishimura, A., Nagasaka, T., Inoue, N., Muroga, T. and Namba, C., "Low Cycle Fatigue Property of Low Activation Ferritic Steel at Room Temperature," presented in 9th International Conference on Fusion Reactor Materials, Colorado Springs, 1999, to be published.

[10] Inoue, N., Muroga, T., Nishimura, A. and Motojima, O., "Correlation between Microstructure and Hardness of a Low Activation Ferritic Steel (JLF-1) Weld Joint," *Journal of Nuclear Materials*, 258-263, 1998, pp. 1248-1252.

[11] Miwa, Y., Jitsukawa, S. and Hishinuma, A., "Development of a Miniaturized Hourglass Shaped Fatigue Specimen," *Journal of Nuclear Materials*, 258-263, 1998, pp. 457-461.

[12] Fischker, U., Möslang, A. and Ivanov, A. A., *Transactions of Fusion Technology*, vol. 35, No.1, 1999, p. 160.

[13] Brillaud, C., Meylogan, T. and Salathe, P., "Use of Laser Extensometer for Mechanical Tests on Irradiated Materials," *Effects of Radiation on Materials: 17th International Symposium, ASTM STP 1270,* David S. Gelles, Randy K. Nanstad, Arvind S. Kumar, and Edward A. Little, Eds., American Society for Testing and Materials, West Conshohocken, PA, 1996, pp. 1144-1153.

[14] Kuroda, M., Proceedings of the IEA Working Group Meeting on Ferritic/Martensitic Steels, Tokyo, 1997.

[15] Hirose, T., Sakasegawa, H., Kohyama, A., Katoh, Y. and Tanigawa, H., "Effect of specimen size on fatigue properties of reduced activation ferritic/martensitic Steels," presented in 9th International Conference on Fusion Reactor Materials, Colorado Springs, 1999, to be published.

[16] Rao, G. R. and Chin, B. A., "Comparison of Small-Scale Bending and Axial Fatigue Specimens", *Effects of Radiation on Materials: 15th International Symposium, ASTM STP 1125,* R. E. Stoller, A. S. Kumar, and D. S. Gelles, Editors, American Society for Testing and Materials, Philadelphia, 1992, pp. 1095-1102.

[17] Rao, G. R. and Chin, B. A., "Small Specimen Test Techniques Applied to Nuclear Reactor Vessel Thermal Annealing and Plant Life Extension" *Small Specimen Test Techniques Applied to Nuclear Reactor Vessel Thermal Annealing and Plant Life Extension* ASTM STP 1204, W. R. Corwin, F. M. Haggag, and W. L. Server, Eds., American Society for Testing and Materials, Philadelphia, 1993, pp. 267-274.

Hideo Sakasegawa,[1] Takanori Hirose,[1] Akira Kohyama,[2] Yutai Katoh,[2] Toshio Harada,[2] and Toshiei Hasegawa[3]

Correlation Between Creep Properties and Microstructure of Reduced Activation Ferritic/Martensitic Steels

---

Reference: Sakasegawa, H., Hirose, T., Kohyama, A., Katoh, Y., Harada, T., and Hasegawa, T., **"Correlation between Creep Properties and Microstructure of Reduced Activation Ferritic/Martensitic Steels,"** *Effects of Radiation on Materials: 20th International Symposium, ASTM STP 1405,* S. T. Rosinski, M. L. Grossbeck, T. R. Allen, and A. S. Kumar, Eds., American Society for Testing and Materials, West Conshohocken, PA, 2001.

Abstract: Reduced activation ferritic steels (RAFs), JLF-1 (9Cr-2W-V,Ta steel) had been studied as one of the reference materials in the IEA RAF R&D activity, because of its superior resistance to high-fluence neutron irradiation up to 100 dpa.

Recently, JLS-series (9Cr-xW-V,Ta steels; x=2.5, 3.0 and 3.5) have been produced as options for RAFs for use at higher temperatures in advanced blanket systems. In this series, improved high temperature mechanical properties were intended by increased tungsten contents.

The creep properties improved with the tungsten contents. A microstructural examination after creep rupture test revealed differences with a increase of tungsten content from 2.0 (JLF-1) to 3.0 wt% (JLS-2). According to the examination of tested specimens (100 MPa, 700°C), more intragranular fine precipitates and more intergranular precipitates formed preferentially on prior austenite boundaries and martensite lath boundaries were found in JLS-2 than in JLF-1, martensite lath widths became narrower and dislocation density increased with the increase of tungsten content.

Such microstructural features are likely to improve the creep properties by inhibiting grain evolution, microstructural recovery and strengthening grain boundaries at elevated temperatures.

Keywords: reduced activation ferritic/martensitic steels, creep property, microstructure, precipitates, carbide, Laves phase

---

[1]Graduate School of Energy Science, Kyoto University, Gokasho, Uji, Kyoto 611-0011, Japan.
[2]Institute of Advanced Energy, Kyoto University, Gokasho, Uji, Kyoto 611-0011, Japan.
[3]Nippon Steel Corporation, Shintomi, Huttsu, Chiba 293-8511, Japan.

Introduction

Reduced activation ferritic/martensitic steels (RAFs) have attracted attention in as a means simplifying the special waste storage of the highly radioactive blanket and first-wall structures from fusion reactors after service. From the viewpoint of decreasing components such as molybdenum, niobium and nickel which transmute to long-lived radioactive nuclides upon irradiation with 14 MeV high-energy neutrons in D-T fusion reactors, molybdenum, niobium and nickel must be severely restricted and molybdenum is replaced with tungsten.

RAFs have the superior swelling resistance, high heat flux capacity and so on. One of the RAFs, it was reported that JLF-1 has been developed and proven to have good resistance against high-fluence neutron irradiation up to 100 dpa and good phase stability [*1*].

As an option of RAFs for use at higher temperatures in advanced blanket systems, JLS-series (9Cr-xW-V,Ta; x=2.5, 3.0 and 3.5) have been prepared. In this series, improved creep and tensile properties at elevated temperatures were intended by increased tungsten contents. Tungsten composition in JLF-1 is 2.0 wt%, but in the JLS-series, it increases to 2.5, 3.0 and 3.5 wt% designated as JLS-1, JLS-2 and JLS-3 respectively.

In the present work, creep properties of JLS-series and JLF-1 were first characterized to obtain fundamental high temperature mechanical properties. Correlations between microstructural features and creep properties were then studied by means of transmission electron microscopy (TEM) and energy dispersive X-ray spectrometry (EDS) on the creep rupture-tested specimens as well as on as-received specimen, normalized, tempered and post weld heat treatment (NT+PWHT) specimens.

Experimental procedure

Materials

Table 1 shows chemical compositions and heat treatments of JLF-1 and JLS-series.

Table 1 –*Chemical composition and heat treatments of JLF-1 and JLS-series*

	C	Cr	W	V	Ta	Mn	Ti	N	Si	Ni	Mo	Al	P	S
JLF-1	0.097	9.04	1.97	0.19	0.700	0.46	0.001	-	<0.1	<0.1	<0.1	-	-	-
JLS-1	0.094	9.19	2.50	0.24	0.810	0.99	-	0.0267	0.050	-	-	0.003	0.0020	0.0004
JLS-2	0.096	8.99	2.97	0.24	0.810	1.00	-	0.0259	0.048	-	-	0.003	0.0020	0.0004
JLS-3	0.097	8.96	3.59	0.25	0.810	0.90	-	0.0270	0.048	-	-	0.003	0.0020	0.0005

Normalized at 1323 K x 3.6 ks followed by air-cooling
Tempered at 1033 K x 3.6 ks followed by air-cooling
Post Weld Heat Treatment (PWHT) at 1013 K x 10.8 ks followed by furnace cooling

The addition of tungsten instead of molybdenum improves creep rupture strength in high chromium steels, but excess tungsten addition leads to formation of δ-ferrite phase. δ-ferrite phase is known to degrade the toughness. Therefore the manganese content, an austenite former, was modulated in JLS-series up to 1.0 wt%. In addition, Laves phase (Fe_2W) tends to form with increasing of tungsten. This is known to have a bad effect on mechanical properties by coarsening at elevated temperature. Therefore the optimum amount of tungsten addition should be defined.

Vanadium addition also improves high temperature creep rupture strength, however excess vanadium addition leads to mechanical property degradation by causing significant carbide coarsening at high temperature. The optimum vanadium concentration is thought to be around 0.2 wt%. Niobium has same effect as vanadium, but it is detrimental element from the standpoint of RAFs. Tantalum has an equivalent effect to niobium, so tantalum is added instead of niobium.

Nitrogen is considered to be a less desirable element from the view of RAFs, but a small amount of nitrogen produces a remarkable improvement of high temperature creep strength by forming very fine nitride precipitates. Therefore, 0.025 wt% nitrogen is added in JLS-series.

The reason that a PWHT is that the heat treatment is performed for all actual structures which contain both weld joints and parent materials.

Specimen and test condition

Creep rupture tests were carried out in air at 650°C and 700°C by Nippon Steel Corporation.

For the microstructural observation, 3 mm TEM disks were cut from tested specimens as close to the fracture surface as possible with the normal to the tensile direction. After polishing mechanically with emery paper to about 0.1mm in thickness, electrolytic polishing was performed in a solution of 10 vol% sulfuric acid and 90 vol% ethyl alcohol.

In order to identify precipitates in creep rupture-tested specimens as well as on NT+PWHT specimens, the carbon extraction replica method was used. Extraction replicas were taken from polished surfaces etched in Villela solution. After carbon evaporation, the replicas were released in Villela solution. The carbon films were investigated with TEM and EDS. Additionally, extracted residues were investigated. X-ray diffraction analysis was carried out by using Cu K_α (λ=0.1545 nm) radiation.

Result and discussion

Starting Material

The heat treatment given to these steels produces a tempered, fully martensitic structure in both JLF-1 and JLS-2. Figure 1 shows the martensite with a high density of transformation dislocations. A narrower martensite lath width and a finer distribution of precipitates were found in JLS-2 than in JLF-1. Most of the precipitates existed along prior austenite grain boundaries and martensite lath boundaries, but few very fine precipitates were found within the grain [2]. Such changes in microstrucural features with the increase of tungsten content are likely to improve mechanical properties.

Micrographs of extraction replicas are shown in Figure 2. From the result of EDS and selected area diffraction analysis, two different type carbides were present in both specimens; $M_{23}C_6$ and MC. Block-like precipitates (>100 nm) were identified as $M_{23}C_6$ type carbides and these carbides were revealed to be Cr-rich.

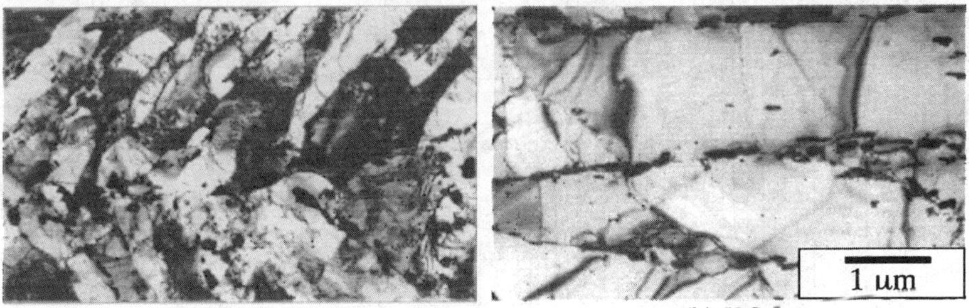

(a) JLF-1 *(b) JLS-2*

Figure 1 – *Microstructure of NT+PWHT specimen*

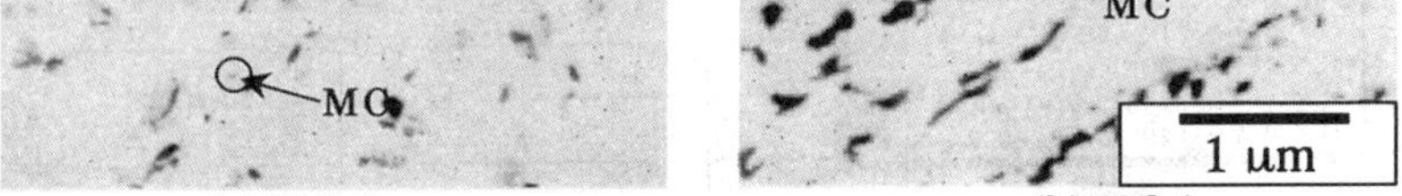

(a) JLF-1 *(b) JLS-2*

Figure 2 – *Micrograph of NT+PWHT specimen's extraction replica*

The results of quantitative analysis by EDS are presented in Table 2. Very fine dot-lik precipitates (<10 nm) were considered to be MC type carbides. These precipitates were (Ta, V)C type carbides. Ta-rich precipitates were dominant, but several were V-rich.

Creep properties

Figure 3 shows the results of creep rupture tests of JLF-1 and JLS-series. The tests are performed in atmosphere at 650°C and 700°C. It can be seen that rupture time become longer with increasing of tungsten content. These creep rupture test results were summarized by Larson-Miller parameter and Figure 4 shows the result. According to it, the improvement of creep life can be seen with increasing tungsten content. JLS-3 (3.5 wt% W) demonstrated the best creep resistance.

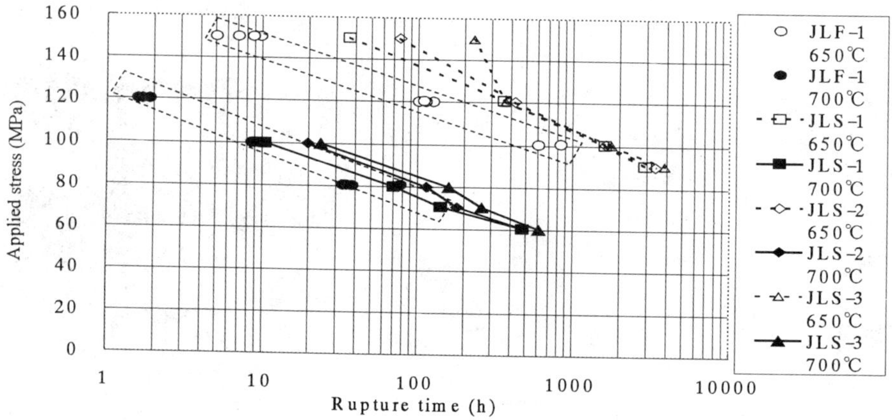

Figure 3 The Result of creep rupture tests

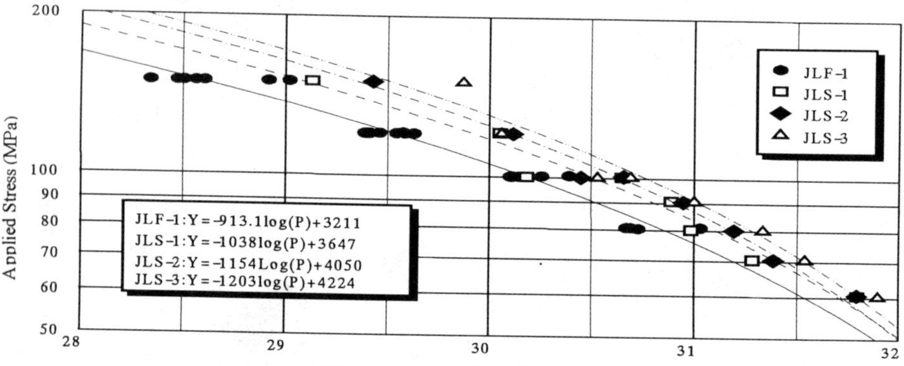

Figure 4 Creep properties summarized by Larson-Miller parameter

Correlation between microstructure and creep properties

650°C Creep rupture tested specimen

Figure 5 shows the result of microstructural observation by TEM in JLF-1 and JLS-2. Its creep test condition is 100MPa, 650°C. Here, the rupture time of JLF-1 was 772 h and that of JLS-2 was 1640 h.

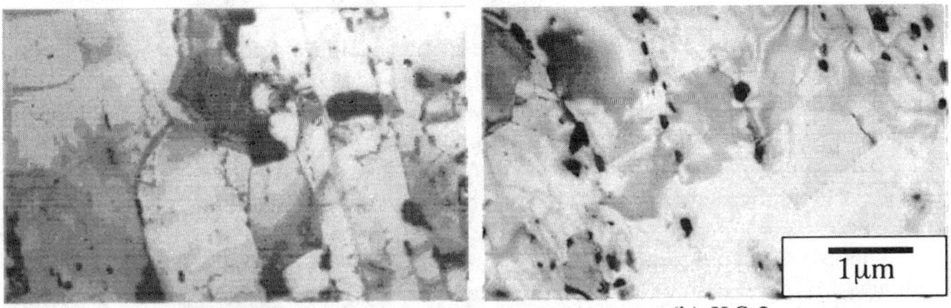

(a) JLF-1 *(b) JLS-2*

Figure 5 – *Microstructure of 650°C crept specimen*

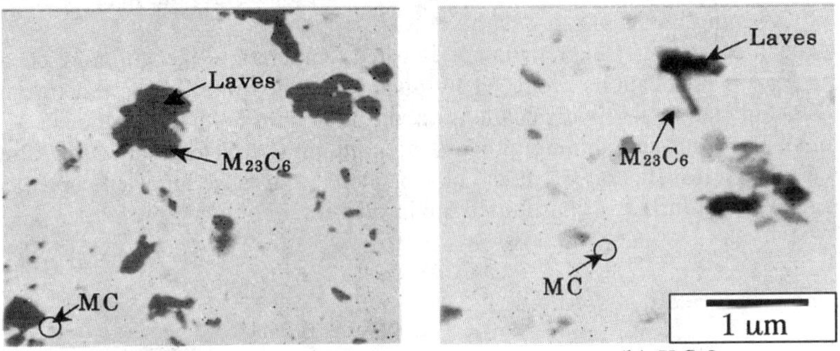

(a) JLF-1 *(b) JLS-2*

Figure 6 – *Micrograph of 650°C crept specimen's extraction replica*

In both specimens, dislocations were recovered and precipitates on grain boundaries were particularly coarsened. Martensite lath structures could not be found.

Very fine intragranular precipitates (<10 nm) and intergranular rod and block-like precipitates were found. Some rod and block-like precipitates (>100 nm) within recovered grain lay along a preferred direction. It seems that they existed on prior austenite grain boundary or martensite lath boundary before creep rupture test. These intragranular precipitates were found in JLS-2 rather than in JLF-1. It can be said that these precipitates hindered prior austenite grain and martensite lath subgrain from evolution at elevated temperatures. Since intragranular fine precipitates are known to strengthen mechanical property, it can be considered that not only such intragranular fine precipitates (<10nm) but also rod and block-like precipitates contributed to the improvement of creep properties with increasing tungsten content.

The results of XRD analysis of extracted residues to identify precipitates are shown in Figure 9. In this condition, $M_{23}C_6$ type carbide was identified only in JLF-1 and JLS-2. M_6C type carbides were not found. It has been reported that block-like and globular M_6C type carbides exist on grain boundaries and affected creep properties similarly to $M_{23}C_6$ type carbide [*3*]. Major precipitates in these specimens can be considered to be $M_{23}C_6$ type carbide, but all of the precipitates could not be defined by the method.

Micrographs of extraction replicas are given in Figure 6 for closer identification to precipitates. Two different type carbides were found: $M_{23}C_6$ and MC type carbides in both specimens. This result was the same as in NT+PWHT specimens. In addition to these carbides, intermetallic compound Laves phase (Fe_2W) was found. Laves phase was found at the same places where $M_{23}C_6$ exists. They aggregated with each other and the size of aggregation was about 1μm. They were found in both JLF-1 and JLS-2 specimens [*4*].

The $M_{23}C_6$ type carbide was identified to be Cr-rich and very fine dot-like precipitates MC type carbide was identified as (Ta, V)C type carbide [*5*]. Most of MC type carbides were Ta-rich, but several were V-rich [*6*]. From the result of quantitative analysis with EDS (in Table 2), chromium, iron and tungsten content in $M_{23}C_6$ type carbide decreased after creep rupture test. This was the result of Laves phase formation. It can be assumed that tungsten, not merely in $M_{23}C_6$ type carbide, but also in the matrix was used by forming Laves phase.

Mechanical property degradation at elevated temperature like softening occurs by recovering excess dislocations produced by quench-induced martensitic transformation. Excess dislocations recover by dislocations climb, interacting with the carbides and other dislocations. The rate-determining process of softening due to dislocation climb is considered to be self-diffusion [*7*]. It can be considered that more tungsten content in the matrix of JLS-2 than in JLF-1 contributed the improvement of creep properties.

700°C Creep rupture tested specimen

Figure 7 shows the result of microstructural TEM observation in JLF-1 and JLS-2. Its creep test condition was 100 MPa at 700°C. Here, the rupture time of JLF-1 was 8.7 h and that of JLS-2 was 19.7 h.

Martensitic structure was recovered somewhat during creep rupture test and lath width increased to more than about 2.0 μm in both specimens.

These structures didn t recover sufficiently as well as structures of the tested specimens at 650°C in spite of higher temperature 700°C. This is the result of creep rupture time, creep rupture time was much longer at 650°C than at 700°C. Holding at elevated temperature affected microstructural recovery more than temperature.

Comparing between JLF-1 and JLS-2, microstructural recovery in JLF-1 advanced further than JLS-2 and many dot-like precipitates in recovered grains were found in JLS-2. These very fine intragranular precipitates can improve creep properties.

In this condition, according to XRD analysis of extracted residues (in Figure 9), $M_{23}C_6$ type carbide was identified in both JLF-1 and JLS-2, which is same as at 650°C. Micrographs of extraction replicas are shown in Figure 8. Two different type carbides were present in both specimens: $M_{23}C_6$ and MC. The result of carbide precipitation was the same NT+PWHT and tested at 650°C.

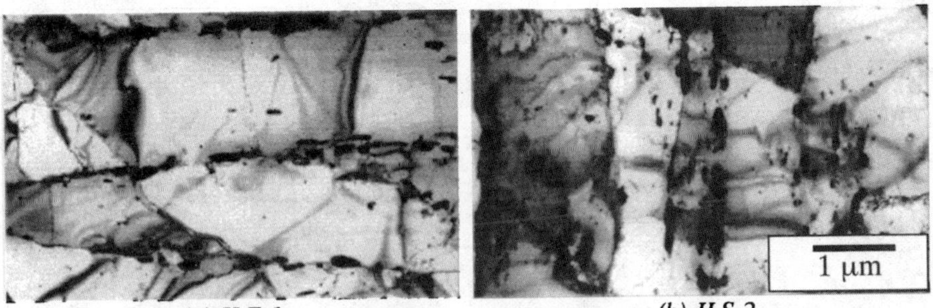

(a) JLF-1 *(b) JLS-2*

Figure 7 – *Microstructure of 700°C crept specimen*

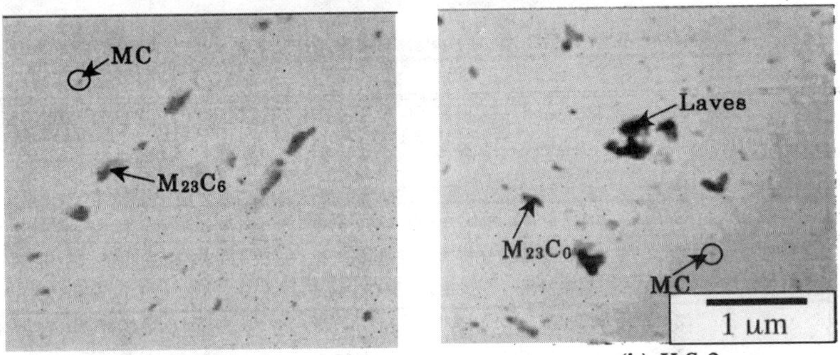

(a) JLF-1 *(b) JLS-2*

Figure 8 – *Micrograph of 700°C crept specimen's extraction replica*

Laves phase was found only in JLS-2. Laves phase was found on the same place where $M_{23}C_6$ type carbide exists. They aggregated with each other and the size of aggregation was about 0.5 μm. This precipitation morphology was the same as at 650°C, but this size in tested at 700°C specimen is smaller than in specimen tested at 650°C. It was influenced by retention time at high temperature as well.

$M_{23}C_6$ type carbide was revealed to be Cr-rich and dot-like very fine precipitates MC type carbide was identified as (Ta, V)C type carbide. Most of MC type carbides were Ta-rich but several were V-rich. These results were the same as in specimens tested at

650°C. It can be assumed that tungsten dissolved in the matrix prevented microstructural recovery by inhibiting iron self-diffusion. Therefore, microstructural recovery in JLF-1 was faster than in JLS-2. Many fine MC type carbides were found within the grains in JLS-2. This precipitation morphology was likely to contribute to the improvement of mechanical properties at elevated temperature.

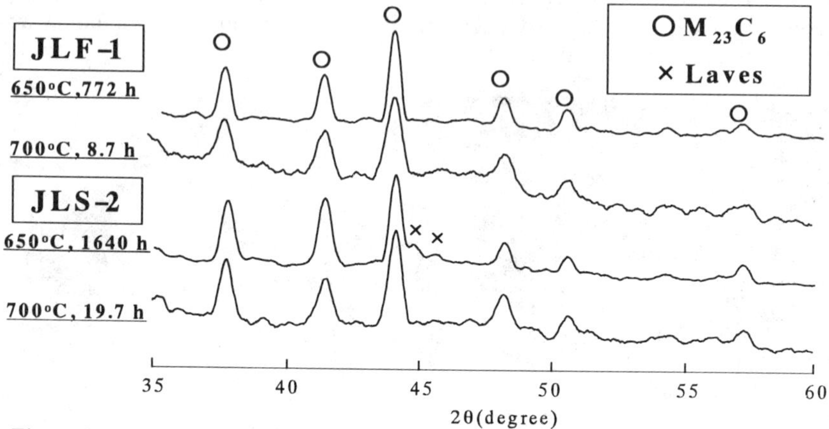

Figure 9 *The result of XRD analysis to extracted residues of crept specimens*

Table 2 –*Chemical composition of precipitates in JLF-1 and JLS-series*

			Fe	Cr	W	Mn	V	Ta
JLF-1	NT+PWHT	$M_{23}C_6$	22.91-25.37	47.14-50.61	11.25-14.94	0.53-1.1	0.49-1.06	10.46-13..42
		MC	0-0.99	1.29-7.33	1.93-5.60	0-0.25	0.60-33.38	56.50-92.50
		Laves	-	-	-	-	-	-
	Creep 650°C	$M_{23}C_6$	15.17-17.84	39.91-46.63	14.02-16.69	0-13.5	0	12.3-15.9
		MC	0.33-1.84	0.42-7.98	1.82-3.70	0-7.24	0.78-39.79	73.09-94.77
		Laves	22.06-21.93	5.14-7.11	50.96-54.20	0	0	18.98-22.02
	Creep 700°C	$M_{23}C_6$	18.96-23.27	47.87-55.64	14.02-16.69	0.13-1.31	0.36-0.89	10.26-14.49
		MC	0.30-1.24	3.83-10.35	3.99-11.34	0-0.99	23.34-37.49	52.45-59.73
		Laves	-	-	-	-	-	-
JLS-2	NT+PWHT	$M_{23}C_6$	17.64-21.46	42.85-47.74	15.38-18.57	1.48-2.52	0.68-4.3	10.94-12.46
		MC	0-0.72	0.53-8.48	0.86-2.38	0-0.47	19.28-56.64	32.42-94.68
		Laves	-	-	-	-	-	-
	Creep 650°C	$M_{23}C_6$	12.21-14.34	42.02-46.86	12.52-14.77	13.63-16.82	0-8.35	9.37-16.22
		MC	0.09-4.27	0.51-8.91	0.58-9.08	0-2.47	17.09-63.60	90.02-93.44
		Laves	17.68-22.49	4.28-4.76	40.26-52.08	0-1.30	0	19.05-35.22
	Creep 700°C	$M_{23}C_6$	18.17-23.27	48.35-52.51	12.52-14.77	13.63-16.82	0-8.35	9.37-16.22
		MC	0.30-1.24	3.83-10.35	3.99-11.06	0-0.99	23.34-29.77	52.45-59.73
		Laves	23.03-25.52	4.68-5.44	46.19-47.15	0.59-0.69	0.15-2.21	21.48-22.88

Consideration of observation on crept specimen

In the present work, microstructural observations were carried out for only the creep rupture tested specimens. For example, in 9Cr-4W steel after tempering at 773 K for 3.6 ks, precipitation of W_2C occurred uniformly in the matrix, which also typical 9Cr-1W and 9Cr-2W steels. W_2C precipitates are most likely responsible for the secondary hardening in 9Cr-xW steels [7]. Such a secondary hardening caused by W_2C during creep rupture test can occur sufficiently in the JLF-1 and JLS-series.

Summary

Creep properties improved with increasing tungsten content. JLS-3 (3.5 wt% W) had the best creep resistance.

In NT+PWHT specimens, differences in microstructure between JLF-1 and JLS-2 were the decrease of martensite lath width and finer distribution of precipitates with increasing tungsten content.

In tested specimens (100 MPa, 650°C), microstructures in both JLF-1 and JLS-2 showed fully recovered structure. Block-like $M_{23}C_6$ type carbide and very fine dot-like MC type carbide were found. $M_{23}C_6$ type carbide was coarsened by holding at high temperature during creep rupture test. Laves phase existed and aggregated with $M_{23}C_6$ type carbide and with each other in both specimens.

In tested specimens (100 MPa, 700°C), $M_{23}C_6$ and MC type carbides were found in both JLF-1 and JLS-2, but Laves phase existed only in JLS-2. Martensite lath width was more than about 2 μm. Many fine intragranular MC type carbides were found in JLS-2.

From the results, microstructural features contributing to the improvement of creep properties with the increase of tungsten content were as follows:

1. The decrease of martensite lath width and finer distribution of precipitation in as-received specimens
2. Finely distributed grain boundary precipitates, such as $M_{23}C_6$ type carbide, decelerated the grain evolution until significant coarsening took place.
3. Tungsten dissolved in the matrix and the fine intragranular precipitates (e.g., MC type carbide) inhibited self-diffusion and dislocation motion.

References

[1] Kohno,Y., Kohyama, A., Yoshino, M., and Asakura, K., Effects of Neutron Irradiation on Microstructural Evolution in Candidate Low Activation Ferritic Steels , *Journal of Nuclear Materials*, 1994, 212-215, pp. 707-713.

[2] Asakura, K. Kohyama, A., and Yamada, T., Mechanical Properties and Microstructure Changes of Low-activation 3Cr-2W-V-Ti Ferritic Steels Developed for Nuclear Applications , *ISIJ International*, 1990, Vol. 30, pp. 947-954

[3] Kawanishi, H., Sekimura, N., Shibata, and M., Ishino, S. Micrpstrucuture Ion-irradiated 9Cr-2Mo Steel (JFMS) , *Journal of Nuclear Materials*, 1985, 133-134, pp. 623-627

[4] Nishiyama, Y., Fukuya, K., Suzuki, M., Eto, M., Irradiatton embrittlement of 2 1/4Cr-1Mo Steel at 400°C and its Electrochemical Evaluation , *Journal of Nuclear Materials*, 1998, 258-263, pp.1187-1192

[5] Shibayama, T., Kimura, A., Kayano, H., The Effect of Small Additional Elements on the Precipitation of Reduced Activation Fe-9Cr-2W Steels , *Journal of Nuclear Materials*, 1996, 233-237, pp. 270-275

[6] Jones, W. B., Effects of Mechanical Cycling on the Substructure of Modified 9Cr-1Mo Ferritic Steel , *ASTM Metals Park*, Ohio 44073, 1983, p.221-235

[7] Abe, F., Araki, H., Noda, T., The Effect of Tungsten on Dislocation Recovery and Precipitation Behavior of Low-Activation Martensitic 9Cr Steels , *Metallurgical Transactions A*, 1991, 22A, pp. 2225-2235

Mychailo B. Toloczko,[1] Benjamin R. Grambau,[2] Frank A. Garner,[3] and Katsunori Abe[1]

Comparison of Thermal Creep and Irradiation Creep of HT9 Pressurized Tubes at Test Temperatures from ~490°C to 605°C

Reference: Toloczko, M. B., Grambau, B. R., Garner, F. A., and Abe, K., **"Comparison of Thermal Creep and Irradiation Creep of HT9 Pressurized Tubes at Test Temperatures from ~490°C to 605°C,"** *Effects of Radiation on Materials: 20th International Symposium, ASTM STP 1405,* S. T. Rosinski, M. L. Grossbeck, T. R. Allen, and A. S. Kumar, Eds., American Society for Testing and Materials, West Conshohocken, PA, 2001.

Abstract: Pressurized creep tubes were used to evaluate the thermal creep behavior and the irradiation creep behavior of several variants of HT9 ferritic-martensitic steel in the temperature range from ~490°C to 605°C. Examination of the thermal creep data as a function of time showed classic thermal behavior at all the test temperatures. A creep transient was followed by a period of steady state creep which, in some instances, was then followed by tertiary creep. The irradiation creep behavior also displayed a transient followed by steady state creep, but no tertiary creep was observed.

While the thermal creep behavior and the irradiation creep behavior were qualitatively similar, the magnitude of both the transient and the steady state creep rate were often different. At ~490°C, the thermal creep specimens exhibited a greater transient strain, while at 600°C, the irradiation creep specimens exhibited a greater transient strain. For both the thermal creep specimens and the irradiation creep specimens, the transient strain was observed to increase with increasing applied stress. The steady state creep rates of both the thermal creep specimens and the irradiation creep specimens were observed to increase with increasing temperature. At lower stresses, the steady state creep rates of the thermal creep specimens and the irradiation creep specimens were often equal. At the higher test stresses, the steady state creep rate of the irradiation creep specimens was always higher.

At ~490°C, the steady state thermal creep rate was observed to have a linear stress dependence up to the highest applied stress of 250 MPa. At 575°C, the stress exponent was between 1.5 and 2 for effective stresses less than 200 MPa. At 605°C, the stress exponent was ~1 for effective stresses less than 70 MPa. For the irradiation creep specimens, a stress exponent between 1.5 and 2 was observed at ~490°C and 550°C. At 600°C, the stress exponent was observed to be ~1.

Keywords: ferritic-martensitic steels, thermal creep, irradiation creep

[1] Research Associate and Professor respectively, Tohoku University, Sendai 980-8579, Japan.
[2] Graduate student, University of Michigan, Ann Arbor, MI 48109, USA.
[3] Fellow Scientist, Pacific Northwest National Laboratory, Richland, WA 99352, USA.

Background

In the open literature, the results of many neutron irradiation creep experiments have been published [1-6]. However, in only a few of these experiments have the irradiation creep data been compared to thermal creep data obtained under otherwise identical experimental conditions [6-10]. In the present work, for temperatures between ~490°C and 605°C, neutron irradiation creep data from several heats of the ferritic-martensitic steel HT9 were compared to corresponding thermal creep data. Pressurized tube capsules were utilized for both the irradiation creep and thermal creep experiments. The data were obtained from the liquid metal reactor (LMR) irradiation creep program conducted at the Pacific Northwest National Laboratory between 1980 and 1994.

Experimental

Materials, Specimens, Stress State, and Deformation

Three different heats of HT9 with very similar compositions were studied. The compositions are shown in Table 1. These materials were formed into a tubular shape by gun drilling of billet stock. After gun drilling, tubing was taken to its final dimensions through repeated drawing and annealing. The final tubing dimensions are shown in Table 2. After the final anneal, the materials were heat treated. Two different heat treatments were used which led to four different combinations of material and heat treatment as shown in Table 2. In the present work, these four combinations will be referred to as HT9-1, HT9-2, HT9-3, and HT9-5.

Pressurized tube capsules were constructed by cutting tube sections from the tube stock, and electron beam welding plugs to the ends of the tube sections. Tube pressurization was accomplished by introducing helium gas through a small entrance hole in one of the plugs. After setting the appropriate pressure, the entrance hole was laser welded shut [11]. The pressurization procedure was performed at room temperature, and thus it was necessary to select a room temperature pressure that would give the desired pressure at the test temperature. Between room temperature and the maximum test temperature utilized in the experiment, helium behaves as an ideal gas behavior [12-14], and thus the required pressure at room temperature was calculated using the ideal gas law.

Table 1 -- *Composition of each of the HT9 variants. Heat number is in parentheses.*

Alloy	Fe	Cr	Ni	Mo	Mn	V	W
HT9-1 (84425)	Bal	11.8	0.51	1.03	0.50	0.33	0.52
HT9-2,3 (91353)	Bal	11.8	0.57	0.94	0.54	0.24	0.52
HT9-5 (92235)	Bal	11.8	0.60	1.06	0.62	0.33	0.52
Alloy	**Si**	**C**	**Ti**	**Al**	**S**	**P**	**N**
HT9-1 (84425)	0.21	0.21	< 0.01	0.03	0.003	0.008	0.006
HT9-2,3 (91353)	0.28	0.17	< 0.01	0.05	0.003	0.007	0.006
HT9-5 (92235)	0.29	0.21	---	0.01	0.002	0.011	---

Table 2 -- *Heat treatments for each of the HT9 variants. Heat number is in parentheses.*

Alloy	Thermomechanical Treatment	Tube Dimensions (cm)
HT9-1 (84425)	1038°C/5 min/air cool, 760°C/30 min/air cool	0.457 outside diameter 0.020 wall thickness 2.24 length
HT9-2 (91353)	1038°C/5 min/air cool, 760°C/30 min/air cool	0.457 outside diameter 0.020 wall thickness 2.24 length
HT9-3 (91353)	1100°C/2 min/air cool, 650°C/2 hr/air cool	0.457 outside diameter 0.020 wall thickness 2.24 length
HT9-5 (CDE Heat, 92235)	1100°C/2 min/air cool, 650°C/2 hr/air cool	0.686 outside diameter 0.055 wall thickness 2.82 length

For each target temperature, a set of capsules with each capsule having a different pressure were constructed. For the present work, a set of capsules shall be called a tube-set. The range of pressures, and thus stresses, was chosen based on the test environment (irradiation or thermal) and the target test temperature. In general, a tube-set consisted of between four and eight capsules. For the tube-sets designated to be irradiated, a "stress-free" capsule was included in each tube-set to monitor swelling.

The uncertainty in the magnitude of the stress in the tube-wall arises mainly from deviations in the actual test temperature (T_{test}) from the target test temperature (T_{target}). First order error analysis shows that the uncertainty in the stress is given approximately by

$$\frac{\Delta\sigma}{\sigma} \sim \frac{\left|T_{test} - T_{target}\right|}{T_{target}} \tag{1}$$

where the temperature is in Kelvin.

The stress state in the wall of a capped thin-wall pressurized tube is well known [15], and for net internal pressure inside the tube, P_g, the principle stresses in the cylindrical coordinate system are

$$\sigma_H = \frac{P_g r}{t} \tag{2}$$

$$\sigma_z = \frac{\sigma_H}{2} \tag{3}$$

$$\sigma_r \approx 0 \tag{4}$$

where r is the inner radius of the tube and t is the tube wall thickness. The "H" subscript is meant to represent the hoop stress. To aid in comparing creep data obtained from pressurized tube capsules with creep data obtained from other specimen geometries, the von Mises effective stress is used. For the biaxial stress state in a pressurized tube capsule, the von Mises stress ($\bar{\sigma}$) is given by

$$\bar{\sigma} = \frac{\sqrt{3}}{2}\sigma_H \tag{5}$$

Due to the 2:1 ratio of the hoop stress to the axial stress, and due to the fact that creep involves only deviatoric deformation, a plane strain deformation condition exists when a pressurized tube undergoes creep. The deviatoric deformation lies in the rθ plane, and there is no deviatoric deformation along the z-axis.

Creep strains are obtained from the change in the outer-wall diameter of a tube. This measured outer-wall strain is converted into a mid-wall strain which is an average of the strain across the tube wall thickness. For the tube geometries utilized in this experiment, the conversion factor is of the order of 1.1 and is essentially constant for outer-wall strains up to 10%. The mid-wall strain (ε_M) is then converted to an effective plastic strain [16]. Assuming plane strain deformation (in the rθ plane), the effective plastic strain is given by

$$\bar{\varepsilon} = \frac{2}{\sqrt{3}}\varepsilon_M \tag{6}$$

Because of the plane strain deformation and the conservation of volume which occurs during creep of a pressurized tube capsule, the pressure and wall thickness both decrease in such a way that the stresses in the tube wall remain nearly constant for uniform strains as large as 10%. This is in contrast to typical uniaxial creep tests where the load remains constant and the stress increases as cross-sectional area of a specimen decreases.

Thermal Creep and Irradiation Creep Environment

For both the thermal creep tests and the irradiation creep tests, the selected temperatures were ~490°C, 550°C, and 600°C. The time that the irradiated specimens spent in reactor was very similar to the time that the thermal creep specimens were held at temperature.

Irradiations were performed at the Fast Flux Test Facility (FFTF) located in Richland, Washington using the Materials Open Test Assembly (MOTA) [17, 18]. Specimen irradiation temperatures were achieved using an actively controlled gas-gap method which permitted temperature control to within ±5°C [18]. In most every instance, capsules in a tube-set were placed side-by-side in the reactor to provide nearly identical irradiation conditions for each capsule in a tube-set. However, a tube-set was not necessarily kept in the same position from one irradiation cycle to the next which occasionally resulted in small temperature differences from one irradiation cycle to the next. Depending on the location in reactor and the particular MOTA cycle, the dose rate for this experiment ranged from ~0.8×10^{-6} dpa/sec to ~1.7×10^{-6} dpa/sec. Diameter measurements were performed at the conclusion of each irradiation cycle.

Results and Discussion

At all three test temperatures, both the thermal creep and the irradiation creep specimens displayed typical creep behavior defined by an initial transient which was followed by a region of steady state creep. Several of the higher stress thermal creep specimens also showed a transition to tertiary creep. Tertiary creep was not observed in the irradiation creep specimens, but this is probably due to the irradiated specimens not being in reactor for a sufficient amount of time. Examples of creep from each test temperature are shown in Figs. 1-3. The neutron irradiation creep data were plotted as a function of "effective full power hours" (EFPH) which represents the time during which the reactor was running at full power during an irradiation cycle. EFPH is very close to the actual irradiation cycle time. Without exception, the irradiation creep data had less scatter. While the data here do not provide an explanation of this behavior, it is reasonable to assume that the homogenous microstructures which result from neutron irradiation caused this behavior. The stress-free capsules which were included in each of the irradiated tube-sets showed no evidence of swelling.

A comparison of the thermal creep data and the irradiation creep data leads to several observations. For the same applied stress, the magnitude of the total creep strain is generally greater for the irradiated specimens. At higher stresses, this difference is greater. As the temperature increases, the difference in the total creep strain between thermal creep specimens and irradiation creep specimens also increases. While the creep strain in the irradiated specimens is generally larger for the range of temperatures in this experiment, the strains are still within the same order of magnitude.

The magnitude of the transient strains are compared in Figs. 4-6. Since the dose rate in the irradiated specimens varied from one tube-set to the next, and irradiation creep is known to be dependent on the dose rate, the dose rates are shown in the figures. At ~490°C, it can be clearly seen that the transient strain in the thermal creep specimens is

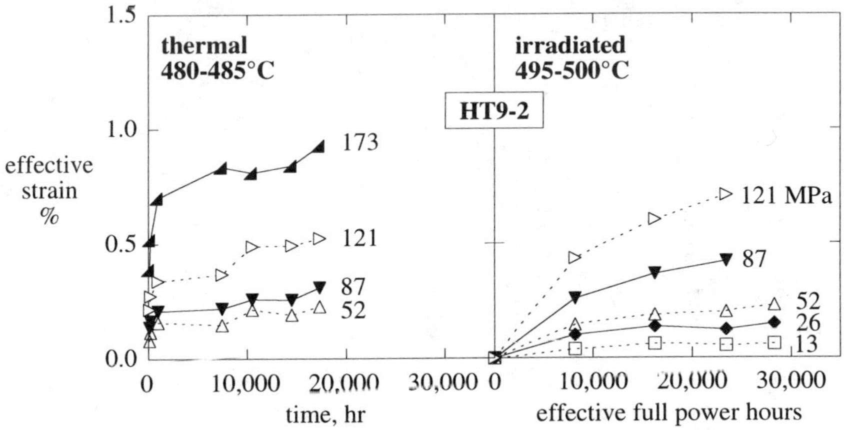

Figure 1 -- *Creep as a function of time for pressurized tubes in a thermal environment or in a neutron irradiation environment at ~490°C.*

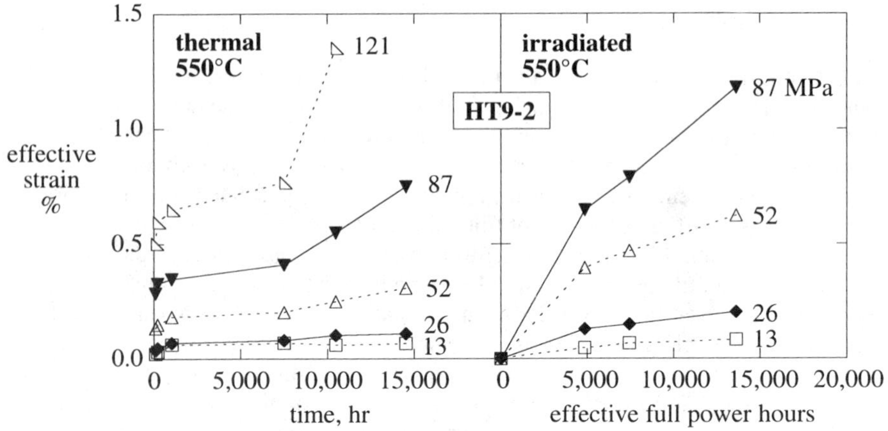

Figure 2 -- *Creep as a function of time for pressurized tubes in a thermal environment or in a neutron irradiation environment at 550°C.*

greater than the transients in the irradiation creep specimens at any given stress. At 550°C, the transient strains are nearly equal, and at 600°C, the transient strains are by far greater for the irradiation creep specimens. Creep transients are generally considered to be due to a relaxation or a re-orientation of the dislocation microstructure which is followed by an increase in dislocation density due to strain hardening [19]. The transient ends when resulting strain hardening reduces dislocation mobility and thus the creep rate. A comparison of precipitate evolution in thermally aged and in irradiated HT9 shows that at lower temperatures (below 600°C) precipitates form in irradiated HT9 that do not form

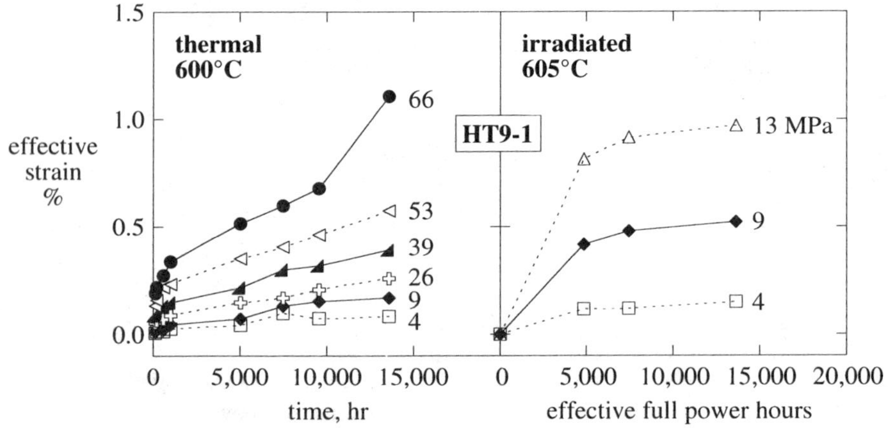

Figure 3 -- *Creep as a function of time for pressurized tubes in a thermal environment or in a neutron irradiation environment at 600°C.*

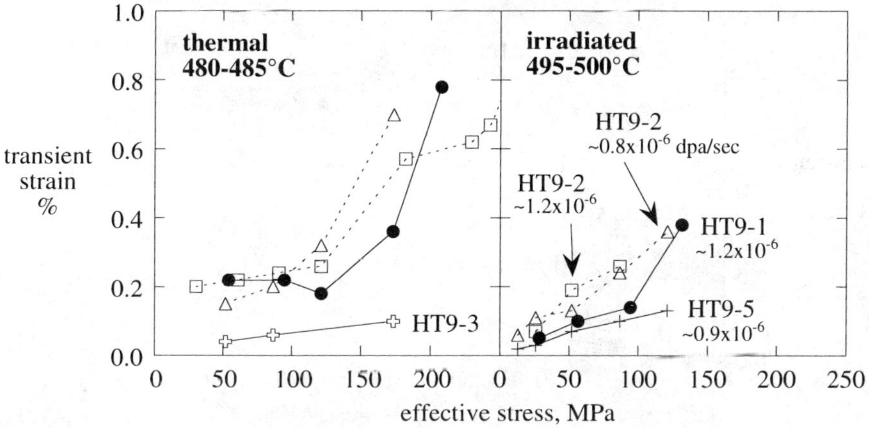

Figure 4 -- *The magnitude of the creep transient as a function of stress for HT9 thermal creep specimens and HT9 neutron irradiation creep specimens at ~490°C.*

during thermal aging at the same temperature (e.g., G-phase, α', and χ) [20, 21]. It also appears that at thermal aging temperatures below 600°C, precipitate formation occurs only after relatively long aging times [21]. If the creep transients are due to dislocation re-organization, then it may be possible that the precipitate formation which occurs in the irradiated materials may be hindering the dislocation re-organization process or perhaps accelerating the increase in the dislocation density. This would at least explain why transients are smaller in the irradiated materials at ~490°C. At 600°C, the difference in precipitate evolution between irradiated and thermally aged ferritic-martensitic steels is

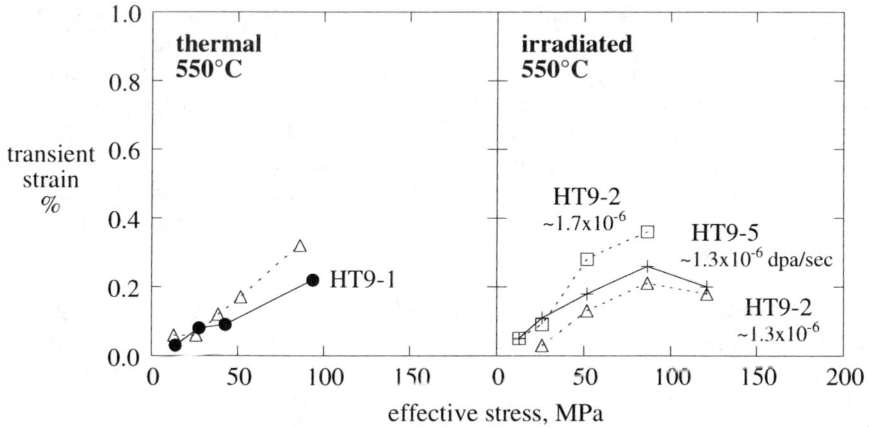

Figure 5 -- *The magnitude of the creep transient as a function of stress for HT9 thermal creep specimens and HT9 neutron irradiation creep specimens at 550°C.*

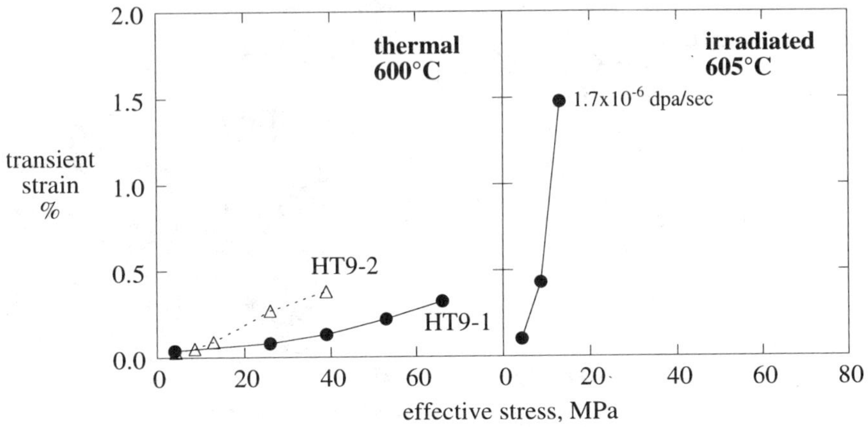

Figure 6 -- *The magnitude of the creep transient as a function of stress for HT9 thermal creep specimens and HT9 neutron irradiation creep specimens at 600°C.*

not as great as at lower temperatures, but this would only suggest that the transients should be similar at this temperature. Thus, some other unidentified mechanism must be contributing to the transient strain behavior. Chin has also performed a comparative study of thermal creep and irradiation creep of HT9 using pressurized tubes [10], and in his study, he observed transient behavior in the thermal creep specimens but not in the irradiation creep specimens. This is in contrast to the results presented here where transients were observed in both the thermal creep and the irradiation creep specimens. The reason for the difference in transient behavior between Chin's irradiation creep specimens and the specimens studied for the present work is not known.

The steady state creep rates as a function of applied stress for the thermal creep and for the irradiation creep specimens are compared in Figs. 7-9. As with the transient data, there is some variability in the steady state creep rate data. For the irradiation creep specimens, it is possible that the difference in dose rate from one tube-set to another contributed to the varied behavior, and in fact, the irradiation creep specimens with the lowest creep rate were irradiated at the lowest dose rate. While there is some variability in the data, the thermal creep data and the irradiation creep data behave differently with the thermal creep data and the irradiation creep data forming different trend bands. At lower stresses, the steady state creep rates of the thermal creep specimens and the irradiation creep specimens overlap, but at the higher stresses, the steady state creep rate of the irradiation creep specimens is greater.

In considering the contributions to creep during irradiation, it should be recognized that the amount of thermal creep which occurs in an irradiated specimen is probably less than which occurs in a thermal environment. Irradiation can cause the formation of microstructures that are very different from the microstructures which occur in a thermal environment. Several ion-irradiation based studies of creep have shown that for materials which form precipitates or other extended defects during irradiation, the post-irradiation steady state thermal creep rate is lower than the steady state creep rate of a corresponding

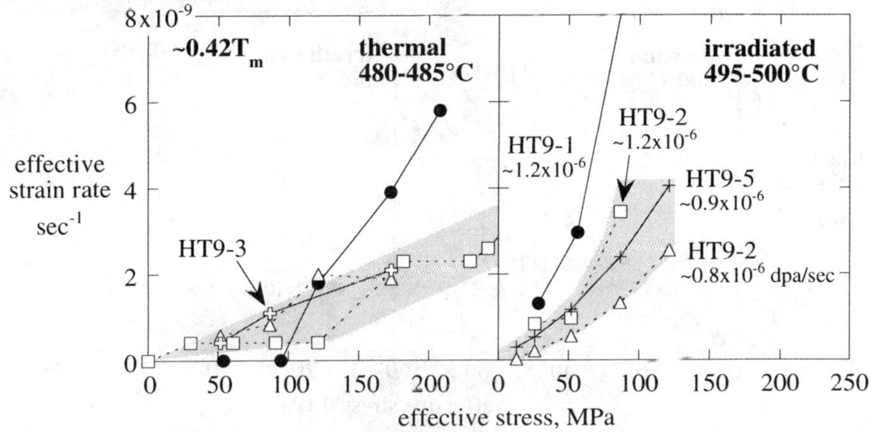

Figure 7 -- *Steady state creep rate of HT9 pressurized tubes as a function of stress for specimens in either a thermal environment or an irradiation environment at ~490°C.*

specimen exposed only to a thermal environment [22, 23]. This has been attributed to the formation of precipitates and dislocation loops which hinder the motion of dislocations. Thus, the amount of thermal creep which occurred during irradiation is likely to be less than that which occurred in the thermal creep specimens, and the majority of the creep in the irradiated specimens may be due mainly to the irradiation-induced increase in point defect concentrations and diffusion. The decrease in the difference between the steady state thermal creep rate and steady state irradiation creep rate at 600°C can be explained

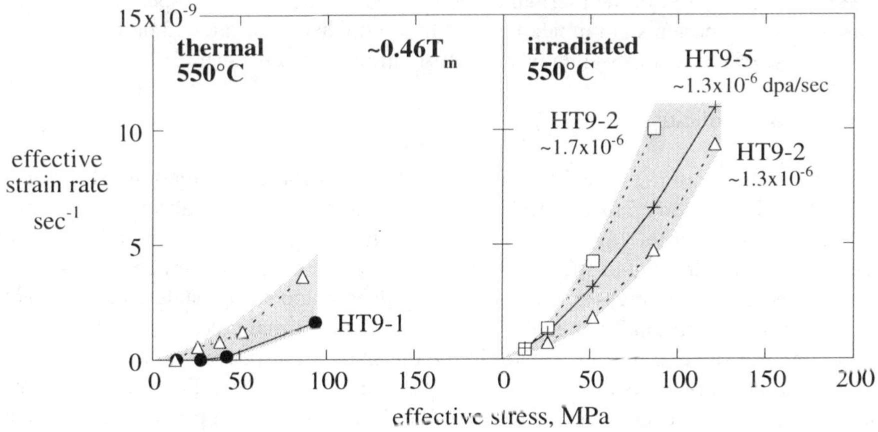

Figure 8 -- *Steady state creep rate of HT9 pressurized tubes as a function of stress for specimens in either a thermal environment or an irradiation environment at 550°C.*

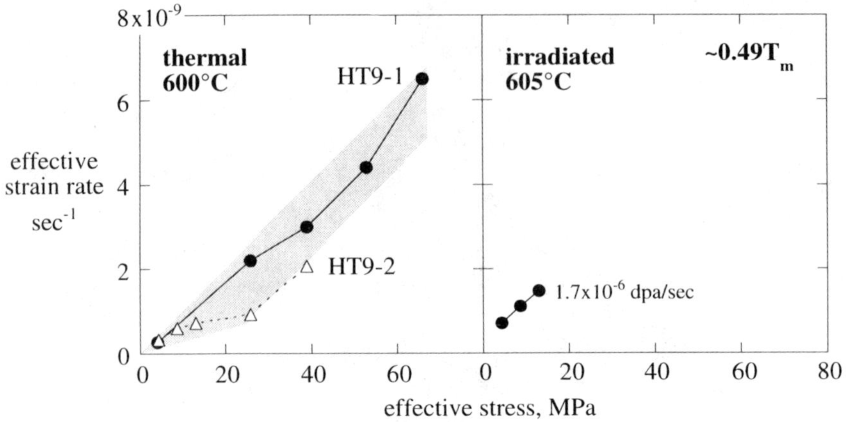

Figure 9 -- *Steady state creep rate of HT9 pressurized tubes as a function of stress for specimens in either a thermal environment or an irradiation environment at 600°C.*

by the collapse in radiation-induced point defect supersaturations which occurs at higher temperatures.

The stress dependence of the thermal creep rate is varied for low to moderate stresses. At 490°C and 600°C, a stress exponent of one provides the best fit, while at 575°C, a stress exponent between 1.5 and 2 provides the best fit. For the irradiated specimens at ~490°C and 550°C, a stress exponent between 1.5 and 2 provides the best fit, and at 600°C, the stress exponent has a value of 1. These results sometimes match and sometimes do not match Chin's study of thermal creep and irradiation creep of HT9 [10]. Within the same range of stresses, Chin observed a stress exponent of 1.75 between 425°C and 600°C for both the thermal creep and the irradiation creep specimens. The observed stress exponents are indicative of either a diffusion dominated deformation mechanism (n=1) or a climb-glide based deformation mechanism (n=2).

Summary and Conclusions

Over the temperature range from ~490°C to 605°C, irradiation creep of HT9 was compared to thermal creep using identical materials and identically fabricated specimens. The data show that for the higher applied stresses, the combined transient and steady state creep strains of the irradiated specimens are greater than the thermal creep strains at all temperatures. At lower stresses, the combined transient and steady state creep strains are comparable for the irradiation creep specimens and the thermal creep specimens. At ~490°C, the transient strains are greater in the thermal creep specimens, whereas at 600°C, the transient strains in the irradiated specimens are greater. For the steady state creep rate, the opposite situation occurs; at ~490°C, the steady creep rate of the irradiated specimens is greater, but at 600°C, the steady state creep rates of the thermally aged specimens and the irradiated specimens are comparable. For both the irradiated specimens and the thermally aged specimens, the stress exponent for the steady state creep rate varied between the values of 1 and approximately 2.

The data suggest that precipitate formation at lower temperatures may be causing the reduced transient strain in the irradiated specimens. It is also probable that precipitate formation in the irradiated specimens is leading to a steady state thermal creep contribution in the irradiated specimens that is lower than what occurs in the thermally aged specimens. The difference in steady state creep rates between the thermal creep specimens and the irradiation creep specimens can be explained by the variation in the radiation induced point defect supersaturation with temperature. The observed stress exponents are indicative of either a diffusion-dominated deformation mechanism (n=1) or a climb-glide based deformation mechanism (n=2).

References

[1] Gilbert, E. R., and Bates, J. F., "Dependence of Irradiation Creep on Temperature and Atom Displacements in 20% Cold-Worked Type 316 Stainless Steel," *Journal of Nuclear Materials*, Vol. 65, 1977, pp. 204-209.

[2] Lewthwaite, G. W., and Mosedale, D., "The Effects of Temperature and Dose-Rate Variations on the Creep of Austenitic Stainless Steels in the Dounreay Fast Reactor," *Journal of Nuclear Materials*, Vol. 90, 1980, pp. 205.

[3] Porter, D. L., and Garner, F. A., "Cessation of Irradiation Creep in AISI 316 Concurrent with High Levels of Swelling," *Influence of Radiation on Material Properties: 13th International Symposium (Part II), ASTM STP 956*, F.A. Garner, C.H. Henager, Jr., and N. Igata, Eds., American Society for Testing and Materials, Philadelphia, 1987, pp. 11-21.

[4] Grossbeck, M. L., and Horak, J. A., "Irradiation Creep in Type 316 Stainless Steel and US PCA with Fusion Reactor He/dpa Levels," *Journal of Nuclear Materials*, Vols. 155-157, 1988, pp. 1001-1005.

[5] Porter, D. L., Hudman, G. D., and Garner, F. A., "Irradiation Creep and Swelling of Annealed Type 304L Stainless Steel at ~390°C and High Neutron Fluence," *Journal of Nuclear Materials*, Vols. 179-181, 1991, pp. 581-584.

[6] Shibahara, I., Ukai, S., Onose, S., and Shikakura, S., "Irradiation Performance of Modified 316 Stainless Steel for Monju Fuel," *Journal of Nuclear Materials*, Vol. 204, 1993, pp. 131-140.

[7] Gilbert, E. R., Kaulitz, D.C., Holmes J. J., and Claudson, T. T., "Fast Reactor Induced Creep in 20% Cold Worked Type 316 Stainless Steel," *Irradiation Embrittlement and Creep in Fuel Cladding and Core Components*, British Nuclear Energy Society, 1972, pp 239-251.

[8] Kenfield, T. A., Busboom, H. J., and Appleby, W.K., "In-Reactor Stress Relaxation in Bending of 20% Cold-Worked 316 Stainless Steel," *Journal of Nuclear Materials*, Vol. 65, 1977, pp. 238-243.

[9] Puigh, R. J., Gilbert, E. R., and Chin, B. A., "An In-Reactor Creep Correlation for 20% Cold Worked AISI 316 Stainless Steel," Effects of Radiation on Materials: Eleventh Conference, ASTM STP 782, 1982, pp. 108-121.

[10] Chin, B. A., "Analysis of the Creep Properties of a 12Cr-1Mo-W-V Steel," Proceedings of: Topical Conference on Ferritic Alloys for use in Nuclear Energy Technologies, Snow Bird, Utah, June 19-23, 1983, The Metallurgical Society of AIME, 1984, pp. 593-599.

[11] Gilbert, E. R., and Chin, B. A., "In-Reactor Creep Measurements," *Nuclear Technology*, Vol. 52, 1981, pp. 273-283.

[12] Dymond, J. H., and Smith, E. B., *The Virial Coefficients of Gases, A Critical Compilation*, Oxford University Press, 1969, p. 177.

[13] Schneider, W. G., and Duffie, J. A. H., "Compressibility of Gases at High Temperatures. II. The Second Virial Coefficient of Helium in the Temperature Range from 0°C to 600°C," *Journal of Chemical Physics*, Vol. 17, 1949, pp. 751-754.

[14] Yntema, J. L., and Schneider, W. G., "Compressibility of Gases at High Temperatures. III. The Second Virial Coefficient of Helium in the Temperature Range from 600°C to 1200°C," *Journal of Chemical Physics*, Vol. 18, 1950, pp. 641-646.

[15] Timoshenko, S. P., and Goodier, J. N., *Theory of Elasticity, Third Edition*, McGraw-Hill Book Company, 1970.

[16] Malvern, L. E., *Introduction to Mechanics of a Continuous Medium*, Prentice-Hall, Inc., New Jersey, 1969.

[17] Irradiation Parameters for the FFTF Materials Open Test Assemblies from 1983 to 1992, WHC-SD-FF-TD-010 Revision 0, August, 1994.

[18] Gilbert, E. R., and Chin, B. A., "Irradiated Materials Measurement Technology," *Effects of Radiation on Materials: Tenth Conference, ASTM STP 725*, David Kramer, H.R. Brager, J.S. Perrin, Eds., American Society for Testing and Materials, Philadelphia, 1981, pp. 665-679.

[19] Barrett, C. R., Nix, W. D., and Tetelman, A. S., *The Principles of Engineering Materials*, Prentice-Hall, New Jersey, 1973.

[20] Hsu, C. Y., Gelles, D. S., and Lechtenberg, T. A., "Microstructural Examination of 12% Cr Martensitic Stainless Steel After Irradiation at Elevated Temperatures in FFTF," *Radiation-Induced Changes in Microstructure: 13th International Symposium (Part I), ASTM STP 955*, F.A. Garner, N.H. Packan, and A.S. Kumar, Eds., American Society for Testing and Materials, Philadelphia, 1987, pp. 545-559.

[21] Pai, R. N., and Wilcox, R. C., "Precipitation Evolution in HT-9 Steel Under Thermal Aging and Stress Rupture Conditions," *Effects of Radiation on Materials: 15th International Symposium, ASTM STP 1125*, R.E. Stoller, A.S. Kumar, and D.S. Gelles, Eds., American Society for Testing and Materials, Philadelphia, 1992, pp. 1190-1205.

[22] Nagakawa, J., Sethi, V. K., and Turner, A. P. L., "Irradiation Creep in Simple Binary Alloys," *Journal of Nuclear Materials*, Vols., 103 & 104, 1981, pp. 1275-1280.

[23] Henager, C. H., Jr., and Simonen, E. P., "Light-Ion Irradiation Creep and Hardening of Model Ferritic Alloys," *Influence of Radiation on Material Properties: 13th International Symposium (Part II), ASTM STP 956*, F.A. Garner, C.H. Henager, Jr., and N. Igata, Eds., American Society for Testing and Materials, Philadelphia, 1987, pp. 69-80.

Proton and Spallation Neutron Sources

Kerry A. Dunn,[1] McIntyre R. Louthan, Jr.,[1] John I. Mickalonis,[1] Stuart Maloy,[2] and Michael R. James[2]

Examination of 304L Stainless Steel to 6061-T6 Aluminum Inertia Welded Transition Joints after Irradiation in a Spallation Neutron Spectrum

Dunn, K. A., Louthan, M. R., Jr., Mickalonis, J. I., Maloy, S., and James, M. R., **"Examination of Irradiated 304L Stainless Steel to 6061-T6 Aluminum Inertia Welded Transition Joints in a Spallation Neutron Spectrum,"** *Effects of Radiation on Materials: ASTM STP 1405,* S. T. Rosinski, M. L. Grossbeck, T. R. Allen, and A. S. Kumar, Eds., American Society for Testing and Materials, West Conshohocken, PA, 2001.

Abstract: Tritium target/blanket assemblies were irradiated for six months at the Los Alamos Neutron Science Center (LANSCE). Stainless steel to aluminum inertia weld transition joints were welded to the target/blanket assemblies to simulate expected transition joints in the Accelerator Production of Tritium (APT) where as many as a thousand SS/Al weld transition joints will be used. No leaking at the inertia-welded interface was observed following irradiation. The galvanic interaction at the 6061 Al to 304L stainless steel inertia-welded interface did not present significant degradation as a result of the irradiation or the cooling water chemistry. General corrosion was evident along the entire length of the aluminum piping and pitting corrosion was present along the aluminum piping and is attributed to the conductivity of and the copper deposits in the cooling water. The reduction or elimination of impurities in the cooling water will aid in corrosion control of the aluminum piping.

Keywords: corrosion, irradiation, inertia weld, weld, pitting, interface, galvanic corrosion, aluminum, stainless steel, spallation

Introduction

The Accelerator Production of Tritium (APT) program is coordinated by Los Alamos National Laboratory (LANL) for the Department of Energy (DOE)/ Defense Programs (DP). As part of this program, the Savannah River Technology Center (SRTC) designed and fabricated tritium target/blanket assemblies which were sent to LANL and irradiated for six months at the Los Alamos Spallation Radiation Effects Facility (LASREF) at the Los Alamos Neutron Science Center (LANSCE). Cooling water was supplied to the assemblies through 2.54 cm. diameter 304L stainless steel (SS) tubing. To attach the

[1]Fellow, Sr. Advisory, and Principal Engineers, respectively, Westinghouse Savannah River Company, 773-41A, Aiken, SC 29808.

[2]Materials Team Leader, Staff Member, respectively, Los Alamos National Laboratory, MS H809 APT-TPO, Los Alamos, NM 87545.

304L SS tubing to the assemblies, a 304L SS to 6061-T6 aluminum (Al) inertia-welded (friction welded) transition joint was used. Four SS/Al weld transition joints were attached to each of the two assemblies called Mark I and Mark II (two inlet and two outlet cooling lines) and two were attached to each of the three "High Flux" Assemblies (one inlet and one outlet). While the interior of the tubing was exposed to the cooling water, the exterior of the tubing was exposed to an ambient air environment.

Several thousand of these inertia welds are currently in the design for the APT target/blanket (T/B). In the APT design, aluminum piping in the target/blanket region houses the $^3He/^3H$ gas. During production, this gas is transferred to the extraction and purification facility, where SS piping is used. The Al/SS inertia weld provides the path for gas flow between these two systems. Materials compatibility between the 304L SS and the 6061-T6 Al in the radiation environment as well as the corrosion associated with the cooling water that will interact with the piping in the APT facility are major concerns.

Examination of the interior of the cooling water piping irradiated at LANSCE indicated that the cooling water chemistry provided a noticeable influence on the corrosion of both the inertia weld and the parent material. The effect of radiation on the inertia weld and surrounding area was evaluated against the cooling water effect. In addition, corrosion on the outside of the piping was investigated. This exterior corrosion was accelerated by the presence of nitric acid, which was formed by the radiolysis products of air and water from other components leaking during the irradiation.

Analyses of the irradiated 304L to 6061-T6 Al inertia welds included a variety of methods including visual examination, microstructural analysis, hardness measurements, optical microscopy, Scanning Electron Microscopy (SEM), Energy Dispersive Spectroscopy (EDS) and electrochemical corrosion testing. The data were used to evaluate the concepts and methods for target/blanket manufacturing, materials and weld metal properties for spallation neutron environments, and joining methods for aluminum and aluminum to stainless steel.

Experimental Design

Several prototypical target/blanket assemblies were fabricated for irradiation testing in the LANSCE facility at LANL. These target/blanket assemblies consisted of lead encapsulated in aluminum. The lead surrounded Al tubes filled with either helium-3 (^3He) gas, or solid aluminum-lithium targets. The lead provided an additional source of neutrons from n,xn spallation reactions. Cooling water was supplied to the target/blanket assemblies in the LANSCE facility via 2.54 cm. diameter 304L SS tubing. In order to transition from the SS cooling water tubing to the Al cooling water piping, a 304L SS to 6061-T6 Al inertia-welded transition joint was utilized [*1*]. The transition joint was machined from an inertia-welded part made by Interface Welding [*2*], a commercial vendor. The inertia-welded part was made by inertia welding solid cylinders of 304L SS and 6061-T6 Al. Certification data sheets,

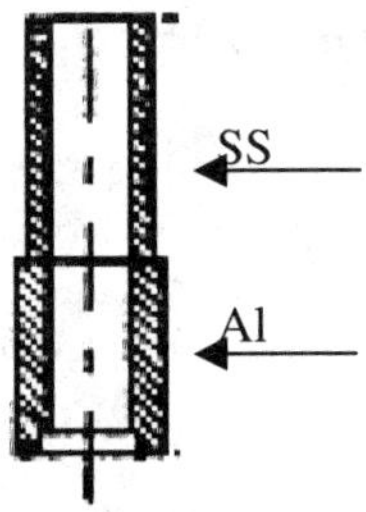

Figure 1. *Schematic of inertia welded piping.*

supplied by the vendor, showed that the solid state welds had a tensile strength equivalent to the 6061-T6 aluminum base metal and a ^{3}He leak rate of approximately 10 ^{9}cc/s. A piping schematic is shown in Figure 1 and the solid inertia-welded piece prior to machining is shown in Figure 2.

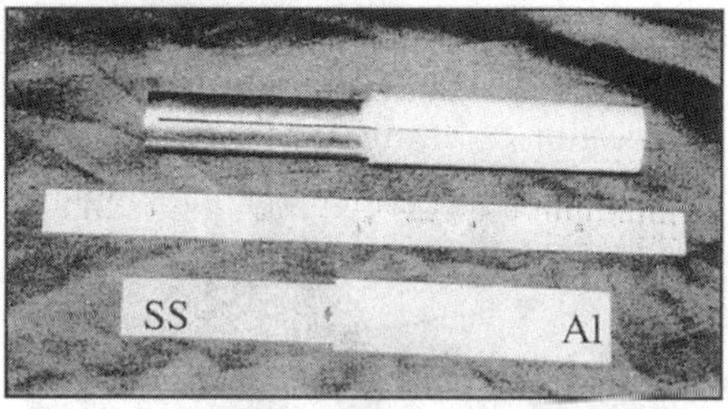

Figure 2. *Solid inertia welded piece prior to machining.*

The SS/Al inertia-welded transition joints were welded to the inlet and outlet cooling lines of all test assemblies. A 304L sleeve was placed over the inertia-welded section and gas tungsten arc welded (GTAW) to the stainless steel to prevent bending stresses during installation of the assembly.

To ensure the appropriate analyses were being conducted, a task plan was developed for the overall target/blanket materials characterization which included the analyses of the irradiated transition joints [*3*].

A fixture and cutting apparatus (Figure 3) was designed, fabricated, and mocked up at SRTC to support the sectioning of the irradiated inertia-welded samples in the Chemical and Metallurgical Research (CMR) facility at LANL. The cutting apparatus was designed to provide ease of operation in the hot cells while working with hand operated remote manipulators. A mock-up cutting operation was performed on non-irradiated inertia-welded transition joints at SRTC to ensure the cutting apparatus would operate smoothly in the hot cells. Once the mock-up operation at SRTC was complete the cutting apparatus was transferred to the LANL CMR hot cells.

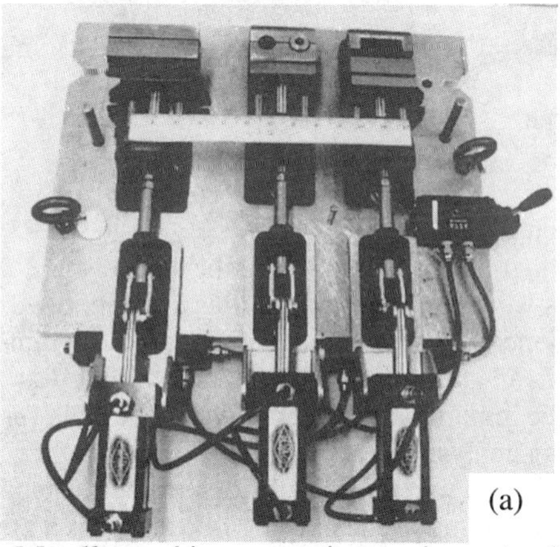

(a)

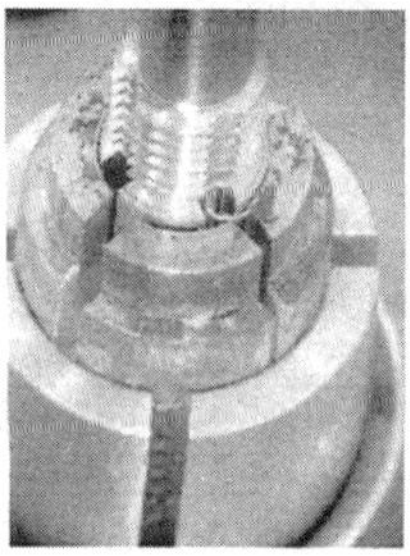

(b)

Figure 3. *(a) Fixture used to section pieces in the hot cell. (b) Sectioned pieces.*

Metallographic preparation equipment, which would be compatible with the hot cell environment, was purchased and mocked up at SRTC prior to transfer to the LANL hot cells. Remote handling of the samples and the limited use of fluids in the hot cells needed to be considered for the operation. Several modifications to the equipment were made to make the equipment compatible with the hot cells. These modifications included the attachment of remote water lines to allow for the introduction of water from outside the hot cells and the displacement of the power operation to the exterior of the hot cells.

Once the modifications and mock-up were completed the equipment was transferred to the LANL hot cells.

Experimental Setup

The prototypical lead/aluminum blanket assemblies were attached to the shielding inserts by welding the stainless steel water cooling lines to just above the transition joints as shown in Figure 4. The inserts were placed in a location downstream of the primary target inserts and outside of the direct proton beam. Over the course of the six-month irradiation, the Mark I and Mark II assemblies saw a neutron fluence of approximately $8x10^{19}$ n/cm^2. The fluence from secondary protons was very small (<$3.0x10^{18}$ p/cm^2). The High Flux assemblies, located closer to the tungsten spallation target, saw approximately $1x10^{20}$ n/cm^2 with a negligible proton fluence as well. The cooling water used for this particular irradiation at LASREF was the XO2 water source, which was a recirculating water system that cooled many areas during the irradiation including a copper beam stop. Thermocouples placed on the outside of the cooling lines for the Mark I and Mark II, as well as the high flux assemblies, indicated that the temperatures were maintained at approximately 30°C.

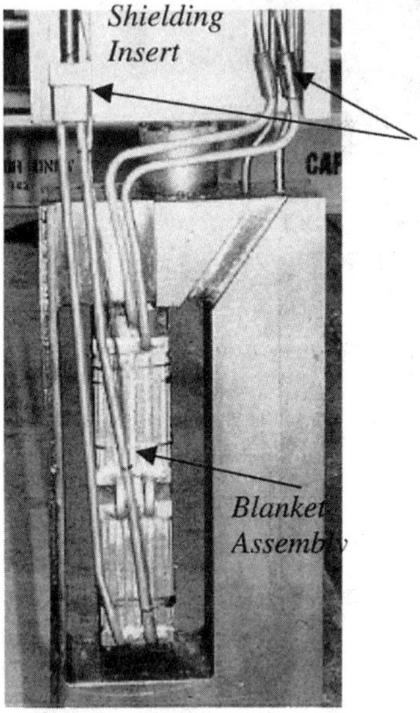

Following irradiation, the assemblies were detached from the shielding inserts and transferred in a cask from LANSCE to the hot cells in the CMR facility at LANL.

The fixtures and cutting apparatus were used to separate the sleeve from the transition joints prior to sectioning the samples for further examination. Once the sleeve was removed, the inertia-welded transition joints were placed in the fixtures and sectioned in half and then one of the cut pieces was sectioned in half again, creating two ¼ sections and one ½ section. After visual examination, a ¼ section of each transition joint was mounted for metallographic preparation. The remaining two pieces were available for additional testing, such as SEM/EDS or x-ray analysis.

Metallographic preparation of these specimens was not straightforward because of the dissimilar metals at the weld interface. Rounding of the aluminum edges was a great concern because Al 6061-T6 is a softer material than stainless steel. Therefore, metallographic preparation of the specimen using standard techniques for aluminum samples would not ensure that the stainless steel microstructure would be properly prepared for examination. An added intricacy was the fact that this metallographic preparation took place in the hot cells at LANL where remote handling was used and a minimum amount of liquids was allowed. Under these conditions, a metallographic technique was developed that provided the optimum operating parameters to ensure a properly prepared sample (Table 1 and Table 2). A baseline characterization of non-irradiated Al-SS inertia-welded transition joints was completed to support the characterization of the irradiated transition joints [*4*].

A sample from each inertia-welded transition joint was mounted and prepared metallographically. In addition, a non-irradiated inertia-welded transition joint (control sample) was prepared metallographically in the hot cells to compare to the irradiated specimens.

Table 1 – *Sample preparation sequence for metallographic analysis using automated grinder/polisher.*

Grinding/Polishing Sequence	Preparation Time (seconds)	Application Force (N)
500, 600, 800, 1200, 2400 grit SiC Papers	90 (Each)	100,150,100 (Each)
6 μm Diamond Nap Cloth	240	200,150,100
1 μm Diamond Nap Cloth	240	200, 150, 100

Table 2 – *Exposure times and compositions of etchants used in metallographic analysis.*

Poulton's Etching Solution (Al etchant)	Oxalic Acid Solution (SS etchant)
50 mL Poulton's Reagent[1] 40 mL chromic acid solution[2] 25 mL HNO_3	10 g oxalic acid 100 mL H_20 (6V applied to solution)
Exposure Time: 10 sec.	Exposure Time: 60 sec.

[1] 12 parts HCl, 6 parts HNO_3, 1 part HF, 1 part H_20
[2] 3 g chromic acid per 10 mL H_2O

Electrochemical corrosion tests were performed to evaluate the corrosion process at the weld interface of non-irradiated samples as a function of water chemistry. The galvanic interactions between Al and SS were suspect. The corrosion tests included potential and galvanic current monitoring and potentiodynamic polarization. Samples for electrochemical testing were obtained from the same as-received inertia-welded slugs that were used for the irradiated sections. Additional test samples included the Al parent metal, the SS parent metal, and the heat affected zone (HAZ) of the Al. The samples were prepared as metallographic mounts with electrical connections for the electrochemical tests.

The water chemistries were simulated conditions for XO2 water, dirty APT water and APT water. For the simulated APT water, laboratory distilled water was used since the impurity levels were low and similar to that anticipated for the APT. The dirty APT water contained 2 ppm Cl^- and 7 ppm SO_4^-. The XO2 water was a derivative of the dirty

APT water with added copper, either 1.0 ppm Cu (XO2b) or 0.1 ppm Cu (XO2a). All the waters were adjusted to a pH 5 with nitric acid and had hydrogen peroxide levels of 0.001 M.

Testing was conducted in a blackened glass cell equipped with gas spargers. When deaerating, nitrogen gas was used. A Ag/AgCl electrode was used as the potential reference and graphite rods were the counter electrodes. A temperature regulated stir/hot plate was used for agitation and heating. All tests were conducted at 30° C.

Experimental Results and Discussion

Corrosion

Over the course of the irradiation some leaks developed from other components in Area A. These leaks produced a spray of the cooling water to the exterior of the target/blanket assemblies. Radiolysis of this water in the air environment created a nitric acid spray on the components. Corrosion on the outside of the assemblies was clearly evident following irradiation and has been attributed, primarily, to the presence of the nitric acid, Figure 5.

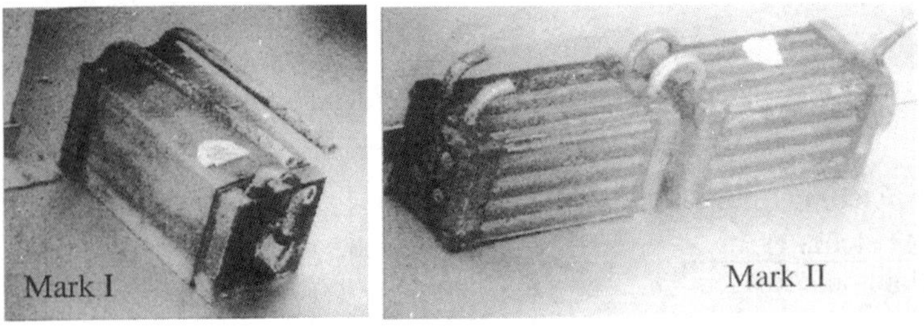

Figure 5. *Mark I and Mark II following irradiation showing corrosion product.*

Visual examination of the sectioned inertia-welded transition joints showed the presence of corrosion on not only the exterior of the cooling water tubing but also on the interior of the tubing. The corrosion on the interior of the tubing was evident both on the aluminum section of the cooling water tubing and at the inertia-welded interface, Figures 6-7. The stainless steel was relatively unaffected. Differences in the appearance of the corrosion on the aluminum metal and at the inertia-welded interface were clearly evident among the different samples. The inside wall of cooling water piping from Mark II assembly had a black corrosion product on the aluminum that appeared to be relatively thick. The inside wall of cooling water piping from the High Flux assemblies had a grayish oxide appearance on the aluminum. These differences are attributed to the different flow rates of the XO2 cooling water through the Mark II and Mark I (7.5 and 7.0 gpm, respectively) and through the high flux assemblies (25.86 gpm). Understanding these results required further characterization through metallographic examination and electrochemical testing.

Figure 6. *Inside wall of cooling line from Mark II assembly. Note black corrosion product on aluminum.*

Figure 7. *Inside wall of cooling line from High Flux Assembly. Note the greyish oxide on aluminum.*

Metallographic Examination

Samples from each assembly (Mark I, Mark II, and High Flux Assemblies) were sectioned, prepared metallographically, and examined in the CMR hot cells. Because Al and SS are dissimilar metals, the inertia-welded interfaces were susceptible to galvanic corrosion. Metallographic examination of the irradiated inertia welds indicated that some of the samples did exhibit degradation at the interface, however because the degradation is equally damaging to the SS as the Al, it is not seen as galvanic corrosion,

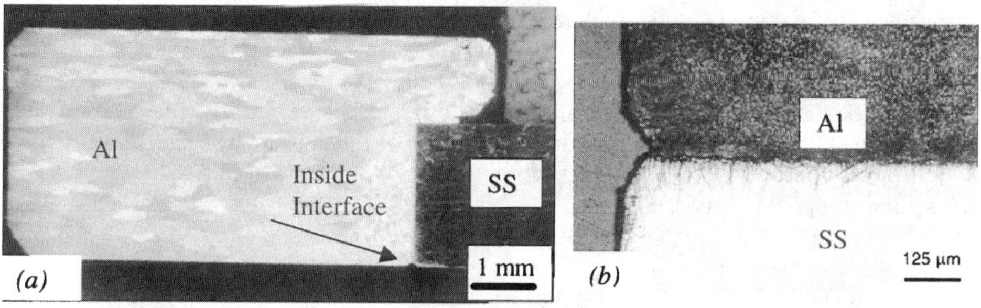

Figure 8. *Slight degradation at inertia welded interface on the cooling water side of the High Flux inlet piece (a) macrograph and (b) higher magnification of inside at interface.*

Figure 8. General corrosion along the length of the aluminum piping and pitting corrosion was more prevalent, Figure 9. The pitting corrosion appeared to be the most severe aspect of corrosion. This pitting corrosion, which was located along the piping away from the interface to a depth of 75 µm, indicated that the effects of the cooling water chemistry, flow rate, and irradiation on corrosion needed a closer analysis than the

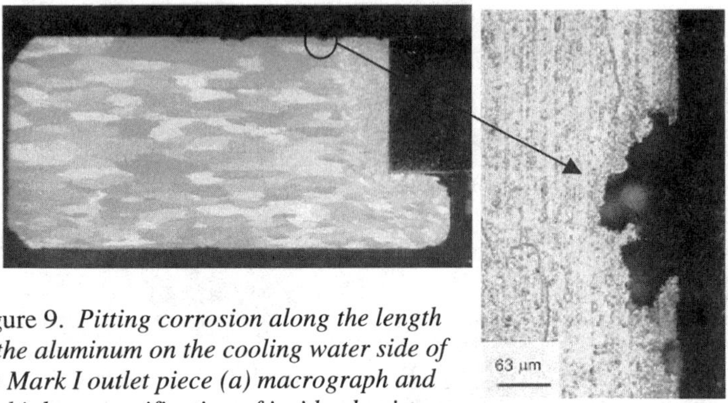

Figure 9. *Pitting corrosion along the length of the aluminum on the cooling water side of the Mark I outlet piece (a) macrograph and (b) higher magnification of inside aluminum wall showing representative pitting.*

effect of the galvanic couple at the SS/Al inertia weld interface. The cooling water used for the Mark I, Mark II and high flux assemblies was provided from the XO2 system. This XO2 system contains more impurities than are expected in the APT facility where the water chemistry will be more stringently controlled. Flow of the XO2 cooling water through the 9A and 9C inserts (Mark II and Mark I) was 7.5 and 7.0 gpm, respectively, while the flow through the 18B insert (high flux assemblies) was 25.86 gpm. As a result of these different cooling water flow rates, general corrosion oxide thickness on the inside wall of the cooling water piping from the 9A and 9C inserts was approximately 50 µm, while the oxide layer on the cooling water piping from the 18B insert was approximately 15 µm, Figure 10. The predominant oxide layer was Bayerite, (β-Al_2O_3*$3H_2O$) because the temperature of the cooling water was maintained at approximately 30°C. Initially, Gibbsite (α-Al_2O_3*$3H_2O$) is formed before a phase change to Bayerite [5]. Fifty microns was the maximum oxide thickness from general corrosion expected for the LANSCE test conditions [6].

Another observation made during the metallographic examination was the effect of the recrystallization of the HAZ in the aluminum at the inertia-welded interface and grain growth in the aluminum parent material. In each of the unirradiated inertia-welded samples prepared for the baseline study, as well as the control inertia-welded sample that was prepared in the CMR hot cells, the HAZ did not show the presence of individual grains in the structure, Figure 11. In addition, the grain structure in the Al parent material of the unirradiated samples showed

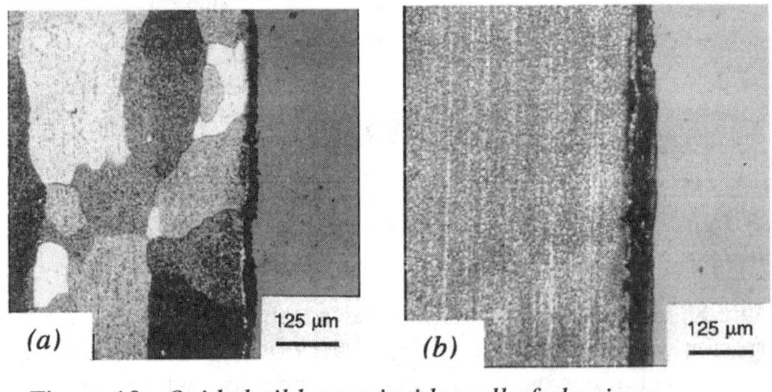

Figure 10. *Oxide buildup on inside wall of aluminum cooling water piping near the inertia weld (a) high flux section (25.86 gpm) (b) Mark I section (7.0gpm).*

small equiaxed grains along the exterior length of the samples with an elongated grain structure near the center of the samples, Figure 12. However, all of the irradiated inertia-welded samples showed grain definition in the HAZ of the aluminum, indicating recrystallization, and grain growth in the Al parent material (Figure 9(a), 10(a) and 13). The grain growth in the parent material resulted in a fairly consistent grain size throughout the microstructure.

Figure 11. *Control sample metallographically prepared in CMR hot cells. Solid inertia welded piece machined to simulate irradiated sections.*

Other inertia welded samples, prepared for tensile testing, were irradiated during the same time frame as these assemblies. Recrystallization and grain growth was not observed with the tensile samples. All of the irradiated samples received a low dose and the temperatures measured during irradiation were approximately 30°C. Therefore, recrystallization and grain growth associated with the test assemblies is best explained by heating during welding prior to irradiation which was done to connect the transition joints to the connector water piping.

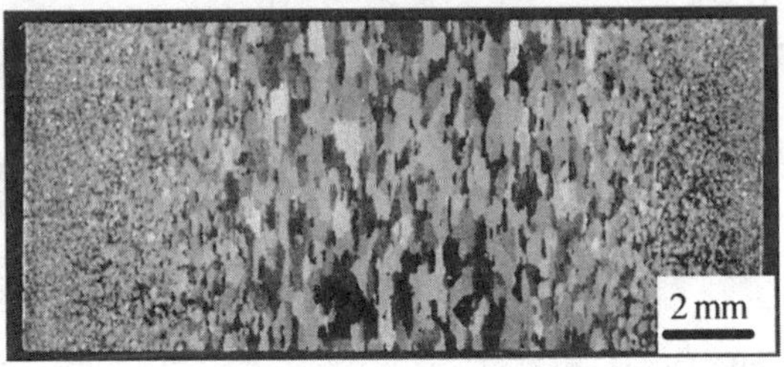

Figure 12. *Grain size variation across solid aluminum parent metal from inertia welded specimens.*

Hardness measurements were performed on the non-irradiated as well as the irradiated inertia-welded samples to evaluate differences as a result of the recrystallization and grain growth. Hardness values on the aluminum side of the non-irradiated samples were higher than the hardness values in the irradiated samples, as shown in Figure 14. These hardness results indicated an overaged structure in the aluminum side of the irradiated samples and, hence, a reduction in strength. Presumably, this reduction in strength occurred during fusion welding. It

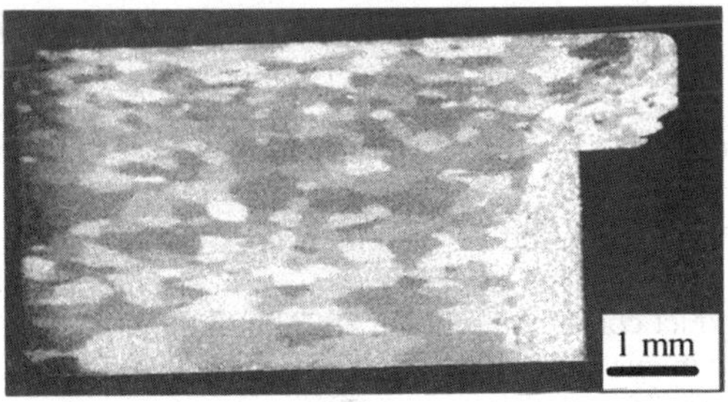

Figure 13. *Irradiated piping from High Flux outlet section showing evidence of recrystallization at inertia welded interface (HAZ) and grain growth across the Al parent metal.*

did not produce a leak at the inertia weld interface during the six-month irradiation. However, the method of attaching the inertia-welded joints to adjacent sections of tubing needs to be considered to avoid obtaining such an over-aged and recrystallized 6061 Al grain structure.

Preliminary measurements of the effects of irradiation on the strengths of inertia-welded tensile specimens with comparable test conditions indicate that irradiation had a minimal effect on the strength or ductility of the joint. These studies showed, however, that yield strengths of the inertia welds were controlled by the behavior of the 6061 Al microstructure (resulting from the inertia welding process) and do not reflect the strength of the starting T6 condition. Inertia welding lowered the yield strength from 275 MPa,

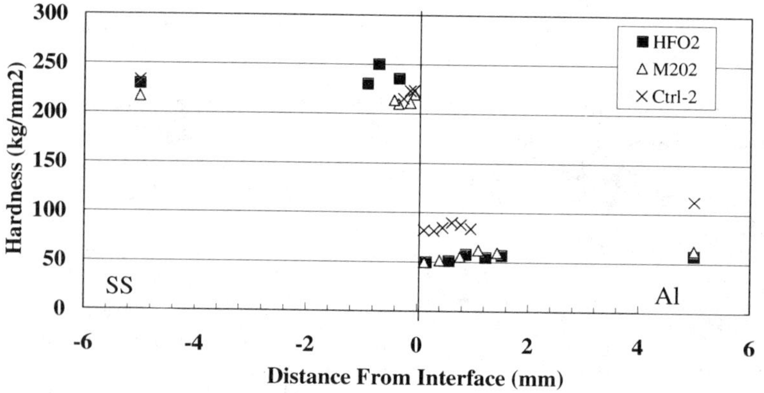

Figure 14. *Hardness vs. Position for Stainless Steel/Aluminum Inertia Welds*

typical of 6061-T6 Al, to approximately 170 MPa. To prevent additional losses in yield strength during attachment of the bi-metallic joint, proper cooling techniques will be needed when joining the inertia-welded 6061-T6 aluminum to the aluminum tubing in the target/blanket region of the APT facility.

Scanning Electron Microscopy and Energy Dispersive Analysis

The SEM and EDS study was used to investigate and characterize the inertia-welded specimens. The exterior of the piping of the samples generally indicated corrosion product consistent with an aluminum oxide, Figure 15. The interior of the inertia-welded specimens showed a high degree of oxidation on the aluminum portion of the specimens, Figure 16, consistent with the previous visual examinations. Several areas observed had portions where the aluminum oxide had flaked off of the surface, exposing another layer below the outer layer, Figure 16, which again is consistent with earlier findings. The presence of a "mud-flat cracking" appearance was evident on several samples, Figure 17. This flaking of the oxide layer and "mud-flat cracking" is consistent with the presence of Bayerite, which does not adhere well to the surface of the aluminum substrate.

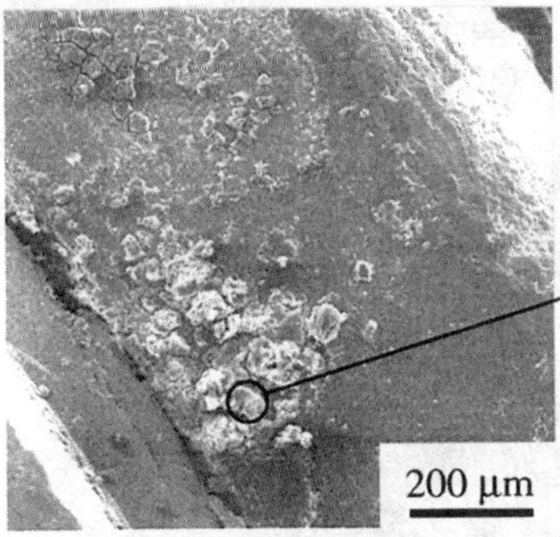

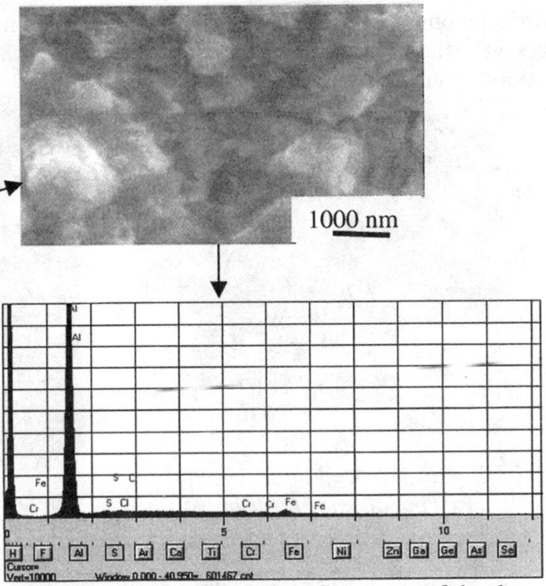

Figure 15. *Mark II inlet piece exposed to atmospheric air during irradiation. Circled area, above, shown in top right and EDS spectrum at bottom right indicates aluminum oxide.*

As discussed in the *"Metallographic Examination"* section, flow rates of the three assemblies placed in the Area A irradiation were different. Flow of the XO2 cooling water through the 9A and 9C inserts (Mark I and Mark II) was 7.5 and 7.0 gpm, respectively, while the flow through the 18B insert (high flux assemblies) was 25.86 gpm. The oxide layer adjacent to the weld interface shown in Figure 16 and 17 (high flux) has flaked off leaving a thin oxide layer while the oxide layer shown in Figure 18 (Mark II) does not indicate flaking. This observation is consistent with the increased agitation from the flow rate with the High Flux assemblies and the observation made visually and with metallography (Figures 6, 7 and 10).

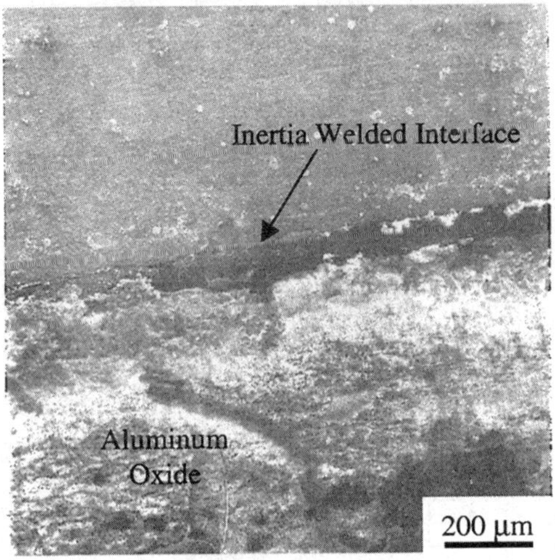

Figure 16. *SEM of inside weld taken from High Flux Assembly on inertia welded interface. Note the region of "flaked-off" corrosion product adjacent to the interface.*

Another observation with the SEM was the presence of copper (Cu) rich deposits on the surface of the stainless steel, Figure 19. No evidence of the copper deposits was observed on the aluminum material. However, the $Al_2O_3*3H_2O$ oxide layer was fairly thick and may have masked the presence of the copper deposits. Copper deposits on aluminum are known to cause pitting, which would explain the pitting corrosion observed in the metallographic examinations, Figure 9. The copper was present in the water system as a result of the corrosion of the copper beam stop, which was also cooled by the XO2 cooling water system.

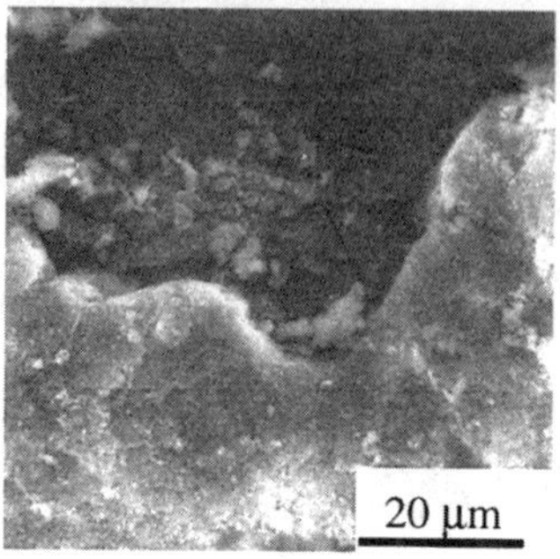

Figure 17. *Mudflat cracking of oxide layer in high flux sample.*

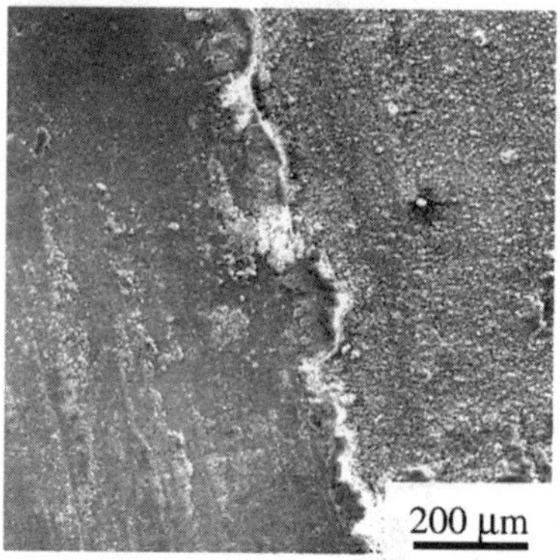

Figure 18. *SS/Al interface of Mark II with Al oxide adhering to the surface.*

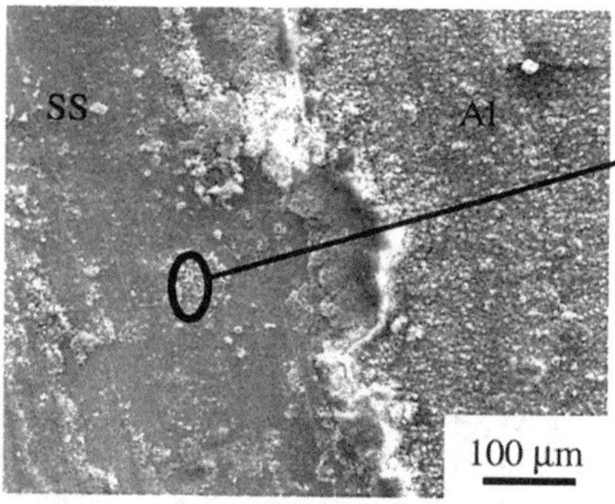

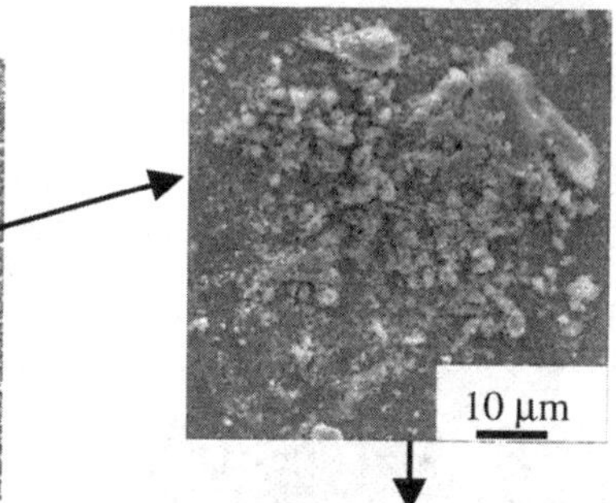

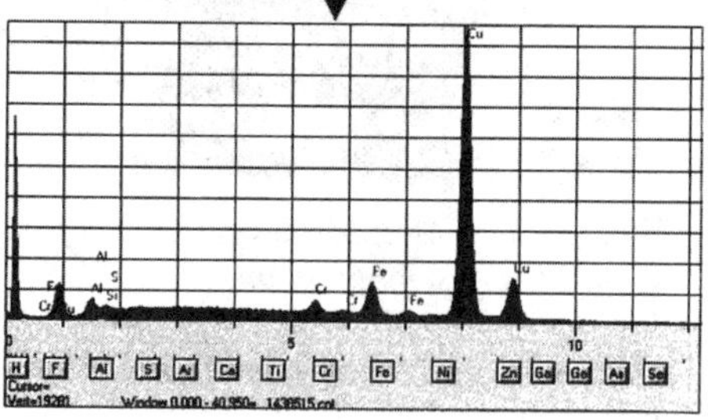

Figure 19. *Copper deposits found on stainless steel portion of the inertia weld.*

Electrochemical Testing

Corrosion processes characterized through the electrochemical testing were found to be consistent with those observed on the irradiated inertia welds from the Mark I, Mark II, and high flux assemblies. Galvanic and pitting corrosion were the primary mechanisms, similar to the irradiated inertia welds. These processes were more aggressive in the water chemistries with higher impurities. XO2 water chemistries produced the most significant pitting and the highest galvanic currents. The test results are summarized in Table 3, which includes the nominal galvanic current of the SS/Al couple and open-circuit potentials. The XO2a and XO2b water chemistries contained 0.1 and 1.0 ppm copper ions, respectively.

Open-circuit potential measurements were made on samples of the parent material as well as a section of the weld. As expected, 304L stainless steel was more electropositive than aluminum. The potential of the weld fell between that of the two parent materials. The potential of aluminum was variable due to the oxide formation dependence on the water chemistry. The presence of copper in the XO2 water chemistry caused the potential of both stainless steel and aluminum to shift to more electropositive values. This shift was attributed to the increase in the cathodic reaction due to the reduction of copper.

As the impurity level of the water increased the galvanic current increased. This large galvanic interaction led to more significant pitting on the aluminum samples. The worst pitting was observed in the XO2 water chemistries. In the XO2 water chemistries, the high degree of pitting was associated with the deposition of copper. The stainless steel sample had the most deposits. Some pitting was also observed on the stainless steel in the XO2b water. These pits were associated with copper deposits on the surface. Few copper deposits were observed on the aluminum sample. The results show that 304L stainless steel is the cathode in the couple where the reduction of copper occurs. These results are consistent with those observed on the irradiated inertia welds.

Table 3 - *Galvanic current and potential measurements*

Water Test Solution	Galvanic Current (μA)	Aluminum potential (V)	Stainless steel potential (V)	Weld potential (V)
Typical APT	0.75	-0.375	0.375	-0.290
Dirty APT	5.5	-0.250	0.330	-0.23
XO2a	7	-0.225	0.400	ND
XO2b	8.75	-0.250	0.410	ND

Summary

No leaking at the inertia-welded interface was observed following irradiation. The galvanic interaction at the 6061 Al to 304L stainless steel inertia-welded interface did not present significant degradation as a result of the irradiation or the cooling water chemistry. General corrosion was evident along the entire length of the aluminum piping after irradiation in a high conductivity water system. Pitting corrosion was present along the aluminum piping and is attributed to the conductivity of and the copper deposits in the cooling water. The reduction or elimination of impurities in the cooling water will aid in corrosion control of the aluminum piping.

Water chemistry and flow rate are important considerations for the APT design. Static or low flow rate of cooling water produced a greater oxide buildup on the aluminum portion of the inertia weld while a higher flow rate induced flaking of the aluminum oxide layer.

Welding of these inertia-welded sections to the aluminum cooling water piping caused recrystallization at the aluminum HAZ and grain growth in the parent material. This recrystallization and grain growth in the Al material resulted in a reduction in hardness of the aluminum material. The quality of the welding used for the design and construction of the APT facility will have to be controlled via procedures that outline appropriate cooling mechanisms for the inertia welds.

Acknowledgements

The authors would like to acknowledge Andy Cheek and Becky Dinkins from SRTC and Tobias Romero and Manuel Lopez from LANL for their expertise and seemingly endless hours working with the irradiated specimens in the hot cells in the CMR facility at LANL. Also, many thanks to Angela Bowser from SRTC for her excellent laboratory work in support of the electrochemical testing and Charlie Davis from LANL for his willingness to spend time on the SEM examining these samples.

References

[1] Peacock, Jr., H. B., and Leader, D. R., "Design and Fabrication of Prototypical APT Target/Blanket Assemblies and Weld Samples for Area-A Test in the LANSCE Facility at LANL," Internal Report WSRC-TR-96-0180, Westinghouse Savannah River Company, Aiken, SC, April, 1997.

[2] Wadleigh, Al S., "Bi-Metal Welding of Aluminum," *Interface Welding*, Carson, CA, 90746.

[3] Dunn, K. A., "Task Technical Plan for Target/Blanket Materials Examination", Internal Report WSRC-RP-98-00038, Westinghouse Savannah River Company, Aiken, SC, May 7, 1998.

[4] Dunn, K. A., Tosten, M. H., Louthan, Jr., and M. R., Birt, M. L., "Microstructural Characterization of 6061 Aluminum to 304L Stainless Steel Inertia Welds," *Conference Proceedings from the 32nd Annual Convention of the International Metallographic Society (Understanding Processing/Structure/Property/Behavior),* November 1999.

[5] Chandler, G. T., Sindelar, R. L., and Lam, P. S., "Evaluation of Water Chemistry on the Pitting Susceptibility of Aluminum," *Corrosion 97, Paper No. 104*, 1997.

[6] Peacock, Jr., H. B., Sindelar, R. L., Lam, P. S., Murphy, T. H., "Evaluation of Corrosion of Aluminum Base Reactor Fuel Cladding Materials During Dry Storage", Internal Report WSRC-TR-95-0345, Westinghouse Savannah River Company, Aiken, SC, November 1995.

B. H. Sencer[1,2], G. M. Bond[1], F. A. Garner[2], S. A. Maloy[3], W. F. Sommer[3] and M. R. James[3]

Microstructural Alteration of Structural Alloys by Low Temperature Irradiation with High Energy Protons and Spallation Neutrons

Reference: Sencer, B. H., Bond, G. M., Garner, F. A., Maloy, S. A., Sommer, W. F. and James, M. R., "**Microstructural Alteration of Structural Alloys by Low Temperature Irradiation with High Energy Protons and Spallation Neutrons,**" *Effects of Radiation on Materials: 20th International Symposium, ASTM STP 1405,* S. T. Rosinski, M. L. Grossbeck, T. R. Allen, and A. S. Kumar, Eds., American Society for Testing and Materials, West Conshohocken, PA, 2001.

Abstract: Several candidate alloys (nickel-based Alloy 718, iron-based 316 L and 304 L stainless steels) have been exposed to mixed proton-neutron particle fluxes and spectra at the Los Alamos Neutron Science Center (LANSCE), with the radiation conditions similar to those expected in the proposed Accelerator Production of Tritium (APT) facility. All alloys developed a high density of small black-spot damage and larger Frank loops, even at the lowest dose examined, producing rapid hardening and concurrent loss of ductility. The radiation-induced evolution of both microstructure and mechanical properties in the 300-series stainless steels is relatively simple, whereas that in Alloy 718 is rather more complex,. Superlattice spots corresponding to the age-hardening precipitate phases γ' and γ'' are lost from the diffraction patterns for Alloy 718 by only 0.6 dpa, the lowest proton/neutron-induced dose level achieved in this experiment. Examination of neutron-irradiated samples to doses of only ~0.1 dpa showed that precipitates are faintly visible in diffraction patterns but are rapidly becoming invisible. It is proposed that the γ' and γ'' first become disordered (by <0.6 dpa), but remain as solute-rich aggregates that still contribute to the hardness at relatively low dpa levels, and then are gradually dispersed at higher doses.

The two 300-series stainless steels (SS) appear to exhibit essentially identical evolution both in their microstructure and mechanical properties. In the 300-series stainless steels, there is an initial strong decrease in uniform elongation at very low dose levels, followed by a second, rather abrupt contribution to ductility loss at higher doses (>3dpa) that is not accompanied by any observed new or enhanced microstructural development. This may be a manifestation of gas-induced loss of ductility resulting from large levels of hydrogen and helium generated by irradiation in these alloys. Although

[1] New Mexico Institute of Mining and Technology, Socorro, NM 87801, USA
[2] Pacific Northwest National Laboratory, Richland, WA 99352, USA.
[3] Los Alamos National Laboratory, Los Alamos, NM 87545, USA.

the retained gas levels approached ~1 at% at the highest exposure levels, no discernible cavities were observed in any of the alloys.

With one significant exception, all observed features of radiation-affected properties can be explained in terms of the observed developments in irradiated but undeformed microstructure. Explanation of the remaining feature will require observation of post-deformation microstructure.

Keywords: structural alloys, proton irradiation, neutron irradiation, Alloy 718, stainless steels, 316 L, 304 L, microstructural evolution, helium, hydrogen

Introduction

An understanding of the response of candidate materials to the proton and spallation neutron irradiation conditions they would experience in the Accelerator Production of Tritium (APT) facility is essential to proper development of APT components [*1*]. Almost no information was previously available for materials applications in such severe environments. Those environments produce high radiation damage levels, concurrent with high rates of helium and hydrogen generation, at relatively low irradiation temperatures. The data already available were derived, in general, from irradiations involving relatively low particle energies, much lower gas (He/H) generation rates, and much lower doses, compared to those expected in APT.

Therefore, specimens of APT candidate materials were irradiated at the 800 MeV 100 mA LANSCE accelerator to relatively low exposures, and to displacement levels as high as ~15 dpa in the direct proton beam in order to simulate APT conditions [*1*]. Nevertheless, even the high exposure data obtained will have to be extrapolated to doses two to three times higher, in order to predict materials performance in APT conditions for the expected service life. A good understanding of the microstructural changes responsible for changes in mechanical behavior is essential, if the mechanical property data are to be extrapolated to higher exposures with confidence.

Changes in tensile strength of the materials irradiated in the mixed proton and spallation neutron spectra at LANSCE have been reported by Maloy and coworkers [2]. The yield stress of Alloy 718 increases rapidly with increasing dose up to ~0.5 dpa, and then decreases slowly thereafter with further increases in dose. The uniform elongation, however, decreases drastically at very low doses (< 0.5 dpa), and does not recover when the alloy softens at higher doses. Both 304 L and 316 L stainless steels also exhibit an initial strong hardening at the low (<0.1 dpa) damage levels attained in the neutron furnaces, where the specimens are far away from the proton beam. At the higher doses the 300-series stainless steels maintain the initial radiation-induced hardening. The uniform elongation of the 300-series stainless steels falls slowly with increasing dose, and then drops precipitously at doses around 3-4 dpa.

Helium and hydrogen measurements on these materials have also been reported [*3*]. The retained helium increases continuously at a rate of ~150 to 180 appm/dpa, depending on the alloy, while the hydrogen retention is ~50-100% higher. For example, in Alloy 718, ~1830 appm helium and ~5000 appm hydrogen are retained at 13.4 dpa.

To investigate the microstructural evolution of the materials in a mixed proton and spallation neutron environment, specimens irradiated between 0.1 and 13.4 dpa were examined by transmission electron microscopy (TEM). The present paper describes the radiation-induced microstructural evolution of these alloys and its role in the concurrent changes in mechanical properties.

Experimental Conditions

The compositions of the Alloy 718 and 300-series stainless steels are given in Table 1. Alloy 718 specimens were electro-discharge-machine (EDM) cut from the as-received material (annealed condition), then wrapped in Nb foil and encapsulated in a quartz tube (evacuated and back-filled with Ar). The specimens were then heat treated with the following steps: annealed at 1065^{0}C for 30 minutes and air cooled, aged at 760^{0}C for 10 hours, furnace cooled to 650^{0}C and held for a total time of 20 hours, with a final air cool to room temperature. The 316 L and 304 L stainless steel samples were also EDM cut from as-received (annealed) material, and received no further heat treatment.

Table 1- *Composition of the materials (in wt.%).*

ELEMENT	ALLOY 718	SS 316L	SS 304L
Al	0.48	----	----
C	0.04	0.019	0.02
Cr	18.13	17.26	18.23
Cu	0.08	0.26	0.38
Fe	Bal.	Bal.	Bal.
Mn	0.13	1.75	1.77
Mo	3.06	2.57	0.33
Ni	53.58	12.16	9.68
P	0.008	0.022	0.026
S	0.001	0.006	0.002
Si	0.11	0.65	0.54
Ti	1.03	---	---
(Nb+Ta)	4.98	---	---

The irradiation was conducted in LANSCE at Los Alamos National Laboratory. Details of the irradiation were published previously [*1*]. Doses were determined by analysis of the isotopes produced in activation foils placed alongside the specimens [*4*]. The irradiation temperatures were determined from thermocouples placed in each reference tube [*5*]. Irradiated specimens were chosen for microstructural evaluation depending on their locations relative to the proton beam. The materials examined, together with the irradiation temperatures and doses, are given Figure 1. Figure 1 also shows which types of energetic particle were primarily responsible for the radiation damage to the materials.

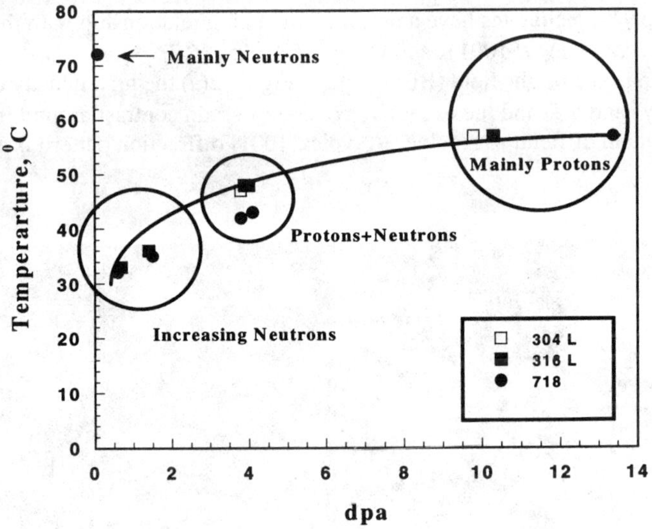

Figure 1- *The materials examined, together with the irradiation temperature, dose and type of energetic particles responsible for radiation damage. At high doses (≥9.6 dpa), the flux is mainly high-energy protons, at medium doses (3-5 dpa) protons and spallation neutrons, and at low doses (<1.4 dpa) increasing amounts of spallation neutrons.*

Thin-foil samples were prepared for transmission electron microscope (TEM) examination by conventional jet electropolishing methods, in 5% perchloric acid, 95% ethyl alcohol at - 25^{0}C and 55V for Alloy 718, and in 5% perchloric acid, 95% methanol at - 20 ^{0}C and 50 V for the stainless steels. TEM examinations were conducted with a JEOL 2010F (FEG-STEM, 200KeV) for high resolution, and a JEOL 2000E (200 KeV) and JEOL 1200EX (120KeV) for the diffraction-contrast microstructural characterization and energy-dispersive x-ray spectrometry (EDS).

Results

Microstructural Evolution of Alloy 718:

Unirradiated - Alloy 718 is a γ'-γ'' age-hardenable alloy. It develops high strength during thermal aging by precipitating finely dispersed Ni_3Nb-γ'' and $Ni_3(Ti,Al)$-γ' in an austenitic γ matrix. The γ' and γ'' phases represent ordered variations of the fcc lattice. γ' has a cubic $L1_2$ ordered structure based on Ni_3Al, with Ti and some Nb substituting on Al sites. γ'' has a tetragonal DO_{22} ordered structure based on Ni_3Nb, with minor Al and Ti substitutions on the Nb sites. From consideration of the Nb, Al, and Ti atom positions, it is apparent that the DO_{22} structure can also be described as having body-centered tetragonal (bct) symmetry, while the $L1_2$ structure can be considered as having cubic symmetry. The γ'' precipitates have a disc-shaped morphology, while γ' particles are spheroidal. The γ'' precipitates have a definite orientation relationship with the austenitic γ matrix: $(001)_{\gamma''} // \{001\}_{\gamma}$ and $[100]_{\gamma''} // <100>_{\gamma}$ [6,7].

As shown in the bright-field (BF) image in Figure 2(a) the high density of precipitates (γ' and γ'') and the excessive coherency strain contrast around individual particles make them difficult to resolve. A typical [001] diffraction pattern is also shown in Figure 2(b).

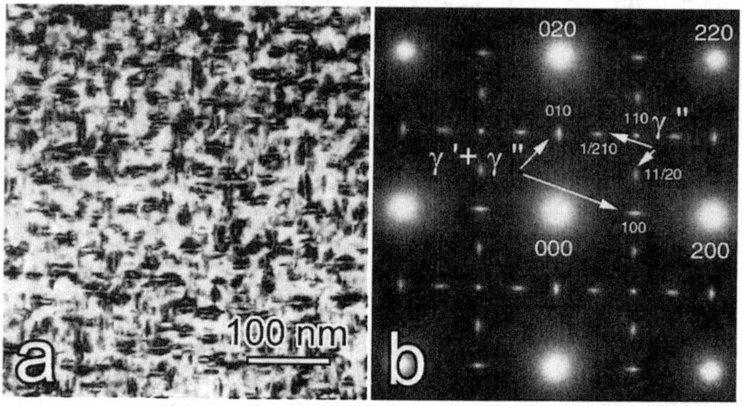

Figure 2 - *(a) Bright-field image of unirradiated Alloy 718, and (b) [001] electron diffraction pattern showing the superlattice reflections.*

The γ'/γ'' precipitates can be clearly shown by dark-field (DF) analysis with specific superlattice reflections. Reflections of the type (100), (010) and (110) could arise from either γ' or specific variants of the γ''. The γ' superlattice reflections in this pattern have been indexed with respect to the fcc matrix. Reflections of the form (11/20) are allowed for the DO_{22} bct phase, but not for γ'. Thus, the observation of such reflections in Figure 2 (b) clearly demonstrates that a bct γ'' precipitate is present.

A systematic dark-field analysis is shown in Figure 3. Here the (010), (1/210), (110), (11/20) and (100) reflections were used. These dark-field images (all from the same region in the foil) are shown along with their corresponding diffraction patterns. The (010), (110) and (100) dark-field images (Figure 3 (b,d,f)) clearly reveal the presence of three different variants of a fine, disc-shaped precipitate. In Figure 3 (b,f), two different variants are lying along the (010) and (100) planes. The third variant, lying in the (001) plane of the foil, is visible in Figure 3 (d). The orientation and selective imaging of the different precipitate variants are consistent with the bct γ''. The (1/210) and (11/20) dark-field images (Figure 3 (c,e)) identify the disc-shaped precipitate as bct γ''. The [100] γ'' variant is imaged with both the (100) and (1/210) reflections (Figure 3 (c,f)) while the [010] variant is imaged with the (010) and (11/20) reflections (Figure 3 (b,e)). Existence of γ' can also be seen by comparison of the dark-field images produced by, for example, (1/210) and (100) superlattice reflections, as shown in Figure 4.

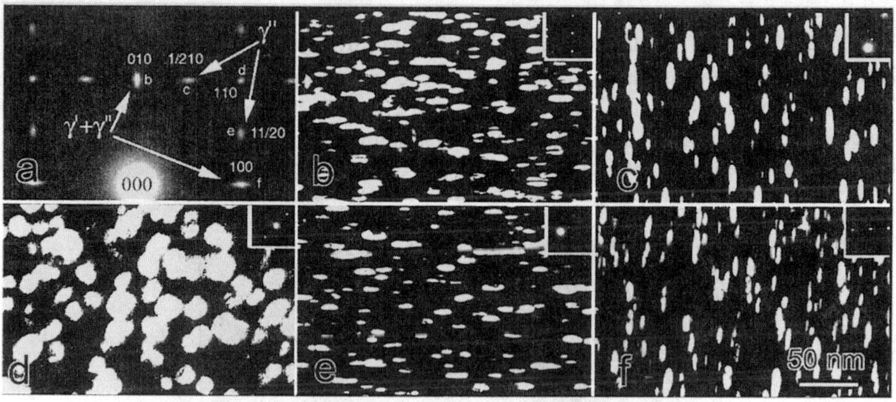

Figure 3 - *Unirradiated 718, [001] orientation. (a) Electron diffraction pattern of [001]; dark-field micrographs with (b) g = (010) ($\gamma'+\gamma''$); (c) g = (1/110) (γ''); (d) g =(110) ($\gamma'+\gamma''$); (e) g = (11/20) (γ''); (f) g = (100) ($\gamma'+\gamma''$).*

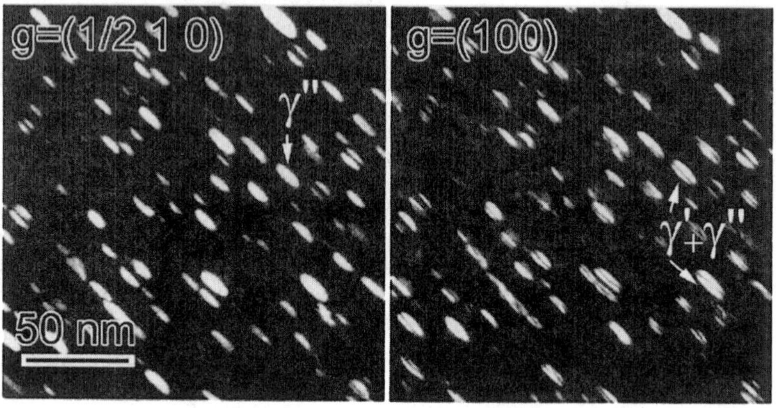

Figure 4 - *Unirradiated 718, dark-field micrographs with g = (1/210) (γ″) and g= (100) (γ′+ γ″) from the same area of the grain.*

Lattice imaging of the unirradiated Alloy 718 confirmed that the intragranular γ′ and γ″ particles occur in pairs, each γ′ particle sharing an interface with a γ″ particle. In the [001] lattice image shown in Figure 5, the γ′ is distinguished from the fcc matrix by its doubled lattice spacing. The γ″ has a different structure, and shows evidence of coherency strain along its interface with the matrix.

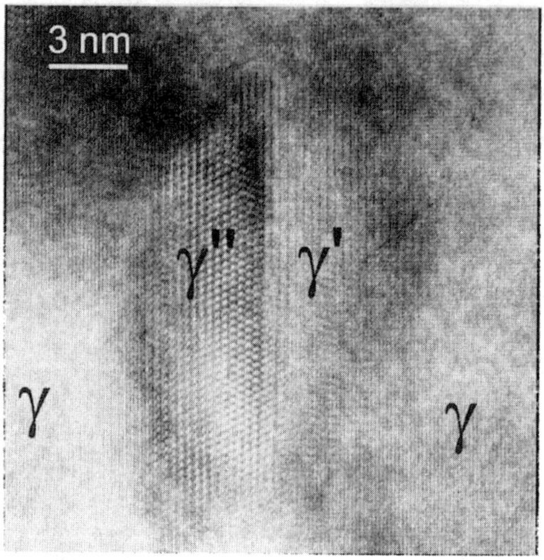

Figure 5 - *High-resolution electron micrograph of unirradiated Alloy 718. [001] lattice image showing γ′– γ″ precipitates.*

TEM-EDS analysis was performed on unirradiated Alloy 718 samples in order to identify the carbides that are also present in the material. Because electropolished thin-foil TEM samples are random sections through material containing embedded precipitate particles with widely varied sizes, the particle analysis generally contains contributions from an unknown thickness of superimposed matrix material. The results of compositional analysis for representative particles that appeared nearly free of matrix contributions indicated that the particles were Nb-rich carbides with an average composition of 80-87%Nb, 7-11%Ti, 2-7%Ni,1-3%Fe, 1-3%Cr, 0-2%Mo, < 0.5% Al, Si in wt.%.

Irradiated Inconel 718

γ' and γ'' Phase Instability - γ' and γ'' phases in Alloy 718 exhibited instabilities under proton irradiation between 32 and 55 ^{0}C. After doses as low as 0.6 dpa, all evidence of the γ' and γ'' particles in the matrix had disappeared from the electron diffraction patterns. As shown in Figure 6, the γ' and γ'' superlattice diffraction spots, clearly evident for unirradiated samples, were absent after irradiation. No other precipitate spots (that would correspond to irradiation-induced phases) appeared in electron diffraction patterns from the irradiated samples.

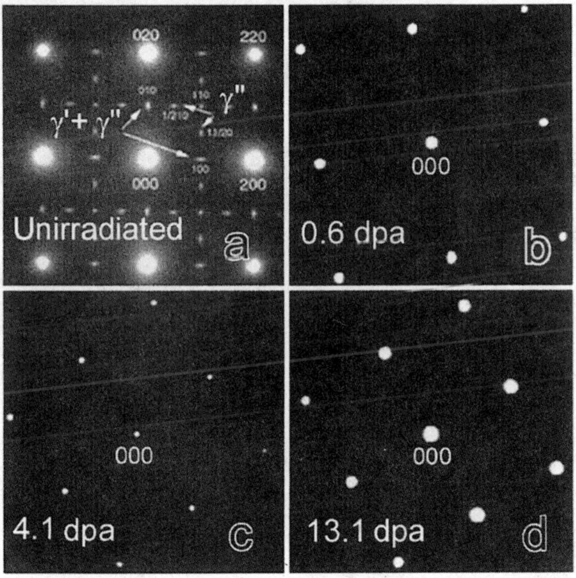

Figure 6 - *[001] Selected-area electron diffraction patterns from Alloy 718 (a)showing γ' and γ'' reflections, (b),(c),(d) showing disappearance of γ' and γ'' reflections after irradiation.*

In order to investigate further the disappearance of the γ' and γ'', samples of Alloy 718 were examined that had been neutron irradiated to a lower dose of ~0.1 dpa at ~72^{0}C. γ' and γ'' reflections were still faintly visible in the diffraction patterns, as demonstrated by Figure 7 in which diffraction patterns from unirradiated and ~0.1 dpa samples are compared. As seen in Figure 8, precipitate dark-field images from these reflections confirmed that the precipitates were still present, although they gave a more patchy, discontinuous contrast.

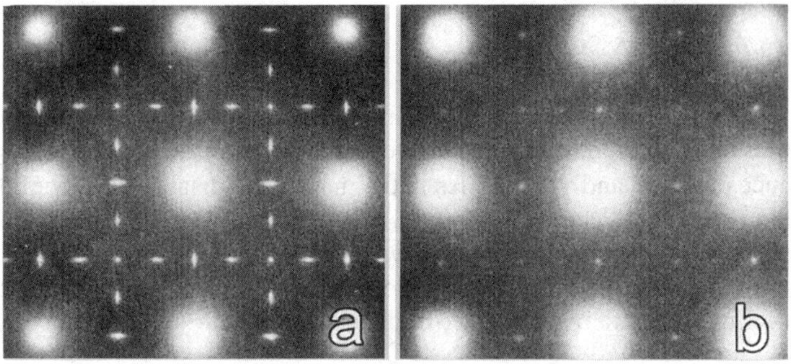

Figure 7 - *[100] electron diffraction patterns of Alloy 718 a) unirradiated and b) ~ 0.1 dpa, 72^{0}C.*

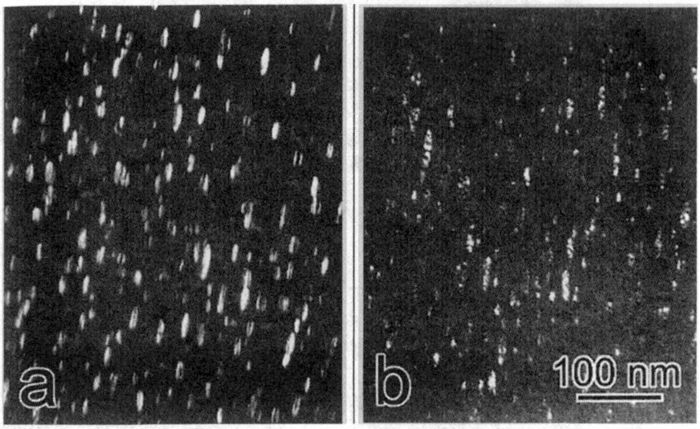

Figure 8 - *Dark-field images from close to [100] zone axis in Alloy 718, g= (010) (γ'+γ'') precipitate dark-field images a) unirradiated and b) ~ 0.1 dpa, 72^{0}C.*

Radiation-Induced Defects - {111}-faulted dislocation loops, i.e. Frank loops, were present at all dpa levels. In the case of neutron-irradiated samples (~0.1 dpa), small black spots were visible in addition to the precipitates (Figure 9). The loops produced distinctive satellite spots around the fundamental matrix spots on diffraction patterns. The satellite spots arise from extended diffraction streaks (relrods) perpendicular to the four sets of {111} planes. Similar intensities of satellite diffraction spots from the four <111> relrods indicated essentially equal Frank-loop populations on all four {111} planes. Dark-field images of the loops taken with such satellite spots revealed faulted Frank loops, as shown in Figure 10. More than 75 loops from dark-field TEM micrographs were measured to obtain an average loop diameter. Average loop diameter was ~ 10.6 nm at 3.8 dpa and ~ 23.7 nm at 13.4 dpa. Figure 11 shows the [011] lattice image of a nearly edge-on faulted Frank loop at 1.5 dpa.

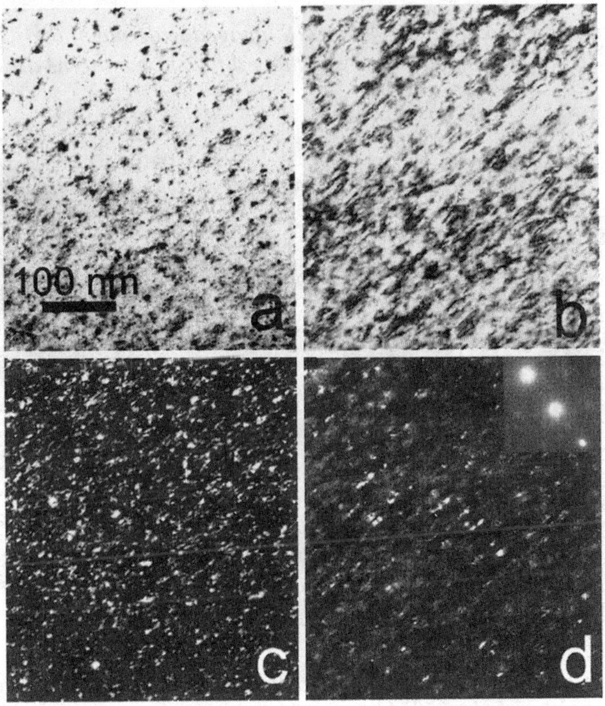

Figure 9 - *Black-spot damage in Alloy 718 (~0.1 dpa) (a) g=200 s>>o bright-field contrast, b) g=200 s>0 bright-field contrast c) g=200 g/4g weak-beam image d) precipitate dark-field contrast. All images are from the same area of the grain.*

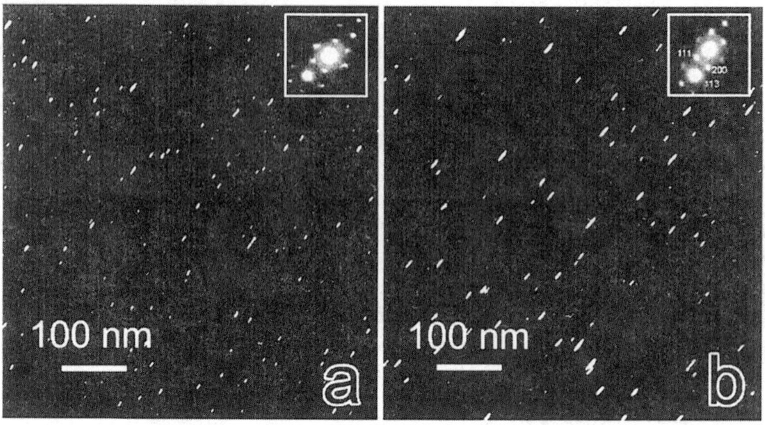

Figure 10 - *Faulted Frank loops in Alloy 718, imaged with one of the <111> relrods, in (a) 3.8 and (b) 13.4 dpa samples. Corresponding diffraction patterns are shown as insets.*

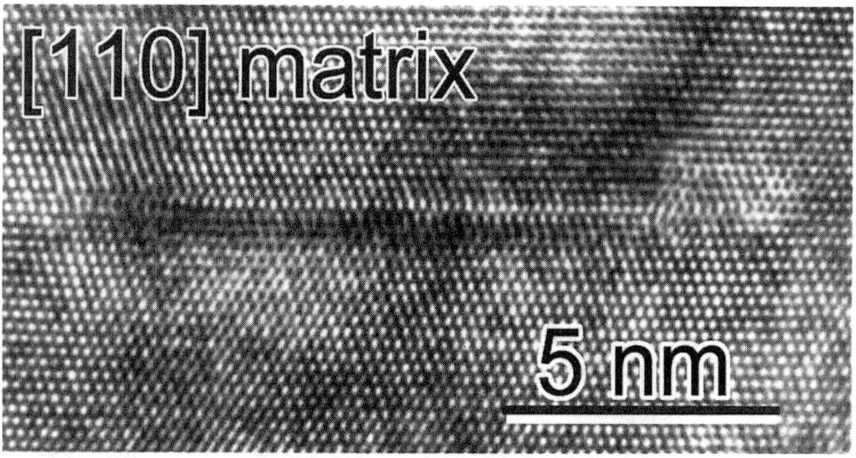

Figure 11 - *High-resolution electron micrograph of radiation-induced Frank loop in irradiated Alloy 718, 1.5 dpa, [011] lattice image.*

The possibility of cavity formation was investigated in near-kinematic conditions (Fresnel contrast). Several through-focal images were taken, as shown in Figure 12. Examination of the samples irradiated to the higher dose (13.4 dpa) showed that no cavity formation had occurred, down to a resolution of ~ 1 nm [*8*].

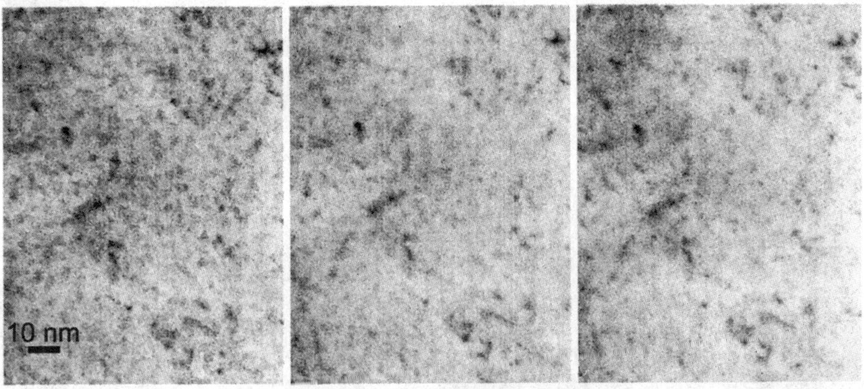

Figure 12 - *Kinematic through-focal image series from 13.4 dpa Alloy 718. From left to right: under-focus, near-focus and over-focus, showing that no cavities are observed in the matrix.*

Microstructural Evolution of 316 L/304 L Stainless Steels

Proton and spallation neutron irradiation produced a significant change in the general microstructure of both 316 L and 304 L stainless steels. Black-spot damage was present at all dpa levels. Figures 13 and 14 show small black-spot damage and larger faulted Frank loops in bright-field (BF) and weak-beam dark-field (WBDF) images of 316 L at 1.4, 3.9 and 10.3 dpa respectively. Relrod images of faulted Frank loops are shown in Figure 15 for 316 L at 1.3, 3.9 and 10.3 dpa. Microstructural evolution of 304 L is shown in bright-field (BF) images in Figure 16, and in weak-beam dark-field images in Figure 17, at 0.7 and 9.8 dpa. Relrod images of faulted Frank loops are shown in Figure 18 for 316 L at 0.7, 3.8 and 9.8 dpa.

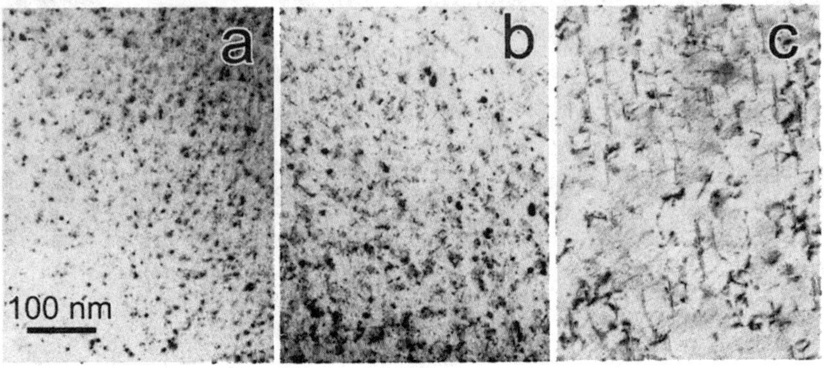

Figure 13 - *Bright-field TEM micrographs of 316L SS, showing small defect clusters and larger faulted Frank loops, g = 200, s>0 at a) 1.3, b) 3.9 and c) 10.3 dpa.*

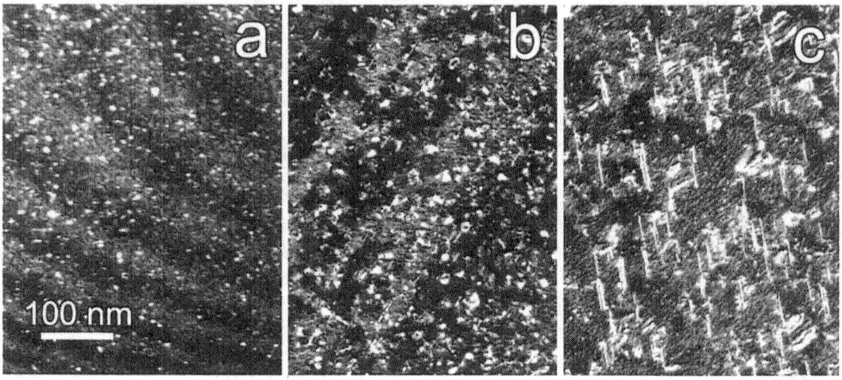

Figure 14 – *WBDF g = 200 (g/4g) images of 316L SS, showing black-spot damage and larger faulted Frank loops, taken with beam direction close to <110> zone axis, at a) 1.4, b) 3.9 and c) 10.3 dpa.*

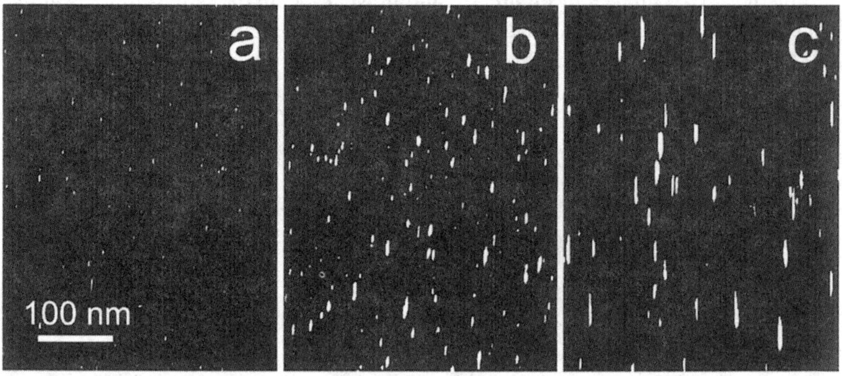

Figure 15 - *TEM micrographs of 316L SS showing faulted Frank loops, taken with beam direction close to <110> zone axis. (a), (b) and (c) are DF images taken from one of the <111> relrods for samples at a) 1.3, b) 3.9 and c)10.3 dpa.*

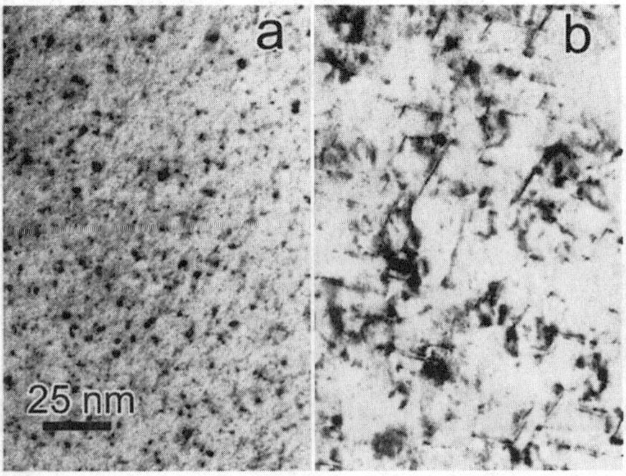

Figure 16 - *Bright-field TEM micrographs of 304L SS, showing small defect clusters and larger faulted Frank loops, g = 200, s>0, at a) 0.7 dpa and b) 9.8 dpa.*

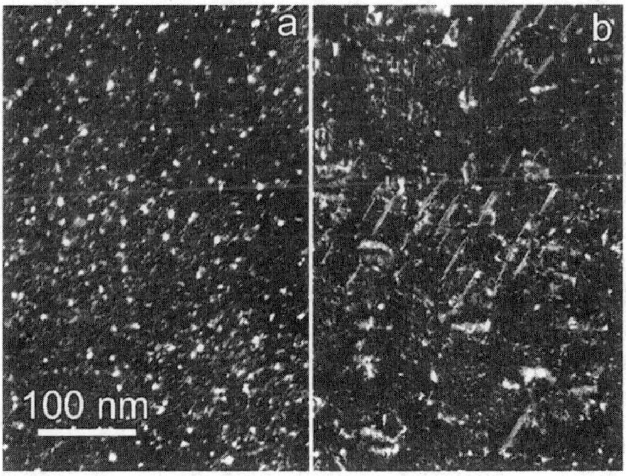

Figure 17 - *WBDF g = 200 (g/4g) images of 304L SS, showing small defect clusters and larger Frank loops, taken with beam direction close to <110> zone axis, at a) 0.7 dpa and b) 9.8 dpa.*

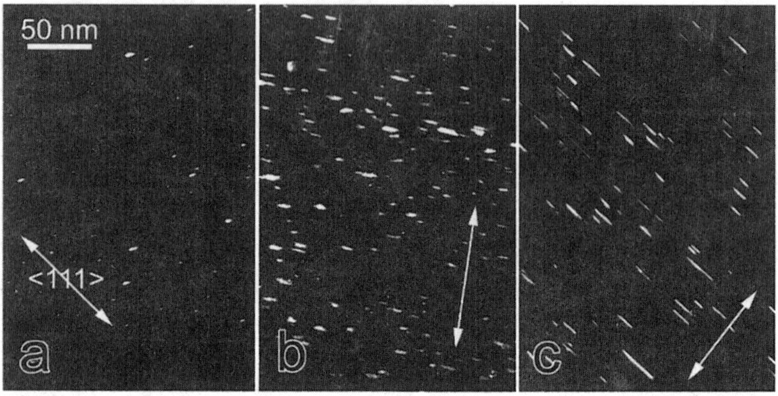

Figure 18 - *TEM micrographs of SS 304L showing faulted Frank loops, taken with beam direction close to <110> zone axis. (a), (b) and (c) are DF images taken from one of the <111> relrods for samples at 0.7, 3.8 and 9.8 dpa respectively.*

Electron diffraction patterns did not show any precipitate reflections. The possibility of cavity formation was investigated in Fresnel contrast. However, no cavity (bubble and/or void) formation was observed in the samples examined between 0.7 and 10.3 dpa.

Dislocation loop size was estimated from the size of the outside loop contrast. The loops tended to appear elliptical, depending on their orientation in the TEM foil. More than 150 loops from high-magnification dark-field TEM micrographs were measured to obtain an average loop diameter. Frank loops were imaged by the relrod technique, which makes it possible to measure Frank loops down to very small sizes (1-2 nm). This does not, however, eliminate the possibility that there may be smaller dislocation loops or defect clusters of other types present. This technique images only the faulted Frank loops, and also there is a resolution limit below which smaller faulted Frank loops are either not visible in the relrod image, or cannot be confidently identified.

Loop densities were determined from several micrographs. The corresponding foil thickness was determined from convergent-beam electron diffraction (CBED) patterns. Quantitative measurements are given in Table 2. Figure 19 shows loop size distributions for 304 L and 316 L. Average loop size and the density of loops, as well as the total dislocation density (associated with the loops) are shown for both steels in Figures 20, 21, and 22, respectively.

Table 2 - *Summary of dislocation loop densities, mean loop diameters, and total dislocation densities for 316 L and 304 L.*

Material	Dose (dpa)	Loop Number Density (m^{-3})	Mean Loop Diameter (nm)	Total Dislocation Density (Loops) (m^{-2})
316 L	1.4	$2.1x10^{22}$	4	$2.6x10^{14}$
316 L	4.1	$4.8x10^{22}$	10.2	$1.54x10^{15}$
316 L	10.3	$1.9x10^{22}$	23.6	$1.41x10^{15}$
304 L	0.7	$1.6x10^{22}$	1.8	$9.0x10^{13}$
304 L	3.8	$5x10^{22}$	9.6	$1.35x10^{15}$
304 L	9.8	$2.1x10^{22}$	20.1	$1.32x10^{15}$

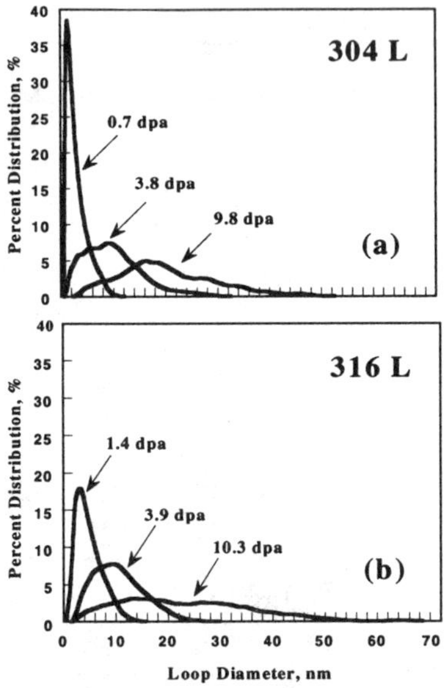

Figure 19 - *(a) Loop size distributions for 304 L samples at 0.7 dpa, 3.8 dpa, 9.8 dpa (b) Loop size distributions for 316 L samples at 1.4 dpa, 3.9 dpa, 10.3 dpa.*

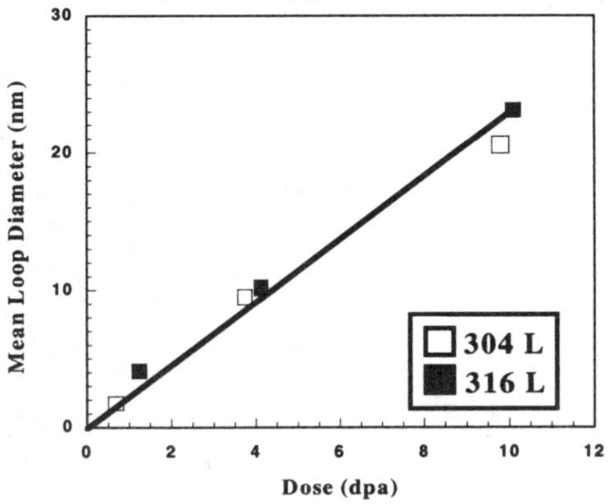

Figure 20 - *Mean loop diameter as a function of dpa for two 300-series stainless steels irradiated with mixed spectra of protons and spallation neutrons.*

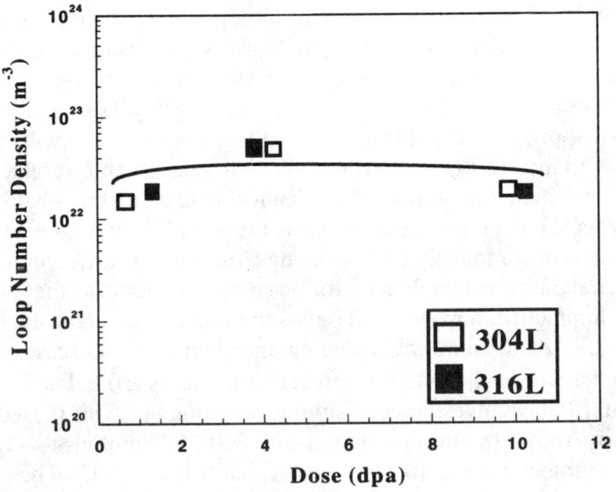

Figure 21 - *Number density of loops as a function of dpa for two 300-series stainless steels.*

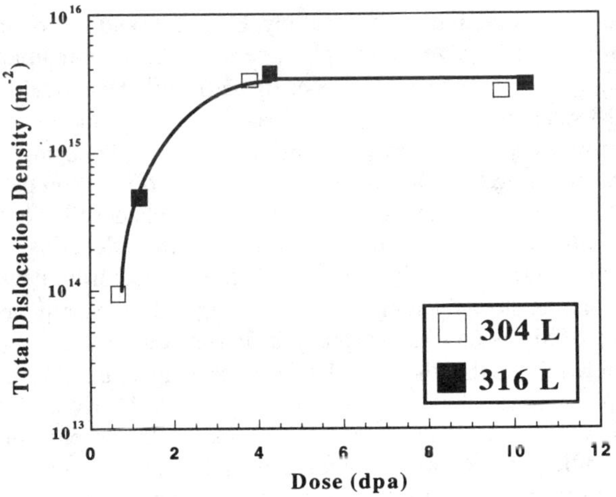

Figure 22 - *Total dislocation density as a function of dpa for two 300-series stainless steels.*

Discussion

Alloy 718 - The most interesting and complex evolution of both microstructure and mechanical properties occurred in Alloy 718. The stability of γ' and γ'' phases in Alloy 718 is affected by irradiation with mixed spectra of protons and spallation neutrons at low temperatures (32-55^{0}C), even at doses as low as 0.6 dpa. All evidence for these phases had disappeared from the electron diffraction patterns after proton-induced doses ranging from 0.6 through 13.4 dpa. The microstructure of irradiated Alloy 718 was dominated instead by small black-spot damage and larger faulted loops at all dose levels, the average loop size increasing with increasing dose. These microstructural trends might, at first thought, suggest a very different evolution of mechanical behavior from what is actually observed in tensile tests [*2*] and microhardness measurements [*9*]. If all the age-hardening precipitates were completely lost by 0.6 dpa, one would expect an initial softening of the material. This would then be followed by hardening, as the loop density and sizes increased. In practice, however, the opposite trend is seen in the evolution of mechanical behavior [*2*]. After an initial hardening, the yield stress decreases with increasing dose; in other words, the material softens for all doses $\geq$ 0.6 dpa.

Carsughi et al. [*9*] have investigated a similar softening behavior (based on microhardness data) in Alloy 718 after irradiation with 800 MeV protons to ~10 dpa at temperatures that were somewhat variable and higher, but below 250^{0}C. The disappearance of γ' and γ'', and the presence of η-phase (which can form as a product of overaging in irradiated or unirradiated 718) were reported. The presence of η-phase was suggested by Carsughi to be a possible cause of softening of the material. In the present study at lower temperatures, however, no evidence of η-phase was found, although faulted Frank loops were imaged with the <111> relrods (see Figure 10). Identification of η-phase in the work of Carsughi et al. [*9*] is thought to be questionable, possibly resulting from misidentification of faulted loops as η-phase.

Between ~0.6 and ~13.4 dpa, in the present study, lack of γ' and γ'' precipitate reflections in diffraction patterns gave the initial impression that the precipitates disappeared at very low ($\leq$0.6 dpa) doses. To investigate further the behavior of these precipitates, Alloy 718 samples irradiated to ~0.1 dpa with only spallation neutrons were examined. In these samples, γ' and γ'' precipitates were found to be losing their superlattice intensity, becoming only faintly visible in electron diffraction patterns, and difficult to image with precipitate reflections. The next available dose for examination was ~0.6 dpa (protons and spallation neutrons). The precipitate reflections were totally absent from diffraction patterns by 0.6 dpa, and no additional spots from new phases were observed. This finding strongly suggests that γ' and γ'' disappeared well before 0.6 dpa. The question then becomes how to interpret this disappearance.

One interpretation would be that the γ' and γ'' have dissolved, but this is inconsistent with the changes in mechanical behavior, as explained above. It is also difficult to envisage complete dissolution occurring at such a low dose, in the absence of significant thermal diffusion. Another interpretation, however, would be that the precipitates have only lost much of their ordered structure, with the local solute concentrations remaining.

The loss of order may result from the displacement cascades that occur during radiation damage [*10*]. At low irradiation temperatures, thermal diffusion is limited, and cascade-induced mixing may become the most important short-range transport mechanism. Under 800 MeV proton irradiation, very large cascades are thought to be produced. As a result of cascade formation, atomic mixing occurs, along with interstitial and vacancy production [*11*]. Such processes might not be detrimental to solution-hardened material, but, in the case of ordered intermetallic phases, extensive mixing and defect formation will lead to extensive disordering. Complete dissolution by 0.6 dpa would lead to a softening that is inconsistent with the observed initial increase in hardening [2]. It is therefore proposed that the precipitates become disordered, but are retained in the range of 0.0 to ~0.6 dpa as regions of locally high solute concentration that can still harden the material similar to GP (Guinier-Preston) zones. Thus, for low doses there is no significant softening associated with the disorder, and the concurrent development of black-spot damage leads to an overall increase in hardness.

Alloy 718 exhibited a partial softening between ~ 0.6 and ~13.4 dpa. This is attributed to gradual redistribution of the solutes as the dose increases over this range. The observed mechanical behavior might be explained as follows. The early study of Silcock et al. [*12*] on GP zones in binary aluminum-copper alloys showed that the main contribution to hardness comes not from the ordering of GP zones, but from segregation of solute atoms to the GP zones. Therefore, by analogy, the disordering of γ' and γ'' alone should not remove the hardening associated with the precipitates. Subsequent dissolution of the solute concentrations could, however, produce substantial softening that might be sufficient to outweigh hardening due to further formation of Frank loops, and hence result in a net softening effect at doses > 0.6 dpa.

A drastic loss of ductility was observed in Alloy 718 at all dose levels examined [2]. The early loss of ductility may be attributed to radiation-induced hardening, as loops are formed. What is perhaps more surprising, however, is that there is no recovery of ductility as softening of the material occurs at higher doses. Some increase in ductility might be expected to accompany the decreases in yield stress that are seen as dose increases beyond 0.6 dpa, but this is not the case. A possible explanation for this is as follows. In the unirradiated condition the γ' and γ'' particles are hard particles that induce Orowan hardening. At low dose, they become disordered, and start to become shearable by dislocations. This allows strain localization (dislocation channeling), which gives considerable loss in uniform elongation and less hardening. Increasing dose reinforces strain localization, and the loss in elongation does not recover. It must also be remembered that substantial amounts of helium and hydrogen accumulated in the samples during the LANSCE irradiations [*3*]. TEM examination revealed a complete absence of cavity formation in Alloy 718, even though levels of ~1830 appm helium and ~5000 appm hydrogen were reached at 13.4 dpa. It appears that there is insufficient mobility of helium atoms in the temperature range 32-72^{0}C to allow nucleation and growth of bubbles or voids. Thus, the helium and hydrogen must be dispersed on a very fine scale within the material. The high gas levels within the lattice may also contribute to the continued loss of ductility of the material, even as the yield strength and hardness decrease. Hydrogen accumulation is initially greater in Inconel 718 than in the 300-series stainless steels [*3*].

316 L and 304 L Stainless Steels - Combined proton and spallation neutron irradiation produced less pronounced, but still significant changes in the microstructure of both 304 L and 316 L. Black-spot damage was the most common feature, together with larger faulted Frank loops. Black-spot damage was present at all dpa levels, while the larger faulted Frank loops became more numerous at higher doses. At 0.7 dpa the predominant microstructural feature is black-spot damage or small Frank loops, almost all of which are smaller than 6 nm. Loops grow larger as a function of dose, reaching 50 nm at 9.8 dpa in 304 L. The average loop diameter at 0.7 dpa is smaller by a factor of ten than that at 9.8 dpa. Neither electron diffraction patterns nor stereo-micrographs revealed any evidence of radiation-induced precipitation in the matrix or at the grain boundaries.

The microstructures of 316 L and 304 L austenitic stainless steels are remarkably similar after irradiation under similar irradiation conditions at LASREF. Both materials essentially contain black-spot damage and larger faulted Frank loops. Comparisons of the average size and density of the Frank loops, and total dislocation densities are shown in Figures 20, 21, and 22.

The density of loops increases quickly at very low doses, and changes only a little with increasing dose thereafter. Dislocation loop densities for both alloys appear to be the highest at about 3 - 4 dpa. At ~ 4 dpa, the densities of loops in both 316 L and 304 L are identical within the scatter of the data. For both alloys, the overall microstructure coarsens as a function of increasing dose.

The microstructural response of the two steels to the irradiation is so similar that it is not surprising to see similar trends in the mechanical behavior [*3*]. If the observed radiation-induced changes in mechanical properties are to be explained in terms of the microstructural alterations induced by radiation, it is important to check whether the hardening observed in the 300-series stainless steels can be explained in terms of only the microstructural alterations observed. According to the theory of hardening based on a hard-sphere model (Orowan model) [*13*], the change in yield stress change due to dislocation loops, $\Delta\sigma_y$, is expressed by

$$\Delta\sigma_y = M\alpha\mu b(Nd)^{1/2},$$

where

M = Taylor factor,
α = barrier strength of obstacles,
μ = shear modulus,
b = Burgers vector,
N = density of dislocation loops,
d = mean diameter of dislocation loops.

The α values required to produce the observed hardening were calculated from the yield stresses [2], loop densities, and loop sizes. Calculated α values (0.3-0.5) between ~4 and 10.3 dpa are close to the expected values [*14*]. Nothing can be said about the possible contribution of the accumulating gas atoms to the hardening, but the fact that the Frank-loop hardening contributions saturate, as does the yield strength, tends to argue that the increasing gas (in either atomistic form or subresolvable clusters) is not contributing significantly to the hardening. There were no precipitates or cavities observed; only black spots and larger Frank loops were formed.

The results of the electron microscopy indicate that the evolution of radiation damage in undeformed specimens can account for all the features observed in the radiation-induced evolution of mechanical properties, except for the second and most precipitous drop of uniform elongation in 304 L and 316 L stainless steels at ~3 dpa. It is proposed that accumulation of gas, especially hydrogen, may be the cause of this abrupt decrease. To address this issue further, it is planned to examine the deformed microstructure of these alloys at two dpa levels, just below and just above the dpa level where the uniform elongation suffers the second round of reduction. Previously tested tensile specimens will be used to prepare the TEM specimens for further investigation.

Summary

In general, it appears that, with only one exeption, the major features of the radiation-induced evolution of mechanical properties of all three alloys can be explained in terms of their individual microstructural evolution. To summarize:

Alloy 718

By ~ 0.1 dpa, images of γ' and γ'' precipitates lose some of their intensity, and show discontinuous contrast, indicating disordering, while γ' and γ'' superlattice reflections are still faintly visible in the electron diffraction patterns. No new phases are precipitated from γ' and γ''. A dose of ~0.1 dpa is thought to be sufficient to induce partial disordering of the precipitates at ~ 72^{0}C.

All evidence of the γ' and γ'' precipitates disappears from the diffraction patterns by 0.6 dpa. If complete dissolution of the precipitates were to occur, the material should soften, but, on the contrary, Alloy 718 hardens up to ~ 0.6 dpa. It is proposed here that the precipitates are essentially disordered by 0.6 dpa, but that the solutes they contained are not yet fully redistributed into the matrix. Thus the precipitates retain the majority of their original hardening contribution.

At low doses, the dominant microstructural features are black-spot damage and remnants of the γ' and γ'' phases. Together, the black-spot damage and remains of the precipitates result in a net hardening up to 0.6 dpa, since both are obstacles to dislocation movement.

At doses between ~ 0.6 and ~13 dpa, the observable microstructure of Alloy 718 is dominated by only black-spot damage and larger faulted Frank loops. No evidence of any new phase was found in the diffraction patterns.

At doses between 0.6 and ~13 dpa, Alloy 718 softens somewhat. The gradual softening is attributed to mixing-induced redistribution of the solutes from the former γ' and γ'' precipitates as the dose increases.

In spite of the large levels of retained helium and hydrogen, no cavities are produced at doses as high as ~ 13 dpa. The gas atoms must be dispersed in the matrix as single atoms or very small complexes.

316 L/304 L Stainless Steels

Proton and spallation neutron irradiation induces the formation of both black-spot damage and larger faulted Frank loops in 316 L and 304 L stainless steels. Radiation-induced precipitation did not occur in either steel.

At ~ 4 dpa, mean loop sizes, loop number densities and total dislocation densities are identical for the two steels, within the scatter of the data. There was no significant difference in the microstructural evolution of 316 L and 304 L under these irradiation conditions.

In spite of the large levels of helium and hydrogen retention, no cavities were found even at the highest dose examined.

The increase in yield stress in both steels is due primarily to formation and evolution of faulted Frank loops, with gas playing no large role either in solution or in cavities.

The loss of ductility observed in both steels is due to a combination of, first, radiation-induced hardening by black-spot damage and larger loops, and, second, to gas accumulation at higher doses (>3dpa).

Acknowledgments

This work was supported by the U.S. Department of Energy under the Accelerator Production of Tritium Program at Los Alamos National Laboratory. Pacific Northwest National Laboratory is operated for USDOE by Battelle Memorial Institute.

References

[1]. Maloy, S. A., Sommer W. F., Brown, R. D., Roberts, J. E., Eddleman, J., Zimmerman E., and Willcutt, G., Proc. Symp. on Materials for Spallation Neutron Sources, eds M. S. Wechler, L. K. Mansur, C. L. Snead, and W. F. Sommer (TMS, Warrendale, PA, 1998), p. 35.

[2] Maloy, S. A., James, M. R., Sommer, W. F., and Hamilton, M. L., "The Effect of High Energy Protons and Neutrons on the Tensile Properties of Materials Selected for the Target and Blanket Components in the Accelerator Production of Tritium Project', *Effects of Radiation on Materials, 20th International Symposium, ASTM STP 1405*, S. T. Rosinski, M. L. Grossbeck, T. R. Allen, and A. S. Kumar, Eds., American Society for Testing and Materials, West Conshohocken, PA, 2002, in press.

[3] Oliver, B. M., Garner, F. A., Maloy, S. A., Sommer, W. F., Ferguson, P. D., and James, M. R., "Retention of Very High Levels of Helium and Hydrogen Generated in Various Structural Alloys by 800 MeV Protons and Spallation Neutrons", *Effects of Radiation on Materials, 20th International Symposium, ASTM STP 1405*, S. T. Rosinski,

M. L. Grossbeck, T. R. Allen, and A. S. Kumar, Eds., American Society for Testing and Materials, West Conshohocken, PA, 2002, in press.

[4] James, M. R., Maloy, S. A., Sommer, W. F., Ferguson, P. D., Fowler, M. M., and Corzine, K., 2nd International Topical Meeting on Nuclear Applications of Accelerator Technology, Gatlinburg, TN, Sept. 20-23, 1998, pp. 605-608.

[5] Willcutt, G. J., Maloy, S. A., James, M. R., Teague, J., Siebe, D. A., Sommer, W. F., Ferguson, P. D., 2nd International Topical Meeting on Nuclear Applications of Accelerator Technology, Gatlinburg, TN, Sept. 20-23, 1998, p. 254.

[6] Paulonis, D. F., Oblak, J. M., and Duvall, D. S., Trans. ASM Vol.62, 1969, p. 611.

[7] Oblak, J. M., Paulonis, D. F., and Duvall, D. S., Metall. Trans. Vol.5, 1974, p.143.

[8] Ruhle, M., and Wilkens, M., Cryst. Lattice Def. Vol.6., 1975, p.129.

[9] Carsughi, F., Derz, H., Ferguson, P., Pott, G., Sommer, W. and Ullmaier, H., J. Nucl. Mater. Vol.264, 1999, p. 78.

[10] Stoller, R. E., Odette, G. R., and Wirth, B. D., J. Nucl. Mater., Vol.251, 1997, p. 49.

[11] Bacon, D. J., Calder, A. F. and Gao, F., J. Nucl. Mater., Vol.251, 1997, p. 1.

[12] Silcock, J. M., Heal, T. J., and Hardy, H. K., J. Inst. Metals, Vol.82, 1953-1954, p. 239.

[13] Seeger, A., Proc. 2nd UN Int. Conf. on Peaceful Uses of Atomic Energy, Geneva, Vol.6, 1958, p.250.

[14] Lucas, G. E., J. Nucl. Mater., Vol.206, 1993, p.287.

B. M. Oliver,[1] F. A. Garner,[1] S. A. Maloy,[2] W. F. Sommer,[2]
P. D. Ferguson,[2] and M. R. James[2]

Retention of Very High Levels of Helium and Hydrogen Generated in Various Structural Alloys by 800 MeV Protons and Spallation Neutrons

Reference: Oliver, B. M., Garner, F. A., Maloy, S. A., Sommer, W. F., Ferguson, P. D., and James, M. R., **"Retention of Very High Levels of Helium and Hydrogen Generated in Various Structural Alloys by 800 MeV Protons and Spallation Neutrons,"** *Effects of Radiation on Materials: 20th International Symposium, ASTM STP 1405,* S. T. Rosinski, M. L. Grossbeck, T. R. Allen, and A. S. Kumar, Eds., American Society for Testing and Materials, West Conshohocken, PA, 2001.

Abstract: A series of irradiations were conducted at relatively low temperatures (<100°C) in the Los Alamos Neutron Science Center (LANSCE) as part of the test program supporting the Accelerator Production of Tritium Program sponsored by USDOE. In this irradiation campaign, a variety of candidate structural alloys were placed in various particle spectra, ranging from 800 MeV protons, to mixed energy distributions of both protons and spallation neutrons, and finally to distributions consisting primarily of high energy neutrons. At proton energies on the order of hundreds of MeV, exceptionally high levels of gas atoms are generated in all elemental constituents of typical iron-based and nickel-based structural alloys, with helium typically on the order of ~150 appm per dpa and hydrogen at approximately an order of magnitude greater. Since both of these gases are considered to exert a negative influence on structural properties of interest, their retention after both energetic recoil loss and subsequent diffusional loss is of strong programmatic interest.

Helium is essentially immobile at all temperatures of nuclear interest, but hydrogen has some limited temperature-dependent mobility, even at <100°C. To assess the degree of generation and retention, each gas was measured in a number of highly irradiated specimens of different alloy compositions and dpa levels. The results show that helium production is relatively insensitive to composition. The retained hydrogen levels, however, are somewhat sensitive to composition, reflecting different levels of diffusional loss, but are still at rather large concentrations.

The measured helium concentrations are larger than calculated using a calculational code optimized for prediction of neutron/proton ratios in the target tungsten source rods. The use of these gas measurements to provide benchmarks for determination of gas production cross sections for typical elements that comprise common structural alloys is also examined.

[1] Staff Scientist and Sr. Staff Scientist, respectively, Pacific Northwest National Laboratory, P.O. Box 999, Richland, WA, 99352, USA.
[2] Team Leader, Team Leader, Technical Staff and Technical Staff, respectively, Los Alamos National Laboratory, Los Alamos, NM, 87545, USA.

Keywords: structural alloys, proton irradiation, neutron irradiation, helium measurements, hydrogen measurements, Accelerator Production of Tritium (APT), hydrogen release, helium release

Introduction

The Accelerator Production of Tritium (APT) project [*1*] was proposed as one of several solutions to the U.S. national need for tritium. In the APT concept, high-energy protons would impinge on a tungsten target producing high-energy spallation neutrons. These neutrons would in turn be multiplied using a lead blanket, then thermalized in water. Tritium production would occur through capture of the thermalized neutrons by ^{3}He gas. An important technical issue that was addressed during the APT design was the potentially strong impact of radiation damage to structural materials resulting from the various mixed high-energy proton and neutron distributions expected in the APT facility environment.

A series of irradiations were conducted in the Los Alamos Neutron Science Center (LANSCE) as part of the experimental test program supporting the APT Program [*2*]. In this irradiation campaign, a variety of candidate structural alloys were placed in various particle spectra, ranging from 800 MeV protons, to mixed energy distributions of both protons and spallation neutrons, and finally to distributions consisting primarily of high energy neutrons. The irradiation temperatures of all specimens in the LANSCE test were 200°C or less, with most below 100°C, conditions expected to be typical of the APT facility.

At proton energies on the order of hundreds of MeV, exceptionally high levels of gas atoms are generated in all elemental constituents of typical structural alloys, with helium typically on the order of ~150 appm per dpa and hydrogen (i.e., protium) at approximately an order of magnitude greater [*3,4*]. Whereas the generation of these gases in typical fission and fusion neutron spectra are very sensitive to elemental composition, especially the nickel content [*5,6*], there is very little difference expected between the rates of gas generation in nickel, iron or chromium at very high proton energies [*4*].

In high energy proton spectra the hydrogen is born in two roughly equal distributions with very different birth energies, the first on the order of ~100 MeV in the internuclear cascade and the second at ~1 MeV in the subsequent nuclear evaporation event. For typical specimen dimensions, the large range of the very high energy protons produced in the internuclear cascade results in their near-total recoil loss from the irradiated volume, while a significant fraction of the evaporation protons are stopped in the specimen. Thus ~50% or less of the total generated hydrogen is expected to come to rest in the specimen. On the other hand, ~100% of the heavier and lower energy helium is retained in specimens of the sizes employed in this program (Actually, since all specimens in this experiment are covered by adjacent specimens and foil covers, fluxes across surfaces of adjacent pieces of metal should cancel, leading to no net loss).

Since both of these gases are considered to exert a negative influence on structural properties of interest, their retention after both recoil and diffusional losses is of strong interest. Helium is essentially immobile in structural metals at all temperatures of nuclear interest, but hydrogen has some limited temperature-dependent mobility. To assess the degree of retention after both energetic and diffusional losses, each gas was measured in a number of highly irradiated specimens of different alloy compositions and dpa levels.

Experimental Conditions

A variety of structural alloys from the APT materials characterization program were analyzed for helium and hydrogen content, including Alloy 718, 300 series stainless steels (304L and 316L), and 9Cr-1Mo. Other alloys based on aluminum will be measured later in the program. There were a variety of specimen types irradiated in this experiment, but the majority of gas measurements were conducted on standard 3 mm disks used for transmission electron microscopy or shear punch studies.

In this paper we will concentrate on specimens irradiated primarily in the proton beam, although these specimens are also simultaneously exposed to lower fluxes of high energy neutrons, with the relative fraction of neutrons increasing toward the edge of the proton beam, as shown in Figure 1. The 1 mA proton beam was roughly Gaussian in distribution, with a 2σ of ~3 cm. The proton energy spectrum varies across the beam. For example, in tube 21, the protons at beam center vary from ~600 MeV to 800 MeV, peaking at ~700 MeV, while at the beam edge, most of the protons exist at 500 to 800 MeV, peaking at ~650 MeV.

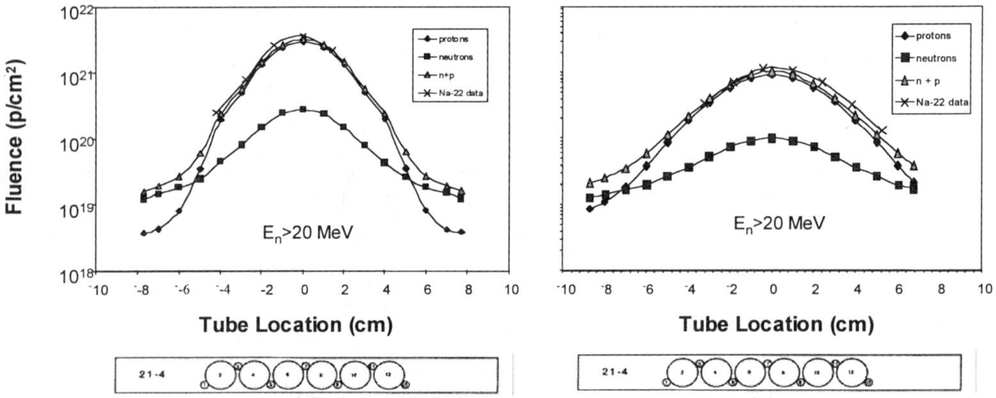

Figure 1 - *Calculated vs. measured particle fluences for representative specimen holders (numbers 1 and 21, left and right respectively) containing samples discussed in this paper. Note that measured profiles were determined using the Na-22 reaction in pure aluminum dosimeters included in the experiment.*

A diagram of the experimental setup in the LANSCE facility is shown in Figure 2. A typical sample holder for the TEM disks is shown in Figure 3. The specimens are mounted in thin foil-covered blades with metal-to-metal contact for optimum cooling, with high velocity water flowing between the various blades. Proton-induced heating leads to somewhat higher temperatures at the higher dpa rates, however. Therefore, there is some coupling between dose and temperature, with the calculated temperature of the samples measured in this study rising from about 30°C at the lower doses to about 65°C in the beam center. This may induce some bias in the hydrogen retention data, tending to reduce it somewhat at higher dpa levels relative to that at the beam edges.

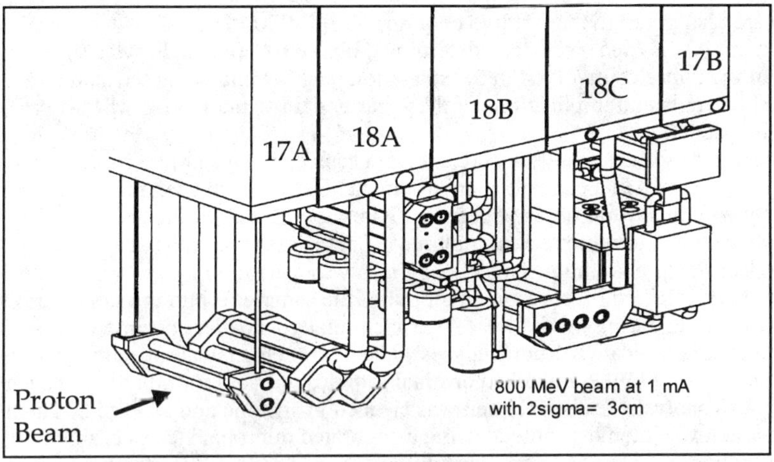

Figure 2 – *APT Mockup in the LANSCE Irradiation Facility. The specimens discussed in this paper were irradiated in assemblies 17A and 18C.*

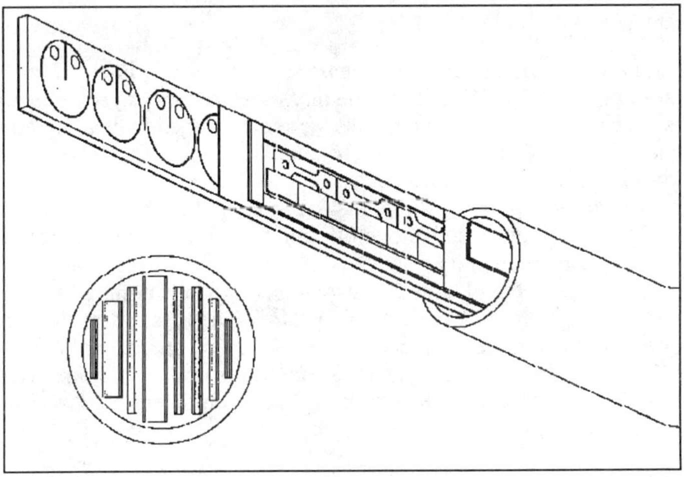

Figure 3 – *Typical specimen holder in the LANSCE irradiation series discussed in this paper. The various materials sit in specimen-shaped holes in 304 stainless steel plates, allowing metal-to-metal contact for good heat transfer. Each of these specimen holders was covered on both sides by thin foils of 304 stainless steel. The various blades were cooled by high velocity flowing water with the inlet and out water temperature measured with thermocouples, allowing calculations of specimen temperature.*

Details on the samples measured for gas content are given in Table 1. Calculated helium and hydrogen contents based on the LAHET code system [7] for each of the samples are also given in the last two columns of the table. The calculated atomic displacement levels (dpa) were also determined by the LAHET code system, averaging over all of the particles involved in the spallation reaction and using the standard NRT model [*8*]. Details and implications of these gas and dpa calculations will be explored later in the paper.

Specimens for gas analysis were cut from each original sample using small diagonal wire cutters in a controlled environment. Before each use, the cutters were cleaned by wiping several times with a dry "Kimwipe". Each of the helium analysis samples was etched to remove a minimum of ~0.013 mm (~0.5 mil) of surface material prior to specimen preparation. This etching step was performed to remove material that may have been affected by α-recoil either out of the sample or into the sample from adjacent materials during irradiation. After etching, two smaller specimens were cut from each sample for duplicate helium analyses. The hydrogen analysis specimens were cut in a similar manner from the un-etched original samples to avoid introduction of hydrogen.

Prior to analysis, each specimen was cleaned in acetone and air-dried. The mass of each specimen was then determined using a calibrated microbalance traceable to the National Institute of Standards and Technology (NIST). Mass uncertainty is estimated to be ±0.002 mg.

Helium Analysis

Helium Analysis System

Helium analyses were conducted by mass spectrometry at Pacific Northwest National Laboratory (PNNL). Details on the mass spectrometry system have been presented elsewhere [*9,10*]. Helium contents were determined by heating and/or vaporizing each sample in a resistance-heated crucible in one of the mass spectrometer system's high-temperature vacuum furnaces. The concentrations of the two helium isotopes were determined either by direct measurements of the mass spectrometer signal for helium-3 or helium-4, or by an isotope-dilution technique where the released helium is compared with a known quantity of added "spike" of the other isotope.

The helium spikes were obtained by expanding and partitioning known quantities of gas through a succession of calibrated volumes[*9*]. The mass spectrometer was calibrated for mass sensitivity during each series of runs by analyzing known mixtures of ^{3}He and ^{4}He. Reproducibility of the analysis system for samples with known homogeneous helium content is ~0.5%. Absolute accuracy has been determined in many previous studies to be generally better than 1%.

Helium Measurements

Helium analyses were conducted on a total of 13 samples from the APT materials tests. Results of the analyses are given in Table 2 and shown graphically in Figures 4 – 7. Helium results in Table 2 are listed as total atoms of ^{3}He and ^{4}He, and total helium concentration in atomic parts per million (10^{-6} atom fraction). Some of the helium measurements on the Alloy 718 material included stepped-anneal analyses conducted to determine the extent of gas removal at temperatures up to ~1200°C. These stepped-anneal analyses were followed by complete vaporization of the sample to completely remove the remaining helium. Because of eutectic formation of the 718 material with the analysis crucible, resulting in premature sample melting, significant levels of helium were removed prior to 1200°C. The helium concentrations in the last two columns represent

Table 1 – *APT sample summary.*

Sample	Material	Tube + No.[a]	Location[b] (mm)	Dose (dpa)	Calculated Gas Content (appm)	
					Helium	Hydrogen
IN66	Alloy 718	21-2-1	-31.2	1.45	87.2	787.9
IN25		21-2-22	-31.2	1.45	87.2	787.9
INE4		21-2-11	10.0	3.87	252.1	2270
IN53		21-2-31	8.0	4.15	272.3	2446
INO1		1-1-7	-12	9.92	757.4	6603
IN43		1-1-10	0	13.77	1097	9544
4121	304L	22-2-12	8.0	3.82	243.3	2151
4077		4-2-11	0.0	9.77	695.1	5967
6138	316L	24-2-22	-31.2	1.34	74.4	657.8
6053		24-2-10	8.0	3.92	237.7	2101
6040		24-2-29	4.1	4.07	249.0	2207
6100		4-2-33	2.0	10.3	730.7	6246
MDC1	9Cr-1Mo	4-2-41	40.0	1.1	55.8	475.8
MD67		4-2-13	3.8	9.55	698.9	5905

[a]Tube-envelope-ID#.
[b]Location from beam centerline.

the sum of the stepped-anneal and vaporization analyses. An example of the helium release profile in Alloy 718 is shown in Figure 4.

Hydrogen Measurements

Hydrogen Analysis System

Hydrogen analyses were conducted using a newly developed analysis system at PNNL. Details of the system have been presented elsewhere [*11*]. The system is based on a low-volume extraction furnace in combination with a quadrupole mass spectrometer, and has a detection limit of ~1 appm for steel. Samples for analysis are loaded into the sample holder carousel located above the extraction furnace. Sample analyses are conducted by dropping the individual specimens into the heated crucible sequentially. Hydrogen release, in terms of current output from the electron multiplier, is measured as a function of time. Total hydrogen released is determined from the integral of the hydrogen release curve and the measured system sensitivity.

Table 2 – *Measured helium in APT alloys.*

Sample	Material	Anal. Type[a]	Specimen Mass[b] (mg)	Measured Helium (10^{13} atoms)		Helium Concentration (appm)[c]			
				^{3}He	^{4}He	^{3}He	^{4}He	Total	Mean[d]
IN66	Alloy 718	V	0.738	1.76	11.17	17.1	147	164	172
		V	1.027	2.59	17.19	18.8	162	181	±12
IN25		A	3.164	5.99	51.40				
		V		-	0.25	18.4	159	177	
INE4		A	1.890	11.03	95.65				
		V		-	0.28	56.7	492	549	548
		A	1.356	8.11	68.38	58.0	490	548	±0
IN53		A	1.714	10.53	90.36				
		V		-	0.62	59.6	512	572	532
		V	1.074	5.731	48.62	51.8	440	492	±56
IN43		V	1.407	32.4	265.0	223	1829	2052	2060
		V	0.900	21.2	170.4	229	1838	2067	±11
4121	304L	V	1.320	90.9	704.8	63.8	494.4	558	
4077		V	1.095	263	1868	220	1565	1785	1813
		V	1.632	404	2869	227	1613	1840	±39
6138	316L	V	0.539	11.3	79.12	19.4	135.9	155	163
		V	0.710	16.4	114.1	21.4	148.8	170	±11
6053		V	0.271	14.4	110.9	49.2	378.9	428	420
		V	0.801	40.9	315.3	47.3	364.5	412	±12
6040		V	0.866	55.0	428.5	58.8	458.2	517	
6100		V	1.451	385	2746	246	1752	1998	1979
		V	1.287	329	2395	237	1723	1960	±27
MDC1	9Cr-1Mo	V	1.682	30.1	213.4	16.6	117.5	134	134
		V	1.551	27.6	195.7	16.5	116.8	133	±1
MD67		V	0.874	206	1462	218	1549	1767	1737
		V	1.181	266	1911	209	1498	1707	±42

[a]Stepped-anneal (A) or vaporization (V) analysis.
[b]Mass uncertainty is ±0.002 mg.
[c]Helium concentration in atomic parts per million (10^{-6} atom fraction) with respect to the total number of atoms in the specimen. Values for samples that were both stepped-anneal and vaporized is the total of the two analyses.
[d]Mean and standard deviation (1σ) of duplicate analyses.

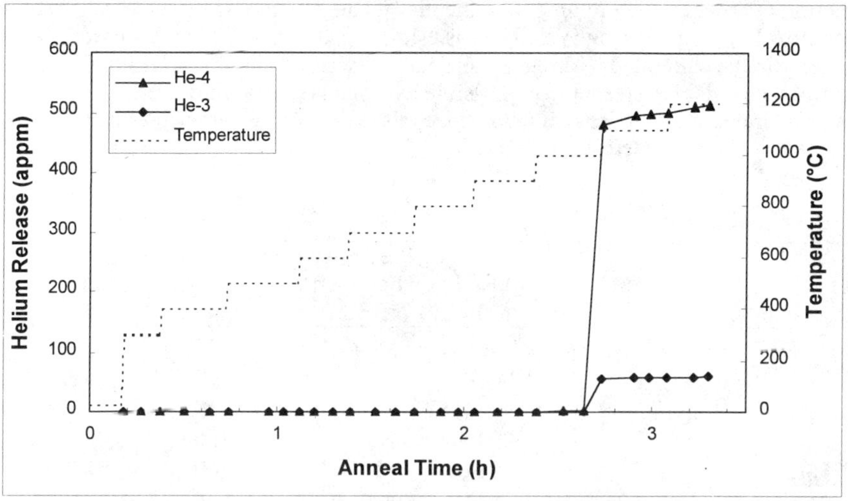

Figure 4 – *Typical stepped-anneal helium release in Alloy 718. Helium release at ~1000°C is attributed to sample melting cause by alloying of the material with the analysis crucible.*

Measurements of hydrogen release with temperature were also made on a few samples of Alloy 718 material. These measurements were made by ramping the crucible temperature in an approximately linear profile from about 250°C to 1200°C over a 400 second time period. The temperature profile was determined following the measurements using a thin-walled K-type thermocouple inserted from the upper sample loading area.

Calibration of the system was accomplished using a hydrogen leak source attached to the vacuum line between the extraction furnace and the detector volume. This calibrated leak has a very small trapped volume, resulting in virtually no lowering of the leak rate with time. Calibration measurements are conducted before and after each sample analysis, and typically show an overall standard deviation of ~2 to 3% for each analysis series. The system has been determined to be linear up to a total hydrogen release of at least 10^{17} atoms, which for a 0.5 mg steel sample, represents a hydrogen concentration of ~20 000 appm.

Measurements are also routinely conducted on specimens of a standard, hydrogen-containing steel maintained in the laboratory. The stated content of the steel is 5.2 ± 0.3 wppm. The average hydrogen content measured in more than 90 specimens ranging in mass from ~2 to ~8 mg is 5.0 wppm with standard deviation of ~20% (1σ). It is speculated that the variability observed in the standard samples is associated with actual heterogeneity in the hydrogen content at this small mass level.

Retained Hydrogen in APT Alloys

Hydrogen measurements were made on a total of 13 samples from the APT materials characterization tests, and on unirradiated control samples of the same materials. The results of the control analyses are given in Table 3; the retained hydrogen measured in the irradiated materials are shown in Table 4. The hydrogen results are shown graphically in Figures 8 – 10. The mean and standard deviation of the replicate measurements are given in the last two columns of the table. The data in the last two columns of the Table 4 represent the "net" hydrogen retained in the samples after subtraction of the measured hydrogen in the unirradiated materials in Table 3. Absolute uncertainty (1σ) in the hydrogen analyses is estimated at ~20%, and is due partly to the uncertainty in the calibrated hydrogen leak source discussed above. Additional uncertainty may also be present from possible hydrogen release from remaining water layers or hydrated metal oxides on the surface of the sample that are subsequently dissociated by the hot crucible.

Table 3 – *Hydrogen in unirradiated alloy material.*

Sample	Material	Mass (mg)[a]	Measured Hydrogen (10^{15} at.)	Hydrogen Concentration (appm)[b]	Hydrogen Concentration Mean[c]
In-blk	Alloy 718	1.931	3.45	174	190
		2.014	4.14	200	±20
304L-blk	304L	3.824	13.3	319	320
		4.940	17.2	320	±0
316L-blk	316L	3.014	11.7	360	448
		3.844	22.2	535	±124
9Cr-1Mo-blk	9Cr-1Mo	2.821	1.94	64	58
		4.089	2.30	52	±8

[a]Mass of specimen for analysis. Mass uncertainty is ±0.002 mg.
[b]Hydrogen concentration in atomic parts per million (10^{-6} atom fraction) with respect to the total number of atoms in the specimen.
[c]Mean and standard deviation (1σ) of duplicate analyses.

Discussion and Conclusions: Helium

Total helium (He-3 and He-4) concentrations measured in the APT alloys ranged from 134 appm to 2060 appm. As is evident in Figures 5-7, helium generation rates relative to dpa rates appear to be the highest for the iron-based alloys (300 series SS, and 9Cr-1Mo) at ~180 appm/dpa. The lines through the measured data in the figures are least square fits assuming linear behavior through the origin. Helium generation in nickel-base Alloy 718 appears to be slightly lower at ~150 appm/dpa. It is important to note, however, that the difference between nickel and iron-base may not be the dominant factor, since 718 also contains a rather high level of high atomic weight elements, unlike the iron-base alloys.

It is important to recognize that the measured helium concentrations are the result of generation rates averaged over a distribution of proton energies, distributions that vary across the proton beam, and also across a varying neutron contribution to gas production at each position and associated dpa level. In spite of this inherent bias in the data set, the concentrations are nearly linear with dpa and roughly independent of composition. This latter point is particularly demonstrated in Figure 11, where the measured helium in all the alloys is summarized on a single plot.

Table 4 – *Retained hydrogen in APT alloys.*

Sample	Material	Mass (mg)[a]	Measured Hydrogen (10^{15} at.)	Hydrogen Concentration (ppm)[b]		
				Measured	Corrected[c]	Mean[d]
IN66	Alloy 718	1.639	30.7	1820	1630	1870
		2.366	56.0	2300	2110	±340
IN25		0.587	10.6	1760	1570	1810
		0.509	9.41	1790	1600	±380
		0.683	17.2	2440	2250	
INE4		1.059	29.7	2720	2530	2600
		0.732	21.6	2860	2670	+100
IN01		0.342	19.7	5590	5400	5200
		0.693	37.9	5310	5120	±180
		0.539	29.2	5260	5070	
IN43		2.386	117	4750	4560	4710
		1.912	103	5250	5060	±300
		1.328	64.3	4700	4510	
4121	304L	1.215	17.3	1300	980	1050
		2.336	37.0	1450	1130	±110
4077		1.806	78.0	3960	3640	3560
		0.955	39.7	3810	3490	±110
6138	316L	1.126	10.7	882	434	459
		1.305	13.1	932	484	±35
6053		1.283	21.6	1560	1110	1460
		1.741	42.4	2260	1810	±490
6040		2.407	40.4	1550	1100	1100
		3.497	58.8	1560	1110	±10
6100		1.391	61.0	4060	3610	3460
		1.030	41.8	3760	3310	±210
MDC1	9Cr-1Mo	1.071	85.0	735	677	687
		1.580	12.9	754	696	±13
MD67		1.910	89.5	4340	4280	4260
		2.050	95.2	4300	4240	±30

[a]Mass of specimen for analysis. Mass uncertainty is ±0.002 mg.
[b]Hydrogen concentration in atomic parts per million (10^{-6} atom fraction) with respect to the total number of atoms in the specimen.
[c]Corrected for hydrogen in unirradiated controls (see Table 3).
[d]Mean and standard deviation (1σ) of duplicate analyses.

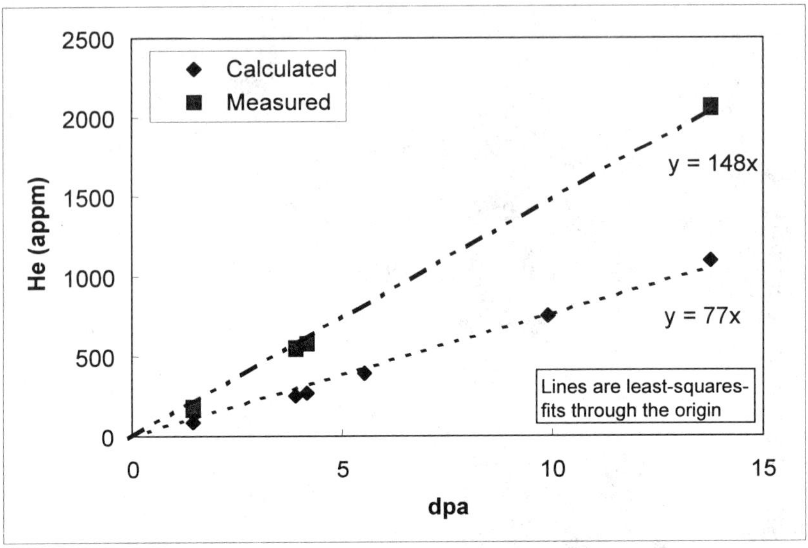

Figure 5 – *Measured helium in Alloy 718 vs. dpa.*

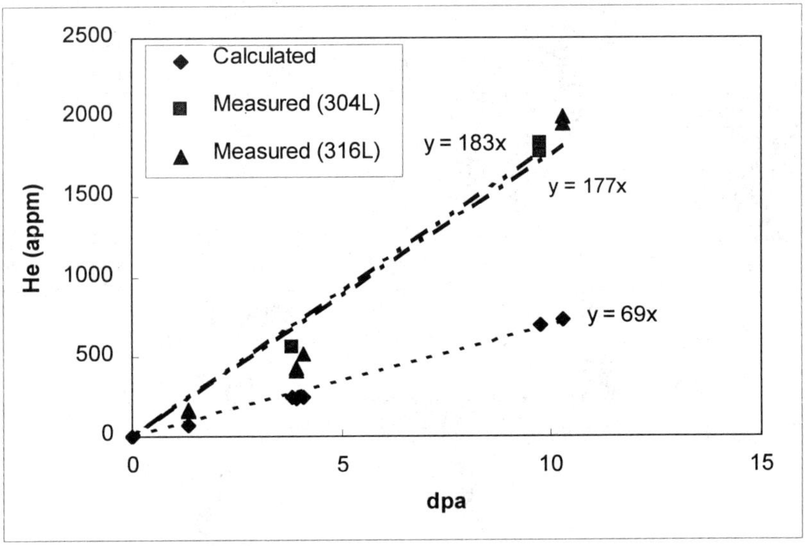

Figure 6 - *Measured helium in 300 stainless steel vs. dpa.*

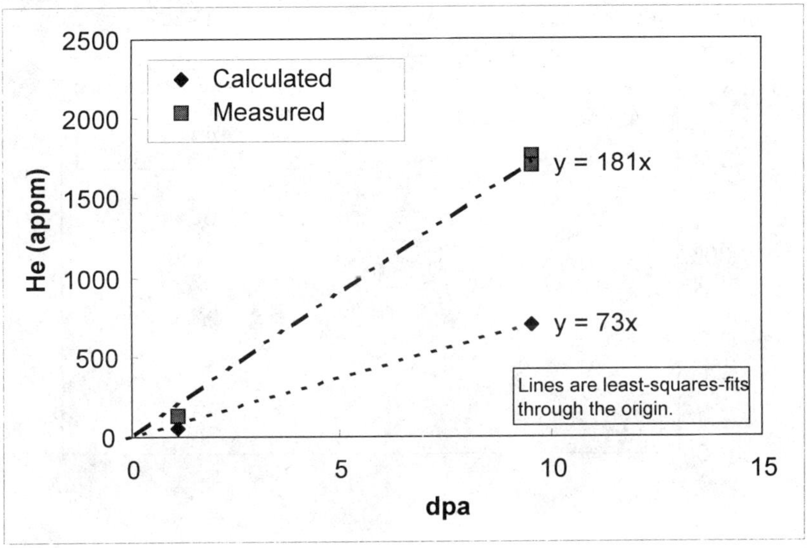

Figure 7 – *Measured helium in 9Cr-1Mo vs. dpa.*

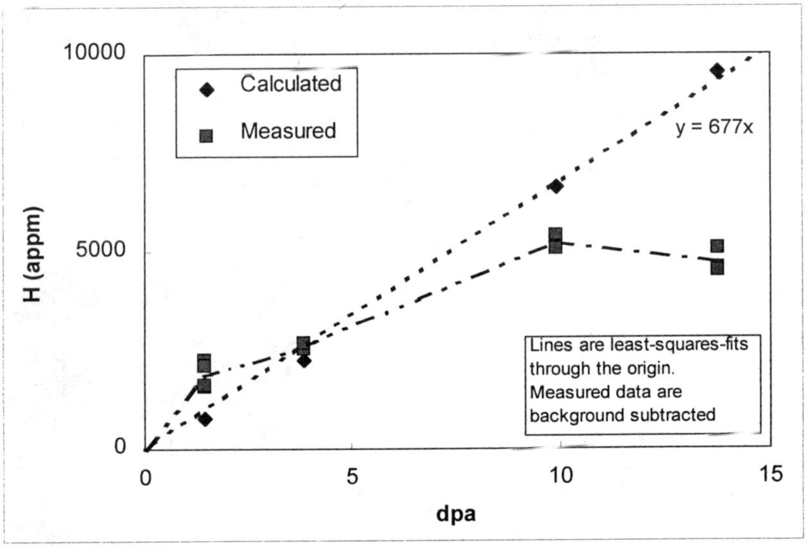

Figure 8 – *Retained hydrogen in Alloy 718 vs. dpa.*

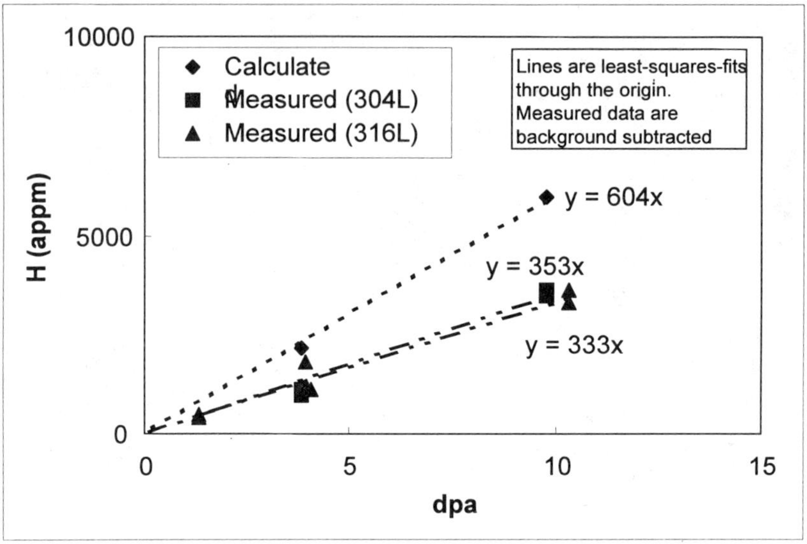

Figure 9 – *Retained hydrogen in 300 stainless steel vs. dpa.*

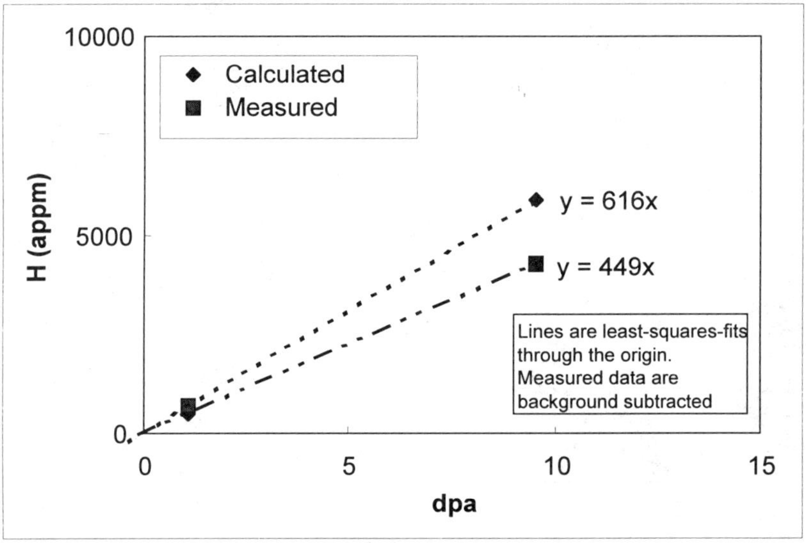

Figure 10 – *Retained hydrogen in 9Cr-1Mo vs. dpa.*

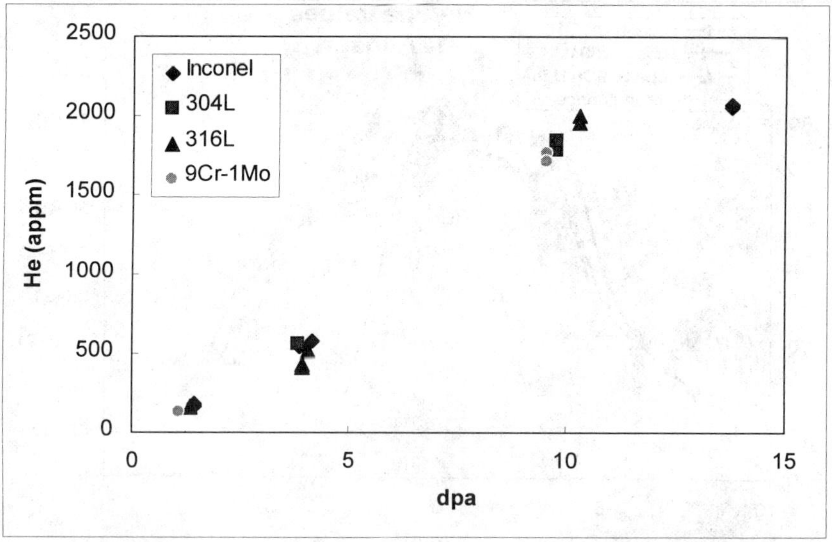

Figure 11 – *Summary plot of measured helium in APT alloys.*

The near-linearity of helium production vs. dpa in spite of the proton energy variation and changing neutron/proton contribution can be understood by noting that over the range of proton energies involved in this experiment, the helium and dpa cross sections exhibit nearly the same energy dependence [*4*]. Thus, the calculated He/dpa ratio arising from protons only remains relatively constant across the beam, as shown in Figure 12, demonstrating that even though the proton fluence falls strongly across the beam and the energy and particle spectra change with position, the He/dpa ratio remains relatively constant and independent of composition until the neutron contribution to the total fluence becomes relatively large near the edges of the beam.

As the relative neutron fluence becomes larger there are two consequences, both of which act to decrease the local He/dpa ratio. First, the spectrum-averaged helium cross section falls, and second, the relative neutron contribution to the dpa rate increases. For this reason, the measured helium levels at low dpa levels tend to lie slightly below the linear lines drawn in Figures 5-7.

In all cases, the measured helium levels are considerably higher than calculated, as shown in Figures 5-7. The calculated productions for both He and H were generated using the LAHET code system (LCS) [*7*], version 2.83. The Bertini intra-nuclear cascade model was used with the pre-equilibrium option turned on. The level density model used was that of Gilbert-Cameron-Cook-Ignatyuk (GCCI) [*12, 13*].

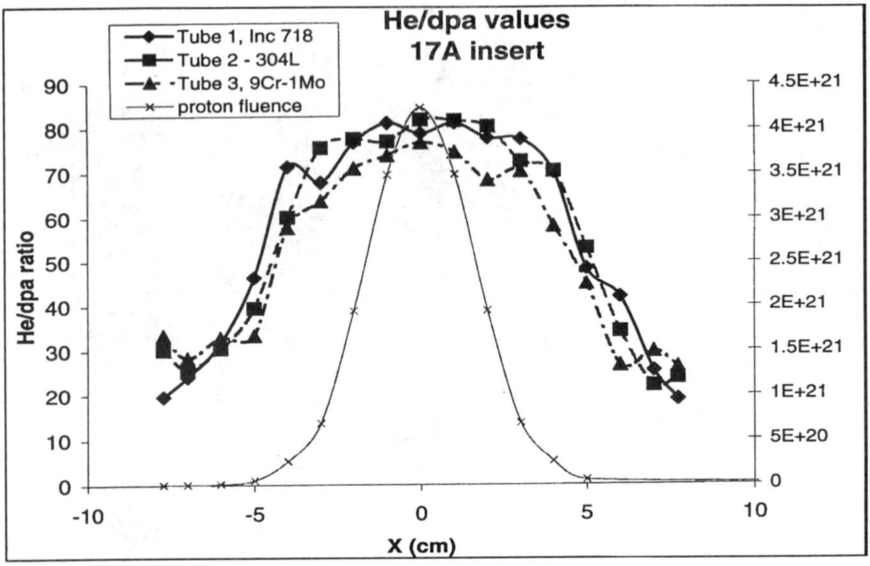

Figure 12 – *Calculated He/dpa ratios for several alloys irradiated in different specimen blades of insert 17A, showing that even though the proton fluence falls strongly and the energy and particle spectra change with position, the He/dpa ratio remains relatively constant until the neutron contribution to the total fluence becomes relatively large near the edges of the proton beam.*

The LCS is the neutronic tool used for the design of APT and the estimation of radiation damage parameters, such as dpa and gas production. The physics options employed in LAHET as the standard APT settings were chosen primarily to provide the proper n/p ratios for high-Z targets such as tungsten. Using these assumptions the accuracy in He production for mid-Z (~20-40) elements is known to be low by about a factor of 2 [*12*]. Interestingly, the current measurements are in relatively good agreement with that of earlier nearly monoenergetic 750 MeV proton irradiations by Green also at LANSCE in pure iron and pure nickel, with He/dpa ratios of 169 and 158 appm/dpa, respectively [*4*]. Unfortunately, it is not known what settings were employed in the then-current version of LCS. Green's results do support a slightly lower He/dpa ratio, however, in agreement with the lower helium generation rate observed in Alloy 718.

As expected, significant levels of ^{3}He were also measured in the various alloys. Very little ^{3}He is produced by neutrons and therefore protons produce most of the ^{3}He. Measured ^{4}He/^{3}He ratios varied from 7.0 to 8.6, and showed little variation with dpa level in the range of 1 to 14 dpa. The helium 4/3 ratio was the highest for Alloy 718, and appears to scale with the level of high-Z components in the material rather than with the major component Fe or Ni. Previous measurements in LANSCE by Green et al. for higher-Z pure materials irradiated with 750 MeV protons [*3*] showed higher helium 4/3 ratios (Fe, 8.66; Ni, 8.16; Cu, 9.44; Mo, 10.1;W,13.7; Au,11.4), indicating that high-Z materials should increase the 4/3 ratio.

Discussion and Conclusions: Hydrogen

The H/dpa ratio remains approximately independent of proton energy and spectral variations across the data set for the same reasons advanced for the He/dpa ratio earlier, as shown in Figure 13. The H/He ratio is also known not to change strongly as a function of proton energy in the range relevant to this experiment [*3*].

Retained hydrogen in the APT alloys ranged from 459 appm to 5200 appm. The data in Figures 8 - 10 are the net retained hydrogen after subtraction of the measured hydrogen in the unirradiated materials. As with helium, the retained hydrogen in the 300 series SS and in the 9Cr-1Mo scaled approximately linearly with dpa, but at a higher rate of ~600 appm/dpa. Remember, however, that the inherent bias in temperature with dpa rate may lower this retention rate per dpa somewhat relative to that in a fully isothermal experiment.

Alloy 718, however, showed a marked nonlinearity in the retained hydrogen, with three trend lines shown in Figure 8 with progressively lower slopes, a relatively higher rate in the first dose interval from 0 up to ~1.5 dpa, a second lower rate from ~1.5 to ~10 dpa, and a third, possibly negative rate above ~10 dpa.

The complex hydrogen behavior observed in Alloy 718 can perhaps be attributed to a rather complex microstructural evolution in the material compared to that of the other alloys. According to Sencer et al. [*14*], the radiation-induced microstructural evolution in the three iron-base alloys was rather simple, involving only the nucleation and growth of small faulted Frank loops and smaller "black spots". Therefore, these microstructural defects probably serve as the only available trapping sites for hydrogen, with the possible exception of He/H complexes in the matrix. (Sencer did not observe any cavities down to a resolution limit of ~1 nm.) In 718, however, Sencer observed that the starting microstructure was composed of a high density of very small, ordered γ' and γ'' precipitates, whose surfaces could serve as trapping sites along with the developing faulted loops and black spots. With increasing damage levels, however, the precipitates were first disordered and then their constituents progressively mixed back into solution. (Again, no resolvable cavities were observed, even though a total gas content approaching 0.7 at% was reached at the highest dpa level). Such a scenario might explain the initially high and then progressively lower hydrogen storage rate with increasing dpa level observed in Alloy 718.

Calculated total hydrogen generation values with dpa, ignoring any diffusional losses, are also shown in Figures 8-10. Contrary to the behavior observed in the helium, the calculated values generally overpredict the measured retained hydrogen content. The possible exception is the initial slope observed in the Alloy 718 up to ~1.5 dpa, where the highest trapping efficiency was postulated. It should be noted that the calculated hydrogen generation is relatively insensitive to alloy composition, so only microstructural differences in diffusion can account for the lower and divergent hydrogen levels. Given the uncertainties in the calculated energetic losses and lack of knowledge of diffusional losses, this over-prediction is not unexpected.

The most significant observation from the hydrogen calculations is that the differences with measured values is not very large if diffusional losses of up to 50% are assumed for the temperatures involved in this experiment. Since proton production in LAHET is a primary channel, one would expect better ability to predict hydrogen generation compared to that of helium, which is considered to be a secondary channel with more inherent calculational uncertainty. Although no attempt has been made here to account for diffusional losses, estimates of the average hydrogen diffusion length for the current irradiation conditions (i.e., ~30 to 100°C for 180 days) would suggest that, in the absence of defect trapping, even at the lower temperatures, most of the generated hydrogen should have been lost. Therefore, trapping of the hydrogen, possibly at helium sites, plays a significant role in the hydrogen retention.

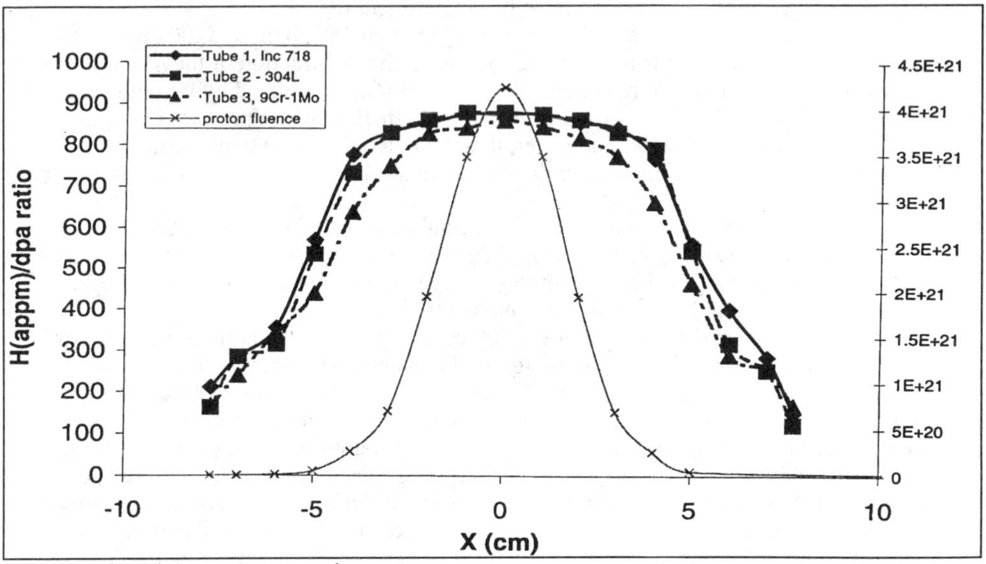

Figure 13 - *Calculated H/dpa ratios for several alloys irradiated in different specimen blades of insert 17A, showing that even though the proton fluence falls strongly and the energy and particle spectra change with position, the H/dpa ratio remains relatively constant until the neutron contribution to the total particle fluence becomes relatively large near the edges of the proton beam.*

Overall Conclusions

In summary, within the respective uncertainties of the gas measurements and the LCS calculations, the following observations can be made.

- Measured He/dpa ratios produced in iron-based and nickel-based alloys measured after irradiation in mixed and varying proton/neutron spectra are quite consistent with earlier studies employing nearly monoenergetic 750 MeV protons on pure iron and nickel.
- As anticipated, the total helium production rate per dpa is not very sensitive to alloy composition. There is some compositional dependence of the helium 4/3 ratio, however, and this difference may provide an opportunity to use these data to adjust and optimize LAHET input parameters for a wide range of atomic weight elements.
- Both helium and hydrogen generation rates are slightly nonlinear with dpa in this experiment due to changing balance of neutron and proton fluxes across the

elements based on this experiment should suffice to address these more diverse spectral environments.

- Essentially all helium generated in the specimens should be retained, but the amount measured is roughly twice that calculated, indicating that calculated helium values for Fe, Cr, Ni require an upward adjustment. This will require that the LCS parameters for tungsten source rods and those for lower-Z materials be maintained separately.
- Retained hydrogen is generally less than predicted but this result is not unexpected, since thermally driven diffusional losses were not included in the calculations. The "relatively" small difference observed between calculated and measured hydrogen levels actually lends confidence that the parameters used to calculate hydrogen in mid-Z elements does not require as large an adjustment as required for helium.
- While the calculated hydrogen production is relatively insensitive to alloy composition, the retained hydrogen is somewhat sensitive. The most complex hydrogen behavior was observed in Alloy 718, which undergoes a rather complex microstructural evolution compared to that of the other alloys. In iron-base alloys with less complicated microstructural evolution, there is much less variability in behavior, both with respect to the accumulated exposure and also the composition.
- Based on the results of this study, it will be possible to confidently predict the gas generation levels in the highest dose and therefore most vulnerable components of the APT facility. The relative energy independence of gas generation rate per dpa allows these 500-800 MeV data to be easily extrapolated to the >1 GeV levels anticipated in the APT.

Acknowledgements

This work was supported by the U.S. Department of Energy under the Accelerator Production of Tritium Program at Los Alamos National Laboratory. Pacific Northwest National Laboratory is operated for USDOE by Battelle Memorial Institute.

References

[1] APT ^{3}He Target/Blanket Topical Report, Los Alamos National Laboratory Report LA-CP-94-27, Rev. 1, Los Alamos National Laboratory, Los Alamos, NM, 1994.

[2] Maloy, S. A. and Sommer, W. F., "Spallation Source Materials Test Program", *Proceedings of the Topical Meeting on Nuclear Applications of Accelerator Technology*, Albuquerque, NM, November 16-20, 1997, pp. 58-61.

[3] Wechsler, M. S. et al., "Calculations of Radiation Damage to Materials at the SNS Facility", 2nd International Topical Meeting on Nuclear Applications of Accelerator Technology (AccApp '98), Gatlinburg, Tennessee, September 20-23, 1998.

[4] Green, S. L., Green, W. V., Hegedus, F. H., Victoria, M., Sommer, W. F., and Oliver, B. M., "Production of Helium by Medium Energy (600 and 800 MeV) Protons," *Journal of Nuclear Materials*, Vols. 155-157, 1988, pp. 1350-1353.

[5] Garner, F. A, Greenwood, L. R. and Oliver B. M., "A Reevaluation of Helium/dpa and Hydrogen/dpa Ratios for Fast Reactor and Thermal Reactor Data Used in Fission-Fusion Correlations," Effects of Radiation on Materials: 18th International Symposium, ASTM STP 1325, R. K. Nanstad, M. L. Hamilton, F. A. Garner and A. S. Kumar, Eds., American Society for Testing and Materials, West Conshohocken, PA,1999, pp. 794-807.

[6] Greenwood, L. R. and Garner, F. A., "Hydrogen Generation Arising from the ^{59}Ni(n,p) Reaction and its Impact on Fission-Fusion Correlations," *Journal of Nuclear Materials*, Vols. 233-237, 1996, pp. 1530-1534.

[7] Prael, R. E. and Lichtenstein, H., "User Guide to LCS: The LAHET Code System," Los Alamos National Laboratory Report LA-UR-89-3014, Los Alamos National Laboratory, Los Alamos, NM, September 1989.

[8] Norgett, J. J., Robinson, J. T., Torrens, I. M., *Nuclear Engineering Design*, Vol. 33, p. 50, 1975.

[9] Farrar, H. and Oliver, B. M., "A Mass Spectrometer System to Determine Very Low Levels of Helium in Small Solid and Liquid Samples," *Journal of Vacuum Science and Technology - A*, Vol. 4, 1986, pp. 1740-1741.

[10] Oliver, B. M., Bradley, J. G., and Farrar, H. IV, "Helium Concentration in the Earth's Lower Atmosphere," *Geochimica et Cosmochimica Acta*, Vol. 48, 1984, pp. 1759-1767.

[11] Oliver, B. M., Garner, F. A., Greenwood, L. R., and Abrefah, J. A., "Quadrupole Mass Spectrometer System for the Determination of Low to High Levels of Hydrogen in Irradiated Materials," Proceeding of the 9th Int. Conf. on Fusion Reactor Materials, Colorado Springs, CO, October 10-15, 1999, *Journal of Nuclear Materials*, Vols. 283-287, 2000, pp. 1006-1010.

[12] Gilbert, A. and Cameron, A. G. W., "A Composite Nuclear-Level Density Formula With Shell Corrections," *Canadian Journal of Physics*, Vol. 43, 1965, pp. 1446-1496.

[13] Ignatyuk, A. V., Smirenkin, G. N., and Tishin, A. S., "Phenomenological Description of the Energy Dependence of the Level Density Parameter," *Soviet Journal of Nuclear Physics*, Vol. 21, 1975, pp. 255-257.

[14] Sencer, B. H., Bond, G. M., Garner, F. A., Maloy, S. A., Sommer, W. F., and James, M. R., "Microstructural Alteration of Structural Alloys by Low Temperature Irradiation with High Energy Protons and Spallation Neutrons," Effects of Radiation on Materials, 20th International Symposium, ASTM STP 1405, S. T. Rosinski, M. L. Grossbeck, T. R. Allen, and A. S. Kumar, Eds., American Society for Testing and Materials, West Conshohocken, PA, 2002.

Scott Lillard[1], Frank Gac[2], Mike Paciotti[2], Phil Ferguson[3], Gordon Willcutt[4], Greg Chandler[5], and Luke Daemen[3]

The Influence of High Energy Proton Irradiation on the Corrosion of Materials

Reference: Lillard, R. S., Gac, F., Paciotti, M., Ferguson, P., Willcutt, G., Chandler, G., and Daemen, L. L., **"The Influence of High Energy Proton Irradiation on the Corrosion of Materials,"** *Effects of Radiation on Materials: 20th International Symposium, ASTM STP 1405,* S. T. Rosinski, M. L. Grossbeck, T. R. Allen, and A. S. Kumar, Eds., American Society for Testing and Materials, West Conshohocken, PA, 2001.

Abstract: This paper presents a summary of our current efforts to characterize the real-time corrosion rates of materials during 800 MeV proton radiation at currents up to 0.4 mA. Specially designed corrosion probes, which incorporated ceramic seals, were mounted in flow tubes on a water manifold that allowed samples to be directly exposed to the proton beam at the Los Alamos Neutron Science Center. The water system that supplied the manifold provided a means for controlling water chemistry, measuring dissolved hydrogen concentration, and measuring the effects of water radiolysis and water quality on corrosion rate. Real-time corrosion rates during proton irradiation increased with average proton beam current. In addition, for any given material type, a trend in corrosion rate with probe location relative to the beam centerline was observed. These results are discussed within the context of particle type, particle flux, and energy deposition.

Keywords: corrosion, proton irradiation, electrochemical impedance spectroscopy, particle flux, energy deposition, Alloy 718, type 316L stainless steel, aluminum alloy 6061, photons, neutrons, protons, EIS

Spallation neutron source target/blanket cooling loops present a unique combination of parameters that may effect the corrosion rates of metals, including high energy protons, high and low energy neutrons, gamma radiation, and water radiolysis products. In previous publications, Lillard and Butt have described a novel method for measuring the corrosion rate of materials in spallation neutron sources *[1, 2]*. Lillard measured the real-time corrosion rate of Alloy 718 with electrochemical impedance spectroscopy (EIS) during irradiation in an 800 MeV proton beam at the Los Alamos Neutron Science Center (LANSCE). In that work, the in-beam Alloy 718 samples were tested in a water system that provided a means for controlling water chemistry, measuring dissolved hydrogen

[1] Materials Corrosion and Environmental Effects Lab, Materials Science and Technology Division, MST-6, Los Alamos National Laboratory, Los Alamos, NM 87545.
[2] APT-TPO, Los Alamos National Laboratory, Los Alamos, NM 87545.
[3] LANSCE-12, Los Alamos National Laboratory, Los Alamos, NM 87545.
[4] TSA-10, Los Alamos National Laboratory, Los Alamos, NM 87545.
[5] Westinghouse Savannah River Company, Aiken, SC 29808.

concentration, and measuring the effects of water radiolysis and water quality on corrosion rate. It was shown that the real-time polarization resistance of Alloy 718 decreased from $1.7x10^3$ ohm·m^2 in the absence of irradiation to 8.2 ohm·m^2 during proton irradiation at an average beam current of 0.4 mA*[3]*. Because the proton beam spot size was smaller than the sample, two methods for determing corrosion rate from polarization resistance were employed. The first method assumed that the distribution of corrosion was uniform across the entire probe surface. With this assumption, the corrosion rate for Alloy 718 was found to vary with average proton beam current from 0.041 μm/yr. at an average current of 0.001 mA to 3.1 μm/yr. at an average current of 0.40 mA. The second method used proton flux as a criterion for determining the area of highest damage. In the proton flux distribution model of corrosion, for an average beam current of 0.4 mA the corrosion rate for Alloy 718 on beam centerline (peak proton flux of $1.77x10^{18}$ p/m^2·s) was determined to be 60.9 μm/yr. The foundation for this flux model was based on thickness measurements from tungsten rods irradiated in a separate cooling water loop at LANSCE. In that work Sommer *et al.* found that the post irradiation thickness profiles of W rods, irradiated at a beam current of 1.0 mA for approximately two months, were Gaussian and corresponded to the Gaussian profile of the beam*[4]*. From these observations of W corrosion during proton irradiation a Gaussian distribution model for corrosion damage appeared to be justified.

Given the short irradiation time in the Lillard and Butt study (several months) and the low corrosion rates of Alloy 718, physical verification of the proton flux model was not possible (via post-irradiation thickness measurements of the probes). To verify the flux model, one would like to measure the real-time polarization resistance of a material as a function of radial distance from the proton beam centerline for each average beam current. The corrosion rate at any given position should be proportional to the Gaussian distribution proton current

$$\phi_o \exp\left\{ {}^{-r^2}\!/_{2\sigma^2} \right\} \tag{1}$$

where r is the radial distance from the proton beam center, σ is the Gaussian distribution of the flux and ϕ_o is given by

$$\phi_o = {}^{I_t}\!/_{2\pi\sigma^2} \tag{2}$$

and I_t is the beam current*[5]*.

In this paper, real-time corrosion rates as a function of average beam current and radial position from the proton beam centerline were measured with EIS for several engineering alloys proposed for spallation neutron target/blanket cooling loops. The probes used in this work were similar in design to those in previous studies. The corrosion data were analyzed within the context of particle flux (proton, neutron, and photon), as well as particle energy deposition (thermal hydraulics).

Experimental

The In-beam Corrosion Loop

All experiments were conducted at the A6 Target Station of LANSCE. A detailed description as well as diagrams of the beam line at A6, the corrosion water loop, and

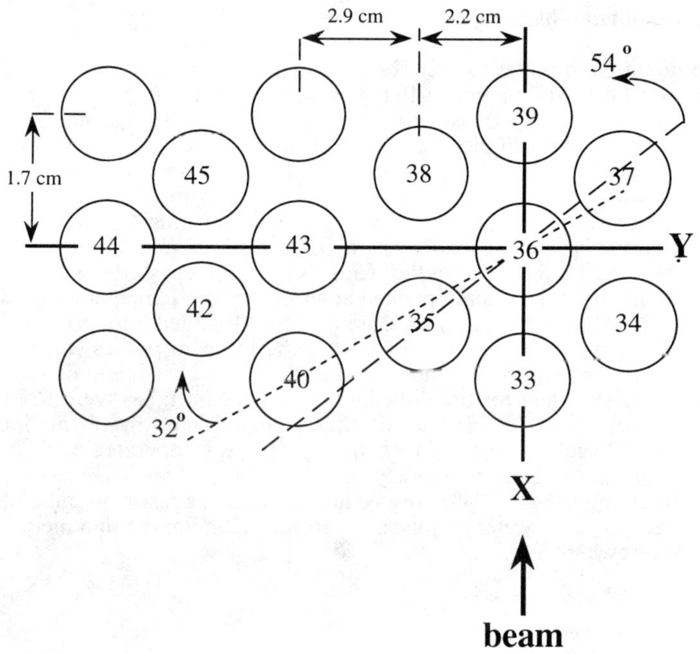

Figure 1 - *A perspective of the tube array irradiated during the FY '99 irradiation at the LANSCE A6 target station from the front, top-down. The path of the proton beam was parallel to the X axis at Y=0. Here we have defined the center of tube #36 as (0,0). Each numbered tube contained a corrosion probe. Alloy 718: 33, 39, and 45; SS 316L: 35, 36, 38, and 40; Al6061: 42, 43, and 44.*

probe design have been presented elsewhere*[2]*. Briefly, the corrosion water loop consisted of a water pumping system that supplied a manifold, which held the corrosion samples in-beam and provided water cooling of the samples. With the exception of the individual corrosion samples, the system was fabricated entirely of type 304 stainless steel (SS). To measure real-time in-beam corrosion rates, it was necessary to isolate electrically the corrosion samples from the stainless steel plumbing system. Conventionally, this can be accomplished with metal-to-glass seals. To overcome the problems associated with metal-to-glass seals in a proton environment, a corrosion probe that employed ceramic to metal seals was employed*[1-3]*. The only significant design difference as compared to previous investigations was the configuration of the water manifold. In this investigation the water manifold (Fig. 1) consisted of fifteen, 15 cm long by 1.73 cm inside diameter flow tubes arranged in close-packed arrays. Each tube contained either an in-beam corrosion probe that was 15.9 cm in length by 1.27 cm in outside diameter or a flow restrictor (unnumbered tubes in Fig. 1). With respect to water flow, tubes 33-39 were in parallel with one another and in series with tubes 40-45 (which were also in parallel with on another). The manifold was welded to the bottom of a 3.4 m supporting insert that not only supported the weight of the manifold but also provided the necessary conduits for electrical and water connections. Thermocouples attached to the front of the manifold verified the position, size, and shape of the proton beam.

Sample Preparation and Water Quality

To provide a fresh metal surface for electrochemical characterization, all samples were ground to 400 p SiC paper. After grinding, the samples were degreased in an ultrasonic bath of acetone. Degreasing was followed by successive sonications in ethanol and de-ionized water. Prior to placing the probes in the water system, the interior of the water system, which included all piping, tanks, and pumps, was rinsed with a mixture of water and ethanol. The water system was then filled with approximately 230 L of deionized (DI) water, operated for several hours, and then flushed. This DI rinse was repeated three times before the final system fill with DI water. The water resistivity varied between 1 $\times 10^6$ ohm·cm (initial) and 8×10^4 ohm·cm (after several weeks of operation). Nominally, the system operated at an inlet water temperature between 18° C (beam off) and 24° C (0.340 mA), a pressure of 1.02 MPa, and a total flow rate of 0.91 L/s. This resulted in a flow rate of 0.13 L/s for the in-beam probes (tubes 33-39) and 0.11 L/s for the near-beam probes (tubes 40-45), and a water velocity of 1.21 m/s. The resulting Reynolds numbers for the in-beam and near-beam tubes were 6214 and 5434 respectively (calculated at 25° C). In an attempt to mitigate the formation of water radiolysis products such as H_2O_2 *[6-9]*, the system was operated with a dissolved hydrogen concentration of approximately 0.40 mole/m^3. This was accomplished by continuously bubbling 6% H_2 - 94% Ar gas into the system's reservoir tank. A thorough discussion of the effects of water radiolysis on corrosion and mitigation methodology has been presented elsewhere*[10]*.

Proton Beam Characteristics

The flux of the incident proton beam had a Gaussian distribution of $2\sigma \approx 3.5$ cm. The energy of this particle beam was 800 MeV. The pulsed beam was characterized by a gate length (macropulse), a macropulse repetition rate, and a fixed peak current (Table 1). Average proton beam currents were controlled by varying the gate length as well as the repetition rate. Nominally, the average proton beam currents were varied between 0.01 and 0.4 mA. Therefore, the irradiation source was a complex pulsed beam and each of its components may play a disproportionate role in the measured corrosion rate. Thus, in extrapolating these results, the beam characteristics must be considered.

Table 1 *Proton beam parameters for corrosion data.*

Average Current (mA)	Peak Current (mA)	Repetition Rate (Hz)	Gate Length (µs)	Duty Cycle (milli-cycles)
0.010	16	3	200	0.600
0.036	16	10	200	2.00
0.10	16	10	625	6.25
0.34	16	36	625	21.6

Electrochemical Measurements

EIS*[11-14]* was used to measure the polarization resistance of each sample as a function of average beam current and irradiation time. To maximize the signal-to-noise ratio, measurements were conducted with a 30 mV peak-to-peak sinusoidal voltage perturbation[6] over the frequency range of 0.005 - 1000 Hz. No applied dc potential was employed; that is, all measurements were conducted at the open circuit potential (OCP). To eliminate the effects of ground loops, a floating ground EIS system was used. In these measurements, the traditional three electrode set-up was used. Here the 304 SS water system acted as the counter electrode. Because a traditional reference electrode was not capable of withstanding the proton / neutron flux at the manifold, one of the in-beam corrosion probes was used as a reference.

All electrochemical measurements were conducted while the proton beam was on.

Radiation Transport Calculations

In an attempt to correlate corrosion rate directly with particle flux, radiation transport calculations, including energy deposition and particle flux calculations, were completed using MCNPX. MCNPX*[15, 16]* is the result of the merger of the Monte Carlo transport codes LAHET*[17]* and MCNP*[18]*. MCNPX™ is a general particle transport code that spans a broad range of particle energies, from sub-MeV to multi-GeV. Charged particles, light ions, and high-energy neutrons use the physics modules of LAHET™ to simulate particle transport and nuclear interactions. As with MCNP™, transport and nuclear interactions of photons and low-energy neutrons use evaluated nuclear data. MCNPX can be used to model arbitrarily complex geometries and multiple material compositions. Uncertainties in MCNPX energy deposition calculations at 800 MeV for in-beam and near-beam samples are generally expected to be within 20% as shown by Corzine, *et al.[19, 20]*, and by Beard and Belyakov-Bodin *[21]*.

Particle flux tallies were completed using the standard flux averaged over the volume of a cell or the new mesh tally in MCNPX for neutral particles. For charged particles, fluxes were calculated on the surface of a cell. Uncertainties in the particle fluxes are more difficult to quantify. Although the secondary particle source is probably well predicted as evidenced by the work of Morgan, et al.*[22]*, there may be errors in the prediction of the angular distribution leading to increasing uncertainty in the calculated particle fluxes as a function of distance from the proton beam centerline. As a guide, an uncertainty of approximately 20% can be assumed for the particle fluxes with larger errors possible in the outer regions. In general, the relative comparisons of the calculations (one calculation vs. another) are more accurate than the absolute comparison of a single calculation.

In addition to the surrounding A6 experiment region, the Monte Carlo model included each probe, the surrounding water layer, and the outer SS 304 tube explicitly. The corrosion probes were assumed to be hollow cylinders as opposed to modeling the rounded ends. Energy deposition calculations were performed along the probe in 1 cm long cells to provide input for temperature calculations as a function of position to allow an analysis of the axial heat flow. In addition, detailed mesh tallies (~3 mm grid spacing) were completed for multiple sections of the geometry with energy deposition separated by particle type. Particle flux tallies were also performed as detailed mesh tallies, as well as averaged over the entire length of the probe. The probe length averages are believed to correspond better with the measurements because the measurements include effects along the length of the probe.

Particle flux and the resulting energy deposition was also used to calculate probe surface temperature (peak and average). For each location, the probe and tube surface

---

[6] Although a 30 mV perturbation is somewhat higher than that typically used in EIS measurements, for metals undergoing passive dissolution the effect of an applied anodic voltage is to thicken the passive film. However this increase is small. Therefore, no appreciable effect of the 30 mV signal sample is anticipated.

™ MCNPX, MCNP, and LAHET are registered trademarks of the University of California.

temperatures were determined using the local water temperature and material power densities from the radiation transport calculations described above. The probe water side surface temperature was determined by adding the film temperature drop to the bulk water temperature, where the film temperature drop was determined by dividing the heat flux by the heat transfer coefficient. The heat flux was caused by the power deposited in the probe wall with the inside of the probe being adiabatic. The surface temperature for the inner side of the probe was determined by adding the temperature increase through the probe wall to the outer surface temperature, where the temperature increase through the wall was obtained from a conduction solution with internal heat generation in the hollow cylindrical wall. A similar calculation is done for the tube wall, except the heat flux was removed from the inner surface and the outer surface was adiabatic. These calculations use one-dimensional assumptions in the radial direction. A separate two-dimensional calculation was done to determine the impact of axial conduction along the probe, and it was shown to be very small. A separate three-dimensional model was used to calculate the temperatures of the thermocouples attached on the outside of the flow tubes. An approximation to this model was then incorporated in the spreadsheet model to calculate the thermocouple temperatures along with the probe and tube wall surface temperatures.

Results

Typical EIS data from a SS 316L in-beam corrosion probe are presented in Fig. 2 in the form of Bode magnitude and phase plots (tube 35, 0.036 mA avg. beam current). The magnitude of the impedance has been normalized for area by multiplying |Z| by the total sample area (approximately 6.34×10^{-3} m^2). To obtain polarization resistance (R_{pol}) from the data, complex non-ljnear least squares (CNLS) fitting of the data was employed. To model the data several electrical equivalent circuits (EC) were used. For SS 316L and Al6061 the electrical equivalent circuit presented in Fig. 3 was used (Warburg model*[23, 24]*) while for Alloy 718 a simplified Randles circuit was used*[3]*. In the Warburg model R_{pol} is the polarization resistance, C_{dl} is the capacitance associated with the double layer, Z_W is a diffusion (Warburg) impedance, and R_{sol} is the geometric solution resistance between the sample and the water system. A CNLS fit of the in-beam data to this EC is also presented in Fig. 2. As seen in this figure, good agreement between the model and the data exists.

From R_{pol} the corrosion rate, CR in µm/yr., was determined from the well known relationship*[25]*

$$CR = \frac{327.6(26/R_{pol})(EW)}{\rho} \tag{3}$$

where ρ is the density of the material in kg/m^3, the units conversion constant 327.6 has units of (kg·µm)/(m·mA·yr.), and EW is the equivalent weight (dimensionless). The constant 26 has units of mV and assumes Tafel slopes of 0.12 V/decade. The error

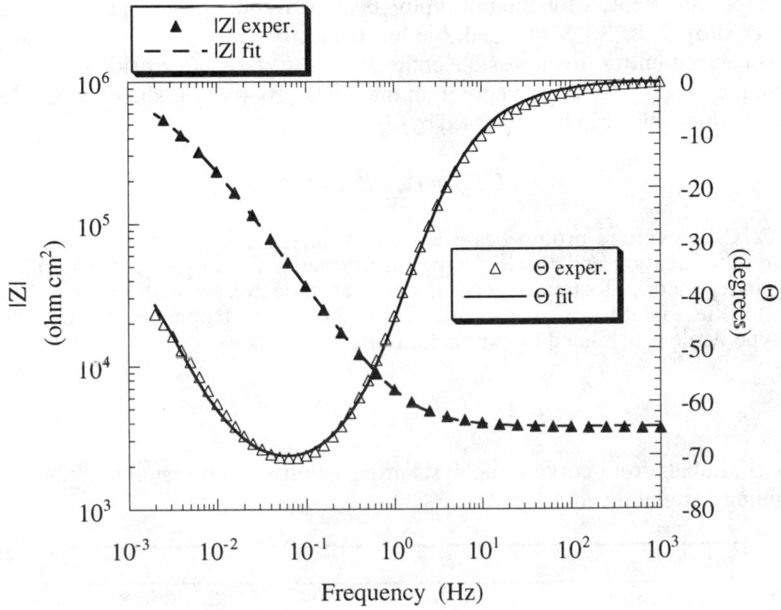

Figure 2 - *Bode magnitude and phase plots for SS 316L in tube 35 during proton irradiation at 0.036 mA. The CNLS fit of the data to the Warburg model in Fig. 3 is also presented. For clarity, not all experimental data are shown.*

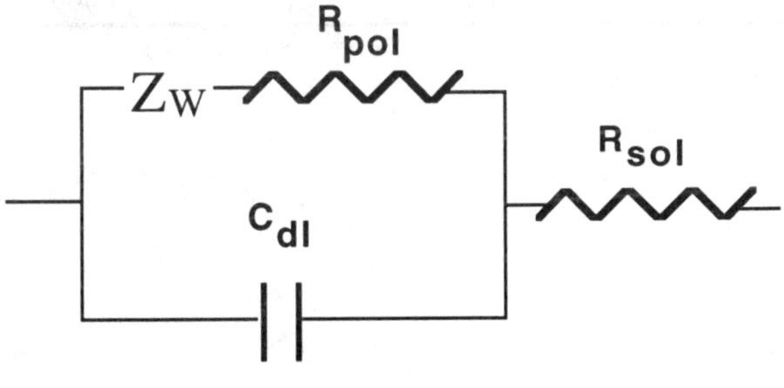

Figure 3 - *Warburg equivalent circuit used to model the SS 316L EIS data. Where:* R_{pol} *is the polarization resistance,* R_{sol} *is the geometric solution resistance,* C_{dl} *is the double layer capacitance, and* Z_W *is a Warburg type impedance (diffusion impedance).*

associated with this assumption is small and has been addressed elsewhere*[13]*. A plot of uniform corrosion rate as a function of proton beam current is presented in Figs. 4a, 4b, and 4c for Alloy 718, SS 316L, and Al6061 respectively. These plots assume that corrosion occurred uniformly across the entire sample surface. Corrosion rate increased as a function of average beam current for all materials. By plotting the data on a log-log scale these relationships can be expressed by*[3]*

$$log(CR) = \alpha + \beta log(BC) \tag{4}$$

where BC is the average proton beam current in mA and α and β are constants and depend on tube location and material type respectively. For any single material β was relatively independent of sample location where as α decreased with distance from the beam centerline for any given material (Table 2). Thus β appears to be related to material type while α is related to sample location.

Table 2 - Variables from curve fit analysis of Equation 4 to corrosion rate data and the corresponding correlation coefficient (r).

tube #	α	β	r
SS 316L: 35	0.99	0.97	0.998
36	1.2	0.95	0.998
38	1.1	0.92	0.999
40	0.48	0.76	0.987
Al6061: 42	1.1	1.4	0.999
43	0.98	1.4	0.989
44	0.35	1.2	0.979
718: 33	0.95	1.0	0.994
39	1.1	1.0	0.992

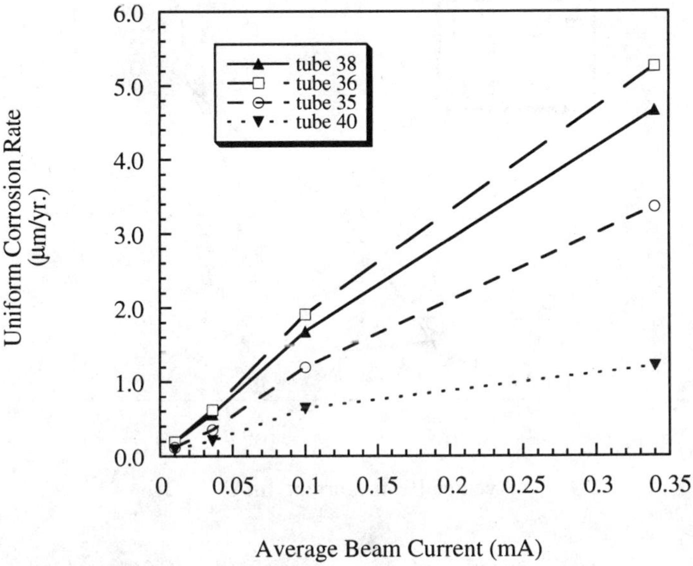

Figure 4a - *Corrosion rate as a function of average beam current for the SS 316L probes.*

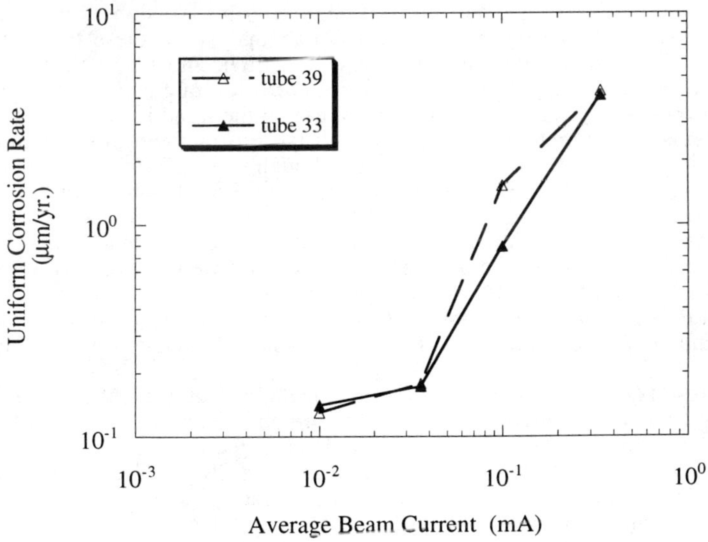

Figure 4b - *Corrosion rate as a function of average beam current for the Alloy 718 probes at the beam centerline.*

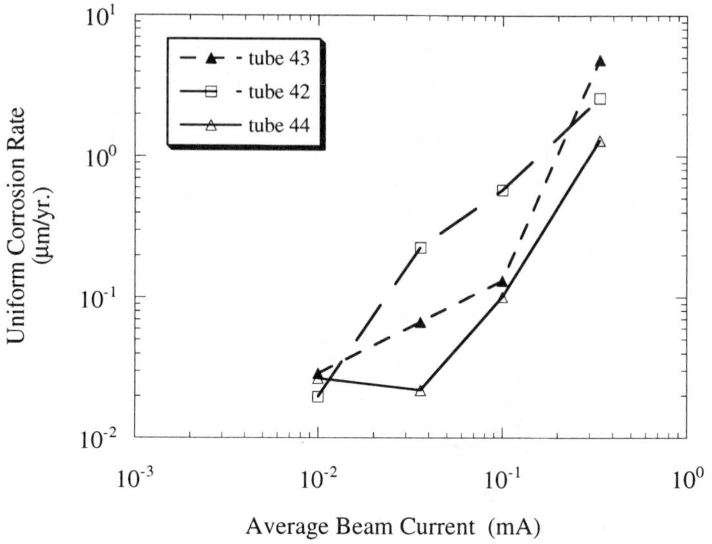

Figure 4c - *Corrosion rate as a function of beam current for the Al6061 probes.*

Discussion

Although the plots in Figs. 4a to 4c assume uniform corrosion across the entire sample surface, this assumption may not be valid as analysis of these figures reveals several trends in corrosion rate with probe location. The most notable trends with position were observed in the SS 316L data where corrosion rate increased in the order (by tube number) 40 < 35 < 38 < 36. It is apparent that the trend in corrosion rate with sample location is not a simple function of radial distance from the beam as predicted by Eq. 1 and previously modeled*[3]*. For example, although tubes 35 and 38 are equidistant from the beam centerline (tube 36) the corrosion rate of SS 316L sample in tube 38 was consistently greater than the corrosion rate of the sample in tube 35. Several potential explanations for the observed differences in corrosion rate between sample location include changes in the passive film and charging at the double layer (the solution/metal interface) owing to interactions with ionizing radiation, water radiolysis, and temperature. To evaluate these mechanisms, a thorough analysis of particle flux from radiation transport calculations, and energy deposition (as it relates to sample temperature) was necessary.

Sample results from the radiation transport calculations are presented in Table 3 for the SS 316L specimens. A direct correlation between both neutron and photon flux with corrosion rate was observed. For neutrons and photons, the flux increased in the order (by tube number) 40 < 35 < 38 < 36. In comparison, the SS 316 L corrosion rate also increased in the order 40 < 35 < 38 < 36. However, as seen in Table 3 proton flux did not predict the relationship between corrosion rate and sample location. Other differences between protons and neutrons and photons also exist. For example, the ratio of the maximum to minimum proton flux was 21 (tube 36 : tube 40), while the maximum to

Table 3 - *Results from radiation transport calculations for tubes containing SS 316L samples. Flux was averaged over the tube length and summed for all particle energies.*

tube #	proton flux $p/m^2 \cdot s \cdot mA$	neutron flux $n/m^2 \cdot p$	photon flux $g/m^2 \cdot p$
35	72.7	38.0	35.2
36	127	52.7	46.2
38	70.9	43.8	37.3
40	6.0	13.5	13.2

minimum photon flux was 3.5. The ratio of the maximum to minimum corrosion rate for SS 316L was 4.3 (tube 36 : tube 40; 0.34 mA beam current). Currently, we are examining these relationships between particle flux and corrosion rate for all sample materials and probe locations. In addition, we are also evaluating energy deposition by particle type both in the sample and the water annulus surrounding the sample as potential explanations for the observed changes in corrosion rate with location and beam current.

Sample results from the thermal hydraulic calculations are presented in Table 4 for SS 316L. Similar trends in measured and calculated TC readings were observed, however, the calculated temperatures were consistently higher than the measured values. This difference likely owed to assumptions made in the modeling of the TC/tube braze. The model assumed that the braze was uniform along the TC shield to the tip of the TC. In practice, the brazes extended beyond the tip and gaps existed between the shield and tube where the braze did not wet. Calculated peak probe water-side surface temperatures were as high as 112° C for the probe on beam centerline for the maximum beam current. At the minimum beam current (0.01 mA) the temperature of this probe was close to water temperature. The measured and calculated temperature trends increased in the same order as proton flux 40 < 38 < 35 < 36. This is not surprising as the relative probe temperatures reflect the differences in total energy deposition which is dominated by proton flux. However, the trend in temperature did not correlate with the measured trends in corrosion rate for SS 316L. In addition, for passive metals the increase in corrosion rate as a function of temperature below 100°C is small (closed systems). For example, for Alloy 718, it has been shown that a 50°C increase in sample temperature produced a 33% increase in corrosion rate*[3]*. Although the change in corrosion rate due to an increase in temperature is small relative to the changes in corrosion rate measured during proton irradiation, the measured corrosion rate must be considered as the sum of the increase due to changes in materials properties (oxide film and double layer changes) as well as temperature.

Table 4 - *Thermocouple (TC) and probe temperatures calculated at a beam current 0.342 mA (SS 316L only, water-side, peak temp). Actual TC temperatures measured at a beam current of 0.34 mA are also presented.*

tube #	TC, Measured °C	TC, Calculated °C	Calculated Probe Temperature, °C
35	48.1	64.2	81.7
36	66.9	94.6	112.2
38	53.4	63.9	79.2
40	33.8	30.8	28.9

Conclusions

Corrosion rate increased with average proton beam current for all materials examined consistent with previous studies on Alloy 718 during proton irradiation. Fitting of the data to a log - log relationship gave a slope that was related to material type and intercept that was proportional to sample location relative to the beam centerline. Transport calculations revealed that the flux of neutrons and/or photons appears to correlate with corrosion rate. Proton flux data from these results did not predict the observed trend in corrosion rate as a function of sample location. Further, the ratio of maximum to minimum proton flux was much larger than the ratio of maximum to minimum corrosion rate. Probe surface temperature calculations based on total energy deposition did not predict the observed trend in corrosion rate as a function of sample location. However, it was concluded that the observed trends in corrosion rate with beam current had a component that was related to sample temperature.

Acknowledgment

Work on this project was performed by the University of California under the auspices of the United States Department of Energy contract W7405-ENG36. The authors would like to thank Laurie Waters and the continued support of the APT Project Office; Walt Sommer, Stuart Maloy, and Mike James for helpful discussions; Richard Werbeck, Michael Baumgartner, and the rest of the LANSCE-7 group for their engineering expertise; R.G. Stone and R.E. Loehman (Sandia National Lab) for manufacturing the corrosion probes. In memory of Russ Kidman (LANL).

References

[1] Lillard, R. S. and Butt, D. P., *Journal of Materials*, vol. 50, 1998, pp. 56.

[2] Lillard, R. S. and Butt, D. P., *Materials Characterization*, vol. 43, 1999, pp. 135.

[3] Lillard, R. S., Willcutt, G. J., Pile, D. L., and Butt, D. P., *Journal of Nuclear Materials*, vol. 277, 2000, pp. 250.

[4] Sommer, W., Maloy, S., and Zaslawsky, M., "Proton Beam Effects on Tungsten Rods, Surface Cooled by Water," presented at 2nd International Workshop on Spallation Materials Technology, Ancona, Italy, 1997, pp. 215.

[5] Sommer, W. F., "APT Materials Safety Experiments Technical Report," Los Alamos National Laboratory, Los Alamos LAUR-93-2850, August 1993.

[6] Indig, M. E. and Weber, J. E., "Mitigation of Stress Corrosion Cracking in an Operating BWR," in Corrosion/83, NACE, Houston, TX, 1983, paper no. 124.

[7] Fox, M., "Water Chemistry and Corrosion in BWRs," in Corrosion/83, NACE, Houston, TX, 1983, paper no. 123.

[8] Fox, M. J., "A Review of Boiling Water Reactor Chemistry: Science, Technology, and Performance," Argonne National Laboratory for the US Nuclear Regulatory Commission NUREG/CR-5115 ANL-88-42, February 1989.

[9] Lin, C. C., Cowan, R. L., and Pathania, R. S., "Evaluation of ECP Measurements in BWR's During HWC Tests," in Corrosion/93, NACE, Houston, TX, 1993, paper no. 619.

[10] Lillard, R. S., Pile, D. L., and Butt, D. P., *Journal of Nuclear Materials*, vol. 278, 2000, pp. 277.

[11] MacDonald, D. D., "Electrochemical Impedance Techniques in Corrosion Science," in *Electrochemical Corrosion Testing, ASTM STP 727*, F. Mansfeld and U. Bertocci, Eds. ASTM, Philadelphia, 1981, pp. 110.

[12] MacDonald, J. R., *Impedance Spectroscopy*. Wiley Publishing, New York, 1987.

[13] Mansfeld, F., "Polarization Measurements," in *Electrochemical Techniques*, R. Baboian, Ed. NACE, Houston, 1986, pp. 67.

[14] Sluyters-Renbach, M. and Sluyters, J. H., *Electroanalytical Chemistry*, vol. 4, no. 1, 1970, pp. 1.

[15] Hughes, H. G., "The MCNPTM/LAHET™ Merger Project," in *Proceedings of the Topical Meeting on Nuclear Applications of Accelerator Technology*, Albuquerque, NM, November 16-20, 1997, pp. 213.

[16] Hughes, H. G., "Recent Developments in MCNPX™," in *Second International Topical Meeting on Nuclear Applications of Accelerator Technology*, Gatlinburg, Tennessee, September 20-23, 1998, pp. 281.

[17] Prael, R. E. and Lichtenstein, H., "User Guide to LCS: The LAHET Code System," Los Alamos National Laboratory LA-UR-89-3014, September, 1989.

[18] Briesmeister, J. F., "MCNP - A General Monte Carlo N-Particle Transport Code," Los Alamos National Laboratory LA-12625-M, Version B, March, 1997.

[19] Corzine, R. K., "MCNPX Benchmark of In-beam Proton Energy Deposition," Los Alamos National Laboratory, Los Alamos, New Mexico LAUR-00-0138, January 2000.

[20] Corzine, R. K., "MCNPX Benchmark of Out-of-Beam Energy Deposition in LiAl," Los Alamos National Laboratory, Los Alamos, New Mexico LAUR-00-0711, February 2000.

[21] Beard, C. A. and Belyakov-Bodin, V. I., *NS&E*, vol. 199, 1995, pp. 87-96.

[22] Morgan, G., "LANL Sunnyside Experiment: Study of Neutron Production in Accelerator-Driven Targets," in *Proceedings of the International Conference on Accelerator-Driven Transmutation Technologies and Applications*, vol. 346. AIP Conference Proceedings, 1994, pp. 682-689.

[23] Armstrong, R. D., Bell, M. F., and Metcalfe, A. A., *Electrochemistry - Specialist Periodical Reports*, vol. 6, 1978, pp. 98.

[24] MacDonald, D. D., "Theoretical Analysis of Electrochemical Impedance," in Corrosion/87, NACE, Houston, TX, 1987, paper no. 479.

[25] Fontana, M. G., *Corrosion Engineering*. McGraw-Hill, New York, 1986.

S.A. Maloy,[1] M.R. James,[1] G.J. Willcutt,[1] W.F. Sommer,[1] W.R. Johnson,[2] M.R. Louthan Jr.,[3] M.L. Hamilton,[4] and F.A. Garner[4]

The Effect of High Energy Protons and Neutrons on the Tensile Properties of Materials Selected for the Target and Blanket Components in the Accelerator Production of Tritium Project

---

Reference: Maloy, S. A., James, M. R., Willcutt, G. J., Sommer, W. F., Johnson, W. R., Louthan, M. R., Hamilton, M. L., Garner, F. A., **"The Effect of High Energy Protons and Neutrons on the Tensile Properties of Materials Selected for the Target and Blanket Components in the Accelerator Production of Tritium Project,"** *Effects of Radiation on Materials: 20th International Symposium, ASTM STP 1405,* S. T. Rosinski, M. L. Grossbeck, T. R. Allen, and A. S. Kumar, Eds., American Society for Testing and Materials, West Conshohocken, PA, 2001.

Abstract: The Accelerator Production of Tritium (APT) project proposes to use a 1.0 GeV, 100 mA proton beam to produce neutrons via spallation reactions in a tungsten target. The neutrons are multiplied and moderated in a lead/aluminum/water blanket and then captured in ^{3}He to form tritium. The materials in the target and blanket region are exposed to protons and neutrons with energies into the GeV range. The effect of irradiation on the tensile properties of candidate APT materials, 316L and 304L stainless steel (annealed), modified (Mod) 9Cr-1Mo steel, and Alloy 718 (precipitation hardened), was measured on tensile specimens irradiated in the Los Alamos Neutron Science Center accelerator, which operates at an energy of 800 MeV and a current of 1 mA. The irradiation temperatures ranged from 50-164°C, prototypic of those expected in the APT target/blanket. The maximum achieved proton fluence was 4.5 x 10^{21} p/cm^2 for the materials in the center of the beam. This maximum exposure translates to a dpa of 12 and the generation of 10,000 appm H and 1,000 appm He for the Type 304L stainless steel tensile specimens.

Specimens were tested at the irradiation temperature of 50-164°C. Less than 1 dpa of exposure reduced the uniform elongation of Alloy 718 (precipitation hardened) and mod 9Cr-1Mo to less than 2%. Approximately 4 dpa of exposure was required to reduce the uniform elongation of the austenitic stainless steels (304L and 316L) to less than 2%. The yield stress of the austenitic steels increased to more than twice its non-irradiated value after less than 1 dpa. These results are discussed and compared with results of similar materials irradiated in fission reactor environments.

Keywords: tensile testing, radiation effects, Accelerator Production of Tritium (APT), stainless steel, alloy 718, Mod 9Cr-1Mo

---

[1] Los Alamos National Laboratory, MS H809, Los Alamos, NM 87545.
[2] General Atomics, LANL, MS H809, Los Alamos, NM 87545.
[3] Westinghouse Savannah River Company, Bldg. 773-41A, Aiken, SC 29808.
[4] Pacific Northwest National Laboratory, Richland, WA 99352.

Introduction

The U.S. Department of Energy is investigating the use of an accelerator to produce tritium for national defense purposes. The Accelerator Production of Tritium, APT, project design uses a 1.0 GeV, 100 mA proton beam to produce neutrons through the spallation of a tungsten target. The spallation neutrons are moderated and multiplied in a surrounding lead/aluminum/water blanket and captured by ^{3}He gas to produce tritium. Therefore, the structural materials in the APT target and blanket systems are exposed to a high energy proton and spallation neutron flux at energies up to 1.0 GeV [1]. This exposure to such high energy particles causes displacement damage and the production and implantation of large quantities of hydrogen and helium in structural materials. The energy of this proton/neutron flux is high compared to the particle energies in a fission reactor. Thus the amounts of hydrogen and helium produced per dpa of exposure in the APT will be significantly greater than the amount of gas produced in a typical fission reactor. The temperature of the APT target/blanket, 50-160°C, is also significantly lower than most fission reactor environments except for mixed spectrum fission reactors such as the High Flux Isotope Reactor (HFIR) and the Advanced Test Reactor in the US and HFR in Europe. The combination of high particle energies, enhanced gas production and low operating temperatures may make irradiation induced degradation of the APT target/blanket materials differ significantly from the degradation previously experienced in fission reactors.

The effect of high energy proton and neutron irradiation on the mechanical properties of APT target and blanket structural materials was determined in an APT project which exposed mechanical test specimens to the 800 MeV, 1mA Gaussian proton beam (where 2σ=3cm) at the Los Alamos Neutron Science Center (LANSCE) [2]. Specimens were held in and near the proton beam in stainless steel envelopes to allow them to be irradiated without direct exposure to the cooling water. Because of the small size and high intensity of the Gaussian beam, specimens had to be small and thin to obtain a uniform flux and achieve prototypic APT temperatures (50-160°C).

Various mechanical properties including tensile, fracture toughness, etc., were measured on specimens after irradiation. This paper summarizes the effect of high energy proton and neutron irradiation on the tensile properties of Alloy 718, 316L and 304L and Mod 9Cr-1Mo. Each of these alloys is a candidate structural material for the APT target and blanket.

Experimental

The S-1 tensile specimens, Figure 1, were electro-discharge machined from two thicknesses of sheet material, 0.25 and 0.75-1.0 mm. Variations in specimen thickness coupled with the thickness effects on energy implantation and heat transfer, provided a technique to irradiate materials to the same flux at different temperatures. The composition of the heats used to make the two thicknesses of material that were irradiated are shown in Table 1. The Mod 9Cr-1Mo specimens were machined from the same heat of material regardless of specimen thicknesses. The as-machined Alloy 718 and Mod 9Cr-1Mo specimens were heat treated by first separately wrapping the specimens of each material in Nb foil packets and placing them in a quartz tube with a titanium ampoule which was backfilled with argon and sealed. The Alloy 718 specimens were annealed at

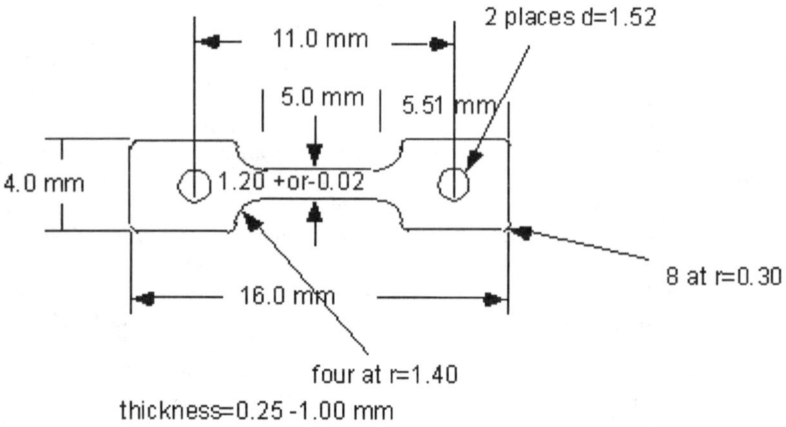

Figure 1. *Schematic showing the dimensions of the S-1 tensile specimen used for measuring the tensile properties of materials after irradiation in a spallation environment.*

1065°C for 30 min. and air cooled; they were then aged at 760°C for 10 hrs., furnace cooled to 650°C and held for a 20 hour total furnace time and finally air cooled. The Mod 9Cr-1Mo specimens were normalized at 1038°C for 1 hr. and air cooled followed by a temper at 760°C for 1 hr. and air cooling. The 304L and 316L materials were received in the annealed form, and no further heat treatment was performed. The Vickers hardness at a 500 g load and grain size measured for each heat of material, Table 2, suggest that the heat-to-heat variation in properties should be minimal. The largest difference in hardness observed was between the two different heats of 304L stainless steel (174 kg/mm^2 at 0.75 mm thick and 159 kg/mm^2 at 0.25 mm thick).

The tensile specimens were irradiated for six months in the LANSCE accelerator. After irradiation, the dose for each specimen was determined from analysis of pure metal activation foils placed next to specimens during irradiation. Details of the dose determination can be found in Reference [3]. The fluence was first calculated using the Los Alamos High Energy Transport (LAHET) Code System [4], [5]. Then, the gamma spectra measured from the activation foils were used with the STAYSL2 code to adjust the calculated fluences. A mathematical fit of the position dependence of fluence was used to determine the fluence at each specimen position. The temperature for each specimen was determined by interpolation of measurements from thermocouples placed near each specimen. The details of this temperature measurement can be found in Reference[6]. Because of beam heating, the calculated irradiation temperature varied from 50-164°C depending on specimen thickness and the location of the specimen with respect to center of the 800 MeV, 1mA proton beam.

The irradiated specimens were tensile tested in air using an Instron machine equipped with a high temperature furnace. Except where noted, materials were tested at the irradiation temperature. Specimens were tested at an initial strain rate of 10^{-4}/s. Load and displacement were measured and converted to stress and strain. The stress/strain curve for each specimen was corrected for machine compliance. The compliance-corrected stress/strain curves were used to determine 0.2% offset yield stress, ultimate

Table 1- *Elemental Composition of Materials used in APT Irradiation in wt. %*

Material	Lot	Al	C	Cr	Cu	Fe	Mn	Mo	Ni	P	S	Si	Ti	Others
718 0.25 mm	E618	.48	.04	18.13	0.08	Bal	0.13	3.06	53.58	0.008	0.001	0.11	1.03	Nb + Ta-4.98
718 0.75 mm	L426	.54	.05	18.13	0.05	Bal	0.21	3.01	52.7	0.005	0.002	0.13	1.06	Nb + Ta-5.07; Co-0.4; B-0.004
316L 0.25 mm	E835		.019	17.26	0.26	Bal	1.75	2.57	12.16	0.022	0.006	0.65		
316L 0.75 mm	L406		.022	16.05		Bal	1.82	2.08	10.11	0.022	.0002	0.48		
304L 0.25 mm	K953		.020	18.23	0.38	Bal	1.77	0.33	9.68	0.026	0.002	0.54		
304L 0.75 mm	L404		.060	18.19	0.4	Bal	1.86	0.34	8.14	0.030	.0003	0.48		
Mod 9Cr-1Mo	10148	.002	.089	9.24	0.08	Bal	0.47	0.96	0.16	0.021	0.006	0.28	0.002	V-0.21; Nb-0.054; Co-0.019; N-0.035; O-0.008

Table 2-*Hardness and Grain Size of Materials used in APT Irradiation*

Material	Thickness (mm)	Micro-hardness (kg/mm^2)	Grain Size (microns)
Alloy 718 Precipitation Hardened	1.0	458	30
	0.25	465	47
Mod 9Cr-1Mo	0.75	283	21
	0.25	278	19
304L	0.75	174	18
	0.25	159	15
316L	0.75	149	21
	0.25	153	16

Table 3- *Reduction of Area measurement on selected tensile specimens of Alloy 718, 316L and 304L Stainless Steel*

Alloy	Reduced area (mm²)	Original dimensions (mm) t	w	Original area (mm²)	RA (%)	Dose (dpa)	UE (%)
718	0.64	1.03	1.20	1.24	48.3	0	16.9
	0.89	1.04	1.20	1.26	29.4	4.0	1.2
	0.16	0.244	1.20	0.29	45.1	11	0.7
	0.82	1.04	1.20	1.26	34.7	1.0	1.4
304L	0.11	0.851	1.18	1.01	89.1	0	57.2
	0.10	0.304	1.18	0.36	72.2	4.0	0.7
	0.22	0.859	1.18	1.02	78.3	2.6	22.7
	0.10	0.296	1.18	0.35	71.5	0.8	30.2
316L	0.04	0.243	1.18	0.29	86.1	0	48.1
	0.04	0.250	1.19	0.30	86.6	0	48.9
	0.08	0.249	1.19	0.30	72.9	1.2	20.1
	0.18	0.878	1.19	1.04	82.7	4.1	0.4
	0.11	0.253	1.19	0.30	63.6	9.2	0.2

stress, uniform elongation and total elongation. Fracture surfaces of selected tensile specimens were examined using a Scanning Electron Microscope (SEM). These SEM micrographs were used to determine reduction of area by measuring the fracture surface area in relation to the area of the gage before testing.

Results

Tensile Testing

The tensile properties of Alloy 718 in the precipitation hardened condition, 316L and 304L in the annealed condition and Mod 9Cr-1Mo in the tempered condition were determined after irradiation to a maximum dose of 12 dpa. Representative stress/strain curves of irradiated Alloy 718, Figure 2, show that the work hardening rate is essentially zero after only 0.09 dpa of exposure and does not recover as dose increases. The yield strength, Figure 3, increases with dose up to 1 dpa and then gradually decreases out to 12 dpa. On the other hand, the uniform elongation drops almost immediately to less than 2% after 0.5 dpa and remains low for irradiation to 12 dpa.

Representative stress/strain curves for 316L/304L stainless steel specimens, Figure 4, irradiated and tested at 50°C, show irradiation-induced increases in yield strength, losses in ductility and reductions in work hardening capacity. Test data, for all the 304L/316L stainless steel specimens, Figure 5, show a sharp irradiation-induced increase in yield strength. The irradiation-induced changes in uniform elongation are more complex, Figure 5. The non-irradiated uniform elongation ranges from 40-60% and sharp decreases in uniform elongation are observed after 0.01 dpa to less than 20% leading to a more gradual decrease out to 3-4 dpa where the uniform elongation decreases rapidly to less than 1%.

The effect of test temperature on the irradiated tensile properties of 316L/304L is shown in Figure 6. Specimens irradiated to the same dose between 1 and 4 dpa were

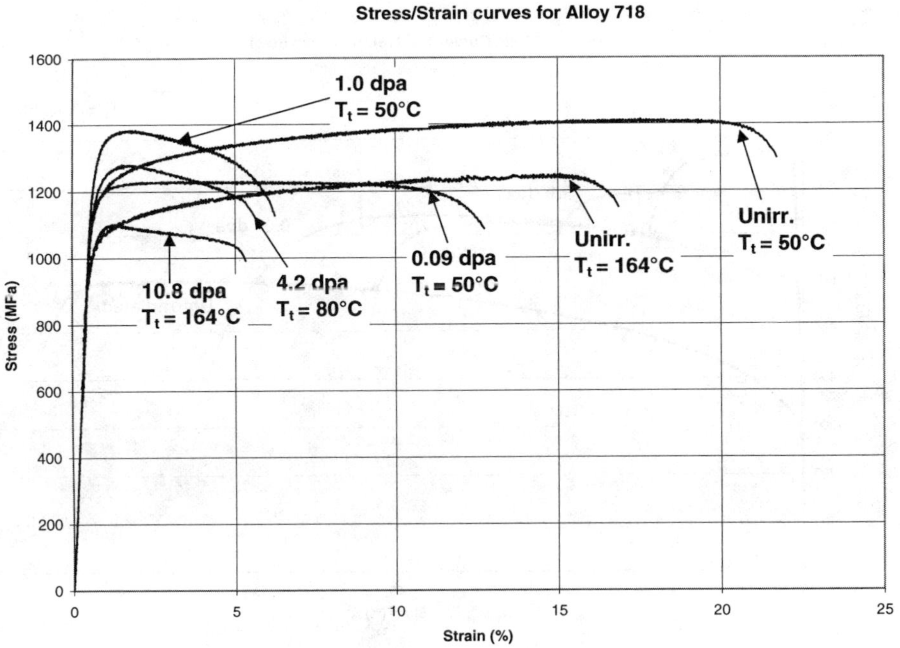

Figure 2. *A representative plot of stress/strain curves measured on Alloy 718 in the precipitation hardened condition after irradiation in a spallation environment.*

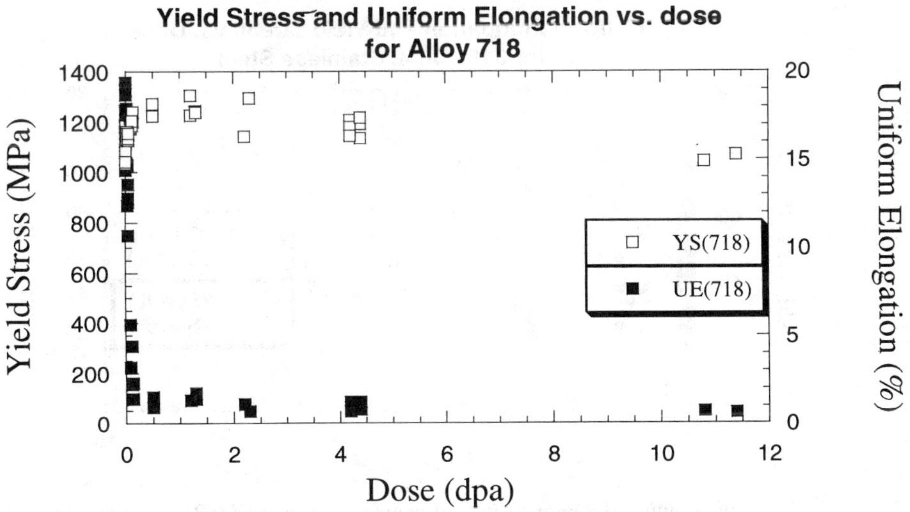

Figure 3. *Graph showing the change in 0.2% offset yield stress and uniform elongation in Alloy 718 after irradiation in a spallation environment.*

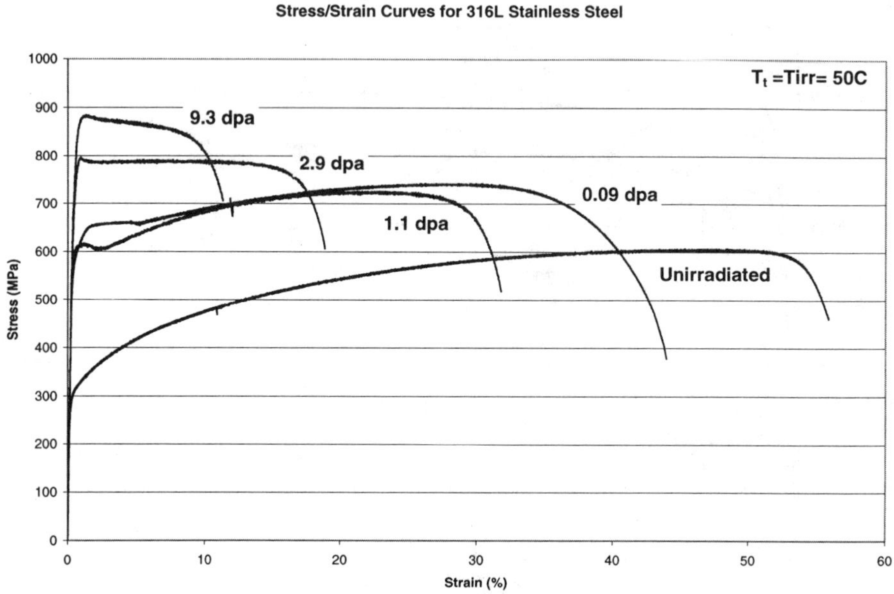

Figure 4. *A representative plot showing stress/strain curves measured on annealed 316L stainless steel after irradiation in a spallation environment.*

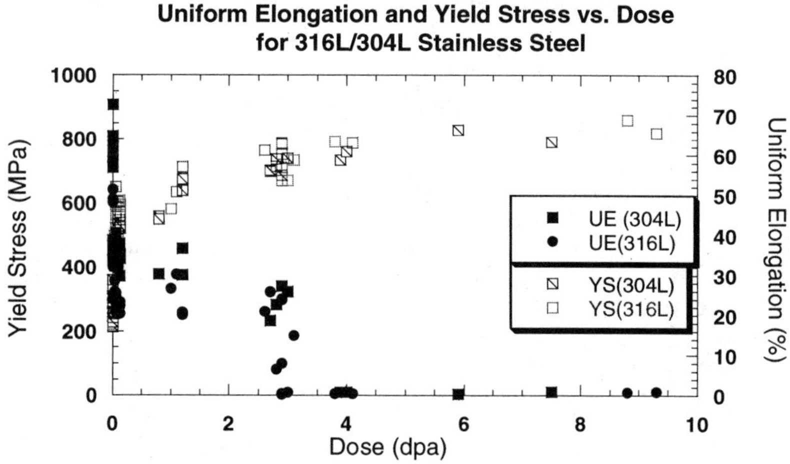

Figure 5. *A graph showing the change in uniform elongation and 0.2% offset yield stress with dose for 304L and 316L stainless steel after irradiation in a spallation environment.*

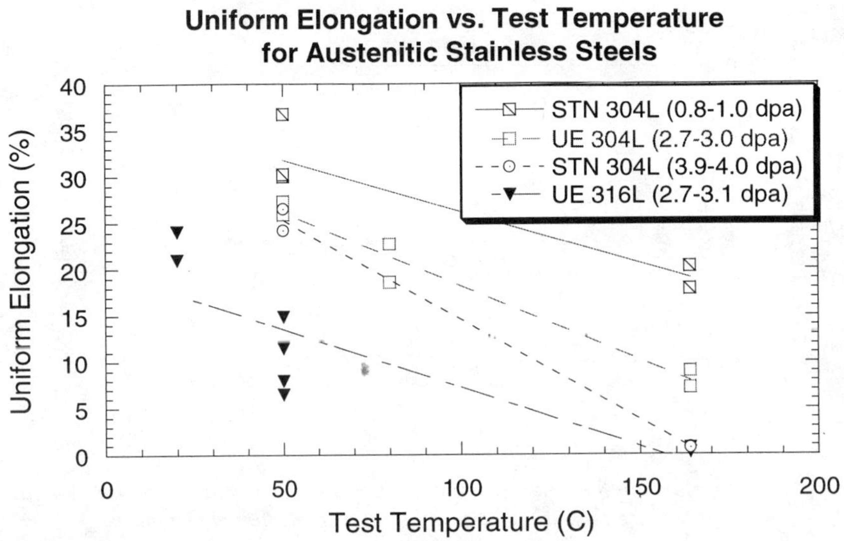

Figure 6. *A graph showing the decrease in uniform elongation or strain-to-necking with increasing test temperature after irradiation to 1-4 dpa in a spallation environment.*

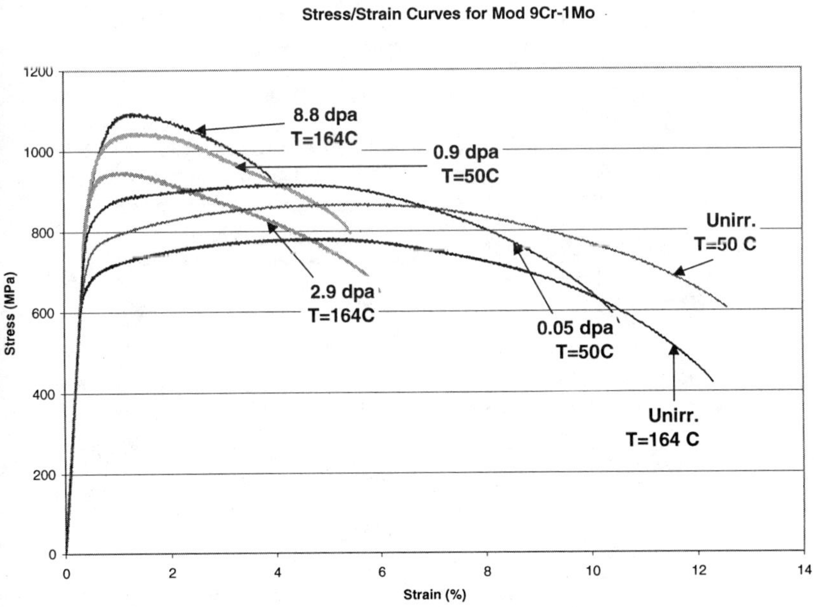

Figure 7. *A representative plot showing stress/strain curves for Mod 9Cr-1Mo after irradiation in a spallation environment*

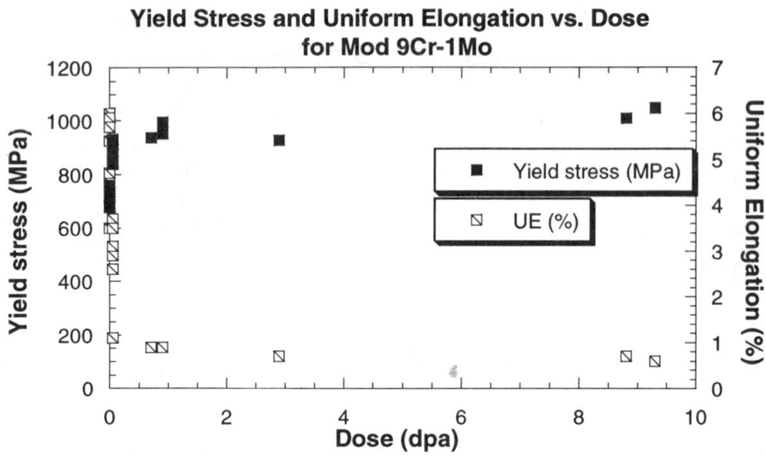

Figure 8. *A graph showing the change in 0.2% offset yield stress and uniform elongation with increasing dose after irradiation in a spallation environment.*

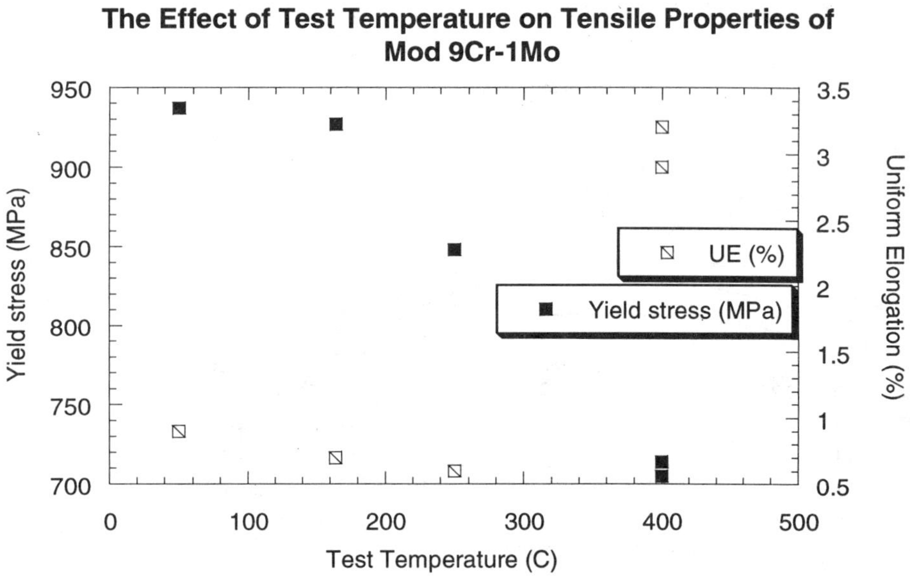

Figure 9. A graph showing the change in 0.2% offset yield stress and uniform elongation with increasing test temperature for specimens irradiated to 3 dpa in a spallation environment.

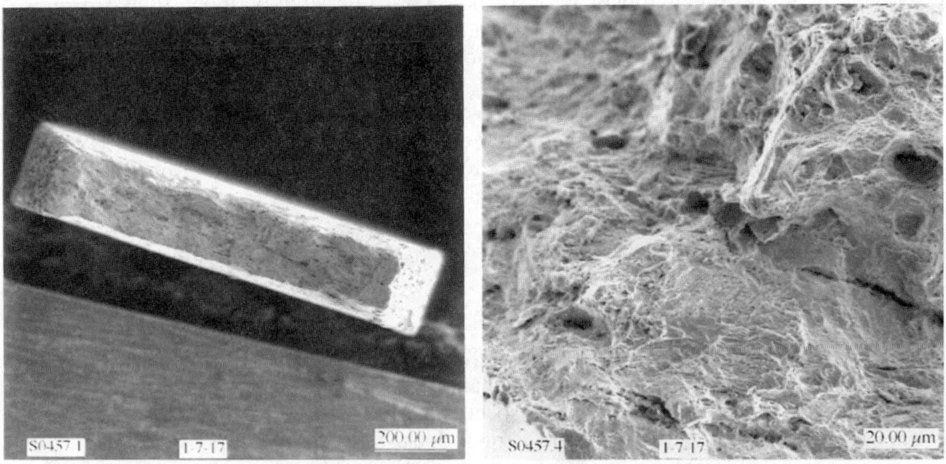

Figure 10. *Scanning Electron Micrographs showing the fracture surface of an Alloy 718 tensile specimen after irradiation to 11 dpa in a spallation environment.*

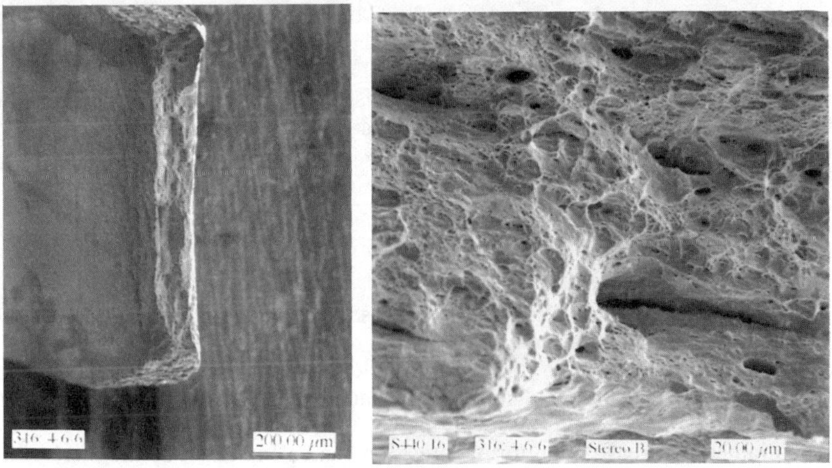

Figure 11. *Scanning Electron Micrographs showing the fracture surface of a 316L tensile specimen after irradiation to 9 dpa in a spallation environment.*

tested at increasing test temperature from 20 to 164°C. Uniform elongation shows a significant decrease with increasing test temperature between 20 and 164°C.

Representative stress/strain curves for Mod 9Cr-1Mo after irradiation are shown in Figure 7. As observed for Alloy 718, the work hardening rate goes to zero after greater than 0.05 dpa of exposure. The data, for all tests performed on Mod 9Cr-1Mo, are plotted in Figure 8 and show that the yield stress sharply increases with dose and the uniform elongation drops precipitously to less than 1% after slightly more than 0.1 dpa of exposure.

The effect of test temperature was also evaluated on the irradiated tensile properties of Mod 9Cr-1Mo as shown in Figure 9. Because Mod 9Cr-1Mo is also a candidate material for the Accelerator Transmutation of Waste program in which the target operates at higher temperatures, test temperatures up to 400°C were investigated. The results show that uniform elongation is low (<1%) and does not significantly increase until the test temperature is raised to 400°C. A significant decrease of yield stress is also observed when testing at 400°C.

SEM Analysis

The fracture surfaces of selected Alloy 718 and 316L and 304L austenitic stainless steel tensile specimens were observed using scanning electron microscopy. Figure 10 shows the fracture surface of an Alloy 718 specimen irradiated to 11 dpa. The fracture still appears ductile although the fracture is quite localized, i.e., the necked/fracture region is confined to a very small portion of the gauge section of the specimen. Reduction of area was also measured on this specimen and on others at lower doses. The results are shown in Table 3. Although significant reductions in uniform elongation are apparent with increasing dose, very little change in reduction of area was observed.

The fracture surface of a 316L stainless steel specimen irradiated to 9 dpa is shown in Figure 11. Fracture is also quite localized and the appearance is more ductile than that observed for Alloy 718. Reduction of area was also measured on this specimen and on other 316L and 304L stainless steel specimens. The results shown in Table 3 reveal very little change in reduction of area for both stainless steels with increasing dose while the change in uniform elongation with dose is quite large.

Discussion

Alloy 718

The tensile properties of Alloy 718 in the precipitation hardened condition show two salient effects with irradiation in a high energy proton and neutron flux. First, the yield stress increases slightly while the uniform elongation decreases dramatically after exposure to 0.5 dpa. Second, the yield stress gradually decreases with increasing dose up to 12 dpa.

The dramatic decrease in uniform elongation can be explained by the increase in the density of irradiation-induced defects with increasing dose, which hardens the material and increases in strength, producing a concomitant decrease in ductility. Black spot damage and Frank loop formation has been observed by Sencer et al. [7] in Alloy 718. Such damage is similar to what has been observed for irradiation of Alloy 718 in a fission reactor[8]. In addition, comparison of the change in uniform elongation with dose obtained in this study to that measured after irradiation in the High Flux Isotope Reactor (HFIR) by Farrell et al.[9] shows very little difference (Figure 12).

The decrease in the yield stress with increasing dose above 1 dpa may be explained by a second microstructural change observed by Sencer et al. [7]. In TEM studies, Sencer observed that the superlattice reflections for the strengthening precipitates, γ'/γ'', disappear after irradiation to doses greater than 0.5 dpa in a high energy proton and neutron flux. Since the irradiation temperature is very low, making complete dissolution of the precipitates unlikely, he interprets the loss of these reflections, and the decrease in yield stress, as the results of irradiation-enhanced disordering of the hardening precipitates.

The fracture surfaces on the Alloy 718 specimens reveal very little change in fracture appearance or reduction of area with increasing dose. On the other hand, the uniform elongation drops significantly. The fracture still appears ductile but the localization of the neck in the tensile specimens, and the shape of the stress/strain curve, suggests a very localized type of plastic flow, resulting in a small uniform elongation. Such strain localization has commonly been caused by both hydrogen and irradiation embrittlement.

316L/304L Stainless Steel

The response of the tensile properties of the austenitic stainless steels 316L and 304L to irradiation in a high energy proton and neutron flux is a more gradual increase in yield stress concomitant with a decrease in uniform elongation with increasing dose, (Figure 5). These changes occurring at low dose are attributed to increases in the density of irradiation-induced defects (black spot damage and Frank loops) with increasing dose as observed by Sencer et al. [7]. Sencer measured total dislocation loop densities of 1.3-1.5 x $10^{15}/m^2$ after irradiation to 4 dpa in a spallation spectrum. With increasing dose above 3 dpa, the uniform elongation dramatically decreases. Comparison of measured values of ductility, in this case, strain to necking, for LANSCE-irradiated vs. fission neutron-irradiated 316 stainless steel at 20-100°C, does not show the same significant ductility decrease at 3 dpa, (Figure 13). Strain to necking values equal uniform elongation values except in very few cases where the maximum load does not coincide with the maximum

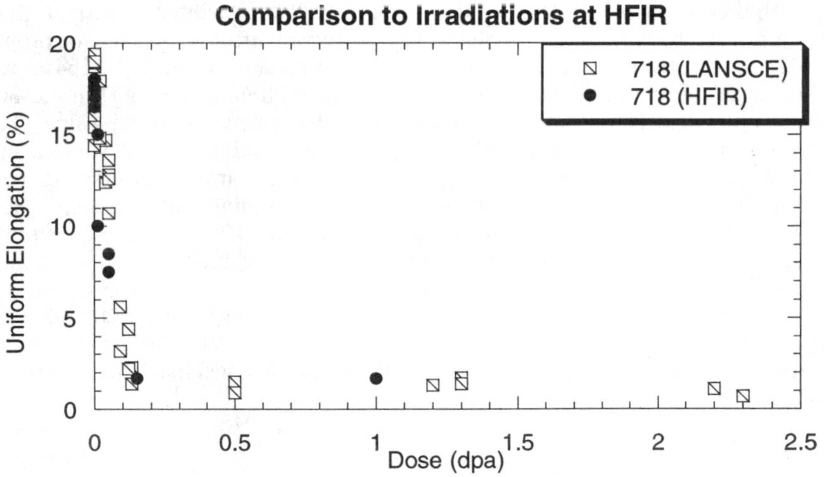

Figure 12. *A graph comparing the uniform elongation measured on Alloy 718 in tension after irradiation in an accelerator to similar doses irradiated in the High Flux Isotope Reactor.*

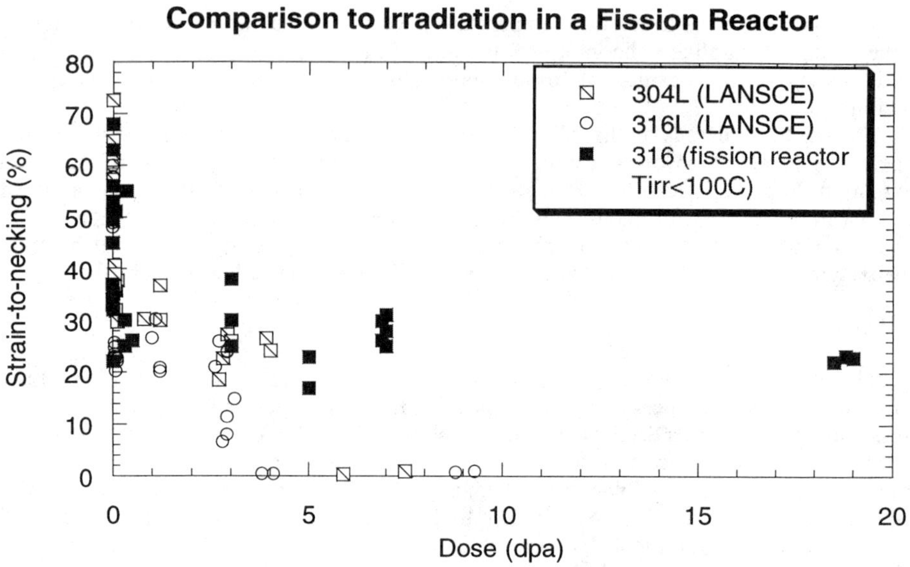

Figure 13. *A graph comparing the strain-to-necking measured on 300 series stainless steels after irradiation in the LANSCE accelerator to those measured after irradiation in a fission reactor[11] at low temperatures (<100°C).*

elongation before necking occurs. Analyses by Oliver et al. [10] show that in addition to the displacement damage in these alloys, they also retain up to 2000 appm He and 2500 appm H after irradiation in a high energy proton/spallation neutron spectrum. Such large amounts of retained gas may contribute to the observed differences in tensile properties.

A second effect is that increasing the test temperature from 20 to 164°C results in decreases in the measured uniform elongations (and strain-to-necking) in the 304L and 316L stainless steels for specimens irradiated to doses between 1 and 4 dpa. A similar effect has been observed by Pawel et al. [11] in 316 stainless steel where a minimum uniform elongation (strain-to-necking) is observed for irradiations/testing performed around 300°C. Our data reported herein suggest that a minimum may appear at a lower temperature for irradiation in a spallation environment. However, more tests would be required at higher irradiation and test temperatures in a high energy proton/spallation neutron environment to confirm this effect.

The fracture surfaces on the 316L/304L stainless steel specimens still show a ductile appearance after irradiation to 9 dpa. Reduction of area also remains high at 9 dpa. Therefore, the low uniform elongation results from very localized fracture in a ductile mode.

Mod 9Cr-1Mo

The effect of increasing proton and neutron fluence on the tensile properties of Mod 9Cr-1Mo in the tempered condition is to decrease the uniform elongation while increasing the yield stress. This can be attributed to the microstructural changes observed by numerous investigators as well as Sencer [12] which show an increase in black spot and Frank loop damage with increasing dose. Our results compare well with the results of Spatig et al.[13] for F82H who measured a $\Delta\sigma$ of 210 MPa at 25°C after a dose of 0.16-0.26 dpa for irradiation in a 590 MeV proton beam while we observe a $\Delta\sigma$ of 200-250 MPa after a dose of 0.12 dpa at 50°C.

The effect of increasing test temperature on the tensile properties after irradiation to 3 dpa is an increase in the uniform elongation and a decrease in the yield stress as the temperature is raised from 250 to 400°C. This could be caused by annealing of radiation-induced defect clusters with increasing temperature.

Conclusions

The tensile properties of Alloy 718 in the precipitation hardened condition, annealed 316L and 304L stainless steel and Mod 9Cr-1Mo in the tempered condition show the following effects after irradiation in a high energy proton/spallation neutron environment:

Alloy 718:

1. The uniform elongation drops to less than 2% after only 0.5 dpa of exposure.
2. The yield stress increases slightly at low doses and then decreases with dose above 1 dpa of exposure.
3. The reduction of area shows very little change with increasing dose to 12 dpa, remaining greater than 30%.

These results show that the tensile properties are severely affected by irradiation in a spallation environment and they are in good agreement with irradiation results in a fission environment at similar temperatures.

316L/304L Stainless Steel:

1. The uniform elongation gradually decreases with increasing dose but then abruptly drops at 3 dpa.
2. The yield stress increases with increasing dose to more than two times its unirradiated value.
3. For specimens irradiated to 1-3 dpa, increasing test temperature from 20-160°C results in decreasing uniform elongation.
4. The reduction of area does not significantly change with increasing dose to 9 dpa, remaining greater than 60%.

These effects observed after irradiation to high doses in a spallation spectrum are not seen for irradiation in a fission neutron spectrum at these low temperatures and may be related to high retained helium and hydrogen produced in the material.

Mod 9Cr-1Mo:

1. The yield stress increases to over 900 MPa while the uniform elongation decreases to less than 2% after only 0.5 dpa exposure.
2. A slight increase in uniform elongation is observed along with a decrease in yield stress when the test temperature is raised to 400°C.

These results show good agreement with those from irradiations of similar material in a fission reactor.

Acknowledgments

The authors would like to thank Manuel Lopez and Tobias Romero of LANL and Gary Whiting of PNNL for their expertise in handling and testing these tensile specimens using manipulators in hot cells.

References

[1] Cappiello, M.W. and Pitcher, E., "Design and Operation of the APT Target/Blanket," *Materials Characterization*, **43**, 1999, p. 73-82.

[2] Maloy, S.A., Sommer, W.F., Brown, R.D., Roberts, J.E., Eddlemen, J., Zimmermann, E., and Willcutt, G., "Progress Report on the Accelerator Production of Tritium Materials Irradiation Program," in *Materials for Spallation Neutron Sources*, The Minerals, Metals & Materials Society, edited by M.S. Wechsler, C.L. Snead, and W.F. Sommer, 1998, p.131-138.

[3] James, M.R., et al., "Determination of Mixed Proton/Neutron Fluences in the LANSCE Irradiation Environment," in *2nd International Topical Meeting on Nuclear Applications of Accelerator Technology*, Gatlinburg, TN, American Nuclear Society, 1998, p. 605-608.

[4] Prael, R.E. and Lichtenstein, H., "User Guide to LCS: The LAHET Code System," Radiation Transport Group, Los Alamos National Laboratory: Los Alamos, NM, 1989.

[5] Prael, R.E. and Madland, D.G., "LAHET Code System Modifications for LAHET 2.8," Los Alamos National Laboratory: Los Alamos, NM, 1995.

[6] Willcutt, G.J., et al., "Thermal Analysis of the APT Materials Irradiation Samples" in *Accelerator Applications '98*, Gatlinburg, TN, American Nuclear Society, 1989, p. 254-259.

[7] Sencer, B.H., et al., "Microstructural Alteration of Structural Alloys by Low Temperature Irradiation with High Energy Protons and Spallation Neutrons," in these proceedings.

[8] Thomas, L.E. and Bruemmer, S.M., "Radiation-Induced Microstructural Evolution and Phase Stability in Nickel-Base Alloy 718," in the *Proc. 8th Int. Symp. Environmental Degradation of Materials in Nuclear Power Systems WaterReactors*, editors S. M. Bruemmer and A. R. McIlree, American Nuclear Society, 1997, p. 772.

[9] Farrell, K., "Comparison of Tensile Properties of Ferritic/Martensitic Steels and Austenitic Alloys after Irradiation in HFIR," in *Third International Workshop on Spallation Materials Technology*, Santa Fe, NM, Los Alamos National Laboratory document number: LA-UR-00-3892, 1999.

[10] Oliver, B.M., et al., "Retention of Very High Levels of Helium and Hydrogen Generated in Various Structural Alloys by 800 MeV Protons and Spallation Neutrons," in these proceedings.

[11] Pawel, J.E., et al., "Irradiation Performance of Stainless Steels for ITER Application," *J. Nuc. Mat.*, **239**: 1996, p. 126-131.

[12] Sencer, B.H., "Structural Alloys Irradiated with Protons and Spallation Neutrons at Low Temperature Under High Helium and Hydrogen Evolution: Microstructural

Evolution and Implications for Mechanical Properties," PhD thesis, New Mexico Institute of Mining and Technology: Socorro, NM., 2000, pp. 161.

[13] Spatig, P., et al., "Evolution of the mechanical properties of the F82H ferritic/martensitic steel after 590 MeV proton irradiation," *J. Nuc. Mat.*, **258-263**, 1998, p. 1345-1349.

Michael R. James,[1] Stuart A. Maloy,[1] Walter F. Sommer,[1] W. Robert Johnson,[1] David A. Lohmeier,[1] and Margaret L. Hamilton[2]

High-Energy Spallation Neutron Effects on the Tensile Properties of Materials for the Target and Blanket Components for the Accelerator Production of Tritium Project

---

Reference: James, M. R., Maloy, S. A., Sommer, W. F., Johnson, W. R., Lohmeier, D. A., and Hamilton, M. L., **"High-Energy Spallation Neutron Effects on the Tensile Properties of Materials for the Target and Blanket Components for the Accelerator Production of Tritium Project,"** *Effects of Radiation on Materials: 20th International Symposium, ASTM STP 1405,* S. T. Rosinski, M. L. Grossbeck, T. R. Allen, and A. S. Kumar, Eds., American Society for Testing and Materials, West Conshohocken, PA, 2001.

Abstract: The Accelerator Production of Tritium (APT) project proposes to use a 1.0 GeV, 100 mA proton beam to produce neutrons by spallation from a tungsten target clad with Alloy 718. The neutrons are multiplied and moderated and then captured in ^{3}He to form tritium. In this process, the materials in the target and blanket (T/B) region are exposed to a wide range of irradiation conditions. To test the effects of spallation neutrons on APT candidate materials, ASTM subsize samples were irradiated in a spallation neutron spectrum created from the interaction of an 800 MeV, 1 mA proton beam on a tungsten target at the Los Alamos Neutron Science Center (LANSCE). Specimens were located in controlled-temperature furnaces that were placed in proximity to the proton beam, but in locations where spallation neutrons would form the bulk of the exposure. The materials irradiated were the candidate APT materials 316L and 304L annealed stainless steels (SS), modified 9Cr-1Mo steel, and Alloy 718 (precipitation hardened). Irradiation temperatures varied from 100 to 280 °C and the approximate proton fluences were 4 x 10^{18} p/cm^2. The neutron fluences were ~5 x 10^{19} n/cm^2 for energies < 0.1 MeV and 6x10^{19} n/cm^2 for energies > 0.1 MeV with the spectra including a high-energy tail up to the proton beam energy (800 MeV). The samples experienced an exposure range of 0.023 to 0.051 dpa.

The tensile tests were conducted in air at ambient temperatures (~20° C). The results indicated that even for low exposures (< 0.05 dpa), there was a decrease in ductility for the all of the alloys. Increases in yield and ultimate strength were observed in the austenitic stainless steels and the 9Cr-1Mo alloy. The Alloy 718 showed no change in yield strength but a marked decrease in the ultimate strength. The results are discussed and compared with results of the in-beam irradiated materials where smaller S-1 type tensile samples were used. Comparison also is made with irradiations of materials in fission environments in a similar dose and temperature environment.

---

[1] Los Alamos National Laboratory, P.O. Box 1663, Los Alamos, NM 87545.
[2] Pacific Northwest National Laboratory, P.O. Box 999, Richland, Washington 99352.

Keywords: proton irradiation, spallation, neutron, 316L, 304L, stainless steel, Alloy 718, 9Cr-1Mo, tensile properties

Introduction

Irradiation experiments were performed at the Los Alamos Spallation Radiation Effects Facility (LASREF) at the Los Alamos Neutron Science Center (LANSCE) to provide data in support of the APT project. Various APT candidate materials were irradiated using an 800 MeV, 1 mA proton beam. Details of the irradiation configuration can be found in Maloy et al. [*1*]. Figure 1 illustrates the targets as they were placed in the target area. In addition to test samples that were placed in the direct proton beam, many samples were placed in temperature-controlled "furnaces" in locations that would be subjected to a strong flux of secondary particles (primarily neutrons) created from spallation reactions. In relation to the targets shown in Fig. 1, these furnaces are located behind the last insert (17B) and off to one side, outside of the proton beam. The work presented here will focus on samples taken from three furnaces (numbered 9, 10, and 11). Samples in furnace 9 saw a fluence of approximately 6.6 x 10^{19} neutrons/cm^2 and 7.3 x 10^{17} protons/cm^2 and the irradiation temperatures averaged 127 ± 40 °C. Furnace 10 experienced nearly identical flux with only slightly more neutrons at an average temperature of 110 ± 15 °C. The samples in furnace 11 experienced a lower fluence with only 5.2 x 10^{19} neutrons/cm^2 but the average temperature was much higher at 180 ± 110 °C. The flux of secondary neutrons was produced from a W target (insert 18A) placed in the proton beam and as well as from the other targets used to irradiate materials in the proton beam. The samples were exposed to approximately 6 months of irradiation in two separate intervals of 2 and 4 months. This work compares the results of tensile tests of ASTM subsize specimens and the smaller S-1 type tensile specimens that were irradiated along side the ASTM subsize samples.

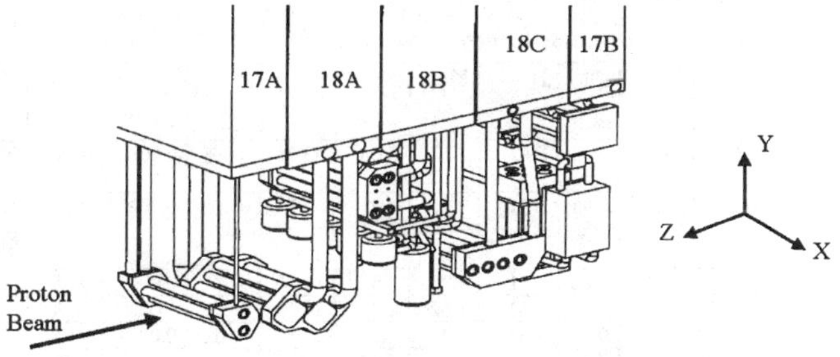

Figure 1. *Schematic of in-beam targets for LASREF 1996-1997 irradiation.*

Experimental Testing

The mechanical test samples were fabricated following ASTM E8 "Standard Test Methods for Tension Testing of Metallic Materials" subsize specification for standard sheet-type samples. Figure 2 presents the actual specimen dimensions used and also shows the dimensions of the smaller S-1 samples that were irradiated simultaneously with the ASTM subsize samples. Testing of the ASTM subsize samples was performed using an

Instron 5567 test stand modified for remote operations in a hot cell. Specimens were tested at an initial strain rate of 10^{-4}/s. The load was measured with a 30 kN load cell which was calibrated in accordance with the ASTM standard E4-94 "Standard Practices for Force Verification of Testing Machines." An extensometer was clipped to the gauge length of each sample to provide strain data. Testing was performed on several APT-relevant structural materials, Alloy 718, 316L SS, 304L SS, and 9Cr-1Mo. The S-1 samples were tested at Pacific Northwest National Laboratory [2] at the same strain rate and temperature as the ASTM subsize testing.

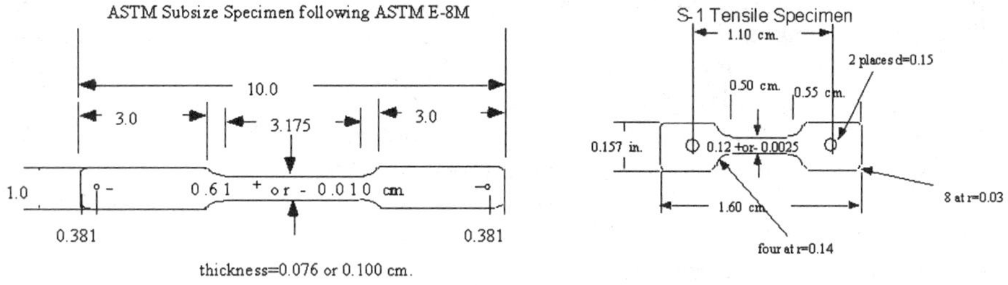

Figure 2. *Specifications for ASTM subsize and S-1 tensile samples. All dimensions in cm.*

Compositions of the alloys are given in Table 1. The first column gives the material type and thickness (in mm) of the sheet from which samples were taken. For the cases of the Alloy 718, 316L SS and 304 L SS, the ASTM subsize samples were fabricated from a different heat of material and a thicker sheet (thicknesses of 0.76 and 1.0 mm) than was used to fabricate the S-1 samples (which were 0.25 mm thick).

Table 1 - *Composition of candidate APT materials (wt %).*

Material (thickness - mm)	Al	C	Cr	Cu	Fe	Mn	Mo	Ni	P	S	Si	Others
Alloy 718 (0.25)	0.48	0.04	18.13	0.08	BAL	0.13	3.06	53.58	0.008	0.001	0.11	Nb + Ta - 4.98; Ti – 1.03
Alloy 718-(1.0)	0.54	0.05	18.13	0.05	BAL	0.21	3.01	52.7	0.005	0.002	0.13	Nb + Ta-5.07; Co - 0.4; B - 0.004; Ti – 1.06
316L (0.25)		0.019	17.26	0.26	BAL	1.75	2.57	12.16	0.022	0.006	0.65	
316L (0.76)		0.022	16.05		BAL	1.82	2.08	10.11	0.022	.0002	0.48	
304L (0.25)		0.020	18.23	0.38	BAL	1.77	0.33	9.68	0.026	0.002	0.54	
304L (0.76)		0.060	18.19	0.4	BAL	1.86	0.34	8.14	0.030	.0003	0.48	
Mod 9Cr-1Mo	0.002	0.089	9.24	0.08	BAL	0.47	0.96	0.16	0.021	0.006	0.28	V – 0.21; Nb - 0.054; Co - 0.19; W - 0.01; N - 0.035; O - 0.008

The testing yielded stress-strain curves for each sample from which the 0.2 % offset yield strength (YS), ultimate tensile strength (UTS), uniform elongation (UE) and total elongation (TE) were determined. For some samples (the irradiated 304L and 316L), a clear yield point was reached. In those cases, the yield point was used rather then the 0.2 % offset for the YS. These properties were measured to assess the effects of the radiation and for comparisons among the sample types and with literature data. Total elongation comparisons that were made between the ASTM subsize samples and the S-1 samples were *not* adjusted for sample geometry effects such as those found in Dieter [*3*] because this adjustment relies on the assumption that difference in total elongation arises entirely from the necking and this was not borne out in the current results.

The Alloy 718 samples were annealed at 1065°C for 30 min. and air-cooled (AC) followed by aging at 760°C for 10 hr and furnace cooled (FC) to 650°C and held for a 20 hour total furnace time and AC. The Mod 9Cr-1Mo specimens were normalized at 1038°C for 1 hr. and AC followed by a temper at 760°C for 1 hr and AC. The 316L and 304L SS were in the annealed condition.

Results

Dosimetry

The quantification of the fluences was accomplished using dosimetry foil stacks that were irradiated in the sample containers. Analysis of the fluences and determination of the irradiation exposure parameters (dpa, H, and He production) can be found in James et al. [*4,5*]. Three different containers (furnaces) held the ASTM subsize and S-1 samples and the doses for these three containers varied depending upon their location in the target area. The samples tested here had doses ranging from 0.023 to 0.051 dpa.

Tensile tests

Alloy 718 -The stress-strain curves from tensile tests on the Alloy 718 are given in Fig. 3. Two tests were performed at each condition but sample-to-sample reproducibility was very good so only one test trace is shown. Stress-strain curves for the unirradiated material as

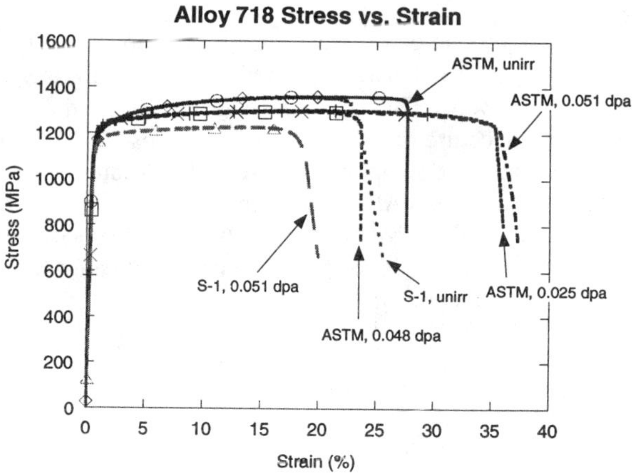

Figure 3. *Stress-strain curves of irradiated Alloy 718 for three different doses for ASTM subsize and S-1 tensile samples under the same conditions.*

well as from samples irradiated to 0.025, 0.048 and 0.051 dpa are shown. The results at these doses show no change in YS, however the UTS drops significantly. Reduction in UE is also observed, although there is no clear trend seen in changes in TE. Also shown in Fig. 3 are the unirradiated and irradiated curves for the S-1 type samples. A comparison of the ASTM subsize results with those from the smaller S-1 tensile samples is shown in Fig. 4. The comparison is made from samples receiving a dose of approximately 0.051 dpa at an irradiation temperature of 105 ± 15 °C. The test temperature was 22 °C. Two samples from each condition were tested; error bars indicate the standard deviations from these tests. Although the absolute values differ because the specimens are from different heats, the S-1 and ASTM subsize specimens are in agreement on direction and magnitude of changes. The stress-strain curves in Fig. 3 also show good agreement.

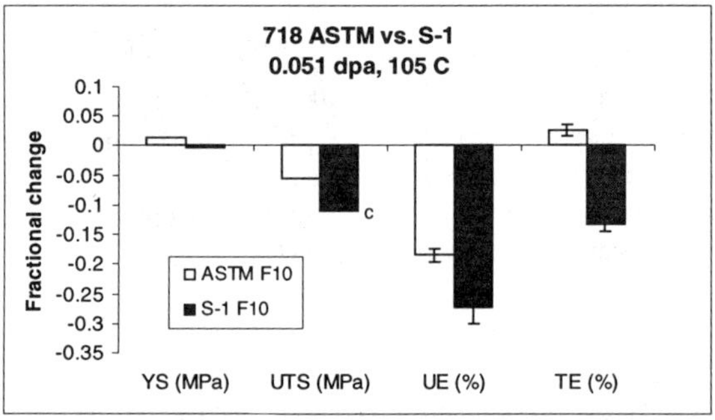

Figure 4. *Comparison of ASTM and S-1 tensile test results of Alloy 718 to 0.051 dpa at an irradiation temperature of 105 C.*

316L SS - The stress-strain curves for the 316L SS are shown in Fig. 5. A small yield drop was observed in all of the irradiated stress/strain curves for the ASTM samples and in most of the S-1 samples as well. This is a typical feature of irradiated stainless steels at doses out to 7 dpa [*6*]. An increase in YS as a function of dose is clearly shown. UTS also increases as a function of dose, although less significantly. Reductions in UE and TE were observed in all irradiated samples. Also shown in Fig. 5 are the S-1 sample results for unirradiated and 0.051 dpa. A direct comparison of the ASTM subsize and S-1 samples are presented in Fig. 6. The changes in irradiated properties between ASTM subsize and S-1 sample types show very good agreement.

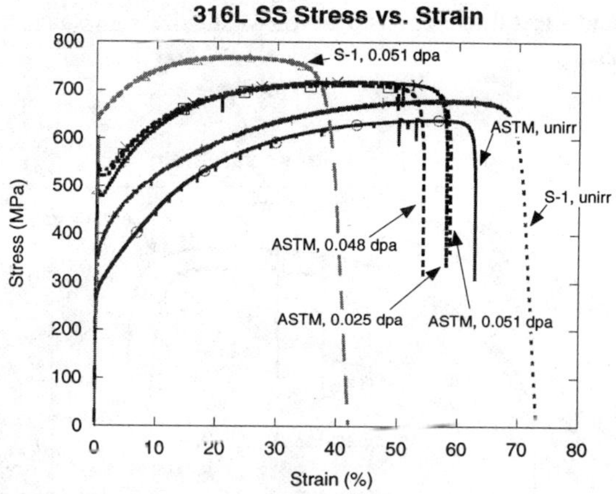

Figure 5. *Stress-strain curves for irradiated 316L SS for three different doses. Also shown are results from S-1 tensile samples under the same conditions.*

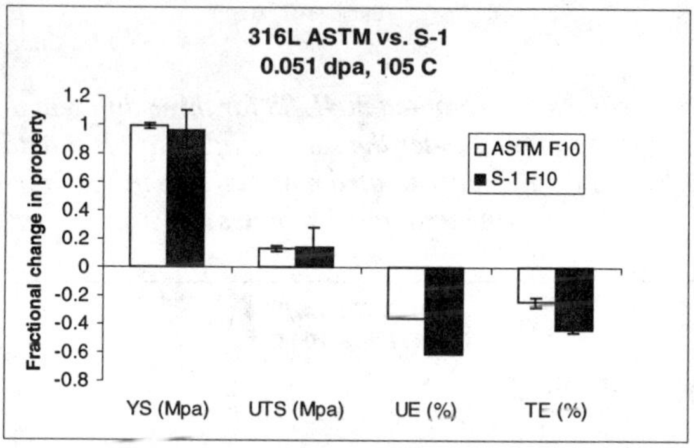

Figure 6. *Comparison of ASTM and S-1 tensile test results from Furnace 10. It should be noted that the YS compares the offset YS for the unirradiated material and the yield point in the case of the irradiated samples.*

304L SS - Stress-strain curves for 304L SS are given in Fig. 7. The yield drops seen in the 316L data are also present for the 304L. Trends in the irradiated 316L SS are also present in the 304L SS such as the changes in strength and reductions in ductility. The comparison for the ASTM subsize and S-1 tensile samples are given in Fig. 8. The results are in fairly good agreement. The one exception is the change in YS observed for the ASTM subsize sample is approximately twice that observed for the S-1 samples. This difference is not fully understood, but may arise from the fabrication of the S-1's from a different heat or a thinner sheet that effected the unirradiated grain size or microstructure. It is unlikely that this discrepancy is attributable to the sample geometry as S-1 samples made from the same heat and sheet thickness as the ASTM subsize samples have better agreement in unirradiated properties.

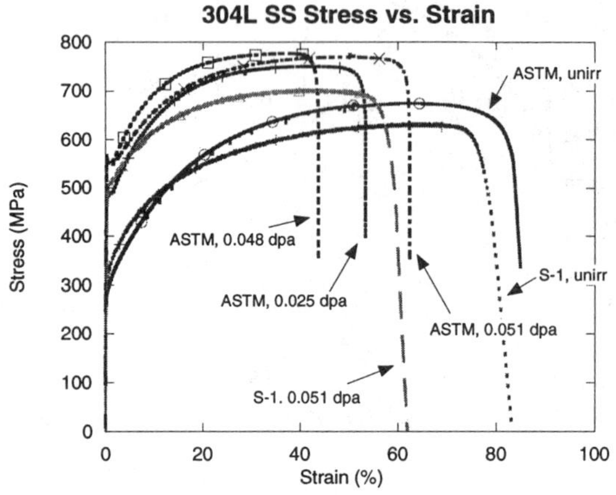

Figure 7. *Stress-strain curves of irradiated 304L SS for three different doses. Also shown are results from S-1 tensile samples under the same conditions. It should be noted that the YS compares the offset YS for the unirradiated material and the yield point in the case of the irradiated samples.*

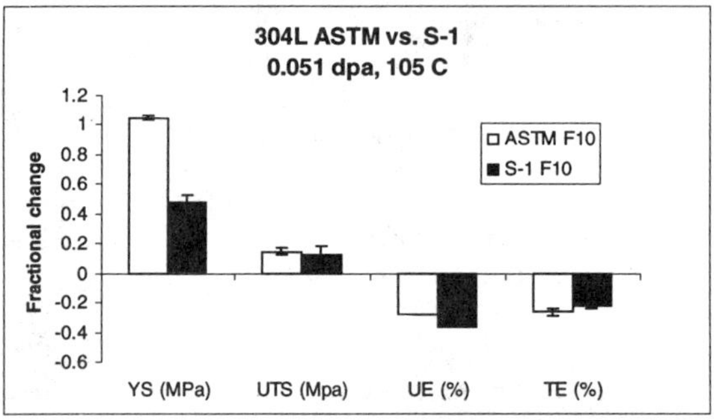

Figure 8. *Comparison of ASTM and S-1 tensile test results from Furnace 10.*

9Cr-1Mo - The stress-strain curves for the 9Cr-1Mo alloy are given in Fig. 9 for both the ASTM subsize and S-1 specimens. A noticeable strengthening from irradiation can be seen in these curves as well as a loss in uniform and total elongation. Comparisons for the ASTM subsize and S-1 sample data are given in Fig. 10. Excellent agreement is seen in the changes of the four material properties (YS, UTS, UE, TE). Note that only one heat of material was used to fabricate all the 9Cr-1Mo samples.

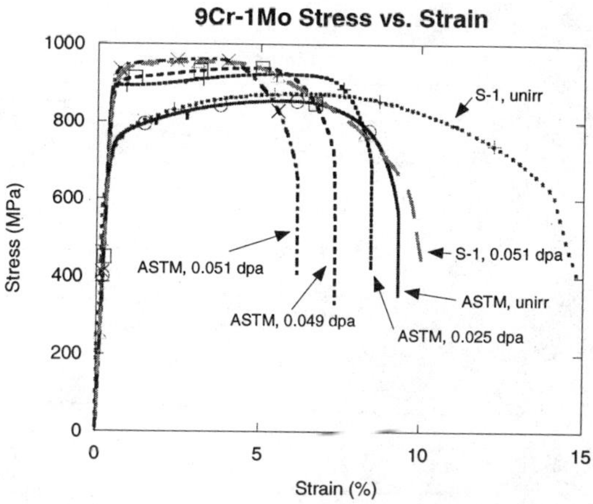

Figure 9. *Stress-strain curves of irradiated 9Cr-1Mo for three different doses. Also shown are results from S-1 tensile samples under the same conditions*

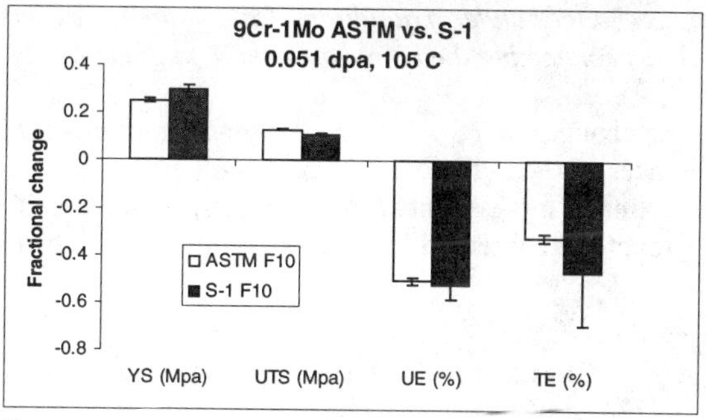

Figure 10. *Comparison of ASTM and S-1 tensile test results from Furnace 10*

Discussion

Comparisons of the measurements between the S-1 and ASTM samples show general agreement in the property changes induced from the radiation exposure. Some notable exceptions occur in the elongation values for the Alloy 718 and 316L SS and the change in YS for the 304L. The Alloy 718 stress/strain curve showed a larger loss in elongation from radiation exposure and a somewhat larger reduction in UTS in the S-1 samples. The reductions in elongation for the 316L SS were not dramatically different between the two sample types, but the S-1 samples clearly showed a greater loss of both UE and TE. Previous work has shown that the S-1 sample type has a tendency to produce larger reductions in the elongation of post-irradiated material so these results are not unexpected [*7,8*]. The discrepancy in the YS for the 304L is likely attributed to the use of different material in fabrication of the S-1 specimens.

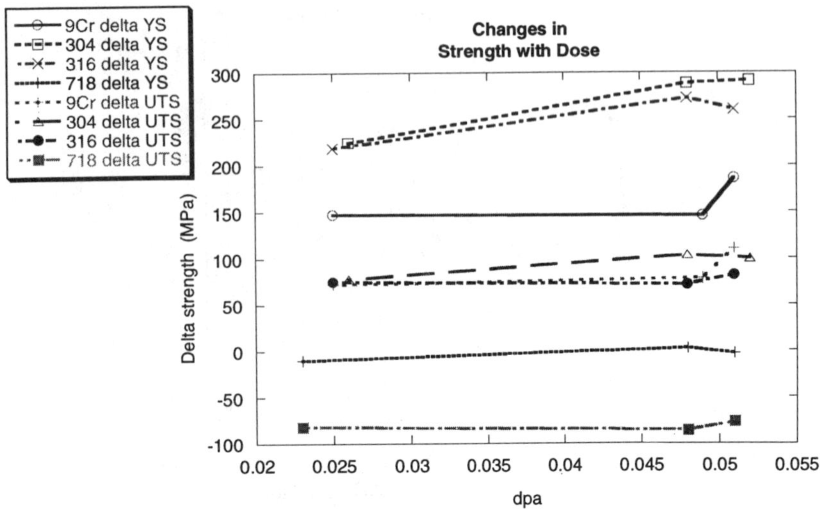

Figure 11. *Strength changes under irradiation. Data from ASTM subsize samples. The lines connecting the points only serve to guide the eye.*

Figure 11 gives the changes in YS and UTS as a function of dpa for the alloys discussed. Although the differences in the radiation dose are small, especially between the two points of highest exposure, trends are evident in the 304LSS, 316LSS and 9Cr-1Mo. In these alloys, changes in strength are observed that increase with increasing dose.

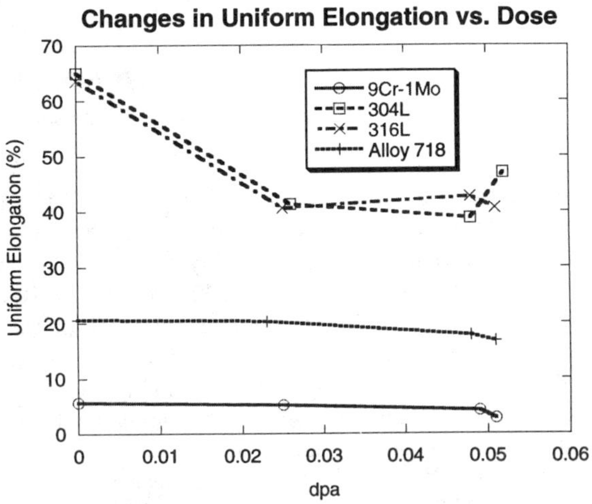

Figure 12. *Changes in UE with dose for a APT candidate materials. Data taken from ASTM subsize tests.*

Figure 12 shows the effects of dose on UE. The results here are not as clear as the strength changes perhaps due to the greater variability in the elongation parameter. Also, given that the irradiation temperatures were highly variable and did not correlate with dose, other conditions may have played a role in the changes in elongation.

The current irradiation-induced changes are comparable to fission-irradiation results as a function of dpa. While the spectrum seen by the samples in this work is harder and the level of gas introduced into these samples is higher than for a typical reactor spectrum, there are no indications of the drastic reduction in uniform elongation seen at the higher doses in high-energy proton irradiations [*9*].

Data from experiments in fission reactor irradiations on 316 type SS are shown in Figs. 13 and 14. The literature data were taken from a summary in Pawel et al. [*6*]. In Fig. 13, the YS data are data from fission irradiations exposed and tested in the range of 25 - 250 °C. Figure 14 contains strain-to-necking (STN)* data from Pawel et al. [*6*] and represents samples irradiated and tested at 100 °C. In both cases, the results from the ASTM subsize samples irradiated at LANSCE agree well with values for reactor irradiated material. It should be noted that the He and H production as a function of dpa is expected to be quite high in the LANSCE-irradiated samples due to the high-energy particles from spallation events. For the Alloy 718, the He production is expected to be approximately 40 appm/dpa. While this number is lower than is typically seen in spallation, Green et al. [*10*], it is still much higher than for fission reactor spectra. Earlier published results of tests on materials in a spallation environment suggest that while strength increases in 316L and 304L seem consistent with fission irradiation work, the ductility changes occur at an accelerated rate, possibly due to the higher levels of spallation-produced gas in the matrix [*9*].

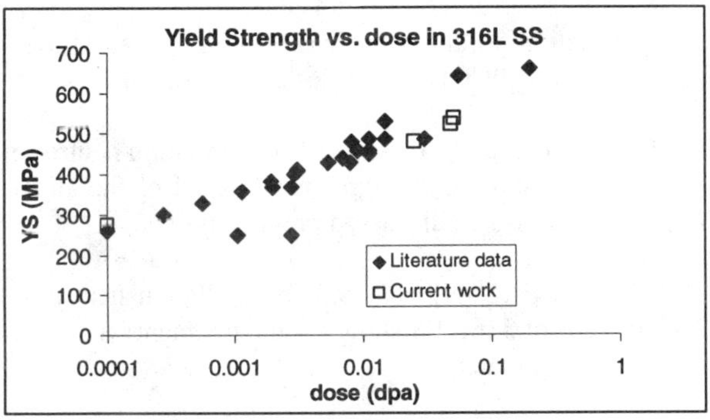

Figure 13. *Yield strength changes in 316L SS taken from literature values* [*6*] *and current work.*

---

* The strain-to-necking as it is applied here is the equivalent of the uniform elongation except in those cases where the sample undergoes a yield drop and then subsequent strain hardening but does not exceed the original yield strength [*5*].

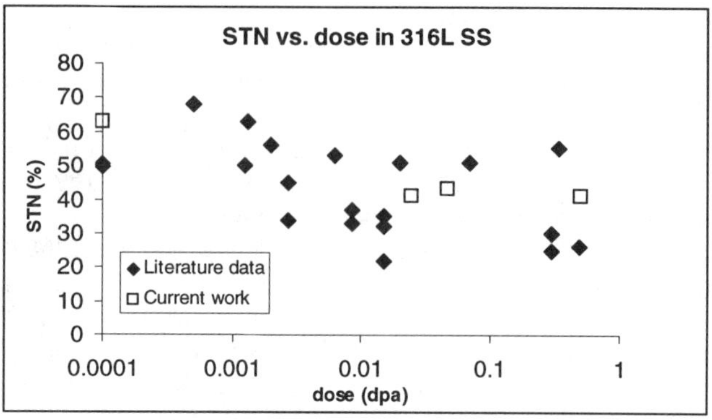

Figure 14. *STN vs. dose for 316L SS showing literature data collected from Pawel [6]. Comparison to the current work is made.*

Conclusions

- Spallation neutron irradiation to low dpa values (up to ~0.05 dpa) produces an increase in YS for 316L and 304L stainless steels and 9Cr-1Mo, but no change in YS for Alloy 718.
- Spallation neutron irradiation produces slight to moderate increases in UTS for the 316L and 304L stainless steels and 9Cr-1Mo, but a small decrease in the Alloy 718.
- Spallation neutron irradiation produces decreases in UE for all alloys, 316L and 304L stainless steels, Alloy 718 and 9Cr-1Mo, for doses to ~0.05 dpa.
- Comparisons of YS, UTS, UE, and TE obtained for spallation neutron-irradiated ASTM subsize and S-1 specimens show good agreement for all of the alloys tested with the exception of the 304L SS YS value discussed previously.
- Comparisons of YS and UE (STN) obtained for spallation neutron-irradiated ASTM subsize and S-1 specimens of 316L SS show good agreement with similar data obtained for fission neutron-irradiated materials.

Acknowledgments

The authors would like to thank Manny Lopez and Toby Romero of LANL and Gary Whiting of PNNL for their expertise in handling and testing these tensile specimens using manipulators in hot cells.

References

[1] Maloy, S M., Sommer, W. F., Brown, R.D., Eddlemen, J., Zimmerman, E., and Willcutt, G., "Progress Report on the Accelerator Production of Tritium Materials Irradiation Program, " *Materials For Spallation Neutron Sources,* Eds., M.S.

Wechsler, L.K. Mansur, C.L. Snead and W.F. Sommer, The Minerals, Metals and Materials Society, 1998, p.131.

[2] Dieter, G.E., "Mechanical Metallurgy," McGraw Hill, New York, NY, 1986, p.294.

[3] Garner, F. A., Hamilton, M. L., Heinisch, H. L., and Kumar, A. S., "Application of Miniature Tension Specimens to Studies of Radiation Damage in Metals," *Small Specimen Test Techniques Applied to Nuclear Reactor Vessel Thermal Annealing and Plant Life Extension, ASTM STP 1204,* W. R. Corwin, F. M. Haggag, and W. L. Server, Eds., American Society for Testing and Materials, Philadelphia, PA, 1993, pp 336-355.

[4] James, M. R., Maloy, S. A., Sommer, W. F., Ferguson, P. D., Fowler, M. M., Mueller, G. E., and Corzine, R. K., "Spectral Unfolding of Mixed Proton/Neutron Fluences in the LANSCE Irradiation Environment," *Reactor Dosimetry, ASTM STP 1398,* John. G. Williams, David. W. Vehar, Frank H. Ruddy and David. M. Gilliam, Eds., American Society for Testing and Materials, West Conshohocken, PA, 2000.

[5] James, M. R., Maloy, S. A., Sommer, W. F., Fowler, M. M., Corzine, R. K., "Determination of Mixed Proton/Neutron Fluences in the LANSCE Irradiation Environment," *2nd International Topical Meeting on Nuclear Applications of Accelerator Technology*, Gatlinburg, TN, Sept. 20-23, 1998, p. 605.

[6] Pawel, J. E., Rowcliffe, A. F., Lucas, G. E., and Zinkle, S. J., "Irradiation Performance of stainless steels for ITER application," *Journal of Nuclear Materials*, 1996, 239, p. 126.

[7] Kohyama, A., Matsui, H., Abe, K., Hamada, K., and Asano, K., "Specimen Size Effects on Mechanical Properties of 14 MeV Neutron Irradiated Metals," *Journal of Nuclear Materials*, 1988, 155-157, pp. 1354-1358.

[8] . Kohyama, A., Asakura, K., and Igata, N., "Mechanical Property Changes in Ferritic Steels by 14 MeV Neutron Bombardment," *Journal of Nuclear Materials*, 1986, 141-143, pp. 921-925.

[9] Maloy, S. A., James, M. R., Sommer, W. F., Ferguson, P. D., Willcutt, G., Alexander, D., Louthan, M. R., Hamilton, M. L., Snead, L., and Sokolov, M. A., "The Change in the Mechanical Properties of Alloy 718, 304L and 316L Stainless Steel and Al6061 after Irradiation in a High-Energy Proton Beam," *3rd International Topical Meeting on Nuclear Applications of Accelerator Technology*, Long Beach, CA, Nov 14-18, 1999, p.541.

[10] Green, S. L., et al., "Production of Helium by Medium Energy Protons," *Journal of Nuclear Materials,* 1988, 155-157, pp. 1350-1353.

Radiation Damage Fundamentals

Marie-Hélène Mathon,[1] Yann de Carlan,[2] Ghita Geoffroy,[3] Xavier Averty,[4] Charles-Henri de Novion,[1] and Ana Alamo[2]

Microstructural Evolution of Reduced Activation and Conventional Martensitic Steels after Thermal Aging and Neutron Irradiation

---

Reference: Mathon, M. H., de Carlan, Y., Geoffroy, G., Averty, X., de Novion, C. H., and Alamo, A., "**Microstructural Evolution of Reduced Activation and Conventional Martensitic Steels after Thermal Aging and Neutron Irradiation,**" *Effects of Radiation on Materials: 20th International Symposium, ASTM STP 1405,* S. T. Rosinski, M. L. Grossbeck, T. R. Allen, and A. S. Kumar, Eds., American Society for Testing and Materials, West Conshohocken, PA, 2001.

Abstract: Small-Angle Neutron Scattering (SANS) allows characterization, in a non-destructive way, of the heterogeneities (precipitates, pores, cavities,...) ranging in size between 1 and 100 nm. The SANS technique has been used in order to study the microstructural evolution of martensitic steels under neutron irradiation at 325°C up to a dose of 0.9 dpa, and after thermal aging between 250°C and 550°C up to 22 000 hours. The investigated materials were Fe-7.5/11Cr WVTa Reduced Activation and Fe-9/12Cr MoWV conventional martensitic steels. SANS experiments were coupled with Transmission Electron Microscopy (TEM) observations. The structural information obtained by SANS and TEM is compared with the evolution of mechanical properties. Three main results have been obtained: (1) above 8.8 at.% Cr in solid solution, the ferritic matrix is unstable at an aging temperature of 400°C and separates into a Cr-depleted α and a Cr-rich α' b.c.c. phase; (2) moreover, a strong radiation-accelerated or induced α' precipitation is observed in alloys irradiated at 325°C; (3) the precipitation of Laves phase $Fe_2(Mo,W)$ is strongly dependent on the Mo and W content in the alloys.

Keywords: 7–12% Cr martensitic steels, reduced activation, small-angle neutron scattering, thermal aging, neutron irradiation, Laves phase, α-α' phase separation

---

Commissariat à l'Energie Atomique, CEA-Saclay, F-91191 Gif-sur-Yvette, France:
[1] Laboratoire Léon Brillouin (Laboratoire commun CEA-CNRS),
[2] DTA/DECM/Service de Recherches Métallurgiques Appliquées,
[3] DSM/DRECAM/Laboratoire des Solides Irradiés,
[4] DRN/DRE/Service d'Etudes des Matériaux Irradiés

Introduction

Reduced Activation (RA) martensitic steels are candidates for internal structures of fusion reactors. The concept of Low Activation steels or Reduced Activation steels was introduced for new materials that offer benefits on maintenance operations and waste management [*1,2,3,4*]. For martensitic/ferritic steels, the main alloying elements such as molybdenum, niobium and nickel present in commercial steels are substituted by elements such as tungsten, vanadium and tantalum. These elements have a similar influence on the structure and the behavior, but exhibit a lower radiological impact. The assessment of potential reduced activation steels requires a good understanding of the microstructural features in correlation with the mechanical behavior. Consequently, much effort is under way to study the microstructural evolution in these materials, occurring during long thermal aging or neutron irradiation, and which may be the origin of hardening and embrittlement. In this context, two phenomena are of particular interest:

- the decomposition of the b.c.c. ferritic solid solution by a phase separation mechanism, resulting in an ultra fine-scale array of Fe-enriched α and Cr-enriched α' coherent phases or in very small α' particles,
- the formation of Laves phases $Fe_2(Mo,W)$.

In order to assess reduced activation materials and to understand the effect of the substitution of W for Mo and the role of Ta on the microstructural evolution during long-term thermal aging or neutron irradiation, several RA martensitic steels of different compositions, with a chromium content between 7 and 12 at.%, were studied by Small Angle Neutron Scattering (SANS) and Transmission Electron Microscopy (TEM). Indeed, SANS allows characterization, in these materials, of the mean size, shape and number density of very small precipitates, even at early aging times when they are not detected by TEM [*5*]. This has been shown recently for nanosized M_2C carbide precipitates responsible for secondary hardening in a Fe-9Cr-1Mo martensitic steel [*6*]. The SANS technique is particularly useful to detect and study the α-α' phase separation, which introduces very weak contrast for X-rays or electron scattering. Furthermore, since the ferrite matrix is ferromagnetic, the A(q) ratio of the magnetic and nuclear SANS contrasts between the matrix and particles gives information about their chemical composition [*7*].

This paper presents the results obtained on the RA materials in comparison with those acquired on two commercial martensitic steels, containing Mo and Ni. The results are discussed with regard to the mechanical behavior.

Experimental

Materials

The chemical compositions of the steels used in this study are listed in Table 1. The materials include two large heats of Reduced Activation steels of Japanese origin (respectively supplied by JAERI-NKK and Nippon Steel Corporation), F82H (Fe-7.5Cr-2W(V,Ta)) and JLF-1 (Fe-9Cr-2W(V,Ta)). In order to evaluate the role of chemical composition, and in particular of Cr, Ta and W contents, five additional RA martensitic steels (the "LA series," from AEA-Culham, UK, described in ref. [*2*]) were studied;

compared to LA12Ta (Fe-9.8Cr-1W(Ta,V)), the LA13Ta steel has a large W content (nearly 3%), the LA4Ta steel has a higher Cr content (11%), LA12LC does not contain Ta, and LA12TaLC has a low carbon content.

In addition, two conventional martensitic steels, already used in the nuclear industry, were also studied; these materials, which contain molybdenum and nickel, with respectively low and high Cr content, are EM10 (Fe-9Cr-1Mo) and HT9 (Fe-12Cr-1Mo-0.5W(V)) (supplied by Aubert et Duval and Sandvik respectively).

Table 1 - *Chemical analysis of experimental steels (wt %)*

Steel	Cr	W	Mo	Ni	Mn	Ta	V	Si	C	N
F82H	7.47	1.96		0.02	0.21	0.023	0.15	0.10	0.087	0.006
JLF-1	8.70	1.91			0.52	0.08	0.18	0.05	0.106	0.028
LA12TaLC	8.80	0.73			1.13	0.10	0.30	0.03	0.090	0.019
LA12LC	8.92	0.73			1.13	0.01	0.30	0.03	0.089	0.035
LA12Ta	9.86	0.84			0.88	0.10	0.28	0.03	0.155	0.043
LA13Ta	8.39	2.79			0.79	0.09	0.24	0.04	0.179	0.048
LA4Ta	11.08	0.72			0.78	0.07	0.23	0.03	0.142	0.041
EM10	8.76	<0.05	1.05	0.18	0.48		0.03	0.37	0.11	0.024
HT9	11.8	0.51	0.99	0.48	0.50		0.29	0.37	0.21	NA*

* Not Analysed

The alloys were austenitized at high temperature, then they were tempered at 750/800°C for 1 hour. The LA materials have been 10% cold rolled; this mechanical treatment has been shown to induce more stable impact properties after thermal ageing in the case of Fe-9Cr-1Mo(V,Nb) martensitic steels [*8*]. The details of the treatments for each material are given in Table 2. After these thermal treatments, a fully martensitic structure is obtained for all the alloys except for HT9, which contains a small amount of delta-ferrite (<5 %) in the martensitic structure.

Thermal aging and irradiation conditions

Samples were thermally aged under vacuum for various temperatures ranging between 250 and 550°C and for durations up to 22 000 hours. The list of thermal aging conditions for each alloy is given in Table 2.

Four materials (F82H, LA12LC, LA4Ta and HT9) were irradiated in the OSIRIS reactor at Saclay under PWR conditions, in the frame of the "Alexandre" irradiation [*9*]. The irradiation conditions were : temperature 325 ± 10 °C; fast neutron flux 2 10^{14} n/cm^2.s (E > 1 MeV); duration 43 days, corresponding to an average displacement dose of 0.9 dpa (displacement per atom).

SANS Experiments

Experimental set-up - The neutron scattering experiments were performed at the Laboratoire Léon Brillouin (Laboratoire Commun CEA-CNRS), Saclay, on PAXY and PAXE small-angle instruments [*10*]. The wavelength λ was 6 Å and sample-to-detector

Table 2 - *Metallurgical conditions before aging, thermal aging and irradiation conditions of LA and conventional martensitic steels*

Steel	Normalisation	Tempering	Cold-work	Thermal ageing	Irradiation
F82H	40 min. - 1040°C	1 h - 750°C	-	2000 h - 13 500 h 250 - 350 - 400 - 550°C	0.9 dpa 325°C
JLF-1	1 h - 1050°C	1 h - 780°C	-	2000 h - 13 500 h 250 - 350 - 400 - 550°C	-
LA12TaLC	40 min. - 1030°C	1 h - 750°C	10%	5000 h 250 - 400 - 450 - 550°C	-
LA12LC	40 min. - 1030°C	1 h - 740°C	10%	5000 h 250 - 400 - 450 - 550°C	0.9 dpa 325°C
LA12Ta	40 min. - 1030°C	1 h - 770°C	10%	2000 h - 10 000 h 250 - 350 - 400 - 550°C	-
LA13Ta	40 min. - 1030°C	1 h - 800°C	10%	2000 h - 10 000 h 250 - 350 - 400 - 550°C	-
LA4Ta	40 min. - 1030°C	1 h - 770°C	10%	2000 h - 10 000 h 250 - 350 - 400 - 550°C	0.9 dpa 325°C
EM10	30 min. - 980°C	0.5 h - 760°C	-	15 000 h 400-450-500°C	-
HT9	30 min. - 1050°C	2.5 h - 780°C	-	22 000 h 400-450-500°C	0.9 dpa 325°C

distance (D) was 2 m, covering a scattering vector (q) range from 0.3 to 1.6 nm^{-1} ($q = 4\pi \sin\theta / \lambda$, where 2θ is the scattering angle). Measurements have been made at room temperature, under saturating magnetic field H = 2T perpendicular to the incident neutron beam direction, in order to separate the magnetic and nuclear scattering cross-sections.

The samples for SANS were cut and chemically polished (in the hot cells of CEA/Saclay in the case of the irradiated steels). They were in the form of platelets, sizes ranging from $10 \times 10 \times 1$ mm^3 (for thermally aged samples) to $7 \times 7 \times 0.5$ mm^3 (for irradiated samples); the transmission value for thermal neutrons ($\lambda = 0.6$ nm) showed that multiple scattering corrections were negligible.

Data treatment - Details of preliminary data treatment (correction, normalization, and calibration) have been reported in [*5,7,11*].

In the case of ferromagnetic materials, the SANS intensity is composed of a nuclear and a magnetic contribution which depend respectively on the difference of composition and of magnetisation between particles and matrix. In terms of cross-section, the SANS intensity can be written as:

$$(d\Sigma / d\Omega)(q) \approx f_p \left[\Delta\rho^2_{nucl} + \Delta\rho^2_{mag} \sin^2 \alpha\right] F(q, h(R), S(q, R))$$

where q is the scattering vector, f_p the precipitated atomic fraction, and F(q, h(R),

S(q,R)) the specific structure factor of the scattering particles, depending on their form factor S(q,R) and on their size distribution h(R). $\Delta\rho_{nucl,mag}$ are the nuclear and magnetic contrasts given by:

$$\Delta\rho_{nucl,mag} = \frac{b^{p}_{nucl,mag}}{v^{p}_{at}} - \frac{b^{m}_{nucl,mag}}{v^{m}_{at}}$$

where b is the nuclear (nucl) or magnetic (mag) mean scattering length in the precipitates (p) or in the matrix (m), and $v^{p,m}_{at}$ is the mean atomic volume of the precipitate (p) and of the matrix (m). α is the angle between the magnetization of the sample and the scattering vector q.

As the magnetic moments are aligned parallel to the applied external magnetic field, the magnetic scattered intensity is zero in this direction and maximum in the perpendicular direction. Some information about chemical composition can be deduced from this anisotropy through the ratio between scattered intensities measured perpendicular and parallel to the sample magnetization, called "A ratio". For homogeneous particles, the A ratio depends only on the chemical compositions and on the atomic density variation between precipitates and the matrix, and is given by:

$$A = \frac{\left(d\Sigma/d\Omega\right)_{\perp\vec{H}}}{\left(d\Sigma/d\Omega\right)_{//\vec{H}}} = \frac{\Delta\rho^2_{nucl} + \Delta\rho^2_{mag}}{\Delta\rho^2_{nucl}} = 1 + \frac{\Delta\rho^2_{mag}}{\Delta\rho^2_{nucl}}$$

In principle, the A ratió can be calculated for a given matrix/precipitate couple, and compared to the experimental value in order to check the precipitate composition. But in fact A depends on several parameters which are not always known precisely, such as the atomic volumes in the matrix and in the precipitate, and the magnetic moments. So, to estimate the A ratio, we used atomic volumes deduced from tabulated unit cell parameters and magnetic moments from literature data on bulk materials. For example, the calculated A ratio in a pure iron matrix for TaC is 10, for VN 3.9 and for VC 2.2.

If the material contains several types of precipitates, the mean A ratio does not depend only of the contrasts $\Delta\rho$, but takes explicitly into account the form factor and the volume fraction of each family of precipitates; in this case, A is q-dependent.

Results

Initial microstructure of the materials

The microstructure and precipitation state of the initial materials was studied by TEM. The chemical compositions of precipitated phases (C and N excluded) was measured by Analytical Electron Microscopy (XEDS : X-ray Emission Dispersive Spectrometry) on carbon extraction replicas. TEM and XEDS results are summarized in Table 3.

In all materials, after normalisation and tempering treatments, the predominant precipitated phase is a chromium-rich $M_{23}C_6$ carbide. These particles are located preferentially along prior austenite grain boundaries and along lath boundaries. The

Table 3 - *Precipitation behaviour in as-received and thermally aged specimens obtained from Analytical Transmission Electron Microscopy.*

Steel	Aging	Type of precipitates	Composition of precipitates (at %)
EM10 9Cr-1Mo [*12*]	As-received	$M_{23}C_6$ M_2X	64Cr 30Fe 6Mo
	Aged 15 000 hours at 400°C	$M_{23}C_6$ (7) M_2X (11)	63Cr 27Fe 6Mo 4Mn 90Cr 9V
	Aged 15 000 hours at 500°C	$M_{23}C_6$ (11) M_2X (7) Laves phase (13)	62Cr 26Fe 7Mo 4Mn 86Cr 8Mo 3Fe 3V 40Fe 30Mo 15Cr 10Si 4Mn
HT9 12Cr-1Mo WV [*9*]	As-received	$M_{23}C_6$	
	Aged 22 000 hours at 450°C	$M_{23}C_6$ Laves phase	
	Aged 22 000 hours at 500°C	$M_{23}C_6$ Laves phase	41Fe 23Cr 19Mo 11Si 5W #
F82H 7.5Cr-2W VTa	As-received	$M_{23}C_6$(8)	60Cr 34Fe 5W 1V
	Aged 13 500 hours at 400°C	$M_{23}C_6$(5)	60Cr 34Fe 5W 1V
	Aged 13 500 hours at 550°C	$M_{23}C_6$(7)	66Cr 27Fe 5W 1V
JLF-1 9Cr-2W VTa	As-received	$M_{23}C_6$(15) M_4X_3 * MX (2)	55Cr 34Fe 8W 2V V rich 90Ta 6V 3.5Fe 2Cr 0.5W
	Aged 13 500 hours at 550°C	$M_{23}C_6$(6) M_4X_3 ** MX (5)	65Cr 29Fe 5W 1V V rich 86Ta 9V 4Fe 1Cr
LA4Ta 11Cr-1W VTa	As-received	$M_{23}C_6$(9) M_4X_3(1) MX (2) M_2X (3)	65Cr 28Fe 4W 3V 57V 32Ta 10Cr 1Fe 93Ta 4V 2Fe 1Cr 80Cr 17 V 2Fe 1W
	Aged 10 000 hours at 400°C	$M_{23}C_6$(9) M_4X_3 (7) MX (1)	66Cr 28Fe 3W 3V 68V 18Ta 13Cr 2Fe 86Ta 7V 5Fe 3Cr
LA12Ta 9.8Cr-1W VTa	As-received	$M_{23}C_6$(11) M_4X_3(4) MX (3)	63Cr 29Fe 4V 4W 77V 11Cr 9Ta 3Fe 90Ta 7V 2Fe 1Cr
	Aged 10 000 hours at 550°C	$M_{23}C_6$(7) M_4X_3(3) MX (1)	63Cr 29Fe 4V 4W 66V 15Ta 15Cr 4Fe 88Ta 6Fe 6V
LA13Ta 8.5Cr-3W VTa	As-received	$M_{23}C_6$(6) M_4X_3(5) MX (3)	59Cr 32Fe 7W 1V 68V 16Ta 13Cr 2.5Fe 80Ta 15V 4Fe 2Cr
	Aged 10 000 hours at 550°C	$M_{23}C_6$ M_4X_3 MX Laves phase (2)	49Fe 24Cr 25W 2V

(numbers in brackets indicate the number of individual particle spectra analysed)
* V rich precipitates, observed on carbon extraction replica
** V rich precipitates, not observed on carbon extraction replica but probably present
# The quantification of W and Si contents in HT9 was done from the W-L and Si-K X-ray emission lines with the software TRACOR TN 5400 from a registered library. However, the Si content value should be considered with caution, because of the difficulty to deconvoluate the Si-K from the W-M line.

chemical composition of the $M_{23}C_6$, determined by XEDS, does not differ for RA and conventional alloys : M ~ 60%Cr-35%Fe-5%W/Mo (in atomic %).

For the F82H steel, no other precipitation has been detected. In the JLF-1 alloy, some particles of Ta(C,N) and V-rich $M_4(C,N)_3$ have been observed. For the LA4Ta, LA12Ta and LA13Ta materials, the TEM observations show that V-rich M_4X_3 and Ta-rich M(C,N) coexist with $M_{23}C_6$; the LA12LC and LA12TaLC have not yet been observed by TEM, but similar precipitation can be expected (but without Ta(C,N) precipitates in LA12LC which is tantalum-free).

In the conventional steel EM10, another carbonitride second phase, $M_2(C,N)$, very rich in chromium, has also been detected by TEM.

Thermal aging: conventional alloys EM10 and HT9

The precipitated phases and their chemical compositions, determined by XEDS before and after aging, are reported in Table 3. The $M_{23}C_6$ particles do not evolve during thermal aging. The main aging effect observed by TEM is the formation at high temperature (450-500°C) of $Fe_2(Mo,W)$ Laves phase. This phase occurs as a continuous thin film on prior austenitic grain boundaries and interlath boundaries (see Figure 1).

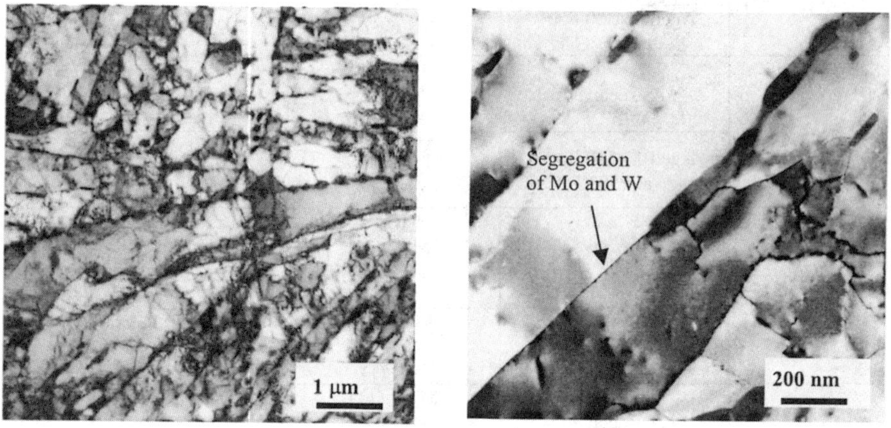

Figure 1 – *Microstructure of the conventional steel HT9 aged 22 000 hours at 450°C, observed by TEM. The presence of the Laves phase, which appears as an enhanced edging at the lath boundaries, has been checked by TEM microanalysis.*

In both materials, after each aging treatment, we observed an increase of the SANS intensity in comparison with the as-received samples. So, we present in Figures 2 and 3 the scattered SANS intensities, after subtraction of the reference (without thermal aging) contribution.

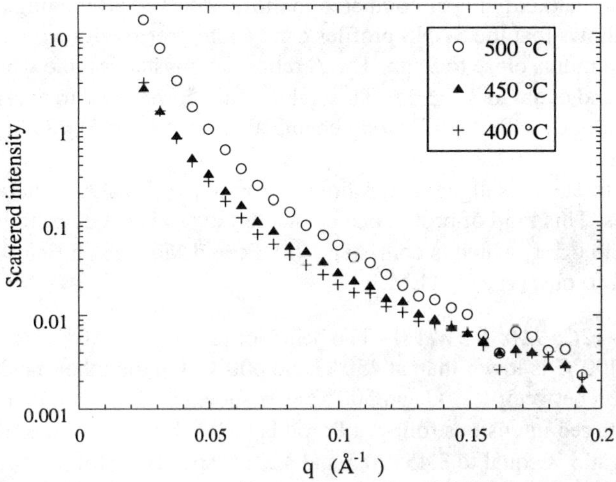

Figure 2 - *Intensities scattered perpendicularly to the applied magnetic field, obtained after subtraction of the reference as-received sample, measured on EM10 samples thermally aged 15 000 h at different temperatures*

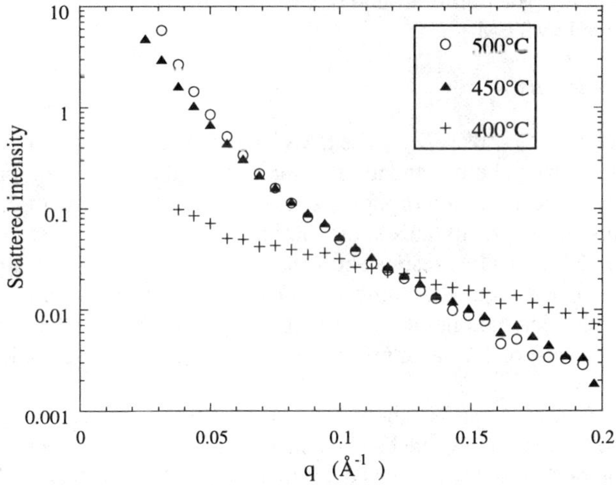

Figure 3 - *Intensities scattered perpendicularly to the applied magnetic field, obtained after subtraction of the reference as-received sample, measured on HT9 samples thermally aged 22 000 h at different temperatures*

EM10 (Fe-9Cr-1Mo) alloy - The scattered intensities after treatment at 400 and 450°C are similar and markedly lower compared to those observed after aging at 500°C. The data treatment shows that the SANS profiles can be interpreted with a distribution of particles with a mean radius close to 3 nm. The A ratios are constant on the whole measured q domain and equal to 3.5 ± 0.6. This value could correspond to several kinds of particles : M_2C carbides with the following chemical content M(at%)= 83Cr, 12Fe, 5Mo, or VC carbides.

At 500°C, the intensity is higher and follows a Porod law ($I \propto A/q^4$) characteristic of large size particles. This kind of profile can be due to Laves phases at grain boundaries. The A ratio is equal to 6 ± 1, which is consistent with Fe and Mo-rich particles such as the Laves phase Fe_2Mo observed by TEM.

HT9 alloy (Fe-12Cr-1Mo-0.5W(V)) - The behavior at 400 and 450°C is different and the intensity at 400°C is lower than at 450°C and 500°C. On the other hand, we note only a small evolution between 450°C and 500°C as is shown on Figure 3. At high temperature, the scattered intensities follow a Porod law ($I \propto A/q^4$) as for the EM10 alloy. Furthermore, the A ratio is equal to 3.45 ± 0.5 and 4 ± 1 respectively for 450°C and 500°C, which is consistent with Fe_2(Mo,W) Laves phase precipitates. Indeed, the substitution of Mo by W in the Laves phase induces an important decrease in the calculated A ratio, compared to the case of EM10 (containing no W) aged at 500°C.

At 400°C, the study of the form factor shows that the radius of the particles is around 1 nm. The associated A ratio is equal to 2 ± 0.4, in agreement with the value expected for α' clusters. With the assumption of α' clusters containing 90% Cr and 10% Fe, the precipitated volume fraction is around 0.25%. For this thermal treatment, no new phase has been observed by TEM.

Thermal aging: RA materials

F82H alloy (Fe-7.5Cr-2W(V,Ta) - The SANS intensities obtained on F82H samples present an anisotropy, before and after aging, even in the absence of a magnetic field. This is probably the result of a morphologic texture in the microstructure, i.e. a preferential orientation of the precipitates in the matrix.

The obtained SANS and TEM results show that, whatever the aging treatment, there is no evolution of the microstructure in comparison with the as-received sample. All the SANS intensities are identical taking into account the error bars. The scattered intensities are high at low q and correspond to large precipitate size, i.e. the $M_{23}C_6$ carbides initially present in the material.

At low q, the measured A ratio is about 4-4.5. Taking into account of the uncertainties discussed above, this value can be considered to be in agreement with the chemical composition of the $M_{23}C_6$ carbides determined by TEM-XEDS (i.e. M = 65%Cr-30%Fe-5%W), from which the calculated A ratio is 3.5.

JLF-1 alloy (Fe-9Cr-2W(V,Ta)) - As in the case of the F82H alloy, the variation of scattered intensity in comparison with the reference sample is too weak to be considered as representative of a microstructure evolution. At 550°C, the SANS intensity seems to increase at low q with aging, but this evolution is too weak to be interpreted.

Furthermore, the A ratio measured at low q, does not change with aging and is about 3-3.5. This value is slightly lower than for the F82H alloy, although the chemical composition of the $M_{23}C_6$ carbides is not significantly different in these two materials. The difference of the A ratio between the steels is significant and probably representative of the presence of additional or different phases. TEM observations have shown that this material contains Ta-rich MX and V-rich M_4X_3 carbonitrides in addition to $M_{23}C_6$ (see Table 3). Qualitatively, only the VC (A = 2.2) or V(C,N) particles can induce a decrease of the mean A ratio. Indeed, the presence of TaC carbides (A = 10) should sharply increase the A factor. So, to explain the results obtained on JLF-1, we have to assume an important contribution of vanadium carbide particles in the matrix.

LA4Ta alloy (Fe-11Cr-1W(Ta,V)) – For this alloy, which presents the highest Cr content among the studied RA materials, SANS data showed that, whatever the aging duration and temperature, the scattered intensity in the aged materials, in all the q range, is always clearly increased compared to the as-received material (Figure 4). This behavior could be explained by a modification of the carbonitride precipitation, observed directly by TEM: a dissolution of the M_2X precipitates, located preferentially near $M_{23}C_6$ (before thermal aging), which could enhance the precipitation of the M_4X_3 V-rich particles.

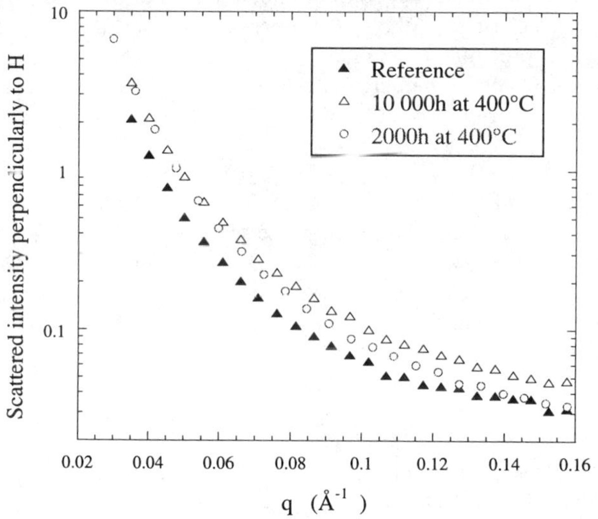

Figure 4 - *Intensities scattered perpendicularly to the applied magnetic field, measured on LA4Ta samples (as-received, aged 2000 and 10 000 h at 400°C)*

On the other hand, aging for 10 000 hours at 400°C, shows a further increase in the scattered intensity for the high q values ($q \geq 0.08$ Å^{-1}, see Figure 4): this must be due to the appearance of some new small particles, which had not precipitated after 2000 h at 400°C. The increment of SANS intensity from 2000 to 10 000 hours aging at 400°C is consistent with spherical nanoparticles of mean radius 1.5 nm. The corresponding A ratio is equal to 2, a value near the theoretical one for the chromium-rich b.c.c. α' particles. The precipitated volume fraction is around 0.1-0.2% if one assumes for α' the composition 90%Cr - 10%Fe.

LA12Ta alloy (Fe-9.8Cr-1W(Ta,V)) – This alloy is similar to the preceding, but with a lower Cr content. No change of the precipitation state was observed by TEM (see Table 3); consistently, no evolution of the SANS intensity was observed after 2000 h aging, whatever the temperature, or after 10 000 h at 550°C. However, a similar increase as for LA4Ta, was observed after thermal aging at $T \leq 400°C$ for times $\geq$ 10 000 hours, for the high q values ($q \geq 0.08$ Å^{-1}). This has been analyzed in the same way as for LA4Ta, as a precipitation of the α' Cr-enriched b.c.c. phase.

LA13Ta (Fe-9Cr-3W) – This steel has the highest W content among all the alloys studied. The comparison between the scattered intensities of thermally aged and as-received samples shows that thermal aging does not induce significant differences in the range 250 to 450°C up to 10 000 hours, in agreement with TEM observations.

This is not the case for thermal aging performed at 500°C. Indeed, in all the q domain, the scattered intensities of samples aged for 10 000 hours, but also for a shorter time (2000 hours), are significantly higher compared to the reference sample. TEM observations on the LA13Ta steel aged for 10 000 hours at 550°C show that the precipitation of Laves phase (Fe_2W type) occurs as a very thin film on grain and lath boundaries [*13*]. This precipitation explains the signal observed by SANS. The chemical composition of precipitates from microanalysis is close to 48Fe-25Cr-2V-25W (at %), in agreement with the value of the A ratio ($A \approx 5$, after subtracting the signal of the as-received sample).

LA12TaLC and LA12LC alloys - As for JLF-1 and F82H materials, we did not observe by SANS any significant variation of the scattered intensity after aging. The A ratio measured at low q is about 2.5 for the LA12LC alloy and 3 for the LA12TaLC alloy. These values are low compared to the other alloys. The A ratio is higher for the Ta rich alloy which is compatible with the presence of TaC in the matrix of LA12TaLC and not in the matrix of LA12LC.

Irradiated steels

Four materials were irradiated by fast neutrons to a dose of 0.9 dpa at 325°C : three RA materials, F82H, LA4Ta and LA12LC, and the conventional alloy HT9.

F82H alloy - The SANS signal is unchanged by the irradiation in the above conditions (see Figure 5) : this shows that, as for thermal aging reported previously, the precipitation state is unchanged by irradiation in this material. Literature data concerning this material are contradictory: the results reported here are in agreement with those of Kohno et al [*14*] who did not detect α' in F82H after irradiation up to 37 dpa between 365 and 750°C, but contradict those of Materna-Morris [*15*], who observed some α' phase and even cavities on the same alloy by TEM, after irradiation at 0.8 dpa at 250 and 400°C.

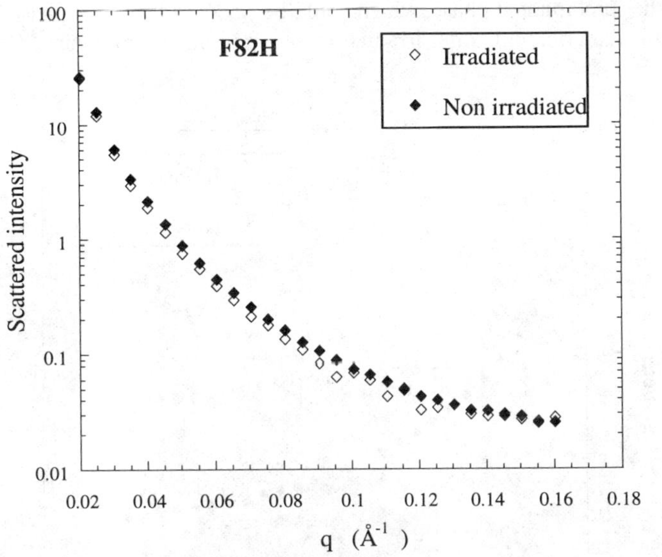

Figure 5 - *Scattered intensities measured on F82H samples, non irradiated and irradiated 0.9 dpa at 325°C.*

LA4Ta alloy - In this RA material, contrary to F82H, neutron irradiation induces a large increase in the SANS intensity at high q values ($q \geq 0.05$ Å^{-1}) (see Figure 6). The increment of SANS signal between irradiated and as-received samples shows a maximum around $q \sim 0.1$ Å^{-1} : this behavior, typical of concentrated alloys, indicates a spatial periodicity in the composition of the material, with a characteristic length $2\pi/q \sim$ 2-3 nm. The A value is in agreement with α-α' phase separation of the ferritic matrix. Two interpretations are possible : (1) isolated Cr-enriched α' precipitates surrounded by a spherical exclusion volume depleted in Cr, or (2) bipercolating α-α' two-phase system, close to the image of a spinodal decomposition. Modelling of the present SANS data is in progress. Detailed work on the thermally aged Fe-Cr binary alloy suggests that for the considered Cr concentrations (< 20%), the exclusion-sphere model is more adequate [*16*].

LA12LC alloy - This RA material shows a small increment of SANS signal after irradiation, indicative of some α' precipitation.

HT9 alloy - This conventional material shows a large increase of scattered intensity after irradiation. Taking into account its high Cr content (11.8%), this must include some α' contribution (the α' phase has been detected by TEM in irradiated HT9, but at higher temperature [*17*]). But, contrary to LA4Ta, the SANS increment induced by irradiation is a monotonically decreasing function of q and shows no maximum; moreover, the A ratio is not compatible with simple α-α' phase separation. Therefore, the material probably contains another precipitated phase, which could be M_6X; indeed, this phase has been observed in this material, only after irradiation, but at higher temperature [*18,19*].

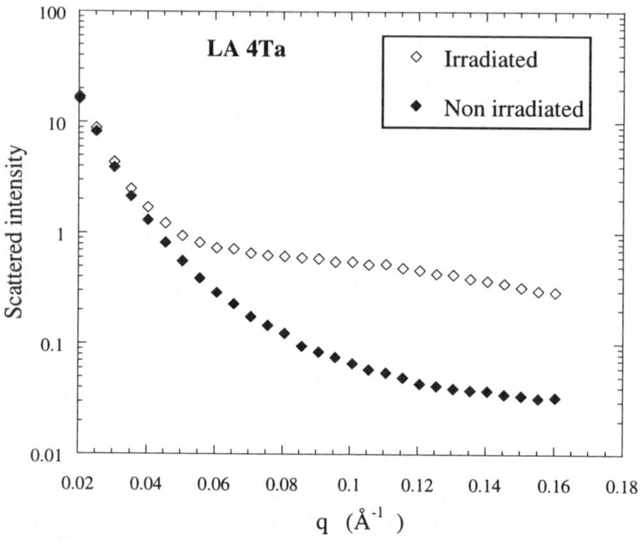

Figure 6 – *Scattered intensities measured on LA4Ta samples, non irradiated and irradiated 0.9 dpa at 325°C*

Discussion

Effect of the chemical composition on the microstructure evolution during thermal aging.

Effect of the Cr content on α - α' phase separation - The interpretation of α-α' phase separation requires the knowledge of the Cr concentration in the matrix, which differs from the bulk Cr concentration because of the presence of precipitates. The "THERMOCALC" software, associated with the SGTE ("Scientific Group Thermodata Europe") thermodynamical data base [*20*], allows to calculate the phase diagram and the chemical composition of each phase in a multicomponent system, by minimizing its free enthalpy. The Cr concentrations in the ferritic matrix of the various materials, obtained from the "THERMOCALC" software, are given in column a of Table 4. These calculations also show that the carbon concentration remaining in solid solution in the matrix after tempering ranges between 50 and 130 at. ppm.

For the martensitic steels containing 7.5 to 12% Cr, the major precipitated phase after tempering is $M_{23}C_6$ carbides, which dissolve a large amount of Cr; indeed, TEM observations show that in first approximation these carbides have the chemical formula $Cr_{15}M'_8C_6$, with M' = Fe, V, W or Mo. If one takes the average value of 100 at. ppm for the C concentration in the matrix, and assumes that all the precipitated carbon is in the form of $M_{23}C_6$ carbides, approximate values of the Cr concentrations in the ferritic matrix of the various materials can be calculated. These, given in column b of Table 4, are found to be very close to those obtained from the "THERMOCALC" software.

The SANS results show that the Cr-rich α' phase appears only around 400°C in the materials with a Cr concentration in the ferrite matrix close to or larger than 8.8 at.% (LA12Ta, LA4Ta and HT9 alloys, see Table 4). For LA4Ta aged 10 000 hours at 400°C, the volume fraction of α' phase estimated by SANS is 0.1-0.2 %, with a mean precipitate radius of 1.5 nm. This corresponds to an average distance between precipitates of approximately 50 nm, which is of the same order as the diffusion distance covered by Cr at this temperature.

Table 4 - *Conditions of appearance of α-α' phase separation in the ferrite matrix of several martensitic steels after thermal aging at 400°C or fast neutron irradiation at 325°C: correlation with the total Cr content and the Cr concentration in the ferritic matrix. The latter is obtained from a "THERMOCALC" calculation (column a) or from an approximation described in the text (column b).*

Alloy	Total Cr content (at.%)	Cr in matrix (at.%)		α/α' phase separation	
		a)	b)	Thermal aging at 400°C	Irradiation at 325°C 0.9 dpa - 1000 hours
LA13Ta	9.1	7.5	7.0	No after 10 000h	
F82H	8.1	7.2	7.1	No after 13 500 h	No
EM10	9.3		8.2	No after 15 000 h	
JLF-1	9.4	8.4	8.2	No after 13 500 h	
LA12TaLC	9.4	8.5	8.4	No after 5000 h	
LA12LC	9.5	8.7	8.5	No after 5000 h	Yes
LA12Ta	10.5	8.9	8.8	Yes after 10 000 h	
LA4Ta	11.8	10.3	10.2	Yes after 10 000 h	Yes
HT9	12.5	10.7	10.2	Yes after 22 000 h	Yes *

* In irradiated HT9 steel, another phase is also present

Effect of Mo and W content on Laves phase formation – The formation of the Laves phase is only observed at high temperature on materials sufficiently charged in W (LA13Ta, 0.8% at. W) or Mo (EM10 and HT9, 0.6 % at. Mo).

One can note that the appearance of Mo-rich Laves phase in conventional steels occurs at lower temperature (450°C for EM10, 400-450°C for HT9) than W-rich Laves phase in LA13Ta (550°C). These results confirm the observations of Tamura et al [*21*] who show, in a TTT diagram, a shift of the "nose" of Laves phase precipitation towards high temperatures for W-containing compared to Mo-containing steels.

Effect of Ta content – The LA12TaLC (0.1% Ta) and LA12LC (Ta-free) steels show similar microstructure stability under thermal aging; so, no effect of tantalum on precipitation evolution has been observed.

Precipitation under irradiation

Radiation-induced α-α' phase separation in model Fe-Cr alloys was observed first by Little et al [22] in samples containing 10%Cr after fast neutron irradiation (30 dpa) at

420°C, and more recently in Fe-10 to 25%Cr after electron irradiation at 300 and/or 200°C [*23*].

In the present martensitic steels, our results show that under irradiation α - α' phase separation in ferrite occurs at lower temperature (325°C) and lower Cr concentration in the matrix (8.5%) than observed by thermal aging. Indeed, the duration of the irradiation experiments reported here, around 1000 hours, does not allow to obtain α - α' phase separation by simple thermal vacancy diffusion mechanism (in these conditions, the thermal diffusion length of Cr at 325°C is of the order of 1 nm). One has therefore to involve a direct effect of irradiation; on one hand, irradiation by fast energetic particles is known to accelerate the nucleation and growth rates for precipitation, by inducing an excess of vacancies and interstitials ("irradiation-enhanced" mechanism); furthermore, from coupling between migrating point defects and solute atoms, it can induce non-equilibrium phases or modify the composition range of existing phases ("irradiation-induced" mechanism) [*24*].

Although our results follow an explanation based on simple irradiation-enhanced diffusion and thermal equilibrium phase separation response, recent work on electron-irradiated model Fe-Cr alloys suggests that the true mechanism of α' precipitation might be more complex [*23*].

Comparison with mechanical properties

After thermal ageing - The materials with high Cr content (LA4Ta, HT9 and to a lesser extent LA12Ta) show a decrease of their toughness after aging at 400°C [*13,25*]: the precipitation of α' phase detected by SANS in these alloys can be used to explain this behavior, which up to now could not be clearly related to a microstructural evolution.

In some of the 9-12%Cr(W,Mo) steels, aging at high temperature (450-550°C) leads to the formation of Laves phase, which is more damaging for the toughness properties than the α' phase. The SANS technique allowed detection of this phase which precipitates, at these temperatures, as films on the grain and lath boundaries. Nevertheless, SANS in this case does not allow determination of the quantity of precipitated Laves phase. In the case of LA13Ta alloy, the SANS signal saturates after 2000 h aging at 550°C, although the Charpy impact tests show that between 2000 and 10 000 h, the toughness properties deteriorate (the Upper-Shelf Energy (USE) decreases by 30 J/cm^2, and the brittle-to-ductile transition temperature increases by 20°C). A similar observation is made for the HT9 alloy, where the SANS signal after 22 000 h aging at 450°C is identical to the one after aging at 500°C, although the toughness properties degrade significantly between these two aging temperatures.

Finally, the SANS measurements on EM10 steel show the formation of an unidentified nanometric phase after aging at 400 and 450°C. The mechanical tests on this material showed a slight hardening, which could be explained by a precipitation of this unidentified phase.

After irradiation - Figure 7 shows the evolution of the yield stress and of the Reduction in Area to Failure (RAF) of the irradiated materials characterized by SANS. Because these materials have not been tested in the same metallurgical state, the comparison is problematic. It appears that the 10% cold-working applied to the "LA series" (e.g. LA4Ta and LA12LC) somewhat masks the effect of chemical composition: whatever the composition, an equivalent irradiation-induced hardening is obtained [9,26].

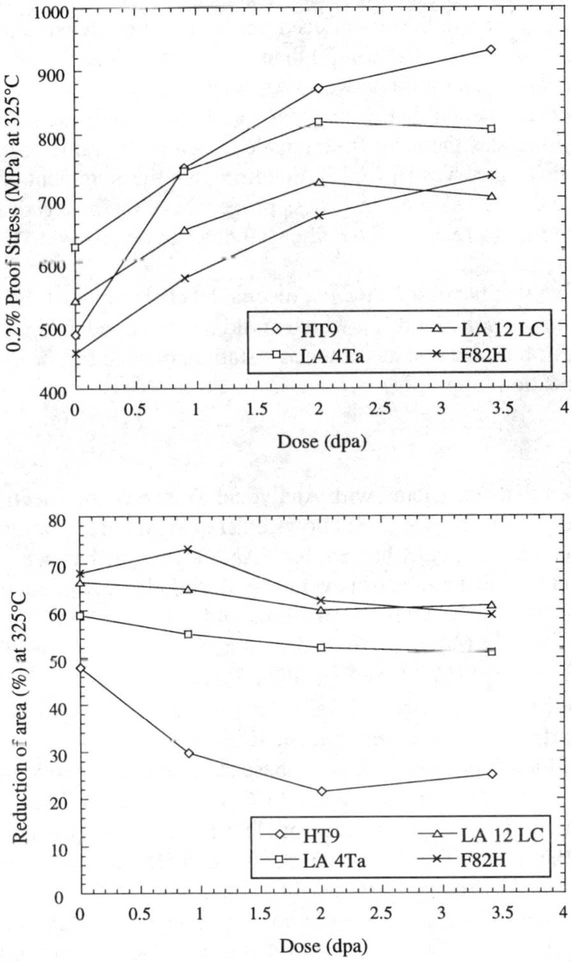

Figure 7 – *Evolution of the yield stress and of the reduction of area at failure versus irradiation dose measured at 325°C for HT9, LA4Ta, LA12LC and F82H*

On the other hand, for the F82H steel which, according to our work, does not show new phases or cavities after irradiation at 0.9 dpa and 325°C, one observes an increase of the yield stress by more than 100 MPa; this seems to be mainly due to the occurrence of radiation defects, such as dislocation loops.

For the HT9 steel, which has also been normalized and tempered but not cold-worked, and which contains more Cr, the hardening after 0.9 dpa neutron irradiation at 325°C is largely more important (250 MPa) and might be explained by the presence, in addition to radiation defects, of the new phases detected by SANS. This confirms a general observation that the irradiation-induced hardening of martensitic steels is more important for Cr-rich materials (10-12% Cr) than for the materials containing 9% Cr, whatever the irradiation temperature between 70 and 370°C [*27*].

It has been recently shown that, for a great variety of martensitic steels, a linear empirical relationship relates the RAF from tensile tests and the value of the USE in the brittle-to-ductile transition curves [*9,13,25*]; therefore, the measurement of the RAF allows us to estimate the decrease of toughness properties of the material. Here again, for the HT9 steel, the strong decrease of RAF after 0.9 dpa can be related to the appearance of new phases in the material.

One should note that beyond 2 dpa, the mechanical behaviour of the four materials studied seems to saturate (see Figure 7), which could be explained either by a saturation of the precipitation, or because one has reached a stationary state for the creation and annihilation of dislocation loops.

Conclusions

We have shown that, associated with Analytical Transmission Electron Microscopy, Small-Angle Neutron Scattering (SANS) allows a detailed characterization of the microstructure of martensitic steels. Indeed, the SANS technique has been used successfully to detect the appearance of new phases during thermal aging or irradiation. In particular, we have been able to determine a "threshold concentration" of Cr in solid solution in the ferrite matrix for the precipitation of nanometric α' phase after thermal ageing at 400°C: for Fe-7.5/12Cr-MoW type alloys, it is close to 8.8 at.%; the α' precipitation, detected up to now only by SANS, allows explanation for the increase in embrittlement of Cr-rich materials after aging at 400°C. We have also shown that SANS allowed detection of the appearance of Laves phase at the interfaces in some materials.

After irradiation at a dose of 0.9 dpa at 325°C, the appearance of α' phase is also detected for the Cr-rich materials; in the case of HT9 steel, it is associated with the precipitation of another phase. For this material, the hardening measured after irradiation could be partly explained by the appearance of these phases. For the other materials, and in particular for the steel F82H, where we did not detect any precipitation or cavities, the hardening must originate from the formation of dislocation loops or point defect clusters, which give only a small (undetected) SANS contribution.

The characterization by SANS of materials irradiated at higher dose (> 1 dpa) has not been made yet, and it might be difficult to realize, because of the residual (radio)activity of the materials. Nevertheless, it would be especially interesting to quantify the precipitation kinetics and to try to separate the effect of irradiation-induced defects, such as dislocation loops, from precipitation phenomena.

Acknowledgments

We wish to thank G. Jaskierowicz (LSI, Palaiseau) and I. Monnet (SRMA, Saclay) for their help during the TEM experiments, and J.C. Brachet (SRMA, Saclay) for useful discussions.

References

[1] Klueh, R.L., Maziasz, P.J., and Corwin, W.R., "Development of Ferritic Steels for Fusion Reactor Applications," *ORNL - 6472*, August 1988.

[2] Dulieu, D., Tupholme, K.W., and Butterworth, G.J., "Development of Low-Activation Martensitic Stainless Steels," *Journal of Nuclear Materials* 141-143, 1986, pp. 1097-1101.

[3] Tamura, M., Hayakawa, H., Tanimura, M., Hishinuma, A., and Kondo, T., "Development of Potential Low Activation Ferritic and Austenitic Steels," *Journal of Nuclear Materials* 141-143, 1986, pp. 1067-1073.

[4] Ehrlich, K., Cierjacks, S.W., Kelzenberg, S., and Möslang, A., "The Development of Structural Materials for Reduced Long-Term Activation," *Effects of Radiation on Materials : 17th. Int. Symposium, ASTM STP 1270*, 1996, pp. 1109-1122

[5] Albertini, G., Carsughi, F., Coppola, R., Rustichelli, F., and Stefanon, M., "Radiation Damage Studies using Small-Angle Neutron Scattering," *Effects of Radiation on Materials : 17th. Int. Symposium, ASTM STP 1270*, 1996, pp. 206-219.

[6] Mathon, M.H., Geoffroy, G., de Carlan, Y., Alamo, A., and de Novion, C.H., "SANS Study of the Microstructural Evolution of Martensitic Steels under Thermal Ageing and Neutron Irradiation," *Physica B* 276-278, 2000, pp. 939-940.

[7] Mathon, M.H., Barbu, A., Dunstetter, F., Maury, F., Lorenzelli, N., and de Novion, C.H., "Experimental Study and Modelling of Copper Precipitation under Electron Irradiation in Dilute *Fe*Cu Binary Alloys," *Journal of Nuclear Materials* 245, 1997, pp. 224-237.

[8] Alamo, A., Brachet, J. C., Castaing, A., and Foucher, C., "Effects of the Initial Metallurgical Condition on the Microstructure and Mechanical Properties of 9Cr-1Mo and 9Cr-1MoVNb Thermal Aged Steels," *Proc. of the Symposium on Microstructure and Mechanical Properties of Aging Materials II"*, The Minerals, Metals & Materials Society, 1996, pp. 121-134.

[9] Brachet, J.C., Averty, X., Lamagnère, P., Alamo, A., Rozenblum, F., Raquet, O., and Bertin, J.L., "Behavior of different Austenitic Stainless Steels and Conventional, Reduced Activation (RA) and ODS Chromium Rich Ferritic-Martensitic Steels

under Neutron Irradiation at 325°C for PWR Environment," *Effects of Radiation on Materials : 20th. Int. Symposium, ASTM STP 1405*, (2002).

[10] "*Equipements expérimentaux*", Ed. by Laboratoire Léon Brillouin, CEA Saclay, 91191 Gif-sur-Yvette, France, 1995, p. 67.

[11] Mathon, M.H., and de Novion, C.H., "De l'intensité à la structure des matériaux," *Journal de Physique IV (France)* 9, 1999, pp. 127-146.

[12] Brachet, J.C., "Alliages Martensitiques 9Cr-1Mo : Effets de l'addition de l'azote, du niobium et du vanadium sur la microstructure, les transformations de phase et les propriétés mécaniques," (in french), CEA-Report R-5581, France, 1991.

[13] de Carlan, Y., Alamo, A., Mathon, M.H., Geoffroy, G., and Castaing, A., "Effect of Thermal Ageing on the Microstructure and Mechanical Properties of 7-11CrW RAF/M Steels," Proc. of *ICFRM 9, Colorado Springs, Colorado, USA, Oct. 11-15 1999,* to be published in the *Journal of Nuclear Materials.*

[14] Kohno, Y., Gelles, D.S., Kohyama, A., Tamura, M., and Hishinuma, A., "Irradiation Response of Reduced Activation Fe-8Cr-2W Martensitic Steel (F82H) after FFTF Irradiation," *Journal of Nuclear Materials* 191-194, 1992, pp. 868-873.

[15] Materna-Morris, E.I., Rieth, M., and Ehrlich, K., "Mechanical Properties and Microstructure of HFR-Irradiated Ferritic/Martensitic Low-Activation Alloys," *Effects of Radiation on Materials : 19th. Int. Symposium, ASTM STP 1366*, 2000, pp. 597-611.

[16] Bley, F., "Neutron Small-Angle Scattering Study of Unmixing in Fe-Cr Alloys," *Acta Metallurgica et Materialia* 40, 1992, pp. 1505-1517.

[17] Kai, J.J., and Klueh, R.L., "Microstructural Analysis of Neutron-Irradiated Martensitic Steels," *Journal of Nuclear Materials* 230, 1996, pp. 116-123.

[18] Dubuisson, P., Gilbon, D. and Séran, J.L., "Microstructural Evolution of Ferritic-Martensitic Steels Irradiated in the Fast Breeder Reactor Phénix," *Journal of Nuclear Materials* 205, 1993, pp. 178-189.

[19] Maziasz, P.J., "Formation and Stability of Radiation-Induced Phases in Neutron-Irradiated Austenitic and Ferritic Steels," *Journal of Nuclear Materials* 169, 1989, pp. 95-115.

[20] Sundman, B., Janssen, B., and Anderson, J.O., "The Thermo-calc databank system" *Calphad* 9 (2), 1985, pp. 153-190.

[21] Tamura, M., Hayakawa, H., Yoshitake, A., Hishinuma, A., and Kondo, T., "Phase Stability of Reduced Activation Ferritic Steel : 8% Cr-2%W-0.2%V-0.04%Ta-Fe," *Journal of Nuclear Materials* 155-157, 1988, pp. 620-625.

[22] Little, E.A. and Stowe, D.A., "Void-Swelling in Irons and Ferritic Steels - II. An Experimental Survey of Materials Irradiated in a Fast Reactor," *Journal of Nuclear Materials* 87, 1979, pp. 25-39.

[23] Jaquet, V., "Effet de l'irradiation sur la demixtion des alliages modèles Fe-Cr autour de 15% de chrome," (in french), PhD thesis, Laboratoire des Solides Irradiés, Ecole Polytechnique, 91128 Palaiseau, France (6 march 2000).

[24] "*Materials under Irradiation*," Ed. Dunlop A., Rullicr-Albenque, F., Jaouen, C., Templier C., and Davenas, J., Trans Tech Publications – Solid state phenomena, Volumes 30 & 31, 1993 - see articles of Bellon, P. and Martin, G., pp. 107-148, and of Barbu, A. and Martin, G., pp. 179-228.

[25] Brachet, J. C., Castaing, A., and Foucher, C., " Effects of the Chemical Composition on the Behavior of 9-12Cr/0.5-1Mo/(W,V,Nb) Thermal Aged Steels," Proc. of *Int. Symposium on Materials ageing and component life extension, Milan, Italy, 10-13 Oct. 1995*, Engineering Materials Advisory Services Ltd., UK, 1995, pp. 75-87.

[26] Alamo, A., Horsten, M., Averty, X., Materna-Morris, E.I., Rieth, M. and Brachet, J.C., "Mechanical Behaviour of Reduced Activation and Conventional Martensitic Steels after Neutron Irradiation in the range 250-450°C," Proc. of *ICFRM 9, Colorado Springs, Colorado, USA, Oct. 11-15 1999,* to be published in the *Journal of Nuclear Materials.*

[27] Horsten, M.G., Van Osch, E.V., Gelles, D.S., and Hamilton, M.L., "Irradiation Behavior of Ferritic - Martensitic 9-12 Cr Steels," *Effects of Radiation on Materials: 19th. Int. Symposium, ASTM STP 1366*, 2000, pp. 579-593.

Alexandre V. Kozlov[1], Irina A. Portnykh[1], Leonid A. Skryabin[1], and Sergey S. Lapin[2]

Dimensional Characteristics of Displacement Cascades in Austenitic Steels under Neutron Irradiation at Cryogenic Temperature

---

Reference: Kozlov, A. V., Portnykh, I. A., Skryabin, L. A., and Lapin, S. S., "**Dimensional Characteristics of Displacement Cascades in Austenitic Steels under Neutron Irradiation at Cryogenic Temperature,**" *Effects of Radiation on Materials: 20th International Symposium, ASTM STP 1405,* S. T. Rosinski, M. L. Grossbeck, T. R. Allen, and A. S. Kumar, Eds., American Society for Testing and Materials, West Conshohocken, PA, 2001.

ABSTRACT: Samples of 16Cr - 15 Ni - 3 Mo - 1Ti steel were irradiated to a fast neutron fluence of $1.5 \cdot 10^{18}$ cm $^{-2}$ and $1.5 \cdot 10^{19}$ cm $^{-2}$ (E > 0.1 MeV) in the IVV - 2M reactor and investigated by transmission electron microscopy methods. Electron microscopy patterns were obtained. It was shown that the size distribution of clusters had a form close to a Gaussian after irradiation to a fluence of $1.5 \cdot 10^{18}$ n·cm $^{-2}$. There were several peaks in the distribution histogram of the sample irradiated to $1.5 \cdot 10^{19}$ n·cm $^{-2}$.

The formation of the low temperature irradiation clusters in austenitic steel was analyzed with a model of displacement cascades overlapping under cryogenic neutron irradiation. All the clusters formed were differentiated into two groups: low-temperature radiation clusters formed by not overlapped displacement cascades (LTRC - 1) and radiation clusters formed by overlapped cascade regions (LTRC - 2). Analytical expressions were obtained to describe the relationships of the average dimensions of LTRC of both groups and the dynamics of cluster accumulation.

Keywords: D01, M05, N01, R01, S05

Introduction

Physical and mechanical properties of austenitic steels are changed essentially under cryogenic neutron irradiation even at rather small damage doses (~0.01 dpa). The yield strength increases considerably, plasticity is twice reduced, electric resistance increases by ~5 %, and Young's modulus similarly decreases [*1*]. These changes are connected with radiation defect formation, the main type of which are low-temperature radiation clusters.

---

[1]Doctor, Head of Post-Irradiation Examination Department, SUSE "SB RDIPE", P.O.Box 29, 21ª Klara-Tsetkin St.-60, Zarechny, 624051, Sverdlovsk region, Russia.

[2]Doctor, Leading scientist, Laboratory of mechanical properties, IMP UB RAS 18 S. Kovallevskaya St., 620219 Ekaterinburg, Russia

The knowledge of characteristics of the defects is important not only for description of their formation, evolution and kinetics of accumulation, but also for the purpose of predicting physical and mechanical property changes.

This paper presents examination results of low-temperature radiation clusters formed in the 16Cr-15Ni-3Mo-1Ti austenitic steel at room temperature after its irradiation in the IVV-2M reactor up to fast neutron fluences $1.5*10^{18}$ cm^{-2} and $1.5*10^{19}$ cm^{-2} (E > 0.1MeV) at 77 K.

Experimental Results

Experimental examinations were performed by transmission electron microscopy methods on JEM-2000 EX. Samples were studied at room temperature. Low-temperature radiation clusters (LTRCA) were observed after warming up to the temperature (annealing up to room temperature). The main feature of low-temperature radiation clusters (LTRC) is lower a vacancy concentration, hence lower contrast in comparison with radiation clusters formed at room temperature and higher [*1*]. In order to observe such objects the two-beam diffraction mode was used [*2*]. LTRCA were objects of "black-and-white" contrast having round- or oval-shaped form (Figure1) and with a direction determined by the *g* vector joining bright diffraction peaks on an electron diffraction pattern.

Figure 1 – *A View of Low-Temperature Radiation Clusters in 16Cr16-15Ni-3Mo-1Ti Steel Irradiated up to 0.016 dpa at 77 K in the IVV-2M Reactor*

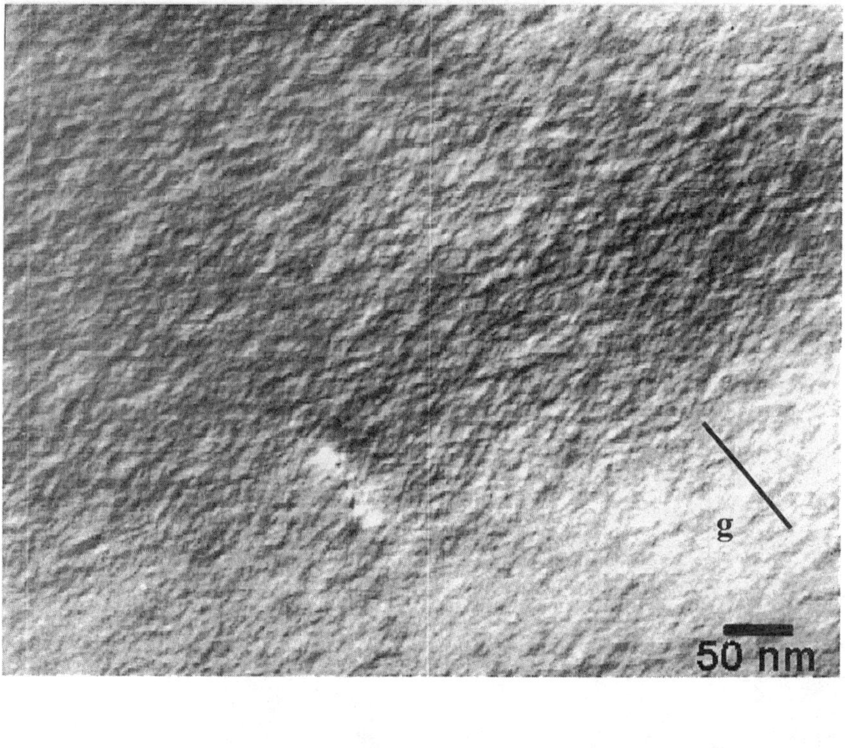

LTRCA identified on electron-microscopy photos were measured and the histograms of LTRCA size distribution were plotted. A histogram was plotted using ~750 measured objects of a tested sample irradiated up to a fast neutron fluence $1.5 \cdot 10^{18}\ n \cdot cm^{-2}$ (E > 0.1 MeV) (or 0.0016 dpa). The size distribution of a sample irradiated up to a fast neutron fluence $1.5 \cdot 10^{19}\ n \cdot cm^{-2}$ (0.016 dpa) was determined by measuring ~430 objects.

A view of the histogram obtained on the sample with lower fluence is shown in Figure 2.

Figure 2 – *Histogram of LTRC Distribution on Sizes in 16Cr16-15Ni-3Mo-1Ti* Steel *Irradiated up to 0.0016 dpa at 77 K in the IVV-2M Reactor (a – One Modal Description, b – Decomposition on Two One-Modal Distributions)*

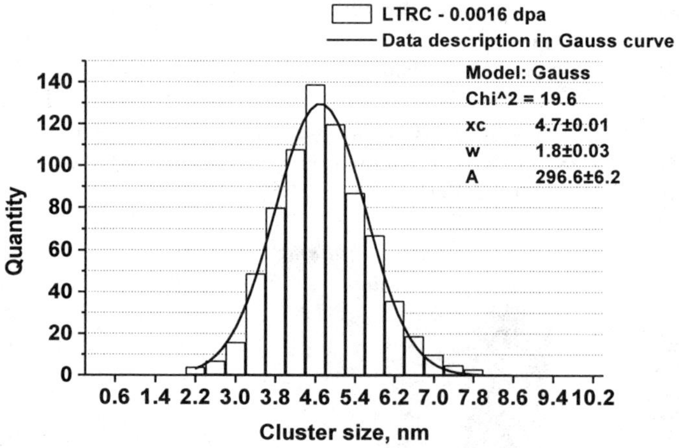

a

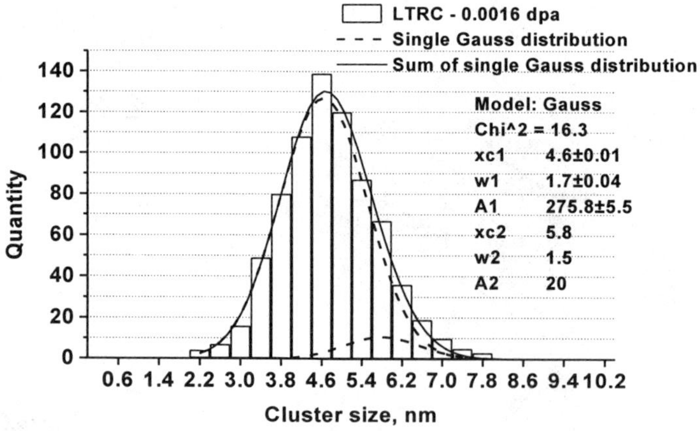

b

This picture presents both density f(x) of a Gaussian LTRCA size distribution and also distribution parameters obtained by optimization methods.

$$f(x) = \frac{A}{w \cdot \sqrt{\pi / 2}} \cdot \exp(-\frac{2(x - x_c)^2}{w^2}) \qquad (1)$$

where

x_c = distribution center
w = dispersion
A = a normalization factor (an area under the curve on the diagram of a density distribution)
Ch^2 = criterion χ^2, characterizing a quality of a description of experimental data with the help of a distribution found out [*3*].

A normal distribution describes the measurement data satisfactory with a distribution center of ~4.7 nm. The value of corresponds to probability ~15 % that the theory does not conflict with the experiment. According to Reference [*3*] it means that such hypothesis is acceptable. One can see a little asymmetry of the histogram. An attempt to describe the histogram as a sum of two one-modal Gaussian distributions. The value χ^2 = 16.3 corresponds to probability more than 30 % [*3*], (Figure 2b). It speaks that LTRCA are probably objects exhibiting two different distributions in this sample: near 93 % of LTRC have average size ~4.6 nm and 7 % have average size ~5.8 nm.

Figure 3 shows results from examination of a sample irradiated up to 0.016 dpa at 77 K. One can see that the histogram is not symmetrical concerning an axis transited through a maximum and therefore it is not described by one-modal distribution.

Figure 3 – *Histogram of LTRC Distribution on Sizes in 16Cr16-15Ni-3Mo-1Ti* Steel *Irradiated up to 0.016 dpa at 77 K in the IVV-2M Reactor*

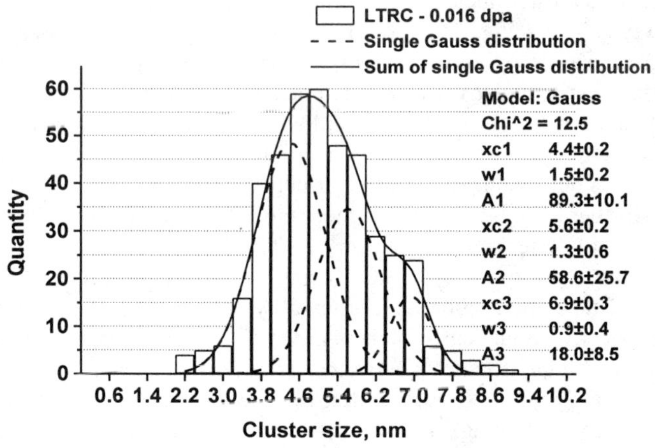

A harmonic analyses allows the observed distribution to be represented as a sum of three one-modal Gaussian distributions with close characteristics of a half-width and different distribution centers. Three types of LTRCA are presented in the sample according to characteristics obtained: 53 % - with average size ~4.4 nm, 35 % - ~5.6 nm, 12 % - ~6.9 nm.

A Model of Cascade Area Overlaps

In order to analyze the formation and evolution of LTRC the model described in Reference [*4*] was used, according to which a low-temperature radiation cluster forming in austenitic steel under cryogenic irradiation is an area enriched by point defects having the following structure. An internal part (nucleus) consists of an excess amount of vacancies (~6 %). Then there is a layer with thickness ~2*b* (*b* – lattice period) containing approximately an equal amount of vacancies and interstitials which did not recombine at 77 K due to their poor mobility (low layer of cluster "atmosphere"). The outside layer is enriched with interstitials (~1 %) which had time to migrate into this area during thermal peak. The whole height of the cluster "atmosphere" is equal to a nucleus radius.

During LTRC formation by subsequent cascade, the new cluster may be superposed on an already existing LTRC. Various possibilities for their relative spacing can occur. We can distinguish two significantly different types of spacing of LTRC interacting among themselves, as shown in Figure 4.

The first type of spacing is as follows, atmospheres containing only superposed interstitials, Figure 4a. In this case LTRC were subjected to no significant changes. The interstitial concentration is doubled in the area of atmosphere, and vacancy concentration is decreased in the area of superposition of a nucleus with atmosphere. These LTRC maintain their individuality and become vacancy clusters after annealing up to room temperature. This case occurs when the distance between LTRC is larger than the diameter of a cluster nucleus.

The second type of spacing, - nuclei of LTRC are superposed against each other, figure 4b. In this case a process of transformation takes place in one area and point defects of one sign (in particular vacancies) interact among themselves in others. Thus oval- or dumb-bells-shaped LTRC are formed, their shape depends on the distance between LTRC centers. Such type of spacing we shall term "overlapping". Arrival of a new cascade area is not accompanied with formation of a new LTRC, only an existing LTRC (further LTRC-1) is transformed into LTRC of other type (further we shall term as "double low-temperature cluster" – LTRC-2).

Let's express an average size of LTRC-2 using the size characteristics of LTRC-1. For this purpose we take into account that probability of formation of a double cluster with the distance r between nucleus centers is proportional to r^2. Probability density of LTRC-2 formation with the distance r between nucleus centers, using normalization requirements, is equal $f(r) = 3r^2/(8\, r_{n1}^3)$, where r_{n1} = an average radius of LTRC-1. From here follows that the average distance between nuclei in LTRC-2 is equal:

$$r_{n2} = 3/(8 \cdot r_{n1}{}^{3}) \int_{0}^{2r_{n1}} r^{3} dr = 1.5 \cdot r_{n1} \qquad (2)$$

Hence an average LTRC –2 has a form close to oval-shaped one, fig. 5a, with a relation of the longitudinal and cross sizes 5.5 : 4, and sizes of its elongated nucleus as 3.5 : 2 respectively.

Figure 4 – Schematic View of Low-Temperature Radiation Clusters (LTRC) Produced under Various Type of Cascade Area Superposition (a – LTRC-1 Maintaining Individuality, b – "Overlapping" of LTRC Takes Place, Resulted in LTRC-2 Formation)

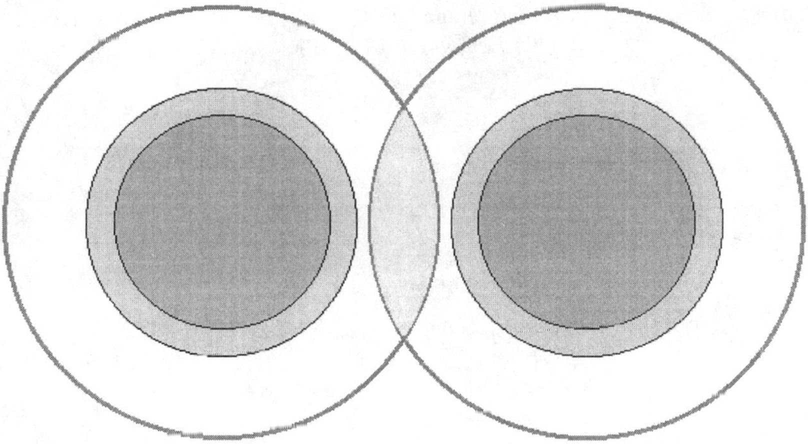

a

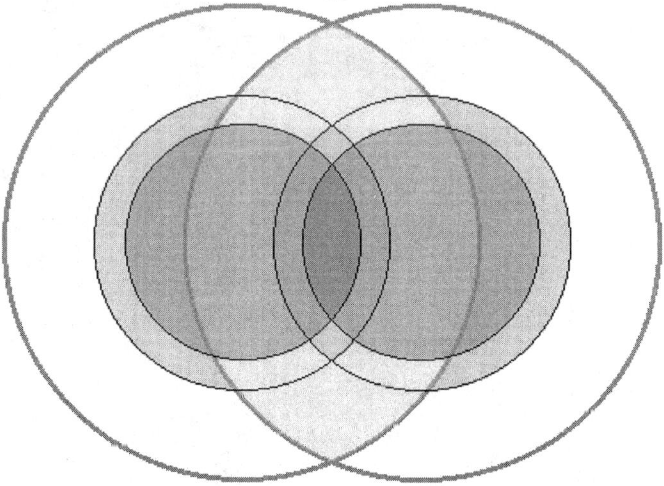

b

It is convenient to use an approximation of average spherical clusters for describing a series of processes of LTRC-2 isotropic spacing, in particular their interaction with dislocations. Thus the sizes of an equivalent spherical LTRC-2 are obtained under condition when it both occupies the same volume and contains the same amount of vacancies in the nuclei and in the low layer of cluster atmosphere as an oval-shaped cluster, Figure 5b. From geometrical relationships it is possible to calculate volumes of overlapping areas and a number of point defects transforming in clusters nucleuses and clusters atmosphere during formation of oval-shaped LTRC-2. Omitting simple but unwieldy evaluations, for example, we present the equations to calculate a volume of overlapping of nuclei between themselves – v_{nn} and a volume of overlapping of nuclei with clusters atmosphere – v_{na}

Figure 5 - Schematic View of Average LTRC-2 (a – a View of Average LTRC-2, *b – Equivalent LTRC-2)*

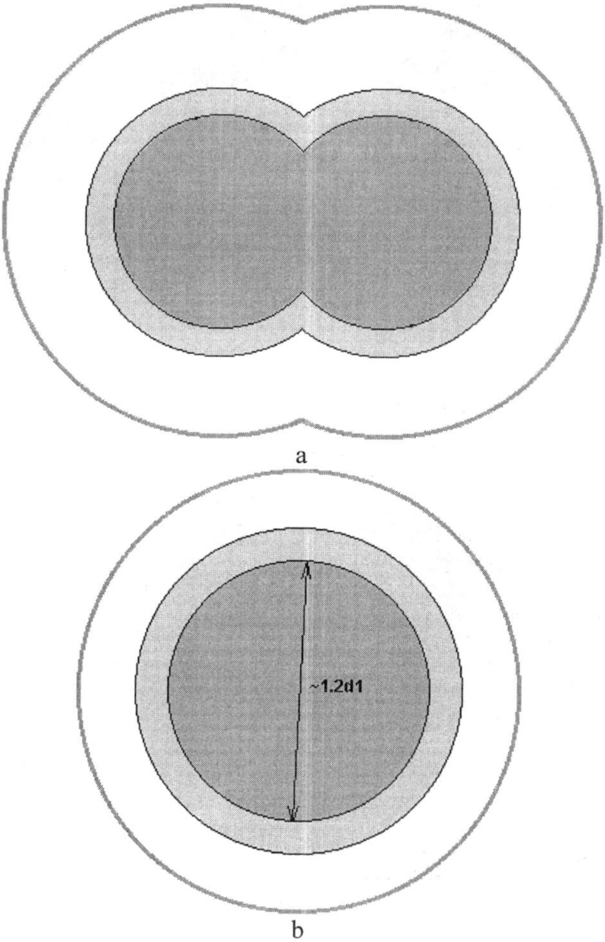

$$v_{nn} = r_{n1}^3 \cdot (8/3 \text{ across } (3/4) - 7\pi/32) \quad (3)$$

$$v_{na} = 2 \cdot (V_1 + V_2 + V_3) \quad (4)$$

where V_1, V_2, V_3 are evaluated from the equations

$$V_1 = 2/3 \cdot \pi \cdot ((\sqrt{(r_{n1} + 2b)^2 - r_{c1}^2}^{\,3} - \sqrt{4r_{n1}^2 - r_{c1}^2}^{\,3}) + ((2r_{n1})^3 - (r_{n1} + 2b)^3)) \quad (5)$$

$$V_2 = 2/3 \cdot \pi \cdot ((\sqrt{(4r_{n1}^2 - r_{c1}^2}^{\,3} - \sqrt{4r_{n1}^2 - r_{c2}^2}^{\,3} + (\sqrt{r_{n1}^2 - r_{c1}^2})^3 - (\sqrt{r_{n1}^2 - r_{c2}^2})^3 + 9/4 \cdot r_{n1} \cdot (r_{c1}^2 - r_{c2}^2)) \quad (6)$$

$$V_3 = 4 \cdot \pi / 3 \cdot (\sqrt{r_{n1}^2 - r_{c2}^2})^3 \quad (7)$$

$$r_{c1} = \sqrt{7/16 \cdot r_{n1}^2 + 2 \cdot r_{n1} \cdot b + 9/2 \cdot b^2 - \frac{16 \cdot (b^2 + 2 \cdot r_{n1} \cdot b) \cdot b^2}{9 \cdot r_{n1}^2}} \quad (8)$$

$$r_{c2} = \frac{\sqrt{15}}{4} \cdot r_{n1} \quad (9)$$

It is possible to calculate size characteristics of LTRC-2 using these and obtained similar equations for evaluation of overlapping atmospheres based on characteristics of unary clusters.

For concentration N_1 (LTRC-1) and concentration N_2 (LTRC-2) we can write down the following system of differential equations neglecting threefold overlapping.

$$dN_1 = (1 - N_1 \cdot \upsilon) \cdot n\, dt - N_1 \cdot \upsilon \cdot n\, dt$$
$$(10)$$
$$dN_2 = N_1 \cdot \upsilon \cdot n\, dt$$

where

n = a number of subcascades formed per second ($n = \lambda \cdot \Phi / t$)
Φ = neutron fluence
λ = a number of subcascades formed under irradiation by single fluence, in our case it is equal to 15.8 s/(m · neutron)
$\upsilon = 4/3 \cdot \pi r^3 = 32/3 \cdot \pi r_{n1}^3$ - an area in which a generation of the primary knock-on atom resulted in a formation of LTRC-2.

The solution of the system are the following equations

$$N_1 = 1/(2\upsilon) \cdot (1 - \exp(-2\upsilon \cdot \lambda \cdot \Phi)) \quad (11)$$

$$N_2 = \lambda \cdot \Phi/2 + 1/(4\upsilon) \cdot (\exp(-2\upsilon \cdot \lambda \cdot \Phi) - 1) \quad (12)$$

The equations can be used as for evaluation of relation between of LTRC-1 and LTRC-2 at different irradiation stages.

Theoretical Analyses

Using the presented above simulations we analyze experimental LTRCA size distributions. Various one-modal distributions describe single, double and threefold LTRC after their warming up. As it is shown in Reference [*4*], there is a transformation of a part of point defects under warming up and thus a vacancy cluster remains (LTRCA) which radius makes ~87 % from a radius of a nucleus of LTRC (before warming up). If we take as a diameter of single low-temperature radiation cluster after warming up (LTRCA-1) the value equal 4.5 nm (an average value among LTRCA-1 distribution characteristics on sizes in both samples), we can calculate both sizes of unary and double LTRC before annealing and a size of LTRCA-2 using the equations stated above. Diameters of LTRCA-1 and LTRCA-2 are exhibited in Table 1.

Table 1 – *Sizes and Relative Amount of Low-Temperature Radiation Clusters in Samples Irradiated up to 0.0016 dpa and 0.016 dpa at 77 K after their Annealing to Room Temperature*

Characteristics	Type of definition	Damage dose, dpa	LTRCA-1	LTRCA-2	LTRCA-3
Average	Test		4.6	5.8	
size, nm	Model	0.0016	4.5	5.5	
A part of the total	Test		0.93	0.07	
amount of LTRCA	Model		0.93	0.07	
Average	Test		4.4	5.6	6.9
size, nm	Model	0.016	4.5	5.5	
A part of the total	Test		0.53	0.35	0.12
amount of LTRCA	Model		0.51	0.49	

Using the obtained earlier values of vacancies (370) in the nuclei and (100) in the low atmosphere respectively for LTRC-1 [*4*] we can calculate an amount of point defects in various areas of LTRC-2. Size characteristics and concentration of point defects in reference areas of LTRC-1 and LTRC-2 are shown in Table 2.

Table 2 – *Calculated Characteristics of LTRC Formed in Austenitic Steels under Cryogenic Neutron Irradiation*

LTRC type	Area	Size, nm	n_v	n_i	c_v, %	c_i, %
LTRC-1	Nucleus (diameter)	5.2	370	0	6.2	
	Lower atmosphere (height)	0.7	100	100	1.3	1.3
	Whole atmosphere (height)	2.6		470		1.1
LTRCA-1	(diameter)	4.5	250		6.4	
LTRC-2	Nucleus (diameter)	6.2	670	0	6.7	
	Lower atmosphere (height)	0.8	100	100	1.1	1.1
	Whole atmosphere (height)	3.1		770		1.1
LTRCA-2	(diameter)	5.5	500		7.1	

As shown in Reference [5], ~ 10 % of generated point defects under irradiation are disappeared during aging at 77 K of samples irradiated by neutrons at 4 K. It permits the amount of point defects contained in radiation clusters at 77 K to be obtain using damage dose calculated. A relative amount of LTRC-1 and LTRC-2 after irradiation up to 0.016 dpa and 0.0016 dpa were calculated using their data, the values from the Table 2 and the equations 11 and 12. One can see that they are in a good agreement with experimental data (Table 1).

Conclusions

Using microstructure results of the 16Cr-15Ni-3Mo-1Ti steel obtained by transmission electron microscopy methods after their irradiation in the BN-600 at 77 K and warming up at room temperature based on analyses performed with the help of presented simulation of overlapping of cascade areas it was obtained the following:

- Sizes of low-temperature radiation clusters formed under cryogenic neutron irradiation are measured at room temperature
- Histograms of LTRC size distributions are plotted for the samples irradiated up to 0.0016 dpa and 0.016 dpa and parameters of their distributions are obtained
- LTRC of various types are revealed with close distribution dispersion and different average sizes in the sample irradiated up to 0.016 dpa
- A model taking into account overlapping of cascade areas was simulated in which both a formation of single and double low-temperature radiation clusters and a connection of their geometrical sizes with accumulation kinetics were described
- Characteristics of LTRC are obtained at 77 K using the model. They are in a good agreement with the experimental data.

References

[1] Kozlov, A.V.and Kirsanov, V.V., Jr., “Radiation Defect Formation and Evolution in C0.03Cr20Ni16Mn6 Steel under Low-Temperature Neutron Irradiation and their Effect on Physical and Mechanical Properties of the Steel,” J. Nucl. Mater., Vols. 233-237, 1996, pp. 1062- 1066.

[2] Goshchitskii, B.N., Sagaradze, V.V., Arbuzov, V.L. and et al., Jr., Nucl. Mater., Vols. 258-263 ,1998, pp. 1681- 1686.

[3] Schenck, H., "Theories of Engineering Experimentation", Peace, Moscow, 1972, pp. 242-253.

[4] Kozlov, A.V., Jr., Metals Physics and Material Science, Vol. 81, 1996, pp. 97-106. (in Russian)

[5] Dimitrov, C., Tenti, M., Dimitrov, O., "Resistivity Recovery in Austenitic Fe-Cr-Ni Alloys Neutron Irradiated at 23 K", J., Phis. F: Met. Phys.11, 1981, pp. 753-765

Wolfgang Schüle[1]

On the α+γ↔γ-Phase Boundary in Nickel and in Manganese Containing Stainless Steel Alloys

Reference: Schüle, W., "**On the $\alpha+\gamma$↔γ-Phase Boundary in Nickel and in Manganese Containing Stainless Steel Alloys,**" *Effects of Radiation on Materials: 20th International Symposium, ASTM STP 1405,* S. T. Rosinski, M. L. Grossbeck, T. R. Allen, and A. S. Kumar, Eds., American Society for Testing and Materials, West Conshohocken, PA, 2001.

Abstract: The α-ferrite-phase in nickel and in manganese containing stainless steel alloys is very brittle. Its formation is connected with a huge decrease in volume, which also causes stresses in the alloys that are very dangerous if the materials cannot accommodate them. The α-ferrite-phase is formed in stainless steel alloys only if nucleation sites are provided. α'-martensites are nucleation sites for the formation of α-ferrite, even at temperatures at which α-ferrite is not stable. If α'-martensite is dissolved during an anneal, no new α-ferrite is formed, and the remaining α-ferrite transforms back into γ-austenite.

In "pure" nickel or in "pure" manganese containing stainless steel alloys the martensitic temperature is above 100°C and thus α'-martensite is always present at ambient temperature in these materials, giving rise to the formation of α-ferrite during a subsequent anneal. It was established in the present work that the γ↔γ+α-phase boundary in iron-chromium-nickel alloys is also almost independent of the temperature as in iron-chromium-manganese alloys. Thus the existing phase diagram for nickel containing stainless steels has to be revised. The various elements added to "pure" stainless steel alloys, as we find them in EUR-316L, US-316L, US-PCA, and in AMCR, cause a drastic decrease of the martensitic temperature so that neither α'-martensite nor α-ferrite is found in these alloys.

However, in all these four alloys α-ferrite is formed readily during irradiation with high energy particles and a tentative γ↔γ+α-phase boundary, valid during irradiation with high energy particles, is derived. The amount of α-ferrite formed during irradiation increases with decreasing irradiation temperature and with decreasing applied stress. The alloys EUR-316L, US-316L, US-PCA, and AMCR do not survive one reactor cycle, if irradiated at 100°C.

Keywords: Phase diagram, nickel and manganese containing stainless steels, nucleation, α-ferrite, α'-martensite, dislocations, volume change, brittleness

[1] Professor, Institut für Angewandte Physik der Johann Wolfgang-Goethe Universität, Robert Mayer Straße 2-4, D-60325 Frankfurt, Germany. Corresponding address: Via Bosco 23, I-21038 Leggiuno(Va), Italy.

Introduction

Many different phases in iron-chromium-nickel alloys of the 316-type and in iron-chromium-manganese alloys of the AMCR-type were identified during the last century (see e.g. references *1-5*). Interest in the constitution of stainless steel alloys increased during the last forty years, when it was found that stainless steel materials can be very resistant against irradiation with high energy particles *[2-6]*. The phases and compounds which are found in e.g. 316-type stainless steel alloys are γ-austenite, α-ferrite, α'-martensite, and Laves-, γ'-, γ''-, δ-, ß-, σ-, χ-, μ-, G-, and R-phase, and compounds such as carbides, nitrides, and sulphides. Most of these phases and compounds are dissolved during a solution anneal at 1100°C, but they are formed again during cooling or annealing after a solution anneal. The formation of all these phases is connected with volume changes, and for this reason it is better to use well aged stainless steel alloys and not materials in the solution annealed state.

The compositions of all the phases mentioned above change during irradiation with high energy particles. Usually the nickel content is increased in all the phases, because nickel is the smallest atom in comparison to iron and chromium in the stainless steel alloys and thus it diffuses faster than the other two component elements *[7]*. Many different new phases, namely radiation-induced phases, e.g. Ni_3Si-phases, are formed during irradiation with high-energy particles by radiation-enhanced diffusion, if the stainless steel alloys contain undersized atoms, e.g. silicon.

It is very important to be aware that the mobilities of the atoms in stainless steel alloys are very similar to those found in pure nickel *[7]*. Also the radiation-enhanced diffusion coefficients in the stainless steel alloys are very similar to those found in pure nickel. The formation of many phases in stainless steel alloys requires nucleation and thus all these phases are often formed in a very sluggish way. However, the dissolution of phases formed during irradiation in stainless steel alloys is as fast as the dissolution of agglomerates of interstitials in pure nickel.

The recovery of work-hardening is also very sluggish in stainless steel alloys, because the many compounds in stainless steel alloys act as obstacles to the annihilation of dislocations.

The mechanical properties of stainless steel alloys seemed to be improved by adding titanium. However, the increase in volume due to the formation of titanium carbide during service at e.g. T = 200°C is so large that the matrix of the stainless steel alloys can not accommodate the volume change and fracture occurs without extra applied stresses *[2-4]*.

The formation of α-ferrite is connected with a decrease in volume, which is so large that the formation of this phase causes changes in the dimensions and in the shape of the stainless steel specimens. In addition, α-ferrite itself is very brittle so that specimens containing very small amounts of this phase break *[10, 11]*.

The formation of α-ferrite during irradiation with high energy particles in nickel and in manganese-containing stainless steel alloys has been observed by investigations of irradiation creep *[9-12]* and by means of electron microscopy *[13-16]*. For studies on the constitution of nickel and of manganese containing stainless steel materials more than fifty different types of stainless steel materials were investigated using electrical resistivity techniques, magnetic response, stored energy, X-ray techniques, electron microscopy, and light microscopy. The following main phases were identified: γ-austenite, α-ferrite, α'-martensite, ε-martensite, $\varepsilon_{def.}$-martensite, dislocations, and α-Mn-phase *[4-6, 9-11]*.

A recent analysis of these data revealed that no α-ferrite is formed during annealing of solution-annealed or cold-worked 316L-type steels, e.g. in the European reference steels EUR-316L, and in US-316L or US-PCA steels. This finding is in contrast to results reported in the literature for "pure" iron-chromium-nickel alloys *[1]*, according to which α-ferrite should be formed in a wide composition range in 316 stainless steels. However, α-ferrite is formed during annealing in solution-annealed or cold-worked pure 316-type steel, which contains only iron, chromium and nickel (or manganese). It is further known that α-ferrite is also formed in the above mentioned 316L-type steels during an anneal after cold-work at very low temperatures *[17]*.

All these findings suggest, that nucleation sites are necessary for the formation of α-ferrite in stainless steel alloys. We will show in the following that at least α'-martensite must be present in stainless steel alloys so that α-ferrite can be formed.

The constitutions of the nickel and of the manganese containing stainless steel alloys are very different. The main handicap in using manganese-containing stainless steel alloys is e.g. the very narrow γ-phase field, which is limited by the α-Mn-phase. This phase is very brittle when it contains chromium. However, the phases formed in the vicinity of the compositions corresponding to that of the 316-type and the AMCR-type alloys are in some respects very similar, especially with respect to the formation of α-ferrite.

Experiments

Materials

The materials investigated in the present work were obtained from Creusot-Loire, France, from Vakuum Schmelze Hanau, Germany, and from the Metals and Ceramics Division of the Oak Ridge National Laboratory, USA. In the context of the present work about fifty alloys were extensively investigated but only the results obtained on seven alloys, the compositions of which are listed in Table 1, are discussed in the present work.

The stainless steel materials were obtained in the solution-annealed state in plates having a cross section of about 15 cm x 3 cm and a length of about 300 cm or in rods having a diameter of about 2.5 cm. Small pieces were cut from these plates and cold-worked to foils having a thickness of about 0.2 mm with many intermediate annealings at high temperatures. The detailed treatments of the materials prior to the investigation including the final degree of cold-work are mentioned in the text.

The annealing treatments of the materials were performed in an argon atmosphere containing 5% or 10% hydrogen and under vacuum ($<10^{-4}$ torr). The evaporation of manganese in some of the investigated materials was noticed on annealing at temperatures higher than 800°C under vacuum ($<10^{-4}$ torr).

Electrical Resistivity Measurements

The measurements of the electrical resistivity were carried out in two different modes *[4, 5]*:

1. By continuous measurements during a very slow heating and cooling (1°C·min^{-1}) between room temperature and 1000°C in either pure argon or in argon and 5% H_2.

2. By measurements at a constant measuring temperature of −196°C (liquid nitrogen), after isochronal (30 min) annealing stages with stepwise (50°C) increasing and decreasing temperatures followed by rapid cooling. These annealing treatments were carried out in either dynamic vacuum or in an argon +5% H_2 atmosphere.

The Matthiessen rule does not hold for the electrical resistivity of stainless steel alloys. This means that the residual electrical resistivity of stainless steel alloys is dependent on the temperature, that is, it decreases with increasing temperature in contrast to the temperature dependent electrical resistivity. Because of the opposite temperature dependencies of the two resistivity contributions (ρ_{res} and ρ_{th}) the resulting temperature coefficient α is relatively small compared to that of pure metals and is itself temperature dependent.

Table 1 --*The composition in wt.% of stainless steel alloys investigated in the present work is listed. The balance is iron.*

Alloy	C	Cr	Mn	Ni	Mo	W	Si	Ti	V
	[w	[w	[wt%	[wt	[wt	[w	[wt%	[wt%	[wt
t%]	t%]	]	%]	%]	t%]	]	]	%]	
AMCR	0.	10	17.5	<0.	<0.	-	0.55	-	<0.
AMCR	0.	14	19.8			-	0.63	-	-
EUR-	0.	17	1.81	12.	2.3	-	0.45	-	-
81/5283	0.	11	6.85		-	-	-	-	-
81/5282	0.	16	-	12.	-	-	-	-	-
US-316	0.	16	2.0	10-	2.3	-	1.3	-	-
US-	0.	13	1.5-	15-	-	-	0.2-	0.2-	-

There is a further peculiarity in that the electrical resistivity in stainless steel alloys varies from alloy to alloy. The electrical resistivity of an annealed and of a cold-worked AMCR 0035 specimen (Table 1), is plotted versus the measuring temperature in Fig. 1. The electrical resistivity of a cold worked specimen of AMCR 0035 is smaller than that of an annealed specimen, if the measuring temperature is below about -100°C. However, the electrical resistance of a deformed specimen is larger than that of an annealed specimen, if it is measured above about -100°C.

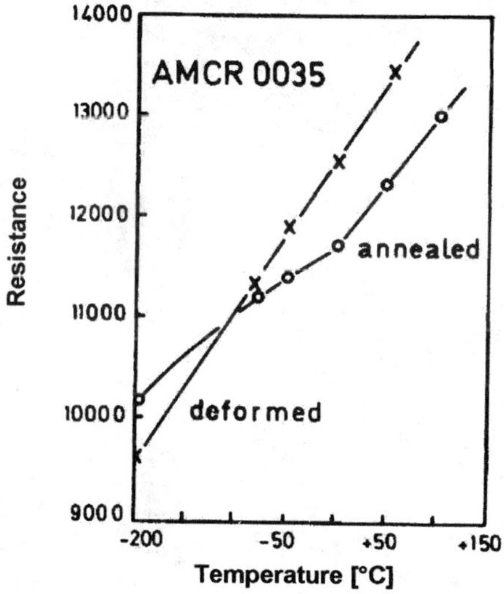

Fig. 1-- The electrical resistivity is plotted versus the measuring temperature for a specimen first annealed at 1000°C in an argon atmosphere containing 10% hydrogen and then 41% cold worked (after references 4 and 5).

The electrical resistance of an annealed and of a cold worked specimen of alloy 81/5283 is plotted versus the temperature in Fig. 2. The resistivity of a cold worked specimen is always larger than that of an annealed specimen in the temperature range between -200 and +100°C. However, the two curves extrapolated to high temperatures cross at about 440°C. This means that the resistivity of an annealed sample is larger than that of a cold worked specimen for temperatures higher than about 440°C.

This means that the formation of α-ferrite in alloy 81/5283 causes a decrease of the electrical resistivity, if measured below about 440°C, and the dissolution of this phase also causes a decrease, if measured above 440°C.

The electrical resistivity is often very sensitive to any microstructural change occurring during or after a thermo-mechanical treatment. However, the electrical resistivity is a non-specific property for microstructural variations. Therefore, almost all the microstructural variations observed in the present work were determined by means of X-ray diffraction spectroscopy (XRD). In addition highly sensitive calorimetric and magnetic response measurements were performed. In the present work only resistivity data are reported and reference is made to results of investigations obtained by other means *[4, 5]*.

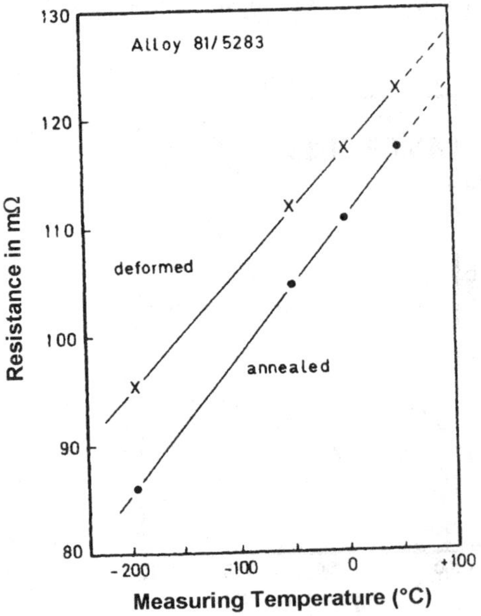

Fig. 2 -- The electrical resistance of a cold worked and of an annealed specimen is plotted versus the measuring temperature. The specimen was 18% cold worked prior to annealing and was annealed in an argon atmosphere containing 5% hydrogen. After references [4] and [5].

Results

In the following we will show results of investigations on the identification of various phases, which were obtained in seven characteristic alloys and which were observed in more than fifty stainless steel alloys of various compositions and additions by means of light and transmission electron microscopy, and by X-ray diffraction spectroscopy, are discussed. In addition, results of highly sensitive calorimetric and magnetic response measurements are included. However, the formation and dissolution of the various phases formed during various thermo-mechanical treatmeants of mainly two types of stainless steel alloys, namely of 316- and AMCR-type materials by means of electrical resistivity techniques, are presented in the present work. The variation of the elongation during irradiation with high energy particles is obtained as a function of the applied stress and the irradiation temperature.

On the Formation of α-Ferrite in Alloy 81/5283

The change of the electrical resistance and the derivative of this quantity with respect to the temperature is plotted for increasing and decreasing temperature in Fig. 3. The variation of the electrical resistance with temperature can be better noticed from the variation of the derivative of the electrical resistance with respect to the temperature, which is a quantity proportional the temperature coefficient of the electrical resistivity.

The sample was about 20% cold worked prior to annealing. It is seen that the electrical resistance decreases with increasing temperature first in two stages at about 300 and 450°C and decreases again in two stages at 550 and 650°C (Fig. 3). On cooling the electrical resistance decreases steadily from 1000°C to about 150°C and increases below this temperature.

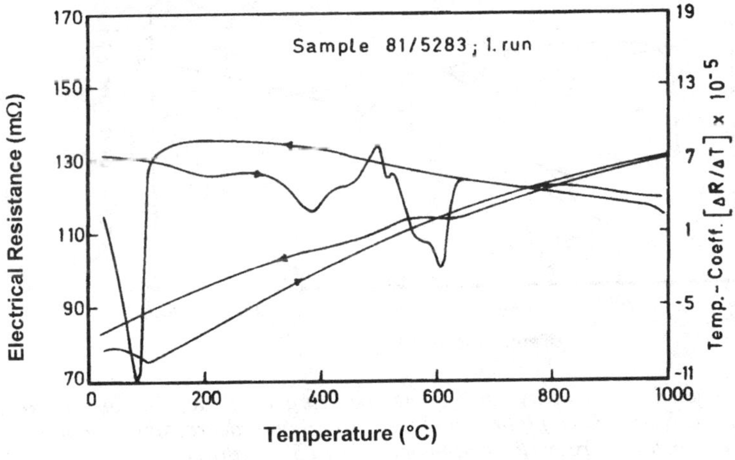

Fig. 3 -- The electrical resistivity and the temperature coefficient of a previously 20% cold worked specimen of alloy 81/5283 is plotted for increasing and decreasing temperature (first run). The anneals were performed in argon with 5% argon. After references 4-6.

The analysis of the microstructure of alloy 81/5283 revealed that the specimen contained only (or mainly) α'-martensite after a 20% cold work. α-ferrite and γ-austenite is formed during subsequent heating between 300 and 450°C. During further heating α-ferrite and the remaining α'-martensite transforms into γ-austenite[2] between 550 and 650°C. Dislocations introduced by 20% cold work prior to annealing are recovered between 550 and 650°C, too.

The microstructure formed during the first heating between 300 and 450°C decreases the electrical resistivity and the dissolution of the microstructure, which was formed between 300 and 450°C and which is dissolved in the temperature range between 550 and 650°C also decreases the electrical resistivity. Thus, the formation of a certain microstructure causes a decrease of the electrical resistivity, when measured at low temperatures, between 300 and 450°C, and the dissolution of this microstructure causes a decrease of the electrical resistivity, also when measured at high temperatures, between 550 and 650°C (Figs. 1 and 2).

During cooling no microstructural change is found between 1000 and 150°C on measuring the electrical resistivity (Fig. 3). This means that no α-ferrite is formed below 650°C. However, a huge increase of the electrical resistivity is found during cooling below 150°C due to the formation of a martensitic phase (Fig. 3), which has been identified as α'-martensite [4-6]. The mobility of the component atoms in stainless steel alloys below 150°C is much too small so that no phases are formed by diffusion at or below 150°C *[7]*.

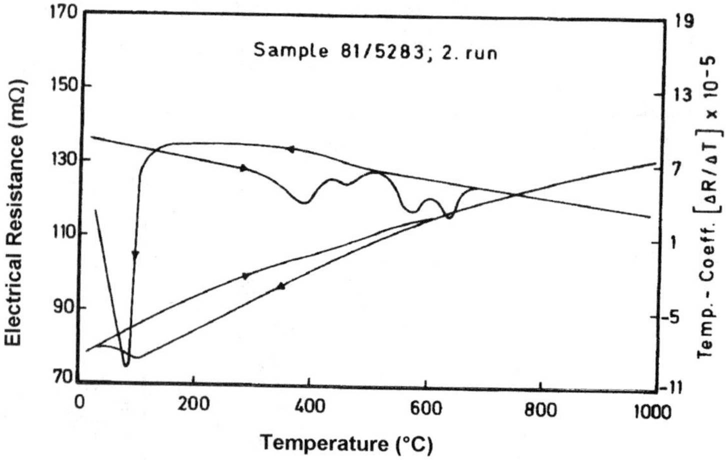

Fig. 4 -- The electrical resistivity and the temperature coefficient of an annealed specimen of alloy 81/5283 is plotted for increasing and decreasing temperature (second run) The anneals were performed in argon with 5% hydrogen.

---

2 The content of e.g. nickel in the γ-austenite depends on the annealing temperature and on annealing time.

This means that α-ferrite, which is formed during the first annealing between 300 and 450°C (Fig. 3) is not in thermodynamic equilibrium at these temperatures in alloy 81/5283 nor is the temperature range between 550 and 600°C the temperature of its dissolution. If α-ferrite would be stable between e.g. 300 and 600°C, then it would be formed during a slow cooling from 1000°C, which can be excluded (Figs. 3 and 4).

The analysis of the microstructure of the slowly cooled specimen revealed that it consists at ambient temperature mainly of α'-martensite. The specimen, which was slowly cooled and cold worked prior to the first annealing, contained mainly α'-martensite and in addition a high concentration of dislocations. This means that the specimen 20% cold-worked prior to annealing and after the first run treatment contained mainly or only α'-martensite.

The variation of the electrical resistivity of the previously slowly cooled specimen (Fig. 3) during a second run with increasing and decreasing temperature is plotted in Fig. 4. This variation is almost identical with that shown in Fig. 3. Thus the interpretation of the curves in Fig. 4, which concerns the formation and dissolution of phases during heating and cooling, is identical with that given previously.

In summary, α-ferrite and γ-austenite is formed between 300 and 450°C in a specimen, which consists only of α'-martensite. The remaining α'-martensite, α-ferrite, and γ-austenite transforms into γ-austenite[2] between 550 and 650°C (Figs. 3 and 4). Above 650°C the specimen consists of only γ-austenite. This phase is maintained during cooling till about 150°C. At ambient temperature almost all the γ-austenite is again transformed into α'-martensite as derived from X-ray diffraction and from measurements of the magnetic response and the stored energy *[4-6]*.

Again α-ferrite is not formed during cooling from 1000°C. The specimen consists after the first and after the second run only of α'-martensite. We conclude that i) the martensitic transition temperature is M_S = 150°C for alloy 81/5283, that ii) α'-martensite acts as nucleation site for the formation of α-ferrite, and that – most important – iii) the α-ferrite-phase is thermodynamically unstable in alloy 81/5283 in the entire investigated temperature range, e.g. between 1000 and ca 200°C. This means that α'-martensite acts as nucleation site for the formation of α-ferrite. Apparently α-ferrite is formed in alloy 81/5283 only when α'-martensite is present even if α-ferrite is thermodynamically unstable at the temperature at which it is formed. In the progress of this work we will show that dislocations alone are very likely not able to act as nucleation sites for the formation of α-ferrite.

The variation of the electrical resistivity after 20% cold work of alloy 81/5283, which has been prepared in the same way as the specimen used previously (Fig. 3) is plotted versus the annealing temperature in Fig. 5. The explanation of the variation of the electrical resistivity is again supported by the results obtained by means of investigations using X-ray spectroscopy, by measurements of the stored energy, and the magnetic response, and by the conclusions, which were obtained from the investigations by means of electrical resistivity techniques, and which are shown in Figs. 3 and 4.

One has to be aware that the electrical resistivity is in this experiment (Fig. 5), after each heat treatment, always measured in liquid nitrogen, i.e. at –196°, that the martensitic temperature of this alloy is ca M_S = 150°C. Consequently, some or most of the γ-austenite, which is formed during annealing, is transformed in α'-martensite during the measurement of the electrical resistivity in liquid nitrogen. α-ferrite, which is formed during annealing, remains α-ferrite during the measurement of the electrical resistivity in liquid nitrogen.

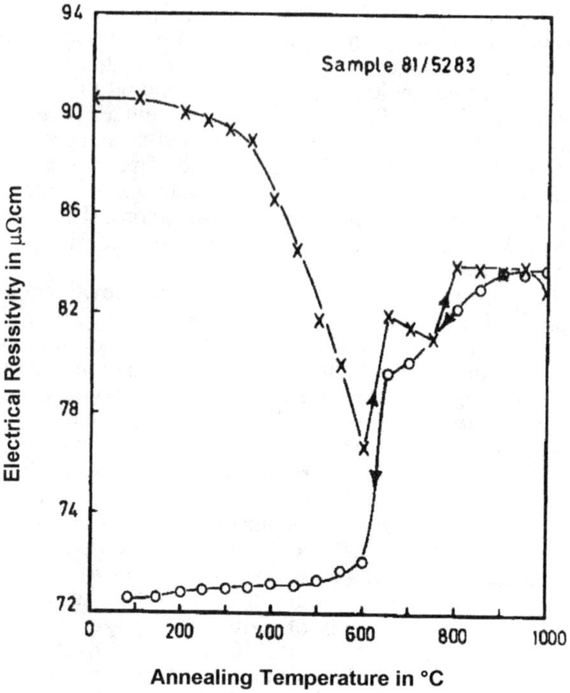

Fig. 5 -- The change of the electrical resistivity of 20% cold worked specimen of alloy 18/5283 is plotted for increasing and decreasing temperature. The measurements of the electrical resistivity were performed in liquid nitrogen (at −196°C). The specimen was annealed for thirty minutes at each indicated temperature in argon with 5% hydrogen.

The specimen consisted prior to the beginning of annealing mainly of α'-martensite and dislocations, which were introduced by plastic deformation prior to annealing. The electrical resistivity decreases between about 400 and 600°C due to the formation of γ-austenite and α-ferrite from α'-martensite (Fig. 5). Then with further increasing temperature the electrical resistivity increases due to the transformation of α-ferrite and γ-austenite, which are formed during heating to the annealing temperature indicated in the figure, and α'-martensite into γ-austenite. On cooling the electrical resistivity decreases somewhat between 1000 and 650°C and drops rapidly between 650 and 600°C. More and more α-ferrite is formed in the temperature region between ca 650 and 600°C, and the sample consists below 600°C mainly of α-ferrite, the content of which is only slightly increasing with further decreasing temperature [3].

In summary the specimen, which is cooled slowly from 1000°C to ambient temperature, consists of mainly α'-martensite (Figs. 3 and 4), and the specimen, which is after each annealing step immersed in liquid nitrogen for the measurement of the

---

[3] At first, we assumed erroneously that the γ↔γ+α-phase boundary would be at 650°C or slightly below this temperature *[4-6]*, and that below this temperature α-ferrite would precipitate from γ-austenite.

electrical resistivity, consists of mainly α-ferrite after the end of the annealing treatment (Fig. 5). We concluded from the results obtained after a slow cooling (Figs. 3 and 4) that α-ferrite is only formed in the presence of α'-martensite acting as nucleation site for the formation of α-ferrite. The validity of this explanation is readily confirmed by the results obtained from the following experiment.

A 20% cold worked specimen, which consists of α'-martensite and dislocations was annealed at T_a = 356°C for several months and the electrical resistivity was continuously measured as a function of the annealing time always at T_a = 356°C. A huge decrease of the electrical resistivity was obtained during the first day of annealing and the specimen consisted of α'-martensite, α-ferrite, and γ-austenite, and contained still many dislocations. This means that the same phases are formed during annealing of one day as during annealing with increasing temperature (Figs. 3 and 4).

After an annealing time of two months always at the same temperature of T_a = 356°C the electrical resistivity of the same specimen increased and the specimen consisted finally of only γ-austenite. This means that α'-martensite and α-ferrite were dissolved, and also the dislocations introduced by cold work prior to annealing were annihilated.

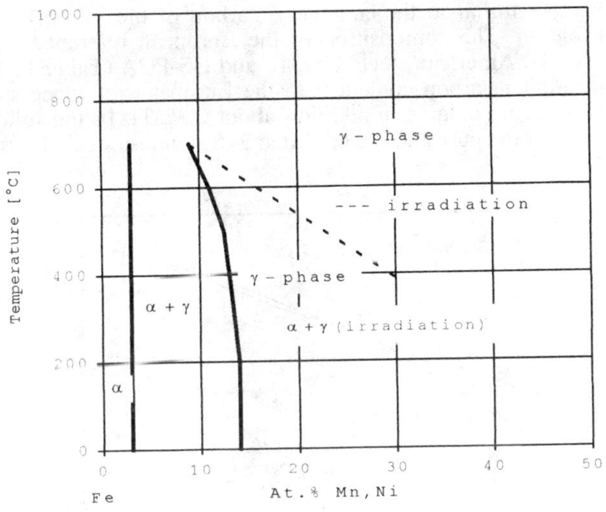

Fig. 6 -- The γ↔γ+α phase boundary for the iron-chromium-manganese and iron-chromium-nickel phase diagrams are shown for a chromium content of about 12%. The dashed curve is the γ↔γ+α phase boundary which is valid during irradiation with high energy particles [9-11].

This result implies that during the first day of the anneal, when α'-martensite was present, α-ferrite and γ-austenite was formed in the specimens. During the following months a kind of quasi-dynamic phase equilibrium among the various phases was established; more and more α'-martensite became dissolved and less α-ferrite was formed. Finally, after an annealing time of two months all the α'-martensite and α-ferrite were transformed into γ-austenite. This means that γ-austenite is the only stable phase in alloy 81/5283, i.e. in pure stainless steel alloys at 356°C.

Therefore the γ↔γ-α-phase boundary shown in Fig. 6 is almost independent of the temperature, in agreement with the γ↔γ-α-phase boundary known for "pure" stainless steel alloys consisting of iron, chromium and manganese *[1]*. However, to establish the precise position of the γ↔γ-α-phase boundary, many more alloys with composition in the vicinity of those discussed in the present work must be investigated.

The mobilities of the main component atoms in stainless steel alloys are so large at 200°C *[7]* that α-ferrite would have been formed at 200°C, if this phase is stable at this temperature.

The γ↔γ+α-Mn phase boundary is not shown in Fig. 6, because a similar phase boundary does not exist in the iron-chromium-nickel phase diagram. The γ-phase field is very narrow in the iron-chromium-manganese alloys, depending on the chromium content. For iron-chromium-manganese alloys containing about 14% Cr the γ↔γ+α-Mn phase boundary lies at about 20% Mn (Fig. 6).

On the Formation of α-Ferrite in Alloy 81/5282 and Alloy EUR-316L

Alloy 81/5282 consists mainly of the main component atoms of stainless steel alloys, namely of iron, chromium, and nickel, and its only addition is carbon, the content of which is very similar to the content of carbon of the European reference steel EUR-316L (Table 1). The composition of the European reference steel EUR-316L is very similar to the American steels US-316 and US-PCA (Table 1). Both US-steels have a slightly higher carbon content than the European reference steel EUR-316L and the US-PCA steel contains in addition about 0.3%Ti. In the following we will discuss the features of the pure and the modified 316 stainless steel alloys.

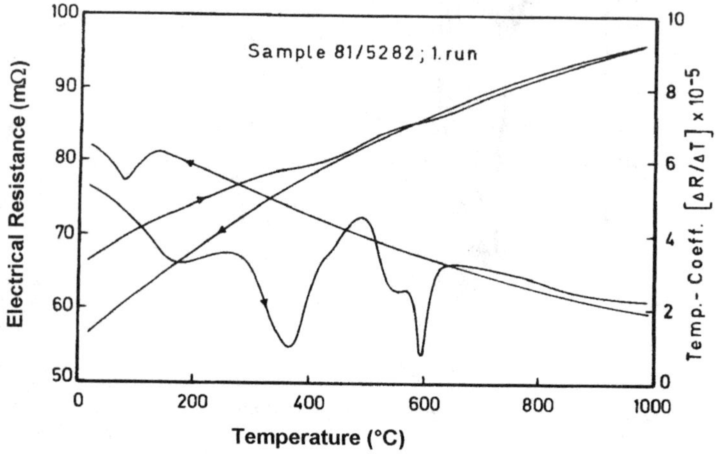

Fig. 7 -- The electrical resistivity and the temperature coefficient of the electrical resistivity of an annealed and cold worked specimen are plotted for increasing and decreasing temperature (first run). The degree of cold work prior to annealing was 37.5%. The anneals were performed in argon with 5% hydrogen.

The following phases were found after a 37.5% cold work in alloy 81/5282, which is considered a "pure" 316L stainless steel: α'-martensite, γ-austenite, and a small amount of $\varepsilon_{def.}$-martensite [4].

During a subsequent anneal $\varepsilon_{def.}$-martensite is transformed into γ-austenite at about 150°C (Fig. 7), and α-ferrite and γ-austenite is formed between 300 and 450°C. With further increasing annealing temperature α'-martensite and α-ferrite are dissolved into γ-austenite between 450 and 650°C (Fig. 7). This means that the microstructural changes obtained in alloy 81/5282 during heating of a cold worked specimen are very similar to those observed during the first run of alloy 81/5283 (Fig. 3).

During cooling, no phase transition can be recognized in the entire temperature range. The amount of α'-martensite, which is formed below 100°C, which is the martensitic temperature of this alloy, is and this is an important difference in the behavior to alloy 81/5283 very small in alloy 81/5282.

Consequently a very small amount of the α-ferrite is formed again (second run) on re-annealing alloy 81/5282 between 350 and 450°C, because the slowly cooled specimen contained only a very small amount of α'-martensite, which can act as nucleation sites for α-ferrite formation (Fig. 8). All the phases present in this alloy are transformed into γ-austenite during heating between 550 and 650°C (Fig. 8).

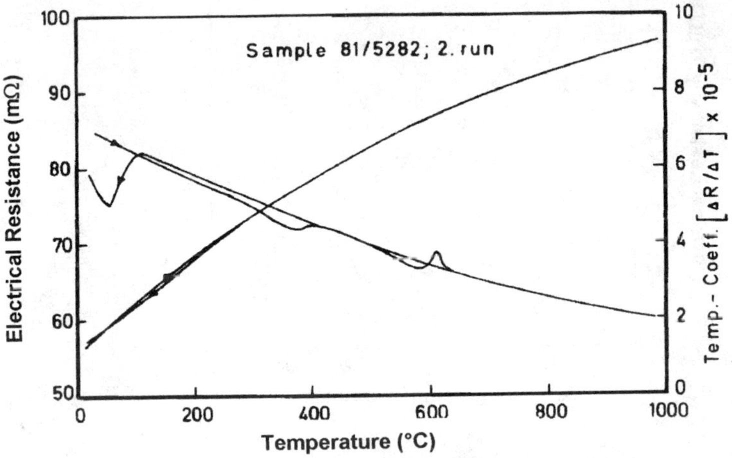

Fig. 8 -- The electrical resistivity and the temperature coefficient of the electrical resistivity of an annealed specimen of alloy 81/5282 are plotted for increasing and decreasing temperature (second run). The anneals were performed in argon with 5% hydrogen.

Another sample of alloy 81/5282, which was only 18.5% cold worked, contained the same phases as the sample which was 37.5% cold worked (Fig. 7), namely α'-

[4] The $\varepsilon_{def.}$-martensite phase has also been recognized in alloy 81/5283, but the amount of this phase formed by plastic deformation is small in this alloy *[4, 5]*.

martensite, γ-austenite, and a small amount of $\varepsilon_{def.}$-martensite [4]. The electrical resistivity of such a material decreases in three steps during a subsequent anneal, when the electrical resistivity is measured after each annealing temperature in liquid nitrogen (Fig. 9). The three steps indicate three different microstructural changes which develop during heating. $\varepsilon_{def.}$-martensite transforms into α-ferrite at about 150°C [4]. In this material α'-martensite is present, too, acting as nucleation site for the formation of α-ferrite and γ-austenite in the temperature region between 300 and 450°C. In the third annealing step, between 450 and 650°C, α'-martensite and α-ferrite is dissolved into γ-austenite. The electrical resistivity increases between 800 and 1000°C due to the transformation of α-ferrite and γ-austenite into γ-austenite.

The electrical resistivity decreases between 1000 and 600°C due to the formation of α-ferrite, which is formed during annealing, because α'-martensite is always formed when measuring the electrical resistivity in liquid nitrogen at –196°C.

Thus the specimens of alloys 81/5282 and 81/5283 consisted of only α-ferrite after cooling below 600°C, when the specimens are put into liquid nitrogen for the measurements of the electrical resistivity after each anneal (Figs. 5 and 9).

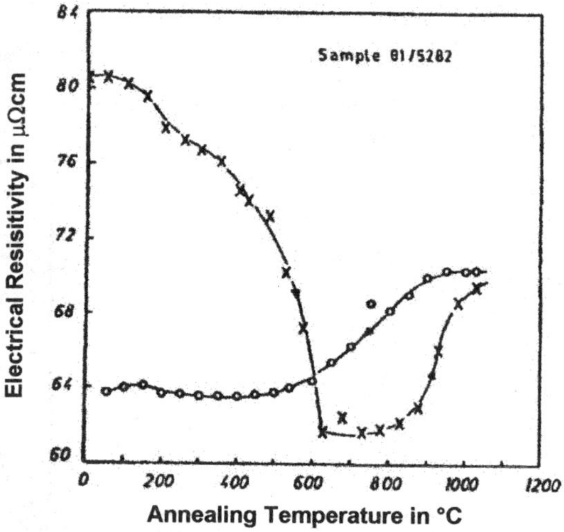

Fig. 9 -- The change of the electrical resistivity is plotted versus increasing and decreasing temperature. The specimen was held for 30 min. at each indicated temperature before it was cooled and put into liquid nitrogen (-196°C) to measure the electrical resistivity. The specimen was 18 % cold worked prior to annealing. The anneals were performed in vacuum (<10-4 torr).

Alloy 81/5282 is a "pure" 316L-type alloy without additions. The various additions added in EUR-316L-, US-316L-, and US-PCA-steels are listed in Table 1. These three steels consist of only γ-austenite, apart from the precipitates and segregations, nitride compounds and hydrides, which were mentioned in the introduction of this work, which may be present in these materials, and which are not of great interest and subject of the present work.

None of the phases, namely $\varepsilon_{def.}$-martensite, α'-martensite, and α-ferrite which were detected after cold work and during subsequent anneals in alloy 81/5282 or alloy 81/5283, were found in EUR-316L-, in US-316L, in US-PCA, and in AMCR [5]. During cooling from 1000°C to ambient temperature no α'-martensite is obtained and on reheating alloy EUR-316L no microstructural change could be detected during a second annealing by measurements of the electrical resistivity.

However, we have to mention that α'-martensite is formed in all commercial 316L-type stainless steel alloys, if these materials are cold-worked at –196°C [17], and we believe that this phase would have also be formed in EUR-316L-, US-316L, and US-PCA-steels, if these materials would have been cold worked at very low temperatures.

The change of the electrical resistivity of alloy EUR-316L after 10% cold work is plotted versus the annealing temperature in Fig. 10. The total change of the electrical resistivity is very small, namely only 3.9%, in contrast to more than 30% found for e.g. alloy 81/5282 (Fig. 9). We repeat again that in EUR-316L only γ-austenite was found after cold work by means of XRD-measurements and none of the other phases found in alloy 81/5282 or alloy 81/5283 could be detected [4-6].

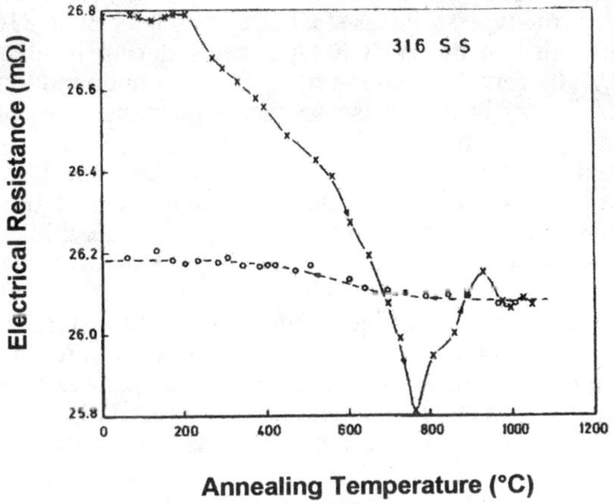

Fig. 10 --The change of the electrical resistivity is plotted versus increasing and decreasing temperature. The specimen was kept for 30 min. at each indicated temperature before it was cooled and put into liquid nitrogen (-196°C) to measure the electrical resistivity. The specimen was 10 % cold worked prior to annealing. The anneals were performed in vacuum (<10-4 torr).

The electrical resistivity decreases in three or even four steps during annealing in alloy EUR-316L (Fig. 10). These decreases can be attributed to the annihilation of point defects, namely to interstitials at about 250°C, and to vacancies at about 400°C [7]. The decrease of the electrical resistivity at 600 and at 700°C is again attributed to

---

[5] The $\varepsilon_{def.}$-martensite phase is present in AMCR-type stainless steel alloys after cold work and the ε-martensite phase is always found below about + 50°C (Fig. 1). The appearance of both phases are not of interest in the context of this work.

the formation of microstructural changes due to the migration of vacancies in thermal equilibrium, e.g. recovery of work hardening and/or to the formation or dissolution of phases or compounds.

This means that the additions in EUR-316L material have the effect of suppressing the formation of α'-martensite during cooling the materials down to temperatures as low as – 196°C. However, in "pure" 316L stainless steel alloys, e.g. in alloys 81/5182 and 81/5283, α'-martensite is formed during cooling these alloys down to –196°C without additional cold work. Consequently α-ferrite is not formed during annealing in most commercially available 316L-type alloys, e.g. in EUR-316L-, US-316L, and US-PCA-steels, because no nucleation sites are present.

The effect of additions of various elements on the stabilization of γ-austenite in stainless steel alloys was also the subject of numerous studies in the past. The present work, however, indicates that α-ferrite is only formed in stainless steels, if α'-martensite is present. Thus, the purpose of the additions to prevent the formation of α-ferrite is to suppress the formation of α'-martensite or to decrease the martensite temperature M_S to such low temperatures that α'-martensite is not formed spontaneously and that this phase is also not formed during cold work at ambient temperature.

Fortunately, it was possible in the past century to develop 316-type stainless steel alloys, in which no α'-martensite is formed and consequently the γ-austenite in these materials is very stable. Unfortunately, α-ferrite is readily formed in all these commercially available 316L-type stainless steel alloys such as EUR-316L-, US-316L, and US-PCA-steels, and also in the AMCR-type steels during irradiation with high energy particles *[9-10]*. The formation of α-ferrite causes a huge reduction in volume. In addition α-ferrite is a very brittle phase so that specimens that containing small amounts of this phase are very brittle.

The results obtained for the iron-chromium-nickel alloy 81/5282 are in many respects identical with those of alloy 81/5283. We conclude that the γ↔γ-α-phase boundary must be almost identical for both types of stainless steel alloys, namely for nickel and for manganese containing stainless steel alloys. However, the γ↔γ-α-phase boundary for "pure" iron-chromium-nickel alloys, known from the literature *[1]* is completely different from that shown in Fig. 6. We believe that α-ferrite formation can be observed at almost any temperature as long as nucleation sites are provided, i.e. also at temperatures, at which α-ferrite is not stable. The presence of α-ferrite at temperatures, at which this phase is not stable in thermal equilibrium led to an erroneous interpretation of experimental results in the past concerning the course of the γ↔γ-α-phase boundary.

On the Formation of α-Ferrite during Irradiation in US-316, US-PCA, EUR-316L, and AMCR 0033

A systematic study of neutron irradiation creep on various stainless steel alloys, mainly on AMCR 0033, EUR-316L, US 316, and US-PCA, has been reported during the last decade *[6-11]*. A negative creep elongation was found at the beginning of an irradiation mainly of 20% cold worked materials. The results for all three alloys were very similar; the results obtained for US-316 are shown in Fig. 11. It is evident that the length change of a material increases due to irradiation creep and that, consequently, the observed negative creep elongation must be attributed to the formation of precipitates during irradiation, which causes a decrease in the volume of the material *[6-11]*.

Results of investigations by means of TEM techniques, which were performed on non-stressed specimens, revealed that α-ferrite is formed during irradiation with high energy particles *[13-16]*. The formation of α-ferrite is connected with a large decrease in volume in both types of stainless steel alloys. In addition only the formation of α-ferrite in stainless steel alloys is connected with a huge decrease in volume in both types of stainless steel alloys, namely in nickel- and in manganese-containing steel alloys. Thus we believe that the negative creep elongation observed during irradiation with high energy particles is due to the formation of α-ferrite.

It is known that α-ferrite is a very brittle phase. Even small amounts of this phase precipitated in γ-austenite cause fracture, when the lattice can not accommodate the large volume change due to the formation of α-ferrite. Many of the AMCR- and 316-type materials broke during irradiation at +100°C before they received a neutron irradiation dose of 0.1 dpa *[11]*.

The γ↔γ-α-phase boundary, which was derived from measurements of the creep elongation during irradiation with high energy particles is shown as a dashed line in Fig. 6 *[10, 11]*. It is not understood why α-ferrite is only formed during irradiation with high energy particles in AMCR 0033, EUR-316L, US 316, and US PCA materials, and that this phase is not obtained without irradiation.

It is possible that α-ferrite is a radiation-induced structure formed during irradiation by radiation-induced interstitials, which are mobile at temperatures as low as room temperature [7]. This explanation is supported by the finding that α-ferrite was identified in foils by means of the transmission electron microscope *[13-16]*.

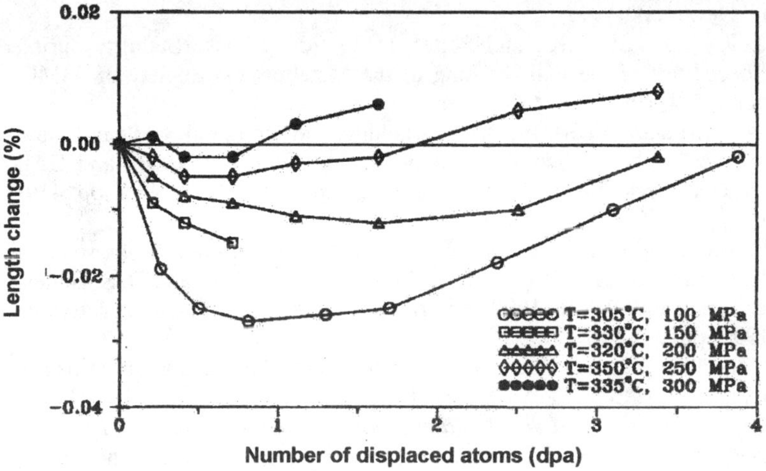

Fig. 11 -- The changes in length of five samples of US 316 are plotted versus the number of displaced atoms. The irradiation temperatures and the respective applied stresses are listed in the figure. The samples were cold rolled prior to irradiation (after reference 10).

We found a very pronounced negative creep elongation in the AMCR 0033, US 316, and US PCA materials only after cold work prior to irradiation. This can mean that dislocations introduced by cold work prior to irradiation act during irradiation with high energy particles as nucleation sites for the formation of α-ferrite, enhancing the formation of this phase with respect to annealed specimens, in which the concentration of dislocations is small.

A further very important result concerning the appearance of the negative creep elongation as a function of the irradiation fluence is shown in Fig. 11, i.e. the formation of α-ferrite during irradiation with high energy particles, is shown in Fig. 11. It is found that this quantity increases with decreasing applied stress for a constant irradiation temperature. This means that the amount of α-ferrite, which is formed during irradiation, increases drastically with decreasing applied stress. The consequence of this finding is twofold: 1) The γ↔γ-α-phase boundary valid during irradiation with high energy particles (Fig. 6) must depend on the applied stress and 2) the amount of α-ferrite formed during irradiation with high energy particles increases with decreasing applied stress.

The results of the present work imply that the behavior of stainless steel alloys and of all structural materials, which may be used in an irradiation environment, have to be tested as function of the irradiation temperature and of the flux of the high energy particle. In such irradiation environments, in which the materials are subject to compression – this concerns many components especially in a fusion reactor – the mechanical tests, e.g. irradiation creep tests have to be performed in compression.

References

[1] Peckner, D. and Bernstein, I. M., Eds., Handbook of Stainless Steels, McGraw Hill Book Company, New York, 1977

[2] Holland, J. R., Mansur, L. K., and Potter, D. I., Eds., "Phase Stability during Irradiation," Proceedings of the Fall Meeting of the Metallurgical Society of AIME, Pittsburgh, October 5-9, 1980

[3] Garner, F. A., "Irradiation Performance of Cladding and Structural Steels in Liquid Metal Reactors," Materials Science Technology, R. W. Cahn, P. Haasen, and E. J. Kramer, Eds., Nuclear Materials, B. R. T. Frost, Ed., Part I, Vol. 10A, 1994, pp. 419-543

[4] Schüle, W. and Lang, E., *Proceedings of the 15th International Symposium of Radiation on Materials, ASTM STP 1125*, R. E. Stoller, A. S. Kumar, and D. S. Gelles, Eds., Nashville, Tenn., June 19-21, 1990, American Society for Testing and Materials, Philadelphia, 1992, pp. 945-957

[5] Schüle, W. and Lang, E., and Panzarasa, A, "Studies of Phase Transformations in Fe-Cr-Mn-Alloys," Report EUR 11756 EN, 1988

[6] Schüle, W., *Proceedings of the 19th International Symposium on the Effect of Radiation on Materials, ASTM STP 1366*, Seattle, Washington, USA, June 16-19, 1998, M. L. Hamilton, A. S. Kumar, S. T. Rosinski, and M. L. Grossbeck, Eds., American Society for Testing and Materials, Philadelphia, USA, 2000, pp. 894-918

[7] Schüle, W., Z. Metallkd. "On Properties of Interstitials and Vacancies in Stainless Steel Alloys," to be published

[8] Hausen, H., Schüle, W., and Cundy, M. R., *Proceedings of the 15th International Symposium on Fusion Technology*, 15th SOFT, 9-13 September, 1988, Utrecht, The Netherlands, Fusion Technology 1988 Vol. 2, 905

[9] Schüle, W. and Hausen, H., *6th International Conference on Fusion Reactor Materials (ICFRM-6)*, Stresa, Lago Maggiore, Italy, September 27-October 1, 1993, D. G. Rickerby, H. Stamm, K. Ehrlich, and M. Victoria, Eds., J. Nucl. Mat., Vols. 212-215, 1994, pp. 388-392

[10] Hausen, H. and Schüle, W., *Proceedings of the 18th International Symposium on the Effect of Radiation on Materials*, ASTM STP 1325, R. K. Nanstad, M. L. Hamilton, F. A. Garner, and A. S. Kumar, Eds., Hyannis, Mass., USA, June 22-26, 1996, American Society for Testing and Materials, West Conshohocken, PA 1999, pp. 742-764

[11] Hausen, H. and Schüle, W., *Proceedings of the 18th International Symposium on the Effect of Radiation on Materials, ASTM STP 1325*, R. K. Nanstad, M. L. Hamilton, F. A. Garner, and A. S. Kumar, Eds., Hyannis, MA., June 22-26, 1996, American Society for Testing and Materials, West Conshohocken, PA, 1999, pp. 830-849

[12] Robertson, J. P., Ioka, I., Rowcliffe, A. F., Grossbeck, M. L., and Jitsukawa, S., *Proceedings of the 18th International Symposium on the Effect of Radiation on Materials, ASTM STP 1325*, R. K. Nanstad, M. L. Hamilton, F. A. Garner, and A. S. Kumar, Eds., Hyannis, MA., June 22-26, 1996, American Society for Testing and Materials, West Conshohocken, PA 1999, pp. 671-688

[13] Maziasz, P. J., *Proceedings on Phase Stability during Irradiation*, J. R. Holland, L. K. Mansur, and D. I. Potter, Eds., Metallurgical Society of AIME, Pittsburgh, PA., 1981, p. 477

[14] Garner, F. A., and Brager, H. R., *Proceedings of the 13th International Symposium on Radiation-Induced Changes in Microstructure, ASTM STP 955*, Part 1, F. A. Garner, N. H. Packan, and A. S. Kumar, Eds. Philadelphia, PA., (1986), p. 195

[15] Garner, F. A. and McCarthy, J. M., *Proceedings of the Symposium on Reduced Reduced Activation Materials for Fusion Reactors, ASTM STP 1047*, R. L. Klueh, D. S. Gelles, M. Okada, N. H. Packan, Eds., Philadelphia, PA., 1990, pp. 19-29

[16] Porter, D. L., Garner, F. A., and Bond, G. L., *Proceedings of the 19th International Symposium on the Effect of Radiation on Materials, ASTM STP* 1366 Seattle, WA., June 16-19, 1998, M. L. Hamilton, A. S. Kumar, S. T. Rosinski, M. L. Grossbeck, F. Quinzi, Eds., American Society for Testing and Materials, West Conhohocken, PA., 2000, pp. 884-893

[17] Angel T., Journal of the Iron and Steel Institute London, Great Britain, Vol. 77, 1954 pp. 165

H. Shibamoto,[1] T. Kitao,[2] H. Matsui,[3] M. Hasegawa,[3] S. Yamaguchi,[3] and A. Kimura[2]

Effect of Nickel on Irradiation Hardening and Microstructure Evolution of Proton Irradiated Fe-Cu Alloys

---

Reference: Shibamoto, H., Kitao, T., Matsui, H., Hasegawa, M., Yamaguchi, S., and Kimura, A., **"Effect of Nickel on Irradiation Hardening and Microstructure Evolution of Proton Irradiated Fe-Cu Alloys,"** *Effects of Radiation on Materials: 20th International Symposium, ASTM STP 1405,* S. T. Rosinski, M. L. Grossbeck, T. R. Allen, and A. S. Kumar, Eds., American Society for Testing and Materials, West Conshohocken, PA, 2001.

Abstract: Hardness measurements and microstructure examinations of Fe(-C)-Cu-Ni model alloys were performed following 1 MeV proton irradiation below 80°C. Microstructural examinations by transmission electron microscope (TEM) were carried out by means of a cross section method. A band of damage structures, parallel to the irradiated surface, was observed at a depth of 6.5 μm in agreement with calculation based on the TRIM code. TEM observation revealed that the band consisted of high density of small black spots, which were considered to be interstitial-type dislocation loops. The amount of irradiation hardening increased with increase in copper concentration. An addition of 0.6wt%Ni to Fe-Cu alloys further increased the hardening, although the effect was reduced with increasing copper concentration. Irradiation hardening of pure iron was also significantly increased by the addition of nickel. The size and number density of the spot-like structures in Fe-Cu alloys decreased and increased, respectively, with addition of nickel. Three recovery stages were found in Fe-Cu-Ni alloys during post-irradiation isochronal annealing to 675°C: the stages are at around 150°C, 400°C and 600°C. The first stage was only observed in nickel-containing alloys, while the third stage was only observed in copper-containing alloys. After annealing to 375°C, the density of spot-like structures decreased in Fe-Cu-Ni alloy but increased in Fe-Cu alloy, while for both the size of spots increased.

Keywords: irradiation hardening, recovery, nickel effects, iron-copper model alloys

---

[1]Scientist, Chubu Electric Power Company, Inc., 1 Higashi-shinmachi, Higashi-ku, Nagoya 461-0006, Japan.

[2]Graduate student and professor, respectively, Institute of Advanced Energy, Kyoto University, Gokasho, Uji, Kyoto 611-0011, Japan.

[3]Professors, Institute for Materials Research, Tohoku University, 2-1-1 Katahira, Aoba-ku, Sendai 980-0812, Japan.

Introduction

In order to understand the mechanism of irradiation embrittlement and hardening of reactor pressure vessel (RPV) steel, it is important to clarify the effects of alloying elements on microstructural evolution under irradiation as well as to know the correlation with irradiation hardening. In RPV steels, copper as an impurity of steel has been considered to play a significant role on irradiation embrittlement and hardening [*1-6*]. In the previous works on the effects of carbon and copper on the irradiation hardening, we reported that irradiation hardening increased with increasing concentration of copper, and addition of carbon further increased the hardening [*7,8*].

It is well known that RPV steel contains a small amount of nickel to improve low temperature fracture toughness. Although there are some studies on the nickel effects on irradiation embrittlement of RPV steels and model alloys, the details of the effects are not yet understood. Buswell et al. investigated the microstructure of thermally aged Fe-Cu-Ni model alloys with atom probe field ion microscopy (APFIM) and small angle neutron scattering (SANS), and they reported that a peak-hardening occurred before reaching a maximum copper precipitate volume fraction, and a bimodal size distribution in the copper precipitates at an over aged condition, suggesting the presence of small BCC copper precipitates at peak-hardening [*9*]. As for the role of nickel, bimodal response became significant in Fe-Cu-Ni alloys. It was also reported that there was a nickel-rich region at the surface of copper precipitates and the concentration of solid-solution copper was reduced by nickel addition for unirradiated conditions [*9,10*]. In copper-containing alloys, very small copper precipitates were induced by neutron irradiation in high density and an addition of nickel reduced the amount of copper precipitates [*11*].

It was also shown that irradiation hardening of RPV steels increased with increasing nickel concentration at low irradiation temperature, and irradiation-induced copper precipitates contained a high concentration of nickel and manganese, and a thermodynamic calculation predicted an interaction between nickel atom and manganese atom so that the concentration of manganese in copper precipitates was influenced by nickel concentration [*12*]. Based on those experimental results, it was considered that nickel and manganese influenced nucleation and growth of copper precipitates [*13,14*]. In contrast, it was reported that for RPV steel weld containing less than 0.4wt% nickel, irradiation embrittlement depended on irradiation temperature with nickel effects only for a certain range of concentration. Miller conducted tensile tests to measure yield strength and tensile strength of Fe-Cu-Ni-P alloys, and showed that even if nickel concentration was increased, the nickel effect was not accumulative [*10*], while he suggested ultra-fine carbides, such as molybdenum and might be nickel-carbides, attribute for the irradiation hardening [*15*].

Although many studies have been done on the role of nickel on the irradiation hardening of Fe-Cu model alloys, it is still unclear that the enhancement of irradiation hardening by nickel addition is due to only effects on copper precipitation related phenomena. The objectives of this research are to correlate the irradiation hardening and microstructure change in Fe-Cu, Fe-Ni and Fe-Cu-Ni alloys and to investigate the effects of nickel addition on the irradiation hardening of Fe-Cu model alloys.

Experimental

Model alloys were prepared by arc-melting method and the alloy compositions are shown in Table 1. Sheet specimens, 1mm wide, 12mm long and 0.25mm thick, were fabricated from the ingots. All the specimens were solution-treated at 880°C for 7.5 hours and tempered at 670°C for 2 hours, followed by aging at 620°C for 22.5hours to simulate the post-weld heat treatment of RPV steels. After the heat treatment, specimens were irradiated with 1MeV protons up to 3 x 10^{17}ions/cm^2 at below 80°C. Micro-Vickers hardness tests were carried out on the as-irradiated surface with a load of 25g. An averaged hardness of 20 indentations on each sample was obtained. Microstructural examinations by transmission electron microscopy were performed by means of a cross section method [*16*]. Proton-irradiated sheet specimens were plated with iron and then nickel to form a cylindrical rod of about 5mm in diameter. A large fraction of the rod was nickel to reduce ferro-magnetic effects on stigmation of the electron beam during TEM operations. The cylindrical rods were cut into slices 0.2mm thick, and TEM disk specimens were punched out so that the highly damaged region was centered. TEM disks were electrolytic-polished by a twin-jet method in a solution of 5vol.% perchloric acid with methanol below –10°C. After the irradiation, isochronal annealings of 30 minutes (and air-cooling) were performed by 25°C step from 100 to 675°C.

Table 1- *Chemical compositions of model alloys.*

wt.%	C	Cu	wt.%	C	Cu	Ni
Fe	0.0028	-	Fe-0.6Ni	0.20	-	0.64
Fe-0.05Cu	0.20	0.057	Fe-0.05Cu-0.6Ni	0.20	0.062	0.75
Fe-0.15Cu	0.21, (0.35)	0.157, (0.157)	Fe-0.3Cu-0.2Ni	0.20	0.30	0.21
Fe-0.3Cu	0.20, (0.35)	0.339, (0.339)	Fe-0.3Cu-0.6Ni	0.20	0.30	0.65

Results and Discussion

Irradiation Hardening

Effects of 0.6wt% nickel addition on the irradiation hardening of Fe-Cu alloys are shown in Fig. 1. No significant alloying effect on the hardness of the model alloys before the irradiation is found in the bottom Fig. ure. In contrast, after irradiation both the alloy elements of copper and nickel addition increase the irradiation hardening, as can be shown in the top Fig. ure. Irradiation hardening, ΔHv, for Fe-Cu alloys increases significantly with increasing copper concentration. The addition of nickel enhances the irradiation hardening of Fe-Cu alloys, but the effect decreases with increasing copper concentration. Note also that the addition of nickel enhances irradiation hardening. The *Hv* of Fe-Ni alloy becomes three times larger than pure iron.

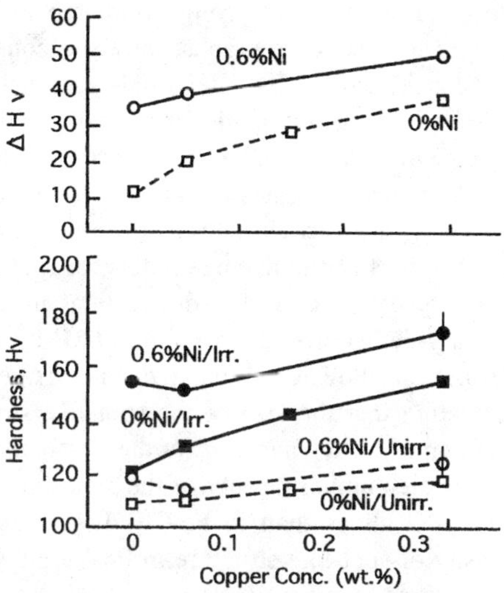

Figure 1-*Effects of irradiation on the change in hardness and base hardness of Fe-Cu and Fe-Cu-Ni alloys.*

Microstructure Observation

Observations by optical microscope revealed that all the alloys consisted of ferrite and cementite particles several microns in diameter. It was found by TEM observation of cross-sections that proton irradiation caused the formation of a damage band in a region about 6 .5 μ m beneath the surface, corresponding to the position for the peak in number density of

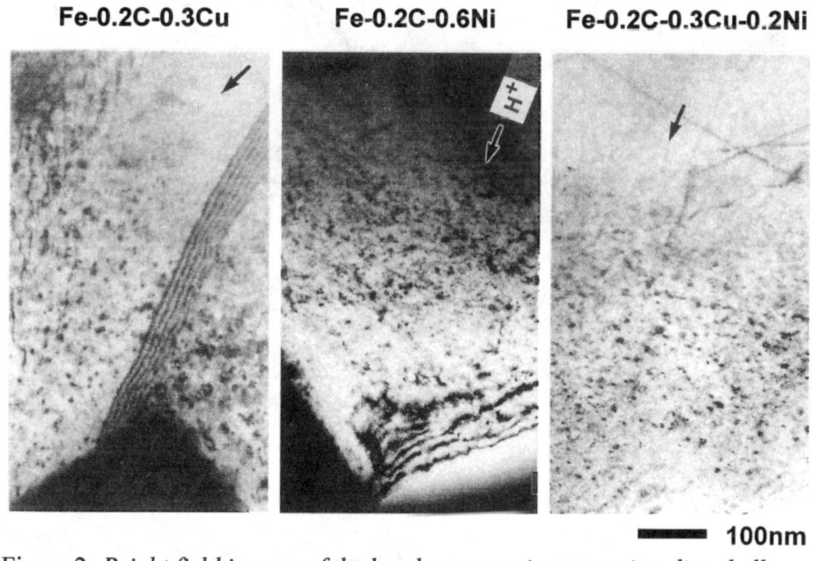

Figure 2- *Bright field images of the band structure in proton-irradiated alloys. (Arrows indicate the beam direction.)*

primary defects calculated by the TRIM code. Figure 2 shows the bright field images of the band structure in proton-irradiated Fe-C-Cu, Fe-C-Ni and Fe-C-Cu-Ni alloys. As shown in these micrographs, the band consists of small black spot-like features, which are considered to be interstitial-type dislocation loops. As for vacancy-type defect clusters, a positron annihilation study (PAS) [7] suggested that microvoids with an average diameter of 0.7nm were formed in the present model alloys by proton irradiation. The large black regions found at the bottom of the micrographs of Fe-C-Cu and Fe-C-Ni are cementite particles. From the micrographs shown in Fig. 2, the depth profiles of number density and average radius for spot-like structures at a given depth were obtained and shown in Fig. 3. The dotted lines in the figures are the calculated depth profile of primary-defect density from the TRIM code. The size of spot-like structures in the Fe-0.15Cu alloy is almost two times larger than those of Fe-Ni and Fe-Cu-Ni alloys, indicating that the size of the spot-like structures in an Fe-Cu alloy is reduced by the addition of nickel, and that for the Fe-Ni alloy is not influenced by the presence of copper. As for the number density of the spot-like structures, it is four times larger in Fe-Ni and Fe-Cu-Ni alloys than in Fe-Cu alloy, indicating that the nickel addition increases the density of spot-like structure in Fe-Cu alloy significantly, but copper has little influence on the density in Fe-Ni alloy.

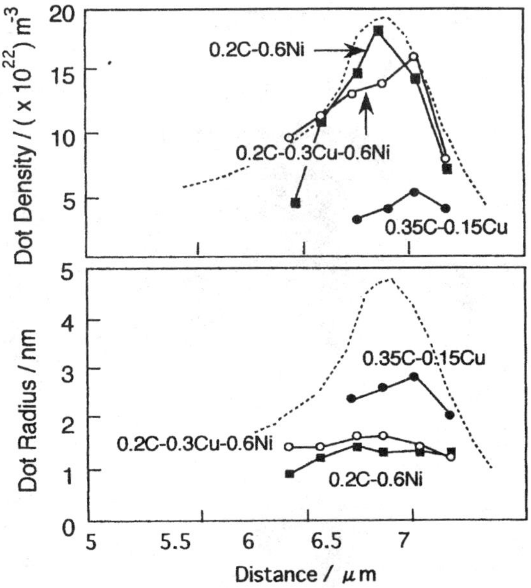

Figure 3--*Depth profiles for number density and average radius of spot-like structures for proton-irradiated model alloys. Dotted lines give calculation results of defect density using the TRIM code.*

Figure 4 shows the microstructures of model alloys near the peak position of the damaged area. Images are obtained with g=110 close to a two beam condition. In all the alloys, there are three types of images, namely, black-white lobe contrast (≈1nm), small black spots (≈3nm), and large black spots (≈15nm). Since the black-white lobe contrast, which has been considered to be due to copper precipitates with spherical strain field symmetry, were also observed in the copper-free alloy, they are considered to stem from not only copper precipitates but also small defect clusters or interstitial loops.

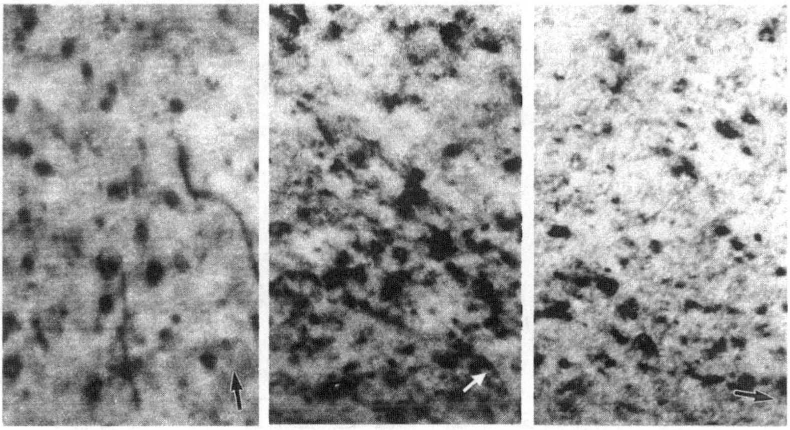

Figure 4--*Microstructures of model alloys at around peak position of the damaged area. Images are obtained with g=110 at near two beam condition. (Arrows indicate the beam direction.)*

Correlation between Hardening and Microstructure

The estimation of irradiation hardening from the microstructure using the Orowan relation seems to be very complex, since the distribution of damage structures is inhomogeneous in the proton-irradiated specimens, and hardness measurements were performed on the irradiated surfaces with loading in the direction of proton beam. The amount of irradiation hardening, ΔH_V, however, can be compared between the alloys. In order to assess the role of microvoids on the irradiation hardening, the irradiation hardening was estimated from positron data [7] obtained for the same alloys studied in this work. Irradiation hardening has been expressed by the following equation as a function of the number density of damage features, *N*, and their diameter, *d*.

$$\Delta H_V \propto \sqrt{(N \times d)} \text{---} (1)$$

Relationships between ΔH_V and $(N \times d)^{1/2}$ of model alloys were derived from both the experimental results, PAS measurements, and TEM observations, as shown in Fig. 5, indicating that there appears to be a linear relationship between ΔH_V and $(N \times d)^{1/2}$ derived from TEM measurements, but there is no such relationship between those derived from PAS measurements. This suggests that irradiation hardening is related to not microvoids but spot-like damage futures, which are considered to be interstitial-type dislocation loops and/or small precipitates.

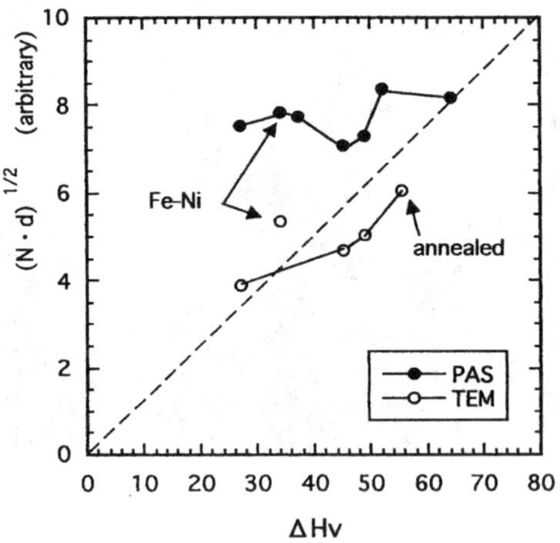

Figure 5-*Correlation between $(N \cdot d)^{1/2}$ and ΔH_V in Fe-Cu-Ni alloys after irradiation.*

Post-irradiation Annealing Behavior

Recovery behavior of irradiation hardening of model alloys during post-irradiation isochronal annealing is shown in Fig. 6. Here, ΔH_V is the difference between the hardness of the isochronally annealed specimen without irradiation and the hardness of the post-irradiation annealed specimen. There are three recovery stages in the annealing behavior of irradiation hardening in Fe-Cu-Ni alloys: the first stage at around 150°C, the second stage at around 400°C, and the third stage at around 600°C. The first stage was only observed in nickel-containing alloys, while the third stage was only observed in copper-containing alloys. As for the mechanism of the first stage, it is considered that the low temperature recovery stage is related with the motion of carbon, and nickel may influence the mobility of carbon atoms. The coauthors [*17*] investigated the effects of nickel on the neutron irradiation hardening of martensitic steels, and found that the steel doped with 1wt%Ni showed large additional irradiation hardening when the irradiation temperature was lower than 180°C, and the additional hardening was recovered by post-irradiation annealing above 220°C. A similar nickel effect was observed in this work. It is expected that the irradiation at 80°C causes the formation of Ni-C pairs or complexes, which may induce large hardening and be unstable above 220°C. From the viewpoint of lifetime estimation for RPV steel, the role of nickel may be important to evaluate the irradiation hardening of RPV steel for which service temperatures are lower than 220°C. The third stage is only observed in copper-containing alloys. According to the thermal aging experiment using the same specimens as in this study [7], age hardening was observed in this temperature region in water quenched Fe-Cu alloys. Since all the specimens were finally annealed at 620°C for 22.5hr before the tests, they were in a solid-solution condition for copper before irradiation, but no age hardening was observed in unirradiated copper-containing alloys during annealing because the aging was short-term. The appearance of the third stage in irradiated copper-containing alloys can be interpreted in terms of irradiation induced copper precipitation. The second stage was observed in all the alloys, suggesting that the recovery was mainly due to annealing out of matrix damage such as dislocation loops.

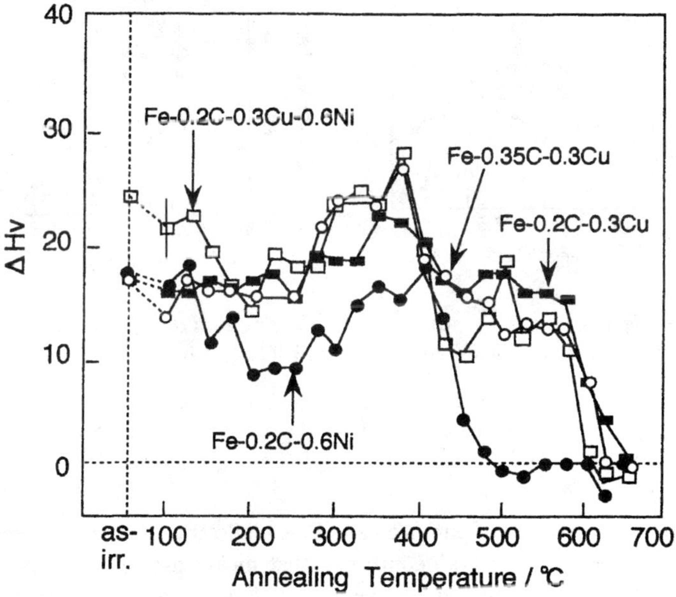

Figure 6—*Recovery behavior of irradiation hardening of each alloy during isochronal annealing to 673°C for 20min. There are three recovery stages: 1) below 200°C, 2) 350-450°C, 3) above 550°C.*

The TEM observation of damage structure of Fe-0.2C-0.3Cu-0.2Ni alloys was also conducted after annealing to 375°C near the peak temperature for anneal-hardening. It is confirmed that interstitial-type dislocation loops still remain after the annealing. From the micrographs, the size distribution and the density of spot-like structures were measured. As shown in Fig. 7, after annealing, the average radius became larger, and the number density decreased slightly in Fe-Cu-Ni, suggesting that interstitial-type dislocation loops grow by Ostwald's ripening during annealing. In the Fe-Cu alloy, however, the density of spot-like structures increased after the annealing. This suggests that the spot-like structures observed in the post-irradiation annealed Fe-Cu alloy consists of both the dislocation loops and copper precipitates that form during the annealing.

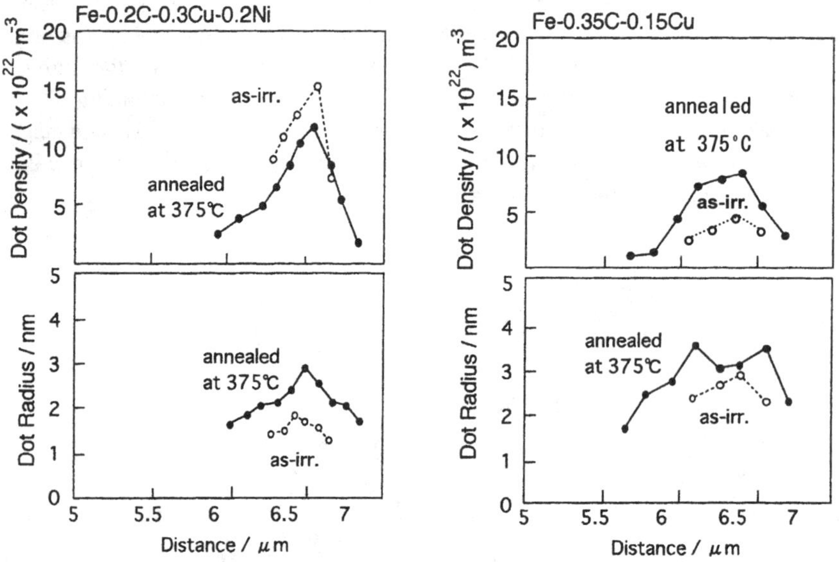

Figure 7—*Effects of annealing at 375°C on the size and density of dot-structures observed in proton-irradiated Fe-Cu-Ni and Fe-Cu alloys.*

Summary

The irradiation hardening of pure iron and Fe-Cu alloys were markedly enhanced by the addition of 0.6wt. %Ni. It was confirmed that nickel as well as copper addition significantly enhanced irradiation hardening. Microstructural observations also confirmed that the spot-like structures, which were considered to be interstitial-type dislocation loops, were formed in Fe-C, Fe-C-Ni, Fe-C-Cu and Fe-C-Cu-Ni alloys by proton irradiation. In Ni doped alloys, the size and density of the structures was smaller and higher, respectively, than Fe-Cu alloys. A good correlation was found between the irradiation hardening and morphology of the spot-like structures observed by TEM.

There are three recovery stages in the annealing behavior of irradiation hardening in Fe-Cu-Ni alloys: the first stage at around 150°C, the second stage at around 400°C and the third stage at around 600°C. The first stage was only observed in nickel-containing alloys, while the third stage was only observed in copper-containing alloys. The first recovery stage is considered to be related to the motion of carbon, and nickel may influence the mobility of carbon atoms. The second stage was observed in all the alloys, suggesting that the recovery was mainly due to annealing out of matrix damage such as dislocation loops. The appearance of the third stage in irradiated copper-containing alloys can be interpreted in terms of irradiation induced copper precipitation.

Acknowledgments

We would like to thank Dr. Nagata and Dr. Takahiro for performing proton irradiations, and to thank Mr. Takahashi for producing the alloys by arc-melting.

References

[1] Odette, G. R. and Lucas, G. E., *"Radiation Embrittlement of Nuclear Reactor Pressure Vessel Steels: An International Review", ASTM STP 909,* 1986, p.206.

[2] Odette, G. R., *"On the Dominant Mechanism of Irradiation Embrittlement", Scripta Metallurgica,* Vol.17, 1983, p.1183.

[3] Buswell, J. T. and Jones, R. B., *"The Modelling of Radiation Hardening and Embrittlement in Magnox Mild Steel Submerged-Arc Welds", Effects of Radiation on Materials: 16th International Symposium, ASTM STP 1175*, Philadelphia, 1993, p.424.

[4] Phythian, W. J., Foremann, A. J. E., English, C. A., Buswell, J. T., Hetherrington, M., Roberts, K. and Pizzini, S., *"The Structure and Hardening Mechanism of Copper Precipitation in Thermally Aged or Irradiated Fe-Cu and Fe-Cu- Ni Model Alloys", Effects of Radiation on Material: 15th International Symposium, ASTM STP 1125*, 1992, P.131.

[5] Beaven, P. A., Frisius, F. R., Kampmann, R. and Wagner, R., *"SANS/TEMS Studies of the Defect Microstructure of Test Reactor Irradiated Fe-Cu Alloys and Cu-Containing RPV Steels", Proceedings of Second International Symposium on Environmental Degradation of Materials in Nuclear Reactors-Water Reactors*, 1986, p.400.

[6] Kampann, R., Frisius, F., Hackbarth, H., Beaven, P. A., Wagner, R. and Hawthorne, J. R., "SANS Analysis of Reactor Vessel Materials In Neutron Irradiated, Annealed and Re-Irradiated Conditions", Proceeding of Fifth Internal Symposium on Environmental Degradation of Materials in Nuclear Reactors-Water Reactors, 1992, p.679.

[7] Shibamoto, H., Koyama, K., Yuya, H., Hasegawa, M., Kimura, A., Matsui, H., and Yamaguchi, S., "Effects of Proton Irradiation on Positron Annihilation and Micro-Vickers Hardness of Fe-C-Cu Model Alloys", Effects of Radiation on Materials: 17th International Symposium, *ASTM STP 1270*, 1996, p.623.

[8] Kimura, A., Shibamoto, H., Yuya, H., Hasegawa, M., Yamaguchi., H., and Matsui, H., *"Damage Structures of Proton Irradiated Fe-Cu Alloys", Effects of Radiation on Materials: 17th International Symposium, ASTM STP 1270*, 1996, p.220.

[9] Buswell, J. T., English, C. A., Hetherington, M. G., Phythian, W. J., Smith, G. D. W. and Worrall, G. M., *"An Analysis of Small Clusters Formed in Thermally Aged and Irradiated FeCu and FeCuNi Model Alloys", Effects of Radiation on Materials: 14th International Symposium, ASTM STP 1046*, 1990, p.127.

[10] Miller, M. K., Hoelzer, D. T., Hawthorne, J. R. and Burke, M. G., *Characterization of Irradiated Model Pressure Vessel Steels", Proceedings of the Third International Symposium on Environmental Degradation of Materials in Nuclear Power Systems Water Reactors*, 1987, p.71.

[11] Williams, T. J., Burch, P. R., English, C. A. and Ray, P. H. N., *"The Effect of Irradiation Dose Rate and Temperature and Copper and Nickel Content on the Irradiation Shift of Low Alloy Steel Submerged Arc Welds" Environmental Degradation of Materials in Nuclear Reactors-Water Reactors*, 1988, p.121.

[12] Lucas, G. E., Mader, E. and Odette, G. R., *"Effect of Metallurgical Variables and Irradiation Conditions on the Kinetics of Post-Irradiation Annealing of Pressure Vessel Steels", Proceedings of the Fourth International Symposium on Environmental Degradation of Materials in Nuclear Power Systems-Water Reactors*, 1989, p.21.

[13] Phythian, W. J., English, C. A. and Buswell, J. T., *"The Microstructural and Microchemical Examination of Reactor Pressure Vessel Steels",Proceedings of the Fifth International Symposium on Environmental Degradation of Materials in Nuclear Power Systems-Water Reactors*, 1991, p.27.

[14] Onchi, T., *"Radiation Damage of Light Water Reactor Materials"*, *Materia Japan*, Vol. 34, No.1, 1995, p.25.

[15] Miller, M. and Burke, M. G., *"Fine Scale Microstructural Characterization of Pressure Vessel Steels and Related Materials using APFIM"*, *Effects on Radiation on Materials: 14th International Symposium (Volume II) ASTM STP 1046*, 1990, P.107.

[16] Odette, G. R. and Lucas, G. E., *"The Effect of Nickel on Irradiation Hardening of Pressure Vessel Steels"*, *Effects on Radiation on Materials: 14th International Symposium (Volume II) ASTM STP 1046*, 1990, P.323.

[17] R. Kasada, T. Morimura, A. Kimura, H. Matsui, M. Narui, *"Two-Step Recovery Process of Irradiation Hardening in 1%Ni doped 9%Cr-2%W Martensitic Steel"*, *19th Symposium on Radiation Effects on Materials, ASTM STP 1366*, 2000, p448.

Other Materials

Nobuyasu Nita,[1] Ken-ichi Fukumoto,[1] and Hideki Matsui[1]

Effect of Final Irradiation Temperature and Frequency of Irradiation Temperature Cycles on Microstructural Evolution of Vanadium Alloys

Reference: Nita, N., Fukumoto, K., and Matsui, H., "**Effect of Final Irradiation Temperature and Frequency of Irradiation Temperature Cycles on Microstructural Evolution of Vanadium Alloys,**" *Effects of Radiation on Materials: 20th International Symposium, ASTM STP 1405,* S. T. Rosinski, M. L. Grossbeck, T. R. Allen, and A. S. Kumar, Eds., American Society for Testing and Materials, West Conshohocken, PA, 2001.

Abstract: Effects of varying temperature on microstructure of materials under irradiation can be very complicated. In this paper, evolution of microstructure in vanadium alloys during temperature changes is closely examined with emphasis on the effect of frequency of the temperature changes and the final temperature.

Pure vanadium, V-5Cr and V-5Nb were prepared. Neutron irradiation was conducted in JMTR at constant temperatures of 200℃, 400℃ and under cyclic temperature changes of 200/400℃. The numbers of temperature cycles were 0.5, 5.5 and 6 cycles. TEM observations, Positron Annihilation Spectroscopy and micro-Vickers hardness test were performed.

By comparison between 0.5 and 5.5 temperature cycle, coarser microstructures are observed in 5.5 cycles. Frequent temperature change is considered to have restricted the defect growth. By comparison between 5.5 and 6 temperature cycle irradiation, which is focused on the effect of the final temperature, coarser microstructures are observed in 6 cycle irradiation.

Keywords: temperature varying irradiation, vanadium alloys, microstructure, neutron irradiation, JMTR

Introduction

Microstructural evolution under irradiation is very sensitive to the irradiation temperature because the defect clustering processes are essentially controlled by the thermal activation processes. In a future fusion environment, temperature of the first wall is expected to change depending on the operating condition. During the start-up and shut-down of the reactor, materials will be exposed to neutrons at lower temperature than expected in the rated operation condition. Therefore, it is necessary to develop an understanding of the effects of temperature change to predict the material behavior in

[1]Graduate student, Research Associate and Professor, respectively, Institute for Material Research, Tohoku University, Katahira, Aoba-ku, Sendai 980-8577, Japan.

fusion reactors.

Several irradiation experiments with temperature changes were conducted to understand the effects of temperature history on microstructural evolution [*1-3*]. From the stepwise temperature change experiments, it has been confirmed that the irradiation temperature changes between the regimes where different species of defects are mobile and stability of the defects are significantly different from each other, microstructural evolution will be greatly affected [*4*]. From the simulation experiment by means of rate theory, it has also been pointed out that vacancy flux to sinks significantly increases immediately after an upward temperature change [*5*]. The stability of small dislocation loops against stepwise temperature change is discussed by conducting heavy ion irradiation experiments [*6*]. That is, unstable interstitial clusters that were formed under pre-irradiation at low temperature shrink during upward temperature change; on the other hand, the clusters that were formed at relatively higher temperature survive during upward temperature change.

It is important to reveal the essential parameters that most affects the microstructure, for example, number of temperature cycles, frequency of temperature cycles, final temperature, fluence under low temperature irradiation, and regimes of temperature change etc.

The objective of the present work is to clarify the effects of the frequency of temperature cycles and the final temperature under neutron irradiation on microstructural evolution of vanadium alloys. Vanadium alloys are selected as the material to be studied, since they have a number of attractive features as candidate materials for the first wall of a fusion reactor.

Experimental Procedures

Pure vanadium and two binary vanadium alloys, V-5at.%Cr and V-5at.%Nb, were prepared. Chromium is an undersized element in the vanadium matrix while niobium is oversized. The alloys were prepared by arc-melting in an argon gas atmosphere, followed by cold rolling into 0.25 mm thickness sheets. Transmission electron microscopy (TEM) disks of 3 mm diameter were punched from the alloy sheets. The resulting TEM disks were wrapped with zirconium getter foil and annealed at 1100 °C in a high vacuum (2×10^{-6} torr).

Neutron irradiation was conducted in Japan Material Testing Reactor (JMTR) at constant temperatures of 200°C and 400°C and cyclic temperature change between 200/400°C. The latter type of irradiation will be referred to below as varying temperature irradiation-200/400 °C, or VTI-200/400°C. The temperature cycles were 0.5, 5.5 and 6 cycles. Here, 0.5 cycle, for example, is defined as stepwise temperature change from 200°C to 400°C. The damage levels are from 0.09 up to 0.21 dpa.

In order to understand the mechanism of microstructure evolution during temperature change, knowledge of simple constant-temperature irradiation is obviously a prerequisite. The effect of frequency of the temperature cycles is focused on by comparing between 0.5 and 5.5 cycles. The effect of the final irradiation temperature is focused on by comparing 5.5 cycles irradiation with 6 cycles irradiation, although the damage level is not precisely the same.

After electro-polishing, the microstructure of alloys was observed by TEM. In order to obtain complementary information relating to invisible small vacancy-type clusters such as microvoids, Positron Annihilation Spectroscopy measurement was carried out. Micro-Vickers hardness test was also carried out in order to understand the effect of temperature change on the mechanical properties of vanadium alloys.

Results

Micro-Vickers hardness test

Vickers hardness changes by irradiation are plotted in Fig.1. The trends are almost the same in all alloys. Approximately the same large amount of irradiation-induced hardening occurred under constant temperature irradiation at 200°C and 400 °C. Temperature changes during irradiation decrease the hardening compared with constant temperature irradiation in this condition. By comparison between 0.5 and 5.5 cycles, the levels of hardness are higher in the latter. Irradiation hardening increases with the frequent temperature change. By comparison between 5.5 and 6 cycles, larger hardening has occurred during 5.5 cycle. Final irradiation at a lower temperature of 200°C decreases the hardening, which is inconsistent with the knowledge of constant temperature irradiation that irradiation hardening is more significant at low temperatures.

Microstructure of pure vanadium

Fig.2 shows the dislocation images of pure vanadium. The high density of tiny dislocation loops of interstitial type is observed in the specimen irradiated at 200 °C. Coarse microstructure of network dislocations is observed in the specimen irradiated at 400 °C. As in the previous study [*4*], temperature of 200 °C is in the regime where nucleation of defect cluster mainly occurs, and 400 °C is in the defect growth regime. Microstructures after temperature change are similar to the microstructure irradiated at

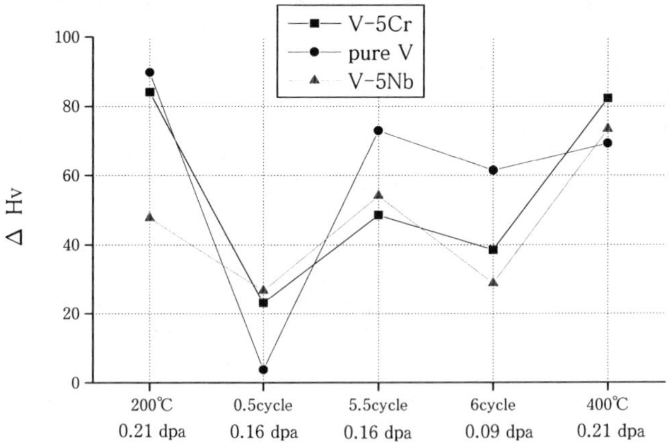

Fig.1 *Micro-Vickers hardness test*

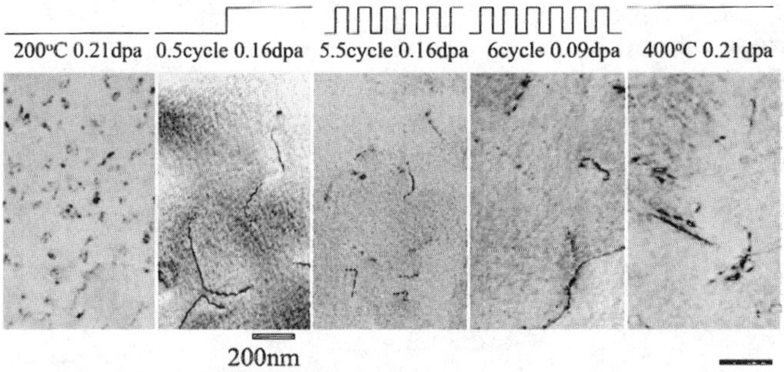

Fig.2 *Dislocation images of pure vanadium*

400 °C, where dislocation loops have grown to network dislocations.

Fig.3 shows the cavity images of pure vanadium. No cavity is observed in the specimen irradiated at 200°C, while a microvoid component is detected by PAS. In other words, tiny vacancy clusters nucleated, but they were unable to grow to observable size by TEM under 200°C irradiation. 30 nm size of large cavities with low density are observed in the specimen irradiated at 400°C. Smaller size and higher densities of cavities are observed in specimens irradiated under varying temperature conditions compared to the 400°C irradiation. Coarser microstructure of cavities is observed in 0.5 cycle irradiation than in 5.5 cycle irradiation. Frequent temperature changes increase the number density and decrease the size of cavity. Cavity mean size is larger in 6 cycle irradiation than in 5.5 cycle irradiation despite the lower final temperature and damage level in the former.

Microstructure of V-5Cr

Fig.4 shows the dislocation images of V-5Cr. The high densities of dot-like contrast observed in the specimen irradiated at 200°C are dislocation loops of interstitial type. Unlike the case of pure vanadium, interstitial clusters retain the shapes of dislocation loops rather than dislocation network in the specimen irradiated at 400°C and VTI-200/400°C. In vanadium alloy, growth of defects is suppressed by the interaction

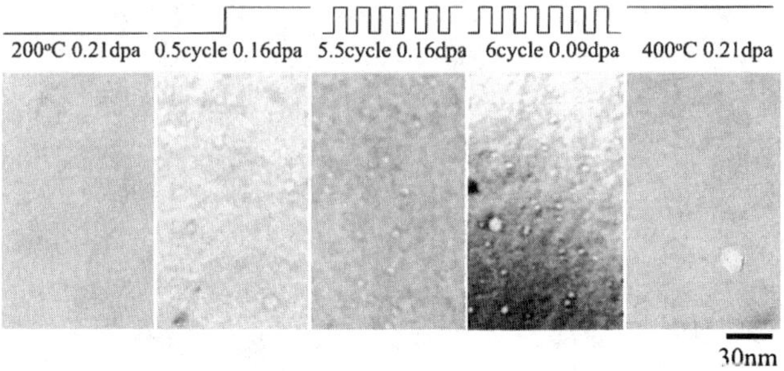

Fig. 3 *Cavity images of pure vanadium*

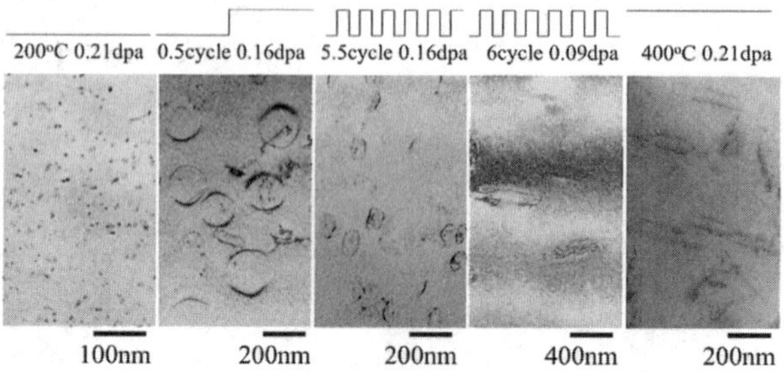

Fig. 4 *Dislocation images of V-5Cr*

between the solute and point defect as shown in a previous study [7]. By comparison between 0.5 and 5.5 cycle irradiation, about 200 nm dislocation loops are observed in the 0.5 cycle specimen, while a high density of small dislocation loops is observed in 5.5 cycle. Apparently, frequent temperature changes suppress the growth of dislocation loops. Large dislocation loops are observed in the specimen irradiated up to 6 cycle in spite of its low damage level. By comparison between 5.5 and 6 cycle, It is seen that a lower final temperature decreases the density of defects and increases the defect size similarly to the microstructural evolution of pure vanadium.

Fig.5 shows the cavity images of V-5Cr. The number density and mean size of cavities are plotted in Fig.6. Tendencies of microstructural evolution are very similar to pure vanadium. In this alloy, almost all the cavities are formed within the compressive strain field of dislocation loops instead of being formed in the matrix. Largest size and lowest density of cavities are observed in the specimen of 0.5 cycle irradiation. Temperature change enhances the growth of cavities in this case. In 6 cycles of irradiation, cavity density is higher than for 5.5 cycle irradiation. Frequent temperature changes suppress the defect growth. Moreover, a microvoid component is detected in 6 cycle irradiation by means of PAS (Fig.6). These microvoids seem to have been formed in the final irradiation sequence at 200°C, since there is no microvoid component in the specimen irradiated at a final temperature of 400°C.

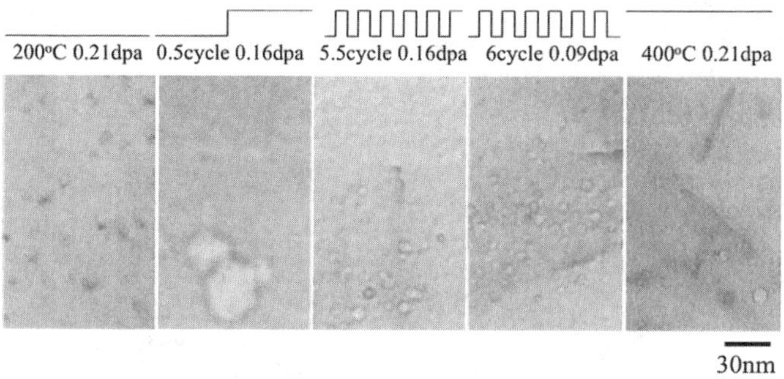

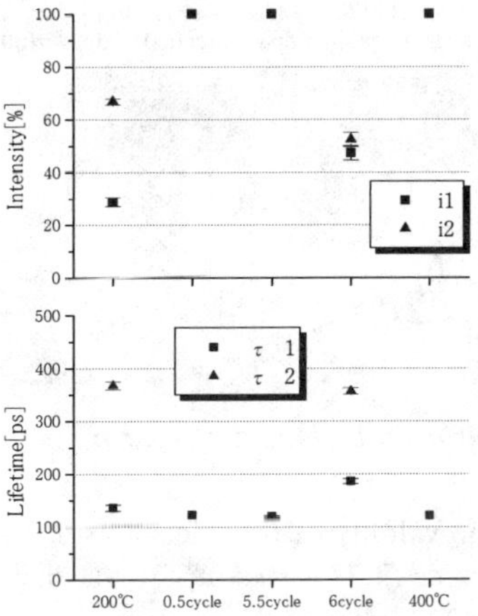

Fig.6 *PAS in V-5Cr*

Microstructure of V-5Nb

Fig.7 shows the dislocation images of V-5Nb. Tendencies of microstructural evolution are almost the same as other materials. Namely, frequent temperature changes suppress the growth of dislocation loops, and coarser microstructure is obtained after a VTI ending with a lower final irradiation temperature. But concerning vacancy clusters, no cavities were observed by TEM in V-5Nb. Moreover, no microvoids were detected by PAS. This is explained in terms of the solute atomic size effect [*7*]; oversized niobium atoms decrease the mobility of vacancy so that vacancies cannot migrate to form microvoids.

Discussion

Frequency of irradiation temperature cycles

Frequent temperature cycles under irradiation suppress the growth of interstitial clusters. There must be a mechanism that shrink and decrease the density of interstitial clusters. One of the mechanisms that is likely to occur is an excess vacancy flux during upward temperature change as described below. At first, a lot of small interstitial loops and tiny vacancy clusters that are invisible by TEM are formed under irradiation at 200°C. During the upward temperature change under irradiation, irradiation enhancement of vacancy clusters occurs, which has been reported in a previous paper [*6*]. Not only the thermal dissociation but also the irradiation-enhanced dissociation of vacancy clusters occurs under VIT-200/600°C [*8*]. In an other annealing study [*9*], which focused on recovery temperature of vacancy clusters, it has been reported that microvoids in vanadium recover in the temperature regime of 400-600°C. There seems to be a possibility of thermal and irradiation-induced dissociation of vacancy clusters under VTI-200/400°C in the present work. The number of point defects in clusters has been

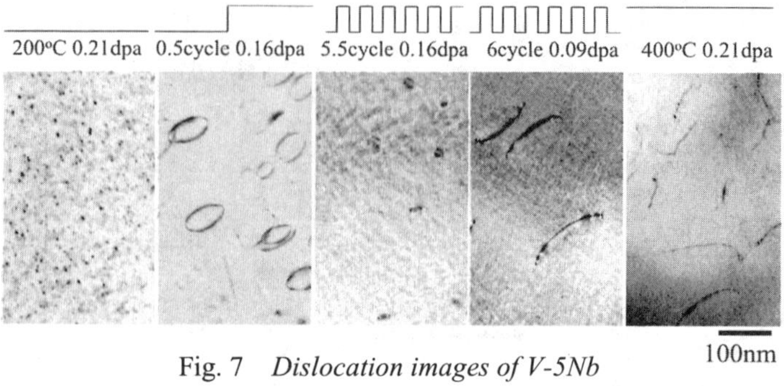

Fig. 7 *Dislocation images of V-5Nb*

roughly estimated for checking the validity of this mechanism. Under irradiation at 200 °C in V-5Cr, for example, microvoids ($3.72\times10^{23}/m^3$, 20 vacancies in a microvoid) and interstitial loops ($2.78\times10^{22}/m^3$, 4038 interstitials in a loop) are formed. If it is assumed that all the microvoids dissociate and the resultant extra vacancies shrink the interstitial loops, the density of survived interstitial loops will be $2.64\times10^{22}/m^3$. This value is much larger than the density of interstitial loops under 0.5 cycle irradiation ($1.50\times10^{20}/m^3$). That is, the number of extra vacancies due to the dissociation of microvoids is too small to decrease the density of interstitial loops. Consequently, this mechanism seems to be ineffective on the excess vacancy flux in the present work.

Another consideration is based on reaction kinetics of point defects. The steady-state concentration of vacancies; C_v, is greater than the concentration of interstitials; C_i, ($C_v > C_i$). The diffusivity of vacancies; D_v is lower than the diffusivity of interstitials; D_i, ($Di > Dv$).

C_v is greater at lower temperatures. ($C_{vlowT} > C_{vhighT}$). A vacancy rich condition should occur just after the upward temperature change, since interstitials migrate to sinks leaving a lot of vacancies in matrix. In other words, unbalance state of point defect flux appears just after an upward temperature change ($D_vC_v > D_iC_i$).

As a result, a surplus vacancy flux shrinks the interstitial clusters. Growth of survived defects occurs in the subsequent irradiation at 400°C. This process takes place every time when the upward temperature change occurs during irradiation with many temperature change cycles. In 5.5 cycle irradiation, the damage level of each sequence at high temperature is so low (0.013 dpa) that the cluster cannot grow compared to 0.5 cycle irradiation in which the damage level during the high temperature phase is 0.08dpa. Especially, this tendency is prominent in V-5Nb where very tiny interstitial loops are observed in 5.5 cycle irradiation. If the dislocation loops are only taken into account as interstitial sinks and the size of dislocation loops is assumed to have a linear dependence to damage level [*10*], the estimated loop sizes after 5.5 cycle irradiation is 14 nm in V-5Nb. This value is in good agreement with experimental data of 21 nm.

As for the microstructural evolution of vacancy clusters, the tendency of microstructural evolution is similar to the case of interstitial clusters. That is, frequent temperature cycles under irradiation suppress the growth of vacancy clusters. In this case, the above mechanism is also suitable. Growth of survived defects occurs in high temperature periods.

This work revealed the prominent influence of varying temperature, that is, defects of small size but high density are formed under irradiation with many temperature cycles. Further experiments are needed in order to clarify the effects of number of cycles under irradiation up to high damage level such as 100dpa.

Final irradiation temperature

Varying temperature irradiation ending with a lower temperature decreases the density of defect clusters and increases the cluster size, as can be seen by comparing 5.5 and 6 cycle irradiations in Fig.4 and 7. One possible mechanism is as follows. The sink strength of point defects is large due to the dense extended defects that were formed during irradiation at the former high temperature of 400°C. Although an unbalance of point defect fluxes appears ($D_vC_v < D_iC_i$), it is assumed that no significant change of point defect flux occurs during the downward temperature change since C_i is much smaller than C_v. Moreover, dissociation of vacancy clusters is unlikely to occur during the downward temperature change. During subsequent irradiation at the low temperature of 200°C, the process of point defect migration to sinks is dominant compared with the mutual recombination of point defects or nucleation of clusters. The extended defects formed during irradiation at high temperature grow absorbing these migrating point defects. Consequently, coarse microstructures of both interstitial and vacancy types are observed.

In the case of downward temperature change, it is emphasized that the sink strength of extended defects that were formed during former irradiation at high temperature determines the microstructure.

Conclusions

A series of Varying Temperature Irradiations-200/400°C was conducted in JMTR in order to understand the effects of the frequency of temperature cycles and the final irradiation temperature on microstructural evolution of vanadium alloys. It has been revealed that the frequency of temperature cycles and the final temperature significantly affect the microstructural evolution of vanadium alloys in this temperature regime.

Mechanical properties

Though the damage levels are slightly higher in constant temperature irradiation, irradiation temperature changes decrease the hardening compared with constant temperature irradiation. Irradiation hardening increases due to the frequent temperature changes and final irradiation at the lower temperature of 200°C decreases the hardening.

Effects of the frequency of temperature cycles

Coarser microstructures are observed in 0.5 cycle than in 5.5 cycle irradiation and the hardness changes induced by irradiation are smaller in 0.5 cycle than in 5.5 cycle. Frequent temperature changes restrict the defect growth because of the following mechanism: A Short period of excess vacancy flux appears immediately after upward temperature change, which shrinks interstitial loops, and finally defects are not able to grow after several temperature changes.

Effects of the final temperature

Coarser microstructures are observed in 6 cycle than in 5.5 cycle irradiation, and the hardness changes induced by irradiation are smaller in 6 cycles than in 5.5 cycle irradiations. Defects grow under lower final temperature irradiation in spite of our understanding that nucleation of defects occurs mainly under irradiation at low temperatures such as 200°C. This is understood in terms of sink strength as follows. Pre-existing defects that were grown during the previous high temperature irradiation act as strong sinks for point defects under low final temperature irradiation. Then, the defects grow at low temperature without new nucleation.

ACKNOWLEDGMENTS

The authors are grateful to the members of Material Irradiation Division of JMTR, Oarai Research Establishment of Japan Atomic Energy Research Institute; without their cooperation this work would not have been performed. The authors would like to thank Professor M. Hasegawa and Dr. T. Yamamoto for valuable discussions.

References

[1] Kiritani, M., "The Need for Improved Temperature Control during Reactor Irradiation", *Journal of Nuclear Materials,* 1988, 160, pp. 135-141.

[2] Kasada, R., Kimura, A., Matsui, H., Hasegawa, M., and Narui, M., "Effects of Varying Temperature Irradiation on the Neutron Irradiation Hardening of Reduced Activation 9Cr-2W Martensitic Steels", *Journal of Nuclear Materials,* 1999, 271-272, pp. 360-364.

[3] Matsuda, M., Yoshiie, N., Muroga, T., and Yoshida, N., "Modification of Microstructures Induced by Temperature Variation during Irradiation with 14 MeV Neutrons", *Journal of Nuclear Materials,* 1991, 179-181, pp. 962-965.

[4] Nita, N., Fukumoto, K., Kimura, A., and Matsui, H., "Effect of Temperature Change on Microstructural Evolution of Vanadium Alloys under Neutron Irradiation in JMTR", *Journal of Nuclear Materials,* 1999, 271-272, pp. 365-369.

[5] Katoh, Y., Stoller, R. E., Kohyama, A., and Muroga, T., "Simulating the Influence of Radiation Temperature Variations on Microstructural Evolution", *Journal of Nuclear Materials,* in press

[6] Nita, N., Fukumoto, K., Iwai, T., and Matsui, H., "Effects of Temperature Change on the Microstructural Evolution of Vanadium Alloys under Ion Irradiation", *Journal of Nuclear Materials,* in press

[7] Candra, Y., et al., "Microstructural Evolution and Hardening of Neutron Irradiated Vanadium Alloys at Low Temperature in JMTR", *Journal of Nuclear Materials,* 1999, 271-272, pp. 301.

[8] Ochiai, K., Watanabe, H., Muroga, T., Yoshida, N., and Matsui, H., " Microstructural Evolution in Vanadium Irradiated during Ion Irradiation at Constant and Varying Temperature", *Journal of Nuclear Materials,* 1999, 271-272, pp. 376-380.

[9] Riewcharoon, T., and Hasegawa, M., *Physical Review B,* in press

[10] Hayashi,T., Fukumoto, K., and Matsui, H., *Journal of Nuclear Materials,* in press

Takuya Nagasaka,[1] Hiroyuki Takahashi,[2] Takeo Muroga,[3] Naoaki Yoshida,[4] and Tetsuo Tanabe[5]

Hardening of Vanadium Doped with Nitrogen by Heavy Ion Irradiation and Post-Irradiation Annealing

---

Reference: Nagasaka, T., Takahashi, H., Muroga, T., Yoshida, N., and Tanabe, T.,"**Hardening of Vanadium Doped with Nitrogen by Heavy Ion Irradiation and Post-Irradiation Annealing,**" *Effects of Radiation on Materials: 20th International Symposium, ASTM STP 1405,* S. T. Rosinski, M. L. Grossbeck, T. R. Allen, and A. S. Kumar, Eds., American Society for Testing and Materials, West Conshohocken, PA, 2001.

Abstract: Copper ion irradiation at room temperature and at 573 K up to 1 dpa was conducted on vanadium doped with controlled levels of nitrogen ranging from 12 to 770 weight ppm (wppm). Micro-indentation test was performed to make a quantitative estimation of hardening of the irradiated surface layer. Microhardness of the unirradiated and as-irradiated vanadium increased with increasing nitrogen content. They showed linear relationship when nitrogen was singly varied. Irradiation hardening was independent of nitrogen content and of irradiation temperatures. This fact agrees with microstructure observations. After post-irradiation annealing at 573 K, radiation anneal hardening (RAH) was observed at the nitrogen contents of 170 wppm and above. In comparison with the previous anneal hardening experiments with cold-worked specimens, stepwise increase in the anneal hardening with nitrogen content at about 100 wppm was commonly derived for cold-worked and irradiated vanadium.

Keywords: Vanadium, nitrogen effect, irradiation hardening, radiation anneal hardening, micro-indentation test

Introduction

Vanadium (V) alloys are regarded as the advanced fusion structural materials because of its superior strength at elevated temperature, low-activation and nonmagnetic properties.

---

[1]Research associate, National Institute for Fusion Science, Oroshi-cho 322-6, Toki-shi, Gifu 509-5292, Japan.

[2]Graduate student, Graduate School of Engineering, Nagoya University, Furo-cho, Chikusa-ku, Nagoya 464-8601, Japan.

[3]Professor, National Institute for Fusion Science, Oroshi-cho 322-6, Toki-shi, Gifu 509-5292, Japan.

[4]Professor, Research Institute for Applied Mechanics, Kyushu University, Kasuga 6-1, Kasuga-shi, Fukuoka 816-8580, Japan.

[5]Professor, Center for Integrated Research in Science and Engineering, Nagoya University, Furo-cho, Chikusa-ku, Nagoya 464-8603, Japan.

It is known that interstitial impurities, such as nitrogen (N), oxygen (O), and carbon (C), strongly affect their workability[*1, 2*], mechanical properties[*3*] and irradiation properties[*4*]. Previous studies have suggested that the interstitial impurities cause the irradiation embrittlement that was associated with irradiation hardening[*5-7*].

Irradiation hardening in vanadium and its alloys is caused by interaction of mobile dislocations with various irradiation-induced defects and their clusters, such as dislocation loops, cavities, precipitates and so on. Therefore, influence of the interstitial impurities on interaction of dislocations with defects and defect clusters should be investigated for research into the impurity effects on radiation embrittlements of low activation vanadium alloys.

Radiation anneal hardening (RAH) is an additional hardening due to post-irradiation annealing. It is a unique phenomenon to body-centered cubic (bcc) metals and their alloys, and was one of the major interests of irradiation studies on vanadium in 1970s[*4*]. RAH is explained by interaction between interstitial impurities and dislocation loops produced by irradiation[*4, 8, 9*]. In those studies, however, nitrogen, oxygen and carbon effects have not been well separated from each other.

Recently, micro-indentation technology has been developed for obtaining mechanical property change in very limited testing volume, and made it possible to measure a hardness change in a quite thin layer produced by heavy ion irradiation[*10, 11*]. The heavy ion irradiation is a valuable technique for radiation damage studies, because it reaches high dose for relatively short period and is capable of wide parametric survey studies for temperature, damage rate and so on[*12*].

The purpose of this study is to apply the micro-indentation technology to vanadium irradiated with heavy ions, and to investigate the effects of nitrogen on irradiation hardening and RAH by heavy ion irradiation using vanadium doped with controlled levels of nitrogen. The results will be compared with those on anneal hardening after plastic deformation (work anneal hardening) for the same set of materials.

Experimental Method

Table 1 shows contents of interstitial impurities in the vanadium specimens used in this study. V-EB-1 was made from electrolytic vanadium by Electron-Beam-refining (EB-refining). V-LI (Vanadium Large Ingot) and V-HI (High-nitrogen Ingot) are made by industrial scale EB-melting. V-HI is a conventional large ingot in Japan. It was purified into V-LI by modification of fabrication process and used to make NIFS-HEAT-1, which is the reference V- 4 wt% chromium- 4 wt% titanium (V-4Cr-4Ti) alloy in Japanese universities[*13, 14*]. V-LI was arc-melted into 30 g buttons with different nitrogen content. They are designated V-LI-R, V-N-1, -2 and -3. The button of V-LI-R was melted in argon (Ar) atmosphere, whereas the other buttons were melted in the atmosphere of mixed Ar with N_2 gas. Then, they were arc-melted twice in 100 % Ar atmosphere to homogenize. In the same way, 30 g buttons of V-HI-R and V-HI-N-1 were made out of V-HI. As a result, nitrogen contents of the specimens ranged from 12 to 770 weight ppm (wppm).

All the specimens were cold-rolled into 0.2 mm-thick sheets. Disks for transmission electron microscopy (TEM) were punched from the sheets, then they were wrapped with tantalum (Ta) and zirconium (Zr) foil and annealed at 1373 K for 3 hours in a higher vacuum than 1.3×10^{-3} Pa.

These specimens were irradiated with 2.0 or 2.4 MeV copper (Cu^{2+}) ions at room

Table 1 - *Contents of interstitial impurities in the vanadium.*

Code	N, wppm	O, wppm	C, wppm	Note
V-EB-1	12	290	33	Electro- and EB-refined V
V-LI	91	43	93	High-purity 25-kg V ingot
V-LI-R	94	55	70	Made out of V-LI
V-N-1	170	51	79	
V-N-2	400	52	72	
V-N-3	540	56	74	
V-HI	540	46	270	Conventional 50-kg V ingot
V-HI-R	610	82	210	Made out of V-HI
V-HI-N-1	770	74	210	

temperature (RT) and at 573 K with Tandem accelerator of Kyushu University. The damage peak for 2.0 and 2.4 MeV Cu ions calculated by TRIM code[*15*] was at 460 nm and at 530 nm from the surface, respectively. The total fluence was about 1 dpa at the damage peak. The average damage of the region between the irradiated surface and the damage peak was about 0.6 dpa. The stopping range of the implanted Cu was 570 and 680 nm from the surface, respectively, where the concentration of the Cu was the maximum and was equivalent to 430 wppm in all the specimens. Effect of the implanted Cu on the hardness is not evaluated, however it is expected to have the same effect on all the specimens because of the same concentration.

Microhardness at the damaged layer was measured by Akashi-MZT-3, which is a microindentation test machine installed in the National Institute for Fusion Science. This indenter has a Berkovich (triangular pyramid) indenter with a semi-apex angle of 68 degrees. Figure 1 shows typical indentation curves for loading process for unirradiated and irradiated specimens. The left vertical axis indicates L (indentation load) /h (indenter displacement). The right vertical axis indicates the damage rate calculated by TRIM code, where h is used as the depth from the irradiated surface to compare the indentation curves with the damage profile. In the case of microindentation test on unirradiated specimens it is known that L/h shows linear relationship with h when h is large enough[*16*]. In this study a good linear relationship was obtained as the solid line when h is larger than about 60 nm. Microhardness of the unirradiated specimen, $M_{Unirrad.}$, is defined as the slope of the solid line in Fig. 1. For irradiated specimens L/h was proportional to h for only the range of h from about 60 to 200 nm. The part of the indentation curve from 60 to 200 nm for h provides the broken line in Fig. 1. The indentation curve begins to deviate from the broken line at about 200 nm in h, which is estimated to be around 1/3 of the depth of the damage peak from a comparison with the damage profile. Before the deviation begins the curves is considered to exhibit the hardness of the damaged region between the irradiated surface and the damage peak, because dislocations promoted by the indenter can interact with the barrier of defects well removed from the apex of the indenter. The deviation is understood to reflect a beginning of an interaction between the dislocations promoted by the indenter and the undamaged region, which lies behind the damage peak. After the beginning of the deviation the slope of the curve will be asymptotic towards that of the unirradiated solid line. Therefore, microhardness, $M_{RT\text{-}irrad.}$, was defined as the slope of the broken line. Considering the damage profile shown in Fig. 1, $M_{RT\text{-}irrad.}$ corresponds to somewhat lower damage than the

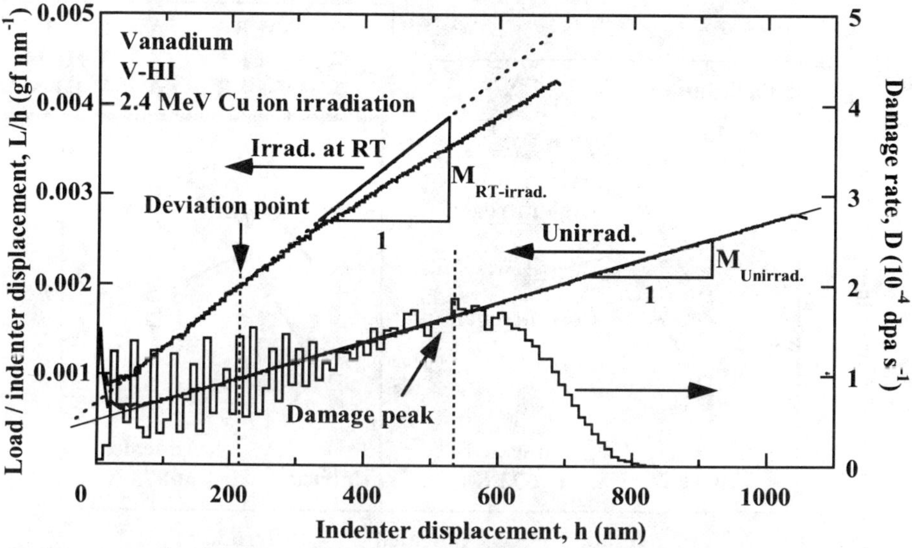

Figure 1 – *Indentation curves for unirradiated and irradiated vanadium. The calculated damage profile is also shown. Microhardness,* $M_{Unirrad.}$ *and* $M_{RT\text{-}irrad.}$, *are defined as the slope of the solid and the broken line, respectively.*

peak damage of 1 dpa.

Post-irradiation annealing was conducted at 573 K and 873 K for 1 hour after the wrapping with Ta and Zr in a higher vacuum than 1.3 X 10^{-3} Pa.

To make a comparison study, the Vickers hardness of unirradiated specimens, which were annealed at 1373 K for 1 hour, was measured. A load condition was 500 g for 30 s.

TEM observations were performed around the damage peak after sectioning and back-thinning, using JEM-2000EXII and JEM-2000FX of Kyushu University.

Results

All the specimens showed irradiation hardening by both RT and 573 K irradiation. Figure 2 shows the irradiation hardening and its recovery behavior by post-irradiation annealing. In the case of RT irradiation, the low-nitrogen specimens, such as V-EB-1, V-LI and V-LI-R showed a reduction in microhardness after post-irradiation annealing at 573 K, while the other specimens showed RAH. On the other hand, post-irradiation annealing at 873 K always caused decrease in microhardness. In the case of 573 K irradiation all the specimens showed irradiation hardening and its recovery after post-irradiation annealing at 873 K.

Figure 3 shows dependence of microhardness on nitrogen content. Some previous data[*17*] of Vickers hardness are also plotted in Fig. 3 (b). Vickers hardness of unirradiated specimens, and microhardness of unirradiated and as-irradiated specimens increased with

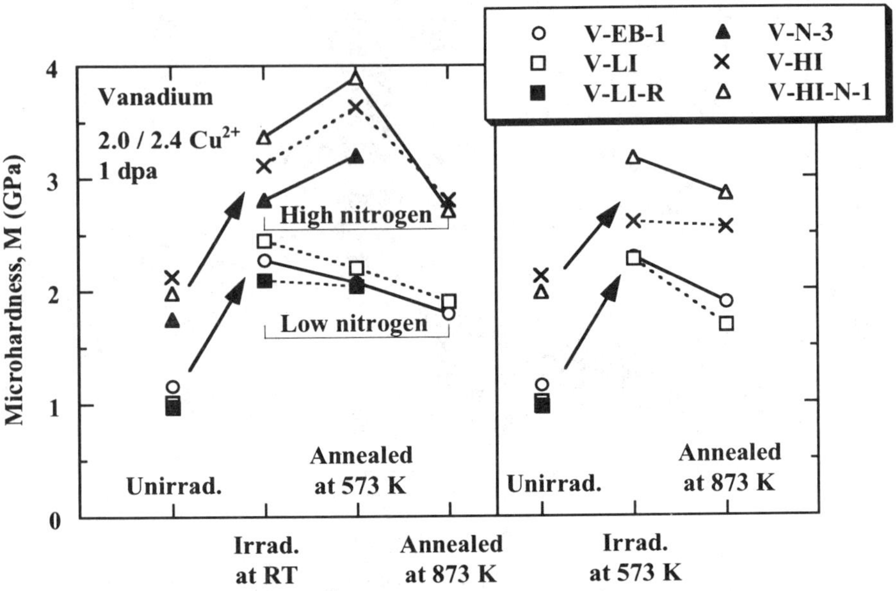

Figure 2 – *Irradiation hardening and its recovery by post-irradiation annealing. All the specimens showed irradiation hardening. After post-irradiation annealing at 573 K, V-EB-1, V-LI and V-LI-R showed hardness recovery, while the other specimens showed RAH.*

increasing nitrogen content. Two solid lines and one broken line were drawn for the data of V-LI, V-LI-R and V-N series. For the these alloys oxygen and carbon contents are close with each other in spite of significant difference in nitrogen content. The broken line indicates the microhardness after the irradiation at RT. When nitrogen is singly varied, as the set of V-LI, V-LI-R and V-N series, Vickers hardness and microhardness for unirradiated and irradiated specimens show a linear relationship with nitrogen content. Irradiation hardening did not show a clear correlation with nitrogen content, and seemed to be constant either at RT and at 573 K. The average of the irradiation hardening at RT, $M_{RT\text{-}irrad.} - M_{Unirrad.}$, was 1.2 GPa, which is indicated in Fig. 3 (a) as $\Delta M_{RT\text{-}irrad.}$. After the irradiation at 573 K, the average of irradiation hardening, $M_{573\ K\text{-}irrad.} - M_{Unirrad.}$, was 1.0 GPa.

Figure 4 shows typical TEM microstructures after the irradiation. Fine black dots and dislocation loops were observed in the bright field images. The dislocation loops were identified to be a/2 <111> type dislocation loops. The micrographs of weak-beam dark field show higher density of dots than that indicated by the bright field images. Most of them were less than 5 nm in diameter. Their images are heavily overlapped with each other so that their number density was estimated roughly to be 6×10^{22} - 1×10^{23} m^{-3} or actually higher. The density seems not to be changed either by the impurity level or irradiation-anneal histories.

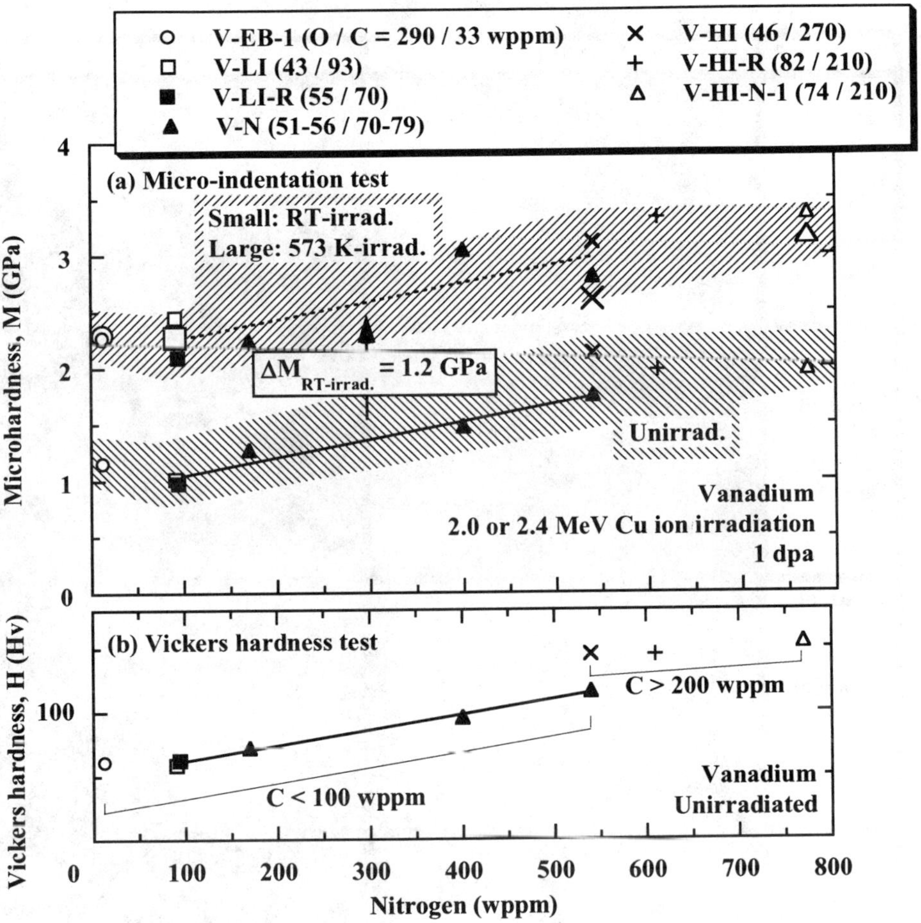

Figure 3 – *Dependence of (a) microhardness and (b) Vickers hardness on nitrogen content. Oxygen contents are indicated in parentheses after the material codes. In Fig. 3 (a) irradiated and unirradiated data are separated each other by hatching. In irradiation data the small symbols indicate RT-irradiation data, while the large ones indicate 573 K-irradiation data. Solid lines and a broken line are drawn after the data on V-LI, V-LI-R and V-N series in which oxygen content varies within only 13 wppm.* $\Delta M_{RT\text{-}irrad.}$ *is the average of irradiation hardening at RT.*

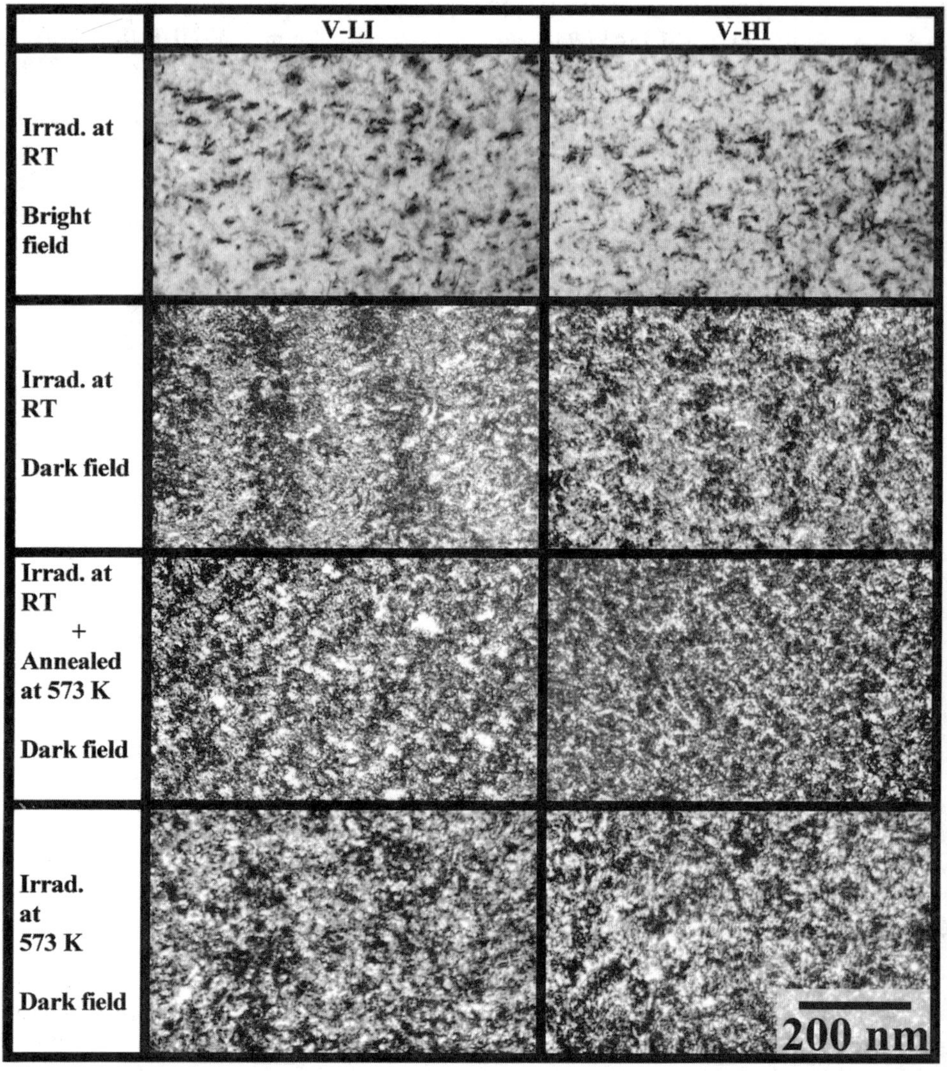

Figure 4 – *TEM microstructure of V-LI and V-HI irradiated to 1 dpa with 2.4 MeV Cu ions. The fine black dots and the small dislocation loops were identified to be a/2 <111> type dislocation loops.*

Discussion

Correlation between Microhardness and Vickers Hardness

In Fig. 3 (b) the Vickers hardness of V-EB-1, V-HI, V-HI-R and V-HI-N-1 was higher than that obtained by extrapolating the solid line. In the case of V-EB-1 this is attributed to its oxygen content of 290 wppm, which is much higher than those of the other specimens. The difference between oxygen content of V-EB-1 and that of the other specimens is about 240 wppm, which is converted into about 14 Hv by using the oxygen hardening coefficient in the previous work[*17*]. In the case of V-HI, V-HI-R and V-HI-N-1 oxygen level is comparable to that of V-LI and its arc-melting products. According to Table 1, higher carbon content of 210 or 270 wppm can explain the gap from the solid line. The Vickers hardness of V-N-3 and V-HI was 116 and 145 Hv, respectively, therefore about 200 wppm carbon caused a hardening of 30 Hv. In the previous tensile study, it was revealed that hardening due to carbon tends to saturate at above 200 wppm[*3*]. It is possible to categorize the specimens including V-EB-1, V-LI, V-LI-R and V-N into a low carbon (C < 100 wppm) group, and the specimens including V-HI, V-HI-R and V-HI-N-1 into a high carbon (C > 200 wppm) one. However, effects of several tens wppm carbon on mechanical properties still remains to be clarified using a vanadium with controlled nitrogen, oxygen and other impurities.

By a comparison among the two solid lines and the other deviated data for unirradiated specimens in Fig. 3, one would know that the dependence of microhardness on nitrogen content reproduces that of Vickers hardness well. Figure 5 shows the correlation

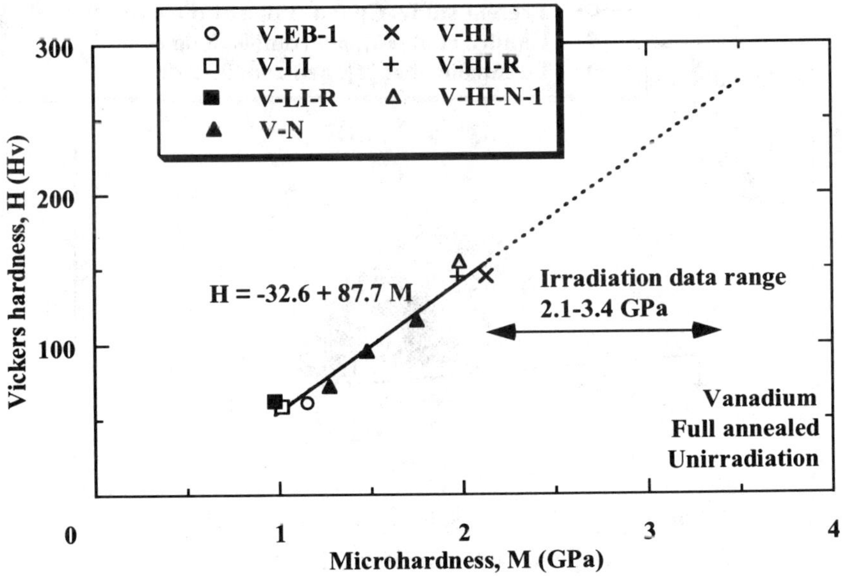

Figure 5 – *Correlation between microhardness and Vickers hardness. The solid line is obtained after the data for unirradiated specimens. The liner relationship is extrapolated to the irradiation data range as shown by the broken line.*

between microhardness and Vickers hardness of unirradiated specimens. This figure shows a linear relationship described by an equation

$$H\,(Hv) = -32.6 + 87.7\,M\,(GPa) \tag{1}$$

The set of vanadium used was uniformly hardened by solid solute and had a grain size larger than 40 μm, which does not influence either Vickers hardness or microhardness, because the indentation size is 140 μm and 10 μm at the maximum, respectively. This equation is useful to convert microhardness into Vickers hardness for the range from 1.0 to 2.1 GPa in vanadium. The equation will not be applicable to the lower hardness region near the origin. On the other hand, in higher hardness region Vickers hardness would be estimated by extrapolation, as shown by the broken line in Fig. 5, unless different deformation mode such as cracking occurs around the indentation.

Irradiation Hardening

As shown in Fig. 3 (a) irradiation hardening is not strongly influenced by either nitrogen content or irradiation temperature within the range investigated. This fact agrees with TEM observations indicating that the microstructures of all irradiated specimens were similar to each other. Assuming no change in deformation mode in irradiated specimens, the average irradiation hardening of 1.2 GPa at RT and of 1.0 GPa at 573 K can be converted into 110 and 88 Hv by using Eq (1), respectively. Figure 6 shows a comparison between the converted values and the irradiation hardening obtained at neutron irradiation. The

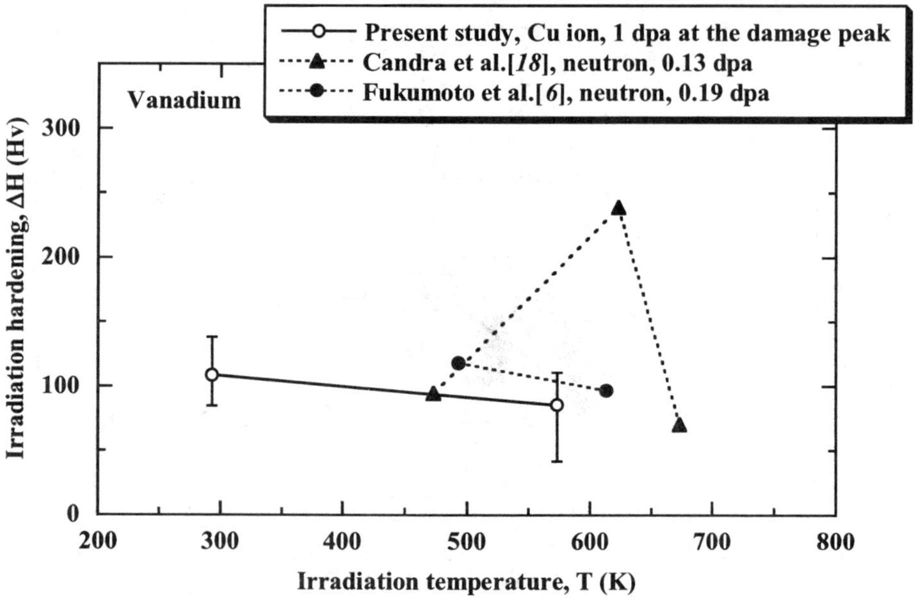

Figure 6 – *Comparison between temperature dependence of irradiation hardening at Cu ion irradiation and that at neutron irradiation. The irradiation hardening at Cu ion irradiation is converted from microhardness by using Eq (1).*

converted values are similar to or smaller than the hardening induced by neutron irradiation at the dose of 0.13 or 0.19 dpa at 473 – 673 K[*6, 18*].

The microstructures by TEM consisted of the a/2<111> type dislocation loops, which are common to neutron-irradiated vanadium[*18-20*]. It should, however, be noted that cavities were formed at 455 K or higher temperature at 1 dpa in the case of neutron irradiation[*19*].

The irradiation hardening and also the microstructure obtained from 573 K irradiation in the present study seem to correspond to that by neutron irradiation at lower dose and at lower temperature. One of the probable explanations for the difference between the heavy ion and neutron irradiation is that the higher dose rate, such as about 10^{-3} dpa/s in heavy ion irradiation, should cause a longer incubation dose for cavity and less defects for hardening than the lower dose rate of about 10^{-6} dpa/s in neutron irradiation, because of enhanced recombination of irradiation defects arising from the overlapping of cascades. In addition it is commonly known that higher irradiation temperature is necessary for cavity formation by heavy ion irradiation compared with neutron irradiation[*21, 22*].

Radiation Anneal Hardening at 573 K

After the post-irradiation annealing at 573 K, RAH was observed at the nitrogen contents of 170 wppm and above. It has been suggested that RAH arises from the decoration of dislocation loops with interstitial impurities[*4, 8, 9*]. The authors have investigated the nitrogen and oxygen effects on work anneal hardening of vanadium. In the study, the effects of nitrogen and oxygen have successfully been separated by using the same set of specimens as this study[*17*]. It has been clarified that the work anneal hardening at 573 K is dominated by nitrogen content. Figure 7 shows a comparison between RAH ($\Delta M_{573\,K}$) and work anneal hardening of vanadium ($\Delta H_{573\,K}$). $\Delta M_{573\,K}$ and $\Delta H_{573\,K}$ are defined as the increase in microhardness due to the post-irradiation annealing at 573 K, and as the increase in Vickers hardness due to annealing at 573 K after deformation to 90 %-cold working, respectively. In Fig. 7 oxygen contents are also indicated in parentheses after the material codes.

The solid line in Fig. 7 (b) indicates that work anneal hardening of vanadium is small at low nitrogen level. The work anneal hardening increases with the nitrogen content in a step-wise manner at around 100 wppm, followed by saturation. Figure 7 (b) suggests that nitrogen is the dominant element for the hardening, because the work anneal hardening of V-EB-1 is nearly zero despite its large oxygen content of 290 wppm. V-O series exhibit similar extent of work anneal hardening to the other specimens, though their oxygen contents varies from 260 to 920 wppm. In all the deformed specimens a dislocation network was the only set of defects observed, which did not change by annealing at 573 K. The dislocation structure was independent of nitrogen and oxygen content.

As shown in Fig. 7 (a), the data itself does not represent a sharp step function, however they does not conflict with the data on the work anneal hardening. In the present study a broken part on the solid line shown in Fig. 7 (a) is suggested, because the correlation between RAH and nitrogen content can be understood in the same way as that derived for the deformed specimens. The microstructures consisted of only dislocation loops, and were not significantly affected either by annealing at 573 K or its nitrogen level. In this case the loops and invisible simple defects and their clusters can be responsible for the RAH. Single vacancies and their small clusters, such as divacancy, is expected to migrate below 573 K,

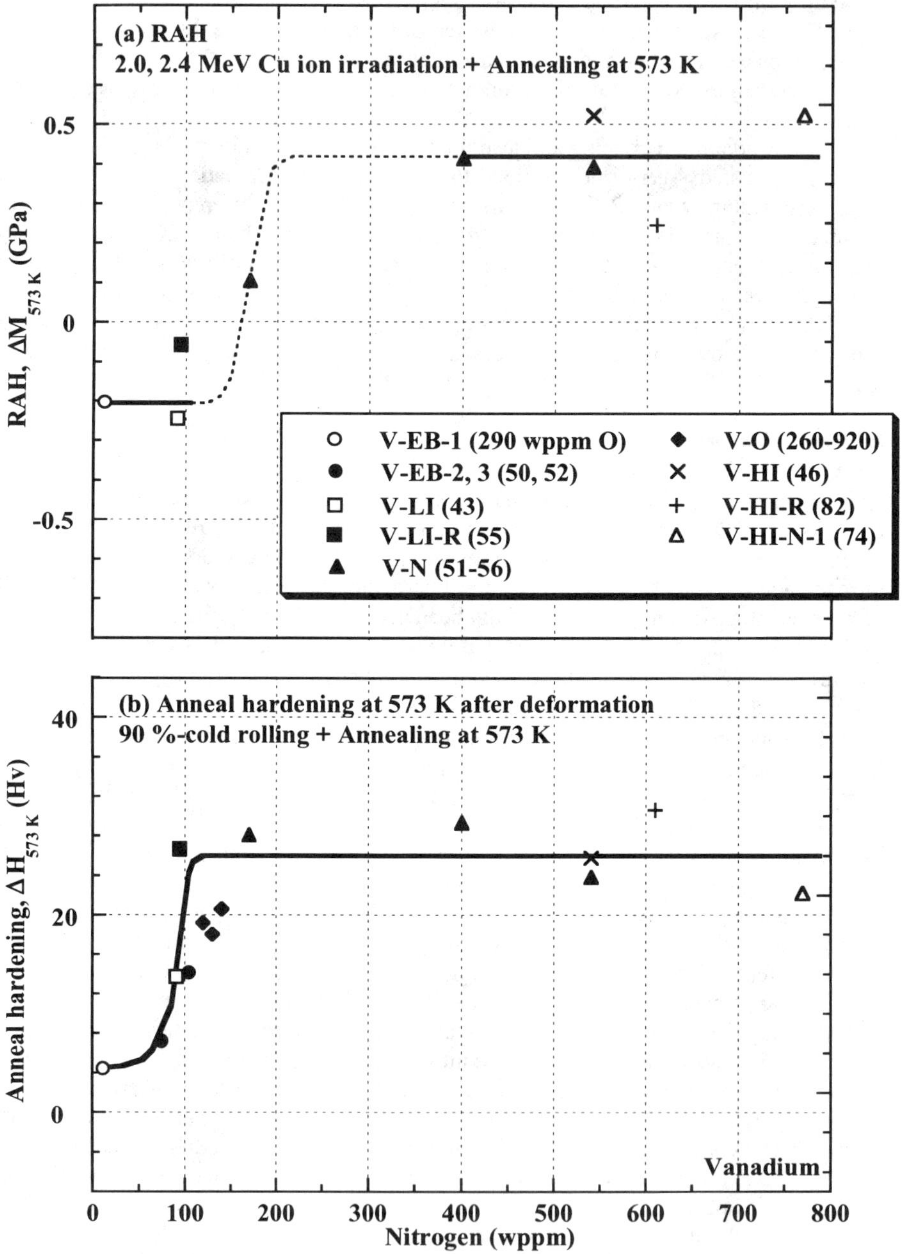

Figure 7 – *Comparison between (a) RAH and (b) anneal hardening at 573 K after 90% cold-rolling. Oxygen contents are indicated in parentheses after the material codes. Detail of the anneal hardening of deformed specimens shown in Figure 3 (b) has been mentioned elsewhere[17].*

according to the previous estimation of their migration energy ranging from 0.5 to 1.2 eV[*23, 24*]. In the low-nitrogen specimens, such as V-EB-1, V-LI and V-LI-R, a slight hardness recovery was observed after post-irradiation annealing at 573 K in Fig. 7 (a). This indicates that the small, unstable and invisible defect clusters produced by irradiation at RT were recovered by the post-irradiation annealing at 573 K. In contrast to the irradiation, such unstable defect clusters were not introduced by deformation. This is the reason why hardness recovery was not observed in the low nitrogen specimens in Fig. 7 (b). RAH occurred in high nitrogen specimens although such small defect clusters were considered to be recovered also in those specimens. Therefore, the dislocation loops should be the responsible defects for RAH. According to the clear correlation between RAH and nitrogen content, RAH should be caused by migration of nitrogen to dislocation loops in irradiated vanadium. In the previous work with Cu ion irradiation up to 0.75 dpa, number density of dislocation loops similar to that in the present study was decreased by two orders during post-irradiation annealing at 873 K[*12*]. The decrease of hardness at 873 K in Fig. 2. is expected to arise from the decrease of dislocation loops and detrapping of nitrogen from dislocation loops.

Microhardness of the specimens showing RAH, M_{RAH}, and Vickers hardness of the specimens showing work anneal hardening, H_{AH}, are separated into three terms as

$$M_{RAH} = M_{Unirrad.} + \Delta M_{RT\text{-}irrad.} + \Delta M_{RAH} \tag{2}$$

$$H_{AH} = H_{FA} + \Delta H_{CW} + \Delta H_{AH} \tag{3}$$

where $M_{Unirrad.}$, $\Delta M_{RT\text{-}irrad.}$ and ΔM_{RAH} are microhardness of full-annealed specimen, irradiation hardening at RT and RAH, respectively. H_{FA} is the Vickers hardness of full-annealed specimens. ΔH_{CW} and ΔH_{AH} are work hardening and the work anneal hardening, respectively. In the present study the average of $\Delta M_{RT\text{-}irrad.}$ is 1.2 GPa as shown in Fig. 3 (a). ΔM_{RAH} is 0.42 GPa, which is given as the upper level indicated in Fig. 7 (a) for the specimens with nitrogen content of 400 wppm or higher. ΔH_{CW}, the difference between the as-rolled and the full-annealed specimens, was shown to be independent of nitrogen content and estimated to be 75 Hv[*17*]. From Fig. 7 (b) the average of ΔH_{AH} is estimated to be 26 Hv, which is the upper level of the solid line drawn after the specimens with 400 wppm nitrogen or higher.

Orowan's dispersed barrier equation for the change in the sheer stress, $\Delta\tau$, can be written as

$$\Delta\tau = \frac{\mu b}{\beta l} \tag{4}$$

in which μ is sheer modulus and b is the magnitude of the Burgers vector[*25*]. The factor β is a measure of the barrier strength. The average barrier spacing for the dislocation loops in the specimens irradiated at RT, $l_{RT\text{-}irrad.}$, is described as

$$l_{RT\text{-}irrad.} = \frac{1}{\sqrt{N d}} \tag{5}$$

in which N and d is the number density and the diameter of the loops, respectively. The spacing for network dislocations in the cold-worked specimens, l_{CW}, is described as

$$l_{CW} = \frac{1}{\sqrt{\rho}} \tag{6}$$

in which ρ is the network dislocation density produced by cold working. $\Delta\tau$ is converted into change in the uniaxial yield stress, $\Delta\sigma_y$, using Taylor factor, M_T, as

$$\Delta\sigma_y = M_T\,\Delta\tau \tag{7}$$

Assuming both changes in Vickers hardness and Microhardness are proportional to $\Delta\sigma_y$, the hardness change in the specimens irradiated at RT, $\Delta M_{RT\text{-}irrad.}$, and that in the cold-worked specimens, ΔH_{CW}, are written as follows by using Eq (4) – (7).

$$\Delta M_{RT\text{-}irrad.} = \alpha_M\,\mu\,b\sqrt{N\,d} \tag{8}$$

$$\Delta H_{CW} = \alpha_H\,\mu\,b\sqrt{\rho} \tag{9}$$

α_M and α_H are measures of the barrier strength of the dislocation loops and the network dislocations. Because TEM showed no change in N, d and ρ after annealing at 573 K, RAH and the work anneal hardening are considered to arise from decorations of dislocations with nitrogen. The decoration is expected to induce an increase in α_M and α_H into $\alpha_M{}'$ and $\alpha_H{}'$, respectively. Therefore, followings are obtained.

$$\Delta M_{RT\text{-}irrad.} + \Delta M_{RAH} = \alpha_M{}'\,\mu\,b\sqrt{N\,d} \tag{10}$$

$$\Delta H_{CW} + \Delta H_{AH} = \alpha_H{}'\,\mu\,b\sqrt{\rho} \tag{11}$$

Using Eq (8)-(11), the change ratios of the barrier strength are given as follows.

$$\frac{\alpha_M{}'}{\alpha_M} = 1 + \frac{\Delta M_{RAH}}{\Delta M_{RT\text{-}irrad.}} = 1 + \frac{0.42\ (GPa)}{1.2\ (GPa)} = 1.350 \tag{12}$$

$$\frac{\alpha_H{}'}{\alpha_H} = 1 + \frac{\Delta H_{AH}}{\Delta H_{CW}} = 1 + \frac{26\ (Hv)}{75\ (Hv)} = 1.347 \tag{13}$$

Thus, the ratios of the barrier strength before to after anneal hardening are very similar for RAH and the work anneal hardening. The similarity supports the idea that the same mechanism is operating to RAH and the work anneal hardening.

In conclusion, RAH and the work anneal hardening in vanadium at 573 K should be accounted for by interaction of nitrogen with dislocation loops and dislocation network, respectively.

Conclusions

Copper ion irradiation at room temperature (RT) and at 573 K up to 1 dpa was conducted on vanadium doped with controlled levels of nitrogen. A microindentation test was performed to make a quantitative estimation of irradiation hardening and radiation anneal hardening (RAH) of the irradiated surface layer. New findings and understandings are obtained as follows.

1. In vanadium, microhardness obtained by the microindentation test has a linear relationship with Vickers hardness at least for the range from 1.0 to 2.1 GPa in microhardness.
2. Irradiation hardening and microstructures are not strongly influenced by nitrogen content and irradiation temperature. The average of irradiation hardening at RT and at 573 K is 1.2 and 1.0 GPa, respectively.
3. RAH appears with 170 wppm nitrogen and above after post-irradiation annealing at 573 K. RAH increases with nitrogen content in stepwise manner and similar to anneal hardening of deformed vanadium.
4. RAH and the anneal hardening of deformed vanadium at 573 K should be accounted for by interaction of nitrogen with dislocation loops and dislocation network, respectively.

Acknowledgments

This study was supported by the inter-university cooperative research program of the Institute for Materials Research, Tohoku University.

The authors are grateful to Professor H. Matsui for his continuous encouragement.

The authors are indebted to T. Takahashi and T. Sugawara of "Core Laboratory of Crystal Science" in Institute for Materials Research (IMR) of Tohoku University for EB-refining and nitrogen doping, and to K. Takada and the staff of "Analytical Research Core for Advanced Materials" for chemical analyses.

References

[1] Smith, D. L., Loomis, B. A., and Diercks, D. R., "Vanadium-Base Alloys for Fusion Reactor Applications – a Review," *Journal of Nuclear Materials*, Vol. 135, 1985, pp. 125-139.

[2] Diercks, D. R., and Loomis, B. A., "Alloying and Impurity Effects in Vanadium-Base Alloys," *Journal of Nuclear Materials*, Vol. 141-143, 1986, pp. 1117-1124.

[3] Harrod, D. L., and Gold, R. E., "Mechanical Properties of Vanadium and Vanadium-Base Alloys," *International Metals Reviews*, T. L. Hughes, Ed., The Metals Society and The American Society for Metals, Nos. 5 and 6, 1980, pp. 163-221.

[4] Gold., R. E., and Harrod, D. L., "Radiation Effects in Vanadium and Vanadium-Base Alloys," *International Metals Reviews*, T. L. Hughes, Ed., The Metals Society and The American Society for Metals, Nos. 5 and 6, 1980, pp. 232-254.

[5] Chung, H. M., Gazda, J., and Smith, D. L., "Irradiation-Induced Precipitation and Mechanical Properties of Vanadium Alloys at <430°C," *Fusion Materials Semiannual Progress Report, DOE/ER-0313/24*, 1998, pp. 49-60.

[6] Fukumoto, K., Matsui, H., Candra, Y., Takahashi, K., Sasanuma, H., Nagata, S., and

Takahiro, K., "Radiation-Induced Precipitation in V-(Cr, Fe)-Ti Alloys Irradiated at Low Temperature with Low Dose during Neutron or Ion Irradiation," *Journal of Nuclear Materials*, Vol. 283-287, 2000, pp. 535-539.

[7] Satou, M., Chuto, T., and Abe, K., "Improvement in Post-Irradiation Ductility of Neutron Irradiated V-Cr-Ti-Si-Al-Y Alloy and the Role of Interstitial Impurities," *Journal of Nuclear Materials*, Vol. 283-287, 2000, pp. 367-371.

[8] Shiraishi, K., Fukuya, K., and Katano, Y., "Radiation and Anneal Hardening in Vanadium," *Journal of Nuclear Materials*, Vol. 44, 1972, pp. 228-238.

[9] Stanley, J. T., Williams, J. M., Brundage, W. E., and Wechsler, M. S., "The Effect of Interstitial Impurities on The Annealing of Neutron-Irradiated Vanadium," *Acta Metallurgica* Vol. 20, 1972, pp. 191-198.

[10] Katoh, Y., Muroga, T., Iwai, T. and Motojima, O., "Hardness Evaluation of MeV-ion Irradiated Materials by Means of Very Low-Load Indentation Technique," *Journal of Japan Institute of Metals*, 61 (1997) 191-198.

[11] Muroga, T., Yasunaga, K., Katoh, Y., Watanabe, H., Yoshida, N., and Noda, N., "Correlation of Hardening and Microstructure of Tantalum Irradiated with Heavy Ions," *Effects of Radiation on Materials: 19th Conference, ASTM STP 1366*, M. L. Hamilton, A. S. Kumar, S. T. Rosinski, and M. L. Grossbeck, Eds., American Society for Testing and Materials, 2000, pp. 1186-1196.

[12] Ochiai, K., Watanabe, H., Muroga, T., Yoshida, N., and Matsui, H., "Microstructural Evolution in Vanadium Irradiated during Ion Irradiation at Constant and Varying Temperature," *Journal of Nuclear Materials*, Vol. 271-272, 1999, pp. 376-380.

[13] Muroga, T., and Nagasaka, T., "Reduction of Impurity Levels of Vanadium and Its Alloys for Fusion Application," *International Journal of Refractory Metals & Hard Materials*, Vol. 18, 2000, pp. 225-230.

[14] Muroga, T., Nagasaka, T., Iiyoshi, A., Kawabata, A., Sakurai, S., and Sakata, M., "NIFS Program for Large Ingot Production of a V-Cr-Ti alloy," *Journal of Nuclear Materials*, Vol. 283-287, 2000, pp. 711-715.

[15] Ziegler, J. F., Biersack, J. P., and Littmark, U., *The Stopping and Range of Ions in Solid*, Vol. 1, Pergamon, Press, London, 1985.

[16] Inamura, M., and Suzuki, T., "Evaluation of Materials Strength by Ultra-Micro-Indentation," *Seisan Kenkyu*, Vol. 42, 1990, pp. 257-260.

[17] Nagasaka, T., Takahashi, H., Muroga, T., Tanabe, T., Matsui, H., "Recovery and Recrystallization Behavior of Vanadium at Controlled Various Nitrogen and Oxygen Levels," " *Journal of Nuclear Materials*, Vol. 283-287, 2000, pp. 816-821.

[18] Candra, Y., Fukumoto, K., Kimura, A., and Matsui, H., "Microstructural Evolution and Hardening of Neutron Irradiated Vanadium Alloys at Low Temperatures in Japan Material Testing Reactors," *Journal of Nuclear Materials*, Vol. 271&272, 1999, pp. 301-305.

[19] Horton, L. L., and Farrell, K., "The Temperature Dependence of The Damage Microstructures in Neutron-Irradiated Vanadium," *Journal of Nuclear Materials*, Vol. 122&123, 1984, pp. 687-692.

[20] Bradley, E. R., and Jones, R. H., "Strength and Microstructure in Vanadium Irradiated with T (d, n) Neutron at 300 to 675 K," *Journal of Nuclear Materials*, Vol. 133&134, 1985, pp. 671-675.

[21] Packan, N. H., Farrell, K., and Stiegler, J. O, "Correlation of Neutron and Heavy-Ion Damage I. The Influence of Dose Rate and Injected Helium on Swelling in Pure

Nickel," *Journal of Nuclear Materials*, Vol. 78, 1978, pp. 143-155.

[22] Mansur, L. K., "Correlation of Neutron and Heavy-Ion Damage II. The Predicted Temperature Shift if Swelling with Changes in Radiation Dose Rate," *Journal of Nuclear Materials*, Vol. 78, 1978, pp. 156-160.

[23] George, B., Janot, C., Ablitzer, D., and Chabre, Y., "A Mechanism for Self-Diffusion and Impurity Diffusion in Vanadium," *Philosophical Magazine A*, Vol. 44, 1981, pp. 763-778.

[24] Schultz, H., "Vacancy Migration in B. C. C. Transition Metals," *Scripta Metallurgica*, Vol. 8, 1974, pp. 721-726.

[25] Bement, Jr., A. L., "Fundamental Materials Problems in Nuclear Reactors," *Proceedings of the Second International Conference on the Strength of Metals and Alloys*, American Sciety for Metals, Vol. 2, 1970, pp. 693-728.

B. M. Oliver,[1] F. A. Garner,[1] M. L. Hamilton,[1] W. F. Sommer,[2]
S. A. Maloy,[2] P. D. Ferguson,[2] and M. R. James[2]

Hydrogen and Helium Gas Formation and their Release Kinetics in Tungsten Rods after Irradiation with 800 MeV Protons

Reference: Oliver, B. M., Garner, F. A., Hamilton, M. L., Sommer, W. F., Maloy, S. A., Ferguson, P. D., and James, M. R., **"Hydrogen and Helium Gas Formation and their Release Kinetics in Tungsten Rods after Irradiation with 800 MeV Protons,"** *Effects of Radiation on Materials: 20th International Symposium, ASTM STP 1405,* S. T. Rosinski, M. L. Grossbeck, T. R. Allen, and A. S. Kumar, Eds., American Society for Testing and Materials, West Conshohocken, PA, 2001.

Abstract: In the Accelerator Production of Tritium Program, thermalized neutrons produced from multiplication and moderation of spallation neutrons will be absorbed in ^{3}He gas to produce tritium. The spallation neutrons will be generated by the interaction of high energy (~1 GeV) protons with solid tungsten rods. An unavoidable byproduct of the spallation reactions is large amounts of helium and hydrogen gas generated in the rods. The release kinetics of these gases during various proposed off-normal scenarios involving loss of coolant and afterheat-induced rises in temperature are of particular interest. In addition, the magnitude of the gas generation cross sections and the fractional retention of these gases is necessary for extrapolation of these data to higher exposures.

Tungsten rods irradiated with 800 MeV protons in the Los Alamos Neutron Science Center (LANCE) to rather high exposures have been sectioned to produce small specimens suitable for measurement of both helium and hydrogen. Hydrogen evolution was measured both by dropping the specimen into a small ceramic crucible at 1200°C and also by subjecting the specimen to a simulated temperature ramp from ~200 to ~1200°C, as has been calculated to be induced by loss of cooling and subsequent afterheat in an APT plant. The latter technique showed four distinct hydrogen release peaks at temperatures of approximately 500, 800, 1000 and 1200°C, suggesting a variety of trapping sites with different binding energies. Helium release and total content were measured by subjecting the specimens to the same temperature excursion, followed later by melting to release the remaining helium. Approximately 99% of the helium was retained until melting occurred. For both gases, release measurements were conducted using mass spectrometric techniques.

The measured amounts of helium agreed well with predictions. The hydrogen measurements were somewhat lower than predicted. This may reflect diffusional losses from the rods or some needed modifications in the evaporation model employed in the LAHET code.

[1] Staff Scientist, Sr. Staff Scientist and Sr. Research Scientist, respectively, Pacific Northwest National Laboratory, P.O. Box 999, Richland, Washington 99352, USA.

[2] Team Leader, Team Leader, Technical Staff and Technical Staff, respectively, Los Alamos National Laboratory, Los Alamos, NM 87545, USA.

Keywords: proton irradiation, helium measurements, hydrogen measurements, Accelerator Production of Tritium (APT), hydrogen release, helium release, tungsten

Introduction

The Accelerator Production of Tritium (APT) project [*1*] was proposed as one of several solutions to the U.S.A. national need for tritium. In the APT concept, high-energy protons would impinge on a tungsten target producing high-energy spallation neutrons. These neutrons would in turn be multiplied using a lead blanket, then thermalized using water. Tritium production would occur through capture of the thermalized neutrons by ^{3}He gas. A main technical issue that was addressed during the APT design was radiation damage to materials in the mixed high-energy proton and neutron environment. A materials irradiation program [*2*] using the 800-MeV proton accelerator at the Los Alamos Neutron Science Center (LANSCE) was therefore developed. To simulate the tungsten neutron source in the proposed APT target, a series of tungsten rods was included in the LANSCE materials irradiation assembly.

While reviewing postulated off-normal conditions for the APT target, the question of gas buildup inside the Alloy 718 clad tungsten rods was raised. Spallation reactions typically result in large amounts of helium and hydrogen being produced in the target material. If enough gas is produced and released from the target material, and then trapped at the tungsten/cladding interface, this gas might result in the failure of the target cladding due to a pressure build-up. Therefore, the tungsten rods included in the LANSCE materials irradiation test were used to address two important questions; (1) is a large fraction of the two gases released from the target material and at what rate, and (2) how accurately can gas production be predicted for the APT target rods?

Tungsten Analysis Samples

Several sets of tungsten samples were prepared at Los Alamos National Laboratory (LANL) for helium and hydrogen measurements. Each sample was in the form of a thin disk cut from selected 3.18 mm diameter tungsten target rods previously irradiated with 800 MeV protons in LANSCE for an exposure time of approximately two months. Proton fluence for the two in-beam samples (W-2 and W-3) and the out-of-beam sample (1Wc) was ~1 x 10^{21} p/cm^2 and ~5 x 10^{19} p/cm^2, respectively. The radial temperature of the rods was estimated to range from ~166°C (surface) to ~175°C (center) at beam center, to ~35°C out of the beam. The tungsten rods were formed by powder metallurgy (pressed, sintered and then drawn) and had an elemental composition of 99.96% tungsten with the major impurities being Mo at 100 μg/g and C and O at 30 μg/g each. The initial grain size was ~15 μm.

As the rods were highly radioactive, sample preparation was performed in a hot cell. Extensive radiometric analysis was conducted on the samples to provide flux and gradient data for the model calculations. A diagram of the experimental setup in the LANSCE facility is shown in Figure 1. Locations of the individual measured tungsten specimens, relative to the center of the proton beam, are shown in Figure 2.

Samples for gas analysis were cut using small diagonal wire cutters. Prior to analysis, each specimen is individually cleaned in acetone and air-dried. The mass of each analyzed specimen was determined using a microbalance with calibration traceable to the National Institute of Standards and Technology (NIST). Mass uncertainty is conservatively estimated to be ±0.002 mg.

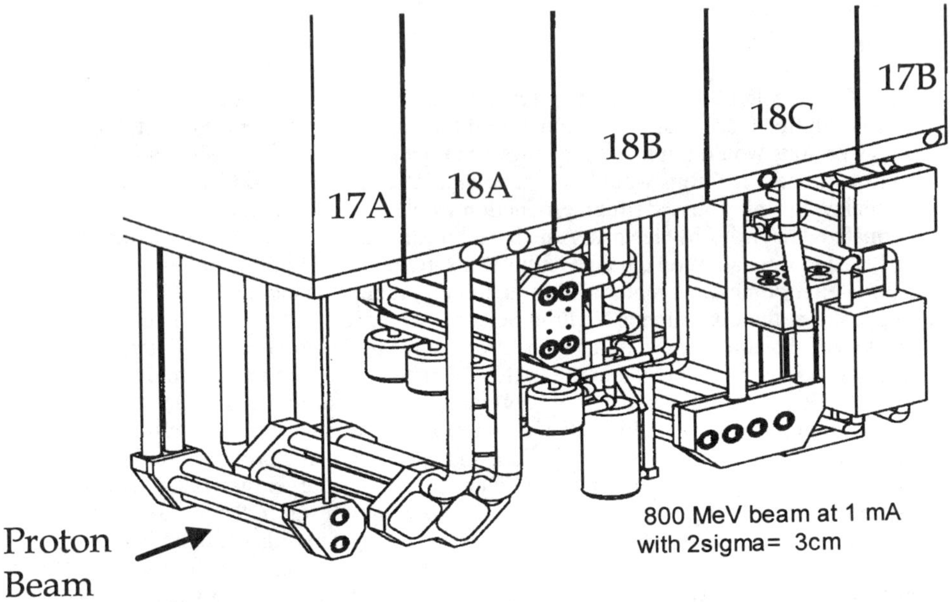

Figure 1 - *LANSCE A-6 irradiation setup.*

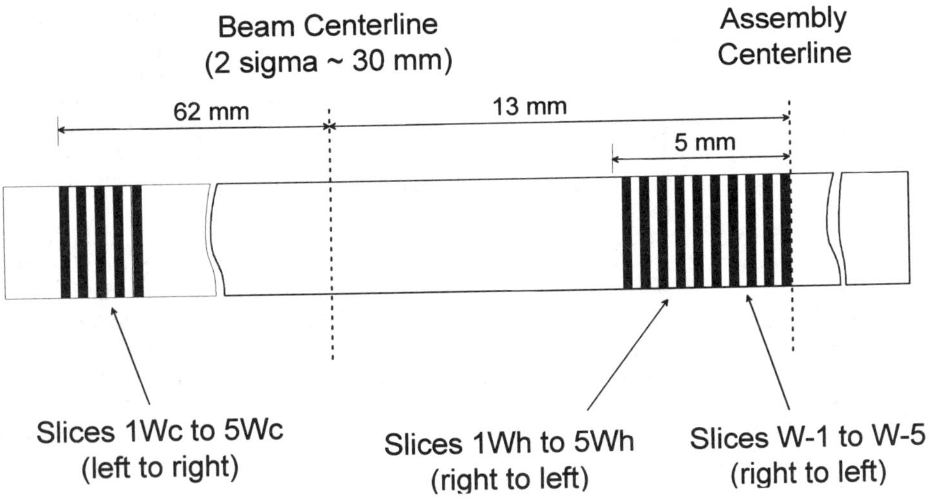

Figure 2 - *Tungsten sample location relative to proton beam centerline.*

Helium Analyses

Helium Analysis System

Helium analyses were conducted by mass spectrometry at Pacific Northwest National Laboratory (PNNL). Details on the mass spectrometry system have been presented elsewhere [*3,4*]. Helium contents were determined by heating and/or vaporizing each sample in a resistance-heated crucible in one of the mass spectrometer system's high-temperature vacuum furnaces. Helium concentrations are generally determined either by direct measurements of the mass spectrometer helium signal, or by an isotope-dilution technique where the released helium isotope (He3 or He4) is compared with a known quantity of added "spike" of the other isotope. The helium isotopic spikes were obtained by expanding and partitioning known quantities of gas through a succession of calibrated volumes [*3*]. The mass spectrometer was calibrated for mass sensitivity during each series of runs by analyzing known mixtures of ^{3}He and ^{4}He. Reproducibility of the analysis system for samples with known homogeneous helium content is ~0.5%. The absolute accuracy is generally better than 1%.

Stepped-Anneal and Total Helium Measurements

Details of the stepped-anneal and total helium measurements have been presented earlier [*5*]. For the stepped-anneal helium release measurements, the specimens cut from the tungsten disks were loaded into the central section of a 0.48 cm diameter graphite crucible. Small graphite plugs were placed at each end of the hole to position the samples as close as possible to the middle of the crucible. Temperature measurements were conducted using either a thermocouple or an optical pyrometer. Temperature uncertainty is estimated to be ±50°C. As both ^{3}He and ^{4}He were expected in the samples, no helium spikes were used in this measurement series.

Results of the stepped-anneal measurements conducted earlier are shown graphically in Figures 3 and 4. Helium concentrations are plotted in atomic parts per million (appm), based on a calculated value of 0.328 x 10^{22} atoms per gram for tungsten. Helium levels were corrected for background helium buildup in the analysis furnace as a function of time, obtained from separate "control" analyses conducted immediately after the sample runs. For sample W-2, helium release is first observed at a temperature of ~800°C. From 800°C up to the final temperature of ~1600°C, the incremental helium release is approximately the same for each temperature step, although there is some indication of a subsequent, slowly increasing release rate up to ~1400°C. At the final temperature of 1600°C, a total of five measurements were made, and these data show a leveling off of the gas release at a value of ~1.4 appm. Similar trends were observed for sample W-3, with the ^{4}He release tending asymptotically to ~1.0 appm at the highest temperature of ~1200°C. For this second sample, however, there is little evidence for an increasing helium release rate with temperature (above ~800°C). The ^{3}He release for both specimens was at or below the detection limit of the analysis system (~10^{12} atoms) for the particular setup used in these tests. Variability in the helium release data are due largely to variability in the subtracted helium background [*5*].

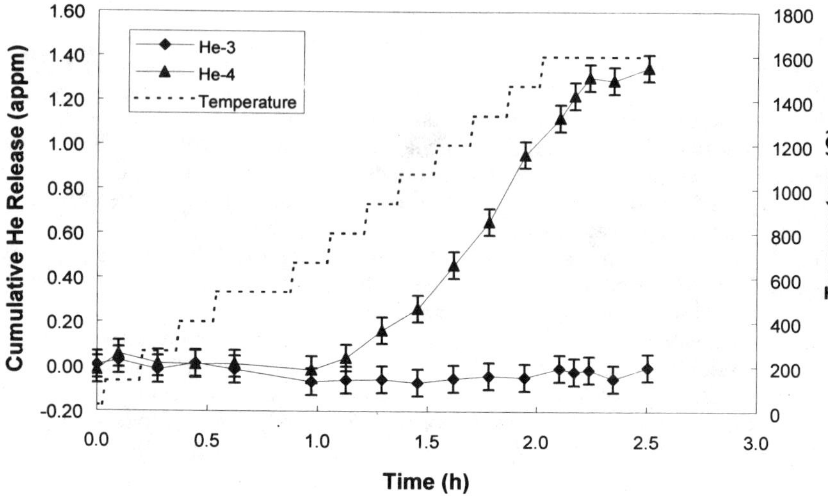

Figure 3 - *Helium release in Sample W-2.*

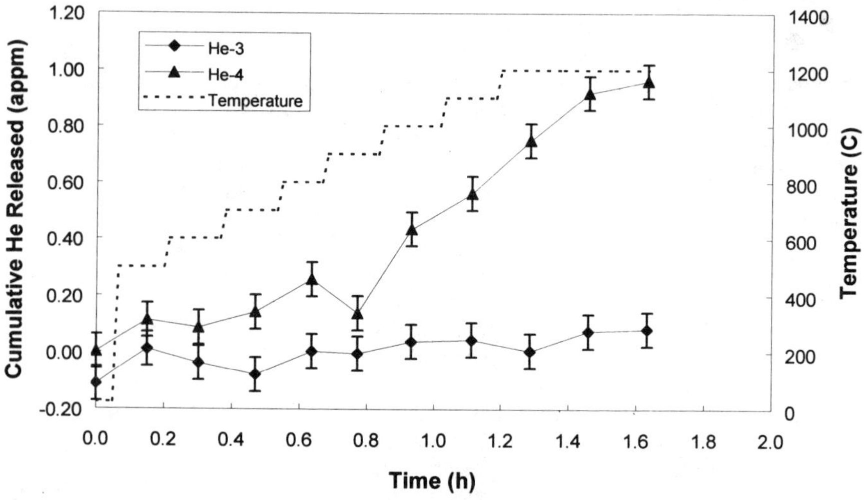

Figure 4 - *Helium release in Sample W-3.*

Following the stepped-anneal measurements, the two tungsten specimens were relocated, in their original crucibles, to different positions in the same furnace for subsequent vaporization analysis. For these measurements, the residual helium contained in the specimens was determined using the isotope-dilution method. Prior to the addition of the spike, however, a first aliquot of the sample gas was taken for a determination of the helium isotopic ratio in the sample itself.

Helium measurements in the two samples after vaporization are summarized in Table 1. Total helium measured was 754 appm for W-2 and 715 appm for W-3, with an average of 733 ± 28 (1σ) appm. The $^4He/^3He$ isotopic ratio observed in the two specimens was 13.9 and 13.7, respectively, with an average value of 13.8 ± 0.1 (1σ). The absolute uncertainty (1σ) in the individual helium contents is estimated to be between 1% and 2%.

As can be seen from Figures 3 and 4, and Table 1, helium release from the tungsten during the initial stepped-anneal tests was small compared to the total helium inventory. For W-2, the total ^4He released at 1600°C was only ~0.2% of the total ^4He. Similarly, for W-3, the ^4He released by 1200°C was only ~0.1% of the total.

Hydrogen Analyses

Hydrogen Analysis System

Hydrogen measurements conducted previously [*5*] on these tungsten samples were performed using one of the analysis furnaces attached to the helium analysis system that had been extensively modified for the task. This earlier system was based on isotope-dilution, static-mode, gas mass spectrometry, similar to the method used for the helium measurements. Although this system provided useful initial data for the APT program, the system could only be used for relatively high hydrogen levels (thousands of appm), and the absolute accuracy of the system was difficult to assess. In particular, background hydrogen levels in the detector were difficult to determine.

In order to improve on the accuracy and sensitivity of the hydrogen measurements, a second development effort was undertaken that involved improvements to both the

Table 1 – *Total helium in APT tungsten.*

Specimen	Analysis Type[a]	Specimen Mass[b] (mg)	Measured Helium (10^{14} atoms) ^3He	Measured Helium (10^{14} atoms) ^4He	Total Helium Concentration (appm)[c] ^3He	Total Helium Concentration (appm)[c] ^4He
W-2-B	Anneal	5.373	0.09	0.2		
	Vap.		8.88	123.4	51.0	702.2
W-3-B	Anneal	3.606	0.03	0.17		
	Vap.		5.74	78.50	48.8	665.9
1Wc-E	Vap.	3.510	0.268	2.570	2.33	22.35
-F	Vap	5.965	0.460	4.043	2.35	20.69

[a]Stepped-anneal or vaporization (Vap.) analysis.
[b]Mass uncertainty is ±0.002 mg.
[c]Total helium concentration in atomic parts per million (10^{-6} atom fraction) with respect to the total number of atoms in the specimen. Values for the W-2 and W-3 represent the total of the stepped-anneal and vaporization analyses.

sample hydrogen extraction furnace and detector components [*6*]. The hydrogen extraction system improvements involved a significant reduction of the chamber volume and a change of materials in the sample furnace that were heated and in direct connection to the mass spectrometer detector. The detector improvements involved a complete redesign of the detector volume to minimize its size and its proximity to the extraction furnace.

A schematic diagram of the hydrogen analysis system is shown in Figure 5. Hydrogen is extracted from the analysis sample by heating to temperatures up to ~1200°C. The extraction furnace shown on the left-hand side consists of a sample turntable assembly located above a ceramic test tube shaped crucible heated on the outside by a resistance-heated tungsten filament. A tungsten heating filament is contained inside a water-cooled copper enclosure connected to its own separate vacuum system. Samples are loaded into the upper section of the turntable and remotely dropped one at a time through the central vertical channel into the crucible using a bellows-sealed push rod.

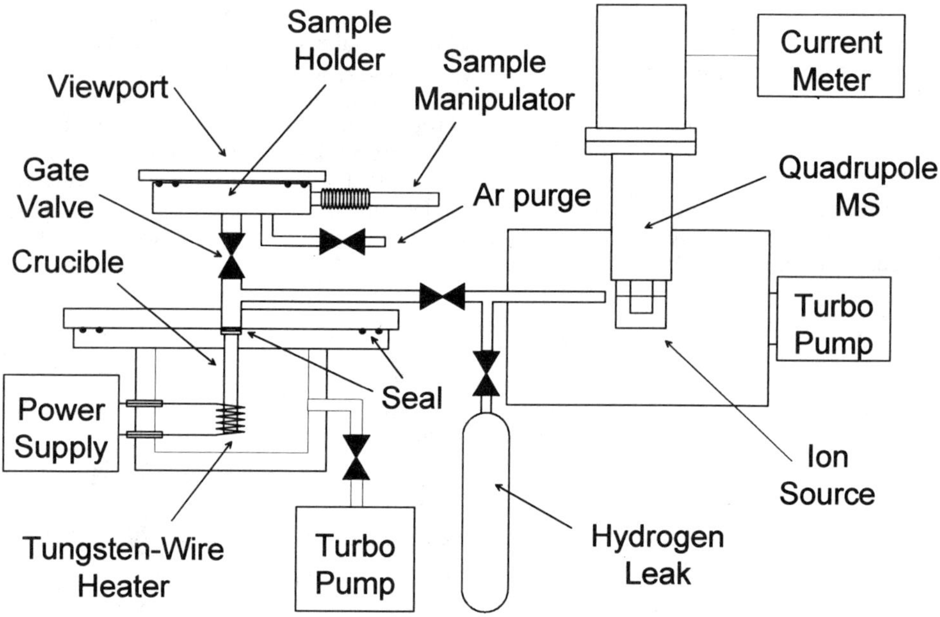

Figure 5 - *Hydrogen analysis system employed in the present measurements.*

The hydrogen detector system is shown in the right-hand side of Figure 5. Sample gas flow from the extraction system travels through the horizontal vacuum line which terminates at a point that is very near the ionization region of the quadrupole detector. The mass spectrometer detector is a quadrupole analyzer with an electron multiplier detector. For the present measurements, the output of the electron-multiplier was fed directly to a digital multimeter (DMM). The mass spectrometer was tuned to a mass of 2 amu for all measurements.

Calibration of the system sensitivity is accomplished using a hydrogen leak source attached to the vacuum line between the extraction furnace and the detector volume. The calibrated leak has a very small trapped volume, resulting in virtually no reduction of the leak rate with time. Calibration measurements are conducted before and after each sample analysis, and show an overall reproducibility of ~2 to 3%.

Measurements are also routinely conducted on specimens of a standard, hydrogen-containing steel maintained in the laboratory. The stated content of the steel is 5.2 ± 0.3 wppm. The average hydrogen content measured in more than 90 of these samples is 5.0 wppm with a standard deviation of ~20% (1σ). It is speculated that at least some of the observed variability in the standard samples is associated with actual heterogeneity in the hydrogen content at this small mass level.

Absolute uncertainty (1σ) in the hydrogen analyses is currently estimated at ~20%, and is due partly to the uncertainty in the calibrated hydrogen leak source. The system has been determined to be linear in response up to a total hydrogen release of at least 10^{17} atoms, which for a 0.5 mg steel sample, represents a hydrogen concentration of ~20,000 appm. Additional uncertainty may also be present from possible hydrogen release from remaining water layers or hydrated metal oxides on the surface of the sample that are subsequently dissociated by the hot crucible. It is hoped that with more experience with the system, this uncertainty will be further reduced.

Retained Hydrogen in Tungsten

Using the new analysis system, hydrogen measurements were conducted on additional specimens from two of the irradiated tungsten samples from the original group of five samples prepared at LANL and reported earlier [*5*], on samples from two additional specimen sets from LANL, and on specimens taken from unirradiated control tungsten material. Except for one sample, which was analyzed using an approximate linear temperature ramp, all of the analyses were conducted at a fixed crucible temperature of ~1200°C, which is the postulated off-normal scenario for the APT design.

Hydrogen measurements are summarized in Table 2. The mean concentrations and 1σ standard deviations in the replicate analyses are given in the last column. As discussed above, the absolute uncertainty (1σ) in the hydrogen analyses is estimated at ~20%.

Hydrogen release measurements during the temperature ramp were also conducted on one of the tungsten specimens and are shown in Figure 6. The thin solid line in the graph is the crucible temperature profile as a function of heating time in seconds. The crucible temperature is given on the right hand y-axis. The thicker line indicates the instantaneous hydrogen release in atoms per second on the left y-axis. Total heating time is given in the bottom axis.

LAHET Calculations of Helium and Hydrogen Content

A detailed Monte Carlo model of the APT materials irradiation has been completed at LANL using the LAHET Code System (LCS) [*7*]. The LCS is the neutronic tool used for the design of APT and the estimation of radiation damage parameters, such as DPA and gas production. Because the proton energy spectrum and beam profile vary throughout the APT target/blanket (with values ranging as high as 1.6 GeV), the model presented is therefore prototypic for only one region of the target system. However,

Table 2 – *Retained hydrogen in APT tungsten.*

Sample	Mass[a] (mg)	Measured Hydrogen (10^{15} atoms)	Hydrogen Concentration (appm)[b] Measured	Corrected[c]	Average[d]
Unirradiated:					
W-blank	7.506	4.7	191	-	210
	10.22	7.8	233	–	±30
Irradiated:					
W-2	9.063	~42	~1400	~1190	1700
	1.526	10.2	2030	1820	±470
	3.282	25.0	2320	2100	
W-3	0.761	9.27	3710	3500	3500
1Wh	4.538	25.1	1680	1470	1990
	6.197	46.4	2280	2070	±480
	4.432	38.3	2630	2420	
1Wc	5.240	6.36	370	160	220
	4.390	7.23	502	292	±70
	3.719	5.18	424	214	

[a]Mass of specimen for analysis. Mass uncertainty is ±0.002 mg.
[b]Helium concentration in atomic parts per million (10^{-6} atom fraction) with respect to the total number of atoms in the specimen.
[c]Corrected for hydrogen measured in unirradiated control material.
[d]Mean and standard deviation (1σ) of duplicate analyses.

using the measurements of helium and hydrogen in the LANSCE Area A test at 800 MeV as a calibration for the LCS will provide some insight into the potential accuracy of gas production predictions throughout the APT target/blanket.

Calculations of the helium and hydrogen (i.e., proton) production in the current LANSCE Area A test were carried out using the LCS [*5*]. Because the measured results are related to the total number of helium and hydrogen atoms that are stopped and retained in the sample volume and not necessarily the total production rate in the sample, the tally option used was IOPT=14 with charged particle transport turned on in LAHET (this is not the default LAHET option). The LCS default level density model was used with a pre-equilibrium model following the intranuclear cascade (not the LCS default). This allows some estimation of the energetic loss of helium atoms and protons, especially the latter, which can be on the order of 50% or larger. It does not account for subsequent thermally induced diffusional losses of hydrogen, however.

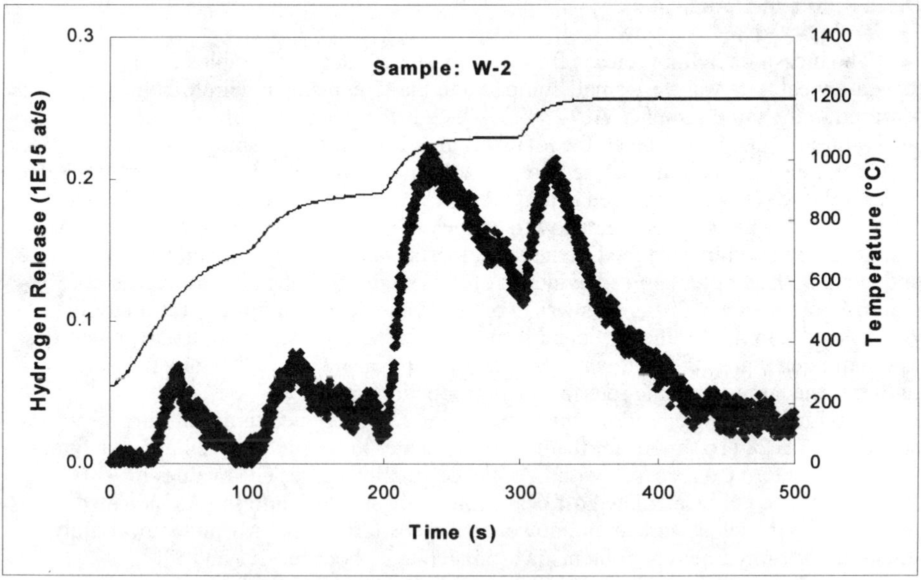

Figure 6 – *Hydrogen release from irradiated tungsten at temperatures between 750°C and 1200°C.*

The results of the generation and retention calculations are summarized in Table 3 along with a comparison to the average measured values. Statistical errors in the LCS calculations were calculated to be less than 3%. Although the statistical errors are small, overall uncertainty is expected to be about 50% for the prediction of radionuclides in general. This uncertainty estimate is based primarily upon the work of Ullmann, et al. [*8*], who measured radioactive spallation products from tungsten and lead targets. The variation of predictive capabilities is relatively large and seems, to some extent, to be a function of the target material [*8,9*].

Table 3 - *Comparison of LCS calculations with measured helium and hydrogen concentrations.*

		Gas Concentrations (appm)					
	Location	Helium			Hydrogen		
Sample	(mm)[a]	Measured[b]	Calculated	C/M[c]	Measured	Calculated	C/M[c]
W-2	12.5	753	760	1.0	1700	4980	2.9
W-3	12.0	715	760	1.1	3500	4980	1.3
1Wh	10.5	-	760	-	1990	4980	2.5
1Wc	62	23.9	TBD	-	220	TBD	-

[a]Sample location relative to proton beam center.
[b]Average total helium ($^3He+^4He$) measured in samples.
[c]Calculated-to-measured ratio.

Discussion and Conclusions

The measured helium release from two irradiated tungsten samples during the stepped-anneal tests was very small compared to the total residual helium inventory measured in the same samples. At 1200°C, which is the postulated off-normal temperature for the APT project, the total helium release for one sample was only ~0.1%. At 1600°C, the total helium release in a second sample was only slightly higher at ~0.2%. Helium release was first observed at about 800°C. One sample showed some evidence for an increasing helium release rate with temperature between 800°C and 1400°C. At the highest temperatures reached in the anneal tests, where multiple helium analyses were conducted, both samples showed evidence of a leveling off in the helium release rate. Helium-3 release for both samples was at or below the detection limit of the analysis system (~10^{12} atoms) for the particular setup used in these tests. Given the low quantities of helium release at 1200°C, it is unlikely that this release will result in any pressure buildup at the target/cladding interface of a production target.

Total helium concentrations measured in the two irradiated tungsten samples averaged 733 ± 28 (1σ) appm for the near beam center locations, and ~24 appm at 65 mm from beam center. Comparison with LAHET calculations gives C/M values of ~1.0 to 1.1 near beam center. Calculated off beam values are still in progress. As indicated previously [*5*] this close agreement, however, may be fortuitous. Absolute uncertainty (1σ) in the measured helium contents is estimated to be between 1% and 2%.

New hydrogen measurements conducted on two unirradiated specimens of as-received tungsten material showed a low hydrogen impurity content of about 200 appm, in good agreement with the preliminary data obtained earlier [*5*]. New analyses of irradiated samples show total hydrogen contents ranging from approximately 1700 up to 3500 appm, also in good agreement with the earlier preliminary data.

Comparison of the hydrogen values with LAHET calculations gives C/M values ranging from ~1.3 to 2.9. Some of the discrepancy between the calculated and measured values is the result of the uncertainties in the measurements and the calculations (about 20% and 50%, respectively). While energetic losses were estimated, thermally-induced diffusional losses were not accounted for in the LAHET calculations. Calculations of effective diffusion length [*10*] for the LANSCE irradiation conditions (~170°C for ~60 days) would indicate a value of about 9 mm, which is considerably larger than the rod radius. Such a large value would suggest that most if not all the generated hydrogen would have diffused out of the sample during irradiation, at least for the in-beam samples W-2 and W-3. The fact that significant hydrogen was in fact retained, strongly supports trapping of the hydrogen in the tungsten matrix. Although LAHET calculations are still to be performed for the out of beam locations, correlation of the retained hydrogen in Sample 1Wc with the relative proton fluence would suggest lower diffusional loss of hydrogen which would be consistent with the lower irradiation temperature for this sample.

Hydrogen release profiles with increasing temperature, in Figure 6, show four release peaks at temperatures of approximately 550, 850, 1100, and 1200°C. Some of release features may be attributable to the nonlinearity in the temperature ramp. Recent calculations [*11*] show a good fit to the release data using a single trap model with a trap energy of 1.4 eV. Both the present measurements and the earlier preliminary measurements suggest complete hydrogen release from the tungsten at temperatures of 1200°C (or lower), within a few minutes. Whether or not this released hydrogen will result in a pressure buildup at the target/cladding interface of a production target will depend largely on whether or not any gaps exist between the tungsten and the cladding. If no gap exists, then the hydrogen will likely diffuse through the Alloy 718 at a rate equal to its release from the tungsten [*12*]. If a gap exists, however, the hydrogen will likely recombine to form molecular H_2, and subsequent decomposition and release through the 718 cladding may be slower, allowing some build up in pressure within the cladding, leading to potential cladding rupture.

Within the respective uncertainties of the gas measurements and calculations, several observations can be made.

- As expected, significant levels of helium and hydrogen were observed in the tungsten target material.
- Helium release at the APT postulated off-normal temperature of 1200°C is small at about 0.1 to 0.2% of the total generated helium. A very small helium release first occurs at ~800°C, and the release rate does not change significantly with temperature.
- Given the small levels of helium released, helium gas buildup should not contribute significantly to any pressure buildup at the target/cladding interface of a production target.
- Hydrogen release was seen to occur in tungsten starting at ~500°C, with release "peaks" at ~550, 850, 1100, and 1200°C, possibly from temperature nonlinearities. Release data are consistent with a single trap model, with a trap energy of 1.4 eV.
- The measurements suggest complete release of generated hydrogen within a few minutes at 1200°C. Whether or not this released hydrogen will contribute to a pressure buildup at the target/cladding interface will depend on whether there is a gap at the interface allowing recombination of the hydrogen into molecular form.
- LAHET calculations of helium generation agreed reasonably well with measurements within estimated uncertainties. Calculation of hydrogen generation were higher than the observed retention, likely from relatively small measurement and somewhat larger calculational uncertainties and possibly from some thermally-induced diffusional losses from the tungsten samples during irradiation.

Acknowledgements

This work was supported by the U.S. Department of Energy under the Accelerator Production of Tritium Program at Los Alamos National Laboratory. Pacific Northwest National Laboratory is operated for USDOE by Battelle Memorial Institute.

References

[1] "APT ^{3}He Target/Blanket Topical Report", Los Alamos National Laboratory Report LA-CP-94-27, Rev. 1, Los Alamos National Laboratory, Los Alamos, NM, 1994.

[2] Maloy, S. A. and Sommer, W. F., "Spallation Source Materials Test Program", *Proceedings of the Topical Meeting on Nuclear Applications of Accelerator Technology*, American Nuclear Society, ISBN: 0-89448-629-2, Albuquerque, NM, November 16-20, 1997, pp. 58-61.

[3] Farrar, H. IV and Oliver, B. M., "A Mass Spectrometer System to Determine Very Low Levels of Helium in Small Solid and Liquid Samples," *Journal of Vacuum Science and Technology - A*, Vol. 4, 1986, pp. 1740-1741.

[4] Oliver, B. M., Bradley, J. G., and Farrar, H. IV, "Helium Concentration in the Earth's Lower Atmosphere," *Geochimica et Cosmochimica Acta*, Vol. 48, 1984, pp. 1759-1767.

[5] Oliver, B. M., Hamilton, M. L., Garner, F. A., Sommer, W. F., Maloy, S. A., and Ferguson, P. D., "Helium/Hydrogen Measurements in High-Energy Proton-Irradiated Tungsten", *Effects of Radiation on Materials, 19th International Symposium, ASTM STP 1366*, M. L. Hamilton, A. S. Kumar, S. T. Rosinski, and M. L. Grossbeck, Eds., American Society for Testing and Materials, West Conshohocken, PA, 1999, pp 1109-1121.

[6] Oliver, B. M., Garner, F. A., Greenwood, L. R., and Abrefah, J. A. "Quadrupole Mass Spectrometer System for the Determination of Low to High Levels of Hydrogen in Irradiated Materials". Proceeding of the 9th Int. Conf. on Fusion Reactor Materials, Colorado Springs, CO, October 10-15, 1999, *Journal of Nuclear Materials*, in press.

[7] Prael, R. E. and Lichtenstein, H., "User Guide to LCS: The LAHET Code System," Los Alamos National Laboratory Report LA-UR-89-3014, Los Alamos National Laboratory, Los Alamos, NM, September 1989.

[8] Ullmann, J. L., Gavron, A. I., King, J. C., Laird, R., Mayo, D. R., Waters, L. S., Zoeller, C., Staples, P., Jagnow, D., Koster, J. E., Lisowski, P. W., Nelson, R. O., Wender, S. A., Butler, G. W., Gritzo, R. E., Yates, M. A., Fowler, Malcolm M., Wilhelmy, J. B., Wilson, W. B., Laird, C. E., and Mullins, D. H., APT Radionuclide Production Experiment Technical Report, LA-UR-95-3327, Los Alamos National Laboratory, Los Alamos, NM, September 1995.

[9] Green, S. L., Green, W. V., Hegedus, F. H., Victoria, M., Sommer, W. F., and Oliver, B. M., "Production of Helium by Medium Energy (600 and 800 MeV) Protons," *Journal of Nuclear Materials*, Vol. 155-157, 1988, pp. 1350-1353.

[10] Frauenfelder, R., "Solution and Diffusion of Hydrogen in Tungsten", *Journal of Vacuum Science and Technology,* Vol. 4, 1969, p. 388.

[11] Oliver, B. M., Garner, F. A., Hamilton, M. L., Venhaus, T.J., Causey, R.A., and Maloy, S. A., "Hydrogen Release From 800-Mev Proton-Irradiated Tungsten Rods", Proceeding of the 5th International Workshop on Hydrogen Isotopes in Solids, May 17-19, 2000, Stockholm, Sweden, *Physica Scripta, in press.*

[12] Oliver, B. M., Garner, F. A., Maloy, S. A., Sommer, W. F., Ferguson, P. D., and James, M. R., "Retention of Very High Levels of Helium and Hydrogen Generated in Various Structural Alloys by 800 MeV Protons and Spallation Neutrons", *Effects of Radiation on Materials, 20th International Symposium, ASTM STP 1405*, S. T. Rosinski, M. L. Grossbeck, T. R. Allen, and A. S. Kumar, Eds., American Society for Testing and Materials, West Conshohocken, PA, 2002.

Hirotatsu Kishimoto,[1] Yutai Katoh,[2] Akira Kohyama,[2] and Masami Ando[3]

The Influence of Temperature, Fluence, Dose Rate, and Helium Production on Defect Accumulation and Swelling in Silicon Carbide

---

Reference: Kishimoto, H., Katoh, Y., Kohyama, A., and Ando, M., **"The Influence of Temperature, Fluence, Dose Rate, and Helium Production on Defect Accumulation and Swelling in Silicon Carbide,"** *Effects of Radiation on Materials: 20th International Symposium, ASTM STP 1405,* S. T. Rosinski, M. L. Grossbeck, T. R. Allen, and A. S. Kumar, Eds., American Society for Testing and Materials, West Conshohocken, PA, 2001.

Abstract: Swelling and microstructure of silicon carbide (SiC) are studied by means of MeV-range ion irradiation. The material used is chemical vapor deposited high purity polycrystalline cubic (3C)-SiC. The swelling behavior is characterized by precision interferometric profilometry following ion bombardment to the diamond-finished surface over a molybdenum micro-mesh. Irradiation was carried out at temperatures up to 873 K, followed by profilometry at room temperature. Microstructural characterization by means of cross-sectional transmission electron microscopy has also been finished for selected materials. Irradiation induced swelling was increased with increasing the displacement damage level up to 0.3 dpa at all evaluated temperatures. At 333 K, the swelling was increased with increasing the damage level up to 1 dpa, and irradiation-induced amorphization was observed over 1.07 dpa. At the higher irradiation temperature, swelling was saturated over 0.3 dpa. The temperature dependence of saturated swelling obtained so far appeared very close to the neutron irradiation data. For the study of the synergism of displacement damage and helium production, a dual-beam experiment was performed up to 100 dpa at 873 K. Swelling of the dual-beam irradiated specimen was larger than that of single-beam irradiated specimen. The result also suggested the onset of unsteady swelling in high He/dpa conditions after "saturated point defect swelling" is once achieved at displacement damage levels over 50 dpa.

Keywords: silicon carbide, swelling, ion irradiation, dual-beam irradiation method

---

[1]Graduate student, Graduate school of Energy Science, Kyoto University, Gokasho, Uji, Kyoto 611-0011 Japan.

[2] Associate Professor and Professor, respectively, Institute of Advanced Energy, Kyoto University, Gokasho, Uji, Kyoto 611-0011 Japan.

[3] Doctor Researcher, Japan Atomic Energy Research Institute, Tokai 319-1195, Japan

Introduction

Silicon carbide (SiC) materials have been considered as the structural material for advanced energy systems, including fusion system, due to their superior mechanical performance and chemical stability at high temperature and low induced radioactivity. For the design of SiC materials as the structural components of nuclear energy systems, irradiation effects such as thermal and mechanical property degradation and swelling cannot be ignored. Because irradiation-produced defect accumulation causes these property changes, it is important to understand the nature and processes of radiation defect evolution in SiC [*1*].

In a fusion environment, the synergistic effect of helium and displacement induced by neutron irradiation is of concern. The most damaging component of radiation will be high-energy neutrons (≈ 14MeV), which cause not only displacement of the atom but also helium production through (n, α) nuclear reactions [*2*]. Since the fusion reactors do not yet exist and only a few high temperature capsules have been irradiatesd in research fission reactors, irradiation effects data under the existence of both displacement damage and helium production at the high temperature in SiC materials are very limited.

A dual-beam ion irradiation study [*3, 4*] is suitable to the evaluation of materials for a fusion system. The objective of this work is to determine the effects of irradiation condition for the influence to SiC microstructure and study the synergistic effects of displacement damage and helium production.

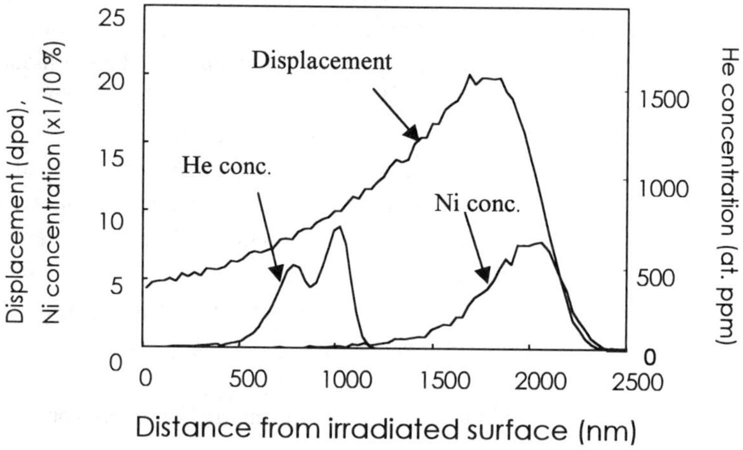

Fig. 1 - *The depth profile calculated by TRIM92 code assuming Ed=35eV, ρ=3.21g/cm^3.*

Experimental

Irradiation and Observation

The material used is chemical vapor deposited high purity polycrystalline cubic (3C)-SiC produced by Morton. For ion-beam irradiation, the CVD-SiC was cut to disk shape and its irradiation surface was polished. The dimensions of specimen were 3.0 mm diameter and 0.25 mm thickness. Ion-beam irradiation was carried out at the High-fluence Irradiation Facility, The University of Tokyo (HIT Facility). The specimens were irradiated with single- ion, only 4 MeV Ni^{3+} ions or dual-ions, 4 MeV Ni^{3+} ions and simultaneously implanted helium ions. 4 MeV Ni^{3+} ions were accelerated by the 1 MeV Tandetron accelerator, and a Van de Graaff accelerator was used for 1 MeV He^{+} ions, which were degraded in energy and slowed down by a thin Ni foil degrader placed between the accelerator and the irradiation chamber. The depth profiles of displacement damage, concentrations of Ni and He were calculated by TRIM-92 code. An example of TRIM92 calculation result is shown in Fig. 1 [*4*]. A sublattice-averaged displacement energy was assumed to 35 eV. The damage level was up to 25 dpa, and irradiation temperature was up to 873 K. Displacement damage rate was 1×10^{-5}, 1×10^{-4} and 1×10^{-3} dpa/s. He/dpa ratio of simultaneously irradiated helium was 60 appm. Irradiation surface was polished by diamond and covered with molybdenum mesh to make unirradiated area for the swelling characterization as shown in Fig. 2.

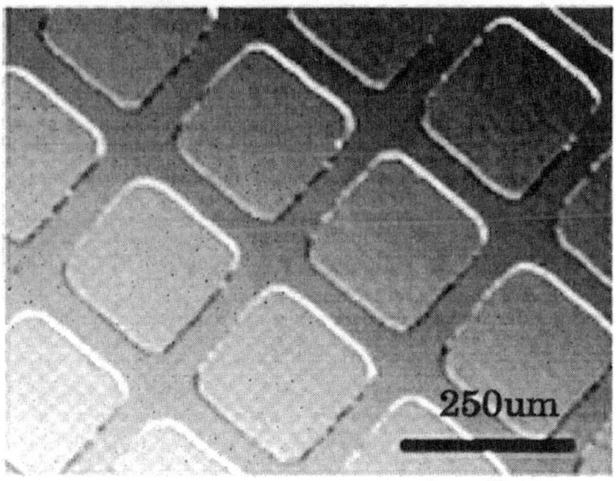

Fig. 2 - *Nomarski micrograph of ion irradiated surface over Mo mesh in CVD-SiC.*

CVD-SiC was sliced about 0.1 - 0.3 mm thick by dicing saw and focused ion beam (FIB) processing [*3*] was performed to make a thin foil for the microstructural investigation by a transmission electron microscopy (TEM). The thin foil was normal to irradiation surface. The microstructural investigation was performed with JEOL JEM-2010 conventional TEM.

Swelling Characterization

The surface profile was characterized by precision interferometric profilometry. Swelling is calculated from the surface profile and TRIM92 code. The interferometric profilometry is possible to analysis about 2nm height. Ion irradiation induces the gradient displacement damage, the swelling at the average displacement damage level, $\Delta V / V(\Phi av)$, is given by

$$\Delta V / V(\Phi\, av) \cong \Delta Z / d \qquad (1)$$

where ΔZ is the step height between irradiated and unirradiated surface measured by interferometric profilometry, d is the depth of irradiated region from the surface, and Φav is the average displacement damage level of all irradiated region. Irradiated aria depth, *d* and average damage level, Φav were calculated from the TRIM92 code and evaluated by EDS analysis.

For the swelling evaluation of dual-beam irradiated specimen, it is necessary to consider the implanted helium depth distribution in CVD-SiC. The swelling of dual-beam irradiated specimen is calculated as following equation.

$$(\Delta V / V^{dual}_{@\Phi_d} - \Delta V / V^{single}_{@\Phi_{av}}) \times Z_d = \Delta Z^{dual} - \Delta Z^{single} \qquad (2)$$

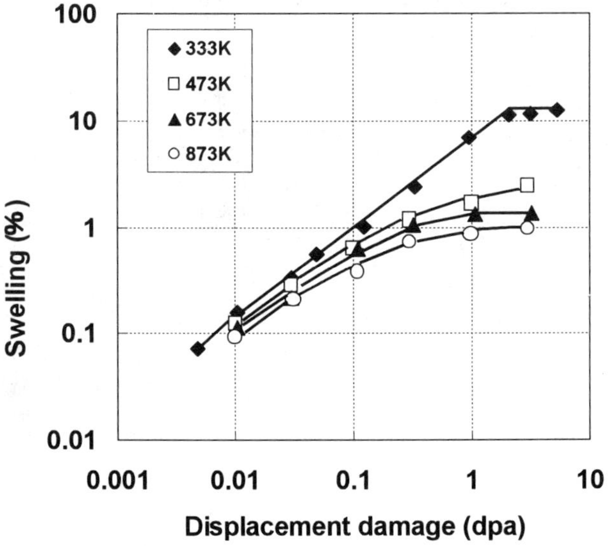

Fig. 3 - *Temperature dependence of ion induced swelling in CVD-SiC.*

$\Delta V / V^{dual}_{@\Phi_d}$ is the swelling of dual-beam irradiated specimen at the Φd, average displacement damage level in helium implanted zone. $\Delta V / V^{single}_{@\Phi_{av}}$ is the swelling of single-beam irradiated specimen at the Φav. The thickness of helium implanted region is Z_d, and the step heights of dual-beam and single-beam irradiated specimen are ΔZ^{dual} and ΔZ^{single}, respectively.

Results

Temperature Dependence of Point-Defect Swelling in CVD-SiC

Temperature dependence of swelling was performed with single-beam irradiation at the temperature up to 873 K and shown in Fig. 3. Displacement damage rate was 1×10^{-3} dpa/s. At 333 K, the swelling increased with increasing average damage level up to 1 dpa, and saturated by amorphization. Fig. 4 shows the cross-sectional TEM image of single-beam irradiated SiC at 1 dpa and 333 K. There are 3 regions that are normal to the surface because of the gradient damage level by ion beam irradiation. The near-surface region,

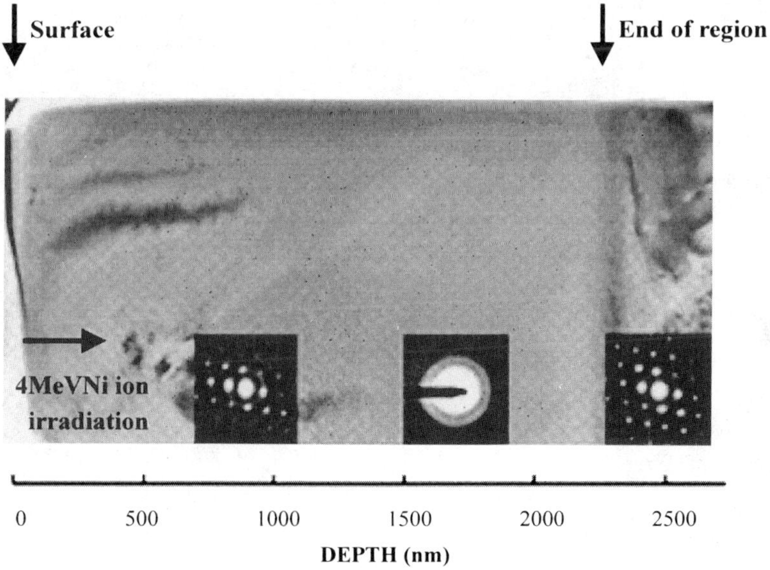

Fig. 4 - *Cross-sectional TEM image of single-ion irradiated CVD-SiC at the temperature 333 K and dose level 1 dpa.*

from the surface to about 1.5 μm depth, was not observed amorphization. Amorphization of SiC occurred about 0.8 μm wide band centered near the peak damage region. The bottom of damaged range is 2.25 μm depth from the surface, unirradiated region is located deeper than the amorphization region. From the TEM evaluation in Fig. 4, amorphization threshold for CVD-SiC at 333 K is estimated at 1.07 dpa [5].

At the higher temperature, swelling increased with increasing the displacement damage level up to ~ 0.3 dpa and saturated without amorphization as shown in Fig. 3. Unsaturated and saturated swelling is induced by point defect accumulation, and depends on the irradiation temperature at 473-873K.

Dose Rate Dependence of Ion Irradiation Induced Swelling in CVD-SiC

Dose rate of ion irradiation by accelerator is much larger than that of neutron irradiation in reactor, it is necessary to evaluate the influence of displacement damage rate in SiC. The evaluated displacement damage rates were 1×10^{-4} and 1×10^{-3} dpa/s at 333 K - 873 K and 1×10^{-5} dpa/s at 333 K with single-beam irradiation. The results of these evaluations are shown in Fig. 5. The error bars represent the

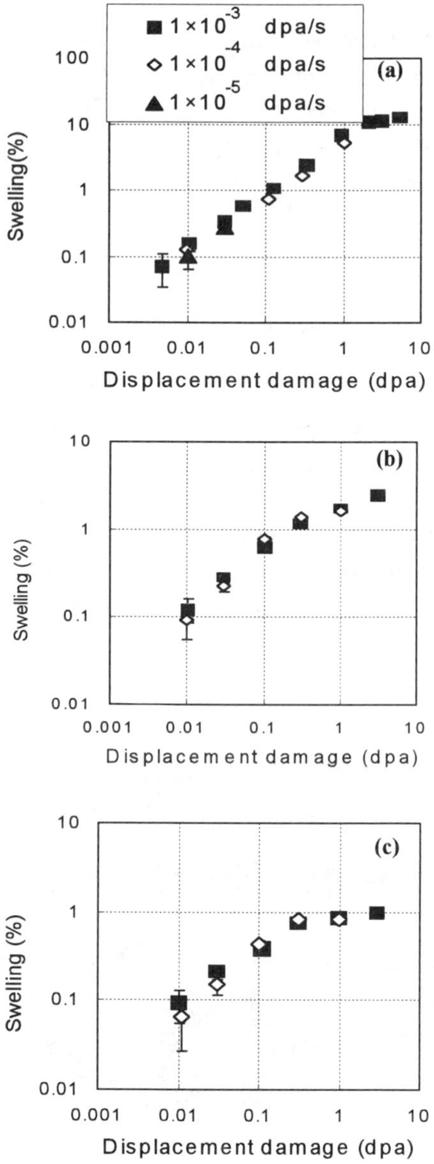

Fig. 5 - *Dose rate dependence of ion induced swelling in CVD SiC (a)333K (b) 473K (C)873K*

96% confidence limits for the Gaussian distribution. It appears no significant effects of displacement damage rate are shown in Fig. 5. At the low dose level, error bars of both displacement damage late results overlapped each other, and the swelling results of both rates were almost equal at the high dose level. Fig. 5 suggests that ion irradiation induced swelling (10^{-3} - 10^{-5} dpa/s) should be comparable with the neutron induced saturation swelling (10^{-7} - 10^{-6} dpa/s).

Effects of Simultaneously Implanted Helium for the Swelling in CVD-SiC

Dual-beam study was performed at the 333 K and 873 K. Displacement damage rate was 1×10^{-3} dpa/s. Fig. 6 shows the comparison of the swelling between dual-beam irradiated specimens and single-beam irradiated one. The swelling of dual-beam irradiated specimens were calculated by Eq. (2). Fig. 6 shows no significant difference due to single or dual irradiation at 333K. At 873K, however, the simultaneously implanted helium affected the swelling behavior. Both the unsaturated swelling at the low dose level and saturation swelling at the high dose level of dual-beam irradiated specimens are larger than that of single-beam irradiated one.

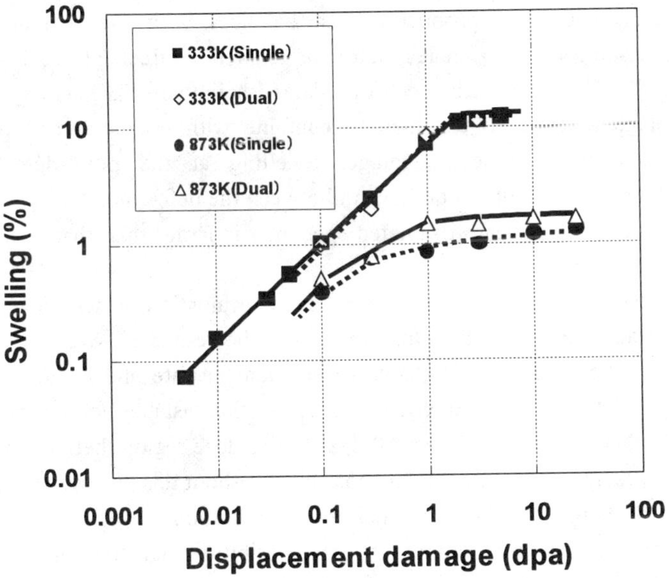

Fig. 6 - *Dual-ion irradiation induced swelling in CVD-SiC*

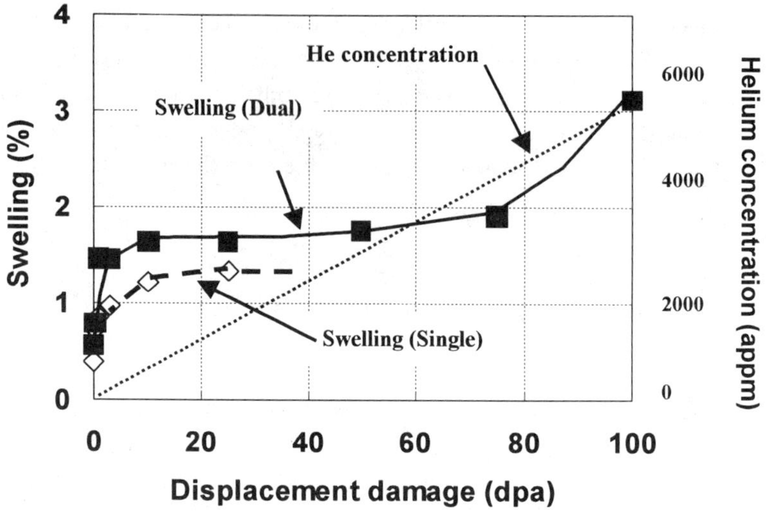

Fig. 7 - *He accumulation and irradiation-induced swelling in dual- and single-ion irradiated CVD-SiC at 873K*

Fig. 6 suggests that the implanted helium affects the defect annihilation rate. Under the irradiation environment at elevated temperature, the two competing processes of damage production and recovery are simultaneously occurring [*6*]. In saturated swelling region, displacement damage production just balances with the relaxation by the recombination of defects and vacancies. Since the relaxation effects at 333 K are smaller than those of at 873 K and the influence of implanted helium at 873 K is larger than that at 333 K, implanted helium is assumed to combine with vacancies and prevent the relaxation of irradiation induced damage. Swelling at the point-defect swelling temperature region is the result of equilibrium between the defect production and damage relaxation, swelling of dual-beam irradiated specimen is larger than that of single-beam irradiated specimen at 873 K.

High fluence examination was performed for the evaluation of helium accumulation by dual-beam irradiation up to 100 dpa at 873 K. The results of swelling and helium concentration are shown in Fig. 7. Displacement damage rate and He/dpa ratios were 1×10^{-3} dpa/s and 60 appm, respectively. Fig. 7 shows the unstable volumetric expansion at high displacement damage level over 50 dpa. In this dose region, helium accumulation increases over 3000 appm. There are some possibilities about this phenomenon, one is the effect of Ni ion. The other is the effect of helium accumulation.

The data in Figs. 6 and 7 suggest one possibility that the implanted helium significantly affects the defect annihilation rate. Under the irradiation environment at elevated temperature, the two competing processes of damage production and recovery are simultaneously occurring [*6*]. In saturated swelling region, displacement damage production just balances with the relaxation by the recombination of defects and vacancies. Since the relaxation effects at 333 K smaller than those at 873 K and the influence of implanted helium at 873 K is larger than that at 333 K, implanted helium is assumed to combine with vacancies and prevent the relaxation of irradiation induced damage. Swelling at the point-defect swelling temperature region is the result of equilibrium between the defect production and damage relaxation, swelling of dual-beam irradiated specimen is larger than that of single-beam irradiated specimen at 873 K.

Discussion

Ion irradiation induced swelling results in the present work are comparable with the neutron irradiation ones published by Price [*6,7,8*], Blackstone [*9*], Snead [*10*]. The results of ion irradiation induced saturated swelling generally agree with the published results of neutron induced volumetric swelling. No significant effects of displacement damage rate for unsaturated and saturated swelling in CVD-SiC observed at least for dose rate of 10^{-5} to 10^{-3} dpa/s, ion beam irradiation results should be directly relevant for investigation of SiC as a potential fusion reactor structural material.

Though ion beam irradiation induced swelling generally agrees with neutron induced swelling, exactly, it is necessary to consider the difference between ion and neutron irradiation. This makes a possible that the ion irradiation induced swelling is somewhat larger than that of neutron irradiation, though it is not appeared in the present work.

Helium production is suspected to prevent the interstitial-vacancy recombination by interacting vacancies. This changes the balance of displacement damage production and relaxation, and dual-beam irradiation swelling become larger than single-beam irradiated one at the elevated temperature. He accumulation over 3000appm in dual-beam irradiation at 873K brings about the unstable expansion. If this phenomenon is the result of helium accumulation, this might be important because SiC structural materials irradiated to damage levels in saturation swelling region (>70 dpa) have a possibility to expand with the excess of helium accumulation in fusion environment.

Conclusion

The present work established the validity of characterizing ion irradiation induced swelling in SiC by using precision the interferometric profilometry method. The results of ion irradiation induced saturated swelling generally agree with the published results of neutron induced volumetric swelling and confirmed the possibility of being comparable to the neutron irradiation behavior.

The amount of ion irradiation swelling depends on irradiation temperature and the displacement damage level. Amorphization occurs over the threshold dose level of 1.07 dpa at 333 K. The swelling is saturated without amorphization over 0.3 dpa at the elevated temperature. Unsaturated and saturated swelling for single ion irradiation condition decrease with increasing irradiation temperature. dose rate does not significantly affect to swelling in the present work condition.

The dual-beam irradiation study is suitable for microstructural evaluation of fusion materials. Simultaneously implanted helium affects the irradiation induced swelling larger at 873 K. Since the helium effect at 333 K is not observed, implanted helium is estimated to combine vacancies and prevent the interstitial-vacancy recombination. High fluence dual beam irradiation achieved unstable expansion at the displacement damage level and helium accumulation over 50 dpa and 3000 appm, respectively. This phenomenon is suspected to be associated with the obstruction of relaxation and suggests the unstable volumetric expansion of structural materials may occur in the fusion environment.

Acknowledgiment

The dual-beam ion irradiation experiment was carried out with the assistance of Drs. H. Shibata、 T. Iwai, and Mr. T. Omata at the Research Center for Nuclear Science and Technology, the University of Tokyo. Dr. H. Tanigawa, Japan Atomic Energy Research Institute, helped with the irradiation experiment and with advice about irradiation etc.

References

[1] Snead, L. L., Jones, R. H., Kohyama, A. and Fenici, P., "Status of Silicon Carbide Composites for Fusion," *Journal of Nuclear Materials*, Vol. 233-237, 1996, pp. 26-36.

[2] Corelli, J. C., Hoole, J., Lazzaro, J. and Lee, C. W., "Mechanical, Thermal, and Microstructural Properties of Neutron-Irradiated SiC," *Journal of American Ceramic Society*, Vol 66, 1983, p. 529.

[3] Katoh, Y., Kohyama, A., Hinoki, T., " Microstructural Evolution of SiC/SiC

Composite under Irradiation," *European Conference on Composite Materials Science -Technologies and applications*, Vol.4, 1998, pp.351-357.

[4] Katoh, Y., Hinoki, T., Kohyama, A., Shibayama, T. and Takahashi, H., "Microstructural Stability of SiC/SiC Composites Under Dual-Beam Ion Irradiation," *Ceramic Engineering and Science Proceedings*, Vol.20, Issue 4, 1999, pp. 325-332.

[5] Snead, L. L., Zinkle, S. J., Hay, J. C., Osbone, M. C., "Amorphization of SiC under ion and neutron irradiation," *Nuclear Instruments and Method in Physics Research B*, vol. 141, 1998, pp. 123-132.

[6] Price, R. J., "Effect of fast-neutron irradiation on Pyrolytic Silicon Carbide," *Journal of Nuclear Materials*, vol. 33, 1969, pp. 17-22.

[7] Price, R. J., "Thermal conductivity of Neutron-Irradiation Pyrolytic β-Silicon Carbide," *Journal of Nuclear Materials*, vol. 46, 1973, pp. 268-272.

[8] Price, R. J., "Neutron Irradiation-induced Voids in β-Silicon Carbide," *Journal of Nuclear Materials*, vol. 48, 1973, pp. 47-57.

[9] Blackstone, R., "The expansion of Silicon Carbide by Neutron Irradiation at High Temperature," *Journal of Nuclear Materials*, vol. 39, 1971, pp. 319-322.

[10] Snead, L. L., Osbone, M. C., Lowden, R. A., Strizak, J., Shinavski, R. J., More, K. L., Eartherly, W. S., Bailey, J., Williams, A. M., "Low dose irradiation performance of SiC interphase SiC/SiC composites," *Journal of Nuclear Materials*, vol. 253, 1998, pp. 20-30.

Yutai Katoh,[1,2] Hirotatsu Kishimoto,[1] Masami Ando,[3] Akira Kohyama,[1,2] Tamaki Shibayama,[4] and Heishichiro Takahashi[4]

Microstructural Stability of SiC/SiC Composites under Dual-Beam Ion Irradiation

Reference: Katoh, Y., Kishimoto, H., Ando, M., Kohyama, A., Shibayama, T., and Takahashi, H., **"Microstructural Stability of SiC/SiC Composites under Dual-Beam Ion Irradiation,"** *Effects of Radiation on Materials: 20th International Symposium, ASTM STP 1405,* S. T. Rosinski, M. L. Grossbeck, T. R. Allen, and A. S. Kumar, Eds., American Society for Testing and Materials, West Conshohocken, PA, 2001.

Abstract: Microstructural evolution of silicon carbide fiber-reinforced silicon carbide matrix composites (SiC/SiC composites) during dual-beam ion irradiation was studied. The composites had been fabricated by means of chemical vapor infiltration (CVI) and polymer impregnation and pyrolysis (PIP) methods using advanced fibers, i.e., Hi-Nicalon™ Type-S and Tyranno™-SA, as well as conventional fibers. Dual-beam ion irradiation was performed to 10 dpa at 873K and 60appmHe/dpa. Composites with advanced low-oxygen near-stoichiometric SiC fiber with high crystallinity and beta-SiC matrix exhibited superior irradiation resistance, in contrast to that amorphous Si-C-O based fibers and conventional polymer-derived matrix exhibited microstructural instability in association with irradiation-induced crystallization. Pyrolytic carbon deposited as the fiber-matrix (F-M) interphase exhibited irradiation-induced amorphization and the helium co-implantation enhanced it. Post-irradiation heat treatment caused significant microstructural changes across the F-M interphases at temperatures as low as the composites are supposed to retain their mechanical strength.

Keywords: SiC/SiC composite, ion irradiation, helium effect, microstructure, SiC fiber, TEM

[1]Associate Professor, Graduate Student and Professor, Institute of Advanced Energy, Kyoto University, Gokasho, Uji, Kyoto 611-0011, Japan
[2]CREST-ACE, Japan Science and Technology Corporation, Kawaguchi, Japan
[3]Japan Atomic Energy Research Institute, Naka-gun, Japan
[4]Research Associate and Professor, Hokkaido University, Sapporo, Japan

Introduction

Silicon carbide fiber-reinforced silicon carbide matrix composites (SiC/SiC composites) are attractive potential materials for advanced energy systems. In addition to superior mechanical performance and chemical stability at elevated temperatures, low induced radioactivity of silicon carbide makes it especially attractive for heavily neutron-loaded applications represented by fusion blanket first wall structures and advanced fission reactor core components [*1,2*]. The improvement in mechanical properties and neutron irradiation resistance, as a result of recent efforts in developing high performance SiC fibers and composites for fusion and advanced gas turbine applications, makes SiC/SiC composites even more attractive [*3-5*].

The most serious issue particular to materials for fusion and nuclear applications is the effect of irradiation on structural integrity [*1*]. Of crucial concern for fusion materials is a synergistic effect of atomic displacement due to energetic neutron bombardment and insoluble gas (i.e. helium) production through (n,α) nuclear reaction, which may cause microstructural, physical and mechanical property degradation. For example, irradiation-induced swelling of crystalline SiC[*6,7*], shrinkage of oxygen-containing SiC fibers due to irradiation-assisted oxidation[*8*], irradiation-induced recrystallization of micro-crystalline fibers[*9*], amorphization[*10*], the fiber-matrix debonding caused by radiation-induced deformation[*11*] and irradiation creep of fibers[*12*] have been observed. SiC fibers with lower oxygen content, reduced free carbon and increased crystallinity are likely to minimize these effects of radiation.

As for the SiC fibers, advanced near-stoichiometric crystalline fibers, i.e., Hi-Nicalon™ Type-S (Nippon Carbon, Tokyo, Japan) and Tyranno™-SA (Ube Industries Ltd., Ube, Japan), are expected to exhibit superior radiation tolerance. A low-fluence neutron irradiation experiment confirmed that Hi-Nicalon Type-S -reinforced composites do not undergo radiation-induced strength degradation that is common for Nicalon-CG and Hi-Nicalon composites [*5*]. In this series of work, radiation response of conventional and advanced fibers are compared.

Matrix densification of SiC/SiC composites are being performed by chemical vapor infiltration (CVI), polymer impregnation and pyrolysis (PIP), reaction bonding (RB, or 'reaction sintering,' RS), liquid phase sintering (LPS) and other methods including various hybrid processes. Among two industrially common methods, namely CVI and PIP, the former has extensively been employed for the studies for nuclear and fusion applications because of its advantages in crystallinity, stoichiometry and homogeneity in the produced matrix. On the other hand, PIP is a widely studied technique for industrial production of SiC/SiC composites, due to the potential for drastic reduction in fabrication cost, to which a continuous impregnation and pyrolysis process is applicable for large scale production. Several disadvantages of conventional PIP-produced SiC/SiC have been pointed out, such as low thermal conductivity, insufficient elastic modulus, poor radiation tolerance and chemical incompatibility with liquid coolants and breeders. These undesirable properties are associated with amorphous structure and excess carbon (and oxygen) contents in the PIP-produced matrix after typical pyrolysis at temperatures of 1473-1673K. In order to obtain a highly crystalline and stoichiometric matrix through a

PIP process, pyrolysis at temperatures as high as 2073K was attempted in the current work. This was enabled by the recent commercialization of highly heat-resistant Tyranno-SA fiber [*13*].

The response of polycrystalline beta-SiC to fast neutron and electron irradiation has been relatively well studied [*14-16*]. However, understanding of the influences of high-fluence irradiation and/or intense helium production is so far very limited for either fibers, matrices or interphases.

In this paper, we will present some results from dual-beam ion irradiation experiments on SiC/SiC composites, that have been reinforced by conventional and advanced SiC fibers and matrix-densified by either CVI, conventional PIP or high temperature PIP process.

Experimental Procedure

The CVI-SiC/SiC composites, Hi-Nicalon(U/D)/PyC/SiC^{CVI} and Hi-Nicalon Type-S(U/D)/PyC/SiC^{CVI}, were fabricated at the National Research Institute of Metals (NRIM) using regular and Type-S Hi-Nicalon fibers. The Type-S fiber contains a reduced amount of oxygen and has near-stoichiometric chemical composition. Representative properties and chemical compositions of the fibers have been compiled in ref. [*17*]. The matrix of a unidirectional fiber-reinforced composite was formed by chemical vapor infiltration (CVI). A pyrolitic carbon (PyC) coating was deposited as an interlayer between the fiber and matrix (interphase, hereafter), on the woven fiber fabrics by CVI prior to the matrix processing. The nominal thickness of the carbon coating on the fiber preforms was 200nm. The high temperature-pyrolyzed PIP composite, Tyranno-SA(U/D)/SiC^{PVS} was fabricated at the Institute of Advanced Energy, Kyoto University. Tyranno-SA fiber was selected as the reinforcement due to its proven strength retention after heat treatment at over 2073K[*18*]. Polyvinylsilane (PVS) was used as the matrix precursor[*19*]. The pyrolyzing temperature was selected to be 2073K, at which an almost complete crystallization of SiC and a significant reduction in excess carbon in argon environment could be expected [*20*]. The number of impregnation and pyrolysis cycles was limited to four, since the matrix porosity suppression was not essential for the purpose of ion irradiation study. As a representative conventional PIP-SiC/SiC, Tyranno-TE(P/W)/SiC^{PCS} composite was also subjected to ion irradiation. The PIP condition was eight cycles times pyrolysis at 1573K for 3600s in nitrogen. A F-M carbon interphase was produced by a thermochemical treatment.

For the dual-beam ion irradiation experiment, the composites were square-cut into a $15.0 \times 2.8 \times 2.0$ mm bar with the 2.0 mm-length parallel to the fiber direction. To enable examination of the depth-dependent irradiated microstructures of the fiber, the matrix and the interphase, the surface normal to the fiber direction was irradiated. The surface was carefully polished and diamond powder-finished prior to irradiation. The ion irradiation was carried out at the dual-beam irradiation target at the High-fluence Irradiation Facility, University of Tokyo (HIT Facility) [*21*]. 4 MeV Ni^{3+} ions accelerated by a Tandetron accelerator operating at 1MV were used to produce atomic displacement damage.

Simultaneous implantation of He^- ions was accomplished using a Van de Graaff accelerator, operating at 1MV. The 1 MeV He^- ions were decelerated by a nickel foil energy degrader.

The calculated depth profiles of displacement damage, stopped nickel and deposited helium are presented in Fig. 1. The calculated depth profile was made using the TRIM-92 code [*22*], by assuming an average displacement threshold energy of 35eV [*23*], stoichiometric chemical composition and 3.1g/cm^3 of mass density. The irradiation temperature, displacement damage rate and total dose were 873K, 1×10^{-3} dpa/s and 10 dpa, respectively. Note that most of the deposited Ni and He atoms occur at different depths in SiC.

Following the ion irradiation, the specimen was sliced into 0.2 mm-thick foil and the irradiated edge region was further thinned to 0.05mm using a dicing saw. For examination by transmission electron microscopy (TEM), the thinned tip was further thinned in a focused ion beam (FIB) micro-processing device. The plane of the thinned film was parallel to both fiber axis and the ion bombardment direction. The details of the specimen processing has been described elsewhere [*24*]. The FIB processing was performed so that a single thin film contained the fiber, matrix and the interphase. Microstructural investigation was performed with a JEOL JEM-2010 conventional TEM and with a JEOL JEM-2010F field emission TEM operating at 200kV.

Some of the irradiated specimens were subjected to post-irradiation annealing at temperatures as high as 1873K, in order to study the evolution of implanted helium and microstructural stability during high temperature exposure following irradiation. The annealing was performed on bulk samples. TEM specimen preparation and examination were perfomed in a same way as the irradiated ones.

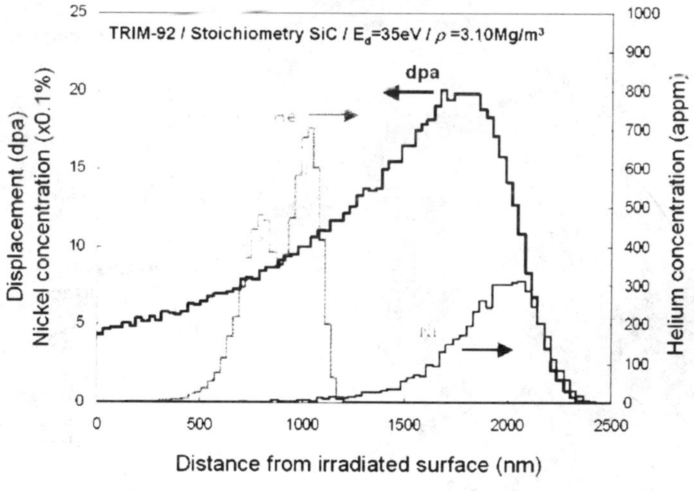

Figure 1 - *Depth profiles of displacement damage, stopped nickel and deposited helium in stoichiometric silicon carbide, calculated by TRIM-92 assuming 35eV of an average displacement threshold energy and 3.10Mg m^3 of mass density.*

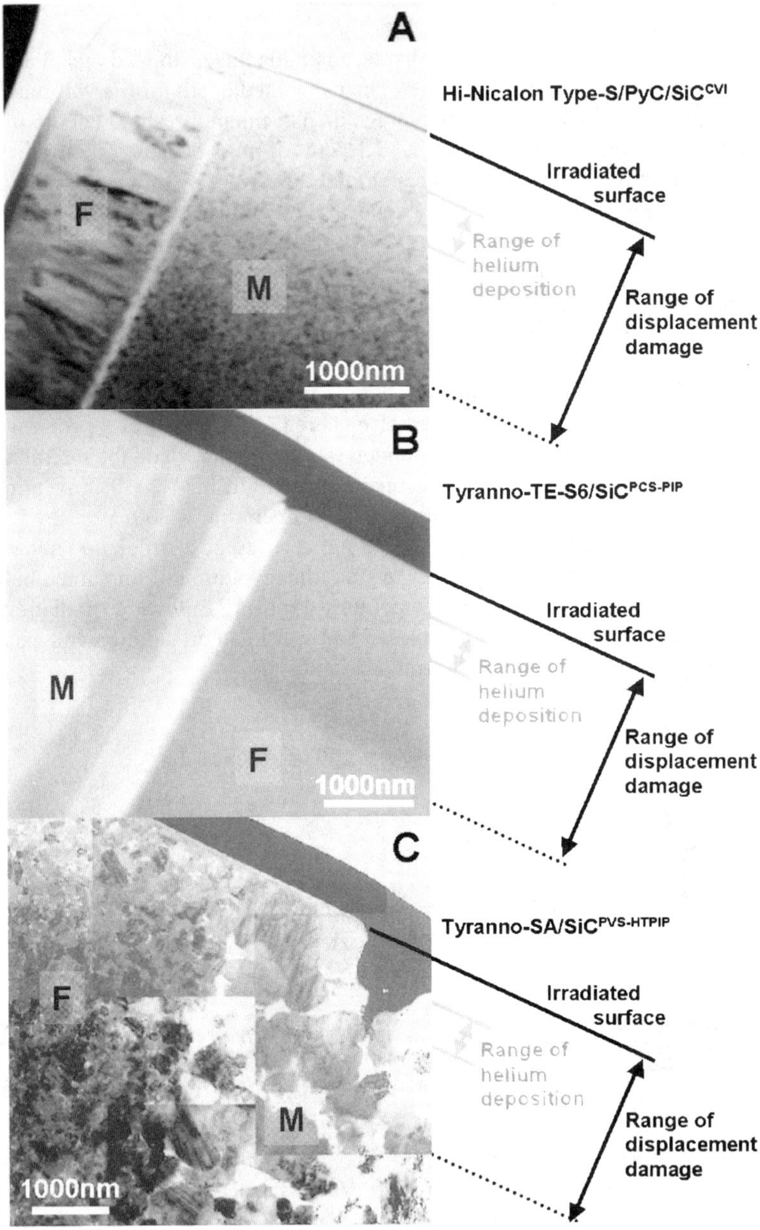

Figure 2 · *Low magnification transmission electron micrographs of irradiated thin foils of (A) Hi-Nicalon(U D) PyC SiC^{CVI}, (B) Tyranno-TE(P W) SiC^{PCS} and (C) Tyranno-SA(U D) SiC^{PVS}.*

Results and Discussion

Low magnification transmission electron micrographs of irradiated thin foils are presented in Figs. 2 (A)-(C). Three different regions of interest were selected from the fiber portion of the thin foils. The first one, 'unirradiated' hereafter, is located about 3~5 μm from the irradiated surface, which is far beyond the maximum range of the implanted ions. The second region, or 'single-beam,' was selected to be approximately 500 nm from the surface, in which both the total deposited concentration of helium and any surface effects during irradiation should be negligible and only atomic displacement due to nickel ion irradiation is expected to contribute to the microstructural modification. In the last one, or 'dual-beam' irradiated region, approximately 1 μm from the surface, both the atomic displacement and the helium deposition rate are significant. The average helium deposition rate in this region was 60 appm (atomic parts per million) / dpa nominal, which corresponds to the typical anticipated helium generation rate in a SiC fusion blanket.

In Fig. 2, the images from fibers are marked 'F' and the matrix 'M'. The pyrolytic carbon interphases typically appear as thin bright stripes in the absorption contrast images. The thin foil's edges generally on the top of micrographs correspond to the irradiated surfaces, covered with thin tungsten layers, which were physically deposited following irradiation to protect the edge region of the thin foil during the FIB processing.

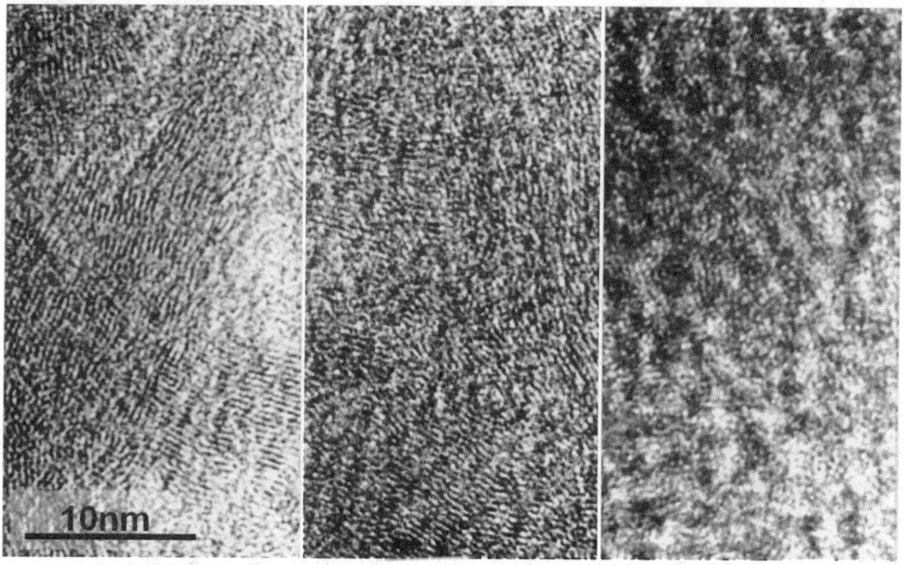

Figure 3 - *High resolution TEM images of unirradiated and single- and dual-beam irradiated pyrolytic carbon deposited as a fiber-matrix interlayer.*

The appearance of Hi-Nicalon/PyC/SiC^{CVI} and Hi-Nicalon Type-S/PyC/SiC^{CVI} was very similar in this magnification, in which no significant influence of irradiation was observed either in fibers or matrices. High resolution transimission electron microscopy (HRTEM) revealed irradiation induced crystallite growth to typically 10 to 15nm in diameter in the Hi-Nicalon fiber, compared to the original crystallite sizes as large as 5nm. No difference was observed between single- and dual-beam irradiated microstructures. Influences of irradiation could not be detected in Hi-Nicalon Type-S fiber and CVI-produced beta-SiC matrix even by HRTEM.

The pyrolytic carbon interphases appeared to be susceptible to irradiation, as anticipated from neutron irradiation data found in the literature [*25*]. As seen in Fig. 3, heavy ion irradiation to 10 dpa-SiC at 873K caused very significant disturbance in graphitic substracture of the glassy carbon. Co-implanted helium in the dual-beam irradiation condition was clearly enhancing this trend so that the resultant microstructure

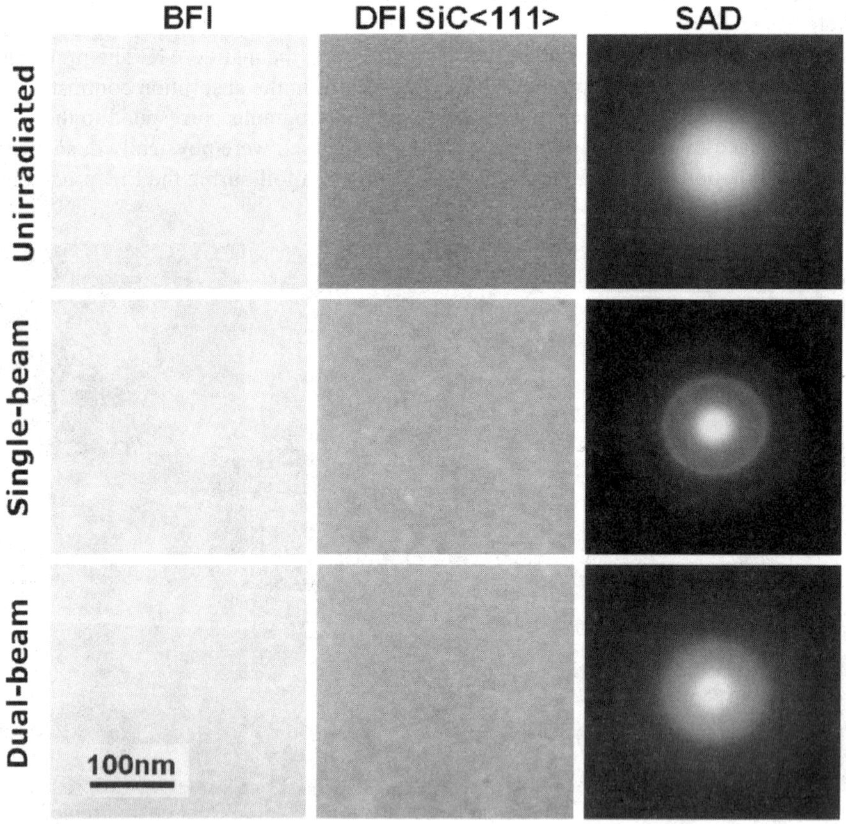

Figure 4 - *Unirradiated and irradiated bright and dark field images and selected area diffraction patterns taken from the PCS-derived SiC matrix.*

could be identified as amorphous. The average basal plane spacings measured on HRTEM micrographs were 0.69, 0.75 and 0.79nm, respectively, for unirradiated, single- and dual-beam irradiated materials. These observations confirm that the ion-irradiatied evolution of the pyrolytic carbon was following the amorphization process of neutron-irradiated graphite and carbon fibers and it was accelerated in the presence of constantly-supplied helium.

In Tyranno-TE(P/W)/SiC^{PCS}, the irradiated area could easily be distinguished by its slightly reduced brightness in the bright field image. The irradiated volume in both fiber

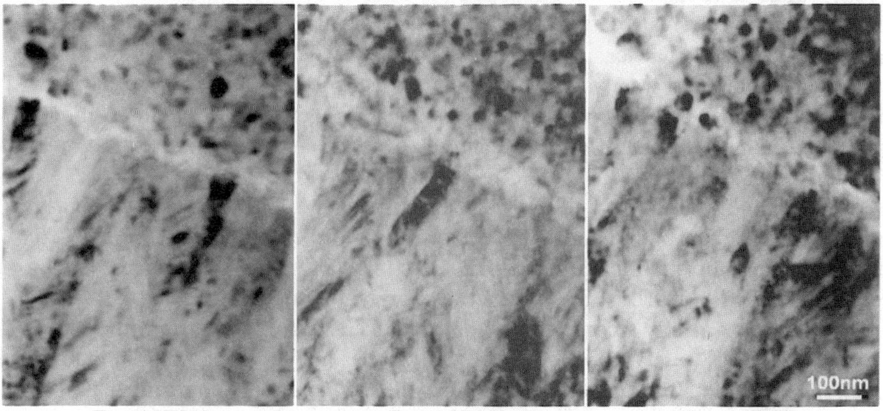

T = 1073K T = 1273K T = 1473K

Figure 5 - *Appearance of fiber-matrix interfacial regions of Hi-Nicalon PyC SiC^{CVI} after annealing for 3600s at the indicated temperature following ion irradiation.*

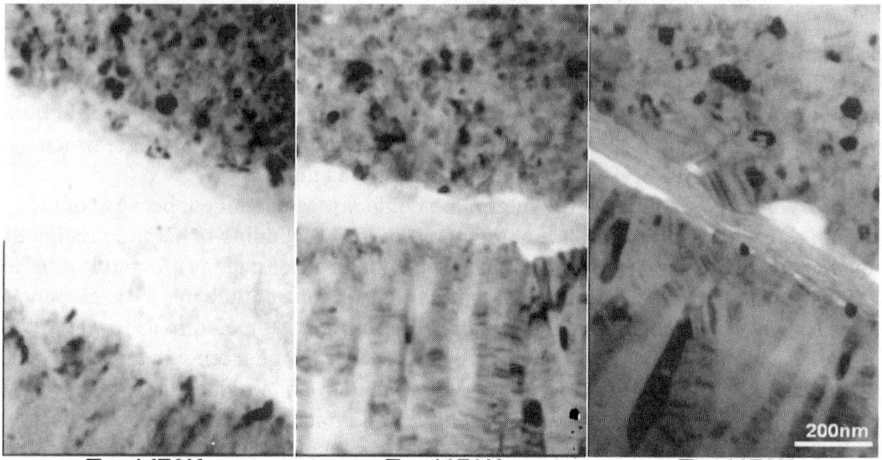

T = 1473K T = 1673K T = 1873K

Figure 6 - *Appearance of fiber-matrix interfacial regions of Hi-Nicalon Type-S PyC SiC^{CVI} after annealing for 3600s at the indicated temperature following irradiation.*

and matrix also shrunk significantly, as clearly demonstrated by morphological changes in the fiber-interphase boundary and the matrix-interphase boundary in Fig.2(B). The surface step between fiber and matrix is not necessarily representing a differential densification, because of potential polishing differences during specimen preparation.

Unirradiated and irradiated bright and dark field images and selected area diffraction (SAD) patterns taken from the PCS-derived SiC matrix are compared in Fig. 4. The dark field images were obtained by picking up a part of beta-SiC <111> diffraction ring or corresponding diffraction angle in case the ring was not clearly observed. Crystallites as large as several nm in diameters were detected at low number density following single-beam irradiation, while none was confirmed in the unirradiated and dual-beam irradiated matrices with a conventional TEM. The set of SAD patterns tells that the original almost totally amorphous matrix tends to crystallize by irradiation and that the co-implanted helium retards the crystallization. It could be concluded that the irradiation-induced densification was due to crystallization of the amorphous structure of PCS-derived SiC matrix.

In Tyranno-SA(U/D)/SiC^{PVS}, the matrix pyrolyzed at 2073K consisted of beta-SiC grains with diameters up to about 1000nm and a small fraction of excess carbon. The volume fraction of crystalline SiC was measured to be 81 percent. The excess carbon was present, mixed with matrix pores, within the intergranular openings. The structure of carbon phase was mostly amorphous, while a small amount of graphitic carbon was detected by an SAD analysis. Effect of irradiation was not apparent either in matrix or in fibers. A very smooth irradiated surface across the F-M boundary confirms that no significant differential volume change had been induced by irradiation.

A post-irradiation annealing experiment was performed using the Hi-Nicalon(U/D)/PyC/SiC^{CVI} and Hi-Nicalon Type-S(U/D)/PyC/SiC^{CVI} composites. A precipitation of implanted helium as bubbles was observed in the matrix at temperatures over 1473K. Most of the bubbles were observed on grain boundaries and triple points in non-spherical irregular shapes. This leads to an interpretation that along-range intragranular migration, that might be associated with trapping/detrapping at vacancies, is determining the condition for helium precipitation and the helium atoms once reached to grain boundaries may rather easily migrate along the boundaries to precipitate bubbles. No helium precipitation was observed interior Hi-Nicalon and Type-S fibers, which consist of fine beta-SiC crystallites and a glassy Si-C-O matrix.

The evolution of helium at the interphase could not be examined, because of the substantial mechanical disturbance at temperatures corresponding to helium precipitation in the matrix. Fig. 5 shows the appearance of F-M interfacial regions after annealing for 3600s at the indicated temperature. A significant disturbance of the pyrolytic carbon interphase was observed after post-irradiation annealing at temperature as low as 1073K in the Hi-Nicalon composite. An annealing at 1273K or a higher temperature, at which Hi-Nicalon composites with CVI matrices usually retain its mechanical properties in an unirradiated condition, resulted in a disruption of interfacial structure. The Hi-Nicalon Type-S composite exhibited a superior tolerance to the heat treatment, where the interphase integrity was maintained at temperatures up to 1473K as presented in Fig. 6. Therefore, it is likely to be related with the chemical stability of fibers and the irradiation effect on it. A closer look at the fiber-interphase and matrix-interphase interfacial structures revealed that the disturbance had been caused by intrusion of the growing SiC crystallites on the interfaces toward the carbon interphase. The crystallite size varied depending on the fiber grade and annealing temperature, and generally larger on the fiber side than on the matrix side of the interphase. The observation of SiC crystallite growth from the fully crystalline stoichiomeric SiC matrix suggests the reaction of evolving silicon with the interphase carbon. The enhanced crystallite growth on the fiber side and its dependence on the fiber grade suggest that the major part of evolving silicon comes from the glassy Si-C-O matrix of the fibers.

Conclusion

SiC/SiC composites reinforced with several kinds of SiC fibers and matrix-densified by CVI and PIP processes were subjected to microstructural examination following dual-beam ion irradiation to 10dpa at 873K and 60appmHe/dpa.

Advanced SiC fibers, namely Hi-Nicalon Type-S and Tyranno-SA, did not develop any detectable irradiation effects, while major microstructural modifications and a consequent densification took place in Hi-Nicalon and Tyranno-TE, respectively.

It was confirmed that a CVI-produced matrix were stable under irradiation. A PIP appeared to qualify as a radiation-resistant matrix consolidation process when pyrolyzed at temperatures high enough to allow a total crystallization of SiC. A conventional PIP matrix suffered from an irradiation-induced crystallization but the dual-beam irradiation retarded it.

Pyrolytic carbon, introduced as a fiber-matrix interphase, exhibited an irradiation-induced amorphization, which appeared to be enhanced by the dual-beam irradiation.

Post-irradiation annealing caused a loss of fiber-matrix interphase integrity of the CVI composites at temperatures lower than expected. The composite with near-stoichiometry SiC fibers exhibited a superior stability.

Acknowledgment

This work is supported by CREST-ACE program for advanced energy material systems. The CVI-processing was performed by Drs. H. Araki and T. Noda at the National Research Institute of Metals. The dual-beam ion irradiation experiment was carried out with the assistance of Drs. H. Shibata and T. Iwai at the Research Center for Nuclear Science and Technology, the University of Tokyo. The authors are grateful to these people for their valuable help.

References

[1] Fenici, P., Frias Rebelo, A. J., Jones, R. H., Kohyama, A., and Snead, L. L., "Current Status of SiC/SiC Composite R&D," *Journal of Nuclear Materials*, 1998, 258-263, pp.215-225.

[2] Kohyama, A., Katoh, Y., Hinoki, T., Zhang, W., and Kotani, M., "Progress in the Development of SiC/SiC Composites for Advanced Energy Systems: CREST-ACE Program," *Proceedings of the 8th European Conference on Composite Materials*, 1998, 4, pp.15-22.

[3] Nakayasu, T., Sato, M., Yamamura, T., Okamura, K., Katoh, Y., and Kohyama, A., "Ceramic Matrix Composites for Improved Thermal Conductivity and Reduced Radioactivation," *Ceramic Engineering and Science Proceedings*, 1999, 20, pp.301-308.

[4] Katoh, Y., Hinoki, T., Kohyama, A., Shibayama, T., and Takahashi, H., "Microstructural Stability of SiC/SiC Composites Under Dual-Beam Ion Irradiation," *Ceramic Engineering and Science Proceedings*, 1999, 20, pp.325-332.

[5] Snead, L. L., Katoh, Y., Kohyama, A., Shinavski, R., Williams, A. M., Bailey, J. L., Vaughn, N. L., and Lowden, R. A., "Evaluation of Neutron Irradiated Near-Stoichiometric Silicon Carbide Fiber Composites," *Journal of Nuclear Materials*, 2000, 283-287, pp.551-555.

[6] Price, R. J., "Properties of Silicon Carbide for Nuclear Fuel Particle Coatings," *Nuclear Technology*, 1979, 35, pp.320-336.

[7] Kishimoto, H., Katoh, Y., Kohyama, A., and Ando, M., "The Influence of Temperature, Fluence, Dose Rate and Helium Production on Defect Accumulation and Swelling in Silicon Carbide," presented at ASTM 20th Symposium on Effects of Radiation on Materials, June 6-8, 2000, Williamsburg.

[8] Snead, L. L., Osborne, M. C., and More, K. L., "Effect of Radiation on SiC-based Nicalon Fibers," *Journal of Materials Research*, 1995, 10, pp.736-747.

[9] Hasegawa, A., Youngblood, G. E., and Jones, R. H., "Effect of Irradiation on Microstructure of Nicalon Fibers," *Journal of Nuclear Materials*, 1996, 231, pp.245-248.

[10] Snead, L. L., and Zinkle, S. J., "Threshold Irradiation Dose for Amorphization of Silicon Carbide," *Materials Research Society Symposium Proceedings*, 1997, 439, pp.595-606.

[11] Snead, L. L., Zinkle, S. J., and Steiner, D., "Radiation Induced Microstructure and Mechanical Property Evolution of SiC/C/SiC Composite Materials," *Journal of Nuclear Materials*, 1992, 191-194, pp.560-565.

[12] Scholz, R.,, "Deuteron Irradiation Creep of Chemically Vapor Deposited Silicon Carbide Fibers," *Journal of Nuclear Materials*, 1998, 254, pp.74-77.

[13] Ishikawa, T., Kohtoku, Y., Kumagawa, K., Yamamura, T., and Nagasawa, T., "High-Strength Alkali-Resistant Sintered SiC Fibre Stable to 2,200°C," *Nature*, 1998, 391 [3669], pp.773-774.

[14] Price, R. J., "Neutron Irradiation Induced-Voids in β-Silicon Carbide," *Journal of Nuclear Materials*, 1973, 48, pp.47-57.

[15] Yano, T., and Iseki, T., "High-Resolution Electron Microscopy of Neutron-Irradiation-Induced Dislocations in SiC," *Philosophical Magazine*, 1990, A62, pp.421-430.

[16] Itoh, H., Yoshikawa, M., Nashiyama, I., Misawa, S., Okumura, H., and Yoshida, S., "Radiation Induced Defects in CVD-Grown 3C-SiC," *IEEE Transactions on Nuclear Science*, 1990, 37, pp.1732-1738.

[17] Takeda, M., Sakamoto, J., Saeki, A., and Ichikawa, H., "Mechanical and Structural Analysis of Silicon Carbide Fiber Hi-Nicalon Type S," *Ceramic Engineering & Science Proceedings*, 1996, 17, pp.35-42.

[18] Ishikawa, T., Kajii, S., and Hisayuki, T., "High Heat-Resistant SiC-Polycrystalline Fiber and Its Fiber-Bonded Ceramic," *Ceramic Engineering and Science Proceedings*, 2000, 21, pp.323-330.

[19] Kohyama, A., Kotani, M., Katoh, Y., Nakayasu, T., Sato, M., Yamamura, T., and Okamura, K., "High Performance SiC/SiC Composites for Fusion by New Precursor Polymers and Improved PIP Process," *Journal of Nuclear Materials*, 2000, 283-287, pp.565-569.

[20] Katoh, Y., Kohyama, A., Yang, W., Hinoki, T., Yamada, R., Suyama, S., Ito, M., Tachikawa, N., Sato, M. and Yamamura, T., "SiC/SiC Thermo-Physical Properties," presented at *International Town Meeting on SiC/SiC Design and Material Issues for Fusion Systems*, January 18-19, 2000, Oak Ridge, http://aries.ucsd.edu/PUBLIC/SiCSiC/Katoh2.pdf.

[21] Kohno, Y., Asano, K., Kohyama, A., Hasegawa, K., and Igata, N., "New Dual-Ion Irradiation Station at the University of Tokyo," *Journal of Nuclear Materials*, 1986, 141-143, pp.794-798.

[22] Ziegler, J. F., "Ion Beam Interactions with Matter," http://www.research.ibm.com/ionbeams/

[23] Heinisch, H. L., "Analysis and Recommendations on DPA Calculations in SiC", *Proceedings of the Second JUPITER/IEA Joint International Symposium on SiC/SiC Ceramic Composites for Fusion Structural Applications*, Kohyama, A., Jones, R. H., and Fenici, P., Eds., 1997, Japanese Society of Materials for Advanced Energy Systems.

[24] Katoh, Y., Kohyama, A., and Hinoki, T., "Microstructural Evolution of SiC/SiC Composite under Irradiation," *Proceedings of the 8th European Conference on Composite Materials Scinence -Technologies and applications*, 1998, 4, pp.351-357.

[25] Burchell, T. D., Eatherly, W. P., and Strizak, J. P., "The Effect of Neutron Irradiation on the Structure and Properties of Carbon-Carbon Composite Materials," *Effects of Radiation on* Materials, ASTM STP 1175, Kumar, A. S., Gelles, D. S., Nanstad, R. K., and Little, E. A., Eds, American Society for Testing and Materials, West Conshohocken, PA, 1993, pp.1266-1282.

Lorenzo Malerba,[1] J. Manuel Perlado,[1] Isabel Pastor,[1] and Tomás Díaz de la Rubia[2]

Molecular Dynamics Simulation of Radiation Damage Production in Cubic Silicon Carbide

Reference: Malerba, L., Perlado, J. M., Pastor, I., and Díaz de la Rubia, T., **"Molecular Dynamics Simulation of Radiation Damage Production in Cubic Silicon Carbide,"** *Effects of Radiation on Materials: 20th International Symposium, ASTM STP 1405,* S. T. Rosinski, M. L. Grossbeck, T. R. Allen, and A. S. Kumar, Eds., American Society for Testing and Materials, West Conshohocken, PA, 2001.

Abstract: Silicon carbide (SiC) is a candidate material for nuclear fusion reactor blankets; hence the importance of investigating its response to irradiation. Molecular dynamics (MD) simulations are a powerful tool to study radiation-damage production from the microscopic standpoint. Results of displacement-cascade MD simulations, conducted using the Tersoff potential to describe the interatomic forces, are presented herein. The number of point-defects produced in the material by silicon- (Si) and carbon- (C) primary knock-on atoms (PKAs) of increasing energy (between 0.25 and, respectively, 8 and 4 keV) is studied systematically. By comparison with standard theoretical models, threshold-displacement-energy (TDE) values of practical usefulness for SiC are derived. The effect of irradiation temperature is also allowed for. Qualitatively, the C sublattice turns out to be more heavily damaged than the Si sublattice. The effect of the irradiation temperature becomes visible only above ≈2000 K.

Keywords: silicon carbide, radiation effects, displacement cascades, point-defects, molecular dynamics simulation

Introduction

Silicon carbide (SiC), a low-activation, low-Z, high-temperature-resistant material, has long been considered for performing structural and protective functions in the first wall and blanket of future nuclear fusion reactors [*1*], both magnetic [*2-4*] and inertial [*5-7*], particularly in the form of SiC-fibre/SiC-matrix composite [*8*]. It is therefore paramount to achieve a deep knowledge of the effects of intense irradiation in this material.

---

[1] PhD student, professor and student, respectively, Instituto de Fusión Nuclear, ETSI Industriales, Universidad Politécnica de Madrid, José Gutiérrez Abascal 2, 28006 Madrid, Spain.

[2] Manager scientist, Chemistry and Materials Division, Lawrence Livermore National Laboratory, L-268, P.O. Box 808, Livermore, CA 94550, USA.

The computational atomistic simulations are today a powerful tool for the study of the different phases of radiation damage production and evolution [*9*], from the microscopic standpoint. More specifically, Molecular Dynamics (MD) has long been identified as the only technique that permits the investigation in detail of the characteristics of the displacement cascades [*10*], which are the true primordial phenomenon of radiation damage production. The short duration of these physical events (a few picoseconds) and their limited spatial extension (a few nanometres) prevent their analysis by direct experimental observation. Through experiments it is only possible to extrapolate information on displacement cascades by observing and measuring their secondary effects. On the contrary, MD, which is essentially an appropriate computational algorithm to solve the equations of motion for an ensemble of atoms, enables naturally the study of the evolution of a displacement cascade [*11*]. The remarkable advances in the knowledge of radiation effects in metals, achieved in the last decade, are mainly due to the impressive power of modern computers, that allow treating by MD a suitably large amount of atoms [*12*].

Relatively few MD studies of displacement cascade dynamics in SiC exist in the literature [*13-22*], all of them preliminary. The main qualitative features emerging from the published work can be synthesised as follows: (1) The duration of the thermal spike in SiC is very short and no evidence exists of any local phase-change to liquid, so that, for instance, no after-cascade amorphous-pocket formation has been observed, as opposed to pure silicon [*13,20*]; (2) Ion-beam mixing is also very limited during displacement cascades in SiC, most displaced atoms ending up at small distances from their initial positions [*13,16*]; (3) Most displaced atoms belong to the C sublattice, owing to both lower threshold displacement energy (TDE) for C atoms and to easier recombination for Si atoms [*13-22*]; (4) Antisites do form, but represent a minority of the total amount of defects [*13-22*].

In this paper, for the first time, some results of a systematic displacement cascade study in SiC are reported. A statistically meaningful collection of events has been gathered, in order to deduce quantitatively significant information on how the number of defects generated per cascade depends on recoil energy. By comparison with standard theoretical models, such as Norgett, Robinson and Torrens' (NRT [*23*]) or Coulter and Parkin's (CP [*24*]), TDE values for practical use have been inferred. Moreover, the effect of irradiation temperature has been analysed in certain detail.

Methodological Details

All simulations herein reported were carried out with the MD parallel program MDCASK, a code optimised for studying the interaction of high-energy ions with crystals [*25*]. The interatomic forces for SiC are described using the many-body, semiempirical potential proposed by Tersoff [*26,27*], merged with a binary *ab initio* repulsive potential, to adequately treat the atomic collisions in the short-distance range (> 0.5 Å) [*28,29*]. The Tersoff potential has been long demonstrated to allow the performance of reliable MD simulations of the behaviour of SiC in a variety of conditions and situations [*30-32*]. The MDCASK code is implemented in the Cray-T3E massively parallel supercomputer of the CIEMAT (Madrid, Spain), which can make available up to 32 processors for a single computational task. The parallel implementation is based on the PVM message-passing library.

The simulations were conducted in the canonical (*NVT*) ensemble, that is, maintaining number of particles (*N*), volume (*V*) and temperature (*T*) constant throughout the duration of the calculation. The temperature was controlled by using the velocity rescaling method, applied to the sole outer box-shell when the recoil atoms were introduced [*11*]. Simulations were performed at 20, 800, 1300, 1800 and 2300 K (SiC is a high-temperature applied material), though most of them were carried out at 300 K. The simulation-box size was chosen in order to guarantee that it contained the whole cascade, thereby avoiding unphysical energy or recoil-atom re-entering, as a result of the periodic boundary conditions. All simulations where the possibility of such re-entering existed were rejected. Globally, three box-sizes were used: $25a_0 \times 25a_0 \times 30a_0$ (150 000 atoms-8 processors), $40a_0 \times 40a_0 \times 40a_0$ (512 000 atoms-16 processors), $40a_0 \times 40a_0 \times 60a_0$ (768 000 atoms-32 processors), with $a_0 = 0.436$ nm. In most cases the smallest box was enough; for higher recoil energies it was necessary to turn to the intermediate box; the largest box, due to the long wait of the process in queue, was only used in extreme cases. All recoils were introduced from the upper part of the box and set in motion along a direction close to $[00\bar{1}]$. The highest recoil energy that could be simulated was 8 keV for Si PKA and 4 keV for C PKA, since on average the range of C recoils almost doubles the range of Si recoils of the same energy, as can be seen in Figure 1. In the same figure the recoil-energy data points produced in this study at 300 K are visible (except for the 8 keV, Si-PKA point). Each point is the result of an average of 5 displacement cascades, though for the very-high-energy range this number dropped to 2 or 3. It should be noticed that the PKA energy coincides with the damage energy, as the dissipation through electron excitation was not simulated.

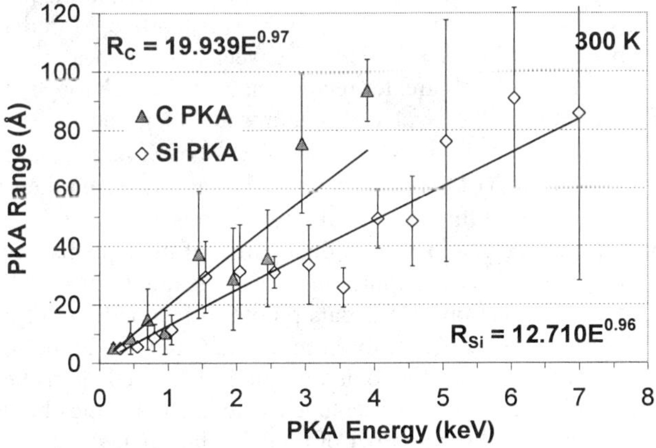

Figure 1 – *Range of Si and C PKAs vs. energy. Though both data points and error bars show high scattering, lineal interpolations are possible, as indicated.*

Each simulation consisted of 4 phases: (1) The system was equilibrated at the desired temperature (1000 timesteps = 1 ps); (2) After introducing the PKA, the cascade reached its peak and got to its end within 2000 timesteps (≤ 2 ps, adaptive timestep duration); (3) The system was relaxed for another 2000 timesteps (2 ps) at the simulation temperature, (4) The defects were counted and the system was visualised using adequate analysis and visualisation programs. The criteria adopted for counting the defects were as follows: (1) An atom is counted as *displaced* whenever it is found at a distance greater than the 1st neighbor distance ($d_1=\sqrt{3}/4a_0$) from its original position; (2) A *replacement* occurs whenever a displaced atom is found at a distance smaller than $d_1/2$ from a *vacancy* left by another displaced atom: if the vacancy corresponds to an equilibrium position for the type of atom displaced, then a *recombination* has taken place; if, on the contrary, the vacancy is of the opposite type, then an *antisite* has formed; (3) Any displaced atom found at a distance greater than $d_1/2$ from all vacancies is counted as an *interstitial* atom; the number of interstitials, that equals the number of vacancies, is referred to as number of *Frenkel pairs*. The count of atoms which, after being displaced, went back to their original position, was carried out separately.

Results

Recoil energy dependence

In Figure 2 the number of atomic displacements (Si atoms, C atoms and total) at the peak and at the end of the cascade, at 300 K, is plotted as a function of Si-recoil energy. For better legibility, the error bars (standard deviation) have been indicated only for the total amount. The lines are simply guides for the eye. The curves stemming from the study of C recoils (not shown) are essentially coincident, though limited to a smaller energy range. As anticipated in previous work (see Introduction), both at the peak and at the end of the cascade most displaced atoms belong to the C sublattice. Moreover, only on the Si sublattice a substantial difference is visible between cascade peak and end. This can be interpreted as a sign of both greater recombination distance for Si atoms and smaller average separation between the displaced Si atoms and their vacancies, compared to C atoms. On average, between ≈30% (in the case of Si recoils) and ≈45% (in the case of C recoils, not shown) of the peak-displaced Si atoms regain their initial position, in contrast to only ≈10% of the peak-displaced C atoms.

Despite the apparent scattering, all series of data points of the type displayed in Figure 2 can be interpolated with excellent approximation by straight lines. In Table 1 the linear laws used for the interpolations of the data points corresponding to final displacements (end of the cascade) are given, for both Si and C recoils. In the same table, the average per cent subdivision of the total amount of displaced atoms between the two sublattices is also indicated. These percentages, calculated as ratios between the number of Si or C displacements and the number of total displacements, are independent of the recoil energy, precisely because both the numerator and the denominator of such ratio depend linearly on the recoil energy. The peak displacements are not given, as they are not to be considered representative of the damage, since, as mentioned, they include a substantial portion of atoms that have not been actually displaced, but only temporarily removed from their position. In Table 1 the following notation is used: ν_{ij}^d is the number of displacements on the j sublattice induced by a recoil of type i; ν_i^d is the total number of displacements generated by a recoil of type i and E is the recoil energy in keV, coincident with the damage energy.

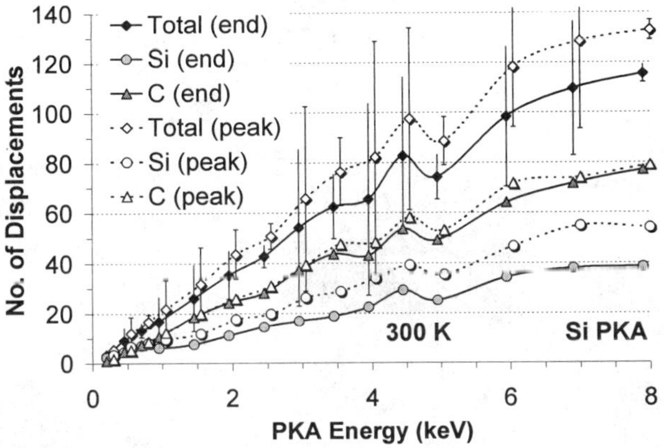

Figure 2 – *Number of atomic displacements per cascade (total, Si and C) as a function of Si-recoil energy, at 300 K.*

Table 1 – *Linear laws interpolating the MD data points for the number of displacements per cascade and relevant per cent subdivisions between the two sublattices, at 300 K.*

Displaced atoms	Si PKA	C PKA
Si atoms	$\nu^d_{Si,Si} = 1+5.20E$ 38.1±10.8 %	$\nu^d_{C,Si} = 4.65E$ 26.1±5.9 %
C atoms	$\nu^d_{Si,C} = 10.53E$ 61.9±10.8 %	$\nu^d_{C,C} = 1+10.87E$ 73.9±5.9 %
Total	$\nu^d_{Si} = 1+15.73E$	$\nu^d_{C} = 1+15.52E$

From Table 1 it can be seen that, although the total amount of atomic displacements is practically independent of the type of recoil, their subdivision is not. As it is, in the case of a C recoil only 1 out of 4 displaced atoms is a Si atom, whereas a Si recoil succeeds in distributing the damage more equally.

Not all of the displaced atoms given by the linear expressions of Table 1 correspond to actual defects. Part of them recombine, without contributing to the damage; a small portion recombine "wrongly," that is, create antisites; the rest of them remain out of place, forming Frenkel pairs. Vacancies, interstitials and antisites can appear in clusters, but this issue, though of fundamental importance, is not addressed in this paper. In Figure 3 the number of Frenkel pairs (Si, C and total) at 300 K is plotted as a function of Si-recoil energy. In Figure 4 the number of recombinations and antisites versus the same independent variable is given. The curves obtained from the study of C recoils (not shown) are very similar.

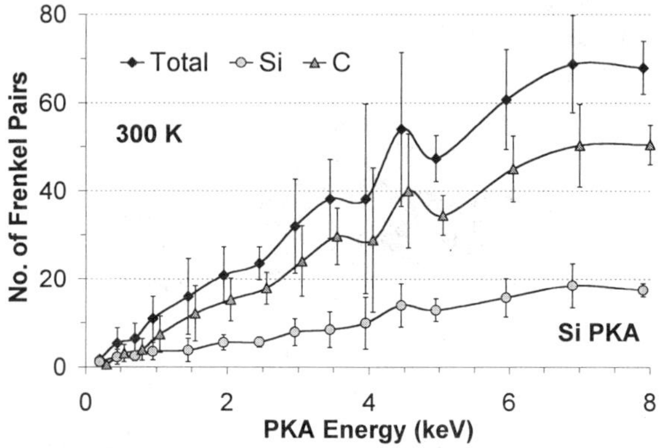

Figure 3 – *Number of Frenkel pairs per cascade (total, Si and C) as a function of Si-recoil energy, at 300 K.*

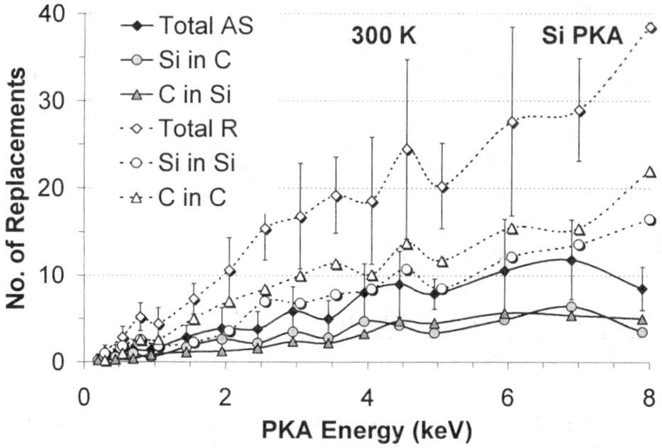

Figure 4 – *Number of replacements per cascade (total antisites, Si_C and C_{Si}; total recombinations, Si_{Si} and C_C) as a function of Si-recoil energy, at 300 K.*

Table 2 is analogous to Table 1, but refers to the number of Frenkel pairs per cascade. In Table 3, similar data are given for the number of replacements per cascade; however, in this case the exponents of E were used as fitting parameters for the interpolations, instead of using straight lines.

Table 2 – *Linear laws interpolating the MD data points for the number of Frenkel pairs per cascade and relevant per cent subdivisions between the two sublattices, at 300 K.*

Frenkel Pairs	Si PKA	C PKA
Si sublattice	$\nu^{FP}_{Si,Si} = 1+2.33E$ 30.6±11.0 %	$\nu^{FP}_{C,Si} = 2.38E$ 20.2±5.6 %
C sublattice	$\nu^{FP}_{Si,C} = 7.22E$ 69.4±11.0 %	$\nu^{FP}_{C,C} = 1+7.59E$ 79.8±5.6 %
Total	$\nu^{FP}_{Si} = 1+9.55E$	$\nu^{FP}_{C} = 1+9.97E$

Table 3 – *Free-exponent laws interpolating the MD data points for the number of replacements per cascade (antisites and recombinations) and relevant per cent subdivisions between the two sublattices, at 300 K.*

Replacements	Si PKA	C PKA
Si_C	$\nu^{Si\text{-}C}_{Si} = 0.61E^{0.96}$ 49.0±7.7 %	$\nu^{Si\text{-}C}_{C} = 0.97E^{0.91}$ 54.7±9.5 %
C_{Si}	$\nu^{C\text{-}Si}_{Si} = 0.64E^{0.94}$ 51.0±7.7 %	$\nu^{C\text{-}Si}_{C} = 0.70E^{1.06}$ 45.3±9.5 %
Total Antisites	$\nu^{AS}_{Si} = 1.26E^{0.95}$	$\nu^{AS}_{C} = 1.71E^{0.97}$
Si_{Si}	$\nu^{Si\text{-}Si}_{Si} = 1.19E^{1.27}$ 33.4±11.3 %	$\nu^{Si\text{-}Si}_{C} = 0.97E^{0.81}$ 46.5±12.2 %
C_C	$\nu^{C\text{-}C}_{Si} = 2.81E^{0.83}$ 66.6±11.3 %	$\nu^{C\text{-}C}_{C} = 2.25E^{1.18}$ 53.5±12.2 %
Total Recombinations	$\nu^{R}_{Si} = 4.16E^{0.94}$	$\nu^{R}_{C} = 5.30E^{0.95}$

Table 2 shows that, as with the atomic displacements, though the total amount of Frenkel pairs is substantially independent of the type of recoil, their subdivision is not. Indeed, only 1 out of 5 interstitials is a Si atom when the recoil is a C, Si recoils succeeding in distributing a little more equally the damage.

Both Figure 4 and Table 3, if compared with the other figures and tables, state clearly that antisites represent a reduced minority of the total number of defects. However, their (slow) growth against recoil energy is practically linear, as revealed by the exponents of E in the upper part of Table 3. Moreover, the two types of antisites appear to be produced with basically the same probability.

As to recombinations, they always represent between ≈25% (C recoils) and ≈30% (Si recoils) of the total amount of displaced atoms. In addition, though no relevant figure is shown here, it should be stressed that, by comparing the number of recombinations and displacements separately on the Si and C sublattice, it turns out that the recombination fraction on the former is almost ≈1.5 times the same fraction on the latter. This is a sign of the greater ease with which displaced Si atoms recombine, compared to displaced C atoms. Finally, it should be noted that the slight deviation from the linear behavior exhibited by the number of Si_{Si} and C_C recombinations (exponents of E in the lower part of Table 3) is only apparent. Indeed, the number of recombinations is seen to be proportional to the number of displacements, which are interpolated by a linear law of the type $\delta_{ij}+aE$, where δ_{ij} is 1 if $i=j$ and 0 if $i \neq j$. The absence of any addendum substituting δ_{ij} in the free-exponent laws used to interpolate the number of recombinations causes the exponents in such analytical expressions of Table 3 to be different from 1, as they have to "make up for" the lack of the mentioned addendum.

Irradiation temperature dependence

In Figure 5 the curves representing the total number of defects per cascade (including recombinations) against irradiation temperature are plotted. Each point is an average of at least 5 cascades induced by a 3-keV Si recoil. Error bars correspond to standard deviation. Lines are simply guides for the eye.

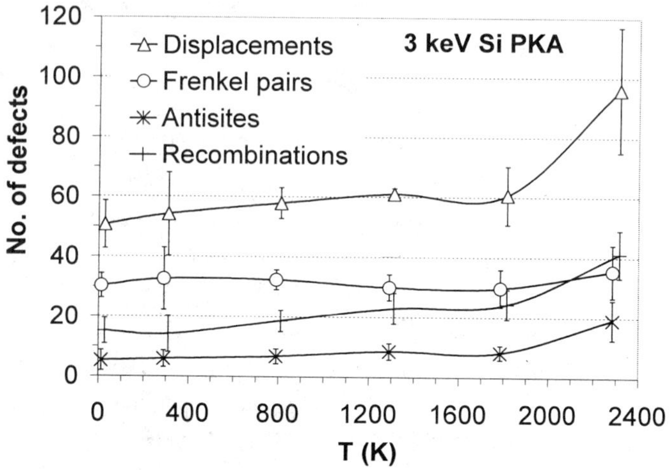

Figure 5 – *Number of defects per cascade (total) as a function of irradiation temperature (3keV Si recoils).*

It can be seen that in a wide range of temperature, spanning form 0 K to ≈2000 K, the number of defects per cascade is substantially constant, except for a slow increase in the number of displacements and replacements. Above ≈2000 K, suddenly there appears to be a substantial increase in the number of displacements, a sign of a diminution in the TDEs. Nonetheless, since this rise is accompanied by a growth in the number of recombinations and, also, antisites per cascade, the net effect is that the total amount of Frenkel pairs generated per cascade remains constant for all temperatures. This behavior is particularly interesting for practical purposes, as it will be seen in the next section.

Discussion

The fact that the number of displacements, replacements and Frenkel pairs per cascade in SiC is a linear function of the recoil energy can be used to make useful predictions. For instance, if the number of displacements is known, then the portions of recombinations, antisites and Frenkel pairs can be estimated, by employing fixed coefficients, valid below 2000 K, obtainable from the data of Tables 1, 2 and 3. As an example, on average it has been seen that the total number of displacements is about 1.6 times the total number of Frenkel pairs. Similarly, the *average number of displacements per unit length and per ion* can be estimated by dividing the total number of displacements by the PKA range (Figure 1), thereby obtaining a constant value of $\approx 1.1 \cdot 10^8$ displ.$\cdot$cm^{-1}/ion (or, applying the factor 1.6, $\approx 0.7 \cdot 10^8$ Frenkel-pairs$\cdot$cm^{-1}/ion).

Another useful application of the linear behavior versus damage energy exhibited by the number of displacements and Frenkel pairs per cascade is that it permits a direct comparison with linear theoretical models such as NRT [*23*] and CP [*24*]. These models, in their simplest form, can be synthesised by the following expressions:

$$\text{NRT} \qquad \nu_i^{d}(E) = 1 + \frac{0.8E}{2E_d} L_i(E) \qquad (1)$$

$$\text{CP:} \qquad \nu_{ij}^{d}(E) = \delta_{ij} + \frac{\gamma_j E}{2E_d^j} L_{ij}(E) \qquad (2)$$

where:

ν_i^d, ν_{ij}^d = respectively, the number of total and j displacements per cascade induced by a recoil of type i;
E – damage energy;
E_d, E_d^j = respectively, the average and the j-species TDE;
δ_{ij} = Kronecker delta (=1 if i=j; =0 if $i \neq j$);
γ_j = j-species-concentration-dependent coefficient (in the present case, $\gamma_j = 0.8c_j = 0.4$, since $c_j = 0.5\ \forall\, j$ in SiC); and
$L_i(E)$, $L_{ij}(E)$ = *damage functions*, which take into account the departure from the linear behavior; in the present case they can be set equal to unity.

Either by comparing the slopes of equations (1) and (2) with the slopes of the corresponding equations in Tables 1 and 2, or by considering each data point of Figures 2 and 3 as the result of applying equations (1) and (2), it is possible to deduce a value for the TDEs. A similar approach was used by Devanathan and Weber, applied to a much smaller set of data [*33*].

If models (1) and (2) are compared with the number of final displacements per cascade (Figure 2, Table 1), then the TDE values thereby obtained are more liable to approach a realistic value. However, such a comparison is not totally justified from the theoretical point of view, as the simple binary NRT model does not include the count of recombinations, whereas MD final displacements do. On the other hand, comparing models (1) and (2) with the number of Frenkel pairs (Figure 3, Table 2), though theoretically more close to the spirit of the NRT model, will give TDE values supposedly higher than the real ones, as atoms which have been removed from their positions, but ended up occupying other vacancies (recombinations), are not taken into account. At any rate, both approaches were followed here and the results are given in Table 4, in which E_d denotes the threshold *displacement* energies and E_{FP} denotes the threshold *Frenkel pair formation* energies.

Table 4 – *Threshold* "displacement" *energies as deduced from the comparison between the results of the MD simulation and the theoretical models NRT and CP.*

Threshold energies (eV)	From *slope* comparison		From *data point* comparison		Average values
	Si PKA	C PKA	Si PKA	C PKA	
From final displacements					
E_d^{Si}	38.5	43.0	36.9±5.1	52.8±16.5	**≈43**
E_d^{C}	19.0	18.4	20.5±8.1	19.4±1.6	**≈19**
E_d	25.4.	25.8	25.7±4.6	27.9±2.8	**≈26**
From Frenkel pairs					
E_{FP}^{Si}	85.8	84.0	99.0±41.6	108.6±40.5	**≈94**
E_{FP}^{C}	27.7	26.0	31.5±14.3	27.0±2.5	**≈28**
E_{FP}	41.9	40.2	47.7±21.2	45.0±5.4	**≈44**

The TDE values for the C sublattice obtained from the different calculation methods utilised are quite homogeneous around the average value. The ≈19 eV found in the final-displacement comparison is very close to the 20 eV recommended by Zinkle and Kinoshita, on the basis of both experimental and computational data available from the literature [*34*]. On the other hand, the ≈28 eV encountered in the Frenkel-pair comparison do not differ so much from the previous value. The small difference is justified by the little per cent of recombinations on the C sublattice, so that the number of C Frenkel pairs is very close to the number of C displacements.

In the case of the Si sublattice the data are not as much homogeneous around the average value. One cause for the large uncertainty can be identified in the much higher Si-atom TDE *seen* by the C recoils. Another cause could be the existence of wide uncertainty bands around the value of the TDE for Si atoms in SiC along different crystallographic directions, as determined by MD in earlier work [*22*]. However, the

≈43 eV found on average in the final-displacement comparison do not differ much from the 40 eV recommended by Zinkle and Kinoshita [*34*]. On the contrary, the ≈94 eV encountered in the Frenkel-pair comparison appear unrealistically high, as a consequence of the significant percentage of recombinations on the Si sublattice.

Probably the most interesting values are the total ones, ≈26 eV for displacements and ≈44 eV for Frenkel pairs. As it is, these numbers, inserted in the simple linear NRT expression, provide a correct estimation of the total number of displacements and Frenkel pairs per cascade in SiC. According to Figure 5, the latter magnitude, is independent of irradiation temperature, whereas the former is not. Moreover, the total number of Frenkel pairs per cascade is a good index of damage level in the material. Hence, it is suggested here that the adoption of the straightforward linear NRT model with a value $E_d \approx$ 40-45 eV could allow an easy and reliable estimation of the total *dpa* (displacements per atom) level in SiC.

To conclude the discussion, it is important to mention that it is likely that the linear behaviour detected in this work is valid only for sufficiently low recoil-energies. By comparison with the behavior of metals, it is expected that a substantial deviation towards sublinearity will take place above a hitherto unknown recoil-energy value, > 8 keV. According to theoretical calculations of total displacement function, the number of displacements per cascade in SiC begins to saturate between 100 and 200 keV and only up to ≈100 keV can the linear approximation be considered roughly valid [*35*]. Although these results cannot be currently confirmed by MD simulation in SiC, because a multi-million-atom simulation box would be necessary for such purpose, they probably give an indication, however crude, of the limit of the approximation proposed in the present paper. Above ≈100-200 keV, the number of displacements per cascade could perhaps be considered constant and equal to the value extrapolated for ≈100 keV.

Conclusions

By systematic MD simulation of a statistically meaningful amount of displacement cascades in SiC, induced by both Si and C recoils, the following features have been deduced:

- The number of defects (atomic displacements, Frenkel pairs, antisites) produced within a displacement cascade in SiC is a linear function of the recoil energy (damage energy), at least below 10 keV and probably up to ≈100 keV. Whether the recoil is a Si atom or a C atom, the total number of defects does not vary significantly. However, the type of recoil does have an influence in determining how the total amount of defects of all sorts is distributed between the two sublattices.
- The C sublattice is always much more damaged than the Si sublattice: only 1 out 4 displaced atoms is a Si atom and only 1 out of 5 Frenkel pairs belong to the Si sublattice, when the recoil is a C atom. Si recoils succeed in distributing a little more equally the damage between the two sublattices. The higher damage level on the C sublattice should be ascribed to both lower threshold displacement energy for C atoms and greater ease in recombining for Si atoms.

- Antisites do form within displacement cascades in SiC, but they represent a minority of the population of defects (<10% of the displaced atoms). C_{Si} and Si_C antisites are generated with basically the same probability.
- Between 25% (C recoils) and 30% (Si recoils) of the atoms displaced in a cascade recombine within the cascade itself. The recombination fraction is about 1.5 times higher for the Si sublattice than for the C sublattice.
- The number of defects per cascade is seen to be insensitive to irradiation temperature in a wide range, spanning from 0 K to ≈2000 K. Above 2000 K the number of displacements and replacements increases. However, the number of Frenkel pairs is the same for all temperatures.
- By comparison of the MD results with theoretical models such as NRT and CP, an average TDE of ≈19 eV and ≈43 eV at 300 K is found for, respectively, C and Si atoms in SiC, in agreement with previous recommendations. However, it is suggested that it may be convenient to adopt a global value ≈40-45 eV to predict the total number of Frenkel pairs per cascade (independent of irradiation temperature) by using the simple linear NRT model. This scheme could be used for *dpa* calculations, at least for recoils of sufficiently low energy. Above ≈100-200 keV the number of displacements per cascade could be considered constant.

Acknowledgments

Work financed by the EU-Commission: Marie Curie Research Training-Grant, Fusion Programme, Contract no. ERB-5004-CT97-5002. Work also performed under the auspices of the US DoE by LLNL under contract W-7405-Eng-48.

References

[1] Rovner, L. H., and Hopkins, G. R., "Ceramic Materials for Fusion," *Nuclear Technology*, 1976, 29, pp. 274-302.

[2] Sharafat, S., Najmabadi, F., and Wong, C. P.C., "ARIES-I Fusion-Power-Core Engineering," *Fusion Engineering & Design*, 1991, 18, pp. 215-222.

[3] Pérez Ramírez, A. S., et al., "TAURO: a Ceramic Composite Structural Material Self-cooled Pb-17Li Breeder Blanket Concept," *Journal of Nuclear Materials*, 1996, 233-237, pp. 1257-1261.

[4] Ueda, S., et al. (DREAM Design Team), "A Fusion Power Reactor Concept Using SiC/SiC Composites," *Journal of Nuclear Materials*, 1998, 258-263, pp. 1589-1593.

[5] McDonnell Douglas Aerospace Team, "PROMETHEUS -L, PROMETHEUS -H: Inertial Fusion Energy Reactor Design Studies - Final Report," *DOE/ER-54101, MDC92E0008*, vols. I, II and III, March 1992.

[6] Badger, B., et al., "HIBALL: A Conceptual Heavy Ion Beam Driver Fusion Reactor Study", *Kernforschungszentrum Karlsruhe Report no. KfK-3202, Madison, Univ. of Wisconsin Fusion Technology Institute Report no. UWFDM-450*, June 1981.

[7] Perlado, J. M., Mima, K., Nakai, S., Alonso, E., Muñoz, E., and Sanz, J., "Activation Analysis and Materials Choice in the Laser Fusion Reactor KOYO," *Journal of Nuclear Materials*, 1996, 233-237, pp. 1523-1529.

[8] Hopkins, G. R., and Chin, J., "SiC-matrix/SiC-fibre Composite: a High-heat Flux, Low-activation, Structural Material," *Journal of Nuclear Materials*, 1986, 141-143, pp. 148-151.

[9] Díaz de la Rubia, T., et al., "Multiscale Modelling of Radiation Effects in FCC and BCC Metals," *Mathematics and Computation, Reactor Physics and Environmental Analysis in Nuclear Applications*, vol. 2, Aragonés, J. M., Ahnert, C., Cabellos, O., Eds., ISBN: 84-699-0943-6, Madrid (Spain), 1999, pp. 1129-1140.

[10] Gibson, J. B., Goland, A. N., Milgram, M., and Vineyard, G. H., "Dynamics of Radiation Damage," *Physical Review*, 1960, 120, pp. 1229-1253.

[11] Allen, M. P., and Tildesley, D. J, "Computer Simulation of Liquids," Clarendon Press, Oxford, 1987.

[12] Averback, R. S., and Díaz de la Rubia, T., "Displacement Damage in Irradiated Metals and Semiconductors," *Solid State Physics*, 1998, 51, pp. 281-402.

[13] Díaz de la Rubia, T., Caturla, M.-J., and Tobin, M., "Molecular Dynamics Studies of Radiation Effects in Silicon Carbide," *Materials Research Society Symposium Proceedings*, 1995, 373, pp. 555-566.

[14] Díaz de la Rubia, T., Perlado, J. M., and Tobin, M., "Radiation Effects in Silicon Carbide: High-energy Cascades and Damage Accumulation at High Temperature," *Journal of Nuclear Materials*, 1996, 233-237, pp. 1096-1101.

[15] Perlado, J. M., "Behaviour and Computer Simulation of SiC under Irradiation with Energetic Particles," *Journal of Nuclear Materials*, 1997, 251, pp. 98-106.

[16] Devanathan, R., Weber, W. J., and Díaz de la Rubia, T., "Atomistic Simulation of Defect Production in β-SiC," *Materials Research Society Symposium Proceedings*, 1998, 504, pp. 45-50.

[17] Devanathan, R., Weber, W. J., and Díaz de la Rubia, T., "Computer Simulation of a 10 keV Si Displacement Cascade in SiC," *Nuclear Instruments & Methods in Physics Research B*, 1998, 141, pp. 118-122.

[18] Hensel, H., and Urbassek, H. M., "Preferential Effects in Low-energy Si Bombardment of SiC," *Nuclear Instruments & Methods in Physics Research B*, 1998, 142, pp. 287-294.

[19] Perlado, J. M., Malerba, L., and Díaz de la Rubia, T., "MD Simulation of High-energy Cascades and Damage Accumulation in β-SiC in Inertial Fusion Conditions," *Fusion Technology*, 1998, 34, pp. 840-847.

[20] Perlado, J. M., Malerba, L., and Díaz de la Rubia, T., "Molecular Dynamics Simulation of Neutron Damage in β-SiC," *Materials Research Society Symposium Proceedings*, 1999, 540, pp. 171-176.

[21] Perlado, J. M., Malerba, L., Sánchez-Rubio, A., and Díaz de la Rubia, T., "Analysis of Displacement Cascades and Threshold Displacement Energies in β-SiC," *Journal of Nuclear Materials*, 2000, 276, pp. 235-242.

[22] Malerba, L., Perlado, J. M., Sánchez-Rubio, A., Pastor, I., Colombo, L., and Díaz de la Rubia, T., "Molecular Dynamics Simulation of Defect Production in Irradiated β-SiC," *Journal of Nuclear Materials*, 2000, in press.

[23] Norgett, M. J., Robinson, M. T., and Torrens, I. M., "A Proposed Method of Calculating Displacement Dose Rates," *Nuclear Engineering & Design*, 1975, 33, pp. 50-54.

[24] Parkin, D. M., Coulter, C. A., *Journal of Nuclear Materials*, 1983, 117, pp. 340.

[25] Díaz de la Rubia, T., and Guinan, M. W., "Progress in the Development of a Molecular Dynamics Code for High-energy Cascade Studies," *Journal of Nuclear Materials*, 1990, 174, pp.151-157.

[26] Tersoff, J., "Modelling Solid-state Chemistry: Interatomic Potentials for Multicomponent Systems," *Physical Review B*, 1989, 39, pp. 5566-5568.

[27] Tersoff, J., "Carbon Defects and Defect Reactions in Silicon," *Physical Review Letters*, 1990, 64, pp. 1757-1760.

[28] Nordlund, K., Runeberg, N. , and Sundholm, D., "Repulsive Interatomic Potentials Calculated Using Hartree-Fock and Density-functional Theory Methods," *Nuclear Instruments & Methods in Physics Research B*, 1997, 132, pp. 45-54.

[29] Devanathan, R., Weber, W. J., and Díaz de la Rubia, T., "Displacement Threshold Energies in β-SiC," *Journal of Nuclear Materials*, 1998, 253, pp. 47-52.

[30] Tang, M., and Yip, S., "Lattice Instability in β-SiC and Simulation of Brittle Fracture," *Journal of Applied Physics*, 1994, 76, pp. 2719-2725.

[31] Tang, M., and Yip, S., "Atomistic Simulation of Thermomechanical Properties of β-SiC," *Physical Review B*, 1995, 52, pp. 15150-15159.

[32] Huang, H. , Ghoniem, N. M., Wong, J. K., and Baskes, M. I., "Molecular Dynamics Determination of Defect Energetics in β-SiC Using Three Representative Empirical Potentials," *Modelling Simulation in Material Science and Engineering*, 1995, 3, pp. 615-627.

[33] Devanathan, R., Weber, W. J., "Displacement Energy Surface in 3C- and 6H-SiC," *Journal of Nuclear Materials*, 2000, 278, pp. 258-265.

[34] Zinkle, S. J., and Kinoshita, C., "Defect Production in Ceramics," *Journal of Nuclear Materials*, 1997, 251, pp. 200-217.

[35] Weber, W. J., Williford, R. E., Sickafus, K. E., "Total Displacement Function for SiC," *Journal of Nuclear Materials*, 1997, 244, pp. 205-211.

Oleg P. Maksimkin[1] and Maxim N. Gusev [1]

Influence of the Reactor and Cyclotron Irradiation on Energy Transformation during Plastic Deformation of Metal Materials

Reference: Maksimkin, O. P. and Gusev, M. N., **"Influence of the Reactor and Cyclotron Irradiation on Energy Transformation during Plastic Deformation of Metal Materials,"** *Effects of Radiation on Materials: 20th International Symposium, ASTM STP 1405,* S T. Rosinski, M. L. Grossbeck, T. R. Allen, and A. S. Kumar, Eds., American Society for Testing and Materials, West Conshohocken, PA, 2001.

Abstract: The processes of energy dissipation and accumulation were studied during plastic deformation. In addition, the characteristics of strength and ductility were obtained for pure metals and structural materials pre-irradiated by neutrons or alternatively by alpha-particles. The temperature dependencies were obtained and analyzed for both strength and ductility characteristics and stored energy. It was found, that for the studied materials in their unirradiated and/or irradiated conditions, the stored energy increase correlates with the applied stress increase, i.e. $E_S \sim \sigma^n$, where n = 2 at early stages of deformation. Due to radiation effects the material, ability to accumulate mechanical energy decreases to a great extent with irradiating fluence increase. At the same time, heat release is increasing during deformation process.

Keywords: metals and alloys, deformation, stored energy, irradiation, martensite, true strength, dissipation of heat

Introduction

The defect structure, formed and developing under irradiation and/or deformation of metal crystals, is a type of a "carrier" of the latent (stored) energy E_S. The quantity E_S is associated with variation in numerous significant (physical, mechanical, chemical) characteristics of metals and alloys, as well as in peculiarities of structural and phase transformations occurring in them [*1, 2*]. It is assumed that the latent energy values are necessary to know when dealing with many aspects of the theory of radiation-induced and fatigue damage, as well as predicting the lifetime for structural materials considered as candidates for fission/fusion reactors. Due to this, investigation of the process of energy absorption in metal materials under deformation is of theoretical and practical interest.

[1] Head of Division, Dr. Sc. (Solid State Physics), Professor and Post graduated student, respectively, Institute of Atomic Energy, National Nuclear Center, Almaty, 480082, Kazakhstan

The objective of this work is to study, by means of calorimetry, the regularities and characteristics of the processes of energy accumulation and heat release in the course of plastic deformation of irradiated metals and alloys.

Materials and Techniques of the Investigations

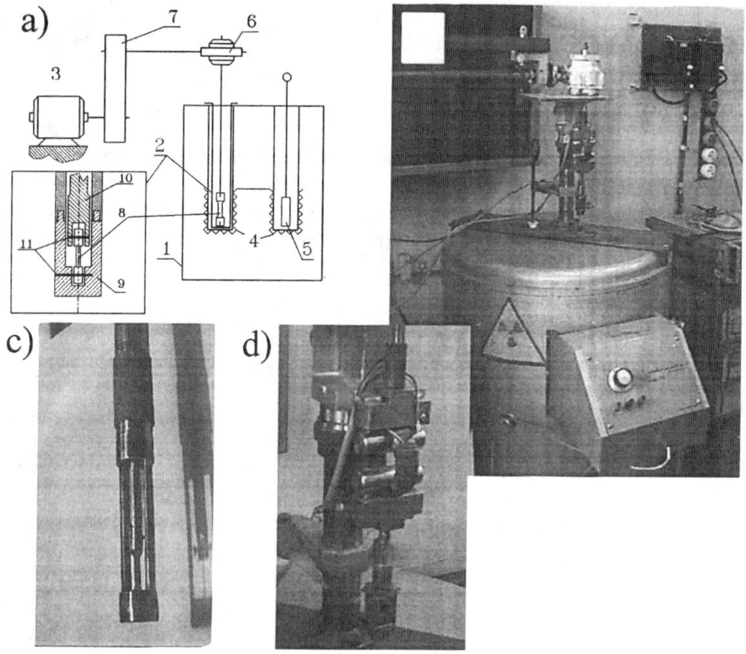

Figure 1. a) Schematic of the facility for study of energy accumulation and dissipation during deformation and failure of irradiated materials. 1 – calorimeter, 2 – assembly for deforming, 3 – power drive, 4 – measuring cells, 5 – sample-bystander, 6 and 7 – reduction gears, 8 – sample under investigation, 9 – receiver of the sample, 10 – tie-rod, 11 - pins for specimen fixing. b)Calorimetric unit. c) Exterior of the assembly for a tensile-test with a specimen fixed inside. d) The unit of load/elongation sensors.

Flat specimens of armco-iron, copper and the 12Cr18Ni10Ti steel with sizes of a working part 10x3.5x0.3 mm have been studied in mechanical tests. These were pre-annealed in vacuum not worse than $6.7 \cdot 10^{-3}$ Pa at 1125 K for 30min – for copper, at 1185K for 2h followed by cooling with the furnace – for iron, and at 1325K for 1h, followed by cooling in water – for the steel 12Cr18Ni10Ti.

The thermally treated specimens have been irradiated by neutrons in the WWR-K reactor core at the temperature not higher than 353K to the maximum fluence of $2 \cdot 10^{24}$ n/m^2 (E>0.1 MeV). Irradiation of the specimens by 50-MeV alpha particles, in view of obtaining uniform helium implantation throughout a specimen volume, has been executed at the isochronous cyclotron U–150 (NNC RK).

In view of the heat release investigation performed directly in a uniaxial tensile test, a special installation was designed. The installation includes a calorimetric unit with a micro-disruption machine, placed inside it (see Figure 1) [*3*]. The standard differential Calvet micro-calorimeter (the manufacturer is the French firm "Setaram") is used as a calorimetric block (Figure 1). The calorimeter sensitivity is to $2 \cdot 10^{-7}$ W, the operational temperature range is from the room temperature to 473 K.

Due to transit-time effects of calorimetric sensors, the typical primary thermogram, usually obtained in experiments, is strongly distorted, making impossible both recovery of the true behaviour of thermal processes accompanying the deformation and study of kinetics of energy accumulation in a material. In order to eliminate this disadvantage and to extract the true chain of events, thermogram reconstruction was produced by means of the regularisation technique by Tikhonov [*4*].

The installation makes possible simultaneous recording of both the tensile diagram in the co-ordinates *"force F – elongation* Δl*"* and the thermogram *"heat release rate* $\partial Q/\partial t$ *– time t"*. As a result, one can obtain the deformation dependence of the heat release $\int \partial Q/\partial t \cdot \partial t$ in experiments for every continuously strained specimen and, as a sequence, compare the effects of the deformation and the thermal processes in the same time scale. On calculating, from the tensile diagram, the work spent on variation in a material shape A ($A=\int Fdl$) versus the residual deformation ε, one can determine experimentally the stored energy $E_s=A-Q$ and the relative stored energy $P=E_s/A$, as well as their dependencies on applied stresses and strain values.

Results and Discussion

Influence of in-reactor irradiation, the extent of cold deformation and the test temperature on the latent energy and the heat release in steel 12Cr18Ni10Ti under deformation

In Figure 2 the typical "force–elongation" diagrams and the corresponding experimental thermograms are represented for unirradiated specimens of the steel 12Cr18Ni10Ti and those irradiated by neutrons to the fluence of $5 \cdot 10^{22}$ n/m^2. The characteristics of strength and ductility, calculated from the diagrams, as well as the values A, Q and E_s, related to the moment of failure, are presented in Table 1. In Figure 3 the A-, Q- and E_s -dependencies versus the strain value ε are shown.

As follows from the presented data, the steel is hardened as a result of irradiation, and the effect of embrittlement is small. Figure 3 demonstrates that the values of A and Q increase too, and E_s tends to saturation, as strain increases. At the same value of ε, the A and Q values are higher and E_s is somewhat lower for irradiated steel than for unirradiated one; i.e., a material capability to store energy in the course of tensile strain is lost as a result of irradiation.

Increase in the temperature from 293 to 473 K leads to reduction of both the strength and plasticity characteristics and to considerable reduction in the work A, the dissipated heat Q and the stored energy E_s.

Table 1. – *The mechanical and energy characteristics of unirradiated and neutron-irradiated steel 12Cr18Ni10Ti versus the test temperature*

T,K	State	Yield stress $\sigma_{0.2}$, MPa	Ultimate stress σ_B, MPa	Uniforme elongation $\varepsilon_{equal.}$, %	Total elongation ε_{total}, %	Work of deformation A, MJ/m^3	Released heat Q, MJ/m^3	Stored energy E_s, MJ/m^3	Relative stored energy P, %
293	Unirradiated	180	650	63	70	410	310	100	24
	After irradiation	440	700	48	60	406	333	73	18
473	Unirradiated	130	475	35	37	144	95	48	33
	After irradiation	320	500	27	33	162	115	47	29

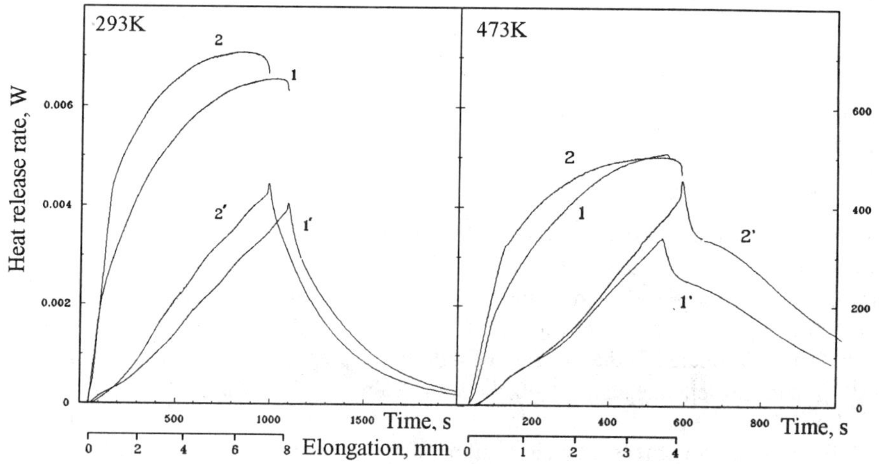

Figure 2. Typical machine «load-elongation» diagrams (1 and 2) and thermograms (1′ and 2′) for an unirradiated specimen 1 from the steel 12Cr18Ni10Ti and neutron-irradiated ones 2. Scales of time and elongation correspond.

One of the reasons for this variation in the properties of the steel 12Cr18Ni10Ti is the temperature suppression of the $\gamma \rightarrow \alpha'$ transformation. It has been found that the temperature close to 373 K is the critical point for the $\gamma \rightarrow \alpha'$ transformation, which isn't observed at the temperatures higher and lower, both in an unirradiated material and an irradiated one, subject to single-axial tension [*5*].

Another factor responsible for potential reduction in the mechanical and energy characteristics is the temperature variation in the stacking fault energy (SFE), which determines an extent of dislocation splitting: the higher the test temperature, the higher is the SFE in austenitic steel [*6*], and the less is the energy stored as a result of deformation.

It follows also from the data presented in Table 1 that the neutron irradiation leads to increase in the dissipated heat Q and reduction of E_s. The increase of Q in particular can be associated with the release of additional energy as a result of interaction between moving dislocations and radiation defects.

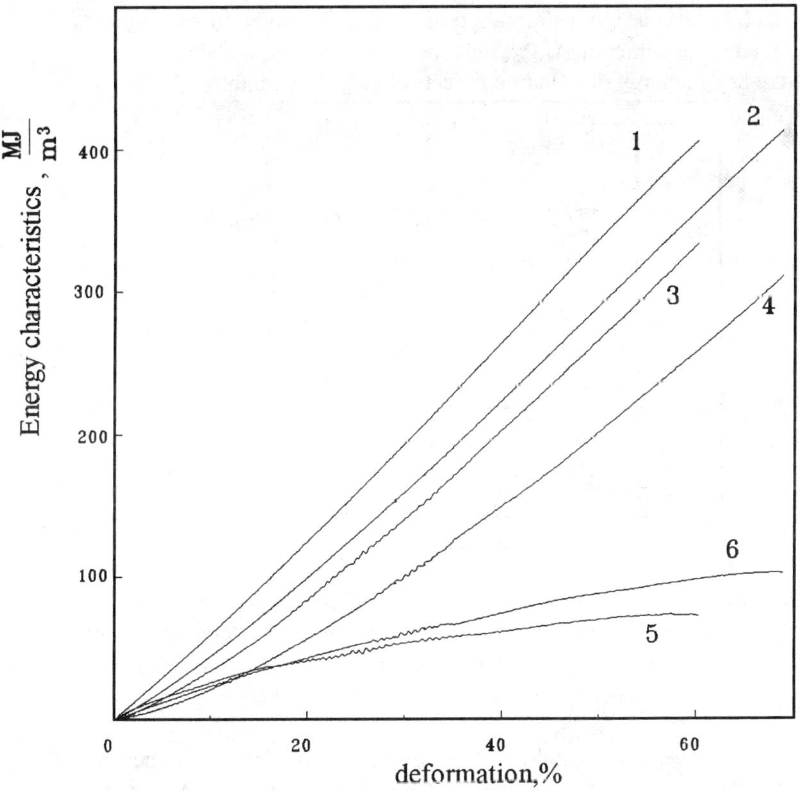

Figure 3 The deformation work A (1 and 2), the dissipated heat Q (3 and 4) and the latent energy Es (5,6) versus the deformation for unirradiated specimens (2,4,6) and irradiated ones (1,3,5) for steel 12Cr18Ni10Ti. Deformation is performed at 293 K.

Role of the latent energy in the martensitic $\gamma \rightarrow \alpha'$ transformation under deformation in the Fe–Cr–Ni irradiated stainless steel

In Reference [7] an attempt was undertaken to consider the phase transition $\gamma \rightarrow \alpha'$ in terms of energy and to represent accumulation of the α' phase in irradiated steel 12Cr18Ni10Ti before and after irradiation by neutrons as a certain function of the latent energy E_s. Earlier we have found for the 18–10-type steel [8] that the values of the relative energy $P=E_s/A$ stored during the deformation of a neutron-irradiated specimen are considerably lower than those of unirradiated specimens throughout the entire range of ε values. On the other hand, in Ref. [9] it was shown that for steel of the same type the critical value of deformation ε_{crit} related to the formation of α' phase also decreases as the fluence grows. These facts can be matched under assumption that a certain critical value of energy $U_{\gamma\rightarrow\alpha}$ exists (see.Fig.4), corresponding to elastic distortions, at which conditions are available for occurrence of the martensitic $\gamma \rightarrow \alpha'$ transition in steel. In this case, transformation of the energy, stored in the course of deformation, needed for achieving $U_{\gamma\rightarrow\alpha}$, must be less for an irradiated material, because in it the latent energy of radiation defects already is available.

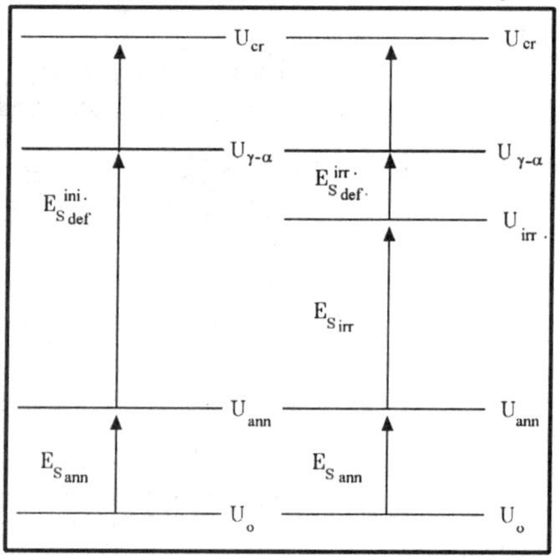

Figure 4. Change of the stored energy in steel after annealing, deformation and irradiation. U_0 – Level of internal energy of the material at T=0K, U_{ann} – internal energy in the material after annealing; U_{irr} – level of the internal energy, after irradiation, U_{cr} – critical level of the internal energy, on achievement of which material will be failured. $U_{\gamma\rightarrow\alpha}$ level of the internal energy, on achievement of which will start $\gamma \rightarrow \alpha'$ transformation.

To evaluate this concept, the in-calorimeter experiments on deforming the samples of the 12Cr18Ni10Ti steel before and after neutron irradiation have been carried out. The dependence of the stored energy in tension has been found with determination, for each case, of the value of $E_{s.def}$ absorbed irreversibly by a sample at the moment of start of the $\gamma \rightarrow \alpha'$ transformation. The absolute values of $E_{s.def}$ have been found to be equal to 60 MJ/м^3 for unirradiated ($Es^{ini}_{def.}$) steel, 40 and 8 MJ/м^3 – for the steel samples irradiated ($Es^{irr.}_{def.}$), respectively, with the fluences $5 \cdot 10^{18}$ and $2 \cdot 10^{19}$ n/cm^2 (E>0.1 MeV). Thus, it has been confirmed experimentally that, at other equal conditions, the value of $E_{s.def}$ reduces indeed as the fluence grows.

In particular in our case the following relations are valid:

$$U_{\gamma\to\alpha.} = U_0 + E_{s\,ann} + Es^{ini}_{def.} \qquad \text{for unirradiated steel} \quad (1)$$

$$U_{\gamma\to\alpha} = U_0 + E_{s\,ann} + E_{s.irr} + Es^{irr.}_{def.} \qquad \text{for irradiated steel} \quad (2),$$

where $E_{s.irr}$ energy of radiation defects, $E_{s.ann}$ value of stored energy after annealing.

We obtain from Eqs. (1) and (2):

$$E_{s\,irr} = Es^{ini}_{def} - Es^{irr}_{def} \qquad (3)$$

Thus, on deforming the unirradiated and irradiated specimens up to the moment of α' phase appearing, one can determine the value of the latent energy related purely to irradiation. On the other hand, with a known value of $E_{s\,irr.}$, one can estimate with ease the relation between the value of cold deformation and the integral dose of irradiation. For instance, it has been found that for the fluence values $5 \cdot 10^{22}$ and $2 \cdot 10^{23}$ n/m^2 the values of the cold work are equal to 5 and 19% respectively.

It follows also from Eq. (3) that, with the assumption that $Es^{irr.}_{def} = 0$, one can determine the fluence value related to direct $\gamma\to\alpha'$ transformation for a given material, occurring as a result of irradiation without extra deformation. On a base of the obtained dependence of latent energy ($E_{s.irr}$) versus the fluence, we can estimate the required integral irradiation dose, comprising ~10^{25} n/m^2 for the steel 12Cr18Ni10Ti.

Thus, relying on the obtained results, one can conclude that the process of the martensitic $\gamma\to\alpha'$ transformation in the unirradiated/neutron-irradiated austenitic stainless steel 12Cr18Ni10Ti subject to deformation correlates with the process of energy storing, and the value of the latent energy is a crucial parameter of the process. The amount of the α' phase obtained under deformation is determined by the amount of the latent energy: $M_f = C\,exp(mE_s)$ [8], where C, m – constant.

Peculiarities of the helium effect on energy accumulation/dissipation during deformation of armco-iron

Investigations of the armco-iron implanted up to the concentration ~10^{-3} at.% He have shown that the character of the dynamical strain ageing process (DSA) varies substantially. The start of DSA is shifted toward higher temperatures and higher deformations. In contrast to an unirradiated armco-iron, in a case of the helium implantation, DSA evolution doesn't lead to abrupt reduction of plasticity. This fact can be related, in particular, to peculiarities in accumulation of the mechanical energy under deformation. For the specimens implanted by helium, in the DSA temperature range the intensity of the Es growth is considerably lower than for unirradiated ones. Besides, it should be noted that in a material implanted by helium the dissipation process intensity *dQ*, in the course of deformation at 423 K (ε ranges from 6 to 25 %) is higher than the deformation work rate dA. This phenomenon, seemingly, is caused by interaction between dislocation assemblies and radiation defects.

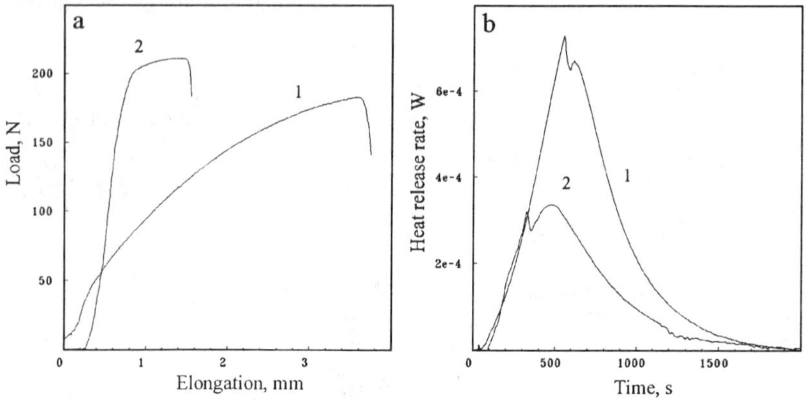

Figure 5 Tension diagrams (a) and heat release curves (b) obtained during deformation for copper (1 – the unirradiated specimen, 2 – the specimen subjected to neutron irradiation).

Peculiarities of heat release during deformation and failure for copper pre-irradiated with neutrons

The samples of the oxygen-free copper, both unirradiated and subjected to neutron irradiation in the WWR-K reactor up to the fluence $2{\cdot}10^{24}$ n/m^2, have been studied. The experimental curves obtained are presented in Figure 5, and the results obtained are given in Table 2.

Table 2 - *Properties of copper, initial and irradiated by neutron*

Condition	Yield stress $\sigma_{0.2}$ MPa	Ultimate stress σ_B MPa	Uniforme elongation $\varepsilon_{equal.}$ %	Total elongation ε_{total} %	Work of deformation A, MJ/m^3	Released heat Q, MJ/m^3	Stored energy Es, MJ/m^3	Relative stored energy P, %
Unirradiated	40	150	30	32	38	30	8	21
Irradiated, $2{\cdot}10^{24}$ n/m^2	115	202	8	9	16	10	6	37

It can be observed that irradiation has led to increases in $\sigma_{0,2}$ and σ_B and considerable reduction of plasticity. The values of A and Q for irradiated samples are also considerably lower. Failure of irradiated material occurs at E_s=6 MJ/m^3, being somewhat less than for the unirradiated material; however, growth of E_s with increase of ε in the neutron-irradiated copper occurs considerably faster. For instance, at ε=8 % $E_s^{irr.}$=6 MJ/m³, and E_s^{ini}=3.5 MJ/m³. One can assume that in copper the formation and the evolution of defects under irradiation have led to formation of the defect structure of the special type which absorbs the deformation energy stronger.

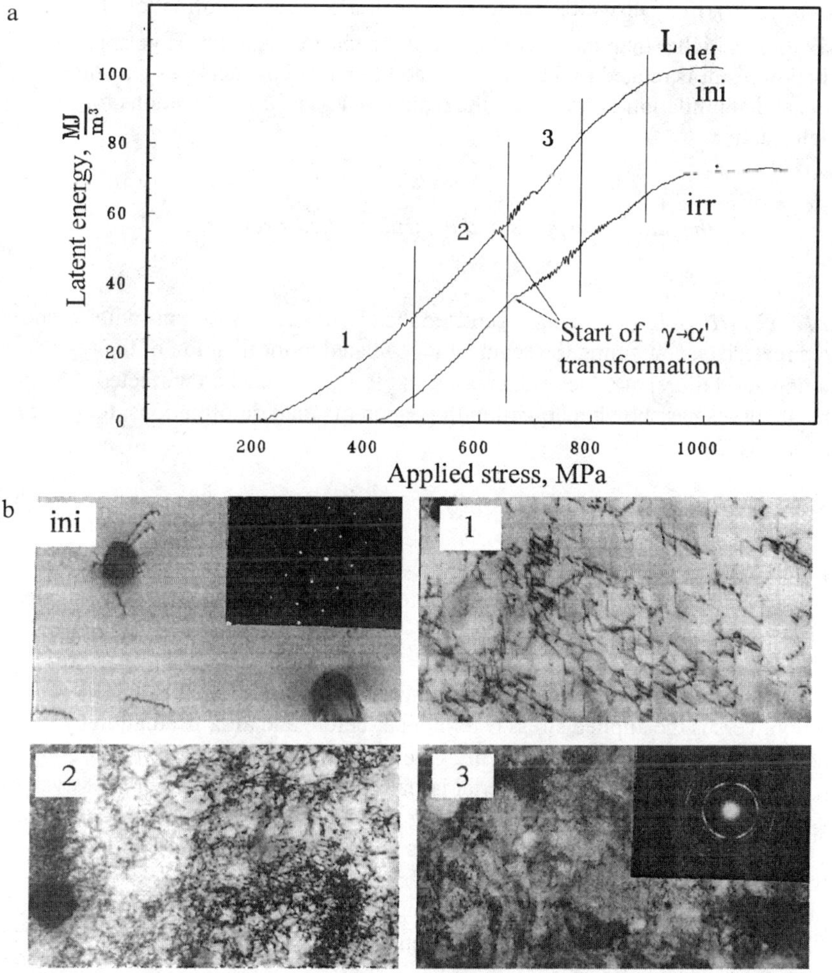

Figure 6 a) The latent energy and the applied stress for the steel 12Cr18Ni10Ti subjected to tensile strain at 293K. ini – unirradiated material, irr- material irradiated by neutron with fluence 5×10²² n/m². 1- a region where the relation (4) is valid, Ldef – a region of strain localization. b) Electron microscopic photographs of the deformed steel, × 25,000 ini – initial structure, 1- 3 – a dominant structure for the segments 1- 3 of the curve «σ-E_s». (1 – FCC-lattice, 3 – FCC+BCC)

It is of special interest to consider the phenomenon of relaxation in a deformed material, when after failure the process of heat release is observed for ~300s with intensity decreasing almost exponentially with time. And, during post-deformation calorimetry at 293K from the failure point till the moment of noticeable heat release stop, the irradiated copper releases, in average, $Q_{rel} \approx 8$ MJ/m^3, that is higher than for an unirradiated material and exceeds the value of latent energy in copper after irradiation ($E_s{}^{irr}_{def}$). In other words, the total thermal effect from deformation, failure and relaxation is higher than the total work A of the tensile machine. This, apparently, contradiction result is related to the release of additional latent energy as a result of migration and annihilation of a fraction the radiation defects on dislocations formed under deformation.

Relation between the latent energy Es and the true applied stress σ

Steel 12Cr18Ni10Ti. In Figure 6 are depicted the «E_s-σ» curves for unirradiated and irradiated materials, containing the segment L_{def}, related to localization of the deformation and failure, at which σ increases and E_s is, practically, unaffected. Analysis of curves has shown that, within the region of small deformations, they obey the relation like this:

$$E_s \sim E_{s0} + \alpha \sigma^2 \tag{4}$$

where α and E_{s0} are constants.

The «E_s-σ» curve for an irradiated material looks like that for the unirradiated one, but is shifted toward the region of lesser E_s by ~20 MJ/m^3.

Copper. Figure 7 shows the latent energy E_s, accumulated during plastic deformation, versus the applied stress σ for copper before and after irradiation by neutrons fluence at the fluence $2 \cdot 10^{24}$ n/m^2. One can observe that the curves are S-shaped. The curve of an irradiated material is shifted to the range of larger stress values and lesser values of E_s. Directly after achieving the yield strength, a parabolic relation between the latent energy and the applied stress takes place (see Figure 7). On achieving certain critical values of E_s and σ, a break occurs in the curve, showing the change in the character of material strengthening. The extent of the parabolic segment on the "E_s-σ" curve for irradiated copper is considerably lower than for unirradiated and failure of the specimen occurs at lesser values of E_s.

Armco-iron. For armco-iron, in contrast to copper and the 12Cr18Ni10Ti steel, a specific segment takes place on the «E_s–σ» curve at 293 K, where the Chernov-Luderse band appears and propagates on the specimen. It should be mentioned that increase in E_s occurs at, practically, unchanged value of σ. As a whole, with the exeption of this feature, the E_s-σ dependence for both primary armco-iron and helium-doped one to certain values of the stress and the latent energy is described rather satisfactorily by Eq. (4).

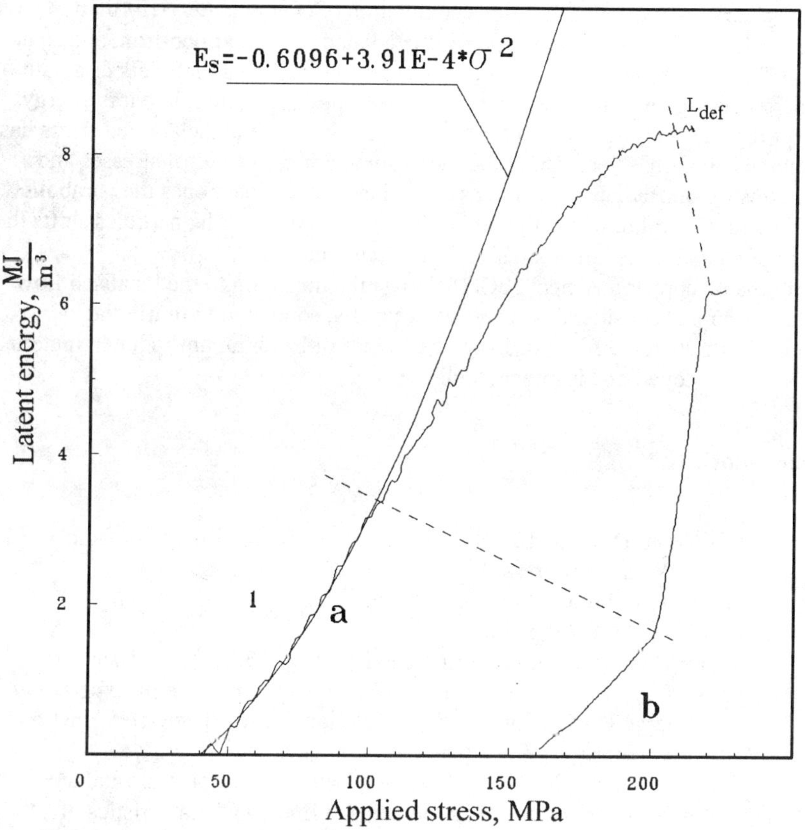

Figure 7. Relation between the applied stresses and the latent energy for copper deformed at 293 K. a – unirradiated material, b- copper irradiated by neutrons at 2×10^{24} n/m^2. The spot 1 is marked; approximating parabola corresponding to spot 1 is shown.

Thus, as for the relation between *Es* and σ, one can conclude the following. The relation like Eq. (1) is satisfied to a certain critical extent of deformation (stress) Then the square law for material hardening is changed. Irradiation either by neutrons or by alpha particles shifts the «E_S-σ» curves toward the larger stresses and lesser values of E_S. For copper and the steel 12Cr18Ni10Ti in a neutron-irradiated state transition to a localised flow and failure occurs at considerably lower values of Es, compared to primary samples. For iron, difference in the critical values of E_s for an unirradiated material and for a helium-doped one is small.

Conclusions

With application of the new calorimetric technique, the value of the stored energy and the regularities of dissipation and accumulation of energy under deformation have been determined for neutron- and α-particle irradiated copper, armco-iron, and steel Cr18Ni10Ti. It was found that in the general case, the irradiated materials dissipate more energy during deformation than unirradiated ones and accumulate less energy.

Except for this, a relationship between the stored energy and the applied stress has been obtained and it has been found that up to some critical values of stresses and deformations for unirradiated specimens as well as for irradiated ones the parabolic law of hardening is valid. Irradiation either by neutrons or by alpha particles shifts the «E_S-σ» curves toward the larger stresses and lesser values of E_S.

For irradiated copper and steel 12Cr18Ni10Ti the transition to the localised flow and failure occurs at considerably lower values of Es, compared to unirradiated specimens. For iron, the difference in critical values of E_s for an unirradiated material and for a helium-doped one is rather small.

Reference Block

[1] Bever, M. B., Holt, D. I., and Titchener, A. L. "The Stored Energy of Cold Work", Oxford etc., Pergamon Press, 1973, 192p.

[2] Panin, V. E. et al., - "Structural Levels of Plastic Deformation and Failure". //Novosibirsk, 1990, 256p. – in Russian

[3] Calvet, E., Prat H., "Microcalorimetrie". //Paris, 1956, 158p. – in French.

[4] Maksimkin, O. P., Astafiev, I. V., "Restoration of Thermogramm in Experiments for Study Energy Dissipation and Accumulation during Plastic Deformation" //*Industry laboratory,* 1, 1994, p.44.

[5] Maksimkin, O. P., Gusev, M. N., "Correlation Between Temperature-Induced Changes in Physicomechanical Properties and Energy Characteristics of Deformed 12Cr18Ni10Ti Steel" // *Physics of Metals and Metallography,* v.84, 1997, 3, pp. 299 - 301.

[6] Latanison, R. M., Ruffl, A. W., "The Temperature Dependence of Stacking Fault Energy Fe–Cr–Ni Alloy" //*Met.Trans*., v.2, N 2, 1971, pp. 505–509.

[7] Astafiev, I. V., Maksimkin, O. P., "About Role of Stored Energy in Martensitic Transformation During Deformation of Irradiated Stainless Fe-Cr-Ni-Steel"// *Physics of Metals and Metallography*, 77, 1994, N3, pp. 90 – 95.

[8] Bolotov, A. V., Ibragimov, Sh., Sh., and Maksimkin, O., P., "Dissipation and Accumulation of Energy during Deformation of Steel 12Cr18Ni10Ti Irradiated by Neutrons"// *Atomic Energy,* v.66, 1989, N6 pp.430-431.

[9] Ibragimov, Sh. Sh., Maksimkin, O. P., and Sadvakasov, D. H., "Martensitic transformation and mechanical properties of stainless steel 12Cr18Ni10Ti irradiated by neutron"// *Physics of Metals and Metallography*, 1990, N7, pp. 199 - 201.

Akihiko Kimura,[1] Takanori Hirose,[1] and Hideki Matsui[2]

Irradiation-Induced Amorphization and Its Recovery Behavior in Cold-Rolled and Aged Ti-Ni Shape Memory Alloys

Reference: Kimura, A., Hirose, T., and Matsui, H., **"Irradiation-Induced Amorphization and Its Recovery Behavior in Cold-Rolled and aged Ti-Ni Shape Memory Alloys,"** *Effects of Radiation on Materials: 20th International Symposium, ASTM STP 1405,* S. T. Rosinski, M. L. Grossbeck, T. R. Allen, and A. S. Kumar, Eds., American Society for Testing and Materials, West Conshohocken, PA, 2001.

Abstract: The materials used were Ti-50.0, 50.5 and 51.0at%Ni alloys which were cold rolled and aged at 400°C for 1 hr. Neutron irradiation was performed in the Japan Materials Test Reactor (JMTR) up to a dose of 1.2 x 10^{24}n/m^2 at cooling water temperature (about 60°C), which meant the specimens were irradiated in the parent phase (B2). After the irradiation, micro-Vickers hardness measurements, microstructure observations by transmission electron microscope (TEM) and positron annihilation spectrometry (PAS) measurements were carried out. Differential scanning calorimetry (DSC) measurements were taken at temperatures between -120°C and 120°C. After the irradiation, all the alloys show extremely high irradiation hardening: the estimated increase in the yield stress of Ti-50Ni alloy is almost 1GPa. Although the hardness of the Ti-Ni alloys depends on the alloy composition in the unirradiated condition, all the alloys reach the same hardness value after the irradiation. DSC measurements revealed that the martensitic transformation was completely suppressed by the irradiation in all the alloys. PAS study revealed that no structural vacancies existed in the alloys before the irradiation, and that vacancies were formed after the irradiation. Microstructural observations indicate that disordered regions, which were considered to be amorphous, were homogeneously distributed in the irradiated parent ordered phase, accompanied by an appearance of the halo ring on the diffraction pattern. The irradiation effects disappeared following a post-irradiation anneal at 250°C for 1 hr. The recovery of the martensitic transformation by post-irradiation annealing is attributed to the migration of vacancies that causes the reordering. It is expected that Ti-Ni alloys are potential self-restorative materials for the irradiation above 250°C where vacancies are mobile.

Keywords: shape memory effects, aged Ti-Ni alloys, neutron irradiation, martensitic transformation

[1]Professor and graduate student, IAE, Kyoto University, Gokasho, Uji 611-0011, Japan.
[2]Professor, IMR, Tohoku University, Sendai 980-0812, Japan.

Introduction

The application of shape memory alloys to nuclear fusion technology has been considered from the viewpoint of security of work in a radiation environment where demands for remote controlling systems for reactor maintenance have been rising. Irradiation effects on shape memory alloys have been studied by many workers. It was found in an in-situ transmission electron microscope (TEM) observation study that electron irradiation caused a broadening of the dislocation and grain boundary images, which was attributed to the preferential irradiation-induced transition from crystalline to amorphous at lattice defects [*1-3*]. The resistivity measurements and differential scanning calorimetry (DSC) technique were used to measure the transformation temperature of the electron and neutron [*4*] irradiated Ti-Ni alloys, and the transformation temperatures were shown to be decreased by the irradiations that might induce disordering [*5*]. Neutron irradiation often suppressed the transformation completely, but the irradiation-induced suppression of the transformation recovered after post-irradiation annealing at 250°C [*6,7*]. Although there are several works on irradiation effects on shape memory alloys, these are the studies on solid solution annealed alloys. The shape memory effect of aged alloys at 400°C is well known as superelasticity.

In this research, the effects of neutron irradiation and post-irradiation annealing on the aged Ti-Ni shape memory alloys have been investigated to make clear the mechanism of irradiation-induced suppression of the martensitic transformation.

Experimental

The materials used were Ti-50.0, 50.5 and 51.0 at.%Ni alloys, which were produced by melting in a high frequency vacuum furnace from high-purity nickel (99.99mass%) and pure titanium (99.85mass%). The ingots were hot-forged and rolled, followed by cold rolling to sheets of 0.3 mm thickness. The specimen preparation conditions of the alloys are summarized in Table 1.

Table 1- *Specimen preparation conditions*

	Cold work ratio	Heat Treatment (1)	Heat Treatment (2)
Ti-50.0Ni	0.20	800°C, 1 hr, WQ	400°C, 1 hr, WQ
Ti-50.5Ni	0.33	800°C, 1 hr, WQ	400°C, 1 hr, WQ
Ti-51.0Ni	0.18	800°C, 1 hr, WQ	400°C, 1 hr, WQ

WQ: water quench

Disk shape specimens were fabricated with the dimensions of 3 mm diameter and 0.15 mm thickness after the heat treatment in a vacuum of 5×10^{-4}Pa. Neutron irradiation was performed in the Japan Materials Testing Reactor (JMTR) to a fluence of 1.2×10^{24}n/m^2

in a water leaking capsule to irradiate specimens at temperature as low as the cooling water or 60°C. The DSC measurements were carried out by scanning in the range of temperatures between -120°C and 120°C at a heating/cooling rate of 15°C/min. Micro-Vickers hardness tests with a load of 0.1kg and PAS measurements were carried out at room temperature. From the PA spectra, average lifetime was obtained. Microstructures of the neutron irradiated 51.0 at.%Ni alloy was observed by TEM.

Results and discussion

DSC Measurement

The typical features of the DSC curves of neutron irradiated and post-irradiation isothermally annealed Ti-Ni alloys are shown in Fig. 1 with those before the irradiation. The upper and bottom curves in each pair are for cooling and heating, respectively. It is clear that in all the alloys, martensitic transformation was suppressed by the irradiation and recovered after the annealing at 250°C for 1 hr. This trend is similar to that of the solution annealed alloys [7].

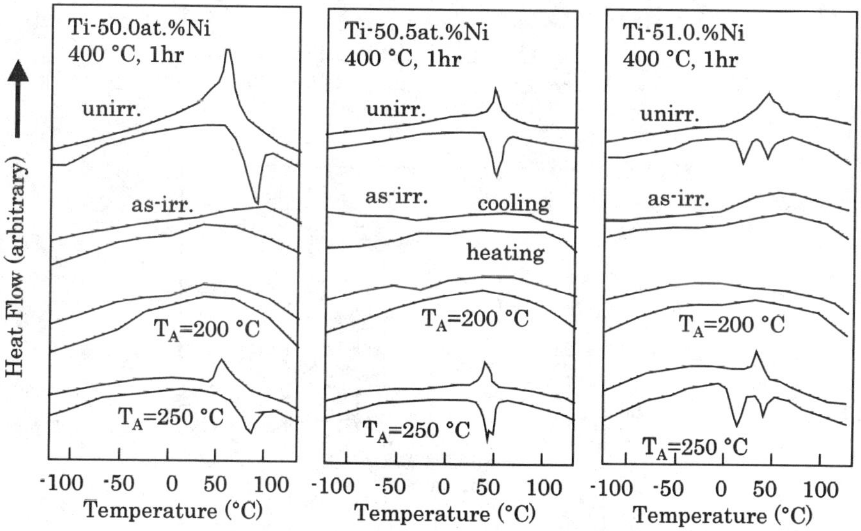

Figure 1- *DSC spectra of aged Ti-Ni alloys before and after the irradiation and after post-irradiation annealing. The upper and bottom curves are of cooling and heating, respectively. Isothermal annealing was carried out for 1 hr at each temperature.*

Micro-Vickers Hardness Test

Age hardening—The change in the hardness during aging is shown in Fig. 2 and 3 for solid solution annealed Ti-Ni alloys. The Ti-51Ni alloy was age hardened by the annealing above 250°C and attained the maximum hardness at 350°C, while no such a significant age hardening was observed in the Ti-50Ni alloys. A small amount of hardening was observed in the Ti-50.5Ni alloy. Overaging was recognized at 400°C for Ti-51Ni alloy. It was reported that the precipitation of intermetallics such as $Ti_{11}Ni_{14}$ and Ti_2Ni_3 was observed in Ti-51Ni alloy after the aging at 400°C [8].

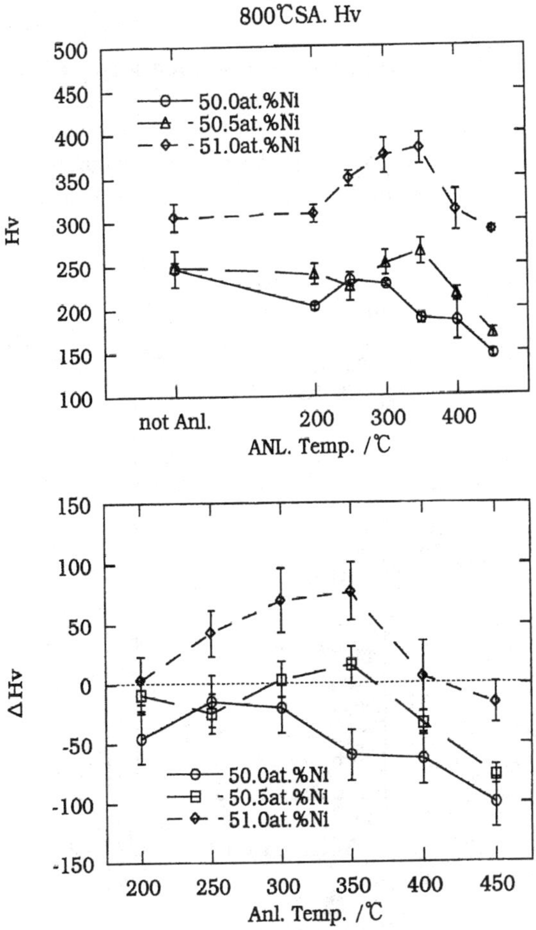

Figures 2 (top) and 3 (bottom)-*Age hardening of solid solution annealed Ti-Ni alloys. Specimens were isochronally annealed for 1 hr at each temperature.*

Irradiation Effects--Figure 4 and 5 show the effects of irradiation and post-irradiation annealing on micro-Vickers hardness of aged Ti-Ni alloys, indicating that the hardness before irradiation and the amount of irradiation hardening depends on the alloy composition significantly. In the unirradiated condition, the hardness of Ti-51Ni alloy is about 1.8 times higher than that of Ti-50Ni alloy. The hardness of the alloys, however, increased to almost the same level following the irradiation, showing the smallest and largest irradiation hardening in Ti-51Ni and Ti-50Ni alloy, respectively. The irradiation hardening again completely recovered after the annealing at 250°C for 1 hr.

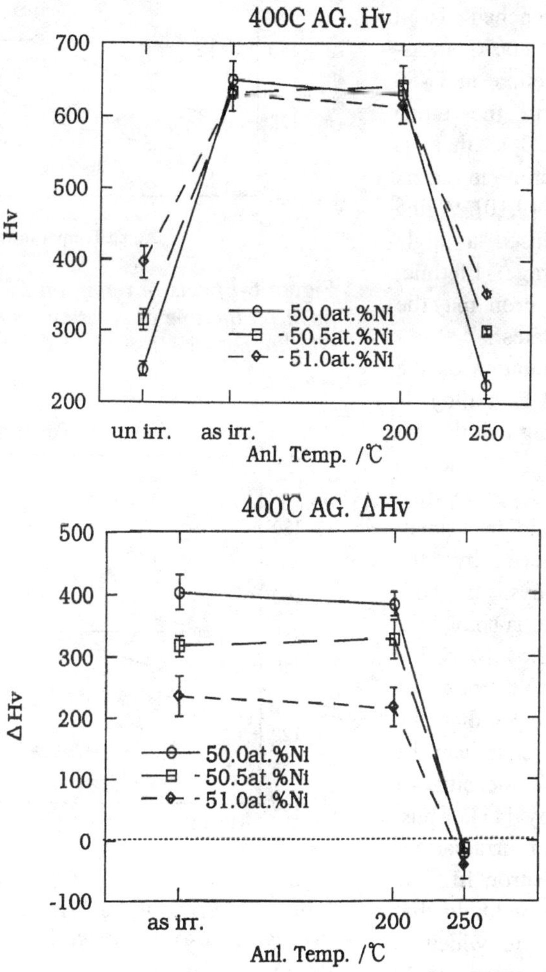

Figures 4 (top) and 5 (bottom)-*Effects of neutron irradiation and post-irradiation isothermal annealing (1 hr) on micro-Vickers hardness of aged Ti-Ni alloys.*

PAS Measurement

Since the structural vacancies have been considered to exist sometimes in off-stoichiometric intermetallic compounds, PAS measurements were performed for solid solution annealed Ti-Ni alloys. Figure 6 shows the average lifetime of positron in water quenched Ti-Ni alloys after annealing at 800°C. In as-quenched alloys, the lifetime in Ti-50 and 51Ni alloys is in the range between 110 and 120ps that is considered to be the lifetime in matrix of many intermetallics [9,10]. Aging at 350 and 400°C caused a slight increase in the average lifetime, suggesting that the positron may be trapped at small precipitates.

The effect of irradiation on the average lifetime in aged TiNi alloys is shown in Fig.7, indicating that the life time is increased by the irradiation to around 160ps. This suggests that single vacancies and/or small microvoids are induced by the irradiation in the alloys. In both Figure 6 and 7, the trend is somewhat different in Ti-50.5Ni alloy. Since the Ti-50.5Ni alloy was cold rolled to 33%, which is much larger than the other alloys, the difference can be attributed to the high density of dislocations in the alloy [11]. Thus, although the cause of irradiation-induced increase in positron lifetime is still unclear, it is very likely that vacancies contribute to it, which is also supported by the previous study on kinetics of the recovery process of neutron irradiation effects in Ti-Ni alloys [7].

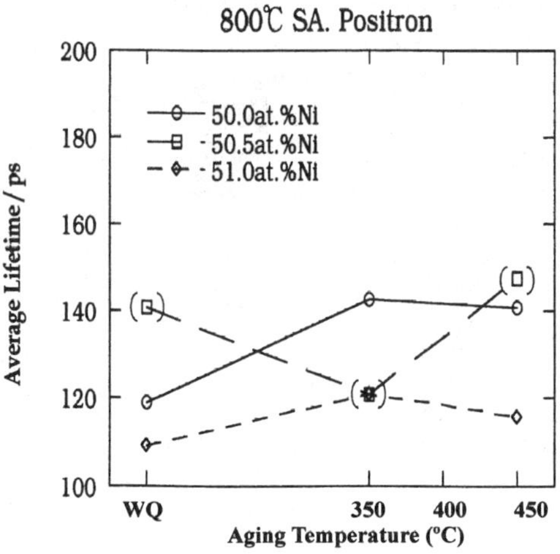

Figure 6-*Effects of aging on the average positron lifetime of solution annealed Ti-Ni alloys.*

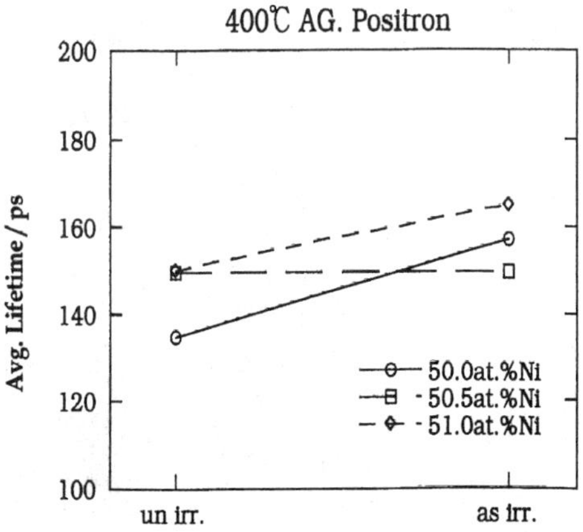

Figure 7-*Effects of the irradiation on the average positron lifetime of aged Ti-Ni alloys.*

TEM Examinations

Figure 8 shows the TEM micrographs of aged Ti-51Ni alloy in the as-irradiated condition (g=110): a) bright field image and b) image taken by a 001 ordered spot, c) [110]B2 diffraction pattern. As indicated by white arrows in the micrographs, black-white contrast becomes reversed between the two, indicating that the black region of the bright field image is the disordered (maybe amorphous) region. There are some black-dot structures which are considered to be dislocation loops. Since the dislocation loops lose their contrast when they were observed with g=112, the Burgers vector of the dislocation loops were identified to be b=a/2<111>, which is different from b=a<001> of deformation-induced dislocations [*12*]. The diffraction pattern shown in Fig.8(c) is observed from <111> direction of the parent phase of the alloys. It is clear that the ordered spots still exist and halo rings are superimposed on the diffraction pattern. This suggests that the irradiation-induced disordered region includes amorphous as well as disordered lattice and defect clusters such as dislocation loops and microvoids. The suppression of the martensitic transformation is considered to be due to high resistance to the motion of martensite domain boundaries caused by the highly disordered region and defect clusters.

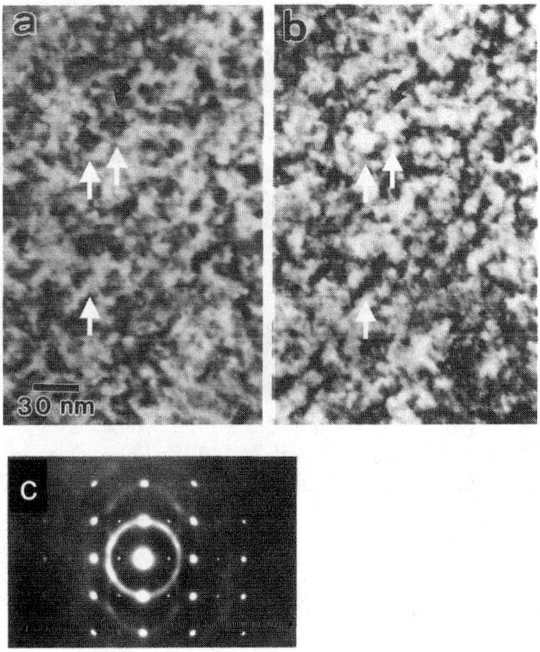

Figure 8-*TEM micrographs of aged Ti-51Ni alloys:* a) bright field image and b) dark field image taken by 001 ordered spot, c) [110]B2 diffraction pattern.

Figure 9 shows the in-situ results obtained by a hot-stage TEM holder, showing the change in the irradiation damage structure by post-irradiation annealing. The same region was observed following the post-irradiation annealing of aged Ti-51Ni specimen at 300°C for 45 min. This annealing resulted in almost complete recovery of the irradiation effects in the alloy. After the annealing, a number of spherical precipitates (several nm in size) were observed. Since atomic diffusion in B2 intermetallic compounds has been known to occur by the so called six-jump mechanism through the motion of vacancies[*13,14*], reordering, which was accompanied by the precipitation, occurred during the annealing where vacancies were mobile. The halo rings observed before the annealing were still observed after the annealing. This suggests that the reordering has not been completed yet.

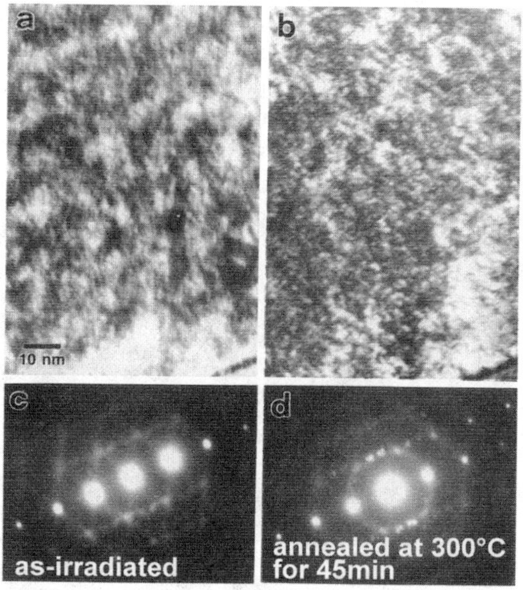

Figure 9-*TEM micrographs of aged Ti-51Ni alloys:* a) as-irradiated, b) post-irradiation annealed at 300°C for 40min.

Appearance of R-phase Transition in Ti-50Ni Alloy

The Ti-51Ni alloy is well known to show superelasticity that relates to two step martensitic transformation, that is from B2 to R-phase and from R to M-phase [15]. The R-transition has been considered to relate to the precipitation of $Ti_{11}Ni_{14}$ [8]. In the stoichiometric Ti-50Ni alloy, no R-phase has been observed so far. However, post-irradiation annealing causes the occurrence of R-transition in Ti-50Ni alloy. Figure 10 shows the [110] diffraction pattern of parent phase in solution annealed Ti-50.0%Ni alloy after post-irradiation annealing at 250°C for 2hr. Observation was carried out at room temperature where the parent phase and R-phase are considered to exist. There are extra spots at 1/3 position in the direction of <112>, which have been considered to stem from R-phase [8]. Even in stoichiometric alloy, during post-irradiation annealing, R-phase relevant precipitates, such as $Ti_{11}Ni_{14}$ and/orTi_2Ni_3, can be formed locally in the parent phase and induce the R-transition.

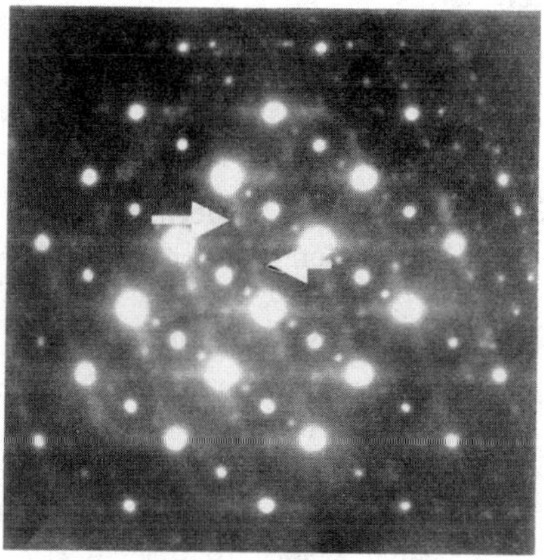

Figure 10--[110] *diffraction pattern of parent phase in solution annealed Ti-50.0%Ni alloy after post-irradiation annealing at 250°C for 2 hr.*

Conclusions

Irradiation effects on the martensitic transformation of aged Ti-50.0, 50.5 and 51.0at.%Ni alloys were investigated after neutron irradiation to 1.2 x 10^{24}n/m^2 at the temperature of cooling water (about 60°C). The neutron irradiation suppressed the martensitic transformation in all the alloys. Irradiation hardening was the largest in Ti-50Ni alloy that showed the lowest hardness before the irradiation. The hardness after the irradiation is almost the same for the three alloys. The irradiation effect recovers in all the alloys after post-irradiation annealing at 250°C for 1 hr.

PAS and TEM examinations suggest the existence of a disordered region that is considered to be amorphous with defect clusters, which suppress the transformation in the alloys. The recovery is considered to be due to migration of vacancies, which causes the reordering accompanied by precipitation of nickel-rich particles. The R-transition occurs in the post-irradiation annealed Ti-50Ni alloy. It is expected that Ti-Ni alloys are potential self-restorative materials for irradiation above 250°C where vacancies are mobile.

Acknowledgment

The authors are grateful to Prof. S. Miyazaki, Tsukuba University, for providing Ti-Ni alloys. Special thanks go to Mr. R. Kasada, graduate student, IAE, Kyoto University, and Prof. M. Hasegawa, Tohoku University, for his assistance in PAS measurements, and Mr. M. Narui, the branch of IMR at Tohoku University, for performing neutron irradiation at cooling water temperature.

References

[1] Mori, H., and Fujita, H., "Temperature Dependence of Electron-Irradiation Induced Amorphization of TiNi Alloys",

[3] Pedraza, D.F., "Mechanisms of the Electron Irradiation-Induced Amorphous Transition in Intermetallic Compounds", J. Mater. Res., 1986, 1, 3,p.425

[4] Hoshiya, T., Shimakawa, S., Ichihashi, Y., Nishikawa, M., and Watanabe, K., "Fast Neutron Irradiation of Ti-Ni Shape Memory Alloys", *J. Nucl. Mater.,* 1991, 179-181, p.1119.

[5] Liou, K.Y., and Wilkes, P., "The Radiation Disorder Model of Phase Stability", J. Nucl. Mater., 1979, 87, p.317

[6] Hoshiya, T., Takada, F. and Ichihashi, Y., "Restoration Phenomena of Neutron-irradiated Ti-Ni Shape Memory Alloys", *Mater. Sci. Eng.,* 1990, A130, p.185.

[7] Kimura, A., Tsuruga, H., Morimura, T., Miyazaki, S. and Misawa, T., "Effects of Post-irradiation Annealing on the Transformation Behavior of Ti-Ni Alloys" *Materials Trans. JIM,* 1993, 34, 11, p1076.

[8] Nishida, M., Wayman, C. M. and Honma, T., "Precipitation Process in Near-Equiatomic TiNi Shape Memory Alloys", *Metallurgical Transaction A*., 1986, 17A, 9, p.1505.

[9] Shirai, Y. and Yamaguchi, M., "Studies of Vacancies and Dislocations in TiAl by Positron Annihilation", *Mater. Sci. Eng.,* 1992, 152, p.173.

[10] Yamaguchi, Y. and Umakoshi, Y., "The Deformation Behavior of Intermetallic Superlattice Compounds", Prog. Mater. Sci., 1990, 34, p.1.

[11] Kamimura, Y., Tsutsumi, T. Kuramoto, E., "Influence of Dislocations on Positron Lifetime in Iron", *J. Phys. Soc. Japan,* 1997, 66, 10, p.3090.

[12] Miyazaki, S., Matsumoto, O. and Ohtsuka, K., "Determination of the Slip System in Ti-Ni Single Crystals", *Proc. 6th JIM Int. Symp. on Intermetallic Compounds*, Japan Inst. Metals, 1991, p.269.

[13] Elcock, E. W. and McCombie, C. W., "Vacancy Diffusion in Binary Ordered Alloys", *Phys. Rev.,* 1958, 109, p.605.

[14] Nakajima, H., Yasuda, H. and Koiwa, M., "Self- and Impurity Diffusion in $L1_2$ Type Intermetallic Compounds Ni_3Ge and Co_3Ti", *Proc. 6th JIM Int. Symp. on Intermetallic Compounds*, Japan Inst. Metals, 1991, p.57.

[15] Hwang, C. M., Meichle, M., Salamon, M. B. and Wayman, C. M., "Transformation Behavior of a $Ti_{50}Ni_{47}Fe_3$ Alloy, *Phil. Mag.*, 1983, 47, 1, A, p.9.

Mitsuhiro Takeda,[1] S. Ohnuki,[1] T. Suda,[1] S. Watanebe,[1] H. Abe,[2] and I. Nashiyama[2]

Effect of Mass and Energy on Preferential Amorphization in Polycrystalline Silicon Film during Ion Irradiation

Reference: Takeda M., Ohnuki S., Suda T., Watanebe S., Abe H., and Nashiyama I., **"Effect of Mass and Energy on Preferential Amorphization in Polycrystalline Silicon Film during Ion Irradiation,"** *Effects of Radiation on Materials: 20th International Symposium, ASTM STP 1405,* S. T. Rosinski, M. L. Grossbeck, T. R. Allen, and A. S. Kumar, Eds., American Society for Testing and Materials, West Conshohocken, PA, 2001.

Abstract: In-situ transmission electron microscopy was applied for clarifying radiation-induced amorphization, the behavior of grain boundaries under ion irradiation. The effect of mass and energy of several ions on preferential amorphization was discussed. The critical fluence for amorphization strongly depended on the temperature, where it increased with increasing temperature. Further, with increasing ion mass and decreasing energy, the critical fluence was reduced. The onset temperature for preferential amorphization increased in the case of heavier mass and lower energy ion irradiation. All of the results imply the importance of the balance between damage production and recovery.

Keywords: polycrystalline silicon, in-situ TEM observation, preferential amorphization, ions mass, energy, irradiation temperature

[1]Department of Material Science, Facility of Engineering, Hokkaido University. N-13, W-8, Kita-ku, Sapporo 060-8628, Japan.

[2]Japan Atomic Energy Research Institute, Takasaki, 1233 Watanuki-machi, Takasaki 370-1292, Japan.

Introduction

Polycrystalline silicon (poly-Si) is a widely used material for thin film transistors (TFTs); active-matrix addressed flat-panel display [*1,2*], static random-access memory [*3*] and image sensors [*4*], for example. However, the grain size obtained by commonly used methods is too small. For achieving much higher integration and performance, excimer laser [*5-7*] and ion-beam [*8,9*] techniques have been applied for increasing the grain size.

Heavy ion irradiation can easily produce the amorphous transition, where the stability of phases, crystalline and amorphous, is strongly dependent on irradiation temperature and ion flux [*10-13*]. It has been reported that the preferential amorphization occurred on grain boundaries (GB) during Xe^+ irradiation in the transformation temperature range [*14*], which was supposed to be a thermodynamically controlled process and to be analogous to the concept of grain boundary melting in the solid at the equilibrium temperature [*15*]. However, the details of this phenomenon have not yet been clarified. Especially, the kinetics of radiation-produced amorphization and crystallization are expected to be very important in controlling the phenomenon.

In this work, in-situ transmission electron microscopy (TEM) was carried out for understanding the details of the process of radiation-induced amorphization and GB behavior. The objectives of this paper are defining the mass and the energy effect on critical fluence for the amorphization, and clarifying the relation between the fluence and preferential amorphization.

Experimental Method

Specimen preparation

Poly-Si of about 200 nm in thickness was grown on oxide silicon by LPCVD (Low Pressure Chemical Vapor Deposition) followed by thermal annealing at 1223 K for 4 hours in an Ar atmosphere. Figure 1 shows the microstructures in cross-sectional and plane-view after annealing. The mean grain size (d) adapted by equation (1) [*16*] was about 230nm.

$$d = \sqrt{ab\pi}/2 \qquad (1)$$

Where a and b are the long and short axes of the ellipsoid, respectively.

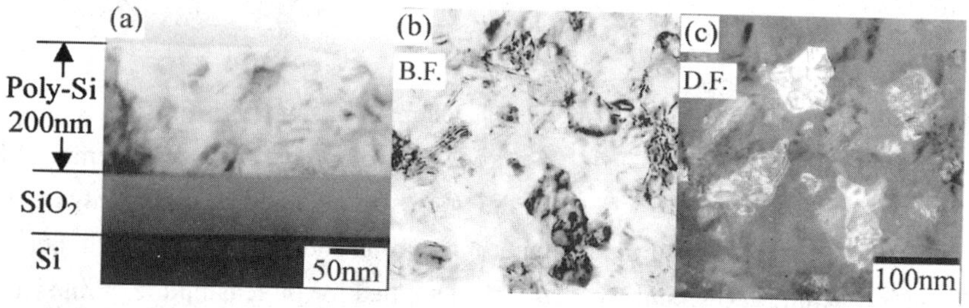

Figure 1-*Grain structure of poly-Si film, (a) cross-sectional view, (b) plan-view at bright field image and (c) plane-view at dark filed image.*

Irradiation experiment

Ion irradiation was performed at TIARA Facility of Japan Atomic Energy Research Institute, Takasaki Laboratory, where 400 kV electron microscopes, JEOL 4000FX, is equipped with 400keV ion accelerator.

Specimens were irradiated with 100 - 300 keV C^+, 100 keV N^+ and 200 – 300 keV Ar^+ with ion fluxes 3.0 - 4.0 x 10^{17} $m^{-2}s^{-1}$. The irradiation temperature was from 100 K to 423 K. The details of the irradiation conditions were shown in Table 1.

For detecting the critical fluence, changes in selected-area electron diffraction (SAED) patterns were used, as shown in Figure 2. SAED pattern of polycrystalline presented clearly diffraction spots. On the other hands, SAED pattern of amorphous phase revealed halo rings. In this work, the critical fluence for amorphization was defined by SAED pattern. SAED spots were completely changed from spots (Figure 3 (a)) to halo ring pattern (Figure 3 (c)) by ion irradiation. The critical fluence was the fluence at the complete halo ring pattern.

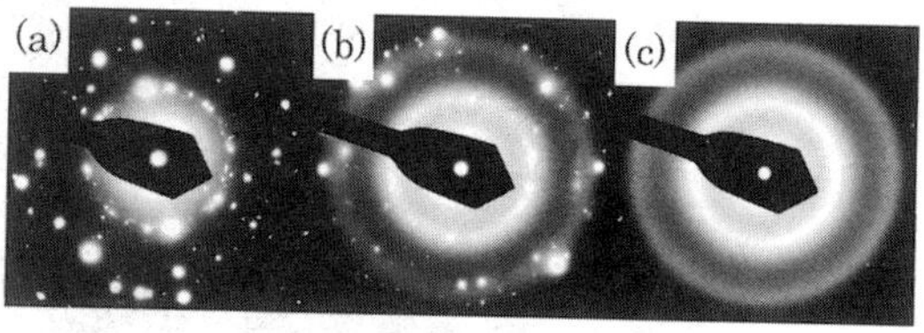

Figure 2-*Selected-area electron diffraction (SAED) pattern during ion irradiation: (a) crystal, (b) defective crystal and (c) amorphous phase.*

Table 1 – *Details of ion irradiation conditions*

Ion/energy	Temperature, K
C^+ 300keV	213 – 333
C^+ 100keV	104 – 453
N^+ 100keV	113 – 453
Ar^+ 300keV	298 – 453
Ar^+ 200keV	323 – 453

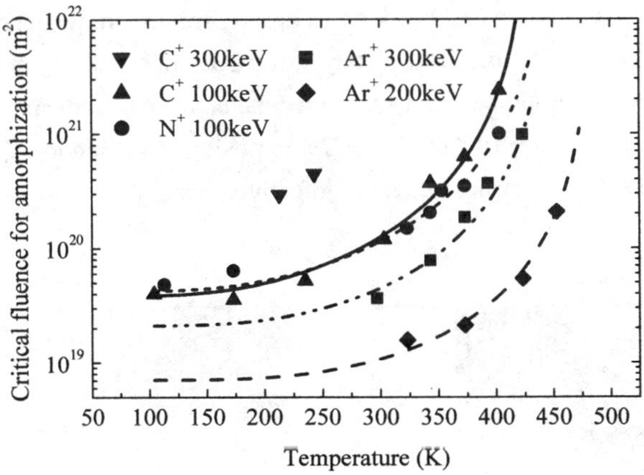

Figure 3 *Temperature dependence of critical fluence at different irradiation condition.*

Results

Effect of mass and energy on the critical fluence

Figure 3 shows the temperature dependence of the critical fluence at each irradiation conditions. The critical fluence showed an exponential function. The critical fluence had little temperature dependence below a room temperature (RT). On the other hands, about RT, it increased with increasing of temperature. In case of the heavy ion and the low energy, the critical fluence was reduced at below a RT. The cut-off temperature, which means crystal is stable even if under irradiation condition, was increased. Those behaviors suggest that the critical temperature depends on both of the production of amorphization and crystallization at the local area.

Minimum temperature of preferential amorphization on GB

The microstructure of crystal-amorphous (c-a) transformation was strongly depended on the irradiation temperature. Figure 4 shows microstructural change of amorphization process during 300 keV Ar^+ irradiation at each temperature.

In case of RT, diffraction spots disappeared with diminishing of irradiation and then turn to halo ring at the fluence of 3.7×10^{19} m^{-2}. In TEM image, the polycrystalline structures composed of white and black areas before irradiation, which depend on Bragg condition. If the crystal becomes completely amorphous phase during the irradiation, the contrast of the image should be turn to intermediate color, which means the amorphous area. At the fluence of 1.3×10^{19} m^{-2}, SAED composing of diffraction spots and the weak halo ring indicates that the structure was a defective crystal.

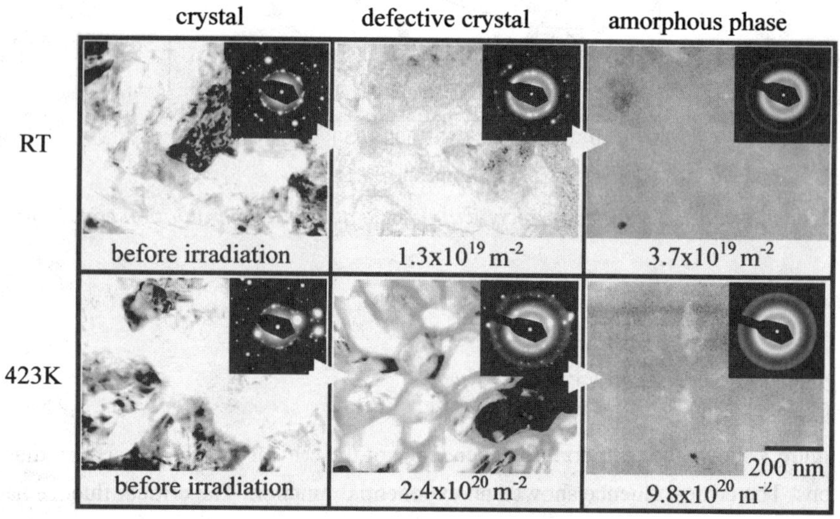

Figure 4 *TEM microstructures irradiated by 300keV Ar^+ at the RT and the 423K*

At higher temperature of 423 K, SAED pattern showed completely halo ring after the irradiation of 9.8×10^{20} m^{-2}. During medium fluence, the preferential amorphization developed on GB occurred and the amorphous layer grew with increasing of fluence. This phenomenon was recognized in other ion species and energies as shown in figure 5.

Table 2 shows the minimum temperature of the onset of preferential amorphization for different ion irradiation condition. In case of heavier ion and the lower energy, the minimum temperature was increased. The data of 1.5 MeV Xe^+ in table 2 was the result obtained by Atwater [*14*].

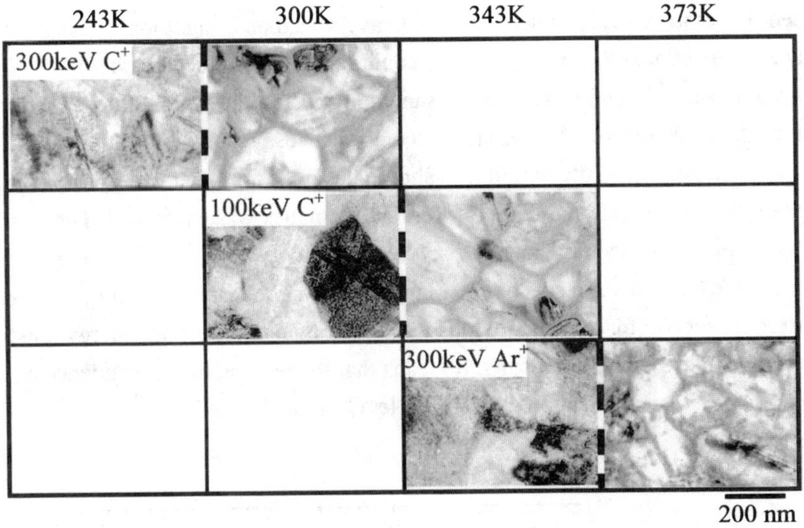

Figure 5 *The structure of polycrystalline silicon above/below the temperature of preferential amorphization during 300 keV C^+, 100 keV C^+ and 300 keV Ar^+ irradiation. Above the temperature indicated at dash line, preferential amorphization occurs. The temperature of preferential amorphization at difference condition shows in Table 2.*

Table 2 – *The minimum temperature of the onset of preferential amorphization on GB*

Ion/energy	Temperature, K
C^+ 300keV	243 – 300
C^+ 100keV	300 – 333
N^+ 100keV	300 – 333
Ar^+ 300keV	343– 373
Ar^+ 200keV	373 – 403
Xe^+ 1.5 MeV	423

Discussion

Effect of the mass and the accelerated energy on critical fluence

With increasing ion mass, the critical fluence reduced as shown in Figure 3. In general the total amount of defects and the size of cascade damage zones produced by an incident ion assumed to be much larger in heavy ions than in light ions. Consequently it is easily considered that the critical fluence could reduce with much heavier ion.

With decreasing of energy for the same ion species, the critical fluence reduced, whereas the crystallization temperature increased. Those results suggest that the damage level at local area especially in thin film should be larger for lower energy. By using TRIM code [17], energy transferred to recoil from ion to matrix was calculated. The condition was as follows: specimen thickness, displacement energy and lattice bonding energy were 200 nm, 16 eV and 2 eV, respectively. Figure 6 shows the result of calculation. Energy transferred to recoil for low energy ions was larger than that for high-energy ions. This corresponds to experimental results. It means that the temperature dependence of the critical fluence is dependent on the damage level in local area.

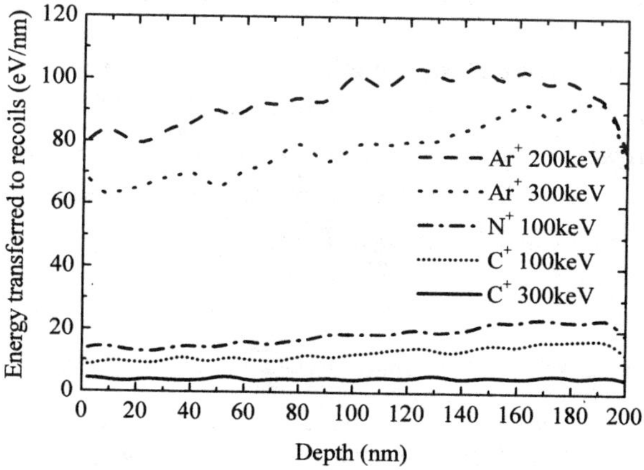

Figure 6 *The result of TRIM calculation. Specimen thickness, displacement energy and lattice bonding energy were 200 nm, 16 eV and 2 eV, respectively.*

Temperature range of preferential amorphization on GB

The temperature of preferential amorphization on GB increased with following order; 300 keV C^+ < 100 keV C^+ < 100 keV N^+ < 300 keV Ar^+ < 200 keV Ar^+. In comparison with the result of TRIM calculation, this temperature shows a positive correlation as the energy transferred to recoil. In the case of large damage energy, it was suggested that the amount of defects introduced by single ion were a lot.

Considering the preferential amorphization on GB region from the point of view of thermodynamics, it is generally known that the Gibbs free energy of GB region (G_{GB}) is larger than that of the free defect crystal (G_{free}). The Gibbs free energy of the whole of crystal is increased due to the introduced defect, which brings about the increase of free energy (G_{delta}). Assuming that amorphization takes place when the Gibbs free energy of crystal exceeds the threshold energy (G_{amor}) for amorphization, it can be considered that GB region become easily to amorphous phase, because of its higher potential for amorphization. Therefore, it can be considered that GB region can be easily transformed to amorphous phase in comparison with the inside of grain. In the growth mechanism of preferential amorphization, the growth of amorphous phase means the continual amorphous nucleation on c-a interface. If the c-a interface has large free energy with a strain in comparison with the inside of grain, amorphous nucleation on c-a interface can be understand as the same as GB region.

In case of lower energy ions, the preferential amorphization temperature was increased: increasing of the energy of damage formation indicates that G_{delta} becomes large. Considering that the crystal can transfer to amorphous phase, if the free energy becomes to $G_{per} + G_{delta} \geq G_{amor}$. In order to nucleate amorphous phase on GB, G_{delta} must decrease, e.g. $G_{per} + G_{delta} < G_{amor}$ and $G_{GB} + G_{delta} \geq G_{amor}$, where G_{delta} depends on amount of defects. If many defects are introduced by the irradiation, the cut-off temperature must be at high for reducing G_{delta}. Therefore, it can be considered that the preferential amorphization temperature shift to high temperature side, in case of higher energy transferred to recoil is large.

Conclusion

We investigated the mass and energy effect on the critical fluence for amorphization, the temperature for preferential amorphization and during ion irradiation.

With increasing mass and decreasing energy, the critical fluence was reduced, whereas the cut-off temperature was increased. It was considered that the amount of the damage should be much larger. The temperature of preferential amorphization was increased with increasing mass and decreasing energy. This result tended to be the same as the cut-off temperature. These results were equivalent to large of energy transferred to recoil calculated by TRIM code. It is confirmed that these phenomena are controlled by the balance between the introduced defects by energetic ion and the recovery of defects by temperature.

References

[1] Tsukada T., "Scaling theory of liquid-crystal displays addressed by thin-film transistors" *IEEE Transaction on Electron Device*, 1998, 45, pp. 387-393.

[2] Aoyama T., Ogawa K., Mochizuki Y. and Konishi N., "Inverse staggered poly-Si and amorphous Si double structure TFT's for LCD panels with peripheral driver circuits integration" *IEEE Transaction on Electron Device*, 1996, 43, pp. 701-705.

[3] Son K.S., Known S.W., Lee Y.J., and Kim D. M., "Metal node contact TFT SRAM cell for high-speed, low-voltage applications" *IEEE Transaction on Electron Device*, 1999, 46, pp. 805-806.

[4] King T.J., Saraswat K.C., "Polycrystalline silicon-germanium thin-film transistors" *IEEE Transaction on Electron Device*, 1994, 41, pp. 1581-1591.

[5] Voutsas A.T., "Enhanced Grain Growth Phenomena in Excimer Laser Annealed Microcrystalline Silicon Films" *Japanese Journal of Applied Physics*, 1998, 37, pp. 338-396.

[6] Rezek B., Nebel C.E., "Polycrystalline Silicon Thin Films Produced by Interference Laser Crystallization of Amorphous Silicon" *Japanese Journal of Applied Physics*, 1999,38, pp. L1083-L1084.

[7] Mariucci L. and Sala D.D., "A Two-Pass Excimer Laser Annealing Process to Control Amorphous Silicon Crystallization" *Japanese Journal of Applied Physics*, 1999, 38, pp. L907-L910.

[8] Yamaychi N. and Reif R., "Polycrystalline silicon thin films processed with silicon ion implantation and subsequent solid-phase crystallization: Theory, experiments, and thin-film transistor applications" *Journal of Applied Physics*, 1994, 75, pp. 3235-3257.

[9] Kong K.T-Y. and Reif R., "Implant-dose dependence of grain size and {110} texture enhancements in polycrystalline Si films by seed selection through ion channeling" *Journal of Applied Physics*, 1986, 59, pp. 2422-2428.

[10] Iverson R. B. and Reif R., "Recrystallization of amorphized polycrystalline silicon films on SiO_2: Temperature dependence of the crystallization parameters" *Journal of*

Applied Physics, 1987, 62, pp. 1675-1681.

[11] Jackson K. A., "A defect model for ion-induced crystallization and amorphization" *Journal of Materials Research*, 1988, 3, pp. 1218-1226.

[12] Carter G., "The effects of flux, fluence and temperature on amorphization in ion implanted semiconductors" *Journal of Applied Physics*, 1996, 79, pp. 8285-8289.

[13] Kinomura A., Williams J. S. and Fujii K., "Mass effects on regrowth rates and activation energies of solid-phase epitaxy induced by ion beams in silicon" *Physical Review B*, 1999, 59, pp. 15214-15224.

[14] Atwater H. A. and Brown W.L., "Grain boundary mediated amorphization in silicon during ion irradiation" *Applied Physics Letters*, 1990, 56, pp. 30-32.

[15] Atwater H. A., Im J. S. and Brown W.L., "Heterogeneous amorphization of Si during ion irradiation: dependence of amorphous Si nucleation kinetics on defect energy and structure" *Nuclear Instruments and Methods in Physics Research Section B*, 1991, 59/60, pp. 386-390.

[16] Lee J.N., Choi Y.W., Lee B.J., and Ahn B. T., "Microwave-induced low-temperature crystallization of amorphous silicon thin films" *Journal of Applied Physics*, 1997, 82, pp. 2918-2921.

[17] Ziegler J.F. and Biersack J.P., "The Stopping and Ranges of Ions in Solids", pergamon press, New York, 1985

Pauline H. Davies,[1] Donald D. Himbeault,[2] Robert S.W. Shewfelt,[3] and Robert R. Hosbons[4]

Crack Growth Resistance of Irradiated Zr-2.5Nb Pressure Tube Material at Low Hydrogen Levels

---

Reference: Davies, P. H., Himbeault, D. D., Shewfelt, R. S. W., and Hosbons, R. R., "**Crack Growth Resistance of Irradiated Zr-2.5Nb Pressure Tube Material at Low Hydrogen Levels,**" *Effects of Radiation on Materials: 20th International Symposium, ASTM STP 1405,* S. T. Rosinski, M. L. Grossbeck, T. R. Allen, and A. S. Kumar, Eds., American Society for Testing and Materials, West Conshohocken, PA, 2001.

Abstract: The primary factors influencing the crack growth resistance of irradiated Zr-2.5Nb pressure tube material at low concentrations of hydrogen/deuterium are reviewed. These factors include the initial characteristics of the material, which have brought about improvements in the toughness, and the operating conditions in reactor. The paper presents an update on the current status of this work using J-R curves. Such curves are determined from curved compact and rising-pressure burst test specimens at 250°C, i.e., the lower end of the operating temperature range. Some of the challenges encountered in assessing the crack growth toughness of this high-strength, thin-walled material are discussed. The role of chlorine, known to be responsible for the presence of Zr-Cl-C particles and preferential decohesion and fissuring, is also highlighted. The results from the curved compact specimens suggest a limiting level of chlorine above which no further significant degradation in crack growth resistance occurs. This level of chlorine is about 3 wt ppm for material having a low concentration of zirconium phosphide (P < 20 wt ppm). Such results require confirmation using burst tests on material with intermediate levels of chlorine and low levels of zirconium phosphide.

Keywords: crack growth resistance (J-R) curve, fracture toughness, Zr-2.5Nb, pressure tube, irradiation, slant fracture

---

[1]Senior research scientist, AECL, Chalk River Laboratories, Chalk River, Ontario K0J 1J0.

[2]Formerly research scientist, AECL, Whiteshell Laboratories, Pinawa, Manitoba R0E 1L0.

[3]Research scientist, AECL, Whiteshell Laboratories, Pinawa, Manitoba R0E 1L0.

[4]Formerly Head, Deformation Technology Branch, AECL, Chalk River Laboratories, Chalk River, Ontario K0J 1J0.

Introduction

The primary containment for the fuel in CANDU® reactors is provided by thin-walled pressure tubes of cold-worked Zr-2.5Nb. These tubes typically have a length of 6.3 m, an inside diameter of 103 mm, and a wall thickness of 4.2 mm. During service heavy water flows through the tubes to cool the fuel. The tubes operate at an internal pressure of between 10 and 11 MPa at temperatures varying from 250-265°C (inlet ends) to 290-315°C (outlet ends) depending upon the reactor unit. Considerable progress has been made recently in understanding the primary factors influencing the crack growth resistance and failure mechanism of the material. This includes the effects of irradiation, deuterium pickup and the role of pre-existing particles. The paper presents an update on the current status of this work for irradiated material having low concentrations of hydrogen/deuterium. The results presented are from crack growth resistance curves obtained from curved compact and burst test specimens tested at 250°C, i.e. the lower end of the operating temperature range.

Material

CANDU reactor pressure tubes are manufactured from extruded and cold-worked (about 26%) Zr-2.5Nb. The results reported here are for material manufactured to the earlier specification shown in Table 1.

Table 1 - *Chemical specification for Zr-2.5Nb pressure tubes for CANDU® reactors.*

Element	Specification (up to 1987)	Current
Niobium	2.4-2.8 wt%	2.5-2.8 wt%
Oxygen	900-1300[a]	1000-1300
Carbon	< 270	≤ 125
Chlorine	------	≤ 0.5
Chromium	≤ 200	≤ 100
Hydrogen	≤ 25[b]	≤ 5
Iron	≤ 1500	≤ 650
Nickel	≤ 70	≤ 35
Nitrogen	≤ 65	≤ 65
Phosphorus	------	≤ 10
Silicon	≤ 120	≤ 100
Tantalum	≤ 200	≤ 100
Zirconium and Other Impurities	Balance	Balance

[a] Concentration of remaining elements given in ppm by weight.
[b] 20 ppm hydrogen for ingot, 25 ppm hydrogen for final tube.

The final microstructure of the material is dual-phase with a network of elongated α–phase (hexagonal-close-packed), containing about 1 wt% Nb in solution, surrounded by a thin film of β-phase (body-centered-cubic), with about 20 wt% Nb [*1*].

Degradation occurs during service as a result of irradiation damage as well as deuterium pickup from the pressurised heavy water. Periodically surveillance tubes are removed from reactor and the mechanical properties assessed to ensure that the tubes remain "fit for service". Although some small specimen studies have been conducted on pre-selected material irradiated in test reactors, the majority of information on the crack growth resistance of irradiated Zr-2.5Nb pressure tube material is from surveillance tubes.

Mechanical test results have been obtained to date on irradiated Zr-2.5Nb pressure tube material sampled from tubes removed from Pickering Nuclear Generating Station (NGS) A Units 3 and 4, Bruce NGS A Units 1 to 4 and Wolsong Nuclear Power Plant (NPP) Unit 1. The tubes sampled operated for up to 18 years in service at a fast neutron flux of between 2.5 and 4×10^{17} $n \cdot m^{-2} \cdot s^{-1}$ to give a maximum fast neutron fluence of 18×10^{25} $n \cdot m^{-2}$ (E >1 MeV). (All neutron fluences in this paper are for E >1 MeV.) All the material was fabricated in the late 1960s and 1970s by early production techniques, e.g. from double-vacuum, arc-melted ingots. The majority of mechanical test results have been obtained at a test temperature of 250°C. This temperature corresponds to the lowest (inlet) irradiation temperature of the surveillance material and was selected to avoid removal of any irradiation damage by annealing. In addition, the maximum total equivalent hydrogen concentration (hydrogen + 0.5 deuterium concentration by weight) of this material is < 30 wt ppm, so that the hydrogen isotopes would all be in solution at test temperatures ≥ 250°C. (The terminal solid solubility for hydride dissolution at 250°C is about 29 wt ppm.)

Unirradiated archive material from the surveillance tubes is also available for testing. The material comes from rings of material that are removed from the front and back ends of a tube before installation. The front end refers to the end that emerged first in the extrusion operation.

Experimental Techniques

Two different techniques are used to assess the crack growth resistance of irradiated pressure tube material. The first is a small-scale specimen method using 17-mm wide curved compact specimens machined directly from the tube material [*2-4*] (see Figure 1). The second is a large-scale specimen method for rising-pressure burst tests using 500-mm long sections with a 55-mm long, axial through-wall starter crack [*5-7*] (see Figure 2). In each case specimens are oriented for crack growth in the axial direction on the radial-axial plane. After fatigue pre-cracking of the starter notch, specimens are loaded at a rate corresponding to an initial rate of increase of stress intensity factor of about 1 $MPa\sqrt{m} \cdot s^{-1}$. Any stable crack growth is monitored using the direct-current potential drop method. The positions of the constant current and voltage leads are selected to ensure that the relationship between the voltage and crack length for a planar crack front is approximately linear.

Curved compact specimens are loaded in displacement control to produce about 3 to 4 mm of crack growth and the load, load point displacement and potential drop signal are all monitored. After unloading, the crack extension area is marked by heat tinting, and the measured (average) crack extension matched to the change in potential drop voltage during the test. This allows direct calibration of the potential drop method for each

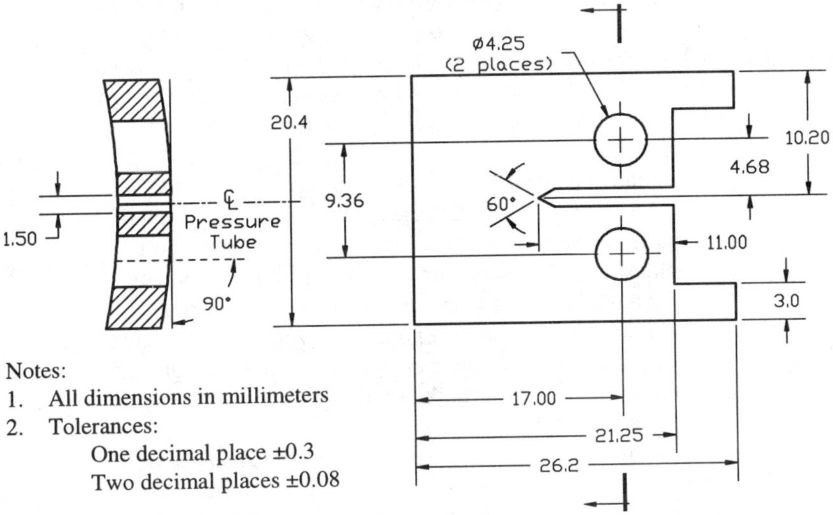

Figure 1 - *Standard (17-mm wide) curved compact specimen configuration.*

Figure 2 - *Burst test specimen with mechanical end caps and potential drop leads.*

specimen on an individual basis. Problems occasionally arise when testing irradiated specimens from tubes of higher toughness. Such material may exhibit a load drop after about 1.5 to 2 mm crack growth which is associated with secondary crack initiation from the back face at about 45° to the radial-axial plane. No correction method for such secondary cracking has been found. Current practice is to use the calibration factor from a specimen of similar toughness (i.e. load, displacement and voltage characteristics before the load drop) which did not exhibit "back-face cracking." The origin of such crack growth behavior is under investigation.

Burst test specimens are internally pressurized to failure with argon gas and the pressure and potential drop voltage monitored. In this case calibration of the potential drop method on an individual basis is not possible, since crack instability occurs just beyond maximum pressure. Studies of burst test specimens interrupted before achieving crack instability at 250°C suggest the potential drop method provides conservative assessments of crack extension during the early stages of crack growth. This is because of contributions to the voltage change from through-thickness yielding. However, given the mixed-mode nature of the fractures observed for both the small- and large-scale specimen geometry (see next section) the true bias of the results is unknown.

The J-integral parameter used for characterizing the crack growth resistance of Zr-2.5Nb pressure tube material may be considered the elastic-plastic equivalent of the linear elastic strain energy release rate, G. This is especially true for the rising-pressure burst test [*5-7*], where J is calculated using a strip yield equation for an axial through-wall crack in a thin shell [*8*], i.e. an elastic equation with a small-scale plasticity correction factor.

In comparison, for curved compact specimens the standard ASTM equation for J for a flat compact specimen is used [*2-4*]. This equation is based on the original non-linear elastic interpretation and analysis of Rice for the deformation J-integral [*9*], but the validity of this approach is questionable after crack initiation, e.g. work by Turner [*10*]. In fact, different values for J are obtained if alternative analyses are used, e.g. the modified J-integral [*11*], the dissipation J-integral [*12*]. Such differences can be significant when crack front tunneling occurs, see recent paper on Zr-2.5Nb pressure tube material by Davies [*13*]. However, for the purposes of surveillance testing, the limitations of the standard ASTM equation based on the deformation J-integral are accepted due to the simplicity of the approach and its current widespread acceptance within the field of elastic-plastic fracture mechanics.

Crack Growth Resistance (J-R) Curves and Fracture Modes

A J-integral versus crack extension (crack growth resistance or J-R) curve may be considered a measure of the elastic-plastic energy expended during crack growth, with the shape reflecting the macroscopic crack growth process in the material. For irradiated Zr-2.5Nb pressure tube material a large range of different shape J-R curves are observed depending upon the characteristics of the individual pressure tube, the irradiation and test conditions as well as the specimen geometry. For example, Figure 3 shows the range of J-R curves obtained from burst tests at 250°C for the original tubes removed from Pickering NGS A Units 3 and 4 [*6*].

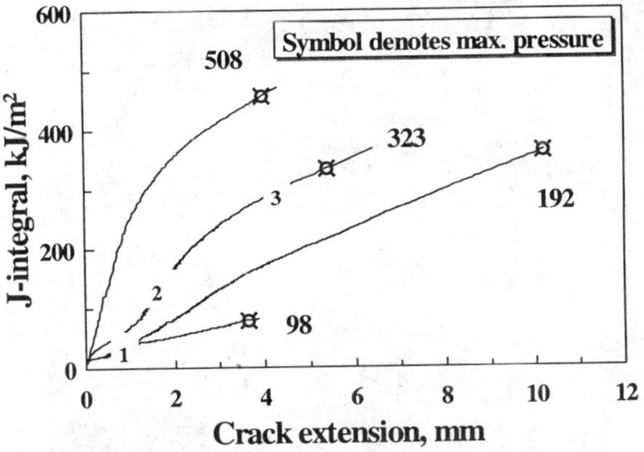

Figure 3 - *Range of crack growth resistance (J-R) curves from rising-pressure burst tests on surveillance tubes removed from Pickering NGS A Units 3 and 4 at 250°C.*

Such curves show evidence of the three stage crack growth behavior observed earlier for small specimens [*14*]. These three stages can be classified as indicated below.

1) Stage 1 (plateau) due to a low energy-absorbing, flat fracture mode developing in the region of highest constraint at the mid-thickness.
2) Stage 2 (increasing slope of J-R curve) due to a high energy-absorbing, transition fracture mode developing in the region of intermediate constraint between the mid-thickness and surface.
3) Stage 3 (decreasing slope of J-R curve) due to a low energy-absorbing slant fracture mode developing from the region of lowest constraint at the surface.

Thus individual irradiated Zr-2.5Nb pressure tubes may be characterized as being of low, intermediate or high toughness depending upon the relative contributions of these different fracture modes.

1) Low toughness: dominated by flat and slant fracture modes, i.e. stage 1 and 3.
2) Intermediate toughness: flat, transition and slant fracture modes all significant i.e. stage 1, 2 and 3.
3) High toughness: dominated by transition and slant fracture, i.e. stage 2 and 3.

An example of a fractograph for a high toughness specimen showing the three different fracture modes is shown in Figure 4.

The relative contributions of the different stages depends upon the relative proportions and energy-absorbing capacities of the different fracture modes, which is determined by the ease of void nucleation, void growth and coalescence ahead of the crack tip. Therefore, the relative proportions and energy-absorbing capacities of the three different fracture modes depends on the following factors.

1) The size, shape and distribution of the particles present since this influences the probability of nucleating voids [*15*].
2) The deformation characteristics of the matrix which influences the local crack-tip stress for void nucleation, growth and coalescence (ligament failure) [*15-17*].
3) The specimen geometry which influences the crack-tip stress and the probability of achieving the stress level required for void nucleation, growth and coalescence [*18,19*].

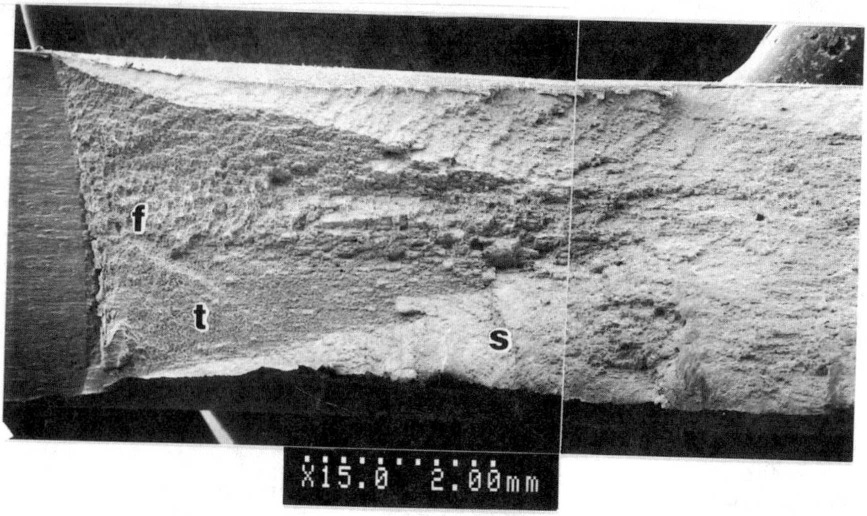

Figure 4 - *Scanning electron fractograph of burst test specimen from a tube of high toughness tested at 250°C showing narrow flat fracture zone at mid-section (f), wide cup-shaped transition zone (t), and slant fracture developing at surfaces (s) (Tilt angle view at 45°.)*

For irradiated Zr-2.5Nb pressure tube material the factors in 1) and 2) are strongly influenced by the initial properties of the as-installed tubes (e.g. ingot material, fabrication variables) as well as the operating conditions of the tubes in a given reactor (e.g. fast neutron fluence and irradiation temperature). Therefore, some discussion of the influence of these variables on the J-R curves is warranted. The majority of this work is based on the results from small curved compact specimens. The effects of specimen geometry, and confirmation of the influence of such variables on the J-R curves from rising-pressure burst tests, is presented afterwards.

Factors Influencing J-R Curves from Small Specimens (Parametric Studies)

Void-Nucleating Particles

Void-nucleating particles may be considered as primary or secondary depending upon whether they are responsible for initial void nucleation ahead of the crack tip at relatively low stress levels or at higher stress levels contributing to ligament failure between the primary voids. The majority of work on Zr-2.5Nb pressure tube material to date has focused on the study (and elimination) of particles responsible for primary void nucleation. In addition to the well known effects of zirconium hydride [*20*], such particles include a Zr-Cl-C (complex carbide), responsible for preferential decohesion and fissuring [*21,22*], zirconium phosphide and carbide [*21-23*]. Quantitative relationships between crack growth toughness and fissure spacing and length (an indirect measure of the density of Zr-Cl-C particles) are given elsewhere for the unirradiated [*21*] and irradiated material [*22*]. Relationships between toughness and density of zirconium phosphide and carbide particles are provided in reference [*23*].

In addition, it has been shown that tubes fabricated from ingots made-up from 100% recycled material (i.e. passed through the production process twice and equivalent to quadruple vacuum-arc-melted material) exhibit the highest toughness [*21-24*]. This is because of the low concentration of chlorine and an absence of Zr-Cl-C particles [*21-23*]. The influence of these pre-existing particles have now been eliminated by careful control of the ingot chemistry, including the use of selected raw materials and quadruple vacuum-arc-melting [*24*].

The effect of chlorine on the crack growth resistance of irradiated Zr-2.5Nb pressure tube material at 250°C is demonstrated in Figures 5 and 6 using the results from small specimens sampled from surveillance tubes. Figure 5 shows the initial crack growth toughness (dJ/da), which is a measure of the "toughening rate" of the material during the early stages of crack growth, and is calculated as the linear regression slope of the J-R curve between the 0.15 and 1.5 mm offset lines. For comparison, the J value based on maximum load-bearing capacity or load, (J_{ml}) is given in Figure 6. The chlorine concentration was determined from unirradiated archive material from the different surveillance tubes using Glow Discharge Mass Spectroscopy (GDMS). For clarity, only results from tubes having a low concentration of zirconium phosphide have been included, i.e. P < 20 wt ppm, since previous work has shown that such particles have little effect below this level [*23*]. Both sets of results demonstrate a sharp reduction in toughness with increasing Cl at concentrations < 3 wt ppm with little further change at higher Cl levels. The results also show that the Cl concentration of the Zr-2.5Nb pressure tube material fabricated in the late 1960s and 1970s from 100% recycled material was < 1 wt ppm. This compares with a level of Cl < 0.2 wt ppm which is now achievable with careful material selection and ingot melting practice.

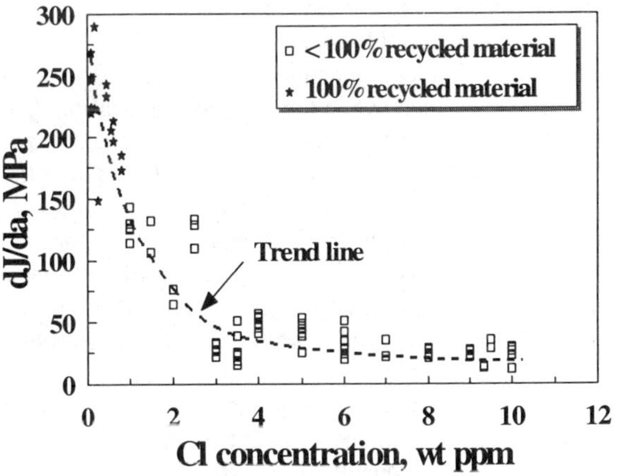

Figure 5 - *Initial crack growth toughness (dJ/da) from curved compact specimens of irradiated material tested at 250°C versus Cl. Results sorted according to source of ingot material. Results selected from material with P < 20 wt ppm. Fast neutron fluence of 4 to 18 x 10^{25} n·m^{-2} and irradiation temperature of 254 to 295°C.*

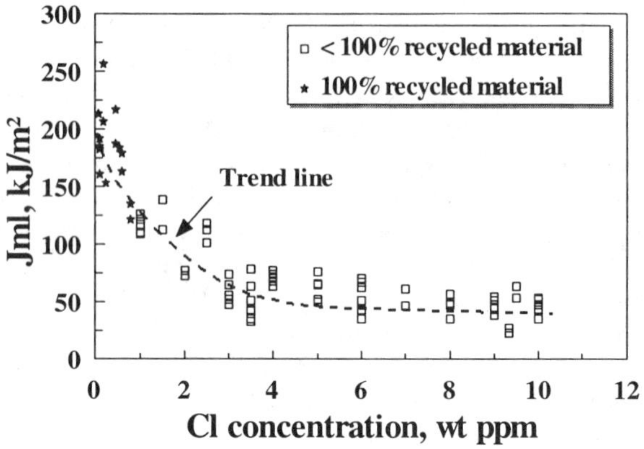

Figure 6 - *J at maximum load (J_{ml}) from curved compact specimens of irradiated material tested at 250°C versus Cl. Results sorted according to source of ingot material. Results selected from material with P < 20 wt ppm. Fast neutron fluence of 4 to 18 x 10^{25} n·m^{-2} and irradiation temperature of 254 to 295°C.*

Deformation Characteristics of Matrix

The deformation characteristics of irradiated Zr-2.5Nb pressure tube material are determined by the propensity for slip, dislocation channeling and twinning at a given test temperature, i.e. by the yield stress and the twinning stress [*14,25,26*]. Such factors influence the tendency for void nucleation and also for void growth and coalescence once the voids are formed, and are sensitive to the operating conditions of a given pressure tube, i.e. the fast neutron fluence and irradiation temperature.

For example, at low fluences (< 3 x 10^{24} n·m^{-2}) irradiation at a temperature of about 250°C increases the transverse yield stress and ultimate tensile strength under static loading conditions at 240°C by about 200 and 180 MPa, respectively, [*27*]. At higher fluences only a small further increase in strength is observed, e.g. about 5 MPa per increase in fluence of 1 x 10^{25} n·m^{-2}) [*27*]. Such results show little difference between the yield stress and ultimate tensile strength after irradiation, a characteristic of a low work-hardening material before the onset of strain localization (work-softening) by twinning and/or dislocation channeling [*14,25,26*]. These changes are primarily associated with an increase in the (**a**-type) dislocation density [*27,28*].

Therefore the main effect of irradiation on crack growth resistance is expected to occur early in the life of a reactor pressure tube, within the first year. This is demonstrated in Figures 7 and 8 which show the initial crack growth toughness (dJ/da) and maximum load toughness (J_{ml}) versus fluence for material from tubes having a lower (< 0.2 wt ppm) and higher (< 4 wt ppm) chlorine concentration. The sharp decrease in toughness at low fluences (< 3 x 10^{24} n·m^{-2}) is associated with an increased susceptibility of the material to void nucleation resulting from irradiation hardening [*22,27*]. At higher fluences there is little evidence of further degradation, behavior consistent with the toughness being governed mainly by the presence of pre-existing particles in addition to the near saturation in the hardening response. However, both sets of results exhibit considerable scatter, the source of which is discussed below.

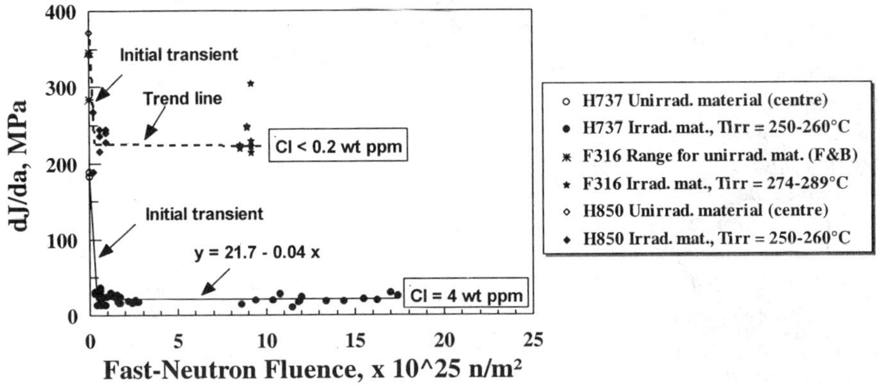

Figure 7 - *Initial crack growth toughness (dJ/da) from curved compact specimens of irradiated material tested at 240/250°C versus fast neutron fluence. Note that tube H737 and H850 were irradiated in test reactors compared with results obtained from the surveillance tube F316.* T_{irr} *is the irradiation temperature in °C. All material with P < 10 wt ppm. F & B refer to unirradiated material obtained from archive rings removed from front and back ends of tube before installation.*

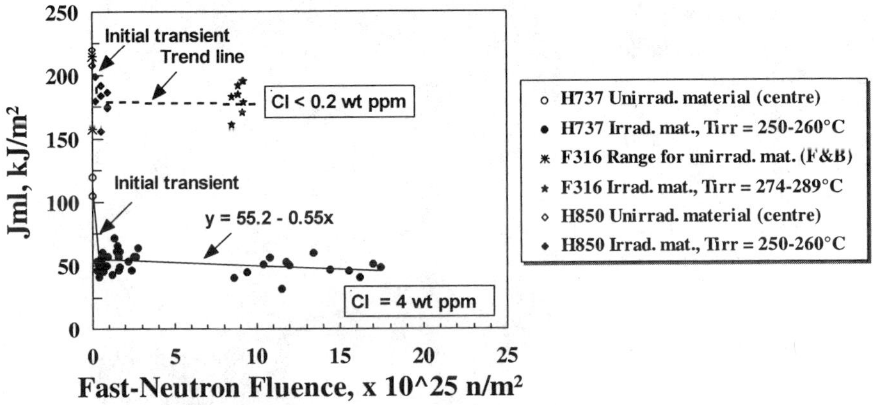

Figure 8 - *J at maximum load (*J_{ml}*) from curved compact specimens of irradiated material tested at 240/250°C versus fast neutron fluence. Note that tube H737 and H850 were irradiated in test reactors compared with results obtained from the surveillance tube F316.* T_{irr} *is the irradiation temperature in °C. All material with P < 10 wt ppm. F & B refer to unirradiated material obtained from archive rings removed from front and back ends of tube before installation.*

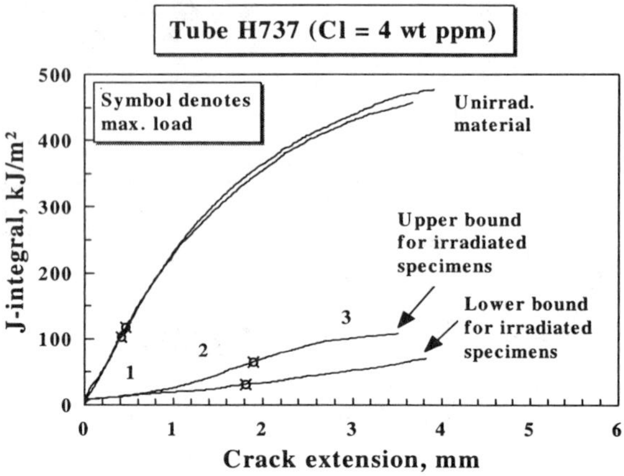

Figure 9 - *Range of J-R curves at 240°C from curved compact specimens of tube H737 (Cl = 4 wt ppm) before and after irradiation in the OSIRIS test reactor, Saclay, France. Fast neutron fluence of 0.4 to 17.4 x 10^{25} $n{\cdot}m^{-2}$ and irradiation temperature about 250°C.*

The effect of irradiation fluence on the J-R curves obtained for the material of lower toughness (Cl = 4 wt ppm) tested at 240°C is shown in Figure 9. Here the sharp reduction in the initial slope and formation of stage 1 plateau is associated with the development of a wide tunnel-shape flat fracture zone at the mid-section, as shown in Figure 10.

Figure 10 - *Optical fractograph showing stable crack growth region of irradiated specimen from tube H737 (Cl = 4 wt ppm) irradiated in OSIRIS test reactor. Features include tunnel-shape, flat fracture zone at mid-section (f), narrow cup-shaped transition zone (t), and wide slant fracture developing at surface (s). (Normal view.)*

Such flat fracture zones are associated with preferential void nucleation and growth at Zr-Cl-C (complex carbide), aligned in the axial or extrusion direction of the tube [*21,22*].. This is manifested by fissures on the radial-axial fracture plane. The large variation in the J-R curves at longer crack extensions for the irradiated material (stage 2 and 3) is related to a tendency of such material to develop slant fracture. The latter may occur preferentially at one or other surface as the crack front tunnels forward at the mid-wall to relieve the high crack-tip constraint, as indicated in Figure 10. Thus specimens with lower J-R curves exhibit not only wider flat fracture zones, but also an increased tendency to develop slant fracture, i.e. a reduction in the width of the transition zones between the regions of flat and slant fracture. In this regard the maximum load toughness (J_{ml}) is a more reliable indicator of such behavior than the initial crack growth toughness (dJ/da), since the latter is a measure of the initial J-R curve slope and does not reflect the crack growth behavior at larger crack extensions.

In comparison, Figure 11 shows the effect of fluence on the J-R curves obtained for a material of higher toughness (Cl = 0.2 wt ppm) tested at 240°C and 250°C.

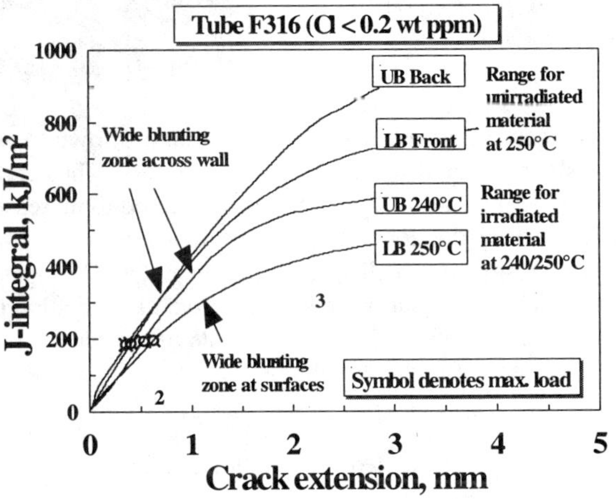

Figure 11 - *Range of J-R curves at 240/250°C from curved compact specimens of surveillance tube F316 (Cl < 0.2 wt ppm). Fast neutron fluence of* 9.2×10^{25} $n{\cdot}m^{-2}$ *and irradiation temperature of 286 to 289°C. Results for unirradiated material obtained from archive rings removed from front and back ends of tube before installation. UB and LB refer to upper and lower bound curve results, respectively.*

In this case the reduction in initial J-R slope after irradiation is not as large as for the material with 4 wt ppm Cl, and there is no evidence of stage 1 crack growth. This is believed to be due to the absence of significant void nucleation sites. The variation in the J-R curves arises mainly from differences in the initial slope which result from variations in size of the crack-tip blunting zone. For example, specimens sampled from unirradiated material exhibit full through-thickness yielding and a wide crack-tip blunting zone extending fully across the wall about 1 mm ahead of the fatigue crack. However, after irradiation the blunting zone does not always extend fully through the thickness, as shown in Figure 12 by the fracture surface for an irradiated specimen tested at 250°C. Here, the blunting zone is narrow at the mid-wall, where the crack-tip stress state (constraint) is high, but extends from the mid-wall as two triangular-shape regions, up to 1 mm ahead of the fatigue crack to the surface where the constraint is low. Therefore, for material of higher toughness, the initial crack growth toughness (dJ/da) does not necessarily reflect a true measure of crack extension, e.g. by microvoid coalescence. In the current case dJ/da reflects a combination of both blunting and crack extension which are difficult to separate.

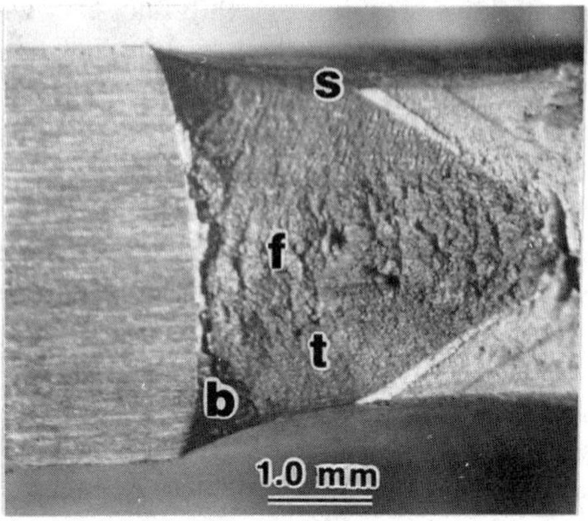

Figure 12 - *Optical fractograph showing stable crack growth region of irradiated specimen from surveillance tube F316 (Cl < 0.2 wt ppm). Features include wide blunting zones at surface (b), tunnel-shape, flat fracture zone at mid-section (f), wide cup-shaped transition zone (t), and narrow slant fracture developing at surface (s). (Normal view.)*

An increase in the irradiation temperature is known to produce a reduction in the **a**-type dislocation density and the hardening [29]. Thus at the same test temperature, increasing the irradiation temperature is expected to increase the crack growth resistance curve due to the reduction in local crack-tip stress.

Limited results on small specimen tests at 250°C suggest that the J-R curves from material of lower toughness are less sensitive to irradiation temperature than those from higher toughness material [7]. An example of the sensitivity of material from a surveillance tube of higher toughness (Cl ≤ 0.2 wt ppm, P ≤ 57 wt ppm) is shown in Figure 13 [7].

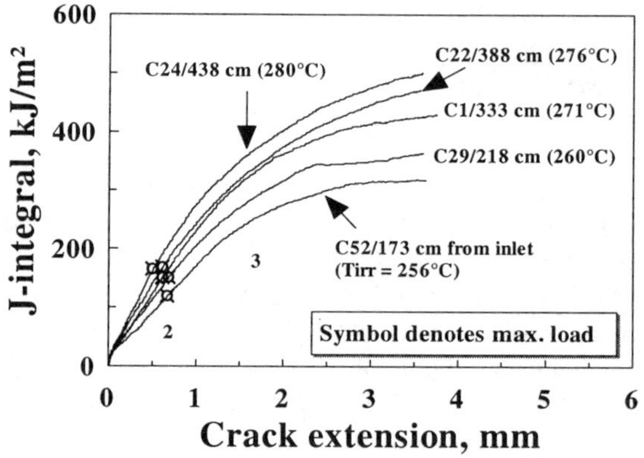

Figure 13 - *Comparison of J-R curves at 250°C from curved compact specimens sampled from* different *axial locations of surveillance tube 508 (Cl ≤ 0.2 wt ppm, P ≤ 57 wt ppm). Fast neutron fluence of 9 to 11 x 10^{25} n·m^{-2}. T_{irr} is the irradiation temperature in °C*

In Figure 13 the J-R curves are from small specimens machined from different axial locations along the main section of the tube where the fast neutron flux was relatively constant, varying from only 9 to 11 x 10^{25} n·m^{-2}. In all cases the specimens were machined from the same circumferential location, i.e. the 9 o'clock position where the 6 o'clock and 12 o'clock positions refer to the bottom and top of the tube, respectively, viewing from the inlet end. Such results clearly show the increase in crack growth resistance (stage 2 and 3) with increasing axial distance from the inlet end of the tube [7].

Figures 14 and 15 show plots of the initial crack growth toughness (dJ/da) and maximum load toughness (J_{ml}), respectively, versus irradiation temperature for the same series of tests. The figures show duplicate small specimen test results from each location, as well as the corresponding values of transverse UTS, and provide confirmation of the inverse relationship between crack growth resistance and transverse strength.

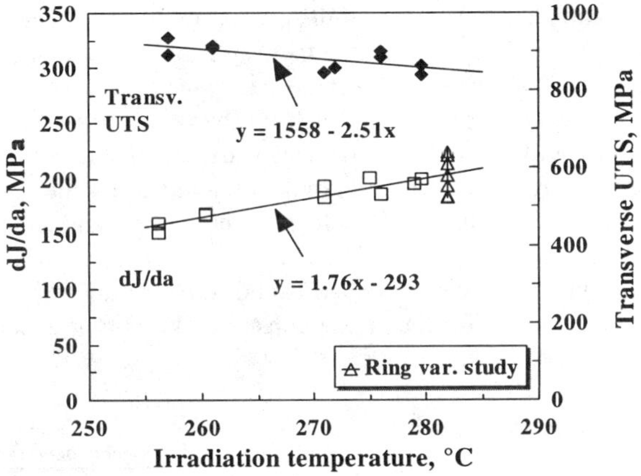

Figure 14 - *Variation in initial crack growth toughness (dJ/da) and transverse UTS at 250°C with irradiation temperature for curved compact specimens sampled from different axial locations of surveillance tube 508 (Cl $\leq$ 0.2 wt ppm, P $\leq$ 57 wt ppm). Fast neutron fluence of 9 to 11 x 10^{25} $n{\cdot}m^{-2}$.*

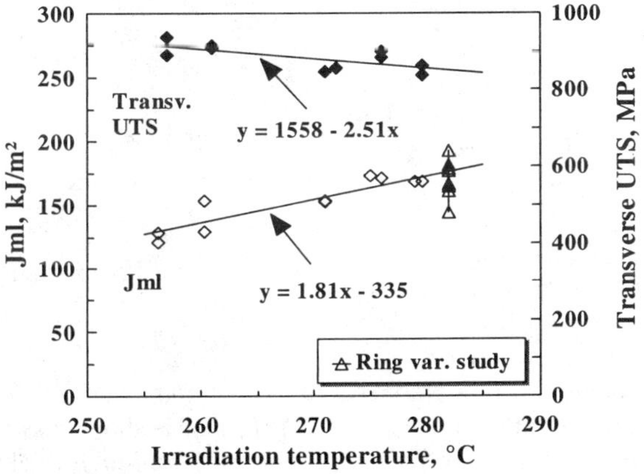

Figure 15 - *Variation in J at maximum load (J_{ml}) and transverse UTS at 250°C with irradiation temperature for curved compact specimens sampled from different axial locations of surveillance tube 508 (Cl $\leq$ 0.2 wt ppm, P $\leq$ 57 wt ppm). Fast neutron fluence of 9 to 11 x 10^{25} $n{\cdot}m^{-2}$.*

For comparison, Figure 14 and 15 include results from a variability study on a ring of material sampled 463 cm from the inlet. For the latter, the lowest and highest toughness results were obtained for locations closest to the top (12 o'clock) and bottom (6 o'clock) positions, respectively, consistent with the lower local temperature towards the top of the tube as a result of more coolant flowing above the fuel. The range in results from the ring variability study are generally in agreement with expected behavior based on those obtained from the different axial locations. In particular, the results imply a temperature difference between the top and bottom of the tube of about 20°C, in good agreement with the predicted behavior of CANDU reactor tubes based on modeling studies [*30*].

Geometry (Constraint) Effects

The previous section has shown the influence of primary variables (particles, deformation characteristics of the matrix) on the J-R curves of irradiated Zr-2.5Nb pressure tube material based on the results of curved compact specimens. However, the majority of J-R curves from small specimens generally lie well below those from burst tests as a result of the higher crack-tip stress (constraint) associated with the former compared with the latter [*5,6*]. This is demonstrated in Figure 16, which compares the J-R curves from a matched set of small- and large-scale specimens for a surveillance tube of lower toughness. Results from both plane and side-grooved curved compact specimens are shown. The shape of the different J-R curves are consistent with a higher proportion of flat fracture (stage 1) and a lower proportion of transition fracture (stage 2) being observed for the higher constraint, bend-type specimen compared with the burst test specimen.

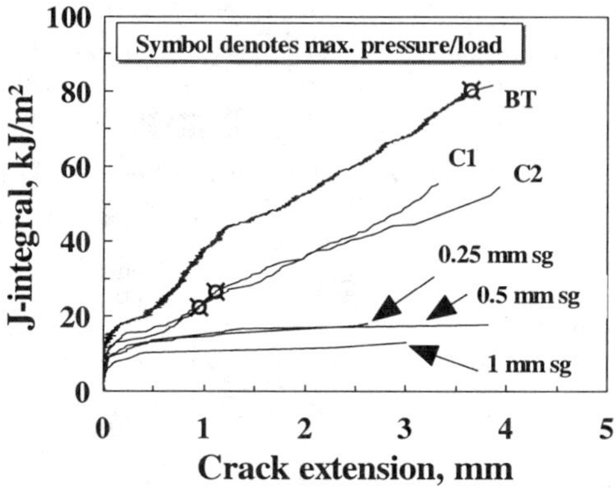

Figure 16 - *Comparison of J-R curves at 250°C from standard burst test and plane and side-grooved curved compact specimens from surveillance tube 98 (Cl = 8 to 10 wt ppm, P = 12 to 15 wt ppm). "sg" refers to depth of side-groove machined on each surface.*

At higher levels of toughness the agreement between the J-R curves from small- and large-scale specimens generally improves due to the dominance of the high-energy absorbing transition fracture mode (stage 2) [*6*]. This is shown in Figure 17, where the use of side-grooves is now less effective in suppressing through-thickness yielding in the curved compact specimens.

However, at all levels of toughness care must be taken in applying the results from small specimens to the actual tube geometry. This is because crack instability in a burst test occurs predominantly by slant fracture (stage 3). Such a fracture mode develops from the two surfaces after the crack-tip stress state is relieved by crack-front tunneling at the mid-section. However, in small specimen testing, such slant fracture may only

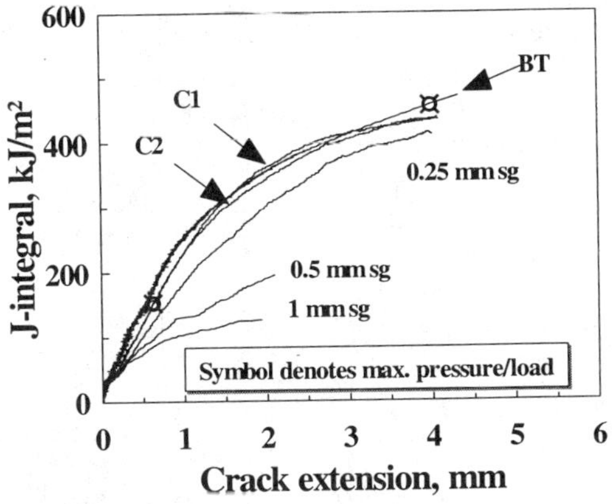

Figure 17 -*Comparison of J-R curves at 250°C from standard burst test and plane and side-grooved curved compact specimens from surveillance tube 508 (Cl ≤ 0.2 wt ppm, P ≤ 57 wt ppm).*

just be in evidence close to the specimen surfaces at the termination of the test. In addition, the out-of-plane bending or bulging associated with a rising-pressure burst test also promotes slant fracture, especially at higher levels of toughness corresponding to wider crack openings. This effect is responsible for the material dependence of crack-size-effects observed recently for material of higher toughness, i.e. a decrease in crack growth resistance for burst tests with long starter cracks (i.e. > 65 mm) such that the J-R curves lie below those of the matching small specimens (see Figure 18) [7].

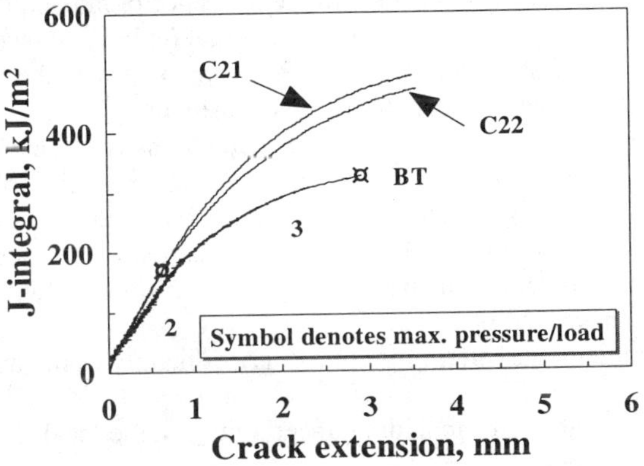

Figure 18 - *Comparison of J-R curves at 250°C from burst test with long initial crack size of 71 mm and curved compact specimens from surveillance tube 508 (Cl ≤ 0.2 wt ppm, P ≤ 57 wt ppm).*

Factors Influencing J-R Curves from Rising-Pressure Burst Tests

The foregoing has demonstrated the importance of confirming the effect of variables on the J-R curves of the irradiated material by means of large-scale tests, such as rising-pressure burst tests, before applying such results to the in-reactor situation.

Figure 19 - *J at maximum pressure toughness based on instantaneous crack size (J_{mpi}) from standard burst tests on surveillance tubes versus Cl. Results sorted according to source of ingot material. P of 6 to 60 wt ppm, fast neutron fluence of 7 to 17 x 10^{25} $n \cdot m^{-2}$ and irradiation temperature of 256 to 287°C.*

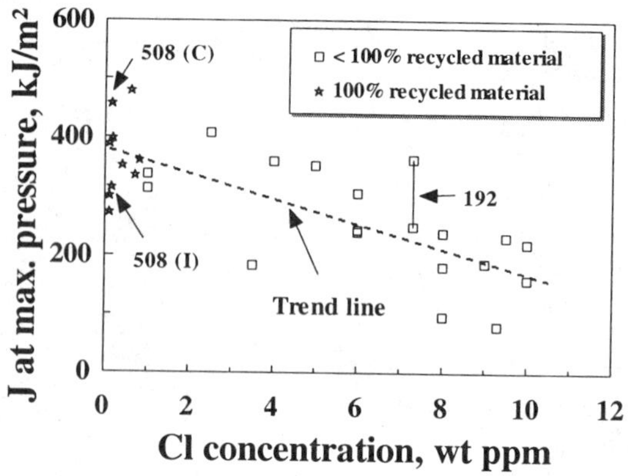

An example is shown in Figure 19 using the results from standard rising-pressure burst tests on surveillance tubes conducted at 250°C. Here the maximum pressure toughness has been calculated based on the instantaneous crack size (J_{mpi}) and plotted versus the chlorine concentration measured from unirradiated archive material. In each case the starter crack was located at the 3 o'clock position to minimize any variations in toughness arising from circumferential variations in irradiation temperature. With the exception of the results for tube 508 (see below), each data point corresponds to the crack instability (maximum pressure) position on the J-R curve for a different tube. The results confirm the influence of bulk chlorine on toughness, with tubes fabricated from 100% recycled material (chlorine concentration < 1 wt ppm) having the highest toughness. However, compared with the small specimen results of J_{ml} versus chlorine in Figure 6, there is considerably more scatter in Figure 19 and insufficient data to determine the true shape of the relationship between J_{mpi} and chlorine concentration. This is in part due to uncertainties in the definition of the instability point, e.g., for tube 192 which exhibited an extended plateau in the pressure versus crack extension curve [*6,7*]. Such uncertainty (and scatter) in the data can be reduced by using a J value based upon the hoop stress at failure and initial crack size (J_{mpo}), as shown in Figure 20.

The remaining scatter in Figure 19 arises mainly from variability in the following factors.

1) Experimental error including uncertainties in the local Cl concentration of the burst test specimen.
2) Size and distribution of zirconium phosphide and carbide particles.
3) Irradiation hardening due to variations in fluence and irradiation temperature.
4) Initial deformation characteristics of the individual tube.

For material with Cl < 1 wt ppm (100% recycled material), variations in the Cl concentration are relatively small, i.e. ≤ 0.1 wt ppm. However, for material with Cl > 1 wt ppm (< 100% recycled material), variations in Cl ≤ 1 wt ppm have been observed in the GDMS results from duplicate samples, as well as from material sampled from the front and back end of the tube. The largest variation in Cl is observed for tubes

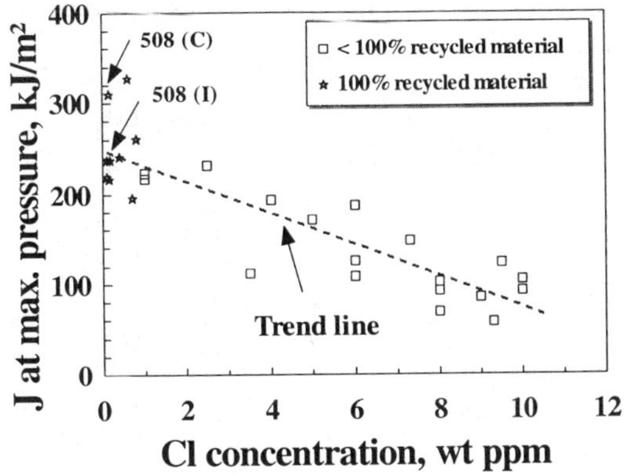

Figure 20: - *J at maximum pressure toughness based on initial crack size (J_{mpo}) from standard burst tests on surveillance tubes versus chlorine concentration. Results sorted according to source of ingot material. P of 6 to 60 wt ppm, fast neutron fluence of 7 to 17 x 10^{25} $n{\cdot}m^{-2}$ and irradiation temperature of 256 to 287°C.*

fabricated from billets sampled from the top of an ingot, i.e. the final solidification zone. Such tubes also tend to have a lower average Cl concentration compared with their "sisters" (tubes from the same ingot) due to the reduction in volatile trace elements towards the top of the ingot.

Zirconium phosphide and carbide particles up to several microns in diameter have been observed on many of the fracture surfaces of the burst test specimens [*6,7*]. This is because the majority of burst tests have been conducted on the original tubes removed from Pickering NGS A Unit 3 and 4 for which the concentration of phosphorus and carbon was generally high, i.e. up to about 80 wt ppm and 180 wt ppm, respectively [*22,23*]. This is well above the level at which zirconium phosphides and carbides have been identified on fractures, i.e. $\geq$ 8 wt ppm and 125 wt ppm, respectively [*22,23*]. However, in comparison to the small specimen results in Figure 6, there are too few burst test results to limit the database, e.g. to tubes with P < 20 wt ppm

The role of irradiation hardening in reducing the J-R curves obtained from burst tests has been confirmed by tests on two sections of tube 508. This tube was fabricated from 100% recycled material and had a phosphorus concentration up to 57 wt ppm. The J-R curve results are shown in Figure 21 for the two sections taken from the center and inlet regions of the pressure tube [*7*]. The fast neutron fluence and irradiation temperature are 11.1 x $10^{25}{\cdot}n{\cdot}m^{-2}$ and 271°C for the former and 9.3 x $10^{25}{\cdot}n{\cdot}m^{-2}$ and 256°C for the latter. For these tests the difference in fast neutron fluence is small and the reduction in crack growth resistance in Figure 21 arises mainly from the variation in transverse strength. The latter was measured after each burst test at the 9 o'clock position (i.e. diametrically opposite the starter flaw), giving values of 852 and 913 MPa for the center and inlet/center test sections, respectively. The relative magnitude of this effect of irradiation hardening on the maximum pressure toughness parameters, J_{mpi} and J_{mpo}, is indicated in Figures 19 and 20, respectively. Such results are very similar to the previous results obtained from the corresponding small specimens (see Figures 13 to 15).

Early attempts to rationalize all the burst test specimen results from the original Pickering 3 and 4 tubes solely on the basis of chlorine concentration and transverse

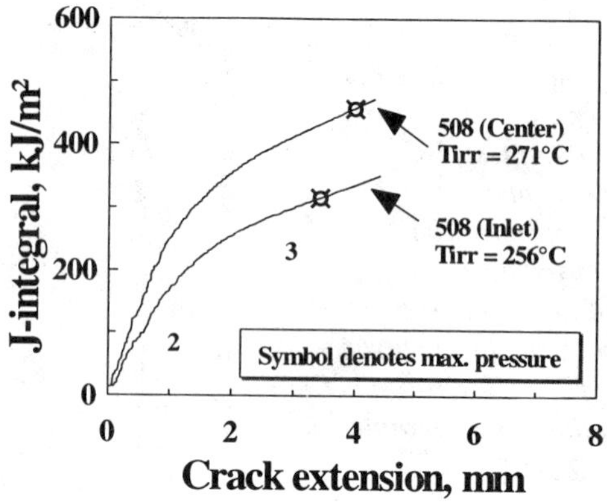

Figure 21 - *Comparison of J-R curves at 250°C from standard burst tests sampled from two different axial locations of surveillance tube 508 (Cl ≤ 0.2 wt ppm, P ≤ 57 wt ppm). T_{irr} is the irradiation temperature in °C. The difference in fast neutron fluence is relatively small and varied from 9 to 11 x 10^{25} n·m^{-2} for material sampled from the center and the inlet end of the pressure tube, respectively.*

strength proved reasonably successful (see Figure 16 in reference 7). However, as results from tubes installed in later reactors have become available, it is now clear that such relationships are sensitive to the initial deformation characteristics of the individual tube, e.g. initial yield stress and twinning stress. For example, variability in yield stress can arise from variations in the oxygen and niobium concentrations [*31*] as well as the initial dislocation density [*31*], grain structure and texture [*32*]. Such variations can result from minor changes in ingot chemistry, fabrication route and extrusion variables [*33-35*].

The sensitivity of the crack growth resistance of irradiated Zr-2.5Nb pressure tube material to the initial deformation properties of the tube at 250°C is due to the similarity of the yield and twinning stresses in the operating temperature regime after irradiation [*26*]. These parameters determine not only the local crack-tip stress for void nucleation, but also the propensity for deformation by dislocation channeling and/or twinning compared with slip, i.e., for strain localization. Such strain localization contributes to the failure of ligaments between neighboring voids (stage 1 and 2 crack growth) and controls the onset of slant fracture at the surface (stage 3). Therefore, removal of void nucleation sites is particularly beneficial. This is because it increases the effective wall thickness for local ligament failure, and also reduces the extents of stable crack growth and tunneling which orient the crack front for easy development of the "sliding-off" mechanism.

Work is now underway to extend the database of burst test results and to determine the quantitative relationship between crack growth resistance and Cl using material with intermediate levels of Cl and low levels of zirconium phosphide. The initial deformation characteristics of the burst test material are also being studied with the aim of elucidating the microstructural factors responsible for the variability in results in Figures 19 and 20. This should allow further rationalization of the results obtained from different reactor units as well as the potential for producing further improvements in crack growth resistance.

Summary

The key factors controlling the crack growth resistance of irradiated Zr-2.5Nb pressure tube material at 250°C at low levels of hydrogen/deuterium have been reviewed using the current database of small- and large-scale specimen results from different CANDU reactor pressure tubes. The review highlights the role of the following factors.

1) The mixed-mode nature of the crack growth process in the irradiated, thin-walled material, i.e. flat, transition and slant fracture modes.
2) The relative proportions and energy-absorbing capacities of the different fracture modes, which is determined by the ease of void nucleation, growth and coalescence ahead of the crack tip.
3) The role of primary void nucleating particles, e.g. Zr-Cl-C (complex carbide), zirconium phosphide and carbide, the beneficial effects of eliminating them. For example, the removal of chlorine and Zr-Cl-C particles by quadruple vacuum-arc-melting.
4) The deformation characteristics of the matrix, e.g. yield stress and twinning stress, which control the tendency for void nucleation, growth, coalescence and strain localization (work-softening).
5) The specimen geometry, e.g. crack size, geometry and material effects in influencing the relationship between the J-R curves obtained from different specimens.

The current results from curved compact specimens, having a low concentration of zirconium phosphide (P < 20 wt ppm), suggest a limiting level of Cl (about 3 wt ppm) above which no further significant degradation in the crack growth resistance occurs. Such results require confirmation using rising-pressure burst tests on material with intermediate levels of Cl and low levels of zirconium phosphide.

The influence of minor variations in tube fabrication on the deformation behavior of Zr-2.5Nb pressure tube material (item 4 above) also requires study. This should allow further rationalization of the results obtained from different reactor units as well as the potential for producing further improvements in crack growth resistance.

Acknowledgments

Many individuals contributed to the production of the results reported in this work over a number of years. In particular, thanks are due to R. Behnke, G. R. Brady, G. R. Kasprick, G. C. Longhurst, A. R. Reich, J. M. Smeltzer, A. K. West, as well as the hot cell staff at the Whiteshell and Chalk River Laboratories for their technical assistance. Special thanks go to R. R. Bawden and J. C. Owens at Chalk River for retrieval and shipment of surveillance material for testing and to G. D. Moan for supplying unirradiated archive material for testing. Funding of this work by the CANDU Owners Group (COG) under Work Packages 3192, 3305, 6506 and 6536 and contract COG-00136t3 is gratefully acknowledged.

References

[1] Cheadle, B. A., Coleman, C. E., and Licht, H., *Nuclear Technology*, Vol. 57, 1982, pp. 413-425.

[2] Chow, C. K. and Simpson, L. A., *Fracture Mechanics: Eighteenth Symposium, ASTM STP 945*, D. T. Read, R. P. Read, Eds., American Society for Testing and Materials, West Conshohocken, PA, 1988, pp. 419-439.

[3] Simpson, L. A., Chow, C. K., and Davies, P. H., CANDU Owners Group Report No. COG-89-110-I, AECL, September 1989.

[4] Himbeault, D. D., and Davies, P. H., CANDU Owners Group Report No. COG-98-161-I, AECL, January 1999.

[5] Davies, P. H. Shewfelt, R. S. W. and Järvine, A. K., *Constraint Effects in Fracture: Theory and Applications, ASTM STP 1244*, M. Kirk and A. Bakker, Eds., American Society for Testing and Materials, West Conshohocken, PA, 1995, pp. 392-424.

[6] Davies, P. H. and Shewfelt, R. S. W., *Zirconium in the Nuclear Industry: Eleventh International Symposium, ASTM STP 1295,* E. R. Bradley and G. P. Sabol, Eds., American Society for Testing and Materials, West Conshohocken, PA, 1996, pp. 492-517.

[7] Davies, P. H. and Shewfelt, R. S. W., *Zirconium in the Nuclear Industry: Twelfth International Symposium, ASTM STP 1354,* G. P. Sabol and G. D. Moan, Eds., American Society for Testing and Materials, West Conshohocken, PA, 2000, pp. 356-376.

[8] Kiefner, J. F., Maxey, W. A., Eiber, R. J. and Duffy, A. R., *Progress in Flaw Growth and Fracture Toughness Testing, ASTM STP 536,* American Society for Testing and Materials, West Conshohocken, PA, 1973, pp. 461-481.

[9] Rice, J. R., *Journal of Applied Mechanics, Transactions of ASME,* Vol. 35, 1968, pp379-386.

[10] Turner, C. E., *Size Effects,* Mechanical Engineering Publications, London, 1986, pp. 25-31.

[11] Ernst, H. A., *Elastic-Plastic Fracture: Second Symposium, Volume I, ASTM STP 803,* C. F. Shih and J. P. Gudas, Eds., American Society for Testing and Materials, 1983, pp. I191-I213.

[12] Turner, C. E., *Fracture Mechanics: Twenty Second Symposium (Volume I), ASTM STP 1131,* H. A. Ernst, A. Saxena and D. L. McDowell, Eds., American Society for Testing and Materials, 1992, pp. 71-92.

[13] Davies, P. H., *Fatigue and Fracture Mechanics: Twenty Eighth Volume, ASTM STP 1321,* J. H. Underwood, B. D. MacDonald and M. R. Mitchell, Eds., American Society for Testing and Materials, West Conshohocken, PA, 1997, pp. 535-561.

[14] Chow, C. K. and Simpson, L. A., *Fracture Mechanics: Nonlinear Fracture Mechanics- Volume II-Elastic-Plastic Fracture, ASTM STP 995,* J. D. Landes, A. Saxena and J. G. Merkle, Eds., American Society for Testing and Materials, West Conshohocken, PA, 1989, pp. 537-562.

[15] Thomason, P. F., *Ductile Fracture of Metals*, Pergamon Press, Oxford, Ch. 1 to 4, 1990, pp. 1-114.
[16] Ritchie, R. O., Server, W. L., and Wullaert, R. A., *Metallurgical Transactions A,* Vol. 10A, 1979, pp. 1557-1570.
[17] Hutchinson, J. W., and Tvergaard, V., *Fracture Mechanics; Perspectives and Directions (Twentieth Symposium), ASTM STP 1020,* R. P. Wei and R. P. Gangloff, Eds., American Society for Testing and Materials, 1989, pp. 61-83.
[18] Shih, C. F., *Journal of Mechanics and Physics of Solids,* Vol. 29, 1981, pp. 305-326.
[19] Dodds, R. H., Shih, C. F., and Anderson, T., *International Journal of Fracture*, Vol. 64, 1993, pp. 101-133.
[20] Wallace, A. C., Shek, G. K., and Lepik, O. E., *Zirconium in the Nuclear Industry: Eighth International Symposium, ASTM STP 1023,* L. F. . Van Swam and C. M. Eucken, Eds., American Society for Testing and Materials, 1989, pp. 66-88.
[21] Aitchison, I. and Davies, P. H., *Journal of Nuclear Materials*, Vol. 203, 1993, pp. 206-220.
[22] Davies, P. H., Hosbons, R. R., Griffiths, M. and Chow, C. K., *Zirconium in the Nuclear Industry: Tenth International Symposium, ASTM STP 1245,* A. M. Garde and E. R. Bradley, Eds., American Society for Testing and Materials West Conshohocken, PA, 1994, pp. 135-167.
[23] Davies, P. H., Aitchison, I., Himbeault, D. D., Järvine, A. K. and Watters, J. F., *Fatigue and Fracture of Engineering Materials and Structures*, Vol. 18, 1995, pp. 789-800.
[24] Theaker, J. R., Choubey, R., Moan, G. D., Aldridge, S. A., Davies, L., Graham, R. A. and Coleman, C. E., *Zirconium in the Nuclear Industry: Tenth International Symposium, ASTM STP 1245,* A. M. Garde and E. R. Bradley, Eds., American Society for Testing and Materials West Conshohocken, PA, 1994, pp. 221-242.
[25] Dutton, R., AECL Report No. 9930, 1989 April.
[26] Himbeault, D. D., Chow, C. K. and Puls, M. P., *Metallurgical and Materials Transactions A*, Vol. 25A, 1994, pp. 135-145.
[27] Hosbons, R. R., Davies, P. H. Griffiths, M. Sagat, S. and Coleman, C. E., *Zirconium in the Nuclear Industry: Twelfth International Symposium, ASTM STP 1354,* G. P. Sabol and G. D. Moan, Eds., American Society for Testing and Materials, West Conshohocken, PA, 2000, pp. 122-138.
[28] Chow, C. K., Coleman, C. E., Hosbons, R. R., Davies, P. H., Griffiths, M., and Choubey, R., *Zirconium in the Nuclear Industry: Ninth International Symposium, ASTM STP 1132,* C. M. Eucken and A. M. Grade, Eds., American Society for Testing and Materials, West Conshohocken, PA, 1992, pp. 246-275.
[29] Sagat, S., Coleman, C. E., Griffiths, M., and Wilkins, B. J. S., *Zirconium in the Nuclear Industry: Tenth International Symposium, ASTM STP 1245,* A. M. Garde and E. R. Bradley, Eds., American Society for Testing and Materials, West Conshohocken, PA, 1994, pp. 35-61.

[30] Tabatabai, M. and Byrne, T., Ontario Hydro Report No. A-NFC-94-34-K, 1994 April.

[31] Winton, J., Murgatroyd, R. A., Watkins, B., and Nichols, R. W., *Transactions of the Japanese Institute of Metals,* Volume 9 Supplement, 1968, pp. 630-636.

[32] Cheadle, B., and Ells, C. E., *Transactions of the Metallurgical Society of AIME,* Vol. 233, 1965, pp. 1044-1052.

[33] Cheadle, B., *Zirconium in the Nuclear Industry, ASTM STP 633,* A. L. Lowe, Jr., G. W. Parry, Eds., American Society for Testing and Materials West Conshohocken, PA, 1977, pp. 457-485.

[34] Holt, R. A. and Aldridge, S. A., *Journal of Nuclear Materials*, Vol. 135, 1985, pp. 246-259.

[35] Choubey, R., Aldridge, S. A., Theaker, J. R., Cann, C. D., and Coleman, C. E., *Zirconium in the Nuclear Industry: Eleventh International Symposium, ASTM STP 1295,* E. R. Bradley and G. P. Sabol, Eds., American Society for Testing and Materials West Conshohocken, PA, 1996, pp. 657-675.

Author Index

J

K

L

M

N

O

P

R

S

Subject Index

O

P

R

S